Lecture Notes in Computer Science 9014

Commenced Publication in 1973
Founding and Former Series Editors:
Gerhard Goos, Juris Hartmanis, and Jan van Leeuwen

T0185357

Yevgeniy Dodis   Jesper Buus Nielsen (Eds.)

Theory of Cryptography

12th Theory of Cryptography Conference, TCC 2015
Warsaw, Poland, March 23-25, 2015
Proceedings, Part I

 Springer

Volume Editors

Yevgeniy Dodis
New York University, Department of Computer Science
251 Mercer Street, New York, NY 10012, USA
E-mail: dodis@cs.nyu.edu

Jesper Buus Nielsen
Aarhus University, Department of Computer Science
Åbogade 34, 8200 Aarhus N, Denmark
E-mail: jbn@cs.au.dk

ISSN 0302-9743   e-ISSN 1611-3349
ISBN 978-3-662-46493-9   e-ISBN 978-3-662-46494-6
DOI 10.1007/978-3-662-46494-6
Springer Heidelberg New York Dordrecht London

Library of Congress Control Number: 2015933013

LNCS Sublibrary: SL 4 – Security and Cryptology

Typesetting: Camera-ready by author, data conversion by Scientific Publishing Services, Chennai, India

Printed on acid-free paper

Springer is part of Springer Science+Business Media (www.springer.com)

Preface

The 2015 Theory of Cryptography Conference (TCC) was held at the Sheraton Warsaw Hotel in Warsaw, Poland, during March 23–25. TCC 2015 was sponsored by the International Association for Cryptologic Research (IACR). The general chair of the conference was Stefan Dziembowski. We would like to thank him for his hard work in organizing the conference.

The conference received 137 submissions, a record number for TCC. Each submission was reviewed by at least three Program Committee (PC) members. Because of the large number of submissions and the very high quality, the PC decided to accept 52 papers, a significant extension of capacity over previous TCCs. Still the PC had to reject many good papers. After the acceptance notification, authors of the accepted papers were given three weeks to revise their paper in response to the reviews. The revisions were not reviewed. The present proceedings consists of the revised versions of the 52 accepted papers.

The submissions were reviewed by a PC consisting of 25 leading researchers from the field. A complete list of the PC members can be found after this preface. Each PC member was allowed to submit two papers. PC-authored papers were held to a higher standard. Initially each paper was given at least three independent reviews (four for PC-authored papers), using external reviewers when appropriate. Following the individual review period there was a discussion phase. This year the review software was extended to allow PC members to interact with authors by sending them questions directly from the discussion page of a submission, and having the answer automatically appear on the discussion page. In order to minimize the chance of making decisions based on misunderstandings, the PC members were strongly encouraged to use this "chat" feature to discuss with the authors all technical issues that arose during the review and the discussion phases. As a result, the feature was used extensively where appropriate, and completely replaced (what we felt was) a much more limited and less effective "rebuttal" phase that was used in recent CRYPTO and Eurocrypt conferences. In particular, this allowed the PC members to spend their effort on the issues that were most likely to matter at the end. We believe that our experiment with the increased interaction with authors was a great success, and that it gives a better quality-to-effort ratio than a rebuttal phase. Thus, we encourage future program chairs to continue increasing the interaction with the authors. This year we also experimented with cross-reviews, letting authors of similar submissions comment on the relation between these submissions. This was less of a success. Although the chance to compare the other submissions was welcomed by some authors, the cross-reviews were found to be controversial by other authors, and it is not clear that the comparisons contributed much more than having a dedicated PC member read all the papers and form an independent opinion.

We would like to thank the PC for their dedication, high standards, and hard work. Indeed, most of the PC members truly went above and beyond. Having such a great PC made it easy to chair the conference. We would also like to thank all the external reviewers who decided to dedicate their time and effort to reviewing for TCC 2015. Your help was indispensable. A list of all external reviewers follows this preface. We apologise for any omissions. Great thanks to Shai Halevi for developing and maintaining the *websubrev* software, which was an invaluable help in running the PC. Thanks in particular for extending the system with a "chat" feature. Tuesday evening the conference had a rump session chaired by Krzysztof Pietrzak from IST Austria. We would like to thank him for his hard work in making it an enjoyable event. We also thank the Warsaw Center of Mathematics and Computer Science (WCMCS) and Google Inc. for contributing to the financing of the conference. Last but not least, thanks to everybody who submitted a paper to TCC 2015!

This year we had two invited speakers. Leonid Reyzin from Boston University talked about "Wyner's Wire-Tap Channel, Forty Years Later" and John Steinberger from Tsinghua University talked about "Block Ciphers: From Practice Back to Theory." We are grateful to both speakers for their interesting contributions to the program.

This was the first year where TCC presented the Test of Time (ToT) award to a paper that has appeared at TCC in yore and has stood the test of time. This year the award was given to Silvio Micali and Leonid Reyzin for the paper "Physically Observable Cryptography," which was presented at TCC 2004. The ToT paper was chosen by a committee selected by the TCC Steering Committee. The ToT committee has the following quotation for the ToT paper:

For pioneering a mathematical foundation of cryptography in the presence of information leakage in physical systems.

The 52 papers selected for this year's TCC testify to the fact that the theory of cryptography community is a thriving and expanding community of the highest scientific quality. We are convinced that this year's TCC program contained many papers that will stand the test of time. Have fun reading these proceedings.

January 2015

Yevgeniy Dodis
Jesper Buus Nielsen

TCC 2015

12th IACR Theory of Cryptography Conference

Sheraton Warsaw Hotel, Warsaw, Poland
March 23–25

Sponsored by *International Association for Cryptologic Research (IACR)*

General Chair

Stefan Dziembowski   University of Warsaw, Poland

Program Chair

Yevgeniy Dodis   New York University, USA
Jesper Buus Nielsen   Aarhus University, Denmark

Program Committee

Joël Alwen	IST Austria, Austria
Benny Applebaum	Tel Aviv University, Israel
Nir Bitansky	Tel Aviv University, Israel
Elette Boyle	Technion, Israel
Kai-Min Chung	Academia Sinica, Taiwan
Nico Döttling	Aarhus University, Denmark
Sebastian Faust	EPFL, Switzerland
Serge Fehr	CWI, Holland
Sanjam Garg	University of California, Berkeley, USA
Shai Halevi	IBM Research, USA
Martin Hirt	ETH Zurich, Switzerland
Dennis Hofheinz	KIT, Karlsruhe, Germany
Thomas Holenstein	ETH Zurich, Switzerland
Yuval Ishai	Technion, Israel
Kaoru Kurosawa	Ibaraki University, Japan
Allison Lewko	Columbia University, USA
Mohammad Mahmoody	University of Virginia, USA
Moni Naor	Weizmann Institute of Science, Israel
Chris Peikert	Georgia Institute of Technology, USA

Phillip Rogaway	UC Davis, USA
Mariana Raykova	SRI International, USA
abhi shelat	University of Virginia, USA
Stefano Tessaro	UC Santa Barbara, USA
Jon Ullman	Harvard University, USA
Daniel Wichs	Northeastern University, USA

External Reviewers

Divesh Aggarwal	Chandan Dubey	Takahiro Matsuda
Shashank Agrawal	Alexandre Duc	Christian Matt
Martin Albrecht	Pooya Farshim	Alexander May
Jacob Alperin-Sheriff	Georg Fuchsbauer	Willi Meier
Prabhanjan Ananth	Benjamin Fuller	Bart Mennink
Frederik Armknecht	Ariel Gabizon	Daniele Micciancio
Gilad Asharov	Peter Gazi	Eric Miles
Nuttapong Attrapadung	Sergey Gorbunov	Payman Mohassel
Jean-Philippe Aumasson	Dov Gordon	Hart Montgomery
Christian Badertscher	Vipul Goyal	Jörn Müller-Quade
Foteini Baldimtsi	Matthew Green	Ryo Nojima
Abhishek Banerjee	Divya Gupta	Adam O'Neill
Carsten Baum	Jan Hazla	Wakaha Ogata
Alexander Belov	Pavel Hubáček	Emmanuela Orsini
Itay Berman	Peter Høyer	Omkant Pandey
Dan Bernstein	Vincenzo Iovino	Omer Paneth
Zvika Brakerski	Abhishek Jain	Dimitrios Papadopoulos
Christina Brzuska	Daniel Jost	Bryan Parno
Ran Canetti	Bhavana Kanukurthi	Anat Paskin-
David Cash	Marcel Keller	Cherniavsky
Nishanth Chandran	Dakshita Khurana	Rafael Pass
Melissa Chase	Susumu Kiyoshima	Le Trieu Phong
Binyi Chen	Venkata Koppula	Krzysztof Pietrzak
Jie Chen	Takeshi Koshiba	Andrew Poelstra
Alessandro Chiesa	Daniel Kraschewski	Antigoni Polychroniadou
Chongwon Cho	Hugo Krawczyk	Raluca Popa
Seung Geol Choi	Sara Krehbiel	Samuel Ranellucci
Aloni Cohen	Robin Kuenzler	Ben Riva
Gil Cohen	Ranjit Kumaresan	Alon Rosen
Ran Cohen	Robin Künzler	Ron Rothblum
Sandro Coretti	Tancrède Lepoint	Andy Rupp
Dana Dachman-Soled	Kevin Lewi	Yusuke Sakai
Bernardo David	Huijia (Rachel) Lin	Carla Rafols Salvador
Gregory Demay	Yehuda Lindell	Jacob Schuldt
Yi Deng	Feng-Hao Liu	Lior Seeman
Itai Dinur	Steve Lu	Gil Segev

Rocco Servedio
Karn Seth
Or Sheffet
Shikha Singh
Adam Smith
Fang Song
Damien Stehlé
John Steinberger
Koutarou Suzuki
Björn Tackmann
Katsuyuki Takashima
Sidharth Telang

Isamu Teranishi
Justin Thaler
Nikolaos Triandopoulos
Mehdi Tibouchi
Daniel Tschudi
Dominique Unruh
Serge Vaudenay
Muthuramakrishnan
    Venkitasubramaniam
Daniele Venturi
Ivan Visconti
Hoeteck Wee

Mor Weiss
Xiaodi Wu
Shota Yamada
Arkady Yerukhimovich
Yu Yu
Mark Zhandry
Haibin Zhang
Vassilis Zikas
Joe Zimmerman
Asaf Ziv

Sponsoring and Co-Financing Institutions

TCC 2015 was co-financed by the Warsaw Center of Mathematics and Computer Science (WCMCS). The conference was also generously sponsored by Google Inc. The conference organizers are grateful for this financial support.

Wyner's Wire-Tap Channel, Forty Years Later (Invited Talk)

Leonid Reyzin

Boston University
Department of Computer Science
Boston, MA 02215, USA

Abstract. Wyner's information theory paper "The Wire-Tap Channel" turns forty this year. Its importance is underappreciated in cryptography, where its intellectual progeny includes pseudorandom generators, privacy amplification, information reconciliation, and many flavors of randomness extractors (including plain, strong, fuzzy, robust, and nonmalleable). Focusing mostly on privacy amplification and fuzzy extractors, this talk demonstrates the connection from Wyner's paper to today's research, including work on program obfuscation.

Table of Contents – Part I

Foundations

Symmetric Key

Multiparty Computation

Concurrent and Resettable Security

Non-malleable Codes and Tampering

Privacy Amplification

Encryption and Key Exchange

Table of Contents – Part II

Pseudorandom Functions and Applications

Proofs and Verifiable Computation

Differential Privacy

Functional Encryption

Obfuscation

On Basing Size-Verifiable One-Way Functions on NP-Hardness

Andrej Bogdanov[1,*] and Christina Brzuska[2,**]

[1] Dept. of Computer Science and Engineering,
The Chinese University of Hong Kong, China
andrejb@cse.cuhk.edu.hk
[2] Microsoft Research, Cambridge, UK
christina.brzuska@gmail.com

Abstract. We prove that if the hardness of inverting a size-verifiable one-way function can be based on NP-hardness via a general (adaptive) reduction, then NP $\subseteq$ coAM. This claim was made by Akavia, Goldreich, Goldwasser, and Moshkovitz (STOC 2006), but was later retracted (STOC 2010).

Akavia, Goldreich, Goldwasser, and Moshkovitz [AGGM06] claimed that if there exists an adaptive reduction from an NP-complete problem to inverting an efficient size-verifiable function, then NP $\subseteq$ coAM. They provided a proof for size-verifiable functions that have polynomial pre-image size as well as a proof for general size-verifiable functions, even if the size of the pre-image can only be approximated. The proof for the latter statement was found to be erroneous and has been retracted [AGGM10].[1] In this note we give a proof of their claim. For motivation about the problem, we refer the reader to the work [AGGM06].

Throughout this paper, we consider efficiently computable functions f with $f(\{0,1\}^n) \subseteq \{0,1\}^{m(n)}$, where m is an injective function on integers. We say an oracle I inverts f if for every $x \in \{0,1\}^*$, $I(f(x))$ belongs to the set $f^{-1}(f(x))$.

We say that f is *size-verifiable* if the decision problem $N_f = \{(y,s) \colon |f^{-1}(y)| = s\}$ is in AM. We say that f is *approximately size-verifiable* if the following promise problem A_f is in AM:

YES instances of A_f: $(y, s, 1^a)$ such that $|f^{-1}(y)| \leq s$

NO instances of A_f: $(y, s, 1^a)$ such that $|f^{-1}(y)| > (1 + 1/a)s$.

---

* This work was partially supported by Hong Kong RGC GRF grant CUHK410113.
** Work done while at Tel Aviv University. Christina Brzuska was supported by the Israel Science Foundation (grant 1076/11 and 1155/11), the Israel Ministry of Science and Technology grant 3-9094), and the German-Israeli Foundation for Scientific Research and Development (grant 1152/2011). Part of this work was done while visiting CUHK.

[1] In the same paper [AGGM06], Akavia et al. also show that the existence of a (randomized) non-adaptive reduction of NP to the task of inverting an arbitrary one-way function implies that NP $\subseteq$ coAM. This result is not affected by the gap found in [AGGM10].

Y. Dodis and J.B. Nielsen (Eds.): TCC 2015, Part I, LNCS 9014, pp. 1–6, 2015.
© International Association for Cryptologic Research 2015

A *reduction* from a decision problem L to inverting f is a randomized oracle algorithm $R^?$ such that for every oracle I that inverts f, R^I decides L with probability at least $2/3$ over the randomness of $R^?$.

Theorem 1. *Let f be an efficiently computable, approximately size-verifiable function. If there exists an efficient reduction from L to inverting f with respect to deterministic inversion oracles, then L is in $\text{AM} \cap \text{coAM}$.*

Corollary 1. *Let f be an efficiently computable, approximately size-verifiable function. There is no efficient reduction from an NP-hard language L to inverting f with respect to deterministic inversion oracles, unless $\text{NP} \subseteq \text{coAM}$.*

We first prove a weaker version of the theorem that relies on two simplifying assumptions. Firstly, we assume that the reduction is correct even with respect to *randomized* inversion oracles. These are oracles that have access to an internal source of randomness when answering their queries. Our inversion oracle will simply sample a uniform pre-image amongst all possible pre-images like an inverter for a distributional one-way function [IL89]. Note that a reduction that works for randomized inversion oracles also works with respect to deterministic oracles, as they are a special case of randomized ones. As we prove a negative result, stronger requirements on the reduction weaken our result. We will thus explain later how to remove this additional requirement on the reduction. Secondly, we assume the function to be size-verifiable rather than approximately size-verifiable. We then adapt the proof to the general case.

Randomized Inversion Oracles. Let $R^?$ be a reduction, I a randomized oracle, and z an input. A *valid transcript* of $R^I(z)$ is a string of the form $(r, x_1, \ldots, x_q)$, where r is the randomness of the reduction and $x_1, \ldots, x_q$ are the oracle answers in the order produced by I. We will assume, without loss of generality, that the length of r and the number of queries q depend only on the length of z.

Consider the randomized inversion oracle U that, on query y, returns an x chosen uniformly at random from the set $f^{-1}(y)$, or the special symbol $\perp$ if this set is empty. Let the set C consist of all tuples $(z, r, x_1, \ldots, x_q, p)$, such that $(r, x_1, \ldots, x_q)$ is an accepting valid transcript of $R^U(z)$ and p is an integer between 1 and $\lceil K/(s(y_1) \cdots s(y_q)) \rceil$. Here,

- y_i is the i-th query of the reduction,
- $s(y)$ is the size of the set of possible answers on query y:

$$s(y) = \begin{cases} |f^{-1}(y)|, & \text{if } f^{-1}(y) \text{ is non-empty} \\ 1, & \text{otherwise,} \end{cases}$$

- and $K = 2 \cdot 2^{q\ell}$, where ℓ is an upper bound on the length of queries $R^?$ makes on inputs of length $|z|$.

Claim. C is in AM.

Proof. On input $(z, r, x_1, \ldots, x_q, p)$, the AM verifier for C runs the reduction on input z with randomness r and checks that for for each query y_i that the reduction makes, the answer x_i is indeed a pre-image of y_i and that the reduction accepts. To see that p is of the right size, we ask the prover to provide $s(y_1), \ldots, s(y_q)$ such that $p \leq K/(s(y_1) \cdots s(y_q))$. We then run the AM verifier for N_f to check that the numbers $s(y_1), \ldots, s(y_q)$ that the prover provided are correct. □

Let $C(z)$ denote the set of all $(r, x_1, \ldots, x_q, p)$ such that $(z, r, x_1, \ldots, x_q, p)$ is in C.

Claim. $C(z)$ has size at least $\frac{2}{3} 2^{|r|} K$ if $z \in L$, and size at most $\frac{1}{2} 2^{|r|} K$ if $z \notin L$.

Proof. Fix the input z. Conditioned on the randomness r, every valid transcript $(r, x_1, \ldots, x_q)$ appears with probability exactly $1/(s(y_1) \ldots s(y_q))$ over the choice of randomness of the inverter. All these probabilities add up to one:

$$\sum_{(x_1, \ldots, x_q)} \frac{1}{s(y_1) \ldots s(y_q)} = 1.$$

If $z \in L$, then at least a $2/3$ fraction of these valid transcripts must be accepting for $R^?(z)$ over the choice of r and so

$$|C(z)| \geq \frac{2}{3} \sum_{\text{valid transcript } (r, x_1, \ldots, x_q)} \left\lceil \frac{K}{s(y_1) \ldots s(y_q)} \right\rceil$$

$$\geq \frac{2}{3} \sum_r K \sum_{(x_1, \ldots, x_q)} \frac{1}{s(y_1) \ldots s(y_q)}$$

$$= \frac{2}{3} 2^{|r|} K.$$

If $z \notin L$, then at most a $1/3$ of the valid transcripts are accepting, and

$$|C(z)| \leq \frac{1}{3} \sum_{\text{valid transcript } (r, x_1, \ldots, x_q)} \left\lceil \frac{K}{s(y_1) \ldots s(y_q)} \right\rceil$$

$$\leq \frac{1}{3} \sum_r (K+1) \sum_{(x_1, \ldots, x_q)} \frac{1}{s(y_1) \ldots s(y_q)}$$

$$\leq \frac{1}{3} \sum_r K \left(\sum_{(x_1, \ldots, x_q)} \frac{1}{s(y_1) \ldots s(y_q)} + \sum_{(r, x_1, \ldots, x_q)} 1 \right)$$

$$\leq \frac{1}{3} 2^{|r|} (K + 2^{q\ell})$$

$$\leq \frac{1}{2} 2^{|r|} K$$

by our choice of K. □

Using the set lower bound protocol of Goldwasser and Sipser [GS86], we conclude that L is in AM. Applying the same argument to the reduction $\overline{R}^?$ that outputs the opposite answer of $R^?$, it follows that L is also in coAM.

Deterministic Inversion Oracles. We now prove Theorem 1 for size-verifiable functions and deterministic inversion oracles. Assume $R^?$ is an efficient reduction from L to inverting f with respect to deterministic inversion oracles. Then, for every inversion oracle I for f, R^I decides L with probability at least 2/3. By averaging, it follows that for every distribution $\mathcal{I}$ on inversion oracles I for f, R^I decides L with probability at least 2/3:

$$\Pr_{r, I \sim \mathcal{I}}[R^I(z; r) = L(z)] \geq \tfrac{2}{3} \quad \text{for every } z.$$

If the oracle U could be written as a probability distribution over deterministic inversion oracles for f, then Theorem 1 would follow immediately from Claims 1 and 1. Unfortunately this is not the case: One reason is that a deterministic oracle sampled from any distribution always produces the same answer to the same query, while the oracle U outputs statistically independent answers. We resolve this difficulty by applying a minor modification to the description of U: The modified oracle U' will choose among the answers to a query y using randomness coming from a random function F applied to y. Specifically, if $x_1, \ldots, x_{s(y)}$ are the possible inverses of y, then $U'(y) = x_{F(y)}$.

Proof of Theorem 1 for size-verifiable functions. Let $\ell(n)$ and $q(n)$ be polynomial, efficiently computable upper bounds on the query length and query complexity of the reduction on inputs of length n, respectively. Let $\mathcal{F} = \{F_m\}$ be a collection of random functions, where F_m takes as input a string $y \in \{0, 1\}^m$ and outputs a number between 1 and $s(y)$. We define the randomized oracle U' as follows:

- **Randomness:** For every query length m, choose a uniformly random F_m, independently of $F_1, \ldots, F_{m-1}$.
- **Functionality:** On input y of length m, output $\perp$ if y is not in the range of f, or $U'(y) = x_{F_m(y)}$ if it is, where $x_1, \ldots, x_{s(y)}$ are the inverses of y under f.

Observe that U' is determined by a product distribution over $F_1, F_2, \ldots$ and any fixing of $F_1, F_2, \ldots$ specifies a deterministic inversion oracle for f. Since, for every z, the event $R^{U'}(z; r) = L(z)$ is measurable both over r and over $(F_1, F_2, \ldots)$, by averaging

$$\Pr_{r, (F_1, F_2, \ldots) \sim \mathcal{F}}[R^{U'}(z; r) = L(z)] \geq \tfrac{2}{3} \quad \text{for every } z.$$

We may now assume, without loss of generality, that $R^{U'}$ never makes the same query twice to the oracle U'. (More formally, we replace $R^?$ by another reduction that memoizes answers to previously made queries, and possibly makes some dummy queries at the end to ensure the number of queries is exactly $q(n)$ on inputs of length n.) We define $C(z)$ as before. Claims 1 and 1 still hold, and so L is in $\text{AM} \cap \text{coAM}$. $\qquad\square$

Extension to Approximately Size-verifiable Functions. Consider the promise problem C', whose YES instances are the same as the YES instances of C, and whose NO instances consist of the $(z, r, x_1, \ldots, x_q, p)$ for which either $(r, x_1, \ldots, x_q)$ is not an accepting valid transcript of $R^{U'}(z)$ or $p > \lceil \frac{6}{5} K / s(y_1) \ldots s(y_q) \rceil$, where $K = \frac{10}{3} 2^{q\ell}$. We now prove the analogues of Claims 1 and 1. We observe that the Goldwasser-Sipser lower bound protocol extends to AM-promise problems and conclude, as before, that L must be in $\mathrm{AM} \cap \mathrm{coAM}$.

Claim. C' is in AM.

Proof. On input $(z, r, x_1, \ldots, x_q, p)$, the AM verifier for C' runs the reduction on input z with randomness r and checks that for for each query y_i that the reduction makes, the answer x_i is indeed a pre-image of y_i and that the reduction accepts. It then asks the prover to provide claims $\hat{s}_i$ for the values $s(y_i), 1 \le i \le q$, runs the AM proof for A_f on input $(y_i, \hat{s}_i, 1^{6q})$, and verifies that $p \le \lceil K / \hat{s}_1 \ldots \hat{s}_q \rceil$. Clearly the verifier accepts YES instances of C'. If $(z, r, x_1, \ldots, x_q, p)$ is a NO instance, then either the transcript is not valid and accepting, or $f(x_i) \ne y_i$ for some i, or $\lceil K / \hat{s}_1 \ldots \hat{s}_q \rceil \ge p > \lceil \frac{6}{5} K / s(y_1) \ldots s(y_q) \rceil$, in which case $s(y_i) > (6/5)^{1/q} \hat{s}_i > (1 + 1/(6q)) \hat{s}_i$ for some i and the verifier for A_f rejects.   □

Let $C'_{\mathrm{YES}}(z)$ and $C'_{\mathrm{NO}}(z)$ consist of those $(r, x_1, \ldots, x_q, p)$ such that $(z, r, x_1, \ldots, x_q, p)$ are YES and NO instances of C', respectively.

Claim. If $z \in L$, then $C'_{\mathrm{YES}}(z)$ has size at least $\frac{2}{3} 2^{|r|} K$. If $z \notin L$, then $\overline{C'_{\mathrm{NO}}(z)}$ has size at most $\frac{1}{2} 2^{|r|} K$, where $\overline{C'_{\mathrm{NO}}(z)}$ denotes all tuples $(z, r, x_1, \ldots, x_q, p)$ that are not in $C'_{\mathrm{NO}}(z)$.

Proof. The proof of the first part is identical to the proof of the first part of Claim 1. For the second part, if $z \notin L$, then by a similar calculation

$$|\overline{C'_{\mathrm{NO}}(z)}| \le \frac{1}{3} \sum_{\text{valid transcript } (r, x_1, \ldots, x_q)} \left\lceil \frac{6K/5}{s(y_1) \ldots s(y_q)} \right\rceil$$
$$\le \frac{1}{3} 2^{|r|} (\tfrac{6}{5} K + 2^{q\ell}) \le \frac{1}{2} 2^{|r|} K.$$

□

Conclusion

In this work we show that counting the number of possible (suitably padded) transcripts from an interaction between a reduction and an inverter for a size-verifiable function is essentially a #P problem. The value of this problem can be approximated in AM using the Goldwasser-Sipser protocol. Alternatively, we can view this protocol as a proof-assisted sampler for an approximately uniformly random transcript.

Akavia et al.'s attempted proof of Theorem 1 is also based on the idea of sampling a transcript from a fixed distribution. Instead of sampling the transcript "globally" as we do, they instantiate a variant of the Goldwasser-Sipser protocol separately for every answer provided by the inverter. Such a protocol would have unbounded round complexity; to obtain a (constant-round) AM proof system, the protocol messages are reordered and grouped. While the samples produced by the Goldwasser-Sipser protocol are close to the desired distribution for each answer, their true distribution is affected by the prover's choices. The adaptive nature of the reduction allows the prover to exercise enough choice to end up with an atypical transcript. In contrast, when the transcript is sampled globally, it is guaranteed to be close to typical.

Acknowledgements. We thank Oded Goldreich and Shafi Goldwasser for helpful comments on the presentation.

References

[AGGM06] Akavia, A., Goldreich, O., Goldwasser, S., Moshkovitz, D.: On basing one-way functions on NP-hardness. In: Proceedings of the Thirty-eighth Annual ACM Symposium on Theory of Computing, STOC 2006, pp. 701–710. ACM, New York (2006)

[AGGM10] Akavia, A., Goldreich, O., Goldwasser, S., Moshkovitz, D.: Erratum for: On basing one-way functions on NP-hardness. In: Proceedings of the Forty-second ACM Symposium on Theory of Computing, STOC 2010, pp. 795–796. ACM, New York (2010)

[GS86]   Goldwasser, S., Sipser, M.: Private coins versus public coins in interactive proof systems. In: Proceedings of the Eighteenth Annual ACM Symposium on Theory of Computing, STOC 1986, pp. 59–68. ACM, New York (1986)

[IL89]   Impagliazzo, R., Luby, M.: One-way functions are essential for complexity based cryptography (extended abstract). In: 30th Annual Symposium on Foundations of Computer Science, Research Triangle Park, North Carolina, USA, October- 30 November 1, pp. 230–235. IEEE Computer Society (1989)

The Randomized Iterate, Revisited - Almost Linear Seed Length PRGs from a Broader Class of One-Way Functions

Yu Yu[1,2], Dawu Gu[1], Xiangxue Li[3], and Jian Weng[4]

[1] Department of Computer Science and Engineering,
Shanghai Jiao Tong University, China
[2] State Key Laboratory of Information Security, Institute of Information Engineering,
Chinese Academy of Sciences, Beijing, China
{yyuu,dwgu}@sjtu.edu.cn
[3] Department of Computer Science and Technology,
East China Normal University, China
xxli@cs.ecnu.edu.cn
[4] College of Information Science and Technology, Jinan University, China
cryptjweng@gmail.com

Abstract. We revisit "the randomized iterate" technique that was originally used by Goldreich, Krawczyk, and Luby (SICOMP 1993) and refined by Haitner, Harnik and Reingold (CRYPTO 2006) in constructing pseudorandom generators (PRGs) from regular one-way functions (OWFs). We abstract out a technical lemma (which is folklore in leakage resilient cryptography), and use it to provide a simpler and more modular proof for the Haitner-Harnik-Reingold PRGs from regular OWFs.

We introduce a more general class of OWFs called "weakly-regular one-way functions" from which we construct a PRG of seed length $O(n \cdot \log n)$. More specifically, consider an arbitrary one-way function f with range divided into sets $\mathcal{Y}_1, \mathcal{Y}_2, \ldots, \mathcal{Y}_n$ where each $\mathcal{Y}_i \overset{\text{def}}{=} \{y : 2^{i-1} \leq |f^{-1}(y)| < 2^i\}$. We say that f is weakly-regular if there exists a (not necessarily efficient computable) cut-off point max such that $\mathcal{Y}_{\max}$ is of some noticeable portion (say n^{-c} for constant c), and $\mathcal{Y}_{\max+1}, \ldots, \mathcal{Y}_n$ only sum to a negligible fraction. We construct a PRG by making $\tilde{O}(n^{2c+1})$ calls to f and achieve seed length $O(n \cdot \log n)$ using bounded space generators. This generalizes the approach of Haitner et al., where regular OWFs fall into a special case for $c = 0$. We use a proof technique that is similar to and extended from the method by Haitner, Harnik and Reingold for hardness amplification of regular weakly-one-way functions.

Our work further explores the feasibility and limits of the "randomized iterate" type of black-box constructions. In particular, the underlying f can have an arbitrary structure as long as the set of images with maximal preimage size has a noticeable fraction. In addition, our construction is much more seed-length efficient and security-preserving (albeit less general) than the HILL-style generators where the best known construction by Vadhan and Zheng (STOC 2012) requires seed length $\tilde{O}(n^3)$.

Y. Dodis and J.B. Nielsen (Eds.): TCC 2015, Part I, LNCS 9014, pp. 7–35, 2015.

1   Introduction

That one-way functions (OWFs) imply pseudorandom generators (PRGs) [13] is one of the central results upon which modern cryptography is successfully founded. The problem dates back to the early 80's when Blum, Micali [2] and Yao [19] independently observed that a PRG (often referred to as the BMY generator) can be efficiently constructed from one-way permutations (OWPs). That is, given a OWP f on n-bit input x and its hardcore predicate h_c (e.g., by Goldreich and Levin [8]), a single invocation of f already implies a PRG $g : x \mapsto (f(x), h_c(x))$ with a stretch[1] of $\Omega(\log n)$ bits and it extends to arbitrary stretch by repeated iterations (seen by a hybrid argument). Unfortunately, the BMY generator does not immediately apply to an arbitrary OWF since the output of f might be of too small amount of entropy to be secure for subsequent iterations.

THE RANDOMIZED ITERATE - PRGS FROM SPECIAL OWFS. Goldreich, Krawczyk, and Luby [7] extended the BMY generator by inserting a randomized operation (using k-wise independent hash functions) into every two applications of f, from which they built a PRG of seed length $O(n^3)$ assuming that the underlying OWF is known-regular[2]. Haitner, Harnik and Reingold [11] further refined the approach (for which they coined the name "the randomized iterate") as in Figure 1 below, where in between every i^{th} and $(i+1)^{th}$ iterations a random pairwise-independent

$$x_1 \xrightarrow{f} y_1 \xrightarrow{h_1} x_2 \xrightarrow{f} y_2 \xrightarrow{h_2} \cdots x_k \xrightarrow{f} y_k \xrightarrow{h_k}$$

Fig. 1. An illustration of the randomized iterate

hash function h_i is applied. Haitner et al. [11] showed that, when f is instantiated with any (possibly unknown) regular one-way function, it is hard to invert any k^{th} iterate (i.e., recovering any x_k s.t. $f(x_k) = y_k$) given y_k and the description of the hash functions. This gives a PRG of seed length $O(n^2)$ by running the iterate $n+1$ times and outputting a hardcore bit at every iteration. The authors of [10] further derandomize the PRG by generating all the hash functions from bounded space generators (e.g., Nisan's generator [17]) using a seed of length $O(n \log n)$. Although the randomized iterate is mostly known for construction of PRGs from regular OWFs, the authors of [10] also introduced many other interesting applications such as linear seed length PRGs from any exponentially hard regular OWFs, $O(n^2)$ seed length PRGs from any exponentially hard OWFs, $O(n^7)$ seed length PRGs from any OWFs, and hardness amplification of regular weakly-OWFs. Dedic, Harnik and Reyzin [3] showed that the amount of secret randomness can be reduced to achieve tighter reductions, i.e., if a regular one-way function f has 2^k images then the amount of secret randomness needed is k (instead of n bits). Yu et al. [21] further

---

[1] The stretch of a PRG refers to the difference between output and input lengths (see Definition 3).

[2] A function $f(x)$ is regular if every image has the same number (say α) of preimages, and it is known- (resp., unknown-) regular if α is efficiently computable (resp., inefficient to approximate) from the security parameter.

reduced the seed length of the PRG (based on any regular OWFs) to $O(\omega(1)\cdot n)$ for any efficiently computable $\omega(1)$.

THE HILL APPROACH - PRGs FROM ANY OWFs. Håstad, Impagliazzo, Levin and Luby (HILL) [13] gave the seminal result that pseudorandom generators can be constructed from any one-way functions. Nevertheless, they only gave a complicated (and not practically efficient) construction of PRG with seed length $\tilde{O}(n^{10})$ and sketched another one with seed length $\tilde{O}(n^8)$, which was formalized and proven in [14]. Haitner, Reingold, and Vadhan [12] introduced the notion of next-block pseudoentropy, and gave a construction of seed length $\tilde{O}(n^4)$. Vadhan and Zheng [18] further reduced the seed length of the uniform construction to $\tilde{O}(n^3)$, which is the current state-of-the-art.

A SUMMARY. The randomized iterate has advantages (over the HILL approach) such as shorter (almost linear) seed length and tighter reductions, but it remains open if the approach can be further generalized[3] (i.e., to go beyond regular one-way functions). In this paper, we answer this question by introducing a more general class of one-way functions and giving a construction based on the randomized iterate that enjoys seed length $O(n \cdot \log n)$ and tighter reductions.

A TECHNICAL LEMMA. First, we abstract out a technical lemma from [10] (see Lemma 1) that, informally speaking, "if any algorithm wins a one-sided game (e.g., inverting a OWF) on uniformly sampled challenges only with some negligible probability, then it cannot do much better (beyond a negligible advantage) in case that the challenges are sampled from any distribution of logarithmic Rényi entropy deficiency[4]". In fact, this lemma was implicitly known in leakage-resilient cryptography. Analogous observations were made in similar settings [1,5,4], where either the game is two-sided (e.g., indistinguishability applications) or the randomness is sampled from slightly defected min-entropy source. Plugging this lemma into [10] immediately yields a simpler proof for the key lemma of [10] (see Lemma 2), namely, "any k^{th} iterate (instantiated with a regular OWF) is hard-to-invert". The rationale is that y_k has sufficiently high Rényi entropy (even conditioned on the hash functions) that is only logarithmically less than the ideal case where y_k is uniform (over the range of f) and independent of the hash functions, which is hard to invert by the one-way-ness assumption.

THE MAIN RESULTS. We introduce a class of one-way functions called weakly-regular one-way functions. Consider an arbitrary OWF $f : \{0,1\}^n \to \{0,1\}^l$ with range divided into sets $\mathcal{Y}_1, \ldots, \mathcal{Y}_n$, where each $\mathcal{Y}_i \stackrel{\text{def}}{=} \{y : 2^{i-1} \leq |f^{-1}(y)| < 2^i\}$ and $|f^{-1}(y)|$ refers to preimage size of y (i.e., the number of images that

---

[3] The randomized iterate handles almost-regular one-way functions as well and this generalization is not hard to see (implicit in [10,21]). Similarly, the construction we introduced in this paper only needs "weakly-almost-regular one-way functions" (of which almost-regular one-way functions fall into a special case). See Remark 1 for some discussions.

[4] The Rényi entropy deficiency of a random variable W over set $\mathcal{W}$ refers to the difference between entropies of $U_\mathcal{W}$ and W, i.e., $\log |\mathcal{W}| - \mathbf{H}_2(W)$, where $U_\mathcal{W}$ denotes the uniform distribution over $\mathcal{W}$ and $\mathbf{H}_2(W)$ is the Rényi entropy of W.

map to y under f). We say that f is weakly-regular if there exists an integer function max $=$ max(n) such that $\mathcal{Y}_{\max}$ is of some noticeable portion (n^{-c} for constant c), and $\mathcal{Y}_{\max+1}$, ..., $\mathcal{Y}_n$ only sum to a negligible fraction $\epsilon(n)$. Note that regular one-way functions fall into a special case for $c = 0$, $\epsilon(n) = 0$ and arbitrary (not necessarily efficient) function max$(\cdot)$. We give a construction that only requires the knowledge about c (i.e., oblivious of max and ϵ). Informally speaking, as illustrated in Figure 1, the main idea is that at each kth round conditioned on $y_k \in \mathcal{Y}_{\max}$ the Rényi entropy of y_k given the hash functions is close to the ideal case where $f(U_n)$ hits $\mathcal{Y}_{\max}$ with noticeable probability (and is independent of the hash functions), which is hard to invert. We have by the pairwise independence (in fact, universality already suffices) of the hash functions that every $y_k \in \mathcal{Y}_{\max}$ is an independent event of probability n^{-c}. By a Chernoff bound, running the iterate $\Delta = n^{2c} \cdot \omega(\log n)$ times yields that with overwhelming probability there is at least one occurrence of $y_k \in \mathcal{Y}_{\max}$, which implies every Δ iterations are hard-to-invert, i.e., for any $j = \mathsf{poly}(n)$ it is hard to predict $x_{1+(j-1)\Delta}$ given $y_{j\Delta}$ and the hash functions. A PRG follows by outputting $\log n$ hardcore bits for every Δ iterations and in total making $\tilde{O}(n^{2c+1})$ calls to f. This requires seed length $\tilde{O}(n^{2c+2})$, and can be pushed to $O(n \cdot \log n)$ bits using bounded space generators [17,16], ideas borrowed from [10] with more significant reductions in seed length (we reduce by factor $\tilde{O}(n^{2c+1})$ whereas [10] saves factor $\tilde{O}(n)$). Overall, our technique is similar in spirit to the hardness amplification of regular weakly-one-way[5] functions introduced by Haitner et al. in the same paper [10]. Roughly speaking, the idea was that for any inverting algorithm A, a weakly one-way function has a set that A fails upon (the failing-set of A), and thus sufficiently many iterations are bound to hit every such failing-set to yield a strongly-one-way function (that is hard-to-invert on an overwhelming fraction). However, in our case the lack of a regular structure for the underlying function and the negligible fraction (i.e., $\mathcal{Y}_{\max+1}$, ..., $\mathcal{Y}_n$) further complicate the analysis (see Remark 2 for some discussions), and we make our best effort to provide an intuitive and modular proof.

ON THE EFFICIENCY, FEASIBILITY AND LIMITS. From the application point of view, known-regular one-way functions may already suffice for the following reasons:

1. If a one-way function behaves like a random function, then it is known(-almost)-regular. In other words, most functions are known(-almost)-regular (see Lemma 8 in Appendix C).
2. In practice, many one-way function candidates turn out to be known-regular or even 1-to-1. For example, Goldreich, Levin and Nisan [9] showed that 1-to-1 one-way functions can be based on concrete intractable problems such as RSA and DLP.

---

[5] We should not confuse "weakly-regular" with "weakly-one-way", where the former "weakly" describes regularity (i.e., regular on a noticeable fraction as in Definition 4) and the latter is used for one-way-ness (i.e., hard-to-invert on a noticeable fraction [19]).

It is folklore (see, e.g., [6,21]) that pseudorandom generators can be constructed almost optimally from known(-almost)-regular one-way functions, i.e., with seed length $O(n \cdot \omega(1))$ and $O(\omega(1))$ non-adaptive OWF calls for any efficiently computable super-constant $\omega(1)$. Despite the aforementioned, the study on minimizing the knowledge required for the underlying one-way functions (and at the same time improving the efficiency of the resulting pseudorandom generator) is of theoretical significance, and it improves our understanding about feasibility and limits of black-box reductions. In particular, Holenstein and Sinha [15] showed that $\Omega(n/\log n)$ black-box calls to an arbitrary (including unknown-regular) one-way function is necessary to construct a PRG, and Haitner, Harnik and Reingold [10] gave an explicit construction (from unknown-regular one-way functions) of seed length $O(n \cdot \log n)$ that matches this bound. In the most general setting, Håstad et al. [13] established the principle feasibility result that pseudorandom generators can based on any one-way functions but the current state-of-the-art [18] still requires seed length $\tilde{O}(n^3)$. We take a middle course by introducing weakly-regular one-way functions that lie in between regular one-way functions and arbitrary ones, and giving a construction of pseudorandom generator with seed length $O(n \cdot \log n)$ and using tighter reductions. We refer to the appendix and the full version of this work [20] for missing details, proofs omitted and a discussion in Appendix C about the gap between weakly one-way functions and arbitrary ones.

2 Preliminaries

2.1 Notations and Definitions

NOTATIONS. We use $[n]$ to denote set $\{1, \ldots, n\}$. We use capital letters (e.g., X, Y) for random variables, standard letters (e.g., x, y) for values, and calligraphic letters (e.g., $\mathcal{Y}$, $\mathcal{S}$) for sets. $|\mathcal{S}|$ denotes the cardinality of set $\mathcal{S}$. We use shorthand $\mathcal{Y}_{[n]} \stackrel{\text{def}}{=} \bigcup_{t=1}^{n} \mathcal{Y}_t$. For function $f : \{0,1\}^n \to \{0,1\}^{l(n)}$, we use shorthand $f(\{0,1\}^n) \stackrel{\text{def}}{=} \{f(x) : x \in \{0,1\}^n\}$, and denote by $f^{-1}(y)$ the set of y's preimages under f, i.e. $f^{-1}(y) \stackrel{\text{def}}{=} \{x : f(x) = y\}$. We use $s \leftarrow S$ to denote sampling an element s according to distribution S, and let $s \stackrel{\$}{\leftarrow} \mathcal{S}$ denote sampling s uniformly from set $\mathcal{S}$, and let $y := f(x)$ denote value assignment. We use U_n and $U_{\mathcal{X}}$ to denote uniform distributions over $\{0,1\}^n$ and $\mathcal{X}$ respectively, and let $f(U_n)$ be the distribution induced by applying function f to U_n. We use $\mathsf{CP}(X)$ to denote the collision probability of X, i.e., $\mathsf{CP}(X) \stackrel{\text{def}}{=} \sum_x \Pr[X = x]^2$, and denote by $\mathbf{H}_2(X) \stackrel{\text{def}}{=} -\log \mathsf{CP}(X)$ the Rényi entropy. We also define conditional Rényi entropy (and probability) of a random variable X conditioned on another random variable Z by

$$\mathbf{H}_2(X|Z) \stackrel{\text{def}}{=} -\log \left(\mathsf{CP}(X|Z) \right) \stackrel{\text{def}}{=} -\log \left(\mathbb{E}_{z \leftarrow Z} \left[\sum_x \Pr[X = x | Z = z]^2 \right] \right)$$

A function $\mathsf{negl} : \mathbb{N} \to [0,1]$ is negligible if for every constant c we have $\mathsf{negl}(n) < n^{-c}$ holds for all sufficiently large n's, and a function $\mu : \mathbb{N} \to [0,1]$ is called noticeable if there exists constant c such that $\mu(n) \geq n^{-c}$ for all sufficiently large n's.

We define the *computational distance* between distribution ensembles $X \overset{\text{def}}{=} \{X_n\}_{n\in\mathbb{N}}$ and $Y \overset{\text{def}}{=} \{Y_n\}_{n\in\mathbb{N}}$, denoted by $\mathsf{CD}_{T(n)}(X,Y) \leq \varepsilon(n)$, if for every probabilistic distinguisher D of running time $T(n)$ it holds that

$$\left| \Pr[\mathsf{D}(1^n, X_n) = 1] - \Pr[\mathsf{D}(1^n, Y_n) = 1] \right| \leq \varepsilon(n) \ .$$

The *statistical distance* between X and Y, denoted by $\mathsf{SD}(X,Y)$, is defined by

$$\mathsf{SD}(X,Y) \overset{\text{def}}{=} \frac{1}{2} \sum_x |\Pr[X = x] - \Pr[Y = x]| = \mathsf{CD}_\infty(X,Y)$$

We use $\mathsf{SD}(X,Y|Z)$ (resp. $\mathsf{CD}_T(X,Y|Z)$) as shorthand for $\mathsf{SD}((X,Z),(Y,Z))$ (resp. $\mathsf{CD}_T((X,Z),(Y,Z))$).

SIMPLIFYING ASSUMPTIONS AND NOTATIONS. To simplify the presentation, we make the following assumptions without loss of generality. It is folklore that one-way functions can be assumed to be length-preserving (see [11] for full proofs). Throughout, most parameters are functions of the security parameter n (e.g., $T(n)$, $\varepsilon(n)$, $\alpha(n)$) and we often omit n when clear from the context (e.g., T, ε, α). By notation $f : \{0,1\}^n \to \{0,1\}^l$ we refer to the ensemble of functions $\{f_n : \{0,1\}^n \to \{0,1\}^{l(n)}\}_{n\in\mathbb{N}}$. As slight abuse of notion, poly might be referring to the set of all polynomials or a certain polynomial, and h might be either a function or its description, which will be clear from the context.

Definition 1 (pairwise independent hashing). *A family of hash functions* $\mathcal{H} \overset{\text{def}}{=} \{h : \{0,1\}^n \to \{0,1\}^m\}$ *is **pairwise independent** if for any $x_1 \neq x_2 \in \{0,1\}^n$ and any $v \in \{0,1\}^{2m}$ it holds that*

$$\Pr_{h \overset{\$}{\leftarrow} \mathcal{H}} [(h(x_1), h(x_2)) = v] = 2^{-2m}$$

or equivalently, $(H(x_1), H(x_2))$ is i.i.d. to U_{2m} where H is uniform over $\mathcal{H}$. It is well known that there are efficiently computable families of pairwise independent hash functions of description length $\Theta(n + m)$.

Definition 2 (one-way functions). *A function $f : \{0,1\}^n \to \{0,1\}^{l(n)}$ is $(T(n),\varepsilon(n))$-one-way if f is polynomial-time computable and for any probabilistic algorithm A of running time $T(n)$*

$$\Pr_{y \leftarrow f(U_n)} [\mathsf{A}(1^n, y) \in f^{-1}(y)] \leq \varepsilon(n).$$

We say that f is a (strongly) one-way function if $T(n)$ and $1/\varepsilon(n)$ are both super-polynomial in n.

Definition 3 (pseudorandom generators [2,19]). *A deterministic function $g : \{0,1\}^n \to \{0,1\}^{n+s(n)}$ ($s(n) > 0$) is a $(T(n),\varepsilon(n))$-secure PRG with stretch $s(n)$ if g is polynomial-time computable and*

$$\mathsf{CD}_{T(n)}(\ g(1^n, U_n)\ ,\ U_{n+s(n)}\ ) \leq \varepsilon(n).$$

We say that g is a pseudorandom generator if $T(n)$ and $1/\varepsilon(n)$ are both super-polynomial in n.

Definition 4 (weakly-regular one-way functions). *Let $f : \{0,1\}^n \to \{0,1\}^l$ be a one-way function. For every $n \in \mathbb{N}$ divide range $f(\{0,1\}^n)$ into sets $\mathcal{Y}_1,\ldots,\mathcal{Y}_n$ (i.e., $\mathcal{Y}_1 \cup \ldots \cup \mathcal{Y}_n = f(\{0,1\}^n)$) where $\mathcal{Y}_j \stackrel{\text{def}}{=} \{y : 2^{j-1} \le |f^{-1}(y)| < 2^j\}$ for every $1 \le j \le n$. We say that f is **weakly-regular** if there exist constant c, integer function $\max = \max(n)$, and negligible function $\epsilon = \epsilon(n)$ such that the following holds for all sufficiently large n's :*

$$\Pr[f(U_n) \in \mathcal{Y}_{\max}] \ge n^{-c} \ , \tag{1}$$

$$\Pr[\, f(U_n) \in (\, \mathcal{Y}_{\max+1} \cup \mathcal{Y}_{\max+2} \cup \ldots \cup \mathcal{Y}_n \,)\,] \le \epsilon \ , \tag{2}$$

Note that $\max(\cdot)$ can be arbitrary (not necessarily efficient) functions and thus regular one-way functions fall into a special case for $c = 0$.

Remark 1 (on further categorization and generalization.). We can further divide the above class of functions into **weakly-known-regular** and **weakly-unknown-regular** one-way functions depending on whether $\max(\cdot)$ is efficiently computable or not. This is however not necessary since our construction needs no knowledge about $\max(\cdot)$ and thus handles any weakly-regular one-way functions. In fact, our construction only assumes that f is **weakly-almost-regular**, i.e., for some $d = d(n) \in O(\log n)$ it holds that

$$\Pr[f(U_n) \in (\mathcal{Y}_{\max-d} \cup \mathcal{Y}_{\max-d+1} \cup \ldots \cup \mathcal{Y}_{\max})] \ge n^{-c}$$

instead of (1), where almost-regular one-way functions become a special case for $c = 0$. We mainly give the proof under the assumption of Definition 4 for neatness, and sketch how to adapt the proof to the weakly-almost-regular case in Remark 3 (see Appendix B).

2.2 Technical Tools

Theorem 1 (Goldreich-Levin Theorem [8]). *Let (X,Y) be a distribution ensemble over $\{\{0,1\}^n \times \{0,1\}^{\mathrm{poly}(n)}\}_{n\in\mathbb{N}}$. Assume that for any PPT algorithm A of running time $T(n)$ it holds that*

$$\Pr[\, \mathsf{A}(1^n, Y) = X \,] \ \le \ \varepsilon(n) \ .$$

Then, for any efficiently computable $m = m(n) \le n$, there exists an efficient function family $\mathcal{H}_c \stackrel{\text{def}}{=} \{h_c : \{0,1\}^n \to \{0,1\}^m\}$ of description size $\Theta(n)^6$, such that

$$\mathsf{CD}_{T'(n)}(\, H_c(X) \, , \, U_m \, | \, Y, H_c) \ \in \ O(2^m \cdot (n \cdot \varepsilon)^{\frac{1}{3}}) \ ,$$

where $T'(n) = T(n) \cdot (\varepsilon(n)/n)^{O(1)}$, and H_c is the uniform distribution over $\mathcal{H}_c$.

---

[6] For example (see [8]), we can use an $m \times n$ Toeplitz matrix $a_{m,n}$ to describe the function family, i.e., $\mathcal{H}_c \stackrel{\text{def}}{=} \{h_c(x) \stackrel{\text{def}}{=} a_{m,n} \cdot x, \ x \in \{0,1\}^n, a_{m,n} \in \{0,1\}^{m+n-1}\}$.

Definition 5 (bounded-width layered branching program - LBP). *An (s, k, v)-LBP M is a finite directed acyclic graph whose nodes are partitioned into $k + 1$ layers indexed by $\{1, \ldots, k + 1\}$. The first layer has a single node (the source), the last layer has two nodes (sinks) labeled with 0 and 1, and each of the intermediate layers has up to 2^s nodes. Each node in the $i \in [k]$ layer has exactly 2^v outgoing labeled edges to the $(i + 1)^{th}$ layer, one for every possible string $h_i \in \{0, 1\}^v$.*

An equivalent (and perhaps more intuitive) model to the above is bounded space computation. That is, we assign labels to graph nodes (instead of associating them with the edges), at each ith layer the program performs arbitrary computation on the current node (labelled by s-bit string) and the current v-bit input h_i, advances (and assigns value) to a node in the $(i+1)^{th}$ layer, and repeats until it reaches the last layer to produce the final output bit.

Theorem 2 (bounded-space generator [17,16]). *Let $s = s(n), k = k(n), v = v(n) \in \mathbb{N}$ and $\varepsilon = \varepsilon(n) \in (0, 1)$ be polynomial-time computable functions. Then, there exist a polynomial-time computable function $q = q(n) \in \Theta(v + (s + \log(k/\varepsilon)) \log k)$ and a generator $BSG : \{0, 1\}^q \rightarrow \{0, 1\}^{k \cdot v}$ that runs in time $\mathrm{poly}(s, k, v, \log(1/\varepsilon))$, and ε-fools every (s, k, v)- LBP M, i.e.,*

$$| \Pr[M(U_{k \cdot v}) = 1] - \Pr[M(BSG(U_n)) = 1] | \leq \varepsilon .$$

3 Pseudorandom Generators from Regular One-Way Functions

3.1 A Technical Lemma

Before we revisit the randomized iterate based on regular one-way functions, we introduce a technical lemma that simplifies the analysis in [10] and is also used to prove our main theorem in Section 4. Informally, it states that if any one-sided game (one-way functions, MACs, and digital signatures) is (T, ε)-secure on uniform secret randomness, then it will be $(T, \sqrt{2^{e+2} \cdot \varepsilon})$-secure when the randomness is sampled from any distribution with e bits of Rényi entropy deficiency.

Lemma 1 (one-sided game on imperfect randomness). *For any $e \leq m \in \mathbb{N}$, let $\mathcal{W} \times \mathcal{Z}$ be any set with $|\mathcal{W}| = 2^m$, let $\mathsf{Adv} : \mathcal{W} \times \mathcal{Z} \rightarrow [0, 1]$ be any (deterministic) real-valued function, let (W, Z) be any joint random variables over set $\mathcal{W} \times \mathcal{Z}$ satisfying $\mathbf{H}_2(W|Z) \geq m - e$, we have*

$$\mathbb{E}[\mathsf{Adv}(W, Z)] \leq \sqrt{2^{e+2} \cdot \mathbb{E}[\mathsf{Adv}(U_{\mathcal{W}}, Z)]} \tag{3}$$

where $U_{\mathcal{W}}$ denotes uniform distribution over $\mathcal{W}$ (independent of Z and any other distributions in consideration).

Proof. For any given δ, define $\mathcal{S}_\delta \stackrel{\text{def}}{=} \{(w, z) : \Pr[W = w|Z = z] \geq 2^{-(m-e)}/\delta\}$.

$$2^{-(m-e)} \geq \sum_z \Pr[Z = z] \sum_w \Pr[W = w|Z = z]^2$$

$$\geq \sum_z \Pr[Z = z] \sum_{w:(w,z)\in\mathcal{S}_\delta} \Pr[W = w|Z = z]\cdot 2^{-(m-e)}/\delta$$

$$\geq (2^{-(m-e)}/\delta) \cdot \Pr[(W, Z) \in \mathcal{S}_\delta] \ ,$$

and thus $\Pr[(W, Z) \in \mathcal{S}_\delta] \leq \delta$. It follows that

$$\mathbb{E}[\mathsf{Adv}(W, Z)] = \sum_{(w,z)\in\mathcal{S}_\delta} \Pr[(W, Z) = (w, z)] \cdot \mathsf{Adv}(w, z)$$

$$+ \sum_{(w,z)\notin\mathcal{S}_\delta} \Pr[Z = z] \cdot \Pr[W = w|Z = z] \cdot \mathsf{Adv}(w, z)$$

$$\leq \sum_{(w,z)\in\mathcal{S}_\delta} \Pr[(W, Z) = (w, z)]$$

$$+ (2^e/\delta) \cdot \sum_{(w,z)\notin\mathcal{S}_\delta} \Pr[Z = z]\cdot 2^{-m} \cdot \mathsf{Adv}(w, z)$$

$$\leq \delta + (2^e/\delta) \cdot \mathbb{E}[\mathsf{Adv}(U_\mathcal{W}, Z)] \ ,$$

and we complete the proof by setting $\delta = \sqrt{2^e \cdot \mathbb{E}[\mathsf{Adv}(U_\mathcal{W}, Z)]}$. $\square$

ON HOW TO USE THE LEMMA. One can think of $\mathsf{Adv}(w, z)$ as the advantage of any specific adversary conditioned on the challenge being w and the additional side information being z (e.g., hash functions that are correlated to the challenges). Thus, the left-hand of (3) gives the adversary's advantage on slightly defected random source in consideration, which is bounded by the ideal case on the right-hand of (3), namely, the advantage on uniformly sampled challenges, such as a uniform random $y \leftarrow f(U_n)$ (for some regular one-way function f) independent of the hash functions.

3.2   The Randomized Iterate

Definition 6 (the randomized iterate [10,7]). *Let $n \in \mathbb{N}$, let $f : \{0,1\}^n \rightarrow \{0,1\}^n$ be a length-preserving function, and let $\mathcal{H}$ be a family of pairwise independent length-preserving hash functions over $\{0,1\}^n$. For $k \in \mathbb{N}$, $x_1 \in \{0,1\}^n$ and vector $\boldsymbol{h}^k = (h_1, \ldots, h_k) \in \mathcal{H}^k$, recursively define the kth randomized iterate by:*

$$y_k = f(x_k), \ x_{k+1} = h_k(y_k)$$

For $k - 1 \leq t \in \mathbb{N}$, we denote the kth iterate by function f^k, i.e., $y_k = f^k(x_1, \boldsymbol{h}^t)$, where $\boldsymbol{h}^t$ is possibly redundant as y_k only depends on $\boldsymbol{h}^{k-1}$.

*The **randomized version** refers to the case where $x_1 \stackrel{\$}{\leftarrow} \{0,1\}^n$ and $\boldsymbol{h}^{k-1} \stackrel{\$}{\leftarrow} \mathcal{H}^{k-1}$.*

The **derandomized version** refers to that $x_1 \overset{\$}{\leftarrow} \{0,1\}^n$, $\boldsymbol{h}^{k-1} \leftarrow BSG(U_q)$, where $q \in O(n \cdot \log n)$, $BSG : \{0,1\}^q \to \{0,1\}^{(k-1)\cdot \log|\mathcal{H}|}$ is a bounded-space generator[7] that 2^{-2n}-fools every $(2n+1, k, \log|\mathcal{H}|)$-LBP, and $\log|\mathcal{H}|$ is the description length of $\mathcal{H}$ (e.g., $2n$ bits for concreteness).

Theorem 3 (PRGs from Regular OWFs [10]). *For $n \in \mathbb{N}, k \in [n+1]$, let f, $\mathcal{H}$, f^k and $BSG(\cdot)$ be as defined in Definition 6, and let $\mathcal{H}_c = \{h_c : \{0,1\}^n \to \{0,1\}\}$ be a family of Goldreich-Levin predicates, where $\mathcal{H}$ and $\mathcal{H}_c$ both have description length $\Theta(n)$. We define $G : \{0,1\}^n \times \mathcal{H}^n \times \mathcal{H}_c \to \{0,1\}^{n+1} \times \mathcal{H}^n \times \mathcal{H}_c$ and $G' : \{0,1\}^n \times \{0,1\}^{q(n)} \times \mathcal{H}_c \to \{0,1\}^{n+1} \times \{0,1\}^{q(n)} \times \mathcal{H}_c$ as below:*

$$G(x_1, \boldsymbol{h}^n, h_c) = (h_c(x_1), h_c(x_2), \ldots, h_c(x_{n+1}), \boldsymbol{h}^n, h_c),$$
$$G'(x_1, u, h_c) = G(x_1, BSG(u), h_c).$$

Assume that f is a regular (length-preserving) one-way function and that $BSG(\cdot)$, $\mathcal{H}$ and $\mathcal{H}_c$ are efficient. Then, G and G' are pseudorandom generators.

PROOF SKETCH OF THEOREM 3. It suffices to prove Lemma 2: for any $1 \le k \le n+1$, given y_k and the hash functions (either sampled uniformly or from bounded space generators), it is hard to recover any x_k s.t. $f(x_k) = y_k$. Then, Goldreich-Levin Theorem yields that each $h_c(x_k)$ is computationally unpredictable given y_k, which (together with $\boldsymbol{h}^n$) efficiently determines all the subsequent $h_c(x_{k+1})$, ..., $h_c(x_{n+1})$. We complete the proof by Yao's "next/previous bit unpredictability implies pseudorandomness" argument [19]. It thus remains to prove Lemma 2 below which summarizes the statements of Lemma 3.2, Lemma 3.4, Lemma 3.11 from [11], and we provide a simpler proof below via Lemma 1. □

Lemma 2 (the kth iterate is hard-to-invert). *For any $n \in \mathbb{N}, k \in [n+1]$, let f, $\mathcal{H}$, f^k be as defined in Definition 6. Assume that f is a (T, ε) regular one-way function, i.e., for every PPT A and A' of running time T it holds that*

$$\Pr [A(f(U_n), \boldsymbol{H}^n) \in f^{-1}(f(U_n))] \le \varepsilon .$$

$$\Pr [A'(f(U_n), U_q) \in f^{-1}(f(U_n))] \le \varepsilon .$$

Then, for every such A and A' it holds that

$$\Pr [A(Y_k, \boldsymbol{H}^n) \in f^{-1}(Y_k)] \le 2\sqrt{k \cdot \varepsilon} , \tag{4}$$

$$\Pr [A'(Y_k', U_q) \in f^{-1}(Y_k')] \le 2\sqrt{(k+1) \cdot \varepsilon} , \tag{5}$$

where $Y_k = f^k(X_1, \boldsymbol{H}^n)$, $Y_k' = f^k(X_1, BSG(U_q))$, X_1 is uniform over $\{0,1\}^n$ and $\boldsymbol{H}^n$ is uniform over $\mathcal{H}^n$.

---

[7] Such efficient generators exist by Theorem 2, setting $s(n) = 2n + 1$, $k(n) = \mathsf{poly}(n)$, $v(n) = \log|\mathcal{H}|$ and $\varepsilon(n) = 2^{-2n}$ and thus $q(n) = O(n \cdot \log n)$.

Proof. To apply Lemma 1, let $\mathcal{W} = f(\{0,1\}^n)$, $\mathcal{Z} = \mathcal{H}^n$, let $(W, Z) = (Y_k, \boldsymbol{H}^n)$, $U_{\mathcal{W}} = f(U_n)$, and define

$$\mathsf{Adv}(y, \boldsymbol{h}^n) \stackrel{\text{def}}{=} \begin{cases} 1, & \text{if } \mathsf{A}(y, \boldsymbol{h}^n) \in f^{-1}(y) \\ 0, & \text{if } \mathsf{A}(y, \boldsymbol{h}^n) \notin f^{-1}(y) \end{cases}$$

where A is assumed to be deterministic without loss of generality[8]. We have by Lemma 3 that

$$\mathbf{H}_2(Y_k \mid \boldsymbol{H}^n) \geq \mathbf{H}_2(f(U_n) \mid \boldsymbol{H}^n) - \log k$$

and thus Lemma 1 yields that

$$\Pr\left[\, \mathsf{A}(Y_k, \boldsymbol{H}^n) \in f^{-1}(Y_k) \,\right] \leq 2\sqrt{k \cdot \Pr\left[\, \mathsf{A}(f(U_n), \boldsymbol{H}^n) \in f^{-1}(f(U_n)) \,\right]}$$
$$\leq 2\sqrt{k \cdot \varepsilon} \ .$$

The proof for (5) is similar except for setting ($W = Y'_k$, $Z = U_q$) and letting $\mathsf{Adv}(y, u) = 1$ iff $\mathsf{A}'(y, u) \in f^{-1}(y)$. We have by Lemma 3 that

$$\mathbf{H}_2(Y'_k \mid U_q) \geq \mathbf{H}_2(f(U_n) \mid U_q) - \log(k+1)$$

and thus we apply Lemma 1 to get

$$\Pr\left[\, \mathsf{A}'(Y'_k, U_q) \in f^{-1}(Y'_k) \,\right] \leq 2\sqrt{(k+1) \cdot \Pr\left[\, \mathsf{A}'(f(U_n), U_q) \in f^{-1}(f(U_n)) \,\right]}$$
$$\leq 2\sqrt{(k+1) \cdot \varepsilon} \ . \qquad \square$$

The proof of Lemma 3 below appeared in [10], and we also include it in the full version [20] for completeness.

Lemma 3 (Rényi entropy conditions [10]). *For the same assumptions as in Lemma 2, it holds that*

$$\mathsf{CP}(f(U_n)) = \mathsf{CP}(f(U_n) \mid \boldsymbol{H}^n) = \mathsf{CP}(f(U_n) \mid U_q) = \frac{1}{|f(\{0,1\}^n)|} \ , \quad (6)$$

$$\mathsf{CP}(Y_k \mid \boldsymbol{H}^n) \leq \frac{k}{|f(\{0,1\}^n)|} \ , \quad (7)$$

$$\mathsf{CP}(Y'_k \mid U_q) \leq \frac{k+1}{|f(\{0,1\}^n)|} \ . \quad (8)$$

4 A More General Construction of Pseudorandom Generators

In this section we construct a pseudorandom generator with seed length $O(n \log n)$ from weakly-regular one-way functions (see Definition 4). We first show how to construct the PRG by running the iterate $\tilde{O}(n^{2c+1})$ times, and thus require large amount of randomness (i.e., $\tilde{O}(n^{2c+2})$ bits) to sample the hash functions. Then, we show the derandomized version where the amount of the randomness is compressed into $O(n \log n)$ bits using bounded space generators.

---

[8] If A is probabilistic, let $\mathsf{Adv}(y, \boldsymbol{h}^n) = \Pr[\mathsf{A}(y, \boldsymbol{h}^n) \in f^{-1}(y)]$, where probability is taken over the internal coins of A.

4.1   The Randomized Version: A PRG with Seed Length $\tilde{O}(n^{2c+2})$

Recall that any one-way function f can be assumed to be length-preserving without loss of generality.

Theorem 4 (the randomized version). *For $n, k \in \mathbb{N}$, assume that f is a weakly-regular one-way function (with c, max and ϵ as defined in Definition 4), let $\mathcal{H}$ and f^k be defined as in Definition 6, and let $\mathcal{H}_c = \{h_c : \{0,1\}^n \rightarrow \{0,1\}^{2\log n}\}$ be a family of Goldreich-Levin hardcore functions. Then, for any efficient $\alpha = \alpha(n) \in \omega(1)$, $\Delta = \Delta(n) = \alpha \cdot \log n \cdot n^{2c}$ and $r = r(n) = \lceil n/\log n \rceil$, the function $g{:}\{0,1\}^n \times \mathcal{H}^{r\Delta-1} \times \mathcal{H}_c \rightarrow \{0,1\}^{2n} \times \mathcal{H}^{r\Delta-1} \times \mathcal{H}_c$ defined as*

$$g(x_1, \boldsymbol{h}^{r\cdot\Delta-1}, h_c) = (\ h_c(x_1), h_c(x_{1+\Delta}), h_c(x_{1+2\Delta}), \ldots, \ h_c(x_{1+r\cdot\Delta}), \boldsymbol{h}^{r\cdot\Delta-1}, h_c\ ) \tag{9}$$

is a pseudorandom generator.
Notice that a desirable property is that a construction assuming a sufficiently large c works with any one-way function whose actual parameter is less than or equal to c.

Proof. The proof is similar to Theorem 3 based on Yao's hybrid argument [19]. Namely, the pseudorandomness of a sequence (with polynomially many blocks) is equivalent to that every block is pseudorandom conditioned on its suffix (or prefix). By the Goldreich-Levin Theorem and Lemma 4 below we know that every $h_c(x_{1+j\Delta})$ is pseudorandom conditioned on h_c, $y_{(j+1)\Delta}$ and $\boldsymbol{h}^{r\Delta-1}$, which efficiently implies all subsequent blocks $h_c(x_{1+(j+1)\Delta}), \ldots, h_c(x_{1+r\Delta})$. This completes the proof. □

Lemma 4 (every Δ iterations are hard-to-invert). *For $n, k \in \mathbb{N}$, let f be a weakly-regular (T,ε)-OWF (with c as defined in Definition 4), and let $\mathcal{H}$, f^k, $\alpha = \alpha(n)$, $\Delta = \Delta(n)$ and $r = r(n)$ be as defined in Theorem 4. Then, for every $j \in [r]$, and for every PPT A of running time $T(n) - n^{O(1)}$ (for some universal constant $O(1)$) it holds that*

$$\Pr_{x_1 \xleftarrow{\$} \{0,1\}^n, \ \boldsymbol{h}^{r\Delta-1} \xleftarrow{\$} \mathcal{H}^{r\Delta-1}} [\ \mathsf{A}(y_{j\cdot\Delta}, \ \boldsymbol{h}^{r\Delta-1}) = x_{1+(j-1)\Delta}\ ] \in O(\ n^{3c/2} \cdot r \cdot \Delta^2 \cdot \sqrt{\varepsilon}\ ). \tag{10}$$

PROOF SKETCH OF LEMMA 4 . Assume towards a contradiction that

$$\exists j^* \in [r], \exists\ \mathrm{PPT}\ \mathsf{A}:\ \Pr[\ \mathsf{A}(Y_{j^*\cdot\Delta}, \ \boldsymbol{H}^{r\Delta-1}) = X_{1+(j^*-1)\Delta}\ ] \geq \varepsilon_{\mathsf{A}} \tag{11}$$

for some non-negligible function $\varepsilon_{\mathsf{A}} = \varepsilon_{\mathsf{A}}(n)$. Then, we build an efficient algorithm M^{A} (see Algorithm 1) that invokes A and inverts f with probablity $\Omega(\varepsilon_{\mathsf{A}}^2/n^{3c} \cdot r^2 \cdot \Delta^4)$ (as shown in Lemma 6), which is a contradiction to the (T,ε)-one-wayness of f and thus completes the proof.

We define the events $\mathcal{E}_k$ and $\mathcal{S}_k$ in Definition 7 below, where $\mathcal{S}_k$ refers to that during the first k iterates no y_t ($1 \leq t \leq k$) hits the negligible fraction region (see Remark 2 in Appendix B for the underlying intuitions), and $\mathcal{E}_k$ defines the desirable event that y_k hits $\mathcal{Y}_{\max}$ (which implies the hard-to-invertness).

Definition 7 (events $\mathcal{S}_k$ and $\mathcal{E}_k$). *For any $n \in \mathbb{N}$ and any $k \leq r\Delta$, define events*

$$\mathcal{S}_k \stackrel{\text{def}}{=} \left((X_1, \boldsymbol{H}^{r\Delta-1}) \in \{(x_1, \boldsymbol{h}^{r\Delta-1}) : \forall t \in [k] \, satisfies \, f^t(x_1, \boldsymbol{h}^{r\Delta-1}) \in \mathcal{Y}_{[\max]} \} \right)$$

$$\mathcal{E}_k \stackrel{\text{def}}{=} \left((X_1, \boldsymbol{H}^{r\Delta-1}) \in \{ (x_1, \boldsymbol{h}^{r\Delta-1}) : y_k \in \mathcal{Y}_{\max}, where \, y_k = f^k(x_1, \boldsymbol{h}^{r\Delta-1}) \} \right)$$

where $\mathcal{Y}_{[\max]} = \mathcal{Y}_1 \cup \ldots \cup \mathcal{Y}_{\max}$ and $(X_1, \boldsymbol{H}^{r\Delta-1})$ is uniform distribution over $\{0,1\}^n \times \mathcal{H}^{r\Delta-1}$. We also naturally extend the definition of collision probability conditioned on $\mathcal{E}_k$ and $\mathcal{S}_k$. For example, $\mathsf{CP}(Y_k \wedge \mathcal{E}_k \wedge \mathcal{S}_k | \boldsymbol{H}^{r\Delta-1}) \stackrel{\text{def}}{=}$
$$\mathbb{E}_{\boldsymbol{h}^{r\Delta-1} \leftarrow \boldsymbol{H}^{r\Delta-1}} \left[\sum_y \Pr[f^k(X_1, \boldsymbol{H}^{r\Delta-1}) = y \wedge \mathcal{E}_k \wedge \mathcal{S}_k | \boldsymbol{H}^{r\Delta-1} = \boldsymbol{h}^{r\Delta-1}]^2 \right]$$
and $\mathsf{CP}(Y_k, \boldsymbol{H}^{r\Delta-1} | \mathcal{E}_k \wedge \mathcal{S}_k) \stackrel{\text{def}}{=} \sum_{(y, \boldsymbol{h}^{r\Delta-1})} \Pr[(f^k(X_1, \boldsymbol{H}^{r\Delta-1}), \boldsymbol{H}^{r\Delta-1}) = (y, \boldsymbol{h}^{r\Delta-1}) | \mathcal{E}_k \wedge \mathcal{S}_k]^2$.

Claim 1. *For any $n \in \mathbb{N}$, let $\mathcal{S}_k$ and $\mathcal{E}_k$ be as defined in Definition 7, assume that f is weakly-regular (with c, ϵ and $\max$ defined as in (1) and (2)). Then, it holds that*

$$\forall k \in [r\Delta] : \Pr[\mathcal{S}_k] \geq 1 - k\epsilon, \quad \Pr[\mathcal{E}_k] \geq n^{-c}, \quad \Pr[\mathcal{E}_k \wedge \mathcal{S}_k] \geq n^{-c}/2 \quad (12)$$

$$\forall k \in \mathbb{N} : \Pr[\mathcal{E}_{k+1} \vee \mathcal{E}_{k+2} \vee \ldots \vee \mathcal{E}_{k+\Delta}] \geq 1 - \exp^{\Delta/n^{2c}} \geq 1 - n^{-\alpha} \quad (13)$$

$$\forall k \in [r\Delta] : \mathsf{CP}(Y_k \wedge \mathcal{E}_k \wedge \mathcal{S}_k | \boldsymbol{H}^{r\Delta-1}) \leq r\Delta \cdot 2^{\max-n+1}, \quad (14)$$

where $Y_k = f^k(X_1, \boldsymbol{H}^{r\Delta-1})$.

Proof. We have that $x_1, x_2 = h_1(y_1), \ldots, x_{r\Delta} = h_{r\Delta-1}(y_{r\Delta-1})$ are all i.i.d. to U_n due to the universality of $\mathcal{H}$. This implies that $\Pr[y_i \in \mathcal{Y}_{[\max]}] \geq 1 - \epsilon$ for every $i \in [k]$ independently, and that $\mathcal{E}_1, \ldots$ and $\mathcal{E}_{r\Delta}$ are i.i.d. events with probability at least n^{-c}. The former further implies

$$\Pr[\mathcal{S}_k] \geq (1 - \epsilon)^k \geq 1 - k \cdot \epsilon,$$

where the second inequality is due to Fact 2 (see Appendix A). Thus, we complete the proof for (12) by

$$\Pr[\mathcal{E}_k \wedge \mathcal{S}_k] \geq \Pr[\mathcal{E}_k] - \Pr[\neg \mathcal{S}_k] \geq n^{-c} - k \cdot \epsilon \geq n^{-c}/2.$$

For every $k \in \mathbb{N}$, $i \in [\Delta]$, define $\zeta_{k+i} = 1$ iff $\mathcal{E}_{k+i}$ occurs (and $\zeta_{k+i} = 0$ otherwise). It follows by a Chernoff-Hoeffding bound that

$$\forall k \in \mathbb{N} : \Pr[(\neg\mathcal{E}_{k+1}) \wedge \ldots \wedge (\neg\mathcal{E}_{k+\Delta})] = \Pr[\sum_{i=1}^{\Delta} \zeta_{k+i} = 0] \leq \exp^{-\Delta/n^{2c}} \leq n^{-\alpha},$$

which yields (13) by taking a negation. For the collision probability in (14), we consider two instances of the randomized iterate seeded with independent x_1 and x_1' and a common random $h^{r\Delta-1}$ and thus:

$$\text{CP}(\,Y_k\,\wedge\,\mathcal{E}_k\wedge\mathcal{S}_k\mid H^{r\Delta-1})\;\leq\;\text{CP}(\,Y_k\,\wedge\,\mathcal{S}_k\mid H^{r\Delta-1})$$

$$\leq\quad\Pr_{x_1,x_1'\xleftarrow{\$}\{0,1\}^n}\;[f(x_1)=f(x_1')\in\mathcal{Y}_{[\max]}]$$

$$+\;\sum_{t=2}^{k}\left(\Pr_{y_{t-1}\neq y_{t-1}',\;h_{t-1}\xleftarrow{\$}\mathcal{H}}\;[f(x_t)=f(x_t')\in\mathcal{Y}_{[\max]}]\right)$$

$$\leq\;r\Delta\sum_{y\in\mathcal{Y}_{[\max]}}\Pr[f(U_n)=y]^2$$

$$\leq\;r\Delta\sum_{i=1}^{\max}\sum_{y\in\mathcal{Y}_i}\Pr[f(U_n)=y]\cdot2^{i-n}\;=\;r\Delta\sum_{i=1}^{\max}\Pr[f(U_n)\in\mathcal{Y}_i]\cdot2^{i-n}$$

$$\leq\;r\Delta\cdot2^{\max-n}(1+2^{-1}+\ldots+2^{-(\max-1)})\;\leq\;r\Delta\cdot2^{\max-n+1}\;,$$

where we omit $\mathcal{E}_k$ in the first inequality (since we are considering upper bound), the second inequality is due to that the collision probability is upper bounded by the sum of events that the first collision occurs on points $y_1, y_2, \ldots, y_k \in \mathcal{Y}_{[\max]}$ respectively, and the third inequality follows from the pairwise independence of $\mathcal{H}$ so that $x_1, x_1', \ldots, x_k$ and x_k' are i.i.d. to U_n. $\square$

Lemma 5. *For any $n \in \mathbb{N}$, with the same assumptions and notations as in Theorem 4, Definition 4 and Definition 7, and let $j^* \in [r]$, A, ε_A be as assumed in (11). Then, there exists $i^* \in [\Delta]$ such that*

$$\Pr[\,\mathsf{A}(Y_{j^*\cdot\Delta},\,H^{r\Delta-1})=X_{1+(j^*-1)\Delta}\,\wedge\,\mathcal{E}_{(j^*-1)\Delta+i^*}\wedge\mathcal{S}_{(j^*-1)\Delta+i^*}\,]\;\geq\;\varepsilon_\mathsf{A}/2\Delta\;. \tag{15}$$

Proof. For notational convenience use shorthand $\mathcal{C}$ for the event $\mathsf{A}(Y_{j^*\cdot\Delta},\,H^{r\Delta-1})$ $= X_{1+(j^*-1)\Delta}$. Then,

$$\sum_{i=1}^{\Delta}\Pr[\mathcal{C}\wedge\mathcal{E}_{(j^*-1)\Delta+i}\wedge\mathcal{S}_{(j^*-1)\Delta+i}]$$

$$\geq\;\sum_{i=1}^{\Delta}\Pr[\mathcal{C}\wedge\mathcal{E}_{(j^*-1)\Delta+i}\wedge\mathcal{S}_{r\Delta}]$$

$$\geq\;\Pr[\,\mathcal{C}\wedge\mathcal{S}_{r\Delta}\wedge(\bigvee_{i=1}^{\Delta}\mathcal{E}_{(j^*-1)\Delta+i})\,]$$

$$\geq\;\Pr[\,\mathcal{C}\,]\;-\;\Pr[\,\neg\mathcal{S}_{r\Delta}]\;-\;\Pr[\neg(\bigvee_{i=1}^{\Delta}\mathcal{E}_{(j^*-1)\Delta+i})]$$

$$\geq\;\varepsilon_\mathsf{A}\;-\;r\Delta\cdot\epsilon\;-\;n^{-\alpha}\;\geq\;\varepsilon_\mathsf{A}/2\;,$$

where the first inequality is due to $\mathcal{S}_{r\Delta} \subseteq \mathcal{S}_{\kappa}$ for any $\kappa \leq r\Delta$, the second inequality is the union bound, and the fourth follows from (12) and (13). We recall that ϵ and $n^{-\alpha}$ are both negligible in n. Thus, there exists i^* (that satisfies (15)) by an averaging argument. □

THE INTUITION FOR M^{A}. Lemma 5 states that there exist some i^* and j^* conditioned on which A inverts the iterate with non-negligible probability. If we knew which i^* and j^*, then we simply replace $y_{(j^*-1)\Delta+i^*}$ with $f(U_n)$, simulate the iterate for the rest iterations and invoke A to invert f. Although the distribution after the replacement will not be identical to the original one, we use Lemma 1 to argue that the Rényi entropy deficiency is small enough and thus the inverting probability will not blow up by more than a polynomial factor. However, we actually do not know the values of i^* and j^*, so we need to randomly sample i and j over $[\Delta]$, $[r]$ respectively. This yields M^{A} as defined in Algorithm 1.

Algorithm 1. M^{A}

Input: $y \in \{0,1\}^n$

Sample $j \overset{\$}{\leftarrow} [r]$, $i \overset{\$}{\leftarrow} [\Delta]$, $\boldsymbol{h}^{r\Delta-1} \overset{\$}{\leftarrow} \mathcal{H}^{r\Delta-1}$;

Let $\tilde{y}_{(j-1)\Delta+i} := y$;

FOR $k = (j-1)\Delta + i + 1$ TO $(j-1)\Delta + \Delta$

   Compute $\tilde{x}_k := h_{k-1}(\tilde{y}_{k-1})$, $\tilde{y}_k := f(\tilde{x}_k)$;

$\tilde{x}_{(j-1)\Delta+1} \leftarrow \mathsf{A}(\tilde{y}_{j\Delta}, \boldsymbol{h}^{r\Delta-1})$;

FOR $k = (j-1)\Delta + 1$ TO $(j-1)\Delta + i - 1$

   Compute $\tilde{y}_k := f(\tilde{x}_k)$, $\tilde{x}_{k+1} := h_k(\tilde{y}_k)$;

Output: $\tilde{x}_{(j-1)\Delta+i}$

Lemma 6 (M^{A} inverts f). *For any $n \in \mathbb{N}$, let A be as assumed in Lemma 5 and let M^{A} be as defined in Algorithm 1. Then, it holds that*

$$\Pr_{y \leftarrow f(U_n); j \overset{\$}{\leftarrow} [r]; i \overset{\$}{\leftarrow} [\Delta]; \boldsymbol{h}^{r\Delta-1} \overset{\$}{\leftarrow} \mathcal{H}^{r\Delta-1}} [\mathsf{M}^{\mathsf{A}}(y; j, i, \boldsymbol{h}^{r\Delta-1}) \in f^{-1}(y)] \geq \frac{\varepsilon_{\mathsf{A}}^2}{2^9 \cdot n^3 c r^2 \Delta^4}.$$

Proof. We know by Lemma 5 that there exist $j^* \in [r]$ and $i^* \in [\Delta]$ satisfying (15), which implies

$\Pr[\mathsf{M}^{\mathsf{A}}(Y_{(j-1)\Delta+i}; j, i, \boldsymbol{H}^{r\Delta-1}) \in f^{-1}(Y_{(j-1)\Delta+i}) \mid (j,i) = (j^*, i^*)$

$\qquad\qquad\qquad\qquad\qquad\qquad \wedge \, \mathcal{E}_{(j-1)\Delta+i} \wedge \mathcal{S}_{(j-1)\Delta+i} \,]$

$\geq \; \Pr[\, \mathsf{A}(Y_{j^* \cdot \Delta}, \boldsymbol{H}^{r\Delta-1}) = X_{1+(j^*-1)\Delta} \mid \mathcal{E}_{(j^*-1)\Delta+i^*} \wedge \mathcal{S}_{(j^*-1)\Delta+i^*} \,]$

$\geq \; \Pr[\, \mathsf{A}(Y_{j^* \cdot \Delta}, \boldsymbol{H}^{r\Delta-1}) = X_{1+(j^*-1)\Delta} \, \wedge \, \mathcal{E}_{(j^*-1)\Delta+i^*} \wedge \mathcal{S}_{(j^*-1)\Delta+i^*} \,]$

$\geq \; \varepsilon_{\mathsf{A}}/2\Delta \; ,$

where the second inequality, in abstract form, is $\Pr[\mathcal{E}_a|\mathcal{E}_b] \geq \Pr[\mathcal{E}_a|\mathcal{E}_b] \cdot \Pr[\mathcal{E}_b] = \Pr[\mathcal{E}_a \wedge \mathcal{E}_b]$. The above is not exactly what we need as conditioned on $\mathcal{E}_{(j^*-1)\Delta+i^*} \wedge$

$\mathcal{S}_{(j^*-1)\Delta+i^*}$, the random variable $(Y_{(j^*-1)\Delta+i^*}, \boldsymbol{H}^{r\Delta-1})$ is not uniform over $\mathcal{Y}_{\max} \times \mathcal{H}^{r\Delta-1}$. However, we show below that it has nearly full Rényi entropy over $\mathcal{Y}_{\max} \times \mathcal{H}^{r\Delta-1}$, i.e.,

$$
\begin{aligned}
& \mathsf{CP}(\ (Y_{(j^*-1)\Delta+i^*}, \boldsymbol{H}^{r\Delta-1}) \mid \mathcal{E}_{(j^*-1)\Delta+i^*} \wedge \mathcal{S}_{(j^*-1)\Delta+i^*}\ ) \\
={}& \frac{\mathsf{CP}(\ (Y_{(j^*-1)\Delta+i^*}, \boldsymbol{H}^{r\Delta-1}) \wedge \mathcal{E}_{(j^*-1)\Delta+i^*} \wedge \mathcal{S}_{(j^*-1)\Delta+i^*})}{\Pr[\mathcal{E}_{(j^*-1)\Delta+i^*} \wedge \mathcal{S}_{(j^*-1)\Delta+i^*}]^2} \\
\leq{}& \frac{\mathsf{CP}(\ Y_{(j^*-1)\Delta+i^*} \wedge \mathcal{E}_{(j^*-1)\Delta+i^*} \wedge \mathcal{S}_{(j^*-1)\Delta+i^*} \mid \boldsymbol{H}^{r\Delta-1})}{(n^{-2c}/4) \cdot |\mathcal{H}|^{r\Delta-1}} \\
\leq{}& \frac{r\Delta \cdot 2^{\max-n+1}}{(n^{-2c}/4) \cdot |\mathcal{H}|^{r\Delta-1}} = \frac{8r\Delta \cdot n^{2c}}{2^{n-\max} \cdot |\mathcal{H}|^{r\Delta-1}} ,
\end{aligned}
$$

where the first equality follows from Fact 1 (see Appendix A) and the two inequalities are by (12) and (14) respectively. Taking a negative logarithm, we get $\mathbf{H}_2(\ (Y_{(j^*-1)\Delta+i^*}, \boldsymbol{H}^{r\Delta-1}) \mid \mathcal{E}_{(j^*-1)\Delta+i^*} \wedge \mathcal{S}_{(j^*-1)\Delta+i^*}\ ) \geq (n - \max + (r\Delta - 1)\log|\mathcal{H}| + 1) - e$, where entropy deficiency $e \leq 2c \cdot \log n + \log r + \log \Delta + 4$. This is due to that the uniform distribution over $\mathcal{Y}_{\max} \times \mathcal{H}^{r\Delta-1}$ has full entropy

$$
\begin{aligned}
\mathbf{H}_2(\ (U_{\mathcal{Y}_{\max}}, \boldsymbol{H}^{r\Delta-1})\ ) &\leq \log(\frac{1}{2^{-n+\max-1}} \cdot |\mathcal{H}|^{r\Delta-1}) \\
&= n - \max + (r\Delta - 1)\log|\mathcal{H}| + 1 .
\end{aligned}
$$

To apply Lemma 1, let $\mathcal{W} = \mathcal{Y}_{\max} \times \mathcal{H}^{r\Delta-1}$, $\mathcal{Z} = \emptyset$, let W be $(Y_{(j^*-1)\Delta+i^*}, \boldsymbol{H}^{r\Delta-1})$ conditioned on $\mathcal{E}_{(j^*-1)\Delta+i^*}$ and $\mathcal{S}_{(j^*-1)\Delta+i^*}$, and define

$$
\mathsf{Adv}(y, \boldsymbol{h}^{r\Delta-1}) \stackrel{\mathrm{def}}{=} \begin{cases} 1, & \text{if } \mathsf{M}^{\mathsf{A}}(y; j^*, i^*, \boldsymbol{h}^{r\Delta-1}) \in f^{-1}(y) \\ 0, & \text{if } \mathsf{M}^{\mathsf{A}}(y; j^*, i^*, \boldsymbol{h}^{r\Delta-1}) \notin f^{-1}(y) \end{cases}
$$

Let $\mathcal{C}_{j^*i^*\max}$ denote the event that $(j, i) = (j^*, i^*) \wedge f(U_n) \in \mathcal{Y}_{\max}$, and we thus have

$$
\begin{aligned}
& \Pr[\mathsf{M}^{\mathsf{A}}(f(U_n); j, i, \boldsymbol{H}^{r\Delta-1}) \in f^{-1}(f(U_n))] \\
\geq{}& \Pr[\mathcal{C}_{j^*i^*\max}] \cdot \Pr[\mathsf{M}^{\mathsf{A}}(f(U_n); j, i, \boldsymbol{H}^{r\Delta-1}) \in f^{-1}(f(U_n)) \mid \mathcal{C}_{j^*i^*\max}] \\
\geq{}& (1/r\Delta n^c) \cdot \mathbb{E}[\ \mathsf{Adv}(U_{\mathcal{Y}_{\max}}, \boldsymbol{H}^{r\Delta-1})\ ]/2 \\
\geq{}& (1/r\Delta n^c) \cdot \frac{\mathbb{E}[\ \mathsf{Adv}(Y_{(j^*-1)\Delta+i^*}, \boldsymbol{H}^{r\Delta-1}) \mid \mathcal{E}_{(j^*-1)\Delta+i^*} \wedge \mathcal{S}_{(j^*-1)\Delta+i^*}\ ]^2}{2^{e+3}} \\
\geq{}& (1/r\Delta n^c) \cdot \frac{\varepsilon_{\mathsf{A}}^2/4\Delta^2}{2^7 \cdot n^{2c}r \cdot \Delta} = \frac{\varepsilon_{\mathsf{A}}^2}{2^9 \cdot n^{3c} \cdot r^2 \cdot \Delta^4} ,
\end{aligned}
$$

where the second inequality is due to Claim 2 (i.e., conditioned on $f(U_n) \in \mathcal{Y}_{\max}$ random variable $f(U_n)$ can be loosely regarded as $U_{\mathcal{Y}_{\max}}$), and the third inequality follows from Lemma 1. □

4.2 The Derandomized Version: A PRG with Seed Length $O(n \cdot \log n)$

The derandomized version uses a bounded-space generator to expand an $O(n \cdot \log n)$-bit u into a long string over $\mathcal{H}^{r\Delta-1}$ (rather than sampling a random element over it).

Theorem 5 (the derandomized version). *For $n, k \in \mathbb{N}$, let f, c, $\mathcal{H}$, $\mathcal{H}_c$, f^k, $\alpha = \alpha(n)$, $\Delta = \Delta(n)$ and $r = r(n)$ be as assumed in Theorem 4, let g be as defined in (9), let*

$$BSG : \{0,1\}^{q(n) \in O(n \cdot \log n)} \to \{0,1\}^{(\alpha \cdot n^{2c+1} - 1) \cdot \log |\mathcal{H}|}$$

be a bounded-space generator that 2^{-2n}-fools every $(2n + 1,\ \alpha \cdot n^{2c+1},\ \log |\mathcal{H}|)$- LBP (see Footnote 7). Then, the function $g' : \{0,1\}^n \times \{0,1\}^q \times \mathcal{H}_c \to \{0,1\}^{2n} \times \{0,1\}^q \times \mathcal{H}_c$ defined as

$$g'(x_1, u, h_c) = g(x_1, BSG(u), h_c) \tag{16}$$

is a pseudorandom generator.

Similar to the randomized version, it suffices to show Lemma 7 (the counterpart of Lemma 4).

Lemma 7. *For the same assumptions as stated in Lemma 4, we have that for every $j \in [r]$, and for every PPT A' of running time $T(n) - n^{O(1)}$ (for some universal constant $O(1)$) it holds that*

$$\Pr_{x_1 \overset{\$}{\leftarrow} \{0,1\}^n,\ u \overset{\$}{\leftarrow} \{0,1\}^q} [\ \mathsf{A}'(y'_{j \cdot \Delta},\ u) = x'_{1+(j-1)\Delta}\ ] \in O(\ n^{3c/2} \cdot r \cdot \Delta^2 \cdot \sqrt{\varepsilon}\ )\ , \tag{17}$$

where $h'^{r\Delta - 1} := BSG(u)$, $y'_k = f^k(x_1, h'^{r\Delta - 1})$ and $x'_{k+1} = h'_k(y'_k)$ for $k \in \mathbb{N}$.

The proof of Lemma 7 follows the steps of that of Lemma 4. We define events $\mathcal{S}'_k$ and $\mathcal{E}'_k$ in Definition 8 (the analogues of $\mathcal{S}_k$ and $\mathcal{E}_k$). Although the events (e.g., $\mathcal{E}'_1, \ldots, \mathcal{E}'_k$) are not independent due to short of randomness, we still have (18), (19) and (20) below. We defer their proofs to Appendix A, where for every inequality we define an LBP and argue that the advantage of the LBP on $H^{r\Delta - 1}$ and $BSG(U_q)$ is bounded by 2^{-2n} and thus (18), (19) and (20) follow from their respective counterparts (12), (13) and (14) by adding an additive term 2^{-2n}.

Definition 8 (events $\mathcal{S}'_k$ and $\mathcal{E}'_k$). *For any $n \in \mathbb{N}$ and any $k \leq r\Delta$, define events*

$$\mathcal{S}'_k \overset{\text{def}}{=} \left((X_1, U_q) \in \{ (x_1, u) : \forall t \in [k] \text{ satisfies } f^t(x_1, BSG(u)) \in \mathcal{Y}_{[\max]} \} \right)$$

$$\mathcal{E}'_k \overset{\text{def}}{=} \left((X_1, U_q) \in \{ (x_1, u) : y'_k \in \mathcal{Y}_{\max}\ , \text{ where } y'_k = f^k(x_1, BSG(u)) \} \right)$$

where (X_1, U_q) is uniform distribution over $\{0,1\}^n \times \{0,1\}^q$. We refer to Definition 9 in Appendix B for the definitions of the collision probabilities in the following proofs.

$$\forall k \in [r\Delta] : \Pr[\mathcal{S}'_k] \geq 1 - k\epsilon - 2^{-2n}\ ,\ \ \Pr[\mathcal{E}'_k] \geq n^{-c} - 2^{-2n},\ \ \Pr[\mathcal{E}'_k \wedge \mathcal{S}'_k] \geq n^{-c}/2 \tag{18}$$

$$\forall k \in [(r-1)\Delta] : \Pr[\mathcal{E}'_{k+1} \vee \mathcal{E}'_{k+2} \vee \ldots \vee \mathcal{E}'_{k+\Delta}] \geq 1 - n^{-\alpha} - 2^{-2n} \tag{19}$$

$$\forall k \in [r\Delta] : \mathsf{CP}(\ Y'_k \wedge \mathcal{E}'_k \wedge \mathcal{S}'_k \mid U_q) \ \leq \ (r\Delta + 1)\cdot 2^{\max - n + 1} \qquad (20)$$

where $Y'_k = f^k(X_1, BSG(U_q))$.

PROOF SKETCH OF LEMMA 7. Assume towards a contradiction that for some non-negligible $\varepsilon_{A'} = \varepsilon_{A'}(n)$ that

$$\exists j^* \in [r], \exists\ \mathsf{PPT}\ \mathsf{A'} : \ \Pr[\ \mathsf{A'}(Y'_{j^*\cdot\Delta},\ U_q) = X'_{1+(j^*-1)\Delta}\ ] \ \geq\ \varepsilon_{A'}\ , \qquad (21)$$

where for $k \in [r\Delta]$ we use notations $\boldsymbol{H}'^{r\Delta - 1} = BSG(U_q)$, $Y'_k = f^k(X_1, \boldsymbol{H}'^{r\Delta - 1})$ and $X'_{k+1} = H'_k(Y'_k)$. Then, we define $\mathsf{M}^{\mathsf{A'}}$ that inverts f with the following probability. Since $\mathsf{M}^{\mathsf{A'}}$ is quite similar to its analogue M^{A} we state it as Algorithm 2 in Appendix B.

$$\Pr_{y \leftarrow f(U_n);\ j \xleftarrow{\$} [r];\ i \xleftarrow{\$} [\Delta];\ u \xleftarrow{\$} \{0,1\}^q} [\ \mathsf{M}^{\mathsf{A'}}(y;\ j, i,\ u) \in f^{-1}(y)\ ] \ \in\ \Omega(\frac{\varepsilon_{A'}^2}{n^{3c}\cdot r^2 \cdot \Delta^4})\ ,$$

$$(22)$$

which is a contradiction to the one-way-ness of f and thus concludes Lemma 7.

PROOF SKETCH OF (22). Denote by $\mathcal{C}'$ the event $\mathsf{A'}(Y'_{j^*\cdot\Delta},\ U_q) = X'_{1+(j^*-1)\Delta}$. Then,

$$\sum_{i=1}^{\Delta} \Pr[\mathcal{C}' \wedge \mathcal{E}'_{(j^*-1)\Delta+i} \wedge \mathcal{S}'_{(j^*-1)\Delta+i}] \ \geq\ \sum_{i=1}^{\Delta} \Pr[\mathcal{C}' \wedge \mathcal{E}'_{(j^*-1)\Delta+i} \wedge \mathcal{S}'_{r\Delta}]$$

$$\geq\ \Pr[\ \mathcal{C}' \wedge \mathcal{S}'_{r\Delta} \wedge (\bigvee_{i=1}^{\Delta} \mathcal{E}'_{(j^*-1)\Delta+i})\ ]$$

$$\geq\ \Pr[\ \mathcal{C}'] \ -\ \Pr[\ \neg \mathcal{S}'_{r\Delta}] \ -\ \Pr[\neg(\bigvee_{i=1}^{\Delta} \mathcal{E}'_{(j^*-1)\Delta+i})]$$

$$\geq\ \varepsilon_{A'} \ -\ r\Delta\cdot\epsilon \ -\ n^{-\alpha} \ -\ 2^{-2n+1} \ \geq\ \varepsilon_{A'}/2\ ,$$

where the first three inequalities are similar to analogues in the proof of Lemma 5 and the fourth inequality is due to (18) and (19). Thus, by averaging we have that there exist $\exists j^* \in [r]$ and $\exists i^* \in [\Delta]$ such that

$$\Pr[\ \mathsf{A'}(Y'_{j^*\cdot\Delta},\ U_q) = X'_{1+(j^*-1)\Delta} \ \wedge \ \mathcal{E}'_{(j^*-1)\Delta+i^*} \ \wedge \ \mathcal{S}'_{(j^*-1)\Delta+i^*}\ ] \ \geq\ \varepsilon_{A'}/2\Delta.$$

The proofs below follow the steps of Lemma 6. We have that (proof of (23) given in Appendix A)

$$\boldsymbol{H}_2(\ (Y'_{(j^*-1)\Delta+i^*}, U_q)\ \mid \mathcal{E}'_{(j^*-1)\Delta+i^*} \wedge \mathcal{S}'_{(j^*-1)\Delta+i^*}\ ) \geq \boldsymbol{H}_2(\ U_{\mathcal{Y}_{\max}}, U_q\ ) \ -\ e\ ,$$

$$(23)$$

where entropy deficiency $e \leq 2c \cdot \log n + \log r + \log \Delta + 5$. Finally, let $\mathcal{W} = \mathcal{Y}_{\max} \times \{0,1\}^q$, $\mathcal{Z} = \emptyset$, let W be $(Y'_{(j^*-1)\Delta+i^*}, U_q)$ conditioned on $\mathcal{E}'_{(j^*-1)\Delta+i^*}$ and $\mathcal{S}'_{(j^*-1)\Delta+i^*}$, and define

$$\mathsf{Adv}(y, u) \overset{\text{def}}{=} \begin{cases} 1, & \text{if } \mathsf{M}^{\mathsf{A'}}(y;\ j^*, i^*,\ u) \in f^{-1}(y) \\ 0, & \text{if } \mathsf{M}^{\mathsf{A'}}(y;\ j^*, i^*,\ u) \notin f^{-1}(y) \end{cases}$$

Let $\mathcal{C}_{j^*i^*\max}$ denote the event that $(j,i) = (j^*, i^*) \wedge f(U_n) \in \mathcal{Y}_{\max}$, and we thus have

$$\Pr[\mathsf{M}^{\mathsf{A}'}(f(U_n); j,i,U_q) \in f^{-1}(f(U_n))]$$

$$\geq \Pr[\mathcal{C}_{j^*i^*\max}] \cdot \Pr[\mathsf{M}^{\mathsf{A}'}(f(U_n); j,i,U_q) \in f^{-1}(f(U_n)) \mid \mathcal{C}_{j^*i^*\max}]$$

$$\geq (1/r\Delta n^c) \cdot \mathbb{E}[\mathsf{Adv}(U_{\mathcal{Y}_{\max}}, U_q)]/2$$

$$\geq (1/r\Delta n^c) \cdot \frac{\mathbb{E}[\mathsf{Adv}(Y'_{(j^*-1)\Delta+i^*}, U_q) \mid \mathcal{E}'_{(j^*-1)\Delta+i^*} \wedge \mathcal{S}'_{(j^*-1)\Delta+i^*}]^2}{2^{e+3}}$$

$$\geq (1/r\Delta n^c) \cdot \frac{\Pr[A'(Y'_{j^*\cdot\Delta}, U_q) = X'_{1+(j^*-1)\Delta} \wedge \mathcal{E}'_{(j^*-1)\Delta+i^*} \wedge \mathcal{S}'_{(j^*-1)\Delta+i^*}]^2}{2^{e+3}}$$

$$\geq (1/r\Delta n^c) \cdot \frac{\varepsilon_{\mathsf{A}'}^2/4\Delta^2}{2^8 \cdot n^{2c}r \cdot \Delta} = \frac{\varepsilon_{\mathsf{A}'}^2}{2^{10} \cdot n^{3c} \cdot r^2 \cdot \Delta^4}.$$

where the inequalities follow the same order as their analogues in the proof of Lemma 6.

Acknowledgement. This research work was supported by the National Basic Research Program of China (Grant 2013CB338004). Yu Yu was supported by the National Natural Science Foundation of China Grant (Nos. 61472249, 61103221). Dawu Gu was supported by the National Natural Science Foundation of China Grant (Nos. 61472250, 61402286), the Doctoral Fund of Ministry of Education of China (No. 20120073110094) and the Innovation Program by Shanghai Municipal Science and Technology Commission (No. 14511100300). Xiangxue Li was supported by the National Natural Science Foundation of China (Nos. 61472472, 61272536, 61172085) and Science and Technology Commission of Shanghai Municipality (Grant 13JC1403500). Jian Weng was supported by the National Science Foundation of China (Nos. 61272413, 61133014, 61472165), the Fok Ying Tung Education Foundation (No. 131066), the Program for New Century Excellent Talents in University (No. NCET-12-0680), and the Research Fund for the Doctoral Program of Higher Education of China (No. 20134401110011).

References

1. Barak, B., Dodis, Y., Krawczyk, H., Pereira, O., Pietrzak, K., Standaert, F.-X., Yu, Y.: Leftover hash lemma, revisited. In: Rogaway, P. (ed.) CRYPTO 2011. LNCS, vol. 6841, pp. 1–20. Springer, Heidelberg (2011)
2. Blum, M., Micali, S.: How to generate cryptographically strong sequences of pseudorandom bits. In: Proceedings of the 23rd IEEE Symposium on Foundation of Computer Science (FOCS 1982), pp. 112–117 (1982)
3. Dedić, N., Harnik, D., Reyzin, L.: Saving private randomness in one-way functions and pseudorandom generators. In: Canetti, R. (ed.) TCC 2008. LNCS, vol. 4948, pp. 607–625. Springer, Heidelberg (2008)
4. Dodis, Y., Pietrzak, K., Wichs, D.: Key derivation without entropy waste. In: Nguyen, P.Q., Oswald, E. (eds.) EUROCRYPT 2014. LNCS, vol. 8441, pp. 93–110. Springer, Heidelberg (2014)

5. Dodis, Y., Yu, Y.: Overcoming weak expectations. In: Sahai, A. (ed.) TCC 2013. LNCS, vol. 7785, pp. 1–22. Springer, Heidelberg (2013)
6. Goldreich, O.: Foundations of Cryptography: Basic Tools. Cambridge University Press (2001)
7. Goldreich, O., Krawczyk, H., Luby, M.: On the existence of pseudorandom generators. SIAM Journal on Computing 22(6), 1163–1175 (1993)
8. Goldreich, O., Levin, L.A.: A hard-core predicate for all one-way functions. In: Proceedings of the 21st Annual ACM Symposium on Theory of Computing (STOC 1989), pp. 25–32 (1989)
9. Goldreich, O., Levin, L.A., Nisan, N.: On constructing 1-1 one-way functions. In: Goldreich, O. (ed.) Studies in Complexity and Cryptography. LNCS, vol. 6650, pp. 13–25. Springer, Heidelberg (2011), http://dx.doi.org/10.1007/978-3-642-22670-0_3
10. Haitner, I., Harnik, D., Reingold, O.: On the power of the randomized iterate. In: Dwork, C. (ed.) CRYPTO 2006. LNCS, vol. CRYPTO 2006, pp. 22–40. Springer, Heidelberg (2006)
11. Haitner, I., Harnik, D., Reingold, O.: On the power of the randomized iterate. SIAM Journal on Computing 40(6), 1486–1528 (2011), http://www.cs.tau.ac.il/ iftachh/papers/RandomizedIteate/RandomIterate.pdf
12. Haitner, I., Reingold, O., Vadhan, S.P.: Efficiency improvements in constructing pseudorandom generators from one-way functions. In: Proceedings of the 42nd ACM Symposium on the Theory of Computing (STOC 2010), pp. 437–446 (2010)
13. Håstad, J., Impagliazzo, R., Levin, L.A., Luby, M.: Construction of pseudorandom generator from any one-way function. SIAM Journal on Computing 28(4), 1364–1396 (1999)
14. Holenstein, T.: Pseudorandom generators from one-way functions: A simple construction for any hardness. In: Halevi, S., Rabin, T. (eds.) TCC 2006. LNCS, vol. 3876, pp. 443–461. Springer, Heidelberg (2006)
15. Holenstein, T., Sinha, M.: Constructing a pseudorandom generator requires an almost linear Number of calls. In: Proceedings of the 53rd IEEE Symposium on Foundation of Computer Science (FOCS 2012), pp. 698–707 (2012)
16. Impagliazzo, R., Nisan, N., Wigderson, A.: Pseudorandomness for network algorithms. In: Proceedings of the 26th ACM Symposium on the Theory of Computing (STOC 1994), pp. 356–364 (1994)
17. Nisan, N.: Pseudorandom generators for space-bounded computation. Combinatorica 12(4), 449–461 (1992)
18. Vadhan, S.P., Zheng, C.J.: Characterizing pseudoentropy and simplifying pseudorandom generator constructions. In: Proceedings of the 44th ACM Symposium on the Theory of Computing (STOC 2012), pp. 817–836 (2012)
19. Yao, A.C.C.: Theory and applications of trapdoor functions (extended abstract). In: Proceedings of the 23rd IEEE Symposium on Foundation of Computer Science (FOCS 1982), pp. 80–91 (1982)
20. Yu, Y., Gu, D., Li, X., Weng, J.: The randomized iterate revisited - almost linear seed length PRGs from a broader class of one-way functions. Tech. Rep. 2014/392, Cryptology e-print archive, full version of this work, http://eprint.iacr.org/2014/392
21. Yu, Y., Li, X., Weng, J.: Pseudorandom generators from regular one-way functions: New constructions with improved parameters. In: Sako, K., Sarkar, P. (eds.) ASIACRYPT 2013, Part II. LNCS, vol. 8270, pp. 261–279. Springer, Heidelberg (2013)

A   Proofs Omitted

Fact 1. *For any $k \in [r\Delta]$, we have*

$$\mathsf{CP}(\ (Y_k, \boldsymbol{H}^{r\Delta-1})\ |\ \mathcal{E}_k \wedge \mathcal{S}_k\ ) = \frac{\mathsf{CP}(\ (Y_k, \boldsymbol{H}^{r\Delta-1})\ \wedge\ \mathcal{E}_k\ \wedge\ \mathcal{S}_k\ )}{\Pr[\mathcal{E}_k \wedge \mathcal{S}_k]^2}$$

$$= \frac{\mathsf{CP}(\ Y_k\ \wedge\ \mathcal{E}_k\ \wedge\ \mathcal{S}_k\ |\ \boldsymbol{H}^{r\Delta-1})}{\Pr[\mathcal{E}_k \wedge \mathcal{S}_k]^2 \cdot |\mathcal{H}|^{r\Delta-1}}$$

Proof of Fact 1. We first have that

$$\mathsf{CP}(\ (Y_k, \boldsymbol{H}^{r\Delta-1})\ |\ \mathcal{E}_k \wedge \mathcal{S}_k\ ) \cdot \Pr[\mathcal{E}_k \wedge \mathcal{S}_k]^2$$

$$= \Pr[\mathcal{E}_k \wedge \mathcal{S}_k]^2 \cdot \sum_{(y, h^{r\Delta-1})} \Pr[\ (Y_k, \boldsymbol{H}^{r\Delta-1}) = (y, h^{r\Delta-1})\ |\ \mathcal{E}_k \wedge \mathcal{S}_k\ ]^2$$

$$= \sum_{(y, h^{r\Delta-1})} \big(\Pr[\ (Y_k, \boldsymbol{H}^{r\Delta-1}) = (y, h^{r\Delta-1})\ |\ \mathcal{E}_k \wedge \mathcal{S}_k\ ] \cdot \Pr[\mathcal{E}_k \wedge \mathcal{S}_k] \big)^2$$

$$= \sum_{(y, h^{r\Delta-1})} \Pr[\ (Y_k, \boldsymbol{H}^{r\Delta-1}) = (y, h^{r\Delta-1})\ \wedge\ \mathcal{E}_k \wedge \mathcal{S}_k\ ]^2$$

$$= \mathsf{CP}(\ (Y_k, \boldsymbol{H}^{r\Delta-1})\ \wedge\ \mathcal{E}_k\ \wedge\ \mathcal{S}_k\ )\ ,$$

and complete the proof by the following

$$\frac{\mathsf{CP}(\ Y_k\ \wedge\ \mathcal{E}_k\ \wedge\ \mathcal{S}_k\ |\ \boldsymbol{H}^{r\Delta-1}\ )}{|\mathcal{H}|^{r\Delta-1}}$$

$$= \frac{\sum_{h^{r\Delta-1}} \Pr[\boldsymbol{H}^{r\Delta-1} = h^{r\Delta-1}] \cdot \sum_y \Pr[Y_k = y \wedge \mathcal{E}_k \wedge \mathcal{S}_k | \boldsymbol{H}^{r\Delta-1} = h^{r\Delta-1}]^2}{|\mathcal{H}|^{r\Delta-1}}$$

$$= \sum_{(y, h^{r\Delta-1})} \big(\Pr[\boldsymbol{H}^{r\Delta-1} = h^{r\Delta-1}] \cdot \Pr[Y_k = y \wedge \mathcal{E}_k \wedge \mathcal{S}_k | \boldsymbol{H}^{r\Delta-1} = h^{r\Delta-1}] \big)^2$$

$$= \sum_{(y, h^{r\Delta-1})} \Pr[(Y_k, \boldsymbol{H}^{r\Delta-1}) = (y, h^{r\Delta-1})\ \wedge\ \mathcal{E}_k\ \wedge\ \mathcal{S}_k\ ]^2$$

$$= \mathsf{CP}(\ (Y_k, \boldsymbol{H}^{r\Delta-1})\ \wedge\ \mathcal{E}_k\ \wedge\ \mathcal{S}_k\ )\ .$$

$\square$

Claim 2. $\mathbb{E}[\ \mathsf{Adv}(f(U_n), \boldsymbol{H}^{r\Delta-1})\ |\ f(U_n) \in \mathcal{Y}_{\max}\ ] \geq \mathbb{E}[\ \mathsf{Adv}(U_{\mathcal{Y}_{\max}}, \boldsymbol{H}^{r\Delta-1})\ ]/2.$

Proof of Claim 2. We recall that $f(U_n)$ is independent of $\boldsymbol{H}^{r\Delta-1}$.

$$\mathbb{E}[\ \mathsf{Adv}(f(U_n), \boldsymbol{H}^{r\Delta-1})\ |\ f(U_n) \in \mathcal{Y}_{\max}\ ]$$

$$= \sum_{(y, h^{r\Delta-1})} \Pr[\boldsymbol{H}^{r\Delta-1} = h^{r\Delta-1}] \cdot \Pr[f(U_n) = y | f(U_n) \in \mathcal{Y}_{\max}] \cdot \mathsf{Adv}(y, h^{r\Delta-1})$$

$$\geq \frac{1}{2} \sum_{(y, h^{r\Delta-1}) \in \mathcal{Y}_{\max} \times \mathcal{H}^{r\Delta-1}} \Pr[\boldsymbol{H}^{r\Delta-1} = h^{r\Delta-1}] \cdot \frac{1}{|\mathcal{Y}_{\max}|} \cdot \mathsf{Adv}(y, h^{r\Delta-1})$$

$$= \mathbb{E}[\ \mathsf{Adv}(U_{\mathcal{Y}_{\max}}, \boldsymbol{H}^{r\Delta-1})\ ]/2\ ,$$

where the inequality is due to for any $y \in \mathcal{Y}_{\max}$ it holds that

$$\Pr[f(U_n) = y \mid f(U_n) \in \mathcal{Y}_{\max}] = \frac{\Pr[f(U_n) = y]}{\sum_{y^* \in \mathcal{Y}_{\max}} \Pr[f(U_n) = y^*]}$$

$$= \frac{1}{\sum_{y^* \in \mathcal{Y}_{\max}} \frac{\Pr[f(U_n)=y^*]}{\Pr[f(U_n)=y]}} \geq \frac{1}{2|\mathcal{Y}_{\max}|} .$$

$\square$

Proof of (18). For any $k \leq r\Delta$, we will define a $(n + 1, r\Delta, \log|\mathcal{H}|)$-LBP M_1 that on input x_1 (at the source node) and $\boldsymbol{h}^{r\Delta-1}$ ($h_i \in \mathcal{H}$ at each ith layer), outputs 1 iff every $t \in [k]$ satisfies $y_t \in \mathcal{Y}_{[\max]}$. The BSG 2^{-2n}-fools M_1, i.e., for any $x_1 \in \{0,1\}^n$

$$| \Pr[M_1(x_1, \boldsymbol{H}^{r\Delta-1}) = 1] - \Pr[M_1(x_1, BSG(U_q)) = 1] |$$
$$= |\Pr[\mathcal{S}_k \mid X_1 = x_1] - \Pr[\mathcal{S}'_k \mid X_1 = x_1]| \leq 2^{-2n}$$

and thus

$$\Pr[\mathcal{S}'_k] \geq \Pr[\mathcal{S}_k] - 2^{-2n} \geq 1 - k\epsilon - 2^{-2n} .$$

The bounded-space computation of M_1 is as follows: the source node input is $(y_1 \in \{0,1\}^n, \mathsf{tag}_1 \in \{0,1\})$, where $y_1 = f(x_1)$ and $\mathsf{tag}_1 = 1$ iff $y_1 \in \mathcal{Y}_{[\max]}$ (or 0 otherwise). At each ith layer up to $i = k$, it computes $x_i := h_{i-1}(y_{i-1})$, $y_i := f(x_i)$ and sets $\mathsf{tag}_i := 1$ iff $\mathsf{tag}_{i-1} = 1$ and $y_i \in \mathcal{Y}_{[\max]}$ ($\mathsf{tag}_i := 0$ otherwise). Finally, M_1 produces tag_k as the final output.

Similarly, we define another $(n + 1, r\Delta, \log|\mathcal{H}|)$-LBP M_2 that on input $(x_1, \boldsymbol{h}^{r\Delta-1})$, outputs 1 iff $y_k \in \mathcal{Y}_{\max}$, and thus

$$\Pr[\mathcal{E}'_k] \geq \Pr[\mathcal{E}_k] - 2^{-2n} \geq n^{-c} - 2^{-2n} .$$

The computation of M_2 is simply to compute $x_i := h_{i-1}(y_{i-1})$ and $y_i := f(x_i)$ at each ith iteration and to output 1 iff $y_k \in \mathcal{Y}_{\max}$. It follows that

$$\Pr[\mathcal{E}'_k \wedge \mathcal{S}'_k] \geq \Pr[\mathcal{E}'_k] - \Pr[\neg\mathcal{S}'_k] \geq n^{-c} - 2^{-2n} - (k\epsilon + 2^{-2n}) \geq n^{-c}/2 .$$

$\square$

Proof of (19). For any $k \in [(r - 1)\Delta]$, consider the following $(n+1, r\Delta, \log|\mathcal{H}|)$-LBP M_3: on source node input $y_1 = f(x_1)$ and layered input vector $\boldsymbol{h}^{r\Delta-1}$, it computes $x_i := h_{i-1}(y_{i-1})$, $y_i := f(x_i)$ at each ith layer. For iterations numbered by $(k + 1){\leq}i \leq (k + \Delta)$, it additionally sets $\mathsf{tag}_i = 1$ iff either $\mathsf{tag}_{i-1} = 1$ or $y_i \in \mathcal{Y}_{\max}$, where tag_k is initialized to 0. Finally, M_3 outputs $\mathsf{tag}_{k+\Delta}$. By the bounded space generator we have

$$| \Pr[M_3(X_1, \boldsymbol{H}^{r\Delta-1}) = 1] - \Pr[M_3(X_1, BSG(U_q)) = 1] |$$
$$= |\Pr[\bigvee_{i=k+1}^{k+\Delta} \mathcal{E}_i] - \Pr[\bigvee_{i=k+1}^{k+\Delta} \mathcal{E}'_i]| \leq 2^{-2n} ,$$

and thus by (13)

$$\Pr[\bigvee_{i=k+1}^{k+\Delta} \mathcal{E}'_i] \geq \Pr[\bigvee_{i=k+1}^{k+\Delta} \mathcal{E}_i] - 2^{-2n} \geq 1 - n^{-\alpha} - 2^{-2n} .$$

$\square$

Proof of (20). For any $k \in [r\Delta]$, consider the following $(2n + 1, r\Delta, \log|\mathcal{H}|)$-LBP M_4: on source node input $(y_1 = f(x_1), y'_1 = f(x'_1), \mathsf{tag}_1 \in \{0, 1\})$, where $\mathsf{tag}_1 = 1$ iff both $y_1, y'_1 \in \mathcal{Y}_{[\max]}$. For $1 \leq i \leq k$, at each ith layer M_4 computes $y_i := f(h_{i-1}(y_{i-1}))$, $y'_i := f(h_{i-1}(y'_{i-1}))$ and sets $\mathsf{tag}_i = 1$ iff $\mathsf{tag}_{i-1} = 1 \wedge y_i \in \mathcal{Y}_{[\max]} \wedge y'_i \in \mathcal{Y}_{[\max]}$. Finally, at the $(k+1)^{th}$ layer M_4 outputs 1 iff $y_k = y'_k \in \mathcal{Y}_{\max}$ (in respect for event $\mathcal{E}_k/\mathcal{E}'_k$) and $\mathsf{tag}_k = 1$ (in honor of $\mathcal{S}_k/\mathcal{S}'_k$). Imagine running two iterates with random x_1, x'_1 and seeded by a common hash function from distribution either $\boldsymbol{H}^{r\Delta-1}$ or $BSG(U_q)$, we have

$$\mathsf{CP}(Y_k \wedge \mathcal{E}_k \wedge \mathcal{S}_k \mid \boldsymbol{H}^{r\Delta-1}) = \Pr_{(x_1,x'_1)\leftarrow U_{2n},\, h^{r\Delta-1}\leftarrow H^{r\Delta-1}}[M_4(x_1, x'_1, h^{r\Delta-1}) = 1]$$

$$\mathsf{CP}(Y'_k \wedge \mathcal{E}'_k \wedge \mathcal{S}'_k \mid BSG(U_q)) = \Pr_{(x_1,x'_1)\leftarrow U_{2n},\, h^{r\Delta-1}\leftarrow BSG(U_q)}[M_4(x_1, x'_1, h^{r\Delta-1}) = 1]$$

and thus

$$\left| \mathsf{CP}(Y_k \wedge \mathcal{E}_k \wedge \mathcal{S}_k \mid \boldsymbol{H}^{r\Delta-1}) - \mathsf{CP}(Y'_k \wedge \mathcal{E}'_k \wedge \mathcal{S}'_k \mid BSG(U_q)) \right|$$

$$\leq \mathbb{E}_{(x_1,x'_1)\leftarrow U_{2n}}\left[\left| \Pr[M_4(x_1, x'_1, \boldsymbol{H}^{r\Delta-1}) = 1] - \Pr[M_4(x_1, x'_1, BSG(U_q)) = 1] \right| \right]$$

$$\leq 2^{-2n} .$$

It follows by (14) that

$$\mathsf{CP}(Y'_k \wedge \mathcal{E}'_k \wedge \mathcal{S}'_k \mid BSG(U_q)) \leq \mathsf{CP}(Y_k \wedge \mathcal{E}_k \wedge \mathcal{S}_k \mid \boldsymbol{H}^{r\Delta-1}) + 2^{-2n}$$
$$\leq (r\Delta + 1) \cdot 2^{\max-n+1} .$$

Note that y_k, $\mathcal{E}'_k$ and $\mathcal{S}'_k$ depend only on x_1 and $h^{r\Delta-1}$, namely, for any h^{k-1} and any $u_1, u_2 \in BSG^{-1}(h^{k-1})$,

$$\mathsf{CP}(Y'_k \wedge \mathcal{E}'_k \wedge \mathcal{S}'_k \mid U_q = u_1) = \mathsf{CP}(Y'_k \wedge \mathcal{E}'_k \wedge \mathcal{S}'_k \mid U_q = u_2)$$
$$= \mathsf{CP}(Y'_k \wedge \mathcal{E}'_k \wedge \mathcal{S}'_k \mid BSG(U_q) = h^{k-1}) .$$

Therefore,

$$\mathsf{CP}(Y'_k \wedge \mathcal{E}'_k \wedge \mathcal{S}'_k \mid U_q) = \mathsf{CP}(Y'_k \wedge \mathcal{E}'_k \wedge \mathcal{S}'_k \mid BSG(U_q)) \leq (r\Delta + 1) \cdot 2^{\max-n+1} .$$

$\square$

Proof of (23). We have that

$$
\begin{aligned}
& \mathsf{CP}(\ (Y'_{(j^*-1)\Delta+i^*}, U_q\ )\ |\ \mathcal{E}'_{(j^*-1)\Delta+i^*} \wedge \mathcal{S}'_{(j^*-1)\Delta+i^*}\ ) \\
&= \frac{\mathsf{CP}(\ (Y'_{(j^*-1)\Delta+i^*}, U_q\ )\ \wedge\ \mathcal{E}'_{(j^*-1)\Delta+i^*} \wedge \mathcal{S}'_{(j^*-1)\Delta+i^*})}{\Pr[\mathcal{E}'_{(j^*-1)\Delta+i^*} \wedge \mathcal{S}'_{(j^*-1)\Delta+i^*}]^2} \\
&\leq \mathsf{CP}(\ Y_{(j^*-1)\Delta+i^*}\ \wedge\ \mathcal{E}_{(j^*-1)\Delta+i^*}\ \wedge \mathcal{S}_{(j^*-1)\Delta+i^*}\ |\ U_q\ )\ \frac{1}{(n^{-2c}/4)\cdot 2^q} \\
&\leq \frac{(r\Delta+1)\cdot 2^{\max-n+1}}{(n^{-2c}/4)\cdot 2^q} \leq \frac{16r\Delta\cdot n^{2c}}{2^{n-\max}\cdot 2^q}\ ,
\end{aligned}
$$

where the equality is similar to that in Fact 1 (by renaming $H^{r\Delta-1}$ to U_q), and the two inequalities are due to (18) and (20) respecitvely and thus

$$\mathbf{H}_2((Y'_{(j-1)\Delta+i^*}, U_q)|\mathcal{E}'_{(j-1)\Delta+i^*} \wedge \mathcal{S}'_{(j-1)\Delta+i^*}) \geq n-\max+q-2c\cdot\log n-\log r\Delta-4\ .$$

The uniform distribution over $\mathcal{Y}_{\max} \times \{0,1\}^q$ has entropy

$$\mathbf{H}_2(\ (U_{\mathcal{Y}_{\max}}, U_q)\ ) \leq \log(\frac{1}{2^{-n+\max-1}}\cdot 2^q) = n-\max+q+1\ ,$$

and thus the entropy deficiency (i.e., the difference of two entropies above) $e \leq 2c\log n + \log r + \log \Delta + 5$. $\qquad\square$

Fact 2. *For any $\delta > -1$ and any positive integer q, it holds that*

$$(1+\delta)^q \geq 1 + q\cdot\delta$$

Proof. We prove by induction. For $q = 1$ the equality holds. Suppose that the above holds for $q = k \in \mathbb{N}$, i.e., $(1+\delta)^k \geq 1+k\cdot\delta$, then for $q = k+1$ we have

$$(1+\delta)^{k+1} \geq (1+k\cdot\delta)(1+\delta) = 1+(k+1)\cdot\delta + k\delta^2 \geq 1+(k+1)\cdot\delta$$

which completes the proof. $\qquad\square$

B   Definitions, Explanations and Remarks

Remark 2 (some intuitions for $\mathcal{S}_k$). Throughout the proofs, we consider the (inverting, collision, etc.) probabilities conditioned on event $\mathcal{S}_k$, which requires that during the first k iterations no y_i ($1 \leq i \leq k$) hits the negligible fraction. This might look redundant as $\mathcal{S}_k$ occurs with overwhelming probability (by (12)). However, our proofs crucially rely on the fact that, as stated in (14), the collision probability of y_k conditioned on $\mathcal{S}_k$ is almost the same (roughly $\tilde{O}(2^{\max-n})$), omitting poly(n) factors) as the ideal case, i.e., the collision probability of $f(U_n)$ conditioned on $f(U_n) \in \mathcal{Y}_{\max}$. This would not have been possible if not being conditioned on $\mathcal{S}_k$ even though $\mathcal{Y}_{\max+1}, \ldots, \mathcal{Y}_n$ only sum to a negligible function

$\mathsf{negl}(n)$. To see this, consider the following simplified case for $k = 1$, the collision probability of y_1 is equal to that of $f(U_n)$, and thus we have

$$\frac{1}{2} \cdot \sum_{i=1}^{n} 2^{i-n} \cdot \Pr[f(U_n) \in \mathcal{Y}_i] \leq \left(\mathsf{CP}(f(U_n)) = \sum_{i=1}^{n} \sum_{y \in \mathcal{Y}_i} \Pr[f(U_n) = y]^2 \right)$$

$$< \sum_{i=1}^{n} 2^{i-n} \cdot \Pr[f(U_n) \in \mathcal{Y}_i] .$$

Suppose that there is some $\mathcal{Y}_t$ such that $t = \max + \Omega(n)$ and $\Pr[f(U_n) \in \mathcal{Y}_t] = \mathsf{negl}(n)$, then the above collision probability is of the order $O(2^{\max - n}(n^{-c} + 2^{\Omega(n)}\mathsf{negl}(n))$. By setting $\mathsf{negl}(n) = n^{-\log n}$, the collision probability blows up by a factor of $2^{\Omega(n)}$ than the desired case $\tilde{O}(2^{\max - n})$, and thus unable to apply Lemma 1. In contrast, conditioned on $\mathcal{S}_1$ the collision probability is $\tilde{O}(2^{\max - n})$.

Definition 9 (Collision probabilities conditioned on $\mathcal{S}'_k$ and $\mathcal{E}'_k$). *In the derandomized version, we will use the following conditional collision probabilities (that are quite naturally extended from the standard collision probabilities):*

$$\mathsf{CP}(\ Y'_k \ \wedge \ \mathcal{E}'_k \wedge \mathcal{S}'_k \ |\ U_q) \ \stackrel{\mathsf{def}}{=} \ \mathbb{E}_{u \leftarrow U_q} \left[\sum_y \Pr[f^k(X_1, H'^{r\Delta-1}) = y \ \wedge \ \mathcal{E}'_k \ \wedge \right.$$

$$\left. \mathcal{S}'_k \ |\ H'^{r\Delta-1} = BSG(u)]^2 \right], \mathsf{CP}(\ Y'_k \wedge \mathcal{E}'_k \wedge \mathcal{S}'_k \ |\ BSG(U_q)\ ) \stackrel{\mathsf{def}}{=} \mathbb{E}_{h'^{r\Delta-1} \leftarrow BSG(U_q)}$$

$$\left[\sum_y \Pr[\ f^k(X_1, H'^{r\Delta-1}) = y \wedge \mathcal{E}'_k \ \wedge \ \mathcal{S}'_k \ |\ H'^{r\Delta-1} = h'^{r\Delta-1}\ ]^2 \right],$$

$$\mathsf{CP}(\ Y'_k, U_q \ |\ \mathcal{E}'_k \wedge \mathcal{S}'_k\ ) \stackrel{\mathsf{def}}{=} \sum_{(y,u)} \Pr[\ f^k(X_1, BSG(U_q)) = y \wedge U_q = u \ |\ \mathcal{E}'_k \wedge \mathcal{S}'_k\ ]^2 .$$

Algorithm 2. $\mathsf{M}^{A'}$

Input: $y \in \{0,1\}^n$

    Sample $j \stackrel{\$}{\leftarrow} [r]$, $i \stackrel{\$}{\leftarrow} [\Delta]$, $u \stackrel{\$}{\leftarrow} \{0,1\}^q$, $h^{r\Delta-1} := BSG(u)$;

    Let $\tilde{y}_{(j-1)\Delta+i} := y$;

    FOR $k = (j-1)\Delta + i + 1$ TO $(j-1)\Delta + \Delta$

        Compute $\tilde{x}_k := h_{k-1}(\tilde{y}_{k-1})$, $\tilde{y}_k := f(\tilde{x}_k)$;

    $\tilde{x}_{(j-1)\Delta+1} \leftarrow A'(\tilde{y}_{j\Delta}, u)$;

    FOR $k = (j-1)\Delta + 1$ TO $(j-1)\Delta + i - 1$

        Compute $\tilde{y}_k := f(\tilde{x}_k)$, $\tilde{x}_{k+1} := h_k(\tilde{y}_k)$;

Output: $\tilde{x}_{(j-1)\Delta+i}$

Remark 3 (On weakening the condition of (1).). In fact, our construction only assumes a weaker condition than (1), i.e., for some constant $c \geq 0$ and $d = d(n) \in O(\log n)$ it holds that

$$\Pr[\ f(U_n) \in (\ \mathcal{Y}_{\max - d} \cup \mathcal{Y}_{\max - d+1} \cup \ldots \cup \mathcal{Y}_{\max}\ )\ ] \geq n^{-c} . \tag{24}$$

We sketch the idea of adapting the proof to the relaxed assumption. By averaging there exists $t \in [0, d]$ such that $\mathcal{Y}_{\max -t}$ has weight at least n^{-c-1}. We thus consider the chance that Y_j hits $\mathcal{Y}_{\max -t}$ (instead of $\mathcal{Y}_{\max}$ as we did in the original proof), and $O(n^{2c+2} \cdot \omega(\log n))$ iterations are bound to hit $\mathcal{Y}_{\max -t}$ at least once. Now we adapt the proof of Lemma 6. Ideally, conditioned on $f(U_n) \in \mathcal{Y}_{\max -t}$ the distribution $(f(U_n), \boldsymbol{H}^{r\Delta-1})$ is uniform over $\mathcal{Y}_{\max -t} \times \mathcal{H}^{r\Delta-1}$ with full entropy

$$\mathbf{H}_2(\ (U_{\mathcal{Y}_{\max -t}}, \boldsymbol{H}^{r\Delta-1})\ ) \leq \log(\frac{1}{2^{-n+\max -t-1}} \cdot |\mathcal{H}|^{r\Delta-1})$$

$$= n - \max +t + (r\Delta - 1)\log|\mathcal{H}| + 1 \ .$$

However, we actually only have that

$$\mathbf{H}_2(\ (Y_{(j-1)\Delta+i^*}, \boldsymbol{H}^{r\Delta-1})\ |\ \mathcal{E}_{(j-1)\Delta+i^*} \wedge \mathcal{S}_{(j-1)\Delta+i^*}\ )$$

$$\geq \left(n - \max +t + (r\Delta - 1)\log|\mathcal{H}| + 1 \right) - e \ ,$$

where entropy deficiency $e \leq t + O(\log n) = O(\log n)$. Then, we apply Lemma 1 and the hard-to-invertness only blows up by a factor of roughly $2^e = n^{O(1)}$ than the ideal ε (and taking a square root afterwards), which does not kill the iterate. Therefore, the iterate is hard to invert for every $O(n^{2c+2} \cdot \omega(\log n))$ iterations. The proof for the derandomized version can be adapted similarly.

C   Regular, Weakly-Regular and Arbitrary OWFs

In this section, we discuss the gap between weakly-regular and arbitrary one-way functions. First, we show that most functions are known-almost-regular and thus weakly-almost-regular as well (see Remark 1), namely, "if a one-way function behaves like a random function, then it is known-almost-regular". More generally, weakly-regular one-way functions cover a wider range of one-way functions (for positive $c \in \mathbb{N}$) than regular ones. We also (attempt to) characterize functions that are not captured by the definition of "weakly-regular". We show that in order not to fall into weakly-regular functions, the counterexamples should be somewhat artificial.

Now, we use probabilistic methods to argue that almost-regularity is a reasonable assumption in the average sense. That is, if the one-way function is considered as randomly drawn from the set of all (not just one-way) functions, then it is very likely to be almost-regular and thus a PRG can be efficiently constructed.

Lemma 8 (A random function is known-almost-regular). *Let $\mathcal{F} = \{f : \{0,1\}^n \to \{0,1\}^m \}$ be the set of all functions mapping n-bit to m-bit strings. For any $0 < d < n$,*

– *if $m \leq n - d$, then it holds that*

$$\Pr_{f \xleftarrow{\$} \mathcal{F}} [\ \mathsf{SD}(\ f(U_n),\ U_m\ ) \leq 2^{-d/4}\ ] \geq 1 - 2^{-d/4} \ ,$$

− *if $m > n − d$, then we have*

$$\Pr_{f \overset{\$}{\leftarrow} \mathcal{F}, \ x \overset{\$}{\leftarrow} \{0,1\}^n} [\ 1 \le |f^{-1}(f(x))| \le 2^{2d+1}\ ] \ge 1 - 2^{-d} \ .$$

Typically, we can set $d \in \omega(\log n)$ so that f will be almost regular except for a negligible fraction. Note that the first bullet gives even stronger guarantee than the second one does.

Proof of Lemma 8. We see $\mathcal{F}$ as a family of universal hash functions and let F be a uniform distribution over $\mathcal{F}$. For $m \le n − d$ we have by the leftover hash lemma that

$$\mathbb{E}_{f \overset{\$}{\leftarrow} \mathcal{F}}[\ \mathsf{SD}(\ f(U_n)\ ,\ U_m\ )\ ] \ = \ \mathsf{SD}(\ F(U_n)\ ,\ U_m \mid F\ ) \ \le \ 2^{-\frac{d}{2}} \ .$$

It follows by a Markov inequality that the above statistical distance is bounded by $2^{-d/2} \cdot 2^{d/4}$ except for a $2^{-d/4}$-fraction of f. We proceed to the case for $m > n − d$ to get

$$\mathsf{CP}(\ F(U_n) \mid F\ ) \ \le \ \mathsf{CP}(U_n) \ + \ \max_{x_1 \ne x_2}\{\ \Pr[F(x_1) = F(x_2)]\ \} \ = \ 2^{-n} + 2^{-m}$$

$$\le \ 2^{-n+d+1}$$

We define $\mathcal{S} \overset{\text{def}}{=} \{(y, f) : |f^{-1}(y)| > 2^{2d+1}\}$ to yield

$$2^{-n+d+1} \ge \mathsf{CP}(F(U_n)|F) = \sum_f \Pr[F = f] \sum_y \Pr[f(U_n) = y]^2$$

$$\ge \ 2^{-n+2d+1} \cdot \sum_f \Pr[F = f] \sum_{y:(y,f)\in\mathcal{S}} \Pr[f(U_n) = y]$$

$$= \ 2^{-n+2d+1} \cdot \Pr[(F(U_n), F) \in \mathcal{S}] \ ,$$

and thus $\Pr[(F(U_n), F) \in \mathcal{S}] \le 2^{-d}$. This completes the proof. Note that $|f^{-1}(y)| \ge 1$ for any $y = f(x)$. □

BEYOND REGULAR FUNCTIONS. We cannot rule out the possibility that the one-way function in consideration is far from regular, namely (using the language of Definition 4), an arbitrary one-way function can have non-empty sets $\mathcal{Y}_i, \ldots,$ $\mathcal{Y}_{i+O(n)}$. Below we argue that Definition 4 is quite generic and any function that fails to satisfy it should be somewhat artificial. As a first attempt, one may argue that if we skip all those $\mathcal{Y}'_j s$ (in the descending order of j) that sum to negligible, the first one that is non-negligible[9] (i.e., not meeting (2)) will satisfy (1) for at least infinitely many n's. In other words, an arbitrary one-way function is weakly-regular (at least for infinitely many n's). This argument

---

[9] Although non-negligible and noticeable are not the same, they are quite close: a non-negligible (resp., noticeable) function $\mu(\cdot)$ satisfies that there exists constant c such that $\mu(n) \ge n^{-c}$ for infinitely many (resp., all large enough) n's.

is unfortunately problematic as (non-)negligible is a property of a sequence of probabilities, rather than a single value. However, we will follow this intuition and provide a remedied analysis below.

Lemma 9 (a necessary condition to be a counterexample). *Let* $f :$ $\{0,1\}^n \to \{0,1\}^{l(n)}$ *be any one-way function and denote* $\mathcal{Y}_j \overset{\text{def}}{=} \{y : 2^{j-1} \leq |f^{-1}(y)| < 2^j\}$, *and let* $\kappa = \kappa(n)$ *be the number of non-empty sets* $\mathcal{Y}_j$ *(that comprise the range of f) for any given n, and write them as* $\mathcal{Y}_{i_1}, \mathcal{Y}_{i_2}, \ldots, \mathcal{Y}_{i_\kappa}$ *with* $i_1 < i_2 < \ldots < i_\kappa$. *For every* $n_0 \in \mathbb{N} \cup \{0\}$, *it must hold that function* $\mu_{n_0}(\cdot)$ *defined as*

$$\mu_{n_0}(n) \overset{\text{def}}{=} \begin{cases} \Pr[f(U_n) \in \mathcal{Y}_{i_{\kappa - n_0}}], & \text{if } \kappa > n_0 \\ 0 & , & \text{if } \kappa \leq n_0 \end{cases} \tag{25}$$

is negligible. Otherwise (if the above condition is not met), there exists constant $c \geq 0$, $\max(n) \in \mathbb{N}$ *and negligible function* $\epsilon(n) \in [0,1]$ *such that (2) holds (for all n's) and (1) holds for infinitely many n's.*

Proof of Lemma 9. If (25) does not hold for every $n_0 \in \mathbb{N} \cup \{0\}$, then there must exist an n_0 such that $\mu_0(\cdot), \ldots \mu_{n_0-1}(\cdot)$ are negligible and $\mu_{n_0}(\cdot)$ is non-negligible. We then define $\max(\cdot)$ as

$$\max(n) \overset{\text{def}}{=} \begin{cases} i_{\kappa(n)-n_0}, & \text{if } \kappa(n) > n_0 \\ \perp & , & \text{if } \kappa(n) \leq n_0 \end{cases}$$

It is easy to see that $\mathcal{Y}_{i_{\kappa-n_0+1}}, \ldots, \mathcal{Y}_{i_\kappa}$ sum to a negligible fraction in n (i.e., the sum of a finite number of negligible functions $\mu_0(\cdot), \ldots \mu_{n_0-1}(\cdot)$ results into another negligible function). Denote by $\mathcal{N}_\perp \overset{\text{def}}{=} \{n \in \mathbb{N} \cup \{0\} : \max(n) = \perp\}$. We have by assumption that for some constant c it holds that $\mu_{n_0}(n) \geq n^{-c}$ for infinitely many $n \in \mathbb{N} \cup \{0\}$, and thus $\mu_{n_0}(n) \geq n^{-c}$ holds also for infinitely many $n \in \mathbb{N} \cup \{0\} \setminus \mathcal{N}_\perp$. This is due to $\mu_{n_0}(n) = 0$ for any $n \in \mathcal{N}_\perp$. Therefore, $\Pr[f(U_n) \in \mathcal{Y}_{\max}]$ is non-negligible, which completes the proof. $\square$

(25) IS A NECESSARY AND STRONG CONDITION. The above lemma formalizes a necessary condition to constitute a counterexample to Definition 4. It is necessary in the sense that any one-way function that does not satisfy it must satisfy Definition 4 (for at least infinitely many n's). Note that the condition is actually an infinite set of conditions by requiring every $\mu_{n_0}(n)$ (for $n_0 \in \mathbb{N} \cup \{0\}$) being negligible. At the same time, it holds unconditionally that all these $\mu_{n_0}(n)$ (that correspond to the weights of all non-empty sets) must sum to unity, i.e., for every n we have

$$\mu_0(n) + \mu_1(n) + \ldots + \mu_{\kappa(n)-1}(n) = 1 .$$

The above might look mutually exclusive to (25) as if every $\mu_{n_0}(n)$ is negligible then the above sum should be upper bounded by $\kappa(n) \cdot \text{negl}(n) = \text{negl}'(n)$ instead of being unity. This intuition is not right in general, as by definition a negligible function only needs to be super-polynomially small for all sufficiently (instead of all) n's. However, it is reasonable to believe that one-way functions satisfying (25) should be quite artificial.

(25) IS NOT SUFFICIENT. Despite seeming strong, (25) is still not sufficient to make a counterexample. To show this, we give an example function that satisfies both (25) (for every $n_0 \in \mathbb{N} \cup \{0\}$) and Definition 4. That is, let f be a one-way function where for every n the non-empty sets of f are

$$\mathcal{Y}_{n/3}, \mathcal{Y}_{n/3+1}, \ldots, \mathcal{Y}_{n/2} \tag{26}$$

with $\Pr[f(U_n) \in \mathcal{Y}_{n/3}] = 1 - n^{-\log n+1}/6$, $\Pr[f(U_n) \in \mathcal{Y}_{n/3+i}] = n^{-\log n}$ for all $1 \le i \le n/6$ and thus $\kappa(n) = n/6 + 1$. It is easy to see that this function satisfies Definition 4 with $\max(n) = n/3$ and $\epsilon(n) = n^{-\log n+1}/6$. In addition, for every $n_0 \in \mathbb{N} \cup \{0\}$ function $\mu_{n_0}(\cdot)$ is negligible as $\mu_{n_0}(n) = n^{-\log n}$ for all $n > 6n_0$. In summary, although an arbitrary one-way function may not be weakly-regular, the counterexamples must be well crafted to satisfy a somewhat artificial (yet still insufficient) condition.

The Power of Negations in Cryptography[*]

Siyao Guo[1], Tal Malkin[2], Igor C. Oliveira[2], and Alon Rosen[3]

[1] Department of Computer Science and Engineering,
Chinese University of Hong Kong, China
syguo@cse.cuhk.edu.hk

[2] Department of Computer Science, Columbia University, USA
{tal,oliveira}@cs.columbia.edu

[3] Efi Arazi School of Computer Science, IDC Herzliya, Israel
alon.rosen@idc.ac.il

Abstract. The study of monotonicity and negation complexity for Boolean functions has been prevalent in complexity theory as well as in computational learning theory, but little attention has been given to it in the cryptographic context. Recently, Goldreich and Izsak (2012) have initiated a study of whether cryptographic primitives can be monotone, and showed that one-way functions can be monotone (assuming they exist), but a pseudorandom generator cannot.

In this paper, we start by filling in the picture and proving that many other basic cryptographic primitives cannot be monotone. We then initiate a *quantitative* study of the power of negations, asking how many negations are required. We provide several lower bounds, some of them tight, for various cryptographic primitives and building blocks including one-way permutations, pseudorandom functions, small-bias generators, hard-core predicates, error-correcting codes, and randomness extractors. Among our results, we highlight the following.

- Unlike one-way functions, one-way permutations cannot be monotone.
- We prove that pseudorandom functions require $\log n - O(1)$ negations (which is optimal up to the additive term).
- We prove that error-correcting codes with optimal distance parameters require $\log n - O(1)$ negations (again, optimal up to the additive term).
- We prove a general result for monotone functions, showing a lower bound on the depth of any circuit with t negations on the bottom that computes a monotone function f in terms of the monotone circuit depth of f. This result addresses a question posed by Koroth and Sarma (2014) in the context of the circuit complexity of the Clique problem.

[*] The first author was partially supported by RGC GRF grant CUHK 410111. The second and the third author were supported in part by NSF grants CCF-116702 and CCF-1423306. The first and the third author did part of this work while visiting IDC Herzliya, supported by the ERC under the European Union's Seventh Framework Programme (FP/2007-2013) Grant Agreement n. 307952. The fourth author was supported by ISF grant no. 1255/12 and by the ERC under the European Union's Seventh Framework Programme (FP/2007-2013) Grant Agreement n. 307952.

Y. Dodis and J.B. Nielsen (Eds.): TCC 2015, Part I, LNCS 9014, pp. 36–65, 2015.

1   Introduction

Why do block ciphers like AES (Advanced Encryption Standard) have so many XOR gates dispersed throughout the levels of its circuit? Can we build a universal hard-core bit alternative to the Goldreich and Levin one [20] that only applies a small (say, constant) number of XORs? Why does the Goldreich, Goldwasser, and Micali [18] construction of a pseudorandom function (PRF) from a pseudorandom generator (PRG) heavily rely on selection functions, and calls the PRG many times? Could there be a monotone construction of a PRF from a PRG?

These are a few of the many fascinating questions related to the negation complexity of cryptographic primitives. The *negation complexity* of a boolean function $f: \{0,1\}^n \rightarrow \{0,1\}$ is the minimum number of negation gates in any fan-in two circuit with AND, OR, and NOT gates computing f. Note that negation gates are equivalent to XOR gates (of fan-in 2), in the sense that any circuit with t negation gates can be transformed into an equivalent circuit with t XOR gates, and vice-versa.[1] A function is *monotone* if and only if its negation complexity is 0.

In this paper, we initiate the investigation of the negation complexity of cryptographic primitives. We take first steps in this study, providing some surprising results, as well as pointing to some basic, intriguing problems that are still open.

This direction fits within the larger program of studying how *simple* basic cryptographic primitives can be, according to various complexity measures such as required assumptions, minimal circuit size, depth, etc (see, e.g., [4]). Exploring such questions helps gaining a deeper theoretical understanding of fundamental primitives and the relationships among them, and may provide the basis for understanding and addressing practical considerations as well.

While the study of monotone classes of functions and negation complexity has been prevalent in circuit complexity ([21,3,38,36,35,7,6,29,28], to name a few) and computational learning theory (see e.g. [8,9,11,32,13]), little attention has been given to it in the cryptographic context.

Recently, Goldreich and Izsak [19] have initiated a study of "cryptography in the monotone world", asking whether basic cryptographic primitives may be monotone. They focus on one-way functions (OWF) and pseudorandom generators, and show an inherent gap between the two by proving: (1) if any OWF exist, then there exist OWFs with polynomial-size monotone circuits, but (2) no monotone function can be a PRG. Quoting from their paper: *these two results indicate that in the "monotone world" there is a fundamental gap between one-way functions and pseudorandom generators; thus, the "hardness-vs-randomness" paradigm fails in the monotone setting.* This raises the following natural question:

> *Can other cryptographic primitives be computed by polynomial-size monotone circuits?*

We consider this question for several primitives and building blocks, showing negative answers for all of them. This may suggest the interpretation

---

[1] $\neg x$ is equivalent to $x \oplus 1$, while $x \oplus y$ is equivalent to $\neg(x \wedge y) \wedge (x \vee y)$.

(or conjecture) that in the "monotone world", there is no cryptography except for one-way functions. We then initiate a *quantitative* study (where our main contributions lie), putting forward the question:

> *How many negations are required (for poly-size circuits) to compute fundamental cryptographic building blocks?*

Markov [27] proved that the negation complexity of any function $h\colon \{0,1\}^n \to \{0,1\}^m$ is at most $\lceil \log(n+1) \rceil$, and Fischer [14] proved that this transformation can be made efficient (see Jukna [22] for a modern exposition). In light of these results, is it the case that all natural cryptographic primitives other than OWFs require $\Omega(\log n)$ negations, or are there primitives that can be computed with, say, a constant number of negations?

We state our results informally in the next section. Since our lower bounds hold for well-known primitives, we postpone their definitions to Section 3.

2   Our Results

Our contributions alongside previously known results are summarized in Figure 1, together with the main idea in each proof (the definition of these primitives can be found in Section 3). We explain and discuss some interesting aspects of these results below, deferring complete details to the body of the paper.

Primitive	Lower Bound	Upper Bound	Ref.	Proof Ideas
OWF	-	(monotone)	[19]	Embedding into middle slice
OWP	**non-monotone**	$\log n + O(1)$	here	Combinatorial and analytic proofs
PRG	non-monotone	$\log n + O(1)$	[19]	AND of one or two output bits
SBG	**non-monotone**	$\omega(1)$	here	Extension of [19]; Parity of Tribes
WPRF	non-monotone	$(\frac{1}{2}+o(1))\log n$	[9]	Weak-learner for monotone functions
PRF	$\log n - O(1)$	$\log n + O(1)$	here	Alternating chains in the hypercube
ECC	$\log n - O(1)$	$\log n + O(1)$	here	Extension of [12]
HCB	$(\frac{1}{2}-o(1))\log n$	$(\frac{1}{2}+o(1))\log n$	here	Low influence and [16]
EXT	$\Omega(\log n)$	$\log n + O(1)$	here	Low noise-sensitivity and [10]

Fig. 1. Summary of the <u>negation complexity</u> of basic cryptographic primitives and building blocks. Boldface results correspond to new bounds obtained in this paper. The $\log n + O(1)$ upper bound is Markov's bound [27] for any Boolean function. Error-correcting codes (ECC) and extractors (EXT) refer to constructions with good distance and extraction parameters.

2.1   Cryptography is Non-monotone

As mentioned above, [19] proved that if OWFs exist, then they can be monotone, while PRGs cannot. We fill in the picture by considering several other cryptographic primitives, and observing that none of them can be monotone (see Figure 1).

A result of particular interest is the lower bound showing that one-way permutations (OWP) *cannot* be monotone. We obtain this result by proving that any monotone permutation f on n variables must satisfy $f(x_1, \ldots, x_n) = (x_{\pi(1)}, \ldots, x_{\pi(n)})$, for some permutation $\pi \colon [n] \to [n]$ (finding the permutation and inverting f can then be done by evaluating f on n inputs). This is surprising in light of the [19] construction for OWFs. In particular, our result can be seen as a separation between OWFs and OWPs in the monotone world.

We provide two proofs of our result. The first is based on analytical methods, and was inspired by an approach used by Goldreich and Izsak [19]. The second is more elementary, and relies on a self-contained combinatorial argument.

2.2 Highly Non-monotone Primitives

We show that many central cryptographic primitives are highly non-monotone. Some of our lower bounds demonstrate necessity of $\log n - O(1)$ negations, which is tight in light of Markov's $\log n + O(1)$ upper bound [27]. For some of the primitives we give less tight $\Omega(\log n)$ lower bounds.

Pseudorandom Functions (PRF). We show that PRFs can only be computed by circuits containing at least $\log n - O(1)$ negations (which is optimal up to the additive term). We prove this by exhibiting an adversary that distinguishes any function that can be implemented with fewer negations gates from a random function. Our result actually implies that for any PRF family $\{F(w, \cdot)\}$, for almost all seeds w, $F(w, \cdot)$ can only be computed by circuits with at least $\log n - O(1)$ negations.[2] The distinguisher we construct asks for the values of the function on a fixed chain from 0^n to 1^n and accept if the alternating number of this chain is large. We note that the distinguisher suceeds for any function that has an implementation with fewer negations than the lower bound, regardless of the specific implementation the PRF designer had in mind. This can be considered as another statistical test to run on proposed candidate PRF implementations.

Error-Correcting Codes (ECC). As shown by Buresh-Oppenheim, Kabanets and Santhanam [12], if an ECC has a monotone encoding function then one can find two codewords that are very close. This implies that there is no monotone ECC with good distance parameters.

We extend this result to show that, given a circuit with t negation gates computing the encoding function, we can find two codewords whose Hamming distance is $O(2^t \cdot m/n)$ (for codes going from n bits to m bits). Consequently, this gives a $\log n - O(1)$ lower bound on the negation complexity of ECC

---

[2] That is, if we consider the circuit computing the PRF family $F(\cdot, \cdot)$ as a single function (with the seed as one of the inputs), then this circuit must have at least logarithmically many negation gates.

with optimal distance parameters.

Hard-core Bits (HCB). Recall that a Boolean function $h\colon \{0,1\}^n \to \{0,1\}$ is a hard-core predicate for a function $f\colon \{0,1\}^n \to \{0,1\}$ if, given $f(x)$, it is hard to compute $h(x)$. We show that general hard-core bit predicates must be highly non-monotone. More specifically, there exists a family of one-way functions f_n for which any hard-core predicate requires $\Omega(\log n)$ negations (assuming one-way functions exist).

Our result follows via the analysis of the *influence* of circuits with few negations, and a corresponding lower bound on hard-core bits due to Goldmann and Russell [16].

(Strong) Extractors (EXT). A strong extractor produces almost uniform bits from weak sources of randomness, even when the truly random seed used for extraction is revealed. We prove that any extractor function $\mathsf{Ext}\colon \{0,1\}^n \times \{0,1\}^s \to \{0,1\}^{100}$ that works for $(n, n^{1/2-\varepsilon})$-sources requires circuits with $\Omega(\log n)$ negations (see Section 3 for definitions).

This proof relies on the analysis of the *noise sensitivity* of circuits containing negations, together with a technique from Bogdanov and Guo [10].

2.3 Non-trivial Upper Bound for Small-Bias Generators

The above lower bounds may suggest the possibility that, with the exception of OWFs, all cryptographic building blocks require $\Omega(\log n)$ negations. We show one example of a primitive – small-bias generator (SBG) – that can be constructed with significantly fewer negations, namely, with any super-constant number of negations (for example, $\log^*(n)$ such gates).

A SBG can be thought of as a weaker version of a PRG, where the output fools linear distinguishers (i.e., it looks random to any distinguisher that can only apply a linear test). Thus, any PRG is also a SBG, but not vice-versa. We construct our SBG with few negations by outputting the input and an additional bit consisting of a parity of independent copies of the Tribes function.

Since SBGs are not quite a cryptographic primitive (these can be constructed unconditionally, and are not secure against polynomial adversaries), one may still conjecture that all "true" cryptographic primitives with the exception of OWFs require $\Omega(\log n)$ negations. We do not know whether this is the case, and it would be interesting to determine whether other primitives not covered in this paper can be monotone.

2.4 Lower Bounds for Boolean Circuits with Bottom Negations

In addition to studying specific primitives, we investigate general structural properties of circuits with negations. We prove a theorem showing that for monotone functions, the depth of any circuit with negations at the bottom (input) level only is lower bounded by the monotone depth complexity of the function minus the number of negations in the circuit. This improves a result by Koroth and

Sarma [25] (who proved a multiplicative rather than additive lower bound), and answers an open problem posed by them in the context of the circuit complexity of the Clique problem. Our proof is inspired by ideas from [25], and relies on a circular application of the Karchmer-Wigderson connection between boolean circuits and communication protocols.

This result suggests that negations at the bottom layer are less powerful and easier to study. In the Appendix we describe some techniques (following results of Blais et al. [8]) that allow one to decompose arbitrary computations into monotone and non-monotone components, and provide further evidence that negations at the bottom are less powerful (see also the discussion in Section 6).

Organization. We provide the definitions for most of the primitives mentioned in this paper in Section 3. Basic results used later in our proofs are presented in Section 4, with some proofs deferred to Appendix A. Our main results appear in Section 5. Finally, Section 6 discusses some open problems motivated by our work.

3   Preliminaries and Notation

In this section, we set up notation and define relevant concepts. We refer the reader to the textbooks [22], [5], [17], and [26] for more background in circuit complexity, computational complexity, theory of cryptography, and communication complexity, respectively.

3.1   Basic Notation

We use $[n]$ to denote the set $\{1, \ldots, n\}$. Given a Boolean string w, we use $|w|$ to denote its length, and $|w|_1$ to denote the number of 1's in w. Unless explicitly stated, we assume that the underlying probability distribution in our equations is the uniform distribution over the appropriate set. Further, we let $\mathcal{U}_\ell$ denote the uniform distribution over $\{0, 1\}^\ell$. We use $\log x$ to denote a logarithm in base 2, and $\ln x$ to refer to the natural base.

Given strings $x, y \in \{0, 1\}^n$, we write $x \preceq y$ if $x_i \leq y_i$ for every $i \in [n]$. A *chain* $\mathcal{X} = (x^1, \ldots, x^t)$ is a monotone sequence of strings over $\{0, 1\}^n$, i.e., $x^i \preceq x^{i+1}$ for every $i \in [1, t-1]$. We say that a chain $\mathcal{X} = (x^1, x^2, \ldots, x^t)$ is k-*alternating* with respect to a function $f: \{0, 1\}^n \to \{0, 1\}$ if there exist indexes $i_0 < i_1 < \ldots < i_k$ such that $f(x^{i_j}) \neq f(x^{i_{j+1}})$, for every $j \in [0, k-1]$. If this is true for every pair of consecutive elements of the chain, we say that the chain is *proper* (with respect to f). We let $a(f, \mathcal{X})$ be the size of the largest set of indexes satisfying this condition. The alternating complexity of a Boolean function f is given by $a(f) \overset{\text{def}}{=} \max_{\mathcal{X}} a(f, \mathcal{X})$, where $\mathcal{X}$ is a chain over $\{0, 1\}^n$. A function $f: \{0, 1\}^n \to \{0, 1\}$ is *monotone* if $f(x) \leq f(y)$ whenever $x \preceq y$. A function $g: \{0, 1\}^n \to \{0, 1\}^m$ is monotone if every output bit of g is a monotone Boolean function. Moreover, we say that a Boolean function $f: \{0, 1\}^n \to \{0, 1\}$ is *anti-monotone* if f is the negation of a monotone Boolean function.

3.2   Boolean Circuits and Negation Gates

Every Boolean circuit mentioned in this paper consists of AND, OR and NOT gates, where the first two types of gates have fan-in two. Recall that a Boolean function $f\colon \{0,1\}^n \to \{0,1\}$ is monotone if and only if it is computed by a circuit with AND and OR gates only.

For convenience, the size of a circuit C will be measured by its number of AND and OR gates, and will be denoted by $\mathsf{size}(C)$. The depth of a circuit C, denoted by $\mathsf{depth}(C)$, is the largest number of AND and OR gates in any path from the output gate to an input variable. The depth of a Boolean function f is the minimum depth of a Boolean circuit computing f. Similarly, the depth of a *monotone* function f, denoted by $\mathsf{depth}^+(f)$, is the minimum depth among all *monotone* circuits computing f. We will also consider multi-output Boolean circuits that compute Boolean functions $f\colon \{0,1\}^n \to \{0,1\}^m$. We stress that whenever we say that a function of this form is computed by a circuit with t negations, it means that there exists a *single* circuit (with multiple output gates) containing at most t negations computing f.

We say that a circuit contains negation gates at the bottom layer only if any NOT gate in the circuit gets as input an input variable x_i, for some $i \in [n]$. We will also say that circuits of this form are DeMorgan circuits. Put another way, a circuit $C(x)$ of size s with t negations at the bottom layer can be written as $D(x, (x \oplus \beta))$, where D is a *monotone* circuit of size s, $\beta \in \{0,1\}^n$ with $|\beta|_1 = t$ encodes the variables that appear negated in C, and $x \oplus \beta \in \{0,1\}^n$ is the string obtained via the bit-wise XOR operation. This notation is borrowed from Koroth and Sarma [25], which refers to β as the orientation vector.

3.3   Complexity Measures for Boolean Functions

Given a Boolean function $f\colon \{0,1\}^n \to \{0,1\}$ and an index $i \in [n]$, we use $I_i(f)$ to denote the *influence* of the i-th input variable on f, i.e., $I_i(f) \overset{\text{def}}{=} \Pr_x[f(x) \neq f(x^{\oplus i})]$, where $x^{\oplus i}$ denotes the string obtained from x by flipping its i-th coordinate. The *influence* of f (also known as *average-sensitivity*) is defined as $I(f) \overset{\text{def}}{=} \sum_{i \in [n]} I_i(f)$. We say that a Boolean function f is *balanced* or *unbiased* if $\Pr_x[f(x) = 1] = 1/2$. We use $\mathsf{NS}_p(f)$ to denote the *noise sensitivity* of f under noise rate $p \in [0, 1/2]$, which is defined as $\Pr[f(X \oplus Y) \neq f(X)]$, where X is distributed uniformly over $\{0,1\}^n$, and Y is the p-biased binomial distribution over $\{0,1\}^n$, i.e., each coordinate of Y is set to 1 independently with probability p.

3.4   Pseudorandom Functions and Weak Pseudorandom Functions

Let $\mathcal{F}_n$ be the set of all Boolean functions on n variables, and $F\colon \{0,1\}^m \times \{0,1\}^n \to \{0,1\}$. We say that F is an (s, ε)-secure *pseudorandom function* (PRF) if, for every (non-uniform) algorithm $\mathcal{A}$ that can be implemented by a circuit of size at most s,

$$\left| \Pr_{w \sim \{0,1\}^m} \left[\mathcal{A}^{F(w,\cdot)} = 1 \right] - \Pr_{f \sim \mathcal{F}_n} \left[\mathcal{A}^f = 1 \right] \right| \leq \varepsilon,$$

where $\mathcal{A}^h$ denotes the execution of $\mathcal{A}$ with oracle access to a Boolean function $h\colon \{0,1\}^n \to \{0,1\}$ (circuits with access to oracle gates are defined in the natural way).

A *weak pseudorandom function* (WPRF) is defined similarly, except that the distinguisher only has access to *random examples* of the form $(x, F(w,x))$, where x is uniformly distributed over $\{0,1\}^n$. In particular, any (s, ε)-secure pseudorandom function is an (s, ε)-secure weak pseudorandom function, while the other direction is not necessarily true.

3.5   Pseudorandom Generators and Small-Bias Generators

A function $G\colon \{0,1\}^n \to \{0,1\}^m$ is an (s, ε)-secure *pseudorandom generator* (PRG) with *stretch* $\ell \stackrel{\text{def}}{=} m - n$ if for every circuit $C(z_1, \ldots, z_m)$ of size s,

$$\left| \Pr_{x \sim \mathcal{U}_n} [C(G(x)) = 1] - \Pr_{y \sim \mathcal{U}_m} [C(y) = 1] \right| \leq \varepsilon.$$

We say that a function $g\colon \{0,1\}^n \to \{0,1\}^m$ is an ε-secure *small-bias generator* (SBG) with stretch $\ell = m - n$ if, for every nonempty set $S \subseteq [m]$,

$$\left| \Pr_{x \sim \mathcal{U}_n,\, y = g(x)} \left[\sum_{i \in S} y_i \equiv 1 \ (\mathrm{mod}\ 2) \right] - \frac{1}{2} \right| \leq \varepsilon.$$

Observe that small-bias generators can be seen as weaker pseudorandom generators that are required to be secure against linear distinguishers only. We refer the reader to Naor and Naor [30] for more information about the constructions and applications of such generators.

3.6   One-Way Functions, One-Way Permutations, and Hard-Core Bits

We say that a function $f\colon \{0,1\}^n \to \{0,1\}^m$ is an (s, ε)-secure *one-way function* (OWF) if for every circuit C of size at most s,

$$\Pr_{x \sim \mathcal{U}_n,\, y = f(x)} [C(y) \in f^{-1}(y)] \leq \varepsilon.$$

If $m = n$, we say that f is *length-preserving*. If in addition f is a one-to-one mapping, we say that f is an (s, ε)-secure *one-way permutation* (OWP).

We say that a function $h\colon \{0,1\}^n \to \{0,1\}$ is an (s, ε)-secure hard-core bit for a function $f\colon \{0,1\}^n \to \{0,1\}^m$ if, for every circuit C of size s,

$$\left| \Pr_{x \sim \mathcal{U}_n} [C(f(x)) = h(x)] - \frac{1}{2} \right| \leq \varepsilon.$$

3.7 Extractors and Error-Correcting Codes

The *min-entropy* of a random variable X, denoted by $H_\infty(X)$, is the largest real number k such that $\Pr[X = x] \leq 2^{-k}$ for every x in the range of X. A distribution X over $\{0,1\}^n$ with $H_\infty(X) \geq k$ is said to be an (n,k)-source. Given random variables X and Y with range $\{0,1\}^m$, we let

$$\delta(X,Y) \overset{\text{def}}{=} \max_{S \subseteq \{0,1\}^m} \big| \Pr[X \in S] - \Pr[Y \in S] \big|$$

denote their *statistical distance*. We say that X and Y are ε-close if $\delta(X,Y) \leq \varepsilon$.

We say that a function $\mathsf{Ext} \colon \{0,1\}^n \times \{0,1\}^s \to \{0,1\}^m$ is a (strong) (k,ε)-*extractor* (EXT) if, for any (n,k)-source X, the distributions $(\mathcal{U}_s, \mathsf{Ext}(X,\mathcal{U}_s))$ and $\mathcal{U}_{s+m}$ are ε-close.[3]

Given strings $y^1, y^2 \in \{0,1\}^m$, we let

$$\Delta(y^1, y^2) \overset{\text{def}}{=} \frac{|\{i \in [m] \mid y_i^1 \neq y_i^2\}|}{m}$$

be their *relative Hamming distance*. Given a function $E \colon \{0,1\}^n \to \{0,1\}^m$, we say that E has *relative distance* γ if for every distinct pair of inputs $x^1, x^2 \in \{0,1\}^n$, we have $\Delta(E(x^1), E(x^2)) \geq \gamma$. As a convention, we will refer to a function of this form as an *error-correcting code* (ECC) whenever we are interested in the distance between its output strings (also known as "codewords").

4 Basic Results and Technical Background

4.1 Karchmer-Wigderson Communication Games

Karchmer-Wigderson games [24] are a powerful tool in the study of circuit depth complexity. We focus here on games for *monotone* functions. Let $f \colon \{0,1\}^n \to \{0,1\}$ be a monotone function, and consider the following deterministic communication game between two players named Alice and Bob. Alice gets an input $x \in f^{-1}(1)$, while Bob receives $y \in f^{-1}(0)$. The goal of the players is to communicate the minimum number of bits (using an interactive protocol), and to output a coordinate $i \in [n]$ for which $x_i = 1$ and $y_i = 0$. The monotonicity assumption on f guarantees that such coordinate always exists.

Proposition 1 (Karchmer and Wigderson [24]). *Let $f \colon \{0,1\}^n \to \{0,1\}$ be a monotone function. There exists a monotone circuit C_f of depth d that computes f if, and only if, there exists a protocol Π_f for the (monotone) KW-game of f with communication cost d.*

---

[3] Two occurrences of the same random variable in an expression refer to the same copy of the variable.

4.2   Markov's Upper Bound

The following result was obtained by Markov [27].

Proposition 2 (Markov [27]). *Let $f\colon \{0,1\}^n \to \{0,1\}^m$ be an arbitrary function. Then f is computed by a (multi-output) Boolean circuit containing at most $\lceil \log(n + 1) \rceil$ negations.*

This result implies that many of our lower bounds are tight up to an additive term independent of n. Some of our proofs also rely on the following relation between negation complexity and alternation.

Proposition 3 (Markov [27]). *Let $f\colon \{0,1\}^n \to \{0,1\}$ be a Boolean function computed by a circuit with at most t negations. Then $a(f) = O(2^t)$.*

4.3   The Flow of Negation Gates

It is useful in some situations to decompose the computation of a function into monotone and non-monotone components. This idea has been applied successfully to obtain almost optimal bounds on the learnability of functions computed with few negation gates (Blais et al. [8]). An important lemma used in their paper can be stated as follows.

Lemma 1 (Blais et al. [8]). *Let $f\colon \{0,1\}^n \to \{0,1\}$ be a Boolean function computed by a circuit C with at most t negations. Then f can be written as $f(x) = h(g_1(x), \ldots, g_T(x))$, where each function g_i is monotone, $T = O(2^t)$, and h is either the parity function, or its negation.*

A drawback of this statement is that the computational complexity of each g_i is not related to the size of C. Roughly speaking, the proof of this result uses a circuit for f in order to gain structural information about f, and then rely on a non-constructive argument. We observe that, by relaxing the assumption on h, we can prove the following effective version of Lemma 1.[4]

Lemma 2. *Let $f\colon \{0,1\}^n \to \{0,1\}$ be a Boolean function computed by a circuit C of size s containing t negation gates. Then f can be written as $f(x) = h(g_1(x), \ldots, g_T(x))$, where each function g_i is computed by a monotone circuit of size at most s, $T = 2^{t+1} - 1$, and $h\colon \{0,1\}^T \to \{0,1\}$ is computed by a circuit of size at most $5T$.*

We state below a more explicit version of Lemma 2 for circuits with a single negation gate and several output bits. The proof of this result follows from the same argument used to derive Lemma 2, whose proof we give in Section A.

---

[4] This result was obtained during a discussion with Clement Canonne, Li-Yang Tan, and Rocco Servedio.

Lemma 3. *Let $f: \{0,1\}^n \to \{0,1\}^u$ be computed by a circuit of size s containing a single negation gate. Assume that the j-th output bit of f is computed by the function $f_j: \{0,1\}^n \to \{0,1\}$. Then, there exist monotone functions $m: \{0,1\}^n \to \{0,1\}$ and $m_{j,\ell}: \{0,1\}^n \to \{0,1\}$, where $j \in [u]$ and $\ell \in \{0,1\}$, which are computed by monotone circuits of size at most s, and a function $h: \{0,1\}^3 \to \{0,1\}$, such that:*

(i) For every $j \in [u]$, $f_j(x) = h(m(x), m_{j,0}(x), m_{j,1}(x))$.
(ii) For every $j \in [u]$ and $x \in \{0,1\}^n$, $m_{j,0}(x) \leq m_{j,1}(x)$.
(iii) The function h is defined as $h(z, y_1, y_0) \stackrel{\text{def}}{=} y_z$.

From a programming perspective, Lemma 3 shows that a single negation gate in a Boolean circuit can be interpreted as an IF-THEN-ELSE statement involving monotone functions. Conversely, the selection procedure computed by h can be implemented with a single negation.

For convenience of the reader, we sketch the proof of these results in Section A, where we also discuss the expressiveness of negations at arbitrary locations compared to negations at the bottom layer of a circuit. Lemmas 1 and 2 can be used interchangeably in our proofs.

4.4   Useful Inequalities

Some of our proofs rely on the following results for Boolean functions.

Proposition 4 (Fortuin, Kasteleyn, and Ginibre [15]). *If $g: \{0,1\}^n \to \{0,1\}$ and $f: \{0,1\}^n \to \{0,1\}$ are monotone Boolean functions, then*

$$\Pr_x[f(x) = 1 \wedge g(x) = 1] \geq \Pr_x[f(x) = 1] \cdot \Pr_x[g(x) = 1].$$

The same inequality holds for anti-monotone functions. In particular, for monotone $f, g: \{0,1\}^n \to \{0,1\}$, following inequality holds

$$\Pr_x[f(x) = 0 \wedge g(x) = 0] \geq \Pr_x[f(x) = 0] \cdot \Pr_x[g(x) = 0].$$

A stronger version of this inequality that will be used in some of our proofs is presented below.

Proposition 5 (Talagrand [37]). *For any pair of monotone Boolean functions $f, g: \{0,1\}^n \to \{0,1\}$, it holds that*

$$\Pr_x[f(x) = 1 \wedge g(x) = 1] \geq \Pr_x[f(x) = 1] \cdot \Pr_x[g(x) = 1] + \psi\left(\sum_{i \in [n]} I_i(f) \cdot I_i(g)\right),$$

where $\psi(x) \stackrel{\text{def}}{=} c \cdot x / \log(e/x)$, and $c > 0$ is a fixed constant independent of n.

Proposition 6 (Kahn, Kalai, and Linial [23]). *Let $f: \{0,1\}^n \to \{0,1\}$ be a balanced Boolean function. Then there exists an index $i \in [n]$ such that $I_i(f) = \Omega\left(\frac{\log n}{n}\right)$.*

Finally, we will make use of the following standard concentration bound (cf. Alon and Spencer [2]).

Proposition 7. *Let $X_1, \ldots, X_m$ be independent $\{0,1\}$ random variables, where each X_i is 1 with probability $p \in [0,1]$. In addition, set $X \overset{\text{def}}{=} \sum_i X_i$, and $\mu \overset{\text{def}}{=} \mathbb{E}[X] = pm$. Then, for any fixed $\zeta > 0$, there exists a constant $c_\zeta > 0$ such that*

$$\Pr[\,|X - \mu| > \zeta\mu\,] < 2e^{-c_\zeta \mu}.$$

5 Main Results

5.1 One-Way Functions versus One-Way Permutations

Goldreich and Izsak [19] proved that if one-way functions exist, then there are *monotone* one-way functions. We show below that this is not true for *one-way permutations*. In other words, one-way permutations are inherently non-monotone. This lower bound follows easily via the following structural result for monotone permutations.

Proposition 8. *Let $f \colon \{0,1\}^n \to \{0,1\}^n$ be a one-to-one function. If f is monotone, then there exists a permutation $\pi \colon [n] \to [n]$ such that, for every $x \in \{0,1\}^n$, $f(x) = x_{\pi(1)} \ldots x_{\pi(n)}$. In particular, there exists a (uniform) polynomial size circuit that inverts f on every input $y = f(x)$.*

Proof. Let $f_i \colon \{0,1\}^n \to \{0,1\}$ be the Boolean function corresponding to the i-th output bit of f. Since f is monotone, each function f_i is monotone. Consider functions f_ℓ and f_k, where $\ell \neq k$. By Talagrand's inequality (Proposition 5),

$$\Pr_x[f_\ell(x) = 1 \wedge f_k(x) = 1]$$

$$\geq \Pr_x[f_\ell(x) = 1] \cdot \Pr_x[f_k(x) = 1] + \psi\Big(\sum_{i \in [n]} I_i(f_\ell) \cdot I_i(f_k) \Big). \quad (1)$$

Since f is a permutation, $\Pr_x[f_\ell(x) = 1 \wedge f_k(x) = 1] = 1/4$ and $\Pr_x[f_\ell(x) = 1] = \Pr_x[f_k(x) = 1] = 1/2$. Consequently, it follows from Equation 1 and the definition of ψ that

$$\sum_{i \in [n]} I_i(f_\ell) \cdot I_i(f_k) = 0.$$

In other words, f_ℓ and f_k depend on a disjoint set of input variables. Since this is true for every pair ℓ and k with $\ell \neq k$, and every output bit of f is non-constant, there exists a permutation $\pi \colon [n] \to [n]$ such that, for every $i, j \in [n]$, if $I_i(f_j) > 0$ then $i = \pi(j)$. Moreover, as f is monotone and one-to-one, we must have $f_j(x) = x_{\pi(j)}$, for every $j \in [n]$. The corresponding permutation can be easily recovered from f by evaluating this function on every indicator string $e^i \in \{0,1\}^n$, where $e^i_j = 1$ if and only if $i = j$. This completes the proof of our result.

We remark that a simple extension of this proof allows us to rule out monotone one-way functions $f: \{0,1\}^n \to \{0,1\}^{n-k}$ where each pre-image set $f^{-1}(x)$ has size exactly 2^k (i.e., regular OWFs), and some relaxations of this notion.

Proposition 8 implies that any circuit computing a one-way permutation contains at least one negation gate. It is not clear how to extend its proof to obtain a stronger lower bound on the negation complexity of one-way permutations, as Talagrand's inequality holds for monotone functions only. Although we leave open the problem of obtaining better lower bounds, we give next an alternative proof of Proposition 8 that does not rely on Talagrand's result.

Proof. Let $S_k \overset{\text{def}}{=} \{x \in \{0,1\}^n \mid |x|_1 = k\}$, where $k \in [0,n]$. In other words, S_k is simply the k-th slice of the n-dimensional Boolean hypercube. Initially, we prove the following claim: For every set S_k, $f(S_k) = S_k$. In other words, f induces a permutation over each set of inputs S_k. We then use this result to establish Proposition 8.

First, observe that $f(0^n) = 0^n$. Otherwise, there exists an input $x \neq 0^n$ such that $f(x) = 0^n$. Since $0^n \preceq x$ and f is monotone, we get that $f(0^n) \preceq f(x)$, which contradicts the injectivity of f. This establishes the claim for S_0. The general case follows by induction on k. Assume the result holds for any $k' < k$, and consider an arbitrary $y \in S_k$. Since f is one-to-one, there exists $x \in S_\ell$ such that $f(x) = y$, where $\ell \geq k$. If $\ell \neq k$, there exists $x' \prec x$ such that $x' \in S_k$. Let $y' \overset{\text{def}}{=} f(x')$. Using that f is monotone and $x' \prec x$, we get that $y' \preceq y$. Since f is one-to-one, $y' \prec y$, thus $y' \in S_{k'}$ for some $k' < k$. This is in contradiction with our induction hypothesis and the injectivity of f, since $f(S_{k'}) = S_{k'}$, $x' \in S_k$, and $y' = f(x') \in S_{k'}$. This completes the induction hypothesis, and the proof of our claim.

Now let $\pi: [n] \to [n]$ be the permutation such that $f^{-1}(e^i) = e^{\pi(i)}$, where $e^j \in \{0,1\}^n$ is the input with 1 at the j-th coordinate only. Clearly, for every $x \in S_0 \cup S_1$, $f(x) = x_{\pi(1)} \ldots, x_{\pi(n)}$. On the other hand, for every $x \in S_k$ with $k > 1$, it follows from the monotonicity of f that

$$\bigvee_{i \,:\, x_i = 1} f(e^i) \preceq f(x),$$

where the disjunction is done coordinate-wise. Finally, it follows from our previous claim that we must also have $f(x) \in S_{|x|_1}$. Therefore,

$$\bigvee_{i \,:\, x_i = 1} f(e^i) = f(x).$$

Consequently, for every $x \in \{0,1\}^n$, it follows that $f(x) = x_{\pi(1)} \ldots x_{\pi(n)}$, which completes the proof.

5.2   Pseudorandom Generators and Small-Bias Generators

In contrast to the situation for one-way functions, Goldreich and Izsak [19] presented an elegant proof that pseudorandom generators cannot be monotone.

More specifically, their result shows that the output distribution of a monotone function $G: \{0,1\}^n \to \{0,1\}^{n+1}$ can be distinguished from random either by the projection of one of its output bits, or via the conjunction of two output bits.

Recall from Section 3 that small-bias generators can be seen as restricted pseudorandom generators that are only required to be secure against linear tests. We prove next that the techniques from [19] can be used to show that there are no $(1/n^{\omega(1)})$-secure monotone small-bias generators with 1 bit of stretch. We observe later in this section that such generators can be constructed with any super-constant number of negation gates.

Proposition 9. *For any monotone function $G: \{0,1\}^n \to \{0,1\}^{n+1}$, there exists a (non-uniform) linear test $D: \{0,1\}^{n+1} \to \{0,1\}$ such that*

$$\left| \Pr_{x \sim \mathcal{U}_n} [D(G(x)) = 1] - \frac{1}{2} \right| = \Omega\left(\frac{1}{n^2}\right).$$

Proof. The proof follows closely the argument in [19], combined with an appropriate application of the FKG inequality (Proposition 4). Let $G_i: \{0,1\}^n \to \{0,1\}$ be the Boolean function corresponding to the i-th output bit of G, where $i \in [n+1]$. Observe that if there exists i such that

$$\left| \Pr_{x \sim \mathcal{U}_n} [G_i(x) = 1] - \frac{1}{2} \right| = \Omega\left(\frac{1}{n^2}\right),$$

then there is a trivial linear distinguisher for G.

Assume therefore that, for every $i \in [n+1]$, G_i is almost balanced. In particular, each function G_i is $\delta(n)$-close under the uniform distribution to an unbiased function $\widetilde{G}_i: \{0,1\}^{n+1} \to \{0,1\}$, where $\delta(n) = o((\log n)/n)$. It follows from Proposition 6 that each function $\widetilde{G}_i$ has an influential variable. More precisely, there exists $\gamma: [n+1] \to [n]$ such that

$$I_{\gamma(i)}(\widetilde{G}_i) = \Omega\left(\frac{\log n}{n}\right),$$

for every $i \in [n+1]$. As each G_i is $\delta(n)$-close to $\widetilde{G}_i$, it follows that $I_{\gamma(i)}(G_i) = \Omega\left(\frac{\log n}{n}\right)$ as well.

By the pigeonhole principle, there exist distinct indexes i and j such that $\gamma(i) = \gamma(j)$. It follows from Proposition 5 that

$$\Pr_x[G_i(x) = 1 \wedge G_j(x) = 1]$$

$$\geq \Pr_x[G_i(x) = 1] \cdot \Pr_x[G_j(x) = 1] + \psi\left(\sum_{k \in [n]} I_k(G_i) \cdot I_k(G_k) \right)$$

$$\geq \Pr_x[G_i(x) = 1] \cdot \Pr_x[G_j(x) = 1] + \Omega\left(\psi\left(\frac{\log^2 n}{n^2}\right)\right)$$

$$\geq \Pr_x[G_i(x) = 1] \cdot \Pr_x[G_j(x) = 1] + \Omega\left(\frac{\log n}{n^2}\right).$$

On the other hand, Proposition 4 implies that

$$\Pr_x[G_i(x) = 0 \wedge G_j(x) = 0] \geq \Pr_x[G_i(x) = 0] \cdot \Pr_x[G_j(x) = 0].$$

Combining both inequalities, and using the assumption that each output bit of G is almost balanced, we get that:

$$\Pr_x[G_i(x) + G_j(x) = 0]$$

$$= \Pr_x[G_i(x) = 1 \wedge G_j(x) = 1] + \Pr_x[G_i(x) = 0 \wedge G_j(x) = 0]$$

$$\geq \Pr_x[G_i(x) = 1] \cdot \Pr_x[G_j(x) = 1] + \Pr_x[G_i(x) = 0] \cdot \Pr_x[G_j(x) = 0] + \Omega\left(\frac{\log n}{n^2}\right)$$

$$\geq \frac{1}{2} - O\left(\frac{1}{n^2}\right) + \Omega\left(\frac{\log n}{n^2}\right)$$

$$= \frac{1}{2} + \Omega\left(\frac{\log n}{n^2}\right).$$

Therefore, the linear function $D(y) \overset{\text{def}}{=} y_i + y_j$ can distinguish the output of G from random with the desired advantage, which completes the proof. $\blacksquare$

In contrast, we show next that there *are* small-bias generators with super-polynomial security that can be computed with *any* super-constant number of negations. Let $\mathsf{Tribes}_{s,t} \colon \{0,1\}^{s \cdot t} \to \{0,1\}$ be the (monotone) Boolean function defined as

$$\mathsf{Tribes}_{s,t}(x_1, \ldots, x_{s \cdot t}) = \bigvee_{i=0}^{s-1} \bigwedge_{j=1}^{t} x_{i \cdot t + j}.$$

Further, we use $\mathsf{Tribes}_m \colon \{0,1\}^m \to \{0,1\}$ to denote the function $\mathsf{Tribes}_{s,t}$, where s is the largest integer such that $1 - (1 - 2^{-t})^s \leq 1/2$, and $t = m/s$ (i.e., we try to make Tribes as balanced as possible as a function over m variables).

Proposition 10. *Let* $f \colon \{0,1\}^n \to \{0,1\}$ *be* $f(x) \overset{\text{def}}{=} \oplus_{i=1}^{k} \mathsf{Tribes}_{n/k}(x^{(i)})$, *where* $x^{(i)}$ *denotes the i-th block of x with length n/k. Let* $1 \leq k(n) \leq n/\log n$, *and* $G \colon \{0,1\}^n \to \{0,1\}^{n+1}$ *be defined by* $G(x) \overset{\text{def}}{=} (x, f(x))$, *Then, there exists a constant $C > 0$ such that, for any linear function* $D \colon \{0,1\}^{n+1} \to \{0,1\}$,

$$\left| \Pr_{x \sim \mathcal{U}_n}[D(G(x)) = 1] - \frac{1}{2} \right| \leq (C \cdot (k/n) \cdot \log(n/k))^k.$$

In particular, when $k = \omega(1)$, we can get a small-bias generator with negligible error that can be computed with roughly $\log k$ negations (via Proposition 2). Interestingly, for $k = 2$ we obtain an SBG computed with a *single negation* and security $\tilde{\Theta}(n^{-2})$, essentially matching the lower bound for *monotone* SBGs given by Proposition 9.

Proof. We assume the reader is familiar with basic concepts from analysis of Boolean functions (cf. O'Donnell [31]). Suppose that $D(y) \overset{\text{def}}{=} \sum_{i \in S} y_i \pmod 2$, where $S \subseteq [n+1]$ is nonempty. If $n+1 \notin S$, using that the first n output bits of G are uniformly distributed over $\{0,1\}^n$, we get that $|\Pr_x[D(G(x)) = 1] - 1/2| = 0$. Assume therefore that $n + 1 \in S$, and let $S' \stackrel{\text{def}}{=} S \backslash \{n + 1\}$. Then,

$$
\left| \Pr_{x \sim \mathcal{U}_n} \left[D(G(x)) = 1 \right] - \frac{1}{2} \right| = \left| \Pr_{x \sim \mathcal{U}_n} \left[f(x) + \sum_{i \in S'} x_i \equiv 1 \ (\text{mod } 2) \right] - \frac{1}{2} \right|
$$

$$
= \left| \Pr_{x \sim \mathcal{U}_n} \left[\sum_{i \in S'} x_i \not\equiv f(x) \ (\text{mod } 2) \right] - \frac{1}{2} \right|
$$

$$
\stackrel{\text{def}}{=} p.
$$

Let $f^- \colon \{-1, 1\}^n \to \{-1, 1\}$ be the corresponding version of f where we map 0 to 1, and 1 to -1, as usual. Observe that, under this correspondence,

$$
\sum_{i \in S'} x_i \not\equiv f(x) \ (\text{mod } 2) \quad \Longleftrightarrow \quad \chi_{S'}(x) \cdot f^-(x) = -1.
$$

Therefore,

$$
p = \left| \left(\frac{1}{2} - \frac{1}{2} \cdot \mathbb{E}_{x \sim \{-1,1\}^n} \left[\chi_{S'}(x) \cdot f^-(x) \right] \right) - \frac{1}{2} \right|
$$

$$
= \left| \left(\frac{1}{2} - \frac{1}{2} \cdot \widehat{f^-}(S') \right) - \frac{1}{2} \right|
$$

$$
= \frac{1}{2} \cdot |\widehat{f^-}(S')|.
$$

In other words, in order to upper bound the distinguishing probability p, it is enough to upper bound $|\widehat{f^-}(S')|$, where $S' \subseteq [n]$. Using that $x^{(i)}$ and $x^{(j)}$ are disjoint for $i \neq j$ and $f^-(x) = \prod_{i \in [k]} \text{Tribes}^-_{n/k}(x^{(i)})$, it follows that $\widehat{f^-}(S')$ is a product of Fourier coefficients of the corresponding Tribes functions. It is known that

$$
\max_{T \subseteq [m]} \left| \widehat{\text{Tribes}^-_m}(T) \right| = O\left(\frac{\log m}{m} \right)
$$

as $m \to \infty$ (see e.g. O'Donnell [31]). Consequently, since we have $m = n/k$, we get that

$$
p = \frac{1}{2} \cdot |\widehat{f^-}(S')| \leq \frac{1}{2} \cdot \max_{T \subseteq [n/k]} \left| \widehat{\text{Tribes}^-_{n/k}}(T) \right|^k \leq (C \cdot (k/n) \cdot \log(n/k))^k,
$$

for an appropriate constant C.

It is possible to use other monotone functions for the construction in Proposition 10, but our analysis provides better parameters with Tribes.

5.3 Pseudorandom Functions

In this section we prove that a pseudorandom function is a highly non-monotone cryptographic primitive. For simplicity, we will not state the most general version of our result. We discuss some extensions after its proof.

Proposition 11. *If $F: \{0,1\}^m \times \{0,1\}^n \to \{0,1\}$ is a $(\mathsf{poly}(n), 1/3)$-secure pseudorandom function, then any Boolean circuit computing F contains at least $\log n - O(1)$ negation gates.*

Proof. Consider the following algorithm D^h that has membership access to an arbitrary function $h: \{0,1\}^n \to \{0,1\}$, and computes as follows. Let $\mathcal{X} \stackrel{\text{def}}{=} (e^0, e^1, \ldots, e^n)$ be the chain over $\{0,1\}^n$ with $e^i \stackrel{\text{def}}{=} 1^i 0^{n-i}$. After querying h on each input $e^0, \ldots, e^n$ and computing $a(h, \mathcal{X})$, D accepts h if and only if $a(h, \mathcal{X}) \geq n/4$. This completes the description of D. Clearly, this algorithm can be implemented in polynomial time.

Observe that if $f \sim \mathcal{F}_n$ is a random Boolean function over n variables, then $\mathbb{E}_f[a(f, \mathcal{X})] = n/2$. In addition, it follows from a standard application of Proposition 7 that $|a(f, \mathcal{X}) - n/2| \leq n/4$ with probability exponentially close to 1. Therefore, under our assumption that F is a $(\mathsf{poly}(n), 1/3)$-secure pseudorandom function,

$$\frac{1}{3} \geq \left| \Pr_{w \sim \{0,1\}^n}[D^{F(w, \cdot)} = 1] - \Pr_{f \sim \mathcal{F}_n}[D^f = 1] \right|$$

$$\geq \left| \Pr_{w \sim \{0,1\}^n}[D^{F(w, \cdot)} = 1] - (1 - o(1)) \right|,$$

which implies in particular that $\Pr[D^{F(w, \cdot)} = 1] \geq 2/3 - o(1)$. Therefore, there must exist some seed w^* for which the resulting function $F_{w^*} \stackrel{\text{def}}{=} F(w^*, \cdot)$ over n-bit inputs satisfies $a(F_{w^*}, \mathcal{X}) \geq n/4$. It follows from Proposition 3 that if C is a circuit with t negations computing F_{w^*}, then

$$n/4 \ \leq \ a(F_{w^*}, \mathcal{X}) \ \leq \ a(F_{w^*}) \ \leq \ c \cdot 2^t,$$

where c is a fixed positive constant. Consequently, $t \geq \log n - O(1)$. Finally, it is clear that any circuit for F also requires $\log n - O(1)$ negations, which completes the proof.

Note that we can replace $1/3$ with any constant in $[0, 1)$. The proof of Proposition 11 also implies that if F is a sufficiently secure pseudorandom function, then for most choices of the seed $w \in \{0,1\}^m$, the resulting function $F(w, \cdot)$ over n input variables requires $\log n - O(1)$ negations. Further, observe that our distinguisher is quite simple, and makes $n + 1$ *non-adaptive* queries.

The same proof does not work for *weak* pseudorandom functions. In this case, most random examples obtained from the oracle are concentrated around the middle layer of the hypercube, and one cannot construct a chain. We remark, however, that weak pseudorandom functions cannot be monotone, as there are weak learning algorithms for the class of monotone functions (cf. Blum, Burch, and Langford [9]). We discuss the problem of obtaining better lower bounds for WPRFs in Section 6. (The upper bound on the negation complexity of WPRFs follows via standard techniques, see Section 5.5 and Blais et al. [8].)

5.4 Error-Correcting Codes

In this section, we show that circuits with few negations cannot compute error-correcting codes with good parameters. The proof generalizes the argument given by Buresh-Oppenheim, Kabanets and Santhanam [12] in the case of *monotone* error-correcting codes.

Proposition 12. *Let* $E\colon \{0,1\}^n \to \{0,1\}^m$ *be an error-correcting code with relative distance* $\gamma > 0$. *If* C *is a circuit with* t *negations that computes* E, *then* $t \geq \log n - \log(1/\gamma) - 1$.

Proof. Assume that $E\colon \{0,1\}^n \to \{0,1\}^m$ is computed by a (multi-output) circuit C^0 with t negation gates, and let $x_1, \ldots, x_n$ be its input variables. For convenience, we write C_i^0 to denote the Boolean function computed by the i-th output gate of C^0. We proceed as in the proof of Lemma 2. More precisely, we remove one negation gate during each step, but here we also inspect the behavior of the error-correcting code on a particular set of inputs of interest. Let $\mathcal{X} \overset{\text{def}}{=} (e^0, e^1, \ldots, e^n)$ be the chain over $\{0,1\}^n$ with $e^i \overset{\text{def}}{=} 1^i 0^{n-i}$.

It follows from an easy generalization of Lemma 3 that there exist functions $f\colon \{0,1\}^n \to \{0,1\}$, $h\colon \{0,1\}^3 \to \{0,1\}$, and $g_{i,b}\colon \{0,1\}^n \to \{0,1\}$, where $i \in [m]$ and $b \in \{0,1\}$, for which the following holds.

- f is monotone;
- h is the addressing function $h(a, d_0, d_1) \overset{\text{def}}{=} d_a$;
- for every $x \in \{0,1\}^n$ and $i \in [m]$,

$$E(x)_i = h(f(x), g_{i,0}(x), g_{i,1}(x)).$$

- there exist (multi-output) circuits $C^{1,0}$ and $C^{1,1}$ over input variables $x_1, \ldots, x_n$ such that, for every $i \in [m]$ and $b \in \{0,1\}$,

$$C^{1,b}(x)_i = g_{i,b}(x).$$

- each circuit $C^{1,b}$ contains at most $t - 1$ negations.

Since $e^0 \prec e^1 \prec \ldots \prec e^n$ and f is monotone, there exists $k \in [0, n]$ such that $f(e^\ell) = 0$ if and only if $\ell < k$. By the pigeonhole principle, f is constant on a (continuous) subchain $\mathcal{X}^1 \subseteq \mathcal{X}$ of size at least $(n+1)/2$, and there exists a constant $b \in \{0,1\}$ such that

$$E(e_i) = g_{1,b}(e^i) \ldots g_{m,b}(e^i),$$

whenever $e^i \in \mathcal{X}^1$. Consequently, there exists a (multi-output) circuit C^1 computed with at most $t - 1$ negations that agrees with E on every $e^i \in \mathcal{X}^1$.

Observe that this argument can be applied once again with respect to $\mathcal{X}^1$ and C^1. Therefore, it is not hard to see that there must exist a chain $\mathcal{X}^t \subseteq \mathcal{X}$ of size $w \geq (n+1)/2^t$ and a *monotone* (multi-output) circuit C^t such that

$$C^t(e^i) = E(e^i),$$

for every $e^i \in \mathcal{X}^t$.

Assume that $\mathcal{X}^t = (e^j, e^{j+1}, \ldots, e^{j+w-1})$, and let $\mathcal{Y} \overset{\text{def}}{=} (y^j, \ldots, y^{j+w-1})$, where $y^i \overset{\text{def}}{=} E(e^i)$. Since C^t is *monotone* and $\mathcal{X}^t$ is a chain over $\{0,1\}^n$, we get that $\mathcal{Y}$ is a chain over $\{0,1\}^m$. By the pigeonhole principle, there exists an index $k \in [j+1, j+w-1]$ for which $y^{j-1} \preceq y^j$ and $|y^j|_1 - |y^{j-1}|_1 \le (m+1)/w$. Now using that E computes an error-correcting code of relative distance at least γ, it follows that

$$\gamma \le \Delta(y^j, y^{j-1}) \le \frac{m+1}{w} \cdot \frac{1}{m} \le \frac{2^t}{n+1} \cdot \frac{m+1}{m},$$

which completes the proof of our result.

It is possible to show via a simple probabilistic construction that there is a sequence of error-correcting codes $E_n \colon \{0,1\}^n \to \{0,1\}^{O(n)}$ with relative distance, say, $\gamma = 0.01$ (see e.g. Arora and Barak [5]). Proposition 12 implies that computing such codes requires at least $\log n - O(1)$ negation gates, which is optimal up to the additive term via Markov's upper bound (Proposition 2).

5.5   Hard-core Bits

We prove in this section that general hard-core predicates must be highly non-monotone. This result follows from a lower bound on the average-sensitivity of such functions due to Goldmann and Russell [16], together with structural results about monotone Boolean functions and Lemma 1. Roughly speaking, our result says that there are one-way functions that do not admit hardcore predicates computed with less than $(1/2) \cdot \log n$ negations (assuming that one-way functions exist).

Proposition 13. *Assume there exists a family $f = \{f_n\}_{n\in\mathbb{N}}$ of $(\text{poly}(n), n^{-\omega(1)})$-secure one-way functions, where each $f_n \colon \{0,1\}^n \to \{0,1\}^n$. Then, for every $\varepsilon > 0$, there exists a family $g^\varepsilon = \{g_n\}_{n\in\mathbb{N}}$ of (length-preserving) $(\text{poly}(n), n^{-\omega(1)})$-secure one-way functions for which the following holds. If $h = \{h_n\}_{n\in\mathbb{N}}$ is a $(\text{poly}(n), n^{-\omega(1)})$-secure hard-core bit for g^ε, then for every n sufficiently large, any Boolean circuit computing h_n contains at least $(1/2 - \varepsilon) \log n$ negations.*

Proof. It follows from the main result of Goldmann and Russell [16] that under the existence of one-way functions, there exists a one-way function family $g^\delta = \{g_n\}_{n\in\mathbb{N}}$ that only admits hard-core bit predicates with average-sensitivity $\Omega(n^{1-\delta})$. Our result follows easily once we observe that the average-sensitivity of Boolean functions computed with t negations is $O(2^t \cdot \sqrt{n})$.[5]

First, if $f \colon \{0,1\}^n \to \{0,1\}$ is a monotone Boolean function, then $I(f) = O(\sqrt{n})$ (see e.g. O'Donnell [31]). On the other hand, it follows from Lemma 1 that any Boolean function $h \colon \{0,1\}^n \to \{0,1\}$ computed by a circuit with t

---

[5] This result is from Blais et al. [8], and we include its short argument here for completeness.

negation gates can be written as $h(x) = P(m_1(x), \ldots, m_T(x))$, where $T = O(2^t)$, each function m_i is monotone, and P is either the parity function or its negation. Therefore, using the definition of influence,

$$I(h) = I(P(m_1, \ldots, m_T)) \leq \sum_{i \in [T]} I(m_i) \leq T \cdot O(\sqrt{n}) = O(2^t \cdot \sqrt{n}),$$

which completes the proof.

This result is almost optimal, as any function $f \colon \{0,1\}^n \to \{0,1\}$ can be $(1/n^{\omega(1)})$-approximated by a Boolean function computed with $(1/2 + o(1)) \log n$ negations (check Blais et al. [8] for more details). More precisely, if h is a hard-core bit for f, its approximator $\tilde{h}$ is also hard-core for f, as the inputs $f(x)$ given to the distinguisher are produced with $x \sim \mathcal{U}_n$.

5.6   Randomness Extractors

In this section, we show in Proposition 14 that strong $(n^{0.5-\varepsilon}, 1/2)$-extractors can only be computed by circuits with $\Omega(\log n)$ negation gates, for any constant $0 < \varepsilon \leq 1/2$. We proceed as follows. First, we argue that such extractors must have high noise sensitivity. The proof of this result employs a technique from Bogdanov and Guo [10]. We then upper bound the noise sensitivity of circuits with few negations. Together, these claims provide a trade-off between the parameters of the extractor, and the minimum number of negations in any circuit computing the extractor.

For convenience, we view the extractor $\mathsf{Ext} \colon \{0,1\}^n \times \{0,1\}^s \to \{0,1\}^m$ as a family of functions

$$\mathcal{H}_{\mathsf{Ext}} \overset{\text{def}}{=} \{h_w \colon \{0,1\}^n \to \{0,1\}^m \mid h_w = \mathsf{Ext}(\cdot, w), \text{ where } w \in \{0,1\}^s\},$$

i.e., the family of functions obtained from the extractor by fixing its seed. Similarly, every such family can be viewed as a strong extractor in the natural way.

Lemma 4. *Let* $0 \leq p \leq 1/2$, $0 \leq \gamma \leq 1/4$, *and* $\mathcal{H} \subseteq \{h \mid h \colon \{0,1\}^n \to \{0,1\}^m\}$ *be a family of functions. Assume that* $\mathsf{NS}_p(h_i) \leq \gamma$ *for every function* $h_i \colon \{0,1\}^n \to \{0,1\}$ *that computes the i-th output bit of some function in* $\mathcal{H}$, *where* $i \in [m]$. *Then there exists a distribution* D *over* $\{0,1\}^n$ *with min-entropy* $H_\infty(D) = n \cdot \log(\frac{1}{1-p})$ *such that the statistical distance between* $(\mathcal{H}, \mathcal{H}(D))$ *and* $(\mathcal{H}, \mathcal{U}_m)$ *is at least* $(1 - 2\sqrt{\gamma} - 2^{-0.1m})(1 - 2\sqrt{\gamma})$.

Proof. For a fixed $y \in \{0,1\}^n$, let D_y denote a random variable distributed according to $y \oplus X$, where X is the p-biased binomial distribution over $\{0,1\}^n$. Since $p \leq 1/2$, observe that the min-entropy of D_y is precisely

$$H_\infty(D_y) = - \log \max_{z \in \{0,1\}^n} \Pr[y \oplus X = z] = - \log \Pr[y \oplus X = y]$$

$$= - \log (1-p)^n = n \cdot \log \left(\frac{1}{1-p} \right).$$

We will need the following result.

Claim. For any fixed $h \in \mathcal{H}$,

$$E_{y \sim \{0,1\}^n} [\delta(h(D_y), \mathcal{U}_m)] \geq (1 - 2\sqrt{\gamma} - 2^{-0.1m})(1 - 2\sqrt{\gamma}). \tag{2}$$

We use this claim to complete the demonstration of Lemma 4, then return to its proof. Observe that, for any fixed $y \in \{0,1\}^n$,

$$\delta((\mathcal{H}, \mathcal{H}(D_y)), (\mathcal{H}, \mathcal{U}_m)) = E_{h \sim \mathcal{H}}[\delta(h(D_y), \mathcal{U}_m)]. \tag{3}$$

It follows from Equation 3 that

$$
\begin{aligned}
E_{y \sim \{0,1\}^n}[\delta((\mathcal{H}, \mathcal{H}(D_y)), (\mathcal{H}, \mathcal{U}_m))] &= E_{y \sim \{0,1\}^n}[E_{h \sim \mathcal{H}}[\delta(h(D_y), \mathcal{U}_m)]] \\
&= E_h[E_y[\delta(h(D_y), \mathcal{U}_m)]] \\
\text{(Using Equation 2)} \quad &\geq E_h[(1 - 2\sqrt{\gamma} - 2^{-0.1m})(1 - 2\sqrt{\gamma})] \\
&= (1 - 2\sqrt{\gamma} - 2^{-0.1m})(1 - 2\sqrt{\gamma}).
\end{aligned}
$$

In particular, there exists $y \in \{0,1\}^n$ such that

$$\delta((\mathcal{H}, \mathcal{H}(D_y)), (\mathcal{H}, \mathcal{U}_m)) \geq (1 - 2\sqrt{\gamma} - 2^{-0.1m})(1 - 2\sqrt{\gamma}),$$

which completes the proof of Lemma 4.

We proceed now to the proof of our initial claim. Fix a function $h \in \mathcal{H}$. By the definition of noise sensitivity and our assumption on $\mathcal{H}$, for every function $h_i \colon \{0,1\}^n \to \{0,1\}$ obtained from a function $h \in \mathcal{H}$ as the projection of the i-th output bit, we have

$$\Pr_y[h_i(D_y) \neq h_i(y)] = \Pr_y[h_i(y \oplus X) \neq h_i(y)] \leq \gamma.$$

Using the linearity of expectation, we obtain

$$E_y[\Delta(h(D_y), h(y))] \leq \gamma.$$

By Markov's inequality,

$$\Pr_y[\Delta(h(D_y), h(y)) \leq 1/4] \geq 1 - 4\gamma.$$

Using an averaging argument, with probability at least $1 - \sqrt{4\gamma}$ over the choice of y, we have that

$$\Pr[\Delta(h(D_y), h(y)) \leq 1/4] \geq 1 - \sqrt{4\gamma}. \tag{4}$$

For any fixed y, consider the following statistical test,

$$T_y \stackrel{\text{def}}{=} \{z \in \{0,1\}^m \mid \Delta(z, h(y)) \leq 1/4\}.$$

The probability that $\mathcal{U}_m \in T_y$ can be upper bounded via a standard inequality by

$$\Pr_{z \sim \mathcal{U}_m}[z \in T_y] \leq \frac{2^{m \cdot H_2(1/4)}}{2^m} \leq 2^{-0.1m}, \tag{5}$$

where $H_2 \colon [0,1] \to [0,1]$ is the binary entropy function, and we use the fact that $H_2(1/4) \leq 0.9$. Combining Equations 4 and 5, we get

$$\Pr_y\left[\left(\Pr_X[h(D_y) \in T_y] - \Pr_{z \sim \mathcal{U}_m}[z \in T_y]\right) \geq 1 - \sqrt{4\gamma} - 2^{-0.1m}\right] \geq 1 - \sqrt{4\gamma},$$

which implies that

$$\Pr_y[\delta(h(D_y), \mathcal{U}_m) \geq 1 - 2\sqrt{\gamma} - 2^{-0.1m}] \geq 1 - 2\sqrt{\gamma}.$$

Finally, since $\delta(\cdot)$ is non-negative and $\gamma \leq 1/4$, it follows that

$$E_y[\delta(h(D_y), \mathcal{U}_m)] \geq (1 - 2\sqrt{\gamma} - 2^{-0.1m})(1 - 2\sqrt{\gamma}),$$

which completes the proof of the claim.

We are now ready to prove a lower bound on the negation complexity of strong extractors.

Proposition 14. *Let $0 < \alpha < 1/2$ be a constant, and $m(n) \geq 100$. Further, suppose that $\mathcal{H} \subseteq \{h \mid h \colon \{0,1\}^n \to \{0,1\}^m\}$ is a family of functions such that each output bit $h_i \colon \{0,1\}^n \to \{0,1\}$ of a function $h \in \mathcal{H}$ is computed by a circuit with t negations. Then, if $\mathcal{H}$ is an $(n^{\frac{1}{2}-\alpha}, 1/2)$-extractor,*

$$t \geq \alpha \log n - O(1).$$

Proof. It is known that for any monotone function $g \colon \{0,1\}^n \to \{0,1\}$ and $p(n) \in (0,1/2)$, $\mathsf{NS}_p(g) = O(\sqrt{n} \cdot p)$ (see e.g. O'Donnell [31]). Using an argument similar to the one employed in the proof of Proposition 13, it follows from Lemma 1 that if $f \colon \{0,1\}^n \to \{0,1\}$ is a Boolean function computed by a circuit with t negations, then

$$\mathsf{NS}_p(f) \leq C_1 \cdot 2^t \sqrt{n} \cdot p \overset{\text{def}}{=} \gamma,$$

where $C_1 > 0$ is a fixed constant. In other words, this upper bound on the noise sensitivity and our assumption on $\mathcal{H}$ allow us to apply Lemma 4 with an appropriate choice of parameters, which we describe next.

We choose a $0 \leq p \leq \frac{1}{2}$ such that $n \cdot \log \frac{1}{(1-p)} = n^{\frac{1}{2}-\alpha}$. Observe that we can take $p \leq C_2 n^{-\frac{1}{2}-\alpha}$, for an appropriate constant $C_2 > 0$. Let C_3 be a sufficiently large constant such that $C_1 C_2 2^{-C_3} < 1/64$, and suppose that $t < \alpha \log n - C_3$. For this setting of parameters, we obtain

$$\gamma = C_1 \cdot 2^t \cdot \sqrt{n} \cdot p < \frac{1}{64}.$$

By Lemma 4, there exists a distribution D of min-entropy $H_\infty(D) = n \log \frac{1}{1-p} = n^{\frac{1}{2}-\alpha}$ for which

$$\delta((\mathcal{H}, \mathcal{H}(D)), (\mathcal{H}, \mathcal{U}_m)) \geq (1 - 2\sqrt{\gamma} - 2^{-0.1m})(1 - 2\sqrt{\gamma})$$
$$> (\frac{3}{4} - 2^{-0.1m}) \cdot \frac{3}{4} \geq \frac{1}{2},$$

which contradicts our assumption that $\mathcal{H}$ is an $(n^{\frac{1}{2}-\alpha}, 1/2)$-extractor. Therefore,

$$t \geq \alpha \log n - C_3 = \alpha \log n - O(1),$$

as desired.

Observe that Proposition 14 provides an almost tight lower bound on the number of negations for extractors with rather weak parameters: in order to extract from reasonable sources only 100 bits that are not ridiculously far from uniform, the corresponding circuits need $\Omega(\log n)$ negations.

5.7 Negations at the Bottom Layer and Circuit Lower Bounds

In this section we solve a problem posed by Koroth and Sarma [25]. Our proof is inspired by ideas introduced in their paper. Our main contribution is the the following general proposition.

Proposition 15. *Let $f: \{0,1\}^n \to \{0,1\}$ be a monotone Boolean function, and C be a circuit computing f with negation gates at the bottom layer only. Then,*

$$\mathsf{depth}(C) + \mathsf{negations}(C) \geq \mathsf{depth}^+(f).$$

Proof. Let $d \overset{\text{def}}{=} \mathsf{depth}(C)$, and $t \overset{\text{def}}{=} \mathsf{negations}(C)$. The idea is to use C, a non-monotone circuit for f, to solve the corresponding monotone Karchmer-Wigderson game of f with communication at most $d + t$. It follows from Proposition 1 that $\mathsf{depth}^+(f) \leq d + t$, which completes the proof. More details follow.

Recall that in the monotone Karchmer-Wigderson game for f, Alice is given a string $x \in f^{-1}(1)$, Bob is given $y \in f^{-1}(0)$, and their goal is to agree on a coordinate i such that $x_i = 1$ and $y_i = 0$. Let $T \subset [n]$ be the set of variables that occur negated in C, where $|T| = t$. Given a string $x \in \{0,1\}^n$, we write x_T to denote the substring of x obtained by concatenating the bits indexed by T. During the first round of the protocol, Alice sends x_T to Bob. If among these coordinates there is an index $i \in T$ for which $x_i = 1$ and $y_i = 0$, the protocol terminates with a correct solution. Otherwise, Bob defines a new input $y' \in \{0,1\}^n$ for him as follows: $y'_j \overset{\text{def}}{=} x_j$ if $j \in T$, otherwise $y'_j \overset{\text{def}}{=} y_j$. For convenience, Alice sets $x' \overset{\text{def}}{=} x$.

It is not hard to see that if there was no good index $i \in T$, then $f(x') = 1$ and $f(y') = 0$. Clearly, $1 = f(x) = f(x')$, since $x = x'$. On the other hand, if there is no good index i, y' is obtained from y simply by flipping some bits of

y from 1 to 0. In other words, $y' \preceq y$, and the monotonicity of f implies that $f(y') \leq f(y) = 0$.

Crucially, the players now have inputs $x', y' \in \{0,1\}^n$ that agree on every bit indexed by T. Therefore, without any communication, they are able to simplify the original circuit C in order to obtain a *monotone* circuit $\tilde{C}$ with input variables indexed by $[n]\backslash T$. Let $\tilde{x} \stackrel{\text{def}}{=} x'_{[n]\backslash T}$ and $\tilde{y} \stackrel{\text{def}}{=} y'_{[n]\backslash T}$ be the corresponding projections of x' and y'. Clearly, $\tilde{C}(\tilde{x}) = C(x') = f(x') = 1$, and $\tilde{C}(\tilde{y}) = C(y') = f(y') = 0$. Furthermore, $\tilde{C}$ computes some monotone function $\tilde{f} \colon \{0,1\}^{[n]\backslash T} \to \{0,1\}$.

Alice and Bob simulate together the standard Karchmer-Wigderson protocol Π granted by Proposition 1, and obtain an index $j \in [n]\backslash T$ for which $\tilde{x}_j = 1$ and $\tilde{y}_j = 0$. Observe that this stage can be executed with communication cost $\mathsf{depth}(\tilde{C}) \leq \mathsf{depth}(C) = d$. However, since x agrees with $\tilde{x}$ on every bit indexed by $[n]\backslash T$, and similarly for y and $\tilde{y}$, it follows that $x_j = 1$ and $y_j = 0$. Put another way, Alice and Bob have solved the monotone Karchmer-Wigderson game for f with communication at most $t + d$, which completes the proof of our result.

An interesting aspect of this proof is that it relies on both directions of the Karchmer-Wigderson connection. Proposition 15 and previous work on monotone depth lower bounds provide a trade-off between circuit depth and negation complexity for DeMorgan circuits solving the clique problem.

Proposition 16 (Raz and Wigderson [33]). *Let k-Clique: $\{0,1\}^{\binom{n}{2}} \to \{0,1\}$ be the Boolean function that is 1 if and only if the input graph G contains a clique of size k. If C is a monotone circuit that computes k-Clique for $k = n/2$, then $\mathsf{depth}(C) \geq \gamma \cdot n$, where $\gamma > 0$ is a fixed constant.*

Corollary 1. *There exists a fixed constant $\gamma > 0$ for which the following holds. If $\delta + \varepsilon \leq \gamma$, then any DeMorgan circuit of depth δn solving the $(n/2)$-Clique problem on n-vertex graphs contains at least εn negation gates.*

This result indicates that negation gates at the bottom layer are much easier to handle from the point of view of complexity theory than negations located at arbitrary positions of the circuit (see also Proposition 17 in Section A).

6 Open Problems and Further Research Directions

While our results provide some strong (indeed, optimal) bounds, they also leave open surprisingly basic questions.

For example, it seems reasonable, in light of our results, to think that most cryptographic primitives require $\Omega(\log n)$ negations. Nevertheless, for a basic primitive like a pseudorandom generator (that cannot be monotone), we leave open the following question: Is there a pseudorandom generator computed with a single negation gate? We stress that our question refers to a single circuit with multiple output bits computing the PRG. If one can use different circuits for distinct output bits, then the work of Applebaum, Ishai, and Kushilevitz [4] provides strong evidence that there are PRGs computed with a constant number of negations.

Having negation gates at the bottom level may be easier to study, and with some work we can show (in results omitted from the current paper) that no function with large enough stretch computed with a single negation at the bottom layer can be a small-bias generator (and thus not a pseudorandom generator either).

Another important open problem relates to the negation complexity of WPRF (weak pseudorandom functions, cf. Akavia et al. [1]), or, viewed from the learning perspective, weak-learning functions computed with a single negation. While for strong PRFs, even non-adaptive ones, we have obtained an $\Omega(\log n)$ lower bound, as far as we know, there may exist WPRFs computed by circuits with a single negation gate. Again, when restricting ourselves to negations at the bottom, we can prove some partial results (it is not hard to prove that a function computed by a circuit with a constant number of negations at the bottom layer cannot be a WPRF).

Finally, we have not imposed additional restrictions on the structure of Boolean functions computing cryptographic primitives. For instance, due to efficiency concerns, it is desirable that such circuits have depth as small as possible, without compromising the security of the underlying primitive. It is known that Markov's upper bound of $O(\log n)$ negations fails under restrictions of this form (cf. Santha and Wilson [34]; see also Hofmeister [21]). In particular, this situation sheds some light into why practical implementations have far more negations (or XORs) when compared to the theoretical lower bounds described in our work. Here we have not investigated this phenomenon, and it would be interesting to see if more specific results can be obtained in the cryptographic context.

Acknowledgments. We would like to thank Ilan Orlov for helpful conversations during an early stage of this work, Rocco Servedio for suggesting an initial construction in Proposition 10, and Andrej Bogdanov for helpful discussions that allowed us to extend some results. We also thank the referees for several suggestions that allowed us to improve the presentation.

References

1. Akavia, A., Bogdanov, A., Guo, S., Kamath, A., Rosen, A.: Candidate weak pseudorandom functions in $AC^0 \circ MOD_2$. In: Innovations in Theoretical Computer Science (ITCS), pp. 251–260 (2014)
2. Alon, N., Spencer, J.H.: The Probabilistic Method. Wiley Interscience (2008)
3. Amano, K., Maruoka, A.: A superpolynomial lower bound for a circuit computing the clique function with at most $(1/6)\log \log n$ negation gates. SIAM J. Comput. 35(1), 201–216 (2005)
4. Applebaum, B., Ishai, Y., Kushilevitz, E.: Cryptography in NC^0. SIAM J. Comput. 36(4), 845–888 (2006)
5. Arora, S., Barak, B.: Complexity Theory: A Modern Approach. Cambridge University Press (2009)

6. Beals, R., Nishino, T., Tanaka, K.: More on the complexity of negation-limited circuits. In: Symposium on Theory of Computing (STOC), pp. 585–595 (1995)
7. Beals, R., Nishino, T., Tanaka, K.: On the complexity of negation-limited boolean networks. SIAM J. Comput. 27(5), 1334–1347 (1998)
8. Blais, E., Canonne, C.C., Oliveira, I.C., Servedio, R.A., Tan, L.-Y.: Learning circuits with few negations (2014) (preprint)
9. Blum, A., Burch, C., Langford, J.: On learning monotone boolean functions. In: Symposium on Foundations of Computer Science (FOCS), pp. 408–415 (1998)
10. Bogdanov, A., Guo, S.: Sparse extractor families for all the entropy. In: Innovations in Theoretical Cmputer Science (ITCS), pp. 553–560 (2013)
11. Nader, H.: Bshouty and Christino Tamon: On the fourier spectrum of monotone functions. J. ACM 43(4), 747–770 (1996)
12. Buresh-Oppenheim, J., Kabanets, V., Santhanam, R.: Uniform hardness amplification in NP via monotone codes. In: Electronic Colloquium on Computational Complexity (ECCC), vol. 13(154) (2006)
13. Dachman-Soled, D., Lee, H.K., Malkin, T., Servedio, R.A., Wan, A., Wee, H.: Optimal cryptographic hardness of learning monotone functions. Theory of Computing 5(1), 257–282 (2009)
14. Michael, J.: Fischer. The complexity of negation-limited networks - A brief survey. In: Automata Theory and Formal Languages, pp. 71–82 (1975)
15. Fortuin, C.M., Kasteleyn, P.W., Ginibre, J.: Correlation inequalities on some partially ordered sets. Communications in Mathematical Physics 22(2), 89–103 (1971)
16. Goldmann, M., Russell, A.: Spectral bounds on general hard-core predicates. In: Reichel, H., Tison, S. (eds.) STACS 2000. LNCS, vol. 1770, pp. 614–625. Springer, Heidelberg (2000)
17. Goldreich, O.: Foundations of Cryptography: Volume 1, Basic Tools. Cambridge University Press (2007)
18. Goldreich, O., Goldwasser, S., Micali, S.: How to construct random functions. J. ACM 33(4), 792–807 (1986)
19. Goldreich, O., Izsak, R.: Monotone circuits: One-way functions versus pseudorandom generators. Theory of Computing 8(1), 231–238 (2012)
20. Goldreich, O., Levin, L.A.: A hard-core predicate for all one-way functions. In: Symposium on Theory of Computing (STOC), pp. 25–32 (1989)
21. Hofmeister, T.: The power of negative thinking in constructing threshold circuits for addition. In: Structure in Complexity Theory Conference, pp. 20–26 (1992)
22. Jukna, S.: Boolean Function Complexity - Advances and Frontiers. Springer, Heidelberg (2012)
23. Kahn, J., Kalai, G., Linial, N.: The influence of variables on Boolean functions. In: Symposium on Foundations of Computer Science (FOCS), pp. 68–80 (1988)
24. Karchmer, M., Wigderson, A.: Monotone circuits for connectivity require superlogarithmic depth. In: Symposium on Theory of Computing (STOC), pp. 539–550 (1988)
25. Koroth, S., Sarma, J.: Depth lower bounds against circuits with sparse orientation. In: Cai, Z., Zelikovsky, A., Bourgeois, A. (eds.) COCOON 2014. LNCS, vol. 8591, pp. 596–607. Springer, Heidelberg (2014)
26. Kushilevitz, E., Nisan, N.: Communication Complexity. Cambridge University Press (1997)
27. Markov, A.A.: On the inversion complexity of a system of functions. J. ACM 5(4), 331–334 (1958)

28. Morizumi, H.: Limiting negations in formulas. In: Albers, S., Marchetti-Spaccamela, A., Matias, Y., Nikoletseas, S., Thomas, W. (eds.) ICALP 2009, Part I. LNCS, vol. 5555, pp. 701–712. Springer, Heidelberg (2009)
29. Morizumi, H.: Limiting negations in non-deterministic circuits. Theoretical Computer Science 410(38-40), 3988–3994 (2009)
30. Naor, J., Naor, M.: Small-bias probability spaces: Efficient constructions and applications. SIAM J. Comput. 22(4), 838–856 (1993)
31. O'Donnell, R.: Analysis of Boolean Functions. Cambridge University Press (2014)
32. O'Donnell, R., Wimmer, K.: Kkl, kruskal-katona, and monotone nets. SIAM J. Comput. 42(6), 2375–2399 (2013)
33. Raz, R., Wigderson, A.: Monotone circuits for matching require linear depth. J. ACM 39(3), 736–744 (1992)
34. Santha, M., Wilson, C.B.: Limiting negations in constant depth circuits. SIAM J. Comput. 22(2), 294–302 (1993)
35. Sung, S.C., Tanaka, K.: An exponential gap with the removal of one negation gate. Inf. Process. Lett. 82(3), 155–157 (2002)
36. Sung, S.C., Tanaka, K.: Limiting negations in bounded-depth circuits: An extension of Markov's theorem. Inf. Process. Lett. 90(1), 15–20 (2004)
37. Talagrand, M.: How much are increasing sets positively correlated? Combinatorica 16(2), 243–258 (1996)
38. Tardos, É.: The gap between monotone and non-monotone circuit complexity is exponential. Combinatorica 8(1), 141–142 (1988)

A   The Flow of Negations in Boolean Circuits

In this section we discuss how to move negations in a Boolean circuit in order to explore different aspects of these gates.

A.1   Moving Negations to the Top of the Circuit

For convenience of the reader, we include here a proof for the following structural result about negation gates.

Lemma 5 (Blais et al. [8]). *Let $f\colon \{0,1\}^n \to \{0,1\}$ be a Boolean function computed by a circuit C with at most t negations. Then f can be written as $f(x) = h(g_1(x), \ldots, g_T(x))$, where each function g_i is monotone, $T = O(2^t)$, and h is either the parity function, or its negation.*

Proof. Recall from Proposition 3 that if a Boolean function f is computed by a circuit with t negations, then $k \stackrel{\text{def}}{=} a(f) \le O(2^t)$. The claimed result follows easily once we "f-slice" the boolean hypercube, as described next. For any $i \in \{0, 1, \ldots, k\}$, let

$$T_i = \{x \in \{0,1\}^n \mid a(f, Y) \le i, \text{ for any chain } Y \text{ starting at } x\}.$$

Observe that T_i is a monotone set: if $x \in T_i$ and $x \preceq y$, then $y \in T_i$. In addition, it is clear that $T_0 \subset T_1 \subset \ldots \subset T_k = \{0,1\}^n$. Finally, for every $i \in \{0, 1, \ldots, k-1\}$, it is not hard to see that $T_{i+1} \setminus T_i \ne \emptyset$, since $k = a(f)$.

Let $S_0 \stackrel{\text{def}}{=} T_0$. Observe that $1^n \in T_0$, hence S_0 is nonempty. For any $j \in \{1, \ldots, k\}$, set $S_j \stackrel{\text{def}}{=} T_j \setminus T_{j-1}$. It follows from the preceding discussion that the family $\mathcal{S} = \{S_0, S_1, \ldots, S_k\}$ is a partition of the n-dimensional boolean cube into nonempty sets.

Next, we prove the following claim: There exists a vector $b = (b_0, b_1, \ldots, b_k) \in \{0, 1\}^{k+1}$ with the following properties:

(i) For each $0 \le i \le k$, and for any $x^i \in S_i$, we have $f(x^i) = b_i$.
(ii) For each $0 < i \le k$, we have $b_i = 1 - b_{i-1}$.
(iii) For each $0 \le i \le k$, given any $x^i \in S_i$, there exist elements $x^0, x^1, \ldots, x^{i-1}$ in $S_0, S_1, \ldots, S_{i-1}$, respectively, such that $Y = (x^i, x^{i-1}, \ldots, x^0)$ is a chain starting at x^i with $a(f, Y) = i$. Further, every proper chain Y starting at x^i with $a(f, Y) = i$ is of this form, i.e., its elements belong to distinct sets S_j, where $j < i$.

The construction of this vector and the proof of these items is by induction on $i \in [0, k]$. For $i = 0$, we set $b_0 = f(1^n)$, and observe that the result is true using the definition of S_0 and T_0. Consider now an element $y^i \in S_i$, where $i > 0$, and assume that items (i), (ii), and (iii) hold for any smaller index. Since $y^i \in S_i$, there exists a proper chain $Y = (y^i, y^{i-1}, \ldots, y^0)$ with $a(f, Y) = i$. Since this chain cannot be extended to a larger chain starting at y^i, it follows from the induction hypothesis that, for every $0 < j \le i$, $f(y^j) = 1 - f(y^{j-1}) = 1 - b_{j-1}$. In particular, we can set $b_i = 1 - b_{i-1}$, since our initial element $y^i \in S_i$ was arbitrary. Finally, the remaining part of item (iii) follows once we consider the subchain $Y' = (y^{i-1}, \ldots, y^0)$, and apply the induction hypothesis.

Finally, we use items (i) and (ii) to prove the lemma. Recall that every family T_i is monotone, where $i \in \{0, 1, \ldots, k\}$. In other words, there exist $k + 1$ monotone functions $g_i \colon \{0, 1\}^n \to \{0, 1\}$ such that $g_i^{-1}(1) = T_i$, where $i \in [0, k]$. As observed before, $S_k = T_k \setminus T_{k-1}$ is nonempty. In particular, $0^n \in S_k$. If $f(0^n) = 0$, we let $h \colon \{0, 1\}^{k+1} \to \{0, 1\}$ be the parity function $\sum_{j=0}^{k} y_i \pmod 2$. Otherwise, we let h be the complement of the parity function. It follows from (i) and (ii) that, for every $x \in \{0, 1\}^n$, we have $f(x) = h(g_0(x), \ldots, g_k(x))$, which completes the proof.

By relaxing the assumption on h, we can prove the following effective version of Lemma 1.

Lemma 6. *Let $f \colon \{0, 1\}^n \to \{0, 1\}$ be a Boolean function computed by a circuit C of size s containing t negation gates, where $t \ge 0$. Then f can be written as $f(x) = h(g_1(x), \ldots, g_T(x))$, where each function g_i is computed by a monotone circuit of size at most s, $T = 2^{t+1} - 1$, and $h \colon \{0, 1\}^T \to \{0, 1\}$ is computed by a circuit of size at most $5T$.*

Proof. The proof is by induction on t. The base case $t = 0$ is trivial. Now let $t \ge 1$, and assume the statement holds for any function computed by circuits with at most $t' < t$ negations. Let $f \colon \{0, 1\}^n \to \{0, 1\}$ be a Boolean function computed by a circuit C of size s that contains t negations. Let $x_1, \ldots x_n, f_1, \ldots, f_s$ be the functions computed at each internal node of C, and $\lceil f_1 \rceil, \ldots, \lceil f_s \rceil$ be the corresponding gates, i.e., each $\lceil f_i \rceil \in \{\mathsf{AND}, \mathsf{OR}, \mathsf{NOT}\}$. Furthermore, assume that this sequence is a topological sort of the nodes of the circuit, in the sense that the inputs of each gate $\lceil f_i \rceil$ are $f_{i_1}(x)$ and $f_{i_2}(x)$, with $i_1, i_2 < i$. Let $i^* \in [s]$ be the index of the first NOT gate in C according to this sequence.

Consider a new circuit C' over $n+1$ variables $x_1, \ldots, x_n, y$, where C' computes exactly as C, except that the output value of f_{i^*} is replaced by the new input y. By construction, C' is a circuit of size at most s containing $t' \overset{\text{def}}{=} t - 1$ negations, and it computes some Boolean function $f' \colon \{0,1\}^{n+1} \to \{0,1\}$. Applying the induction hypothesis, we get that

$$f'(x, y) = h'(g_1'(x, y), \ldots, g_{T'}'(x, y)), \tag{6}$$

where each g_i' is computed by a monotone circuit of size at most s, $T' \leq 2^{t'+1} - 1$, and $h' \colon \{0,1\}^{T'} \to \{0,1\}$ admits a circuit of size $5T'$. In addition, notice that

$$f(x) = \begin{cases} f'(x, 1) & \text{if } f_{i^*}(x) = 1, \\ f'(x, 0) & \text{otherwise.} \end{cases} \tag{7}$$

Let f_i be the input wire of $\lceil f_{i^*} \rceil$. Since $\lceil f_{i^*} \rceil = \mathsf{NOT}$, we obtain using Equation 7 that

$$f(x) = \tilde{h}(f_i(x), f'(x, 0), f'(x, 1)), \tag{8}$$

where $\tilde{h}(z, y_1, y_0) \overset{\text{def}}{=} y_z$ is a function over three input bits that is computed by a circuit of size at most 5. Furthermore, combining Equations 6 and 8, it follows that

$$\begin{aligned} f(x) &= \tilde{h}(f_i(x), h'(g_1'(x, 0), \ldots, g_{T'}'(x, 0)), h'(g_1'(x, 1), \ldots, g_{T'}'(x, 1))) \\ &= h(f_i(x), g_{1,0}'(x), \ldots, g_{T',0}'(x), g_{1,1}'(x), \ldots, g_{T',1}'(x)), \end{aligned}$$

where $g_{j,b}'(x) \overset{\text{def}}{=} g_j'(x, b)$, for $j \in [T']$ and $b \in \{0,1\}$, and $h \colon \{0,1\}^{2T'+1} \to \{0,1\}$ is the function obtained by setting

$$h(v_0, v_1, \ldots, v_{T'}, v_{T'+1}, \ldots, v_{2T'}) \overset{\text{def}}{=} \tilde{h}(v_0, h'(v_1, \ldots, v_{T'}), h(v_{T'+1}, \ldots, v_{2T'})).$$

Using our assumption on i^*, it follows that f_i is computed by a monotone circuit of size s. It is also clear that each $g_{j,b}'$ admits a monotone circuit of size s. Further, observe that

$$2T' + 1 \leq 2(2^{t'+1} - 1) + 1 = 2(2^t - 1) + 1 = 2^{t+1} - 1 \overset{\text{def}}{=} T.$$

Finally, using the induction hypothesis and the upper bound on the circuit size of $\tilde{h}$, we get that h is computed by a circuit of size at most

$$5 + 5T' + 5T' = 5(2T' + 1) = 5T,$$

which completes the proof of Lemma 2.

It is possible to show that the upper bound on T in the statement of Lemma 7 is essentially optimal. This follows from the connection between the number of negation gates in a Boolean circuit for a function f and the alternation complexity of f, as discovered by Markov [27] (see e.g. Blais et al. [8] for further details).

A.2   Moving Negations to the Bottom of the Circuit

We recall the following basic fact about negations.

Fact 1. *Let C be a Boolean circuit of size s containing a negation gate at depth $d \geq 1$. Then C can be transformed into an equivalent circuit C' of size s without this negation gate that contains at most 2^{d-1} additional negations at the bottom layer.*

Proof. The result is immediate from the application of DeMorgan rules for Boolean connectives.

We observe below that this result is optimal. Put another way, a negation gate at an arbitrary location can be more powerful than a linear number of negations at the bottom layer.

Proposition 17. *There exists an explicit Boolean function $f\colon \{0,1\}^n \to \{0,1\}$ that admits a linear size circuit C containing a single negation gate, but for which any equivalent circuit C' with negation gates at the bottom layer only requires n negations.*

Proof. Let $f(x) = 1$ if and only if $x = 0^n$. Clearly, f can be computed by a circuit with a single negation, since this function is the negation of a monotone function. The lower bound follows using an argument from [25]. Assume that $f(x) = D(x, (x \oplus \beta))$, where D is a monotone circuit. We need to prove that $\beta_i = 1$ for every $i \in [n]$. Consider inputs $z \stackrel{\text{def}}{=} 0^n$ and $e_i \stackrel{\text{def}}{=} 0^{i-1}10^{n-i}$. By definition, $f(z) = 1$ and $f(e_i) = 0$, thus $D(0^n, 0^n \oplus \beta) = D(0^n, \beta) = 1$ and $D(e_i, e_i \oplus \beta) = D(e_i, \beta^{\oplus i}) = 0$. If $\beta_i = 0$, then $(0^n, \beta) \prec (e_i, \beta^{\oplus i})$, and since D is monotone, we get $D(0^n, \beta) \leq D(e_i, \beta^{\oplus i})$. However, this is in contradiction with the value of f on z and e_i, which implies that $\beta_i = 1$.

From Weak to Strong Zero-Knowledge
and Applications*

Kai-Min Chung[1],[**], Edward Lui[2], and Rafael Pass[2],[***]

[1] Academia Sinica, Taiwan
kmchung@iis.sinica.edu.tw
[2] Cornell University, USA
{luied,rafael}@cs.cornell.edu

Abstract. The notion of *zero-knowledge* [20] is formalized by requiring that for every malicious efficient verifier V^*, there exists an efficient simulator S that can reconstruct the view of V^* in a true interaction with the prover, in a way that is indistinguishable to *every* polynomial-time distinguisher. *Weak zero-knowledge* weakens this notions by switching the order of the quantifiers and only requires that for every distinguisher D, there exists a (potentially different) simulator S_D.

In this paper we consider various notions of zero-knowledge, and investigate whether their weak variants are equivalent to their strong variants. Although we show (under complexity assumption) that for the standard notion of zero-knowledge, its weak and strong counterparts are not equivalent, for meaningful variants of the standard notion, the weak and strong counterparts are indeed equivalent. Towards showing these equivalences, we introduce new non-black-box simulation techniques permitting us, for instance, to demonstrate that the classical 2-round graph non-isomorphism protocol of Goldreich-Micali-Wigderson [18] satisfies a "distributional" variant of zero-knowledge.

Our equivalence theorem has other applications beyond the notion of zero-knowledge. For instance, it directly implies the *dense model theorem* of Reingold et al (STOC '08), and the leakage lemma of Gentry-Wichs (STOC '11), and provides a modular and arguably simpler proof of these results (while at the same time recasting these result in the language of zero-knowledge).

---

* A full version of this paper is available at https://eprint.iacr.org/2013/260
** Chung is supported in part by NSF Award CNS-1217821, NSF Award CCF-1214844, Pass' Sloan Fellowship, and Ministry of Science and Technology MOST 103-2221-E-001-022-MY3; part of this work was done while being at Cornell University.
*** Pass is supported in part by an Alfred P. Sloan Fellowship, Microsoft New Faculty Fellowship, NSF CAREER Award CCF-0746990, NSF Award CCF-1214844, NSF Award CNS-1217821, AFOSR YIP Award FA9550-10-1-0093, and DARPA and AFRL under contract FA8750-11-2-0211.

Y. Dodis and J.B. Nielsen (Eds.): TCC 2015, Part I, LNCS 9014, pp. 66–92, 2015.

1   Introduction

The notion of *zero-knowledge*, and the *simulation-paradigm* used to define it, is of fundamental importance in modern cryptography—most definitions of protocol security rely on it. In a zero-knowledge protocol, a prover P can convince a verifier V of the validity of some mathematical statement $x \in L$, while revealing "zero (additional) knowledge" to V. This zero-knowledge property is formalized by requiring that for every potentially malicious efficient verifier V^*, there exists an efficient simulator S that, without talking to P, is able to "indistinguishably reconstruct" the view of V^* in a true interaction with P. The traditional way of defining what it means to "indistinguishably reconstruct" is to require that the output of S cannot be distinguished (with more than negligible probability) from the true view of V^* by *any* efficient distinguisher D; that is, we have a *universal* simulator that works for *all* distinguishers D.

A seemingly weaker way to define the zero-knowledge property is to require that for every distinguisher D, there exists a "distinguisher-dependent" simulator S_D such that the output of S_D cannot be distinguished from the true view of V^* by the *particular* distinguisher D; following [12], we refer to this weaker notion of zero-knowledge as *weak zero-knowledge*.

The main question addressed in this paper is whether this switch in the order of the quantifiers yields an equivalent notion. More specifically, we consider various notions of zero-knowledge, and investigate whether their weak (distinguisher-dependent simulator) variants are equivalent to their strong (universal simulator) variants. Towards addressing this question, we introduce new non-black-box simulation techniques permitting us, for instance, to demonstrate that the classical 2-round graph non-isomorphism protocol of Goldreich-Micali-Wigderson [18] satisfies a "distributional" variant of zero-knowledge. Our results also reveal deep connections between the notion of zero-knowledge and the *dense model theorem* of Reingold et al [28] (which in turn is related to questions such as the XOR Lemma [32] and Szemeredi's regularity lemma [15]; see [30] for more details).

1.1   From Weak to Strong Zero-Knowledge

Our first result shows that (under plausible complexity-theoretic assumptions) for the standard definition of zero-knowledge, weak zero-knowledge is a strictly weaker requirement than (strong) zero-knowledge.

Theorem 1 (Informally stated). *Assume the existence of "timed commitments" and "timed one-way permutations". Then, there exists an interactive proof for a language $L \in$ NP that is weak zero-knowledge but not (strong) zero-knowledge.*

Motivated by this separation, we turn to consider relaxed notions of zero-knowledge. We first consider a concrete security variant of the notion of zero-knowledge. Roughly speaking, we call a protocol (t, ϵ)-zero-knowledge if the zero-knowledge property holds with respect to all $t(n)$-time bounded distinguishers (as opposed to all polynomial-time distinguishers), and we require that the distinguishability gap is bounded by $\epsilon(n)$ (as opposed to being negligible), where n is the length of the statement x being proved. Weak (t, ϵ)-zero-knowledge is defined analogously (by again switching the order of the quantifiers).

Note that if (P, V) is (t, ϵ)-zero-knowledge (resp. weak (t, ϵ)-zero-knowledge) for some super-polynomial function t and some negligible function ϵ, then (P, V) is zero-knowledge (resp. weak zero-knowledge) in the classic sense. We here consider a slightly relaxed notion where we only require (P, V) to be (t, ϵ)-zero-knowledge for all polynomials t and all inverse polynomials ϵ. (Note that this is weaker than the standard definition of zero-knowledge since now the running-time of the simulator may depend on the bounds t and ϵ.) Perhaps surprisingly, we show that for this relaxed notion of zero-knowledge, the weak and strong versions lead to an equivalent definition.

Theorem 2 (Informally stated). *If an interactive proof (P, V) is weak (t, ϵ)-zero knowledge for every polynomial t and every inverse polynomial ϵ, then (P, V) is also (t', ϵ')-zero knowledge for every polynomial t' and every inverse polynomial ϵ'.*

We highlight that the "universal" simulator S constructed in the proof of Theorem 2 makes use of the malicious verifier V^* in a non-black-box way. On a very high-level (and significantly oversimplifying), the idea behind Theorem 2 is to rely on Von Neumann's minimax theorem to obtain the universal simulator from the "distinguisher-dependent" simulators; the non-black-box nature of the universal simulator comes from the fact that defining the "utility function" we use with the minimax theorem requires knowing the auxiliary inputs received by V^*, and thus we make non-black-box use of V^*.

Implementing this approach becomes quite non-trivial since we require the existence of a *uniform* polynomial-time simulator for every uniform polynomial-time verifier—the minimax theorem only guarantees the existence of a *distribution* over polynomial-time machines that simulates the view of the verifier, but it is not clear if this distribution can be computed in uniform polynomial time. We overcome this issue by instead relying on a multiplicative weights algorithm to appropriately implement an approximate minimax strategy; see Section 1.4 for more details.

1.2 From Super-Weak to Strong Distributional Zero-Knowledge

Note that although in the definition of weak zero-knowledge the simulator may depend on the distinguisher, we still require that the probability that the distinguisher outputs 1 when given the output of the simulator is *close* to the probability that the distinguisher outputs 1 when given the true view of the malicious verifier V^*. An even weaker condition (considered in [21]) only requires that the simulator manages to make the distinguisher output 1 with *at least as high probability* (minus some "small" gap) as the probability that the distinguisher outputs 1 when given a true view of V^*. That is, we only consider "one-sided" indistinguishability. We refer to such a zero-knowledge property as *super-weak zero-knowledge*.

It is not hard to see that super-weak (t, ϵ)-zero-knowledge is not equivalent to weak (t, ϵ)-zero-knowledge (see the full version of this paper for the proof). Thus, we here consider an alternative "distributional" notion of zero-knowledge (a la [17]) where indistinguishability of the simulation is only required for any distribution over statements (and auxiliary inputs), and the simulator as well as the distinguisher can depend on the distribution. Additionally, we here model both the distinguisher and the simulator as non-uniform polynomial-time algorithms (as opposed to uniform ones). (The combination of these variants was previously considered by [12].[1]) We refer to such a notion of zero-knowledge as *distributional zero-knowledge*, and analogously define *distributional (t, ϵ)-zero-knowledge* as well as *weak (resp. super-weak) distributional (t, ϵ)-zero-knowledge*. Roughly speaking, distributional zero-knowledge captures the intuition that proofs of "random" statements do not provide the verifier with any new knowledge (beyond the statement proved). Perhaps surprisingly, we show that super-weak distributional (t, ϵ)-zero-knowledge is equivalent to (strong) distributional (t, ϵ)-zero-knowledge if we consider all polynomials t and all inverse polynomials ϵ.

Theorem 3 (Informally stated). *If an interactive proof (P, V) is super-weak distributional (t, ϵ)-zero-knowledge for every polynomial t and every inverse polynomial ϵ, then (P, V) is also distributional (t', ϵ')-zero knowledge for every polynomial t' and every inverse polynomial ϵ'.*

In contrast to Theorem 2, the proof of Theorem 3 follows from a rather direct use of the minimax theorem; see Section 1.4 for more details. We also show that any protocol where the prover is "laconic" [19]—that is, it sends only $O(\log n)$ bits in total, is super-weak (distributional) zero-knowledge; combining this result with Theorem 3 thus yields the following theorem.

Theorem 4 (Informally stated). *Let (P, V) be an interactive proof with a laconic prover for a language L. Then (P, V) is distributional (t, ϵ)-zero-knowledge for every polynomial t and every inverse polynomial ϵ.*

Given Theorem 3, the proof of Theorem 4 is very straight-forward: to show that laconic proofs are super-weak zero-knowledge, have the simulator simply enumerate all possible prover messages and keep the one that the distinguisher "likes the most" (i.e., makes the distinguisher output 1 with as high probability as possible); note that we here rely crucially on the fact that we only need to achieve "one-sided" indistinguishability.

Theorem 4 may seem contradictory. An interactive proof with a laconic prover (i.e., with small prover communication complexity) can reveal, say, the first $\log n$ bits of the witness w to the statement x proved, yet Theorem 4 states that such a protocol satisfies a notion of zero-knowledge. But if we leak something specific about the witness, how can we expect the protocol to be "zero-knowledge"? The key point here is that (as shown in Theorem 4), for *random* statements x, the

---

[1] More specifically, the notion of "ultra-weak zero-knowledge" of [12] considers both of these relaxations, but relaxes the notion even further.

information revealed about the witness can actually be efficiently generated. In other words, the *whole* process where the prover first picks the statement (at random), and then provides the proof, is zero-knowledge.

Despite the simplicity of the proof of Theorem 4, it has many (in our eyes) intriguing corollaries. The first one is that the classic two-round graph non-isomorphism protocol of [18] (which is only known to be "honest-verifier" zero-knowledge) is distributional (t, ϵ)-zero-knowledge for every polynomial t and every inverse polynomial ϵ.[2] In fact, by the complete problem for SZK [29], we can show that every language in SZK has a 2-round interactive proof that is distributional (t, ϵ)-zero-knowledge for every polynomial t and every inverse polynomial ϵ.

Theorem 5 (Informally stated). *For every language $L \in SZK$ and every polynomial p, there exists a 2-round interactive proof (P, V) for L with completeness $1 - \text{negl}(\cdot)$ and soundness error $\frac{1}{p(\cdot)}$, and is distributional (t, ϵ)-zero-knowledge for every polynomial t and every inverse polynomial ϵ.*

We proceed to outline two other applications of Theorem 4.

Leakage Lemma of Gentry-Wichs. Roughly speaking, the "Leakage Lemma" of Gentry-Wichs [16] states that for every joint distribution $(X, \pi(X))$, where $|\pi(x)| = O(\log |x|)$ (π should be thought of as leakage on X), and for every distribution Y that is indistinguishable from X, there exists some leakage $\widetilde{\pi}$ such that the joint distributions $(X, \pi(X))$ and $(Y, \widetilde{\pi}(Y))$ are indistinguishable. As we now argue, this lemma (and in fact, a stronger version of it) is a direct consequence of Theorem 4.

In the language of zero-knowledge, let X be a distribution over statements, and consider a one-message interactive proof where $\pi(x)$ denotes the distribution over the prover's message when the statement is x. By Theorem 4, this protocol is distributional zero-knowledge, and thus there exists an *efficient* simulator S that can simulate the interaction (i.e, $(X, S(X))$ is indistinguishable from $(X, \pi(X))$). By the indistinguishability of Y and X (and the efficiency of S), it directly follows that $(Y, S(Y))$ is indistinguishable from $(X, \pi(X))$. Thus we have found $\widetilde{\pi} = S$.

Let us note that our proof of the leakage lemma yields an even stronger statement—namely, we have found an efficient simulator $\widetilde{\pi}$; such a version of the leakage lemma was recently established by Jetchev and Pietrzak [23]. (As an independent contribution, our proof of Theorem 4 is actually significantly simpler than both the proof of [16] and [23].) Additionally, since our result on zero-knowledge applies also to *interactive* protocols, we directly also get an interactive version of the leakage lemma.

Dense Model Theorem. Roughly speaking, the Dense Model Theorem of [28,30] states that if X is indistinguishable from the uniform distribution over

---

[2] Recall that in the classic Graph Non-Isomorphism protocol the prover sends just a single bit and thus is very laconic.

n-bits, U_n, and R is δ-dense[3] in X, then there exists a "model-distribution" M that is (approximately) δ-dense in U_n such that M is indistinguishable from R. Again, we show that this lemma is a direct consequence of Theorem 4. (Furthermore, our proof of Theorem 4 is arguably simpler and more modular than earlier proofs of the dense model theorem.)

Let us first translate the statement of the dense model theorem into the language of zero-knowledge. Let X be a distribution over statements x, and consider some distribution R that is δ-dense in X, i.e., there exists a joint distribution $(X, B(X))$ with $\Pr[B(X) = 1] \geq \delta$ such that $R = X|(B(X) = 1)$. Define a one-bit proof where the prover sends the bit $B(x)$, where x is the statement. By Theorem 4, there exists a simulator S for this interactive proof; let $M = U_n|(S(U_n) = 1)$. By the indistinguishability of the simulation, $(X, S(X))$ is indistinguishable from $(X, B(X))$, and thus by indistinguishability of X and U_n, $(U_n, S(U_n))$ is indistinguishable from $(X, B(X))$. It follows that M is (approximately) δ-dense in U_n, and M is indistinguishable from R.

1.3 A Note on Our Non-Black-Box Simulation Technique

The universal simulators in Theorem 3, 4, and 5 are indirectly obtained via the minimax theorem used in the proof of Theorem 3, and again we make non-black-box usage of the verifier V^*. We remark that our non-black-box usage of V^* is necessary (assuming standard complexity-theoretic assumptions): We show that black-box simulation techniques cannot be used to demonstrate distributional (t, ϵ)-zero-knowledge for 2-round proof systems for languages that are hard-on-average.

Theorem 6 (Informally stated). *Let L be any language that is hard-on-average for polynomial-size circuits, and let (P, V) be any 2-round interactive proof (with completeness $2/3$ and soundness error $1/3$) for L. Then, there exists a polynomial t such that for every $\epsilon(n) < 1/12$, (P, V) is not black-box distributional (t, ϵ)-zero-knowledge*

As as consequence we have that as long as SZK contains a language that is hard-on-average, our non-black-box techniques are necessary (otherwise, Theorems 5 and 6 would contradict each other). As far as we know, the above yields the first example where a non-black-box simulation technique can be used to analyze "natural" protocols (e.g., the classic graph non-isomorphism protocol) that were not "tailored" for non-black-box simulation, but for which black-box simulation is not possible. This stands in sharp contrast to the non-black-box technique of Barak [2] and its follow-ups (see e.g., [26,25,27,3,11,5,10,6]), where non-black-box simulation is enabled by a very specific protocol design. This gives hope that non-black-box techniques can be used to analyze simple/practical protocols.

---

[3] R is said to be δ-*dense* in X if for every r, $\Pr[R = r] \leq (1/\delta)\cdot\Pr[X = r]$; equivalently, R is δ-dense in X if there exists a joint distribution $(X, B(X))$ with $\Pr[B(X) = 1] \geq \delta$ such that $R = X|(B(X) = 1)$.

Let us finally remark that in our non-black-box technique, we only need to make non-black-box use of the malicious verifier V^*'s auxiliary input z and its running-time t, but otherwise we may treat V^*'s Turing machine as a black-box. Although the non-black-box simulation technique of Barak [2] also makes non-black-box usage of V^*'s Turing machine, it is not hard to see that also this technique can be modified to only make non-black-box usage of z and t (but not its Turing machine)—since the description of V^*'s Turing machine is of constant length the non-black-box simulator can simply enumerate all possible Turing machines in the protocol of Barak.

1.4 Our Techniques

As mentioned, both Theorem 2 and 3 rely on the minimax theorem from game theory. Recall that the minimax theorem states that in any finite two-player zero-sum game, if for every distribution over the actions of Player 1, there exists some action for Player 2 that guarantees him an expected utility of v, then there exists some (universal) distribution of actions for Player 2 such that no matter what action Player 1 picks, Player 2 is still guaranteed an expected utility of v. For us, Player 1 will be choosing a distinguisher, and Player 2 will be choosing a simulator; roughly speaking, Player 2's utility will be "high" if the simulation is "good" for the distinguisher chosen by Player 1. Now, by the weak zero-knowledge property, we are guaranteed that for every distinguisher chosen by Player 1, there exists some simulator for Player 2 that guarantees him a high utility. Thus intuitively, by the minimax theorem, Player 2 should have a simulator that yields him high utility with respect to any distinguisher.

There are two problems with this approach. First, to apply the minimax theorem, we require the existence of a good "distinguisher-dependent" simulator for every *distribution* over distinguishers. Secondly the minimax theorem only guarantees the existence of a distribution over simulators that works well against every distinguisher. We resolve both of these issues in quite different ways for Theorem 3 and Theorem 2.

In the context of Theorem 3, since we model both the simulator and distinguisher as non-uniform machines, we can use standard techniques to "derandomize" any distribution over simulators/distinguishers into a single simulator/distinguisher that gets some extra non-uniform advice: we simply approximate the original distribution by sufficiently many samples from it, and these samples can be provided to a single machine as non-uniform advice. (Such "derandomization" techniques originated in the proof of the hard-core lemma [22].)

In the context of Theorem 2, the situation is more difficult since we need both the distinguisher and the simulator to be uniform. In particular, we are only guaranteed the existence of a good distinguisher-dependent simulator for every *uniform* distinguisher and not necessarily for non-uniform ones. Here, we instead try to efficiently and uniformly find the "minimax" distribution over simulator strategies. If this can be done, then we do have a single uniform (and efficient) simulator algorithm. Towards this, we use a *multiplicative weights algorithm*, which can be used to approximately find the minimax strategies of two-player zero-sum games (e.g., see [14]). The multiplicative weights algorithm roughly works as follows. In the first round, Player 1 chooses the uniform distribution over the set of all $t(n)$-time Turing machines with description size $\leq \log n$ (note that any $t(n)$-time uniform distinguisher will be a member of this set for sufficiently large n), and then Player 2 chooses a "good simulator" that yields high payoff with respect to Player 1's distribution (note that since Player 1's distribution is uniformly and efficiently computable, we can view the process of sampling from it, and next running the sampled distinguisher, as a single uniform and efficient distinguisher, and thus we may rely on the weak zero-knowledge definition to conclude that a good simulator exists). In the next round, Player 1 updates its distribution using a multiplicative update rule that depends on Player 2's chosen simulator in the previous round; Player 2 again chooses a simulator that yields high payoff with respect to Player 1's new distribution, etc. By repeating this procedure for polynomially many rounds, Player 2 obtains a sequence of simulators such that the uniform distribution over the multiset of simulators yields high payoff no matter what distinguisher Player 1 chooses.

There are some issues that need to be resolved. In each round, we need to pick a simulator that works well against a (uniformly and efficiently computable) distribution over $t(n)$-time distinguishers. Although the running-time of the underlying distinguishers is bounded by $t(n)$, the time needed to sample from this distribution could be growing (exponentially) in each round, which in turn could potentially lead to an exponential growth in the running-time of the simulator. Thus after polynomially many rounds, it is no longer clear that the simulator or the distribution over distinguishers is polynomial-time.[4] To deal with this issue, we rely on the "good" distinguisher-dependent simulator for a single universal distinguisher that receives as auxiliary input the code of the actual distinguisher it is running; we can then at each step approximate the distribution over distinguishers and feed this approximation as auxiliary input to the universal distinguisher.

Another important issue to deal with is the fact that to evaluate the "goodness" of a simulation w.r.t. to some distinguisher (i.e., to compute the utility function), we need to be able to sample true views of the malicious verifier in an interaction with the honest prover—but if we could do this, then we would already be done! Roughly speaking, we overcome this issue by showing that the goodness of a simulation w.r.t. a particular distinguisher D can be approximated by using the distinguisher-dependent simulator S_D for D.

We remark that in both of the above proofs, the reason that we work with a (t, ϵ)-notion of zero-knowledge is that the running-time of the simulator we construct is polynomial in t and $1/\epsilon$.

1.5   Related Work

As mentioned above, the notion of weak zero-knowledge was first introduced by Dwork, Naor, Reingold and Stockmeyer [12]. Dwork et al also considered non-

---

[4] A similar issue appeared in a recent paper by us in the context of forecast testing [9], where we used a related, but different, technique to overcome it.

uniform versions and distributional versions of zero-knowledge; distributional versions of zero-knowledge were first considered by Goldreich [17] in a uniform setting (called uniform zero-knowledge).

The minimax theorem from game-theory has been applied in various contexts in complexity theory (e.g., [22,4,28,30,31]) and more recently also in cryptography (e.g., [28,13,8,16,31,23]). The proof of Theorem 4 is related to the approaches taken in these previous works, and most closely related to the approach taken in [30]. However, as far as we know, none of the earlier results have applied the minimax theorem in the context of zero-knowledge. Nevertheless, as we mentioned above, our Theorem 4 implies some of these earlier results (and shows that they can be understood in the language of zero-knowledge).

In a recent paper [31], Vadhan and Zheng proved a uniform minimax theorem, but our usage of the multiplicative weights algorithm cannot be simplified by using their uniform minimax theorem. One reason is that in our setting, the payoff (utility) function of the zero-sum game cannot be efficiently computed, and thus we have to approximate it. The uniform minimax theorem of [31] does not handle the usage of an approximate payoff function (their theorem does allow the usage of approximate KL projections in the algorithm, but from what we can see, this is not sufficient for handling our approximate payoff function).

1.6   Overview

In Section 2, we show that weak zero-knowledge is not equivalent to zero-knowledge (Theorem 1 above). In Section 3, we show that weak and strong (t, ϵ)-zero-knowledge are equivalent (Theorem 2 above). In Section 4, we show that super-weak and strong distributional (t, ϵ)-zero-knowledge are equivalent (Theorem 3 above), and interactive proofs with a laconic prover are distributional zero-knowledge (Theorem 4 above), and we also describe applications of this result. In the full version of this paper, we separate the notion of super-weak and weak (t, ϵ)-zero-knowledge.

2   Separation of Weak and Strong Zero-Knowledge

Given a prover P, a verifier V^*, and $x, z \in \{0,1\}^*$, let $Out_{V^*}[P(x) \leftrightarrow V^*(x,z)]$ denote the output of $V^*(x,z)$ after interacting with $P(x)$. We now state the definition of zero-knowledge for convenient reference.

Definition 1 (zero-knowledge). *Let (P,V) be an interactive proof system for a language L. We say that (P,V) is* zero-knowledge *if for every PPT adversary V^*, there exists a PPT simulator S such that for every PPT distinguisher D, there exists a negligible function $\nu(\cdot)$ such that for every $n \in \mathbb{N}$, $x \in L \cap \{0,1\}^n$, and $z \in \{0,1\}^*$, we have*

$$|\Pr[D(x, z, Out_{V^*}[P(x) \leftrightarrow V^*(x,z)]) = 1] - \Pr[D(x, z, S(x,z)) = 1]| \leq \nu(n).$$

Remark 1. If L is a language in NP with witness relation R_L, we usually require the prover P to be efficient, but on common input x, we also give any witness $y \in R_L(x)$ to the prover P. We refer to such a notion as *efficient prover* zero-knowledge. More formally, in the definition of zero-knowledge above, we would change "$x \in L \cap \{0,1\}^n$, and $z \in \{0,1\}^*$" to "$x \in L \cap \{0,1\}^n$, $y \in R_L(x)$, and $z \in \{0,1\}^*$", and we would change $P(x)$ to $P(x,y)$ and require P to be efficient. All subsequent definitions can be extended to an efficient prover setting in an obvious way.

One can relax the definition of zero-knowledge by switching the order of the quantifiers $\exists S$ and $\forall D$ so that the simulator S can depend on the distinguisher D. We call the relaxed definition *weak zero-knowledge* (following [12]).

Definition 2 (weak zero-knowledge). *Let (P, V) be an interactive proof system for a language L. We say that (P, V) is* weak zero-knowledge *if for every PPT adversary V^* and every PPT distinguisher D, there exists a PPT simulator S and a negligible function $\nu(\cdot)$ such that for every $n \in \mathbb{N}$, $x \in L \cap \{0,1\}^n$, and $z \in \{0,1\}^*$, we have*

$$|\Pr[D(x, z, Out_{V^*}[P(x) \leftrightarrow V^*(x,z)]) = 1] - \Pr[D(x, z, S(x,z)) = 1]| \leq \nu(n).$$

We now show that weak zero-knowledge is not equivalent to zero-knowledge if we assume the existence of two-round "timed" commitment schemes and "timed" worst-case weak one-way permutations satisfying certain properties. More precisely, we assume that there exists a polynomial $p(\cdot)$ such that for sufficiently large $n \in \mathbb{N}$, the following hold:

- There exists a collection of two-round "timed" commitment schemes $\{\mathsf{Com}_i\}_{i \in [\ell]}$, where $\ell = \log^2 n$, such that Com_i is hard to break in $p(n)^{i-1}$ steps, but can always be broken in $p(n)^i$ steps to obtain the committed value (e.g., one can get such timed commitment schemes from a timed commitment scheme in [7]).
- There exists a collection of "timed" worst-case weak one-way permutations $\{f_i\}_{i \in [\ell]}$, where $\ell = \log^2 n$, such that f_i is somewhat hard to invert in $p(n)^{i+1}$ steps in the worst case (i.e., an adversary running in $p(n)^{i+1}$ steps will fail to invert some instance $f_i(x')$ with probability at least $1/poly(n)$), but can always be inverted in $p(n)^{i+2}$ steps.

Theorem 7. *Assume the existence of two-round "timed" commitment schemes and "timed" worst-case weak one-way permutations as described above. Then, there exists an interactive proof system (P, V) for an NP language L such that (P, V) is weak zero-knowledge but not zero-knowledge.*

Proof (Proof sketch). The proof roughly works as follows. Let L be the trivial NP language $\{0,1\}^*$ with witness relation $R_L(x) = \{(f_1^{-1}(x), \ldots, f_{\log^2 |x|}^{-1}(x))\}$.

Let $(P(x,y), V(x))$ be the following interactive proof, where $x \in \{0,1\}^*$, $n = |x|$, $\ell = \log^2 n$, and $y = (f_1^{-1}(x), \ldots, f_\ell^{-1}(x))$:

1. The verifier V generates and sends ρ_i for $i = 1, \ldots, \ell$ to the prover, where ρ_i is the first message of an execution of Com_i.

2. The prover P sends $\mathsf{Com}_i(f_i^{-1}(x), \rho_i)$ for $i = 1, \ldots, \ell$ to the verifier, where $\mathsf{Com}_i(v, r)$ denotes the commitment of v using Com_i with first message r.
3. The verifier V accepts (i.e., outputs 1).

To see that (P, V) is weak zero-knowledge, consider any PPT verifier V^* and any PPT distinguisher D, and let $T(n)$ be a polynomial that bounds the combined running time of V^* and D. Then, a simulator S can compute the smallest positive integer j such that $p(n)^{j-1} > T(n)$, and then break $f_1^{-1}(x), \ldots, f_{j-1}^{-1}(x)$ in polynomial time. Then, the simulator S can simulate the protocol except that for $i = j, \ldots, \ell$, the simulator S sends $\mathsf{Com}_i(0^n, \rho_i)$ to V^* since S does not know $f_i^{-1}(x)$. By the hiding property of $\mathsf{Com}_j, \ldots, \mathsf{Com}_\ell$, the distinguisher D cannot distinguish between the output of the verifier V^* (in a true interaction with P) and the output of the simulator S, since D and V^* (combined) cannot break any of the commitment schemes $\mathsf{Com}_j, \ldots, \mathsf{Com}_\ell$ (since D and V^* do not run long enough).

Intuitively, (P, V) is not zero-knowledge because the existence of a (universal) simulator S would allow us to invert a worst-case weak one-way permutation f_j with overwhelming probability and in less time than what is specified in our hardness assumption for f_j. To see this, consider a PPT distinguisher D that, given x and a view of V, runs longer than S and breaks a commitment $\mathsf{Com}_j(w_j, \rho_j')$ from the view of V such that the time needed to break f_j is much longer than the running time of the simulator S, and then verifies whether or not $f(w_j) = x$. The fact that the simulator S works for the distinguisher D will ensure that with overwhelming probability, the output of $S(x)$ will contain a commitment $\mathsf{Com}_j(w_j, \rho_j')$ of some w_j such that $f_j(w_j) = x$. Thus, we can now construct an adversary A that inverts $f_j(w_j)$ with overwhelming probability by running the simulator S on input $f_j(w_j)$ and breaking the commitment $\mathsf{Com}_j(w_j, \rho_j')$ in the output of S. Since breaking f_j takes longer time than running the simulator S and breaking the commitment $\mathsf{Com}_j(w_j, \rho_j')$, the adversary A contradicts our hardness assumption for f_j. □

See the full version of this paper for the full proof of Theorem 7.

3 From Weak to Strong (t, ϵ)-Zero-Knowledge

From Theorem 7, we know that zero-knowledge and weak zero-knowledge are not equivalent. Thus, we now consider relaxed notions of zero-knowledge. We first consider a concrete security variant of the notion of zero-knowledge.

Definition 3 ((t, ϵ)-zero-knowledge). *Let (P, V) be an interactive proof system for a language L. We say that (P, V) is (t, ϵ)-zero-knowledge if for every PPT adversary V^*, there exists a PPT simulator S such that for every t-time distinguisher D, there exists an $n_0 \in \mathbb{N}$ such that for every $n \geq n_0$, $x \in L \cap \{0, 1\}^n$, and $z \in \{0, 1\}^*$, we have*

$$|\Pr[D(x, z, Out_{V^*}[P(x) \leftrightarrow V^*(x, z)]) = 1] - \Pr[D(x, z, S(x, z)) = 1]| \leq \epsilon(n).$$

Similar to before, we can relax the definition of zero-knowledge by switching the order of the quantifiers $\exists S$ and $\forall D$ so that the simulator S can depend on the distinguisher D. We call the relaxed definition *weak (t, ϵ)-zero-knowledge*.

Definition 4 (weak (t, ϵ)-zero-knowledge). *Let (P, V) be an interactive proof system for a language L. We say that (P, V) is weak (t, ϵ)-zero-knowledge if for every PPT adversary V^* and every t-time distinguisher D, there exists a PPT simulator S and an $n_0 \in \mathbb{N}$ such that for every $n \geq n_0$, $x \in L \cap \{0, 1\}^n$, and $z \in \{0, 1\}^*$, we have*

$$| \Pr[D(x, z, Out_{V^*}[P(x) \leftrightarrow V^*(x, z)]) = 1] - \Pr[D(x, z, S(x, z)) = 1]| \leq \epsilon(n).$$

Note that if (P, V) is (t, ϵ)-zero-knowledge (resp. weak (t, ϵ)-zero-knowledge) for some super polynomial function t and some negligible function ϵ, then (P, V) is zero-knowledge (resp. weak zero-knowledge) in the classic sense. We now show that (t, ϵ)-zero-knowledge and weak (t, ϵ)-zero-knowledge are equivalent if we consider all polynomials t and inverse polynomials ϵ.

Theorem 8. *Let (P, V) be an interactive proof system for a language L. Then, (P, V) is weak (t, ϵ)-zero-knowledge for every polynomial t and inverse polynomial ϵ if and only if (P, V) is (t', ϵ')-zero-knowledge for every polynomial t' and inverse polynomial ϵ'.*

Proof. The "if" direction clearly holds by definition. We will now prove the "only if" direction. Suppose (P, V) is weak (t, ϵ)-zero-knowledge for every polynomial t and inverse polynomial ϵ. Let t' be any polynomial, and let ϵ' be any inverse polynomial.

Let V^* be any PPT adversary, and let $T_{V^*}(\cdot)$ be any polynomial that bounds the running time of V^*. It is not hard to see that without loss of generality, we can assume that the auxiliary input $z \in \{0, 1\}^*$ in the definition of (t', ϵ')-zero-knowledge is exactly $C \cdot (T_{V^*}(n) + t'(n))$ bits long, where C is some constant ≥ 1.[5] Furthermore, it is easy to see that without loss of generality, we can also remove the absolute value $| \cdot |$ and change $\epsilon'(n)$ to $O(\epsilon'(n))$. Thus, it suffices to construct a PPT simulator S such that for every t'-time distinguisher D, there exists an $n_0 \in \mathbb{N}$ such that for every $n \geq n_0$, $x \in L \cap \{0, 1\}^n$, and $z \in \{0, 1\}^*$ with $|z| = C \cdot (T_{V^*}(n) + t'(n))$, we have

$$\Pr[D(x, z, Out_{V^*}[P(x) \leftrightarrow V^*(x, z)]) = 1] - \Pr[D(x, z, S(x, z)) = 1] \leq O(\epsilon'(n)).$$

We will now construct the required PPT simulator S for V^*.

High-level Description of the Simulator S: We first give a high-level description of the simulator S. The simulator S uses the multiplicative weights algorithm described in [14]. The simulator S, on input (x, z) with $n := |x|$, first runs a multiplicative weights algorithm to find a "good set" of simulator

---

[5] This follows from standard padding techniques and the fact that the adversary V^* and the distinguisher D cannot read any of the bits after the first $T_{V^*}(n) + t'(n)$ bits of z.

machines $\{S_1, \ldots, S_L\}$; then, the simulator S randomly and uniformly chooses one of the simulator machines in $\{S_1, \ldots, S_L\}$ to perform the simulation, i.e., S runs the chosen simulator machine on input (x, z) and outputs whatever the simulator machine outputs.

Before we describe the multiplicative weights algorithm run by the simulator S, let us introduce some notation. Given a simulator S' and a distinguisher D', let the "payoff" of S' (with respect to D') be

$$\mu(S', D') := \Pr[D'(x, z, S'(x, z)) = 1] - \Pr[D'(x, z, Out_{V^*}[P(x) \leftrightarrow V^*(x, z)]) = 1].$$

Given a simulator S' and a distribution $\mathcal{D}^{(i)}$ over distinguishers, let

$$\mu(S', \mathcal{D}^{(i)}) := \mathbb{E}_{D' \sim \mathcal{D}^{(i)}}[\mu(S', D')] = \sum_{D' \in Supp(\mathcal{D}^{(i)})} \mathcal{D}^{(i)}(D') \cdot \mu(S', D').$$

We note that we want to design the simulator S so that for every t'-time distinguisher D, we have $\mu(S, D) \geq -O(\epsilon'(n))$.

Let $D_1, D_2, D_3, \ldots$ be an enumeration of the set of all (uniform) distinguishers, and let $D'_1, D'_2, D'_3, \ldots$ be the corresponding sequence where D'_j is the same as D_j except that after $t'(n)$ steps, D'_j stops and outputs 0. We note that each fixed t'-time distinguisher D will eventually appear in the set $\{D'_1, \ldots, D'_n\}$ as n gets larger.

We now describe the multiplicative weights algorithm run by S. In the multiplicative weights algorithm, S simulates L rounds (repetitions) of a zero-sum game between a "simulator player" Sim and a "distinguisher player" Adv, where the payoff function for Sim is the function $\mu(\cdot, \cdot)$ defined above. In each round i, Adv chooses a mixed strategy (i.e., a distribution) $\mathcal{D}^{(i)}$ over its set of pure strategies $\{D'_1, \ldots, D'_n\}$ (a set of distinguishers), and then Sim chooses a simulator machine $S_i := S_i(\mathcal{D}^{(i)})$ that hopefully "does well" against Adv's mixed strategy $\mathcal{D}^{(i)}$, i.e., Sim's (expected) payoff $\mu(S_i, \mathcal{D}^{(i)})$ is high.

In the first round, Adv chooses the uniform distribution $\mathcal{D}^{(1)}$ over $\{D'_1, \ldots, D'_n\}$. After each round i, Adv updates its mixed strategy to get $\mathcal{D}^{(i+1)}$ in a manner similar to the multiplicative weights algorithm described in [14], which involves the payoff function μ. However, Adv cannot compute μ efficiently, since μ involves the prover P, which may be inefficient (or has a witness y that Adv does not have). Thus, Adv uses an approximation $\hat{\mu}$ of the payoff function μ. In particular, given a distinguisher D', Adv can approximate $\mu(S_i, D')$ by approximating $Out_{V^*}[P(x) \leftrightarrow V^*(x, z)]$ with the output of a simulator $S_{D'}$ that is good w.r.t. the distinguisher D'; the existence of such a simulator is guaranteed by the weak zero-knowledge property of (P, V). There are still some issues: Adv might not be able to find $S_{D'}$ efficiently and uniformly, and $S_{D'}$ only works well for sufficiently large n. We resolve these issues by using a "universal" distinguisher that essentially takes a description of a distinguisher D' as auxiliary input and runs D', and we use a simulator that is good w.r.t. this universal distinguisher.

Using an analysis similar to that in [14], we will show that if Sim manages to choose a simulator machine S_i that does well against Adv's mixed strategy $\mathcal{D}^{(i)}$ in every round $i \in [L]$, then the uniform mixed strategy over the set

$\{S_1, \ldots, S_L\}$ of chosen simulator machines does well against all the distinguishers in $\{D'_1, \ldots, D'_n\}$. To choose a simulator machine S_i that does well against Adv's mixed strategy $\mathcal{D}^{(i)}$, Sim makes use of the weak zero-knowledge property of (P, V), which guarantees that for every distinguisher D, there exists a simulator S_D that does well against D. However, there are some complications: (1) $\mathcal{D}^{(i)}$ is a *mixture* of distinguishers, not a single distinguisher; (2) Sim might not be able to *efficiently and uniformly* find the distinguisher-dependent simulator; and (3) even if Sim can efficiently and uniformly find the distinguisher-dependent simulator, the simulator depends on the mixed strategy $\mathcal{D}^{(i)}$, and the time needed to sample from $\mathcal{D}^{(i)}$ could be growing (exponentially) in each round, which in turn can potentially lead to an exponential growth in the running time of the distinguisher-dependent simulator as more rounds are performed.

Sim overcomes these problems by (also) using a "universal" distinguisher D_U that takes the weights (i.e., probability masses) of a distribution $\mathcal{D}$ over $\{D'_1, \ldots, D'_n\}$ as auxiliary input, samples a distinguisher from the distribution $\mathcal{D}$, and then runs the sampled distinguisher. Let S_{D_U} be the simulator that is good w.r.t. D_U; again, the existence of such a simulator is guaranteed by the weak zero-knowledge property of (P, V). Sim chooses S_i to be the simulator machine that runs S_{D_U} with the weights of the distribution $\mathcal{D}^{(i)}$ provided as auxiliary input. We now give a formal description of the simulator S.

The Simulator S: Let $D_1, D_2, D_3, \ldots$ be an enumeration of the set of all (uniform) distinguishers, and let $D'_1, D'_2, D'_3, \ldots$ be the corresponding sequence where D'_j is the same as D_j except that after $t'(n)$ steps, D'_j stops and outputs 0.

The simulator S, on input (x, z) with $n := |x|$, proceeds as follows:

1. Let $T_{D_U}(n) = O((T_{V^*}(n) + t'(n) + n)^2)$.
   Given a distribution $\mathcal{D}$ over $\{D'_1, \ldots, D'_n\}$, let $\boldsymbol{p}_\mathcal{D}$ denote the vector of weights (i.e., probability masses) representing $\mathcal{D}$, i.e., $\boldsymbol{p}_\mathcal{D} = (\mathcal{D}(D'_1), \ldots, \mathcal{D}(D'_n))$.
   Let D_U be a "universal" distinguisher that, on input (x, z', v), first parses z' as $z' = z \| \boldsymbol{p}_\mathcal{D}$, where $\boldsymbol{p}_\mathcal{D}$ is a vector the weights representing some distribution $\mathcal{D}$ over $\{D'_1, \ldots, D'_n\}$; then, D_U samples a distinguisher D'_j from the distribution $\mathcal{D}$, and then runs D'_j on input (x, z, v), but D_U always stops after $T_{D_U}(n)$ steps regardless of whether or not D'_j finishes running.
   Let S_{D_U} be the PPT simulator for D_U that is guaranteed by the weak (T_{D_U}, ϵ')-zero-knowledge property of (P, V).
2. Let $L = \Theta(\frac{\log n}{\epsilon'(n)^2})$ and $\beta = \frac{1}{1 + \sqrt{(2 \ln n)/L}}$. ($L$ is the number of rounds we will run the multiplicative weights algorithm for, and β is used in the multiplicative update rule.)
3. **Multiplicative weights algorithm:**
   Let $\mathcal{D}^{(1)}$ be the uniform distribution over $\{D'_1, \ldots, D'_n\}$. (The probability mass $\mathcal{D}^{(1)}(D'_j)$ for D'_j can be thought of as the "weight" for D'_j.)
   For $i = 1, \ldots, L$ do:
   (a) **Choosing a simulator machine S_i that does well against $\mathcal{D}^{(i)}$:**
       Let S_i be a simulator machine that, on input (x, z), outputs $S_{D_U}(x, z \| \boldsymbol{p}_{\mathcal{D}^{(i)}})$.
   (b) **Weight update:**

Compute the distribution $\mathcal{D}^{(i+1)}$ from $\mathcal{D}^{(i)}$ by letting

$$\mathcal{D}^{(i+1)}(D'_j) \sim \beta^{\widehat{\mu}(S_i, D'_j)} \cdot \mathcal{D}^{(i)}(D'_j)$$

for every $D'_j \in \{D'_1, \ldots, D'_n\}$ (and renormalizing), where

$$\widehat{\mu}(S_i, D'_j) := \mathrm{freq}_k[D'_j(x, z, S_i(x, z))] - \mathrm{freq}_k[D'_j(x, z, S_{D_U}(x, z\|\boldsymbol{p}_{D'_j}))],$$

where $\mathrm{freq}_k[D'_j(x, z, S_i(x, z))]$ and $\mathrm{freq}_k[D'_j(x, z, S_{D_U}(x, z\|\boldsymbol{p}_{D'_j}))]$ are approximations of $\Pr[D'_j(x, z, S_i(x, z)) = 1]$ and $\Pr[D'_j(x, z, S_{D_U}(x, z\|\boldsymbol{p}_{D'_j}))$ $= 1]$ by taking $k := \Theta(\frac{\log(nL/\epsilon'(n))}{\epsilon'(n)^2})$ samples, respectively, and computing the relative frequency in which 1 is outputted.

The function $\widehat{\mu}$ should be viewed as being an approximation of the payoff function μ.

    End for

4. Choose $S_i \in \{S_1, \ldots, S_L\}$ uniformly at random.
5. Run the simulator S_i on input (x, z) and output $S_i(x, z)$.

We now continue with the formal proof. It can be easily verified that S runs in time $poly(n, t'(n), \frac{1}{\epsilon'(n)})$. Let D be any distinguisher whose running time is bounded by $t'(n)$. Fix an integer n that is sufficiently large so that the distinguisher D appears in $\{D_1, \ldots, D_n\}$ and S_{D_U} works for the distinguisher D_U on input size n for x. We note that the distinguisher D also appears in $\{D'_1, \ldots, D'_n\}$, since the running time of D is bounded by $t'(n)$. Fix $x \in L \cap \{0,1\}^n$ and $z \in \{0,1\}^*$ with $|z| = C \cdot (T_{V^*}(n) + t'(n))$. To prove the theorem, it suffices to show that

$$\mu(S, D) \geq -O(\epsilon'(n)).$$

To show this, we will proceed as follows: (1) We first show that if, in every round i the chosen simulator S_i does well against the distribution $\mathcal{D}^{(i)}$ with respect to our approximation $\widehat{\mu}$ of μ, then the simulator S does well against D with respect to $\widehat{\mu}$; this is the first lemma below; (2) We then show that the first lemma holds even if we replace $\widehat{\mu}$ with μ; this is the second lemma below; (3) Finally, we show that in each round i, the chosen simulator S_i indeed does well against the distribution $\mathcal{D}^{(i)}$ with respect to μ.

We now proceed with the proof. For $i = 1, \ldots, L$, let

$$\widehat{\mu}(S_i, \mathcal{D}^{(i)}) := \mathbb{E}_{D' \sim \mathcal{D}^{(i)}}[\widehat{\mu}(S_i, D')] = \sum_{k=1}^{n} \mathcal{D}^{(i)}(D'_k) \cdot \widehat{\mu}(S_i, D'_k).$$

One should view $\widehat{\mu}(S_i, \mathcal{D}^{(i)})$ as an approximation of $\mu(S_i, \mathcal{D}^{(i)})$.

Lemma 1. *For every distinguisher $D'_j \in \{D'_1, \ldots, D'_n\}$, if we run the simulator $S(x, z)$, then (with probability 1) $S(x, z)$ generates $\mathcal{D}^{(1)}, \ldots, \mathcal{D}^{(L)}$ and $S_1, \ldots, S_L$ such that*

$$\frac{1}{L} \sum_{i=1}^{L} \widehat{\mu}(S_i, D'_j) \geq \frac{1}{L} \sum_{i=1}^{L} \widehat{\mu}(S_i, \mathcal{D}^{(i)}) - O(\epsilon'(n)).$$

The proof of Lemma 1 is essentially the same as a lemma found in [9], whose proof is very similar to the analysis of the multiplicative weights algorithm found in [14]. In [14], the multiplicative weights algorithm updates the weights of $\mathcal{D}^{(i)}$ using the exact value of $\mu(S_i, D'_j)$; here, we only have an approximation $\hat{\mu}(S_i, D'_j)$ of $\mu(S_i, D'_j)$, but with minor changes, the analysis in [14] can still be used to show Lemma 1. We provide a proof of Lemma 1 in the full version of this paper.

We now show that we can essentially replace the $\hat{\mu}$ in Lemma 1 with μ.

Lemma 2. *For every* $D' \in \{D'_1, \ldots, D'_n\}$, *if we run the simulator* $S(x, z)$, *then with probability* $1 - O(\epsilon'(n))$ *over the random coins of* S, $S(x, z)$ *generates* $\mathcal{D}^{(1)}, \ldots, \mathcal{D}^{(L)}$ *and* $S_1, \ldots, S_L$ *such that*

$$\frac{1}{L} \sum_{i=1}^{L} \mu(S_i, D') \geq \frac{1}{L} \sum_{i=1}^{L} \mu(S_i, \mathcal{D}^{(i)}) - O(\epsilon'(n)).$$

The proof of Lemma 2 roughly works as follows. We take Lemma 1 and show that each time we approximate μ via $\hat{\mu}$, the approximation is good with high probability; this follows from Chernoff bounds and the fact that S_{D_U} is a simulator for V^* that is good with respect to the "universal" distinguisher D_U. Lemma 2 then follows from the union bound. See the full version of this paper for the proof of Lemma 2.

To complete the proof of Theorem 8, we will now show that $\mu(S, D) \geq -O(\epsilon'(n))$. We first show that for every $i \in [L]$, we always have $\mu(S_i, \mathcal{D}^{(i)}) \geq -O(\epsilon'(n))$. Fix $i \in [L]$. Now, we observe that

$$\mu(S_i, \mathcal{D}^{(i)})$$
$$= \sum_{j=1}^{n} \mathcal{D}^{(i)}(D'_j) \cdot (\Pr[D'_j(x, z, S_i(x, z)) = 1] - \Pr[D'_j(x, z, Out_{V^*}[P(x) \leftrightarrow V^*(x, z)]) = 1])$$
$$= \Pr[D_U(x, z \| \boldsymbol{p}_{\mathcal{D}^{(i)}}, S_i(x, z)) = 1] - \Pr[D_U(x, z \| \boldsymbol{p}_{\mathcal{D}^{(i)}}, Out_{V^*}[P(x) \leftrightarrow V^*(x, z)]) = 1]$$
$$= \Pr[D_U(x, z \| \boldsymbol{p}_{\mathcal{D}^{(i)}}, S_{D_U}(x, z \| \boldsymbol{p}_{\mathcal{D}^{(i)}})) = 1]$$
$$\quad - \Pr[D_U(x, z \| \boldsymbol{p}_{\mathcal{D}^{(i)}}, Out_{V^*}[P(x) \leftrightarrow V^*(x, z \| \boldsymbol{p}_{\mathcal{D}^{(i)}})]) = 1]$$
$$\geq -\epsilon'(n), \tag{2}$$

where the second equality follows from the definition of D_U, the third equality follows from the definition of S_i and the fact that $V^*(x, z) = V^*(x, z \| \boldsymbol{p}_{\mathcal{D}^{(i)}})$ (since $|z| \geq T_{V^*}(n)$), and the last inequality follows from the fact that S_{D_U} is a simulator for D_U in the weak (t', ϵ')-zero-knowledge property of (P, V).

Now, combining Lemma 2 and (2), we have that with probability $1 - O(\epsilon'(n))$ over the randomness of S, $S(x, z)$ generates $S_1, \ldots, S_L$ such that

$$\frac{1}{L} \sum_{i=1}^{L} \mu(S_i, D) \geq -O(\epsilon'(n)). \tag{3}$$

Now, recall that after generating $S_1, \ldots, S_L$, the simulator $S(x, z)$ chooses a uniformly random $S_i \in \{S_1, \ldots, S_L\}$ and runs $S_i(x, z)$. Thus, conditional on

$S(x, z)$ generating a particular sequence $S_1, \ldots, S_L$, we have $\mu(S, D) = \sum_{i=1}^{L} \frac{1}{L} \cdot \mu(S_i, D)$. Combining this with (3) (which holds with probability $1 - O(\epsilon'(n))$ over the randomness of S), we get

$$\mu(S, D) \geq -O(\epsilon'(n)) - O(\epsilon'(n)) = -O(\epsilon'(n)),$$

as required. This completes the proof of Theorem 8.   □

4   From Super-Weak to Strong Distributional (T, t, ϵ)-Zero-Knowledge

In this section we consider a "super-weak" notion of zero-knowledge, where not only do we allow the simulator to depend on the distinguisher, but also, we only require that the simulator manages to make the distinguisher output 1 with *at least as high probability* (minus some "small" gap) as the probability that the distinguisher outputs 1 when given a true view of V^*. That is, we only consider "one-sided" indistinguishability. (Such a notion was previously considered in [21].)

In the full version of this paper, we show that super-weak (t, ϵ)-zero-knowledge is not equivalent to weak (t, ϵ)-zero-knowledge. Thus, we here consider an alternative "distributional" notion of zero-knowledge (a la [17]) where indistinguishability of the simulation is only required for any distribution over statements (and auxiliary inputs), and the simulator as well as the distinguisher can depend on the distribution. Additionally, we here model both the distinguisher and the simulator as non-uniform algorithms (as opposed to uniform ones). (The combination of these variants was previously considered by [12].) For concreteness, we also add a parameter T to the definition and require that the simulator is of size at most $T(n)$, and thus we also bound the size of the malicious verifier V^* by $t(n)$.

Definition 5 (distributional (T, t, ϵ)-zero-knowledge). *Let (P, V) be an interactive proof system for a language L. We say that (P, V) is distributional (T, t, ϵ)-zero-knowledge if for every $n \in \mathbb{N}$, every joint distribution (X_n, Y_n, Z_n) over $(L \cap \{0,1\}^n) \times \{0,1\}^* \times \{0,1\}^*$, and every randomized $t(n)$-size adversary V^*, there exists a randomized $T(n)$-size simulator S such that for every randomized $t(n)$-size distinguisher D, we have*

$$|\Pr[D(X_n, Z_n, Out_{V^*}[P(X_n, Y_n) \leftrightarrow V^*(X_n, Z_n)]) = 1] - \Pr[D(X_n, Z_n, S(X_n, Z_n)) = 1]|$$
$$\leq \epsilon(n).$$

In the above definition, if L is an NP-language, then we require (i.e., assume) Y_n to be a witness of X_n (this also applies to the corresponding definition below). *Weak distributional (T, t, ϵ)-zero-knowledge* can be defined in an analogous way by switching the ordering of the quantifiers $\exists S$ and $\forall D$. We now turn to define *super-weak distributional (T, t, ϵ)-zero-knowledge*.

Definition 6 (super-weak   distributional (T, t, ϵ)-zero-knowledge). *Let (P, V) be an interactive proof system for a language L. We say that (P, V) is* super-weak distributional (T, t, ϵ)-zero-knowledge *if for every $n \in \mathbb{N}$, every joint distribution (X_n, Y_n, Z_n) over $(L \cap \{0,1\}^n) \times \{0,1\}^* \times \{0,1\}^*$, every randomized $t(n)$-size adversary V^*, and every randomized $t(n)$-size distinguisher D, there exists a randomized $T(n)$-size simulator S such that*

$$\Pr[D(X_n, Z_n, Out_{V^*}[P(X_n, Y_n) \leftrightarrow V^*(X_n, Z_n)]) = 1] - \Pr[D(X_n, Z_n, S(X_n, Z_n)) = 1]$$
$$\leq \epsilon(n).$$

We may consider an even weaker notion of super-weak distributional zero-knowledge—let us refer to it as *super-weak* distributional zero-knowledge*—where we only require indistinguishability to hold against *deterministic* distinguishers D that may output a real value in $[0, 1]$ (such a distinguisher can easily be converted to a randomized distinguisher by simply first computing the output p of the deterministic one and then sampling a decision bit $b = 1$ with probability p).

We now show that super-weak distributional (T, t, ϵ)-zero-knowledge is equivalent to distributional (T, t, ϵ)-zero-knowledge if we consider all polynomials for T and t and all inverse polynomials for ϵ. In fact, we prove a more general theorem that also describes the loss in the parameters T, t, and ϵ.

Theorem 9. *Let (P, V) be an interactive proof system for a language L, and suppose (P, V) is super-weak distributional (T, t, ϵ)-zero-knowledge. Then, (P, V) is also distributional $(T', t', 2\epsilon)$-zero-knowledge, where $t'(n) = \Omega(\epsilon(n)\sqrt{t(n)} - n)$ and $T'(n) = O(\frac{t'(n)\ln(n+t'(n))}{\epsilon(n)^2}) \cdot T(n)$.*

Proof. Let $n \in \mathbb{N}$, let (X_n, Y_n, Z_n) be any joint distribution over $(L \cap \{0,1\}^n) \times \{0,1\}^* \times \{0,1\}^*$, and let V^* be any $t(n)$-size adversary. It is easy to see that w.l.o.g., we can assume that the length of Z_n is always bounded by $t'(n)$, and we can remove the absolute value $|\cdot|$ in the definition of distributional $(T', t', 2\epsilon)$-zero-knowledge. Thus, it suffices to show the following claim:

Claim. There exists a $T'(n)$-size simulator S such that for every $t'(n)$-size distinguisher D,

$$\Pr[D(X_n, Z_n, S(X_n, Z_n)) = 1] - \Pr[D(X_n, Z_n, Out_{V^*}[P(X_n, Y_n) \leftrightarrow V^*(X_n, Z_n)]) = 1]$$
$$\geq -2\epsilon(n).$$

We now proceed to showing the above claim. We define a two-player zero-sum game between a "simulator player" Sim and a "distinguisher player" Adv. The set $Strat_{\mathsf{Sim}}$ of pure strategies for Sim is the set of all $T(n)$-size simulators, and the set $Strat_{\mathsf{Adv}}$ of pure strategies for Adv is the set of all $t'(n)$-size distinguishers. The payoff for Sim when Sim chooses a simulator $S \in Strat_{\mathsf{Sim}}$ and Adv chooses a distinguisher $D \in Strat_{\mathsf{Adv}}$ is

$$\mu_n(S, D)$$
$$:= \Pr[D(X_n, Z_n, S(X_n, Z_n)) = 1] - \Pr[D(X_n, Z_n, Out_{V^*}[P(X_n, Y_n) \leftrightarrow V^*(X_n, Z_n)]) = 1].$$

For mixed strategies (i.e., distributions) $\mathcal{S}$ over $Strat_{\mathsf{Sim}}$, and $\mathcal{D}$ over $Strat_{\mathsf{Adv}}$, we define
$$\mu_n(\mathcal{S}, \mathcal{D}) := \mathbb{E}_{S \leftarrow \mathcal{S}, D \leftarrow \mathcal{D}}[\mu_n(S, D)].$$

The following simple lemma states that any distribution over circuits can be approximated by a small randomized circuit, obtained by taking an appropriate number of samples from the original distribution. This proof technique was used in [1] and [24] for obtaining sparse approximations to randomized strategies in two-player zero-sum games. A fact similar to our lemma was implicitly used by Impagliazzo [22] and several subsequent works, but we find it useful to explicitly formalize it as a lemma (that we hope will be useful also in other contexts).

Lemma 3 (Approximating a distribution over circuits by a small circuit obtained via sampling). *Let X and A be finite sets, let Y be any random variable with finite support, let C be any distribution over s-size randomized circuits of the form $C : X \times Supp(Y) \to A$, and let U be any finite set of randomized circuits of the form $u : X \times Supp(Y) \times A \to \{0, 1\}$. Then, for every $\epsilon > 0$, there exists a randomized circuit $\widehat{C}$ of size $T = O(\frac{\log |X| + \log |U|}{\epsilon^2} \cdot s)$ such that for every $u \in U$ and $x \in X$, we have*

$$|\mathbb{E}_{C \leftarrow \mathcal{C}}[u(x, Y, C(x, Y))] - \mathbb{E}[u(x, Y, \widehat{C}(x, Y))]| \leq \epsilon.$$

Additionally, there exists a deterministic circuit $\widetilde{C}$ of size T such that for all inputs x, y, $\widetilde{C}(x, y) = \Pr[\widehat{C}(x, y) = 1]$.

The lemma follows easily from a Chernoff bound and a union bound; see the full version of this paper for the proof. This proof of the main theorem now follows from three relatively simple steps:

Step 1. We first show that for any mixed strategy $\mathcal{D}$ for Adv (i.e., any distribution over $t'(n)$-size distinguishers), there exists a $T(n)$-size simulator $S_{\mathcal{D}} \in Strat_{\mathsf{Sim}}$ such that $\mu_n(S_{\mathcal{D}}, \mathcal{D}) \geq -3\epsilon(n)/2$. By Lemma 3, we can approximate $\mathcal{D}$ by a $t(n)$-size distinguisher $\widehat{D}$, and then use the super-weak distributional (T, t, ϵ)-zero-knowledge property of (P, V) to get a $T(n)$-size simulator $S_{\widehat{D}}$ for $\widehat{D}$ such that $\mu_n(S_{\widehat{D}}, \widehat{D}) \geq -\epsilon(n)$. Since $\widehat{D}$ approximates $\mathcal{D}$ to within $\epsilon(n)/2$, we have $\mu_n(S_{\widehat{D}}, \mathcal{D}) \geq -3\epsilon(n)/2$, as required.

Step 2. We now apply the minimax theorem to the result of Step 1 to get a mixed strategy $\mathcal{S}$ for Sim (i.e., a distribution over $T(n)$-size simulators) such that for every $t'(n)$-size distinguisher $D \in Strat_{\mathsf{Adv}}$, we have $\mu_n(\mathcal{S}, D) \geq -3\epsilon(n)/2$.

Step 3. By Lemma 3, we can approximate $\mathcal{S}$ (from Step 2) by a $T'(n)$-size simulator $\widehat{S}$ so that $\mu_n(\widehat{S}, D) \geq -2\epsilon(n)$ for every $t'(n)$-size distinguisher $D \in Strat_{\mathsf{Adv}}$.

The result of Step 3 shows Claim 4, which completes the proof of the theorem. We now provide the details for Steps 1 and 3.

Details of Step 1.
By Lemma 3 (in the statement of the lemma, we let $X = Supp(X_n) \times Supp(Z_n) \times \{0, 1\}^{t'(n)}$, $A = \{0, 1\}$, $Y = 0$, $\mathcal{C} = \mathcal{D}$, U be a set containing only the circuit $(x, y, a) \mapsto a$, and $\epsilon = \epsilon(n)/2$), there exists a distinguisher $\widehat{D}$ of size $O((n +$

$t'(n))^2/\epsilon(n)^2) = t(n)$ such that for every $x \in X_n$, $z \in Z_n$, and $v \in \{0,1\}^{t'(n)}$, we have $|\Pr_{D \leftarrow \mathcal{D}}[D(x,z,v) = 1] - \Pr[\widehat{D}(x,z,v) = 1]| \leq \epsilon(n)/2$. Since (P,V) is super-weak distributional (T,t,ϵ)-zero-knowledge, there exists a $T(n)$-size simulator $S_{\widehat{D}}$ such that $\mu_n(S_{\widehat{D}}, \widehat{D}) \geq -\epsilon(n)$. From the result above and the definition of μ_n, we have $|\mu_n(S_{\widehat{D}}, \mathcal{D}) - \mu_n(S_{\widehat{D}}, \widehat{D})| \leq \epsilon(n)/2$, so $\mu_n(S_{\widehat{D}}, \mathcal{D}) \geq -3\epsilon(n)/2$, as required.

Details of Step 3.
By Lemma 3, there exists a simulator $\widehat{S}$ of size $O((\log |Strat_{\mathsf{Adv}}|/\epsilon(n)^2) \cdot T(n))$ such that for every $t'(n)$-size distinguisher $D \in Strat_{\mathsf{Adv}}$, we have
$|\Pr_{S \leftarrow \mathcal{S}}[D(X_n, Z_n, S(X_n, Z_n)) = 1] - \Pr[D(X_n, Z_n, \widehat{S}(X_n, Z_n)) = 1]| \leq \epsilon(n)/2$,
which implies $|\mu_n(\mathcal{S}, D) - \mu_n(\widehat{S}, D)| \leq \epsilon(n)/2$. Combining this with the result of Step 2, we have $\mu_n(\widehat{S}, D) \geq -2\epsilon(n)$ for every $t'(n)$-size distinguisher $D \in Strat_{\mathsf{Adv}}$. Furthermore, the simulator $\widehat{S}$ has size at most $T'(n)$, since there are at most $O(q(n)+t'(n))^{O(t'(n))}$ circuits of size $t'(n)$ on $q(n)$ input bits, so $|Strat_{\mathsf{Adv}}| \leq O(n + t'(n))^{O(t'(n))}$. $\qquad\square$

We note that by the "additional" part of Lemma 3, the above proof actually directly shows equivalence also between super-weak* distributional zero-knowledge and distributional zero-knowledge:

Theorem 10. *Let (P,V) be an interactive proof system for a language L, and suppose (P,V) is super-weak* distributional (T,t,ϵ)-zero-knowledge. Then, (P,V) is also distributional $(T',t',2\epsilon)$-zero-knowledge, where $t'(n) = \Omega(\epsilon(n)\sqrt{t(n)} - n)$ and $T'(n) = O(\frac{t'(n)\ln(n+t'(n))}{\epsilon(n)^2}) \cdot T(n)$.*

4.1   Laconic Prover Implies Distributional (T, t, ϵ)-Zero-Knowledge

In this section, we first use Theorem 10 to show that an interactive proof with short prover communication complexity implies distributional (T,t,ϵ)-zero-knowledge. We then describe applications of this result.

Theorem 11. *Let (P,V) be an interactive proof system for a language L, and suppose that the prover P has communication complexity $\ell(n)$, i.e., the total length of the messages sent by P is $\ell(n)$, where n is the length of the common input x. Then, for every function $t'(n) \geq \Omega(n)$ and $\epsilon'(n)$, (P,V) is distributional (T', t', ϵ')-zero-knowledge, where $T'(n) = O\left(2^{\ell(n)} \cdot \frac{t'(n)^3 \ln(t'(n))}{\epsilon'(n)^4}\right)$.*

Proof. By Theorem 10, it suffices to show that (P,V) is super-weak* distributional $(T, t, \epsilon'/2)$-zero-knowledge, where $t(n) = \Theta(\frac{t'(n)^2}{\epsilon'(n)^2})$ and $T(n) = O(2^{\ell(n)} \cdot \frac{t'(n)^2}{\epsilon'(n)^2})$. Let $n \in \mathbb{N}$, let (X_n, Y_n, Z_n) be a joint distribution over $(L \cap \{0,1\}^n) \times \{0,1\}^* \times \{0,1\}^*$, let V^* be any randomized $t(n)$-size adversary, and let D be any *deterministic* $t(n)$-size distinguisher outputting a real value in $[0,1]$. Consider some inputs x, z and randomness r for the verifier V^*. For any sequence of messages $(m_1, \ldots, m_k)$, let $(m_1, \ldots, m_k) \leftrightarrow V_r^*(x,z)$ denote the protocol where the prover sends the message m_i to V^* in round i, where the randomness of V^* is fixed to r.

Let S be the simulator that, on input (x, z) and given randomness r, enumerates each of the $2^{\ell(n)}$ possible sequences of messages $(m_1, \ldots, m_k)$ of total length $\ell(n)$ (that the prover P may possibly send) and picks the sequence of messages that maximizes $D(x, z, Out_{V^*}[(m_1, \ldots, m_k) \leftrightarrow V_r^*(x, z)])$. By construction it follows that for every random tape r, $D(x, z, Out_{V^*}[P(x) \leftrightarrow V_r^*(x, z)]) \leq D(x, z, S_r(x, z))$ and thus

$$\Pr[D(x, z, Out_{V^*}[P(x) \leftrightarrow V^*(x, z)]) = 1] - \Pr[D(x, z, S(x, z)) = 1] \leq 0.$$

Furthermore, we note that the size of the simulator S is $O(2^{\ell(n)} \cdot t(n)) = T(n)$. Thus, (P, V) is super-weak* distributional $(T, t, 0)$-zero-knowledge, which completes the proof. □

Let us now provide a few corollaries of Theorem 11. The first two are new proofs of old theorems (with some new generalizations). The third one is a new result on 2-round zero-knowledge.

Application 1: Leakage Lemma of Gentry-Wichs. Roughly speaking, the "Leakage Lemma" of Gentry-Wichs [16] states that for every joint distribution $(X, \pi(X))$, where $|\pi(x)| = O(\log |x|)$ (π should be thought of as leakage on X), and for every distribution Y that is indistinguishable from X, there exists some leakage $\widetilde{\pi}$ such that the joint distributions $(X, \pi(X))$ and $(Y, \widetilde{\pi}(Y))$ are indistinguishable. We now show that this result follows as a simple corollary of Theorem 11.

Two distributions X and Y are (s, ϵ)-*indistinguishable* if every s-size circuit C can only distinguish X from Y by at most ϵ, i.e., $|\Pr[C(X) = 1] - \Pr[C(Y) = 1]| \leq \epsilon$.

Corollary 1 (The leakage lemma of Gentry-Wichs [16]). *Let $(X, \pi(X))$ be any joint distribution, where $|\pi(X)| \leq \ell$. Let Y be any distribution that is (s, ϵ)-indistinguishable from X. Then, there exists a joint distribution $(Y, \widetilde{\pi}(Y))$ such that $(X, \pi(X))$ and $(Y, \widetilde{\pi}(Y))$ are $(s', 2\epsilon)$-indistinguishable, where $s' = \Omega\left(\sqrt[3]{\frac{\epsilon^4 \cdot s}{2^\ell \cdot \ln(s)}}\right)$.*

Proof. Let $L = \{0, 1\}^*$ be the trivial language with the trivial witness relation $R_L(x) = \{0, 1\}^*$. Let (P, V) be an interactive proof system for L where the prover P, on input a statement x with witness y, simply sends the first ℓ bits of y to the verifier V, who simply always accepts. By Theorem 11, (P, V) is distributional (T, s', ϵ)-zero-knowledge, where $T \leq s/2$. By considering the statement distribution X with witness distribution $\pi(X)$, it follows that there exists a T-size simulator S such that $(X, \pi(X))$ and $(X, S(X))$ are (s', ϵ)-indistinguishable. Also, $(X, S(X))$ and $(Y, S(Y))$ are $(s/2, \epsilon)$-indistinguishable, since X and Y are (s, ϵ)-indistinguishable and $T \leq s/2$. It follows that $(X, \pi(X))$ and $(Y, S(Y))$ are $(s', 2\epsilon)$-indistinguishable, so letting $\widetilde{\pi} = S$ yields the result. □

Let us note that our proof of the leakage lemma yields an even stronger statement—namely, we have found an efficient simulator $\tilde{\pi}$; such a version of the leakage lemma was recently established by Jetchev and Pietrzak [23]. (As an independent contribution, our proof of Theorem 4 is actually significantly simpler than both the proof of [16] and [23].) Additionally, since our result on zero-knowledge applies also to *interactive* protocols, we directly also get an interactive version of the leakage lemma.

Application 2: Dense Model Theorem We proceed to show that the *dense model theorem* (e.g., see [28,30,13]) follows as a corollary of Theorem 11. A distribution R is δ-*dense* in a distribution X if for every r, $\Pr[R = r] \leq \frac{1}{\delta} \Pr[X = r]$. Equivalently, R is δ-dense in X if there exists a joint distribution $(X, B(X))$ with $\Pr[B(X) = 1] \geq \delta$ such that $R = X|(B(X) = 1)$. Let U_n be the uniform distribution over $\{0,1\}^n$.

Corollary 2 (The dense model theorem). *Let X be any distribution over $\{0,1\}^n$ that is (s, ϵ)-indistinguishable from U_n, and suppose R is δ-dense in X. Then, there exists a distribution M that is $(\delta - 2\epsilon)$-dense in U_n, and M and R are $(s', \frac{2\epsilon}{\delta})$-indistinguishable, where $s' = \Omega\left(\sqrt[3]{\frac{\epsilon^4 \cdot s}{\ln(s)}}\right)$.*

Proof. Since R is δ-dense in X, there exists a joint distribution $(X, B(X))$ with $\Pr[B(X) = 1] \geq \delta$ such that $R = X|(B(X) = 1)$. Without loss of generality, we can assume that $B(X)$ is always either 0 or 1. Let $L = \{0,1\}^*$ be the trivial language with the trivial witness relation $R_L(x) = \{0,1\}^*$. Let (P, V) be an interactive proof system for L where the prover P, on input a statement x with witness y, simply sends the first bit of y to the verifier V, who simply always accepts. By Theorem 11, (P, V) is distributional $(T, 2s', \epsilon)$-zero-knowledge, where $T \leq s/2$. By considering the statement distribution X with witness distribution $B(X)$, it follows that there exists a T-size simulator S such that $(X, B(X))$ and $(X, S(X))$ are $(2s', \epsilon)$-indistinguishable. Also, $(X, S(X))$ and $(U_n, S(U_n))$ are $(s/2, \epsilon)$-indistinguishable, since X and U_n are (s, ϵ)-indistinguishable and $T \leq s/2$. It follows that $(X, B(X))$ and $(U_n, S(U_n))$ are $(2s', 2\epsilon)$-indistinguishable. Thus, $\Pr[S(U_n) = 1] \geq \delta - 2\epsilon$ (since $\Pr[B(X) = 1] \geq \delta$), so $U_n|(S(U_n) = 1)$ is $(\delta - 2\epsilon)$-dense in U_n. Also, $X|(B(X) = 1)$ and $U_n|(S(U_n) = 1)$ are $(s', 2\epsilon/\delta)$-indistinguishable, so letting $M = U_n|(S(U_n) = 1)$ yields the result. $\square$

Application 3: 2-Round ZK A final corollary of Theorem 11 is that the classic two-round graph non-isomorphism protocol (which is only known to be honest-verifier zero-knowledge) is also distributional (T, t, ϵ)-zero-knowledge for $T(n) = poly(t(n), \frac{1}{\epsilon(n)})$.[6] In fact, by using the complete problem for SZK (the class of promise problems having a statistical zero-knowledge proof for an honest verifier) by Sahai and Vadhan [29], we can show that every language in SZK has a 2-round distributional (T, t, ϵ)-zero-knowledge proof for $T(n) = poly(t(n), \frac{1}{\epsilon(n)})$.

---

[6] Recall that in the classic GNI protocol the prover sends just a single bit.

Theorem 12. *For every language $L \in SZK$ and every function $\delta(n) \geq \frac{1}{2^{poly(n)}}$, there exists a two-round interactive proof (P, V) for L with completeness $1 - negl(n)$ and soundness error $\delta(n)$ such that for every function t and ϵ, (P, V) is distributional (T, t, ϵ)-zero-knowledge, where $T(n) = poly(\frac{1}{\delta(n)}, t(n), \frac{1}{\epsilon(n)})$.*

Proof. From [29], there exists a two-round interactive proof (P', V') for a complete problem L_{SZK} for SZK with completeness negligibly close to 1 and soundness error negligibly close to $\frac{1}{2}$, and the prover P' only sends a single bit to the verifier V'. By repeating the proof in parallel $O(\log \frac{1}{\delta(n)})$ times, we get a two-round interactive proof for L_{SZK} with completeness negligibly close to 1 and soundness error $\delta(n)$, and the prover only sends $O(\log \frac{1}{\delta(n)})$ bits to the verifier. Then, by Theorem 11, this interactive proof for L_{SZK} is distributional (T, t, ϵ)-zero-knowledge, where $T(n) = poly(\frac{1}{\delta(n)}, t(n), \frac{1}{\epsilon(n)})$. Since L_{SZK} is a complete problem for SZK, the theorem follows.   □

In Theorem 12, if we choose $\delta(n) = \frac{1}{n^{\log n}}$, $t(n) = n^{\log n}$, and $\epsilon(n) = \frac{1}{n^{\log n}}$, then every language in SZK has a 2-round "quasi-polynomial-time simulatable" distributional zero-knowledge proof (i.e., $T(n)$ is a quasi-polynomial) with completeness $1 - negl(n)$ and negligible soundness error. Alternatively, if we choose $\delta(n) = \frac{1}{poly(n)}$, $t(n) = poly(n)$, and $\epsilon(n) = \frac{1}{poly(n)}$, then every language in SZK has a 2-round "polynomial-time simulatable" (T, t, ϵ)-distributional zero-knowledge proof (i.e., $T(n)$ is a polynomial) with completeness $1 - negl(n)$ and soundness error $\frac{1}{poly(n)}$.

4.2   Necessity of Non-Black-Box Simulation

The universal simulator in Theorem 12 is obtained via Theorem 11, which uses Theorem 9, so the universal simulator makes non-black-box usage of V^*. We remark that this non-black-box usage is also necessary (assuming standard complexity theoretic assumptions): We will show that black-box simulation techniques cannot be used to demonstrate distributional (T, t, ϵ)-zero-knowledge for 2-round proof systems for languages that are hard-on-average. Thus, as long as SZK contains a problem that is hard-on-average, our non-black-box techniques are necessary. Let us first give the definition of black-box distributional (T, t, ϵ)-zero-knowledge.

Definition 7 (black-box distributional (T, t, ϵ)-zero-knowledge). *Let (P, V) be an interactive proof system for a language L. We say that (P, V) is black-box distributional (T, t, ϵ)-zero-knowledge if for every $n \in \mathbb{N}$ and every joint distribution (X_n, Y_n, Z_n) over $(L \cap \{0, 1\}^n) \times \{0, 1\}^* \times \{0, 1\}^*$, there exists a $T(n)$-size simulator S such that for every $t(n)$-size adversary V^* and every $t(n)$-size distinguisher D, we have*

$$| \Pr[D(X_n, Z_n, Out_{V^*}[P(X_n, Y_n) \leftrightarrow V^*(X_n, Z_n)]) = 1]$$
$$- \Pr[D(X_n, Z_n, S^{V^*(X_n, Z_n)}(X_n, Z_n)) = 1]| \leq \epsilon(n).$$

where $S^{V^*(X_n, Z_n)}$ means that S is given oracle access to the verifier $V^*(X_n, Z_n)$.

For any language L and any $x \in \{0,1\}^*$, let $L(x) = 1$ if $x \in L$, and $L(x) = 0$ otherwise. We now show that any 2-round interactive proof for a language L with "hard-on-average" instances is not black-box distributional zero-knowledge.

Theorem 13. *Let L be any language with hard-on-average instances, i.e., there exists an ensemble $\{X_n\}_{n\in\mathbb{N}}$ of distributions X_n over $\{0,1\}^n$ such that for every non-uniform PPT algorithm A and for sufficiently large $n \in \mathbb{N}$, we have $\Pr[A(X_n) = L(X_n)] \leq \frac{1}{2} + \epsilon(n)$, where ϵ is any function such that $\epsilon(n) < \frac{1}{12}$ for sufficiently large $n \in \mathbb{N}$.*

Then, there exists a polynomial t such that any 2-round interactive proof (P,V) for L with completeness $\frac{2}{3}$ and soundness error at most $\frac{1}{3}$ is not black-box (T, t, ϵ)-distributional zero-knowledge for any polynomial T.

Proof. Let $t(n) = O(T_V(n))$, where $T_V(n)$ is a polynomial bound on the running time of V on instances x of length n. To obtain a contradiction, suppose (P, V) is black-box (T, t, ϵ)-distributional zero-knowledge for some polynomial T. Let $n \in \mathbb{N}$, let X_n' be X_n conditioned on the event $X_n \in L$, let X_n'' be X_n conditioned on the event $X_n \notin L$, let Y_n always be the empty string, and let Z_n be the uniform distribution over $\{0,1\}^{t(n)}$. Then, there exists a polynomial-size simulator S such that for every $t(n)$-size adversary V^* and every $t(n)$-size distinguisher D, we have

$$| \Pr[D(X_n', Z_n, Out_{V^*}[P(X_n', Y_n) \leftrightarrow V^*(X_n', Z_n)]) = 1]$$
$$- \Pr[D(X_n', Z_n, S^{V^*(X_n', Z_n)}(X_n')) = 1]| \leq \epsilon(n). \tag{1}$$

Let V^* be the verifier that, on input (x, z), runs the honest verifier $V_z(x)$ with random tape z to interact with the prover, and then outputs the message a received from the prover. Let D be the distinguisher that, on input (x, z, a), outputs 1 if $V_z(x, a) = 1$, and 0 otherwise, where $V_z(x, a)$ represents the output of $V(x)$ with random tape z and with message a received from the prover.

Claim. $\Pr[D(X_n', Z_n, S^{V^*(X_n', Z_n)}(X_n')) = 1] \geq \frac{2}{3} - \epsilon(n)$.

Proof (of claim). Since (P, V) has completeness $\frac{2}{3}$, we have

$$\Pr[D(X_n', Z_n, Out_{V^*}[P(X_n', Y_n) \leftrightarrow V^*(X_n', Z_n)]) = 1]$$
$$= \Pr[Out_V[P(X_n', Y_n) \leftrightarrow V(X_n')] = 1]$$
$$\geq \frac{2}{3}.$$

Now, combining this with (1), we have

$$\Pr[D(X_n', Z_n, S^{V^*(X_n', Z_n)}(X_n')) = 1] \geq \frac{2}{3} - \epsilon(n),$$

as required. This completes the proof of the claim. $\square$

Claim. $\Pr[D(X_n'', Z_n, S^{V^*(X_n'', Z_n)}(X_n'')) = 0] \geq \frac{2}{3} - \epsilon(n)$.

Proof (of claim). To obtain a contradiction, suppose $\Pr[D(X_n'', Z_n, S^{V^*(X_n'', Z_n)}(X_n'')) = 0] < \frac{2}{3} - \epsilon(n)$. We note that the event $D(X_n'', Z_n, S^{V^*(X_n'', Z_n)}(X_n'')) = 0$ occurs if and only if the event $V_{Z_n}(X_n'', S^{V^*(X_n'', Z_n)}(X_n'')) = 0$ occurs, where $V_{Z_n}(X_n'', S^{V^*(X_n'', Z_n)}(X_n''))$ represents the output of $V(X_n'')$ with random tape Z_n and with message $S^{V^*(X_n'', Z_n)}(X_n'')$ received from the prover. Thus, we have $\Pr[V_{Z_n}(X_n'', S^{V^*(X_n'', Z_n)}(X_n'')) = 0] < \frac{2}{3} - \epsilon(n)$.

Now, consider an adversarial prover P^* that, on input x and upon receiving a message c from the verifier V, simulates $S(x)$ while responding to oracle queries with the message c, and then sends the output of $S(x)$ to V. Now, we note that the event $V_{Z_n}(X_n'', S^{V^*(X_n'', Z_n)}(X_n'')) = 0$ occurs if and only if the event $Out_V(P^*(X_n'', Y_n) \leftrightarrow V_{Z_n}(X_n'')) = 0$ occurs. Thus, we have

$$\Pr[Out_V(P^*(X_n'', Y_n) \leftrightarrow V_{Z_n}(X_n'')) = 0] < \frac{2}{3} - \epsilon(n),$$

and since we always have $X_n'' \notin L$, this contradicts the assumption that (P, V) has soundness error at most $\frac{1}{3}$. This completes the proof of the claim. □

Now, using the polynomial-size simulator S and the $t(n)$-size distinguisher D, we will construct a non-uniform PPT algorithm A that contradicts the assumption that L has hard-on-average instances, i.e., for infinitely many $n \in \mathbb{N}$, we have

$$\Pr[A(X_n) = L(X_n)] > \frac{1}{2} + \epsilon(n).$$

Let A be the non-uniform PPT algorithm that, on input $x \in \{0, 1\}^n$, samples a uniformly random z from Z_n, computes $S^{V^*(x,z)}(x)$ (while simulating the oracle $V^*(x, z)$ for $S(x)$) and outputs $D(x, z, S^{V^*(x,z)}(x))$. Then, for infinitely many $n \in \mathbb{N}$, we have

$$
\begin{aligned}
&\Pr[A(X_n) = L(X_n)] \\
={}& \Pr[D(X_n, Z_n, S^{V^*(X_n, Z_n)}(X_n)) = L(X_n)] \\
={}& \Pr[X_n \in L] \cdot \Pr[D(X_n', Z_n, S^{V^*(X_n', Z_n)}(X_n')) = 1] \\
&+ \Pr[X_n \notin L] \cdot \Pr[D(X_n'', Z_n, S^{V^*(X_n'', Z_n)}(X_n'')) = 0] \\
\geq{}& \Pr[X_n \in L] \cdot (2/3 - \epsilon(n)) + \Pr[X_n \notin L] \cdot (2/3 - \epsilon(n)) \\
={}& \frac{2}{3} - \epsilon(n),
\end{aligned}
$$

where the inequality follows from the two claims above. This contradicts the assumption that L has hard-on-average instances. This completes the proof. □

Acknowledgments. We thank Krzysztof Pietrzak for pointing out a mistake in the parameters in Theorem 11 in an earlier version of this paper.

References

1. Althfer, I.: On sparse approximations to randomized strategies and convex combinations. Linear Algebra and its Applications 199(suppl.1), 339–355 (1994)
2. Barak, B.: How to go beyond the black-box simulation barrier. In: FOCS, pp. 106–115 (2001)
3. Barak, B., Sahai, A.: How to play almost any mental game over the net - concurrent composition via super-polynomial simulation. In: FOCS, pp. 543–552 (2005)
4. Barak, B., Shaltiel, R., Wigderson, A.: Computational analogues of entropy. In: Arora, S., Jansen, K., Rolim, J.D.P., Sahai, A. (eds.) RANDOM 2003 and APPROX 2003. LNCS, vol. 2764, pp. 200–215. Springer, Heidelberg (2003)
5. Bitansky, N., Paneth, O.: From the impossibility of obfuscation to a new non-black-box simulation technique. In: FOCS (2012)
6. Bitansky, N., Paneth, O.: On the impossibility of approximate obfuscation and applications to resettable cryptography. In: STOC (2013)
7. Boneh, D., Naor, M.: Timed commitments. In: Bellare, M. (ed.) CRYPTO 2000. LNCS, vol. 1880, pp. 236–254. Springer, Heidelberg (2000)
8. Chung, K.-M., Kalai, Y.T., Liu, F.-H., Raz, R.: Memory delegation. In: Rogaway, P. (ed.) CRYPTO 2011. LNCS, vol. 6841, pp. 151–168. Springer, Heidelberg (2011), http://dl.acm.org/citation.cfm?id=2033036.2033048
9. Chung, K.M., Lui, E., Pass, R.: Can theories be tested? A cryptographic treatment of forecast testing. In: ITCS, pp. 47–56 (2013)
10. Chung, K.M., Pass, R., Seth, K.: Non-black-box simulation from one-way functions and applications to resettable security. In: STOC. ACM (2013)
11. Deng, Y., Goyal, V., Sahai, A.: Resolving the simultaneous resettability conjecture and a new non-black-box simulation strategy. In: 50th Annual IEEE Symposium on Foundations of Computer Science, FOCS 2009, pp. 251–260. IEEE (2009)
12. Dwork, C., Naor, M., Reingold, O., Stockmeyer, L.: Magic functions: In memoriam: Bernard m. dwork 1923–1998. J. ACM 50(6), 852–921 (2003)
13. Dziembowski, S., Pietrzak, K.: Leakage-resilient cryptography. In: 49th FOCS, pp. 293–302. IEEE Computer Society Press (2008)
14. Freund, Y., Schapire, R.E.: Adaptive game playing using multiplicative weights. Games and Economic Behavior 29(1), 79–103 (1999)
15. Frieze, A., Kannan, R.: Quick approximation to matrices and applications. Combinatorica 19(2), 175–220 (1999)
16. Gentry, C., Wichs, D.: Separating succinct non-interactive arguments from all falsifiable assumptions. In: Proceedings of the 43rd Annual ACM Symposium on Theory of Computing, STOC 2011, pp. 99–108. ACM, New York (2011)
17. Goldreich, O.: A uniform-complexity treatment of encryption and zero-knowledge. Journal of Cryptology 6, 21–53 (1993)
18. Goldreich, O., Micali, S., Wigderson, A.: Proofs that yield nothing but their validity or all languages in np have zero-knowledge proof systems. J. ACM 38(3), 690–728 (1991), http://doi.acm.org/10.1145/116825.116852
19. Goldreich, O., Vadhan, S.P., Wigderson, A.: On interactive proofs with a laconic prover. In: Orejas, F., Spirakis, P.G., van Leeuwen, J. (eds.) ICALP 2001. LNCS, vol. 2076, pp. 334–345. Springer, Heidelberg (2001), http://dl.acm.org/citation.cfm?id=646254.684254
20. Goldwasser, S., Micali, S., Rackoff, C.: The knowledge complexity of interactive proof-systems. In: Proceedings of the Seventeenth Annual ACM Symposium on Theory of Computing, STOC 1985, pp. 291–304. ACM (1985)

21. Halpern, J., Pass, R.: Game theory with costly computation: formulation and application to protocol security. In: Proceedings of the Behavioral and Quantitative Game Theory: Conference on Future Directions, BQGT 2010, p. 89:1. ACM, New York (2010), http://doi.acm.org/10.1145/1807406.1807495

22. Impagliazzo, R.: Hard-core distributions for somewhat hard problems. In: Proceedings of the 36th Annual Symposium on Foundations of Computer Science, FOCS 1995, pp. 538–545. IEEE Computer Society (1995)

23. Jetchev, D., Pietrzak, K.: How to fake auxiliary input. In: Lindell, Y. (ed.) TCC 2014. LNCS, vol. 8349, pp. 566–590. Springer, Heidelberg (2014)

24. Lipton, R.J., Young, N.E.: Simple strategies for large zero-sum games with applications to complexity theory. In: Proceedings of the Twenty-sixth Annual ACM Symposium on Theory of Computing, STOC 1994, pp. 734–740. ACM (1994)

25. Pass, R.: Bounded-concurrent secure multi-party computation with a dishonest majority. In: STOC 2004. pp. 232–241 (2004)

26. Pass, R., Rosen, A.: Bounded-concurrent secure two-party computation in a constant number of rounds. In: FOCS, pp. 404–413 (2003)

27. Pass, R., Rosen, A.: New and improved constructions of non-malleable cryptographic protocols. In: STOC 2005, pp. 533–542 (2005)

28. Reingold, O., Trevisan, L., Tulsiani, M., Vadhan, S.: Dense subsets of pseudorandom sets. In: Proceedings of the 2008 49th Annual IEEE Symposium on Foundations of Computer Science, FOCS 2008, pp. 76–85 (2008)

29. Sahai, A., Vadhan, S.: A complete problem for statistical zero knowledge. J. ACM 50(2), 196–249 (2003)

30. Trevisan, L., Tulsiani, M., Vadhan, S.: Regularity, boosting, and efficiently simulating every high-entropy distribution. In: Proceedings of the 2009 24th Annual IEEE Conference on Computational Complexity, CCC 2009, pp. 126–136. IEEE Computer Society (2009)

31. Vadhan, S., Zheng, C.J.: A uniform min-max theorem with applications in cryptography. In: Canetti, R., Garay, J.A. (eds.) CRYPTO 2013, Part I. LNCS, vol. 8042, pp. 93–110. Springer, Heidelberg (2013)

32. Yao, A.C.: Theory and application of trapdoor functions. In: Proceedings of the 23rd Annual Symposium on Foundations of Computer Science, SFCS 1982, pp. 80–91 (1982)

An Efficient Transform from Sigma Protocols to NIZK with a CRS and Non-programmable Random Oracle*

Yehuda Lindell

Dept. of Computer Science
Bar-Ilan University, Israel
lindell@biu.ac.il

Abstract. In this short paper, we present a Fiat-Shamir type transform that takes any Sigma protocol for a relation R and outputs a non-interactive zero-knowledge proof (not of knowledge) for the associated language L_R, in the common reference string model. As in the Fiat-Shamir transform, we use a hash function H. However, zero-knowledge is achieved under standard assumptions in the common reference string model (without any random oracle), and soundness is achieved in the *non-programmable* random oracle model. The concrete computational complexity of the transform is only slightly higher than the original Fiat-Shamir transform.

1 Introduction

Concretely Efficient Zero Knowledge. Zero knowledge proofs [20,16] play an important role in many fields of cryptography. In secure multiparty computation, zero-knowledge proofs are used to force parties to behave semi-honestly, and as such are a crucial tool in achieving security in the presence of malicious adversaries [17]. This use of zero-knowledge proofs is not only for proving feasibility as in [17], and efficient zero-knowledge proofs are used widely in protocols for achieving concretely efficient multiparty computation with security in the presence of malicious adversaries; see [30,22,26,31,23,24] for just a few examples. Efficient zero knowledge is also widely used in protocols for specific problems like voting, auctions, anonymous credentials, and more.

Most efficient zero knowledge proofs known are based on Sigma protocols [8] and [21, Ch. 7]. Informally, Sigma protocols are honest-verifier perfect zero-knowledge interactive proof systems with some very interesting properties. First, they are three round public-key protocols (meaning that the verifier's single message—or challenge—is just a random string); second, if the statement is not

---

* This work was funded by the European Research Council under the European Union's Seventh Framework Programme (FP/2007-2013) / ERC consolidators grant agreement n. 615172 (HIPS), and under the European Union's Seventh Framework Program (FP7/2007-2013) grant agreement n. 609611 (PRACTICE).

Y. Dodis and J.B. Nielsen (Eds.): TCC 2015, Part I, LNCS 9014, pp. 93–109, 2015.

in the language, then for every first prover message there exists just a single verifier challenge that can be answered; third, there exists a simulator that is given the statement and verifier challenge and generates the exact distribution over the prover's messages with this challenge. Although seemingly specific, there exist Sigma protocols for a wide variety of tasks like proving that a tuple is of the Diffie-Hellman type, that an ElGamal commitment is to a certain value, that a Paillier encryption is to zero, and many more. It is also possible to efficiently combine Sigma protocols to prove compound statements [5]; e.g., $(x_1 \in L \wedge y_1 \in L) \vee (x_2 \in L \wedge y_2 \in L)$. Finally, it is possible to efficiently compile a Sigma protocol to a zero-knowledge proof (resp., zero-knowledge proof of knowledge) with just one additional round (resp., two additional rounds) [21, Ch. 7].

The Fiat-Shamir Transform and NIZK. The Fiat-Shamir transform [13] is a way of transforming any *public-coin zero-knowledge proof* into a *non-interactive zero-knowledge proof* [2,11].[1] The transform is very simple, and works by having the prover compute the verifier's (random) messages by applying an "appropriate" hash function to the previous prover messages. The security of this transform was proven in the random oracle model [29]. This means that if the hash function is modeled as an external random function, then the result of applying the Fiat-Shamir transform to a public-coin zero-knowledge proof is a non-interactive zero-knowledge proof. However, it was also shown that it is not possible to prove this statement for any concrete instantiation of the hash function. Rather, there exist public-coin zero-knowledge proofs for which every concrete instantiation of the hash function in the Fiat-Shamir transform yields an insecure scheme [19].

When applying the Fiat-Shamir transform to a Sigma protocol, the result is an extraordinarily efficient non-interactive zero-knowledge proof. We remark that this is not immediate since Sigma protocols are only *honest-verifier* zero knowledge. Thus, the Fiat-Shamir transform both removes interaction and guarantees zero knowledge for malicious verifiers.

The Fiat-Shamir transform is very beneficial in obtaining efficient protocols since it saves expensive rounds of communication without increasing the computational complexity of the original protocol. In addition, it is very useful in settings where the non-interactive nature of the proof is essential (e.g., in anonymous credentials). However, as we have seen, this reliance on the Fiat-Shamir is only sound in the random-oracle model. This leads us to the following question:

> *Can we construct a Fiat-Shamir type transformation, that is highly efficient and is secure in the standard model (without a random oracle)?*

In this paper, we take a first step towards answering this question.

---

[1] The Fiat-Shamir transform was designed to construct signature schemes from public-coin zero-knowledge proofs, and later works also studied its security as a signature scheme. However, the results are actually the same for non-interactive zero knowledge.

The Random-Oracle Saga. Reliance on the random oracle model is controversial, with strong advocates on one side and strong opponents on the other. However, it seems that almost all agree that proving security without reliance on the random oracle model is preferable. As such, there has been a long line of work attempting to prove security of existing schemes without reliance on a random oracle, and to construct new schemes that are comparable to existing ones (e.g., with respect to efficiency) but don't require a random oracle. In the case of the Fiat-Shamir transform, there is no chance of proving it secure in general without a random oracle, due to the impossibility result of [19]. Thus, the aim is to construct a transform that is comparable to Fiat-Shamir in terms of efficiency, but can be proven secure in the standard model.

An interesting development regarding the random oracle is that not all random oracles are equal. In particular, Nielsen introduced the notion of a *non-programmable* random oracle [25], based on the observation that many proofs of security—including that of the Fiat-Shamir transform—rely inherently on the ability of the simulator (or security reduction) to "program" the random oracle and fix specific input/output pairs. In contrast, a non-programmable random oracle is simply a random function that all parties have access to. In some sense, reliance on a non-programmable random oracle seems conceptually preferable since it more closely models the intuition that "appropriate hash functions" behave in a random way. Formally, of course, this makes no sense. However, proofs of security that do not require programmability are preferable in the sense that they rely on less properties of the random oracle and can be viewed as a first step towards removing it entirely.

Our Results. In this paper, we present a Fiat-Shamir type transform from Sigma protocols to non-interactive zero knowledge proofs (that are *not* proofs of knowledge). The transform is extremely efficient; for example, under the Decisional Diffie-Hellman assumption, the cost of transforming a Sigma protocol to a non-interactive zero-knowledge proof is just 4 exponentiations, and the transmission of a single number in $\mathbb{Z}_q$ (where q is the order of the group). Our transform achieves two advantages over the Fiat-Shamir transform:

1. The zero-knowledge property holds in the *standard model* and does not require any random oracle at all. This is in contrast to the standard Fiat-Shamir transform when applied to Sigma protocols, for which the only known proof uses a (fully programmable) random oracle. Our transform utilizes the common reference string model, which is inherent since one-round zero-knowledge protocols do not exist for languages not in $\mathcal{BPP}$ [18].

2. The soundness property holds when the hash function is modeled as a *non-programmable* random oracle.

The fact that zero knowledge holds without any random oracle implies that the difficulties regarding zero knowledge composition that arise in the random oracle model [32] are not an issue here. It also implies that the random oracle is not needed for any simulation, and one only needs it to prove soundness.

Our Transform. The technique used in our transform is very simple. We use a two-round equivocal commitment scheme for which there exists a trapdoor with which commitments can be decommitted to any value. One example of such a scheme is that of Pedersen's commitment [28]. Specifically, let g and h be two random generators of a group in which the discrete log problem is assumed to be hard. Then, $c = \mathsf{Com}_{g,h}(x) = g^r \cdot h^x$, where $r \leftarrow \mathbb{Z}_q$ is random. This scheme is perfectly hiding, and it can be shown to be computationally binding under the discrete log assumption. However, if the discrete log of h with respect to g is known, then it is possible to decommit to any value (if $h = g^\alpha$ and α is known to the committer, then it can define $c = g^y$ and then for any x it simply sets $r = y - \alpha \cdot x$).

We define a common reference string (CRS) that contains the first message of the commitment scheme. Thus, when the simulator chooses the CRS, it will know the trapdoor, thereby enabling it to equivocate. Let Σ be a Sigma protocol for some language, and denote the messages of the proof that $x \in L$ by (a, e, z). Then, in the Fiat-Shamir transform, the prover uses the verifier challenge $e = H(x, a)$. In our transform, the prover first computes a commitment to a using randomness r, denoted $c = \mathsf{Com}(a; r)$, and sets $e = H(x, c)$. Then, the proof contains (a, r, z), and the verifier computes $c = \mathsf{Com}(a; r)$, $e = H(x, c)$ and verifies that (a, e, z) is an accepting proof that $x \in L$. Intuitively, since c is a commitment to a, soundness is preserved like in the original Fiat-Shamir transform. However, since the simulator can choose the common reference string, and so can know the trapdoor, it can equivocate to any value it likes. Thus, the simulator can generate a commitment c that can be opened later to anything. Next, it computes $e = H(x, c)$. Finally, it runs the Sigma protocol simulator with the verifier challenge e already known, in order to obtain an accepting proof (a, e, z). Finally, it finds r such that $c = \mathsf{Com}(a; r)$ to "explain" c as a commitment to a. This reverse order of operations is possible since the simulator can equivocate; soundness is preserved since the real prover cannot.

As appealing as the above is, the proof of soundness is problematic since the commitment is only computational binding and the reduction would need to construct an adversary breaking the binding from any adversary breaking the soundness. However, since a cheating prover outputs a single proof, such a reduction seems problematic, even in the random oracle model.[2] We therefore use a *dual-mode commitment* (or hybrid trapdoor commitment [4]) which means that there are two ways to choose the common reference: in one way the commitment is perfectly binding, and in the other it is equivocal. This enables us to prove soundness when the commitment is perfectly binding, and zero knowledge when it is equivocal. We construct dual-mode commitments from any "hard" language with an associated Sigma protocol (see Section 3.2). Thus, the security of our transform for such languages requires no additional assumptions. We also

---

[2] It may be possible to prove by relying on the extractability of the random oracle, meaning that it is possible to "catch" the cheating prover's queries to the random oracle. We do not know how to do this in this context. In addition, our solution is preferable since we do not even require extractability of random oracle queries.

demonstrate a concrete instantiation of our construction that is secure under the DDH assumption, and requires only 4 exponentiations to generate a commitment. Our DDH instantiation of this primitive appeared in [4] (for different applications); we present the construction here in any case for completeness.

Open Questions. The major question left open by our work is whether or not it is possible to prove the security of our transform or a similar one using a (concretely efficient) hash function whose security is based on a *standard* cryptographic assumption. Note that even achieving a falsifiable assumption is difficult, and this has been studied by [1] and [10]. However, we have the additional power of a CRS, and this may make it easier.

Related Work. Damgård [7] used a very similar transform to obtain 3-round concurrent zero knowledge in the CRS model. Specifically, [7] uses a trapdoor commitment applied to the first prover message, as we do. This enables simulation without rewinding and thus achieves concurrent zero knowledge. However, as we have described, it seems that in our setting a regular tradpoor commitment does not suffice since there is *no interaction* and thus no possibility of rewinding the adversary (note that in the context of concurrent zero knowledge it is problematic to rewind a cheating verifier when proving zero knowledge, but there is no problem rewinding a cheating prover in order to prove soundness, and this is indeed what [7] do).

The problem of constructing zero knowledge in the non-programmable random oracle model was first considered by [27] with extensions to the UC setting in [9]. However, their constructions are not completely non-interactive and require two messages. This is due to the fact that their aim is to solve the problem of deniability and transferability of NIZK proofs, and so some interaction is necessary (as proven in [27]). We also remark that the transform from Σ-protocols to Ω-protocols used in their construction requires repeating the proof multiple times, and so is far less efficient.

2   Definitions

2.1   Preliminaries

Let R be a relation; we denote the associated language by L_R. That is, $L_R = \{x \mid \exists w : (x, w) \in R\}$. We denote the security parameter by n. We model a random oracle simply as a random length-preserving function $\mathcal{O} : \{0, 1\}^n \to \{0, 1\}^n$. In our work, we use the random oracle only to prove soundness, and there is therefore no issue of "programmability". When S is a set, $x \leftarrow S$ denotes choosing x from S with a uniform distribution.

2.2   Sigma Protocols and NIZK

For the sake of completeness, we define Sigma protocol and adaptive non-interactive zero knowledge (NIZK). Our formulation of non-interactive zero knowledge is both adaptive (meaning that statements can be chosen as a function of the common reference string) and considers the case where many proofs are given.

Sigma Protocols. We briefly define Sigma protocols. For more details, see [8] and [21, Ch. 7]. Let R be a binary polynomial-bounded relation. A Σ protocol $\pi = (P_1, P_2, V_\Sigma)$ is a 3-round public-coin protocol: the prover's first message is denoted $a = P_1(x)$; the verifier's message is a random string $e \in_R \{0,1\}^n$, and the prover's second message is denoted $z = P_2(x, a, e)$. We write $V_\Sigma(x, a, e, z) = 1$ if and only if the verifier accepts, and in this case we say tha transcript (a, e, z) is accepting for x. We now formally define the notion of a Sigma-protocol:

Definition 1. *A protocol* $\pi = (P_1, P_2, V_\Sigma)$ *is a* Sigma-protocol *for relation R if it is a three-round public-coin protocol, and the following requirements hold:*

- **Completeness:** *If P and V follow the protocol on input x and private input w to P where $(x, w) \in R$, then V always accepts.*
- **Special soundness:** *There exists a polynomial-time algorithm A that given any x and any pair of accepting transcripts $(a, e, z), (a, e', z')$ for x, where $e \neq e'$, outputs w such that $(x, w) \in R$.*
- **Special honest verifier zero knowledge:** *There exists a probabilistic polynomial-time simulator S_Σ such that*

$$\left\{ S_\Sigma(x, e) \right\}_{x \in L; e \in \{0,1\}^n} \equiv \left\{ \langle P(x, w), V(x, e) \rangle \right\}_{x \in L; e \in \{0,1\}^n}$$

where $S_\Sigma(x, e)$ denotes the output of simulator M upon input x and e, and $\langle P(x, w), V(x, e) \rangle$ denotes the output transcript of an execution between P and V, where P has input (x, w), V has input x, and V's random tape (determining its query) equals e.

Adaptive Non-interactive Zero-Knowledge. In the model of non-interactive zero-knowledge proofs [2], the prover and verifier both have access to a public common reference string (CRS). We present the definition of adaptive zero knowledge, meaning that both the soundness and zero-knowledge hold when statements can be chosen as a function of the CRS. We also consider the unbounded version, meaning that zero knowledge holds for any polynomial number of statements proven. We present the definition directly, and refer to [14, Section 4.10] for motivation and discussion. We define soundness in the non-programmable random oracle model, since this is what we use in our construction.

Definition 2. (adaptive non-interactive unbounded zero-knowledge): *A triple of probabilistic polynomial-time machines* (GenCRS, P, V) *is called an* adaptive non-interactive unbounded zero-knowledge argument system *for a language $L \in \mathcal{NP}$ with an NP-relation R_L, if the following holds:*

- Perfect completeness: *For every $(x, w) \in R_L$, $\Pr[V(x, \rho_n, P(x, w, \rho_n)) = 1] = 1$ where ρ_n is randomly sampled according to* GenCRS(1^n).

- Adaptive computational soundness with a non-programmable random oracle: *For every probabilistic polynomial-time function $f: \{0,1\}^{\mathrm{poly}(n)} \to \{0,1\}^n \setminus L$ and every probabilistic polynomial-time (cheating) prover $\mathcal{B}$,*

$$\Pr\left[V^{\mathcal{O}}(f(\rho_n), \rho_n, \mathcal{B}^{\mathcal{O}}(\rho_n)) = 1\right] < \mu(n)$$

where ρ_n is randomly sampled according to $\mathsf{GenCRS}(1^n)$ and $\mathcal{O}: \{0,1\}^* \to \{0,1\}^*$ is a random length-preserving function.

- Adaptive unbounded zero knowledge: *There exists a probabilistic polynomial-time simulator $\mathcal{S}_{\mathsf{zk}}$ such that for every probabilistic polynomial-time function*

$$f: \{0,1\}^{\mathrm{poly}(n)} \to \{0,1\}^n \times \{0,1\}^{\mathrm{poly}(n)} \cap R_L,$$

every polynomial $p(\cdot)$ and every probabilistic polynomial-time distinguisher D, there exists a negligible function μ such that for every n,

$$\left|\Pr\left[D\left(\mathcal{R}^f(\rho_n, P^f(1^{n+p(n)}))\right) = 1\right] - \Pr\left[D\left(\mathcal{S}_{\mathsf{zk}}^{f}(1^{n+p(n)}))\right) = 1\right]\right| \leq \mu(n)$$

where ρ_n is randomly sampled according to $\mathsf{GenCRS}(1^n)$, f_1 and f_2 denote the first and second outputs of f respectively, and $\mathcal{R}^f(\rho_n, P^f(1^{n+p(n)}))$ and $\mathcal{S}_{\mathsf{zk}}^{f}(1^{n+p(n)})$ denote the output from the following experiments:

Real proofs $\mathcal{R}^f(\rho_n, P^f(1^{n+p(n)}))$**:**

1. *$\rho \leftarrow \mathsf{GenCRS}(1^n)$: a common reference string is sampled*
2. *For $i = 1, \ldots, p(n)$ (initially $\boldsymbol{x}$ and $\boldsymbol{\pi}$ are empty):*
   (a) *$x_i \leftarrow f_1(\rho_n, \boldsymbol{x}, \boldsymbol{\pi})$: the next statement x_i to be proven is chosen.*
   (b) *$\pi_i \leftarrow P(f_1(\rho_n, \boldsymbol{x}, \boldsymbol{\pi}), f_2(\rho_n, \boldsymbol{x}, \boldsymbol{\pi}), \rho_n)$: the ith proof is generated.*
   (c) *Set $\boldsymbol{x} = x_1, \ldots, x_i$ and $\boldsymbol{\pi} = \pi_1, \ldots, \pi_i$*
3. *Output $(\rho_n, \boldsymbol{x}, \boldsymbol{\pi})$.*

Simulation $\mathcal{S}_{\mathsf{zk}}^{f}(1^{n+p(n)})$**:**

1. *$\rho \leftarrow \mathcal{S}_{\mathsf{zk}}(1^n)$: Simulator $\mathcal{S}_{\mathsf{zk}}$ (upon input 1^n) outputs a reference string ρ*
2. *For $i = 1, \ldots, p(n)$ (initially $\boldsymbol{x}$ and $\boldsymbol{\pi}$ are empty):*
   (a) *$x_i \leftarrow f_1(\rho_n, \boldsymbol{x}, \boldsymbol{\pi})$: the next statement x_i to be proven is chosen.*
   (b) *$\pi_i \leftarrow \mathcal{S}_{\mathsf{zk}}(x_i)$: Simulator $\mathcal{S}_{\mathsf{zk}}$ generates a simulated proof π_i that $x_i \in L$.*
   (c) *Set $\boldsymbol{x} = x_1, \ldots, x_i$ and $\boldsymbol{\pi} = \pi_1, \ldots, \pi_i$*
3. *Output $(\rho, \boldsymbol{x}, \boldsymbol{\pi})$.*

Adaptive NIZK proof systems can be constructed from any (doubly) enhanced trapdoor permutation [11]; see [14, Appendix C.4.1] and [15] regarding the assumption.

3 Dual-Mode Commitments

We use a commitment scheme in the CRS model with the property that it is *perfectly binding* given the correctly constructed CRS, but is *equivocal* to a simulator who generates the CRS in an alternative but indistinguishable way. Stated differently, the simulator can generate the CRS so that it looks like a real one, but a commitment can be decommitted to any value. We show how to construct this from any "hard" NP-relation with a Sigma protocol (to be defined below). This construction has the advantage that we obtain non-interactive zero knowledge for such relations under no additional assumptions. This construction is based on the commitment scheme from Sigma protocols that appeared in [6]. However, [6] constructed a standard commitment scheme, and we show how the same ideas can be used to achieve a dual commitment scheme. Following this, we show a concrete instantiation under the DDH assumption which is extremely efficient.

Such a commitment was called a **hybrid trapdoor commitment** in [4], who studied this primitive in depth and presented a number of constructions. In particular, the DDH-based construction in [4] is identical to ours. We repeat it here for the sake of completeness.

3.1 Definition

Before we show how to construct such commitments, we provide a formal definition.

Definition 3. *A* dual-mode commitment scheme *is a tuple of probabilistic polynomial-time algorithms* $(\mathsf{GenCRS}, \mathsf{Com}, \mathcal{S}_{\mathsf{com}})$ *such that*

- $\mathsf{GenCRS}(1^n)$ *outputs a common reference string, denoted* ρ,
- $(\mathsf{GenCRS}, \mathsf{Com})$: *When* $\rho \leftarrow \mathsf{GenCRS}(1^n)$ *and* $m \in \{0,1\}^n$, *the algorithm* $\mathsf{Com}_\rho(m; r)$ *with a random* r *is a non-interactive perfectly-binding commitment scheme,*
- $(\mathsf{Com}, \mathcal{S}_{\mathsf{com}})$: *For every probabilistic polynomial-time adversary* $\mathcal{A}$ *and every polynomial* $p(\cdot)$, *the output of the following two experiments is computationally indistinguishable:*

REAL$_{\mathsf{Com},\mathcal{A}}(1^n)$	SIMULATION$_{\mathcal{S}_{\mathsf{com}}}(1^n)$
1. $\rho \leftarrow \mathsf{GenCRS}(1^n)$	*1.* $\rho \leftarrow \mathcal{S}_{\mathsf{com}}(1^n)$
2. For $i = 1, \ldots, p(n)$:	*2. For* $i = 1, \ldots, p(n)$:
(a) $m_i \leftarrow \mathcal{A}(\rho, \boldsymbol{c}, \boldsymbol{r})$	(a) $c_i \leftarrow \mathcal{S}_{\mathsf{com}}$
(b) $r_i \leftarrow \{0,1\}^{\mathrm{poly}(n)}$	(b) $m_i \leftarrow \mathcal{A}(\rho, \boldsymbol{c}, \boldsymbol{r})$
(c) $c_i = \mathsf{Com}_\rho(m_i; r_i)$	(c) $r_i \leftarrow \mathcal{S}_{\mathsf{com}}(m_i)$
(d) *Set* $\boldsymbol{c} = c_1, \ldots, c_i$	(d) *Set* $\boldsymbol{c} = c_1, \ldots, c_i$
and $\boldsymbol{r} = r_1, \ldots, r_i$	*and* $\boldsymbol{r} = r_1, \ldots, r_i$
3. Output $\mathcal{A}(\rho, m_1, r_1, \ldots, m_{p(n)}, r_{p(n)})$	*Output* $\mathcal{A}(\rho, m_1, r_1, \ldots, m_{p(n)}, r_{p(n)})$

3.2   Membership-Hard Languages with Efficient Sampling

Intuitively, a *membership-hard language* L is one for which it is possible to sample instances of the problem in a way that it is hard to detect if a given instance is in the language or not. In more detail, there exists a sampling algorithm S_L that receives for input a bit b and outputs an instance in the language together with a witness w if $b = 0$, and an instance not in the language if $b = 1$. The property required is that no polynomial-time distinguisher can know which bit S_L received. We let S_L^x denote the instance part of the output (without the witness, in the case that $b = 0$). We now define this formally.

Definition 4. *Let L be a language. We say that L is* membership-hard *with* efficient sampling *if there exists a probabilistic polynomial-time sampler S_L such for every probabilistic polynomial-time distinguisher D there exists a negligible function $\mu(\cdot)$ such that*

$$\left| \Pr[D(S_L^x(0, 1^n), 1^n) = 1] - \Pr[D(S_L(1, 1^n), 1^n) = 1] \right| \leq \mu(n)$$

Such languages can be constructed from essentially any cryptographic assumption. Specifically, if one-way functions exist then there exists a pseudorandom generator $G : \{0, 1\}^n \to \{0, 1\}^{2n}$. Now, define L to be the language of all images of G; i.e., $L = \{G(s) \mid s \in \{0, 1\}^*\}$, and define $S_L(0, 1^n) = (G(U_n), U_n)$, and $S_L(1, 1^n) = U_{2n}$, where U_k is a uniformly distributed string of length k. It is clear that this language is membership-hard with efficient sampling.

Nevertheless, we will be more interested in such languages that have efficient Sigma protocols associated with them. One simple such examples is the language of Diffie-Hellman tuples (where $S_L(0, 1^n)$ outputs a random Diffie-Hellman tuple (g, h, g^a, h^a) together with a, and $S_L(1, 1^n)$ outputs a random non Diffie-Hellman tuple (g, h, g^a, h^b), where a and b are random).

We remark that Feige and Shamir [12] consider the notion of an invulnerable generator for a language. Their notion considers a *relation* for which it is possible to generate an instance such that it is hard to find the associated witness. In contrast, our notion relates to languages and not relations, and on deciding membership rather than finding witnesses.

3.3   Dual-Mode Commitments from Membership-Hard Languages with Sigma Protocols

We now construct a dual-mode commitment scheme from any language L that is membership hard, and has an associated Sigma protocol. Recall that the verifier message of a Sigma protocol is always a uniformly distributed $e \in_R \{0, 1\}^n$. We denote the first and second prover messages of the Sigma protocol on common input x (and witness w for the prover) by $a = P_1(x, w)$ and $z = P_2(x, w, a, e)$, respectively. We denote by $\mathcal{S}_\Sigma$ the simulator for the Sigma protocol. Thus, $\mathcal{S}_\Sigma(x, e)$ outputs (a, z).

PROTOCOL 1 (Dual-Mode Commitment (General Construction))

- **Regular CRS generation (perfect binding):** Run the sampler S_L for the language L with input $(1, 1^n)$, and receive back an x (recall that $x \notin L$). The CRS is $\rho = x$.
- **Commitment Com:** To commit to a value $m \in \{0,1\}^n$, set $e = m$, run $S_\Sigma(x, e)$ and obtain (a, z). The commitment is $c = a$.
- **Decommitment:** To decommit, provide e, z and the receiver checks that $V_\Sigma(a, e, z) = 1$.
- **Simulator S_{com}:**
  1. Upon input 1^n, simulator S_{com} runs the sampler S_L for the language L with input $(0, 1^n)$, and receives back (x, w) (recall that $x \in L$ and w is a witness to this fact). Then, S_{com} computes $a = P_1(x, w)$, sets $c = a$ and $\rho = x$, and outputs (c, ρ).
  2. Upon input $m \in \{0,1\}^n$, simulator S_{com} sets $e = m$ and outputs $z = P_2(x, w, a, e)$.

The fact that the commitment scheme is *perfectly binding* in the regular CRS case holds since $x \notin L$ and thus for every a, there exists a *single* e, z for which (a, e, z) is an accepting proof. In contrast, in the alternative CRS generation case, $x \in L$ and the simulator knows the witness w. Thus, it can generate a "commitment" $a = P_1(x, w)$, and then for any $m \in \{0,1\}^t$ chosen later, it can decommit to m by setting $e = m$, computing $z = P_2(x, e)$ and supplying (e, z). Since (a, e, z) is a valid proof, and the Sigma protocol simulator is perfect, the only difference between this and a real commitment is the fact that $x \in L$. However, by the property of the sampler S_L, this is indistinguishable from the case that $x \notin L$.

Theorem 2. *Let L be a membership-hard language, and let (P_1, P_2, V_Σ) be a Sigma protocol for L. Then, Protocol 1 is a dual-mode commitment scheme.*

Proof. The fact that $\mathsf{Com}_\rho(m; r)$ is perfectly binding when $\rho \leftarrow \mathsf{GenCRS}(1^n)$ follows from the fact that when $x \notin L$ it holds that for every a there exists a *single* e such that $V_\Sigma(x, a, e, z) = 1$.

We now show that the outputs of $\mathrm{REAL}_{\mathsf{Com}, \mathcal{A}}(1^n)$ and $\mathrm{SIMULATION}_{S_{\mathsf{com}}}(1^n)$ (as in Definition 3) are computationally indistinguishable. (We prove this first since we will use it later to prove the computational hiding of $(\mathsf{GenCRS}, \mathsf{Com})$.) We begin by modifying the $\mathrm{REAL}_{\mathsf{Com}, \mathcal{A}}(1^n)$ experiment to $\mathrm{HYBRID}_{\mathsf{Com}, \mathcal{A}}(1^n)$, where the only difference is that the CRS is generated by running $S_L(0, 1^n)$ in the way that S_{com} generates it, instead of running $S_L(1, 1^n)$. Apart from this, everything remains exactly the same. (Observe that since Com runs the Sigma protocol simulator, it makes no difference if $x \in L$ or $x \notin L$.) By the assumption that $S_L^x(0, 1^n)$ is computationally indistinguishable from $S_L(1, 1^n)$, it follows that the outputs of $\mathrm{REAL}_{\mathsf{Com}, \mathcal{A}}(1^n)$ and $\mathrm{HYBRID}_{\mathsf{Com}, \mathcal{A}}(1^n)$ are computationally indistinguishable. Next, we show that $\mathrm{HYBRID}_{\mathsf{Com}, \mathcal{A}}(1^n)$ and $\mathrm{SIMULATION}_{S_{\mathsf{com}}}(1^n)$ are identically distributed. There are two differences between them. First, in $\mathrm{SIMULATION}$ the real

Sigma-protocol prover is used instead of the simulator; second, in SIMULATION the value c_i is generated before m_i is given, in every iteration. Regarding the first difference, the distributions are identical by the perfect zero-knowledge property of $\mathcal{S}_\Sigma$. Regarding the second difference, once the real prover is used, it makes no difference if c_i is given before or after, since the distribution over a_i is identical. We conclude that $\text{HYBRID}_{\text{Com},\mathcal{A}}(1^n)$ and $\text{SIMULATION}_{\mathcal{S}_{\text{com}}}(1^n)$ are identically distributed, and thus $\text{REAL}_{\text{Com},\mathcal{A}}(1^n)$ and $\text{SIMULATION}_{\mathcal{S}_{\text{com}}}(1^n)$ are computationally indistinguishable.

It remains to show that $(\mathsf{GenCRS}, \mathsf{Com})$ is computationally hiding as a commitment scheme. In order to see this, observe that $\text{SIMULATION}_{\mathcal{S}_{\text{com}}}(1^n)$ is *perfectly hiding*. Intuitively, since it is computationally indistinguishable from a real commitment, this proves computational hiding. More formally, for any pair m_0, m_1 of the same length, the output of REAL with m_0 is computationally indistinguishable from the output of SIMULATION with m_0, and the output of REAL with m_1 is computationally indistinguishable from the output of SIMULATION with m_1. (It is straightforward to modify the experiments to have a fixed message, or to have $\mathcal{A}$ output a pair and choose one at random.) Since the commitment in SIMULATION is perfectly hiding, it follows that the output of SIMULATION with m_0 is identical to the output of SIMULATION with m_1. This implies computational indistinguishability of the output of REAL with m_0 from the output of REAL with m_1.

3.4 A Concrete Instantiation from DDH

In this section, we present a dual-mode commitment scheme from the DDH assumption. This can be used for any transform, and may be more efficient if the Sigma protocol for the language being used is less efficient. The complexity is 4 exponentiations for a commitment (by the prover), and 4 exponentiations for a decommitment (by the receiver).

Let $\mathcal{G}$ be the "generator algorithm" of a group in which the DDH assumption is assumed to hold. We denote the output of $\mathcal{G}(1^n)$ by $(\mathbb{G}, q, g, h)$ where $\mathbb{G}$ is the description of a group of order $q > 2^n$ with two random generators g, h.

PROTOCOL 3 (Dual-Mode Commitment from DDH)

- **Regular CRS generation (perfect binding):** Run $\mathcal{G}(1^n)$ to obtain $(\mathbb{G}, q, g, h)$. Choose $\rho_1, \rho_2 \in_R \mathbb{Z}_q$ and compute $u = g^{\rho_1}$ and $v = h^{\rho_2}$. The CRS is $(\mathbb{G}, q, g, h, u, v)$.
- **Alternative CRS generation (equivocal):** As above, except choose a single $\rho \in_R \mathbb{Z}_q$ and compute $u = g^\rho$ and $v = h^\rho$.
- **Commitment:** To commit to a value $m \in \{0, 1\}^n$, choose a random $z \in_R \mathbb{Z}_q$ and compute $a = g^z / u^m$ and $b = h^z / v^m$. The commitment is $c = (a, b)$.
- **Decommitment:** To decommit to $c = (a, b)$, provide m, z and the receiver checks that $g^z = a \cdot u^m$ and $h^z = b \cdot v^m$.

The fact that the commitment scheme is *perfectly binding* in the regular CRS case holds since (g, h, u, v) is *not* a Diffie-Hellman tuple. Thus, by the property of the DH Sigma Protocol, for *every* (a, b) there exists a *unique* e for which there exists a value z such that $g^z = a \cdot u^e$ and $h^z = b \cdot v^e$. In contrast, in the alternative CRS generation case, (g, h, u, v) *is* a Diffie-Hellman tuple and the simulator knows the witness ρ. Thus, it can generate $a = g^r$ and $b = h^r$ and then for any $m \in \{0,1\}^n$ chosen later, it can decommit to m by computing $z = r + m\rho$ and supplying (m, z). Since $u = g^\rho$ and $v = h^\rho$ it follows that $g^z = g^{r+m\rho} = g^r \cdot (g^\rho)^m = a \cdot u^m$ and $h^z = h^{r+m\rho} = h^r \cdot (h^\rho)^m = b \cdot v^m$, as required.

The proof of the following theorem follows directly from Theorem 2 and the fact that the language of Diffie-Hellman tuples is membership hard, under the DDH assumption.

Theorem 4. *If the Decisional Diffie-Hellman assumption holds relative to $\mathcal{G}$, then Protocol 3 is a dual-mode commitment scheme.*

4   The Non-interactive Zero-Knowledge Transformation

We denote by P_1, P_2 the prover algorithms for a Sigma protocol for the relation R. Thus, a proof of common statement x with witness w (for $(x, w) \in R$) is run by the prover sending the verifier the first message $a = P_1(x, w)$, the verifier sending a random query $e \leftarrow \{0,1\}^t$, and the prover replying with $z = P_2(x, w, e)$. We denote the verification algorithm by $V_\Sigma(x, a, e, z)$.

PROTOCOL 5 (NIZK from Sigma Protocol for Relation R)

- **Inputs:** common statement x; the prover also has a witness w such that $(x, w) \in R$
- **Common reference string:** the (regular) CRS ρ of a dual-mode commitment scheme, and a key s for a hash function family H.
- **Auxiliary input:** 1^n, where $n \in \mathbb{N}$ is the security parameter
- **The prover algorithm $P(x, w, \rho)$:**
  1. Compute $a = P_1(x, w)$
  2. Choose a random value $r \in \{0,1\}^{\mathrm{poly}(n)}$ and compute $c = \mathsf{Com}_\rho(a; r)$, where $\mathsf{Com}_\rho(a; r)$ is the dual-mode commitment to a using randomness r and CRS ρ
  3. Compute $e = H_s(x, c)$
  4. Compute $z = P_2(x, w, a, e)$
  5. Output a proof $\pi = (x, a, r, z)$
- **The verifier algorithm $V(x, \rho, a, r, z)$:**
  1. Compute $c = \mathsf{Com}_\rho(a; r)$
  2. Compute $e = H_s(x, c)$
  3. Output $V_\Sigma(x, a, e, z)$

The intuition behind the transformation has been described in the introduction. We therefore proceed directly to prove its security.

4.1 Zero Knowledge

Lemma 1. *Let $\Sigma = (P_1, P_2, V_\Sigma)$ be a Sigma protocol for a relation R and let Com be a dual-mode commitment. Then, Protocol 5 with Σ is zero-knowledge for the language L_R in the common reference string model.*

Proof. We construct a simulator $\mathcal{S}_{zk}$ (as in Definition 2) for Protocol 5 as follows:

- Upon input 1^n, $\mathcal{S}_{zk}$ runs $\mathcal{S}_{com}(1^n)$ for the dual-mode commitment scheme and obtains the value ρ. In addition, $\mathcal{S}_{zk}$ samples a key s for the hash function. $\mathcal{S}_{zk}$ outputs the CRS (ρ, s).
- Upon input x (for every $x_1, \ldots, x_{p(n)}$), simulator $\mathcal{S}_{zk}$ runs $\mathcal{S}_{com}$ to obtain some c. Then, $\mathcal{S}_{zk}$ computes $e = H_s(x, c)$ and runs the simulator $\mathcal{S}_\Sigma$ for the Sigma protocol upon input (x, e). Let the output of the simulator be (a, z). Then, $\mathcal{S}_{zk}$ runs $\mathcal{S}_{com}(a)$ from the dual-mode commitment to obtain r such that $c = \mathsf{Com}_\rho(a; r)$. Finally, $\mathcal{S}_{zk}$ outputs (x, a, r, z).

Intuitively, the difference between a simulated proof and a real one is in the dual-mode commitment. Note also that $\mathcal{S}_{zk}$ uses the Sigma protocol simulator. However, by the property of Sigma protocols, these messages have an identical distribution. Thus, we prove the zero-knowledge property by reducing the security to that of the dual-commitment scheme, as in Definition 3.

First, we construct an alternative simulator $\mathcal{S}'$ who in every iteration (for $i = 1, \ldots, p(n)$) receives (x, w); i.e., $\mathcal{S}'$ receives both $f_1(\rho, \boldsymbol{x}, \boldsymbol{\pi})$ and $f_2(\rho, \boldsymbol{x}, \boldsymbol{\pi})$ and so also receives the witness for the fact that $x \in L$. In the first stage of the simulation, $\mathcal{S}'$ works exactly like $\mathcal{S}_{zk}$ to generate the CRS (ρ, s). In addition, $\mathcal{S}'$ generates c by running $\mathcal{S}_{com}$, just like $\mathcal{S}_{zk}$. However, in order to generate (a, z), $\mathcal{S}'$ uses (x, w) and works as follows. It first computes $e = H_s(x, c)$ exactly like $\mathcal{S}_{zk}$. However, $\mathcal{S}'$ runs $P_1(x, w)$ to obtain a (instead of running $\mathcal{S}_\Sigma$), and then runs $P_2(x, w, a, e)$ to obtain z. Finally, $\mathcal{S}'$ runs $\mathcal{S}_{com}(a)$ to obtain r such that $c = \mathsf{Com}_\rho(a; r)$. The only difference between $\mathcal{S}_{zk}$ and $\mathcal{S}'$ is how the values a, z are obtained. Since for every e, $\mathcal{S}_\Sigma$ outputs (a, z) that are distributed identically as in a real proof with e, it holds that the output distributions of $\mathcal{S}_{zk}$ and $\mathcal{S}'$ are identical.

We now proceed to show that the output distribution of $\mathcal{S}'$ is computationally indistinguishable to a real proof. Formally, let $f = (f_1, f_2)$ be the function choosing the inputs as in Definition 2. We construct an adversary $\mathcal{A}$ for the dual-mode commitments of Definition 3, with input 1^n:

1. $\mathcal{A}$ receives ρ, chooses a key s for the hash function family H, and sets the CRS to be (ρ, s).
2. For $i = 1, \ldots, p(n)$ ($\boldsymbol{x}$ and $\boldsymbol{\pi}$ are initially empty):
   (a) $\mathcal{A}$ receives $(\rho, \boldsymbol{c}, \boldsymbol{r})$ and knows $m_1, \ldots, m_{i-1}$ and $x_1, \ldots, x_{i-1}$ (since these were generated by $\mathcal{A}$ in previous iterations).

    (b) For every $j = 1, \ldots, i-1$, $\mathcal{A}$ sets $a_j = m_j$, $e_j = H_s(x_j, c_j)$, $z_j = P_2(x_j, w_j, a_j, e_j)$, and $\pi_j = (x_j, a_j, r_j, z_j)$. Finally $\mathcal{A}$ sets the vectors $\boldsymbol{x} = (x_1, \ldots, x_{i-1})$ and $\boldsymbol{\pi} = (\pi_1, \ldots, \pi_{i-1})$.

    (c) $\mathcal{A}$ computes $(x_i, w_i) = f((\rho, s), \boldsymbol{x}, \boldsymbol{\pi})$

    (d) $\mathcal{A}$ outputs $m_i = a_i = P_1(x_i, w_i)$, as in Step 2(a) of the REAL experiment in Definition 3

3. $\mathcal{A}$ receives $((\rho, s), m_1, r_1, \ldots, m_{p(n)}, r_{p(n)})$ and works as follows:

    (a) For every $i = 1, \ldots, p(n)$, $\mathcal{A}$ sets $a_i = m_i$, computes the values $e_i = H_s(x_i, \mathsf{Com}(m_i; r_i))$, $z_i = P_2(x_i, a_i, e_i)$, and defines $\pi_i = (x_i, a_i, r_i, z_i)$.

    (b) $\mathcal{A}$ outputs $((\rho, s), x_1, \ldots, x_{p(n)}, \pi_1, \ldots, \pi_{p(n)})$

Now, if $\mathcal{A}$ interacts in the "real commitment" experiment REAL for dual-mode commitments, then its output is exactly the same output as in the real proofs experiment $\mathcal{R}^f$ in Definition 2. This is because the CRS is generated according to the dual commitment scheme, and the algorithm run by $\mathcal{A}$ to compute all the (x_i, a_i, r_i, z_i) is exactly the same as the honest prover $P(x_i, w_i, \rho_i)$. The only difference is that $\mathcal{A}$ receives (r_i, c_i) externally. However, $c_i = \mathsf{Com}_\rho(a_i; r_i)$ and r_i is uniformly distributed in this experiment. Thus, it is exactly the same as P in Protocol 5.

In contrast, if $\mathcal{A}$ interacts in the "simulation" experiment SIMULATION for dual-mode commitments, then its output is distributed identically to $\mathcal{S}'$. This is because the CRS ρ is computed as $\rho \leftarrow \mathcal{S}_{\mathsf{com}}(1^n)$, as too are $c_i \leftarrow \mathcal{S}_{\mathsf{com}}$ and $r_i \leftarrow \mathcal{S}_{\mathsf{com}}(m_i)$ in the dual-commitment simulation experiment, exactly as computed by $\mathcal{S}'$.

Thus, by Definition 3, the output of $\mathcal{S}'$ is computationally indistinguishable from a real proof. This implies that the output of $\mathcal{S}_{\mathsf{zk}}$ is computationally indistinguishable from a real proof, as required by Definition 2.

4.2 Interactive Argument (Adaptive Soundness)

We now prove that Protocol 5 is a non-interactive argument system. In particular, it is computationally (adaptively) sound.

Lemma 2. *Let $\Sigma = (P_1, P_2, V_\Sigma)$ be a Sigma-protocol for a relation R, let Com be a perfectly-binding commitment, and let H be a non-programmable random oracle. Then, Protocol 5 with Σ is a non-interactive argument system for the language L_R in the common reference string model.*

Proof. Completeness is immediate. We proceed to prove adaptive soundness, as in Definition 2. We will use the fact that for any function g, the relation $\mathcal{R} = \{(x, g(x)\}$ is evasive on pairs $(x, \mathcal{O}(x))$, where $\mathcal{O}$ is a (non-programmable) random oracle. This means that, given oracle access to $\mathcal{O}$, it is infeasible to find a string x so that the pair $(x, \mathcal{O}(x)) \in \mathcal{R}$ [3].

Assume $x \notin L$. Then, by the soundness of the Sigma protocol, we have that for every a there exists a *single* $e \in \{0,1\}^n$ for which (a, e, z) is accepting, for some z. Define the function $g(x, c) = e$, where there exist a, r, z such that $c = \mathsf{Com}(a; r)$ and $V_\Sigma(x, a, e, z) = 1$. We stress that since $x \notin L$ and since c is perfectly binding, there exists a *single* value e that fulfills this property. Thus, it follows that g is a *function*, as required.

Since g is a function, it follows that the relation $\mathcal{R} = \{((x,c), g(x,c))\}$ is evasive, meaning that no polynomial-time machine $\mathcal{A}$ can find a pair (x,c) so that $\mathcal{O}(x,c) = g(x,c)$, with non-negligible probability. Assume now, by contradiction, that there exists a probabilistic polynomial-time function f and a probabilistic polynomial-time cheating prover $\mathcal{B}$ such that $V(f(\rho_n), \rho_n, \mathcal{B}(\rho_n)) = 1$ with non-negligible probability (where $\rho_n \leftarrow \mathsf{GenCRS}(1^n)$).

We construct a probabilistic polynomial-time adversary $\mathcal{A}$ as follows. $\mathcal{A}$ runs the regular generation of the dual-mode commitment scheme to obtain ρ_n. Then, $\mathcal{A}$ runs $\mathcal{B}(\rho_n)$ and obtains a tuple (x,a,e,z). If $V(f(\rho_n), \rho_n, (a,r,z)) = 1$, then $\mathcal{A}$ outputs $(x, \mathsf{Com}(a;r))$ and halts. According to the contradicting assumption, $V(f(\rho_n), \rho_n, (a,r,z)) = 1$ with non-negligible probability. This implies that with non-negligible probability, it holds that $V_\Sigma(x, a, \mathcal{O}(x, \mathsf{Com}(a;r)), z) = 1$. However, there is just a single value e for which $V_\Sigma(x, a, \mathcal{O}(x, \mathsf{Com}(a;r)), z) = 1$. Thus, this implies that $\mathcal{O}(x, \mathsf{Com}(a;r)) = e$, with non-negligible probability. Stated differently, this implies that $\mathcal{O}(x, \mathsf{Com}(a;r)) = g(x, \mathsf{Com}(a;r))$ with non-negligible probability, in contradiction to the fact that any function g is evasive for a (non-programmable) random oracle.

4.3 Summary

Combining Lemmas 1 and 2 with the fact that the dual-mode commitment scheme is perfectly binding when the CRS is chosen correctly, we have:

Corollary 1. *Let L be a language with an associated Sigma protocol. If dual-mode commitments exist, then there exists a non-interactive zero-knowledge argument system for L in the non-programmable random-oracle model. Furthermore, zero-knowledge holds in the standard model.*

In Theorem 2 we showed that dual-mode commitment schemes exist for every membership-hard language with a Sigma protocol. Combining this with the above corollary, we have:

Corollary 2. *Let L be a membership-hard language with an associated Sigma protocol. Then, there exists a non-interactive zero-knowledge interactive proof system for L, in the non-programmable random oracle model. Furthermore, zero-knowledge holds in the standard model.*

Acknowledgements. We thank Ben Riva, Nigel Smart and Daniel Wichs for helpful discussions.

References

1. Barak, B., Lindell, Y., Vadhan, S.: Lower Bounds for Non-Black-Box Zero Knowledge. The Journal of Computer and System Sciences 72(2), 321–391 (2003), (An extended abstract appeared in FOCS 2003)

2. Blum, M., Feldman, P., Micali, S.: Non-interactive Zero-Knowledge and its Applications. In: 20th STOC, pp. 103–112 (1988)
3. Canetti, R., Goldreich, O., Halevi, S.: The Random Oracle Methodology, Revisited. In: The 30th STOC, pp. 209–218 (1998)
4. Catalano, D., Visconti, I.: Hybrid Commitments and their Applications to Zero-Knowledge Proof Systems. Theoretical Computer Science 374(1-3), 229–260 (2007)
5. Cramer, R., Damgård, I.B., Schoenmakers, B.: Proofs of Partial Knowledge and Simplified Design of Witness Hiding Protocols. In: Desmedt, Y.G. (ed.) Advances in Cryptology - CRYPTO 1994. LNCS, vol. 839, pp. 174–187. Springer, Heidelberg (1994)
6. Damgård, I.B.: On the Existence of Bit Commitment Schemes and Zero-Knowledge Proofs. In: Brassard, G. (ed.) Advances in Cryptology - CRYPTO 1989. LNCS, vol. 435, pp. 17–27. Springer, Heidelberg (1990)
7. Damgård, I.B.: Efficient Concurrent Zero-Knowledge in the Auxiliary String Model. In: Preneel, B. (ed.) EUROCRYPT 2000. LNCS, vol. 1807, pp. 418–430. Springer, Heidelberg (2000)
8. Damgård, I.: On Σ Protocols, http://www.daimi.au.dk/~ivan/Sigma.pdf
9. Dodis, Y., Shoup, V., Walfish, S.: Efficient Constructions of Composable Commitments and Zero-Knowledge Proofs. In: Wagner, D. (ed.) CRYPTO 2008. LNCS, vol. 5157, pp. 515–535. Springer, Heidelberg (2008)
10. Dodis, Y., Ristenpart, T., Vadhan, S.: Randomness Condensers for Efficiently Samplable, Seed-Dependent Sources. In: Cramer, R. (ed.) TCC 2012. LNCS, vol. 7194, pp. 618–635. Springer, Heidelberg (2012)
11. Feige, U., Lapidot, D., Shamir, A.: Multiple Non-Interactive Zero-Knowledge Proofs Under General Assumptions. SIAM Journal on Computing 29(1), 1–28 (1999)
12. Feige, U., Shamir, A.: Witness Indistinguishable and Witness Hiding Protocols. In: The 22nd STOC, pp. 416–426 (1990)
13. Fiat, A., Shamir, A.: How to Prove Yourself: Practical Solutions to Identification and Signature Problems. In: Odlyzko, A.M. (ed.) Advances in Cryptology - CRYPTO 1986. LNCS, vol. 263, pp. 186–194. Springer, Heidelberg (1987)
14. Goldreich, O.: Foundation of Cryptography, Volume II. Cambridge University Press (2004)
15. Goldreich, O.: Basing Non-Interactive Zero-Knowledge on (Enhanced) Trapdoor Permutation: The State of the Art. Technical Report (2009), http://www.wisdom.weizmann.ac.il/~oded/PSBookFrag/nizk-tdp.ps
16. Goldreich, O., Micali, S., Wigderson, A.: How to Prove all NP-Statements in Zero-Knowledge, and a Methodology of Cryptographic Protocol Design. In: Odlyzko, A.M. (ed.) Advances in Cryptology - CRYPTO 1986. LNCS, vol. 263, pp. 171–185. Springer, Heidelberg (1987)
17. Goldreich, O., Micali, S., Wigderson, A.: How to Play any Mental Game – A Completeness Theorem for Protocols with Honest Majority. In: 19th STOC, pp. 218–229 (1987)
18. Goldreich, O., Oren, Y.: Definitions and Properties of Zero-Knowledge Proof Systems. Journal of Cryptology 7(1), 1–32 (1994)
19. Goldwasser, S., Kalai, Y.: On the (In)security of the Fiat-Shamir Paradigm. In: The 44th FOCS, pp. 102–113 (2003)
20. Goldwasser, S., Micali, S., Rackoff, C.: The Knowledge Complexity of Interactive Proof Systems. SIAM Journal on Computing 18(1), 186–208 (1989)

21. Hazay, C., Lindell, Y.: Efficient Secure Two-Party Protocols – Techniques and Constructions. Springer (October 2010)
22. Jarecki, S.: Efficient Two-Party Secure Computation on Committed Inputs. In: Naor, M. (ed.) EUROCRYPT 2007. LNCS, vol. 4515, pp. 97–114. Springer, Heidelberg (2007)
23. Lindell, Y., Pinkas, B.: Secure Two-Party Computation via Cut-and-Choose Oblivious Transfer. Journal of Cryptology 25(4), 680–722 (2011), (Extended abstract appeared in Ishai, Y. (ed.) TCC 2011. LNCS, vol. 6597, pp. 329–346. Springer, Heidelberg (2011))
24. Lindell, Y.: Fast Cut-and-Choose Based Protocols for Malicious and Covert Adversaries. In: Canetti, R., Garay, J.A. (eds.) CRYPTO 2013, Part II. LNCS, vol. 8043, pp. 1–17. Springer, Heidelberg (2013)
25. Nielsen, J.B.: Separating Random Oracle Proofs from Complexity Theoretic Proofs: The Non-committing Encryption Case. In: Yung, M. (ed.) CRYPTO 2002. LNCS, vol. 2442, pp. 111–126. Springer, Heidelberg (2002)
26. Nielsen, J.B., Orlandi, C.: LEGO for Two-Party Secure Computation. In: Reingold, O. (ed.) TCC 2009. LNCS, vol. 5444, pp. 368–386. Springer, Heidelberg (2009)
27. Pass, R.: On Deniability in the Common Reference String and Random Oracle Model. In: Boneh, D. (ed.) CRYPTO 2003. LNCS, vol. 2729, pp. 316–337. Springer, Heidelberg (2003)
28. Pedersen, T.P.: Non-interactive and Information-Theoretic Secure Verifiable Secret Sharing. In: Feigenbaum, J. (ed.) CRYPTO 1991. LNCS, vol. 576, pp. 129–140. Springer, Heidelberg (1992)
29. Pointcheval, D., Stern, J.: Security Proofs for Signature Schemes. In: Maurer, U.M. (ed.) EUROCRYPT 1996. LNCS, vol. 1070, pp. 387–398. Springer, Heidelberg (1996)
30. Schoenmakers, B., Tuyls, P.: Practical Two-Party Computation Based on the Conditional Gate. In: Lee, P.J. (ed.) ASIACRYPT 2004. LNCS, vol. 3329, pp. 119–136. Springer, Heidelberg (2004)
31. shelat, A., Shen, C.-h.: Two-Output Secure Computation with Malicious Adversaries. In: Paterson, K.G. (ed.) EUROCRYPT 2011. LNCS, vol. 6632, pp. 386–405. Springer, Heidelberg (2011)
32. Wee, H.: Zero Knowledge in the Random Oracle Model, Revisited. In: Matsui, M. (ed.) ASIACRYPT 2009. LNCS, vol. 5912, pp. 417–434. Springer, Heidelberg (2009)

On the Indifferentiability of Key-Alternating Feistel Ciphers with No Key Derivation[*]

Chun Guo[1,2], and Dongdai Lin[1,**]

[1] State Key Laboratory of Information Security,
Institute of Information Engineering, Chinese Academy of Sciences, China
[2] University of Chinese Academy of Sciences, China
{guochun,ddlin}@iie.ac.cn

Abstract. Feistel constructions have been shown to be indifferentiable from random permutations at STOC 2011. Whereas how to properly mix the keys into an un-keyed Feistel construction without appealing to domain separation technique to obtain a block cipher which is provably secure against known-key and chosen-key attacks (or to obtain an ideal cipher) remains an open problem. We study this, particularly the basic structure of NSA's SIMON family of block ciphers. SIMON family takes a construction which has the subkey xored into a halve of the state at each round. More clearly, at the i-th round, the state is updated according to

$$(x_i, x_{i-1}) \mapsto (x_{i-1} \oplus F_i(x_i) \oplus k_i, x_i)$$

For such key-alternating Feistel ciphers, we show that 21 rounds are sufficient to achieve indifferentiability from ideal ciphers with $2n$-bit blocks and n-bit keys, assuming the n-to-n-bit round functions $F_1, \ldots, F_{21}$ to be random and public and an identical user-provided n-bit key to be applied at each round. This gives an answer to the question mentioned before, which is the first to our knowledge.

Keywords: Block cipher, ideal cipher, indifferentiability, key-alternating cipher, Feistel cipher.

1 Introduction

Block Ciphers, and the Security Notions. Block ciphers are among the most important primitives in cryptography. For a block cipher, the standard security notion is the indistinguishability from a random permutation when the key is fixed to some unknown random values. Such pseudorandomness captures the security in traditional single secret key setting. However, block ciphers find numerous and essential uses beyond encryption. For instance, block ciphers have been used to build hash functions and message authentication codes. These applications require the security in the *open key model*, where the adversary knows

---

[*] A full version of this paper is available [16].
[**] Corresponding Author.

Y. Dodis and J.B. Nielsen (Eds.): TCC 2015, Part I, LNCS 9014, pp. 110–133, 2015.

or even chooses the keys. To assess such stronger-than-pseudorandomness security, the *indifferentiability framework* has to be employed. As a generalization of the indistinguishability notion, the indifferentiability framework provides a formal way to assess the security of idealized constructions of block ciphers and hash functions. It can be used to evaluate the "closeness" of a block cipher construction to an *ideal cipher*[1]. Despite the uninstantiability of idealized models [12,23,9], such indifferentiability proofs are widely believed to be able to show the nonexistence of generic attacks which do not exploit the inner details of the implementations of the underlying building blocks.

Feistel Constructions. Existing block cipher designs can be roughly split into two families, namely Feistel-based ciphers and substitution-permutation networks (SPNs). Starting from the seminal Luby-Rackoff paper [20], Feistel constructions have been extensively studied. Most of the provable security works fall in the Luby-Rackoff framework [20], in which the round functions are idealized as being uniformly random (and secret). Such works covered indistinguishability/provable security in the single secret key model (e.g. [24,22,25]), provable security in the open key model (Mandal *et al.* [21] and Andreeva *et al.* [3]), and provable security under related-key attacks (Manuel *et al.* [5]). A recent series of works studied the indifferentiability from random permutations of Luby-Rackoff construction, including the works of Coron *et al.* [14], Seurin [27], and Holenstein *et al.* [17], and the number of rounds required was finally fixed to 14 by Holenstein *et al.* [17].

Our Problem: How to Mix the Key into Feistel. In this paper, we consider the problem that *how to mix the key material into a Feistel construction by a popular approach to obtain a block cipher indifferentiable from an **ideal cipher***. Since an un-keyed Feistel construction is indifferentiable from a random permutation, a Feistel-based cipher indifferentiable from an ideal cipher can be trivially obtained through domain separation. However, such a result tells us nothing about how to concretely mix the keys into the state – in fact, none of the works mentioned before addressed this problem. To our knowledge, domain separation technique is seldom used in existing block cipher designs. Existing designs usually inserts keys via efficient group operations, e.g. xor and modular addition; therefore, this problem has practical meanings.

A natural candidate solution to this problem is the construction called *key-alternating Feistel cipher* (KAF for short) analyzed by Lampe *et al.* [19]. The KAF cipher they considered has the round keys xored before each round function, as depicted in Fig. 1 (left). Lampe *et al.* studied the indistinguishability of KAF in a setting where the underlying round functions are random and *public* (in contrast to the classical Luby-Rackoff setting) and the keys are fixed and secret; this is also the only provable security work on KAF.

However, due to the well known complementation property, there exist obstacles when trying to achieve an indifferentiability proof for KAF (detailed

---

[1] See Sect. 2 for the formal definitions of indifferentiability and the ideal cipher model.

discussions are deferred to the full version [16]). This motivates us to turn to another candidate construction, which has the round key xored into the halve of the state after the round functions. Due to the similarity between the two constructions, we denote the latter construction by KAF^* to follow the convention of Lampe *et al.* while making a distinction. For KAF^*, the $2n$-bit intermediate state s_i is split to two halves, i.e. $s_i = (x_{i+1}, x_i)$ where $i \in \{0, 1, \ldots, r\}$, and at the i-th round, the state value is updated according to

$$(x_i, x_{i-1}) \mapsto (x_{i-1} \oplus F_i(x_i) \oplus k_i, x_i),$$

as depicted in Fig. 1 (right). KAF^* can be seen as the basic structure of NSA's SIMON family of block ciphers.

Clearly, the proof for KAF^* with no cryptographically strong assumption about the key derivation functions is more attractive, since such key derivations are more relevant to practice than random oracle modeled ones. Whereas KAF^* with independent round keys cannot resist related-key attacks (the case is similar to Even-Mansour ciphers). Hence we consider KAF^* with an identical user-provided n-bit key applied at each round, and call such ciphers *single-key* KAF^* ($SKAF^*$ for short). The 21-round $SKAF^*$ is depicted in Fig. 2. With the discussions above, we focus on the question that *whether it is possible for $SKAF^*$ with sufficiently many rounds to be indifferentiable from ideal ciphers.*

Our Results. We show 21-round $SKAF^*$ to be indifferentiable from ideal ciphers, thus for the first time giving a solution to the problem *how to mix keys into Feistel in the open-key model.*

Theorem. *The 21-round key-alternating Feistel cipher $SKAF^*_{21}$ with all round functions $\mathbf{F} = (F_1, \ldots, F_{21})$ being 21 independent n-to-n-bit random functions and an identical (user-provided) n-bit key k applied at each round is indifferentiable from an ideal cipher with $2n$-bit blocks and n-bit keys.*

To our knowledge, this paper is also the first to study the indifferentiability/provable security of key-alternating Feistel ciphers – in particular, with no key derivation – in the open key model.

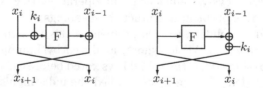

Fig. 1. Mixing the key into: (left) the input of the round function – KAF; (right) the halve of the state after the round function – KAF^*

From a practical point of view, our results suggest a possible choice to resist complementing attack and its extensions (see [7]) when designing Feistel ciphers[2]. KAF with random oracle modeled key derivation functions may also have such resistance. However, practical key derivation algorithms are usually designed to be "lightweight" and moderately complex, and KAF with such moderately complex key derivations may still be broken in hash mode (the example is Camellia, in [7]). Hence we think our results have its own interest. Meanwhile, since publicly released in June 2013, the SIMON family of block ciphers [6] designed by NSA has attracted considerable attention due to its simple structure, high flexibility, and remarkable performance [4,8,1,28,11]. SIMON family is based on KAF^*. Our results may be seen as a first step towards understanding the underlying reasons.

Remark. We heavily borrow the techniques used by Holenstein *et al.* [17] and Lampe *et al.* [18] (see the next paragraph). We stress that our main constructions consist of the indifferentiability result for $SKAF^*$ and the analyzes of $SKAF^*$.

Overview of Techniques. We reuse and adapt the *simulation via chain-completion* technique introduced by Coron *et al.* [14], while the overall framework is very close to that used by Holenstein *et al.* [17] and Lampe *et al.* [18]. This framework consists of constructing a simulator which works by detecting and completing *partial chains* created by the queries of the distinguisher. To ensure consistency in the answers while avoiding exponentially many chain completions, each of the rounds in the construction is assigned a unique and specific role needed in the proof, including *chain detection, uniformness ensuring,* and *chain adaptation* (see Fig. 2). By this, the simulator first detects new partial chains when the associated values have "filled" the chain detection zone; then fills in the corresponding computation chain by both querying the ideal primitive and simulating the other necessary function values, until only the values of the round functions in the chain adaptation zone remain (possibly) undefined; and finally defines these values to complete the whole chain so that the answers of the ideal primitive are consistent with the function values simulated by the simulator.

Adaptations in this Work. To fit into the $SKAF^*$ context, the framework has to be adapted. Note that in the $SKAF^*$ context, each complete chain corresponds to a unique pair of input and output where the input consists of an n-bit key and a $2n$-bit block; hence the entropy of each chain is $3n$ bits, and it is necessary and sufficient to uniquely specify a chain by the queries to 3 round functions (recall that for un-keyed Feistel, the entropy of each chain is only $2n$ bits). Another consequence of this property is that in the $SKAF^*$ context, two different chains may collide at two successive rounds, i.e. for two different chains $(x_0, x_1, \ldots)$ and $(x'_0, x'_1, \ldots)$, it may hold that $x_j = x'_j \wedge x_{j+1} = x'_{j+1}$ for some j. As a comparison, consider the un-keyed Feistel context: in this context, for

---

[2] This idea is not new, as it has been used by XTEA (see [10]). However, this paper provides the first security proof.

two different chains, it is impossible to find j such that $x_j = x'_j \wedge x_{j+1} = x'_{j+1}$, otherwise we will have $x_i = x'_i$ for any i and the two chains are not different.

With these in mind, we introduce the following adaptations: first, we increase the number of rounds used for chain detection to 3, so that given the queries x_i, x_{i+1}, and x_{i+2} to these round functions, a chain can be uniquely specified with the associated key $k = x_i \oplus F_{i+1}(x_{i+1}) \oplus x_{i+2}$, after which it is possible to move forward and backward along the computation path (and do some "completion").

Second, we increase the number of rounds used to ensure randomness. Surrounding each adaptation zone with 2 *always-randomly-defined* buffer rounds is a key point of this framework. The buffer rounds are expected to protect the adaptation zone in the sense that the simulator does not define the values in the 2 buffer rounds while completing other chains. This idea works well in previous contexts. However, in the $SKAF^*$ context, if we continue working with 2 buffer rounds, then since two different chains are possible to collide at two successive rounds, such an expectation may be broken. More clearly, when a chain is to be adapted, the corresponding function values in the buffer rounds may have been defined (this can be shown by a simple operation sequence with only 5 queries; see Appendix A). In such a case, we find it not easy to achieve a proof. To get rid of this, we increase the number of buffer rounds to 4 – more clearly, 2 buffer rounds at each side of each adaptation zone (and in total 8 for the whole construction). We then prove that unless an improbable event happens, the simulator does not define the function values in the buffer rounds *exactly next to* the adaptation zones when completing other chains, and then all chains can be correctly adapted.

Another evidence for the necessity of increasing the number of buffer rounds is that the additional buffer rounds actually play an important role in the proof (see Lemma 2).

At last, to show the indistinguishability of the systems, we combine the *randomness mapping argument* [17] (RMA for short) and its *relaxed* version [2] (RRMA for short). This allows us to bypass the intermediate system composed of the idealized construction and the simulator.

Organization. Sect. 2 presents preliminaries. Sect. 3 contains the main theorem. Sect. 4 gives the simulator. Finally, Sect. 5 sketches the proof. Some additional notations will be introduced later, when necessary.

2   Preliminaries

The Ideal Cipher Model (ICM). The ICM is a widespread model in which all parties have access to a random primitive called ideal cipher $\mathbf{E} : \{0,1\}^n \times \{0,1\}^\kappa \to \{0,1\}^n$. $\mathbf{E}$ is taken randomly from the set of $(2^n!)^{2^\kappa}$ block ciphers with key space $\{0,1\}^\kappa$ and plaintext and ciphertext space $\{0,1\}^n$. ICM finds enormous applications, for instance, the analysis of blockcipher based hash functions (e.g. [13]).

Indifferentiability. The indifferentiability framework was introduced by Maurer, Renner, and Holenstein, at TCC 2004 [23]. It is applicable in settings where the underlying primitives and parameters are exposed to the adversary. Briefly speaking, for a construction C^G from an idealized primitive G (hopefully simpler), if C^G is indifferentiable from another ideal primitive T, then C^G can safely replace T in most "natural" settings[3]. A formal definition is recalled as follows.

Definition 1. *A primitive C^G with oracle access to an ideal primitive G is said to be statistically and strongly (q, σ, ε)-indifferentiable from an ideal primitive T if there exists a simulator $\mathbf{S}^T$ s.t. $\mathbf{S}$ makes at most σ queries to T, and for any distinguisher D which issues at most q queries, it holds that*

$$\left| Pr[D^{C^G,G} = 1] - Pr[D^{T,\mathbf{S}^T} = 1] \right| < \varepsilon$$

Since then, indifferentiability framework has been applied to various constructions, including variants of Merkle-Damgård [13], sponge construction, Feistel [14,17], and iteratated Even-Mansour ciphers [2,18].

3 Indifferentiability for 21-Round Single-Key KAF^*

The main theorem is presented as follows.

Theorem 1. *For any q, the 21-round single-key key-alternating Feistel cipher $SKAF_{21}^*$ with all round functions $\mathbf{F} = (F_1, \ldots, F_{21})$ being 21 independent n-to-n-bit random functions and an identical (user-provided) n-bit key k applied at each round is strongly and statistically (q, σ, ε)-indifferentiable from an ideal cipher $\mathbf{E}$ with $2n$-bit blocks and n-bit keys, where $\sigma = 2^{11} \cdot q^9$ and $\varepsilon \leq \frac{2^{19} \cdot q^{15}}{2^{2n}} + \frac{2^{222} \cdot q^{30}}{2^n} + \frac{2^{34} \cdot q^6}{2^{2n}} = O(\frac{q^{30}}{2^n})$.*

To show it, in the following sections we first give the simulator, then sketch the proof. The full formal proof is deferred to the full version [16].

4 The Simulator

To simplify the proof, we take a strategy introduced by Holenstein *et. al* [17], that is, making the randomness taken by the simulator $\mathbf{S}$, the cipher $\mathbf{E}$ (in the simulated world), and the random functions $\mathbf{F}$ (in the real world) *explicit* as random tapes. The simulator's random tape is an array of tables $\varphi = (\varphi_1, \ldots, \varphi_{21})$, where each φ_i maps entries $x \in \{0,1\}^n$ to uniform and independent values in $\{0,1\}^n$. The cipher's random tape is a table η which encodes an ideal cipher with $2n$-bit blocks and n-bit keys. More clearly, η is selected uniformly at random from all tables with the property of mapping entries $(\delta, k, z) \in \{+, -\} \times \{0,1\}^n \times \{0,1\}^{2n}$

---

[3] Restrictions on the indifferentiability composition theorem have been exhibited in [26,15]. However, indifferentiability has been sufficient in most "natural" settings (see [15]).

to uniform values $z' \in \{0,1\}^{2n}$ such that $\eta(+, k, z) = z'$ iff. $\eta(-, k, z') = z$. The random functions $\mathbf{F}$ have access to the array of tables $f = (f_1, \ldots, f_{21})$ where each f_i maps entries $x \in \{0,1\}^n$ to uniform and independent values in $\{0,1\}^n$. We denote the constructions/primitives which take randomness from the tapes φ, η, and f by $\mathbf{S}(\varphi)$, $\mathbf{E}(\eta)$, and $\mathbf{F}(f)$ respectively. Among the three, $\mathbf{E}(\eta)$ and $\mathbf{F}(f)$ simply relay the values in η and f. As argued by Andreeva *et al.* [2], such a strategy does not reduce the validity of the simulation, since access to such tapes can be efficiently simulated by uniformly sampling.

We now describe the simulator. $\mathbf{S}(\varphi)$ provides an interface $\mathbf{S}(\varphi).F(i, x)$ to the distinguisher for querying the simulated random function F_i on value x, where $i \in \{1, \ldots, 21\}$ and $x \in \{0,1\}^n$. For each i, the simulator maintains a hash table G_i that has entries in the form of pairs (x, y), which denote pairs of inputs and outputs of $\mathbf{S}(\varphi).F(i, x)$. Denote the fact that x is a preimage in the table G_i by $x \in G_i$, and $G_i(x)$ the corresponding image when $x \in G_i$.

Receiving a query $\mathbf{S}(\varphi).F(i, x)$, $\mathbf{S}(\varphi)$ looks in G_i, returns $G_i(x)$ if $x \in G_i$. Otherwise $\mathbf{S}(\varphi)$ accesses the tap φ_i to draw the answer $\varphi_i(x)$ and adds the entry $(x, \varphi_i(x))$ to G_i, and then, if i belongs to the set $\{3, 10, 11, 12, 19\}$, the *chain detection* mechanism and subsequent *chain completion* mechanism of $\mathbf{S}(\varphi)$ will be triggered. These two mechanisms help in ensuring that the answers of the random functions simulated by $\mathbf{S}(\varphi)$ are consistent with the answers of the ideal cipher $\mathbf{E}(\eta)$. Depending on i, there are three case:

1. when $i = 3$, for each newly generated tuple $(x_1, x_2, x_3, x_{20}, x_{21}) \in G_1 \times G_2 \times G_3 \times G_{20} \times G_{21}$, the simulator computes $k := x_1 \oplus G_2(x_2) \oplus x_3$, $x_0 := x_2 \oplus G_1(x_1) \oplus k$, and $x_{22} := x_{20} \oplus G_{21}(x_{21}) \oplus k$. It then calls an inner procedure $\mathbf{S}(\varphi).Check((x_1, x_0), (x_{22}, x_{21}), k)$, which checks whether $\mathbf{E}(\eta).Enc(k, (x_1, x_0)) = (x_{22}, x_{21})$ (i.e. $\eta(+, k, (x_1, x_0)) = (x_{22}, x_{21})$) holds, and returns true if so. Whenever this call returns true, the simulator enqueues a 5-tuple $(x_1, x_2, x_3, 1, 6)$ into a queue *ChainQueue*. In the 5-tuple, the 4-th value 1 informs $\mathbf{S}(\varphi)$ that the first value of the tuple is x_1, and the last value 6 informs $\mathbf{S}(\varphi)$ that when completing the chain $(x_1, x_2, x_3, 1)$, it should set entries in G_6 and G_7 to "adapt" the chain and ensure consistency.
2. when $i = 19$, the case is similar to the previous one by symmetry: for each newly generated tuple $(x_1, x_2, x_{19}, x_{20}, x_{21}) \in G_1 \times G_2 \times G_{19} \times G_{20} \times G_{21}$, the simulator computes $k := x_{19} \oplus G_{20}(x_{20}) \oplus x_{21}$, $x_0 := x_2 \oplus G_1(x_1) \oplus k$, $x_{22} := x_{20} \oplus G_{21}(x_{21}) \oplus k$, and $x_3 := x_1 \oplus G_2(x_2) \oplus k$, makes a call to $\mathbf{S}(\varphi).Check((x_1, x_0), (x_{22}, x_{21}), k)$, and enqueues the 5-tuple $(x_1, x_2, x_3, 1, 15)$ into *ChainQueue* whenever this call returns true.
3. when $i \in \{10, 11, 12\}$, for each newly generated tuple $(x_{10}, x_{11}, x_{12}) \in G_{10} \times G_{11} \times G_{12}$, the simulator enqueues the 5-tuple $(x_{10}, x_{11}, x_{12}, 10, l)$ into the queue *ChainQueue*, where $l = 6$ if $i = 10$ or 11, and $l = 15$ if $i = 12$. The sketch of the whole strategy is illustrated in Fig. 2.

After having enqueued the newly generated tuples, $\mathbf{S}(\varphi)$ immediately takes the tuples out of *ChainQueue* and completes the associated partial chains. More clearly, $\mathbf{S}(\varphi)$ maintains a set *CompletedSet* for the chains it has completed. For each chain C dequeued from the queue, if $C \notin CompletedSet$ (i.e. C has not been completed), $\mathbf{S}(\varphi)$ completes it, by evaluating in the corresponding $SKAF^*$ computation chain both forward and backward (defining the necessary but undefined $G_i(x_i)$ values), and querying $\mathbf{E}.Enc$ or $\mathbf{E}.Dec$ once to "wrap" around, until it reaches the value x_l (when moving forward) and x_{l+1} (when moving backward). Then $\mathbf{S}(\varphi)$ "adapts" the entries by defining $G_l(x_l) := x_{l-1} \oplus x_{l+1} \oplus k$ and $G_{l+1}(x_{l+1}) := x_l \oplus x_{l+2} \oplus k$ to make the entire computation chain consistent with the answers of $\mathbf{E}(\eta)$. This defining action may overwrite values in G_l or G_{l+1} if $x_l \in G_l$ or $x_{l+1} \in G_{l+1}$ before it happens, however we will show the probability to be negligible. $\mathbf{S}(\varphi)$ then adds $(x_1, x_2, x_3, 1)$ and $(x_{10}, x_{11}, x_{12}, 10)$ to $CompletedSet$, where the two chains correspond to C.

During the completion, the values in G_j newly defined by $\mathbf{S}(\varphi)$ also trigger the chain detection mechanism and chain completion mechanism when $j \in \{3, 10, 11, 12, 19\}$. $\mathbf{S}(\varphi)$ hence keeps dequeuing and completing until $ChainQueue$ is empty again. $\mathbf{S}(\varphi)$ finally returns $G_i(x)$ as the answer to the initial query.

Pseudocode of the Simulator. We present the pseudocode of the simulator $\mathbf{S}(\varphi)$ and a modified simulator $\widetilde{S}(\varphi)$ (will be introduced in Sect. 5). When a line has a boxed statement next to it, $\mathbf{S}(\varphi)$ uses the original statement, while $\widetilde{S}(\varphi)$ uses the boxed one.

1: **Simulator $\mathbf{S}(\varphi)$:** $\boxed{\text{Simulator } \widetilde{S}(\varphi):}$
2: **Variables**
3:     hash tables $\{G_i\} = (G_1, \ldots, G_{21})$, initially empty
4:     queue $ChainQueue$, initially empty
5:     set $CompletedSet$, initially empty
        The procedure $F(i, x)$ provides an interface to the distinguisher.
6: **public procedure** $F(i, x)$
7:     $y := F^{inner}(i, x)$
8:     **while** $ChainQueue \neq \emptyset$ **do**
9:         $(x_j, x_{j+1}, x_{j+2}, j, l) := ChainQueue.Dequeue()$
10:        **if** $(x_j, x_{j+1}, x_{j+2}, j, l) \notin CompletedSet$ **then** // Complete the chain
11:            $k := x_j \oplus G_{j+1}(x_{j+1}) \oplus x_{j+2}$
12:            $(x_{l-4}, x_{l-3}, x_{l-2}, l-4) := EvalForward(x_j, x_{j+1}, x_{j+2}, j, l-4)$
13:            $(x_{l+3}, x_{l+4}, x_{l+5}, l+3) := EvalBackward(x_j, x_{j+1}, x_{j+2}, j, l+3)$
14:            $Adapt(x_{l-4}, x_{l-3}, x_{l-2}, x_{l+3}, x_{l+4}, x_{l+5}, l)$
15:            $(x_1, x_2, x_3, 1) := EvalForward(x_j, x_{j+1}, x_{j+2}, j, 1)$
16:            $(x_{10}, x_{11}, x_{12}, 10) := EvalForward(x_1, x_2, x_3, 1, 10)$
17:            $CompletedSet := CompletedSet \cup \{(x_1, x_2, x_3, 1), (x_{10}, x_{11}, x_{12}, 10)\}$
18:        **return** y
        The procedure $Adapt$ randomly sets the "missed" values if necessary
        and adds entries to G_l and G_{l+1} to make the chain match the computation.
19: **private procedure** $Adapt(x_{l-4}, x_{l-3}, x_{l-2}, x_{l+3}, x_{l+4}, x_{l+5}, l)$
20:    $k := x_{l-4} \oplus G_{l-3}(x_{l-3}) \oplus x_{l-2}$
21:    $y_{l-2} := F^{inner}(l-2, x_{l-2})$
22:    $x_{l-1} := x_{l-3} \oplus y_{l-2} \oplus k$
23:    $y_{l-1} := F^{inner}(l-1, x_{l-1})$

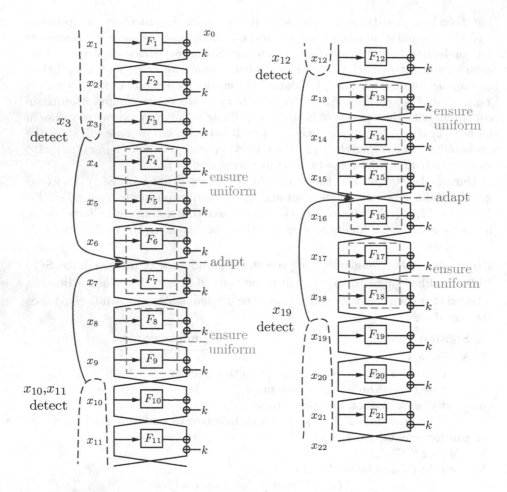

Fig. 2. The 21-round $SKAF^*$ cipher with the zones where the simulator detects chains and adapts them0

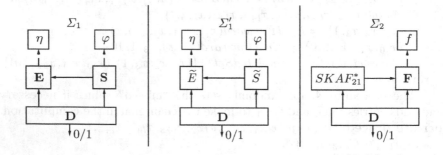

Fig. 3. Systems used in the proof in this paper

24:   $x_l := x_{l-2} \oplus y_{l-1} \oplus k$

25:   $y_{l+3} := F^{inner}(l+3, x_{l+3})$

26:   $x_{l+2} := x_{l+4} \oplus y_{l+3} \oplus k$

27:   $y_{l+2} := F^{inner}(l+2, x_{l+2})$

28:   $x_{l+1} := x_{l+3} \oplus y_{l+2} \oplus k$

29:   $ForceVal(x_l, x_{l-1} \oplus x_{l+1} \oplus k, l)$

30:   $ForceVal(x_{l+1}, x_l \oplus x_{l+2} \oplus k, l+1)$

31: **private procedure** $ForceVal(x, y, l)$

32:   $G_l(x) := y$ // May overwrite the entry $G_l(x)$

       The procedure F^{inner} draws answers from the table G_i, or the tape φ_i if the answers have not been defined in G_i, and enqueue chains when necessary.

33: **private procedure** $F^{inner}(i, x)$

34:   **if** $x \notin G_i$ **then**

35:     $G_i(x) := \varphi_i(x)$

36:     **if** $i \in \{3, 10, 11, 12, 19\}$ **then**

37:       $EnqueueNewChains(i, x)$

38:   **return** $G_i(x)$

39: **private procedure** $EnqueueNewChains(i, x)$

40:   **if** $i = 3$ **then**

41:     **for all** $(x_1, x_2, x_3, x_{20}, x_{21}) \in G_1 \times G_2 \times \{x\} \times G_{20} \times G_{21}$ **then**

42:       $k := x_1 \oplus G_2(x_2) \oplus x_3$

43:       $chk\_pa := ((x_1, G_1(x_1) \oplus x_2 \oplus k), (x_{20} \oplus G_{21}(x_{21}) \oplus k, x_{21}), k)$

44:       $flag := Check(chk\_pa)$ $\boxed{flag := \widetilde{E}.Check(chk\_pa)}$

45:       **if** $flag = true$ **then**

46:         $ChainQueue.Enqueue(x_1, x_2, x_3, 1, 6)$

47:   **else if** $i = 19$ **then**

48:     **for all** $(x_1, x_2, x_{19}, x_{20}, x_{21}) \in G_1 \times G_2 \times \{x\} \times G_{20} \times G_{21}$ **do**

49:       $k := x_{19} \oplus G_{20}(x_{20}) \oplus x_{21}$

50:       $chk\_pa := ((x_1, G_1(x_1) \oplus x_2 \oplus k), (x_{20} \oplus G_{21}(x_{21}) \oplus k, x_{21}), k)$

51:       $flag := Check(chk\_pa)$ $\boxed{flag := \widetilde{E}.Check(chk\_pa)}$

52:       **if** $flag = true$ **then**

53:         $x_3 := x_1 \oplus G_2(x_2) \oplus k$

54:         $ChainQueue.Enqueue(x_1, x_2, x_3, 1, 15)$

55:   **else if** $i = 10$ **then**

56:     **for all** $(x_{10}, x_{11}, x_{12}) \in \{x\} \times G_{11} \times G_{12}$ **do**

57:       $ChainQueue.Enqueue(x_{10}, x_{11}, x_{12}, 10, 6)$

58:   **else if** $i = 11$ **then**

59:     **for all** $(x_{10}, x_{11}, x_{12}) \in G_{10} \times \{x\} \times G_{12}$ **do**

60:       $ChainQueue.Enqueue(x_{10}, x_{11}, x_{12}, 10, 6)$

61:   **else if** $i = 12$ **then**

62:     **for all** $(x_{10}, x_{11}, x_{12}) \in G_{10} \times G_{11} \times \{x\}$ **do**

63:       $ChainQueue.Enqueue(x_{10}, x_{11}, x_{12}, 10, 15)$

The *Check* procedure queries $\mathbf{E}$ to verify whether the inputs are valid pairs of plaintext and ciphertext of $\mathbf{E}$. Note that $\widetilde{S}$ does not own *Check* procedure; instead $\widetilde{S}$ calls the *Check* procedure of a modified cipher $\widetilde{E}$.

64: **private procedure** $Check(x, y, k)$ // $\widetilde{S}$ does not own such a procedure
65:     **return** $\mathbf{E}.Enc(k, x) = y$

The procedures *EvalForward* (and *EvalBackward*, resp.) takes a partial chain $(x_j, x_{j+1}, x_{j+2}, j)$ as input, and evaluate forward (and backward, resp.) in $SKAF^*$ until obtaining the tuple (x_l, x_{l+1}, x_{l+2}) of input values for G_l, G_{l+1}, and G_{l+2} for specified l.

66: **private procedure** $EvalForward(x_j, x_{j+1}, x_{j+2}, j, l)$
67:     $k := x_j \oplus G_{j+1}(x_{j+1}) \oplus x_{j+2}$ // By construction $x_{j+1} \in G_{j+1}$ holds
68:     **while** $j \neq l$ **do**
69:         **if** $j = 20$ **then**
70:             $(x_1, x_0) := \mathbf{E}.Dec(k, (x_{22}, x_{21}))$ $\boxed{(x_1, x_0) := \widetilde{E}.Dec(k, (x_{22}, x_{21}))}$
71:             $x_2 := x_0 \oplus F^{inner}(1, x_1) \oplus k$
72:             $j := 0$
73:         **else**
74:             $x_{j+3} := x_{j+1} \oplus F^{inner}(j+2, x_{j+2}) \oplus k$
75:             $j := j + 1$
76:     **return** $(x_l, x_{l+1}, x_{l+2}, l)$
77: **private procedure** $EvalBackward(x_j, x_{j+1}, x_{j+2}, j, l)$
78:     $k := x_j \oplus G_{j+1}(x_{j+1}) \oplus x_{j+2}$
79:     **while** $j \neq l$ **do**
80:         **if** $j = 0$ **then**
81:             $(x_{22}, x_{21}) := \mathbf{E}.Enc(k, (x_1, x_0))$ $\boxed{(x_{22}, x_{21}) := \widetilde{E}.Enc(k, (x_1, x_0))}$
82:             $x_{20} := x_{22} \oplus F^{inner}(21, x_{21}) \oplus k$
83:             $j := 20$
84:         **else**
85:             $x_{j-1} := x_{j+1} \oplus F^{inner}(j, x_j) \oplus k$
86:             $j := j - 1$
87:     **return** $(x_l, x_{l+1}, x_{l+2}, l)$

5   Proof of the Indifferentiability: Sketch

Denote by $\Sigma_1(\mathbf{E}(\eta), \mathbf{S}(\varphi))$ the simulated system composed of the ideal cipher $\mathbf{E}$ with tape η and the simulator $\mathbf{S}$ with tape φ, and denote by $\Sigma_2(SKAF_{21}^*, \mathbf{F}(f))$ the real system composed of $SKAF_{21}^*$ and the random functions $\mathbf{F}(f)$. Then, for any fixed, deterministic, and computationally unbounded distinguisher $\mathbf{D}$, we show the following two to establish the indifferentiability:

(i) $\Sigma_1(\mathbf{E}(\eta), \mathbf{S}(\varphi))$ and $\Sigma_2(SKAF_{21}^*, \mathbf{F}(f))$ are indistinguishable.
(ii) With overwhelmingly large probability, $\mathbf{S}(\varphi)$ runs in polynomial time.

Note that the underlying ideas used in Sect. 5.2 and Sect. 5.3 are originally used by Coron *et al.* [14] (also used in [17,18]). As stressed in Introduction, the main novelties of this part are in Sect. 5.4. To keep Sect. 5.4 clear and simple, we only present the sketch in the main body, while extracting the lemmas corresponding to the core step (the simulator overwrites with negligible probability) and listing them in Appendix B.

5.1   An Intermediate System Σ'_1

We use an intermediate system $\Sigma'_1(\widetilde{E}(\eta), \widetilde{S}(\varphi))$, which consists of a modified ideal cipher $\widetilde{E}(\eta)$ and a slightly modified simulator $\widetilde{S}(\varphi)$. $\widetilde{E}(\eta)$ maintains a table E, which contains entries of the form $((+, k, x), y)$ and $((-, k, y), x)$, and is initially empty. $\widetilde{E}(\eta)$ provides an additional interface $Check(x, y, k)$. Once being queried on $Enc(k, x)$ or $Dec(k, y)$, $\widetilde{E}(\eta)$ adds the corresponding entries in η to E and returns them as answers. Once being called on $Check(x, y, k)$, $\widetilde{E}(\eta)$ looks in the table E to check whether $E(+, k, x) = y$ and returns the answer. On the other hand, the differences between $\widetilde{S}(\varphi)$ and $\mathbf{S}(\varphi)$ consist of two aspects:

- the cipher queried by them: $\widetilde{S}(\varphi)$ queries $\widetilde{E}(\eta)$ while $\mathbf{S}(\varphi)$ queries $\mathbf{E}(\eta)$;
- the owner of the $Check$ procedure called by them: $\widetilde{S}(\varphi)$ calls $\widetilde{E}(\eta).Check$ while $\mathbf{S}(\varphi)$ calls $\mathbf{S}(\varphi).Check$;

Denote by $\widetilde{E}(\eta).E^+$ the set of entries in $\widetilde{E}(\eta).E$ of the form $((+, \cdot, \cdot), \cdot)$. The pseudocode of $\widetilde{E}(\eta)$ is deferred to the full version [16], while the pseudocode of $\widetilde{S}(\varphi)$ is presented along with $\mathbf{S}(\varphi)$, in Sect. 4, captured by the boxed statements. The three systems are depicted in Fig. 3. Σ'_1 mostly helps in bounding the complexity of $\mathbf{S}(\varphi)$.

5.2   Bounding the Complexity of $\widetilde{S}(\varphi)$ in Σ'_1

The simulator $\widetilde{S}(\varphi)$ in Σ'_1 runs in polynomial time: each time $\widetilde{S}(\varphi)$ dequeues a tuple of the form $(x_1, x_2, x_3, 1, l)$ for which $(x_1, x_2, x_3, 1) \notin CompletedSet$ must correspond to an entry in $\widetilde{E}(\eta).E^+$ previously added during a query issued by $\mathbf{D}$, since $(x_1, x_2, x_3, 1, l)$ can be enqueued only when the corresponding call to $\widetilde{E}(\eta).Check$ returns true. Hence such dequeuing happens at most q times. Based on this, the size of G_i and $\widetilde{E}(\eta).E^+$ is upper bounded to $10q^3$, and the number of queries to $\widetilde{E}(\eta).Check$ issued by $\widetilde{S}(\varphi)$ is upper bounded to $2 \cdot (10q^3)^5$.

5.3 Σ_1 to Σ'_1

To show this, we define a bad event to capture the difference between Σ_1 and Σ'_1. The core difference between the two lies in that the procedure $\mathbf{S}(\varphi).Check$ actually answers queries according to the content of the table/tape η, while the procedure $\widetilde{E}(\eta).Check$ answers according to a much smaller table $\widetilde{E}(\eta).E$: this forms the idea of the bad event. During an execution $\mathbf{D}^{\Sigma_1(\mathbf{E}(\eta), \mathbf{S}(\varphi))}$, the bad event **BadCheck** happens if $\exists(x, y, k)$ s.t. all the following hold:

(i) $\mathbf{S}(\varphi)$ makes a call $\mathbf{S}(\varphi).Check(x, y, k)$.

(ii) $\eta(+, k, x) = y$.

(iii) Before the call in (i), neither $\mathbf{E}(\eta).Enc(k, x)$ nor $\mathbf{E}(\eta).Dec(k, y)$ has been issued (i.e. if the call is made in Σ_1', then $(+, k, x) \notin \widetilde{E}(\eta).E$ before the call).

For some fixed (η, φ), assume that $\widetilde{S}(\varphi)$ makes q_1' calls to $\widetilde{S}(\varphi).Check$ during $\mathbf{D}^{\Sigma_1'(\widetilde{E}(\eta), \widetilde{S}(\varphi))}$. Then if during $\mathbf{D}^{\Sigma_1(\mathbf{E}(\eta), \mathbf{S}(\varphi))}$, **BadCheck** does not happen in the first q_1' calls to $\mathbf{E}(\eta).Check$ – the probability is at least $1 - 2q_1'/2^{2n}$, $\mathbf{D}^{\Sigma_1(\mathbf{E}(\eta), \mathbf{S}(\varphi))}$ and $\mathbf{D}^{\Sigma_1'(\widetilde{E}(\eta), \widetilde{S}(\varphi))}$ behave the same way, and $\mathbf{D}^{\Sigma_1(\mathbf{E}(\eta), \mathbf{S}(\varphi))} = \mathbf{D}^{\Sigma_1'(\widetilde{E}(\eta), \widetilde{S}(\varphi))}$. Since $q_1' \leq 2 \cdot (10q^3)^5$, we have:

Lemma 1. *For any distinguisher* $\mathbf{D}$ *which issues at most* q *queries, we have:*

$$\left| Pr[\mathbf{D}^{\Sigma_1(\mathbf{E}(\eta), \mathbf{S}(\varphi))} = 1] - Pr[\mathbf{D}^{\Sigma_1'(\widetilde{E}(\eta), \widetilde{S}(\varphi))} = 1] \right| \leq \frac{2^{19} \cdot q^{15}}{2^{2n}}.$$

Proof. See the full version [16]. $\square$

This bound on advantage along with the bound on complexity of $\widetilde{S}(\varphi)$ show that with probability at least $1 - \frac{2^{19} \cdot q^{15}}{2^{2n}}$, the complexity of $\mathbf{S}(\varphi)$ is the same as that of $\widetilde{S}(\varphi)$, and can be upper bounded to $2^{11} \cdot q^9$ queries to $\mathbf{E}(\eta)$.

5.4 Σ_1' to Σ_2: The Relaxed Randomness Mapping Argument

We use an RRMA to fill in this part. First, we specify the domain of the randomness map; second, we complete the argument.

Specifying the Domain of the Map. The domain of the map should include overwhelmingly many Σ_1' executions, and these executions should have the same behaviors as the Σ_2 executions in the view of $\mathbf{D}$. Hence we figure out the difference between the two systems first. Consider Σ_1' and Σ_2. In the former, the answers to F-queries are simulated by $\widetilde{S}(\varphi)$, and when $\widetilde{S}(\varphi)$ is forced to overwrite some entries (in $\{G_i\}$), the consistency in the answers will be broken. On the other hand, such inconsistency never appears in Σ_2: this forms the difference. We will take the Σ_1' executions during which $\widetilde{S}$ does not overwrite any entry to specify the domain (later we will show that such Σ_1' executions are the same as the Σ_2 executions in the view of $\mathbf{D}$). For this, we first define an additional bad event **BadHit**, then show that **BadHit** happens with negligible probability, and finally show that during a Σ_1'-execution, if **BadHit** does not happen, then $\widetilde{S}$ does not overwrite, so that the domain we specify covers overwhelmingly many Σ_1' executions as expected. During an execution $\mathbf{D}^{\Sigma_1'(\widetilde{E}(\eta), \widetilde{S}(\varphi))}$, the event **BadHit** happens if the n-bit value read from the tape φ – or either of the two n-bit halves of the value read from the tape η – equals the xor of 9 or less values in the history $\mathcal{H}$, where $\mathcal{H}$ is the set of all the n-bit values – or halves – extracted from the tables $\{G_i\}$ and $\widetilde{E}(\eta).E$ right before the tape accessing action happens. With the bound on the size of the tables, we calculate $Pr[\mathbf{BadHit}] \leq \frac{2^{88} \cdot q^{30}}{2^n}$.

We then show that during the *good* executions $\mathbf{D}^{\Sigma_1'(\widetilde{E}(\eta),\widetilde{S}(\varphi))}$ (during which **BadHit** does not happen), $\widetilde{S}(\varphi)$ never overwrites any entry in $\{G_i\}$. As mentioned in Sect. 1, the reason is that right before any call to *Adapt*, the function values in the two buffer rounds exactly next to the adaptation zone must have not been defined. Then the two values will be set to uniformly random values, which implies that the probability for $\widetilde{S}(\varphi)$ to overwrite is negligible. To illustrate more clearly, we define the 4-tuple $(x_i, x_{i+1}, x_{i+2}, i)$ as *partial chain* for $i \in \{0, \ldots, 20\}$, and borrow the helper functions *next*, *prev*, val_l^+, val_l^-, and the notions *equivalent* partial chains, *table-defined* partial chains from [17,18]. The two helper functions *next* and *prev* take a partial chain C as input and return the partial chain obtained by moving respectively one step forward or backward in $SKAF_{21}^*$ according to the given tables $\widetilde{E}(\eta).E$ and $\{G_i\}$ (wrapping around through $\widetilde{E}(\eta).E$ if necessary), or empty value $\perp$ when some values are necessary for the computation but have not been defined in the tables. The two functions val_l^+ and val_l^- take a partial chain as input and evaluate forward and backward respectively (also according to the given $\widetilde{E}.E$ and $\{G_i\}$) until obtaining and returning the corresponding x_l, or returning $\perp$ when some necessary values have not been defined in the tables. The notions *equivalent* and *table-defined* partial chains are as follows: (i) two partial chains C and D are *equivalent*, if they belong to the same computation chain; (ii) a partial chain $C = (x_i, x_{i+1}, x_{i+2}, i)$ is *table-defined* if all the three values x_i, x_{i+1}, and x_{i+2} have been added to their corresponding tables.

Then, we have the following non-overwriting lemma.

Lemma 2. *In a good execution* $\mathbf{D}^{\Sigma_1'(\widetilde{E}(\eta),\widetilde{S}(\varphi))}$, *before any two successive calls to* $ForceVal(x_l, y_l, l)$ *and* $ForceVal(x_{l+1}, y_{l+1}, l + 1)$, $x_l \notin G_l \wedge x_{l+1} \notin G_{l+1}$ *must hold.*

Proof. A formal proof – along with the lemmas that support the proof – are presented in Appendix B; here we only sketch it. Consider any such two calls $ForceVal(x_l, y_l, l)$ and $ForceVal(x_{l+1}, y_{l+1}, l + 1)$, and suppose that they are triggered by a chain C. Note that $C \notin CompletedSet$ when C is dequeued, otherwise the two calls will not happen. Then the sketch consists of four stages:

First, denote by $Path_C$ the whole computation path that C belongs to. Then before C is enqueued, $val_{l-2}^+(C) = \perp \wedge val_{l+3}^-(C) = \perp$ must hold. Otherwise the values of $Path_C$ must have "filled" another chain detection zone, after which $Path_C$ would be completed, and C would have been added to $CompletedSet$, a contradiction.

Second, since being enqueued, C must have been equivalent to a table-defined chain. Then, during the completion of some other chain D (D is not equivalent to C and is completed after C being enqueued), the subsequent calls to $ForceVal$ cannot affect $val_i^\delta(C)$ for any valid i and $\delta \in \{+, -\}$. The underlying reason is strongly relevant to the number of buffer rounds we arrange. Briefly speaking, for previous calls to $ForceVal$ (triggered by D) to change $val_i^\delta(C)$, C and D must agree on either of the two rounds l' and $l'+1$ where D is supposed to be adapted. However since we arrange two buffer rounds at each side of the adaptation zone, we find that two inequivalent partial chains (C and D) either: (i) cannot collide at three consecutive rounds, and as a result, $val^+_{l'-1}(C) \neq val^+_{l'-1}(D)$ ($val^-_{l'+2}(C) \neq val^-_{l'+2}(D)$, resp.); or: (ii) cannot collide at round $l'-1$ ($l'+2$, resp.) to avoid the bad event **BadHit**. By this, C and D cannot agree on any of the two rounds l' and $l'+1$, and $val^\delta_i(C)$ can only be changed by tape accessing and table entry setting actions, and to avoid **BadHit**, $val^+_{l-2}(C) \notin G_{l-2} \wedge val^-_{l+3}(C) \notin G_{l+3}$ immediately holds after such actions make them two non-empty.

Third, by carefully analyzing all possibilities, we show that any chain completed between the point when C is enqueued and the point when C is dequeued cannot add $val^+_{l-1}(C)$ to G_{l-1} (even if $val^+_{l-1}(C)$ has previously been made non-empty during the completion of some other chains). Similarly, $val^-_{l+2}(C)$ cannot be added to G_{l+2} during this period. Hence $val^+_{l-1}(C) \notin G_{l-1} \wedge val^-_{l+2}(C) \notin G_{l+2}$ keeps holding till C being dequeued.

Finally, after C is dequeued, $G_{l-1}(val^+_{l-1}(C))$ and $G_{l+2}(val^-_{l+2}(C))$ will be defined to values from φ tape, and to avoid **BadHit**, $x_l = val^+_l(C) \notin G_l \wedge x_{l+1} = val^-_{l+1}(C) \notin G_{l+1}$ must hold after these assignments. $\qquad\square$

Completing the RRMA. Fix a distinguisher **D**. Consider a distinguisher $\overline{D}$ which runs **D** and then completes all the chains for each query to $\widetilde{E}(\eta)$ made by **D**. During $\overline{D}^{\Sigma'_1(\widetilde{E}(\eta),\widetilde{S}(\varphi))}$, many entries in the tapes (η, φ) may not be accessed, and those that are really accessed compose *footprints*. Clearly with respect to $\overline{D}$, there is a bijection between the footprint set and the Σ'_1 execution set.

Then, consider the Σ'_1 executions $\overline{D}^{\Sigma'_1(\widetilde{E}(\eta),\widetilde{S}(\varphi))}$ during which $\widetilde{S}(\varphi)$ does not overwrite any entry. By Lemma 2, the probability for such Σ'_1 executions to occur is at least $1 - \frac{2^{222} \cdot q^{30}}{2^n}$. Taking the set of all possible footprints of such Σ'_1 executions as the domain, we define $\tau(\alpha) = f = (f_1, \ldots, f_{21})$ as the exact copies of the tables $(G_1, \ldots, G_{21})$ standing at the end of the execution: $\tau(\alpha) \equiv \{G_i\}$. The Σ'_1 and Σ_2 executions linked by τ have the same behaviors in the view of $\overline{D}$ because the answers in them two are consistent with $\tau(\alpha)$ and $\{G_i\}$; and the probabilities for the tapes (η, φ) and f to respectively agree with the preimage and the image are close – for $22q$-query $\overline{D}$, the ratio of the two probabilities lies in the interval $[1 - \frac{(10(22q)^3)^2}{2^{2n}}, 1]$. By these, with a *nearly* standard process, we upper bound the advantage of distinguishing Σ'_1 and Σ_2 to $\frac{2^{222} \cdot q^{30}}{2^n} + \frac{2^{34} \cdot q^6}{2^{2n}}$.

Lemma 3. *For any distinguisher* **D** *which issues at most q queries, we have:*

$$\left| Pr[\mathbf{D}^{\Sigma'_1} = 1] - Pr[\mathbf{D}^{\Sigma_2} = 1] \right| \leq \frac{2^{222} \cdot q^{30}}{2^n} + \frac{2^{34} \cdot q^6}{2^{2n}}$$

Proof. See the full version [16]. $\qquad\square$

Acknowledgements. We are deeply grateful to the anonymous referees of TCC 2015 for their useful comments. We are also particularly grateful to Jianghua Zhong. This paper would have not been possible without her help.

This work is partially supported by National Key Basic Research Project of China (2011CB302400), National Science Foundation of China (61379139) and the "Strategic Priority Research Program" of the Chinese Academy of Sciences, Grant No. XDA06010701.

References

1. Abed, F., List, E., Lucks, S., Wenzel, J.: Differential cryptanalysis of round-reduced simon and speck. In: Fast Software Encryption 2014. LNCS. Springer, Heidelberg (2014) (to appear)
2. Andreeva, E., Bogdanov, A., Dodis, Y., Mennink, B., Steinberger, J.P.: On the indifferentiability of key-alternating ciphers. In: Canetti, R., Garay, J.A. (eds.) CRYPTO 2013, Part I. LNCS, vol. 8042, pp. 531–550. Springer, Heidelberg (2013)
3. Andreeva, E., Bogdanov, A., Mennink, B.: Towards understanding the known-key security of block ciphers. In: Moriai, S. (ed.) FSE 2013. LNCS, vol. 8424, pp. 348–366. Springer, Heidelberg (2014)
4. Aysum, A., Gulcan, E., Schaumont, P.: Simon says, break the area records for symmetric key block ciphers on fpgas. Tech. rep., Cryptology ePrint Archive, Report 2014/237 (2014), http://eprint.iacr.org
5. Barbosa, M., Farshim, P.: The related-key analysis of feistel constructions. In: Fast Software Encryption 2014. LNCS. Springer, Heidelberg (2014) (to appear)
6. Beaulieu, R., Shors, D., Smith, J., Treatman-Clark, S., Weeks, B., Wingers, L.: The simon and speck families of lightweight block ciphers
7. Biryukov, A., Nikolić, I.: Complementing feistel ciphers. In: Moriai, S. (ed.) FSE 2013. LNCS, vol. 8424, pp. 3–18. Springer, Heidelberg (2014)
8. Biryukov, A., Roy, A., Velichkov, V.: Differential analysis of block ciphers simon and speck. In: Fast Software Encryption 2014. LNCS. Springer, Heidelberg (2014) (to appear)
9. Black, J.A.: The ideal-cipher model, revisited: An uninstantiable blockcipher-based hash function. In: Robshaw, M. (ed.) FSE 2006. LNCS, vol. 4047, pp. 328–340. Springer, Heidelberg (2006)
10. Bouillaguet, C., Dunkelman, O., Leurent, G., Fouque, P.-A.: Another look at complementation properties. In: Hong, S., Iwata, T. (eds.) FSE 2010. LNCS, vol. 6147, pp. 347–364. Springer, Heidelberg (2010)
11. Boura, C., Naya-Plasencia, M., Suder, V.: Scrutinizing and improving impossible differential attacks: Applications to CLEFIA, Camellia, LBlock and SIMON. In: Sarkar, P., Iwata, T. (eds.) ASIACRYPT 2014, PART I. LNCS, vol. 8873, pp. 179–199. Springer, Heidelberg (2014)
12. Canetti, R., Goldreich, O., Halevi, S.: The random oracle methodology, revisited. J. ACM 51(4), 557–594 (2004)
13. Coron, J.-S., Dodis, Y., Malinaud, C., Puniya, P.: Merkle-damgård revisited: How to construct a hash function. In: Shoup, V. (ed.) CRYPTO 2005. LNCS, vol. 3621, pp. 430–448. Springer, Heidelberg (2005)
14. Coron, J.-S., Patarin, J., Seurin, Y.: The random oracle model and the ideal cipher model are equivalent. In: Wagner, D. (ed.) CRYPTO 2008. LNCS, vol. 5157, pp. 1–20. Springer, Heidelberg (2008)
15. Demay, G., Gaži, P., Hirt, M., Maurer, U.: Resource-restricted indifferentiability. In: Johansson, T., Nguyen, P.Q. (eds.) EUROCRYPT 2013. LNCS, vol. 7881, pp. 664–683. Springer, Heidelberg (2013)

16. Guo, C., Lin, D.: On the indifferentiability of key-alternating feistel ciphers with no key derivation. Cryptology ePrint Archive, Report 2014/786 (2014), http://eprint.iacr.org/

17. Holenstein, T., Künzler, R., Tessaro, S.: The equivalence of the random oracle model and the ideal cipher model, revisited. In: Proceedings of the Forty-third Annual ACM Symposium on Theory of Computing, STOC 2011, pp. 89–98. ACM, New York (2011)

18. Lampe, R., Seurin, Y.: How to construct an ideal cipher from a small set of public permutations. In: Sako, K., Sarkar, P. (eds.) ASIACRYPT 2013, Part I. LNCS, vol. 8269, pp. 444–463. Springer, Heidelberg (2013)

19. Lampe, R., Seurin, Y.: Security analysis of key-alternating feistel ciphers. In: Fast Software Encryption 2014. LNCS. Springer, Heidelberg (2014) (to appear)

20. Luby, M., Rackoff, C.: How to construct pseudorandom permutations from pseudorandom functions. SIAM Journal on Computing 17(2), 373–386 (1988)

21. Mandal, A., Patarin, J., Seurin, Y.: On the public indifferentiability and correlation intractability of the 6-round feistel construction. In: Cramer, R. (ed.) TCC 2012. LNCS, vol. 7194, pp. 285–302. Springer, Heidelberg (2012)

22. Maurer, U., Pietrzak, K.: The security of many-round luby-rackoff pseudorandom permutations. In: Biham, E. (ed.) EUROCRYPT 2003. LNCS, vol. 2656, pp. 544–561. Springer, Heidelberg (2003)

23. Maurer, U.M., Renner, R.S., Holenstein, C.: Indifferentiability, impossibility results on reductions, and applications to the random oracle methodology. In: Naor, M. (ed.) TCC 2004. LNCS, vol. 2951, pp. 21–39. Springer, Heidelberg (2004)

24. Patarin, J.: Pseudorandom permutations based on the D.E.S. scheme. In: Charpin, P., Cohen, G. (eds.) EUROCODE 1990. LNCS, vol. 514, pp. 193–204. Springer, Heidelberg (1991)

25. Patarin, J.: Security of random feistel schemes with 5 or more rounds. In: Franklin, M. (ed.) CRYPTO 2004. LNCS, vol. 3152, pp. 106–122. Springer, Heidelberg (2004)

26. Ristenpart, T., Shacham, H., Shrimpton, T.: Careful with composition: Limitations of the indifferentiability framework. In: Paterson, K.G. (ed.) EUROCRYPT 2011. LNCS, vol. 6632, pp. 487–506. Springer, Heidelberg (2011)

27. Seurin, Y.: Primitives et protocoles cryptographiques àsécurité prouvée. Ph.D. thesis, PhD thesis, Université de Versailles Saint-Quentin-en-Yvelines, France (2009)

28. Sun, S., Hu, L., Wang, P., Qiao, K., Ma, X., Song, L.: Automatic security evaluation and (Related-key) differential characteristic search: Application to SIMON, PRESENT, lBlock, DES(L) and other bit-oriented block ciphers. In: Sarkar, P., Iwata, T. (eds.) ASIACRYPT 2014, PART I. LNCS, vol. 8873, pp. 158–178. Springer, Heidelberg (2014)

A   Surrounding Each Adaptation Zone with Two Buffer Rounds – the Broken Expectations

If we increase the number of rounds used for chain detection to 3, while continue surrounding each adaptation zone with two buffer Rounds – exactly same as done in the previous works [17,18] – then we are working on $3 + 1 + 2 + 1 + 3 + 1 + 2 + 1 + 3 = 17$ rounds ($SKAF_{17}^*$). For the modified simulator, the buffer rounds are round 4, 7, 11, and 14, while the first adaptation zone consists of round 5 and 6, the second consists of round 12 and 13. Then the following operation sequence shows that when a chain is to be adapted, the function values in the buffer rounds next to the adaption zone may have been defined:

 (i) arbitrarily chooses x_3, x_2, and x_2';
 (ii) issues queries $G_2(x_2)$ and $G_2(x_2')$ to the simulator;
(iii) arbitrarily chooses k and calculate $k' := k \oplus x_2 \oplus x_2'$;
(iv) calculates $x_1 := x_3 \oplus G_2(x_2) \oplus k$ and $x_1' := x_3 \oplus G_2(x_2') \oplus k'$;
 (v) issues queries $G_1(x_1)$ and $G_1(x_1')$;
(vi) issues queries $G_3(x_3)$;

The last query $G_3(x_3)$ enqueues two chains $(x_1, x_2, x_3, 1)$ and $(x_1', x_2', x_3, 1)$, and whatever value is assigned to $G_3(x_3)$, for the two chains we have $x_4 = x_2 \oplus G_3(x_3) \oplus k = x_2' \oplus G_3(x_3) \oplus k' = x_4'$. When the later one is dequeued, we have $x_4 \in G_4$; this breaks the expectation that *the simulator does not define the values in the buffer rounds while completing other chains*. The underlying reason for this lies in the fact that in the $SKAF^*$ context, it is possible to make two different chains collide at two successive rounds (as already discussed in Introduction). The operation sequence mentioned before indeed takes advantage of this property.

Our at current time, we are not clear whether 17-round single-key $SKAF^*_{17}$ can achieve indifferentiability or not.

B   The Formal Proof for $\widetilde{S}$ Not Overwrites

To give the formal proof, we introduce two notions *key-defined* and *key-undefined* chains and a helper function k, as follows: a partial chain $C = (x_i, x_{i+1}, x_{i+2}, i)$ is called *key-defined* if $x_{i+1} \in G_{i+1}$, otherwise is called *key-undefined*; and $k(C)$ returns the associated key when C is key-defined, while returning $\bot$ otherwise. Moreover, we borrow two additional notions *safe call to Adapt*, and *non-overwriting call to ForceVal* from [17,18]: (i) a call to $Adapt(x_{l-4}, x_{l-3}, x_{l-2}, x_{l+3}, x_{l+4}, x_{l+5}, l)$ is *safe* if the following holds before the call:

$$(((x_{l-2} \notin G_{l-2}) \vee (x_{l-2} \in G_{l-2} \wedge x_{l-3} \oplus G_{l-2}(x_{l-2}) \oplus k(B) \notin G_{l-1}))$$
$$\wedge ((x_{l+3} \notin G_{l+3}) \vee (x_{l+3} \in G_{l+3} \wedge x_{l+4} \oplus G_{l+3}(x_{l+3}) \oplus k(D) \notin G_{l+2}))),$$

where $B = (x_{l-4}, x_{l-3}, x_{l-2}, l-4)$ and $D = (x_{l+3}, x_{l+4}, x_{l+5}, l+3)$; (ii) a call to $ForceVal(x, y, l)$ is *non-overwriting* if $x \notin G_l$ before the call.

Then, we have Lemma 11, which claims that $\widetilde{S}(\varphi)$ does not overwrite in good Σ_1' executions. Before presenting this main lemma, we list some properties of the good Σ_1' executions, as follows. They consist of the foundation of Lemma 11. To save space while highlighting the features of $SKAF^*$, we summarize the properties that are *almost the same* as those owned by un-keyed Feistel [17] and single-key iterated Even-Mansour [18] as Lemma 5.

First, in the good executions, each random tape accessing and the subsequent entry setting action can only extend the key-defined chains one round. Compared to the previous results in [17,18], Lemma 4 only focuses on *key-defined* chains.

Lemma 4. *The following hold in a good execution* $\mathbf{D}^{\Sigma'_1(\widetilde{E}(\eta),\widetilde{S}(\varphi))}$:

(i) *For any key-defined partial chain C, if $next(C) = \bot$ (prev$(C) = \bot$, resp.) before a random tape accessing and subsequent entry setting action on either $\widetilde{E}(\eta).E$ or $\{G_i\}$, then if C is table-defined after the action, it holds that $next^2(C) = \bot$ (prev$^2(C) = \bot$, resp.).*

(ii) *For any key-defined partial chain C and each $\delta \in \{+, -\}$, a random tape accessing and entry setting action $G_j(x_j) := \varphi_j(x_j)$ can only change at most one of the values $val_i^\delta(C)$; and if such change happens, then:*
  - *the value is changed from $\bot$ to some non-empty values.*
  - *if $\delta = +$, $i = j + 1$; if $\delta = -$, $i = j - 1$.*
  - *$val_j^\delta(C) = x_j$ before the assignment.*
  - *after the action, if C is table-defined, then $val_i^\delta(C) \notin G_i$.*

Proof. See the full version [16]. □

Lemma 5. *Informally speaking, during a good execution* $\mathbf{D}^{\Sigma'_1(\widetilde{E}(\eta),\widetilde{S}(\varphi))}$, *we have:*

(i) *the relation $\equiv$ between partial chains is an equivalence relation;*

(ii) *the relation $\equiv$ between table-defined chains is invariant before and after the random tape accessing and subsequent entry setting action;*

(iii) *if a chain C is dequeued such that $C \notin CompletedSet$, then when C was enqueued, no chain equivalent to C had been enqueued.*

Proof. See *Lemma 9, Lemma 10*, and *Lemma 13* in the full version [16]. □

The following lemma claims that two inequivalent chains cannot collide at two consecutive rounds when they are extended by the random tape accessing and entry setting actions.

Lemma 6. *Fix a point in a good execution* $\mathbf{D}^{\Sigma'_1(\widetilde{E}(\eta),\widetilde{S}(\varphi))}$ *and suppose all calls to ForceVal to be non-overwriting up to this point. Assume that a random tape accessing and entry setting action $G_i(x_i) := \varphi_i(x_i)$ happens right after this point, then for any two key-defined partial chains C and D, any $l \in \{3, \ldots, 19\}$, and any $\delta \in \{+, -\}$, the following four cannot be simultaneously fulfilled:*

(i) *before the action, C is not equivalent to D;*

(ii) *before the action, $val_l^\delta(C) = \bot$ or $val_l^\delta(D) = \bot$;*

(iii) *after the action, C and D are table-defined;*

(iv) *after the action, $(val_l^\delta(C) = val_l^\delta(D) \neq \bot) \wedge (val_{l-1}^\delta(C) \oplus k(C) = val_{l-1}^\delta(D) \oplus k(D))$ when $\delta = +$, or $(val_l^\delta(C) = val_l^\delta(D) \neq \bot) \wedge (val_{l+1}^\delta(C) \oplus k(C) = val_{l+1}^\delta(D) \oplus k(D))$ when $\delta = -$;*

Proof. Briefly speaking, once statement (ii), (iii), and (iv) are fulfilled, then either $C \equiv D$, or **BadHit** happens. See [16] for the formal proof. □

If all the previous calls to *ForveVal* were non-overwriting, then the calls to *ForceVal* triggered by safe calls to *Adapt* do not affect the values in previously defined chains, nor the equivalence relation. As mentioned in Introduction, this property is one of the key points of the proof, and is similar to those exhibited in [17] and [18]; the difference is brought in by the increased buffer rounds.

Lemma 7. *Consider a safe call* $Adapt(x_{l-4}, x_{l-3}, x_{l-2}, x_{l+3}, x_{l+4}, x_{l+5}, l)$ *in a good execution* $\mathbf{D}^{\Sigma_1'(\widetilde{E}(\eta), \widetilde{S}(\varphi))}$, *and suppose all the previous calls to Adapt to be safe, then:*

(i) *Right before the subsequent call to* $F^{inner}(l-1, x_{l-1})$, $x_{l-1} \notin G_{l-1}$; *right before the subsequent call to* $F^{inner}(l+2, x_{l+2})$, $x_{l+2} \notin G_{l+2}$;

(ii) *The subsequent calls to* $ForceVal$ *are non-overwriting;*

(iii) *If a chain* C *is table-defined before this call to Adapt and is not equivalent to the chain which is being completed, then for any* $i \in \{1, \ldots, 21\}$, $val_i^+(C)$ *and* $val_i^-(C)$ *are invariant before and after both calls to* $ForceVal$.

(iv) *The relation* $\equiv$ *between table-defined chains is invariant before and after the subsequent calls to* $ForceVal$.

Proof. See the full version [16]. □

Lemma 8. *Consider a good execution* $\mathbf{D}^{\Sigma_1'(\widetilde{E}(\eta), \widetilde{S}(\varphi))}$. *Let* C *be a chain which is dequeued and to be adapted at position* l *s.t.* $C \notin CompletedSet$. *Then the subsequent call to Adapt is safe, if the following holds when* C *is dequeued:*

$$(((val_{l-2}^+(C) \notin G_{l-2}) \vee (val_{l-2}^+(C) \in G_{l-2} \wedge val_{l-1}^+(C) \notin G_{l-1}))$$
$$\wedge((val_{l+3}^-(C) \notin G_{l+3}) \vee (val_{l+3}^-(C) \in G_{l+3} \wedge val_{l+2}^-(C) \notin G_{l+2}))).$$

Proof. See the full version [16]. □

For the following discussions, we introduce a tuple set $KUDCS^4$, as the set of 5-tuples $(x_{10}, x_{11}, x_{12}, 10, 6)$ which is enqueued by a call to $F^{inner}(11, x_{11})$. The tuples in this set are special in the sense that before the call to F^{inner} which leads to they being enqueued, the partial chains correspond to them were *key-undefined*.

Then, the following two lemmas show that the assumptions of Lemma 8 hold in a good execution. Lemma 9 shows them to hold before the chains are enqueued, while Lemma 11 shows them to hold till the chains are dequeued (so that all calls to $ForceVal$ are non-overwriting). Lemma 10 is a helper lemma for Lemma 11.

Lemma 9. *Consider a good execution* $\mathbf{D}^{\Sigma_1'(\widetilde{E}(\eta), \widetilde{S}(\varphi))}$. *Let* C *be a partial chain which is enqueued at some time and to be adapted at position* l. *Suppose that no chain equivalent to* C *was enqueued before* C *is enqueued. Then:*

(i) $val_{l-2}^+(C) = \bot$ *and* $val_{l+3}^-(C) = \bot$ *before the call to* $F^{inner}(i, x)$ *which led to* C *being enqueued.*

(ii) *right after* C *is enqueued,* $val_{l-2}^+(C) \notin G_{l-2} \wedge val_{l+3}^-(C) \notin G_{l+3}$.

Proof. See the full version [16]. □

Lemma 10. *Consider a good execution* $\mathbf{D}^{\Sigma_1'(\widetilde{E}(\eta), \widetilde{S}(\varphi))}$. *Let* $C = (x_{10}, x_{11}, x_{12}, 10, 6) \in KUDCS$ *be a partial chain which is enqueued at some time such that no chain equivalent to* C *was enqueued before* C *is enqueued. Then for any chain* D *which is dequeued before* C *is dequeued, the following two hold;*

---

[4] The term is short for *key-undefined chain set*.

(i) *it cannot be* $val_4^+(C) \neq \perp \wedge val_4^+(C) = val_4^+(D) \wedge val_3^+(C) \oplus k(C) = val_3^+(D) \oplus k(D)$;

(ii) *it cannot be* $val_9^-(C) \neq \perp \wedge val_9^-(C) = val_9^-(D) \wedge val_{10}^-(C) \oplus k(C) = val_{10}^-(D) \oplus k(D)$;

Proof. By the assumption, D must have been enqueued before C is enqueued. Consider proposition (i). After the call to $F^{inner}(11, x_{11})$ which led to C being enqueued, we have:

(i) $val_4^+(C) = \perp$ (follows from Lemma 9);

(ii) C is table-defined, and D is equivalent to some table-defined chain D_{td}, since they have been enqueued (hence C and D_{td} are also key-defined).

After this point in the execution, since C has been table-define, $val_4^+(C)$ can only be changed to non-empty by the tape accessing and entry setting actions (by Lemma 7 (iii)) on $\{G_i\}$ (by Lemma 4 (ii)). Then proposition (i) is established by Lemma 6 (note that $val_i^\delta(D) = val_i^\delta(D_{td})$).

Consider proposition (ii). After the call to $F^{inner}(11, x_{11})$, we have:

(i) $val_9^-(C) \neq \perp \wedge val_9^-(C) \notin G_9$ (also follows from Lemma 9);

(ii) C and D_{td} ($D \equiv D_{td}$) are table-defined (and key-defined);

Depending on $val_9^-(D)$, we distinguish the following cases. First, if $val_9^-(D) \neq \perp$ before the call to $F^{inner}(11, x_{11})$, then D must have been enqueued before this call. By this, for some sufficiently large j, we have $(x'_{10}, x'_{11}, x'_{12}, 10) = prev^j(D)$ where all the three values have been in corresponding tables and $x'_{11} \neq x_{11}$. Then after the call, $val_9^-(C) = val_9^-(D)$ is not possible (and will never be possible in future) since it implies **BadHit**.

Second, if $val_9^-(D) = \perp$ before and after the call to $F^{inner}(11, x_{11})$, then similarly to the argument for proposition (i), $val_9^-(C) \neq \perp \wedge val_9^-(C) = val_9^-(D) \wedge val_{10}^-(C) \oplus k(C) = val_{10}^-(D) \oplus k(D)$ cannot be simultaneously fulfilled.

Finally, if $val_9^-(D) = \perp$ before the call to $F^{inner}(11, x_{11})$ while $val_9^-(D) \neq \perp$ after it, then the only possible case is $D = (x'_{10}, x_{11}, x'_{12}, 10)$ and D is also enqueued by the call to $F^{inner}(11, x_{11})$. In this case, assume that $val_9^-(C) \neq \perp \wedge val_9^-(C) = val_9^-(D) \wedge val_{10}^-(C) \oplus k(C) = val_{10}^-(D) \oplus k(D)$ simultaneously hold; then it necessarily be $x_{12} = x'_{12}$ and $G_{10}(x_{10}) \oplus x_{10} = G_{10}(x'_{10}) \oplus x'_{10}$. By construction, $G_{10}(x_{10})$ and $G_{10}(x'_{10})$ are defined to be $\varphi_{10}(x_{10})$ and $\varphi_{10}(x'_{10})$ respectively (since the 10-th round is not in the adaptation zone), hence the one defined later implies **BadHit**.

Having excluded all possibilities, we establish proposition (ii).   □

Lemma 11. *In a good execution* $\mathbf{D}^{\Sigma_1'(\widetilde{E}(\eta), \widetilde{S}(\varphi))}$*, all calls to Adapt are safe, and all calls to ForceVal are non-overwriting.*

Proof. Suppose that the lemma does not hold, and let C be the first chain during the completion of which the call to *Adapt* is not safe. Clearly $C \notin CompletedSet$ when C is dequeued, and since all calls to *Adapt* before C is dequeued were safe, by Lemma 5 (iii) we know when C was enqueued, no chain equivalent to C had been enqueued. Hence, Lemma 9 implies that $val_{l-2}^+(C) \notin G_{l-2} \wedge$ $val_{l+3}^+(C) \notin G_{l+3}$ immediately holds after C was enqueued. We show that when C is dequeued, $val_{l-1}^+(C) \notin G_{l-1} \wedge val_{l+2}^+(C) \notin G_{l+2}$; this implies that the subsequent call to $Adapt$ is safe (by Lemma 8), so that the calls to $ForceVal$ are non-overwriting (by Lemma 7 (ii)). *Wlog* consider $val_{l-2}^+(C)$ and $val_{l-1}^+(C)$. If $val_{l-2}^+(C) = \perp$ after C was enqueued, we show that $val_{l-2}^+(C) = x_{l-2} \notin G_{l-2}$ immediately holds after $val_{l-2}^+(C) \neq \perp$ holds. Consider the last table entry setting action before $val_{l-2}^+(C) \neq \perp$ holds. Recall that C has been equivalent to a table-defined chain C_{td} since being enqueued; then by Lemma 7 (iii), $val_{l-2}^+(C) = val_{l-2}^+(C_{td})$ cannot be changed by previous calls to $ForceVal$. Hence it was changed by a tape accessing and entry setting action, and we have $val_{l-2}^+(C) = x_{l-2} \notin G_{l-2}$ after this action (Lemma 4 (ii)).

Now assume $val_{l-1}^+(C) \in G_{l-1}$ when C is dequeued. Then during the period between the point C was enqueued and the point C is dequeued, the following two actions must have been induced by the completion of some other chains D:

(i) $G_{l-2}(val_{l-2}^+(C))(= G_{l-2}(x_{l-2}))$ was defined;
(ii) after action (i), $G_{l-1}(val_{l-1}^+(C))$ was defined;

We show it to be impossible to show that $val_{l-1}^+(C) \notin G_{l-1}$ holds when C is dequeued. If the two happen, then for (either of) them two to be defined during the completion of D, we must have $val_{l-2}^+(D) = val_{l-2}^+(C)$ **or** $val_{l-1}^+(D) = val_{l-1}^+(C)$. We then show that for a chain D which is completed in this period,

- during the completion of D, if $val_{l-2}^+(C) = val_{l-2}^+(D)$, then $val_{l-1}^+(C) \neq val_{l-1}^+(D)$ (hence $G_{l-1}(val_{l-1}^+(C))$ cannot be defined).
- during the completion of D, $G_{l-1}(val_{l-1}^+(C))$ can be defined only if $val_{l-2}^+(C) = val_{l-2}^+(D)$ $(val_{l-1}^+(C) = val_{l-1}^+(D) \Rightarrow val_{l-2}^+(C) = val_{l-2}^+(D))$.

Gathering the two claims yields that $G_{l-1}(val_{l-1}^+(C))$ cannot be defined during this period and the call to $Adapt$ will be safe.

For the first claim, assume otherwise, i.e. $val_{l-2}^+(D) = val_{l-2}^+(C)$, and right after $G_{l-2}(val_{l-2}^+(D))$ was defined, $val_{l-1}^+(D) = val_{l-1}^+(C)$ holds. This means that before $G_{l-2}(val_{l-2}^+(D))$ was defined, the following two hold:

(i) $val_{l-2}^+(D) = val_{l-2}^+(C) \neq \perp$
(ii) $val_{l-3}^+(D) \oplus k(D) = val_{l-3}^+(C) \oplus k(C)$

Consider the last table entry setting action before the above two hold. After this action, we have $val_{l-2}^+(D) \neq \perp$ and $val_{l-2}^+(C) \neq \perp$; then after this action, C must have been enqueued (because by Lemma 9 (i), before C was enqueued, $val_{l-2}^+(C)$ should be $\perp$), and D has been enqueued even earlier, hence C and D are equivalent to some table-defined chains C_{td} and D_{td} respectively. Then, if $C \in KUDCS$, a contradiction is directly reached by Lemma 10; if $C \notin KUDCS$, for the action, we exclude possibility for each case:

(i) This cannot have been a tape accessing and table entry setting action on $\{G_i\}$. To illustrate this, assume otherwise. Then this action must be the one or posterior to the one which leads to C being enqueued, and the following five hold simultaneously, which contradicts Lemma 6:
- before the action, both C_{td} and D_{td} are key-defined.
- before the action, C_{td} is not equivalent to D_{td};
- before the action, $val_{l-2}^+(C_{td}) = \bot$ or $val_{l-2}^+(D_{td}) = \bot$;
- after the action, C_{td} and D_{td} are table-defined;
- after the action, $val_{l-2}^+(D_{td}) = val_{l-2}^+(C_{td}) \neq \bot$ and $val_{l-3}^+(D_{td}) + k(D_{td}) = val_{l-3}^+(C_{td}) + k(C_{td})$.

(ii) This cannot have been an entry setting action on E, since such actions cannot change $val_{l-2}^+(C_{td})$ nor $val_{l-2}^+(D_{td})$ (by Lemma 4 (ii));

(iii) This cannot have been because of a previous call to $ForceVal$. For this, assume otherwise; as already discussed before, after this call to $ForceVal$, C and D are enqueued and equivalent to some table-defined chains C_{td} and D_{td} respectively. Then it must be either of the following two cases:
  (a) C has been enqueued before this call to $ForceVal$. Then by Lemma 7 (iii), none of the previous calls to $ForceVal$ affects $val_i^+(D) = val_i^+(D_{td})$ and $val_i^+(C) = val_i^+(C_{td})$, a contradiction.
  (b) C is enqueued by this call to $ForceVal$. This is impossible.

Hence the first claim holds.

For the second claim, assume otherwise, then we know that before the entry setting action on $G_{l-1}(val_{l-1}^+(C))$, the following two hold:

(i) $val_{l-2}^+(C) \in G_{l-2}$, $val_{l-2}^+(D) \in G_{l-2}$, and $val_{l-2}^+(C) \neq val_{l-2}^+(D)$
(ii) $val_{l-1}^+(C) = val_{l-1}^+(D) \notin G_{l-1}$

Consider the last table entry setting action before the above two hold. By Lemma 9 (ii), $val_{l-2}^+(C) \notin G_{l-2}$ immediately holds after C is enqueued; hence this action must happen after C is enqueued, and C, D (enqueued earlier that C) must have been equivalent to some table-defined chains C_{td} and D_{td} respectively, as discussed before. Then, since none of the previous calls to $ForceVal$ affects $val_i^+(D) = val_i^+(D_{td})$ and $val_i^+(C) = val_i^+(C_{td})$ (by Lemma 7 (iii)), the last action before the above two hold must be a tape accessing and entry setting action. Moreover, since $val_{l-2}^+(C_{td}) \notin G_{l-2}$ and C_{td} is table-defined (and $val_{l-2}^+(D_{td}) \notin G_{l-2}$ and D_{td} is table-defined) immediately hold after C (D, resp.) is enqueued, and then this action changed $val_{l-1}^+(C_{td})(= val_{l-1}^+(C))$ and $val_{l-1}^+(D_{td})(= val_{l-1}^+(D))$ from $\bot$ to non-empty values, this action must have been a defining action on either $G_{l-2}(val_{l-2}^+(C_{td}))$ or $G_{l-2}(val_{l-2}^+(D_{td}))$ (by Lemma 4 (ii)). However neither is possible: *wlog* assume the action to be $G_{l-2}(val_{l-2}^+(C_{td})) := \varphi_{l-2}(val_{l-2}^+(C_{td}))$, then after this action, the following holds (by $val_{l-1}^+(C_{td}) = val_{l-1}^+(D_{td}) \notin G_{l-1}$):

$$val_{l-3}^+(C_{td}) \oplus \varphi_{l-2}(val_{l-2}^+(C_{td})) \oplus k(C_{td})$$
$$= val_{l-3}^+(D_{td}) \oplus G_{l-2}(val_{l-2}^+(D_{td})) \oplus k(D_{td})$$

Suppose $C_{td} = (c_i, c_{i+1}, c_{i+2}, i)$ and $D_{td} = (d_j, d_{j+1}, d_{j+2}, j)$, then we have

$$\varphi_{l-2}(val^+_{l-2}(C_{td})) = val^+_{l-3}(C_{td}) \oplus c_i \oplus G_{i+1}(c_{i+1}) \oplus c_{i+2}$$
$$\oplus val^+_{l-3}(D_{td}) \oplus G_{l-2}(val^+_{l-2}(D_{td})) \oplus d_j \oplus G_{j+1}(d_{j+1}) \oplus d_{j+2}$$

which implies an occurrence of **BadHit**. Therefore the claim that $G_{l-1}(val^+_{l-1}(C))$ $(= G_{l-1}(val^+_{l-1}(C_{td})))$ can be defined only if $val^+_{l-2}(C) = val^+_{l-2}(D)$ holds.

Having excluded all possibilities we show $val^+_{l-1}(C) \notin G_{l-1}$ to hold when C is dequeued. The reasoning for $val^+_{l+1}(C) \notin G_{l+1}$ is similar by symmetry. Hence the subsequent call to *Adapt* will be safe; and by Lemma 7 (ii), the subsequent calls to *ForceVal* will be non-overwriting. □

A Little Honesty Goes a Long Way

The Two-Tier Model for Secure Multiparty Computation

Juan A. Garay[1], Ran Gelles[2,*], David S. Johnson[3],
Aggelos Kiayias[4,**], and Moti Yung[5]

[1] Yahoo Labs, USA
garay@yahoo-inc.com
[2] Princeton University, USA
rgelles@cs.princeton.edu
[3] Columbia University, USA
dstiflerj@gmail.com
[4] National and Kapodistrian University of Athens, Greece
aggelos@kiayias.com
[5] Google Inc. and Columbia University, USA
moti@cs.columbia.edu

Abstract. A fundamental result in secure multiparty computation (MPC) is that in order to achieve full security, it is necessary that a majority of the parties behave honestly. There are settings, however, where the condition of an honest majority might be overly restrictive, and there is a need to define and investigate other plausible adversarial models in order to circumvent the above impossibility.

To this end, we introduce the *two-tier model* for MPC, where some small subset of servers is guaranteed to be honest at the beginning of the computation (the *first tier*), while the corruption state of the other servers (the *second tier*) is unknown. The two-tier model naturally arises in various settings, such as for example when a service provider wishes to utilize a large pre-existing set of servers, while being able to trust only a small fraction of them.

The first tier is responsible for performing the secure computation while the second tier serves as a disguise: using novel anonymization techniques, servers in the first tier remain undetected to an adaptive adversary, preventing a targeted attack on these critical servers. Specifically, given n servers and assuming αn of them are corrupt at the onset (where $\alpha \in (0,1)$), we present an MPC protocol that can withstand an optimal amount of less than $(1-\alpha)n/2$ *additional* adaptive corruptions, provided the first tier is of size $\omega(\log n)$. This allows us to perform MPC in a fully secure manner even when the total number of corruptions exceeds $n/2$ across both tiers, thus evading the honest majority requirement.

---

* Work partially done while a student at University of California, Los Angeles.
** Research supported by ERC project CODAMODA.

Y. Dodis and J.B. Nielsen (Eds.): TCC 2015, Part I, LNCS 9014, pp. 134–158, 2015.

1   Introduction

A technically interesting and practically relevant configuration for performing secure multiparty computation (MPC) [GMW87] is the commodity-based *client-server* approach, in which the vast part of the computation is delegated from one or more clients to one or more servers [Bea97]. Indeed, these settings have plenty of practical value, as demonstrated for example by the implementation and deployment of an auction system in the Danish sugar-beet market [BCD+09], and, more generally, in the emerging secure cloud computing paradigm.

Security in MPC is commonly formulated via the following properties: privacy (parties learn only what they should learn); correctness (the honest parties' outputs are correct, despite the disruptive behavior of the corrupt parties); independence of inputs (parties' inputs are independent of other parties'); fairness (either all parties get their output, or none does); and guaranteed output delivery (all honest parties are guaranteed to obtain their outputs). Note that guaranteed output delivery also implies fairness. Achieving all these properties is called *full security*. A fundamental result in MPC with actively malicious participants is that in order to be able to compute any function with full security in the computational setting, an honest majority of the parties is necessary (and sufficient) [Cle86, GMW87, CFGN96].

Indeed, when half or more of the parties are corrupted, fairness might be compromised and guaranteed output delivery cannot be achieved [Cle86]. There are settings, however, where the honest majority requirement might be too costly or unattainable in practice (e.g., the resource-constrained service provider scenario elaborated on below). Thus, it is important to investigate models where it is possible to carry out any computation and obtain full security, even if the number of malicious participants is potentially *higher* than the number of honest parties. Clearly, in order to circumvent the above impossibility, the model in use must be relaxed.[1]

In this paper we put forth a new model for performing client-server-based MPC which we call the *two-tier model* for MPC. In this model, m servers are guaranteed to be properly functioning at the onset of the computation (such servers are identified by the set $\mathcal{P}_1$), while the remaining $n - m$ servers (the set $\mathcal{P}_2$) are of dubious trustworthiness. In addition, it is assumed that $m \ll n$. We call $\mathcal{P}_1$ the *first-tier* servers and $\mathcal{P}_2$ the *second-tier* servers. The objective is to run MPC withstanding a number of corruptions greater than the majority of the overall number of servers—in particular greater than $n/2$. We stress that the adversary may be *adaptive*, i.e., choose which servers to corrupt on the fly.

At first sight, it might seem unlikely that the two-tier setting could provide any advantage in circumventing the honest-majority requirement. Suppose αn servers are initially corrupted (and thus, $m < (1 - \alpha)n$). If we simply run the MPC protocol utilizing all the n servers indiscriminately then the number of additional corruptions (beyond the initial αn) the protocol would withstand is bounded by $\max\{0, (\frac{1}{2} - \alpha)n\}$. On the other hand, if we execute the MPC protocol

---

[1] See Section 1.3 for a comparison of several MPC variants and the security guarantees they offer when a majority of the parties are corrupted.

utilizing only the first-tier servers (and ignoring all the other servers), then the number of additional corruptions is bounded from above by $m/2$. Furthermore, any server allocation strategy that would utilize any arbitrary fraction of the second tier servers along with any fraction the first tier servers would be inferior to one of the above strategies.[2] We thus conclude that applying standard MPC in the two-tiered setting achieves at best tolerance of $\max\{m/2, (\frac{1}{2}-\alpha)n\}$ malicious participants (in addition to the initially αn corrupted parties), which equals $m/2$ for the interesting case of an initial dishonest majority ($\alpha \geq 1/2$).

However, had we known which second-tier servers are honest at the onset of the computation and which are corrupt, we could have (at least in principle) beaten the above bound by executing the MPC over those servers along with all the first-tier servers. The bound on the number of additional corruptions in this case is $(1 - \alpha)n/2$, which surpasses $m/2$ (recall that $m < (1 - \alpha)n$). In fact, if such a protocol was at all feasible, it would imply that the total number of dishonest parties would be $\alpha n + (1 - \alpha)n/2$, which is larger than $n/2$, for any $\alpha > 0$. Note that this is the best possible one could achieve given that there are $(1 - \alpha)n$ honest servers across the two tiers.

Somewhat surprisingly, we show how to construct a protocol that exactly withstands the above maximal number of corruptions, without knowing the status of the second-tier servers, under the assumption that the uncorrupted servers from the two tiers can be made indistinguishable in the view of the adversary. Effectively, this enables our protocol to take advantage of *all* the honest second-tier servers, even in settings where an (unknown) overwhelming majority of them are corrupted. Specifically, we show the following:

Theorem 1 (Informal). *Let $\alpha \in (0,1)$ and let $\mathcal{P} = \mathcal{P}_1 \cup \mathcal{P}_2$ be a set of n servers such that an unknown α-fraction of them are initially corrupted, yet the servers in $\mathcal{P}_1$ are guaranteed to be honest. Then, for any $\epsilon > 0$ there is a two-tier fully secure MPC protocol against any adversary adaptively corrupting up to $(1 - \epsilon) \cdot \frac{1-\alpha}{2} \cdot n$ additional servers, assuming $|\mathcal{P}_1| = \omega(\log n)$ and that the two tiers are indistinguishable to the adversary.*

1.1   How to Obtain Two-Tiers: The Corruption/Inspection Game

The above result is predicated on being able to establish a subset $\mathcal{P}_1$ of honest parties, and that $\mathcal{P}_1$ and $\mathcal{P}_2$ can be made indistinguishable. Theorem 1 says that a super-logarithmic number of $\mathcal{P}_1$ servers would be sufficient to harness the maximal resiliency of the system in terms of number of corrupted servers that can be tolerated. However, it seems challenging to obtain a set $\mathcal{P}_1$ where all its servers are honest, and still keep them hidden within the remaining servers. For example, one cannot form $\mathcal{P}_1$ simply by adding new "trusted" servers into a preexisting pool of

---

[2] To see this, note that if we utilize l servers from $\mathcal{P}_1$ and k servers from $\mathcal{P}_2$ randomly chosen for some values $l \leq m$ and $k \leq n-m$, then the expected number of additional corruptions is bounded by $\frac{l}{2} + \frac{k}{2} - (\alpha\frac{n}{n-m})k$, where $(\alpha\frac{n}{n-m})$ is the probability of picking a corrupt server when choosing a random server from $\mathcal{P}_2$. This function is maximized by taking $l = m$ and is clearly bounded by $\max\{m/2, (\frac{1}{2}-\alpha)n\}$ for any α.

1. $\mathcal{A}$ corrupts $\alpha \cdot n$ of the servers for a parameter $\alpha \in (0,1)$. Distinguishing corrupted from uncorrupted servers is undetectable at this stage (for the service provider S).
2. S inspects $\beta \cdot n$ servers and if they are corrupted it returns them to a clean state. β is the *inspection rate* of the service provider.
3. S opens the service by choosing a subset of the n servers to be tier-1 and the remaining servers tier-2; each server performs a designated protocol specific to its tier. Once the service is activated, $\mathcal{A}$ may adaptively corrupt an additional $\gamma \cdot n$ servers. γ is called the *adaptive corruption rate*.
4. We say that $\mathcal{A}$ wins the game, if it succeeds to corrupt at least half of the tier-1 servers.

Fig. 1. The corruption/inspection game

servers, as those would easily be identified by the adversary (whose existence in the pool of servers precedes the event of the introduction of the new servers). To address this, we now illustrate a realistic setting where two tiers naturally arise.

Assume that there is a single pool of machines out of which an α fraction is corrupted. Furthermore, assume we are allowed to *inspect* some of the servers, say, β-fraction of them, and *restore* them into a safe state if found corrupt. We assume here that corrupting a server means altering its operating program. Therefore, "inspecting" a server means comparing its loaded program with a clean version of the program, and "restoring" a server can be done by simply restoring the original program ("format and reinstall"). Once restored, the machine should be considered as any other honest machine; in particular, it may be corrupted again just like any other machine.

As a motivating example, consider a cloud service with several thousands of machines. As time goes by, some of the machines get hacked. On the other hand, the IT department performs regular maintenance on the servers, possibly restoring compromised machines. Since the IT department has limited resources, it cannot perform a daily maintenance on thousands of machines, but it does service a small fraction (where every day different machines are due for maintenance). Thus, at any given time when a client wishes to utilize a service using the above cloud, we can assume that the above (α, β)-corruption/inspection scenario holds.

We can now define the set $\mathcal{P}_1$ to consist of all the servers that were inspected and found *clean* (i.e., uncorrupted). Note that the restored servers cannot be in $\mathcal{P}_1$, as those would not be indistinguishable from the other honest servers, since the adversary may be aware that it is no longer controlling them. We let $\mathcal{P}_2$ denote all the remaining servers.

For a given rate of corruption α and rate of inspection β at the onset, the question now is what is the maximal possible fraction of active faults γ we can still withstand when running an MPC protocol. In Figure 1 we formalize the above as a "corruption/inspection game" between a service provider S and an adversary $\mathcal{A}$.

The problem posed by the above game is that for a fixed α, the service provider S wants to maximize γ while minimizing β. In the general case, one wants to maximize γ for any given α and β. Observe that, theoretically speaking,

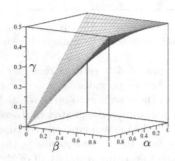 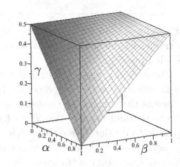

Fig. 2. The maximum adaptive corruption rate γ given α, β in the corruption/inspection game

the maximum value of γ that can be attained is $(1 - \alpha + \alpha\beta)/2$ (see Figure 2), which corresponds to (more than) half the honest servers among the ones originally clean plus (more than) half the ones that were reset to a clean state.

For the special case of $\beta \to 0$, the theoretical maximum is $\gamma = (1 - \alpha)/2$. Indeed, Theorem 1 implies that the service provider can examine a vanishing fraction of the servers and still run a successful MPC protocol amongst those inspected servers that were found clean, given that the adversary's corruption rate γ is below $(1 - \alpha)/2$. However, this still does not show how to obtain the maximal γ for *any* choice of α, β, when the service provider is unaware of the identity of the $(1 - \alpha)n$ honest servers.

Note that the above course of action for the service provider, where the first tier consists of only inspected servers which were uncorrupted, takes no advantage of the servers that were restored to a clean state. As mentioned above, those restored servers cannot be part of the first tier since such servers would be detected by the adversary and hence the required indistinguishability between tiers would be violated. However, by performing a more sophisticated selection of servers which also exploits a small random subset of the restored servers, we can improve on the number of adaptive corruptions obtained by Theorem 1, and maximize γ to its optimal value for any choice of α, β. Specifically:

Theorem 2 (Informal). *In the corruption/inspection game, for any constants $\alpha, \beta \in (0, 1)$ and any constant $\epsilon > 0$, there exists a two-tier fully secure MPC protocol tolerating adaptive corruption rate $\gamma \leq (1 - \epsilon)\frac{1-\alpha+\alpha\beta}{2}$.*

In other words, for any constants α, β, we achieve the maximal theoretical corruption rate of almost half the honest parties across the two tiers. This means our protocol tolerates a total corruption rate arbitrarily close to $\frac{1-\alpha+\alpha\beta}{2} + \alpha(1-\beta)$ across both tiers. Such a corruption rate is above $1/2$ for any $\alpha > 0$, surpassing the maximal corruption rate that can be tolerated in the plain model.

1.2 Our Techniques

The main idea behind our constructions is to have all the servers take part in the protocol, albeit in a way that only tier-1 servers perform the actual computation (as in [Bea97, DI05, ALZ13]), while the tier-2 servers' role is to keep the identities of the tier-1 servers hidden. This way we effectively employ MPC with "honest majority" among tier-1 servers and achieve full security, even though a majority of the overall number of parties may be corrupted.

Hiding the identities of tier-1 servers is done by utilizing a novel message delivery mechanism we describe below, which has the net effect that the adversarial view of the MPC protocol transcript is hidden in a traffic analysis-resistant way amidst a large set of irrelevant (but indistinguishable) messages. Performing MPC with a hidden, anonymous set of servers raises many interesting cryptographic questions; in particular:

> How can first-tier servers run an MPC protocol amongst themselves, while any specific server (whether first- or second-tier) remains oblivious to other servers' identities?

We solve this apparent contradiction by introducing the notion of *Anonymous yet Authentic Communication* (AAC), which allows a party to send a message to any other party in an anonymous *and* oblivious way. Despite being anonymous, the delivery is *authenticated*, that is, only the certified party will be able to send a valid message, and only the certified recipient will be able to correctly learn the message. We believe that such a primitive might be of independent interest.

In more detail, in an AAC message delivery the sender will reveal to the recipient only his "virtual" protocol identity, but not his real identity. At the same time, the sender will remain oblivious to the real identity of the recipient, which will only be specified by its protocol identity. We show how to implement AAC message delivery by utilizing an *anonymous broadcast* protocol [Cha88], which allows parties to broadcast messages without disclosing the real identity of the sender of each message, and composing it with a suitable authentication mechanism. Finally, by substituting point-to-point channels with AAC activations in an (adaptively secure) MPC protocol, we achieve our desired two-tier MPC functionality. The fundamental observation in the security proof is that the usage of the AAC message delivery mechanism effectively transforms any adaptive corruption strategy of the adversary against the MPC protocol into a *random* corruption strategy. Given this observation, we apply a probabilistic analysis using the tail bounds of the hypergeometric distribution to establish Theorem 1.

The proof of Theorem 2 is slightly more complex than Theorem 1's, as we now have to account for the fact that some information is "leaked": the adversary can distinguish the restored servers from the remaining ones, and can therefore "cluster" servers around those two disjoint subsets. Nevertheless, we are able to apply a similar analysis as in Theorem 1 by observing that any adaptive corruption strategy effectively amounts to a partially random corruption strategy: the adversary can control which specific cluster it corrupts, but within a specific cluster, which parties he corrupts are effectively chosen at random.

1.3   Related Work

As mentioned above, fully secure MPC in the cryptographic setting can only be achieved in the standard model against static and adaptive corruptions in the case of an honest majority [GMW87, CFGN96]; in the case of a corrupt majority, the weaker notion of *security with abort* can be achieved as in, e.g., [GMW87, BG89, GL91, CFGN96, GL02, CLOS02, KOS03].

Our two-tier model is inspired by recent work on "resource-based corruptions" [GJKY13], in which corrupting a party (server) carries a (computational) cost to the adversary. Different parties may have different corruption costs, and this information is hidden from the resource-bounded adversary (in [GJKY13], this is termed *hidden diversity*). Due to being uninformed of such costs, the adversary is then "forced" to waste his budget on servers whose corruption cost is high. For a fixed adversarial budget, robustness in the hidden-cost model greatly outweighs robustness in the setting in which all parties have the same corruption cost.

Full security with dishonest majority is also achievable in the case of *incentive-driven adversaries* [GKM+13], which considers a rational type of adversary who is assumed to receive a certain payoff/utility for breaking certain security properties, such as privacy and/or correctness. Intuitively, low payoffs allow for security against corrupt majorities. In contrast to [GJKY13] and [GKM+13], our adversary is the standard cryptographic adversary.

Maybe closest to our work is the work by Asharov, Lindell and Zarosim [ALZ13], which defines a model with a "reputation system." In this model, the service provider knows in advance the probability r_i that a party i remains honest throughout the protocol (in other words, the adversary corrupts each party i independently with probability $1 - r_i$). This allows the service provider to find a subset of the parties, over half of which is guaranteed to remain honest with high probability. In contrast, in our model the service provider only knows that some specific parties are honest *at the onset of the computation*, but has no control or knowledge on whether they remain honest, nor is he aware of the "reputation" of the other parties. In addition, our model also tolerates adaptive corruptions, while the adversary in [ALZ13] only statically corrupts parties according to the reputation system. In more detail, our technique has the property that the best adversarial corruption strategy becomes the random one, that is, we *force* the adversary to corrupt parties at random. Therefore, although our adversary is fully adaptive and is only restricted in the number of parties to corrupt, he is effectively restricted to a uniform corruption pattern. Such a strategy induces a reputation vector which is in the feasible region for secure multiparty computation according to [ALZ13]; yet, we achieve a fully secure MPC without restricting the adversary in advance to a specific corruption pattern.

Table 1 summarizes the current state of the art of MPC with dishonest majority.

Finally, our anonymous message transmission notion, AAC, is related to (but distinct from) the notion of "secret handshakes" [BDS+03, CJT04]. Similarly to this notion, we work in a setting where a certain special action takes place between two parties if and only if they are both members of a hidden subset. If it happens that one party is not a member of the hidden subset, then it cannot infer the membership status of the other party. Our work is, to the best

Table 1. Circumventing the impossibility of full security [Cle86]: a comparison

Paper	"Standard" Adversary	Adaptive Corruption	Dishonest Majority	Security	GOD[a]
[GMW87], [BG89], [GL91], [CFGN96], [GL02], [CLOS02], [KOS03], etc.	✓	✓ (some)	✓	with abort	—
[GKM⁺13]	—[b]	✓	✓	full	✓
[GJKY13]	—[c]	✓	—[d]	full	✓
[ALZ13]	✓	—	✓	full	✓
(this paper)	✓	✓	✓	full	✓

[a] Guaranteed Output Delivery
[b] Incentive-driven adversary; restricted to some utility functions
[c] Resource-based adversary; with different corruption cost per party (unknown to the adversary)
[d] The adversary has enough resources to corrupt a majority of the parties, yet the parties' hidden corruption costs prevent the adversary from doing so

of our knowledge, the first application of such "covert subset" techniques in a setting where anonymity is not the prime objective. In fact, our work shows how anonymity can be effectively used to increase the resiliency (specifically, the number of tolerated corruptions) of MPC, and makes yet another demonstration of the power of such tools [FGMO05, IKOS06].

1.4 Organization of the Paper

The balance of the paper is organized as follows. Notation, definitions, and the two-tier (TT) model for MPC are presented in Section 2. The TT MPC protocol, as well as the AAC (Anonymous yet Authentication Communication) notion and construction it relies on, are presented in Section 3. Finally, the analysis yielding the selection of the two tiers allowing to tolerate the maximal corruption rate appears in Section 4. Some complementary material, including auxiliary definitions and constructions, is presented in the appendix.

2   Model and Definitions

2.1   Notation and Preliminaries

We let κ be the security parameter, and assume that any function, set size or running time implicitly depends on this parameter (especially when we write negl to describe a negligible function in κ—i.e., $\mathsf{negl} < 1/\mathsf{poly}(\kappa)$ for large enough κ). For any ε, we say that two distribution ensembles $\{X_\kappa\}_{\kappa\in\mathbb{N}}$, $\{Y_\kappa\}_{\kappa\in\mathbb{N}}$ are ε-*indistinguishable*, denoted $\{X_\kappa\} \approx_\epsilon \{Y_\kappa\}$, if for any probabilistic polynomial-time (PPT) algorithm C, for large enough κ,

$$|\Pr[C(1^\kappa, X_\kappa) = 1] - \Pr[C(1^\kappa, Y_\kappa) = 1]| < \epsilon + \mathsf{negl}(\kappa).$$

We say that X and Y are *computationally indistinguishable*, denoted $\{X_\kappa\} \approx \{Y_\kappa\}$, if they are ε-indistinguishable with $\varepsilon = 0$. We now proceed to describe some of the cryptographic notions and building blocks that we use throughout the paper.

Security of Multiparty Protocols. For defining security of a multiparty protocol for computing an n-ary function f, we follow the standard simulation-based approach [GMW87, Can00], in which the protocol execution is compared to an ideal protocol where the parties send their inputs to a trusted party who computes f and returns the designated output to each party. Commonly, the trusted-party activity for computing the function f is captured via a so-called ideal functionality $\mathcal{F}_f$.

Let $\mathsf{EXEC}_{\pi,\mathcal{A},\mathcal{Z}}(\kappa, \boldsymbol{x})$ denote an execution of the n-party protocol π with an adversary $\mathcal{A}$ and an environment $\mathcal{Z}$, with $\boldsymbol{x} = x_1, \ldots, x_n$ being the vector of inputs of the parties. In the same manner, define $\mathsf{IDEAL}_{\mathcal{F}_f,\mathcal{S},\mathcal{Z}}(\kappa, \boldsymbol{x})$ to be an execution in the ideal-model, where the ideal functionality is described by $\mathcal{F}_f$, $\mathcal{S}$ is the adversary (commonly known as *simulator*), $\mathcal{Z}$ is the environment, and $\boldsymbol{x}$ defined as above. We say that π *securely realizes* the functionality $\mathcal{F}_f$ if for every polynomial-time real-model adversary $\mathcal{A}$ and any PPT environment $\mathcal{Z}$, there is a polynomial time ideal-model simulator $\mathcal{S}$ such that for any input vector $\boldsymbol{x}$,

$$\{\mathsf{EXEC}_{\pi,\mathcal{A},\mathcal{Z}}(\kappa, \boldsymbol{x})\}_{\kappa \in \mathbb{N}} \approx_\varepsilon \{\mathsf{IDEAL}_{\mathcal{F}_f,\mathcal{S},\mathcal{Z}}(\kappa, \boldsymbol{x})\}_{\kappa \in \mathbb{N}}$$

where ε is a negligible function in the security parameter κ. Throughout this paper, we assume n and κ are polynomially related.

We refer the reader to Appendix A for additional standard definitions and other building blocks we use, and move on to describe the basics of the two-tier (TT) model for MPC.

2.2   The Two-Tier (TT) Model for Secure Multiparty Computation

There are n parties (servers) $\mathcal{P} = \{P_1, P_2, \ldots, P_n\}$, each of them identified by a name P_i, referred to as its *real* identity, and a "virtual" name from $\mathcal{P}^* = \{P_1^*, \ldots, P_n^*\}$, referred to as its *protocol pseudonym*, which identifies them as participants in the MPC protocol; all are probabilistic polynomial-time (PPT) machines. We assume a bijection $\nu : \mathcal{P} \to \mathcal{P}^*$ which maps a real identity P_i to its protocol pseudonym $\nu(P_i) \in \mathcal{P}^*$. The parties are assumed to know both their real name and pseudonym, but they do not know the specific ν.

We are interested in secure function evaluation [GMW87] performed by the servers in $\mathcal{P}$. The inputs to the computation are assumed to be held by a set of clients, who are assumed to be outside the set $\mathcal{P}$. Each such client has an input x_i, and the goal is to compute a joint function f of the clients' inputs. Servers do not have an input of their own and they expect no output from the computation—their sole purpose is to carry out the computation and deliver the output back to clients.

As in the standard MPC setting, parties are connected by pair-wise authentic and reliable channels, which are identified by the *real* names of the two connected parties/servers. Accessing this communication channel does not mandate the disclosure of the protocol pseudonyms of the communicating parties. We assume a synchronous communication model where a party can send a message to multiple parties at the same time [Can00].

The set of servers is divided into two disjoint sets $\mathcal{P} = \mathcal{P}_1 \cup \mathcal{P}_2$—the first and second tier servers respectively. Our communication model assumes that the two tiers are indistinguishable at the communication (real name) layer. As mentioned in Section 1, the two tiers are subject to different adversarial capabilities with respect to corruption. Among the servers in $\mathcal{P}_2$, an unlimited number t_s of *static* corruptions are allowed. The servers in $\mathcal{P}_1$, on the other hand, are assumed to be uncorrupted at the onset of the computation. During the course of the computation, *all* servers are subject to adaptive corruptions; we denote the number of such corruptions by t_a. We assume a threshold corruption model, in which the adversary is restricted to corrupting at most t ($= t_s + t_a$) of the parties overall. At each step, the adversary may choose a party $P_i \in \mathcal{P}$ and corrupt it, as long as the total number of corrupted parties does not exceed his "budget" t. Once P_i gets corrupted, the adversary learns its internal state, including its tier level and protocol pseudonym $\nu(P_i)$.

We assume a standard public-key infrastructure (PKI) setup, in which each party P_i, $i \in \{1, \ldots, n\}$, is given *two* pairs of public/secret keys (pk_i, sk_i), (pk_j^*, sk_j^*) corresponding to its real name and protocol pseudonym, as well as the public keys of all other users (in a certified way) in the form $\{(pk_k, P_k)\}_{k \neq i}$ and $\{(pk_k^*, P_k^*)\}_{k \neq j}$. Note that the correspondence between names and protocol pseudonyms is not revealed. More formally, we express this as the parties having access to two instances of an ideal PKI functionality, denoted by $\mathcal{F}_{\mathsf{PKI}}^{\mathcal{P}}$ and $\mathcal{F}_{\mathsf{PKI}}^{\mathcal{P}^*}$ (see [Can05] for definition of an ideal PKI functionality). If $\nu : \mathcal{P} \to \mathcal{P}^*$ maps between real and protocol identities, we shorthand these two functionalities by $\mathcal{F}_{\mathsf{PKI}}^{\nu} = (\mathcal{F}_{\mathsf{PKI}}^{\mathcal{P}}, \mathcal{F}_{\mathsf{PKI}}^{\nu(\mathcal{P})=\mathcal{P}^*})$.

3 Secure Multiparty Computation in the Two-Tier Model

In this section we present our MPC protocol in the two-tier model, obtaining Theorem 1. As mentioned above, exploiting the indistinguishability between the two tiers requires new cryptographic tools that enable anonymous communication among servers. To this end, our construction assumes a communication capability which allows parties to communicate messages in an authenticated way but without compromising their real identity, which we term *Anonymous yet Authentic Communication* (AAC). Specifically, AAC allows entities to communicate with each other in an authenticated fashion at the protocol

---

Functionality $\mathcal{F}_{\mathsf{AAC}}^{\nu}$

$\mathcal{F}_{\mathsf{AAC}}^{\nu}$ is parameterized by a security parameter κ and a set of n parties with real names $\mathcal{P}$, protocol pseudonyms $\mathcal{P}^*$, and a bijection $\nu : \mathcal{P} \to \mathcal{P}^*$; it assumes a message space $\mathcal{M} = \mathcal{M}(\kappa)$ and proceeds as follows, running with parties $P_1, ..., P_n \in \mathcal{P}$ and an adversary $\mathcal{S}$:

- Upon receiving $(\mathsf{Send}, sid, P_i, \mu, P_j^*)$ from P_i, record this tuple. Once a message is recorded from all honest parties send to the adversary $\mathcal{S}$ the sequence of tuples $(\mathsf{SendLeak}, sid, \nu(P_i), \mu, P_j^*)$ lexicographically ordered.
- Upon receiving $(\mathsf{Deliver}, sid)$ from $\mathcal{S}$, check that a Send message was recorded on behalf of all senders, and if so, for any recorded tuple of the form $(\mathsf{Send}, sid, P_i, \mu, P_j^*)$, deliver $(\mathsf{Sent}, sid, \nu(P_i), \mu)$ to party $\nu^{-1}(P_j^*)$.
- Upon receiving (Abort, sid, A) from $\mathcal{S}$ check that A is a non-empty subset of corrupted parties and forward this message to all honest parties.
- Upon receiving $(\mathsf{Corrupt}, P_i)$ from $\mathcal{S}$ mark P_i as corrupted and return $\nu(P_i)$ to $\mathcal{S}$.

If P_i is corrupted then $\mathcal{S}$ is allowed to submit $(\mathsf{Send}, sid, P_i, \mu, P_j^*)$ on behalf of P_i.

Fig. 3. Ideal functionality for anonymous yet authentic communication (AAC)

(application) layer, yet anonymously at the network (real-name) layer; the latter property comes from the fact that the correspondence between real and protocol names is hidden from the adversary and the functionality does not reveal it. We now define the ideal functionality of such a communication channel, and construct a protocol that securely realizes it.

3.1   The $\mathcal{F}_{\mathsf{AAC}}^{\nu}$ Ideal Functionality

In the ideal world, the sender delivers to the functionality the message μ along with the protocol pseudonym of the intended receiver. The adversary is notified of this event and receives the pseudonyms of the two communicating entities. However, the real names of the two entities remain hidden. The functionality is parameterized by a mapping ν that gives the correspondence between names and pseudonyms. When the adversary instructs the functionality to deliver the message, the functionality recovers the real identity of the receiving entity and writes the message on its network tape along with the protocol pseudonym of the sender. We formally describe the functionality in Figure 3.

We now show how this functionality can be securely realized assuming an *anonymous broadcast channel* functionality (cf. [Cha88]) tolerating an arbitrary number of corrupted parties[3]. Recall that such functionality can be thought of as a bulletin board on which any party can post messages without revealing its

---

[3] We remark that performing an AAC message delivery means that in case the AAC protocol terminates with abort, the protocol is repeated with a subset of parties currently not marked as corrupt.

Functionality $\mathcal{F}_{\mathsf{ABC}}$

The functionality assumes a message space $\mathcal{M} = \mathcal{M}(\kappa)$ with κ being the security parameter, and works as follows, running with n parties $P_1, \ldots, P_n$ and an adversary $\mathcal{S}$:

- Upon receiving (AnonBcast, sid, P_i, μ) from P_i record this tuple. Once a message is recorded for all honest parties send to the adversary $\mathcal{S}$ the message (AnonBcastLeak, sid, M) where M is the (lexicographically ordered) set of messages μ from all recorded tuples of the form (AnonBcast, sid, P_i, μ).
- Upon receiving (Deliver, sid) from $\mathcal{S}$, ignore further AnonBcast messages, and deliver (AnonBcastDeliver, sid, M') to all parties $P_1, \ldots, P_n$ where M' is the set of messages μ (lexicographically ordered) from all recorded tuples of the form (AnonBcast, sid, P_i, μ).
- Upon receiving (Abort, sid, A) from $\mathcal{S}$ check that A is a non-empty subset of corrupted parties and forward this message to all honest parties.
- Upon receiving (Corrupt, P_i) from $\mathcal{S}$ mark P_i as corrupted and return the recorded (AnonBcast, sid, P_i, μ) to $\mathcal{S}$.

If P_i is corrupted then $\mathcal{S}$ is allowed to submit (or substitute existing) (AnonBcast, sid, P_i, μ) messages on behalf of P_i.

Fig. 4. Ideal anonymous broadcast channel functionality (ABC)

identity. This is modeled as the ideal functionality $\mathcal{F}_{\mathsf{ABC}}$ in Figure 4, which we later on show how to implement assuming a PKI setup.[4]

Using the ideal functionality $\mathcal{F}_{\mathsf{ABC}}$ and the PKI setting described in Section 2, we now describe a secure realization of $\mathcal{F}_{\mathsf{AAC}}^{\nu}$. The protocol makes use of an existentially unforgeable digital signature scheme (see Appendix A). The implementation is rather straightforward: the sender uses the (protocol layer) PKI to sign the message and anonymously broadcast it. Any party receiving the message checks whether it is the intended protocol-layer recipient, and if so, it verifies the signature and decrypts the message. This approach prevents impersonation at the protocol layer while still hiding the correspondence between protocol names and real names.

The AAC protocol is described in Figure 5 and operates in the $(\mathcal{F}_{\mathsf{PKI}}^{\nu}, \mathcal{F}_{\mathsf{ABC}})$-hybrid world.

Theorem 3. *Let $n \in \mathbb{N}$, parties $\mathcal{P}$ with protocol pseudonyms $\mathcal{P}^*$ and a bijection $\nu : \mathcal{P} \to \mathcal{P}^*$. The AAC protocol from Figure 5 securely realizes $\mathcal{F}_{\mathsf{AAC}}^{\nu}$ against an adaptive adversary corrupting $t < n$ parties in the $(\mathcal{F}_{\mathsf{PKI}}^{\nu}, \mathcal{F}_{\mathsf{ABC}})$-hybrid model.*

Proof (sketch). Consider a PPT adversary $\mathcal{A}$ and a PPT environment $\mathcal{Z}$. We use the notation $\mathcal{M}$ to denote the space of all messages. We construct a simulator $\mathcal{S}$ so that for every vector of inputs $\boldsymbol{x} = x_1, \ldots, x_n$ with $x_i \in \{(\mathsf{Send}, sid, P_i, \mu, P_j^*) \mid j \in \{1, \ldots, n\}, \mu \in \mathcal{M}\}$, for $i = 1, \ldots, n$, the following holds:

$$\mathsf{EXEC}_{\mathsf{AAC}, \mathcal{A}, \mathcal{Z}}^{\mathcal{F}_{\mathsf{PKI}}^{\nu}, \mathcal{F}_{\mathsf{ABC}}}(\kappa, \boldsymbol{x}) \approx \mathsf{IDEAL}_{\mathcal{F}_{\mathsf{AAC}}^{\nu}, \mathcal{S}, \mathcal{Z}}(\kappa, \boldsymbol{x})$$

---

[4] We note that (most) security proofs in Canetti's synchronous model [Can00] carry over to the *Universal Composability* framework [Can05], given that certain functionalities are available to the protocol [KMTZ13].

Protocol AAC

Setup: Let κ be the security parameter, and $(\mathsf{GenS}, \mathsf{Sig}, \mathsf{Ver})$ be an existentially unforgeable signature scheme. The PKI setup delivers real-layer keys (pk_i, sk_i) and protocol-layer keys (pk_i^*, sk_i^*), as described in Section 2.2. Each pair of keys is generated using $\mathsf{GenKey}(1^\kappa) = (\mathsf{GenE}(1^\kappa), \mathsf{GenS}(1^\kappa))$. We assume real names $\mathcal{P}$ and protocol pseudonyms $\mathcal{P}^*$ are known to all entities (but not ν).

Send message: On input $(\mathsf{Send}, sid, P_i, \mu, P_j^*)$, party P_i sends $(\mathsf{AnonBcast}, sid, P_i, (P_i^*, P_j^*, \mu, \sigma))$ to $\mathcal{F}_{\mathsf{ABC}}$ where $\sigma \leftarrow \mathsf{Sig}_{sk_i^*}(P_j^*, \mu, sid)$.

Receive message: P_j, $1 \leq j \leq n$, upon receiving $(\mathsf{AnonBcast}, sid, M)$ from $\mathcal{F}_{\mathsf{ABC}}$, if it holds that $P_j^* = B$ for some $(A, B, \mu, \sigma) \in M$ (i.e., P_j is the intended protocol level receiver of that message), P_j checks $\mathsf{Ver}_{pk_i^*}(P_j^*, \mu, sid, \sigma)$, where i is such that $A = P_i^*$, and provided Ver returns 1, it records (P_i^*, μ). The action terminates by returning all recorded tuples.

Abort: If $\mathcal{F}_{\mathsf{ABC}}$ returns (Abort, sid, A) then terminate and return A.

Fig. 5. A protocol realizing $\mathcal{F}_{\mathsf{AAC}}^\nu$

where $\nu : \mathcal{P} \to \mathcal{P}^*$ is a random bijection. As a setup step, $\mathcal{S}$ generates keys for all the identities in $\mathcal{P}^*$, and gives $\mathcal{A}$ all the public keys. The simulator maintains a list M which is empty at initialization.

The simulation is straightforward: when $\mathcal{S}$ receives $(\mathsf{SendLeak}, sid, P_i^*, \mu, P_j^*)$ from $\mathcal{F}_{\mathsf{AAC}}^\nu$ it generates a signature $\sigma = \mathsf{Sig}_{sk_i^*}(P_j^*, \mu, sid)$ and updates the list $M = M \cup (P_i^*, P_j^*, \mu, \sigma)$. Once $\mathcal{S}$ processes the $\mathsf{SendLeak}$ message for all the honest parties, it sends $(\mathsf{AnonBcastLeak}, sid, M)$ over to $\mathcal{A}$. If $\mathcal{A}$ issues an Abort message, $\mathcal{S}$ forwards the abort to $\mathcal{F}_{\mathsf{AAC}}^\nu$. Otherwise, $\mathcal{A}$ issues a $\mathsf{Deliver}$ message which is also forwarded to $\mathcal{F}_{\mathsf{AAC}}^\nu$.

When the adversary requests to corrupt some party P_i, the simulator forwards the request to $\mathcal{F}_{\mathsf{AAC}}^\nu$ and learns P_i's protocol pseudonym $\nu(P_i)$. Next, it forms the inner state of P_i accordingly (that contains the signing key of pseudonym $\nu(P_i)$), and delivers this information to $\mathcal{A}$. It is clear that the honest parties' output is identically distributed between the real and ideal executions, with the exception of the event that the adversary $\mathcal{A}$ (or the environment $\mathcal{Z}$) forges a signature on behalf of an honest party. In this case the simulator will fail, but this will happen with negligible probability based on the security of the underlying digital signature scheme. $\qquad \square$

There are several possible ways to realize $\mathcal{F}_{\mathsf{ABC}}$ so that up to $t < n$ corruptions can be tolerated assuming our setup configuration (PKI). We consider some alternatives in Appendix B.

3.2 Pseudonymity and Random Corruptions

With foresight, the approach we will follow is to replace every communication in an (adaptively secure) MPC protocol for the standard setting with an invocation to $\mathcal{F}_{\mathsf{AAC}}^{\nu}$. We now show that if a protocol π that operates at the pseudonym layer is unaware of the real/protocol name correspondence, then the approach does not reveal any information about the mapping ν. (In addition, it is straightforward to verify that the modified protocol would remain correct, i.e., it produces the same outputs as π.)

Let π be a protocol defined over the "pseudonym" protocol layer, i.e., running with parties $P_1^*, \ldots, P_n^*$. Further, π operates in (synchronous) communication rounds. Normally, in an execution of π with an adversary $\mathcal{A}$, an environment $\mathcal{Z}$ and parties $P_1^*, \ldots, P_n^*$, $\mathcal{A}$ is capable of issuing $(\mathsf{Corrupt}, P_i^*)$ messages when it wants to corrupt party P_i^*.[5] We consider a stronger notion of execution, denoted rcEXEC, in which the adversary is allowed to issue $(\mathsf{Corrupt})$ requests to a corruption oracle, upon which a *randomly* chosen honest party gets corrupted. We call this an execution with *random* corruptions. Note that $\mathsf{rcEXEC}_{\pi,\mathcal{A},\mathcal{Z}}$ is the ensemble of views over the adversary's (and environment's) coin tosses, the parties' coin tosses *and* the randomness of the corruption oracle.

Now consider the setting where the communication is handled by a lower "physical" layer where each party has a physical (real) identity $P_1, \ldots, P_n$ and there is a mapping $\nu : \mathcal{P} \to \mathcal{P}^*$ that corresponds protocol identities to communication identities (real names). Given any protocol π that operates in rounds, we can easily obtain a protocol $\tilde{\pi}^{\mathcal{F}_{\mathsf{AAC}}^{\nu}}$ that runs with parties $P_1, \ldots, P_n$ and whenever π, acting on behalf of P_i^*, wishes to send a message μ to party P_j^* the $\tilde{\pi}$ protocol delivers $(\mathsf{Send}, sid, \nu^{-1}(P_i^*), \mu, P_j^*)$ to $\mathcal{F}_{\mathsf{AAC}}^{\nu}$. Thus, each communication round of π is equivalent to a single instantiation of $\mathcal{F}_{\mathsf{AAC}}^{\nu}$.

Next, we show that $\tilde{\pi}^{\mathcal{F}_{\mathsf{AAC}}^{\nu}}$ with a randomly chosen ν is simulatable in the random-corruptions setting. For ease of notation, we identify a bijection $\nu : \mathcal{P} \to \mathcal{P}^*$ with a permutation on n elements.

Lemma 4. *Let π and $\tilde{\pi}^{\mathcal{F}_{\mathsf{AAC}}^{\nu}}$ be as above. For any PPT adversary $\mathcal{A}$ and environment $\mathcal{Z}$, and for any input vector $\boldsymbol{x}$, there exist a PPT simulator $\mathcal{S}$ such that*

$$\left\{ \mathsf{EXEC}_{\tilde{\pi},\mathcal{A},\mathcal{Z}}^{\mathcal{F}_{\mathsf{AAC}}^{\nu}}(\kappa, \boldsymbol{x}^{\nu}) \right\}_{\nu \in_R \mathrm{Perm}(n)} \approx \mathsf{rcEXEC}_{\pi,\mathcal{S},\mathcal{Z}}(\kappa, \boldsymbol{x})$$

where $\mathrm{Perm}(n)$ is the set of all the possible permutations on n elements.

Proof. Consider the following simulator. At first it fixes a randomness tape for $\mathcal{A}$ and follows the computation, replacing each "communication round" with a $\mathcal{F}_{\mathsf{AAC}}^{\nu}$ simulation. Namely, after each round of communication, $\mathcal{S}$ gathers all the messages sent in this round, and provides the adversary with a lexicographical list whose entries are of the form $(\mathsf{SendLeak}, sid, P_i^*, \mu, P_j^*)$ matching the case where P_i^* sent P_j^* the message μ. Note that we assume π runs in rounds, so each party sends exactly one message at each communication round.

---

[5] We emphasize that since π exists only in the pseudonym layer, the parties' identifiers are $\mathcal{P}^*$, and the adversary corrupts by specifying a certain P_i^*. However, when running in our TT model setup, the identities of the parties are $\mathcal{P}$, and the adversary corrupts a party by specifying a certain P_i.

When $\mathcal{A}$ issues (corrupt, P_i), the simulator issues (corrupt) and as a result, P_j^* gets corrupted, for a random j (out of all the parties that are still honest). The simulator sets $\nu(i) = j$ and simulates the inner state of P_j^* so it would correspond to the real identity P_i in a straightforward way.

Note that at the end of the simulation, the simulator has defined a partial mapping ν. The output of this simulation is exactly the same as the output of any instance of the left-hand side experiment, running with the same adversary (set to the same randomness tape), for any mapping ν' that agrees with ν on the identities of all corrupted parties. It easily follows that the two ensembles are identically distributed.   □

3.3   The Two-Tier MPC Protocol

Recall our setting in which n parties (servers) with real names $\mathcal{P} = \{P_1, P_2, \ldots, P_n\}$, are split into two tiers $\mathcal{P} = \mathcal{P}_1 \cup \mathcal{P}_2$, and where the computation is effectively carried out only by servers in the first tier, who then distribute the output to the clients as needed (see also [DI05, ALZ13]). Let $|\mathcal{P}_1| = m$ and $|\mathcal{P}_2| = n - m$. In addition, we assume there are $c \in \mathbb{N}$ clients, each holding a private input x_i; let $\boldsymbol{x} = x_1, \ldots, x_c$. The clients wish to compute some function f of their inputs, described as the c-party functionality $\mathcal{F}_f(\boldsymbol{x})$.

We now describe the two-tier MPC protocol performed by the servers, *assuming they have already (verifiably) secret-shared*[6] *the clients' inputs.* This operation is in fact easy to achieve using standard techniques and without the need for AAC communication. For example, one may assume that the i-th client computes an $(m, \lceil m/2 \rceil - 1)$-verifiable secret sharing of x_i using the adaptively secure VSS scheme of Abe and Fehr [AF04]. Then, the client broadcasts a signed copy of the j-th share encrypted with P_j^*'s public key. (Recall that protocol identities, and in particular those corresponding to servers in $\mathcal{P}_1$, are public.) As a result of the computation, the servers obtain a *share* of $\mathcal{F}_f(\boldsymbol{x})$'s output—we denote this modified functionality by $\mathcal{F}_f^{\mathsf{vss}}(\boldsymbol{x})$; the shares are then sent to the clients.[7]

We now explain how the servers carry out the actual computation of $\mathcal{F}_f(\boldsymbol{x})$. The two-tiered MPC protocol operates in the $(\mathcal{F}_{\mathsf{PKI}}^\nu, \mathcal{F}_{\mathsf{AAC}}^\nu)$-hybrid world and is presented in Figure 6.

It is immediate that the protocol in Figure 6 securely realizes $\mathcal{F}_f^{\mathsf{vss}}$ as long as the adversary does not corrupt a majority of the tier-1 servers. Formally,

Proposition 5. *Let $n, m, c \in \mathbb{N}$. For any given c-ary functionality $\mathcal{F}_f$ and for any bijection $\nu : \mathcal{P} \to \mathcal{P}^*$, the protocol of Figure 6 operating in the $(\mathcal{F}_{\mathsf{PKI}}^\nu, \mathcal{F}_{\mathsf{AAC}}^\nu)$-hybrid world securely realizes $\mathcal{F}_f^{\mathsf{vss}}$ conditioned on the event that the adversary corrupts at most $\lceil m/2 \rceil - 1$ servers.*

---

[6] Refer to Appendix A for the definition of VSS.

[7] We note that in case the identities of first-tier servers need to remain hidden (say, for the continuation of the service in a forthcoming MPC execution), the output delivery should be anonymous as well. This can be achieved, for example, by extending the AAC mechanism to include both servers and clients at the protocol layer.

MPC in the Two-Tier Model

Assume n parties with real names $\mathcal{P} = \{P_1, P_2, \ldots, P_n\}$, split into two disjoint subsets $\mathcal{P} = \mathcal{P}_1 \cup \mathcal{P}_2$, where $|\mathcal{P}_1| = m$. Parameters n and m are public. Furthermore, assume a c-ary functionality $\mathcal{F}_f(\boldsymbol{x})$ to be securely computed on inputs $\boldsymbol{x} = x_1 \ldots x_c$, where each x_i is $(m, \lceil m/2 \rceil - 1)$-VSS'd in $\mathcal{P}_1$.

Trusted setup. Public and secret keys, as well as protocol identities are handed to each party by $\mathcal{F}_{\mathsf{PKI}}^\nu$, as described in Section 2.2.

Computation phase.

- Let $\mathcal{F}_f^{\mathsf{vss}}$ be the m-party functionality that performs the same task as the c-party functionality $\mathcal{F}_f$, assuming that each of the m parties holds a share of each of the c inputs.
  The output of $\mathcal{F}_f^{\mathsf{vss}}$ is a $(m, \lceil m/2 \rceil - 1)$-VSS share of each of the c outputs of $\mathcal{F}_f$.
- The parties in $\mathcal{P}_1$ adaptively securely compute $\mathcal{F}_f^{\mathsf{vss}}$ amongst themselves (for example, via [CFGN96]). During the execution, messages between any two parties are sent invoking $\mathcal{F}_{\mathsf{AAC}}^\nu$ (Fig. 3).

Fig. 6. Computation phase of the TT MPC protocol

Next, we prove a combinatorial lemma, showing that for any $\epsilon > 0$ and t_{s} initial static corruptions among the tier-2 servers, if an adversary adaptively corrupts up to $(1 - \epsilon)\frac{n - t_{\mathsf{s}}}{2}$ parties without knowing the two-tier partition, then the probability of corrupting a majority of $\mathcal{P}_1$ servers is negligible in $|\mathcal{P}_1|$.

Lemma 6. *Assume n parties $\mathcal{P}$, m of which are in $\mathcal{P}_1$ and $t_{\mathsf{s}} \leq n - m$ of $\mathcal{P}_2$ are initially corrupted. Assume that the adversary is bounded to adaptively corrupting t_{a} parties with $t_{\mathsf{a}} \leq (1 - \epsilon)\frac{n - t_{\mathsf{s}}}{2}$, for some constant $\epsilon > 0$, where $\epsilon m \geq 4$. Furthermore, assume that by corrupting a party $P_i \in \mathcal{P}$, the adversary learns its tier level (but not the tier level of other parties). Then, the probability that adversary corrupts at least $m/2$ parties from $\mathcal{P}_1$ is at most $2^{-\Omega(m)}$.*

Proof. Let K be the random variable describing the number of $\mathcal{P}_1$ servers that were corrupted, assuming the adversary corrupts additional $t_{\mathsf{a}} = (1 - \epsilon)\frac{n - t_{\mathsf{s}}}{2}$ servers (i.e., on top of statically-corrupting t_{s} parties). K is distributed according to the hypergeometric distribution (see Appendix C) with parameters $(n - t_{\mathsf{s}}, m, t_{\mathsf{a}})$, and we denote $K \sim \mathsf{HypGeo}_{n - t_{\mathsf{s}}, m, t_{\mathsf{a}}}$. We get that

$$\mathbb{E}[K] = (1 - \epsilon)\frac{n - t_{\mathsf{s}}}{2} \cdot \frac{m}{n - t_{\mathsf{s}}} = (1 - \epsilon)\frac{m}{2}.$$

Assuming that m is odd (the case of even m is similar) we can use the tail bound of Lemma 10 to bound the probability that more than $m/2$ servers get corrupted.

$$\Pr[K > m/2] = \Pr[K - \mathbb{E}[K] > \epsilon m/2]$$
$$< e^{-2\frac{n-t_s+2}{4(m+1)(n-t_s-m+1)}(\epsilon^2 m^2 - 1)}$$
$$\leq 2^{-\Omega(\frac{n-t_s}{n-t_s-m+1}m)} = 2^{-\Omega(m)},$$

since in our case $\alpha_{n,m,t}$ of Lemma 10 satisfies $\alpha_{n,m,t} \geq \frac{n+2}{(m+1)(n-m+1)}$, and assuming $\epsilon m \geq 4$. $\square$

Theorem 1. *Assume $m = \omega(\log n)$ and $\epsilon > 0$. For any given c-ary functionality $\mathcal{F}_f$, there exists a two-tier MPC protocol in the $(\mathcal{F}_{\mathsf{PKI}}^{\nu}, \mathcal{F}_{\mathsf{AAC}}^{\nu})$-hybrid world that securely realizes $\mathcal{F}_f$ against any PPT adversary with $t_a \leq (1-\epsilon)\frac{n-t_s}{2}$ and $t_s \leq n - m$.*

Proof. Observe that (i) the MPC protocol is secure as long as a majority of $\mathcal{P}_1$ servers are honest (Proposition 5); (ii) given that the adversary learns the protocol pseudonym and tier-level of a party only when this party is corrupt, when restricted to $(1-\epsilon)\frac{n-t_s}{2}$ corruptions, it has only an exponentially small probability (in m) to corrupt a majority of $\mathcal{P}_1$ (Lemma 6); (iii) by Lemma 4 an adaptive adversary learns only negligible information about ν (for uncorrupt parties), that is, it does not have an advantage in learning the protocol identity (i.e., tier-level) of uncorrupted parties from the transcript. Therefore, an adaptive adversary has exponentially small probability (in m) to break the protocol of Figure 6. Setting $m = \omega(\log n)$ makes the adversary's success probability negligible in n. $\square$

4   Optimal Strategy for the Corruption/Inspection Game

In this section we present the analysis for the corruption/inspection game (Figure 1). We obtain, for any parameters (α, β), a strategy that maximizes γ up to the theoretical limit. In the previous sections we demonstrated that, *given a two-tier model*, MPC can be realized to resist as much corruptions as less than half the amount of the still-honest parties. However, it is left to be shown *how* to split the n servers into two tiers so that (i) the two tiers are indistinguishable and (ii) the tier-1 servers are honest at the onset of the computation.

As mentioned in Section 1, one possible strategy for the service provider S is to set as tier-1 all the servers that were inspected and found clean. However, S cannot use the servers which were found corrupt, as these are no longer indistinguishable from the honest servers. This strategy leads to a non-optimal adaptive corruption rate of $\gamma = \frac{1-\alpha}{2}$. Thus, better strategies should be sought in order to utilize the "restored" machines. Next, we show a strategy for the service provider which maximizes his utility in the corruption/inspection game. Specifically, we prove the following:

Theorem 2. *For any constants $\alpha, \beta \in (0,1)$, and for any $\varepsilon > 0$, there exists a two-tier MPC protocol in the $(\mathcal{F}^{\nu}_{\mathsf{PKI}}, \mathcal{F}^{\nu}_{\mathsf{AAC}})$-hybrid world, and a winning strategy for a service provider in the corruption/inspection game, such that the protocol is adaptively secure against any PPT adversary with corruption rate $\gamma \leq (1 - \varepsilon)(1 - \alpha + \alpha\beta)/2$.*

We begin by showing a strategy for the service provider that beats any adversary who learns the tier-level of honest parties only by corrupting them. The idea of the strategy is to use two sets of servers as tier-1. One set comprises all the servers that were inspected and found clean, while the second one is a small random subset of the servers that were restored to a clean state. Note that, from the point of view of the adversary, the first set is hidden within all the uncorrupt servers, while the second set is hidden within all the servers that were restored to a clean state. We set the size of the second group so that in both these sets, the ratio of tier-1 servers to the size of the set it is hidden within, is the same.

Lemma 7. *Assume S and $\mathcal{A}$ play the corruption/inspection game with some constants $\alpha, \beta \in (0,1)$ and a small constant $\varepsilon > 0$. Furthermore, assume that when a server becomes corrupt (and only then), the adversary learns its tier level. Then, there exists a strategy for S for choosing tier-1 servers, such that given a corruption rate $\gamma \leq (1 - \varepsilon)(1 - \alpha + \alpha\beta)/2$ the adversary has negligible probability to corrupt half (or more) of tier-1 servers.*

Proof. S will choose the tier-1 servers as a subset of the βn inspected servers. We distinguish between two groups of inspected servers according to their state before the inspection: servers that were uncorrupt before the inspection (denoted G_1), and servers that were corrupt but recovered to a safe state by the inspection (G_2). From the point of view of $\mathcal{A}$, The first group is 'hidden' within the set $\widehat{G_1}$ of size $(1 - \alpha)n$ of the uncorrupt servers at the onset. The second group is fully known to the adversary ($\widehat{G_2} = G_2$ with $\alpha\beta n$ servers[8]). S will pick a small subset of servers in G_2 as tier-1; these will be hidden within the entire $\widehat{G_2}$. Note that the adversary knows which servers belong in $\widehat{G_1}$ and which are in $\widehat{G_2}$, but does not know the tier level of each party within each set. That way, the indistinguishability requirement between tier-1 and tier-2 servers still holds, yet separately in $\widehat{G_1}$ and $\widehat{G_2}$.

Specifically, S chooses tier-1 servers in the following way: all the $(1 - \alpha)\beta n$ servers in G_1 are chosen as tier-1 in addition to a random subset of servers in G_2. We equate the fraction of tier-1 servers in both groups (with respect to the group it is hidden within). Thus, out of the $\alpha\beta n$ servers in G_2, S randomly picks $y = \alpha\beta^2 n$ servers to be tier-1, so that

$$\frac{y}{\alpha\beta n} = \frac{(1 - \alpha)\beta n}{(1 - \alpha)n}.$$

---

[8] The sizes of the groups are only their *expected* value. However for large enough n (and especially, for our asymptotical analysis where $n \to \infty$), with high probability the real size will be very close to the expected value and we treat those sets as having sizes exactly $(1 - \alpha)n$ and $\alpha\beta n$.

We allow the adversary to corrupt at most $t = (1-\varepsilon)(1-\alpha+\alpha\beta)n/2$ servers out of the uncorrupt servers $\widehat{G_1} \cup \widehat{G_2}$. Assume the adversary splits his budget so that it corrupts t_1 servers from $\widehat{G_1}$, and t_2 servers from $\widehat{G_2}$, where $t_1 + t_2 = t$.[9]

Let $r = t_1/t$ (thus, $1 - r = t_2/t$); observe that the adversary cannot spend more budget than the population of each set so $t_1 \leq (1-\alpha)n$ and $t_2 \leq \alpha\beta n$, hence $1 - \frac{\alpha\beta}{t} \leq r \leq \frac{1-\alpha}{t}$. Let K_1, K_2 be the random variables that describe the number of servers $\mathcal{A}$ adaptively corrupts out of $\widehat{G_1}$ and $\widehat{G_2}$ with budget t_1, t_2, respectively. It is clear that

$$K_1 \sim \mathsf{HypGeo}_{(1-\alpha)n,(1-\alpha)\beta n, t_1}, \qquad K_2 \sim \mathsf{HypGeo}_{\alpha\beta n, \alpha\beta^2 n, t_2}.$$

In order to win the game, $\mathcal{A}$ needs to corrupt at least half of the tier-1 servers, where some can be in $\widehat{G_1}$ and the rest in $\widehat{G_2}$. However, no matter how $\mathcal{A}$ splits its budget, $\mathcal{A}$ corrupts more than half of the overall tier-1 servers with only a negligible probability. To that end, we again use the tail bound of Lemma 10. Specifically, the probability that the adversary corrupts, out of $\widehat{G_1}$, at least an r-fraction of *half* of all the tier-1 servers, is negligible:

$$\Pr\left[K_1 > \tfrac{r}{2}((1-\alpha)\beta + \alpha\beta^2)n\right] = \Pr\left[K_1 > t_1\beta\frac{(1-\alpha+\alpha\beta)n}{2t}\right]$$

$$= \Pr\left[K_1 > \frac{1}{1-\varepsilon}\mathbb{E}[K_1]\right]$$

$$= \Pr\left[K_1 > (1+\varepsilon')\mathbb{E}[K_1]\right]$$

$$< e^{-\Omega(\beta t)} = e^{-\Omega(n)},$$

where the second equality follows from $\mathbb{E}[K_1] = t_1\frac{(1-\alpha)\beta n}{(1-\alpha)n} = t_1\beta$.

In a similar way for $\widehat{G_2}$, the probability that $\mathcal{A}$ corrupts more than a $1-r$ fraction of half of tier-1 servers is negligible:

$$\Pr\left[K_2 > \tfrac{1-r}{2}((1-\alpha)\beta + \alpha\beta^2)n\right] = \Pr\left[K_2 > t_2\beta(1-\alpha+\alpha\beta)n/2t\right]$$

$$= \Pr\left[K_2 > \frac{1}{1-\varepsilon}\cdot\mathbb{E}[K_2]\right]$$

$$= \Pr\left[K_2 > (1+\varepsilon')\mathbb{E}[K_2]\right]$$

$$< e^{-\Omega(n)}.$$

It follows that there is a negligible probability for the adversary to corrupt at least $(r + (1-r))\cdot\frac{1}{2}((1-\alpha)\beta + \alpha\beta^2)n$ tier-1 servers, and since the total number of tier-1 servers is $((1-\alpha)\beta + \alpha\beta^2)n$, S wins the game. □

Since the tier-1 servers are now split into two separate sets, we need to extend Lemma 4 to the case where ν is not uniform over $\mathsf{Perm}(n)$. Specifically, we

---

[9] While we assume fixed values t_1 and t_2, in general the attack might be of any arbitrary distribution among the two sets. However, for any such attack we can repeat the analysis with t_1 being the expected number of servers corrupted out of $\widehat{G_1}$, and the two analyses differ with negligible probability when $n \to \infty$.

assume now that $\{P_1, \ldots, P_n\}$ are partitioned into r disjoint sets, $\mathcal{P}_1, \ldots, \mathcal{P}_r$, with respective sizes $s_1, \ldots, s_r$, such that $\sum_{i=1}^{r} s_i = n$. Additionally, assume the protocol pseudonyms $\mathcal{P}^*$ are also partitioned into r disjoint sets $\mathcal{P}_1^*, \ldots, \mathcal{P}_r^*$ where for every $1 \leq i \leq r$, $|\mathcal{P}_i| = |\mathcal{P}_i^*|$. We assume that the mapping ν is composed of r independent uniform permutations on the specific partitions. That is $\nu = (\nu_1, \ldots, \nu_r)$ where $\nu_i : \mathcal{P}_i \to \mathcal{P}_i^*$. For notational convenience, we also treat ν_i as a permutation on $\{1, \ldots, s_i\}$.

We show that even in this setting, where the adversary has some partial knowledge on ν, his best corruption strategy is equivalent to corrupting a random party. To that end, we re-define rcEXEC to be such that the simulator is allowed to choose the set from which the next party will be corrupted. That is, $\mathcal{S}$ may issue (corrupt, i) in which a random honest party in $\mathcal{P}_i$ will get corrupted. We denote an execution of this model as $\mathrm{rc}_r\mathrm{EXEC}$.

Lemma 8. *Let π and $\tilde{\pi}^{\mathcal{F}_{\mathsf{AAC}}}$ be as above. Assume the parties are divided into r sets $\mathcal{P}_1, \ldots, \mathcal{P}_r$ of sizes $s_1, \ldots, s_r$. For any PPT adversary $\mathcal{A}$ and environment $\mathcal{Z}$, and for any input vector $\boldsymbol{x}$, there exist a PPT simulator $\mathcal{S}$ such that*

$$\left\{ \mathsf{EXEC}_{\tilde{\pi}, \mathcal{A}, \mathcal{Z}}^{\mathcal{F}_{\mathsf{AAC}}^{\nu}}(\kappa, \boldsymbol{x}^{\nu}) \right\}_{\substack{\nu = (\nu_1, \ldots, \nu_r) \\ \in_R (\mathrm{Perm}(s_1), \ldots, \mathrm{Perm}(s_r))}} \approx \mathrm{rc}_r\mathsf{EXEC}_{\pi, \mathcal{S}, \mathcal{Z}}(\kappa, \boldsymbol{x}),$$

where $\mathrm{Perm}(k)$ is the set of all the possible permutations on k elements.

Proof. The simulation works similarly to the one of Lemma 4, with the following exception. When the adversary issues (corrupt, P_i), $\mathcal{S}$ will issue (corrupt, k) for the set k such that $P_i \in \mathcal{P}_k$. Assume that as a result P_j^* becomes corrupt, then $\mathcal{S}$ sets $\nu_k(i) = j$ and continues as before. Once again, the output of the simulation in this case is identical to any instance of $\mathsf{EXEC}_{\tilde{\pi}, \mathcal{A}, \mathcal{Z}}^{\mathcal{F}_{\mathsf{AAC}}^{\nu'}}$ running with the same adversary and a mapping ν' that agrees with the partial mapping ν defined by the simulator. $\qquad\square$

Given the above lemmas, the proof of Theorem 2 now follows.

Proof. (**Theorem 2**) The service provider will pick tier-1 servers according to the strategy described in Lemma 7. That is, the service provider will choose as tier-1 all the inspected servers that were found clean and a random β-fraction of the inspected servers that were found corrupt and then restored to a clean state. Then, the service provider runs the MPC scheme described in Figure 6. Similarly to the proof of Theorem 1 we observe the following:

1. The MPC protocol is secure as long as a majority of tier-1 servers are honest (Proposition 5).
2. Given that the adversary learns the protocol pseudonym and tier-level of a party only when this party is corrupt, when restricted to $\gamma \leq (1 - \varepsilon)(1 - \alpha + \alpha\beta)/2$ corruptions, it has only a exponentially small probability in $m = O(n)$ to corrupt a majority of tier-1 servers (Lemma 7).
3. Lemma 8 shows that an adaptive adversary learns only negligible information about ν (for uncorrupt parties).

Therefore, the computation is secure against the above adaptive adversary, except with negligible probability in n.

Observe that the parties are divided into three sets: the set of clean servers after step (1) of the corruption/inspection game (denoted by $\widehat{G_1}$ in Lemma 7); the set of servers that were restored to a clean state ($\widehat{G_2}$); and the rest of the servers. Setting $r = 3$, it is easy to see that Lemma 8 applies to our case by denoting those sets as $\mathcal{P}_1, \mathcal{P}_2$ and $\mathcal{P}_3$, respectively, and setting the protocol pseudonyms $\mathcal{P}_1^*, \mathcal{P}_2^*$ and $\mathcal{P}_3^*$ such that the number of tier-1 servers in each set matches the strategy of the service provider (e.g, β-fraction of the servers in each of the first two sets are tier-1, and no tier-1 servers in the third set).   □

References

[AF04]   Abe, M., Fehr, S.: Adaptively secure Feldman VSS and applications to universally-composable threshold cryptography. In: Franklin, M. (ed.) CRYPTO 2004. LNCS, vol. 3152, pp. 317–334. Springer, Heidelberg (2004)

[AKS83]   Ajtai, M., Komlós, J., Szemerédi, E.: An $O(n \log n)$ sorting network. In: Proceedings of the Fifteenth Annual ACM Symposium on Theory of Computing, STOC 1983, pp. 1–9. ACM, New York (1983)

[ALZ13]   Asharov, G., Lindell, Y., Zarosim, H.: Fair and efficient secure multiparty computation with reputation systems. In: Sako, K., Sarkar, P. (eds.) ASIACRYPT 2013, Part II. LNCS, vol. 8270, pp. 201–220. Springer, Heidelberg (2013)

[BCD+09]   Bogetoft, P., et al.: Secure multiparty computation goes live. In: Dingledine, R., Golle, P. (eds.) FC 2009. LNCS, vol. 5628, pp. 325–343. Springer, Heidelberg (2009)

[BdB90]   Bos, J.N.E., den Boer, B.: Detection of disrupters in the DC protocol. In: Quisquater, J.-J., Vandewalle, J. (eds.) Advances in Cryptology - EUROCRYPT 1989. LNCS, vol. 434, pp. 320–327. Springer, Heidelberg (1990)

[BDS+03]   Balfanz, D., Durfee, G., Shankar, N., Smetters, D.K., Staddon, J., Wong, H.-C.: Secret handshakes from pairing-based key agreements. In: IEEE Symposium on Security and Privacy, pp. 180–196 (2003)

[Bea97]   Beaver, D.: Commodity-based cryptography (extended abstract). In: Proceedings of the Twenty-ninth Annual ACM Symposium on Theory of Computing, STOC 1997, pp. 446–455. ACM, New York (1997)

[BG89]   Beaver, D., Goldwasser, S.: Multiparty computation with faulty majority. In: IEEE Annual Symposium on Foundations of Computer Science, pp. 468–473 (1989)

[Can00]   Canetti, R.: Security and composition of multiparty cryptographic protocols. J. Cryptology 13(1), 143–202 (2000)

[Can05]   Canetti, R.: Universally composable security: A new paradigm for cryptographic protocols. IACR Cryptology ePrint Archive, p. 67 (2005)

[CFGN96]   Canetti, R., Feige, U., Goldreich, O., Naor, M.: Adaptively secure multiparty computation. Tech. rep., Massachusetts Institute of Technology, Cambridge, MA, USA (1996)

[Cha88]   Chaum, D.: The dining cryptographers problem: Unconditional sender and recipient untraceability. Journal of Cryptology 1, 65–75 (1988)

[CJT04]   Castelluccia, C., Jarecki, S., Tsudik, G.: Secret handshakes from CA-oblivious encryption. In: Lee, P.J. (ed.) ASIACRYPT 2004. LNCS, vol. 3329, pp. 293–307. Springer, Heidelberg (2004)

[Cle86]   Cleve, R.: Limits on the security of coin flips when half the processors are faulty (extended abstract). In: Hartmanis, J. (ed.) STOC, pp. 364–369. ACM (1986)

[CLOS02]   Canetti, R., Lindell, Y., Ostrovsky, R., Sahai, A.: Universally composable two-party and multi-party secure computation. In: Proceedings of the Thiry-fourth Annual ACM Symposium on Theory of Computing, STOC 2002, pp. 494–503. ACM, New York (2002)

[DI05]   Damgård, I.B., Ishai, Y.: Constant-round multiparty computation using a black-box pseudorandom generator. In: Shoup, V. (ed.) CRYPTO 2005. LNCS, vol. 3621, pp. 378–394. Springer, Heidelberg (2005)

[FGMO05]   Fitzi, M., Garay, J.A., Maurer, U.M., Ostrovsky, R.: Minimal complete primitives for secure multi-party computation. J. Cryptology 18(1), 37–61 (2005)

[GJ04]   Golle, P., Juels, A.: Dining cryptographers revisited. In: Cachin, C., Camenisch, J.L. (eds.) EUROCRYPT 2004. LNCS, vol. 3027, pp. 456–473. Springer, Heidelberg (2004)

[GJKY13]   Garay, J., Johnson, D., Kiayias, A., Yung, M.: Resource-based corruptions and the combinatorics of hidden diversity. In: Proceedings of the 4th Conference on Innovations in Theoretical Computer Science, ITCS 2013, pp. 415–428. ACM, New York (2013)

[GKKZ11]   Garay, J.A., Katz, J., Kumaresan, R., Zhou, H.-S.: Adaptively secure broadcast, revisited. In: Gavoille, C., Fraigniaud, P. (eds.) PODC, pp. 179–186. ACM (2011)

[GKM+13]   Garay, J., Katz, J., Maurer, U., Tackmann, B., Zikas, V.: Rational protocol design: Cryptography against incentive-driven adversaries. In: 2013 IEEE 54th Annual Symposium on Foundations of Computer Science (FOCS), pp. 648–657 (2013)

[GL91]   Goldwasser, S., Levin, L.A.: Fair computation of general functions in presence of immoral majority. In: Menezes, A., Vanstone, S.A. (eds.) Advances in Cryptology - CRYPTO 1990. LNCS, vol. 537, pp. 77–93. Springer, Heidelberg (1991)

[GL02]   Goldwasser, S., Lindell, Y.: Secure computation without agreement. In: Malkhi, D. (ed.) DISC 2002. LNCS, vol. 2508, pp. 17–32. Springer, Heidelberg (2002)

[GMW87]   Goldreich, O., Micali, S., Wigderson, A.: How to play any mental game. In: Proceedings of the Nineteenth Annual ACM Symposium on Theory of Computing, STOC 1987, pp. 218–229. ACM, New York (1987)

[HS05]   Hush, D., Scovel, C.: Concentration of the hypergeometric distribution. Statistics & Probability Letters 75(2), 127–132 (2005)

[IKOS06]   Ishai, Y., Kushilevitz, E., Ostrovsky, R., Sahai, A.: Cryptography from anonymity. In: 47th Annual IEEE Symposium on Foundations of Computer Science, FOCS 2006, pp. 239–248 (2006)

[KMTZ13]   Katz, J., Maurer, U., Tackmann, B., Zikas, V.: Universally composable synchronous computation. In: Sahai, A. (ed.) TCC 2013. LNCS, vol. 7785, pp. 477–498. Springer, Heidelberg (2013)

[KOS03]   Katz, J., Ostrovsky, R., Smith, A.: Round efficiency of multi-party computation with a dishonest majority. In: Biham, E. (ed.) EUROCRYPT 2003. LNCS, vol. 2656, pp. 578–595. Springer, Heidelberg (2003)

[PW92]   Pfitzmann, B., Waidner, M.: Unconditionally untraceable and fault-tolerant broadcast and secret ballot election. Hildesheimer Informatik Berichte (1992)

[PW96]   Pfitzmann, B., Waidner, M.: Information-theoretic pseudosignatures and byzantine agreement for t ≥ n/3. IBM Research Report RZ 2882 (#90830) (1996)

A   Additional Definitions and Building Blocks

Signature Schemes. A public-key signature scheme consists of three PPT algorithms $(\mathsf{GenS}, \mathsf{Sig}, \mathsf{Ver})$ such that $(sk, pk) \leftarrow \mathsf{GenS}(1^\kappa)$ generates a key; $sig \leftarrow \mathsf{Sig}_{sk}(m)$ generates a signature for $m \in \mathcal{M}$ and $b \in \{0, 1\} \leftarrow \mathsf{Ver}_{pk}(m, sig)$ verifies a signature. For (sk, pk) generated by GenS, it holds that $\mathsf{Ver}_{pk}(m, \mathsf{Sig}_{sk}(m)) = 1$.

We say that a signature scheme is *existentially unforgeable* if any PPT adversary has only negligible advantage (in κ) in winning the following game running with a challenger:

SETUP: The challenger runs $(pk, sk) \leftarrow KeyS(1^\kappa)$. It gives the adversary the resulting public key pk and keeps the private key sk to itself.

QUERIES: The adversary issues signature queries $m_1, \ldots, m_q$. To each query m_i, the challenger computes $sig_i \leftarrow \mathsf{Sig}_{sk}(m_i)$ and sends sig_i back to the adversary. Note that m_i may depend on previous signatures (adaptive queries).

CHALLENGE: The adversary outputs a pair (m, sig), where $m \neq m_i$ for any m_i queried during the previous step. The adversary wins if $\mathsf{Ver}_{pk}(m, sig) = 1$.

Verifiable Secret Sharing (VSS). A (n, t)-VSS scheme is a protocol between a dealer and n parties $P_1, \ldots, P_n$, which extends a standard secret sharing. It consists of a SHARING PHASE where the dealer initially holds a value σ and finally, each party holds a private share v_i; and a RECONSTRUCTION PHASE in which the parties reveal their shares (a dishonest party may reveal $v_i' \neq v_i$) and a value σ' is reconstructed out of the shares $\sigma' = \mathsf{REC}(v_1', \ldots, v_n')$. Assuming an adversary that corrupts up to t parties, the following holds.

PRIVACY: If the dealer is honest, then the adversary's view during the sharing phase reveals no information about σ. More formally, the adversary's view is identically distributed under all different values of σ.

CORRECTNESS: If the dealer is honest, then the reconstructed value equals to σ.

COMMITMENT: After the sharing phase, a unique value σ^* is determined which will be reconstructed in the reconstruction phase; i.e., $\sigma^* = \mathsf{REC}(v_1', \ldots, v_n')$ regardless of the views provided by the dishonest players.

B   Realizing Anonymous Broadcast

First, one may realize ABC via standard adaptively secure multyparty computation techniques [CFGN96]. This construction shows how multiple parties can securely compute any given circuit, which in the case of $\mathcal{F}_{ABC}$ is a lexicographic sorting of the inputs. An asymptotically optimal sorting circuit is given in [AKS83] using $O(n \log n)$ comparators with depth $O(\log n)$.

Assuming the size of each field element is $O(\kappa)$ bits, a field-element comparator can be constructed out of binary gates in a tree fashion in size $O(\kappa)$ and depth $O(\log \kappa)$, or in a pipeline fashion with depth and size $O(\kappa)$. These constructions yield sorting circuits of size $O(\kappa n \log n)$ and depths $O(\log \kappa \log n)$ and $O(\kappa + \log n)$, respectively.

Note that the AAC protocol incurs only one call of ABC (i.e., there are no concurrent instances). Thus, invoking Canetti's modular composition theorem [Can00, Can05], such a construction gives adaptive security (with identifiable abort) against any number $t < n$ of corruptions. Observe that in the case of an abort, the only information that the adversary learns is the output, which is broadcast to all parties, and the security of the construction is not affected.

We refer to this protocol as ABC_{CFGN}; the next corollary immediately follows from [CFGN96].

Corollary 9. *The protocol* ABC_{CFGN} *securely realizes* $\mathcal{F}_{ABC}$ *with round complexity* $O(\min\{\log \kappa \log n, \kappa + \log n\})$, *and total communication* $O(\kappa^2 n \log n)$, *assuming non-committing encryption is used to implement point-to-point secure communication between parties.*

Although the above realization of $\mathcal{F}_{ABC}$ is sufficient for our purposes, we now discuss other alternatives, hoping for higher efficiency. First, we note that Golle and Juels [GJ04] present a scheme for honest-majority anonymous broadcast which uses bilinear maps, assuming the hardness of the Decisional Bilinear Diffie-Hellman problem (DBDH). Besides the honest-majority requirement, the construction does not consider "collisions," a common problem which arises in DC-nets in the selection of message positions; while the first shortcoming could be addressed by a player-elimination technique, addressing the second seems problematic, short of an MPC-type approach.

In [PW92, PW96], Pfitzmann and Waidner give an information theoretically secure sender-anonymous broadcast, based on Chaum's *DC-nets* [Cha88]. Their scheme assumes a *pre-computation* step during which a reliable broadcast is guaranteed. In our setting, we can replace the pre-computation reliable broadcast demand with an adaptively secure broadcast scheme, assuming a PKI setup [GKKZ11].

At a high level, the Pfitzmann-Waidner protocol consists of performing a many-to-many, corruption-detectable variant of a DC-net [BdB90], in which each user begins with a private input x_i and, if all parties behave as expected, ends with the multiset of inputs $\{x_i\}_i$ without being able to relate an input to its source. If some party deviates from the protocol, the other users notice this event (with high probability) and begin an 'investigation' in which each party should publicly reveal its messages and secret state, along with its private input. The parties can now check for consistency and (locally) identify the cheaters.

The resulting scheme, in the $\mathcal{F}_{\mathsf{PKI}}$-hybrid world, however, is less efficient than the generic construction requiring $O(n^4)$ rounds with $O(\kappa n^2)$ communication per round.

C   The Hypergeometric Distribution

We recall the hypergeometric distribution and some of its properties. The Hypergeometric distribution with parameters n, m, t describes the probability to draw k 'good' items out of an urn that contains n items out of which m are good, when one is allowed to draw t items overall. The probability is given by

$$\mathsf{HypGeo}_{n,m,t}(k) = \binom{m}{k}\binom{n-m}{t-k}/\binom{n}{t}.$$

The expectation of a random variable $K \sim \mathsf{HypGeo}_{n,m,t}$ is given by $\mathbb{E}[K] = t\frac{m}{n}$.

In our setting and terminology, $\mathsf{HypGeo}_{n,m,t}(k)$ describes the probability of corrupting k tier-1 servers, if there are n servers out of which m are tier-1, and the adversary is allowed to corrupt up to t servers altogether (assuming that the adversary learns the tier level of a specific server only when it gets corrupt).

A useful tool is a tail bound on the hypergeometric distribution, derived by Hush and Scovel [HS05]:

Lemma 10. *Let* $K \sim \mathsf{HypGeo}_{n,m,t}$ *be a random variable distributed according to the Hypergeometric distribution with parameters* n, m, t. *Then,*

$$\Pr[K - \mathbb{E}[K] > \delta] < e^{-2\alpha_{n,m,t}(\delta^2 - 1)}$$

where

$$\alpha_{n,m,t} = \max\left(\left(\frac{1}{t+1} + \frac{1}{n-t+1}\right), \left(\frac{1}{m+1} + \frac{1}{n-m+1}\right)\right)$$

and assuming $\delta > 2$.

Topology-Hiding Computation

Tal Moran[1,*], Ilan Orlov[1], and Silas Richelson[2]

[1] Efi Arazi School of Computer Science, IDC Herzliya, Israel
{talm,iorlov}@idc.ac.il
[2] Department of Computer Science, UCLA, USA
SiRichel@ucla.edu

Abstract. Secure Multi-party Computation (MPC) is one of the foundational achievements of modern cryptography, allowing multiple, distrusting, parties to jointly compute a function of their inputs, while revealing nothing but the output of the function. Following the seminal works of Yao and Goldreich, Micali and Wigderson and Ben-Or, Goldwasser and Wigderson, the study of MPC has expanded to consider a wide variety of questions, including variants in the attack model, underlying assumptions, complexity and composability of the resulting protocols.

One question that appears to have received very little attention, however, is that of MPC over an underlying communication network whose structure is, *in itself*, sensitive information. This question, in addition to being of pure theoretical interest, arises naturally in many contexts: designing privacy-preserving social-networks, private peer-to-peer computations, vehicle-to-vehicle networks and the "internet of things" are some of the examples.

In this paper, we initiate the study of "topology-hiding computation" in the computational setting. We give formal definitions in both simulation-based and indistinguishability-based flavors. We show that, even for fail-stop adversaries, there are some strong impossibility results. Despite this, we show that protocols for topology-hiding computation can be constructed in the semi-honest and fail-stop models, if we somewhat restrict the set of nodes the adversary may corrupt.

1 Introduction

Secure multi party computation (MPC) has occupied a central role in cryptography since its inception in the '80s. The unifying question can be stated simply:

Can mutually distrusting parties compute a function of their inputs, while keeping their inputs private?

Classical feasibility results [26,18,4,23] paved the way for a plenitude of research which over time simplified, optimized and generalized the original foundational constructions. Some particularly rich lines of work include improving

---

* Supported by ISF grant no. 1790/13 and by the European Union Seventh Framework Programme (FP7/2007-2013) under grant agreement no. 293843

Y. Dodis and J.B. Nielsen (Eds.): TCC 2015, Part I, LNCS 9014, pp. 159–181, 2015.

the complexity (round/communication/computation) of MPC protocols (e.g., [3,15,12,11] and many more) and striving to achieve security in the more diffi-cult (but realistic) setting where the adversary may execute many instantiations of the protocol along with other protocols (e.g., [22,20,8,6] and many more).

Common to nearly all prior work, however, is the assumption that the par-ties are all capable of exchanging messages with one another. That is to say, most work in the MPC literature assumes that the underlying network topol-ogy is that of a complete graph. This is unrealistic as incomplete or even sparse networks are much more common in practice. Moreover, the comparably small body of MPC work that deals with incomplete networks concerns itself with the classical goal of hiding the parties' inputs. In light of the growing impact of net-working on today's world, this traditional security goal is insufficient. Consider, for example, the graph representing a social network: nodes representing people, edges representing relationships. Most computation on social networks today is performed in a centralized way—Facebook, for example, performs computations on the social network graph to provide popular services (e.g., recommendations that depend on what "similar" people liked). However, in order to provide these services Facebook must "know" the entire graph.

One could imagine wanting to perform such a computation in a *distributed* manner, where each user communicates only with their own friends, without revealing any additional information to third parties (there is clearly wide inter-est in this type of service—Diaspora*, a project that was expressly started to provide "Facebook-like" functionality in a more privacy-preserving manner [2], raised over $200,000 in 40 days via Kickstarter).

Another motivating example is the recent push by US auto safety regulators towards vehicle-to-vehicle communication [1], which envisions dynamic networks of communicating vehicles; many "global" computations seem to be interesting in this setting (such as analysis of traffic patterns), but leaking information about the structure of this network could have severe privacy implications.

The rise of the "internet of things", connected by mesh networks (networks of nodes that communicate locally with each other) is yet another case in which the topology of the communication network could reveal private information that users would prefer to hide.

It is with such applications in mind that we initiate the study of *topology-hiding MPC* in the computational setting. We consider the fundamental question:

> *Can an MPC protocol computationally hide the underlying network topol-ogy?*

1.1   Our Contributions

Formally Defining Topology-Hiding MPC: In keeping with tradition we give both an indistinguishability game-based definition and a simulation-based one. Very briefly, in the game-based definition the adversary corrupts $A \subset V$ and sends two network topologies G_0, G_1 on vertices V. These graphs must be so that the neighborhoods of A are the same. The challenger then picks G_b at random and returns the collective view of the parties in A resulting from the execution of the protocol on G_b. The adversary outputs b' and wins if $b' = b$. We say a protocol is *secure against chosen topology attack* (IND-CTA−secure) if no PPT adversary can win the above game with probability negligibly greater than if it simply guesses b'.

We then give a simulation-based definition of security using the UC framework. We define an ideal functionality $\mathcal{F}_{\text{graph}}$ and say that a protocol is "topology hiding" if it is UC secure in the $\mathcal{F}_{\text{graph}}$−hybrid model. The functionality $\mathcal{F}_{\text{graph}}$ models a network with private point-to-point links (private in the sense that the adversary does not know the network topology). It receives G as input, and outputs to each party a description of its neighborhood. It then acts as an "ideal communication channel" allowing neighbors to send messages to each other. For more details on $\mathcal{F}_{\text{graph}}$ and the motivations behind our definition see the discussion below. Finally, we relate the two new notions by proving that simulation-based security implies game-based security.

Feasibility of topology-hiding MPC against semi-honest adversary:

Theorem 1. *Assume trapdoor permutations exist. Let G be the underlying network graph and d a bound on the degree of every vertex in G. Then every multiparty functionality may be realized by a topology hiding MPC protocol which is secure against a semi-honest adversary who does not corrupt all parties in any k−neighborhood of the underlying network graph where k is such that $d^k = \text{poly}(n)$.*

We point out that many naturally occurring graphs satisfy $d^D = poly(n)$ where D is the diameter. Examples of such graphs include binary trees, hypercubes, expanders, and generally graphs with relatively high connectivity such as those occurring from social networks. For such graphs theorem 1 is a feasibility result against a general semi-honest adversary.

Impossibility in fail-stop model:

Theorem 2. *There exists a functionality $\mathcal{F}$ and a network graph G such that realizing $\mathcal{F}$ while hiding G is impossible.*

Our proof uses the ability of the adversary to disconnect G with his aborts; we then prove this is inherent.

Sufficient conditions in fail-stop model:

Theorem 3. *Assume TDP exist. Every multiparty functionality may be realized by a topology hiding MPC protocol which is secure against a fail-stop adversary who does not corrupt all parties in any neighborhood of the underlying network graph and who's aborts do not disconnect the graph.*

1.2 Related Work

MPC on Incomplete Network Topologies One line of work which is in exception to the above began with Dolev's paper [13] proving impossibility of Byzantine agreement on incomplete network topologies with too low connectivity. Dwork et. al. [14] coined the term "almost everywhere Byzantine agreement" to be a relaxed variant of Byzantine agreement where agreement is reached by *almost* all of the honest parties. Garay and Ostrovsky [16] used this to achieve almost everywhere (AE) MPC. Recently [9] gave an improved construction of AE Byzantine agreement translating to an improved feasibility result for AE MPC. These works are all in the information theoretic setting. We refer the curious reader to [9] and the references therein for more details.

Another recent line of work is that of Goldwasser et. al. [5] who consider MPC while minimizing the communication *locality*, the number of parties each player must exchange messages with. Their work is in the cryptographic setting and they give a meaningful upper bound on the locality and overall communication complexity. Their work does not address the notion of hiding the graph. Moreover they employ techniques such as leader election which seem inherently *not* to hide the graph.

Finally, we mention the two classical techniques of *mix-net* and *onion routing*. The mix-net technique introduced by Chaum [10] uses public key encryption to implement a "message transmit" scheme allowing a sender and receiver to using in a message transmit using an additional shuffling mechanism. The onion routing technique [25,24] and its extensions is a useful technique for anonymous communication over the Internet. Its basic idea is establishing paths of entities called proxies that know the topology in order to transmit massages.

Topology-Hiding MPC: While most of the cryptographic MPC literature disregards the interplay between multiparty computation and networking, the above works give a relatively satisfactory view of the landscape. Hiding the topology of the network in secure computation, on the other hand, is somewhat of a novel goal. The only work in the MPC literature of which we are aware that has considered this question is that of Hinkelmann and Jakoby [19] who focused on the information theoretic setting. Their main observation can be summarized:

> If vertices v and w are not adjacent in G then P_v cannot send a message to P_w without some intermediate P_z learning that it sits between P_v and P_w.

They use this observation to prove that any MPC protocol in the information theoretic setting must inherently leak information about G to an adversary. They do, however, prove a nice positive result: given some minimal amount of network information to leak (formalized as a routing table of the network), one can construct an MPC protocol which leaks no further information.

Their work leaves open the interesting possibility that, using cryptographic techniques, one could construct an MPC protocol which (computationally) hides the network topology. In this work we explore this possibility.

Organization of the Rest of the Paper: The rest of the paper is organized as follows. In section 2 we go over the background and general definitions which are required to understand the technical portions which follow. In section 3 we formally define our new notions of "topology hiding" computation. This includes our game-based and simulation-based definitions as well as a proof that simulation-based security implies game-based security. In section 4 we consider achieving topology-hiding MPC against a semi-honest adversary. Our basic protocol is secure as long as the adversary does not corrupt any whole neighborhoods of the network graph. We then lessen this requirement showing how to transform a protocol which is secure against an adversary who doesn't corrupt an entire $k-$neighborhood into one secure as long as $\mathcal{A}$ does not corrupt any $(k+1)-$neighborhood. This proves theorem 1. In section 5 we consider a fail-stop adversary and give a somewhat complete picture of the landscape in this setting, proving theorem 2 and theorem 3.

2   Preliminaries

2.1   General Definitions

We model a network by a directed graph $G = (V, E)$ that is not fully connected. We consider a system with $m = \mathrm{poly}(n)$ parties, denoted $P_1, \ldots, P_m$. We often implicitly identify V with the set of parties $\{P_1, \ldots, P_m\}$. We consider a static and computationally bounded (PPT) adversary that controls some subset of parties. That is, at the beginning of the protocol, the adversary corrupts a subset of the parties and may instruct them to deviate from the protocol according to the corruption model. Through this work, we consider mostly semi-honest and fail-stop adversaries, though we discuss the implications of our fail-stop impossibility result on the hope of achieving topology-hiding MPC in the malicious model. In addition, we assume that the adversary is rushing; that is, in each round the adversary sees the messages sent by the honest parties before sending the messages of the corrupted parties for this round. For general MPC definitions including descriptions of the adversarial models we consider see [17].

2.2   Definitions of Graph Terms

Let $G = (V, E)$ be an undirected graph. For $v \in V$ we let $N(v) = \{w \in V : (v, w) \in E\}$ denote the *neighborhood of v* by; and similarly, the *closed neighborhood of v*, $\mathrm{N}[v] = N(v) \cup \{v\}$. We sometimes refer to $\mathrm{N}[v]$ as the closed $1-$neighborhood of v, and for $k \geq 1$ we define the $k-$neighborhood of v as

$$N^{k+1}[v] = \bigcup_{w \in N^k(v)} \mathrm{N}[w].$$

2.3   UC Security

We employ the UC model [7] in order to abstract away many of the implementation details and get at the core of our definition. Our protocol for hiding the topology in MPC is local in nature, and our final protocol is the result of composing many local subprotocols together. This motivates the need for using subprotocols which are secure under some form of composition. UC security offers strong composability guarantees and thus is well suited to our setting. One of the appealing features of our definition is that it fits entirely within the existing UC framework, hence the UC composition theorem can be applied directly.

The downside of the UC model is that it requires setup [7] and opponents argue that it is "unrealistic". We have two responses to this. First, we encapsulate our setup into realizing the $\mathcal{F}_{graph}$ functionality. This functionality (defined formally in the next section) models the underlying communication network and so we think of the setup required in order to realize it as an implementation issue. Second, we point out that our subroutines need only be secure against bounded self-composition in order to obtain stand-alone security in the $\mathcal{F}_{graph}$–hybrid model, corresponding to a stand-alone variant of topology hiding security. This allows us to instantiate our protocol in the plain model on top of (for example) [20] in order to obtain stand-alone topology hiding MPC.

3   Our Model of Security

3.1   Topology Hiding—The Simulation-Based Definition

In this section, we propose a simulation-based definition for topology hiding computation in the UC framework. Generally, in the UC model, the communication between all parties passes through the environment, so it seems the environment implicitly knows the structure of the underlying communication network. We get around this by working in the $\mathcal{F}_{graph}$–hybrid model. The $\mathcal{F}_{graph}$ functionality (shown in Figure 1) takes as input the network graph from a special "graph party" P_{graph} and returns to each other party a description of their neighbors. It then handles communication between parties, acting as an "ideal channel" functionality allowing neighbors to communicate with each other without this communication going through the environment.

In a real-world implementation, $\mathcal{F}_{graph}$ models the actual communication network; i.e., whenever a protocol specifies a party should send a message to one of its neighbors using $\mathcal{F}_{graph}$, this corresponds to the real-world party directly sending the message over the underlying communication network.

Since $\mathcal{F}_{graph}$ provides local information about the graph to all corrupted parties, any ideal-world adversary must have access to this information as well (regardless of the functionality we are attempting to implement). To capture this, we define the functionality $\mathcal{F}_{graphInfo}$, that is identical to $\mathcal{F}_{graph}$ but contains only the initialization phase. For any functionality $\mathcal{F}$, we define a "composed" functionality $(\mathcal{F}_{graphInfo}||\mathcal{F})$ that adds the initialization phase of $\mathcal{F}_{graph}$ to $\mathcal{F}$. We can now define topology-hiding MPC in the UC framework:

Definition 1. *We say that a protocol Π securely realizes a functionality $\mathcal{F}$ hiding topology if it UC-realizes $(\mathcal{F}_{graphInfo}||\mathcal{F})$ in the $\mathcal{F}_{graph}$-hybrid model.*

Participants/Notation:
This functionality involves all the parties $P_1, \ldots, P_m$ and a special graph party P_{graph}.

Initialization Phase:
Inputs: $\mathcal{F}_{\text{graph}}$ waits to receive the graph $G = (V, E)$ from P_{graph}.
Outputs: $\mathcal{F}_{\text{graph}}$ outputs $N_G[v]$ to each P_v.

Communication Phase:
Inputs: $\mathcal{F}_{\text{graph}}$ receives from a party P_v a destination/data pair (w, m) where $w \in N(v)$ and m is the message P_v wants to send to P_w.
Output: $\mathcal{F}_{\text{graph}}$ gives output (v, m) to P_w indicating that P_v sent the message m to P_v.

Fig. 1. The functionality $\mathcal{F}_{\text{graph}}$

Note that this definition can also capture protocols that realize functionalities depending on the graph (e.g., find a shortest path between two nodes with the same input, or count the number of triangles in the graph).

3.2 Topology Hiding - The Indistinguishability-Based Security Definition

In this section, we propose another definition for topology-hiding security that is not restricted to secure multi-party computation. The definition is formalized using a security game between an adversary $\mathcal{A}$ and a challenger $\mathcal{C}$. In addition, we prove that this definition is implied by the simulation-based definition from subsection 3.1. The basic structure of the game fits several types of adversarial behaviors, e.g., semi-honest, fail-stop, and malicious, thus, we do not emphasize the exact behavior of the adversary during the execution of the protocol.

- Setup: Let $\mathcal{G}$ be a set of graphs. Let Π be a protocol capable of running over any of the communication graphs in $\mathcal{G}$ according to the adversarial model of $\mathcal{A}$ (semi-honest, fail-stop, or malicious). Each P_i gets an input $x_i \in X_i$.
- $\mathcal{A}$ chooses a corrupt subset S, inputs x_j for the corrupted parties $P_j \in S$ and, for $i \in \{0, 1\}$, two graphs $G_i = (V_i, E_i) \in \mathcal{G}$, such that $S \subset V_0 \cap V_1$ and $N_{G_0}[S] = N_{G_1}[S]$ (equality of graphs). It outputs $(S; G_0, G_1; \{x_j\})$. If $S \not\subset V_0 \cap V_1$ or if some input x_j is invalid $\mathcal{C}$ wins automatically.
- Now $\mathcal{C}$ chooses a random $b \in \{0, 1\}$ and runs Π in the communication graph G_b, where each honest P_i gets x_i and each dishonest party gets the input prescribed by $\mathcal{A}$. $\mathcal{A}$ receives the collective view of all parties in S during the protocol execution. [1]
- Finally $\mathcal{A}$ must output $b' \in \{0, 1\}$. If $b' = b$ we say that $\mathcal{A}$ wins the security game. Otherwise $\mathcal{A}$ loses.

---

[1] In the semi-honest model, the joint view of the corrupted parties is given to $\mathcal{A}$ by the end of the execution of P_i, while in the active models such as fail-stop and malicious, $\mathcal{A}$ sends instructions during the execution of Π and can deviate from the prescribed protocol.

Definition 2. We say that an MPC protocol Π is *Indistinguishable under Chosen Topology Attack* (*IND-CTA secure*) over $\mathcal{G}$ if for any PPT adversary $\mathcal{A}$ there exists negligible function $\mu(\cdot)$, such that for every n it holds

$$\left| \Pr(\mathcal{A} \text{ wins}) - \frac{1}{2} \right| \leq \mu(n).$$

We prove below that IND-CTA security is weaker than security with respect to the simulation-based security definition (Definition 1); thus, our impossibility results (in subsection 5.1) imply impossibility of the simulation based definition as well.

Claim. For every functionality $\mathcal{F}$ that does not depend on the communication graph structure, if Π securely realizes $\mathcal{F}$ with topology-hiding security (under Definition 1) then Π is IND-CTA secure.

Proof (Proof Sketch). Let Π be a topology-hiding secure-computation protocol with respect to Definition 1 and let G_0 and G_1 be two graphs. We consider two specific executions of Π on network topologies G_0 and G_1 with corrupt parties given the same inputs. We define random variables $(\text{HYBRID}_{G_0}, \text{IDEAL}_{G_0})$ and $(\text{HYBRID}_{G_1}, \text{IDEAL}_{G_1})$ as usual. We observe that IDEAL_{G_0} are IDEAL_{G_1} are identically distributed, as in both cases the adversary gets the same final output, in addition to the same set of local closed neighborhoods. It follows that if Π realizes $\mathcal{F}$ with topology hiding security then HYBRID_{G_0} and HYBRID_{G_1} are indistinguishable. It follows that $\mathcal{A}$ cannot win the IND-CTA game with probability that is noticeably better than $1/2$. So Π meets also the IND-CTA security definition.

4   Topology Hiding MPC against a Semi-honest Adversary

In this section we describe a protocol for topology-hiding MPC against a semi-honest adversary. This construction is the heart of the main positive result in the paper:

Theorem 4. *Let d be a bound on the degree of any vertex in G. Then for every k satisfying $d^k = \text{poly}(n)$, there exists a protocol Π that securely realizes the broadcast functionality hiding topology against a semi-honest adversary $\mathcal{A}$ that does not corrupt all parties in any closed $k-$neighborhood of G.*

Note that this gives us security against a general semi-honest adversary when the graph has constant degree and a logarithmic bound on the diameter (by setting k to be anything larger than the graph diameter). We point out that many natural families of graphs are of this sort, including binary trees, hypercubes, expanders and more. Theorem 1 follows from theorem 4 by standard methods for compiling broadcast into MPC.

4.1   High-Level Protocol Overview of Our Basic Protocol

Below we give a top-down description of our basic broadcast protocol: one that is secure against adversaries that do not corrupt any complete 1-neighborhood in the graph (i.e., in every star there is at least one honest node). This basic protocol can then be used to construct a broadcast protocol that tolerates larger corrupt neighborhoods (more details of this transformation appear in section 4.6).

A Naïve Broadcast Protocol. To understand the motivation for the construction, first consider a naïve broadcast protocol for a single bit:

1. In the first round, the broadcaster sends the broadcast bit b to all of its neighbors. Every other party sends 0 to all of their neighbors.
2. In each successive round, every party computes the OR of all bits received from their neighbors in the previous round, and sends this bit to all neighbors.

After j rounds, every party at distance j or less from the broadcaster will be sending the bit b (this can be easily shown by induction); after $diam(G)$ rounds all parties will agree on the bit b. This protocol realizes the broadcast functionality, but it is not topology-hiding: a party can tell how far it is from the broadcaster by counting the number of rounds until it receives a non-zero bit (assume $b = 1$ for this attack). It can also tell in which direction the broadcaster lies by noting which neighbor first sent a non-zero bit.

Using Local MPC to Hide Topology. Our construction hides the sensitive information by secret-sharing it among the nodes in a local neighborhood. Essentially, our basic protocol replaces each node in the naïve protocol above with a secure computation between the node and all of its direct neighbors in the graph.

In order to communicate a bit between one local neighborhood and another, without revealing the bit to the vertex connecting the two neighborhoods, each local neighborhood generates a public/private key pair, for which the private key is secret-shared between the parties in the neighborhood. The input to each local MPC includes the private key shares. The output to each party is encrypted under the public key of the neighborhood represented by that party (i.e., of which the party is the center node).

Since no local neighborhood is entirely corrupted, the adversary does not learn any of the plaintext bits. In the final round (at which point the broadcast bit has "percolated" to all the neighborhoods in the graph). a secure computation is used to decrypt the bits and output the plaintext to all the parties.

The protocol is formally specified as two separate functionalities, each instantiated using a local secure computation: the KeyGen functionality ($\mathcal{L}_{\text{KeyGen}}$), handles the generation and distribution of the public/private key-pair shares in each local neighborhood, and the "broadcast-helper" functionality ($\mathcal{L}_{\text{bc-helper}}$), handles the encryption/decryption and ORing of the bits. The details of the construction are in section 4.4.

Implementing Local MPC. To implement $\mathcal{L}_{\text{KeyGen}}$ and $\mathcal{L}_{\text{bc-helper}}$, the basic protocol uses a general MPC functionality, $\mathcal{L}_{\text{MPC}}$, that allows the local neighborhoods to perform secure computation protocols (i.e., among parties connected in a star graph). Realizing $\mathcal{L}_{\text{MPC}}$ ammounts to constructing a compiler which transforms a standard MPC protocol which runs on a complete graph into one which runs on a star graph. We achieve this by having players in the star who are not connected pass messages to each other through the center. The messages are encrypted to ensure privacy. One subtle point is that the protocol must not leak how many players are in the local neighborhood, as parties are not supposed to learn the degrees of their neighbors. We sidestep this issue by having the center node "invent" fake nodes so that parties learn only that the degree is at most d, some public upper bound on the degree of any node in G. The functionality $\mathcal{L}_{\text{MPC}}$ is shown in Figure 2.

4.2   Formal Protocol Construction and Proofs of Security

Below, we give the formal protocol definitions and sketch their proofs of security in the UC framework.

Notation. The protocols and functionalities in the remainder of this section involve parties $P_1, \ldots, P_m$ and a special graph party P_{graph} whose role is always simply to pass his input, G, to $\mathcal{F}_{\text{graph}}$. For $v \in V$ we let P_v be the player corresponding to v. Many of these protocols/functionalities begin with a KeyGen phase which uses a public key encryption scheme $(\text{Gen}, \text{Enc}, \text{Dec})$. Finally, we will make repeated use of "local" MPCs which are executed by the parties in a local neighborhood of G. We will use repeated parallel executions of local MPCs to realize global functionalities. We reserve the letter $\mathcal{L}$ for local functionalities realized by local MPCs, and use $\mathcal{F}$ for global functionalities. For simplicity when describing a local functionality or local MPC protocol, we will describe only the singular execution running in $\text{N}[v]$ (involving P_v and $\{P_w\}_{w \in N(v)}$), even though the same process is occurring in every closed neighborhood in G simultaneously.

4.3   Realizing $\mathcal{L}_{\text{MPC}}$ in the $\mathcal{F}_{\text{graph}}$−Hybrid Model

The local MPC functionality $\mathcal{L}_{\text{MPC}}$ is shown in Figure 2. As we have already mentioned, it is sufficient to securely realize message passing between all parties in $\text{N}[v]$ in the $\mathcal{F}_{\text{graph}}$−hybrid model. This is because, once all parties can send messages to each other, they can simply run their favorite UC secure MPC protocol as if the network topology is that of a complete graph. Note that as we are in the semi-honest model here, UC secure MPC does not require setup. We will use the constant round, protocol of [21], as it is UC secure against a semi-honest adversary (against general adversaries it is bounded concurrent secure).

For simplicity we describe the protocol allowing P_w to securely send a message to P_u for $w, u \in \text{N}[v]$.

Graph Entry Phase:
   Input: $\mathcal{L}_{\text{MPC}}$ receives the graph G from P_{graph}.
   Output: $\mathcal{L}_{\text{MPC}}$ outputs $N[v]$ to P_v.
MPC Phase:
   Input: $\mathcal{L}_{\text{MPC}}$ receives from P_v a d−party protocol Π and input x_w from P_w, for each $w \in N[v]$.
   Computation: $\mathcal{L}_{\text{MPC}}$ simulates Π with inputs $\{x_w\}_w$ obtaining outputs $\{y_w\}_w$.
   Output: $\mathcal{L}_{\text{MPC}}$ gives y_w to each P_w.

Fig. 2. The functionality $\mathcal{L}_{\text{MPC}}$

1. P_u generates a key pair and sends the public key to P_w through P_v;
2. P_w encrypts its message and sends the ciphertext back to P_u through P_v.

Such a protocol naturally extends to allow all parties in $N[v]$ to exchange messages with each other (as long as P_v invents enough nodes to ensure that his neighbors do not learn his degree, but just the preselected bound d). As we mention above, this is sufficient for securely realizing $\mathcal{L}_{\text{MPC}}$ in the $\mathcal{F}_{\text{graph}}$−hybrid model.

Security of $\mathcal{L}_{MPC}$. The proof is very simple so we suffice it to briefly describe $\mathcal{S}$, and leave checking that it accurately emulates $\mathcal{A}$'s view in the real world to the reader. Since Π is a UC secure MPC protocol on a complete graph, there exists a simulator $\mathcal{S}'$ who can replicate any adversary $\mathcal{A}$'s real-world view in the ideal world. The only difference between the view $\mathcal{S}'$ outputs and the view we need to output is that we must take into account that our messages are encrypted and passed through P_v. Therefore, $\mathcal{S}$ generates key pairs $\{(\text{pk}_{w,u}, \text{sk}_{w,u})\}_{w,u \in N[v]}$, where P_w will use $\text{pk}_{w,u}$ to send messages to P_u and computes encryptions of the messages in the view output by $\mathcal{S}$, and distributes them accordingly to the players. Security follows from the security of the encryption scheme.

Participants/Notation: This protocol involves players P_v, P_w, P_u, for $w, u \in N[v]$, and allows P_w to send the message msg to P_u. Let $(\text{Gen}, \text{Enc}, \text{Dec})$ be a public key encryption scheme.
Input: P_v and P_u use no input, P_w uses input msg.
Protocol for Message Passing:
   - P_u chooses a key pair $(\text{pk}, \text{sk}) \leftarrow \text{Gen}(1^n)$ and sends $(v; (w, \text{pk}))$ to $\mathcal{F}_{\text{graph}}$; P_v receives $(u; (w, \text{pk}))$.
   - P_v sends $(w; (u, \text{pk}))$ to $\mathcal{F}_{\text{graph}}$; P_w receives $(v; (u, \text{pk}))$.
   - P_w computes encryption $y = \text{Enc}_{\text{pk}}(\text{msg})$ and sends $(v; (u, y))$ to $\mathcal{F}_{\text{graph}}$; P_v receives $(w; (u, y))$.
   - P_v sends $(u; (w, y))$ to $\mathcal{F}_{\text{graph}}$; P_u receives $(v; (w, y))$.
   - P_u decrypts and outputs $\text{msg} = \text{Dec}_{\text{sk}}(y)$.

Fig. 3. The $\mathcal{F}_{\text{graph}}$−hybrid protocol $\Pi_{\text{msg-transmit}}$

4.4 The Functionalities $\mathcal{L}_{\text{KeyGen}}$ and $\mathcal{L}_{\text{bc-helper}}$

The functionality $\mathcal{L}_{\text{MPC}}$ of the previous section is a general functionality that compiles an MPC protocol Π on a complete graph into an analogous one which can be executed by the parties in $\text{N}[v]$, without compromising the security of Π, and also without leaking any information about the topology. We will be interested in two specific local functionalities, $\mathcal{L}_{\text{KeyGen}}$ and $\mathcal{L}_{\text{bc-helper}}$. These can be securely realized in the $\mathcal{F}_{\text{graph}}$−hybrid model by simply instantiating $\mathcal{L}_{\text{MPC}}$ with two specific MPC protocols.

Recall that our underlying idea is to replace the role of P_v in a usual broadcast protocol with an MPC to be performed by the parties in P_v's neighborhood. This will hide each player's distance from the broadcaster because even though the bit might have been received by P_v's neighborhood, it will not be known to any individual player. Our first functionality, $\mathcal{L}_{\text{KeyGen}}$, is useful towards this end. Intuitively, it generates a key pair (pk, sk) for the neighborhood $\text{N}[v]$ and gives pk to P_v and distributes secret shares of the secret key among P_v's neighbors. Our second functionality, $\mathcal{L}_{\text{bc-helper}}$ will allow the broadcast bit to spread from neighborhood to neighborhood once the neighborhoods have keys distributed according to $\mathcal{L}_{\text{KeyGen}}$. The functionalities $\mathcal{L}_{\text{KeyGen}}$ and $\mathcal{L}_{\text{bc-helper}}$ are shown in Figure 4 and Figure 5, respectively. Let $\mathcal{L}_{\text{KeyGen}}(v)$ and $\mathcal{L}_{\text{bc-helper}}(v)$ denote the copies of $\mathcal{L}_{\text{KeyGen}}$ and $\mathcal{L}_{\text{bc-helper}}$, respectively, which take place in $\text{N}[v]$

Participants/Notation: Let $(\text{Gen}, \text{Enc}, \text{Dec})$ be a public key encryption scheme.

Graph Entry Phase: same as in $\mathcal{L}_{\text{MPC}}$

KeyGen Phase:

- $\mathcal{L}_{\text{KeyGen}}$ generates a key pair $(\text{pk}, \text{sk}) \leftarrow \text{Gen}(1^n)$.
- $\mathcal{L}_{\text{KeyGen}}$ computes random shares $\{\text{sk}_w\}_{w \in \text{N}[v]}$ such that $\bigoplus_w \text{sk}_w = \text{sk}$.

Output: $\mathcal{L}_{\text{KeyGen}}$ gives outputs (pk, sk_v) to P_v, and sk_w to each P_w such that $w \in \text{N}[v]$.

Fig. 4. The functionality $\mathcal{L}_{\text{KeyGen}}$

4.5 Realizing $\mathcal{F}_{\text{broadcast}}$ in $\mathcal{L}_{\text{MPC}}$−hybrid Model

Our $\mathcal{L}_{\text{MPC}}$−hybrid protocol for broadcast, $\Pi_{\text{broadcast}}$ uses the ideal functionalities $\mathcal{L}_{\text{KeyGen}}$ and $\mathcal{L}_{\text{bc-helper}}$ described above. As mentioned in the previous section, these functionalities are obtained from $\mathcal{L}_{\text{MPC}}$ by instantiating $\mathcal{L}_{\text{MPC}}$ with specific MPC protocols. A description of $\Pi_{\text{broadcast}}$ is given in Figure 11. Note that $\Pi^r_{\text{broadcast}}$ is correct as long $r > \text{diam}(G)$, the diameter of the network graph G. Our statement and proof of security are below.

Participants/Notation: For $w \in N[v]$, let pk^w be the public key output to P_w by $\mathcal{L}_{\text{KeyGen}}(w)$. Let sk^v_w denote P_w's share of sk^v (the secret key corresponding to pk^v), given as output by $\mathcal{L}_{\text{KeyGen}}(v)$.

Graph Entry Phase: same as in $\mathcal{L}_{\text{MPC}}$

Main Phase:

Input: $\mathcal{L}_{\text{bc-helper}}$ receives inputs:
- $\alpha_w \in \{\text{"cipher"}, \text{"plain"}\}$ from each P_w;
- (pk^w, sk^v_w) from each P_w;
- encryptions $\{x_w\}_{w \in N[v]}$ from P_v, where $x_w = \text{Enc}_{pk^v}(b_w)$ for a bit $b_w \in \{0, 1\}$.

The first input α_w is a tag which determines whether $\mathcal{L}_{\text{bc-helper}}$ outputs ciphertexts or plaintexts. If all parties do not agree on α_w, $\mathcal{L}_{\text{bc-helper}}$ halts giving no output.

Computation:
- $\mathcal{L}_{\text{bc-helper}}$ reconstructs the secret key $sk^v = \bigoplus_{w \in N[v]} sk^v_w$;
- $\mathcal{L}_{\text{bc-helper}}$ decrypts the bits $b_w = \text{Dec}_{sk^v}(x_w)$;
- $\mathcal{L}_{\text{bc-helper}}$ computes $b = \bigvee_{w \in N[v]} b_w$.

Output:
- If $\alpha_w = \text{"cipher"}$ for all $w \in N[v]$ then $\mathcal{L}_{\text{bc-helper}}$ outputs $y_w = \text{Enc}_{pk^w}(b)$ to each P_w.
- If $\alpha_w = \text{"plain"}$ for all $w \in N[v]$ then $\mathcal{L}_{\text{bc-helper}}$ outputs b to each P_w.

Fig. 5. The functionality $\mathcal{L}_{\text{bc-helper}}$

Claim. The protocol $\Pi^r_{\text{broadcast}}$ UC securely realizes $\mathcal{F}_{\text{broadcast}}$ in the $\mathcal{F}_{\text{graph}}$−hybrid model as long as the network topology graph G is such that

1. $\text{Diameter}(G) < r$;
2. $\mathcal{A}$ does not corrupt any entire closed neighborhood of G.

Simulator. Consider a corrupt party P_v. $\mathcal{S}$ simulates P_v's view as follows:

1. **KeyGen:** $\mathcal{S}$ generates $(pk^v, sk^v) \leftarrow \text{Gen}(1^n)$. When the parties call $\mathcal{L}_{\text{KeyGen}}$, $\mathcal{S}$ returns pk^v to P_v and random strings r^w_v to each P_w such that $w \in N[v]$, instead of shares of P_w's secret key.
2. **Main Computation:** As output to each of the first $r - 1$ calls to $\mathcal{L}_{\text{bc-helper}}$, $\mathcal{S}$ gives output $\{x^c_{v,w}\}_{w,c}$ to P_v, where $x^c_{v,w} = \text{Enc}_{pk^v}(0^n)$ to P_v. To compute the output of the last call of $\mathcal{L}_{\text{bc-helper}}$, $\mathcal{S}$ inputs b_v and all other corrupt parties' input bits to $\mathcal{F}_{\text{broadcast}}$ receiving b^* which it returns to P_v.

Hybrid Argument.

H_0 − This is the real execution of $\Pi^r_{\text{broadcast}}$. Namely, each environment first runs $\mathcal{L}_{\text{KeyGen}}$, after which each P_v has key data $(pk^v, \{sk^v_v\}_{w \in N[v]})$. Then parties enter the loop, running $\mathcal{L}_{\text{bc-helper}}$ r times. Initially, parties enter their secret bit and the key data received from $\mathcal{L}_{\text{KeyGen}}$. In each subsequent call to $\mathcal{L}_{\text{bc-helper}}$, the output from the previous call is also given as input.

Input: P_{graph} inputs the graph G, each P_v inputs a bit $b_v \in \{0,1\}$.

KeyGen: Parties call $\mathcal{L}_{\text{KeyGen}}$ and each P_v receives N[v] and $(\text{pk}^v, \text{sk}^v_v)$ from $\mathcal{L}_{\text{KeyGen}}(v)$ and skv^w from $\mathcal{L}_{\text{KeyGen}}(w)$ for each $w \in \text{N}[v]$.

Main Computation:

- Each P_v sets $x^0_{v,w} = \text{Enc}_{\text{pk}^v}(b_v)$ for each $w \in \text{N}[v]$.
- For $c = 1, \ldots, r-1$, parties call $\mathcal{L}_{\text{bc-helper}}$:
  * P_v gives input $\{\text{``cipher''}; (\text{pk}^v, \text{sk}^v_v); \{x^{c-1}_{v,w}\}_{w \in \text{N}[v]}\}$ to $\mathcal{L}_{\text{bc-helper}}(v)$;
  * For each $w \in \text{N}[v]$, P_v gives input $\{\text{``cipher''}; (\text{pk}^v, \text{sk}^w_v)\}$ to $\mathcal{L}_{\text{bc-helper}}(w)$;
  * P_v receives output $x^c_{v,w}$ from $\mathcal{L}_{\text{bc-helper}}(w)$ for all $w \in \text{N}[v]$.
- Finally, parties call $\mathcal{L}_{\text{bc-helper}}$:
  * P_v gives input $\{\text{``plain''}; (\text{pk}^v, \text{sk}^v_v); \{x^{r-1}_{v,w}\}_{w \in \text{N}[v]}\}$ to $\mathcal{L}_{\text{bc-helper}}(v)$;
  * For each $w \in \text{N}[v]$, P_v gives input $\{\text{``plain''}; (\text{pk}^v, \text{sk}^w_v)\}$ to $\mathcal{L}_{\text{bc-helper}}(w)$;
  * P_v receives the bit $b^*_{v,w}$ as output from $\mathcal{L}_{\text{bc-helper}}(w)$.

Output: P_v outputs $b^*_v = \bigvee_w b^*_{v,w}$.

Fig. 6. The $(\mathcal{L}_{\text{KeyGen}} \| \mathcal{L}_{\text{bc-helper}})$-hybrid protocol $\varPi^r_{\text{broadcast}}$

Finally, P_v receives many copies of the same bit b^*_v as output from the last call to $\mathcal{L}_{\text{bc-helper}}$, and P_v outputs this bit. The view of P_v therefore consists of the following:

1. input $b_v \in \{0,1\}$, output $b^*_v \in \{0,1\}$;
2. key data $\left(\text{pk}^v, \{(\text{sk}^w_v)\}_{w \in \text{N}[v]}\right)$;
3. encryptions $\left\{x^c_{v,w}\right\}^{c=0,\ldots,r-1}_{w \in \text{N}[v]}$.

Let $B \subset V$ be the set of bad parties corrupted by $\mathcal{A}$. The view of the adversary is

$$\left\{\left(b_v, b^*_v; \text{pk}^v, \left\{\text{sk}^w_v\right\}_{w \in \text{N}[v]}; \left\{x^c_{v,w}\right\}_{w,c}\right)\right\}_{v \in B}.$$

H_1 – This is the same as the above experiment except the secret key shares are replaced by random strings. The resulting view is

$$\left\{\left(b_v, b^*_v; \text{pk}^v, \{r^w_v\}_w; \{x^c_{v,w}\}_{w,c}\right)\right\}_{v \in B}.$$

As the secret key sk^v is secret shared among N[v] using a $|\text{N}[v]|$–out–of–$|\text{N}[v]|$ secret sharing scheme, and $\mathcal{A}$ does not corrupt all of N[v], we have that $H_1 \approx H_0$.

H_2 – This is identical to H_1 except that all of the encryptions $x^c_{v,w}$ are changed to encryptions of 0. The resulting view is exactly the view of the ideal world adversary, and is indistinguishable from the view in H_1 by semantic security of the encryption scheme.

4.6   Allowing for Corruption of Whole Neighborhoods

Our protocol $\Pi_{\text{broadcast}}^r$ from the previous section successfully realizes the broadcast functionality while hiding the topology of the graph so long as $\mathcal{A}$ does not corrupt any entire neighborhood of G. If $\mathcal{A}$ were to corrupt $N[v]$ for some v, our protocol immediately becomes insecure, as $\mathcal{A}$ would possess all of the shares of sk^v and so could simply decrypt all of the encrypted bits P_v receives and learn when the broadcast bit reaches P_v. In this section, we show how, given a protocol Π that is secure as long as $\mathcal{A}$ does not corrupt all parties in a $k-$neighborhood, one can construct another protocol Π' for the same functionality as Π, but is secure as long as $\mathcal{A}$ does not corrupt an entire $(k+1)-$neighborhood. The round complexity of Π' will be a constant times the round complexity of Π and so one can only repeat this process logarithmically many times.

The main ideas of this section are essentially the same as those in the previous section; showing that the technique of using local MPC to hide information as it spreads to all parties in the graph is quite general. Like $\Pi_{\text{broadcast}}$, our protocol Π' will be given in the $\mathcal{L}_{\text{MPC}}-$hybrid model, where we will use the ideal functionality $\mathcal{L}_{\text{KeyGen}}$. However, instead of using $\mathcal{L}_{\text{bc-helper}}$, we will use a similar but different local functionality, $\mathcal{L}_\Pi$, shown in Figure 7. Essentially, $\mathcal{L}_\Pi$ allows the role of P_v in Π to be computed using a local MPC by all of the parties in $N[v]$. Then the protocol Π' uses $\mathcal{L}_\Pi$ to execute Π except that each party's role in Π is computed using local MPC by its local neighborhood in Π'. This ensures that if Π is such that any adversary wishing to attack Π must corrupt an entire $k-$neighborhood, then any adversary wishing to attack Π' must corrupt an entire $(k+1)-$neighborhood.

Our hybrid protocol Π' is described in Figure 8. Our statement and construction of simulator are below. We leave out the hybrid argument as it is very similar to the one in subsection 4.5

Claim. The protocol Π' realizes the same functionality as Π. Moreover if Π realizes the functionality UC securely in the $\mathcal{F}_{\text{graph}}-$hybrid model as long as $\mathcal{A}$ does not corrupt an entire $k-$neighborhood of G, then Π' is UC secure in the $\mathcal{F}_{\text{graph}}-$hybrid model as long as $\mathcal{A}$ does not corrupt an entire $(k+1)-$neighborhood of G.

Simulator. We construct a simulator $\mathcal{S}'$ which will make use of the simulator $\mathcal{S}$ for Π. Consider a corrupt party P_v. $\mathcal{S}'$ simulates P_v's view as follows:

1. **KeyGen:** $\mathcal{S}'$ generates $(\text{pk}^v, \text{sk}^v) \leftarrow \text{Gen}(1^n)$ and random shares $\{\text{sk}_w^v\}_{w \in N[v]}$ such that $\bigoplus_w \text{sk}_w^v = \text{sk}^v$. When the parties call $\mathcal{L}_{\text{KeyGen}}$, $\mathcal{S}'$ returns $(\text{pk}^v, \text{sk}_v^v)$ to P_v and sk_w^v to P_w.

2. **Main Computation:** In order to simulate P_v's view we consider two cases:
    Case 1$-(P_v$ has at least one honest neighbor): In this case $\mathcal{S}'$ simulates P_v's view by replacing all the messages P_v would receive with encryptions of 0^n.

Participants/Notation: For $w \in N[v]$, let pk^w be the public key output to P_w by $\mathcal{L}_{\mathrm{KeyGen}}(w)$. Let sk_w^v denote P_w's share of sk^v (the secret key corresponding to pk^v), given as output by $\mathcal{L}_{\mathrm{KeyGen}}(v)$.

Graph Entry Phase: same as in $\mathcal{L}_{\mathrm{MPC}}$

Main Phase:

Input: $\mathcal{L}_\Pi$ receives inputs:
* a round number $c_w \in \{1, \ldots, r\}$ from each $w \in N[v]$;
* $(\mathrm{pk}^w, \mathrm{sk}_w^v)$ from each P_w such that $w \in N[v]$;
* an encrypted transcript so far $\hat{T}_v^{c-1} = \left(x_v, \sigma_v; \{\hat{y}_{v,w}^\ell\}_w^{\ell \leq c-1}\right)$ from P_v, where x_v and σ_v are P_v's input and randomness and $\hat{y}_{v,w}^\ell = \mathrm{Enc}_{\mathrm{pk}^v}(y_{v,w}^\ell)$ is an encryption of the message P_w sent to P_v in the ℓ−th round of Π.

If all parties don't agree on the round number, $\mathcal{L}_\Pi$ halts giving no output.

Computation:
* $\mathcal{L}_\Pi$ reconstructs the secret key $\mathrm{sk}^v = \bigoplus_{w \in N[v]} \mathrm{sk}_w^v$;
* $\mathcal{L}_\Pi$ decrypts $y_{v,w}^\ell = \mathrm{Dec}_{\mathrm{sk}^v}(\hat{y}_{v,w}^\ell)$ for all $w \in N[v]$ and $\ell \leq c-1$;
* $\mathcal{L}_\Pi$ computes the next message function of Π,

$$F_{v,c}^\Pi\left(x_v, \sigma_v, \{y_{v,w}^\ell\}_{w,\ell}\right) = \begin{cases} \{y_{w,v}^c\}_w, & c \leq r-1 \\ z_v, & c = r \end{cases}$$

Output:
* If $c \leq r-1$ then each P_w with $w \in N[v]$ receives $\hat{y}_{w,v}^c = \mathrm{Enc}_{\mathrm{pk}^w}(y_{w,v}^c)$ from $\mathcal{L}_\Pi$.
* If $c = r$ then $\mathcal{L}_\Pi$ outputs z_v to P_v.

Fig. 7. The functionality $\mathcal{L}_\Pi$

Input: P_{graph} inputs the graph G, each P_v inputs x_v, their input to Π.

KeyGen: Parties call $\mathcal{L}_{\mathrm{KeyGen}}$ and each P_v receives $N[v]$ and $\left(\mathrm{pk}^v, \{\mathrm{sk}_v^w\}_{w \in N[v]}\right)$.

Main Computation:
* P_v initializes $\hat{T}_v^0$ to $(x_v, \sigma_v; \emptyset)$.
* For $c = 1, \ldots, r$, parties call $\mathcal{L}_\Pi$:
  · P_v gives input $\left\{c; (\mathrm{pk}^v, \mathrm{sk}_v^v); \hat{T}_v^{c-1}\right\}$ to $\mathcal{L}_{\mathrm{bc\text{-}helper}}(v)$;
  · For each $w \in N[v]$, P_v gives input $\left\{c; (\mathrm{pk}^v, \mathrm{sk}_v^w)\right\}$ to $\mathcal{L}_{\mathrm{bc\text{-}helper}}(w)$;
  · P_v receives output $\hat{y}_{v,w}^c$ from $\mathcal{L}_{\mathrm{bc\text{-}helper}}(w)$ for all $w \in N[v]$.
  · If $c \leq r-1$, P_v updates $\hat{T}_v^c$ to include the messages $\{\hat{y}_{v,w}^c\}_j$ he just received.

Output: When $c = r$, P_v receives z_v from $\mathcal{L}_\Pi(v)$, which it outputs.

Fig. 8. The $(\mathcal{L}_{\mathrm{KeyGen}} \| \mathcal{L}_\Pi)$-hybrid protocol Π'

Case 2−(all of $N[v]$ is corrupt): In this case $\mathcal{A}$ can reconstruct sk^v and so will be able to distinguish if $\mathcal{S}'$ sends encryptions of zero. However, $\mathcal{A}$ does not corrupt an entire $(k+1)$−neighborhood of G which means the set $\{v \in V : N[v]$ is corrupt$\}$ does not contain any $k-$neighborhood. Moreover, since each neighborhood in Π' plays the role of a player in Π, we can simulate the view of such P_v using the simulator $\mathcal{S}$ for Π. Specifically, $\mathcal{S}'$ internally runs $\mathcal{S}$ in order to simulate P_v's view in Π, and encrypts with pk^v to obtain P_v's view in Π'.

5 Topology Hiding MPC against a Fail-Stop Adversary

In this section we consider a stronger adversarial model: the fail-stop adversary. A party controlled by a fail-stop adversary must follow the honest protocol exactly, except that they may abort if the adversary instructs them to.

We have two main results in this section. In section 5.1 we give a general impossibility result, showing that any protocol that implements even a weak version of the broadcast functionality is not IND-CTA secure against fail-stop adversaries. Our proof crucially relies on the ability of the adversary to disconnect the communication graph by aborting with well-placed corrupt parties. In Section 5.2 we show that this is inherent by transforming our broadcast protocol from the previous section into one which is secure against a fail-stop adversary who does not disconnect the graph with his aborts, and who does not corrupt (even semi-honestly) any $k-$neighborhood. We give a high level overview of our techniques of this section before proceeding to the details.

In section 5.1 we consider a protocol Π realizing the broadcast functionality being executed on a line. The proof of the impossibility result is based on two simple observations. First, if some party aborts early in the protocol then honest parties' outputs cannot depend on b. Clearly, if P^* aborts before the information about b has reached him, then no information about b will reach the honest parties on the other side of P^*. This means that the outputs of *all* honest parties must be independent of b, otherwise an adversary would be able to corrupt another party P_{det} to act as a detective. Namely, $\mathcal{A}$ will instruct P_{det} to play honestly and based on P_{det}'s output, $\mathcal{A}$ will be able to guess which side of P^* P_{det} is on. Second, if P^* aborts near the end of the protocol then all parties (other than P^*'s neighbors) must ignore this abort and output what they would have output had nobody aborted. Indeed, if P^* aborts with only k rounds remaining in the protocol, then there simply isn't time for honest parties of distance greater than k from P^* to learn of this abort. Therefore, all honest parties' outputs must be independent of the fact that P^* aborted, lest an $\mathcal{A}$ would be able to employ P_{det} to detect whether is within distance k of P^* or not. This difference in honest parties' outputs when P^* aborts early versus late means there is a round i^* such that the output distribution of P_{det} when P^* aborts in round i^* is distinguishable from P_{det}'s output distribution when P^* aborts in round $i^* + 1$. We take advantage of this by having two aborters P_1^* and P_2^* who abort in rounds i^* and $i^* + 1$. We prove that $\mathcal{A}$ will be able to distinguish the cases from when P_{det} is to the left of P_1^* with the case when he is to the right of P_2^* allowing $\mathcal{A}$ to win the IND-CTA game with non-negligible advantage.

In Section 5.2 we modify our broadcast protocol of section 4 to be secure against a fail-stop adversary who does not disconnect the graph with his aborts.

The idea is to run the semi-honest protocol $2m - 1$ times. Since the adversary can corrupt and abort with at most $m - 1$ parties we are guaranteed that the majority of the executions have no aborts. We ensure that $\mathcal{A}$ learns nothing from the outputs of the executions with aborts by holding off on giving any output until all $2m-1$ executions have occurred. Then we use a final local MPC protocol to compute all outputs, select the majority and output this to all parties.

5.1   Impossibility Result

Definition 3. *We say that a protocol Π weakly realizes the broadcast functionality if Π is such that when all parties execute the protocol honestly, all parties output $\bigvee x_i$ where x_i is P_i's input.*

Note that in weak broadcast, there are no guarantees on the behavior of honest parties if any of the parties deviates from the honest protocol.

Theorem 5. *There does not exist an IND-CTA secure protocol Π that weakly realizes the broadcast functionality in the fail-stop model.*

Let G be a line with m vertices. Namely, $G = (V, E)$ with $V = \{P_1, \ldots, P_m\}$ and $E = \{(P_i, P_{i+1})\}_{i=1,\ldots,m-1}$. Let Π be a protocol executed on G that weakly realizes the broadcast functionality where P_1 (the left most node) is the broadcaster (P_1 has input b, and the inputs to all other nodes is 0). Suppose Π has r rounds. We will show that Π cannot be IND-CTA secure.

Claim. Let $H_{v,b}$ be the event that P_v's output after executing Π matches the broadcast bit b. Let E_i be the event that the first abort occurs in round i. Then either Π is not IND-CTA secure, or there exists a bit $b \in \{0, 1\}$ such that

$$\left| \Pr(H_{v,b} | E_{r-1}) - \Pr(H_{v,b} | E_1) \right| \geq \frac{1}{2} - \mathrm{negl}(n)$$

for all honest P_v whose neighbors do not abort.

Proof. If some P^* aborts during the first round of Π then he disconnects the graph, making it impossible for the parties separated from P_1 to learn about b. These parties' outputs therefore must be independent of b, which implies that there exists a $b \in \{0, 1\}$ such that $\Pr(H_b | E_1) \leq \frac{1}{2}$. If Π is to be IND-CTA secure then it must be that this inequality holds (with possibly a negligible error) for all honest parties. Otherwise an adversary could use the correlation between b and a party's output to deduce that this party is in the same connected component as P_1.

Formally, consider a fail-stop adversary $\mathcal{A}$ who corrupts three parties: the broadcaster P_1, aborter $P^* = P_{\lfloor \frac{m}{2} \rfloor}$ and detective $P_{\det}$. $\mathcal{A}$ then submits (G_0, G_1, S), to the challenger where $G_0 = G$ and G_1 is constructed from G by exchanging the labels of the nodes (P_3, P_4, P_5) with (P_{m-2}, P_{m-1}, P_m). That is, in G_1, the nodes P_3, P_4, P_5 appear at the end of the line. We call $P_{\det} = P_4$ the "detective" node.

The set S consists of the nodes P_1, P^* and $P_{\det}$. Note that $\mathcal{A}$'s neighborhoods are the same in G_0 and G_1 (for $m \geq 12$).

$\mathcal{A}$ instructs P^* to abort during the first round and observes $P_{\det}$'s output. Since P_{m-1}'s output must be independent of b, if P_4's output depends in a non-negligible way on b, this will translate into an advantage for $\mathcal{A}$ in the CTA game.

Finally, note that $\Pr\left(H_{v,b}|E_{r-1}\right) = \Pr\left(H_{v,b}|\text{ no aborts}\right) = 1$ for all P_v which are not neighbors of P^*. The claim follows.

Proof (Proof of Theorem 5).

It follows from Claim 5.1 that there exists a pair $(i^*, b) \in \{1, \ldots, r\} \times \{0, 1\}$ such that

$$\left|\Pr\left(H_{v,b}|E_{i^*}\right) - \Pr\left(H_{v,b}|E_{i^*+1}\right)\right| \geq \frac{1}{2r} - \text{negl}(n). \tag{1}$$

for all honest P_v who do not have an aborting neighbor. Furthermore, assume without loss of generality that $\Pr\left(H_{v,b}|E_{i^*}\right) > \Pr\left(H_{v,b}|E_{i^*+1}\right)$. We construct a fail-stop adversary $\mathcal{A}$ who can leverage this fact to win the CTA game with non-negligible advantage.

Our adversary $\mathcal{A}$ corrupts four parties: the broadcaster P_1, two aborters $\left(P_L^*, P_R^*\right) = \left(P_{\lfloor \frac{m}{2} \rfloor - 1}, P_{\lfloor \frac{m}{2} \rfloor + 1}\right)$, and the detective $P_{\det}$. $\mathcal{A}$ then submits (G_0, G_1, S) to the challenger where $G_0 = G$, G_1 is constructed from G by exchanging (P_3, P_4, P_5) with (P_{m-2}, P_{m-1}, P_m) and $S = \{P_1, P_L^*, P_R^*, P_{\det} = P_4\}$. These graphs are shown in Figure 9. Note that these adversary structures have identical neighborhoods for $m \geq 14$.

Now $\mathcal{A}$ guesses $(i^*, b) \in \{1, \ldots, r\} \times \{0, 1\}$. With non-negligible probability, (i^*, b) is such that inequality (1) is satisfied. $\mathcal{A}$ gives b as input to P_1 and instructs P_L^* to abort on round i^*, P_R^* to abort on round $i^* + 1$. Notice that since the two aborting parties are at distance 2 from each other, the information about P_L^*'s abort does not reach P_R^* by the time he aborts one round later. Therefore, the information about P_L^*'s abort does not reach any of the parties to the right of P_R^* at any point during the protocol. This means that if G_0 was chosen by the challenger, $P_{\det}$'s output will be consistent with E_{i^*} whereas if G_1 was chosen, $P_{\det}$'s output will be consistent with E_{i^*+1}. $\mathcal{A}$ concludes by comparing $P_{\det}$'s output bit to the broadcast bit b. If they are equal, $\mathcal{A}$ sends 0 to the challenger, otherwise he sends 1. The noticeable difference in output distributions ensured by i^* translates to a noticeable advantage for $\mathcal{A}$.

5.2   Feasibility Result

In this section we show how to modify the broadcast protocol from section 4 which is secure against a semi-honest adversary who doesn't corrupt any $k-$neighborhood into one which is secure against a fail stop adversary, who doesn't corrupt a $k-$neighborhood and whose aborts don't disconnect the graph. For simplicity in describing the protocol we take k to be 1, and point out what would have to be changed to accomodate $k = \mathcal{O}(\log n)$. Our protocol, $\Pi_{\text{fstop-bcast}}$, is shown in Figure 11,

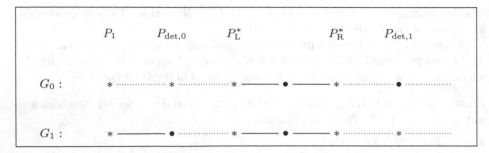

Fig. 9. Graphs used by $\mathcal{A}$ in proof of theorem 5

it makes use of another local functionality $\mathcal{L}_{\mathrm{maj}}$ shown in Figure 12. $\Pi_{\mathrm{fstop\text{-}bcast}}$ realizes the fail-stop broadcast functionality shown in Figure 10

The main idea of our protocol is to run the semihonest protocol many times, ensuring that the majority of the executions contain no aborts. The correctness of the semi-honest protocol guarantees that these executions with no aborts result in correct output. However, to prevent parties from learning which executions contain an abort, we change our protocol so that the outputs of the individual executions are given to the parties in encrypted form, and only after all of them have been completed, $N[v]$ runs a local MPC realizing $\mathcal{L}_{\mathrm{maj}}$ to compute the majority of the outputs it has received. This ensures that all parties will receive the correct output.

We comment that an adversary who aborts during the final majority MPC will stop each local neighborhood he is a part of from being able to reconstruct the output. However, since $\mathcal{A}$ knows which parties he is connected to, forcing them to output $\bot$ does not tell him anything about the graph. If a corrupt party aborts during the main part of the protocol then he ruins the local MPCs running in all closed neighborhoods to which he belongs, but does not affect anything else. Neighbors of the aborting party will get an output of $\bot$ for the current run of the semi-honest protocol, but now that this corrupt party has aborted he cannot ruin any other repitions. The majority at the end will ensure that this abort does not upset the output of $\Pi_{\mathrm{fstop\text{-}bcast}}$.

Input: $\mathcal{F}_{\mathrm{bc\text{-}failstop}}$ receives inputs:
- the graph G from P_{graph};
- a bit $b_v \in \{0, 1\}$ from each P_v;
- a value $\alpha_v \in \{\mathsf{complete}, \mathsf{abort\_early}, \mathsf{abort\_late}\}$ from each corrupt P_v.

Output: $\mathcal{F}_{\mathrm{bc\text{-}failstop}}$ gives outputs:
- $\bot$ to all adversarial P_v such that $\alpha_v \in \{\mathsf{abort\_early}, \mathsf{abort\_late}\}$;
- $\bot$ to all honest P_v who are adjacent to an adversarial P_w such that $\alpha_w = \mathsf{abort\_late}$;
- $\bigvee_v b_v$ to all other parties.

Fig. 10. The fail-stop broadcast functionality $\mathcal{F}_{\mathrm{bc\text{-}failstop}}$

Input: P_{graph} inputs the graph G, each P_v inputs a bit $b_v \in \{0, 1\}$.

KeyGen: Parties call $\mathcal{L}_{\text{KeyGen}}$ and each P_v receives $N[v]$, $(\text{pk}^v, \text{sk}_v^v)$ from $\mathcal{L}_{\text{KeyGen}}(v)$ and sk_v^w from $\mathcal{L}_{\text{KeyGen}}(w)$ for all $w \in N[v]$. If a party aborts during this phase then start again, ignoring that party for the remainder of the protocol.

Main Computation For $j = 1, \ldots, 2m - 1$ do:

- Each P_v sets $\hat{x}_{v,w}^0 = \text{Enc}_{\text{pk}^v}(b_v)$ for each $w \in N[v]$.
- For $i = 1, \ldots, r$, parties call $\mathcal{L}_{\text{bc-helper}}$:
  * P_v gives input $\left\{ \text{``cipher''}; \ (\text{pk}^v, \{\text{sk}_v^w\}_w); \ \{\hat{x}_{v,w}^{i-1}\}_w \right\}$;
  * P_v receives output $\{\hat{x}_{v,w}^i\}_w$.
- P_v picks $w \in N[v]$ randomly and sets $\hat{y}_v^j = \hat{x}_{v,w}^r$.
- If a party aborts during this computation all of his neighbors set $\hat{y}_v^j = \text{Enc}_{\text{pk}^v}(\bot)$ and ignore the aborting party for the remainder of the protocol.

Output:

- Parties call $\mathcal{L}_{\text{maj}}$:
  * P_v gives input $\left\{ (\text{pk}^v, \{\text{sk}_v^w\}_w); \ \{\hat{y}_v^j\}_j \right\}$;
  * P_v receives output b_v^*
- Each P_v outputs b_v^*.

Fig. 11. The protocol $\Pi_{\text{fstop-bcast}}$ (in the $\mathcal{L}_{\text{MPC}}$−hybrid model)

Graph Entry Phase: same as in $\mathcal{L}_{\text{MPC}}$

Main Phase:

Input: $\mathcal{L}_{\text{maj}}$ receives inputs:

- $(\text{pk}^v, \{\text{sk}_v^w\}_{w \in N[v]})$ − the data P_v receives from $\mathcal{L}_{\text{maj}}$
- $\{\hat{y}_v^1, \ldots, \hat{y}_j^{2m-1}\}$ − encryptions $\hat{y}_v^j = \text{Enc}_{\text{pk}^v}(b_v^j)$ of values $b_v^j \in \{0, 1\} \cup \{\bot\}$

from each P_v.

Computation:

- $\mathcal{L}_{\text{maj}}$ reconstructs the secret key $\text{sk}^v = \bigoplus_{w \in N[v]} \text{sk}_w^v$ for each $v \in V$.
- For each $v \in V$, $\mathcal{L}_{\text{maj}}$ decrypts the bits $b_v^j = \text{Dec}_{\text{sk}^v}(\hat{y}_v^j)$ for $j = 1, \ldots, 2m - 1$.
- $\mathcal{L}_{\text{maj}}$ computes $b_v = \text{MAJ}_j(b_v^j)$, where MAJ is the majority function.

Output: $\mathcal{L}_{\text{maj}}$ gives P_v the output b_v.

Fig. 12. The functionality $\mathcal{L}_{\text{maj}}$

Finally, we comment that if the local MPC run at the end to compute the majority is executed by the parties in $N[v]$, then the resulting protocol will only be secure if $\mathcal{A}$ does not corrupt any closed neighborhood in G. However, we can compile a protocol which is secure against a semi-honest $\mathcal{A}$ who does not corrupt all parties in a k−neighborhood of G into one that is secure against a fail-stop $\mathcal{A}$ simply by having the last MPC be computed by all the parties in the k−neighborhood of v. This involves implementing a message passing protocol between all parties in the k−neighborhood of v which may be done similarly to $\Pi_{\text{msg-transmit}}$.

6   Discussion and Open Questions

Malicious Model. The most basic question we leave open in this work is the situation in the malicious model. Clearly, our impossibility results for the fail-stop model also apply here. Our positive results do not carry over, however. This is because a malicious adversary can "pretend" to be connected to an entire graph; in this fake graph, the adversary can corrupt *any* size neighborhood, violating our security assumptions.

General Graphs in the Semi-Honest Model. A second natural question that arises from this work is whether the restriction to graphs of logarithmic diameter is a necessary one, even in the semi-honest model. Does there exist a protocol for topology-hiding secure computation in arbitrary graphs?

Hiding the Identities of Neighbors. Another open problem we leave is whether topology-hiding security can be realized while hiding from P_v the identities of his neighbors. This would involve changing the $\mathcal{F}_{\text{graph}}$ functionality to give as output a local identity for each of the parties in $\mathcal{P}_v$'s neighborhood, and this identity would differ in other closed neighborhoods. One interesting application would be that adversaries in the same local neighborhood would not learn that they are distance 2 from each other. Even our semi-honest protocol revealed this information as parties at distance 2 had to communicate in local MPCs.

References

1. http://www.its.dot.gov/research/v2v.htm
2. http://www.nytimes.com/2010/05/12/nyregion/12about.html
3. Beaver, D., Micali, S., Rogaway, P.: The round complexity of secure protocols (extended abstract). In: STOC, pp. 503–513. ACM (1990)
4. Ben-Or, M., Goldwasser, S., Wigderson, A.: Completeness theorems for non-cryptographic fault-tolerant distributed computation (extended abstract). In: STOC, pp. 1–10. ACM (1988)
5. Boyle, E., Goldwasser, S., Tessaro, S.: Communication locality in secure multi-party computation - how to run sublinear algorithms in a distributed setting. In: Sahai, A. (ed.) TCC 2013. LNCS, vol. 7785, pp. 356–376. Springer, Heidelberg (2013)
6. Canetti, R.: Universally composable security: A new paradigm for cryptographic protocols. Cryptology ePrint Archive, Report 2000/067 (2000), http://eprint.iacr.org/2000/067
7. Canetti, R.: Universally composable security: A new paradigm for cryptographic protocols. In: FOCS, pp. 136–145. IEEE Computer Society (2001)
8. Canetti, R., Lindell, Y., Ostrovsky, R., Sahai, A.: Universally composable two-party and multi-party secure computation. Cryptology ePrint Archive, Report 2002/140 (2002), http://eprint.iacr.org/2002/140
9. Chandran, N., Garay, J., Ostrovsky, R.: Edge fault tolerance on sparse networks. In: Czumaj, A., Mehlhorn, K., Pitts, A., Wattenhofer, R. (eds.) ICALP 2012, Part II. LNCS, vol. 7392, pp. 452–463. Springer, Heidelberg (2012)

10. Chaum, D.: Untraceable electronic mail, return addresses, and digital pseudonyms. Commun. ACM 24(2), 84–88 (1981)
11. Damgård, I., Ishai, Y., Krøigaard, M.: Perfectly secure multiparty computation and the computational overhead of cryptography. In: Gilbert, H. (ed.) EUROCRYPT 2010. LNCS, vol. 6110, pp. 445–465. Springer, Heidelberg (2010)
12. Damgård, I.B., Nielsen, J.B.: Scalable and unconditionally secure multiparty computation. In: Menezes, A. (ed.) CRYPTO 2007. LNCS, vol. 4622, pp. 572–590. Springer, Heidelberg (2007)
13. Dolev, D.: The byzantine generals strike again. J. Algorithms 3(1), 14–30 (1982)
14. Dwork, C., Peleg, D., Pippenger, N., Upfal, E.: Fault tolerance in networks of bounded degree. SIAM J. Comput. 17(5), 975–988 (1988)
15. Franklin, M.K., Yung, M.: Communication complexity of secure computation (extended abstract). In: STOC, pp. 699–710. ACM (1992)
16. Garay, J.A., Ostrovsky, R.: Almost-everywhere secure computation. In: Smart, N.P. (ed.) EUROCRYPT 2008. LNCS, vol. 4965, pp. 307–323. Springer, Heidelberg (2008)
17. Goldreich, O.: Foundations of Cryptography: Basic Applications, vol. 2. Cambridge University Press, New York (2004)
18. Goldreich, O., Micali, S., Wigderson, A.: How to play any mental game or a completeness theorem for protocols with honest majority. In: STOC, pp. 218–229. ACM (1987)
19. Hinkelmann, M., Jakoby, A.: Communications in unknown networks: Preserving the secret of topology. Theor. Comput. Sci. 384(2-3), 184–200 (2007)
20. Pass, R.: Bounded-concurrent secure multi-party computation with a dishonest majority. In: Babai, L. (ed.) STOC, pp. 232–241. ACM (2004)
21. Pass, R.: Bounded-Concurrent Secure Multi-Party Computation with a Dishonest Majority. In: Proceedings of the 36th Annual ACM Symposium on Theory of Computing, STOC 2004, pp. 232–241 (2004)
22. Pass, R., Rosen, A.: Bounded-concurrent secure two-party computation in a constant number of rounds. In: FOCS, pp. 404–413. IEEE Computer Society (2003)
23. Rabin, T., Ben-Or, M.: Verifiable secret sharing and multiparty protocols with honest majority (extended abstract). In: STOC, pp. 73–85. ACM (1989)
24. Reed, M.G., Syverson, P.F., Goldschlag, D.M.: Anonymous connections and onion routing. IEEE Journal on Selected Areas in Communications 16(4), 482–494 (1998)
25. Reiter, M.K., Rubin, A.D.: Anonymous web transactions with crowds. Commun. ACM 42(2), 32–38 (1999)
26. Yao, A.C.-C.: Protocols for secure computations (extended abstract). In: FOCS, pp. 160–164. IEEE (1982)

Secure Physical Computation Using Disposable Circuits

Ben A. Fisch[1], Daniel Freund[2], and Moni Naor[3,*]

[1] Columbia University, USA
benafisch@gmail.com
[2] Cornell University, USA
freund90@mac.com
[3] Weizmann Institute of Science, Israel
moni.naor@weizmann.ac.il

Abstract. In a *secure physical computation*, a set of parties each have physical inputs and jointly compute a function of their inputs in a way that reveals no information to any party except for the output of the function. Recent work in CRYPTO'14 presented examples of physical *zero-knowledge* proofs of physical properties, a special case of secure physical two-party computation in which one party has a physical input and the second party verifies a boolean function of that input. While the work suggested a general framework for modeling and analyzing physical zero-knowledge protocols, it did not provide a general theory of how to prove any physical property with zero-knowledge. This paper takes an orthogonal approach using *disposable circuits* (DC)—cheap hardware tokens that can be completely destroyed after a computation—an extension of the familiar tamper-proof token model. In the DC model, we demonstrate that two parties can compute any function of their physical inputs in a way that leaks at most 1 bit of additional information to either party. Moreover, our result generalizes to any multi-party physical computation. Formally, our protocols achieve unconditional UC-security with input-dependent abort.

1 Introduction

In a *secure two-party computation* (2PC), parties A and B each have secret inputs x_A and x_B respectively, and they jointly compute a function $f(x_A, x_B)$ such that each party receives the output of the function, but no further information about the other party's secret input. A *secure multi-party computation* (MPC) extends 2PC to an arbitrary number of parties each with private inputs to a multi-party function. As early as the 1980s, various results demonstrated that any two-party or multi-party function can be securely computed under standard cryptographic assumptions [21,9].

* Incumbent of the Judith Kleeman Professorial Chair. Research supported in part by grants from the Israel Science Foundation, BSF and Israeli Ministry of Science and Technology and from the I-CORE Program of the Planning and Budgeting Committee and the Israel Science Foundation.

Y. Dodis and J.B. Nielsen (Eds.): TCC 2015, Part I, LNCS 9014, pp. 182–198, 2015.

In the present work, we consider secure computation in a physical context where each party's input is a physical entity. For example, suppose parties A and B wish to securely compute a function of their DNA (e.g. whether they are both carriers of a certain recessive gene). Unless the parties trust each other to honestly supply their genetic information as digital inputs, they cannot use standard secure 2PC to solve this task. They also may not trust one another to directly measure each other's genetic information and potentially learn more than the necessary information. To solve this task, the parties require a *secure physical computation* that is guaranteed to return to both parties the correct function output of the physical inputs and simultaneously prevent either party from learning more than the function output.

A special case of physical secure computation is a *physical zero-knowledge proof*, where a Prover must prove to a Verifier that a physical input object satisfies a physical property without revealing any further information about the input. Informal examples of physical zero-knowledge have appeared throughout the literature [19,20]. Recent work of [4] introduced formal definitions and an analysis framework for physical zero-knowledge protocols, and gave examples of protocols for comparing DNA profiles and neutron radiographs of objects. Earlier work of [7,8] used physical zero-knowledge techniques to demonstrate that two nuclear warheads have the same design without revealing further information on the design. The necessity for such a protocol arises in nuclear disengagement, where one country must prove to the other that it is destroying authentic nuclear weapons without revealing sensitive information on that weapon's design. Previous techniques relied on *information barriers*, or trusted devices that would measure the input objects and display only a red (reject) or green (accept) light. However, in absence of trust, information barriers are problematic as they may reveal too much information to the inspecting party, or may display false information. The protocols in [4] and [7] avoid information barriers and instead use techniques that physically manipulate the inputs and enable the inspecting party to verify physical properties without ever recording any sensitive information. Such protocols seem to be inherently problem-specific, and there is no general zero-knowledge protocol presented for proving any arbitrary physical property of any input.

We revisit the idea of an information barrier using special devices we call *disposable circuits*: instead of avoiding ever recording sensitive information, we suggest using cheap computing devices such as smart cards that can be destroyed or have their memory completely wiped after the computation. Each party will create a disposable device that can investigate the physical inputs and perform the necessary computations. Since neither party can trust the other party's device to act as a true information barrier, we can only allow party A's device to directly supply output to party B and party B's device to directly supply output to party A. Each party must remain isolated from its device during the computation, and the opposing party should be able to destroy or memory-wipe the device after the computation is complete. On the other hand, each party can only trust the correctness of its own device's output. Thus, the two parties need a secure protocol to verify the authenticity of the outputs received.

Many previous works have explored the use of physical hardware and physical separation in cryptographic protocols. Ben-Or et. al. [1] introduced the *multi-prover interactive proof system* (MIP) model, and showed that with two physically isolated provers it is possible to achieve unconditionally secure ZK proofs for NP. In addition, they showed similar results for bit commitment and Rabin-OT. Goldreich and Ostrovsky [10] demonstrated uses of tamper-proof hardware for the purpose of software protection. Katz [15] initiated a long line of works that used tamper-proof hardware to achieve universally composable (UC) security in multi-party computation protocols. In particular, Goyal et. al. [12] showed that tamper-proof hardware could be used to achieve unconditional UC-secure non-interactive multi-party computation.

Our model of disposable circuits is an extension of the standard stateful tamper-proof token model, which was first formalized by Katz [15]. The tokens in our model have the additional capability to measure physical inputs. In particular, a party in our model has the ability to create a token that will directly probe a specific physical input held by another party. We assume that both parties can reference the same object by some uniquely identifiable physical information even if all further physical characteristics of that object are secret. For example, the physical input might be a box held by the second party whose contents are secret, and yet both parties can identify the box by its exterior characteristics and physical location. The issuer of a token could certify that the token is investigating a specified object simply through physical observation. Alternatively, the token itself could be programmed to recognize the pre-specified characteristics used to reference the object. The motivation for this functionality is to remove any need for parties to trust each other to supply correct physical inputs or honest descriptions of physical inputs to the tokens.

Another nuance in our model is the assumption that any user of a token can destroy or memory-wipe that token. Katz's original model did not make any assumptions on the destruction of tokens, but several subsequent works considered tokens with a self-destruction capability. For example, Goldwasser et. al. [11] introduced *one-time memory* (OTM) tokens that immediately self-destruct after delivering output. Goyal et. al. used OTM tokens as well in their construction of unconditionally secure MPC. However, in these works, the self-destruct behavior was only used to make the hardware token tamper-proof against its user, and not to prevent its creator from recovering information. In fact, an important property of the standard tamper-proof token model is that it does not require mutual trust—each party only needs to trust the physical assumptions about its own hardware. This distinguishes the tamper-proof token model from similarly powerful models that require the parties to agree upon a trusted setup, such as the *common reference string* (CRS) model. Our disposable circuit model similarly only requires the device creator to trust a device's full functionality and tamper-proof properties, but we additionally require the user to trust one physical assumption: that the device can be destroyed.

We show that in this disposable circuit model, a prover party can prove any physical property of an arbitrary physical input to a verifier party in a way that leaks at most a single bit about the input. The high-level idea is to allow the verifier to create a tamper-proof device that will investigate the object in isolation. To restrict the communication between the device and verifier, the prover party will mediate all communication between the device and the verifier. The device will conditionally reveal to the prover a secret string after checking the validity of the physical property. The prover could send this string back to the verifier as evidence of the property's validity, but would need to ensure that this string does not reveal any further information to the verifier. To accomplish this, the prover and verifier will execute a secure 2PC protocol that verifies whether both parties "know" the same secret. The device can always communicate an arbitrary bit to the verifier because it can cause the protocol to abort dependent on the physical input. Crucially, the device is destroyed or memory-wiped at the end of the protocol so that it cannot reveal any further information.

More generally, we show that it is possible for two parties to evaluate any function of their physical inputs in a way that leaks at most one bit of information about the inputs. Each party can issue a token to investigate the opposing party's physical input and output a *message authentication code* (MAC) signature on the input description to the opposing party. The two parties then invoke a secure 2PC functionality that verifies the MAC signatures and computes the desired function. Moreover, we show that this approach can be easily extended for physical multi-party computation (MPC). Each party creates multiple tokens, one for each of the other parties. Each token investigates a single party's input and outputs a MAC signature on the input description to that party. The parties each collect the MAC signatures they have received from the other parties' tokens, and input these signatures along with their input descriptions to an MPC functionality that verifies the signatures and computes the desired function.

Building on the result that MPC can be unconditionally realized in the tamper-proof token model [12], our protocols are unconditionally secure (i.e. they do not rely on computational assumptions). We will analyze the security of our protocols using the notion of *security with input-dependent abort* introduced in [14], which cleanly captures the idea that an adversary can learn at most 1 bit about another party's input by causing the protocol to abort conditioned on that party's input.

Lastly, we present in Section 5 a protocol called *isolated circuit secure communication* that allows the two isolated disposable circuits to communicate secretly with each other without leaking any information to A or B and discuss its potential applications. We conclude in Section 6 with a question left open by our work.

2   Preliminaries

In this section we provide further details and formal specifications of the underlying definitions and concepts we address in this paper. We provide formal specifications in terms of *ideal functionalities* (see Canetti [2]). The ideal functionality definition of a protocol or token describes its target behavior in an *ideal world* using a trusted entity.

Disposable Circuits Model. Our formalization of the disposable circuits (DC) model is based on the stateful tamper-proof token model of Katz [15]. Katz formally defines a "wrapper" functionality $\mathcal{F}_{WRAP}$ to model a real world system in which any party can construct a hardware device encapsulating an arbitrary software program and pass it to a user party, who may only interact with the embedded software in a black-box manner. Formally, $\mathcal{F}_{WRAP}$ takes two possible commands: a "creation" command, which initializes a new token T_ρ implementing a stateful interactive polynomial-time program ρ (e.g., an ITM), and an "execution" command, which causes ρ to run on a supplied input and return output. Additionally, the token T_ρ may be partially isolated so that it cannot communicate with its creator and only interacts with a specified user. This is captured in $\mathcal{F}_{WRAP}$ by having the creator party P specify a user party P', whereafter $\mathcal{F}_{WRAP}$ stores (P, P', ρ) and only accepts subsequent "execution" commands from P' to run T_ρ.

We define the ideal functionality $\mathcal{F}_{DC}$ (see Figure 1) as an extension of $\mathcal{F}_{WRAP}$. We model the ability of tokens to directly measure physical inputs in the environment following the paradigm in [4]. Every real world physical object x is assigned a unique identifier id_x, which captures the public meta information that the parties in the real world system use to reference the same object. Upon receiving the identifier id_x, $\mathcal{F}_{DC}$ queries an ideal world oracle for a physical measurement $\mathcal{M}$ with id_x, which returns the logical output of the real world physical measurement $\mathcal{M}(x)$. However, without loss of generality, we may assume that $\mathcal{M}$ is an identity function (outputting a canonical description of x and all its physical characteristics), and hence we eliminate $\mathcal{M}$ from the formal description of $\mathcal{F}_{DC}$ for simplicity. Instead, upon receiving id_x as input to a token implementing ρ, $\mathcal{F}_{DC}$ directly computes $\rho(\bar{x})$ where $\bar{x}$ is the canonical description of x. We also model the capability of the token creator to certify that the token will be used only to investigate a specific object by optionally including id_x in the "creation" command. Additionally, we add a "destroy" command to $\mathcal{F}_{DC}$ that deletes the tuple representing an active token and outputs $\bot$ to the issuer of the command. This models the real world ability of users to destroy or memory-wipe hardware tokens at any point in time as well as certify that the device has been successfully destroyed.

Secure Physical Property Verification (SPV). SPV involves two parties, a prover P and a verifier V. The prover holds a physical input x, and will allow the verifier to verify a physical property π of x, i.e. certify that $\pi(x) = 1$. The ideal functionality for SPV is described below in Figure 2, and only slightly modifies the physical zero-knowledge ideal functionality $\mathcal{F}_{ZK}^{\Pi}$ definition described in [4].

Functionality $\mathcal{F}_{DC}$

$\mathcal{F}_{DC}$ is parametrized by a polynomial $p(\cdot)$ and an implicit security parameter k.

"Creation" Upon receiving (create, $sid, P, P', id_x, mid, \rho$) from a party P where ρ is the description of a deterministic program (e.g., an interactive Turing machine), mid is a machine id, and id_x either uniquely references a physical object x or is set to $\bot$:

1. Send (create, $sid, P, P', id_x, mid, \rho$) to P'.
2. Store $(P, P', id_x, mid, \rho, \emptyset)$.

"Execution" Upon receiving (run, sid, P, mid, msg) from party P':

1. Find the unique tuple $(P, P', id_x, mid, \rho, state)$. If no such tuple is stored, then do nothing.
2. If $id_x \neq \bot$ and $msg \neq id_x$, then do nothing.
3. If $msg = id_x$, set input $= \bar{x}$, the canonical description x.
4. If $id_x = \bot$, then set input $= msg$.
5. Run $\rho(\text{input}, state)$ for at most $p(k)$ steps. Let $(out, state')$ denote the result. If ρ does not terminate in $p(k)$ steps, then set $out = \bot$ and $state' = state$.
6. Send (sid, P, mid, out) to P', store $(P, P', id_x, mid, \rho, state')$ and erase $(P, P', id_x, mid, \rho, state)$.

"Destroy" Upon receiving (destroy, sid, mid, P) from party P', erase any tuple of the form $(P, P', *, mid, *, *)$ and return $\bot$ to P'.

Fig. 1. Ideal world disposable circuit functionality

Secure Physical Two-Party Computation (2PC). Secure physical 2PC involves two parties, A and B, who each hold physical inputs x and y respectively. A function of x and y has two components: a pair of physical measurements $(\mathcal{M}_1, \mathcal{M}_2)$ and a mathematical function $f(\mathcal{M}_1(x), \mathcal{M}_2(y))$. However, in defining the ideal functionality $\mathcal{F}_{\text{Physical2PC}}$, we may assume without loss of generality that $\mathcal{M}_1$ and $\mathcal{M}_2$ are "identity" measurements that each output a sufficiently detailed description of their physical inputs and relevant physical properties. Thus, our definition instead allows $\mathcal{F}_{\text{Physical2PC}}$ to query an oracle $\mathcal{O}$ that outputs canonical descriptions of physical objects. The ideal functionality is formally defined in Figure 3.

Secure Physical Multi-Party Computation (MPC). Secure physical MPC naturally extends the definition of physical 2PC to N parties with N physical inputs $x_1, ..., x_N$. The ideal functionality $\mathcal{F}_{\text{Physical}MPC}$ receives an input identifier id_{x_i} from each ith party, it derives the canonical description $\mathcal{O}(id_{x_i}) = \bar{x}_i$ for each i, computes a function $f(\bar{x}_1, ..., \bar{x}_N)$, and outputs this value to all parties.

Functionality $\mathcal{F}_{SPV}^{\Pi}$

The functionality is parametrized by a physical property Π. We model the physical measurement $\mathcal{M}^{\Pi}$ for verifying Π as an ideal functionality $\mathcal{F}_{\mathcal{M}}^{\Pi}$. We assume any physical input x has a unique identifier id_x in the ideal world.

- Upon receiving $(id_x, \mathsf{Prover}, \mathsf{Verifier})$ from the party Prover, $\mathcal{F}_{SPV}^{\Pi}$ queries $\mathcal{F}_{\mathcal{M}}^{\Pi}$ to compute $\Pi(x)$, and sends $(\Pi(x), id_x, \Pi)$ to $\mathsf{Verifier}$. Finally, if $\mathsf{Verifier}$ receives $\Pi(x) = 1$, it outputs ACCEPT. Otherwise $\mathsf{Verifier}$ outputs REJECT.

Fig. 2. Ideal world Secure Property Verification

Functionality $\mathcal{F}_{\mathsf{Physical2}PC}$

Upon initiation, the functionality is supplied with a function $f : \{0,1\}^* \times \{0,1\}^* \to \{0,1\}^*$ and access to an oracle $\mathcal{O}$, which on input id_x identifying a physical object x outputs a canonical description including physical properties of x.

- Upon receiving id_x from the party A and id_y from the party B, query $\mathcal{O}$ and record $\bar{x} = \mathcal{O}(id_x)$ and $\bar{y} = \mathcal{O}(id_y)$. Output $f(\bar{x}, \bar{y})$ to both parties A and B.

Fig. 3. Ideal world physical 2PC

Unconditional One-Time MAC. We denote by (MAC, VF) an unconditionally secure one-time message authentication code scheme. A message m is signed with a key k as $(m, MAC_k(m))$. The verification algorithm VF computes $VF_k(m, \sigma) = 1$ if and only if σ is a correct signature of m with key k. An example of an unconditional one-time MAC is the map $x \mapsto a \cdot x + b$ over a finite field, where the key $k = (a, b)$ is used to sign m as $a \cdot m + b$.

Proving Security of Protocols. Our security analysis uses the UC-framework [3]. We compare the execution of a *real* protocol ρ (defined in a given computational environment) with a static malicious adversary $\mathcal{A}$ corrupting a subset of the parties to an *ideal* process where the parties only interact with the ideal functionality $\mathcal{F}_\rho$ for ρ and an ideal world adversary $\mathcal{S}$, also called the simulator. When $\mathcal{A}$ (resp. $\mathcal{S}$) corrupts a party, it learns that party's entire state, and takes control of its communication. We define an environment $\mathcal{Z}$ that sets the parties' inputs in both executions and observes their outputs. Additionally, $\mathcal{Z}$ may communicate freely with the adversary $\mathcal{A}$ in the real protocol and $\mathcal{S}$ in the ideal protocol. Let $\mathrm{REAL}_{\rho, \mathcal{A}, \mathcal{Z}}$ denote the output of $\mathcal{Z}$ after interacting with the real protocol and let $\mathrm{IDEAL}_{\mathcal{F}_\rho, \mathcal{A}, \mathcal{Z}}$ denote its output after interacting with the ideal process.

Definition 1. *(Realizing an ideal functionality) A protocol ρ* **securely realizes** *the ideal functionality $\mathcal{F}_\rho$ if for any real world adversary $\mathcal{A}$ there exists an ideal adversary $\mathcal{S}$ such that $\mathrm{REAL}_{\rho, \mathcal{A}, \mathcal{Z}} \sim \mathrm{IDEAL}_{\mathcal{F}_\rho, \mathcal{S}, \mathcal{Z}}$ for every environment $\mathcal{Z}$.*

We will prove that our protocols satisfy a slightly relaxed definition of security called *security with input-dependent abort* [14], which allows an adversary to learn at most 1 bit of information by causing the protocol to abort dependent on another party's inputs. Formally, given any multi-party functionality $\mathcal{F}$, we define the functionality $\mathcal{F}^{\mathsf{IDA}}$ as follows. $\mathcal{F}^{\mathsf{IDA}}$ receives inputs from the parties and runs an internal copy of $\mathcal{F}$. In addition, it receives the description of a predicate ϕ from the adversary. Before delivering any output from $\mathcal{F}$, it evaluates ϕ with all the inputs. If ϕ outputs 1, then $\mathcal{F}^{\mathsf{IDA}}$ replaces the outputs to the honest parties with abort. Otherwise, $\mathcal{F}^{\mathsf{IDA}}$ delivers the correct outputs from $\mathcal{F}$ to both parties. We make only one modification to this definition to account for functionalities with physical inputs: if id_x denotes a physical input identifier, then ϕ is evaluated on the canonical description $\bar{x}$ of the object identified by id_x (along with the other inputs).

Definition 2. *(Security with input-dependent abort) A protocol ρ implementing the ideal functionality $\mathcal{F}_\rho$ is* **secure with input-dependent abort** *if ρ securely realizes $\mathcal{F}_\rho^{\mathsf{IDA}}$.*

3 Secure Property Verification

In this section, we present a solution for SPV that is secure with input-dependent abort. The general setup is as follows. The input to the protocol is the physical object x and a physical property Π. We express Π as a boolean function so that $\Pi(x) \in \{0,1\}$. The verifier, denoted V, programs a DC token C_V to examine x, and compute $\Pi(x)$. In the formal description, C_V is replaced with the ideal functionality $\mathcal{F}_{DC}$, as described in Section 2. The token C_V must then communicate $\Pi(x)$ to V via an indirect channel through the prover, denoted P, who is mediating the communication. The prover's task is to limit this communication to a single bit.

It is straightforward for C_V to authenticate any message it sends to the verifier via the prover by signing the message using a MAC and a secret key it shares with V. In fact, since the message only consists of a single bit, it suffices for C_V to either send the secret key itself or nothing. More precisely, let C_V and V share a secret string sk of length λ (a security parameter). After investigating x, C_V sends sk to P if and only if $\Pi(x) = 1$. P must demonstrate to V that it "knows" sk as "evidence" that $\Pi(x) = 1$. P could simply send sk to V, but would need to ensure that V cannot learn any further information from sk. Namely, P should be convinced that V already "knows" sk as well.

Solutions from Bit Commitment. A natural way to convince P that V already knows sk is to use a commitment protocol in which V sends P a commitment to sk that C_V can later de-commit (without involving or even communicating with V). We discuss briefly a few solutions for this task that we can easily derive from existing protocols in the literature. One suitable protocol is the "commit and reveal" protocol of [1] for two isolated provers in the MIP model, where V acts as one prover and C_V as the second. The UC-secure commitment protocol of Moran and Segev [18] is also applicable, but their protocol uses V to send both the commitment and de-commitment while C_V is used for a validity check. It is easy to modify the protocol so that the token C_V de-commits the secret instead of V, although to maintain UC-security we would need to involve a third isolated token for the validity check (the simulator can extract the committed value by rewinding this token whereas it cannot rewind V). Another option is to adapt the standard construction of bit commitment from Rabin-OT [16]. In the generalized version of this construction using error-correcting codes (e.g., [6,5]), the sender computes a randomized encoding enc(s) of the input string s to be committed, and obliviously transmits to the receiver a subset of the bits of enc(s) via Rabin-OT. To de-commit, the sender reveals s, and the receiver is able to check that enc(s) is consistent with the previously transmitted bits. In our setting, P would simply request from V to see a random subset of the bits of enc(s), and C_V would de-commit by sending s.

All of the above solutions would be unconditionally secure in our model, and we imagine it would not be hard to obtain UC-security as well. However, it is simpler to present and prove our feasibility result by making black-box use of previously defined functionalities that are known to be unconditionally UC-secure in the tamper-proof token model (and the disposable-circuit model by extension). The above solutions could not use the standard ideal commitment functionality $\mathcal{F}_{COM}$ in a black-box way because $\mathcal{F}_{COM}$ does not separate the party that sends the commitment from the party (or token) that reveals the commitment, which seems crucial in our application.

Solution from 2PC. In an effort to present and prove our result in the simplest possible way, we offer a solution using general 2PC, which can be realized with unconditional UC-security in the tamper-proof token model [12]. V and P will run a secure 2PC functionality that takes inputs x and y from each party, and either outputs 1 if $x = y$ and otherwise outputs abort. If both parties are honest, they will both supply the input sk. If the honest V receives the output 1, then it is convinced that P knows sk, which constitutes a proof that C_V verified $\Pi(x) = 1$ (up to the negligible probability that P guessed sk). More formally, given any UC-secure protocol implementing this 2PC functionality, if the protocol execution outputs 1 when V supplies sk, then there exists an efficient simulator that can extract sk from P. Even if V is dishonest, it can learn at most 1 bit from the protocol by causing the 2PC subprotocol to either abort or succeed dependent on the input that the honest P supplies. We formally describe the protocol in Figure 4 and prove that it securely realizes SPV with input-dependent abort.

Claim 1: *Protocol 1 securely realizes $\mathcal{F}_{SPV}^{\Pi}$ with input-dependent abort (unconditionally).*

Protocol 1 - SPV

Setup: Let λ denote a security parameter. The Verifier chooses a random password string sk in $\{0,1\}^{\lambda}$. Let $\bar{x}$ denote a canonical description of the physical input x that the prover supplies. Let Π be a boolean function so that $\Pi(\bar{x}) \in \{0,1\}$. Recall that $\mathcal{F}_{DC}$ is able to derive $\bar{x}$ given id_x, a unique identifier for the object. The prover claims that $\Pi(\bar{x}) = 1$.

1. (*V* **initiates a token**) Let ρ define a program that on input $\bar{x}$ outputs sk if $\Pi(\bar{x}) = 1$ and outputs $\bot$ otherwise. V sends $(\text{create}, sid, V, P, id_x, mid, \rho)$ to $\mathcal{F}_{DC}$.
2. (*P* **queries the token**) P submits $(\text{run}, sid, V, mid, id_x)$ to $\mathcal{F}_{DC}$ and receives in response $\rho(\bar{x})$.
3. (*P* **destroys the token**) P submits $(\text{destroy}, sid, mid, V)$ to $\mathcal{F}_{DC}$ and waits to receive $\bot$ in response.
4. (**2PC equality check**) P inputs $\rho(\bar{x})$ and V inputs sk to a protocol that unconditionally UC-realizes the following ideal functionality:
   - EqualityChecker: On input (msg, sid, P, V) from party P and (msg', sid, V, P) from party V, output 1 to both parties if $msg = msg'$, and otherwise output abort.
5. If V receives the output 1 from EqualityChecker, then it outputs Accept. If either party receives abort, then it outputs abort.

Fig. 4. SPV from secure 2PC

Proof. Let π denote the real protocol according to the description of Protocol 1. We define the ideal world simulator $\mathcal{S}$ given any real world adversary $\mathcal{A}$ and consider separately the cases where $\mathcal{A}$ corrupts Prover and $\mathcal{A}$ corrupts Verifier. We may assume without loss of generality that $\mathcal{A}$ is a proxy for $\mathcal{Z}$. In other words, $\mathcal{Z}$ controls $\mathcal{A}$'s outgoing communication and $\mathcal{A}$ forwards all its incoming messages back to $\mathcal{Z}$.

$\mathcal{A}$ **corrupts** V. $\mathcal{S}$ runs a simulated copy of $\mathcal{A}$. $\mathcal{S}$ intercepts the command $(\text{create}, sid, V, P, id_x, mid, \rho)$ that $\mathcal{A}$ sends to $\mathcal{F}_{DC}$ as well as the value s that $\mathcal{A}$ submits to EqualityChecker. $\mathcal{S}$ defines the unary predicate ϕ_s on strings $m \in \{0,1\}^*$ so that $\phi_s(m) = 0$ when $\rho(m) = s$ and $\phi_s(m) = 1$ when $\rho(m) \neq s$. $\mathcal{S}$ then submits this predicate ϕ_s to $\mathcal{F}_{SPV}^{IDA}$. Recall that $\mathcal{F}_{SPV}^{IDA}$ will proceed identically to $\mathcal{F}_{SPV}$ on input id_x unless $\phi_s(\bar{x}) = 1$ (i.e. $\rho(\bar{x}) \neq s$). Thus, $\mathcal{S}$ will receive the output $(\Pi(x), id_x, 1)$ from $\mathcal{F}_{SPV}^{IDA}$ if and only if $\mathcal{A}$ would receive the output 1 from EqualityChecker. In this case, $\mathcal{S}$ forwards 1 to $\mathcal{Z}$ and sets V's output to Accept. Likewise, $\mathcal{S}$ will receive the output abort if and only if $\mathcal{A}$ would receive abort from EqualityChecker. In this case, $\mathcal{S}$ sends abort to $\mathcal{Z}$ and sets V's output to abort as well. In the former case, the honest P has no output in either the ideal or real processes. In the latter case, the honest party P also outputs abort in both processes. Thus, $\mathcal{Z}$ has identical views in both processes.

$\mathcal{A}$ **corrupts** P. $\mathcal{S}$ runs a simulated $\mathcal{F}_{DC}$ with a simulated $\mathcal{A}$ just as V would, except that it chooses its own random password string sk' and defines ρ' to output sk' when $\Pi(x) = 1$. $\mathcal{S}$ records the value out that $\mathcal{A}$ receives back from

$\mathcal{F}_{DC}$, and the value s' that the simulated $\mathcal{A}$ would submit to EqualityChecker. If either out $= \perp$ or out $\neq s'$, then $\mathcal{S}$ sends the constant predicate $\phi = 1$ to $\mathcal{F}_{SPV}^{IDA}$, causing both P and V to output abort, and sends abort directly to $\mathcal{Z}$ as well. Otherwise, if out $= sk' = s'$, then $\mathcal{S}$ sends $(\Pi(x), id_x, 1)$ to $\mathcal{F}_{SPV}^{IDA}$, causing V to output Accept.

In the case that the simulated $\mathcal{A}$ receives out $= sk'$, the real $\mathcal{A}$ would receive sk from $\mathcal{F}_{DC}$. Let s denote the value that the real $\mathcal{A}$ would send as input to the EqualityChecker. Since sk and sk' are identically distributed random strings, it follows that (sk, s) and (sk', s') must be identically distributed. Conditioned on out $= sk'$, the probability that $s' \neq sk'$ is identical to the probability that $s \neq sk$, which both cause P and V to output abort in the real and ideal processes respectively. In this case, the view of $\mathcal{Z}$ is distributed identically in the real and ideal processes.

In the case that the simulated $\mathcal{A}$ receives out $= \perp$, the real $\mathcal{A}$ also receives $\perp$. Here, $\mathcal{S}$ immediately causes an abort. The only way that $\mathcal{Z}$ might see different values in the real process is if $\mathcal{A}$ avoids an abort by guessing sk. This occurs with probability negligible in λ over V's random choice of sk.

We conclude that in both cases, $\text{REAL}_{\pi, \mathcal{A}, \mathcal{Z}} \sim \text{IDEAL}_{\mathcal{F}_{SPV}^{IDA}, \mathcal{S}, \mathcal{Z}}$.

4   Secure Physical Two-Party Computation

We now have two parties, Party A and Party B, who respectively hold physical inputs x_A and x_B, and wish to evaluate a function of their inputs. Party A creates a DC token C_A to investigate x_B and party B creates a DC token C_B to investigate x_A. We assume that the function can be computed by measuring x_A and x_B separately in isolation, and then evaluating a logical function f of the measured values.[1] The parties exchange circuits so that C_A is isolated with x_B and only a communication channel to B while C_B is isolated with x_A and only a communication channel to A. Figure 5 depicts a schematic of the setup after the circuits are exchanged.

A high level description of the protocol is as follows. Each token computes a canonical description of the input it investigates, and outputs this description signed with an unconditional one time MAC to the opposing party. Let k_A and k_B denote secret keys that A shares with C_A and B shares with C_B respectively. C_A (investigating x_B) derives the description $\bar{x}_B$ and sends $(\bar{x}_B, MAC_{k_A}(\bar{x}_B))$ to B. C_B (investigating x_A) derives the description $\bar{x}_A$ and sends $(\bar{x}_A, MAC_{k_B}(\bar{x}_A))$ to A. Parties A and B also derive the canonical descriptions of their own inputs to check that the descriptions they receive from the tokens are correct. Next, A and B execute a secure 2PC functionality that verifies the MAC signatures and computes the function f. A supplies the inputs $(\bar{x}_A, MAC_{k_B}(\bar{x}_A), k_A)$ and B supplies the inputs $(\bar{x}_B, MAC_{k_A}(\bar{x}_B), k_B)$. The 2PC functionality outputs abort to both parties if either MAC verification fails, and otherwise outputs $f(\bar{x}_A, \bar{x}_B)$ to both parties. We describe the details of the protocol formally in Figure 6.

---

[1] We do not posit that this covers all physical functions (e.g, in the quantum domain).

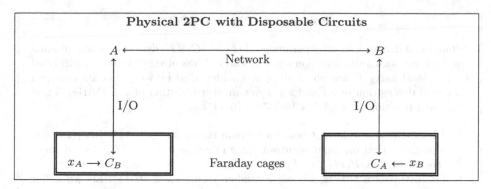

Fig. 5. Schematic of the physical setup for secure physical 2PC using disposable circuits (Protocol 2)

Alternatively, we could use C_A and C_B to locally compute the outputs first, send the outputs signed with a MAC to the opposing party. Then A and B could execute a 2PC functionality that both verifies the signatures and checks that the outputs are equal, outputting 1 if these checks pass. This should be a more efficient method in general since f could be an arbitrarily complex function and costly to evaluate inside the 2PC functionality. This modified protocol uses the same general technique as the *dual-execution garbled circuits* (DualEx) protocol for 2PC [17,13], although the goal of DualEx was to achieve a tradeoff between security and efficiency in standard 2PC. Nonetheless, we find the former method simpler to present and prove secure.

Claim 2: *Protocol 2 securely realizes $\mathcal{F}_{\text{Physical2PC}}$ with input-dependent abort.*

Proof. Let π denote the real protocol according to the description of Protocol 1. We define the ideal world simulator $\mathcal{S}$ given any real world adversary $\mathcal{A}$. It suffices to analyze the case where $\mathcal{A}$ corrupts A, as the case in which $\mathcal{A}$ corrupts B is symmetrical. As before, we may assume without loss of generality that $\mathcal{A}$ is a proxy for $\mathcal{Z}$.

$\mathcal{S}$ runs a simulated copy of $\mathcal{A}$. $\mathcal{S}$ intercepts id_{x_B} and ρ_A from $\mathcal{A}$'s create message to $\mathcal{F}_{DC}$. In addition, $\mathcal{S}$ runs a simulated $\mathcal{F}_{DC}$. $\mathcal{S}$ chooses a random secret key r, defines $\rho_{\mathcal{S}}(y) = (y, MAC_r(y))$, and sends $(\text{create}, sid_{\mathcal{S}}, \mathcal{S}, A, id_{x_B}, mid_{\mathcal{S}}, \rho_{\mathcal{S}})$ to $\mathcal{F}_{DC}$. Next, $\mathcal{S}$ intercepts id_{x_A} from $\mathcal{A}$'s run command to $\mathcal{F}_{DC}$. $\mathcal{S}$ sends $(\text{run}, sid_{\mathcal{S}}, \mathcal{S}, mid_{\mathcal{S}}, id_{x_A})$ to the simulated $\mathcal{F}_{DC}$, and forwards the resulting output $(\bar{x}_A, MAC_r(\bar{x}_A))$ to $\mathcal{A}$ and to $\mathcal{Z}$. $\mathcal{S}$ intercepts the input (m', σ', k') that $\mathcal{A}$ submits to VFChecker. If $VF_r(m', \sigma') \neq 1$ then $\mathcal{S}$ sets $\phi = 1$. Otherwise, $\mathcal{S}$ sets the predicate ϕ such that $\phi(x, y) = 1$ if and only if $VF_{k'}(\rho_A(y)) \neq 1$. Finally, $\mathcal{S}$ sends id_{x_A} and ϕ to $\mathcal{F}_{\text{Physical2PC}}^{IDA}$. $\mathcal{S}$ forwards the output it receives, whether abort or $f(\bar{x}_A, \bar{x}_B)$, to $\mathcal{Z}$ and sets A's output accordingly.

In the real protocol, $\mathcal{Z}$ sees $(\bar{x}_A, MAC_{k_B}(\bar{x}_A))$ and in the ideal world it sees $(\bar{x}_A, MAC_r(\bar{x}_A))$ where k_B is the random value chosen by A and r is the random value chosen by $\mathcal{S}$. These values are identically distributed. Thus, the probability $\mathcal{A}$ responds in the real world with (m, σ, k) such that the verification of the honest

---

Protocol 2 - Physical 2PC

Setup: Let λ denote a security parameter. Let (MAC, VF) denote an unconditional one-time message authentication scheme. Party A has physical input x_A identified by id_{x_A} and party B has physical input x_B identified by id_{x_B}. Let $\bar{x}_A$ denote a canonical description of x_A and $\bar{x}_B$ a canonical description of x_B. Parties A and B share a function $f : \{0,1\}^* \times \{0,1\}^* \to \{0,1\}^*$.

1. (*A* **initiates a token**) Choose a random string $k_A \in \{0,1\}^\lambda$. Let ρ_A define a program that on input y outputs (y, σ_A) where $\sigma_A = MAC_{k_A}(y)$. A sends (create, $sid_A, A, B, id_{x_B}, mid_A, \rho_A$) to $\mathcal{F}_{DC}$.
2. (*B* **initiates a token**) Choose a random string $k_B \in \{0,1\}^\lambda$. Let ρ_B define a program that on input y outputs (y, σ_B) where $\sigma_B = MAC_{k_B}(y)$. B sends (create, $sid_B, B, A, id_{x_A}, mid_B, \rho_B$) to $\mathcal{F}_{DC}$.
3. (*A* **queries *B*'s token**) A submits (run, $sid_B, B, mid_B, id_{x_A}$) to $\mathcal{F}_{DC}$, receives in response (out$_B, \sigma_B$), and checks that out$_B = \bar{x}_A$.
4. (*B* **queries *A*'s token**) A submits (run, $sid_A, A, mid_A, id_{x_B}$) to $\mathcal{F}_{DC}$, receives in response (out$_A, \sigma_A$), and checks that out$_A = \bar{x}_B$.
5. (**The parties destroy the tokens**) A submits (destroy, sid_B, mid_B, B) to $\mathcal{F}_{DC}$ and waits to receive $\perp$ in response. B submits (destroy, sid_A, mid_A, A) to $\mathcal{F}_{DC}$ and waits to receive $\perp$ in response.
6. (**2PC check signatures and compute *f***) A submits inputs $(\bar{x}_A, \sigma_B, k_A)$ and B submits inputs $(\bar{x}_B, \sigma_A, k_B)$ to a protocol that unconditionally UC-realizes the following ideal functionality:
   - VFChecker: On input (m, σ, k_A) from party A and (m', σ', k_B) from party B, check that $VF_{k_A}(m', \sigma') = 1$ and check that $VF_{k_B}(m, \sigma) = 1$. If these checks pass, compute and output $f(m, m')$ to both parties. Otherwise, output abort.
7. If VFChecker outputs abort, then both A and B output abort. Otherwise, VFChecker outputs $f(\bar{x}_A, \bar{x}_B)$ to both A and B, who each locally output this value as well.

Fig. 6. Secure physical 2PC leaking at most 1 bit

party's MAC fails, i.e. $VF_{k_B}(m, \sigma) \neq 1$, is identical to the probability that the simulated $\mathcal{A}$ responds with (m', σ', k') such that $VF_r(m', \sigma') \neq 1$. Likewise, the outcome of the verifying the adversary's MAC in the real world, i.e. $VF_{k'}(\rho_A(\bar{x}_B))$, is identically distributed to $VF_k(\rho_A(\bar{x}_B))$. Thus, the probability that one of the verifications fails in the real world is identical to the probability that one of the verifications fails in the simulation. This event causes the same outcome in the real process and ideal process, namely abort.

On the other hand, when both MAC verifications pass, in the ideal world the output is always the true output $f(\bar{x}_A, \bar{x}_B)$, whereas in the real world the output is $f(m, \bar{x}_B)$ possibly for $m \neq \bar{x}_A$. Nonetheless, when $m \neq \bar{x}_A$, the probability that verifying the honest party's MAC passes, i.e. $VF_{k_B}(m, \sigma) = VF_{k_B}(m, MAC_{k_A}(\bar{x}_A)) = 1$, is negligible by reduction to the security of (MAC, VF).

We conclude that $\text{REAL}_{\pi, \mathcal{A}, \mathcal{Z}} \sim \text{IDEAL}_{\mathcal{F}_{\text{Physical2PC}}^{\text{IDA}}, \mathcal{S}, \mathcal{Z}}$

5   Secure Physical Multi-party Computation

Our protocol for secure physical 2PC can be generalized for n parties computing a multi-party functionality of their n physical inputs $x_1, ..., x_n$. We will sketch the protocol without going into low-level detail, as the extension is quite natural. In the first phase of the protocol, each party P_i creates $n - 1$ tokens $\{T_j^i\}_{j \neq i}$, and transfers the token T_j^i to party P_j. The token T_j^i investigates the input x_j and outputs to P_j the canonical description $\bar{x}_j$ of x_j and a MAC signature using P_i's key k_i: $(\bar{x}_j, MAC_{k_i}(\bar{x}_j))$. At the end of the first phase, each party P_i has received $n - 1$ MAC signed descriptions of its input x_i from the other parties' tokens. Each P_i checks that the $n - 1$ descriptions it has received are all the same, and also match the correct canonical description $\bar{x}_i$ that it has derived on its own. In the second phase, each party P_i inputs $(\bar{x}_i, k_i, \{MAC_{k_j}(\bar{x}_i)\}_{i \neq j})$ to a secure MPC protocol implementing a functionality that verifies all the MAC signatures and either outputs abort (when a signature is invalid) or otherwise returns $f(\bar{x}_1, ..., \bar{x}_n)$ to each party. The resulting protocol is secure with input-dependent abort, as an adversary and its tokens can at best cause the MPC protocol to either succeed or fail depending on the inputs they investigate and the adversary's input. The MPC protocol can be unconditionally UC-securely realized in the DC model (as an extension of the tamper-proof token model) following the result of Goyal et. al. [12].

6   Isolated Circuit Secure Communication

In this section we show a simple protocol we call *isolated circuit secure communication* (ICSC) that enables the two isolated circuits C_A and C_B in the setup depicted in Figure 5 to send messages to each other in a way that keeps the messages hidden from both A and B. The protocol is *malleable*, meaning that A and B can sabotage any message sent between the circuits without detection. However, all messages that A and B see during the protocol reveal no information (are uniformly distributed) even if A colludes maliciously with C_A or B colludes maliciously with C_B.

Security (Sketch). We do not precisely formalize the security of the ICSC protocol (e.g., in the UC-framework), but we sketch here the security intuition. First, it is easy to see the the output of the protocol to C_A and C_B respectively is correct when all parties are honest. Second, the views of A and B are indistinguishable from random. Consider the view of an adversary $\mathcal{A}$ corrupting A (the case for B is symmetric). In the first round, $\mathcal{A}$ receives $k_B^0 \oplus m$ for some m and in the second round it receives $k_B^1 \oplus m'$ for some m'. Given that k_B^0 and k_B^1 are independent uniformly distributed random strings, the simulator can replace both of $\mathcal{A}$'s messages with independent random strings. The argument can be extended to prove computational security in the multi-round setting via a standard hybrid argument, replacing the PRF generated keys with uniform random strings.

Protocol 3 - ICSC

C_A and A share secret key k_A^0 and k_A^1; C_B and B share a secret keys k_B^0 and k_B^1.

1. C_A *sends message* m_0 *to* C_B:
   - C_A to B: send $u_0 = k_A^0 \oplus m_0$
   - B to A: send $v_0 = k_B^0 \oplus u_0$
   - A to C_B: send $w_0 = k_A^0 \oplus v_0 = k_B^0 \oplus m_0$
   - C_B: output $m_0 = k_B^0 \oplus w_0$

2. C_B *sends message* m_1 *to* C_A:
   - C_B to A: send $v_1 = k_B^1 \oplus m_1$
   - A to B: send $u_1 = k_A^1 \oplus v_1$
   - B to C_A: send $w_1 = k_B^1 \oplus u_1 = k_A^1 \oplus m_1$
   - C_A: output $m_1 = k_A^1 \oplus w_1$

Multiple rounds: Using a pseudorandom function or any KDF, on round i party A uses a key pair $\text{KDF}(k_A^0; k_A^1; i)$, and party B uses a key pair $\text{KDF}(k_B^0; k_B^1; i)$.

Fig. 7. Isolated circuits C_A and C_B exchange secret messages indirectly through A and B

Applications. ISCS enables two tokens in isolation to execute arbitrary two-party interactive (digital) functionalities without leaking information to the two parties involved (other than 1 bit by aborting). In theory, this is no more powerful than the protocol we presented for physical two-party computation. However, we can imagine ISCS could have practical import. For example, we could use ISCS to efficiently compile any interactive protocol that leaks up to k bits into one that leaks at most 1 bit of the initial k bits. The isolated tokens could first run the interactive protocol to learn the output and at most k additional bits. Next, the tokens would communicate their outputs to the two parties using the DualEx paradigm (they send their MAC signed output to the opposing party, and the parties run secure 2PC to verify the MACs and equality of the outputs).

7   Further Directions

Open Question: *Is there a protocol for physical SPV or 2PC/MPC in the DC model that is fully secure against malicious adversaries, or that provably does not leak any information?*

Acknowledgements. We would like to thank the anonymous reviewers of this paper for their many helpful comments and suggestions.

References

1. Ben-Or, M., Goldwasser, S., Kilian, J., Wigderson, A.: Multi-prover interactive proofs: How to remove intractability assumptions. In: 20th ACM STOC, pp. 113–131. ACM Press (May 1988)

2. Canetti, R.: Security and composition of multiparty cryptographic protocols. Journal of Cryptology 13(1), 143–202 (2000)
3. Canetti, R.: Universally composable security: A new paradigm for cryptographic protocols. In: 42nd FOCS, pp. 136–145. IEEE (October 2001), Full version at Cryptology ePrint Archive, Report 2001/055, http://eprint.iacr.org/2001/055
4. Fisch, B., Freund, D., Naor, M.: Physical zero-knowledge proofs of physical properties. In: Garay, J.A., Gennaro, R. (eds.) CRYPTO 2014, Part II. LNCS, vol. 8617, pp. 313–336. Springer, Heidelberg (2014)
5. Frederiksen, T.K., Jakobsen, T.P., Nielsen, J.B., Nordholt, P.S., Orlandi, C.: MiniLEGO: Efficient secure two-party computation from general assumptions. In: Johansson, T., Nguyen, P.Q. (eds.) EUROCRYPT 2013. LNCS, vol. 7881, pp. 537–556. Springer, Heidelberg (2013)
6. Garay, J.A., Ishai, Y., Kumaresan, R., Wee, H.: On the complexity of UC commitments. In: Nguyen, P.Q., Oswald, E. (eds.) EUROCRYPT 2014. LNCS, vol. 8441, pp. 677–694. Springer, Heidelberg (2014)
7. Glaser, A., Barak, B., Goldston, R.J.: A new approach to nuclear warhead verification using a zero-knowledge protocol. Presented at 53rd Annual INMM (Institute of Nuclear Materials Management) Meeting (2012)
8. Glaser, A., Barak, B., Goldston, R.J.: A zero-knowledge protocol for nuclear warhead verification. Nature 510, 497–502 (2014)
9. Goldreich, O., Micali, S., Wigderson, A.: How to play any mental game, or a completeness theorem for protocols with honest majority. In: Aho, A. (ed.) 19th ACM STOC, pp. 218–229. ACM Press (May 1987)
10. Goldreich, O., Ostrovsky, R.: Software protection and simulation on oblivious RAMs. J. ACM 43(3), 431–473 (1996)
11. Goldwasser, S., Kalai, Y.T., Rothblum, G.N.: One-time programs. In: Wagner, D. (ed.) CRYPTO 2008. LNCS, vol. 5157, pp. 39–56. Springer, Heidelberg (2008)
12. Goyal, V., Ishai, Y., Sahai, A., Venkatesan, R., Wadia, A.: Founding cryptography on tamper-proof hardware tokens. In: Micciancio, D. (ed.) TCC 2010. LNCS, vol. 5978, pp. 308–326. Springer, Heidelberg (2010)
13. Huang, Y., Evans, D., Katz, J.: Quid-Pro-Quo-tocols: Strengthening semi-honest protocols with dual execution. In: IEEE Symposium on Security and Privacy, pp. 272–284 (2012)
14. Ishai, Y., Kushilevitz, E., Ostrovsky, R., Prabhakaran, M., Sahai, A.: Efficient non-interactive secure computation. In: Paterson, K.G. (ed.) EUROCRYPT 2011. LNCS, vol. 6632, pp. 406–425. Springer, Heidelberg (2011)
15. Katz, J.: Universally composable multi-party computation using tamper-proof hardware. In: Naor, M. (ed.) EUROCRYPT 2007. LNCS, vol. 4515, pp. 115–128. Springer, Heidelberg (2007)
16. Kilian, J.: Founding cryptography on oblivious transfer. In: 20th ACM STOC, pp. 20–31. ACM Press (May 1988)
17. Mohassel, P., Franklin, M.K.: Efficiency tradeoffs for malicious two-party computation. In: Yung, M., Dodis, Y., Kiayias, A., Malkin, T. (eds.) PKC 2006. LNCS, vol. 3958, pp. 458–473. Springer, Heidelberg (2006)
18. Moran, T., Segev, G.: David and Goliath commitments: UC computation for asymmetric parties using tamper-proof hardware. In: Smart, N.P. (ed.) EUROCRYPT 2008. LNCS, vol. 4965, pp. 527–544. Springer, Heidelberg (2008)
19. Naor, M., Naor, Y., Reingold, O.: Applied kid cryptography or how to convince your children you are not cheating. Journal of Craptology (1) (April 1999)

20. Quisquater, J.-J., Guillou, L.C., Berson, T.A.: How to explain zero-knowledge protocols to your children. In: Brassard, G. (ed.) Advances in Cryptology - CRYPT0 1989. LNCS, vol. 435, pp. 628–631. Springer, Heidelberg (1990)
21. Yao, A.C.-C.: How to generate and exchange secrets (extended abstract). In: FOCS, pp. 162–167 (1986)

Complete Characterization of Fairness in Secure Two-Party Computation of Boolean Functions[*]

Gilad Asharov[1], Amos Beimel[2], Nikolaos Makriyannis[3], and Eran Omri[4]

[1] The Hebrew University of Jerusalem, Jerusalem, Israel
[2] Ben Gurion University of the Negev, Be'er Sheva, Israel
[3] Universitat Pompeu Fabra, Barcelona, Spain
[4] Ariel University, Ariel, Israel

Abstract. Fairness is a desirable property in secure computation; informally it means that if one party gets the output of the function, then all parties get the output. Alas, an implication of Cleve's result (STOC 86) is that when there is no honest majority, in particular in the important case of the two-party setting, there exist Boolean functions that cannot be computed with fairness. In a surprising result, Gordon et al. (JACM 2011) showed that some interesting functions can be computed with fairness in the two-party setting, and re-opened the question of understanding which Boolean functions can be computed with fairness, and which cannot.

Our main result in this work is a complete characterization of the (symmetric) Boolean functions that can be computed with fairness in the two-party setting; this settles an open problem of Gordon et al. The characterization is quite simple: A function can be computed with fairness *if and only if* the all one-vector or the all-zero vector are in the affine span of either the rows or the columns of the matrix describing the function. This is true for both deterministic and randomized functions. To prove the possibility result, we modify the protocol of Gordon et al.; the resulting protocol computes with full security (and in particular with fairness) all functions that are computable with fairness.

We extend the above result in two directions. First, we completely characterize the Boolean functions that can be computed with fairness in the multiparty case, when the number of parties is constant and at most half of the parties can be malicious. Second, we consider the two-party setting with asymmetric Boolean functionalities, that is, when the output of each party is one bit; however, the outputs are not necessarily the same. We provide both a sufficient condition and a necessary condition for fairness; however, a gap is left between these two conditions. We then consider a specific asymmetric function in this gap area, and by designing a new protocol, we show that it is computable with fairness. However, we do not give a complete characterization for all functions that lie in this gap, and their classification remains open.

Keywords: Secure computation, fairness, foundations, malicious adversaries.

[*] The first author is supported by the Israeli Centers of Research Excellence (I-CORE) Program (Center No. 4/11). The second author is partially supported by ISF grant 544/13 and by the Frankel Center for Computer Science. The forth author is partially supported by ISF grant 544/13.

Y. Dodis and J.B. Nielsen (Eds.): TCC 2015, Part I, LNCS 9014, pp. 199–228, 2015.

1　Introduction

Secure multiparty computation is one of the gems of modern cryptography. It enables a set of mutually distrusting parties to securely compute a joint function of their inputs in the presence of an adversarial behaviour. The security requirements of such a computation include privacy, correctness, independence of inputs, and *fairness*. Informally, fairness means that if one party gets the output of the function, then all parties get the output. For example, when two parties are signing a contract, it is reasonable to expect that one party signs the contract if and only if the second party signs the contract.

The study of secure multiparty protocols (MPC) started with the works of Yao [15] for the two-party setting and Goldreich, Micali, and Wigderson [10] for the multiparty setting. When a strict majority of honest parties can be guaranteed, protocols for secure computation provide full security, i.e., they provide all the security properties mentioned above including fairness. However, this is no longer the case when there is no honest majority, and in particular in the case of two-party computation where one of the parties may be corrupted. In these settings, protocols for secure computation provide a weaker notion of security, which is known as "security with abort". Specifically, these protocols still provide important security requirements such as correctness and privacy, but can not guarantee fairness – and the adversary can get its output while preventing the honest parties from getting their output. Relaxing the security requirement when there is no honest majority is unavoidable as was shown by Cleve [8].

To elaborate further, Cleve [8] proved that there exist two-party functions that cannot be computed with fairness when there is no honest majority. In particular, he showed that the coin-tossing functionality, where two parties toss an unbiased fair coin, cannot be computed with complete fairness. He proved that in any two-party coin-tossing protocol there exists an adversary that can bias the output of the honest party. A ramification of Cleve's impossibility is that any function that implies coin tossing cannot be computed with full security without an honest majority. An example for such a function is the exclusive-or (XOR) function; a fully secure coin-tossing protocol can be easily constructed assuming the existence of a fair protocol for XOR.

For years, the common interpretation of Cleve's impossibility was that no interesting functions can be computed with full security without an honest majority. In a recent surprising result, Gordon et al. [11] showed that this interpretation is inaccurate as there exist interesting functions that can be computed with full security in the two-party setting, e.g., the millionaires' problem with polynomial size domain. This result re-opened the question of understanding which Boolean functions can be computed with fairness and which cannot.

In more detail, Gordon et al. [11] showed that all functions with polynomial size domain that do not contain an embedded XOR can be computed with full security in the two-party setting; this class of functions contains the AND function and Yao's millionaires' problem. They also presented a protocol, later referred to as the GHKL protocol, which computes with full security a large class of Boolean functions containing embedded XORs. However, the analysis of this protocol is rather involved, and the exact class of function that can be computed using this protocol was unclear until recently.

In this paper, we focus on the characterization of fairness for Boolean functions, and provide a *complete* characterization of Boolean two-party functions that can be computed with full security.

1.1 Previous Works

As mentioned above, Cleve [8] proved that coin tossing cannot be computed with full security without an honest majority, and in particular, in the two-party setting. A generalization of Cleve's result was given recently by Agrawal and Prabhakaran [1]. They showed that any non-trivial sampling functionality cannot be computed with fairness and correctness in the two-party setting, where a non-trivial sampling functionality is a randomized functionality in which two parties, with no inputs, sample two correlated bits.

Gordon et al. [11] re-opened the question of characterizing fairness in secure two-party and multiparty computation. This question was studied in a sequence of works [2,4,14]. Asharov, Lindell, and Rabin [4] focused on the work of Cleve [8] and fully identified the functions that imply fair coin tossing, and are thus ruled out by Cleve's impossibility. Later, Asharov [2] studied the functions that can be computed with full security using the GHKL protocol. He identified three classes of functions: Functions that can be computed using the GHKL protocol, functions that cannot be computed with full security using this specific protocol, and a third class of functions that remained unresolved. Intuitively, [2] showed that if a function is somewhat asymmetric, in the sense that, in the ideal model, one party exerts more influence on the output compared to the other party, then the function can be computed with full security.

Makriyannis [14] has recently shown that the class of functions that by [2] cannot be compute fairly using the GHKL protocol is inherently unfair. He showed a beautiful reduction from sampling functionalities, for which fair computation had already been ruled out in [1], to any function in this class. Indeed, in this class of functions, the influence that each party exerts over the output in the ideal world is somewhat the same. However, the works of [2,14] left the aforementioned third class of functions unresolved. Specifically, this class of functions is significantly different than the class of functions that was shown to be fair, and it contains functions where the parties exert the same amount of influence over the output in the ideal model and yet do not imply sampling, at least not by the construction of [14].

The characterization of fairness was studied in scenarios other than symmetric two-party Boolean functions where the output of the parties is the same. In particular, Gordon and Katz [12] considered fully-secure computation in the multiparty setting without an honest majority and constructed a fully-secure three-party protocol for the majority function and an m-party protocol for the AND of m bits. Asharov [2] also studied asymmetric functions where the parties' outputs are not necessarily the same, as well as functions with non-Boolean outputs, and showed some initial possibility results for these classes of functions.

1.2 Our Results

In this work, we study when functions can be computed with full security without an honest majority. Our main result in this work is a complete characterization of the Boolean functions that can be computed with full security in the two-party setting. This solves an open problem of [11]. We focus on the third class of functions that was left unresolved by [2, 14], and show that *all* functions in this class can be computed with full security (thus, with fairness). This includes functions where the parties' influences on the output in the ideal world are completely equalized, showing that the classification is more subtle than one might have expected.

In order to show possibility for this class of functions, we provide a new protocol (based on the protocol of GHKL), and show that it can compute all the functions in this class. We note that the GHKL protocol had to be modified in order to show this possibility; In particular, we show that there exist functions in this class that cannot be computed using the GHKL protocol for any set of parameters, but can be computed using our modified protocol (see Theorem 1.2). In addition, this protocol computes with full security all functions that were already known to be fair. Combining the result with the impossibility result of [14], we obtain a quite simple characterization for fairness:

Theorem 1.1 (informal). *A Boolean function* $f : X \times Y \to \{0, 1\}$ *can be computed with full security* if and only if *the all one-vector or the all-zero vector are an affine combination of either the rows or the columns of the matrix describing the function.*

We recall that an affine-combination is a linear-combination where the sum of coefficients is 1 (see Theorem 1.2).

The above informally stated theorem is true for both deterministic and randomized functions. Alternatively, our characterization can be stated as follows: either a function implies non-trivial sampling, thus cannot be computed with fairness, or the function can be computed with full security (and, in particular, with complete fairness), assuming the existence of an oblivious transfer protocol.

Example 1.2. The following function is a concrete example of a function that was left as unresolved in [2, 14].

	y_1	y_2	y_3	y_4
x_1	0	0	0	1
x_2	0	0	1	1
x_3	0	1	1	0
x_4	1	1	0	1

We show that the GHKL protocol is susceptible to an attack for this particular function, however, we show that it can be computed using our modified protocol. In this example, the all-one vector is an affine combination of the rows (taking the first row with coefficient -1, the second and the forth with coefficient

1, and the third row with coefficient 0; the sum of the coefficients is 1, as required by an affine combination). Thus, this function can be computed with full security by our protocol.

We extend the above result in two directions. First, we completely characterize the Boolean functions that can be computed with full security when the number of parties is constant and at most half of the parties can be malicious. We show that a function can be computed with full security when at most half of the parties can be malicious if and only if every partition of the parties' inputs into two equal sets results in a fully-secure two-party function. Second, we consider the two-party setting with asymmetric Boolean functionalities, that is, when the output of each party is one bit, but the outputs are not necessarily the same. We generalize our aforementioned protocol for symmetric functions to handle asymmetric functions, and conclude with a sufficient condition for fairness. In addition, we provide a necessary condition for fairness; however, a gap is left between these two conditions. For the functions that lie in the gap, the characterization remains open. We then consider a specific function in this gap area and, by designing a new protocol, we show that it is computable with fairness. This new protocol has some different properties than the generalized protocol, which may imply that the characterization of Boolean asymmetric functions is more involved than the symmetric case.

2   Preliminaries

In this paper all vectors are column vectors over $\mathbb{R}$. Vectors are denoted by bold letters, e.g., $\mathbf{v}$ or $\mathbf{1}$ (the all-one vector). Let $\mathcal{V} \subseteq \mathbb{R}^\ell$ be a set of vectors. A vector $\mathbf{w} \in \mathbb{R}^\ell$ is an *affine combination* of the vectors in $\mathcal{V}$ if there exist scalars $\{a_\mathbf{v} \in \mathbb{R}\}_{\mathbf{v} \in \mathcal{V}}$ such that $\mathbf{w} = \sum_{\mathbf{v} \in \mathcal{V}} a_\mathbf{v} \cdot \mathbf{v}$ and $\sum_{\mathbf{v} \in \mathcal{V}} a_\mathbf{v} = 1$. Let M be an $\ell \times k$ real matrix. A vector $\mathbf{w} \in \mathbb{R}^\ell$ is in the image of M, denoted $\mathbf{w} \in \text{im}(M)$, if there exists a vector $\mathbf{u} \in \mathbb{R}^k$ such that $M\mathbf{u} = \mathbf{w}$. A vector $\mathbf{v} \in \mathbb{R}^k$ is in the kernel of M, denoted $\mathbf{v} \in \ker(M)$, if $M\mathbf{v} = \mathbf{0}_\ell$.

The affine hull of a set $\mathcal{V} \in \mathbb{R}^\ell$, denoted affine-hull$(\mathcal{V})$, is the smallest affine set containing $\mathcal{V}$, or equivalently, the intersection of all affine sets containing $\mathcal{V}$. This is equivalent to the set of all affine combinations of elements in $\mathcal{V}$, that is, affine-hull$(\mathcal{V}) = \{\sum_{\mathbf{v} \in \mathcal{V}} a_\mathbf{v} \cdot \mathbf{v} : \sum_{\mathbf{v} \in \mathcal{V}} a_\mathbf{v} = 1\}$.

Proposition 2.1. *The vector $\mathbf{1}_\ell$ is not a linear combination of the columns of M iff the vector $(\mathbf{0}_k)^T$ is an affine combination of the rows of M.*

Proof. $\mathbf{1}_\ell$ is not a linear combination of the columns of M iff $\text{rank}((M|\mathbf{1})) = \text{rank}(M) + 1$ (where $(M|\mathbf{1})$ is the matrix M with the extra all-one column) iff $\dim(\ker((M|\mathbf{1})^T)) = \dim(\ker(M^T)) - 1$ iff there exists a vector $\mathbf{u} \in \mathbb{R}^\ell$ such that $M^T\mathbf{u} = \mathbf{0}_k$ and $\mathbf{1} \cdot \mathbf{u} = 1$ iff $(\mathbf{0}_k)^T$ is an affine combination of the rows of M. $\square$

Definition 2.2. *Given two $\ell \times k$ matrices matrices A and B, we say that $C = A * B$ if C is the entry-wise (Hadamard) product of the matrices, that is, $C_{i,j} = A_{i,j} \cdot B_{i,j}$.*

Secure Multiparty Computation. We assume that the reader is familiar with the definitions of secure computation, and with the ideal-real paradigm. We refer to [7, 9] and to the full version of this paper [3] for formal definitions. We distinguish between security-with-abort, for which the adversary may receive outputs while the honest parties might not (security without fairness), and full security (security with fairness), where all parties receive outputs (this is similar to security with respect to honest majority as in [7, 9], although we do not have a honest majority).

3   A Fully-Secure Protocol for Boolean Two-Party Functions

In this section we present our protocol that securely computes any function for which the all-one or all-zero vector is an affine combination of either the rows or the columns of its associated matrix. In Section 3.1, we show that our protocol can compute functions that cannot be computed using the original GHKL protocol.

Let $f : X \times Y \to \{0, 1\}$ be a finite Boolean function with the associated $\ell \times k$ matrix M. Without loss of generality, $X = \{1, \ldots, \ell\}$ and $Y = \{1, \ldots, k\}$. In Figure 1 we describe Protocol FAIRTWOPARTY$_\sigma$, a modification of the GHKL protocol [11]. Protocol FAIRTWOPARTY$_\sigma$ is described using an on-line dealer.

Remark 3.1. We show that our protocol computes the function with full security in the presence of an on-line dealer, even though each party can abort this on-line dealer at any point of the execution. We remark that this implies full security in the plain model. In order to see this, we first remark that one can use direct transformation from the on-line dealer to the plain model, as discussed in [2]. An alternative transformation is the following, which is by now standard and is based on [11].

We next sketch the transformation. Protocol FAIRTWOPARTY$_\sigma$ is composed of a fixed number r of rounds (which depends on the security parameter and the function computed by the protocol), the dealer prepares $2r$ values $a_1, \ldots, a_r$ and $b_1, \ldots, b_r$ and in round i it first gives a_i to P_1, if P_1 does not abort it gives b_i to P_2. The values a_i and b_i are called backup outputs and the respective value is outputted by a party if the other party aborts. If one of the parties aborts in round i, then the dealer aborts; otherwise it continues to round $i + 1$. one can transform this protocol to a protocol with an off-line dealer that is only invoked in the initial round of the protocol. This off-line dealer first chooses $a_1, \ldots, a_r$ and $b_1, \ldots, b_r$ in the same way the on-line dealer chooses them, computes authenticated 2-out-of-2 shares of $a_1, \ldots, a_r$ and $b_1, \ldots, b_r$, and gives one share to P_1 and the other to P_2. The protocol now proceeds in rounds, where in round i, party P_2 sends to P_1 its authenticated share of a_i and then party P_1 sends to P_2 its authenticated share of b_i. Assuming that a secure protocol for OT exists, this protocol with an off-line dealer can be transformed into a protocol in the plain model where the parties compute this one-time dealer using security with abort [9]. Alternatively, we can use Kilian's protocol [13] that computes the off-line functionality in the OT-hybrid model providing information-theoretic security-with-abort.

Protocol FAIRTWOPARTY$_\sigma$

1. The parties P_1 and P_2 hand their inputs, denoted x and y respectively, to the dealer.[a]
2. The dealer chooses $i^* \geq 2$ according to the geometric distribution with probability α.
3. The dealer computes out $= f(x,y)$, and for $0 \leq i \leq r$

$$a_i = \begin{cases} f(x, \widetilde{y}^{(i)}) \text{ where } \widetilde{y}^{(i)} \in_U Y & \text{if } i < i^* \\ \text{out} & \text{otherwise} \end{cases}$$

and

$$b_i = \begin{cases} f(\widetilde{x}^{(i)}, y) \text{ where } \widetilde{x}^{(i)} \in_U X & \text{if } i < i^* - 1 \\ \sigma & \text{if } i = i^* - 1 \\ \text{out} & \text{otherwise.} \end{cases}$$

4. The dealer gives b_0 to P_2.
5. For $i = 1, \ldots, r$,
   (a) The dealer gives a_i to P_1. If P_1 aborts, then P_2 outputs b_{i-1} and halts.
   (b) The dealer gives b_i to P_2. If P_2 aborts, then P_1 outputs a_i and halts.

---

[a] If x is not in the appropriate domain or P_1 does not hand an input, then the dealer sends $f(\hat{x}, y)$ (where $\hat{x}$ is a default value) to P_2, which outputs this value and the protocol is terminated. The case of an inappropriate y is dealt analogously.

Fig. 1. Protocol FAIRTWOPARTY$_\sigma$ for securely computing a function f, where σ is either 0 or 1

The main difference in Protocol FAIRTWOPARTY$_\sigma$ compared to the GHKL protocol is in Step 3, where $b_{i^*-1} = \sigma$ (compared to $b_{i^*-1} = f(x, \widetilde{y})$ for a random $\widetilde{y}$ in the GHKL protocol). For some functions f we choose $\sigma = 0$ and for some we choose $\sigma = 1$; the choice of σ depends only on the function and is independent of the inputs. This seemingly small change enables to compute with full security a larger class of functions, i.e., all Boolean functions that can be computed with full security. We note that this change is somewhat counter intuitive as we achieve security against a malicious P_1 by giving P_2 less information. The reason why this change works is that it enables the simulator for P_1 to choose its input to the trusted party in a clever way. See Section 3.1 for more intuition explaining why this change works.

Remark 3.2. There are two parameters in Protocol FAIRTWOPARTY$_\sigma$ that are unspecified – the parameter α of the geometric distribution and the number of rounds r. We show that there exists a constant α_0 (which depends on f) such that taking any $\alpha \leq \alpha_0$ guarantees full security (provided that f satisfies some conditions). As for the number of rounds r, even if both parties are honest the protocol fails if $i^* > r$ (where i^* is chosen with geometric distribution). The probability that $i^* > r$ is $(1 - \alpha)^r$. So, if $r = \alpha^{-1} \cdot \omega(\log n)$ (where n is the security parameter), the probability of not reaching i^* is negligible.

Theorem 3.3. *Let M be the associated matrix of a Boolean function f. If $(\mathbf{0}_k)^T$ is an affine combination of the rows of M, then there is a constant $\alpha_0 > 0$ such that Protocol* FairTwoParty$_0$ *with $\alpha \leq \alpha_0$ is a fully-secure protocol for f.*

Proof. It is easy to see that the protocol is secure against a corrupted P_2, using a simulator similar to [11]. Intuitively, this follows directly from the fact that P_2 always gets the output after P_1.

We next prove that the protocol is secure against a corrupted P_1 by constructing a simulator for every real world adversary $\mathcal{A}$ controlling P_1. The simulator we construct has only black-box access to $\mathcal{A}$ and we denote it by $\mathcal{S}^{\mathcal{A}}$. The simulation is described in Figure 2. It operates along the same lines as in the proof of [11]. The main difference is in the analysis of the simulator, where we prove that there exist distributions $(\mathbf{x}_x^{(a)})_{x \in X, a \in \{0,1\}}$, which are used by the simulator to choose the input that P_1 gives to the trusted party.

The simulator $\mathcal{S}^{\mathcal{A}}$ for Protocol FairTwoParty$_0$

- The adversary $\mathcal{A}$ gives its input x to the simulator.[a]
- The simulator chooses $i^* \geq 2$ according to the geometric distribution with probability α.
- For $i = 1, \ldots, i^* - 1$:
  - The simulator gives $a_i = f(x, \widetilde{y}^{(i)})$ to the adversary $\mathcal{A}$, where $\widetilde{y}^{(i)}$ is chosen according to the uniform distribution.
  - If $\mathcal{A}$ aborts, then the simulator chooses an input x_0 according to a distribution $\mathbf{x}_x^{(a_i)}$ (which depends on the input x and the last bit that was chosen), gives x_0 to the trusted party, outputs the bits $a_1, \ldots, a_i$, and halts.
- At round $i = i^*$, the simulator gives x to the trusted party and gets the output $a = f(x, y)$.
- For $i = i^*, \ldots, r$: The simulator gives $a_i = a$ to the adversary $\mathcal{A}$, if $\mathcal{A}$ aborts, then the simulator outputs the bits $a_1, \ldots, a_i$ and halts.
- The simulator outputs the bits $a_1, \ldots, a_r$ and halts.

---

[a] If the adversary gives an inappropriate x (or no x), then the simulator sends some default $\hat{x} \in X$ to the trusted party, outputs the empty string, and halts.

Fig. 2. The simulator $\mathcal{S}^{\mathcal{A}}$ for Protocol FairTwoParty$_0$

We next prove the correctness of the simulator, that is, if $(\mathbf{0}_k)^T$ is an affine combination of the rows of M, then the output of the simulator and and the output of P_2 in the ideal world are distributed as the adversary's view and the output of the honest P_2 in the real world. The simulator generates its output as the view of P_1 in the execution of the protocol in the real world. First, it chooses i^* as in Protocol FAIRTWOPARTY$_0$, i.e., with geometric distribution. Up to round $i^* - 1$, the backup outputs are uncorrelated to the input of the honest party. That is, for all $i < i^*$ the output $a_i = f(x, \widetilde{y})$ is chosen with a uniformly random $\widetilde{y}$. Starting with round $i = i^*$, the backup outputs are correlated to the true input of the honest party, and are set as $a_i = f(x, y)$, exactly as in the real execution.

As a result of the adversary seeing the same view, $\mathcal{A}$ aborts in the simulation if and only if it aborts in the protocol. Furthermore, if the adversary aborts after round i^*, the output of P_2 in the protocol and in the simulator is identical – $f(x, y)$. The only difference between the simulation and the protocol is the way that the output of P_2 is generated when the adversary aborts in a round $i \leq i^*$. If the adversary aborts in round i^*, the output of P_2 in Protocol FAIRTWOPARTY$_0$ is $b_{i^*-1} = 0$, while the the output of P_2 in the simulation is $f(x, y)$. To compensate for this difference, in the simulation the output of P_2 if the adversary aborts in a round $1 \leq i \leq i^* - 1$ is $f(x_0, y)$, where x_0 is chosen according to a distribution $\mathbf{x}_x^{(a_i)}$ to be carefully defined later in the proof.

To conclude, we only have to compare the distributions $(\mathrm{View}_{\mathcal{A}}, \mathrm{Out}_{P_2})$ in the real and ideal worlds given that the adversary aborts in a round $1 \leq i \leq i^*$. Thus, in the rest of the proof we let i be the round in which the adversary aborts and assume that $1 \leq i \leq i^*$. The view of the adversary in both worlds is $a_1, \ldots, a_i$. Notice that in both worlds $a_1, \ldots, a_{i-1}$ are equally distributed and are independent of $(a_i, \mathrm{Out}_{P_2})$. Thus, we only compare the distribution of $(a_i, \mathrm{Out}_{P_2})$ in both worlds.

First, we introduce the following notation

$$(1/\ell, \ldots, 1/\ell)M = (s_1, \ldots, s_k), \qquad M \begin{pmatrix} 1/k \\ \vdots \\ 1/k \end{pmatrix} = \begin{pmatrix} p_1 \\ \vdots \\ p_\ell \end{pmatrix}. \tag{1}$$

For example, s_y is the probability that $f(\widetilde{x}, y) = 1$, when $\widetilde{x}$ is uniformly distributed. Furthermore, for every $x \in X$, $y \in Y$, and $a \in \{0, 1\}$, define $q_x^{(a)}(y) \triangleq \Pr[f(x_0, y) = 1]$, where x_0 is chosen according to the distribution $\mathbf{x}_x^{(a)}$. That is, $q_x^{(a)}(y)$ is the probability that the output of P_2 in the simulation is 1 when the the adversary aborts in round $i < i^*$, the input of P_1 is x, the input of P_2 is y, and $a_i = a$. Finally, define the column vector $\mathbf{q}_x^{(a)} \triangleq (q_x^{(a)}(y))_{y \in Y}$. Using this notation,

$$M^T \mathbf{x}_x^{(a)} = \mathbf{q}_x^{(a)}, \tag{2}$$

where we represent the distribution $\mathbf{x}_x^{(a)} = (\mathbf{x}_x^{(a)}(x_0))_{x_0 \in X}$ by a column vector. Next, we analyze the four options for the values of $(a_i, \mathrm{Out}_{P_2})$.

First Case: $(a_i, \mathrm{Out}_{P_2}) = (0, 0)$. In the real world $(a_i, \mathrm{Out}_{P_2}) = (0, 0)$ if one of the following two events occurs:

- $i < i^*$, $a_i = f(x, \widetilde{y}) = 0$, and $\mathrm{Out}_{P_2} = b_{i-1} = f(\widetilde{x}, y) = 0$. The probability of this event is $(1 - \alpha)(1 - p_x)(1 - s_y)$.
- $i = i^*$, $a_{i^*} = f(x, y) = 0$, and $\mathrm{Out}_{P_2} = b_{i^*-1} = 0$. Recall that always $b_{i^*-1} = 0$ in Protocol FAIRTWOPARTY$_0$. The probability of this event is $\alpha \cdot (1 - f(x, y)) \cdot 1$ (that is, it is 0 if $f(x, y) = 1$ and it is α otherwise).

Therefore, in the real world $\Pr\left[(a_i, \mathrm{Out}_{P_2}) = (0, 0)\right] = (1 - \alpha)(1 - p_x)(1 - s_y) + \alpha(1 - f(x, y))$. On the other hand, in the ideal world $(a_i, \mathrm{Out}_{P_2}) = (0, 0)$ if one of the following two events occurs:

- $i < i^*$, $a_i = f(x, \widetilde{y}) = 0$, and $\mathrm{Out}_{P_2} = f(x_0, y) = 0$. The probability of this event is $(1 - \alpha)(1 - p_x)(1 - q_x^{(0)}(y))$.
- $i = i^*$, $a_{i^*} = f(x, y) = 0$, and $\mathrm{Out}_{P_2} = f(x, y) = 0$. The probability of this event is $\alpha \cdot (1 - f(x, y))$.

Therefore, in the ideal world $\Pr\left[(a_i, \mathrm{Out}_{P_2}) = (0, 0)\right] = (1 - \alpha)(1 - p_x)(1 - q_x^{(0)}(y)) + \alpha(1 - f(x, y))$. To get full security we need that these two probabilities in the two worlds are the same, that is,

$$(1-\alpha)(1-p_x)(1-s_y)+\alpha(1-f(x,y)) = (1-\alpha)(1-p_x)(1-q_x^{(0)}(y))+\alpha(1-f(x,y)),$$

i.e.,

$$q_x^{(0)}(y) = s_y. \tag{3}$$

As this is true for every y, we deduce, using Equation (1) and Equation (2), that

$$M^T \mathbf{x}_x^{(0)} = \mathbf{q}_x^{(0)} = M^T \left(1/\ell, \ldots, 1/\ell\right)^T. \tag{4}$$

Thus, taking the uniform distribution, i.e., for every $x_0 \in X$

$$\mathbf{x}_x^{(0)}(x_0) = 1/\ell, \tag{5}$$

satisfies these constraints.

Second Case: $(a_i, \mathrm{Out}_{P_2}) = (0, 1)$. In the real world $\Pr\left[(a_i, \mathrm{Out}_{P_2}) = (0, 1)\right] = (1 - \alpha)(1 - p_x)s_y$ (in the real world $\mathrm{Out}_{P_2} = 0$ when $i = i^*$). On the other hand, in the ideal world $\Pr\left[(a_i, \mathrm{Out}_{P_2}) = (0, 1)\right] = (1 - \alpha)(1 - p_x)q_x^{(0)}(y)$ (in the ideal world $a_{i^*} = \mathrm{Out}_{P_2} = f(x, y)$). The probabilities in the two worlds are equal if Equation (3) holds (i.e., Equation (5) holds) for every x.

Third Case: $(a_i, \mathrm{Out}_{P_2}) = (1, 0)$. In the real world $\Pr\left[(a_i, \mathrm{Out}_{P_2}) = (1, 0)\right] = (1 - \alpha)p_x(1 - s_y) + \alpha \cdot f(x, y) \cdot 1$. On the other hand, in the ideal world we have that $\Pr\left[(a_i, \mathrm{Out}_{P_2}) = (1, 0)\right] = (1 - \alpha)p_x(1 - q_x^{(1)}(y))$. The probabilities in the two worlds are equal when

$$(1 - \alpha)p_x(1 - s_y) + \alpha f(x, y) = (1 - \alpha)p_x(1 - q_x^{(1)}(y)).$$

If $p_x = 0$, then $f(x, y) = 0$ and we are done. Otherwise, this equality holds iff

$$q_x^{(1)}(y) = s_y - \frac{\alpha f(x, y)}{(1 - \alpha)p_x}. \tag{6}$$

As this is true for every y, we deduce, using Equation (1) and Equation (2), that

$$M^T \mathbf{x}_x^{(1)} = \mathbf{q}_x^{(1)} = M^T \left(1/\ell, \dots, 1/\ell\right)^T - \frac{\alpha}{(1-\alpha)p_x}(\mathrm{row}_x)^T$$

$$= M^T \left(\left(1/\ell, \dots, 1/\ell\right)^T - \frac{\alpha}{(1-\alpha)p_x} \mathbf{e_x} \right), \tag{7}$$

where row_x is the row of M labeled by the input x, and $\mathbf{e_x}$ is the x-th unit vector. Before analyzing when there exists a probability vector $\mathbf{x}_x^{(1)}$ solving Equation (6), we remark that if the equalities hold for the first 3 cases of the values for $(a_i, \mathrm{Out}_{P_2})$, the equality of the probabilities must also hold for the case $(a_i, \mathrm{Out}_{P_2}) = (1,1)$. In the rest of the proof we show that there exist probability vectors $\mathbf{x}_x^{(1)}$ for every $x \in X$ solving Equation (6).

Claim 3.4. *Fix $x \in X$ and let α be a sufficiently small constant. If $(\mathbf{0}_k)^T$ is an affine combination of the rows of M, then there exists a probability vector $\mathbf{x}_x^{(1)}$ solving Equation (6).*

Proof. By the conditions of the claim, there exists a vector $u \in \mathbb{R}^\ell$ such that $M^T u = \mathbf{0}_k$ and $(1, \cdots, 1)\, u = 1$. Consider the vector $\mathbf{y} = \left(1/\ell, \dots, 1/\ell\right)^T + \frac{\alpha}{(1-\alpha)p_x}(u - \mathbf{e_x})$. The vector $\mathbf{y}$ is a solution to Equation (6) since $M^T u = \mathbf{0}_k$. We need to show that it is a probability vector. First,

$$(1, \cdots, 1)\, \mathbf{y} = (1, \cdots, 1) \left(1/\ell, \dots, 1/\ell\right)^T + (1, \cdots, 1)\, \frac{\alpha}{(1-\alpha)p_x}(u - \mathbf{e_x})$$

$$= 1 + \frac{\alpha}{(1-\alpha)p_x}(1-1) = 1.$$

Second, $y_i \geq 1/\ell - \frac{\alpha}{(1-\alpha)p_x}(1 + |u_i|) \geq 1/\ell - 2k\alpha(1 + \max\{|u_j|\})$ (recall that $p_x \geq 1/k$ as $p_x > 0$ and p_x is a multiple of $1/k$). Thus, by taking $\alpha \leq 1/(2k\ell(1 + \max\{|u_j|\}))$, the vector $\mathbf{y}$ is non-negative, and, therefore, it is a probability vector solving Equation (6).[1] $\square$

To conclude, we have showed that if $(\mathbf{0}_k)^T$ is an affine combination of the rows of M, then there are probability vectors solving Equation (6). Furthermore, these probability vectors can be efficiently found. Using these probability vectors as distributions in the simulator we constructed proves that the protocol FAIRTWOPARTY$_0$ is fully secure for f. $\square$

We provide an alternative proof of Theorem 3.4 in the full version of this paper. Furthermore, in the full version we prove the converse of Theorem 3.4. This converse claim shows that the condition that $(\mathbf{0}_k)^T$ is an affine combination of the rows of M, which is sufficient for our protocol to securely compute f, is necessary for our simulation of protocol FAIRTWOPARTY$_0$.

---

[1] As the size of the domain of f is considered as a constant, the vector $\mathbf{u}$ is a fixed vector and α is a constant.

Changing the Constant. In the above proof, we have fixed b_{i^*-1} (i.e., P_2's backup output at round $i^* - 1$) to be 0. By changing this constant and setting it to 1 (that is, by executing Protocol FAIRTWOPARTY$_1$), we can securely compute additional functions.

Corollary 3.5. *Let M be the associated matrix of a Boolean function f. If $(1_k)^T$ is an affine combination of the rows of M, then there is a constant $\alpha_0 > 0$ such that Protocol FAIRTWOPARTY$_1$ with $\alpha \le \alpha_0$ is a fully-secure protocol for f.*

Proof. Let $\overline{M} = (1) - M$. Executing Protocol FAIRTWOPARTY$_1$ on f is equivalent to executing Protocol FAIRTWOPARTY$_0$ with the function $\bar{f}(x,y) = 1 - f(x,y)$ (whose associated matrix is $\overline{M}$), and flipping the output. Thus, Protocol FAIRTWOPARTY$_1$ is fully secure for f if $(0_k)^T$ is an affine combination of the rows of $\overline{M}$. By the conditions of the corollary, there is an affine combination of the rows of M that equals $(1_k)^T$ and the same affine combination applied to the rows of $\overline{M}$ equals $(0_k)^T$, thus, Protocol FAIRTWOPARTY$_0$ is a fully secure protocol for $\bar{f}$. □

Corollary 3.6. *Assume that there is a secure protocol for OT. A Boolean two-party function f with an associated matrix M is computable with full security if:*

- *either 0_ℓ or 1_ℓ is an affine combination of the columns of M, or*
- *either $(0_k)^T$ or $(1_k)^T$ is an affine combination of the rows of M.*

Proof. By Theorem 3.3, a function can be computed with full security by protocol FAIRTWOPARTY$_0$ if $(0_k)^T$ is an affine combination of the rows of M. Likewise, by Theorem 3.5, a function can be computed with protocol FAIRTWOPARTY$_1$ if $(1_k)^T$ is an affine combination of the rows of M. By changing the roles of the parties in the protocol (in particular, P_2 gets the correct output before P_1), we obtain the other two possibilities in the corollary.

Assuming that there exists a secure protocol for OT, we can transform protocol FAIRTWOPARTY$_\sigma$ to a protocol without a dealer providing full security. □

3.1 Limits of the GHKL Protocol

We consider the function that was given in Example 1.2 and show that it cannot be computed using the GHKL protocol. In fact, we show a concrete attack on the protocol, and construct an adversary that succeeds to influence the output of the honest party. We then explain how the modified protocol is not susceptible to this attack.

Assume that P_2 chooses its input from $\{y_1, y_2, y_4\}$ uniformly at random (each with probability $1/3$), and if the input y_1 was chosen, it flips the output that was received. If the protocol computes f with full security, then P_2 receives a bit that equals 1 with probability $2/3$, no matter what input distribution (malicious) P_1 may use.

Now, assume that the adversary $\mathcal{A}$ corrupts P_1, always uses input x_1, and follows the following strategy: If $a_1 = 1$ it aborts immediately (in which case, P_2 outputs b_0). Otherwise, it aborts at round 2 (in which case, P_2 outputs b_1).

We now compute the probability that P_2 outputs 1 when it runs the GHKL protocol with $\mathcal{A}$. We have the following cases:

1. If $i^* \neq 1$, then both b_0 and b_1 are random values chosen by the online-dealer, and thus both possible values of a_1 yield the same result. In this case, the output of P_2 is 1 with probability $2/3$.

2. If $i^* = 1$, then the value a_1 is the correct output (i.e., $f(x,y)$ where $y \in \{y_1, y_2, y_4\}$). Moreover, the output of P_2 depends on a_1: If $a_1 = 1$, then it outputs b_0, which is random value chosen by the online-dealer. If $a_1 = 0$, it outputs b_1, which is the correct output.

   Since $\mathcal{A}$ uses input x_1, the case of $a_1 = 1$ may occur only if the input of P_2 is y_4, and thus $b_0 = 1$ with probability $3/4$. On the other hand, if $a_1 = 0$ then P_2's input is either y_1 or y_2, and it receives correct output b_1. It outputs 1 only in the case where its input was y_1.

We therefore conclude:

$$\Pr\left[\mathrm{Out}_2 = 1\right] = \Pr\left[i^* \neq 1\right] \cdot \Pr\left[\mathrm{Out}_2 = 1 \mid i^* \neq 1\right] + \Pr\left[i^* = 1\right] \cdot \Pr\left[\mathrm{Out}_2 = 1 \mid i^* = 1\right]$$

$$= (1-\alpha) \cdot \frac{2}{3} + \alpha \cdot \left(\Pr\left[\mathrm{Out}_2 = 1 \wedge a_1 = 0 \mid i^* = 1\right] + \Pr\left[\mathrm{Out}_2 = 1 \wedge a_1 = 1 \mid i^* = 1\right] \right)$$

$$= (1-\alpha) \cdot \frac{2}{3} + \alpha \cdot \left(\frac{2}{3} \cdot \frac{1}{2} + \frac{1}{3} \cdot \frac{3}{4} \right) = \frac{2}{3} - \frac{1}{12}\alpha < \frac{2}{3} .$$

Protocol FAIRTWOPARTY$_1$. Since $\mathbf{1}_k^T$ is an affine combination of the rows of M, we use the protocol FAIRTWOPARTY$_1$, that is, use the backup output at round $i^* - 1$ to be 1. Assume that the adversary corrupts P_1 and instructs him to conduct the same attack in some round $i \geq 2$ (as $i^* \geq 2$ such attack in round 1 is meaningless in our protocol) . Now, if $i = i^*$, the value of b_{i^*-1} is always 1, and the value of b_{i^*} is correct. If $i < i^*$, then b_i is some random value chosen by the dealer. In the former case ($i^* = i$), since P_2 flips its output in case its input was y_1, its output is 1 with probability $2/3$. In the latter case, P_2 outputs 1 with probability $2/3$. We get that the final output of P_2 is 1 with probability $2/3$, and conclude that the protocol is not susceptible to the attack described above.

We next give some intuition why the GHKL protocol is susceptible to this attack whereas our protocol is immune to it. In the GHKL protocol, for each input y the distributions of b_{i^*-1} is different. At round i^*, the value a_{i^*} is "correct", and leaks information about the distribution of b_{i^*-1}. This gives the adversary the ability to bias the output of P_2 once it guesses correctly the round i^*. In contrary, in our protocol, we detach the correlation between a_{i^*} and b_{i^*-1}; although a_{i^*} leaks information about the input of P_2, all distributions of b_{i^*-1} are exactly the same for all possible inputs y, and this attack is bypassed.

4 Characterization of Fairness for Boolean Two-Party Functions

In this section we provide a complete characterization of the Boolean functions that can be computed with full security. To prove this characterization, recall the definition of *semi-balanced* functions given in [14]:

Definition 4.1. *A function $f : X \times Y \to \{0,1\}$ with an associated $\ell \times k$ matrix M is right semi-balanced if $\exists \mathbf{p} \in \mathbb{R}^k$ such that $M\mathbf{p} = \mathbf{1}_\ell$ and $\sum_{i=1}^{k} p_i \neq 1$. Similarly, f is left semi-balanced if $\exists \mathbf{q} \in \mathbb{R}^\ell$ such that $M^T\mathbf{q} = \mathbf{1}_k$ and $\sum_{i=1}^{\ell} q_i \neq 1$. A function f is semi-balanced if it is right semi-balanced and left semi-blanced.*

Makriyannis [14] proved that semi-unbalanced functions are inherently unfair, by showing that a fully-secure protocol for a semi-balanced function implies fair-sampling. We claim that if f is not semi-balanced then f is computable with full security by Protocol FAIRTWOPARTY$_\sigma$.

Lemma 4.2. *If f is not right semi-balanced, then either $(\mathbf{0}_k)^T$ is an affine combination of the rows of M or $\mathbf{1}_\ell$ is an affine combination of the columns of M.*

If f is not left semi-balanced, then either $(\mathbf{1}_k)^T$ is an affine combination of the rows of M or $\mathbf{0}_\ell$ is an affine combination of the columns of M.

Proof. We show only the case where f is not right semi-balanced, the case of left semi-balanced is proven analogously. If f is not right semi-balanced, then one of the following is implied: Either $\mathbf{1}_\ell \notin \text{im}(M)$, which, by Theorem 2.1, implies that $(\mathbf{0}_k)^T$ is an affine combination of the rows of M. Alternatively, it can be that $\mathbf{1}_\ell \in \text{im}(M)$. In this case the vector $\mathbf{p}$ for which $M \cdot \mathbf{p} = \mathbf{1}_\ell$ satisfies $\sum_{i=1}^{\ell} p_i = 1$ (since f is not right semi-balanced). This in particular implies that $\mathbf{1}_\ell$ is an affine combination of the columns of M. □

Theorem 4.3. *Assume an that there is a secure protocol for OT. Let f be a Boolean two-party function f with an associated matrix M. The function f can be computed with full-security if and only if f is not semi-balanced if and only if at least one of the following conditions holds*

I. *$(\mathbf{0}_k)^T$ is an affine combination of the rows of M or $\mathbf{1}_\ell$ is an affine combination of the columns of M,*

II. *$(\mathbf{1}_k)^T$ is an affine combination of the rows of M or $\mathbf{0}_\ell$ is an affine combination of the columns of M.*

Proof. If f is semi-balanced, then by [14] it cannot be computed with complete fairness, hence, it cannot be computed with full security.

If f is not semi-balanced, then, by Lemma 4.2, at least one of the conditions (I) or (II) holds. By Corollary 3.6, conditions (I) or (II) imply (assuming that there is a secure protocol for OT) that f can be computed with full security. □

4.1 Extensions

First we consider the OT-hybrid model. As explained in Theorem 3.1, our protocol can be executed in the OT-hybrid model providing information-theoretic security (without any dealer). Furthermore, the impossibility result of [1] holds in the OT-hybrid model and the reduction of [14] is information-theoretic secure. Thus, our characterization remains valid in the OT-hybrid model.

Corollary 4.4. *Let f be a Boolean two-party function f with an associated matrix M. The function f can be computed with full-security in the OT-hybrid model with information-theoretic security if and only if f is not semi-balanced if and only if at least one of the following conditions holds*

I. *$(\mathbf{0}_k)^T$ is an affine combination of the rows of M or $\mathbf{1}_\ell$ is an affine combination of the columns of M,*

II. *$(\mathbf{1}_k)^T$ is an affine combination of the rows of M or $\mathbf{0}_\ell$ is an affine combination of the columns of M.*

Second, consider randomized functionalities. Let $f : X \times Y \to \Delta(\{0,1\})$ be a randomized finite Boolean function and define associated $\ell \times k$ matrix M such that for all i,j, $M_{i,j} = \Pr\left[f(x_i, y_j) = 1\right]$. One can modify Protocol FAIRTWOPARTY$_\sigma$ such that the dealer now computes randomized outputs, we note that the analysis above still holds. In particular, the following theorem is true.

Theorem 4.5. *Assume that there is a secure protocol for OT. A randomized Boolean two-party function f is computable with complete security if and only if it is not semi-balanced.*

4.2 A Geometric Interpretation of the Characterization

We review the geometric representation of our characterization, which may shed some light on on our understanding of full security. We start by linking between semi-balanced functions and linear hyperplanes.

A *linear hyperplane* in $\mathbb{R}^m$ is an $(m-1)$-dimensional affine subspace of $\mathbb{R}^m$, and is defined as all the points $\boldsymbol{x} = (x_1, \ldots, x_m) \in \mathbb{R}^m$ that are a solution of some linear equation $a_1 x_1 + \ldots + a_m x_m = b$, for some constants $\mathbf{a} = (a_1, \ldots, a_m) \in \mathbb{R}^m$ and $b \in \mathbb{R}$. We denote this hyperplane by $\mathcal{H}(\mathbf{a}, b) \triangleq \{X \in \mathbb{R}^m \mid \langle X, \mathbf{a} \rangle = b\}$. We show alternative representations for the semi-balanced property:

Claim 4.6. *Let f be a Boolean function, let $X_1, \ldots, X_\ell \in \mathbb{R}^k$ denote the rows of the matrix M, and let $Y_1, \ldots, Y_k \in \mathbb{R}^\ell$ denote the columns of M. The following are equivalent:*

1. *The function is semi-balanced.*
2. *$\mathbf{0}_k, \mathbf{1}_k \notin$ affine-hull$\{X_1, \ldots, X_\ell\}$ and $\mathbf{0}_\ell, \mathbf{1}_\ell \notin$ affine-hull$\{Y_1, \ldots, Y_k\}$.*
3. *There exists an hyperplane $\mathcal{H}(\mathbf{q}, 1)$ that contains all the rows $X_1, \ldots, X_\ell$, and there exists yet another hyperplane $\mathcal{H}(\mathbf{p}, 1)$ that contains all the columns $Y_1, \ldots, Y_k$. In addition, $\mathbf{1}_k, \mathbf{0}_k \notin \mathcal{H}(\mathbf{q}, 1)$ and $\mathbf{1}_\ell, \mathbf{0}_\ell \notin \mathcal{H}(\mathbf{p}, 1)$.*

The proof can be found in the full version of this paper [3]. The following theorem divides the Boolean functions to three categories. Each category has some different properties which we will discuss in the following.

Theorem 4.7. *Let f be a Boolean function. Then:*

1. **The function is balanced:**
   There exists an hyperplane $\mathcal{H}(\mathbf{q}, \delta_1)$ that contains all the rows $X_1, \ldots, X_\ell$, and there exists yet another hyperplane $\mathcal{H}(\mathbf{p}, \delta_2)$ that contains all the columns $Y_1, \ldots, Y_k$. Then:
   (a) *If $\mathbf{0}_\ell, \mathbf{1}_\ell \notin \mathcal{H}(\mathbf{p}, \delta_1)$ and $\mathbf{0}_k, \mathbf{1}_k \notin \mathcal{H}(\mathbf{q}, \delta_2)$, then the function is semibalanced and cannot be computed fairly.*
   (b) *If either $H(\mathbf{p}, \delta_1)$ or $\mathcal{H}(\mathbf{q}, \delta_2)$ contains one of the vectors $\mathbf{1}_k, \mathbf{0}_k, \mathbf{1}_\ell, \mathbf{0}_\ell$, then f can be computed with full-security.*
2. **The function is unbalanced (full-dimensional):**
   If the rows do not lie on a single hyperplane, or the columns do no lie on a single hyperplane, then the function is unbalanced and can be computed with full-security.

We remark that case 1b was left unresolved in [2,14], and is finally proven to be possible here.

Intuition. Consider a single (ideal, fair) execution of some function f where P_1 chooses its input according to some distribution $\mathbf{a} = (a_1, \ldots, a_\ell)$, i.e., chooses input x_i with probability a_i, and P_2 chooses its input according to distribution $\mathbf{b} = (b_1, \ldots, b_k)$. Then, the parties invoke the function and receive the same output simultaneously. The vector $\alpha^T \cdot M_f = (w_1, \ldots, w_k)$ is the output vector of P_2, where w_j represents the probability that the output of P_2 is 1 when it uses input y_j in this execution. Similarly, we consider also the output vector $M_f \cdot \mathbf{b}$.

Intuitively, fairness is achievable in functions which are more "unbalanced" and is impossible to achieve in more "balanced" functions. In our context, the term "balanced" relates to the question of whether there exists a party that can influence and bias the output of the other party in a single (and fair) invocation of the function, as mentioned above.

More concretely, for balanced functions (class 1a), we use the fact that in a single execution no party can bias the output of the other in the reduction to sampling [14]. Specifically, the reduction works by considering a single (fair) execution of the function, where each party chooses its input randomly according to some cleverly chosen distributions $\mathbf{a}'$ and $\mathbf{b}'$. The geometric representation of the functions in this class may explain why these functions are reducible to sampling. The fact that all the rows lie on a single hyperplane $\mathcal{H}(q, \delta_1)$ implies that all their convex combinations also lie on that hyperplane. Therefore, no matter what input distribution $\mathbf{a}'$ a malicious P_1 may choose (i.e., what convex combination of the rows), the resulting vector $\mathbf{a}' \cdot M_f = (w_1, \ldots, w_k)$ lies on the hyperplane $\mathcal{H}(\mathbf{q}, \delta_1)$ as well, and therefore satisfies the relation $\langle \mathbf{a}', \mathbf{q} \rangle = \delta_1$. This guarantees that there is the exact *same* correlation between the possible outputs of P_2 no matter what input P_1 may use, and this enables P_2 to deduce a coin from this function. We also have a similar guarantee for the output of P_1 and malicious P_2, and thus this function is reducible to sampling.

The other extreme case is class 2, the class of unbalanced functions. In this class of functions, one party has significant more power over the other, and can manipulate its output. In this case, the single invocation of the function that we consider is the process of the ideal execution: the honest party and the simulator send their inputs to the trusted party, who computes the function correctly and sends the outputs back simultaneously. What enables successful simulation is the fact that that the simulator can actually manipulate the output of the honest party. In particular, in the proof of Theorem 3.3 the simulator chooses the input distribution $\mathbf{x}_x^{(b)}$ cleverly in order to succeed in the simulation. Geometrically, since the inputs of the honest party do not lie on any hyperplane, its outputs are not correlated and the simulator has enough freedom to manipulate them.

Additional interesting geometric properties that show the differences between the two class of functions (classes 1a and 2) are the following. First, one can show that for each function in class 1a, the two hyperplanes that contain the rows and resp. the columns are *unique*. This implies that the *affine dimensions* of the affine hulls of the rows and the affine-hull of the columns are equal. On the other hand, in class 2, the affine dimensions of these two affine-hulls are always distinct. Moreover, almost all functions that satisfy $|X| = |Y|$ are in class 1a, whereas almost all functions that satisfy $|X| \neq |Y|$ are in class 2.

Class 1b. The third class of functions is where the things become less clearer, and may even contradict the intuition mentioned above. This class contains functions that are totally symmetric and satisfy $M_f^T = M_f$ (see, for instance, Example 1.2), and thus both parties have the exact same influence on the output of the other party in a single invocation of the function, the affine dimensions of the rows and the columns are the same, and also in most cases $|X| = |Y|$. Yet, somewhat surprisingly, fairness is possible. Overall, since all the rows and columns lie on hyperplanes, all the functions in this class are reducible to sampling; however, the sampling is a trivial one (where the resulting coins of the parties are uncorrelated). In addition, although the influence that a party may have on the output of the other is significantly less than the case of class 2, it turns out that simulation is still possible, and the simulator can "smudge" the exact advantage that the real world adversary has in the real execution. This case is much more delicate, and the GHKL protocol fails to work for some functions in this class. Nevertheless, as we show, the limited power that the simulator has in the ideal execution suffices.

5 On the Characterization of Fairness for Asymmetric Functions

In this section, we study asymmetric Boolean functions. Namely, we consider functions $f = (f_1, f_2)$, where $f_i : X \times Y \to \{0, 1\}$ for each $i \in \{1, 2\}$. As two-party functionalities, P_1 and P_2's input domains correspond to sets X and Y

respectively, like the symmetric case, however, their outputs now are computed according to f_1 and f_2 respectively. In other words, the parties are computing different functions on the same inputs.

Our goal is to extend the characterization of full security to this class of functions. In particular, we show how the feasibility result (Theorem 3.3) and the impossibility result (the reduction from non-trivial sampling) translate to the asymmetric case. Unfortunately, while these two results provide a tight characterization for symmetric functionalities, the same cannot be said about asymmetric ones. As a first step toward the characterization the latter functions, we consider a particular function that lies in the gap and describe a new protocol that computes this function with full security. While this protocol may be considered as another variant of the GHKL-protocol, we believe that it departs considerably from the protocols we have considered thus far. Consequently, it seems that the characterization for asymmetric functionalities is more involved than the symmetric case.

5.1 Sufficient Condition for Fairness of Two-Party Asymmetric Functionalities

We analyze when Protocol $\text{FAIRTWOPARTY}_\sigma$, described in Figure 1, provides full security for Boolean asymmetric functionalities $f = (f_1, f_2)$. We modify Protocol $\text{FAIRTWOPARTY}_\sigma$ such that the backup values of P_i are computed with f_i; we call the resulting protocol $\text{FAIRTWOPARTYASYMM}_\sigma$. For the next theorem recall that $M_1 * M_2$ is the entry-wise product of the matrices.

Theorem 5.1. *Let f_1, f_2 be functions with associated matrices M_1 and M_2 respectively. If $\mathbf{0}_k$ is an affine combination of the rows of M_2, and all the rows of $M_1 * M_2$ are linear combinations of the rows of M_2, then there is a constant $\alpha_0 > 0$ such that Protocol $\text{FAIRTWOPARTYASYMM}_0$ with $\alpha \leq \alpha_0$ is a fully-secure protocol for (f_1, f_2).*

In Theorem 5.1, we require that all the rows of the matrix $M_1 * M_2$ are in the row-span of M_2, and that the vector $\mathbf{0}_k$ is an affine combination of the rows of M_2. Note that when the function is symmetric, and thus, $M_1 = M_2$, the first requirement always holds and the only requirement is the second, exactly as in Theorem 3.3.

Proof (of TheoremTheorem 5.1). We only discuss the required changes in the proof of Protocol $\text{FAIRTWOPARTYASYMM}_0$ compared to the proof of Protocol FAIRTWOPARTY_0. We use the same simulator for P_1. The difference are in its proof. As in the proof of Theorem 3.3, we only need to compare the distribution of (a_i, Out_{P_2}) in both worlds given that the adversary aborts in round $i \leq i^*$.

First, we introduce the following notation

$$(1/\ell, \ldots, 1/\ell)M_2 = (s_1, \ldots, s_k), \qquad M_1 \begin{pmatrix} 1/k \\ \vdots \\ 1/k \end{pmatrix} = \begin{pmatrix} p_1 \\ \vdots \\ p_\ell \end{pmatrix}. \qquad (8)$$

For example, s_y is the probability that $f_2(\widetilde{x}, y) = 1$, when $\widetilde{x}$ is uniformly distributed. Furthermore, for every $x \in X$, $y \in Y$, and $a \in \{0, 1\}$, define $q_x^{(a)}(y)$ as the probability that the output of P_2 in the simulation is 1 (given that the adversary has aborted in round $i \leq i^*$) when the input of P_1 is x, the input of P_2 is y, and $a_i = a$, that is $q_x^{(a)}(y) \triangleq \Pr\left[f(x_0, y) = 1\right]$, where x_0 is chosen according to the distribution $\mathbf{x}_x^{(a)}$. Finally, define the column vector $\mathbf{q}_x^{(a)} \triangleq (q_x^{(a)}(y))_{y \in Y}$. Using this notation,

$$M_2^T \mathbf{x}_x^{(a)} = \mathbf{q}_x^{(a)}, \tag{9}$$

where we represent the distribution $\mathbf{x}_x^{(a)} = (\mathbf{x}_x^{(a)}(x_0))_{x_0 \in X}$ by a column vector.

We next analyze the four options for the values of $(a_i, \mathrm{Out}_{P_2})$.

First Case: $(a_i, \mathrm{Out}_{P_2}) = (0, 0)$. In the real world $\Pr\left[(a_i, \mathrm{Out}_{P_2}) = (0, 0)\right] = (1 - \alpha)(1 - p_x)(1 - s_y) + \alpha(1 - f_1(x, y))$. On the other hand, in the ideal world $\Pr\left[(a_i, \mathrm{Out}_{P_2}) = (0, 0)\right] = (1 - \alpha)(1 - p_x)(1 - q_x^{(0)}(y)) + \alpha(1 - f_1(x, y))(1 - f_2(x, y))$. To get full security we need that these two probabilities in the two worlds are the same, that is,

$$(1 - \alpha)(1 - p_x)(1 - s_y) + \alpha(1 - f_1(x, y))$$
$$= (1 - \alpha)(1 - p_x)(1 - q_x^{(0)}(y)) + \alpha(1 - f_1(x, y))(1 - f_2(x, y)). \tag{10}$$

If $p_x = 1$, then $f_1(x, y) = 1$ and (10) holds. Otherwise, (10) is equivalent to

$$q_x^{(0)}(y) = s_y - \frac{\alpha \cdot (1 - f_1(x, y))f_2(x, y)}{(1 - \alpha)(1 - p_x)}. \tag{11}$$

As this is true for every y, we deduce, using Equations (2) and (1), that

$$M_2^T \mathbf{x}_x^{(0)} = \mathbf{q}_x^{(0)} = M_2^T \left(1/\ell, \ldots, 1/\ell\right)^T - \frac{\alpha}{(1 - \alpha)(1 - p_x)}(\overline{M_1} * M_2)^T \mathbf{e_x}. \tag{12}$$

Claim 5.2. *Fix $x \in X$ and let α be a sufficiently small constant. If $(\mathbf{0}_k)^T$ is an affine combination of the rows of M_2 and all the rows of $M_1 * M_2$ are linear combinations of the rows of M_2, then there exists a probability vector $\mathbf{x}_x^{(0)}$ solving Equation (12).*

Proof. Let $\mathbb{1}$ denote the $\ell \times k$ all-one matrix, and let $\lambda_\alpha = \frac{\alpha}{(1-\alpha)(1-p_x)}$. We get that $(\overline{M_1} * M_2)^T \mathbf{e_x} = ((\mathbb{1} - M_1) * M_2)^T \mathbf{e_x} = (M_2 - M_1 * M_2)^T \mathbf{e_x} = M_2^T \mathbf{e_x} - (M_1 * M_2)^T \mathbf{e_x}$. Literally, this is the subtraction of some row x in matrix M_2, and the row x in the matrix $M_1 * M_2$. However, from the conditions in the statement, this is just a vector in the linear-span of the rows of M_2, and can be represented as $M_2^T \cdot \mathbf{v}$ for some vector $\mathbf{v} \in \mathbb{R}^\ell$. We therefore are looking for a probability vector $\mathbf{x}_x^{(0)}$ solving:

$$M_2^T \mathbf{x}_x^{(0)} = M_2^T \left(1/\ell, \ldots, 1/\ell\right)^T - \lambda_\alpha \cdot M_2^T \mathbf{v}. \tag{13}$$

Let $\beta = \langle \mathbf{1}, \mathbf{v} \rangle$, and recall that $\mathbf{0}_k$ is an affine combination of the rows of M_2, and thus there exists a vector $\mathbf{u} \in \mathbb{R}^\ell$ such that $M^T \cdot \mathbf{u} = \mathbf{0}_k$ and $\langle \mathbf{1}, \mathbf{u} \rangle = 1$.

Consider the vector $\mathbf{y} = (1/\ell, \ldots, 1/\ell)^T - \lambda_\alpha \mathbf{v} + \lambda_\alpha \beta \mathbf{u}$. This vector is a solution for Eq. (13) since $M_2^T \cdot \mathbf{u} = \mathbf{0}_k$. Moreover, it sums-up to 1 since:

$$\langle \mathbf{1}, \mathbf{y} \rangle = \langle \mathbf{1}, (1/\ell, \ldots, 1/\ell)^T \rangle - \lambda_\alpha \langle \mathbf{1}, \mathbf{v} \rangle + \lambda_\alpha \beta \langle \mathbf{1}, \mathbf{u} \rangle = 1 - \lambda_\alpha \beta + \lambda_\alpha \beta \cdot 1 = 1.$$

Finally, for appropriate choice of α, all the coordinates of $\mathbf{y}$ are non-negative, and thus $\mathbf{y}$ is a probability vector solving Eq. (13).   □

Second Case: $(a_i, \mathrm{Out}_{P_2}) = (0, 1)$. In the real world $\Pr[(a_i, \mathrm{Out}_{P_2}) = (0, 1)] = (1 - \alpha)(1 - p_x)s_y$ (in the real world $\mathrm{Out}_{P_2} = b_{i^*-1} = 0$ when $i = i^*$). On the other hand, in the ideal world $\Pr[(a_i, \mathrm{Out}_{P_2}) = (0, 1)] = (1 - \alpha)(1 - p_x)q_x^{(0)}(y) + \alpha \cdot (1 - f_1(x, y))f_2(x, y)$ (in the ideal world $a_{i^*} = f_1(x, y)$ and $\mathrm{Out}_{P_2} = f_2(x, y)$). The probabilities in the two worlds are equal if (11) holds.

Third Case: $(a_i, \mathrm{Out}_{P_2}) = (1, 0)$. In the real world $\Pr[(a_i, \mathrm{Out}_{P_2}) = (1, 0)] = (1 - \alpha)p_x(1 - s_y) + \alpha \cdot f_1(x, y) \cdot 1$. On the other hand, in the ideal world $\Pr[(a_i, \mathrm{Out}_{P_2}) = (1, 0)] = (1 - \alpha)p_x(1 - q_x^{(1)}(y)) + \alpha \cdot f_1(x, y)(1 - f_2(x, y))$. The probabilities in the two worlds are equal when

$$(1 - \alpha)p_x(1 - s_y) + \alpha f_1(x, y) = (1 - \alpha)p_x(1 - q_x^{(1)}(y)) + \alpha f_1(x, y)(1 - f_2(x, y)).$$

If $p_x = 0$, then $f_1(x, y) = 0$ and we are done. Otherwise, this equality holds iff

$$q_x^{(1)}(y) = s_y - \frac{\alpha f_1(x, y) f_2(x, y)}{(1 - \alpha)p_x}. \tag{14}$$

As this is true for every y, we deuce, using Equations (9) and (8), that

$$M_2^T \mathbf{x}_x^{(1)} = \mathbf{q}_x^{(1)} = M^T (1/\ell, \ldots, 1/\ell)^T - \frac{\alpha}{(1 - \alpha)p_x}(M_1 * M_2)^T \mathbf{e}_x, \tag{15}$$

Analogically to case $(0, 0)$, there exists a probability vector $\mathbf{x}_x^{(1)}$ solving Equation (15) if each row of $M_1 * M_2$ is in the row span of M_2, and $\mathbf{0}_k$ is in the affine hull of the rows of M_2.

If the equalities hold for the first 3 cases of the values for $(a_i, \mathrm{Out}_{P_2})$, the equality of the probabilities must also hold for the case $(a_i, \mathrm{Out}_{P_2}) = (1, 1)$.

To conclude, the conditions of the theorem imply that there are probability vectors solving Equations (12) and (15) for every x. Using these probability distributions in the simulator we constructed, we conclude that protocol FAIRTWOPARTYASYMM$_0$ is fully secure for $f = (f_1, f_2)$.   □

Changing the Matrices. In the symmetric setting, we have showed that by flipping the matrix M associated with the function (i.e., by taking the matrix $\overline{M} = (1) - M$) we can construct protocols with full security for a richer class of functions. In the asymmetric setting we can go even further: P_1 and P_2 can flip some of the rows of M_1 and obtain a matrix $\hat{M}_1$ and flip some of the columns of M_2 and obtain a matrix $\hat{M}_2$. The parties now execute FAIRTWOPARTYASYMM$_0$

on the flipped matrices and the parties obtain outputs a and b respectively. If the input of P_1 corresponds to a row that was flipped, then P_1 outputs $1 - a$, otherwise, it outputs a. Similarly, if the input of P_2 corresponds to a column that was flipped, then P_2 outputs $1 - b$, otherwise, it outputs b. Call the resulting protocol FairTwoPartyAsymm'. Thus, we obtain the following corollary.

Corollary 5.3. *Let M_1, M_2 be the associated matrices of f_1 and f_2 respectively. Assume that $\hat{M}_1$ is computed from M_1 by flipping some of its rows and $\hat{M}_2$ is computed from M_2 by flipping some of its columns. If $\mathbf{0}_k$ is an affine-combination of the rows of $\hat{M}_2$, and all the rows of $\hat{M}_1 * \hat{M}_2$ are in the linear span of the rows of $\hat{M}_2$, then then there is a constant $\alpha_0 > 0$ such that the Protocol FairTwoPartyAsymm' with $\alpha \leq \alpha_0$ is a fully-secure protocol for $f = (f_1, f_2)$.*

5.2 Necessary Condition

In this section, we provide a necessary condition for fully secure computation of asymmetric functionalities. It consists of a natural generalization of the semi-balanced criterion for symmetric functionalities. Namely, we show that certain functions imply (non-private) non-trivial sampling and are thus unfair. Informally, the next theorem states that if both parties have some distribution over their inputs that "cancels out" the other party's choice of input, then, assuming the resulting bits are statistically dependent, the function cannot be computed with complete fairness.

Theorem 5.4. *Let f_1, f_2 be functions with associated matrices M_1 and M_2 respectively. If there exist $\mathbf{p} \in \mathbb{R}^\ell$, $\mathbf{q} \in \mathbb{R}^k$ such that $\mathbf{p}^T M_1 = \delta_1 \cdot \mathbf{1}_k^T$, $M_2 \mathbf{q} = \delta_2 \cdot \mathbf{1}_\ell$ and $\mathbf{p}^T (M_1 * M_2) \mathbf{q} \neq \delta_1 \delta_2$, then the functionality $f(x, y) = (f_1(x, y), f_2(x, y))$ implies (non-private) non-trivial sampling.*

Proof. Suppose there exist $\mathbf{p} \in \mathbb{R}^\ell$, $\mathbf{q} \in \mathbb{R}^k$ such that $\mathbf{p}^T M_1 = \delta_1 \cdot \mathbf{1}_k^T$, $M_2 \mathbf{q} = \delta_2 \cdot \mathbf{1}_\ell$ and $\mathbf{p}^T (M_1 * M_2) \mathbf{q} \neq \delta_1 \delta_2$. Further assume that $\sum_i |p_i| = \sum_i |q_j| = 1$, and define $\delta_{1,2} = \mathbf{p}^T (M_1 * M_2) \mathbf{q}$. Consider the following protocol Π in the hybrid model with ideal access to f (with full security). This protocol achieves non-trivial sampling.

- **Inputs:** Empty for both parties.
- **Invoke trusted party:** Parties choose x_i, y_j according to probability vectors $|\mathbf{p}|$ and $|\mathbf{q}|$ respectively, and invoke the trusted party, write a and b for the bits received by the first and second party.
- **Outputs:** P_1 outputs a if $p_i \geq 0$ and $1 - a$ otherwise. P_2 outputs b if $q_j \geq 0$ and $1 - b$ otherwise.

Claim 5.5. *Let Out_1, Out_2 for the outputs of P_1 and P_2 in the above protocol. In an honest execution of Π, the parties' outputs satisfy $\Pr[\mathrm{Out}_1 = 1] = \delta_1 + p^-$, $\Pr[\mathrm{Out}_2 = 1] = \delta_2 + q^-$ and $\Pr[\mathrm{Out}_1 = 1 \wedge \mathrm{Out}_2 = 1] = \delta_{1,2} + p^- \delta_2 + q^- \delta_1 + p^- q^-$, where $p^- = \sum_{p_i < 0} |p_i|$ and $q^- = \sum_{q_j < 0} |q_j|$.*

Proof. We begin the proof by introducing some notation. Let $\text{row}_{1,i}$, $\text{row}_{2,i}$ denote the i-th row of M_1 and M_2 respectively, and let $\text{col}_{1,j}$, $\text{col}_{2,j}$ denote the j-th column of M_1 and M_2 respectively. Construct matrices $\hat{M}_1$ and $\hat{M}_2$ such that

- the i-th row of $\hat{M}_1$ is equal to $\text{row}_{1,i}$ if $p_i \geq 0$ and $\mathbf{1}_k^T - \text{row}_{1,i}$ otherwise,
- the j-th column of $\hat{M}_2$ is equal to $\text{col}_{2,j}$ if $q_j \geq 0$ and $\mathbf{1}_\ell - \text{col}_{2,j}$ otherwise.

Let $\hat{\text{row}}_{1,i}$, $\hat{\text{col}}_{1,j}$ and $\hat{\text{row}}_{2,i}$, $\hat{\text{col}}_{2,j}$ denote the rows and columns of $\hat{M}_1$ and $\hat{M}_2$ respectively. The proof consists of a straightforward computation of each probability.

$$\Pr\left[\text{Out}_1 = 1 \mid y = y_j\right] = |\mathbf{p}^T|\hat{M}_1 \mathbf{e}_{y_j} = \left(\sum_{p_i < 0} |p_i| \hat{\text{row}}_{1,i} + \sum_{p_i \geq 0} p_i \hat{\text{row}}_{1,i} \right) \mathbf{e}_{y_j} \qquad (16)$$

$$= \left(\sum_{p_i < 0} |p_i|(\mathbf{1}_k^T - \text{row}_{1,i}) + \sum_{p_i \geq 0} p_i \text{row}_{1,i} \right) \mathbf{e}_{y_j} = \left(\sum_{p_i < 0} |p_i| \mathbf{1}_k^T + \mathbf{p}^T M_1 \right) \mathbf{e}_{y_j} = \delta_1 + p^-.$$

The output of P_2 is obtained in a similar fashion. Next,

$$\Pr\left[(\text{Out}_1, \text{Out}_2) = (1,1)\right] = |\mathbf{p}^T|(\hat{M}_1 * \hat{M}_2)|\mathbf{q}| = |\mathbf{p}^T| \left(\sum_j (\hat{\text{col}}_{1,j} * \hat{\text{col}}_{2,j})|q_j| \right)$$

$$= |\mathbf{p}^T| \left(\sum_{q_j \geq 0} (\hat{\text{col}}_{1,j} * \text{col}_{2,j})q_j + \sum_{q_j < 0} (\hat{\text{col}}_{1,j} * (\mathbf{1}_\ell - \text{col}_{2,j}))|q_j| \right)$$

$$= |\mathbf{p}^T| \left((\hat{M}_1 * M_2)\mathbf{q} + \sum_{q_j < 0} \hat{\text{col}}_{1,j}|q_j| \right). \qquad (17)$$

Now, since by (17), $|\mathbf{p}^T|\hat{\text{col}}_{1,j} = |\mathbf{p}^T|\hat{M}_1 \mathbf{e}_{y_j} = \delta_1 + p^-$, we deduce that (17) is equal to

$$\left(\sum_{p_i \geq 0} p_i(\text{row}_{1,i} * \text{row}_{2,i}) + \sum_{p_i < 0} |p_i|((1 - \text{row}_{1,i}) * \text{row}_{2,i}) \right) \mathbf{q} + (\delta_1 + p^-)q^-$$

$$= \mathbf{p}^T(M_1 * M_2)\mathbf{q} + \sum_{p_i < 0} |p_i|(\text{row}_{2,i}\mathbf{q}) + (\delta_1 + p^-)q^- = \delta_{1,2} + p^-\delta_2 + q^-\delta_1 + p^-q^-.$$

$\square$

It remains to show that protocol Π is a secure realization of (non-private) non-trivial sampling. First, we note that the parties' outputs above are statistically dependent. This follows from the fact that two bits are independent if and only if $\Pr\left[\text{Out}_1 = 1\right] \cdot \Pr\left[\text{Out}_2 = 1\right] = \Pr\left[\text{Out}_1 = 1 \wedge \text{Out}_2 = 1\right]$. Since, by assumption, $\delta_{1,2} \neq \delta_1 \delta_2$, we deduce that in an honest execution the parties' outputs are statistically dependent. To conclude, note that no matter how the adversary (say controlling P_2) chooses his input (or does not send one at all), by (17), the probability that the output of P_1 is 1 is equal to $\delta_1 + p^-$. Hence, by [1], we get a contradiction. $\square$

5.3   Special Round Protocol with a Twist

Define an asymmetric functionality $f_{sp}(x, y) = (f_1(x, y), f_2(x, y))$, where f_1, f_2 are given by the following matrices

$$M_1 = \begin{pmatrix} 1 & 1 & 1 & 0 \\ 0 & 0 & 0 & 1 \\ 1 & 0 & 0 & 1 \\ 0 & 1 & 0 & 1 \end{pmatrix}, \quad M_2 = \begin{pmatrix} 1 & 0 & 1 & 0 \\ 1 & 0 & 0 & 1 \\ 1 & 0 & 0 & 0 \\ 0 & 1 & 1 & 1 \end{pmatrix}.$$

In this section we show that the above function can be computed with full security. First, note that f_{sp} does not satisfy the hypothesis of Theorem 5.4. On the other hand, both the GHKL protocol as well as FAIRTWOPARTYASYMM$_\sigma$ are susceptible to fail-stop attacks for this particular function, as explained below.

On the limits of the known protocols for the function f_{sp}. Suppose that the parties execute protocol FAIRTWOPARTYASYMM$_0$ for computing f_{sp}. In addition, suppose that P_2 chooses his input uniformly at random from $\{y_1, y_2\}$. If the protocol computes f with full security, then this particular choice of inputs should result in P_2 obtaining an unbiased random bit as an output, regardless of the actions of P_1. We claim that a corrupt P_1 can bias P_2's output toward zero and thus protocol FAIRTWOPARTYASYMM$_0$ does not compute f with full security. Consider an adversary $\mathcal{A}$ that quits immediately upon receiving a_2. Let's compute the probability that P_2's output is equal to 1 under the action of $\mathcal{A}$, and assuming that $y \in_U \{y_1, y_2\}$:

$$\Pr[\text{out}_2 = 1] = \Pr[b_1 = 1]$$
$$= \Pr[b_1 = 1 \wedge i^* = 2] + \Pr[b_1 = 1 \wedge i^* \neq 2]$$
$$= 0 + \frac{1}{2} \cdot (1 - \alpha).$$

In summary, knowing that $b_{i^*-1} = 0$ and that $\mathcal{A}$ can guess i^* with non-negligible probability ($\mathcal{A}$ is betting that $i^* = 2$) we deduce that the above attack will result in P_2 outputting a bit that does not satisfy the prescribed probability distribution. If we revert to the original GHKL protocol, i.e., $i^* \geq 1$, and $b_{i^*-1} = f_2(\tilde{x}^{(i^*-1)}, y)$ where $\tilde{x}^{(i^*-1)}$ is chosen uniformly at random, then a very similar attack will produce the same result:

- The adversary instructs P_1 to use x_3 and execute the protocol.
- At the first round, if $a_1 = 1$ quit. Otherwise, quit upon receiving a_2.

Again, let's compute the probability that P_2's output is equal to 1 under the action of $\mathcal{A}$ assuming that $y \in_U \{y_1, y_2\}$:

$$\Pr[\text{out}_2 = 1] = \Pr[\text{out}_2 = 1 \wedge i^* \neq 1] + \Pr[\text{out}_2 = 1 \wedge i^* = 1]$$
$$= (1 - \alpha)\frac{1}{2} + \alpha \cdot \left(\Pr[a_1 = 1 \wedge b_0 = 1 \,|\, i^* = 1] + \Pr[a_1 = 0 \wedge b_1 = 1 \,|\, i^* = 1] \right)$$
$$= (1 - \alpha)\frac{1}{2} + \alpha \cdot \left(\frac{1}{4} + 0 \right) = \frac{1}{2} - \alpha\frac{1}{4}.$$

Once again, $\mathcal{A}$ can guess i^* with non-negligible probability ($\mathcal{A}$ is betting that $i^* = 1$), and since P_1 obtains a_{i^*} prior to P_2 obtaining b_{i^*}, the adversary can successfully bias P_2's output. We note that an adversary corrupting P_1 can bias P_2's output regardless of how we choose to describe the function. In other words, flipping some of the rows of M_1 and/or some of the columns of M_2 does not offer any solution, since the attacks above can be easily modified to successfully bias P_2's output. Nor is switching the players' roles helpful, i.e., considering (M_2^T, M_1^T), since $M_1 = M_2^T$.

To remedy this, we propose a new protocol, named FAIRTWOPARTYSPECIAL (described in Figure 3), which foils the attacks described above and provides full security for f_{sp}. Our protocol is the same as the original GHKL protocol for every round, except special round i^*. Recall that in the GHKL protocol (as well as all other protocols we have considered), at round i^*, party P_1 obtains $a_{i^*} = f_1(x, y)$ followed by P_2 who obtains $b_{i^*} = f_2(x, y)$. We can reverse the player's roles, however, there is a party that always gets the correct output before the other party. We now make the following modification for the protocol computing f_{sp}: if $x \in \{x_1, x_2\}$, then $a_{i^*} = f_1(x, y)$, otherwise $a_{i^*} = f_1(x, \widetilde{y}^{(i^*)})$, where $\widetilde{y}^{(i^*)}$ is chosen uniformly at random. However, always $b_{i^*} = f_2(x, y)$. Thus, for certain inputs, the second player P_2 effectively obtains the output first. We claim that this modification suffices to compute f_{sp} with full security. and we dedicate the rest of the section to the proof this claim.

Theorem 5.6. *For every $\alpha \le 1/9$, protocol* FAIRTWOPARTYSPECIAL *is fully secure for f_{sp}.*

Proof. The proof follows the ideas of the symmetric case with two significant modifications. First, we need two distinct simulations for P_1 depending on the choice of the input. In particular, if $\mathcal{A}$ hands x_1 or x_2 to $\mathcal{S}^{\mathcal{A}}$, then the simulation is exactly the same as in the symmetric case (or the equivalent asymmetric protocol FAIRTWOPARTYASYMM$_0$). If not, i.e., if $\mathcal{A}$ hands x_3 or x_4, then we simulate P_1 as if he were P_2 in the symmetric case. On the other hand, regarding the second player, we note that a difficulty arises due to the fact that P_2 obtains the output first depending on P_1's choice of input. Nevertheless, we claim that by simulating P_2 as if he were P_1 in the symmetric case (or Protocol FAIRTWOPARTYASYMM$_0$), i.e., as if he *always* gets the output first, results in the correct simulator. Thus, we will first consider the case of a corrupt P_1 when the adversary hands either x_1 or x_2 to the simulator, followed by the security analysis of a corrupt P_2.

We only discuss the required changes in the security proof of Protocol FAIRTWOPARTYSPECIAL compared to the proof of Protocol FAIRTWOPARTYASYMM$_0$. As in the proofs of Theorem 3.3 and Theorem 5.1, we only need to compare the distribution of $(a_i, \mathrm{Out}_{P_2})$ in both worlds given that the adversary aborts in round $i \le i^*$. Similarly to the previous proofs, we compute vectors $\mathbf{q}_x^{(0)}$ and $\mathbf{q}_x^{(1)}$ where $q_x^{(a)}(y)$ is the desired probability that the output of P_2 in the simulation (given that the adversary has aborted in round $i < i^*$) is 1 when the input of P_1 is x, the input of P_2 is y and $a_i = a$. In contrast to the proofs of Theorem 3.3 and Theorem 5.1, we only need to consider $x \in \{x_1, x_2\}$, since we have a separate simulation for $x \in \{x_3, x_4\}$. However, we emphasize that the simulator

---

Protocol FAIRTWOPARTYSPECIAL

1. The parties P_1 and P_2 hand their inputs, denoted x and y respectively, to the dealer.[a]
2. The dealer chooses $i^* \geq 1$ according to a geometric distribution with probability α.
3. The dealer computes $\text{out}_1 = f_1(x,y)$, $\text{out}_2 = f_2(x,y)$ and for $0 \leq i \leq r$

$$a_i = \begin{cases} f_1(x, \widetilde{y}^{(i)}) \text{ where } \widetilde{y}^{(i)} \in_U Y & \text{if } i < i^* \\ f_1(x, \widetilde{y}^{(i)}) \text{ where } \widetilde{y}^{(i)} \in_U Y & \text{if } i = i^* \text{ and } x \in \{x_3, x_4\} \\ \text{out}_1 & \text{otherwise} \end{cases}$$

$$b_i = \begin{cases} f_2(\widetilde{x}^{(i)}, y) \text{ where } \widetilde{x}^{(i)} \in_U X & \text{if } i < i^* \\ \text{out}_2 & \text{otherwise.} \end{cases}$$

4. The dealer gives b_0 to P_2.
5. For $i = 1, \ldots, r$,
   (a) The dealer gives a_i to P_1. If P_1 aborts, then P_2 outputs b_{i-1} and halts.
   (b) The dealer gives b_i to P_2. If P_2 aborts, then P_1 outputs a_i and halts.

---

[a] If x is not in the appropriate domain or P_1 does not hand an input, then the dealer sends $f_2(\hat{x}, y)$ (where $\hat{x}$ is a default value) to P_2, which outputs this value and the protocol is terminated. The case of an inappropriate y is dealt analogously.

Fig. 3. Protocol FAIRTWOPARTYSPECIAL for securely computing a function f_{sp}

can choose any input from the entire domain to send to the trusted party. By considering the four possible values of (a_i, Out_{P_2}), we deduce that

$$\mathbf{q}_x^{(0)}(y) = s_y + \frac{\alpha}{(1-\alpha)(1-p_x)}(1 - f_1(x,y))(s_y - f_2(x,y))$$

$$\mathbf{q}_x^{(1)}(y) = s_y + \frac{\alpha}{(1-\alpha)p_x} f_1(x,y)(s_y - f_2(x,y))$$

and thus,

$$\mathbf{q}_{x_1}^{(0)} = \begin{pmatrix} 3/4 \\ 1/4 \\ 1/2 \\ 1/2 \end{pmatrix} + \frac{\alpha}{(1-\alpha)(1-p_{x_1})}\begin{pmatrix} 0 \\ 0 \\ 0 \\ 1/2 \end{pmatrix}, \quad \mathbf{q}_{x_1}^{(1)} = \begin{pmatrix} 3/4 \\ 1/4 \\ 1/2 \\ 1/2 \end{pmatrix} + \frac{\alpha}{(1-\alpha)p_{x_1}}\begin{pmatrix} -1/4 \\ 1/4 \\ -1/2 \\ 0 \end{pmatrix},$$

$$\mathbf{q}_{x_2}^{(0)} = \begin{pmatrix} 3/4 \\ 1/4 \\ 1/2 \\ 1/2 \end{pmatrix} + \frac{\alpha}{(1-\alpha)(1-p_{x_2})}\begin{pmatrix} -1/4 \\ 1/4 \\ 1/2 \\ 0 \end{pmatrix}, \quad \mathbf{q}_{x_2}^{(1)} = \begin{pmatrix} 3/4 \\ 1/4 \\ 1/2 \\ 1/2 \end{pmatrix} + \frac{\alpha}{(1-\alpha)p_{x_2}}\begin{pmatrix} 0 \\ 0 \\ 0 \\ -1/2 \end{pmatrix}.$$

We conclude that $M_2^T \mathbf{x}_x^{(a)} = \mathbf{q}_x^{(a)}$ for the following vectors $\mathbf{x}_x^{(a)}$:

$$\mathbf{x}_{x_1}^{(0)} = 1/4 \cdot (1,1,1,1)^T + \frac{\alpha}{(1-\alpha)(1-p_{x_1})} 1/2 \cdot (0,1,-1,0)^T$$

$$\mathbf{x}_{x_1}^{(1)} = 1/4 \cdot (1,1,1,1)^T + \frac{\alpha}{(1-\alpha)p_{x_1}} 1/4 \cdot (-3,-1,3,1)^T$$

$$\mathbf{x}_{x_2}^{(0)} = 1/4 \cdot (1,1,1,1)^T + \frac{\alpha}{(1-\alpha)(1-p_{x_2})} 1/4 \cdot (1,-1,-1,1)^T$$

$$\mathbf{x}_{x_2}^{(1)} = 1/4 \cdot (1,1,1,1)^T + \frac{\alpha}{(1-\alpha)p_{x_2}} 1/2 \cdot (0,-1,1,0)^T.$$

It remains to show that for some $\alpha \in (0,1)$ the above vectors become probability vectors – they already sum to 1, we just need them to be positive. A straightforward computation yields that any $\alpha \leq 1/9$ will do.

Corrupt P_2. We now consider the case of an adversary $\mathcal{A}$ corrupting P_2. As mentioned above, we construct a simulator $\mathcal{S}^{\mathcal{A}}$ given a black-box to $\mathcal{A}$ that is completely analogous to P_1's simulator in protocol FAIRTWOPARTYASYMM$_\sigma$. Namely, say that $\mathcal{A}$ hands $y \in Y$ to $\mathcal{S}^{\mathcal{A}}$ for the computation of f. The simulator chooses i^* according to a geometric distribution with parameter α and,

- for $i = 0, \ldots, i^* - 1$, the simulator hands $b_i = f(\widetilde{x}^{(i)}, y)$ to $\mathcal{A}$, where $\widetilde{x}^{(i)} \in_U X$. If $\mathcal{A}$ decides to quit, $\mathcal{S}^{\mathcal{A}}$ sends y_0 according to distribution $\mathbf{y}_y^{(b_i)}$ (to be defined below), outputs $(b_0, \ldots, b_i)$ and halts.
- for $i = i^*$, the simulator sends y to the trusted party, receives $b = f_2(x, y)$ and hands $b_{i^*} = b$ to $\mathcal{A}$. If $\mathcal{A}$ decides to quit, $\mathcal{S}^{\mathcal{A}}$ outputs $(b_0, \ldots, b_{i^*})$ and halts.
- for $i = i^* + 1, \ldots, r$, the simulator hands $b_i = b$ to $\mathcal{A}$. If $\mathcal{A}$ decides to quit, $\mathcal{S}^{\mathcal{A}}$ outputs $(b_0, \ldots, b_i)$ and halts.
- If $\mathcal{A}$ has not quitted yet, $\mathcal{S}^{\mathcal{A}}$ outputs $(b_0, \ldots, b_r)$ and halts.

For reasons mentioned in the proof of Theorem 3.3, we only need to compare the distribution of $(b_i, \mathrm{Out}_{P_1})$ in both worlds assuming that $i \leq i^*$. Now, for every $x \in X$, $y \in Y$, and $b \in \{0,1\}$, define $q_y^{(b)}(x) \triangleq \Pr[f_1(x, y_0) = 1]$, where y_0 is chosen according to the distribution $\mathbf{y}_y^{(b)}$, and define the column vector $\mathbf{q}_y^{(b)} \triangleq (q_y^{(b)}(x))_{x \in X}$. Using this notation,

$$M_1 \mathbf{y}_y^{(b)} = \mathbf{q}_y^{(b)}, \tag{18}$$

where we represent the distribution $\mathbf{y}_y^{(b)} = (\mathbf{y}_y^{(b)}(y_0))_{y_0 \in Y}$ by a column vector. We now analyze the four options for the values of $(b_i, \mathrm{Out}_{P_1})$. Bare in mind that we need to carefully distinguish between $x \in \{x_1, x_2\}$ and $x \in \{x_3, x_4\}$ in the real world.

First Case: $(b_i, \mathrm{Out}_{P_1}) = (0,0)$. In the real world, if $x \in \{x_1, x_2\}$, then we have $\Pr[(b_i, \mathrm{Out}_{P_1}) = (0,0)] = (1-\alpha)(1-p_x)(1-s_y) + \alpha(1-f_1(x,y))(1-f_2(x,y))$. Otherwise, $\Pr[(b_i, \mathrm{Out}_{P_1}) = (0,0)] = (1-\alpha)(1-p_x)(1-s_y) + \alpha(1-p_x)$

$(1 - f_2(x, y))$. On the other hand, in the ideal world $\Pr\left[(b_i, \mathrm{Out}_{P_1}) = (0,0)\right] = (1 - \alpha)(1 - s_y)(1 - q_y^{(0)}(x)) + \alpha(1 - f_1(x,y))(1 - f_2(x,y))$. To get full security we need that these two probabilities in the two worlds are the same, that is,

$$(1 - \alpha)(1 - s_y)(1 - q_y^{(0)}(x)) + \alpha(1 - f_1(x,y))(1 - f_2(x,y)) \tag{19}$$
$$= \begin{cases} (1 - \alpha)(1 - p_x)(1 - s_y) + \alpha(1 - f_1(x,y))(1 - f_2(x,y)) & \text{if } x \in \{x_1, x_2\} \\ (1 - \alpha)(1 - p_x)(1 - s_y) + \alpha(1 - p_x)(1 - f_2(x,y)) & \text{if } x \in \{x_3, x_4\} \end{cases}.$$

Eq. (19) is equivalent to

$$\mathbf{q}_y^{(0)}(x) = \begin{cases} p_x & \text{if } x \in \{x_1, x_2\} \\ p_x + \dfrac{\alpha(p_x - f_1(x,y))(1 - f_2(x,y))}{(1 - \alpha)(1 - s_y)} & \text{if } x \in \{x_3, x_4\} \end{cases}. \tag{20}$$

Let $\lambda = \alpha/(1 - \alpha)$; we now compute the vectors $\mathbf{q}_y^{(0)}$:

$$\mathbf{q}_{y_1}^{(0)} = \begin{pmatrix} 3/4 \\ 1/4 \\ 1/2 \\ 1/2 \end{pmatrix} + \frac{\lambda}{1 - s_{y_1}}\begin{pmatrix} 0 \\ 0 \\ 0 \\ 1/2 \end{pmatrix}, \quad \mathbf{q}_{y_2}^{(0)} = \begin{pmatrix} 3/4 \\ 1/4 \\ 1/2 \\ 1/2 \end{pmatrix} + \frac{\lambda}{1 - s_{y_2}}\begin{pmatrix} 0 \\ 0 \\ 1/2 \\ 0 \end{pmatrix},$$

$$\mathbf{q}_{y_3}^{(0)} = \begin{pmatrix} 3/4 \\ 1/4 \\ 1/2 \\ 1/2 \end{pmatrix} + \frac{\lambda}{1 - s_{y_3}}\begin{pmatrix} 0 \\ 0 \\ 1/2 \\ 0 \end{pmatrix}, \quad \mathbf{q}_{y_4}^{(0)} = \begin{pmatrix} 3/4 \\ 1/4 \\ 1/2 \\ 1/2 \end{pmatrix} + \frac{\lambda}{1 - s_{y_4}}\begin{pmatrix} 0 \\ 0 \\ -1/2 \\ 0 \end{pmatrix}.$$

and deduce that that $M_1 \mathbf{y}_y^{(0)} = \mathbf{q}_y^{(0)}$ for the following vectors $\mathbf{y}_y^{(0)}$:

$$\mathbf{y}_{y_1}^{(0)} = 1/4 \cdot (1,1,1,1)^T + \frac{\lambda}{1 - s_{y_1}} 1/2 \cdot (0,1,-1,0)^T$$

$$\mathbf{y}_{y_2}^{(0)} = 1/4 \cdot (1,1,1,1)^T + \frac{\lambda}{1 - s_{y_2}} 1/2 \cdot (1,0,-1,0)^T$$

$$\mathbf{y}_{y_3}^{(0)} = 1/4 \cdot (1,1,1,1)^T + \frac{\lambda}{1 - s_{y_3}} 1/2 \cdot (1,0,-1,0)^T$$

$$\mathbf{y}_{y_4}^{(0)} = 1/4 \cdot (1,1,1,1)^T + \frac{\lambda}{1 - s_{y_4}} 1/2 \cdot (-1,0,1,0)^T.$$

Note that, to obtain probability vectors, any $\alpha \leq 1/9$ will do.

We refer to the full proof in [3] for cases $(b_i, \mathrm{Out}_{P_1}) = (0,1)$ and $(b_i, \mathrm{Out}_{P_1}) = (1,0)$. $\qquad\square$

6   Characterization of Fairness When Half of the Parties are Honest

In this section, we extend our discussion to the case of any constant number of parties, where at most half of the parties are corrupted. Using the characterization of symmetric two-party functionalities from Section 4, we fully characterize the functions that can be computed with full security in this setting. In this section, we only consider symmetric functions, i.e., where all parties receive the same output. The main result of this section is stated in Theorem 6.2 below.

Let $X = X_1 \times X_2 \times \cdots \times X_m$ and let $f : X \to R$ be some function. For a subset $\emptyset \subset I \subset [m]$ let $\bar{I} = [m] \setminus I$ and let X_I be the projection of X on I. For an input $\mathbf{x} \in X$, let $\mathbf{x}_I$ be the projection of $\mathbf{x}$ on I. We define the function $f_I : X_I \times X_{\bar{I}} \to R$, by $f_I(\mathbf{w}, \mathbf{z}) = f(\mathbf{x})$, where $\mathbf{x}$ is such that $\mathbf{x}_I = \mathbf{w}$ and $\mathbf{x}_{\bar{I}} = \mathbf{z}$.

The full proof of the theorem appears in the full version of this work (see, Section 5 in [3]). The proof of the feasibility part of Theorem 6.2 is proved by describing a protocol (with an on-line dealer), in which the dealer runs many two-party protocols simultaneously. More specifically, for each subset $I \subset [m]$ of size k, the dealer runs the appropriate two-party protocol Π_I that securely computes the two-party function f_I. The description of Protocol FAIRMULTIPARTY, proving the feasibility, appears in Figure 4. The proof of its security as well as the explanation of how the on-line dealer is eliminated are deferred to the full version of this paper. Our proofs draw on ideas from [5,6].

Protocol FAIRMULTIPARTY

1. The parties $P_1, \ldots, P_m$ hand their inputs, denoted $\mathbf{x} = x_1, \ldots, x_m$, respectively, to the dealer. If a party P_j does not send an input, then the dealer selects $x_j \in X_j$ uniformly at random. If half of the parties do not send an input, then the dealer sends $f(x_1, \ldots, x_m)$ to the honest parties and halts.
2. The dealer computes for every $I \subset [m]$, such that $|I| = k = m/2$, and for every $0 \le i \le r$ the backup outputs $a_i^{(I)}$ and $b_i^{(I)}$ using Π_I.
3. The dealer shares $b_0^{(I)}$ among the parties of $S_I^{(2)}$ for each I as above, in a k-out-of-k Shamir secret-sharing scheme.
4. For $i = 1, \ldots, r_f$,
    (a) The dealer shares $a_i^{(I)}$ among the parties of $S_I^{(1)}$ for each I as above, in a k-out-of-k Shamir secret-sharing scheme. If *all* the parties of some subset $S_I^{(1)}$ abort, then the parties in $S_I^{(2)}$ reconstruct b_{i-1}^I, output it and halt.
    (b) The dealer shares $b_i^{(I)}$ among the parties of $S_I^{(2)}$ for each I as above, in a k-out-of-k Shamir secret-sharing scheme. If *all* the parties of some subset $S_I^{(2)}$ abort, then the parties in $S_I^{(1)}$ reconstruct a_i^I, output it and halt.
5. All subsets $S_I^{(1)}$ and $S_I^{(2)}$ (for all sets I as above) reconstruct a_r^I and b_r^I, respectively, and output it (by correctness, with all but negligible probability in the security parameter – all successfully reconstructed values are equal).

Fig. 4. Protocol FAIRMULTIPARTY for securely computing a function f, if at least half of the parties are honest

We next introduce some of the notation that is necessary for understanding the description of Protocol FAIRMULTIPARTY described in Figure 4. We assume without loss of generality that there exists an integer r such that each Π_I is an

r-round protocol with an on-line dealer for securely computing the two-party function f_I. We show in the full version that this is without loss of generality.

Notation 6.1. *Fix $I \subset [m]$, such that $1 \in I$ and $|I| = k$. Let A_I be the party that plays the role of A in Π_I, and let B_I be the other party in this protocol. Denote by $a_i^{(I)}$ (resp. $b_i^{(I)}$) the backup output that party A_I (resp. B_I) receives in round i of Π_I. In addition, let $S_I^{(1)}$ be the set of parties whose inputs correspond to the input of A_I, that is, $S_I^{(1)} = \{P_i : i \in I\}$; let $S_I^{(2)}$ be the remaining parties (i.e., whose inputs correspond to that of party B).*

Theorem 6.2. *Let $m = 2k$ be a constant, let $X = X_1 \times X_2 \times \cdots \times X_m$ be a finite domain, and let $f : X \to R$ be a deterministic function. Then, f is computable with full security in the multiparty setting against an adversary that can corrupt up to k parties if and only if for every subset $I \subset [m]$, with $|I| = k$, the function f_I is computable with full security in the two-party setting. The same is true if f is a randomized Boolean function.*

7 Open Problems

As a conclusion, we provide a short list of questions that are left unanswered by our paper. First, regarding asymmetric functionalities, it would be interesting to know if a generalized version of Protocol FAIRTWOPARTYSPECIAL offers any significant improvement toward bridging the gap in the asymmetric case. In Protocol FAIRTWOPARTYSPECIAL, for certain inputs P_1 gets the output first, and for others P_2 gets the output first. In our protocol this partition depends only on the input of P_1. The question is which asymmetric functions can be computed with full security when the protocol uses an arbitrary partition of the parties' inputs.

On the other hand, some functions that lie in the gap might be outright impossible to compute with full security. If true, then these functions do not imply non-trivial sampling by the construction[2] of [14], and thus, any impossibility result would require a new argument. We provide an example of a function that lies in the gap, and whose fairness does not seem to derive from the work of the present paper.

$$M_1 = \begin{pmatrix} 0 & 1 & 1 & 0 \\ 1 & 0 & 1 & 1 \\ 1 & 0 & 0 & 0 \\ 0 & 1 & 0 & 1 \end{pmatrix}, \quad M_2 = \begin{pmatrix} 1 & 1 & 1 & 0 \\ 1 & 0 & 1 & 1 \\ 0 & 1 & 0 & 1 \\ 1 & 1 & 0 & 0 \end{pmatrix}.$$

Another direction of inquiry would be the computation of Boolean functions in the multi-party setting, where the corrupted parties form a strict majority. In particular, there may be room to generalize protocol FAIRMULTIPARTY to handle strict majorities of corrupted parties. However, several difficulties arise in view of the fact that the adversary would have access to the backup outputs of multiple

---

[2] The construction is also presented in the proof of Theorem 5.4 in Section 5.2.

partial functions, and not just one. Furthermore, the characterization of functions with arbitrary output domains, a subject already touched upon by [2], seems like a hard problem to tackle.

Our protocols assume that the size of the input domain is a constant independent of the security parameter. It can be shown that in our protocols for two parties, if the size of the input domains is $\log n$ (where n is the security parameter), then the number of rounds and the computation are still polynomial in n. It would be interesting to construct protocols for families of functions $\{f_n\}_{n \in \mathbb{N}}$, where the size of the domain of f_n is polynomial or even exponential in n.

References

1. Agrawal, S., Prabhakaran, M.: On fair exchange, fair coins and fair sampling. In: Canetti, R., Garay, J.A. (eds.) CRYPTO 2013, Part I. LNCS, vol. 8042, pp. 259–276. Springer, Heidelberg (2013)
2. Asharov, G.: Towards characterizing complete fairness in secure two-party computation. In: Lindell, Y. (ed.) TCC 2014. LNCS, vol. 8349, pp. 291–316. Springer, Heidelberg (2014)
3. Asharov, G., Beimel, A., Makriyannis, N., Omri, E.: Complete characterization of fairness in secure two-party computation of boolean functions. Cryptology ePrint Archive, Report 2014/1000 (2014), http://eprint.iacr.org/
4. Asharov, G., Lindell, Y., Rabin, T.: A full characterization of functions that imply fair coin tossing and ramifications to fairness. In: Sahai, A. (ed.) TCC 2013. LNCS, vol. 7785, pp. 243–262. Springer, Heidelberg (2013)
5. Beimel, A., Lindell, Y., Omri, E., Orlov, I.: 1/p-secure multiparty computation without honest majority and the best of both worlds. In: Rogaway, P. (ed.) CRYPTO 2011. LNCS, vol. 6841, pp. 277–296. Springer, Heidelberg (2011)
6. Beimel, A., Omri, E., Orlov, I.: Protocols for multiparty coin toss with dishonest majority. In: Rabin, T. (ed.) CRYPTO 2010. LNCS, vol. 6223, pp. 538–557. Springer, Heidelberg (2010)
7. Canetti, R.: Security and composition of multiparty cryptographic protocols. J. of Cryptology 13(1), 143–202 (2000)
8. Cleve, R.: Limits on the security of coin flips when half the processors are faulty. In: 18th STOC, pp. 364–369 (1986)
9. Goldreich, O.: Foundations of Cryptography, Voume II Basic Applications. Cambridge University Press (2004)
10. Goldreich, O., Micali, S., Wigderson, A.: How to play any mental game. In: 19th STOC, pp. 218–229 (1987)
11. Gordon, S.D., Hazay, C., Katz, J., Lindell, Y.: Complete fairness in secure two-party computation. J. of the ACM 58(6), Article No. 24 (2011)
12. Gordon, S.D., Katz, J.: Complete fairness in multi-party computation without an honest majority. In: Reingold, O. (ed.) TCC 2009. LNCS, vol. 5444, pp. 19–35. Springer, Heidelberg (2009)
13. Kilian, J.: Basing cryptography on oblivious transfer. In: 20th STOC, pp. 20–31 (1988)
14. Makriyannis, N.: On the classification of finite boolean functions up to fairness. In: Abdalla, M., De Prisco, R. (eds.) SCN 2014. LNCS, vol. 8642, pp. 135–154. Springer, Heidelberg (2014)
15. Yao, A.C.: How to generate and exchange secrets. In: 27th FOCS, pp. 162–167 (1986)

Richer Efficiency/Security Trade-offs in 2PC

Vladimir Kolesnikov*, Payman Mohassel, Ben Riva, and Mike Rosulek**

[1] Bell Labs
kolesnikov@research.bell-labs.com
[2] University of Calgary
pmohasse@cpsc.ucalgary.ca
[3] Bar-Ilan University
benr.mail@gmail.com
[4] Oregon State University
rosulekm@eecs.oregonstate.edu

Abstract. The dual-execution protocol of Mohassel & Franklin (PKC 2006) is a highly efficient (each party garbling only one circuit) 2PC protocol that achieves malicious security apart from leaking an *arbitrary, adversarially-chosen* predicate about the honest party's input. We present two practical and orthogonal approaches to improve the security of the dual-execution technique.

First, we show how to greatly restrict the predicate that an adversary can learn in the protocol, to a natural notion of "only computation leaks"-style leakage. Along the way, we identify a natural security property of garbled circuits called *property-enforcing* that may be of independent interest.

Second, we address a complementary direction of reducing the probability that the leakage occurs. We propose a new dual-execution protocol — with a very light cheating-detection phase and each party garbling $s+1$ circuits — in which a cheating party learns a bit with probability only 2^{-s}. Our concrete measurements show approximately 35% reduction in communication for the AES circuit, compared to the best combination of state of the art techniques for achieving the same security notion.

Combining the two results, we achieve a rich continuum of practical trade-offs between efficiency & security, connecting the covert, dual-execution and full-malicious guarantees.

1 Introduction

Garbled circuits were initially conceived as a technique for secure computation protocols [23]. Now they are recognized as fundamental and useful to a wide range of cryptographic applications (see the survey in [3]). By themselves, garbled circuits are generally only useful for achieving semi-honest security. For example, in the setting of two-party secure computation, malicious security costs approximately 40 times more than semi-honest security using current techniques.

* Supported by the Office of Naval Research (ONR) contract N00014-14-C-0113.
** Supported by NSF award CCF-1149647.

Y. Dodis and J.B. Nielsen (Eds.): TCC 2015, Part I, LNCS 9014, pp. 229–259, 2015.

The majority of this extra overhead is to mitigate the effects of adversarially crafted garbled circuits.

In many circumstances, because of the high cost of achieving full malicious security, a slight security relaxation may be acceptable, in return for performance improvements. This was the motivation for the k-leaked model of Mohassel and Franklin [17], covert security of Aumann and Lindell [2], and others. This work falls into the line of research aiming to get as much security as possible, while bringing the required resources down to, ideally, that of the semi-honest model.

Dual execution. The **dual-execution** 2PC protocol of Mohassel & Franklin [17] is a natural starting point in this line; it works as follows. The parties run two separate instances of Yao's semi-honest protocol, so that each party is the "sender" in one instance and "receiver" in the other. Each party evaluates a garbled circuit to obtain their garbled output. Then the two parties run a (much simpler and smaller) fully-secure "equality test" protocol to check whether their outputs match; each party inputs the garbled output they computed along with output wire labels of the garbled circuit they generated. If the outputs don't match, then the parties abort.

The protocol is not secure against malicious adversaries. An honest party executes an adversarially crafted garbled circuit and then uses the garbled output in the equality-test sub-protocol. However, since the equality test has only one bit of output, it can be shown that the dual-execution protocol leaks at most one (adversarially chosen) bit about the honest party's input.

1.1 Our Results

In this section we summarize our results. In Section 2, we present at the high level the main motivation, intuition and insights of our constructions, as well as put all the pieces in the unifying perspective. In the subsequent corresponding sections, we present formalizations, complete constructions, proofs and performance analysis of each individual contribution.

Our theme is to explore and reduce the leakage allowed by the dual-execution protocol. We develop new techniques for restricting the kinds of predicates that the adversary can learn, as well as for reducing the probability that the adversary succeeds in his attack. Combining the two approaches results in a more efficient continuum of cost-security trade-offs, connecting the covert, dual-execution and full-malicious guarantees.

Limiting Leakage Functions in Dual-Execution. The original security notion introduced in the dual-execution paper [17], and the follow-up [7] allows the adversary to learn an arbitrary predicate of the player's input. We show how to significantly limit this leakage to a conjunction of what we call "gate-local" queries, i.e., boolean queries that only operate on the input wires to a single gate of the original circuit. In our formalization, we follow the framework of [3] and introduce the notion of Property-Enforcing Garbling Schemes (PEGS), which may be of independent interest.

Reducing Leakage Probability in Dual-Execution. In a complementary research direction, using the ϵ-CovIDA security notion of [18], we develop new techniques for reducing the *probability* of leakage in the Dual-Execution framework. We improve on their construction by achieving 2^{-s} security with only s circuits for each party, similar to state-of-the-art results of Lindell [14] and Huang, Katz and Evans [8] from the malicious setting. However, we replace the "cheating-recovery" computation of [14] and repeated dual-execution mechanism of [8] with a much more lightweight procedure based on Private Set Intersection (PSI) that provides significant gains in computation and bandwidth. We note that the protocol of [8] has a similar high-level idea to ours: each party sends (approximately) s circuits, then the parties run a fully-secure processing phase. However, their protocol does not achieve ϵ-CovIDA security. In particular, their protocol performs separate equality checks of wire labels for each output bit of the circuit. Hence, in the event that an adversary successfully passes the circuit-check phase (with probability 2^{-s}), she can learn more than one bit.[1]

Our concrete measurements (see Figure 8) show that our techniques yield 35% reduction in overall communication for the AES circuit (compared to protocol of [18] augmented with Lindell's circuit reduction techniques).

Putting it Together: A Richer and Cheaper Security/Efficiency Trade-offs. Restricting the leakage functions a successful adversary may evaluate, and further limiting the probability of his success, allows for a fine-grained practical trade-offs between security guarantees and efficiency of 2PC. This work can be viewed as interpolating between the guarantees of covert, dual execution (i.e. the k-leaked model [17]) and fully-malicious models. Indeed, setting $s = 2$, our protocols correspond to an improved hybrid of covert and k-leaked models and protocols. We guarantee probability of $1/2$ of catching the cheating adversary, and at the same time limit the leakage to an "only computation leaks" one-bit functions. On the opposite end of the spectrum, setting $s = 40$ gives fully-secure 2PC, which, while having better latency (since parties can work in parallel and due to a cheaper cheating recovery) than [14], should be seen as less efficient than [14] as it sends $2s$ total circuits. However, in the extremely important (in practice) set of security parameters/associated costs of $s \in \{1, \ldots, 20\}$, our protocols provide the best "value". Indeed, the guarantee of covert 2PC can be unacceptable in such scenarios as a successful adversary may learn the entire input of the honest party with a non-negligible probability (e.g., the set of long-term keys) making [14] less suitable, while our protocols remain attractive. Further, as noted in Section 6, the parameter s may differ between the two players to reflect different risk/trust assumptions.

1.2   Related Work

To the best of our knowledge, only a handful of prior work consider trading off security for better efficiency in the context of fully-malicious 2PC. Mohassel and

---

[1] Concretely, suppose Alice passes the circuit-check phase with malicious circuits that compute an arbitrary (multi-bit) function $g(x, y)$. Then Alice will learn the length of the longest common prefix of $g(x, y)$ and the correct output $f(x, y)$.

Franklin [17] introduced the notion of k-leaked model and showed through their dual-execution protocol, that leaking a single bit of information can yield major improvement in efficiency. The follow-up work of [7] implemented/enhanced their protocol, confirming the efficiency gains. The work of [9] also considers leakage of information with the goal of designing more efficient non-interactive secure computation protocols. In fact, they also propose a construction where the leakage function is restricted to disjunction of intermediate wire values in the computation. However, what mainly separates our construction from theirs is that we focus on concrete efficiency and small constant factors for fast implementation, while their work is focused on optimal asymptotic complexity, and results in a construction with noticeably larger constant factors (to the best of our knowledge the exact constants have not been worked out).

In a complementary direction, Aumann and Lindell [2] introduced the notion of covert security where one trades the probability of deterring malicious behavior (i.e., making the probability non-negligible) for more efficient 2PC. The recent work of [18] introduces the notion of ϵ-CovIDA security which can be seen as a strengthening of both the covert and the k-leaked models for 2PC. We adopt their security definition for reducing probability of leakage in dual execution.

Application specifics and hard performance requirements sometimes drive the trade-offs. In recent works on practical private DB [10,19] some of the execution patterns are revealed to the adversary as a trade-off for efficient sublinear execution time. Similarly, in the setting of searchable encryption (e.g., [5]), access patterns can often be leaked, yielding significant improvements in performance.

Besides allowing a bit of leakage, other methods have been suggested for relaxing standard security guarantees: input-indistinguishable MPC [15] that allows for better composability, one-sided/two-sided non-simulatability (e.g. see [1]), superpolynomial-time simulation [20,21,4], and security against uniform adversaries [12].

2   Overview of Our Approach and Constructions

Following our focus — reducing leakage power and probability in dual-execution, — we now go a little deeper into each of our results and present their intuition and insights.

2.1   Limiting Leakage Functions in Dual-Execution

Garbled circuits & property-enforcement. It is clear that an adversary can learn an *arbitrary* predicate in the dual-execution protocol, as long as the honest party evaluates *any* garbled circuit given by the other party. To limit leakage in any way therefore requires two things:

1. The parties must perform some check of the garbled circuit they receive in the protocol. In the extreme, parties could demand a zero-knowledge proof of correctness of the garbled circuit, but in that case dual-execution is not even

needed — the protocol would already be fully-secure in the malicious setting. In our setting it makes sense to consider only lightweight checks on the garbled circuits. In their treatment of the dual-execution protocol, Huang, Katz and Evans [7] mention briefly that a party could perform a "sanity check" of the garbled circuit it received. At the same time, most simple checks, such as verifying the circuit topology and XOR gates placement, seem helpful, but bring little guarantees and can exclude few leakage functions due to a simple attack we describe below.

2. Since the checks on the garbled circuit cannot guarantee complete correctness, we are still in a setting involving possibly malicious garbled circuits. Hence, we need some way of reasoning about what information is leaked when a honest party evaluates a malicious garbled circuit. This natural problem has not been investigated before, to the best of our knowledge. Previous work considered all-or-nothing security from a garbled circuit (i.e., either it is correct or not). We suspect that our conceptual approach regarding malicious garbled circuits may be useful in many other settings, given how ubiquitous and powerful the garbled circuit technique has become throughout cryptography.

To address these needs, we introduce a new security notion for garbled circuits called **property-enforcing** garbling schemes (PEGS). Roughly speaking, in a property-enfocring garbling scheme an honest user can locally verify that a garbled circuit F indeed computes a function with a certain property.

More formally, let prop be some property of (plain) circuits: size, topology, the circuit itself, etc. A property-enforcing garbling scheme has additional procedures Prop and Extract. We require that, for all possibly malicious garbled circuits F, if $\mathsf{Extract}(F) \to f$ (where f is a plain circuit) then (1) $\mathsf{Prop}(F) = \mathsf{prop}(f)$ and (2) F produces garbled outputs in direct correspondence with the output of f. That is, the logic of F is "explained by" a plain circuit f with $\mathsf{prop}(f) = \mathsf{Prop}(F)$. See Section 3.2 for the formal definitions. In our actual definition, Extract requires extra information typically only available to the simulator, whereas Prop can be computed publicly.

Suppose we use a property-enforcing garbling scheme in the dual-execution protocol. Both parties would ensure that $\mathsf{Prop}(F) = \mathsf{prop}(f)$ for the garbled circuit F they receive and the objective function f that is being computed. We show that with this modification, the adversary cannot learn *arbitrary* predicates of the honest party's input, but rather only predicates (roughly) of the form $\tilde{f}(x) \overset{?}{=} c$ where $\mathsf{prop}(\tilde{f}) = \mathsf{prop}(f)$.

Achieving topology-enforcement. Intuitively, it seems that classical/standard garbling schemes already give the receiver some guarantees along the lines of property-enforcement. An honest party seems to *enforce* the circuit's topology in how it evaluates the garbled circuit; hence, standard garbling schemes should be topology-enforcing at the least. Interestingly, this is not quite the case.

Imagine a classical garbled circuit for a single gate. This garbled circuit consists of four ciphertexts:

$$\mathsf{Enc}_{A_0,B_0}(C_1) \qquad \mathsf{Enc}_{A_0,B_1}(C_2) \qquad \mathsf{Enc}_{A_1,B_0}(C_3) \qquad \mathsf{Enc}_{A_1,B_1}(C_4)$$

Here A_0, A_1, B_0, B_1 are wire labels of input wires. The garbler is supposed to choose $C_1, \ldots, C_4$ from among two possiblities (i.e., the two wire labels of the output wire). Yet there is no way for the evaluator to check that these four ciphertexts encode only two values. It is trivial to let $C_1, \ldots, C_4$ be distinct. In that case, the garbled output of this circuit reveals the entire input. The behavior of this garbled circuit cannot be "explained" by a single-gate circuit, so the scheme is not topology-enforcing.

Our intuition about standard/classical garbling schemes is thus not quite right, but it is not far off either. These schemes do enforce topology in some sense, but they do not enforce the "information bandwidth" on each wire. In an extreme example, one can make a garbled circuit with just one output wire, but whose garbled output reveals the entire input (in the sense that all distinct inputs give different garbled outputs).

To achieve topology-enforcement in a more reasonable sense (i.e., the behavior of a malicious garbled circuit can always be described by a *boolean* circuit of the advertised topology), it suffices to simply limit the wire bandwidth to 2.[2] In our construction, the sender includes a hash of the two wire labels on each wire. When evaluating a garbled circuit, the receiver checks its wire labels at each step against these hashes. We prove that this construction enforces the topology of the circuit, when the hash function is modeled as a random oracle. Furthermore, the construction remains very practical.

Only computation leaks. "Only computation leaks" (OCL) [16] refers to a paradigm for defining information leakage. In short, OCL means that an adversary cannot leak jointly on two internal values x and y unless they are both computed on simultaneously at some point. Armed with topology-enforcing garbling schemes, we are able to restrict leakage in the dual-execution protocol to OCL-style leakage at the level of gates in the circuit.

More precisely, say that a query is *gate-local* if the query can be expressed as a function of the two input wires to some *single gate* in the circuit. We are able to restrict the dual-execution adversary to learn only a *conjunction* of gate-local queries (with respect to the function f being computed).

The main idea in our construction is as follows. Dual-execution allows parties to (roughly) check the equality of *outputs* of their garbled circuits. To keep a malicious circuit from "building up" a complicated leakage expression (i.e., more complicated than a gate-local query), we try to apply dual execution to check equality of *all* intermediate values in the computation. Hence, we modify the original circuit so that every intermediate wire is secret shared, with each party

---

[2] Actually, we can only limit the wires to contain 0, 1, or an error, where all errors are guaranteed to propogate forward. This turns out to be sufficient for the dual-execution protocol.

receiving one of the shares. The parties then use the dual-execution mechanism to ensure that their shares (hence, the intermediate values of the computation) all agree.

Formalizing and proving this intuition requires some care, and we also need to extend the dual-execution paradigm to cope with outputs known to only one of the parties.

Overall, our modifications to the dual-execution protocol — adding topology-enforcement to the garbling scheme, and adding secret-sharing gadgets to the ciruict — remain quite practical. The resulting protocol has very limited 1-bit leakage but is still much less expensive than fully malicious-secure 2PC. Exploring the continuum of trade-offs between plain dual-execution and full security is an interesting direction, which we address in combination with our next contribution.

2.2 Reducing Leakage Probability in Dual-Execution

As discussed earlier, an alternative to restricting the leakage function is to restrict the probability of occurrence of leakage. This is indeed the notion of ϵ-CovIDA Security recently introduced in [18] which augments the notion of Covert Security of [2]. Essentially, this notion requires that if a player is trying to cheat, the other players can catch him with probability $1-\epsilon$, but even if he is not caught (i.e., with probability ϵ) the cheater can only learn a *single bit* of extra information about the other players' inputs, and the correctness of the output is still guaranteed. In other words, the leakage of the single bit of information only occurs with probability ϵ. The 2^{-s}-CovIDA security is particularly attractive for low and medium values of s (e.g. $1 \leq s \leq 20$) since it provides a *much* stronger guarantee than covert 2PC, and is at the same time more efficient than fully-malicious 2PC where $s \geq 40$.

[18] presents two protocols that are secure in this model, requiring about $3s$ garbled circuits from each player (a total of $6s$) to obtain 2^{-s}-CovIDA security. We observe that it is possible to combine their dual-execution approach with the underlying ideas of Lindell [14], in order to obtain a 2^{-s}-CovIDA 2PC protocol using only $2s$ garbled circuits, as opposed to $6s$ of [18]. The main observation that makes this possible is that the garbler's input consistency check of [22] can be extended to enforce equality of a party's input not only in the circuits he garbles but also in those garbled by his counterparts. However, simply running [14] as a dual execution still results in high overhead as it requires each party to execute a relatively expensive "cheating recovery" phase. We note that we are not aware of the above observation having been published elsewhere, but we consider it a natural combination of ideas in [14] and [18], and the input-consistency check technique of [22].

Protocol Overview. We propose a new approach for designing a 2^{-s}-CovIDA 2PC, wherein we replace the cheating recovery phase with a private set intersection protocol on the outputs. The high level idea of the protocol is as follows. We follow the insight of Lindell [14] for achieving 2^{-s} security with s circuits, namely that cheater only can cheat if *all* of the evaluated circuits are incorrect, and not just the majority. At the same time we avoid the expensive cheating-punishment setup and execution. So, our Alice and Bob perform a simplified version of the protocol of Lindell [14], where they skip all the steps associated with the cheating recovery. Instead, Alice and Bob then switch roles (i.e. Bob becomes the garbler) and perform the same steps *à la* dual execution. All the circuits generated by each party have the same output labels, and we use the universal hashing circuit of [22] in both sets of circuits in order to enforce equality of a player's inputs not only in the circuits he generated but also those generated by his counterpart. There are two points to consider when using the universal hashing approach in the two executions. First, we need to use the same hash function in both sets of executions, and second, we need to commit both the garbler and the evaluator to their inputs before generating the hash function (in standard 2PC, only the garbler needed to commit to his input). To achieve the latter, the garbler commits to his garbled inputs, and parties execute the oblivious transfer for the evaluator's input before choosing the hash function at random.

Let's denote the output labels for the s circuits created by Alice (resp. by Bob) by $(\mathsf{out}_{A,0}^1, \mathsf{out}_{A,1}^1), \ldots, (\mathsf{out}_{A,0}^m, \mathsf{out}_{A,1}^m)$ (resp. $(\mathsf{out}_{B,0}^1, \mathsf{out}_{B,1}^1), \ldots, (\mathsf{out}_{B,0}^m, \mathsf{out}_{B,1}^m)$), where m is the number of output wires in the circuit. At the end of the two executions up to the opening phase, Alice and Bob each create initially empty sets T_A and T_B. For every circuit evaluated by Alice, if the output is valid and equal to, say, $z^A = (z_1^A, \ldots, z_m^A)$, Alice computes $q = \mathsf{out}_{A,z_1^A}^1 \oplus \mathsf{out}_{B,z_1^A}^1 \oplus \cdots \oplus \mathsf{out}_{A,z_m^A}^m \oplus \mathsf{out}_{B,z_m^A}^m$, and lets $T_A = T_A \cup \{q\}$. Bob does a symmetric computation. Each party then adds enough dummy random values to its set until its size is the same as the number of evaluated circuits (note that the expected size of T_A will be $s/2$, the size of the evaluation set).

The idea is to have the parties run a fully secure two-party private set intersection (PSI) protocol computing $T_A \cap T_B$. If the intersection is empty each party aborts, and otherwise, it uses the translation table to compute the final output from the labels in the intersection (note that the intersection can at most be of size one, since the circuits created by the honest party all evaluate to the same correct output). The intuition is that as long as the malicious party did not cheat for just one of the garbled circuits he created, the correct output of that circuit will be the unique value in $T_A \cap T_B$.

There are several issues to resolve for this approach to work: if we perform the PSI before the opening/checking phase, then the output of the PSI can leak extra information to a malicious party. For example, this leakage can happen with probability one in case of a malicious party that garbles the same bad circuit s times, while we want to reduce the probability of leakage to 2^{-s}. If we perform the PSI after the opening phase, on the other hand, we fix the above. But we encounter a different problem, that at the end of the opening, the output labels are revealed, and this allows a malicious party to modify his input to the PSI and hence trick the honest party to learn an incorrect output. We address this dilemma by using a two-stage PSI (see Section 6.2) where in the first stage parties commit their input sets but learn nothing while in the second stage, one of the parties learn the intersection. We then perform the first stage of the PSI before the opening phase, while postponing the second stage until after the openings.

This almost works except that all existing PSI protocol with security against malicious adversaries only let one of the parties learn the intersection. Simply sending the result to the other party is not secure since a malicious party can lie and provide a wrong answer (note that since this step takes place after the opening lying about the output is not hard). We solve this problem by having each party randomly permute its set and commit to each element in the permuted set, before the two-stage PSI is invoked. Then, in the final stage of exchanging the output, in order to prove to the other party that the output is correct (i.e. the output of one of the evaluated circuits), each party also opens the commitment to the PSI input corresponding to the intersection (note that these commitments where issued before the opening phase when it was not possible to forge any output value not returned by the evaluated circuits).

The intuition behind 2^{-s}-CovIDA security of the protocol is that with probability $1 - 2^{-s}$, at least one of the outputs evaluated by the honest party is "the correct output" and hence included in his set. On the other hand, before the opening phase, the malicious party only learns the output labels for the correct output and hence can only commit to the correct output in the first stage of the PSI. Hence, with probability $1 - 2^{-s}$, either the honest party aborts, or the computed intersection of the two sets will be the correct output and among those inputs that parties committed to, before the opening phase.

With probability 2^{-s}, however, a malicious party can cheat in all the evaluated circuits and not get caught. In this scenario, whether the output of the intersection is empty or not leaks one bit of additional information to the malicious player. In either case, the correctness is guaranteed since the honest party cannot be tricked into accepting an incorrect output.

3   Property-Enforcing Garbling Schemes

3.1   Garbling Schemes

Bellare, Hoang, and Rogaway [3] introduce the notion of a garbling scheme as a cryptographic primitive. We refer the reader to their work for a complete treatment and give a brief summary here.[3] A garbling scheme consists of the following algorithms: Garble takes a circuit f as input and outputs (F, e, d) where F is a garbled circuit, e is encoding information, and d is decoding information. Encode takes an input x and encoding information e and outputs a garbled input X. Eval takes a garbled circuit F and garbled input X and outputs a garbled output Y. Finally, Decode takes a garbled output Y and decoding information d and outputs a plain circuit-output (or an error $\perp$).

---

[3] Their definitions apply to any kind of garbling, but we specify the notation for *circuit* garbling.

3.2   Property Enforcing

We extend the definition of garbling schemes as follows. Let prop be a property of circuits; e.g., prop(f) might output the topology of a circuit f.

A garbling scheme is prop-enforcing if it meets the following additional requirements:

- The property prop is extended to *garbled* circuits. That is, when F is a garbled circuit, anyone can publicly compute a value Prop(F).
- There is a deterministic procedure Extract that can "explain" any (possibly adversarially generated) garbled circuit F as a plain circuit f' satisfying prop(f') = Prop(F).

More formally, Extract(F, e) either outputs $\perp$ or a pair (f', d') where d' is a simple mapping of values to wire labels. We define the following security game:

Initialize:
$b \leftarrow \{0, 1\}$

Finalize(b'):
return $b \overset{?}{=} b'$

Query(F, e, x):
$(f', d') \leftarrow$ Extract(F, e)
if $b = 0$ then
   $Y :=$ Eval(F, Encode(e, x))
else $\tilde{y} := f'(x)$
   $Y := d'_{1, \tilde{y}_1} \| \cdots \| d'_{m, \tilde{y}_m}$
return Y

Hence, the garbled output Y contains no more information about x than $f'(x)$, a circuit satisfying property Prop(F).

3.3   Construction: Enforcing Topology

Definition 1. *A* **circuit-with-abort** *is a standard circuit with ternary values* $\{0, 1, \perp\}$ *on the wires, where* $\perp$ *values cascade. That is, for every gate G in the circuit, $G(\perp, \cdot) = G(\cdot, \perp) = \perp$.*

Throughout this section, we use the term "circuit" to refer to circuits with abort.

Construction. For simplicity we make minimal additions to the "Garble2" construction of [3]. Our modification to achieve property-enforcement is rather simple. In Garble2, each wire i is associated with two wire labels X_i^0 and X_i^1. The garbled circuit then simply contains the values $C[i, lsb(X_i^0)] = H(X_i^0)$ and $C[i, lsb(X_i^1)] = H(X_i^1)$, where H is a random oracle. It is straightforward that these new values do not compromise the standard security properties.

When evaluating a gate g, the evaluator obtains a visible wire label X_g, and now checks whether it is valid. By *valid*, we mean that $H(X_g) = C[g, b]$, where b is the select bit of wire label X_g. If this is not the case, then the evaluator aborts.

The Extract procedure maintains the invariant that there exist at most two valid wire labels for each wire. This is true for the input wires by definition.

Provided that the invariant is true for the input wires of a gate, there are at most 4 wire label combinations that Extract needs to try to extract the logic of this gate. If there are more than 2 valid output wire labels for this gate, then we have obtained an explicit collision under H: an event that happens only with negligible probability. Otherwise, the invariant holds at this gate as well.

Note that in the unmodified Garble2 scheme, the number of possible wire labels can be made to grow exponentially at each level of the circuit. Hence, the garbled values can encode more than 1 bit of information on a wire. The key idea here is to limit the "bandwidth" of each wire to a single bit.

The full details of our construction are provided in Figure 1.

Garble($1^k, f$):

$(n, m, q, A', B', G) \leftarrow f$
for $i \in \{1, \ldots, n+q\}$ do
$\quad t \xleftarrow{\$} \{0,1\}; \quad X_i^0 \xleftarrow{\$} \{0,1\}^{k-1}t$
$\quad X_i^1 \xleftarrow{\$} \{0,1\}^{k-1}\bar{t}$
$\star \quad C[i,t] \leftarrow H(X_i^0); \quad C[i,\bar{t}] \leftarrow H(X_i^1)$
for $(g,i,j) \in \{n+1, \ldots, n+q\} \times \{0,1\} \times \{0,1\}$
$\quad a \leftarrow A'(g); \quad b \leftarrow B'(g)$
$\quad A \leftarrow X_a^i; \quad a \leftarrow lsb(A); \quad B \leftarrow X_b^j$
$\quad b \leftarrow lsb(B)$
$\quad T \leftarrow g\|a\|b; \quad P[g,a,b] \leftarrow E_{A,B}^T(X_g^{G_g(i,j)})$
$\star \quad F \leftarrow (n, m, q, A', B', P, C)$
$e \leftarrow (X_1^0, X_1^1, \ldots, X_n^0, X_n^1)$
$d \leftarrow (X_{n+q-m+1}^0, X_{n+q-m+1}^1, \ldots, X_{n+q}^0, X_{n+q}^1)$
return (F, e, d)

Extract(F, e):

$(n, m, q, A', B', P, C) \leftarrow F$
$(X_1^0, X_1^1, \ldots, X_n^0, X_n^1) \leftarrow e$
for $(g,i,j) \in \{n+1, \ldots, n+q\} \times \{0,1\} \times \{0,1\}$
$\quad a \leftarrow A'(g); \quad b \leftarrow B'(g)$
$\quad$ skip loop iteration if X_a^i or X_b^j undefined
$\quad A \leftarrow X_a^i; \quad a \leftarrow lsb(A); \quad B \leftarrow X_b^j$
$\quad b \leftarrow lsb(B)$
$\quad \tilde{X} \leftarrow D_{A,B}^T(P[g,a,b]); \quad x \leftarrow lsb(\tilde{X})$
$\quad$ if $C[g,x] = H(\tilde{X})$ then:
$\quad\quad$ if X_g^x already defined then return $\bot$
$\quad\quad X_g^x \leftarrow \tilde{X}; \quad G_g(a,b) \leftarrow x$
$\quad$ else $G_g(a,b) \leftarrow \bot$
$d' \leftarrow (X_{n+q-m+1}^0, X_{n+q-m+1}^1, \ldots, X_{n+q}^0, X_{n+q}^1)$
$f' \leftarrow (n, m, q, A', B', G)$
return (f', d')

Encode(e, x):

$(X_1^0, X_1^1, \ldots, X_n^0, X_n^1) \leftarrow e$
$x_1 \cdots x_n \leftarrow x$
$X \leftarrow (X_1^{x_1}, \ldots, X_n^{x_n})$

Decode(d, Y):

$(Y_1, \ldots, Y_m) \leftarrow Y$
$(Y_1^0, Y_1^1, \ldots, Y_m^0, Y_m^1) \leftarrow d$
for $i \in \{1, \ldots, m\}$ do:
$\quad$ if $Y_i = Y_i^0$ then $y_i \leftarrow 0$
$\quad$ else if $Y_i = Y_i^1$ then $y_i \leftarrow 1$
$\quad$ else return $\bot$
return $y \leftarrow y_1 \cdots y_m$

Eval(F, X):

$\star \quad (n, m, q, A', B', P, C) \leftarrow F$
$(X_1, \ldots, X_n) \leftarrow X$
for $g \leftarrow n+1$ to $n+q$ do:
$\quad a \leftarrow A'(g), b \leftarrow B'(g)$
$\quad A \leftarrow X_a; \quad a \leftarrow lsb(A)$
$\quad B \leftarrow X_b; \quad b \leftarrow lsb(B)$
$\quad T \leftarrow g\|a\|b$
$\quad X_g \leftarrow D_{A,B}^T(P[g,a,b])$
$\star \quad$ if $H(X_g) \neq C[g, lsb(X_g)]$:
$\star \quad\quad$ return $\bot$
return $(X_{n+q-m+1}, \ldots, X_{n+q})$

Fig. 1. Topology-enforcing garbling scheme construction. H denotes a random oracle, and E denotes a dual-key cipher, following [3]. "$\star$" denotes differences from the Garble2 construction of [3] (the entire Extract procedure is new).

Theorem 1. *The construction in Figure 1 is a secure garbling scheme (in random oracle model) satisfying privacy, authenticity, obliviousness, and prop-enforcement, when prop denotes the topology of the circuit.*

Proof. We focus on the proof of prop-enforcement. First, observe that Extract can output $\perp$ with only negligible probability, since it only outputs $\perp$ when it explicitly finds a collision under the random oracle H. But, as we will argue, the two branches of the security game are identical except for the possibility of Extract outputting $\perp$.

Extract works by identifying at most 2 "valid" wire labels X_g^0, X_g^1 for each wire g. For input wires, these valid wire labels are given as the encoding information e. For a gate g with input wires i & j, Extract produces a gate with logic G_g such that evaluating the corresponding gate in F with wire labels X_g^a, X_g^b yields $X_g^{G_g(a,b)}$ (or $\perp$ if $G_g(a,b) = \perp$). Hence it follows by induction that the output of Eval(F, Encode(e, x)) is exactly characterized by the choice of wire labels $f'(x)$.

4   Applications to Dual Execution

In this section, we write a functionality as $f(x_A, x_B) = (y_A, y_B, y_{AB})$ where y_A, y_B, y_{AB} denote outputs for Alice only, Bob only, and both parties, respectively. We must augment the existing dual-execution protocol and proofs of [17,7] to account for functionalities that give different inputs to the two parties.

$\mathcal{L}$-leaked model. In the $\mathcal{L}$-leaked model for computing function f, where $\mathcal{L} = (\mathcal{L}^A, \mathcal{L}^B)$, Alice provides input x_A and Bob provides input x_B to the functionality. The functionality then computes $(y_A, y_B, y_{AB}) \leftarrow f(x_A, x_B)$. Let (y_A, y_{AB}) denote Alice's potential output and let (y_B, y_{AB}) denote Bob's potential output..

The functionality delivers the corrupt party's potential output. If Alice is corrupt, then the adversary supplies a leakage function $L \in \mathcal{L}^A$ to the functionality. If Bob is corrupt, the adversary supplies a leakage function $L \in \mathcal{L}^B$. The functionality evaluates $\ell = L(x_A, x_B)$. If $\ell = 0$ then the functionality delivers $\perp$ to the honest party; otherwise waits for instruction from the adversary before delivering the honest party's potential output.

Dual execution protocol. Given a functionality f, we let $f(\cdot, x)$ and $f(x, \cdot)$ denote residual circuits with one input hard-coded. We assume that for all possible inputs x we have prop($f(x, \cdot)$) = prop($f(0^n, \cdot)$), and prop($f(\cdot, x)$) = prop($f(\cdot, 0^n)$).

Given a functionality f a garbling scheme $\mathcal{G}$ we define the dual execution protocol DualEx$^f[\mathcal{G}]$ as follows:

1. Alice has input x_A and Bob has input x_B. Alice does $(F_A, e_A, d_A) \leftarrow$ Garble $(f(x_A, \cdot))$. Bob similarly does $(F_B, e_B, d_B) \leftarrow$ Garble($f(\cdot, x_B)$).
2. Alice commits to F_A, Bob commits to F_B (if the garbling scheme is adaptively secure then the garbled circuits can be sent in the clear here).
3. Using instances of OT, Alice acts as sender with inputs e_A and Bob acts as receiver with input x_B. Bob receives garbled input X_B. Likewise, the parties use OT for Alice to obtain X_A.

4. The parties open their commitments to the garbled circuits. Alice aborts if $\mathsf{Prop}(F_B) \neq \mathsf{prop}(f(\cdot, 0^n))$. Similarly Bob aborts if $\mathsf{Prop}(F_A) \neq \mathsf{prop}(f(0^n, \cdot))$.
5. Alice does $(Y_A, Y_B, Y_{AB}) = \mathsf{Eval}(F_B, X_A)$. Bob also computes values (Y_A, Y_B, Y_{AB}). If either party obtains $\perp$ from executing Eval, then it continues below using randomly chosen values for these garbled outputs (Y_A, Y_B, Y_{AB}).
6. Alice can decode Y_A and Y_{AB} to obtain plain outputs y_A and y_{AB}. She can use d_A to compute $\tilde{Y}_A, \tilde{Y}_{AB}$, which are garbled output encodings of y_A and y_{AB} according to d_A. She sends $C = \tilde{Y}_A \| Y_B \| Y_{AB} \| \tilde{Y}_{AB}$ to the equality test functionality.
7. Similarly, Bob can decode Y_B and Y_{AB} to obtain plain outputs y_B and y_{AB}. He computes $\tilde{Y}_B$ and $\tilde{Y}_{AB}$, garbled output encodings of these values according to d_B. He sends $C = Y_A \| \tilde{Y}_B \| \tilde{Y}_{AB} \| Y_{AB}$ to the equality test functionality.
8. If the equality test returns false, then the parties abort; otherwise Alice outputs (y_A, y_{AB}) and Bob outputs (y_B, y_{AB}).

Leakage functions. Let $f(x_A, x_B) = (y_A, y_B, y_{AB})$ be a 2-party functionality as above. Define:

$$L^A_{f,f',\tilde{y}}(x_A, x_B) = \begin{cases} 1 & \text{if } f'(x_B) = (\tilde{y}, y_B, y_{AB}), \text{ where } (y_A, y_B, y_{AB}) \leftarrow f(x_A, x_B) \\ 0 & \text{otherwise} \end{cases}$$

$$L^B_{f,f',\tilde{y}}(x_A, x_B) = \begin{cases} 1 & \text{if } f'(x_A) = (y_A, \tilde{y}, y_{AB}), \text{ where } (y_A, y_B, y_{AB}) \leftarrow f(x_A, x_B) \\ 0 & \text{otherwise} \end{cases}$$

Then define $\mathcal{L}^f_{\mathsf{prop}} = ((\mathcal{L}^f_{\mathsf{prop}})^A, (\mathcal{L}^f_{\mathsf{prop}})^B)$, where:

$$(\mathcal{L}^f_{\mathsf{prop}})^A = \{L^A_{f,f',\tilde{y}} \mid \mathsf{prop}(f') = \mathsf{prop}(f(0^n, \cdot)) \text{ and } \tilde{y} \in \{0,1\}^*\}$$
$$(\mathcal{L}^f_{\mathsf{prop}})^B = \{L^B_{f,f',\tilde{y}} \mid \mathsf{prop}(f') = \mathsf{prop}(f(\cdot, 0^n)) \text{ and } \tilde{y} \in \{0,1\}^*\}$$

Intuitively, the $\mathcal{L}^f_{\mathsf{prop}}$-leaked model allows the adversary to choose a circuit f' such that $\mathsf{prop}(f')$ has the "expected value" and learn whether f', when evaluated on the honest party's input, equals $f(x_A, x_B)$. In addition, for the output of f' that is not revealed to the honest party, the adversary can check that this output of f' is any fixed value of the adversary's choice.

Recall that f' may be a circuit with abort. In the case that f' aborts, these leakage functions will always return 0 (since the main equality condition will not hold).

Security. In Appendix A we prove the following:

Theorem 2. *The dual-execution protocol* $\mathsf{DualEx}^f[\mathcal{G}]$ *is secure in the* $\mathcal{L}^f_{\mathsf{prop}}$*-leaked model when* $\mathcal{G}$ *satisfies* prop-*enforcing, authenticity, and privacy/obliviousness.*

We point out that our simulator chooses f' before seeing the output of the functionality, and can choose only $\tilde{y}$ after seeing the output. Hence, one could slightly strengthen the definition of the $\mathcal{L}^f_{\mathsf{prop}}$-leaked model. For simplicity, we choose not to.

5   Achieving "Only Computation Leaks" with Dual Execution

Micali & Reyzin [16] proposed a model of leakage, one of whose axioms was that *"computation, and only computation, leaks information"* ("only computation leaks", or OCL, for short). One can think of decomposing a large computation into smaller "atomic" steps. Each step does not use all of the information in the system. The OCL axiom restricts us to leakage that is a function of the information used in a single atomic step. If two values are never used in the same step, then OCL leakage precludes (directly) leaking a *joint* function of those two values.

In a circuit model, the smallest "atomic" steps are gates. Hence, we consider leakage on the information available to a single gate (i.e., its two input wire values). By extension, it is natural to consider leaking on the information available to several gates, but only *separately* and not *jointly*. Only when two wires are inputs to a common gate can the leakage be a *joint* function of those two wires' values.

We formalize this kind of leakage as follows:

Gate-local queries. Let f be a circuit. We say that a leakage query $L(x_A, x_B)$ is *gate-local* if there exists a gate g in f such that $L(x_A, x_B)$ can be expressed as a function of the input wires of g in the computation of $f(x_A, x_B)$. We define $\mathcal{L}_{\text{ocl}}^f$ to be the set of conjunctions of gate-local queries; that is:

$$\mathcal{L}_{\text{ocl}}^f = \{L = L_1 \wedge \cdots \wedge L_k \mid \text{each of } L_1, \ldots, L_k \text{ is gate-local for } f\}$$

Circuit transformation. Let f be a circuit and define $\widehat{f}$ as follows: For each gate g in f, we add input bits $r_{g,A}$ for Alice and $r_{g,B}$ for Bob. We add output bits $s_{g,A}$ for Alice only and $s_{g,B}$ for Bob only. We then perform the following transformation for each gate:

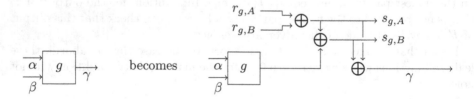

Intuitively, we additively secret share the output of g into shares $s_{g,A}$ and $s_{g,B}$, so that each party learns one share. Then the shares are re-assembled to again form the output of g that is used elsewhere in the circuit.

In Appendix B we prove the following:

Theorem 3. *Let* prop *be a property that includes the circuit topology and let* f, $\widehat{f}$ *be as above. We define Π to be a protocol for f in the $\widehat{f}$-hybrid, $\mathcal{L}_{\text{prop}}^{\widehat{f}}$-leaked model. In Π, parties simply send their inputs along with random values* *for* $\{r_{g,A}, r_{g,B}\}_g$ *to the ideal* $\widehat{f}$. *Then* Π *is a secure realization of* f *in the* $\mathcal{L}_{ocl}^f$-*leaked model.*

More concretely, if we would like to compute f restricting adversaries to $\mathcal{L}_{ocl}^f$ leakage, then we need to simply run the dual execution protocol on $\widehat{f}$ with a topology-enforcing garbling scheme.

Generalizations. One can also define "only computation leaks" at a higher level than individual gates. Indeed, this is more in line with most work on OCL, which does not consider circuit computations. Also, doing so leads to efficiency improvements in our protocols and transformations.

Consider partitioning a circuit into well-defined *components*. Then a *component-level* leakage query is one that can be written as a function of the inputs to a single component. Finally, let $\mathcal{L}$ denote the set of all conjunctions of component-level functions.

Then our results of the previous section can be easily applied to yield a protocol that is secure in the $\mathcal{L}$-leaked model.

We sketch the important differences:

- One would need a garbling scheme which enforces only the topology connecting of *components*, but need not enforce anything about the internals of each component. Our (gate-)topology-enforcing construction of Section 3.3 adds two hashes to each wire. To preserve topology of components, one need only add these hashes to the wires connecting different components. Concretely, this may result in a significantly smaller overhead than gate-topology-enforcement.
- Recall our transformation from a circuit f to a circuit $\widehat{f}$. It replaces each gate g with some gadget of 4 gates. However, the construction and proof go through verbatim with respect to *components*, if one interprets our diagram to let g be a larger component, each line to represent a *bundle* of wires, and the XOR gates to be string-XOR gates.

  Concretely, instead of adding 3 XOR gates for each gate, we add 3 XOR gates for each wire connecting different components. Similarly, the number of additional inputs/outputs is related to the number of "component-connecting wires", not the total size of the circuit.

6   Reducing the Probability of Leakage in Dual Execution

In Section 2.2 we gave a high level overview of our ϵ-CovIDA protocol. Here, we describe the protocol in detail, and evaluate its asymptotic and concrete efficiency. We start with a brief review of the two sub-protocols we use, i.e., committing-OTs and two-stage PSI. (For completeness, we provide a formal definition of CovIDA security in Appendix C.)

6.1 Committing-OTs

Oblivious Transfer (OT) protocol implements securely the following functionality; A sender inputs two tuples $[K_1^0, K_2^0, \ldots, K_s^0], [K_1^1, K_2^1, \ldots, K_s^1]$ and a receiver inputs a bit b. Then, the receiver learns the tuple $[K_1^b, K_2^b, \ldots, K_s^b]$.

A stronger variant of OT, called committing-OT, is one in which the sender is also *committed* to his inputs, meaning, the sender cannot claim in retrospect that his inputs were different than the ones he entered to the OT in the beginning. In other words, if the sender is asked to show what was his inputs $[K_1^0, K_2^0, \ldots, K_s^0], [K_1^1, K_2^1, \ldots, K_s^1]$, he cannot answer with different inputs without being caught.

For simplicity, from now on we abstract out the details of the committing-OT and just work with the following notation: We denote by $\mathrm{COT}_1(b)$ the message sent by the receiver to the sender (where b is the receiver's input bit) and similarly use $\mathrm{COT}_2([K_1^0, K_2^0, \ldots, K_s^0], [K_1^1, K_2^1, \ldots, K_s^1], \mathrm{COT}_1(b))$ to denote the message sent by the sender to the receiver. When we say that the sender *decommits* his input, we refer to the operation in which he reveals $[K_1^0, K_2^0, \ldots, K_s^0], [K_1^1, K_2^1, \ldots, K_s^1]$ and proves that these are the correct inputs he had used in the protocol.

Committing-OTs can be realized in several ways, and even be efficiently extended for specific implementations (see [18]). Throughout this work we will assume that the cost of committing-OT is $\mathcal{O}(s)$ exponentiations. (The exact constant can be computed as done in [14], but we prefer stating our efficiency claims for general committing-OT constructions.)

6.2 Two-Stage Private Set Intersection

In standard two-party private set intersection (PSI), player P_i holds his input set S_i and the goal is for one or both parties to learn the intersection $S_1 \cap S_2$.

The two-stage variant of PSI, which we denote by the functionality $\mathcal{F}_{2PSI}$, is split the protocol to two stages in order to emulate a commitment on the inputs before revealing the result (we will use this property in our constructions). I.e., in the first stage, players submit their input sets and learn nothing, and in the second stage, one of the parties ask for the output and obtains the result. The reason we formalize the functionality with only one party receiving the output is that all of the realizations we are aware of are this form. In our constructions we need both parties to learn the intersection, thus we address this issue directly as part of our 2PC constructions. We define the functionality $\mathcal{F}_{2PSI}$ in Figure 6.2.

Realizing two-stage PSI. There are several two-round, fully-simulatable PSI with security against malicious adversaries in the literature (e.g. see [11,6], both in the random oracle model). These protocols do not automatically realize $\mathcal{F}_{2PSI}$ but can be modified at little cost to do so. In particular, two properties that are shared by these constructions are that (1) only one party learns the intersection (denoted by P_1), and (2) P_1 does not learn any information before receiving the second (and last) message of the protocol from P_2.

First Stage
    Inputs: P_1 inputs S_1 and P_2 inputs S_2 (both of size l).
    Outputs: Both players receive Inputs Accepted.
Second Stage
    Inputs: P_2 inputs Reveal.
    Outputs: P_1 obtains $(S_1 \cap S_2)$.

Fig. 2. $\mathcal{F}_{2PSI}$

Given a fully-simulatable PSI that has these two properties, all we need in order to realize the $\mathcal{F}_{2PSI}$ is to execute the protocol until the step in which P_2 is supposed to send his last message. Instead of sending the last message, we modify the protocol so that P_2 only sends a commitment on that message. The commitment in use should be equivocal and extractable. (Such a commitment can be constructed in the random oracle model, using $H(m, r)$, or in the standard model, e.g., [13].) This completes the first stage. When P_2 wants to reveal the intersection (i.e. the second stage), he decommits his last message and P_1 completes the protocol. The intuition is that the simulation for the two-stage PSI is done by calling the original PSI simulators and replacing the last message with a commitment to it. Note that after the commitment to the last message is sent, both players cannot change their inputs (obviously, P_2 cannot decommit to a different message, and P_1 cannot change his inputs since otherwise the original PSI would be insecure). We defer a more formal treatment to the full version of the paper.

Using the above transformation on the protocol of [6], for instance, yields an efficient two-stage PSI that requires only a linear number (in the number of sets) of public key operations by each party.

6.3 The Protocol

A detailed description of the protocol is in Figures 3 and 5. In Appendix D we prove the following theorem:

Theorem 4. *Assume that the committing-OT, the PSI protocol, the commitment, and the garbling scheme are secure. Then, the protocol from Figures 3 and 5 is a 2^{-s+1}-CovIDA secure realization of f.*

Input-consistency check. The consistency of the players' inputs is handled using the technique of [22], where a universal hash function (UH) is evaluated on the input of each player, and the players verify that the outputs of this function are the same in all circuits. A player's input is padded with a short random string r in order to increase its entropy and by that, reduce the amount of information that can be learnt about the input from the output of the UH. Let l be the input length and s be a security parameter. [22] shows a matrix of dimensions $s \times (l + 2s + \log s)$ that can be used as a UH, where the evaluation consists of multiplying this matrix with the input vector (and getting a vector of length s).

In our protocol we require an additional property from this matrix: Given a matrix M and an output vector v, we require that for any input vector v_i, it is easy to find a vector v_r such that $M \times (v_i \| v_r)^T = v^T$. Interestingly, we propose a solution meeting this property that is simpler and more efficient than the solution of [22]. In particular, we generate a random matrix in $\{0,1\}^{s \times l}$ and concatenate it with the identity matrix of size s, resulting in matrix of dimensions $s \times (l + s)$. Evaluation consists of multiplying this matrix with the input vector which is the l bits of real input and s random bits. The construction is a UH for reasons identical to the construction of [22], and knowing the output of UH reveals nothing about the real input since the output vector will be uniformly random (give the s random bits).

Efficiency comparison. We briefly discuss efficiency of our protocol compared to the best alternative using existing techniques. As discussed earlier, the best alternative (referred to as Best Previous in the Figures), is to use the construction of [18] augmented with the technique of [14] in order to reduce the number of garbled circuits. In particular, in this potential solution (not published elsewhere) we run the 2PC once in each direction (with careful incorporation of the input consistency checks), run the cheating-recovery computation at the end of each, and perform a maliciously secure equality-check to compare the two outputs.

We initially focus on the overhead in computation and communication beyond what is required to garble s circuits by each party and the associated input-consistency checks, since those are part of any know solution for ϵ-CovIDA secure 2PC. Ignoring the cost of the equality-check, the overhead here consists of the two cheating-recovery executions. This is included in Figure 6 based on the numbers given in [14] (see Section 3.1).

The overhead of our protocol is simply to run a two-stage PSI with malicious security where both parties learn the output, and where each set is of expected size $s/2$. This requires s commitments and a standard maliciously secure PSI for which we use the concrete numbers given in [6]. As can be seen in the table, the overhead in our construction is significantly smaller, i.e. a factor of 10 or more in communication, and a factor of 200 or more in computation even for input size of 1. This improvement further increases as the input size grows since the overhead in our construction is independent of the input while the cost of cheating recovery linearly grows with it.

While this improvement is only in the "overhead" cost, we stress that the overhead can be a significant portion of the overall cost in small circuits. In Figure 7 we compare the overall costs for the two approaches where to estimate the cost of garbling and the input-consistency checks needed for dual-execution, we use the numbers given in [14] (multiplied by two) both for our solution and the "Best Previous" (with the exception that we assume the use of 2-row reduction techniques when measuring communication for both, but this only effects the bandwidth column). We note that this comparison is on the conservative side, and should be seen as the minimum improvement since more optimized options are available in the RO model (specially for input-consistency checks), and those would highlight our improvements in the overhead even further.

Alice's input: $x_A \in \{0,1\}^\ell$. **Bob's input:** $x_B \in \{0,1\}^\ell$.

Common input: Alice and Bob agree on the description of a circuit C, where $C(x_A, x_B) = f(x_A, x_B)$, and a collision resistance hash function $H : \{0,1\}^* \to \{0,1\}^\ell$. Let $\mathsf{Commit}(\cdot)$ be an extractable and equivocal commitment.

s is a statistical security parameter that represents the bound on the cheating probability. L is a computational security parameter, so, for example, each key label is L-bits long. s' is a statistical security parameter associated with the input-consistency matrix.. Let $\ell' = \ell + m$, and $m' = m + 2s'$.

Output: Both players learn an m-bit string $f(x_A, x_B)$.

Below, we describe the protocol for the case where Alice is the garbler and Bob is the evaluator. But the protocol is symmetric and each step is performed simultaneously in the other direction as well (where Bob is the garbler and Alice is the evaluator) before moving to the next step. In what follows, we slightly abuse the notations we introduced for a garbling scheme. In particular, we feed input/output labels as inputs to the Garble algorithm while in previous section, they were the output of Garble. This is compatible with all existing instantiations.

Garbler's input preparation.

1. Alice chooses s PRF seeds $sd_1^A, \ldots, sd_s^A$, and commits on them using $\mathsf{Commit}(sd_1^A), \ldots, \mathsf{Commit}(sd_s^A)$. All the randomness Alice will use for generating the ith garbled circuit and its input labels will be derived from sd_i^A.

2. Alice chooses $r_A \in_R \{0,1\}^t, r_A' \in_R \{0,1\}^m$ and sets $x_A' = x_A \| r_A \| r_A'$. She will be using x_A' as her input to the circuits instead of x_A. We denote the jth bit of x_A' by $x_{A,j}'$.

3. Alice chooses $\mathsf{in}_b^{A,i,j}, \mathsf{in}_b^{B,i,j} \in_R \{0,1\}^L$ for $b \in \{0,1\}$, $1 \leq i \leq s$ and $1 \leq j \leq \ell'$. $\mathsf{in}_b^{A,i,j}$ would be the b-key for Alice's jth input wire in the ith garbled circuit. ($\mathsf{in}_b^{B,i,j}$ is defined similarly with respect to Bob.) Using the garbling schemes notation we have $e_A^i = (\mathsf{in}_0^{A,i,1}, \mathsf{in}_1^{A,i,1}, \ldots, \mathsf{in}_0^{A,i,\ell'}, \mathsf{in}_1^{A,i,\ell'}, \mathsf{in}_0^{B,i,1}, \mathsf{in}_1^{B,i,1}, \ldots, \mathsf{in}_0^{B,i,\ell'}, \mathsf{in}_1^{B,i,\ell'})$.

4. Alice sends $\mathsf{Commit}(H(\mathsf{in}_{x_{A,1}'}^{A,i,1} \| \cdots \| \mathsf{in}_{x_{A,\ell'}'}^{A,i,l'}))$ for $1 \leq i \leq s$, i.e. commitments to encoding of her inputs.

Oblivious transfer for evaluator's input.

1. Alice and Bob engage in ℓ' OTs, where in the jth OT, Bob sends $q_j = \mathrm{COT}_1(x_{B,j}')$ and Alice answers with $\mathrm{COT}_2([\mathsf{in}_0^{B,1,j}, \ldots, \mathsf{in}_0^{B,s,j}], [\mathsf{in}_1^{B,1,j}, \ldots, \mathsf{in}_1^{B,s,j}], q_j)$

Fig. 3. 2^{-s}-CovIDA 2PC via PSI

Continued

Circuit Preparation.

1. Alice and Bob jointly choose matrices $M_A, M_B \in_R \{0,1\}^{s' \times \ell'}$, and concatenate each with a $s' \times s'$ identity matrix to obtain M'_A, M'_B respectively. Let $C'(x'_A, x'_B) = (C(x_A, x_B) \oplus r'_A \oplus r'_B, M'_A \cdot x'_A, M'_B \cdot x'_B)$.

2. Alice chooses m' random label pairs and sets $d_A = (\mathsf{out}_0^{A,1}, \mathsf{out}_1^{A,1}) \ldots (\mathsf{out}_0^{A,m'}, \mathsf{out}_1^{A,m'})$.

3. For $1 \le i \le s$, Alice computes $\mathsf{GC}_i^A \leftarrow \mathsf{Garble}(C', e_A^i, d_A)$. Note that unlike standard garbling, here the input and output labels are fixed and fed as input to the garbling algorithm, and note that the same output label is used for all s circuits.

4. Alice computes the output decoding table $\mathsf{GDec}^A = \left\{ [i, H(\mathsf{out}_0^{A,i}), H(\mathsf{out}_1^{A,i})] \right\}_{i=1}^{m'}$. (Note that this is different from d_A and the same table is used for all garbled circuits.)

5. Alice sends garbled circuits $\mathsf{GC}_1^A, \ldots, \mathsf{GC}_s^A$ and the output decoding table GDec^A.

Challenge Generation.

1. Alice chooses $\alpha_1^{(A)}, \alpha_2^{(A)} \in_R \{0,1\}^s$. Similarly, Bob chooses $\alpha_1^{(B)}, \alpha_2^{(B)} \in_R \{0,1\}^s$

2. Alice sends $\mathsf{Commit}(\alpha_1^{(A)})$ and Bob sends $\mathsf{Commit}(\alpha_2^{(B)})$.

3. Alice sends $\alpha_2^{(A)}$ and Bob sends $\alpha_1^{(B)}$.

4. Both players decommit their commitments and set $\alpha_1 = \alpha_1^{(A)} \oplus \alpha_1^{(B)}$ and $\alpha_2 = \alpha_2^{(A)} \oplus \alpha_2^{(B)}$. If one of those values is all zeros, or all one, they go back to step 1.

5. We define the evaluation set $\mathbb{E}_A$ such that $i \in \mathbb{E}_A$ if and only if ith bit of α_1 is one (Similarly, $\mathbb{E}_B$ would be generated using α_2).

Garbled Circuit Evaluation.

1. Alice sends $\mathsf{in}_{x'_{A,j}}^{A,i,j}$ for $i \in \mathbb{E}_A$ and $1 \le j \le \ell'$, and decommits $\mathsf{Commit}(H(\mathsf{in}_{x'_{A,1}}^{A,i,1} \| \cdots \| \mathsf{in}_{x'_{A,l'}}^{A,i,l'}))$.

2. For $i \in \mathbb{E}_A$, Bob evaluates GC_i^A and gets the garbled output Z_i^A. Let z_i^A be the actual output resulted from decoding Z_i^A using GDec^A. If any of the decoded bits is $\bot$, Bob sets z_i^A to $\bot^{m'}$.

Committing to PSI input sets.

1. For all $i \in \mathbb{E}_A$, if $z_i^A \ne \bot^{m'}$, Bob parses $z_i^A = z_{i,1}^A \cdots z_{i,m'}^A$. He then computes $q_i = ((\mathsf{out}_{z_{i,1}^A}^{A,1} \oplus \mathsf{out}_{z_{i,1}^A}^{B,1}) \oplus \cdots \oplus (\mathsf{out}_{z_{i,m'}^A}^{A,m'} \oplus \mathsf{out}_{z_{i,m'}^A}^{B,m'}))$. If $z_i^A = \bot^{m'}$, on the other hand, Bob sets q_i to be a random $(L + m')$-bit value. If $q_i \in T_B$, he modifies q_i to be a random $(L + m')$-bit value. He adds q_i to T_B.

2. Alice and Bob call the first stage of $\mathcal{F}_{2PSI}$ with their inputs T_A and T_B.

3. For all $i \in \mathbb{E}_A$, Alice commits to q_i using $\mathsf{Commit}(\cdot)$ and sends these commitments in a random order. (Bob does not need to follow this step.)

Fig. 4. 2^{-s}-CovIDA 2PC via PSI

Continued

Opening.

1. Alice decommits sd_i^A for all $i \notin \mathbb{E}_A$. She also reveals her OT inputs corresponding to the opened circuits (the OTs for Bob to learn his input labels).
2. Bob aborts if any of the following occurs:
   - $\exists i \in \mathbb{E}_A$ such that Alice's garbled inputs are invalid.
   - $\exists i \notin \mathbb{E}_A$ in which $\mathsf{GC}_i^A \neq \mathsf{Garble}(C_A, sd_i^A, d_A)$, or, the OTs are not consistent with GC_i^A. (Note that once some sd_i^A is revealed, Bob can compute d_A by himself.)
   - Some of the output labels $(\mathsf{out}_{A,0}^1, \mathsf{out}_{A,1}^1) \ldots (\mathsf{out}_{A,0}^{m'}, \mathsf{out}_{A,1}^{m'})$ are not properly constructed or not consistent with GDec^A.

Output generation.

1. Alice and Bob perform the second stage of $\mathcal{F}_{2PSI}$ in order for Alice to learn $I = T_A \cap T_B$.
2. Alice aborts if $I = \emptyset$. Else, she decommits $\mathsf{Commit}(q_i)$ corresponding to the single element in I (note that the intersection is guaranteed to be of size at most one). Bob aborts if the decommitment is invalid or if the decommitted value is not in set T_B.
3. Recall that the computation output is masked by r_A' and r_B'. Alice sends r_A' and the labels that correspond to r_A' in GC_i^B for $i = \min(\mathbb{E}_A)$. (These labels are used for authenticating r_A'.) If the labels are invalid, Bob aborts.
4. Both players unmask the output from q_i and output the result.

Fig. 5. 2^{-s}-CovIDA 2PC via PSI

Finally, for concreteness, in Figure 8 we look at the overall cost of the two protocols for the AES circuit with 6800 non-XOR gates, input size $\ell = 128$, the symmetric ciphertext size $n = 128$ and group element size 220 bits. We have combined the communication cost into one column named bandwidth which is in bits. We express all costs in terms of parameter s. For instance, our new protocol reduces overall bandwidth by 35% and the number of exponentiations by more than 50%.

Different deterrence value for each player. Besides its better efficiency, an additional advantage of our new protocol is that each party has a separate challenge generation and uses a different challenge set $\mathbb{E}_i$, to determine which circuits to check and which ones to evaluate. This allows us to use a different number of garbled circuits for each party and as a result achieve different deterrence factors for each. This variant can be quite useful in practice where different participants in a protocol may have different reputations (to protect) or different levels of tolerance for risk. One can take these real-world factors into account when adjusting the deterrence factor for each party.

Construction	Fixed-based Exponent.	Regular Exponent.	Symmetric Encryption	Group elements sent	Symmetric communication
Best Previous	$18s\ell + 1080s$	$960s$	$78s\ell$	$42s\ell$	$24sn\ell$
Ours	$4.5s$	s	$2s$	$4s$	sn

Fig. 6. Comparison of overhead cost of our CovIDA 2PC protocol with the best alternative using state of the art techniques. s is the number of circuits, n is length of symmetric-key ciphertext, and ℓ is the input sizes.

Construction	Fixed-based Exponent.	Regular Exponent.	Symmetric Encryption	Group elements sent	Symmetric communication
Best Previous	$42s\ell + 1080s$	$7s\ell + 36\ell + 960s$	$26s|C| + 78s\ell$	$52s\ell$	$4ns|C| + 28sn\ell$
Ours	$24s\ell + 4.5s$	$7s\ell + 36\ell + s$	$26s|C| + 2s$	$10s\ell + 4s$	$4ns|C| + sn$

Fig. 7. Comparison of overall cost of our CovIDA 2PC protocol with the best alternative using state of the art techniques. s is the number of circuits, n is length of symmetric-key ciphertext, ℓ is the input sizes, and $|C|$ is the circuit size.

Construction	Fixed-based	Regular	Symmetric	Bandwidth (bits)
Best Previous	$6456s$	$1856s + 4608$	$186784s$	$5404672s$
Ours	$3076s$	$897 s + 4608$	$176802 s$	$3483012s$

Fig. 8. Concrete comparison of overall costs for the AES. s is the number of circuits, $n = 128$ is length of symmetric-key ciphertext, $\ell = 128$ is the input sizes, and $|C| = 6800$ is the number of non-XOR gates in the AES circuit.

References

1. Aiello, W., Ishai, Y., Reingold, O.: Priced oblivious transfer: How to sell digital goods. In: Pfitzmann, B. (ed.) EUROCRYPT 2001. LNCS, vol. 2045, pp. 119–135. Springer, Heidelberg (2001)
2. Aumann, Y., Lindell, Y.: Security against covert adversaries: Efficient protocols for realistic adversaries. In: Vadhan, S.P. (ed.) TCC 2007. LNCS, vol. 4392, pp. 137–156. Springer, Heidelberg (2007)
3. Bellare, M., Hoang, V.T., Rogaway, P.: Foundations of garbled circuits. In: Yu, T., Danezis, G., Gligor, V.D. (eds.) ACM CCS 2012, pp. 784–796. ACM Press (October 2012)
4. Canetti, R., Lin, H., Pass, R.: Adaptive hardness and composable security in the plain model from standard assumptions. In: 51st FOCS, pp. 541–550. IEEE Computer Society Press (October 2010)
5. Curtmola, R., Garay, J.A., Kamara, S., Ostrovsky, R.: Searchable symmetric encryption: improved definitions and efficient constructions. In: Juels, A., Wright, R.N., Vimercati, S. (eds.) ACM CCS 2006, pp. 79–88. ACM Press (October/November 2006)
6. De Cristofaro, E., Kim, J., Tsudik, G.: Linear-complexity private set intersection protocols secure in malicious model. In: Abe, M. (ed.) ASIACRYPT 2010. LNCS, vol. 6477, pp. 213–231. Springer, Heidelberg (2010)
7. Huang, Y., Katz, J., Evans, D.: Quid-Pro-Quo-tocols: Strengthening semi-honest protocols with dual execution. In: 2012 IEEE Symposium on Security and Privacy, pp. 272–284. IEEE Computer Society Press (May 2012)

8. Huang, Y., Katz, J., Evans, D.: Efficient secure two-party computation using symmetric cut-and-choose. In: Canetti, R., Garay, J.A. (eds.) CRYPTO 2013, Part II. LNCS, vol. 8043, pp. 18–35. Springer, Heidelberg (2013)

9. Ishai, Y., Kushilevitz, E., Ostrovsky, R., Prabhakaran, M., Sahai, A.: Efficient non-interactive secure computation. In: Paterson, K.G. (ed.) EUROCRYPT 2011. LNCS, vol. 6632, pp. 406–425. Springer, Heidelberg (2011)

10. Jarecki, S., Jutla, C.S., Krawczyk, H., Rosu, M.-C., Steiner, M.: Outsourced symmetric private information retrieval. In: Sadeghi, A.-R., Gligor, V.D., Yung, M. (eds.) ACM CCS 2013, pp. 875–888. ACM Press (November 2013)

11. Jarecki, S., Liu, X.: Fast secure computation of set intersection. In: Garay, J.A., De Prisco, R. (eds.) SCN 2010. LNCS, vol. 6280, pp. 418–435. Springer, Heidelberg (2010)

12. Lin, H., Pass, R., Venkitasubramaniam, M.: A unified framework for concurrent security: Universal composability from stand-alone non-malleability. In: Mitzenmacher, M. (ed.) 41st ACM STOC 2009, pp. 179–188. ACM Press (May/June 2009)

13. Lindell, Y.: Highly-efficient universally-composable commitments based on the DDH assumption. In: Paterson, K.G. (ed.) EUROCRYPT 2011. LNCS, vol. 6632, pp. 446–466. Springer, Heidelberg (2011)

14. Lindell, Y.: Fast cut-and-choose based protocols for malicious and covert adversaries. In: Canetti, R., Garay, J.A. (eds.) CRYPTO 2013, Part II. LNCS, vol. 8043, pp. 1–17. Springer, Heidelberg (2013)

15. Micali, S., Pass, R., Rosen, A.: Input-indistinguishable computation. In: 47th FOCS, pp. 367–378. IEEE Computer Society Press (October 2006)

16. Micali, S., Reyzin, L.: Physically observable cryptography (extended abstract). In: Naor, M. (ed.) TCC 2004. LNCS, vol. 2951, pp. 278–296. Springer, Heidelberg (2004)

17. Mohassel, P., Franklin, M.K.: Efficiency tradeoffs for malicious two-party computation. In: Yung, M., Dodis, Y., Kiayias, A., Malkin, T. (eds.) PKC 2006. LNCS, vol. 3958, pp. 458–473. Springer, Heidelberg (2006)

18. Mohassel, P., Riva, B.: Garbled circuits checking garbled circuits: More efficient and secure two-party computation. In: Canetti, R., Garay, J.A. (eds.) CRYPTO 2013, Part II. LNCS, vol. 8043, pp. 36–53. Springer, Heidelberg (2013)

19. Pappas, V., Krell, F., Vo, B., Kolesnikov, V., Malkin, T., Choi, S.G., George, W., Keromytis, A., Bellovin, S.: Blind seer: A scalable private DBMS. In: Security and Privacy, Oakland (2014)

20. Pass, R.: Simulation in quasi-polynomial time, and its application to protocol composition. In: Biham, E. (ed.) EUROCRYPT 2003. LNCS, vol. 2656, pp. 160–176. Springer, Heidelberg (2003)

21. Prabhakaran, M., Sahai, A.: New notions of security: Achieving universal composability without trusted setup. In: Babai, L. (ed.) 36th ACM STOC, pp. 242–251. ACM Press (June 2004)

22. Shelat, A., Shen, C.-H.: Fast two-party secure computation with minimal assumptions. In: Sadeghi, A.-R., Gligor, V.D., Yung, M. (eds.) ACM CCS 2013, pp. 523–534. ACM Press (November 2013)

23. Yao, A.C.-C.: How to generate and exchange secrets (extended abstract). In: 27th FOCS, pp. 162–167. IEEE Computer Society Press (October 1986)

A   Proof of Theorem 2

Proof. We sketch only the proof for corrupt Alice: the other case is symmetric.

In the real execution, Alice provides F_A as input to the commitment scheme, e_A as input to the OTs (as sender), and x_A as input to the OTs (as receiver). Finally, Alice sends a string to the equality test functionality. An honest Bob provides honest F_B and causes Alice to receive honest garbled input X_B. An honest Bob further calculates his input to the equality test as a direct result of the garbled output $\mathsf{Eval}(F_B, \mathsf{Encode}(e_A, x_B))$.

We consider a sequence of hybrid interactions, taking hybrid H0 to be the real execution of the protocol:

H1: The simulator extracts $(f', d') \leftarrow \mathsf{Extract}(F_A, e_A)$, and Bob's effective garbled output is instead computed as $\mathsf{Encode}(d', f'(x_B))$.[4] This hybrid is indistinguishable by the prop-enforcing guarantee of $\mathcal{G}$. Provided that Bob does not abort in step 4, we also have $\mathsf{prop}(f') = \mathsf{prop}(f(0^n, \cdot))$.

H2: We focus on the values Y_B and Y_{AB} that Alice provides to the equality test. These values are compared to values that Bob computes by encoding $\tilde{y}_B$ and $\tilde{y}_{AB}$ under encoding d_B, where $(\tilde{y}_A, \tilde{y}_B, \tilde{y}_{AB}) = f'(x_B)$. Importantly, Bob provides *valid* garbled outputs (under encoding d_B). By the authenticity guarantee of $\mathcal{G}$, Alice cannot guess any valid garbled outputs besides the ones she is prescribed via $\mathsf{Eval}(F_B, X_A)$.[5] By the correctness of the garbling scheme, these prescribed garbled outputs encode the "correct" values y_B and y_{AB} computed from $f(x_A, x_B)$. Thus the simulator in this hybrid returns false for the equality test if $f'(x_B)$ and $f(x_A, x_B)$ disagree in their y_B or y_{AB} components, or if Alice provides Y_B or Y_{AB} different from those prescribed via $\mathsf{Eval}(F_B, X_A)$. This change is indistinguishable by the authenticity property of $\mathcal{G}$.

H3: The simulator uses Alice's prescribed output (y_A, y_{AB}) to generate a simulated garbled circuit F_A and garbled input X_A. This hybrid is indistinguishable by the privacy/obliviousness guarantee of $\mathcal{G}$.[6]

H4: We focus on the other values $\tilde{Y}_A$ and $\tilde{Y}_{AB}$ that Alice provides to the equality test. These are compared to Bob's value that is determined from $\mathsf{Encode}(d', f'(x_B))$. The simulator can easily check whether $\tilde{Y}_A, \tilde{Y}_{AB}$ are valid encodings under d' (i.e., possible outputs of $\mathsf{Encode}(d', \cdot)$). If not, then the equality test will always return false and the simulator can also do so. Otherwise, the simulator can easily determine $\tilde{y}_A$ and $\tilde{y}_{AB}$ such that

---

[4] For simplicity, we are glossing over the case where f' outputs $\bot$ (corresponding to the event that Bob's execution of Eval results in $\bot$). In this event, Bob will run the equality test on random inputs, and the equality test will result in false. Looking ahead, this matches the semantics of $L_{f,f',\tilde{y}_A}$ that will be chosen as the leakage function in the ideal world. For the rest of the proof we can therefore condition on f' not outputting $\bot$.

[5] Since F_B and X_A are designated by Bob, this execution of Eval will not abort.

[6] We use obliviousness since the simulated garbled circuit contains no information about y_B.

$\tilde{Y}_A\|\cdots\|\tilde{Y}_{AB} = \mathsf{Encode}(d', \tilde{y}_A\|\cdot\|\tilde{y}_{AB})$. Then the equality check involving these $\tilde{Y}_A, \tilde{Y}_{AB}$ values is logically equivalent to $(\tilde{y}_A, \cdot, \tilde{y}_{AB}) \overset{?}{=} f'(x_B)$.

We see that in the hybrid labeled H4, the equality test outcome is determined by the following logic:

$$f'(x_B) = (\tilde{y}_A, \tilde{y}_B, \tilde{y}_{AB}) \text{ and } f(x_A, x_B) = (y_A, y_B, y_{AB}) \text{ and } \tilde{y}_B = y_B \text{ and } \tilde{y}_{AB} = y_{AB}$$

Indeed, the outcome of the equality test is precisely $L_{f,f',\tilde{y}_A}(x_A, x_B)$ where $\tilde{y}_A$ is the value that the simulator extracts as described above. Overall, we have described a simulator that is indistinguishable from the real execution; it needs to know only Alice's prescribed output (y_A, y_{AB}), and the answer to a $\mathcal{L}_{\mathsf{prop}}^f$-leakage query described above.

B   Proof of Theorem 3

Proof. For simplicity, suppose f gives all of its output to both parties (there is no output given to just one of the parties). Then $\hat{f}$ has syntax $\hat{f}((x_a, \{r_{g,A}\}_g), (x_B, \{r_{g,B}\}_g)) = (\{s_{g,A}\}_g, \{s_{g,B}\}_g, y = f(x_A, x_B))$.

First, consider the case where a corrupt Alice attacks the Π protocol. She provides input $(x_A, \{r_{g,A}\}_g)$ to $\hat{f}$, then receives output $(\{s_{g,A}\}_g, y)$ from $\hat{f}$. She then chooses a legal leakage function $L_{\hat{f},h,\tilde{y}}^A \in \mathcal{L}_{\mathsf{prop}}^{\hat{f}}$ and learns the result. Bob aborts if the leakage function evaluates to zero.

In the simulation, the simulator picks outputs $\{s_{g,A}\}_g$ uniformly at random. It remains to show how the simulator simulates the outcome of the leakage function given only $\mathcal{L}_{\mathsf{ocl}}^f$ leakage to an ideal f.

Alice chooses a leakage function $L_{\hat{f},h,\tilde{y}}^A$ where $\mathsf{prop}(h) = \mathsf{prop}(\hat{f}(0^n, \cdot))$. Since prop includes the circuit topology, we can naturally talk about a correspondence between the topology of h and $\hat{f}$.

The analysis proceeds one gate a time, in a topological order. Suppose we are considering some gate g in f, which corresponds to the larger gadget in $\hat{f}$, described above. Then h has a similar gadget, in which some of the gate logic may be changed:

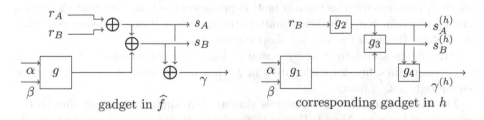

gadget in $\hat{f}$ corresponding gadget in h

By our inductive hypothesis, we'll assume that the values of α and β agree with the corresponding values in $\hat{f}$ (and hence in f). By construction, the same r_B value is used as input for both circuits. Hence we use the same variable names for these values in the above diagram.

The leakage function $L^A_{\widehat{f},h,\tilde{y}}$ simply performs a string equality related to the outputs of $\widehat{f}$ and h. It is helpful to think of this string comparison as a conjunction of single-bit comparisons (taken in the same order as our gate-by-gate analysis), which will "short circuit" to return 0 as soon as a mismatch is encountered.

The current gadget in $\widehat{f}$ and h includes outputs which the leakage function checks in the following way. The string $\tilde{y}$ (a parameter of the leakage function, chosen by Alice) includes a single position whose value we call $\tilde{s}$. The leakage function checks the two bit-comparisons $s_A^{(h)} = \tilde{s}$ and $s_B^{(h)} = s_B$.

We rewrite these two conditions as follows:

$$(s_A^{(h)} = \tilde{s}) \wedge (s_B^{(h)} = s_B)$$
$$\Longleftrightarrow (s_A^{(h)} = \tilde{s}) \wedge (g_3(s_A^{(h)}, g_1(\alpha,\beta)) = s_B)$$
$$\Longleftrightarrow (s_A^{(h)} = \tilde{s}) \wedge (g_3(\tilde{s}, g_1(\alpha,\beta)) = s_B)$$
$$\Longleftrightarrow (s_A^{(h)} = \tilde{s}) \wedge (g_3(\tilde{s}, g_1(\alpha,\beta)) = g(\alpha,\beta) \oplus s_A)$$
$$\Longleftrightarrow (g_2(r_B) = \tilde{s}) \wedge (g_3(\tilde{s}, g_1(\alpha,\beta)) = g(\alpha,\beta) \oplus s_A)$$
$$\Longleftrightarrow (g_2(r_A \oplus s_A) = \tilde{s}) \wedge (g_3(\tilde{s}, g_1(\alpha,\beta)) = g(\alpha,\beta) \oplus s_A)$$

Observe that the gates $g, g_1, \ldots, g_3$ and values $s_A, r_A, \tilde{s}$ are known to the simulator. Hence, this condition is a function of α, β alone — it is a *gate-local* constraint in f! A simulator only needs to know the result of this gate-local function to simulate the corresponding bit-comparisons in the leakage function $L^A_{\widehat{f},h,\tilde{y}}$.

Conditioned on the constraint being true, we examine the output $\gamma^{(h)}$ of the gadget in h. We have:

$$\gamma^{(h)} = g_4(s_A^{(h)}, s_B^{(h)}) = g_4(\tilde{s}, s_A \oplus g(\alpha,\beta)) = g_4(\tilde{s}, s_A \oplus \gamma)$$

Let $\pi(\cdot) = g_4(\tilde{s}, s_A \oplus \cdot)$. Clearly the simulator can extract the unary function π. All gates downstream of $\gamma^{(h)}$ will receive the value of $\pi(\gamma)$, where γ is the "correct" value that leaves this gadget in $\widehat{f}$. But this is equivalent to sending γ along this wire and modifying a downstream gate g' to have logic $g'(\pi(\cdot), \cdot)$ instead. This modified circuit has the same topology as h, so our analysis is not affected. Furthemore, this transformation preserves the invariant that the inputs to all gadgets in h match their counterparts in $\widehat{f}$ (conditioned on the event that the leakage function has not yet short-circuited).

Overall, the simulator will only need to know the conjunction of many gate-local constraints, one for each gate g in f. Hence, the simulator can succeed by asking only a $\mathcal{L}^f_{\text{ocl}}$ query.

The case where Bob is corrupt is similar, but slightly different due to the asymmetry between Alice & Bob in the gadgets. In this case, the gadget is only slightly different (now it is Alice's input r_A which goes to gate g_2):

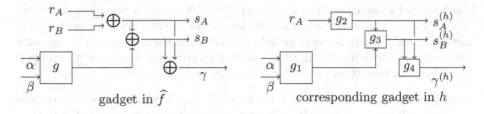

gadget in $\widehat{f}$ corresponding gadget in h

Also, the leakage function now checks the complementary condition $(s_A^{(h)} = s_A) \wedge (s_B^{(h)} = \tilde{s})$. The key to manipulating this expression is the observation that $s_A = g(\alpha, \beta) \oplus s_B$.

We obtain:

$$(s_A^{(h)} = s_A) \wedge (s_B^{(h)} = \tilde{s})$$
$$\iff (s_A^{(h)} = s_A) \wedge (g_3(s_A^{(h)}, g_1(\alpha, \beta)) = \tilde{s})$$
$$\iff (s_A^{(h)} = s_A) \wedge (g_3(s_A, g_1(\alpha, \beta)) = \tilde{s})$$
$$\iff (s_A^{(h)} = s_A) \wedge (g_3(g(\alpha, \beta) \oplus s_B, g_1(\alpha, \beta)) = \tilde{s})$$
$$\iff (s_A^{(h)} = g(\alpha, \beta) \oplus s_B) \wedge (g_3(g(\alpha, \beta) \oplus s_B, g_1(\alpha, \beta)) = \tilde{s})$$
$$\iff (g_2(r_A) = g(\alpha, \beta) \oplus s_B) \wedge (g_3(g(\alpha, \beta) \oplus s_B, g_1(\alpha, \beta)) = \tilde{s})$$
$$\iff (g_2(g(\alpha, \beta) \oplus s_B \oplus r_B) = g(\alpha, \beta) \oplus s_B) \wedge (g_3(g(\alpha, \beta) \oplus s_B, g_1(\alpha, \beta)) = \tilde{s})$$

As before, all gates $g, g_1, \ldots, g_3$ and values $s_B, r_B, \tilde{s}$ are known to the simulator, making this expression a gate-local function of α, β alone.

Conditioned on the above expression being true, we also have:

$$\gamma^{(h)} = g_4(s_A^{(h)}, s_B^{(h)}) = g_4(s_A, \tilde{s}) = g_4(\gamma \oplus s_B, \tilde{s})$$

As before, $\gamma^{(h)}$ is a fixed unary function of the "correct" value γ, and the rest of the argument goes through analogously.

C   CovIDA Security Definition

The following definitions for ϵ-CovIDA security are taken from [18].

Real-model execution. The real-model execution of protocol Π takes place between players (P_1, P_2), at most one of whom is corrupted by a non-uniform probabilistic polynomial-time machine adversary $\mathcal{A}$. At the beginning of the execution, each party P_i receives its input x_i. The adversary $\mathcal{A}$ receives an auxiliary information aux and an index that indicates which party it corrupts. For that party, $\mathcal{A}$ receives its input and sends messages on its behalf. Honest parties follow the protocol.

Let $\mathrm{REAL}_{\Pi, \mathcal{A}(aux)}(x_1, x_2)$ be the output vector of the honest party and the adversary $\mathcal{A}$ from the real execution of Π, where aux is an auxiliary information and x_i is player P_i's input.

Ideal-model execution. Let $f : (\{0,1\}^*)^2 \to \{0,1\}^*$ be a two-party function-ality. In the ideal-model execution, all the parties interact with a trusted party that evaluates f. As in the real-model execution, the ideal execution begins with each party P_i receiving its input x_i, and $\mathcal{A}$ receives the auxiliary information aux. The ideal execution proceeds as follows:

Send inputs to trusted party: Each party P_1, P_2 sends x_i' to the trusted party, where $x_i' = x_i$ if P_i is honest and x_i' is an arbitrary value if P_i is controlled by $\mathcal{A}$.

Abort option: If any $x_i' = \mathsf{abort}$, then the trusted party returns abort to all parties and halts.

Attempted cheat option: If P_i sends $\mathsf{cheat}_i(\epsilon')$, then:

- If $\epsilon' > \epsilon$, the trusted party sends $\mathsf{corrupted}_i$ to all parties and the adversary $\mathcal{A}$, and halts.
- Else, with probability $1 - \epsilon'$ the trusted party sends $\mathsf{corrupted}_i$ to all parties and the adversary $\mathcal{A}$ and halts.
- With probability ϵ',
  - The trusted party sends $\mathsf{undetected}$ and $f(x_1', x_2')$ to the adversary $\mathcal{A}$.
  - $\mathcal{A}$ responds with an arbitrary boolean (polynomial) function g.
  - The trusted party computes $g(x_1', x_2')$. If the result is 0 then the trusted party sends abort to all parties and the adversary $\mathcal{A}$ and halts. (i.e. $\mathcal{A}$ can learn $g(x_1', x_2')$ by observing whether the trusted party aborts or not.)

  Otherwise, the trusted party sends $f(x_1', x_2')$ to the adversary.

Second abort option: The adversary sends either abort or $\mathsf{continue}$. In the first case, the trusted party sends abort to all parties. Else, it sends $f(x_1', x_2')$.

Outputs: The honest parties output whatever they are sent by the trusted party. $\mathcal{A}$ outputs an arbitrary function of its view.

Let $\mathrm{IDEAL}_{f,\mathcal{A}(aux)}^{\epsilon}(x_1, x_2)$ be the output vector of the honest party and the adversary $\mathcal{A}$ from the execution in the ideal model.

Definition 2. *A two-party protocol Π is secure with input-dependent abort in the presence of covert adversaries with ϵ-deterrent (ϵ-CovIDA) if for any non-uniform probabilistic polynomial-time adversary $\mathcal{A}$ in the real model, there exists a non-uniform probabilistic polynomial time adversary $\mathcal{S}$ in the ideal model such that*

$$\left\{ \mathrm{REAL}_{\Pi,\mathcal{A}(aux)}(x_1, x_2) \right\}_{x_1, x_2, aux \in \{0,1\}^*} \overset{c}{\approx} \left\{ \mathrm{IDEAL}_{f,\mathcal{S}(aux)}^{\epsilon}(x_1, x_2) \right\}_{x_1, x_2, aux \in \{0,1\}^*}$$

for all $|x_1| = |x_2|$ and aux.

D   Proof Sketch of Theorem 4

We only present the proof for the case when Alice is corrupted. The case of corrupted Bob is symmetric. The probability that at least one of the evaluated garbled gates that were generated by Alice was constructed properly (and consistently with the COT) is $1 - 2^{-s}$ (because of the cut-and-choose and the fact that Bob checks both the garbled circuits and their corresponding COT inputs). On the other hand, Alice is forced (because of the OTs) to use only one input for the garbled circuits generated by Bob. If that input is different than the one she has used for the garbled circuits that Bob evaluates, then with good probability the outputs of the input-consistency check will be different, causing the PSI to return an empty set (since in order to know one element in Bob's set, Alice has to guess output wire labels which were not revealed to her in her evaluations). Therefore, if the PSI returns at least one element, that element is indeed the result of correct evaluations of valid garbled circuits done by both players, and since Bob is honest, this is the right output. Furthermore, since Bob is honest, Alice will get only a single output from all her evaluations, and will have to use random elements for the rest of her PSI inputs. Since Bob's inputs to the PSI include information that must be learned from the output wire labels chosen by Bob, the only element in the intersection could be the right output.

More formally, let $\mathcal{A}$ be an adversary controlling Alice in the execution of the protocol. We describe a simulator $\mathcal{S}$ that runs $\mathcal{A}$ internally and interacts with the trusted third party (TTP) that computes f. $\mathcal{S}$ does the following:

1. Emulates a honest Bob with input 0 until the end of "Circuit Preparation" Stage.
2. During the emulation, $\mathcal{S}$ extracts the seeds Alice committed on, and learns all her inputs to the OTs (including her input x^A in the OTs for her to learn her input labels for Bob's circuits). We say that a seed, a garbled circuit and its COT inputs constitute a good set if they are consistent and properly generated, and a bad set otherwise. For each of Alice's circuits, $\mathcal{S}$ determines if it is a good or a bad set. Note that unless all Alice's circuits are bad, at this stage $\mathcal{S}$ can compute the output labels of Alice's circuits.
   Let r_B, r'_B be the values chosen by the emulated Bob and let r_A, r'_A be the values chosen by Alice in the OTs. Alice receives only the output $f(x_A, 0) \oplus r'_A \oplus r'_B$ from her evaluated circuits and can compute only a single valid value q_i. (She can also guess other valid values with a negligible in L probability.) $\mathcal{S}$ can also compute this value at this stage. Denote this valid q_i by Q.
   If all sets are good, $\mathcal{S}$ sends x_A to the TTP, receives the output z, and continues the emulation of Bob. If Alice enters Q to the PSI, she receives the same value as the intersection, but otherwise she receives an empty set. Next, if Alice decommits to a value different than Q in the "Output generation" stage, $\mathcal{S}$ aborts. Else, it sends $R'_B = r'_B \oplus z \oplus f(x_A, 0)$ to Alice and outputs whatever she does. (This causes the output of Alice to be z since she received $f(x_A, 0) \oplus r'_A \oplus r'_B$ from the evaluations.) Note that $\mathcal{S}$ knows the corresponding valid labels for R'_B from the OTs.
   If some of the sets are bad, let $c \in \{0, 1\}^s$ be a bitstring such that $c_i = 1$ if set i is good, and $c_i = 0$ otherwise. Also, let e_b be the number of bad sets. If $e_b = s$, $\mathcal{S}$ simulates Bob aborting and outputs whatever Alice does. This is due to the fact that we generate the challenge set such that at least one circuit is always checked.

If $0 < e_b < s$, then

- Let $p = \frac{2^{s-e_b}-1}{2^s-2} - \frac{1}{2^s-2}$. With probability p, S extracts $\alpha_1^{(A)}$ and chooses a random $\alpha_1^{(B)}$ such that all bad sets are in $\mathbb{E}_A$, but also at least one good set is also in it. It calls the TTP with x_A and receives the output z and proceeds with the emulation until the output unmasking. As before, if Alice does not send Q to the PSI, it receives an empty set, and if she does not decommit Q afterwards, S aborts. Last, S sends r_A' so that the unmasked output would be z, and outputs whatever Alice outputs.
- With probability $1 - p$, S calls the TTP with x_A and then sends the message $\mathsf{cheat}(1/(2^s - 2)/(1-p))$. (Note that $(1/2^s - 2)/(1-p) \le 2^{-s+1}$ when $e_b \ge 1$.)

If the TTP returns a corrupted message, S chooses a random $\alpha_1^{(B)}$ such that some bad circuits will be checked, and continues the emulation until Bob aborts once the bad circuits are checked. Otherwise, i.e., in case the TTP returns undetected and the output z, S causes $\mathbb{E}_A$ to be the set of all bad garbled circuits, and sends to the TTP the function that has hardcoded the bad garbled circuits, their COT inputs, all output labels (chosen by both players), the values r_B, r_B' and $R_B' = r_B' \oplus z \oplus f(x_A, 0)$, and the value Q. (This part is similar to [7,18].) The function takes Bob's real input x_B, finds R_B such that $M_B \cdot (0^l \| R_B \| R_B') = M_B \cdot (0^l \| r_B \| r_B')$, emulates a honest Bob with inputs x_B, R_B, R_B' that receives its input labels from the OTs, evaluates the garbled circuits, and checks if any of the outputs would give $q_i = Q$. In high level, the function emulates what a honest Bob would get from the evaluation using Bob's real and output mask R_B' that makes sure both players receive the same output. If the TTP responds with abort, S emulates Bob aborting after Alice decommits the output of the intersection (or before in case the emulated Bob aborts). If the TTP does not respond with abort, as before, if Alice inputs Q to the PSI she receives it back, or empty set otherwise, and if she decommits to a different value than Q, then S emulates Bob Aborting. In case no abort happens until the end of the protocol, S sends R_B' as Bob's mask. (Note that the COT for the bits of R_B' must be fine since otherwise the function that emulated Bob by the TTP would have failed producing the value Q.)

We now analyse the probabilities of the different cases: (1) If all sets are good, then S simply retrieves the output and unmask the output accordingly. The simulation looks the same as the real execution except for Bob's inputs which are hidden because of the security of the COT and the garbling scheme; (2) if all sets are bad, then S will emulate Bob aborting and outputs what Alice does. This is identical to the real world since at least once circuit is always checked and hence Alice caught. (3) If some sets (but not all) are bad then there are three possibilities:

- Alice is caught cheating - Happens with probability $(1-p) \times (1 - \frac{1/(2^s-2)}{1-p}) = 1-p-1/(2^s-2) = \frac{2^{s-e_b}-1}{2^s-2}$.
- The protocol ends without accusing Alice of cheating - Happens with probability $p = \frac{2^{s-e_b}-1}{2^s-2} - \frac{1}{2^s-2}$.
- Alice successfully cheats - Happens with probability $(1-p) \times \frac{1/(2^s-2)}{1-p} = 1/(2^s-2)$.

Note that the soundness of the protocol is 2^{-s+1} since we call the TTP with the message cheat$\frac{1}{2^s-1}/(1-p)) \le 2^{-s+1}$ for the p we have. We stress that the actual cheating probability is only $1/(2^s - 1)$. (This "gap" is a because the adversary is not always accused of cheating, even if it gets caught.)

We remark that the adversary can guess one of Bob's output labels with a negligible in L probability. Since it can guess several values and enter them as inputs to the PSI, the probability that at least one of them would be valid is $|\mathbb{E}_A| \cdot neg(L)$. (This affects only the parameter L, and not s.)

Round-Efficient Concurrently Composable Secure Computation via a Robust Extraction Lemma

Vipul Goyal[1], Huijia Lin[2], Omkant Pandey[3,5,*,**], Rafael Pass[4,***],
and Amit Sahai[5,*]

[1] Microsoft Research India, Bangalore, India
vipul@microsoft.com
[2] University of California Santa-Barbara
rachel.lin@cs.ucsb.edu
[3] University of Illinois at Urbana Champaign
omkant@uiuc.edu
[4] Cornell University
rafael@cs.cornell.edu
[5] UCLA and Center for Encrypted Functionalities
sahai@cs.ucla.edu

Abstract. We consider the problem of constructing protocols for secure computation that achieve strong concurrent and composable notions of security in the plain model. Unfortunately UC-secure secure computation protocols are impossible in this setting, but the Angel-Based Composable Security notion offers a promising alternative. Until now, however, under standard (polynomial-time) assumptions, only protocols with polynomially many rounds were known to exist.

In this work, we give the first $\tilde{O}(\log n)$-round secure computation protocol in the plain model that achieves angel-based composable security in the concurrent setting, under standard assumptions. We do so by constructing the first $\tilde{O}(\log n)$-round CCA-secure commitment protocol. Our CCA-secure commitment protocol is secure based on the minimal assumption that one-way functions exist.

A central tool in obtaining our result is a new *robust concurrent extraction lemma* that we introduce and prove, based on the minimal

___

* Research supported in part from a DARPA/ONR PROCEED award, NSF Frontier Award 1413955, NSF grants 1228984, 1136174, 1118096, and 1065276, a Xerox Faculty Research Award, a Google Faculty Research Award, an equipment grant from Intel, and an Okawa Foundation Research Grant. This material is based upon work supported by the Defense Advanced Research Projects Agency through the U.S. Office of Naval Research under Contract N00014-11- 1-0389. The views expressed are those of the author and do not reflect the official policy or position of the Department of Defense, the National Science Foundation, or the U.S. Government.

** Work done in part while visiting Microsoft Research India (Bangalore).

*** Pass is supported in part by an Alfred P. Sloan Fellowship, Microsoft New Faculty Fellowship, NSF CAREER Award CCF-0746990, NSF Award CCF-1214844, NSF Award CNS-1217821, AFOSR YIP Award FA9550-10-1-0093, and DARPA and AFRL under contract FA8750-11-2-0211.

Y. Dodis and J.B. Nielsen (Eds.): TCC 2015, Part I, LNCS 9014, pp. 260–289, 2015.
© International Association for Cryptologic Research 2015

assumptions that one-way functions exist. This robust concurrent extraction lemma shows how to build concurrent extraction procedures that work even in the context of an "external" protocol that cannot be rewound by the extractor. We believe this lemma can be used to simplify many existing works on concurrent security, and is of independent interest. In fact, our lemma when used in conjunction with the concurrent-simulation schedule of Pass and Venkitasubramaniam (TCC'08), also yields a constant round construction based additionally on the existence of quasi-polynomial time $(\mathcal{PQT})$ secure one-way functions.

1 Introduction

The notion of *secure multi-party computation* protocols is central to cryptography. Introduced in the seminal works of [Yao86, GMW87], secure multi-party computation allows a group of (mutually) distrustful parties $P_1, \ldots, P_n$, with private inputs $x_1, \ldots, x_n$, to jointly compute any functionality f in such a manner that the honest parties obtain correct outputs yet no group of malicious parties learn anything beyond their inputs and the prescribed outputs. These early results on secure computation [Yao86, GMW87], along with a rich body of followup works that further refined and developed the concept [GL90, GMW91, Bea91, MR91, Can00, PW01, Can01, Gol04], demonstrated that the delicate task of designing secure protocols can be captured by general secure computation.

Much of the early literature on secure computation only considered the *stand-alone* setting where security holds only if a single execution of the protocol takes place, in isolation with no other cryptographic activity in the system. We call this security *stand-alone security*. While stand-alone security may be sufficient for basic purposes, it does not suffice in today's more complex networked environments where other cryptographic protocols might be running in the system simultaneously.

Concurrent Security. To deal with more complex systems, the last decade has seen a push towards obtaining protocols that have strong concurrent *composability* properties. For example, we could require concurrent self-composability: the protocol should remain secure even when there are multiple copies executing concurrently. The framework of *universal composability* (UC) was introduced by Canetti [Can01] to capture a more general security requirement for a protocol that may be executed concurrently with not only several copies of itself but also with other protocols in an arbitrary manner.

Unfortunately, strong impossibility results have been shown ruling out the existence of secure protocols in the concurrent setting. UC secure protocols for most functionalities of interest have been ruled out in [CF01, CKL03]. Protocols in even less demanding settings of concurrent security were ruled on in [Lin04, BPS06, AGJ+12, Goy12]. We stress that, in fact, the latest sequence of these impossibility results provide an *explicit attack* in the concurrent setting using which the adversary may even fully recover the input of an honest party (see, e.g., the chosen protocol attack in [BPS06]). Hence, designing secure protocols in the concurrent setting is a question of great theoretical interest as well as practical motivation.

To overcome these impossibility results, UC secure protocols were proposed based on various "trusted setup assumptions" such as a common random string that is published by a *trusted party*[CF01, CLOS02, BCNP04, CPS07, Kat07, CGS08]. Nevertheless, a driving goal in cryptographic research is to eliminate the need to trust other parties. The main focus of this paper is to obtain concurrently-secure protocols in the *plain model*.

Relaxing the Security Notion. To address the problem of concurrent security for secure computation in the plain model, a few candidate definitions have been proposed, the most well studied one being that of *super-polynomial simulation* [Pas03, PS04, BS05]. The notion of security with *super-polynomial simulators* (SPS) [Pas03, PS04, BS05] is one where the adversary in the ideal world is allowed to run in (fixed) super-polynomial time. Very informally, SPS security guarantees that any polynomial-time attack in the real execution can also be mounted in the ideal world execution, albeit in super-polynomial time. This is directly applicable and meaningful in settings where ideal world security is guaranteed statistically or information-theoretically (which would be the case in most "end-user" functionalities that have been considered, from privacy-preserving data mining to electronic voting).

Angel-based UC security. To formalize the notion of SPS security in a way that allows modular analysis and provides composability, Prabhakaran and Sahai [PS04] put forward the notion of *angel-based composable security*. Very roughly, in the angel based security notion, the parties (including the simulator and the adversary) are all polynomial time but have access to an angel which will perform certain *specific* super-polynomial time computations. This angel-based definition is in contrast to the case where the simulator is given direct access to super-polynomial computation power: in this case, the resulting security notion is *not* closed under composition and thus does not permit a modular protocol design in the concurrent setting[1]. A construction for concurrently secure computation in the angel based composable security model were given in [PS04, BS05], but only based on non-standard super-polynomial hardness assumptions.

Very recently, Canetti, Lin, and Pass [CLP10] obtained the first secure computation protocol that achieves angel-based composable security based on *standard polynomial-time assumptions*. Unfortunately, however, the improvement in terms of assumptions comes at the cost of the round complexity of the protocol. Specifically, the protocol of [CLP10] incurs *polynomial-round complexity*. A follow up work of Lin and Pass [LP12] considers the problem of designing *black-box* constructions for secure computation in the concurrent setting. They propose a protocol making only a black-box use of oblivious transfer satisfying the angel-based composable security notion. However the round complexity of their protocol continues to remain polynomial.

---

[1] However we note that according to this weaker SPS security notion, concurrently secure protocols in constant rounds are now known [GGJS12].

We note that the latency of sending messages back and forth has been shown to often be the dominating factor in the running time of cryptographic protocols [MNPS04, BNP08]. Indeed, round complexity has been the subject of a good deal of research in cryptography. For example, in the context of concurrent zero knowledge (ZK) proofs, round complexity was improved in a sequence of works [RK99, KP01, PRS02] from polynomial to slightly super-logarithmic (that nearly matches the lower bound w.r.t. black-box simulation [CKPR01]). The round complexity of non-malleable commitments in the stand-alone and concurrent settings has also been studied in several works[DDN91, Bar02, PR05b, PR05a, LP09, Wee10, Goy11, LP11], improving the round complexity from logarithmic rounds to constant rounds under minimal assumptions. We observe that for the setting of concurrently secure computation protocols with angel-based composable security, the situation is worse since the only known protocols that achieves angel-based composable security based on standard assumptions incurs polynomial-round complexity [CLP10, LP12]. This raises the following natural question:

"Do there exists round-efficient protocols in the concurrent setting satisfying the angel-based composable notion of security based on standard assumptions?"

Our Results. We answer the above question in the affirmative and provide a $\widetilde{O}(\log n)$ round construction of concurrently secure computation in the plain model. Our construction satisfies the angel-based composable notion of security [PS04, CLP10]. To obtain our result, we construct a "CCA-secure commitment" protocol in $\widetilde{O}(\log n)$ rounds, based only the assumption that one-way functions exist. CCA secure commitments were introduced in [CLP10]; roughly speaking, a commitment protocol is CCA-secure if it remains hiding even when the adversary is given an oracle that can open all commitment values (except the commitment given as a challenge to the adversary). In [CLP10], Canetti et al. show how to construct a protocol that securely realizes any functionality—under the angel-based composable notion of security—given an (appropriate) protocol for CCA secure commitments (see full version of [CLP10]). Prior to our work, the best known construction for CCA secure commitments required n^ϵ rounds [CLP10, LP12]. In contrast, the round complexity of our protocol matches that of the best known constructions for concurrent extractable commitment schemes [PRS02, MOSV06].

A Robust Concurrent Extraction Lemma. A key technical tool that we introduce is a lemma that allows *robust* extraction of secrets from an adversarial committer A^* in the *concurrent* setting. We call this lemma, the *robust concurrent extraction lemma*, which is of independent interest. Roughly speaking, the lemma is a strengthening of the concurrent extraction mechanism for the PRS preamble [PRS02] (we shall call this the PRS commitment), and states that concurrent extraction can be performed even in the presence of an external protocol which cannot be rewound by the "simulator."

More precisely, consider an adversarial committer A^* who commits to multiple values in *concurrent* sessions of the PRS commitment to honest receivers; let us label these sessions as the **right** sessions. Simultaneously, A^* participates in a **left** execution of an arbitrary k-round protocol, denoted $\Pi := \langle B, A \rangle$. Then, the robust concurrent-extraction lemma states that for every A^* there exists a simulator S which, *without rewinding the external party B in its execution of Π*, extracts the values committed to by A^* in every session of the PRS commitment. Furthermore, if ℓ is the round complexity of the PRS preamble, and T is the running time of A^*, then S only fails with probability that is exponentially small in $\ell - O(k \cdot \log T)$.

In order to capture correctness of extraction, we formulate our lemma by considering a "real-world" experiment in which A^* receives the values committed to in every valid PRS commitment from an exponentially powerful party $\mathcal{E}$, called the "online extractor." The extraction of values are provided by $\mathcal{E}$ as soon as a PRS session ends. Our lemma states, intuitively speaking, that both the adversary A^* and the exponentially powerful party $\mathcal{E}$ can be replaced with a polynomial-time simulator, that still interacts with the external party B in the external protocol Π. We remark that formulating the lemma in a generic way, so that it can handle as general usage of the PRS commitment as we have seen in the literature, is a delicate task. Nevertheless, we show that it is possible to precisely capture the concurrent-extraction property of the PRS commitment in a generic way without referring to any specific protocol that uses it.

An immediate benefit of our formulation is that when the PRS commitment is used inside a larger protocol, the task of "concurrent extraction" can be *formally* isolated from other parts of the protocol. This allows one to design hybrid experiments without having to worry about the extraction. We provide two procedures for this purpose—a simulator S and an online extraction $\mathcal{E}$—and demonstrate their use in the security proof of our protocol.

We also wish to remark that the ability to extract without rewinding B turns out to be a very useful tool during concurrent security proofs, and we expect this will have significant applications elsewhere. This flexibility simplifies security proofs (of even previous works) to a great extent. For example, a situation similar to our lemma arises in previous works on concurrent non-malleable zero-knowledge [BPS06, LPTV10], In these works, the problem is solved in an arguably ad-hoc fashion, which stops rewindings after a certain point in the simulation during certain hybrids. This, overall, leads to a rather delicate analysis, and the order of hybrid experiments becomes important.

By using our robust-extraction lemma, this problem can be avoided almost directly. We note however, that in our case, it is crucial that the rewindings not be stopped. This is because, in our situation, one needs to implement the super-polynomial angel from the beginning in each hybrid experiment. Hence, every session in every hybrid requires online extraction—which is possible either by using (all) rewindings or by using super-polynomial simulation.

Technical Overview. As mentioned before, the starting point of our construction is the robust extraction lemma. The basic problem encountered in proving the lemma is that given the entire transcript of interaction between A^* and the honest right parties, there are various "breakpoints" in the transcript (representing messages of the left protocol) which cannot be rewound. In addition, during rewindings (or look-ahead threads), the breakpoints can change their location and can even come earlier than expected (at which point the current thread must be discontinued and another one started). At a high level, we start by considering the necessary modification of the KP/PRS simulator so that when a breakpoint is encountered during the execution of a look-ahead thread, the look-ahead thread is stopped and abandoned.

Our proof shows that even with this modification, the simulation still succeeds. The key observation is that each breakpoint can "spoil" at most a $d = \log T$ number of "recursive blocks" used in the swapping argument of [PRS02]. Since there are k breakpoints, this incurs an additional loss of $k \cdot \log T$ blocks. However, to execute this proof strategy, it becomes crucial to use the information learned during a "look-ahead block" in certain special sibling blocks. This new feature of our analysis must be done carefully to maintain the correctness of our swapping argument.

We note that we strive to obtain the best possible bounds for the round complexity ℓ of PRS that is necessary to extract for a given value of k (rounds of the left protocol). For this reason, we choose to re-analyze the proof presented in [PRS02] (see also [PTV08])[2].

For the goal of constructing CCA secure commitments, a direct application of the robust extraction lemma will not be sufficient. This is because the number of rounds in the left and the right interaction will be the same, and the robust extraction lemma cannot work with respect to protocols with the same round complexity as the PRScommitment! To construct CCA secure commitments (which is our main technical goal), we instead build upon techniques from prior work on concurrent non-malleable zero-knowledge (CNMZK) [BPS06, GJO10, LPTV10]. At a high level, our protocol is simple: commit to the value using a regular commitment scheme, and then, prove the validity of the commitment using a concurrent zero-knowledge protocol that is also *simulation sound* [Sah99]. We note that we design our own protocol for this task since we strive to achieve a construction based on *one-way functions* only (for CCA secure commitments). Using techniques from [BPS06, LPTV10] is either not possible or results in stronger assumptions such as the existence of collision-resistant hash functions. Our protocol is presented in section 3.

A Constant Round Protocol from $\mathcal{PQT}$ One-way Functions. Our techniques can be seen as a general method which reduce the task of concurrent-extraction to that of concurrent-simulation. The method requires only a constant-

---

[2] For example, one can consider the approach of applying a pigeonhole principle argument to argue that $\omega(\log n)$ slots must occur between some two breakpoints, and then trying to apply the PRS analysis simply to these slots. However, note that even if no breakpoints occur during these slots, *look-ahead threads* that are started during these slots can still encounter breakpoints, since the adversary can choose the scheduling adaptively. Dealing with this analytically would require further loss, and result in a worse asymptotic bound than ours for super-constant values of k. Our more direct approach shows how to amortize the gains made over all slots, even if only a few slots occur between some breakpoints.

factor blow up in the round-complexity (of a given concurrent-simulation method) and a polynomial-factor blow up in the running time of the (given) simulator. By applying our method to the concurrent-simulator of [PV08], we obtain a constant round protocol for CCA-secure commitments based only on the existence of one-way functions secure against adversaries running in time super quasi-polynomial time ($\mathcal{PQT}$) (see full version of this work [GLP+15]).

Assuming more complex, and somewhat non-standard assumptions—namely the existence of adaptive one-way functions [PPV08]—a constant round protocol for CCA secure commitments is already known [PPV08]. Recent progress on program obfuscation [GGH+13] has given rise to new non-black-box simulation techniques for fully concurrent zero-knowledge [PPS15, IPS15, CLP14]; it would be interesting to explore if these techniques can also lead to improved constructions of CCA secure commitments.

Related Works. The work of Garg, Goyal, Jain, and Sahai [GGJS12] is closely related to our work, who provide a constant round protocol under the non-composable SPS notion (instead of angel-based composable security). Their work requires the existence of statistically hiding commitment. Independently of [GGJS12], a recent work of Pass, Lin, and Venkitasubramaniam [PLV12] also provides a constant round protocol achieving non-composable SPS security, using very different techniques.

Other security notions that deal with concurrent security were presented in [MPR06, GGJS12] who propose the notion of *input indistinguishable computation*, and in [GS09, GJO10] who considered a modified ideal world that allows the adversary to make more output queries to the ideal functionality (than just one) per session.

2   Robust Concurrent Extraction

In this section, we will prove the robust extraction lemma. We use standard notation. In particular, $A(x; r)$ denotes the process of evaluating (randomized) algorithm A on input x with random coins r, and $A(x)$ the process of sampling a uniform r and then evaluating $A(x; r)$. We define $A(x, y; r)$ and $A(x, y)$ analogously. The set of natural numbers is represented by $\mathbb{N}$. Unless specified otherwise, $n \in \mathbb{N}$ represents the security parameter available as an implicit input. when necessary. All inputs are assumed to be of length at most polynomial in n.

For two probability distributions D_1 and D_2, we write $D_1 \stackrel{c}{\equiv} D_2$ to mean that D_1 and D_2 are computationally indistinguishable. For two interactive Turing machines (ITM) A and B, we write $\mathsf{OUT}_B [A(1^n, x) \leftrightarrow B(1^n, y)]$, the output of B after an interaction with A where their inputs in the interaction are y and x respectively, and their random tapes are independent and uniform.

We assume familiarity with commitment schemes. Without loss of generality, we will be using commitment schemes with non-interactive reveal phase—i.e., the committer sends a single message (v, d) to decommit. For a commitment scheme

$\langle C, R \rangle$, we denote by $\mathsf{open}_{\langle C,R \rangle}(c, v, d)$ the decommitment function. That is, the receiver accepts v as the value committed to in the commitment-transcript c if $\mathsf{open}_{\langle C,R \rangle}(c, v, d)$ outputs 1, and rejects otherwise. For statistically binding commitments, v is uniquely determined given c with high probability.

2.1 The PRS Preamble

The robust extraction lemma deals with the commitment preamble of Prabhakaran, Rosen, and Sahai [PRS02]. This preamble has been used in many prior works, and is often referred to as the PRS preamble. The preamble uses an underlying commitment scheme Com. Roughly speaking, the committer first commits to many shares of the value v to be committed using Com. This is followed by a several rounds where in each round, the receiver sends a random challenge, and the committer responds with appropriate decommitments. Each round is called a *slot*.

Essentially, the PRS preamble is an interactive commitment scheme, which is statistically binding (resp. hiding) if the underlying scheme Com is statistically binding (resp., hiding). The formal description of the PRS preamble is given in figure 1. As before, we write $\mathsf{open}_{\mathsf{PRS}}(c, v, \rho) = 1$, to formally mean that there exists randomness ρ such that c is the transcript of the PRS preamble, executed between the honest committer with input v and randomness ρ and the honest receiver with randomness (equal to its challenges) appearing in c.

The security parameter is n, the value to be committed is $v \in \{0,1\}^n$, and the round-parameter is $\ell := \ell(n)$.

Commitment. The committer and the receiver execute the following steps.

1. The committer chooses $n\ell$ pairs of n-bit random strings $(v_{i,j}^0, v_{i,j}^1)$ for $i \in [n], j \in [\ell]$ such that (for every i,j): $v_{i,j}^0 \oplus v_{i,j}^1 = v$. It commits to strings v, $v_{i,j}^b$ using the commitment scheme Com, for every $b \in \{0,1\}, i \in [n], j \in [\ell]$.
2. For $j = 1$ to ℓ:
   (a) the receiver sends a n-bit challenge string $r_j = r_{1,j}, \ldots, r_{n,j}$
   (b) the committer responds by sending a decommitment to strings $v_{1,j}^{r_{1,j}}, \ldots, v_{n,j}^{r_{n,j}}$

Decommitment. The committer decommits to all remaining strings which were not opened in the commit phase.

Fig. 1. The PRS Preamble based on Com

2.2 The Extraction Lemma

In this section, we present the robust extraction lemma. We will consider an adversary A^* who interacts in many sessions of the PRS preamble; simultaneously, A^* also participates in a single execution of a two party computation protocol Π. The running time of A^* is not necessarily polynomial in n. However, the lemma becomes trivial if the hiding/binding of the underlying commitment Com can be broken in $\text{poly}(n) \cdot T^2$ time, where $T = T(n)$ is the maximum number of PRS preambles A^* initiates.

SIMPLIFYING ASSUMPTION. We assume, for the clarity of presentation, that the commitment scheme Com underlying the PRS preamble is *statistically binding*. Later, we will present the general form which deals with both kinds of Com, as well as varying round complexity of the preamble.

PROTOCOL Π. Let $\Pi := \langle B, A \rangle$ be an arbitrary two-party computation protocol. We assume w.l.o.g. that both B and A receive a parameter $n \in \mathbb{N}$ as their first input. In addition, for a fixed $n \in \mathbb{N}$, let $\text{dom}_B(n)$ denote the domain of valid (second) input for algorithm B, and $k := k(n)$ denote the round complexity of Π.

The Robust-Concurrent Attack. Let A^* be an interactive Turing machine, called the adversary, $n \in \mathbb{N}$ the security parameter, and $x \in \text{dom}_B(n)$ an input. In the robust-concurrent attack, A^* interacts with a special, not necessarily polynomial time, party $\mathcal{E}$ called the "online extractor." Party $\mathcal{E}$ simultaneously participates in one execution of the protocol Π, and several executions of the PRS preamble with A^*. Party $\mathcal{E}$ follows the (honest) algorithm $B(1^n, x)$ in the execution of Π with A^*. Further, it follows the (honest) receiver algorithm in each execution of the PRS preamble. If A^* successfully completes a PRS preamble s, $\mathcal{E}$ sends a string α_s to A^*, together with a special message END_s, to mark the completion of the preamble.

The scheduling of all messages in all sessions—Π as well as PRS preambles— is controlled by A^* including starting new sessions and finishing or aborting existing sessions. We adopt the following conventions [Ros04, PRS02]:

1. When A^* sends a round i message of session s, it immediately receives the next—i.e., $(i + 1)$-st message of s; this is without loss of generality,[3] and holds for messages of Π as well.
2. If a session s has been aborted, A^* does not schedule any further messages of s.
3. If A^* starts a PRS preamble s, it also sends a special message, denoted START_s, immediately *after the last message* of step 1 of this preamble is completed (see figure 1). Message START_s indicates that the challenge-response phase is about to start.

---

[3] This is because the next message can be stored and delivered whenever needed during the attack.

At some point, A^* halts. We say that A^* *launches* the robust-concurrent attack.

For $n \in \mathbb{N}, x \in \mathrm{dom}_B(n), z \in \{0,1\}^*$, let $\mathsf{REAL}_{\mathcal{E},\Pi}^{A^*}(n,x,z)$ denote the output of the following probabilistic experiment: on input 1^n and auxiliary input z, the experiment starts an execution of A^*. Adversary A^* launches the robust-concurrent attack by interacting with the special party $\mathcal{E}$ throughout the experiment, as described above. When A^* halts, the experiment outputs the view of A^* which includes: all messages sent/received by A^* to/from $\mathcal{E}$, the auxiliary input z, and the randomness of A^*.   □

We are now ready to present the robust extraction lemma. Informally speaking, the lemma states that there exists an interactive Turing machine—a.k.a the robust simulator—whose output is statistically close to $\mathsf{REAL}_{\mathcal{E},\Pi}^{A^*}(n,x,z)$ *even if* the final response of $\mathcal{E}$ at the end of a successful PRS session is actually the value A^* commits to in that session. Further, the robust simulator does not "rewind" B, and runs in time polynomial in total sessions opened by A^*.

Lemma 1 (Robust Concurrent Extraction). *There exists an interactive Turing machine S ("robust simulator"), such that for every A^*, for every $\Pi :=$ $\langle B, A \rangle$, there exists a party $\mathcal{E}$ ("online extractor"), such that for every $n \in \mathbb{N}$, for every $x \in \mathrm{dom}_B(n)$, and every $z \in \{0,1\}^*$, the following conditions hold:*

1. **Validity constraint.** *For every output ν of $\mathsf{REAL}_{\mathcal{E},\Pi}^{A^*}(n,x,z)$, for every PRS preamble s (appearing in ν) with transcript τ_s, if there exists a unique value $v \in \{0,1\}^n$ and randomness ρ such that $\mathsf{open}_{\mathsf{PRS}}(\tau_s, v, \rho) = 1$, then:*

$$\alpha_s = v,$$

   where α_s is the value $\mathcal{E}$ sends at the completion of preamble s.
2. **Statistical simulation.** *If $k = k(n)$ and $\ell = \ell(n)$ denote the round complexities of Π and the PRS preamble respectively, then the statistical distance between distributions $\mathsf{REAL}_{\mathcal{E},\Pi}^{A^*}(n,x,z)$ and $\mathsf{OUT}_s\left[B(1^n, x) \leftrightarrow S^{A^*}(1^n, z)\right]$ is given by:*

$$\Delta(n) \leq 2^{-\Omega(\ell - k \cdot \log T(n))},$$

   where $T(n)$ is the maximum number of total PRS preambles between A^* and $\mathcal{E}$.[4] Further, the running time of S is $\mathrm{poly}(n) \cdot T(n)^2$.

We prove this lemma by presenting an explicit simulator S and a corresponding party $\mathcal{E}$. The explicit constructions appear in subsection 2.3, and the full proof of the lemma is given in the full version [GLP+15]. We now make some important remarks about the lemma.

---

[4] The lemma allows for exponential $T(n)$ as well. However, if it is too large—e.g., $T(n) = 2^{2n}$, the PRS preamble should be modified suitably. For example, the value v as well as the challenges in each slot, must be of length at least $n + 2 \log T(n)$.

Remarks

1. The special party $\mathcal{E}$ is not completely defined by the lemma. In particular, when a PRS preamble s does not commit to a *unique and valid* value v, the value α_s sent by $\mathcal{E}$ to A^* is not defined. This can happen, e.g., when not all shares committed to in step 1 XOR to v. In such situations, $\mathcal{E}$ can choose whatever value α_s it wants. The only requirements on $\mathcal{E}$ are that it uses honest algorithms during the robust-concurrent attack *and* that for each successfully completed PRS preamble it satisfies the validity constraint. Every $\mathcal{E}$ satisfying these requirements is called a *valid* $\mathcal{E}$.

2. The PRS preamble is used in a variety of complex ways. For example, some protocols require opening the committed PRS-value (e.g., [PRS02, BPS06, OPV10], whereas some others may never open this value (e.g., [LPTV10, GJO10]). To be able to capture such uses generically, we do not enforce any consistency requirements on the PRS preambles. The choice of not fully defining $\mathcal{E}$ when PRS preamble is not valid provides sufficient flexibility to capture such generic uses of the preamble.

3. When the preamble is used in a larger protocol, a "main" simulator is used to prove the security of the larger protocol. Typically, the main simulator employs the rewinding strategy of [KP01, PRS02] to extract the PRS-values and simultaneously deals with other details of the protocol. Our lemma separates the task of extracting PRS-values from other necessary actions of the main simulator. This makes the overall proofs simpler. Party $\mathcal{E}$ then only acts as mechanism to transfer the extracted PRS-values back to the main simulator. The main simulator takes upon the role of A^* to receive extracted values from $\mathcal{E}$, while only dealing with other details of the larger protocol.

4. A consequence of the above two remarks is that when a PRS preamble is not consistent, we do not know what value α_s actually gets extracted. Our choice of the order of quantifiers allows $\mathcal{E}$ to depend on A^* as well as S. This essentially allows $\mathcal{E}$ to extract and supply the same value α_s (by running S internally) that a typical "main" simulator would extract for inconsistent PRS preambles.

5. Requirements 1 and 2 of the lemma imply that if we sample an output of the simulator and consider a PRS preamble s with transcript τ_s which contains a *unique* and *valid* value v, and receives α_s as $\mathcal{E}$'s response in the end, then except with probability $\Delta(n)$, it holds that $\alpha_s = v$.

2.3   A Robust Simulator and an Online Extractor

In this section, we present an explicit construction of a robust simulator S, and the (online extractor) party $\mathcal{E}$ for which (the robust extraction) lemma 1 holds. The simulator is a slight modification of [KP01, PRS02], to also deal with messages of Π, without rewinding them. We start by defining a few terms first.

The States of A^*. Recall that the scheduling of messages in the robust-concurrent attack is controlled by A^*, and when A^* sends the i-th message of a session s (either PRS or Π), it immediately receives the next message of s, namely the $(i + 1)$-st message. The *state* of A^* at any given point during the attack consists of its *view up to that point*: it includes all messages sent/received by A^*, its auxiliary input z and its randomness. The *starting* (or *original*) state of A^*—denoted throughout by st_0—is its state before it receives the first message. If st denotes the state of A^* at some point during the robust-concurrent attack, the set of all PRS preambles which have not completed yet is denote by $\mathsf{LIVE}(\mathsf{st})$.

The Robust Simulator S. The simulator receives as input an auxiliary string $z \in \{0,1\}^*$, and the security parameter n. The simulator participates with an external party B of Π. Let $x \in \mathrm{dom}_B(n)$ and γ denote the input and uniformly chosen randomness of B. The simulator incorporates the adversary A^* as a black-box; let $T = T(n)$ define the maximum number of PRS preambles that A^* can open during the robust-concurrent attack.

Simulator S starts by setting $(1^n, z)$ on A^*'s input tape, and a sufficiently long uniform string on its random tape. The simulator then initiates a helper procedure recurse as follows:

$$(\mathsf{st}, \mathcal{T}) \leftarrow \mathsf{recurse}(T, \mathsf{st}_0, \emptyset, 1, \emptyset, 0).$$

Throughout its execution, messages of recurse are forwarded back and forth to $B(1^n, x; \gamma)$ and (the black-box) A^* as appropriate. Finally, the output of S is the first output of recurse, namely st; the output st is also known as the *main thread*. Procedure recurse is given in figure 2, each execution of recurse is called a block, and has a unique name denoted by id. $\qquad\square$

The Online Extractor $\mathcal{E}$. Formally, an execution of $\mathcal{E}$ begins during a robust-concurrent attack. Let (γ, ρ) denote the random tape of $\mathcal{E}$, and (n, x, z) denote its inputs. $\mathcal{E}$ incorporates the program of A^*, and performs the following *internal steps* before it sends out its first message in the robust-concurrent attack,

1. $\mathcal{E}$ proceeds identically to the robust simulator algorithm S, using ρ as its random tape and $(1^n, z)$ as its inputs. To successfully proceed in this step, $\mathcal{E}$ uses (x, γ) to simulate the honest algorithm $B(1^n, x; \gamma)$, as well as black-box access to A^*.[5] However, $\mathcal{E}$ differs from S in its actions only when a PRS preamble s completes, in the manner described below.
2. Let s be a successfully completed PRS preamble; at this point S either extracts a value μ_s or reaches an ExtractFail. When this happens, $\mathcal{E}$ neither sends μ_s nor aborts the simulation; instead it proceeds as follows. First, $\mathcal{E}$ attempts to extract the *actual value* committed to in the preamble by inverting all instances of the underlying commitment Com. Then, it decides the value α_s, to be sent, as follows. If a valid and *unique* value v_s exists, set $\alpha_s = v_s$. Otherwise, it has following cases:

---

[5] Observe that although $\mathcal{E}$ depends on A^* here, it still uses A^* only as a black-box, as remarked earlier.

procedure recurse(t, st, $\mathcal{T}$, f, aux, id):

1. If $t = 1$, **repeat**:
   (a) If the next message is START, start a new session s.
      – send $r \leftarrow \{0,1\}^n$ as the challenge of the first slot of s.
      – add entry $(s : 1, r, \_)$ to $\mathcal{T}$.
   (b) If the next message is the slot-i challenge of an existing session s.
      – send $r \leftarrow \{0,1\}^n$ as the slot-i challenge of s.
      – add entry $(s : i, r, \_)$ to $\mathcal{T}$.
   (c) If the next message is the slot-i *response*, say β, of an existing session s.
      – If β is a valid message:
         – update entry $(s : i, r_i, \_)$ to $(s : i, r_i, \beta)$.
         – if $i = \ell$, i.e., it is the last slot, send (END, extract(s, id, $\mathcal{T}$, aux)).
      – Otherwise, if $\beta = \bot$, abort session j, and add $(s : \bot, \bot, \bot)$ to $\mathcal{T}$.
      – Update st to be the current state of A^*
      – **return** (st, $\mathcal{T}$).
   (d) If the next message is a *response* from A^* for the external protocol Π.
      – If f $= 0$, i.e., it is a look-ahead block, then **return** (st, $\mathcal{T}$);
      – If f $= 1$, i.e., it is the *main thread*), do the following:
         – send A^*'s message to the external party of Π, return the response to A^*.
      – Update st to be the current state of A^*
      – For every live session $s \in$ LIVE(st), do the following:
         – $\times_{s,\text{id}} = $ true,
         – for every block id$'$ that contains the block id, set: $\times_{s,\text{id}'} = $ true.
2. If $t > 1$,
   # Rewind the first half twice:
   (a) $(\text{st}_1, \mathcal{T}_1) \leftarrow$ recurse($t/2$, st, $\mathcal{T}$, 0, aux, id $\circ$ 1)   [look-ahead block C']
   (b) Let aux$_2 = (\text{aux}, \mathcal{T}_1 \setminus \mathcal{T})$,
      $(\text{st}_2, \mathcal{T}_2) \leftarrow$ recurse($t/2$, st, $\mathcal{T}$, f, aux$_2$, id $\circ$ 2)   [main block C]

   # Rewind the second half twice:
   (c) $(\text{st}_3, \mathcal{T}_3) \leftarrow$ recurse($t/2$, st, $\mathcal{T}^*$, 0, aux, id $\circ$ 3)   [look-ahead block D']
   (d) Let $\mathcal{T}^* = \mathcal{T}_1 \cup \mathcal{T}_2$ and aux$_4 = (\text{aux}, \mathcal{T}_3 \setminus \mathcal{T}^*)$,
      $(\text{st}_4, \mathcal{T}_4) \leftarrow$ recurse($t/2$, st, $\mathcal{T}^*$, f, aux$_4$, id $\circ$ 4)   [main block D]

   (e) **return** (st$_4$, $\mathcal{T}_3 \cup \mathcal{T}_4$).

---

procedure extract(s, id, $\mathcal{T}$, aux):

1. Attempt to extract a value for s from $\mathcal{T}$.
2. If extraction fails, consider every block id_1 for which $\times_{s,\mathsf{id}_1} = \mathsf{true}$.
   - Let id_1' be the sibling of id_1, with input/output tables $\mathcal{T}_{\mathsf{in}}, \mathcal{T}_{\mathsf{out}}$ respectively.
   - Attempt to extract from $\mathsf{aux}_{\mathsf{id}_1'} := \mathcal{T}_{\mathsf{out}} \setminus \mathcal{T}_{\mathsf{in}}$; (included in aux).
3. If all attempts fail, abort the simulation and **return** ExtractFail. Otherwise **return** the extracted value.

Fig. 2. Procedures recurse and extract used by the robust simulator S

(a) If there are more than one valid v_s, proceed as follows: if μ_s equals to any of them, set $\alpha_s = \mu_s$, otherwise, set α_s to be one of them chosen at random.

(b) If no valid v_s exists, then proceed as follows: if $\mu_s = \mathsf{ExtractFail}$, set α_s to be a random value; otherwise, set $\alpha_s = \mu_s$.

3. After reaching the end of the simulation, $\mathcal{E}$ internally stores the randomness ρ_s and the values α_s for every PRS preamble s appearing on the *main thread*.

Having completed the steps above,[6] $\mathcal{E}$ is now ready to interact with the (outside) A^* launching the robust-concurrent attack, and proceeds as follows.

- If A^* sends a message intended for the (only) session of Π, $\mathcal{E}$ interacts with A^* by following actions of $B(1^n, x; \gamma)$. Likewise, if A^* sends a messages for a PRS preamble s, A^* follows the honest receiver algorithm of PRS preamble with *randomness* ρ_s already computed internally.
- If A^* successfully completes a PRS preamble s, $\mathcal{E}$ sends the already stored value α_s to A^*. $\qquad\qquad\qquad\qquad\qquad\qquad\qquad\qquad\qquad\square$

3   CCA Secure Commitments in $\widetilde{O}(\log n)$ Rounds

In this section we apply our robust extraction lemma to construct a $\widetilde{O}(\log n)$-round protocol for CCA secure commitments. We will need the generalized version of the lemma which allows for *statistically hiding* PRS preamble as well; the general version appears in 4. We will also use a non-malleable commitment scheme, denoted NMCom, that is *robust* w.r.t. constant round protocols [LP09]. Constant round constructions for such protocols are now known (see the non-malleable commitments section in the full version [GLP+15]).

---

[6] We insist that all the steps above are *internal* to $\mathcal{E}$ and that as of now it has not sent any external message in the robust-concurrent attack. Further, it can successfully complete these steps since it has all the required inputs.

The protocol has a very intuitive "commit-and-prove" structure where the committer commits to the value v using a PRS preamble, and then proves its consistency using what resembles a "concurrent simulation-sound" protocol. We start by recalling the notion of CCA secure commitments from [CLP10].

3.1   CCA Secure Commitments

Let $\langle C, R \rangle$ denote a statistically binding and computationally hiding commitment scheme. We assume w.l.o.g. that $\langle C, R \rangle$ has a non-interactive reveal phase— i.e., the committer simply sends (v, d). The decommitment is verified using a function $\mathsf{open}(c, v, d)$; that is, the receiver accepts v as the value committed in the commitment-transcript c if $\mathsf{open}(c, v, d)$ outputs 1, and rejects otherwise. A tag-based commitment scheme with $l(n)$-bit identities [PR05b, DDN91] is a commitment-scheme where in addition to 1^n, C and R also receive a "tag" (or *identity*) of length $l(n)$ as common input. We will consider schemes which are *efficiently checkable*: meaning that if R *accepts* in the interaction (with transcript c) then there exists a decommitment pair (v, d) such that $\mathsf{open}(c, v, d) = 1$.

In CCA-secure commitments, we consider an adversarial receiver A, who has access to an oracle $\mathcal{O}$, called the "decommitment oracle." The oracle participates with A in many *concurrent* sessions of (the commit phase of) $\langle C, R \rangle$, using tags of length $l(n)$, chosen adaptively by A. At the end of each session, if the session is accepting, the oracle returns the (unique) value committed by A in that session; otherwise it returns $\perp$. (In case there is more than one possible decommitment, $\mathcal{O}$ returns any one of them.)[7]

Roughly speaking, we say that a tag-based scheme $\langle C, R \rangle$ is CCA-secure if there exists a decommitment oracle $\mathcal{O}$ for $\langle C, R \rangle$, such that the hiding property of the scheme holds even for adversaries A with access to $\mathcal{O}$. Formally, let $\mathsf{IND}_b(\langle C, R \rangle, \mathcal{O}, A, n, z)$ denote the output of the following probabilistic experiment: on common input 1^n and auxiliary input z, the PPT adversary $A^{\mathcal{O}}$ (adaptively) chooses a pair of challenge values $(v_0, v_1) \in \{0, 1\}^n$ and a tag $\mathsf{id} \in \{0, 1\}^{l(n)}$, and receives a commitment to v_b using the tag id (note that A interacts with $\mathcal{O}$ throughout the experiment as described before); finally, when $A^{\mathcal{O}}$ halts, the experiment returns the output y of $A^{\mathcal{O}}$; y is replaced by $\perp$ if during the experiment, A sends $\mathcal{O}$ any commitment using the tag id.

Definition 1 (CCA-secure Commitments, [CLP10]). *Let $\langle C, R \rangle$ be a tag-based commitment scheme with tag-length $l(n)$, and $\mathcal{O}$ be a decommitment oracle for it. We say that $\langle C, R \rangle$ is CCA-secure w.r.t. $\mathcal{O}$, if for every PPT ITM A, every $z \in \{0, 1\}^*$, and every sufficiently large n, it holds that:*

$$\mathsf{IND}_0(\langle C, R \rangle, \mathcal{O}, A, n, z) \overset{c}{\equiv} \mathsf{IND}_1(\langle C, R \rangle, \mathcal{O}, A, n, z).$$

---

[7] Note that since $\langle C, R \rangle$ is efficiently checkable, and the session is accepting, such a valid decommitment always exists. In addition, note that since we only have statistical binding, this value is unique except with negligible probability.

We say that $\langle C, R \rangle$ is CCA-secure if there exists a decommitment oracle $\mathcal{O}'$ such that $\langle C, R \rangle$ is CCA-secure w.r.t. $\mathcal{O}'$.

An analogous version of the definition which considers many concurrent executions on left (instead of just one), is known to be equivalent to the current definition (via a simple hybrid argument). Also note that, for this reason, this definition implies concurrent non-malleable security for commitments [PR05a].

3.2 Our Protocol for CCA Secure Commitments

We are now ready to present our CCA secure commitment protocol, denoted CCA-Com. The protocol employs the PRS preamble in *both* directions (similar to [GJO10, LPTV10]). However, our protocol is much simpler, and admits an easier security proof.

We now provide a quick overview of our protocol. Our protocol will also use a constant-round non-malleable commitment scheme—NMCom—that is robust w.r.t. 4 rounds. The formal description of our protocol appears in figure 3.

Let $v \in \{0, 1\}^n$ be the value to be committed. Our commitment protocol, CCA-Com, consists of five phases. In the first phase, the committer C commits to v using a statistically-binding PRS preamble. We denote this instance of the preamble by PRS_1. In phase 2, the receiver R uses a statistically-*hiding* PRS preamble to commit to a random value $\sigma \in \{0, 1\}^n$; we denote this instance of the PRS preamble by PRS_2.

In phase 3, C commits to 0^n using the *robust* NMCom scheme. In phase 4, R decommits to the value σ (of phase PRS_2). Finally, in phase 5, C uses a witness-indistinguishable (WI) *proof* system to prove that: "either there exists a valid value v in phase 1 or the value committed in NMCom is σ." It can use, for example, Blum's 3-round protocol repeated in parallel n times [Blu87]. We remark that it is necessary to use a proof system which ensures soundness against unbounded provers.

To decommit, C sends decommitments corresponding to the first phase, namely PRS_1. The round-complexity of both PRS preambles is $\ell \in \omega(\log n)$. We have the following theorem, whose proof is given in the full version [GLP+15].

Theorem 1. *Assuming the existence of collision-resistant hash functions, protocol CCA-Com presented in figure 3 is a $\widetilde{O}(\log n)$-round CCA secure commitment scheme for identities of length n.*

3.3 Construction Based on One-Way Functions

Protocol CCA-Com of the previous section requires the use of a statistically-hiding phase PRS_2. If we change the protocol so that PRS_2 is statistically-binding, we will not be able to prove that the right sessions of CCA-Com remain statistically binding. This is because of the presence of the super-polynomial time oracle $\mathcal{O}$. Indeed, without the oracle we can make PRS_2 statistically-binding and

---

Protocol CCA-Com. The committing algorithm is denoted by C. The receiver algorithm is denoted by R. The common input is the security parameter n and an identity id of length n. The private input of C is the value $v \in \{0,1\}^n$ to be committed. The protocol proceeds in following five phases:

- *Phase 1:* C commits to v using a statistically-binding PRS preamble. This instance of the preamble is denoted by PRS_1, and let τ_1 be the commitment-transcript.
- *Phase 2:* R commits to random $\sigma \leftarrow \{0,1\}^n$ using a statistically-hiding PRS preamble. This instance of the preamble is denoted by PRS_2.
- *Phase 3:* C commits to the all-zero string 0^n using NMCom, with common identity id. Let τ_3 be the commitment-transcript.
- *Phase 4:* R decommits to σ, by sending the appropriate decommitment strings.
- *Phase 5:* C proves to R, using a public coin, constant round WI *proof* system (e.g., n parallel repetitions of Blum's protocol), that either:

  (a) $\exists\, v \in \{0,1\}^n$ s.t. τ_1 is a valid PRS-commitment to v; OR
  (b) τ_3 is a valid commitment to σ as per NMCom.

Decommitment oracle $\mathcal{O}$. The oracle extracts the value committed to in transcript τ_1 of the first phase, and returns it.

Fig. 3. Protocol CCA-Com

have a protocol based on one-way functions; but in the presence of oracle, it may happen that A^* is able to ensure that for some right session s, $\widetilde{u}_s = \widetilde{\sigma}_s$.

To avoid this problem, we must somehow remove PRS_2 yet still be able to later modify the left PRS slot-by-slot. The key observation is that in case of commitments, there is only one left session. Therefore, we can use a different simulation strategy which guarantees "soundness" w.r.t. unbounded committers as well.

Based on this observation, our one-way functions based protocol does not use PRS_2. Instead it uses a CoinFlip protocol in which no unbounded prover can succeed in "setting up a trapdoor" but a rewinding party can. Our new protocol, denoted CCA-Com*, appears in figure 4.

Theorem 2. *Assuming the existence of one way functions, protocol* CCA-Com* *presented in figure 4 is a $\widetilde{O}(\log n)$-round CCA secure commitment scheme for identities of length n.*

For the proof of this theorem see appendix ??. Although the proof is very similar to that of theorem 1, one crucial point is that we need to re-use the idea of robust-simulation, and be careful about how we apply it.

> **Protocol CCA-Com*.** The committing algorithm is denoted by C. The receiver algorithm is denoted by R. The common input is the security parameter n and an identity id of length n. The private input of C is the value $v \in \{0,1\}^n$ to be committed. Com is a statistically-binding scheme.
>
> **Round parameters.** Let $q := q(n)$ be a fixed function in $\omega(1)$, and let $\ell := \ell(n) = q(n) \cdot \log n \cdot \omega(1)$. Let $k \in O(1)$ be the round complexity of NMCom.
>
> - *Phase 1:* C commits to v using the PRS preamble with parameter $\ell \in \omega(\log n)$. This instance of the preamble is denoted by PRS_1, and let τ_1 be the commitment-transcript.
> - *Phase 2:* C and R execute the following CoinFlip protocol. In round $i \in [q]$:
>   (a) C commits a "short" string $u_i \in \{0,1\}^{\log n}$ using Com,
>   Let c_i be the commitment-transcript, and d_i a decommitment-string.
>   (b) R sends a "short" random string $\sigma_i \in \{0,1\}^{\log n}$
>   Define $u = (u_1, \ldots, u_q)$, $\sigma = (\sigma_1, \ldots, \sigma_q)$, and $d = (d_1, \ldots, d_q)$.
> - *Phase 3:* C commits to string (u,d) using NMCom andidentity id. Let τ_3 be the commitment-transcript.
> - *Phase 4:* C proves to R, using a public coin, constant round WI proof—e.g., n parallel repetitions of Blum's protocol—that either:
>   (a) $\exists\, v \in \{0,1\}^n$ s.t. τ_1 is a valid PRS-commitment to v; OR
>   (b) τ_3 is a valid commitment to (u,d) as per NMCom, such that $\forall i \in [q]$:
>   - (i) $\mathsf{open}_{\mathsf{Com}}(c_i, u_i, d_i) = 1$, *and*
>   - (ii) $u_i = \sigma_i$.
>
> **Decommitment oracle $\mathcal{O}$.** The oracle extracts the value committed to in transcript τ_1 of the first phase, and returns it.

Fig. 4. Protocol CCA-Com* based on one-way functions

4 Generalized Version of the Robust Extraction Lemma

In the generalized version of the lemma, we allow A^* to open sessions of the PRS preambles that are *statistically hiding*. At the start of the preamble, it is already understood whether it is statistically-binding or statistically-hiding. Since statistically-hiding preambles are only computationally-binding, we require that A^* is a PPT machine. Furthermore, we make following new adjustments to the robust-concurrent attack:

At the successful completion of a *statistically-hiding* PRS preamble s, when A^* receives the string α_s from $\mathcal{E}$, it can choose to respond with an opening of the committed value v_s. It does so by sending appropriate decommitment strings (v_s, d_s). Furthermore, the scheduling of this message is decided by A^*.

Other than this, the attack remains unchanged. We now present the generalized version of the lemma. The essence of the lemma is still the same as before. We only need to add conditions to deal with the statistically-hiding preambles. For such preambles, the validity constraint requires that α_s be equal to the opened value v_s (if any), except with negligible probability.

Lemma 2 (Robust Extraction: General Version). *There exists an interactive Turing machine S ("robust simulator"), such that for every PPT A^*, for every $\Pi := \langle B, A \rangle$, there exists a party $\mathcal{E}$ ("online extractor"), such that for every $n \in \mathbb{N}$, for every $x \in \mathrm{dom}_B(n)$, and every $z \in \{0,1\}^*$, the following conditions hold:*

1. **Validity constraint.** *For every output ν of $\mathrm{REAL}^{A^*}_{\mathcal{E},\Pi}(n, x, z)$, we have:*
   (a) *for every statistically-binding preamble s (appearing in ν) with transcript τ_s, if there exists a unique value $v \in \{0,1\}^n$ in the commitment-transcript τ_s, then $\alpha_s = v$,*
   (b) *for every statistically-hiding preamble s (appearing in ν) with transcript τ_s, if there exists a valid opening (v_s, d_s) in the view ν, then $\alpha_s = v_s$,*
   where α_s is the value $\mathcal{E}$ sends at the completion of s.
2. **Statistical simulation.** *If $k = k(n)$ and $\ell = \ell(n)$ denote the round complexities of Π and the PRS preamble respectively, then the statistical distance between distributions $\mathrm{REAL}^{A^*}_{\mathcal{E},\Pi}(n, x, z)$ and $\mathrm{OUT}_s\left[B(1^n, x) \leftrightarrow S^{A^*}(1^n, z)\right]$ is given by:*
$$\Delta(n) \leq 2^{-\Omega(\ell - k \cdot \log T(n))},$$
   where $T(n)$ is the maximum number of total PRS preambles between A^* and $\mathcal{E}$.[8] Further, the running time of S is $\mathrm{poly}(n) \cdot T(n)^2$.

The proof of this general version of the lemma is identical to the proof of the original lemma and appears in the full version [GLP+15]. We only need to show that condition 1(b) also holds. We show that if it does not then we can violate the computational binding of the statistically-hiding PRS. Suppose that Com_{sh} is the underlying commitment scheme of the statistically hiding PRS. Then, if α_s is not equal to v_s, look at the two opened challenges in the execution of recurse whose XOR results in α_s; let they belong to slot i of this session. Both of these strings must have been decommitted to correctly (for the scheme Com_{sh}). Further, since opening v_s requires opening of all slots of PRS such that pairs in each slot XOR to the same v_s, we have that all pairs of slot i must have been correctly opened to strings that are different from what recurse learned. Therefore, we must have an instance of Com_{sh} with correct openings to two different values. The details are standard and omitted. We note that the value of $\Delta(n)$ in the second condition does not change since both $\mathcal{E}$ and S extract value α_s identically for the statistically-hiding PRS.

---

[8] The lemma allows for exponential $T(n)$ as well. However, if it is too large—e.g., $T(n) = 2^{2n}$, the PRS preamble should be modified suitably. For example, the value v as well as the challenges in each slot, must be of length at least $n + 2 \log T(n)$.

5   Proof of Security for CCA-Com

We start by noting that the scheme is statistically-binding even against an un-bounded cheating committer algorithm C^*. This is because of the following. PRS_2 is statistically-hiding for σ, and therefore except with negligible probability, the value committed to in NMCom (which is statistically-binding) does not equal σ. Then, from the soundness of the WI *proof*, it follows that C^* can succeed only when the PRS_1 is consistent. It will be important to have statistical-binding w.r.t. unbounded $C*$ since we will be dealing with the super-polynomial time oracle $\mathcal{O}$ in various hybrid experiments.

We now proceed to demonstrate the CCA security of our scheme. To do so, we will directly construct a PPT simulator SIM, which simulates the view of any CCA adversary $A^{*\mathcal{O}}$. SIM will not have access to the decommitment oracle $\mathcal{O}$, and uses A^* only as a black-box. Recall that $\mathcal{O}$ extracts the value from the transcript of PRS_1. SIM only receives 1^n and auxiliary-input z for A^*, as its own inputs.

Lemma 3. *There exists a strict polynomial time machine SIM such that for every PPT ITM machine A^*, every $z \in \{0,1\}^*$, every sufficiently large n, and every $b \in \{0,1\}$, it holds that:*

$$\mathsf{SIM}^A(1^n, z) \stackrel{c}{\equiv} \mathsf{IND}_b(\langle C, R\rangle, \mathcal{O}, A, n, z).$$

Proof. Recall that during its execution A^* opens one session of CCA-Com := $\langle C, R \rangle$ on left, while simultaneously interacting in multiple concurrent sessions of CCA-Com with $\mathcal{O}$, called the right sessions. Let $T = T(n)$ be a polynomial upper-bounding the number of right sessions of A^*.

Algorithm SIM will use the robust simulator S guaranteed by the (general version of) the robust-extraction lemma (see sections 2 and 4), allowing it to extract the values committed to by A^* in all PRS preambles. It then uses these values to simulate the answers of $\mathcal{O}$, as well as to succeed in WI proof (by committing the extracted value in NMCom of the left session).

SIM uses two helper procedures: the robust simulator S and an "interface" algorithm I to be described shortly. Procedure I essentially "decouples" the PRS preambles from the rest of the CCA-Com protocol. It incorporates A^* as a black-box, and handles all messages of all sessions of CCA-Com internally and honestly, *except* for all the PRS preambles in which A^* acts as the *committer*. All these preambles are forwarded to outside PRS-receivers. That is, I participates in a robust-concurrent attack, interacting with the party $\mathcal{E}$. It executes PRS preambles with $\mathcal{E}$, and at the end of each PRS preamble s, I expects to receive a value α_s from $\mathcal{E}$. This value will be used internally.

SIM is a polynomial time machine without access to any super-polynomial time helper. Therefore, to run I, it runs the robust-simulator S providing it black-box access to the "adversary" procedure I. SIM outputs whatever S outputs. Formal descriptions follow.

Algorithm SIM($1^n, z$). Return the output of $S^I(1^n, z)$, where procedure I—which has black-box access to adversary A^*—is described below.

Procedure $I(1^n, z)$. Procedure I launches the robust-concurrent attack, by committing in several PRS sessions to external receivers denoted $R_1, \ldots, R_T$. At the end of each preamble, it expects to receive a string α_s. I incorporates the CCA adversary A^* internally, as a black-box. I initiates an execution of A^*, simulating various sessions of CCA-Com that A^* opens as follows.

1. If A^* starts a new session s of CCA-Com on *right*, I starts by initiating a new session of the statistically-binding PRS_1 with an external receiver R_s. Then, messages of phase-1 of s are relayed between A^* and R_s. I simulates all other phases of s internally by following the honest algorithm for each phase.

2. If A^* starts a new session s of CCA-Com on *left*, I initiates a new session of the statistically-*hiding* preamble to be used as PRS_2 of s with an external receiver R'_s. I completes various phases of s as enumerated below.

   (a) *Phase 1:* I commits to an all zero string to A^*.
   (b) *Phase 2:* I simply relays messages between A^* and R'_s.
   (c) *Phase 3:* I commits to value α_s using NMCom (instead of 0^n). Value α_s was received from outside at the end of PRS_2 of sessions s.
   (d) *Phase 4:* If A^* correctly opens the value in PRS_2 of session s, I checks that the opened value is equal to the "fake witness" α_s. If not, it outputs a special symbol ExtractFail, and halts. Otherwise it continues the execution.
   (e) *Phase 5:* I uses α_s and the randomness used in phase 3 (NMCom) to complete the WI proof in phase 5.[9]

3. *Oracle answers:* If A^* successfully finishes a session s of CCA-Com on right, I sends the (already extracted) value α_s to A^*.

When A^* halts, I outputs the view of A^*, and halts.   □

Proving Indistinguishability. We are now ready to prove our lemma. We will prove this by using a series of hybrid experiments. Our hybrid experiments will be designed my making step-by-step changes to how I communicates with A^* internally in various phases of the protocol. For $i \in [7]$, we denote by ν^i the output of hybrid H_i.

Hybrid H_0: This hybrid is identical the experiment $\mathsf{IND}_b(\langle C, R \rangle, \mathcal{O}, A, n, z)$. Recall that in this experiment, $A^{*\mathcal{O}}$ receives a commitment to value $v_b \in \{0, 1\}^n$ from the honest committer C, while interacting with the oracle $\mathcal{O}$. The output of the experiment consists of the view of A^*, which is also the output of H_0.

---

[9] Note that, technically, I can use the valid witness corresponding to the PRS_1 phase s. This is since it committed to a valid value, namely 0^n. However, we choose to use α_s so that SIM will in fact be a valid simulator even for our concurrent non-malleable zero-knowledge protocols.

Hybrid H_1: This hybrid is identical to H_0, except for the following differences:

(a) H_1 does not forward the right sessions to $\mathcal{O}$. Instead, it executes all right sessions on its own, by playing the honest receiver strategy R. Denote by R_s the instance of R executed in session s.

(b) When A^* successfully completes the session R_s, H_1 queries a different oracle $\mathcal{O}'$, which acts as follows. The query to $\mathcal{O}'$ consists of only part of the commitment-transcript PRS_1 that belongs for the value to be committed. All other messages (e.g., commitments of shares and slot-messages) are not part of the query. In particular, if Com is non-interactive, the query will be commitment of the value $\widetilde{v}_s$. If there is a unique value defined by the query, $\mathcal{O}'$ extracts that value, and returns it as the answer. In all other cases, it behaves exactly as $\mathcal{O}$.

Hybrid H_2: This hybrid is identical to H_1; we use it to set up some notation. Define a procedure I_2, which is identical to procedure I (defined above), except that it it executes (all internal phases of) the left session honestly and that it uses the oracle $\mathcal{O}'$ as in H_1. That is, it only forwards the PRS preambles received from A^* outside, but executes all other phases internally and honestly. Formally, I_2 is identical is identical to I except:

(a) I_2 receives the value v_b to be committed in left sessions of CCA-Com, as an input. Furthermore, I_2 commits to value v_b in PRS_1 of the left session (instead of 0^n, step 2(a)).

(b) When a session s of PRS_1 ends, I_2 expects to receive a value α_s from outside. If s is a statistically-binding sessions, then I_2 gives α_s to A^*. If s is statistically-hiding, it *ignores* the value s—there is only one such session corresponding to the left session.

(c) I_2 uses the valid witness, v_b and the randomness of PRS_1 to complete its WI proof in the left session (step 2(e)). I_2 does not check the validity of the "fake witness" (step 2(d)).

Observe that I_2 is essentially launching a robust-concurrent attack. It does not perform any rewindings. Hybrid H_2 simply runs the procedure $I_2^{A^*}$, and simulates PRS-receivers for it exactly as H_1 does; furthermore, when a statistically-binding session s finishes, H_2 forwards its commitment to $\mathcal{O}'$ and returns the oracles answer, denoted α_s, to I. On the other hand, if the statistically-hiding session of PRS ends, H_2 sends a random string α_s (which, by construction, is ignored by I_2, see point (b)). The output of H_2 is the view of I_2 (which in turn is the view of A^*).

Hybrid H_3: For procedure I_2, let $\mathcal{E}$ be our online extractor as defined in section 2.3 (with addendum from section 4 to deal with statistically-hiding PRS as well). Recall that $\mathcal{E}$ is a super-polynomial time machine, which facilitates the robust-concurrent attack. In particular, it acts honest receivers in all PRS preambles, as well as provides extractions for each one of them when they finish.

On input $(1^n, v_b, z)$, hybrid H_3 starts an execution of $I_2(1^n, v_b, z)$, making it interact with party $\mathcal{E}$. The output of H_3 is the output of the robust-concurrent attack, which in turn consists of the view of I_2 (and hence A^*). Therefore, H_3 differs significantly from H_2: it does not run PRS receivers and does not have access to $\mathcal{O}'$, since these are automatically done by $\mathcal{E}$ for H_3.

From hereon, we will only be making changes to the interface procedure I_2. All future hybrids, except for the last one, differ from H_2 in only that they use a modified version of I_2. Furthermore, changes are made to the phases of the *left session* only, of which there is only one.

Hybrid H_4: This hybrid is identical to H_3 except that instead of using I_2, it uses procedure I_4. Let α_s be the value that I_2 receives from $\mathcal{E}$ at the end of statistically-hiding PRS_2 of the *left session*. Furthermore, let σ_s be the value opened by A^* which appears internally, in the *left* session of $\mathsf{CCA\text{-}Com}$ (simulated for A^* by I_2).

Procedure I_4 is identical to I_2 except that when A^* sends σ_s along with valid openings, I_4 tests that $\alpha_s = \sigma_s$; it aborts the entire simulation if this test fails, and outputs $\mathsf{BindingFail}$. Recall that I_2 simply ignores α_s.

Hybrid H_5: This hybrid is identical to H_4 except that instead of using I_4, it uses a modified procedure I_5. Procedure I_5 is identical to I_4 except that instead of committing to 0^n, it commits to α_s in protocol NMCom of the *left* session s.

Hybrid H_6: Identical to H_5 except that instead of using I_5, it uses a modified procedure I_6. Procedure I_6 is identical to I_5 except that it uses the "fake witness" in the WI proof of the left session s. Recall that the "fake witness" consists of the value α_s and the randomness used in NMCom.

Hybrid H_7: Identical to H_6 except that instead of using I_6, it uses a modified procedure I_7. Procedure I_7 is identical to I_6 except it does not receive the input v_b, and commits to 0^n in the first phase PRS_1 of the left session. Observe that I_7 is in fact identical to I.

Hybrid H_8: This hybrid differs from H_7 in two crucial places. First, it does not receive the value v_b as input. Therefore, its only inputs are $(1^n, z)$. In addition, it does not use the party $\mathcal{E}$. Instead, H_8 simply runs the robust simulator S with inputs $(1^n, z)$ and black-box I_7. It outputs whatever S outputs. Observe that H_8 is in fact our original simulator SIM, presented earlier. $\square$

Notation. Let ν^0 denote the output of H_0. Let u^0 be the variable denoting the value committed by C in NMCom in the left session, and let σ^0 be the variable denoting PRS-value opened by A^* in (phase-4 of) the left session. Since there is only one left session, we will not use any subscripts. Analogously, define variables $\widetilde{u}_s^0$ and $\widetilde{\sigma}_s^0$ for the s-th right session in hybrid H_0. Finally, we denote by $\widetilde{v}_s^0$ the value A^* commits in PRS_1 of the s-th right session in H_0. For $i \in \{0, \ldots, 8\}$ define values $\nu^i, \widetilde{u}_s^i, \widetilde{\sigma}_s^i$ and $\widetilde{v}_s^i$ w.r.t. the hybrid H_i analogously. Identity of the left session is referred to by id, and that of the s-th right sessions by $\widetilde{\mathsf{id}}_s$ without mention of the hybrid. Further, unless specified otherwise, index s is in $[T]$, and $\widetilde{\mathsf{id}}_s$ is not equal to id, and the probability is taken over the randomness of the hybrid in consideration.

We start by noting that ν^0 is identical to $\mathsf{IND}_b(\langle C, R \rangle, \mathcal{O}, A, n, z)$. Next, we claim the following.

Claim. For every hybrid $i \in \{0, \ldots, 4\}$ and for every right session $s \in [T]$,

$$\nu^0 \stackrel{\mathrm{s}}{\equiv} \nu^1 \equiv \nu^2 \stackrel{\mathrm{s}}{\equiv} \nu^3 \stackrel{\mathrm{s}}{\equiv} \nu^4 \tag{1}$$

$$\Pr\left[\widetilde{u}_s^i = \widetilde{\sigma}_s^i\right] \leq \mathrm{negl}(n) \tag{2}$$

PROOF. It is seen by way of construction of hybrids H_1 and H_2, that $\nu^0 \stackrel{\mathrm{s}}{\equiv} \nu^1 \equiv \nu^2$.[10] In case of hybrid H_3 party $\mathcal{E}$, which also simulates PRS sessions exactly as H_2 does (since $\mathcal{E}$ is valid). Furthermore, $\mathcal{E}$ extracts the values α_s in the (first message) of the PRS preamble exactly as $\mathcal{O}'$ does in H_2. If a unique and valid α_s exists, the value α_s returned to I_2 is the same in both hybrids. However, when the value is not unique, $\mathcal{E}$ uses a different decision procedure to decide the value of α_s. Nevertheless, statistical-binding of PRS ensures that this happens with only negligible probability. It follows that the distribution of answers α_s in both hybrids is statistically close, and hence $\nu^2 \stackrel{\mathrm{s}}{\equiv} \nu^3$.

Finally, in H_4 the only difference is that I_4 verifies that $\alpha_s = \sigma_s$ (i.e., the fake witness is correct). From the validity constraint 1(b) on $\mathcal{E}$ (see lemma 2, section 4) this condition fails with only negligible probability. Therefore, $\nu^3 \stackrel{\mathrm{s}}{\equiv} \nu^4$.

We first prove the second equation for hybrid H_0. Fix any right-session s of H_0. Observe that PRS_2 of s is statistically-hiding. Therefore, except with negligible probability, we have that value $\widetilde{\sigma}_s^0$ is not defined until after the completion of NMCom of session s. Since NMCom is statistically-binding, and there are exponentially many possible values for $\widetilde{\sigma}_s^0$, the claim follows (for hybrid H_0). Now, observe that the same argument applies for hybrids H_1 and H_2 as well. In case of H_3 and H_4, since $\mathcal{E}$ simulates PRS receivers honestly, the same argument applies to these hybrids as well. □

A corollary of the second equation is that in each of these hybrids, for every right session s that is accepting, there exists a *unique* and *valid*[11] value $\widetilde{v}_s^i$ to which A^* is committed to (except with negligible probability). This is because if the second equation holds, from the soundness of WI proof against unbounded adversaries,[12] phase PRS_1 must be consistent defining a unique and valid value.

---

[10] First two hybrids are not identical since the oracle changes: $\mathcal{O}$ extracts from the full PRS transcript (and therefore always checks for consistency), whereas $\mathcal{O}'$ simply extracts from the (first) committing message of PRS. Statistical-binding ensures that this difference happens only in negligible cases.

[11] Recall that v is valid if all shares XOR to v; formally, there exists randomness ρ such that $\mathsf{open}_{\mathsf{PRS}}(\tau_1, v_s, \rho) = 1$ where τ_1 is the commitment-transcript of PRS_1 of s.

[12] We need this since A^* does have access to super-polynomial computations via $\mathcal{O}$, and we do not know what he might be learning.

Therefore, in all future hybrids, we will continue to maintain the second equation as an invariant.

Claim. We have that, $\nu^4 \overset{c}{\equiv} \nu^5$; and $\forall s : \ \Pr \left[\tilde{u}_s^5 = \tilde{\sigma}_s^5 \right] \leq \mathrm{negl}(n)$.

PROOF. Both hybrids H_4 and H_5 use party $\mathcal{E}$ which is super-polynomial time. Therefore, to prove the claim, we need to first need to eliminate the use of $\mathcal{E}$. Observe that this can be done by employing the robust simulator S. However, since we aim to reduce the claim to the non-malleability of NMCom, we would like one left execution and one right execution of NMCom that the robust-simulator S does not rewind.

Recall that the only difference between H_4 and H_5 is that they use different procedures: I_4 and I_5 respectively. The only difference between I_4 and I_5 is that the first one commits to $x = 0^n$ whereas the second one commits to $x = \alpha_s$ in phase NMCom of the left session. Recall that α_s is the value sent by $\mathcal{E}$ at the conclusion of PRS$_2$ of the left session.

We design an intermediate procedure I_*, which is identical to I_4 except for the following differences:

1. When the NMCom phase of the left session is about to begin, I_* sends $(0^n, \alpha_s)$ to an external committer.
2. I_* does not execute the NMCom of the left session internally. Instead, it expects to receive it from an external honest committer, denoted C_{NM}, who either commits to 0^n or α_s.
3. For a randomly chosen right session j, I_* does not execute the the NMCom of session s internally. Instead, it forwards the messages of this NMCom to an external honest receiver, denoted R_{NM}^j.

Therefore, I_* is executed several PRS preambles, and at the same time acting as a man-in-the-middle for protocol NMCom by receiving a commitment and making a commitment at the same time. Let B denote the party who runs algorithm C_{NM} as well as R_{NM}^j for I_*, both *honestly*. Observe that the number of rounds of interaction between B and I_* are $2k$ where k is the round-complexity of NMCom. The input x to B consists of the value to be committed by C_{NM}.

Viewed this way, I_* is an adversary who launches the robust-concurrent attack with respect to party B, and the party $\mathcal{E}$. Furthermore, if B commits to $x = 0^n$ then the execution is identical to that of H_4 with I_4; on the other hand, if $x = \alpha_s$, the execution is identical to that of H_5 with procedure I_5. That is,

$$\nu^4 \equiv \mathrm{REAL}_{\mathcal{E}, \Pi}^{I_*}(n, 0^n, z),$$
$$\nu^5 \equiv \mathrm{REAL}_{\mathcal{E}, \Pi}^{I_*}(n, \alpha_s, z).$$

where protocol $\Pi = (B, P_2)$, and P_2 is "converse of B—that is P_2 acts as a receiver in one NMCom in which B is acting as a committer and vice-versa in the other.

Now, we can remove $\mathcal{E}$ and instead use the robust-simulator S to sample statistically close views. That is, suppose that we run S with I_* and consider the output $\mathsf{OUT_S}(x) := \mathsf{OUT_S}\left[B(1^n, 0^n) \leftrightarrow S^{I_*}(1^n, z)\right]$. By applying the robust concurrent extraction lemma, we have that statistical distance between $\mathsf{OUT_S}(x)$ and $\mathsf{REAL}^{I_*}_{\mathcal{E},\Pi}(n, x, z)$ is at most:

$$\Delta(n) \leq 2^{-\Omega(\ell - 2k \cdot \log T)} \leq \mathsf{negl}(n),$$

for every $x \in \{0^n, \alpha_s\}$ since $\ell \in \omega(\log n)$, k is a constant, and T is at most a polynomial. Using this with the equations above, we get:

$$\nu^4 \overset{\mathrm{s}}{\equiv} \mathsf{OUT_S}(0^n) \tag{3}$$

$$\nu^5 \overset{\mathrm{s}}{\equiv} \mathsf{OUT_S}(\alpha_s) \tag{4}$$

By construction, $\mathsf{OUT_S}(x)$ is the output of S^{I_*} in an interaction with B on input x. Algorithm S^{I_*} is a PPT man-in-the-middle adversary for NMCom who receives a commitment to x from C_{NM}, and commits a value, say $\widetilde{u}_j(x)$, to R^j_{NM}. From non-malleability of NMCom w.r.t. to itself, we have that:

$$(\widetilde{u}_j(0^n), \mathsf{OUT_S}(0^n)) \overset{\mathrm{c}}{\equiv} (\widetilde{u}_j(\alpha_s), \mathsf{OUT_S}(\alpha_s)) \tag{5}$$

It immediately follows from (3), (4), and (5), that $\nu^4 \overset{\mathrm{c}}{\equiv} \nu^5$. Furthermore, suppose that the other part of the claim is false, so that for some session $s \in [T]$ value $\widetilde{u}^5_s = \widetilde{\sigma}^5_s$ with noticeable probability p. Define this to be event bad_5 for hybrid H_5, and analogously define bad_4 for H_4.

Now observe that since j was chosen uniformly from T sessions, $j = s$ with probability $1/T$. Therefore, value $\widetilde{u}^5_s$ appears as the variable $\widetilde{u}_j(\alpha_s)$ with probability $1/T$. In addition, value $\widetilde{\sigma}_j(\alpha_s)$ is a part of $\mathsf{OUT_S}(\alpha_s)$, and therefore, event bad_5 is efficiently observable given both $(\widetilde{u}_j(\alpha_s), \mathsf{OUT_S}(\alpha_s))$; and it occurs with probability p/T which is noticeable. By an analogous argument, event bad_4 is also efficiently observable; furthermore bad_4 must occur with noticeable probability as well due to equation (5). This contradicts equation (2). Hence the claim. □

Claim. We have that, $\nu^5 \overset{\mathrm{c}}{\equiv} \nu^6$; and $\forall s : \ \Pr\left[\widetilde{u}^6_s = \widetilde{\sigma}^6_s\right] \leq \mathsf{negl}(n)$.

PROOF. The proof of this claim is almost identical to the proof of claim 5. The only difference is that instead of using non-malleability w.r.t. itself property of NMCom, we use the fact that NMCom is robust w.r.t. every *interesting* 3-round protocol (i.e., one that "hides" its input, see the non-malleable commitments definition in the full version [GLP+15]). Since the only difference between H_5 and H_6 is in the WI part, the proof follows; we omit the details. □

Claim. We have that, $\nu^6 \overset{\mathrm{c}}{\equiv} \nu^7$; and $\forall s : \ \Pr\left[\widetilde{u}^7_s = \widetilde{\sigma}^7_s\right] \leq \mathsf{negl}(n)$.

PROOF. Observe that in H_6, since the "fake witness" is being used in the WI part, the PRS_1 phase of the left session (which commits to input value v_b) does not have to be consistent. Therefore, we proceed as follows:

1. Design $\ell+1$ intermediate hybrids $H_{6:i}$ for $i = \{0, \ldots, \ell-1\}$ where $H_{6:0} = H_6$, $H_{6:\ell} = H_7$.
2. Hybrid $H_{6:i}$ is the same as $H_{6:i+1}$ except that in slot-i of PRS_1 phase of the left session, $H_{6:i+1}$ commits to shares of an all-zero string.
3. Now, following the proof of claim 5 and by using the robustness of NMCom w.r.t. the 3-round protocols, we conclude that $\nu^{6:i} \stackrel{c}{\equiv} \nu^{6:i+1}$ and that $\forall s : \Pr\left[\widetilde{u}_s^{6:i} = \widetilde{\sigma}_s^{6:i+1}\right] \leq \mathsf{negl}(n)$, where variables are analogously defined. The details of this part are repetitive, and omitted.

The claim now follows. $\qquad\qquad\qquad\qquad\qquad\qquad\qquad\qquad\qquad\qquad\quad$ $\square$

Completing the Proof. Note that the output of H_8, is statistically-close to ν^7 due to the robust-concurrent-extraction lemma. Therefore, combining all the equations we have that $\nu^0 \stackrel{c}{\equiv} \nu^8$. Since H_8 is identical tot he simulator, the lemma follows. $\qquad\qquad\qquad\qquad\qquad\qquad\qquad\qquad\qquad\qquad\qquad\qquad\qquad\quad$ $\square$

References

[AGJ+12] Agrawal, S., Goyal, V., Jain, A., Prabhakaran, M., Sahai, A.: New impossibility results for concurrent composition and a non-interactive completeness theorem for secure computation. In: Safavi-Naini, R., Canetti, R. (eds.) CRYPTO 2012. LNCS, vol. 7417, pp. 443–460. Springer, Heidelberg (2012)

[Bar02] Barak, B.: Constant-round coin-tossing with a man in the middle or realizing the shared random string model. In: FOCS (2002)

[BCNP04] Barak, B., Canetti, R., Nielsen, J.B., Pass, R.: Universally composable protocols with relaxed set-up assumptions. In: Proc.45th FOCS, pp. 186–195 (2004)

[Bea91] Beaver, D.: Foundations of secure interactive computing. In: Feigenbaum, J. (ed.) Advances in Cryptology - CRYPT0 1991. LNCS, vol. 576, pp. 377–391. Springer, Heidelberg (1992)

[Blu87] Blum, M.: How to prove a theorem so no one else can claim it. In: Proceedings of the International Congress of Mathematicians, pp. 1444–1451 (1987)

[BNP08] Ben-David, A., Nisan, N., Pinkas, B.: Fairplaymp: A system for secure multi-party computation. In: ACM Conference on Computer and Communications Security, pp. 257–266 (2008)

[BPS06] Barak, B., Prabhakaran, M., Sahai, A.: Concurrent non-malleable zero knowledge. In: FOCS, pp. 345–354 (2006), Full version available on eprint arhive

[BS05] Barak, B., Sahai, A.: How to play almost any mental game over the net - concurrent composition using super-polynomial simulation. In: Proc. 46th FOCS (2005)

[Can00] Canetti, R.: Security and composition of multiparty cryptographic protocols. Journal of Cryptology: The Journal of the International Association for Cryptologic Research 13(1), 143–202 (2000)

[Can01]   Canetti, R.: Universally composable security: A new paradigm for cryp-
           tographic protocols. In: Werner, B. (ed.) Proc. 42nd FOCS, pp. 136–147
           (2000), Preliminary full version available as Cryptology ePrint Archive Re-
           port 2000/067
[CF01]   Canetti, R., Fischlin, M.: Universally composable commitments. Report
           2001/055, Cryptology ePrint Archive (July 2001), Extended abstract ap-
           peared in CRYPTO 2001
[CGS08]   Chandran, N., Goyal, V., Sahai, A.: New constructions for UC secure
           computation using tamper-proof hardware. In: Smart, N.P. (ed.) EURO-
           CRYPT 2008. LNCS, vol. 4965, pp. 545–562. Springer, Heidelberg (2008)
[CKL03]   Canetti, R., Kushilevitz, E., Lindell, Y.: On the limitations of univer-
           sally composable two-party computation without set-up assumptions. In:
           Biham, E. (ed.) EUROCRYPT 2003. LNCS, vol. 2656, Springer,
           Heidelberg (2003)
[CKPR01]   Canetti, R., Kilian, J., Petrank, E., Rosen, A.: Black-box concurrent zero-
           knowledge requires (almost) logarithmically many rounds. In: STOC (2001)
[CLOS02]   Canetti, R., Lindell, Y., Ostrovsky, R., Sahai, A.: Universally composable
           two-party computation. In: Proc. 34th STOC, pp. 494–503 (2002)
[CLP10]   Canetti, R., Lin, H., Pass, R.: Adaptive hardness and composable security
           in the plain model from standard assumptions. In: FOCS, pp. 541–550
           (2010), Full version:
           http://www.cs.cornell.edu/~rafael/papers/ccacommit.pdf
[CLP14]   Chung, K.-M., Lin, H., Pass, R.: Constant-round concurrent zero-
           knowledge from indistinguishability obfuscation. Cryptology ePrint
           Archive, Report 2014/991 (2014), http://eprint.iacr.org/
[CPS07]   Canetti, R., Pass, R., Shelat, A.: Cryptography from sunspots: How to use
           an imperfect reference string. In: FOCS, pp. 249–259 (2007)
[DDN91]   Dolev, D., Dwork, C., Naor, M.: Non-malleable cryptography (extended
           abstract). In: STOC, pp. 542–552 (1991)
[GGH+13]   Garg, S., Gentry, C., Halevi, S., Raykova, M., Sahai, A., Waters, B.: Can-
           didate indistinguishability obfuscation and functional encryption for all
           circuits. In: FOCS, pp. 40–49 (2013)
[GGJS12]   Garg, S., Goyal, V., Jain, A., Sahai, A.: Concurrently secure computation
           in constant rounds. In: Pointcheval, D., Johansson, T. (eds.) EUROCRYPT
           2012. LNCS, vol. 7237, pp. 99–116. Springer, Heidelberg (2012)
[GJO10]   Goyal, V., Jain, A., Ostrovsky, R.: Password-authenticated session-key
           generation on the internet in the plain model. In: Rabin, T. (ed.)
           CRYPTO 2010. LNCS, vol. 6223, pp. 277–294. Springer, Heidelberg (2010),
           http://research.microsoft.com/en-us/um/people/vipul/pke.pdf
[GL90]   Goldwasser, S., Levin, L.A.: Fair computation of general functions in pres-
           ence of immoral majority. In: Menezes, A., Vanstone, S.A. (eds.) Advances
           in Cryptology - CRYPTO 1990. LNCS, vol. 537, pp. 77–93. Springer,
           Heidelberg (1991)
[GLP+15]   Goyal, V., Lin, H., Pandey, O., Pass, R., Sahai, A.: Round-efficient con-
           currently composable secure computation via a robust extraction lemma.
           In: Dodis, Y., Nielsen, J.B. (eds.) TCC 2015, Part I. LNCS, vol. 9014, pp.
           260–289. Springer, Heidelberg (2015); Full version of this work available as
           IACR Eprint Report 2012/652
[GMW87]   Goldreich, O., Micali, S., Wigderson, A.: How to play ANY mental game.
           In: ACM (ed.) Proc. 19th STOC, pp. 218–229 (1987), See [Gol04, ch. 7]
           for more details

[GMW91]   Goldreich, O., Micali, S., Wigderson, A.: Proofs that yield nothing but their validity or all languages in NP have zero-knowledge proof systems. Journal of the ACM 38(3), 691–729 (1991), Preliminary version in FOCS 1986

[Gol04]   Goldreich, O.: Foundations of Cryptography: Basic Applications. Cambridge University Press (2004)

[Goy11]   Goyal, V.: Constant round non-malleable protocols using one way functions. In: STOC, pp. 695–704 (2011), See full version available on http://research.microsoft.com/en-us/people/vipul/nmcom.pdf

[Goy12]   Goyal, V.: Positive results for concurrently secure computation in the plain model. In: FOCS, pp. 695–704 (2012)

[GS09]   Goyal, V., Sahai, A.: Resettably secure computation. In: Joux, A. (ed.) EUROCRYPT 2009. LNCS, vol. 5479, pp. 54–71. Springer, Heidelberg (2009)

[IPS15]   Ishai, Y., Pandey, O., Sahai, A.: Public-coin differing-inputs obfuscation and its applications. In: Dodis, Y., Nielsen, J.B. (eds.) TCC 2015, Part II. LNCS, vol. 9015, pp. 668–697. Springer, Heidelberg (2015)

[Kat07]   Katz, J.: Universally composable multi-party computation using tamper-proof hardware. In: Naor, M. (ed.) EUROCRYPT 2007. LNCS, vol. 4515, pp. 115–128. Springer, Heidelberg (2007)

[KP01]   Kilian, J., Petrank, E.: Concurrent and resettable zero-knowledge in polylogarithm rounds. In: Proc. 33th STOC, pp. 560–569 (2000), Preliminary full version published as cryptology ePrint report 2000/013

[Lin04]   Lindell, Y.: Lower bounds for concurrent self composition. In: Naor, M. (ed.) TCC 2004. LNCS, vol. 2951, pp. 203–222. Springer, Heidelberg (2004)

[LP09]   Lin, H., Pass, R.: Non-malleability amplification. In: STOC, pp. 189–198 (2009)

[LP11]   Lin, H., Pass, R.: Constant-round non-malleable commitments from any one-way function. In: STOC (2011)

[LP12]   Lin, H., Pass, R.: Black-box constructions of composable protocols without set-up. In: Safavi-Naini, R., Canetti, R. (eds.) CRYPTO 2012. LNCS, vol. 7417, pp. 461–478. Springer, Heidelberg (2012)

[LPTV10]   Lin, H., Pass, R., Tseng, W.-L.D., Venkitasubramaniam, M.: Concurrent non-malleable zero knowledge proofs. In: Rabin, T. (ed.) CRYPTO 2010. LNCS, vol. 6223, pp. 429–446. Springer, Heidelberg (2010)

[MNPS04]   Malkhi, D., Nisan, N., Pinkas, B., Sella, Y.: Fairplay - secure two-party computation system. In: USENIX Security Symposium, pp. 287–302 (2004)

[MOSV06]   Micciancio, D., Ong, S.J., Sahai, A., Vadhan, S.P.: Concurrent zero knowledge without complexity assumptions. In: Halevi, S., Rabin, T. (eds.) TCC 2006. LNCS, vol. 3876, pp. 1–20. Springer, Heidelberg (2006)

[MPR06]   Micali, S., Pass, R., Rosen, A.: Input-indistinguishable computation. In: FOCS, pp. 367–378 (2006)

[MR91]   Micali, S., Rogaway, P.: Secure computation. In: Feigenbaum, J. (ed.) CRYPTO 1991. LNCS, vol. 576, pp. 392–404. Springer, Heidelberg (1992)

[OPV10]   Ostrovsky, R., Pandey, O., Visconti, I.: Efficiency preserving transformations for concurrent non-malleable zero knowledge. In: Micciancio, D. (ed.) TCC 2010. LNCS, vol. 5978, pp. 535–552. Springer, Heidelberg (2010)

[Pas03]   Pass, R.: Simulation in quasi-polynomial time, and its application to protocol composition. In: Biham, E. (ed.) EUROCRYPT 2003. LNCS, vol. 2656, pp. 160–176. Springer, Heidelberg (2003)

[PLV12] Pass, R., Lin, H., Venkitasubramaniam, M.: A unified framework for UC from only OT. In: Wang, X., Sako, K. (eds.) ASIACRYPT 2012. LNCS, vol. 7658, pp. 699–717. Springer, Heidelberg (2012)

[PPS15] Pandey, O., Prabhakaran, M., Sahai, A.: Obfuscation-Based Non-black-box Simulation and Four Message Concurrent Zero Knowledge for NP. In: Dodis, Y., Nielsen, J.B. (eds.) TCC 2015, Part II. LNCS, vol. 9015, pp. 638–667. Springer, Heidelberg (2015)

[PPV08] Pandey, O., Pass, R., Vaikuntanathan, V.: Adaptive one-way functions and applications. In: Wagner, D. (ed.) CRYPTO 2008. LNCS, vol. 5157, pp. 57–74. Springer, Heidelberg (2008)

[PR05a] Pass, R., Rosen, A.: Concurrent non-malleable commitments. In: FOCS (2005)

[PR05b] Pass, R., Rosen, A.: New and improved constructions of non-malleable cryptographic protocols. In: STOC (2005)

[PRS02] Prabhakaran, M., Rosen, A., Sahai, A.: Concurrent zero knowledge with logarithmic round-complexity. In: FOCS (2002)

[PS04] Prabhakaran, M., Sahai, A.: New notions of security: achieving universal composability without trusted setup. In: STOC, pp. 242–251 (2004)

[PTV08] Pass, R., Tseng, W.-L.D., Venkitasubramaniam, M.: Concurrent zero knowledge: Simplifications and generalizations (2008) (manuscript), http://hdl.handle.net/1813/10772

[PV08] Pass, R., Venkitasubramaniam, M.: On constant-round concurrent zero-knowledge. In: Canetti, R. (ed.) TCC 2008. LNCS, vol. 4948, pp. 553–570. Springer, Heidelberg (2008)

[PW01] Pfitzmann, B., Waidner, M.: A model for asynchronous reactive systems and its application to secure message transmission. In: IEEE Symposium on Security and Privacy (2001)

[RK99] Richardson, R., Kilian, J.: On the concurrent composition of zero-knowledge proofs. In: Stern, J. (ed.) EUROCRYPT 1999. LNCS, vol. 1592, pp. 415–432. Eurocrypt 1999, Heidelberg (1999)

[Ros04] Rosen, A.: The Round-Complexity of Black-Box Concurrent Zero-Knowledge. PhD thesis, Department of Computer Science and Applied Mathematics, Weizmann Institute of Science, Rehovot, Israel (2004)

[Sah99] Sahai, A.: Non-malleable non-interactive zero knowledge and adaptive chosen-ciphertext security. In: Proc. 40th FOCS, pp. 543–553 (1999)

[Wee10] Wee, H.: Black-box, round-efficient secure computation via non-malleability amplification. In: FOCS (2010)

[Yao86] Yao, A.C.-C.: How to generate and exchange secrets. In: Proc. 27th FOCS, pp. 162–167 (1986)

An Alternative Approach to Non-black-box Simulation in Fully Concurrent Setting

Susumu Kiyoshima

NTT Secure Platform Laboratories, Japan
kiyoshima.susumu@lab.ntt.co.jp

Abstract. We give a new proof of the existence of public-coin concurrent zero-knowledge arguments for $\mathcal{NP}$ in the plain model under standard assumptions (the existence of one-to-one one-way functions and collision-resistant hash functions), which was originally proven by Goyal (STOC'13).

In the proof, we use a new variant of the non-black-box simulation technique of Barak (FOCS'01). An important property of our simulation technique is that the simulator runs in a straight-line manner in the fully concurrent setting. Compared with the simulation technique of Goyal, which also has such a property, the analysis of our simulation technique is (arguably) simpler.

1 Introduction

Zero-knowledge Proofs and Non-black-box Simulation. *Zero-knowledge* (ZK) *proofs* [13], with which the prover can convince the verifier of the correctness of a mathematical statement without providing any additional knowledge, have played fundamental roles in cryptography. In particular, ZK protocols[1] have been used as building blocks in many cryptographic protocols, and techniques developed for them have been used in a variety of fields of cryptography.

Traditionally, the security of all ZK protocols was proven via *black-box simulation*. That is, their zero-knowledge property was proven by showing a simulator that uses the adversary only in a black-box way. Although such a simulator can get advantage only through the rewinding of the adversary, black-box simulation is known to be powerful enough to construct ZK protocols with a variety of additional properties, security, and efficiency.

Black-box simulation has, however, inherent limitations. For example, let us consider *public-coin* ZK *protocols*—the ZK protocols such that in each round the verifier sends only the outcome of its coin-tossing—and *concurrent* ZK *protocols*—the ZK protocols such that the zero-knowledge property holds even when the adversary concurrently interacts with many provers in an arbitrary schedule. It was shown that public-coin ZK protocols and concurrent ZK protocols can be constructed with black-box simulation techniques [12,22,15,21]. However, it was also shown that neither of them can be constructed with black-box simulation technique if we additionally require round efficiency. Concretely, it was shown that no *constant-round* public-coin ZK protocol and no $o(\log n/ \log \log n)$-*round* concurrent ZK protocol can be proven secure via

---

[1] We use "ZK protocols" to denote ZK proofs and arguments.

Y. Dodis and J.B. Nielsen (Eds.): TCC 2015, Part I, LNCS 9014, pp. 290–318, 2015.

black-box simulation [11,5]. Furthermore, it was also shown that no public-coin concurrent ZK protocol can be proven secure via black-box simulation irrespective to its round complexity [20].

Because of these impossibility results on black-box simulation, developing *non-black-box simulation* techniques is an important research direction. Developing non-black-box simulation techniques is however considered to be a significantly hard task since non-black-box simulation seems to require the "reverse engineering" of the adversary.

The first non-black-box simulation technique was proposed in a groundbreaking work of Barak [1]. The simulation technique of Barak is completely different from previous ones. In particular, in the simulation technique of Barak, the simulator runs in a "straight-line" manner, i.e., it does not rewind the adversary. With his non-black-box simulation technique, Barak showed that we can go beyond the black-box simulation barrier; in particular, Barak constructed the first constant-round public-coin ZK protocol, which cannot be proven secure via black-box simulation.

Non-black-box Simulation in the Concurrent Setting. Since we can overcome the black-box impossibility result of constant-round public-coin ZK protocols by using Barak's non-black-box simulation technique, it is natural to expect that we can also overcome other black-box impossibility results by using Barak's technique. In particular, since Barak's simulation technique works in a straight-line manner and therefore completely removes the issue of *recursive rewinding* [9], it is natural to think that we can overcome the black-box impossibility results of $o(\log n/ \log \log n)$-round concurrent ZK protocols and public-coin concurrent ZK protocols by using Barak's simulation technique in the concurrent setting.

Unfortunately, Barak's non-black-box simulation technique does not work in the concurrent setting. Although Barak's simulation technique can be extended so that it can handle *bounded-concurrent execution* [1] (i.e., a concurrent execution such that there is an a-priori upper-bound on the number of concurrent sessions) and *parallel execution* [19], it had been open for years to extend Barak's simulation technique so that it can handle fully concurrent execution.

Recently, several works showed that with a trusted setup or non-standard assumptions, Barak's simulation technique can be extended so that it can handle fully concurrent execution. These works then showed that, with their extended simulation techniques, we can overcome the black-box impossibility results of $o(\log n/ \log \log n)$-round concurrent ZK protocols and public-coin concurrent ZK protocols. For example, Canetti, Lin, and Paneth [6] constructed a public-coin concurrent ZK protocol in the *global hash function (GHF) model*, where a single hash function is used in all concurrent sessions. Also, Chung, Lin, and Pass [7] constructed a constant-round concurrent ZK protocol by assuming the existence of $\mathcal{P}$-*certificates* (i.e., "succinct" non-interactive proofs/arguments for $\mathcal{P}$), and Pandey, Prabhakaran, and Sahai [17] constructed a constant-round concurrent ZK protocols by assuming the existence of *differing-input indistinguishability obfuscations*.

Additionally, Goyal [14] showed that even in the plain model under standard assumptions, Barak's non-black-box simulation technique can be extended so that it can handle fully concurrent execution. With his simulation technique, then, Goyal constructed the first public-coin concurrent ZK protocol in the plain model under standard assumptions (the existence of a family of collision-resistant hash functions). Like the original simulation technique of Barak and many of its variants, the simulation technique of Goyal has a straight-line simulator; thus, in the simulation technique of Goyal the simulator performs straight-line concurrent simulation. Because of this straight-line concurrent simulation property, the simulation technique of Goyal has huge potential. In fact, Goyal notes in [14] that his technique can be used to obtain new results on concurrently secure multi-party computation and blind signatures.

Thus, we currently have several good positive results on non-black-box simulation in the concurrent setting, and in particular we have a one that has a straight-line concurrent simulator even in the plain model under standard assumptions [14].[2] However, the state-of-the-art is still not satisfactory and there are many open problems to be addressed. (For example, the simulation technique of Goyal [14] requires the protocol to have $O(n^\epsilon)$ rounds, where $\epsilon > 0$ is an arbitrary constant. Thus, the problem of constructing $o(\log n / \log \log n)$-round concurrent ZK protocols in the plain model under standard assumptions is still open.) Thus, studying more on non-black-box simulation and developing new non-black-box simulation techniques in the concurrent setting is an important research direction.

1.1   Our Result

The main result of this paper is the new non-black-box simulation that we develop to give a new proof of the following theorem, which was originally proven by Goyal [14].

Theorem. *Assume the existence of one-to-one one-way functions and a family of collision resistant hash functions. Then, for any constant $\epsilon > 0$, there exists an $O(n^\epsilon)$-round public-coin concurrent zero-knowledge argument of knowledge.*

Like the simulation technique of Goyal [14], our simulation technique is based on Barak's simulation technique, can handle fully concurrent execution in the plain model under standard assumptions, and has a simulator that runs in a straight-line manner even in the fully concurrent setting. Our non-black-box simulation technique requires the same hardness assumption and the same round complexity as that of Goyal, and therefore it does not lead to immediate improvement over the result of Goyal. Nonetheless, our simulation technique is meaningful since the simulation technique of ours is different from that of Goyal and the analysis of our simulation technique is (in our opinion) simpler than the analysis of Goyal's technique. Since there is only a limited number of non-black-box simulation techniques that can handle fully concurrent execution (and in particular there is only one straight-line concurrent simulation technique in the plain model under standard assumptions), constructing a new non-black-box simulation technique in the concurrent setting is meaningful even when there is no improvement. We hope that our technique leads to further study on non-black-box simulation in the concurrent setting.

---

[2] Also, in their groundbreaking works [3,4], Bitansky and Paneth showed a non-black-box simulation technique that is *not* based on Barak's simulation technique.

Brief Overview of Our Technique. Our public-coin concurrent ZK protocol is based on the public-coin concurrent ZK protocol of Canetti, Lin, and Paneth (CLP) [6], which is secure in the global hash function model. Below, we give a brief overview of our technique, assuming familiarity with Barak's non-black-box simulation technique and the techniques of CLP. In Section 2, we give a more detailed overview of our technique, including the explanation of the techniques of Barak and CLP.

The protocol of CLP is similar to the ZK protocol of Barak except that it has multiple "slots" (i.e., pairs of a prover's commitment and a receiver's random-string message). In any of these slots, the simulator can generate a PCP-proof as a trapdoor witness for the universal argument (UA). Thus, with multiple slots, the simulator can choose which slot to use in the generation of the PCP-proof, and therefore by using a good "proving strategy" that determines which slot to use, the simulator can avoid the blow-up of its running time and can generate PCP-proofs in all sessions in polynomial time even in the concurrent setting. The proving strategy that CLP uses is similar in spirit to the *oblivious rewinding strategy* of [15,21] (in which black-box concurrent ZK protocols are constructed). In particular, in the proving strategy of CLP, the transcript is recursively divided into blocks and then PCP-proofs are generated only at the end of the blocks.

A problem that CLP encountered is that there is only one opportunity for the simulator to give a UA-proof in each session and therefore the simulator need to remember all previously generated PCP-proofs during the simulation. Because of this problem, the length of the PCP-proofs can be rapidly blowing up in the concurrent setting and therefore the size of the simulator cannot be bounded by a polynomial. In [6], CLP solved this problem in the global hash function model by cleverly using the global hash function in UA.

To solve this problem in the plain model, we modify the protocol of CLP so that the simulator also has multiple opportunities to give UA-proofs. We then show that, by using a good proving strategy that also determines which opportunity the simulator takes to give UA-proofs, the simulator can avoid the blow-up of its size as well as its running time. (Our proving strategy works so that a PCP-proof that is generated at the end of a block is used only until the end of the "parent block" of this block; thus, the simulator need to remember PCP-proofs only for a limited time, and therefore the length of the PCP-proofs does not blow up.) This proving strategy is the core of our simulation technique and the main deference between the simulation technique of ours and that of Goyal [14]. (The simulator of Goyal also has multiple opportunities to give UA-proofs, and it determines which opportunity to take by using a proving strategy that is different from ours.) Interestingly, the strategy that we use here is *deterministic* (whereas the strategy that Goyal uses is probabilistic). Because of the use of this deterministic strategy, when we show that every session is successfully simulated, we need to use only a simple counting argument. Because of this, the analysis of our simulation technique is quite simple.

2   Overview of Our Technique

As briefly described in Section 1.1, our protocol is based on the protocol of Canetti et al. [6], which in turn is based on Barak's non-black-box zero-knowledge protocol [1]. Below, we first recall the protocols of [1,6] and then give an overview of our protocol.

2.1 Known Techniques

Barak's Protocol. Roughly speaking, Barak's non-black-box zero-knowledge protocol BarakZK proceeds as follows.

Protocol BarakZK

1. The verifier V chooses a hash function $h \in \mathcal{H}_n$ and sends h to the prover P.
2. P sends $c \leftarrow \mathsf{Com}(0^n)$ to V, where Com is a statistically binding non-interactive commitment scheme. Then, V sends random string r to P. In the following, the pair (c, r) is called a *slot*.
3. P proves the following statement by using a witness-indistinguishable argument.
   - $x \in L$, or
   - $(h, c, r) \in \Lambda$, where $(h, c, r) \in \Lambda$ holds if and only if there exists a machine Π such that c is a commitment to $h(\Pi)$ and Π outputs r in $n^{\log \log n}$ steps.[3]

Note that the statement proven in Step 3 is not in $\mathcal{NP}$. Thus, P proves this statement by a witness-indistinguishable *universal argument* (WIUA), with which P can prove any statement in $\mathcal{NEXP}$.

Intuitively, the security of BarakZK is proven as follows. The soundness is proven by showing that $\Pi(c) \neq r$ holds with overwhelming probability when a cheating prover P^* commits to $h(\Pi)$ for a machine Π. The zero-knowledge property is proven by using a simulator that commits to $h(\Pi)$ such that Π is a machine that emulates the cheating verifier V^*; since $\Pi(c) = V^*(c) = r$ holds from the definition, the simulator can give a valid proof in WIUA. This simulator runs in polynomial time since, from the property of WIUA, the running time of the simulator during WIUA is bounded by $\mathsf{poly}(t)$, where t is the running time of $\Pi(c)$.

Barak's Protocol in the Concurrent Setting. The proof of the ZK property of BarakZK does not work in the concurrent setting. In particular, the above simulator does not work in the concurrent setting since we have $V^*(c) \neq r$ when V^* receives messages during a slot (i.e., when V^* receives messages in other sessions before sending r).

A potential approach for achieving concurrent ZK property with BarakZK is to use a simulator S that commits to a machine that emulates S itself. A key observation behind this approach is that although V^* can receive unbounded number of messages during a slot, all of these messages are generated by S. Thus, if the committed machine Π emulates S from the point that V^* receives c to the point that V^* sends r, Π can output r even when V^* receives many messages during a slot.

This approach, however, causes a problem in simulator's running time. For example, let us consider the following "nested concurrent sessions" schedule (Fig. 1).

- The i-th session is executed so that the $(i + 1)$-th session is completely contained in the slot of the i-th session. That is, the $(i + 1)$-th session starts after V^* receives c in the i-th session, and the $(i + 1)$-th session ends before V^* sends r in the i-th session.

---

[3] Here, $n^{\log \log n}$ can be replaced with any super-polynomial function. We use $n^{\log \log n}$ for concreteness.

Let m be the number of sessions, and let t be the running time of S during the simulation of the m-th session. Then, to simulate the $(m - 1)$-th session, S need to run at least $2t$ steps—t steps for simulating the slot (which contains the m-th session) and t steps for simulating WIUA. Then, to simulate the $(m - 2)$-th session, S need to run at least $4t$ steps—$2t$ steps for simulating the slot and $2t$ steps for simulating WIUA. In general, to simulate the i-th session, S need to run at least $2^{m-i}t$ steps. Thus, the running time of S becomes super-polynomial when $m = \omega(\log n)$.

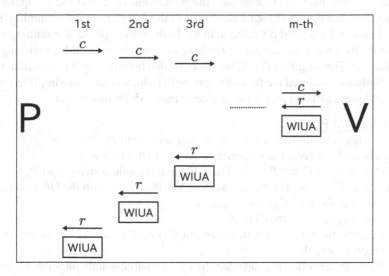

Fig. 1. The "nested concurrent sessions" schedule

Protocol of Canetti et al. [6]. To avoid the blow-up of the simulator's running time, Canetti, Lin, and Paneth (CLP) [6] used the "multiple slots" approach that was originally used in the black-box concurrent zero-knowledge protocols of [22,15,21]. The idea is that with many sequential slots, S can choose any of them as a witness in WIUA, and therefore with a good *proving strategy* that determines which slot to use as a witness, S can avoid the nested computation in WIUA. To implement this approach, CLP first observed that the four-round public-coin UA of [2], from which WIUA can be constructed, can be divided into the *offline phase* and the *online phase* such that all heavy computation is done in the offline phase. As explained later, this online/offline property enables S to perform all heavy computations only at specific points during the simulation, which is crucial to avoid the nested computation in WIUA. Concretely, the UA of [2] is divided as follows. Let $x \in L$ be the statement to be proven in UA and w be a witness for $x \in L$.

Offline/online UA
- Offline Phase:
  1. V sends a random hash function $h \in \mathcal{H}_n$ to P.
  2. P generates a PCP-proof π of statement $x \in L$ by using w as a witness. Then, P computes $\mathsf{UA}_2 := h(\pi)$. In the following, (h, π, UA_2) is called the *offline proof*.

- Online Phase:
  1. P sends UA_2 to V.
  2. V chooses randomness ρ for the PCP-verifier and sends $UA_3 := \rho$ to P.
  3. P computes queries Q by executing the PCP-verifier with statement $x \in L$ and randomness ρ.[4] Then, P computes the replies for queries Q and sends them to V. We denote these replies by UA_4.
  4. V verifies the correctness of the replies by executing the PCP-verifier.

Note that the only heavy computation—the generation of π and the computation of $h(\pi)$—is performed in the offline phase. Thus, in the online phase, the running time of P can be bounded by a fixed polynomial in n.[5] In the offline phase, the running time of P is bounded by a fixed polynomial in t, where t is the time needed for verifying $x \in L$ with witness w. The length of the offline proof is also bounded by a polynomial in t.

CLP [6] then considered the following protocol (which is an over-simplified version of their final protocol). Let N_{slot} be a parameter that is determined later.

Protocol BasicCLP
Stage 1. V chooses a hash function $h \in \mathcal{H}_n$ and sends h to P.
Stage 2. For each $i \in [N_{slot}]$ in sequence, P and V do the following.
- P sends $c_i \leftarrow \mathsf{Com}(0^n)$ to V. Then, V sends a random string r_i to P.
Stage 3. P and V execute the special-purpose WIUA of [18] with the UA system of [2] being used as the underlying UA system.
  1. P sends $c_{UA} \leftarrow \mathsf{Com}(0^n)$ to V.
  2. V sends the third UA message UA_3 to P (i.e., V sends a random string of appropriate length).
  3. P proves the following statement by using a witness-indistinguishable proof of knowledge (WIPOK).
     - $x \in L$, or
     - there exists $i \in [N_{slot}]$ and the second and the fourth UA messages UA_2, UA_4 such that UA_2 is the committed value of c_{UA} and (h, UA_2, UA_3, UA_4) is an accepting proof of the statement $(h, c_i, r_i) \in \Lambda$.

Recall that the idea of the multiple-slot approach is that S avoids the nested computation in WIUA by using a proving strategy that determines which slot to use as a witness. Thus, roughly speaking, the simulation proceeds as follows: First, S commits to a machine Π in each slot and then computes an offline proof (in particular, a PCP-proof) w.r.t. a slot chosen according to a proving strategy; then, S commits to the second UA message (i.e., the hash of the PCP-proof) in Stage 3-1 and gives a valid WIPOK proof in Stage 3-3. As a proving strategy that determines which slot to use as a witness, CLP considered a strategy that is similar in spirit to the oblivious rewinding strategy of [15,21]. In this strategy, the entire transcript of all sessions is recursively divided into blocks. Let M be the total number of messages across the sessions, and let q be a parameter called the *splitting factor*. Assume for simplicity that M is a power of q, i.e., $M = q^d$ for $d \in \mathbb{N}$. Then, the entire transcript is divided into blocks as follows.

---

[4] Recall that the PCP-verifier performs the verification by making a few queries to the PCP-proof.
[5] Here, P is assumed to have random access to π.

- The level-d block is the entire transcript of all sessions. Thus, the level-d block contains $M = q^d$ messages.
- Then, the level-d block is divided into q sequential blocks, where each block contains q^{d-1} messages. These blocks are called the level-$(d-1)$ blocks.
- Similarly, each level-$(d-1)$ block is divided into q sequential blocks, where each block contains q^{d-2} messages. These blocks are called the level-$(d-2)$ blocks.
- In this way, each block is continued to be divided into q blocks until level-0 blocks are obtained. A level-0 block contains only a single message.

Then, at the end of each block of each level, S computes offline proofs w.r.t. all slots that are contained in this block. Note that when $q = n^\epsilon$ for a constant ϵ, the maximum level of blocks (i.e., d) is constant. Thus we have at most constant level of nesting in the execution of WIUA. Furthermore, it was shown by CLP that when $N_{slot} = \omega(q) = \omega(n^\epsilon)$, the simulator does not "get stuck," i.e., at least one offline proof is computed before Stage 3 begins in every session except with negligible probability.

The protocol BasicCLP is, however, not concurrent zero-knowledge in the plain model since the size of S's state can become super-polynomial. Recall that in the simulation, S generates an offline proof in Stage 2 and uses it in Stage 3. Then, since V^* can choose any concurrent schedule (and therefore can delay the execution of Stage 3), S need to remember all previously generated offline proofs during its execution. Thus, each committed machine need to contain all previously generated offline proofs, and therefore an offline proof w.r.t. a slot (which is generated by using a machine committed in this slot as a witness) is as long as the total length of all offline proofs that are generated before this slot. Thus, the length of offline proofs can be rapidly blowing up and therefore the size of S's state cannot be bounded by a polynomial.

A key observation by CLP [6] is that this problem can be solved in the *global hash model*, in which a global hash function is shared by all protocol executions. Roughly speaking, CLP avoids the blow-up of the simulator's size by considering machines that contain only the hash of the offline proofs; then, to guarantee that the simulation works with such machines, they modified BasicCLP so that P proves in WIUA that $x \in L$ or the committed machine outputs r given an access to the *hash-inversion oracle*; in the simulation, S commits to a machine that emulates S by recovering offline proofs from the hash value with the hash-inversion oracle. In this modified protocol, the soundness is proven by using the fact that the same hash function is used across all sessions.

In this way, CLP [6] obtained a public-coin concurrent zero-knowledge protocol in the global hash model. Since $q = n^\epsilon$ and $N_{slot} = \omega(q)$, the round complexity is $O(n^{\epsilon'})$ for a constant ϵ'. (Since ϵ is an arbitrary constant, ϵ' can be an arbitrary small constant.) CLP also showed that by modifying the protocol further, the round complexity can be reduced to $O(\log^{1+\epsilon} n)$.

2.2 Our Techniques

We obtain an $O(n^\epsilon)$-round protocol by removing the use of a global hash function from the protocol of CLP [6]. Recall that in the protocol of CLP, a global hash function is used to avoid the blow-up of the simulator's state size. In particular, a global hash function is used so that the simulation works even when the committed machines do not contain previously computed offline proofs. Below, to obtain our protocol, we first modify the machines to be committed by the simulator (and slightly modify BasicCLP and the simulator accordingly). The modified machines do not contain previously generated offline proofs and therefore their sizes are bounded by a fixed polynomial. We then modify BasicCLP and the simulator so that the simulation works even when the simulator commits to the modified machines.

In the following, we set $q := n^\epsilon$ and $N_{\text{slot}} := \omega(q)$.

Modification on the Machines to be Committed. We modify machines so that they emulate the simulator *not from the start of a slot but from a more prior point of the simulation*; thus, the modified machines emulate more part of the simulation than before. Intuitively, if a machine emulates more part of the simulation, it potentially generates more offline proofs by itself, and therefore more likely to be able to output r even when it contains no offline proof. For example, let us consider an extreme case that each committed machine emulates the simulator from the beginning of the simulation. In this case, each committed machine generates every offline proofs by itself, and therefore it can output r even when it contains no offline proof. Unfortunately, in this case the running time of the simulator becomes super-polynomial since the running time of each committed machine is too long. Thus, we need to consider machines that do not emulate too much of the simulation.

Concretely, we consider a machine that emulates the simulator *from the beginning of a block*. In particular, for each $i \in [n]$, we consider the following machine Π_i.

- Π_i emulates the simulator from the beginning of the level-i block that contains the commitment in which Π_i is committed. Π_i does not contain any previously generated offline proofs, and if the emulation fails due to the lack of the offline proofs, Π_i terminates and outputs fail.

(Recall that the maximum level of the blocks is $d < n$.) Then, we modify BasicCLP so that P gives n commitments in parallel in each slot, and let the simulator commit to Π_i in the i-th commitment. More precisely, the simulator does the following. In the interaction with V^*, for each $i \in [d]$, we say that a level-i block is the *current level-i block* if it will contain the next-scheduled message. In each slot, we call the i-th commitment the i-th *column*.

- In each slot, in the i-th column for each $i \in [n]$, the simulator commits to machine Π_i, which emulates the simulator from the beginning of the current level-i block.
- At the end of each block in each level, for each slot that is contained in this block, the simulator generates the offline proof w.r.t. this slot by using a machine that emulates the simulator from the beginning of this block. Note that such a machine must have been committed in each slot.

(A machine that emulates S from the beginning of a block is already considered in [6] for a different purpose. In [6], such a machine is used to reduce the round complexity. Here, we use such a machine to avoid the blow-up of the simulator's size.)

When the simulator commits to these machines, the running time of the simulator can be bounded by a polynomial in n as follows. First, since each committed machine contains no offline proofs, the size of each committed machine is bounded by a fixed polynomial. Then, let t_i be the maximum time spent by the simulation of a level-i block. (Thus, at the end of a level-i block, each offline proof can be computed in time poly(t_i).) Then, since a level-i block contains q level-$(i-1)$ blocks, and since at most $m := \mathsf{poly}(n)$ offline proofs are generated at the end of each level-$(i-1)$ block, we have

$$t_i \leq q \cdot (t_{i-1} + m \cdot \mathsf{poly}(t_{i-1})) \leq \mathsf{poly}(t_{i-1}) \ .$$

Then, since the maximum level $d = \log_q M$ is constant and since we have $t_0 = \mathsf{poly}(n)$, we have $t_d = \mathsf{poly}(n)$. Thus, the running time of the simulator is bounded by a polynomial in n.

We note that although the above machines do not contain any previously generated offline proofs, they do contain all previously generated witnesses of WIPOK (i.e., UA_2 and UA_4).[6] As explained below, allowing the machines to contain all previously generated witnesses is crucial to obtain a protocol and a simulator with which the simulation works even when the modified machines are committed.

Modifications on the Protocol and the Simulator. When the above machines are committed, the simulation may fail since the committed machines can output fail. In particular, the simulation fails if there exists a block in which the simulator uses an offline proof that are generated before this block starts. (If such a block exists, the machines that are committed in this block output fail since they do not contain the necessary offline proof.) Thus, to guarantee successful simulation, we need to make sure that in each block, the simulator uses only the offline proofs that are generated in this block. Of course, we also need to make sure that the simulator does not "get stuck," i.e., we need to guarantee that in each session, the simulator computes a valid witness of WIPOK before WIPOK starts.

To avoid the simulation failure, we first modify BasicCLP as follows. As noted in the previous paragraph, we need to construct a simulator such that in each block, the simulator uses only the offline proofs that are generated in this block. In BasicCLP, it is hard to construct such a simulator since an offline proof may be used long after it is generated. (Recall that during the simulation, offline proofs are generated in Stage 2 and they are used in Stage 3 to compute witnesses of WIPOK.) Thus, we modify BasicCLP so that the simulator can use offline proofs soon after generating them. In particular, we modify BasicCLP so that the simulator can use the offline proofs in Stage 2. Toward this end, we first observe the following.

- The simulator can compute a witness of WIPOK from the offline proof anytime after Stage 3-2.
- The pair of Stage 3-1 and Stage 3-2 is syntactically the same as a slot: P sends a commitment in Stage 3-1 and V sends a random string in Stage 3-2. Thus, we can merge Stage 3-1 and Stage 3-2 into sequential slots.[7][8]

---

[6] Since the length of the witnesses of WIPOK is bounded by a fixed polynomial, the size of the machines does not blow up even when they contain all previously generated witnesses of WIPOK.

[7] The idea of merging a part of special purpose WIUA into slots is also used in [8] for different purpose. In [8], this idea is used to reduce the round complexity.

[8] Alternatively, we can also think of executing the pair of Stage 3-1 and Stage 3-2 in parallel with each slot.

Following these observations, we modify BasicCLP and obtain the following protocol. (As stated before, we also modify BasicCLP so that P gives parallel commitments in each slot.)

Protocol OurZK

Stage 1. V chooses a hash function $h \in \mathcal{H}_n$ and sends h to P.

Stage 2. For each $i \in [N_{\text{slot}}]$ in sequence, P and V do the following.

- P sends $c_{i,1} \leftarrow \text{Com}(0^n), \ldots, c_{i,n} \leftarrow \text{Com}(0^n)$ to V. Then, V sends a random string r_i to P.

Stage 3. P proves the following statement with WIPOK.

- $x \in L$, or
- there exist $i_1, i_2 \in [N_{\text{slot}}]$, $j \in [n]$, and the second and the fourth UA message UA_2 and UA_4 such that UA_2 is the committed value of $c_{i_2,j}$ and $(h, \text{UA}_2, r_{i_2}, \text{UA}_4)$ is an accepting proof of the statement $(h, c_{i_1,j}, r_{i_1}) \in \Lambda$.

In OurZK, a witness of WIPOK can be computed in a session if there are two slots such that (i) a machine is committed in a slot and (ii) an offline proof w.r.t. this slot is committed in the other slot. The computation of the WIPOK witness can be done anytime after such two slots, and after that, the offline proof will never be used.

We next modify the simulator as follows. Recall that, as noted above, we need the simulator such that (i) each committed machine does not output fail due to the lack of offline proofs, and (ii) the simulator does not get stuck.

Roughly speaking, our simulator does the following (see Fig. 2). Recall that for each $i \in \{0, \ldots, d-1\}$, a level-$(i+1)$ block is divided into q level-i blocks. Then, in each level-$(i+1)$ block, for each session, our simulator first tries to obtain a level-i block that contains a slot in which a machine is committed in the i-th column. If it succeeds in obtaining such a level-i block, our simulator computes an offline proof w.r.t. this slot. Next, our simulator tries to obtain a level-i block that contains a slot in which this offline proof is committed in the i-th column. If it succeeds in obtaining such a level-i block, our simulator computes a witness of WIPOK from this offline proof.

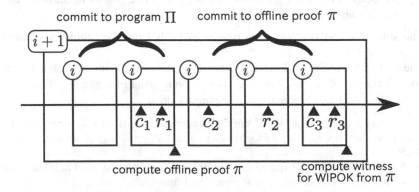

Fig. 2. Our simulator's strategy, when splitting factor is $q = 5$

More precisely, we consider the following simulator. In what follows, for each $i \in \{0, \ldots, d-1\}$, we say that two level-i blocks are *sibling* if they are contained by the same level-$(i+1)$ block.

- In each slot of each session, in the i-th column for each $i \in [n]$, the simulator commits to a machine that emulates the simulator from the beginning of the current level-i block *if no sibling of the current level-i block contains a slot of this session*; if there exists a sibling that contains a slot of this session, an offline proof must have been computed at the end of this sibling (see below), and the simulator commits to this offline proof.
- At the end of each level-i block for each $i \in \{0, \ldots, d-1\}$, the simulator does the following for every slot that is contained in this block: if a machine is committed in the i-th column of this slot, the simulator computes an offline proof by using the committed machine; if an offline proof is committed in the i-th column of this slot, the simulator computes a witness of WIPOK from this offline proof.
- When WIPOK starts, the simulator does the following: if the simulator have a valid witness, the simulator give a valid proof with this witness; if the simulator does not have a valid witness, the simulator aborts with output stuck.

Note that the simulator can compute a witness of WIPOK if there exists a block in which the simulator obtains two lower-level blocks such that each of them contains a slot.

We first note that each committed machine does not fail due to the lack of offline proofs. This follows immediately from the fact that in each block, the simulator uses only offline proofs that are generated in this block.

Thus, it remains to show that the simulator does not get stuck, i.e., the simulator has a valid witness when each WIPOK starts. Below, we use the following terminology.

- A block is *good* w.r.t. a session if it contains a slot of this session and does not contain the first message of WIPOK of this session.
- For each $i \in [d]$, we say that a level-$(i-1)$ block is a *child* of a level-i block if the former is contained by the latter. (Thus, each block has q children.)

From the construction, the simulator does not get stuck if for each session that reaches WIPOK in the simulation, there exists a block that has at least two children that are good w.r.t. this session. Thus, we show that for each session that reaches WIPOK in the simulation, there exists a block that has at least two children that are good w.r.t. this session. To prove this, it suffices to show that for each session that reaches WIPOK in the simulation, there exists a block such that it has at least three children that contain a slot of this session. (This is because at most one child contains the first message of WIPOK.) Assume for contradiction that there exists a session such that it reaches WIPOK and every block has at most two children that contain a slot of this session. Let $C(i)$ be the maximum number of slots that are contained by a level-i block. Then, since in each block there are at most $q-1$ slots that are contained by the block but do not contained by its children, we have

$$C(i) \leq 2C(i-1) + q - 1 \ .$$

Then, since $C(0) = 0$ and since the maximum level d is constant, we have

$$C(d) \leq 2^d C(0) + \sum_{i=0}^{d-1} 2^i (q - 1) = O(q) .$$

This means that in the entire transcript there are at most $O(q)$ slots of this session. Since $N_{slot} = \omega(q)$, this contradicts to the assumption that this session reaches WIPOK. Thus, for each session that reaches WIPOK, there exists a block that has at least two children that are good w.r.t. this session. Thus, the simulator does not get stuck.

Since $q = O(n^{\epsilon})$ and $N_{slot} = \omega(q)$, the round complexity of our protocol is $O(n^{\epsilon'})$ for a constant $\epsilon' > \epsilon$. Since ϵ is an arbitrary constant, ϵ' can be an arbitrary small constant.

Toward the Final Protocol. To obtain a formal proof of security, we need to add a slight modification to the above protocol. In particular, as pointed out in previous works [14,6,7,17], when the code of the simulator is committed in the simulation, we have to take special care to the randomness of the simulator.[9] Fortunately, the techniques used in the previous works can also be used here to overcome this problem. If we use the technique of [6,7], which requires only one-way functions, we can remove the requirement of one-to-one one-way functions from our result. However, to simplify the analysis, in this paper we use the technique of [14], which requires one-to-one one-way functions.

2.3 Comparison with the Simulation Technique of Goyal [14]

In this section, we compare the simulation technique of ours with that of Goyal [14], which is the only known simulation technique that realizes straight-line concurrent simulation in the plain model under standard assumptions.

Since both the simulation technique of ours and that of Goyal are based on Barak's non-black-box simulation technique, there are many similarities between them: For example, the simulator commits to a machine that emulates itself; the protocol is modified so that it has multiple slots; the simulator is given multiple opportunities to give UA proofs[10]; the blocks are used to determine which opportunity the simulator takes to give UA proofs.

However, there are also differences between them. A notable difference is how the simulator determines which opportunity to take to give UA proofs. Recall that, in the simulation technique of ours, the strategy that the simulator uses to determine whether it embeds a UA message in a slot is *deterministic* (the simulator checks whether a sibling of the current block contains a slot; see Fig. 2 in Section 2.2). In contrast, in the

---

[9] When the code of the simulator is committed, the randomness used for generating this commitment is also committed; thus, if a protocol is designed naively, we need a commitment scheme such that the committed value is hidden even when it contains the randomness used for the commitment.

[10] In our work, the simulator is given multiple opportunities to give UA proofs by modifying the protocol so that the encrypted UA is merged into the sequential execution of slots. In [14], the simulator is given multiple opportunities to give UA proofs by modifying the protocol so that the encrypted UA is explicitly executed multiple times.

simulation technique of Goyal, the strategy that the simulator uses is *probabilistic* (the simulator uses a probabilistic procedure that performs the "marking" of the blocks and the UA messages). Since in the simulation technique of ours the simulator uses a deterministic strategy, the analysis of our simulator is simple: We use only a simple counting argument (and no probabilistic argument) to show that the simulator will not get stuck.

3 Preliminary

We assume familiarity to the definition of basic cryptographic protocols, such as interactive proofs of knowledge, witness-indistinguishable proofs, and commitment schemes.

3.1 Notations

We use n to denote the security parameter. For any $k \in \mathbb{N}$, let $[k] \overset{\text{def}}{=} \{1, \ldots, k\}$. For any randomized algorithm Algo, we use $\mathsf{Algo}(x; r)$ to denote the execution of Algo with input x and randomness r. We use $\mathsf{Algo}(x)$ to denote the execution of Algo with input x and uniformly chosen randomness.

3.2 Assumptions

We assume the existence of a family of collision-resistant hash functions $\mathcal{H} = \{h_\alpha\}_{\alpha \in \{0,1\}^*}$. Let $\mathcal{H}_n \overset{\text{def}}{=} \{h_\alpha \in \mathcal{H} : \alpha \in \{0,1\}^n\}$. Then, we require that $\mathcal{H}_n$ satisfies the following properties.

- For any $h \in \mathcal{H}_n$, the domain of h is $\{0,1\}^*$ and the range of h is $\{0,1\}^n$.
- For any $h \in \mathcal{H}_n$, $x \in \{0,1\}^{\mathsf{poly}(n)}$, and $i \in \{1, \ldots, |x|\}$, after computing $h(x)$, we can compute a short certificate $\sigma_i \in \{0,1\}^{n^2}$ for the fact that the i-th bit of x is x_i.

We can achieve these properties by using the Merkle hash tree.

We also assume the existence of one-to-one one-way function f. Recall that from one-to-one one-way function f, we can construct perfectly binding non-interactive commitment scheme Com, where $\mathsf{Com}(b; r) = (f(r), \mathsf{hc}(r) \oplus b)$ for $b \in \{0,1\}$ and the hard-core bit hc of f. (See, e.g., [10].) Note that each valid commitment of Com has a unique decommitment.

3.3 Concurrent Zero-Knowledge

We recall the definition of *concurrent zero-knowledge*. For any polynomial $m(\cdot)$, *m-session concurrent cheating verifier* is a PPT Turing machine V^* such that on input $(1^n, x, z)$, V^* concurrently interacts with $m(n)$ independent copies of P. The interaction between V^* and each copy of P is called *session*. There is no restriction on how V^* schedules messages among sessions, and V^* can abort some sessions. Let $\mathsf{view}_{V^*}\langle P(w), V^*(z)\rangle(1^n, x)$ be the view of V^* in the above concurrent execution, where 1^n and $x \in L$ are the common inputs, $w \in \mathbf{R}_L(x)$ is the private input of P, and z is the auxiliary input of V^*.

Definition 1 (Concurrent Zero-Knowledge). *An interactive proof or argument $\langle P, V \rangle$ for language L is* concurrent zero-knowledge *if for every polynomial $m(\cdot)$ and every m-session concurrent cheating verifier V^*, there exists a PPT simulator S such that following are computationally indistinguishable.*

- $\{\mathsf{view}_{V^*}\langle P(w), V^*(z)\rangle(1^n, x)\}_{n \in \mathbb{N}, x \in L \cap \{0,1\}^{\mathrm{poly}(n)}, w \in R_L(x), z \in \{0,1\}^*}$
- $\{S(1^n, x, z)\}_{n \in \mathbb{N}, x \in L \cap \{0,1\}^{\mathrm{poly}(n)}, w \in R_L(x), z \in \{0,1\}^*}$

$\Diamond$

3.4 PCP and Universal Argument

We recall the definitions of *probabilistically checkable proof* (PCP) systems and *universal argument* system.

Universal Language $L_{\mathcal{U}}$. For simplicity, we show the definitions of PCPs and universal arguments that prove only the membership of a single "universal" language $L_{\mathcal{U}}$. For triplet $y = \langle M, x, t \rangle$, we have $y \in L_{\mathcal{U}}$ if non-deterministic machine M accepts x within t steps. Let $R_{\mathcal{U}}$ be the witness relation of $L_{\mathcal{U}}$, i.e., $R_{\mathcal{U}}$ is a polynomial-time decidable relation such that $y = \langle M, x, t \rangle \in L_{\mathcal{U}}$ if and only if there exists $w \in \{0,1\}^{\leq t}$ such that $(y, w) \in R_{\mathcal{U}}$. Note that every language $L \in \mathcal{NP}$ is linear-time reducible to $L_{\mathcal{U}}$ via mapping $x \mapsto \langle M_L, x, 2^{|x|} \rangle$, where M_L is any fixed non-deterministic polynomial-time machine that decides L. Thus, a proof system for $L_{\mathcal{U}}$ allows us to handle all $\mathcal{NP}$ statements.[11]

PCP System. Roughly speaking, a PCP system is a PPT verifier that can decide the correctness of a statement $y \in L_{\mathcal{U}}$ given access to an oracle π that represents a proof in a redundant form. Typically, the verifier reads only few bits of π in the verification.

Definition 2 (PCP system—basic definition). *A probabilistically checkable proof (PCP) system (with a negligible soundness error) is a PPT oracle machine V (called verifier) that satisfies the following:*

- **Completeness:** *For every $n \in \mathbb{N}$ and every $y \in L_{\mathcal{U}} \cap \{0,1\}^{\mathrm{poly}(n)}$, there exists an oracle π such that*

$$\Pr\left[V^\pi(1^n, y) = 1\right] = 1 \ .$$

- **Soundness:** *For every $n \in \mathbb{N}$, every $y \in \{0,1\}^{\mathrm{poly}(n)} \setminus L_{\mathcal{U}}$, and every oracle π, there exists a negligible function $\mathsf{negl}(\cdot)$ such that*

$$\Pr\left[V^\pi(1^n, y) = 1\right] < \mathsf{negl}(n) \ .$$

$\Diamond$

---

[11] In fact, every language in $\mathcal{NEXP}$ is polynomial-time reducible to $L_{\mathcal{U}}$.

In this paper, we use PCP systems as a building block in the universal argument UA of [2]. To be used in UA, PCP systems need to satisfy four auxiliary properties: relatively efficient oracle construction, non-adaptive verifier, efficient reverse sampling, and proof of knowledge. To understand this paper, the definitions of the first two properties are required; for the definitions of other properties, see [2].

Definition 3 (PCP system—auxiliary properties). *Let V be a PCP-verifier.*

- *Relatively efficient oracle construction: There exists an algorithm P (called prover) such that, given any $(y, w) \in R_{\mathcal{U}}$, algorithm P outputs an oracle π_y that makes V always accepts (i.e., as in the completeness condition). Furthermore, there exists a polynomial $p(\cdot)$ such that on input (y, w), the running time of P is $p(|y| + |w|)$.*
- *Non-adaptive verifier: The verifier's queries are determined based only on the input and its internal coin tosses, independently of the answers given to previous queries. That is, V can be decomposed into a pair of algorithms Q and D such that on input y and random tape r, the verifier makes the query sequence $Q(y, r, 1), Q(y, r, 2), \ldots, Q(y, r, p(|y|))$, obtains the answers $b_1, \ldots, b_{p(|y|)}$, and decides according to $D(y, r, b_1 \cdots b_{p(|y|)})$, where p is some fixed polynomial.*

$\Diamond$

Universal Argument. Universal arguments [2], which are closely related to the notion of CS proofs [16], are "efficient" arguments of knowledge for proving the membership in $L_{\mathcal{U}}$. For $y = \langle M, x, t \rangle \in L_{\mathcal{U}}$, let $T_M(x, w)$ be the running time of M on input x with witness w, and let $R_{\mathcal{U}}(y) \stackrel{\text{def}}{=} \{w : (y, w) \in R_{\mathcal{U}}\}$.

Definition 4 (Universal argument). *A pair of interactive Turing machines $\langle P, V \rangle$ is a universal argument system if it satisfies the following properties:*

- *Efficient verification: There exists a polynomial p such that for any $y = \langle M, x, t \rangle$, the total time spent by (probabilistic) verifier strategy V on inputs 1^n and y is at most $p(n + |y|)$.*
- *Completeness by a relatively efficient prover: For every $n \in \mathbb{N}$, $y = \langle M, x, t \rangle \in L_{\mathcal{U}} \cap \{0, 1\}^{\text{poly}(n)}$, and $w \in R_{\mathcal{U}}(y)$, it holds that*

$$\Pr\left[\langle P(w), V \rangle (1^n, y) = 1\right] = 1 .$$

Furthermore, there exists a polynomial q such that the total time spent by P, on input $(1^n, y, w)$, is at most $q(n + |y| + T_M(x, w)) \leq q(n + |y| + t)$.
- *Computational Soundness: For every PPT Turing machine P^*, there is a negligible function $\text{negl}(\cdot)$ such that for every $n \in \mathbb{N}$, $y = \langle M, x, t \rangle \in \{0, 1\}^{\text{poly}(n)} \setminus L_{\mathcal{U}}$, and $z \in \{0, 1\}^*$, it holds that*

$$\Pr\left[\langle P^*(z), V \rangle (1^n, y) = 1\right] < \text{negl}(n) .$$

- *Weak Proof of Knowledge: For every polynomial $p(\cdot)$ there exists a polynomial $p'(\cdot)$ and a PPT oracle machine E such that the following holds: For every PPT*

Turing machine P^*, every sufficiently large $n \in \mathbb{N}$, every $y = \langle M, x, t \rangle \in \{0, 1\}^{\mathsf{poly}(n)}$, and every $z \in \{0, 1\}^*$, if $\Pr\left[\langle P^*(z), V \rangle(1^n, y) = 1\right] > 1/p(n)$, then

$$\Pr_r\left[\exists w = w_1 \cdots w_t \in \mathbf{R}_{\mathcal{U}}(y) \text{ s.t. } \forall i \in [t], E_r^{P^*(1^n, y, z)}(1^n, y, i) = w_i\right] > \frac{1}{p'(n)}$$

where $E_r^{P^*(1^n, y, z)}(\cdot, \cdot, \cdot)$ denotes the function defined by fixing the randomness of E to equal r, and providing the resulting E_r with oracle access to $P^*(1^n, y, z)$.   ◇

The weak proof-of-knowledge property of universal arguments only guarantees that each individual bit w_i of some witness w can be extracted in probabilistic polynomial time. Given an input 1^n and $y = \langle M, x, t \rangle \in L_{\mathcal{U}} \cap \{0, 1\}^{\mathsf{poly}(n)}$, since the witness $w \in \mathbf{R}_{\mathcal{U}}(y)$ is of length at most t, it follows that there exists an extractor running in time polynomial in $\mathsf{poly}(n) \cdot t$ that extracts the whole witness; we refer to this as the *global proof-of-knowledge property* of a universal argument.

In this paper, we use the public-coin four-round universal argument system UA of [2] (Fig. 3). As observed in [6], the construction of UA can be separated into an expensive *offline stage* and an efficient *online stage*. In the online stage, the running time of the prover is polynomial in n, and in the offline stage, the running time of the prover is $\mathsf{poly}(n + |y| + T_M(x, w))$. In this paper, the third message UA_3 satisfies $|\mathsf{UA}_3| = n \cdot \mathsf{poly}(\log |y|) \le n^2$ and the fourth message UA_4 satisfies $|\mathsf{UA}_4| = \mathsf{poly}(n)$.

- **Input:** The common input of the prover P and the verifier V is $y = \langle M, x, t \rangle \in L_{\mathcal{U}}$. The private input of P is $w \in \mathbf{R}_{\mathcal{U}}(y)$.
- **Offline Phase:**
  1. V sends a random hash function $h \in \mathcal{H}_n$ to P.
  2. P generates a PCP-proof π of statement $y \in L_{\mathcal{U}}$ by using w as a witness. Then P computes $\mathsf{UA}_2 := h(\pi)$. The tuple (h, π, UA_2) is called the *offline proof*.
- **Online Phase:**
  1. P sends UA_2 to V.
  2. V chooses randomness $\rho \in \{0, 1\}^{n^2}$ for the PCP-verifier and sends $\mathsf{UA}_3 := \rho$ to P.
  3. P computes queries Q by executing the PCP-verifier with statement $y \in L_{\mathcal{U}}$ and randomness ρ. Then, P sends $\mathsf{UA}_4 := \{(i, \pi_i, \sigma_i)\}_{i \in Q}$ to V, where π_i is the i-th bit of π and σ_i is a certificate that the i-th bit of π is indeed π_i.
  4. V verifies the correctness of all certificates, and checks whether the PCP-verifier accepts on input $(y, \{(i, \pi_i)\}_{i \in Q})$ with randomness ρ.

Fig. 3. Online/offline UA system of [2,6]

---

4   Our Public-Coin Concurrent Zero-Knowledge Argument

Theorem 1. *Assume the existence of one-to-one one-way functions and a family of collision resistant hash functions. Then, for any constant $\epsilon > 0$, there exists an $O(n^\epsilon)$-round public-coin concurrent zero-knowledge argument of knowledge* cZKAOK.

Proof. cZKAOK is shown in Fig. 4, where the following building blocks are used in cZKAOK.

- Perfectly binding non-interactive commitment scheme Com such that each valid commitment of Com has a unique decommitment. (As noted in Section 3.2, such a commitment scheme can be constructed from one-to-one one-way functions.)
- Constant-round public-coin witness-indistinguishable proof of knowledge WIPOK.
- Four-round public-coin universal argument UA of [2] (Fig. 3 in Section 3.4).

Clearly, cZKAOK is public-coin and its round complexity is $O(n^\epsilon)$. Thus, Theorem 1 follows from the following lemmas.

Lemma 1. cZKAOK *is concurrently zero-knowledge.*

Lemma 2. cZKAOK *is argument of knowledge.*

Lemma 1 is proven in Section 4.1 and Lemma 2 is proven in Section 4.2.   □

Remark 1. The languages Λ_2 in Fig. 5 is slightly over-simplified and will make cZKAOK work only when $\mathcal{H}$ is collision resistant against $\mathsf{poly}(T(n))$-time adversaries. To make it work assuming collision resistance against polynomial-time adversaries, one should use a "good" error-correcting code ECC (i.e., with constant relative distance and with polynomial-time encoding and decoding), and replacing the condition $C = \mathsf{Com}(h(\Pi); R)$ with $C = \mathsf{Com}((|\mathsf{ECC}(\Pi)|, h(\mathsf{ECC}(\Pi))); R)$.

Input: The input of the prover P is (x, w), where $x \in L$ and $w \in \mathbf{R}_L(x)$. The input of the verifier V is x.

Parameter: An integer $N_{\mathrm{slot}} = O(n^\epsilon)$.

Stage 1: The verifier V chooses a random hash function $h \in \mathcal{H}_n$ and sends h to the prover P.

Stage 2: For each $i \in [N_{\mathrm{slot}}]$ in sequence, P and V do the following.

    1. P computes $C_{i,j} \leftarrow \mathsf{Com}(0^n)$ for each $j \in [n]$. Then, P sends $C_i = (C_{i,1}, \ldots, C_{i,n})$ to V.

    2. V sends random $r_i \in \{0, 1\}^{n^2}$ to P.

Stage 3: P proves the following by using WIPOK.

    - $x \in L$, or

    - $\langle h, C_1, r_1, \ldots, C_{N_{\mathrm{slot}}}, r_{N_{\mathrm{slot}}} \rangle \in \Lambda_1$, where language Λ_1 is shown in Fig. 5.

Fig. 4. Public-coin concurrent zero-knowledge argument cZKAOK

4.1 Concurrent Zero-knowledge Property

Proof (of Lemma 1). Let V^* be a cheating verifier. Without loss of generality, we assume that V^* is deterministic. Let $m(\cdot)$ be a polynomial such that V^* invokes $m(n)$ concurrent sessions. Let $q \overset{\text{def}}{=} n^{\epsilon/2}$. We assume without loss of generality that in the interaction between V^* and provers, the total number of messages across all sessions is always the power of q (i.e., it is q^d for an integer d). Note that since the number of messages is polynomially bounded, we have $d = \log_q(\mathsf{poly}(n)) = O(1)$.

Language Λ_1: (statement for WIPOK) $\langle h, C_1, r_1, \ldots, C_{N_{slot}}, r_{N_{slot}} \rangle \in \Lambda_1$ if and only if there exist

- $i_1, i_2 \in [N_{slot}]$ and $j \in [n]$ such that $i_1 < i_2$
- the second and fourth UA messages $\mathsf{UA}_2 \in \{0, 1\}^n$ and $\mathsf{UA}_4 \in \{0, 1\}^{\mathsf{poly}(n)}$
- randomness $R \in \{0, 1\}^{\mathsf{poly}(n)}$ for Com

such that

- $C_{i_2,j} = \mathsf{Com}(\mathsf{UA}_2; R)$, and
- $(h, \mathsf{UA}_2, r_{i_2}, \mathsf{UA}_4)$ is an accepting proof of $\langle h, C_{i_1,j}, r_{i_1} \rangle \in \Lambda_2$.

Language Λ_2: Let $T(\cdot)$ be a slightly super-polynomial function (say, $T(n) = n^{\log \log n}$). Then, $\langle h, C, r \rangle \in \Lambda_2$ if and only if there exist

- an oracle machine Π such that $|\Pi| \leq T(n)$
- a string τ such that $|\tau| \leq T(n)$
- randomness $R \in \{0, 1\}^{\mathsf{poly}(n)}$ for Com
- a string y such that $|y| \leq T(n)$

such that

- $C = \mathsf{Com}(h(\Pi); R)$, and
- r is a substring of τ, and
- Π^O outputs τ within $T(n)$ steps, where O is an oracle that receives a commitment of Com and returns the (unique) decommitment of this commitment, and
- In the execution of Π^O, for every query $\tilde{\rho}$ from Π to O, there exists $(\tilde{\rho}, \rho, r) \in y$ such that $\tilde{\rho} = \mathsf{Com}(\rho; r)$ (i.e., (ρ, r) is the decommitment of $\tilde{\rho}$).

Fig. 5. Languages used in cZKAOK

Simulator S

Before describing our simulator S, we first introduce *blocks*. The level-d block is defined to be the entire transcript of all sessions. Then, for $\ell \in \{0, \ldots, d - 1\}$, level-$\ell$ blocks are defined by dividing each level-$(\ell + 1)$ block into q sequential blocks of equal length. Thus, for every $\ell \in \{0, \ldots, d\}$, a level-$\ell$ block contains q^ℓ messages.

We next introduce a subroutine SOLVE, which generates a simulated transcript by recursively executing itself. (On input $\ell \in [d]$, SOLVE generates a simulated transcript of a level-ℓ block by executing itself q times on input $\ell - 1$.) The details of SOLVE is described below.

We give SOLVE an oracle access to O, where O is the oracle that is defined in the definition of Λ_2 (Fig. 5). Roughly speaking, we give SOLVE an access to O to avoid the issue of randomness sketched in Section 2.2.[12] Specifically, we give SOLVE only "encrypted randomness" $\tilde{\rho}$, which is a Com commitment to which true randomness ρ is committed. SOLVE computes the committed value ρ from $\tilde{\rho}$ by using O and then uses

---

[12] This technique is borrowed from [14].

ρ as randomness in the simulation of Com and WIPOK. (When SOLVE is used in $\mathcal{S}$, oracle O is emulated by $\mathcal{S}$ in polynomial time.)

We also give SOLVE the following input $(x, z, \ell, \text{trans}, \mathsf{V}, \mathsf{W}, \tilde{\rho})$:

- x and z are the input of V^*.
- $\ell \in \{0, \ldots, d\}$ is an execution level.
- trans $\in \{0, 1\}^{\mathsf{poly}(n)}$ is a partial transcript that was simulated so far. The goal of SOLVE is to simulate subsequent q^ℓ messages after trans.
- $\mathsf{V} = \{v_{s,j}\}_{s \in [m], j \in [n]}$ is the values to be committed in Com in the simulation. (In a slot of the s-th session, $v_{s,1}, \ldots, v_{s,n}$ are committed.) Each $v_{s,j}$ is $\perp$ if the value to be committed has not been determined yet.
- $\mathsf{W} = \{w_s\}_{s \in [m]}$ is the WIPOK witnesses that are computed so far. Each w_s is $\perp$ if the witness of the s-th session has not been computed yet.
- $\tilde{\rho} = (\tilde{\rho}_1, \ldots, \tilde{\rho}_{q^d})$ is a vector of "encrypted" randomness (i.e., commitments of Com), where the randomness decrypted from $\tilde{\rho}_i$ is used to simulate the i-th messages.

The output of SOLVE is $(\text{trans}', \mathsf{W}', \Pi^{(\cdot)})$, where trans' is the simulated messages that are generated by this execution of SOLVE, W' is the updated table of the WIPOK witnesses, and $\Pi^{(\cdot)}$ is a machine that emulates this execution of SOLVE and outputs trans'.

Given the above input, SOLVE does the following. Below, we use a function car that takes a triple (a, b, c) as input and outputs a.

$\text{SOLVE}^O(x, z, \ell, \text{trans}, \mathsf{V}, \mathsf{W}, \tilde{\rho})$

- When $\ell = 0$ (the base case), do the following.
  - If the next-scheduled message msg is a verifier message, feed (x, z, trans) to V^* and receive msg from V^*.
  - If the next-scheduled message msg is a prover message, do the following. Let init $\in [q^d]$ be the index of msg across all sessions (thus, trans contains init $- 1$ messages). Parse $(\tilde{\rho}_1, \ldots, \tilde{\rho}_{q^d}) \leftarrow \tilde{\rho}$ and compute the committed value ρ_{init} of $\tilde{\rho}_{\mathsf{init}}$ by using O. Then, compute msg as follows with randomness ρ_{init}.
    * If msg is a message of Com in the s-th session for $s \in [m]$, compute the following for each $j \in [n]$.

$$C_j \leftarrow \begin{cases} \mathsf{Com}(v_{s,j}) & \text{if } v_{s,j} \neq \perp \\ \mathsf{Com}(0^n) & \text{otherwise} \end{cases}$$

    Then, set $\text{msg} := (C_1, \ldots, C_n)$.
    * If msg is the first message of WIPOK in the s-th session for $s \in [m]$, honestly compute msg by using witness w_s. (If w_s is not a valid witness, aborts with output stuck.) If msg is another message of WIPOK, honestly compute msg by reconstructing the prover state of WIPOK from trans, W, and $\tilde{\rho}$.
  - Output $(\text{msg}, \mathsf{W}, \Pi^{(\cdot)})$, where $\Pi^{(\cdot)}$ is a machine that computes $\text{car}(\text{SOLVE}^{(\cdot)}(x, z, \ell, \text{trans}, \mathsf{V}, \mathsf{W}, \tilde{\rho}))$.

– When $\ell > 0$, do the following.

Step 1: Updating values to be committed. Let $\Pi^{\langle \cdot \rangle}$ be a machine that computes car(SOLVE$^{\langle \cdot \rangle}(x, z, \ell, \text{trans}, \mathsf{V}, \mathsf{W}, \tilde{\rho})$). Then, for every $s \in [m]$ such that $v_{s,\ell} = \bot$ holds and the s-th session has already started in trans, update V by setting $v_{s,\ell} := h_s(\Pi)$, where h_s is the hash function used in the s-th session.

Step 2: Initializing temporary variables. Let $\text{ctr}_s := 0$ and $\text{tmp}_s := \bot$ for every $s \in [m]$.

Step 3 to Step $2q + 2$: For each $k \in [q]$, do the following:

Step $2k + 1$: Executing the k-th child-block. Compute

$$(\text{trans}_k, \mathsf{W}_k, \Pi_k) \leftarrow \text{SOLVE}^O(x, z, \ell - 1, \text{trans}, \mathsf{V}, \mathsf{W}, \tilde{\rho}) .$$

Then, update $\text{trans} := \text{trans} \,\|\, \text{trans}_k$ and $\mathsf{W} := \mathsf{W}_k$. Let y_k be the set of all query-answer pairs with O during this recursive execution.

Step $2k + 2$: For each $s \in [m]$ such that (i) trans_k includes a slot sl of the s-th session and (ii) the s-th session has started before trans_k, do the following.[13]

Case 1. When $\text{ctr}_s = 0$, do the following.

1. Let i_1 be the *slot-index* of slot sl (i.e., $i_1 \in [N_{\text{slot}}]$ s.t. slot sl is the i_1-th slot of the s-th session). Let (C_{i_1}, r_{i_1}) denote slot sl, where $C_{i_1} = (C_{i_1,1}, \ldots, C_{i_1,n})$.

2. **Computing offline proof.** By using $\tilde{\rho}$ and O, compute the randomness R_1 used for $C_{i_1,\ell-1}$. Then, compute PCP-proof π_s of statement $\langle h_s, C_{i_1,\ell-1}, r_{i_1} \rangle \in \Lambda_2$ with witness $\langle \Pi_k, \text{trans}_k, R_1, y_k \rangle$ and compute $\mathsf{UA}_2 := h_s(\pi_s)$.

   COMMENT: SOLVE *and* S *are designed so that the committed value of* $C_{i_1,\ell-1}$ *is* Π_k, *which outputs* trans_k *by emulating the recursive execution of* SOLVE *of Step* $2k + 1$.

3. **Updating values to be committed.** Update V by setting $v_{s,\ell-1} := h_s(\pi_s)$.

4. Update $\text{tmp}_s := (i_1, \pi_s, \mathsf{UA}_2)$.

5. Update $\text{ctr}_s := \text{ctr}_s + 1$.

Case 2: When $\text{ctr}_s = 1$, do the following.

1. Let i_2 be the slot-index of slot sl and let (C_{i_2}, r_{i_2}) denote slot sl, where $C_{i_2} = (C_{i_2,1}, \ldots, C_{i_2,n})$.

2. **Completing UA.** Parse $(i_1, \pi_s, \mathsf{UA}_2) \leftarrow \text{tmp}_s$. Then, compute the fourth UA message UA_4 from offline proof $(h_s, \pi_s, \mathsf{UA}_2)$ and the third UA message r_{i_2}.

3. Let R_2 be the randomness used for $C_{i_2,\ell-1}$. Then, update W by setting $w_s := \langle i_1, i_2, \ell - 1, \mathsf{UA}_2, \mathsf{UA}_4, R_2 \rangle$.

   COMMENT: SOLVE *and* S *are designed so that the committed value of* $C_{i_2,\ell-1}$ *is* $\mathsf{UA}_2 = h_s(\pi_s)$.

4. Update $\text{ctr}_s := \text{ctr}_s + 1$.

Step $2q + 3$: Output $(\text{trans}_1 \,\|\, \text{trans}_2 \,\|\, \cdots \,\|\, \text{trans}_q, \mathsf{W}, \Pi^{\langle \cdot \rangle})$.

With SOLVE, the simulator S is defined as follows.

---

[13] We note that if trans_k includes a slot, we have $|\text{trans}_k| \geq 1$ and thus we have $\ell \geq 2$.

$\underline{S(1^n, x, z)}$

1. For every $i \in [q^d]$, choose random $\rho_i \in \{0, 1\}^n$ and $r \in \{0, 1\}^{\text{poly}(n)}$, and compute $\tilde{\rho}_i := \text{Com}(\rho_i; r)$. Let $\tilde{\rho} := (\tilde{\rho}_1, \ldots, \tilde{\rho}_{q^d})$.
2. Let $V := \{v_{s,j}\}_{s \in [m], j \in [n]}$ and $W := \{w_s\}_{s \in [m]}$, where $v_{s,j} = w_s = \perp$ for every $s \in [m]$ and $j \in [n]$.
3. Compute $(\text{trans}', W', \Pi^{(\cdot)}) \leftarrow \text{SOLVE}^{(\cdot)}(x, z, d, \varepsilon, V, W, \tilde{\rho})$, where ε is an empty string. When SOLVE queries $\tilde{\rho}$ for O, find i such that $\tilde{\rho} = \text{Com}(\rho_i; r_i)$ and return (ρ_i, r_i) to SOLVE.[14]
4. Output trans'.

Running Time of S

Lemma 3. $S(x, z)$ *runs in polynomial time.*

Proof. We first show that in each execution of SOLVE, for any $k \in [q]$, the size of Π_k in Step $2k + 1$ is bounded by a fixed polynomial in n. From the construction of SOLVE, Π_k is a machine that computes $\text{car}(\text{SOLVE}^{(\cdot)}(x, z, \ell - 1, \text{trans}, V, W, \tilde{\rho}))$. Then, since the length of $(x, z, \ell - 1, V, W, \text{trans}, \tilde{\rho})$ is bounded by a fixed polynomial in m, the size of Π_k is bounded by a fixed polynomial in m. Thus, the size of Π_k is bounded by a fixed polynomial in n.

We then bound the running time of S as follows. Note that from the constructions of S and SOLVE, each execution of SOLVE can be uniquely identified by the value of ℓ and init $\overset{\text{def}}{=} |\text{trans}| + 1$. For $\ell \in \{0, \ldots, d\}$ and init $\in [q^d]$, we use $\text{SOLVE}_{\ell, \text{init}}$ to denote the execution of SOLVE with input ℓ and init. Let $t_{\ell, \text{init}}$ be the running time of $\text{SOLVE}_{\ell, \text{init}}$, and let $t_\ell \overset{\text{def}}{=} \max_{\text{init}}(t_{\ell, \text{init}})$. Note that in every $\text{SOLVE}_{\ell, \text{init}}$, the running time of the recursive execution of SOLVE in Step $2k + 1$ is at most $t_{\ell-1}$; thus, in Step $2k + 2$, PCP-proof π_s can be computed in time $\text{poly}(t_{\ell-1})$ and the length of π_s is at most $\text{poly}(t_{\ell-1})$ for every $s \in [m]$. Note also that every computation in $\text{SOLVE}_{t, \text{init}}$ can be performed in fixed polynomial time in n except for the following computations:

Type-1 computation. The recursive executions of SOLVE.
Type-2 computation. The generations of the offline proofs (i.e., PCP-proofs and their hash values).

Each type-1 computation can be performed in time $t_{\ell-1}$, and each type-2 computation can be performed in time $\text{poly}(t_{\ell-1})$. Then, since for each $k \in [q]$ there are a single type-1 computation and m type-2 computations, we have

$$t_\ell \le q \cdot (t_{\ell-1} + m \cdot \text{poly}(t_{\ell-1}) + \text{poly}(n)) \le \text{poly}(t_{\ell-1})$$

for any $\ell \in [d]$. Then, since we have $d = O(1)$ and $t_0 = \text{poly}(n)$, we have $t_d = \text{poly}(n)$. Thus, S runs in polynomial time. □

---

[14] From the construction of SOLVE, there must exist such i.

Indistinguishability of Views

Lemma 4. *The output of $S(x,z)$ is computationally indistinguishable from the view of V^*.*

Proof. We prove this lemma by considering a sequence of hybrid experiments. Let H_0 be the real execution of V^* and honest provers. Then, for each $i \in [q^d]$, we consider the following hybrids. In what follows, we use $\mathsf{SOLVE}_{\ell,\mathsf{init}}$ to denote the execution of SOLVE with input ℓ and init (see the proof of Lemma 3).

Hybrid H_i proceeds identically to the execution of S until the end of the execution of $\mathsf{SOLVE}_{0,i}$. At this point, the view of V^* is simulated up until the i-th message across all sessions (inclusive). Let $(\mathsf{msg}, \mathsf{W}, \Pi^{(\cdot)})$ be the output of $\mathsf{SOLVE}_{0,i}$, and let trans be the simulated view of V^* at this point (including msg). Then, after $\mathsf{SOLVE}_{0,i}$ is executed, the view of V^* is simulated from trans as follows: Every message is computed with true randomness (instead of "decrypted" randomness), every commitment is generated by committing to 0^n, every WIPOK that starts after trans is executed with witness for $x \in L$, and every WIPOK that starts in trans is executed as in SOLVE (i.e., by reconstructing the prover state). The output of H_i is the simulated view of V^*.

Note that the output of H_{q^d} is identical to that of S.

To show the indistinguishability between the output of H_i and that of H_{i-1} for each $i \in [q^d]$, we consider a sequence of intermediate hybrid experiments in which H_i is gradually changed to H_{i-1} as follows.

Hybrid $H_{i:1}$ is the same as H_i except that in the execution of $\mathsf{SOLVE}_{0,i}$, the next message msg is computed with true randomness (instead of the one decrypted from $\tilde{\rho}_i$).

Hybrid $H_{i:2}$ is the same as $H_{i:1}$ except that in the execution of $\mathsf{SOLVE}_{0,i}$, if the next message is Com commitments, then the commitments are computed as in the honest prover (i.e., $C_j \leftarrow \mathsf{Com}(0^n)$ for every $j \in [n]$).

Hybrid $H_{i:3}$ is the same as $H_{i:2}$ except that in the execution of $\mathsf{SOLVE}_{0,i}$, if the next message is the first message of WIPOK, then subsequently all messages in this WIPOK are computed with witness for $x \in L$.

Before showing the indistinguishability among these intermediate hybrids, we show the following claim.

Claim 1. *In H_{q^d}, S does not output stuck.*

Proof. We first introduce notation. Recall that in hybrid H_{q^d}, SOLVE is recursively executed many times. We use *block* to denote an execution of SOLVE. A block is in level ℓ if the corresponding SOLVE is executed with input ℓ. For each block, the *child-blocks* of this block are blocks that are recursively executed by this block; thus, each block has q child-blocks. For any slot of any session, we say that a block *contains* this slot if the corresponding execution of SOLVE outputs a transcript that includes this slot (both the prover and the verifier message). A block is *good* w.r.t. a session if this block

(i) contains a slot of this session, (ii) begins after this session begins, and (iii) does not contain the first message of WIPOK of this session.[15]

From the construction, S does not output stuck if for every session that reaches Stage 3, there exists a block such that two of its child-blocks are good. (If there exists such a block, Case 2 of Step $2k + 2$ is executed in this block for this session and therefore a witness of WIPOK is computed.)

Thus, it remains to show that for every session that reaches Stage 3, there exists a block that has two good child-blocks. To show this, it suffices to show that for every session that reaches Stage 3, there exists a block that has four child-blocks that contain slots of this session. (If four child-blocks contain slots, two of them are good since they begin after this session begins and they do not contain the first message of WIPOK.) Assume for contradiction that there exists a session such that every block has at most three child-blocks that contain slots of this session. For $\ell \in \{0, \ldots, d\}$ and init $\in [q^d]$, let $C_{\mathsf{init}}(\ell)$ be the number of slots of this session that are contained by the block corresponding to $\mathsf{SOLVE}_{\ell,\mathsf{init}}$, and let $C(\ell) \overset{\text{def}}{=} \max_{\mathsf{init}}(C_{\mathsf{init}}(\ell))$. Then, since each block contains at most three child-blocks that contain slots, and since in each block there are at most $q - 1$ slots that are contained by this block but do not contained by any of its child-block, we have

$$C(\ell) \le 3C(\ell - 1) + q - 1 \ .$$

Thus, we have

$$
\begin{aligned}
C(d) &\le 3C(d - 1) + q - 1 \\
&\le 3^2 C(d - 2) + 3(q - 1) + q - 1 \\
&\le \cdots \le 3^d C(0) + \sum_{i=0}^{d-1} 3^i (q - 1) \\
&= 3^d C(0) + \frac{1}{2}(3^d - 1)(q - 1) \ .
\end{aligned}
$$

From $d = O(1)$ and $C(0) = 0$, we have $C(d) = O(q)$. Then, since S outputs the view of V^* that was generated by a block of level d, there are at most $O(q) = O(n^{\epsilon/2})$ slots in the simulated view. Then, since we have $N_{\mathsf{slot}} = O(n^\epsilon)$, this contradicts to the assumption that the session reaches Stage 3. $\square$

Now, we are ready to show the indistinguishability among the intermediate hybrids.

Claim 2. *For every $i \in [q^d]$, the output of $H_{i:1}$ is computationally indistinguishable from that of H_i.*

Proof. Recall that $H_{i:1}$ differs from H_i in that ρ_i is replaced with true randomness that is independent of $\tilde{\rho}$. Note that if we replace ρ_i with true randomness, we can no

---

[15] To simplify the description of our simulator, we use a definition of the good block that is slightly different from that given in the overview of our technique (Section 2). (Here, we additionally require that the block begins after the session begins.) We can also use the definition given Section 2 if we modify SOLVE so that the machine that emulates itself is committed even in the sessions that start after it begins.

longer compute PCP-proofs w.r.t. $\mathsf{SOLVE}_{0,i}$ and any execution of SOLVE that contains $\mathsf{SOLVE}_{0,i}$. However, since in both H_i and $H_{i:1}$ no PCP-proof is computed after $\mathsf{SOLVE}_{0,i}$, this causes no problem. Thus, the indistinguishability follows from the hiding property of Com. $\qquad\square$

Claim 3. *For every* $i \in [q^d]$, *the output of* $H_{i:2}$ *is computationally indistinguishable from that of* $H_{i:1}$.

Proof. It suffices to consider the case that the next message msg in $\mathsf{SOLVE}_{0,i}$ is Com commitments. Note that both in $H_{i:1}$ and $H_{i:2}$, the Com commitments are generated with true randomness; furthermore, the committed values of Com and the randomness used in Com are not used anywhere else (in particular, not used in the PCP generations and WIPOK). Thus, the indistinguishability follows from the hiding property of Com. $\quad\square$

Claim 4. *For every* $i \in [q^d]$, *the output of* $H_{i:3}$ *is computationally indistinguishable from that of* $H_{i:2}$.

Proof. It suffices to consider the case that the next message msg in $\mathsf{SOLVE}_{0,i}$ is the first message of WIPOK. Note that both in $H_{i:2}$ and $H_{i:3}$, WIPOK are executed with true randomness that is not used anywhere else; furthermore, from Claim 1, a valid witness is used both in $H_{i:2}$ and $H_{i:3}$. (Recall that $H_{i:2}$ and $H_{i:3}$ proceed identically with H_{q^d} until $\mathsf{SOLVE}_{0,i}$ starts.) Thus, the indistinguishability follows from the witness indistinguishability of WIPOK. $\qquad\square$

Claim 5. *For every* $i \in [q^d]$, *the output of* H_{i-1} *is identically distributed to that of* $H_{i:3}$.

Proof. Note that in $H_{i:3}$, the next message msg in $\mathsf{SOLVE}_{0,i}$ is computed in exactly the same way as in H_{i-1}. Thus, the claim follows. $\qquad\square$

From Claims 2, 3, 4, and 5, the output of H_0 and that of H_{q^d} are computationally indistinguishable. This completes the proof of Lemma 4. $\qquad\square$

This completes the proof of Lemma 1. $\qquad\square$

4.2 Argument of Knowledge Property

As noted in Remark 1, the language Λ_2 shown in Fig. 5 is slightly over-simplified; in particular, the argument-of-knowledge property of cZKAOK can be proven only when $\mathcal{H}$ is collision resistant against $\mathsf{poly}(T(n))$-time adversaries.

Below, we prove the argument-of-knowledge property assuming that $\mathcal{H}$ is collision resistant against $\mathsf{poly}(T(n))$-time adversaries. By using a trick shown in [2], it is easy to extend this proof so that it works under the assumption that $\mathcal{H}$ is collision resistant only against polynomial-time adversaries.

Proof (of Lemma 2, when $\mathcal{H}$ is collision resistant against $\mathsf{poly}(T(n))$-time adversaries).
For any cheating prover P^*, let us consider the following extractor E.

– Given oracle access to P^*, E interacts with P^* as a honest verifier until the start of Stage 3. Then, E uses the extractor of WIPOK to extract a witness w.

To show that E outputs a witness of $x \in L$, it suffices to show that w is a witness of $\langle h, C_1, r_1, \ldots, C_{N_{\text{slot}}}, r_{N_{\text{slot}}} \rangle \in \Lambda_1$ with at most negligible probability. In the following, we use the word "fake witness" to denote a witness for $\langle h, C_1, r_1, \ldots, C_{N_{\text{slot}}}, r_{N_{\text{slot}}} \rangle \in \Lambda_1$. Then, we say that P^* is *bad* if E outputs a fake witness with non-negligible probability. In the following, we show that if there exists a bad cheating prover, we can break the collision resistance of $\mathcal{H}$.

We first show the following claim.

Claim 6. *For any ITM P, let us consider an experiment $\mathsf{Exp}_1(n, P)$ in which P interacts with a verifier V as follows.*

1. **Interactively generating statement.** *First, V sends random $h \in \mathcal{H}_n$ to P. Next, P sends a commitment C of Com to V, and V sends a random $r_1 \in \{0, 1\}^{n^2}$ to P.*
2. **Generating UA proof.** *P sends to V the second UA message UA_2 of statement $\langle h, C, r_1 \rangle \in \Lambda_2$, and V sends to P random $r_2 \in \{0, 1\}^{n^2}$. Then, P sends to V the fourth UA message UA_4.*
3. *We say that P wins in the experiment if $(h, \mathsf{UA}_2, r_2, \mathsf{UA}_4)$ is an accepting UA proof for $\langle h, C, r_1 \rangle \in \Lambda_2$.*

Then, if there exists a bad P^*, there exists PPT ITM P^{**} that wins in $\mathsf{Exp}_1(n, P^{**})$ with non-negligible probability.

Proof. From the assumption that P^* is bad, for infinitely many n we can extract a fake witness from P^* with probability at least $\delta(n) \overset{\text{def}}{=} 1/\mathsf{poly}(n)$. In the following, we fix any such n. Then, from an average argument, there exist $i_1^*, i_2^* \in [N_{\text{slot}}]$ and $j^* \in [n]$ such that with probability at least $\delta'(n) \overset{\text{def}}{=} \delta(n)/n(N_{\text{slot}})^2 > \delta(n)/n^3$, we can extract a fake witness $\langle i_1, i_2, j, \ldots \rangle$ such that $(i_1, i_2, j) = (i_1^*, i_2^*, j^*)$. Then, we consider the following P^{**} that participates in Exp_1.

1. P^{**} internally invokes P^* and interacts with P^* as a honest verifier of cZKAOK with the following exceptions:
   - In Stage 1, P^{**} forwards h from the external V to P^*.
   - In the i_1^*-th slot of Stage 2, P^{**} forwards $C_{i_1^*, j^*}$ from P^* to the external V and forward r_1 from the external V to P^*.
   - In Stage 3, P^{**} extracts a witness w from P^* by using the extractor of WIPOK.
2. If w is not a fake witness of the form $\langle i_1^*, i_2^*, j^*, \ldots \rangle$, P^{**} aborts with output fail. Otherwise, parse $\langle i_1^*, i_2^*, j^*, \mathsf{UA}_2, \mathsf{UA}_4, R \rangle \leftarrow w$. Then, P^{**} sends UA_2 to the external V and receives r_2.
3. P^{**} rewinds the internal P^* to the point that P^* had sent Com in the i_2^*-th slot. Then, P^{**} sends r_2 to P^* as the verifier message of the i_2^*-th slot. Then, P^{**} interacts with P^* as a honest verifier and extracts a witness w' in Stage 3.
4. If w' is not a fake witness of the form $\langle i_1^*, i_2^*, j^*, \ldots \rangle$, P^{**} aborts with output fail. Otherwise, parse $\langle i_1^*, i_2^*, j^*, \mathsf{UA}_2', \mathsf{UA}_4', R' \rangle \leftarrow w'$. Then, P^{**} sends UA_4' to the external V.

To analyze the probability that P^{**} wins in $\mathsf{Exp}_1(n, P^{**})$, we first observe the following. Let trans be the prefix of a transcript of cZKAOK up until the prover-message of the

i_2^*-th slot (inclusive). Then, we say that trans is *good* if under the condition that a prefix of the transcript is trans, a fake witness of the form $\langle i_1^*, i_2^*, j^*, \ldots \rangle$ is extracted from P^* with probability at least $\delta'/2$. From an average argument, the prefix of the transcript is good with probability at least $\delta'/2$ when P^* interacts with a honest verifier of cZKAOK. Then, since a transcript of cZKAOK is perfectly emulated in Step 1 of P^{**}, the prefix of the internally emulated transept is good with probability at least $\delta'/2$.

We next observe that under the condition that the prefix of the internally emulated transcript is good in Step 1 of P^{**}, P^{**} wins in $\mathsf{Exp}_1(n, P^{**})$ with probability at least $(\delta'/2)^2 - \mathsf{negl}(n)$. This follows from the following. First, from the definition of good prefix, both w and w' are fake witnesses of the form $\langle i_1^*, i_2^*, j^*, \ldots \rangle$ with probability at least $(\delta'/2)^2$. Next, when w and w' are fake witnesses, both UA_2 and UA_2' are the committed values of $C_{i_2^*, j^*}$ and thus we have $\mathsf{UA}_2 = \mathsf{UA}_2'$ except with negligible probability; thus, when w and w' are fake witnesses, $(h, \mathsf{UA}_2, r_2, \mathsf{UA}_4')$ is an accepting UA proof except with negligible probability.

Thus, by combining the above two observations, we conclude that the probability that P^{**} wins in $\mathsf{Exp}_1(n, P^{**})$ is at least

$$\frac{\delta'}{2} \left(\left(\frac{\delta'}{2}\right)^2 - \mathsf{negl}(n) \right) \geq \frac{1}{\mathsf{poly}(n)} .$$

$\square$

Next, we show the following claim.

Claim 7. *For any ITM E^*, let us consider an experiment $\mathsf{Exp}_2(n, E^*)$ in which E^* interacts with a verifier V as follows.*

1. ***Interactively generating statement.** This step is the same as Exp_1, where E^* plays as P. Let $\langle h, C, r_1 \rangle$ be the interactively generated statement.*
2. ***Outputting witness.** E outputs $w = \langle \Pi, \tau, R, y \rangle$. We say that E wins in the experiment if w is a valid witness of $\langle h, C, r_1 \rangle \in \Lambda_2$.*

Then, if there exists PPT ITM P^{**} that wins in $\mathsf{Exp}_1(n, P^{**})$ with non-negligible probability, there exists $\mathsf{poly}(T)$-time ITM E^* that wins in $\mathsf{Exp}_2(n, E^*)$ with non-negligible probability.

Proof. We first note that the extractor of UA works even when the statement is interactively generated after h is sent. This is because the extractor of UA extracts a witness by first emulating honest execution till the end and then restarting the execution from the second verifier message. Thus, by simply using the extractor of UA for P^{**}, we obtain E^* that outputs a valid witness of $\langle h, C, r \rangle \in \Lambda_2$ with non-negligible probability. Note that since the running time of the global extractor of UA is $\mathsf{poly}(T(n))$, the running time of E^* is also $\mathsf{poly}(T(n))$. $\square$

Finally, we reach a contradiction by showing that given E^* described in Claim 7, we can break the collision-resistance property of $\mathcal{H}$.

Claim 8. *If there exists* $\mathsf{poly}(T)$*-time ITM* E^* *that wins in* $\mathsf{Exp}_2(n, E^*)$ *with non-negligible probability, there exists* $\mathsf{poly}(T)$*-time machine* $\mathcal{A}$ *that breaks the collision-resistance property of* $\mathcal{H}$.

Proof. We consider the following $\mathcal{A}$.

1. Given $h \in \mathcal{H}$, $\mathcal{A}$ internally invokes E^* and emulates $\mathsf{Exp}_2(n, E^*)$ for E^* perfectly except that $\mathcal{A}$ forwards h to E^* in Step 1. Let $\langle h, C, r \rangle$ and w be the statement and the output of E^* in this emulated experiment.
2. If w is not a valid witness of $\langle h, C, r \rangle \in \Lambda_2$, $\mathcal{A}$ aborts with output fail. Otherwise, parse $\langle \Pi, \tau, R, y \rangle \leftarrow w$.
3. $\mathcal{A}$ rewinds E^* to the point that E^* had sent C, and from this point $\mathcal{A}$ emulates $\mathsf{Exp}_2(n, E^*)$ again with flesh randomness. Let $\langle h, C, r_1' \rangle$ be the statement and w' be the witness in this emulated experiment.
4. If w' is not a valid witness of $\langle h, C, r_1' \rangle \in \Lambda_2$, $\mathcal{A}$ aborts with output fail. Otherwise, let $\langle \Pi', \tau', R', y' \rangle \leftarrow w'$.
5. Then, $\mathcal{A}$ outputs (Π, Π') if $\Pi \neq \Pi'$ and $h(\Pi) = h(\Pi')$. Otherwise, $\mathcal{A}$ outputs fail.

We first show that both w and w' are valid witnesses with non-negligible probability. Note that for infinitely many n, E^* outputs a valid witness in $\mathsf{Exp}_2(n, E^*)$ with probability $\epsilon \stackrel{\text{def}}{=} 1/\mathsf{poly}(n)$. In the following, we fix any such n. Let trans be the prefix of a transcript of $\mathsf{Exp}_2(n, E^*)$ up until E^* sends C (inclusive). Then, we say that trans is *good* if under the condition that a prefix of the transcript is trans, E^* outputs a valid witness with probability at least $\epsilon/2$. Then, from an average argument, the prefix of the internally emulated transcript is good with probability at least $\epsilon/2$. Thus, w and w' are valid witnesses with probability at least $(\epsilon/2)(\epsilon/2)^2 = (\epsilon/2)^3$.

Next, we show that when $\mathcal{A}$ obtains two valid witnesses $w = \langle \Pi, \tau, R, y \rangle$ and $w' = \langle \Pi', \tau', R', y' \rangle$, we have $\Pi \neq \Pi'$ and $h(\Pi) = h(\Pi')$ except with negligible probability. First, from the binding property of Com, we have $h(\Pi) = h(\Pi')$ except with negligible probability. (Recall that from the condition that w and w' are valid witnesses, we have $\mathsf{Com}(h(\Pi); R) = \mathsf{Com}(h(\Pi'); R') = C$.) Next, since r is a substring of τ and r' is a substring of τ', we have $\tau \neq \tau'$ except with negligible probability. (If $\tau = \tau'$ holds, r' is a substring of τ' with probability at most $T(n)/2^n = \mathsf{negl}(n)$ since r' is chosen at random after τ is determined.) Then, since Π^O always outputs τ and Π'^O always outputs τ', we conclude that we have $\Pi \neq \Pi'$ except with negligible probability. (Recall that for each query to O, the reply is uniquely determined.)

From the above two observations, we conclude that $\mathcal{A}$ breaks the collision-resistance property of $\mathcal{H}$. $\quad\square$

From Claims 6, 7, and 8, we conclude that there exists no bad P^*. Thus, the extractor E outputs a witness of $x \in L$ except with negligible probability. $\quad\square$

Acknowledgment. I greatly thank the anonymous reviewers of TCC 2015 for pointing out an error I made in the earlier version of this paper. Their comments also help me to improve the presentation of this paper.

References

1. Barak, B.: How to go beyond the black-box simulation barrier. In: FOCS, pp. 106–115 (2001)
2. Barak, B., Goldreich, O.: Universal arguments and their applications. SIAM J. Comput. 38(5), 1661–1694 (2008)
3. Bitansky, N., Paneth, O.: From the impossibility of obfuscation to a new non-black-box simulation technique. In: FOCS, pp. 223–232 (2012)
4. Bitansky, N., Paneth, O.: On the impossibility of approximate obfuscation and applications to resettable cryptography. In: STOC, pp. 241–250 (2013)
5. Canetti, R., Kilian, J., Petrank, E., Rosen, A.: Black-box concurrent zero-knowledge requires (almost) logarithmically many rounds. SIAM J. Comput. 32(1), 1–47 (2002)
6. Canetti, R., Lin, H., Paneth, O.: Public-coin concurrent zero-knowledge in the global hash model. In: Sahai, A. (ed.) TCC 2013. LNCS, vol. 7785, pp. 80–99. Springer, Heidelberg (2013)
7. Chung, K.M., Lin, H., Pass, R.: Constant-round concurrent zero knowledge from P-certificates. In: FOCS, pp. 50–59 (2013)
8. Chung, K.-M., Ostrovsky, R., Pass, R., Venkitasubramaniam, M., Visconti, I.: 4-round resettably-sound zero knowledge. In: Lindell, Y. (ed.) TCC 2014. LNCS, vol. 8349, pp. 192–216. Springer, Heidelberg (2014)
9. Dwork, C., Naor, M., Sahai, A.: Concurrent zero-knowledge. J. ACM 51(6), 851–898 (2004)
10. Goldreich, O.: Foundations of Cryptography: Volume 1, Basic Tools. Cambridge University Press (August 2001)
11. Goldreich, O., Krawczyk, H.: On the composition of zero-knowledge proof systems. SIAM J. Comput. 25(1), 169–192 (1996)
12. Goldreich, O., Micali, S., Wigderson, A.: Proofs that yield nothing but their validity or all languages in NP have zero-knowledge proof systems. J. ACM 38(3), 691–729 (1991)
13. Goldwasser, S., Micali, S., Rackoff, C.: The knowledge complexity of interactive proof systems. SIAM J. Comput. 18(1), 186–208 (1989)
14. Goyal, V.: Non-black-box simulation in the fully concurrent setting. In: STOC, pp. 221–230 (2013)
15. Kilian, J., Petrank, E.: Concurrent and resettable zero-knowledge in poly-loalgorithm rounds. In: STOC, pp. 560–569 (2001)
16. Micali, S.: Computationally sound proofs. SIAM J. Comput. 30(4), 1253–1298 (2000)
17. Pandey, O., Prabhakaran, M., Sahai, A.: Obfuscation-based non-black-box simulation and four message concurrent zero knowledge for NP. Cryptology ePrint Archive, Report 2013/754 (2013), http://eprint.iacr.org/
18. Pass, R., Rosen, A.: New and improved constructions of non-malleable cryptographic protocols. In: STOC, pp. 533–542 (2005)
19. Pass, R., Rosen, A., Tseng, W.L.D.: Public-coin parallel zero-knowledge for NP. J. Cryptology 26(1), 1–10 (2013)
20. Pass, R., Tseng, W.-L.D., Wikström, D.: On the composition of public-coin zero-knowledge protocols. In: Halevi, S. (ed.) CRYPTO 2009. LNCS, vol. 5677, pp. 160–176. Springer, Heidelberg (2009)
21. Prabhakaran, M., Rosen, A., Sahai, A.: Concurrent zero knowledge with logarithmic round-complexity. In: FOCS, pp. 366–375 (2002)
22. Richardson, R., Kilian, J.: On the concurrent composition of zero-knowledge proofs. In: Stern, J. (ed.) EUROCRYPT 1999. LNCS, vol. 1592, pp. 415–431. Springer, Heidelberg (1999)

General Statistically Secure Computation with Bounded-Resettable Hardware Tokens

Nico Döttling[1,*,**], Daniel Kraschewski[2,***], Jörn Müller-Quade[3],
and Tobias Nilges[3]

[1] Aarhus University, Denmark
[2] TNG Technology Consulting GmbH, Munich, Germany
[3] Karlsruhe Institute of Technology, Germany

Abstract. Universally composable secure computation was assumed to require trusted setups, until it was realized that parties exchanging (untrusted) tamper-proof hardware tokens allow an alternative approach (Katz; EUROCRYPT 2007). This discovery initialized a line of research dealing with two different types of tokens. Using only a single *stateful* token, one can implement general statistically secure two-party computation (Döttling, Kraschewski, Müller-Quade; TCC 2011); though all security is lost if an adversarial token receiver manages to physically reset and rerun the token. *Stateless* tokens, which are secure by definition against any such resetting-attacks, however, do provably not suffice for statistically secure computation in general (Goyal, Ishai, Mahmoody, Sahai; CRYPTO 2010).

We investigate the natural question of what is possible if an adversary can reset a token at most a bounded number of times (e.g., because each resetting attempt imposes a significant risk to trigger a self-destruction mechanism of the token). Somewhat surprisingly, our results come close to the known positive results with respect to non-resettable stateful tokens. In particular, we construct polynomially many instances of statistically secure and universally composable oblivious transfer, using only a constant number of tokens. Our techniques have some abstract similarities to previous solutions, which we grasp by defining a new security property for protocols that use oracle access. Additionally, we apply our techniques to zero-knowledge proofs and obtain a protocol that achieves the same properties as bounded-query zero-knowledge PCPs (Kilian, Petrank, Tardos; STOC 1997), even if a malicious prover may issue *stateful* PCP oracles.

* Supported by European Research Commission Starting Grant no. 279447.
** The authors acknowledge support from the Danish National Research Foundation and The National Science Foundation of China (under the grant 61061130540) for the Sino-Danish Center for the Theory of Interactive Computation, within which part of this work was performed; and also from the CFEM research center (supported by the Danish Strategic Research Council) within which part of this work was performed.
*** Work done while at Technion, Israel. Supported by the European Union's Tenth Framework Programme (FP10/2010-2016) under grant agreement no. 259426 – ERC Cryptography and Complexity.

Y. Dodis and J.B. Nielsen (Eds.): TCC 2015, Part I, LNCS 9014, pp. 319–344, 2015.

1   Introduction

The model of untrusted tamper-proof hardware was introduced by Katz [38] to circumvent trusted setup assumptions and has proven to be a strong tool for creating cryptographic protocols, especially in the context of universally composable (UC-secure) [6] multi-party computation. In the tamper-proof hardware model, (possibly malicious) parties can create tokens and send them to other parties, who then can interact with the tokens but not access any internal secrets.

In this line of research, there are two different types of tokens considered: *stateful* and *stateless/resettable* tokens. Studies of the latter are usually motivated by so-called *resetting-attacks*, meaning that an adversarial receiver could physically reset a token's internal state (e.g., by cutting off the power supply). Döttling et al. [21] implemented multiple instances of statistically UC-secure oblivious transfer (OT), using only a single stateful token. On the downside, Goyal et al. [28] showed that with any number of stateless tokens statistical OT is impossible, even if one only goes for stand-alone security. In fact, only a few statistically secure protocols based on stateless tokens have been proposed, namely stand-alone secure commitments [28] and a UC-secure variant [16], the latter using bidirectional exchange of polynomially many tokens for each commitment. These positive results are complemented again by [28], showing that unconditional non-interactive commitments cannot be performed by using only stateless tokens. Since all known approaches based on stateful tokens completely break down if only a single resetting attempt is successful, and strong impossibility results hold with respect to arbitrarily resettable tokens, it seems a natural question to ask what is still possible if an a priori bound for successful resettings is known. Therefore, similar in nature to the well-studied problems of bounded leakage [39,24], bounded-resettable zero-knowledge [40,42,47,3], and bounded-query zero-knowledge PCPs [41,35], we propose a bounded-resettable hardware model.

The new model can also be seen as a variant of the PCP model [25,1] or interactive PCP model [37], depending on whether a considered protocol contains direct interaction between the token issuer and the token receiver. The difference to the (interactive) PCP model is that maliciously issued tokens/oracles can be stateful. This seems reasonable, since it is hard to verify that a malicious token is stateless. We show that this weakened version of the PCP model still allows non-interactive zero-knowledge with $O(1)$ rounds of oracle queries and even general (interactive) secure computation. [28] also considered stateful malicious PCP oracles, though without an a priori query bound. They constructed in that model interactive zero-knowledge proofs with non-constant round complexity and showed impossibility of general statistically secure computation. Moreover, the result on ZK-PCPs by Kilian et al. [41] can be made robust against malicious stateful oracles straightforwardly at the cost of issuing polynomially many oracles and having the verifier query each oracle only once. It is not clear, however, if the same result can be obtained by using only a constant amount of oracles.

Our Results. We define a *bounded-resettable hardware model* and achieve in this model to a large extent the known positive results for stateful tokens. This is surprising, because the corresponding stateful-token protocols from the literature are all susceptible to resetting-attacks. We construct

- multiple commitments, based on a single token issued by the commitment sender,
- a single string-commitment, based on one token issued by the commitment receiver,
- multiple OT instances, based on $O(1)$ tokens issued by the OT-sender, and
- a bounded-resettable zero-knowledge proof of knowledge, based on $O(1)$ tokens issued by the prover.

All protocols are statistically UC-secure and efficient. The first commitment protocol can be made non-interactive, sacrificing UC-security against a corrupted sender, remaining statistically binding. The zero-knowledge protocol can be implemented such that the verifier does not communicate with the prover but only with tokens sent by the prover. Moreover, if we assume that even malicious provers can only issue stateless tokens, then all token functionalities in the zero-knowledge protocol can be combined on a single token and we end up with the same result as Kilian et al. [41].

Our Techniques. The main technical difficulty we have to deal with is that a malicious token issuer can store an arbitrarily complicated function an a token. We enforce (to some extent) honest programming by a simple challenge-response protocol. The domain of allowed token functionalities is chosen such that it is a linear space. The token receiver announces a random linear projection and the token issuer has to reveal the token functionality under this projection. The receiver can then check if the token reacts consistently, while learning only part of the function parameters. We put forward an abstract notion of this technique, which we call *oracle validation*. It has previously been used in a more ad-hoc manner by [21,11], though their space of token functions is quite different from ours: They use affine functions that map length-n bit-vectors to $(n{\times}n)$-matrices, whereas we use higher-degree univariate polynomials that operate on a large finite field.

The composability proof for one of our commitment schemes also requires some constructive algebraic geometry, namely efficient uniform sampling from large finite varieties [9].

Related Work. The notion of resettable zero-knowledge was introduced by [7]. In this model, a malicious verifier is allowed to reset the prover arbitrarily and rerun the protocol. Constant-round black-box zero-knowledge protocols with resettable provers were only achieved in various public key models where the verifier has to register a public key, like Bare Public Key Model [7,14,46,19], Upperbounded Public Key (UPK) Model [47], Weak Public Key (WPK) Model [42] and Counter Public Key Model [15]. UPK and WPK assume that the amount of resets is a priori bounded, similar to our model. Barak et al. [2] provided the first construction of a (non-black-box) resettably-sound zero-knowledge argument system, where soundness for a resettable verifier is achieved. This work was later improved [5,13] and generalized to simultaneously resettable zero-knowledge protocols [18,17,10,5,12]. Since then resettability has found its way into general multi-party computation [31,30].

Early works concerning tamper-proof hardware made computational assumptions and assumed stateful tokens [38,27]. This was later relaxed to resettable or stateless tokens [8,23,11] and/or unconditional security [43,29,28,21,16].

2   Preliminaries

2.1   The UC-Framework

We state and prove the security of our protocols in the Universal Composability (UC) framework of Canetti [6]. In this framework security is defined by comparison of a *real model* and an *ideal model*. The protocol of interest Π is running in the former, where an adversary $\mathcal{A}$ coordinates the behavior of all corrupted parties. We assume static corruption, i.e., the adversary $\mathcal{A}$ cannot adaptively change corruption during a protocol run. In the ideal model, which is secure by definition, an ideal functionality $\mathcal{F}$ implements the desired protocol task and a simulator $\mathcal{S}$ tries to mimic the actions of $\mathcal{A}$. An environment $\mathcal{Z}$ is plugged either to the ideal or the real model and has to guess which model it is actually plugged to. Denote the random variable representing the output of $\mathcal{Z}$ when interacting with the real model by $\mathsf{Real}_{\mathcal{A}}^{\Pi}(\mathcal{Z})$ and when interacting with the ideal model by $\mathsf{Ideal}_{\mathcal{S}}^{\mathcal{F}}(\mathcal{Z})$. Protocol Π is said to UC-implement $\mathcal{F}$, if for every adversary $\mathcal{A}$ there exists a simulator $\mathcal{S}$, such that for all environments $\mathcal{Z}$ the distributions of $\mathsf{Real}_{\mathcal{A}}^{\Pi}(\mathcal{Z})$ and $\mathsf{Ideal}_{\mathcal{S}}^{\mathcal{F}}(\mathcal{Z})$ are indistinguishable. Since we aim at statistical security, all entities are computationally unbounded. However, the (expected) runtime complexity of the ideal model has to be polynomial in the runtime complexity of the real model.

2.2   Definitions and Notations

We write $\Delta(x,y)$ for the statistical distance between x and y. The inner product of x,y is denoted as $\langle x \mid y \rangle$ and their concatenation as $x\|y$. By $\mathbb{F}_q$ we denote the finite field with q elements.

We canonically extend the notion of polynomials over a field $\mathbb{F}$ as follows. By $\mathbb{F}^n[X]$ we denote the set of all n-tuples of polynomials $p_1,\ldots,p_n \in \mathbb{F}[X]$. Each polynomial $p := (p_1,\ldots,p_n) \in \mathbb{F}^n[X]$ can be seen as a function $\mathbb{F} \to \mathbb{F}^n$, $x \mapsto \big(p_1(x),\ldots,p_n(x)\big)$, whose degree is $\deg(p) := \max_{i=1}^{n}(\deg(p_i))$. We treat $\mathbb{F}^n[X]$ as an $\mathbb{F}$-linear vector space in the natural way.

All (close to) standard ideal functionalities for the UC-framework can be found in Appendix A.

3   Query-Once Oracle Validation

We introduce now our abstract notion of enforcing honest token programming. Consider a scenario consisting of an honest receiver party, a (possibly) malicious

---

Protocol $\Pi_{k\text{-ind}}^{\text{val}}$

Implicitly parametrized by a finite field $\mathbb{F}$ and a dimension $n \in \mathbb{N}$. The sender's input domain consists of all polynomials $p \in \mathbb{F}^n[X]$ of degree at most $k - 1$. The security parameter is $\ell := \log |\mathbb{F}|$.

1. Sender: Let $p \in \mathbb{F}^n[X]$ be the sender's input. Pick $p' \in \mathbb{F}^n[X]$ of degree at most $k - 1$ uniformly at random. Program the oracle such that on input $x \in \mathbb{F}$ it outputs $(p(x), p'(x))$.
2. (a) Receiver: Pick $\lambda \in \mathbb{F}$ uniformly at random and send it to the sender.
   (b) Sender: Compute $\tilde{p} := \lambda \cdot p + p'$ and send it to the receiver.
3. Receiver: Let $x \in \mathbb{F}$ be the receiver's input. Input x into the oracle; let (y, y') denote the response.
4. Receiver: Verify that $\deg(\tilde{p}) \leq k - 1$ and $\lambda \cdot y + y' = \tilde{p}(x)$. If so, output y; otherwise reject.

Fig. 1. Construction of a query-once validation scheme for a k-wise independent oracle

sender party, and an oracle which is arbitrarily programmable by the sender in a setup phase. All entities are computationally unbounded. The security feature we aim at is that the sender has to choose the oracle functionality from some predefined class and otherwise is caught cheating, even though the receiver queries the oracle only once. If this is achieved, we speak of a *query-once oracle validation scheme*. More particularly, such a scheme consists of four stages (for a concrete example protocol, where the domain of allowed functions consists of bounded degree polynomials over a finite field, see Figure 1):

1. The sender programs the oracle.
2. Sender and receiver run an interactive protocol which is independent of the receiver's input.
3. The receiver chooses his input and queries the oracle.
4. The receiver either rejects or produces some output.

Let g denote the sender input and x the receiver input. We require the following properties.

Efficiency: All computations by honest entities have polynomial complexity.

Correctness: If the sender is honest, then the receiver always outputs $g(x)$ and never rejects.

Privacy: The receiver does not learn anything else about g than $g(x)$.

Extractability: Even if the sender is corrupted, an extractor Ext with access to the oracle program T^* and the message transcript τ of Stage 2 can compute a valid sender input g such that a receiver R with uniformly random input x with overwhelming probability (taken over the randomness of x and all of R's and Ext's random choices) either rejects or outputs $g(x)$. The extractor has to be efficient in the sense that its expected runtime on any input (τ, T^*)

is asymptotically bounded by $(\ell \cdot |\mathsf{T}^*|)^{O(1)} \cdot \rho^{-1}$, where ℓ is a security parameter, $|\mathsf{T}^*|$ is the size of the oracle program, and ρ is R's accept probability conditioned on τ (still with random x).

Note that $g(x')$ can be information-theoretically reconstructed from the receiver's view for any input x' that matches his oracle query. It follows by the privacy property that his input x must be uniquely determined by his message to the oracle. Thus, w.l.o.g. he just sends x to the oracle.

Next, we show that the oracle validation scheme $\Pi^{\mathrm{val}}_{k\text{-ind}}$ is indeed extractable—efficiency, correctness, and privacy are straightforward to see. The extractor construction is the main ingredient for our upcoming UC proofs.

Lemma 1. *Figure 1 describes an oracle validation scheme. In particular, there exists an extractor* Ext, *such for every pair* $(\mathsf{S}^*, \mathsf{T}^*)$ *of a corrupted sender* S^* *and a corrupted oracle* T^* *it holds:*

- *Provided arbitrarily rewindable access to* T^* *and given a transcript* $\tau = (\lambda, \tilde{p})$ *of the messages between* S^* *and an honest receiver* R *(i.e., with uniformly random* $\lambda \in \mathbb{F}$*),* Ext *computes a polynomial* $p \in \mathbb{F}^n[X]$ *of degree at most* $k - 1$.
- *If* R*'s input* x *is uniformly random, then with some overwhelming probability* $1 - \rho'$ *(taken over the randomness of* λ, x, *and* Ext*'s random tape),* R *either rejects or outputs* $p(x)$*. In particular, we have a failure probability* $\rho' \leq |\mathbb{F}|^{-\Omega(1)}$.
- *For every possible transcript* τ, *the expected number of queries from* Ext *to* T^* *is* $k \cdot \rho^{-1}$, *where* ρ *is* R*'s accept probability conditioned on* τ *and averaged over all inputs* $x \in \mathbb{F}$. *The rest of* Ext*'s calculations have an overall time complexity which is polynomial in* $n, k, \log |\mathbb{F}|$.

Proof (Sketch). The extractor Ext runs a simple trial-and-error approach. It repeatedly samples a uniformly random oracle input $x \in \mathbb{F}$, until it has found inputs $x_1, \ldots, x_k$ such that the corresponding oracle outputs $(y_i, y_i') := \mathsf{T}^*(x_i)$ pass the consistency checks $\lambda \cdot y_i + y_i' \stackrel{?}{=} \tilde{p}(x_i)$. Then, Ext computes and outputs the minimal-degree interpolation polynomial $p \in \mathbb{F}^n[X]$ with $p(x_i) = y_i$ for $i = 1, \ldots, k$. Note that we do not enforce pairwise distinctness of $x_1, \ldots, x_k$.

There are two things to show. Firstly, we have to show that R on random input x basically either rejects or produces output $p(x)$. Secondly, we have to estimate the expected number of queries from Ext to T^*. We start with the latter. The sampling of each x_i is a stochastic process with geometric distribution of the number of oracle queries: Given that ρ is R's accept probability conditioned on some transcript τ, the expected number of queries for sampling one x_i is $\sum_{j=1}^{\infty} j \cdot (1 - \rho)^{j-1} \cdot \rho = \rho^{-1}$. The sampling of $x_1, \ldots, x_k$ hence requires $k \cdot \rho^{-1}$ queries to T^* on average.

Next, we turn to the question of how well the extracted polynomial p approximates the functionality of a real protocol run. W.l.o.g., S^* follows a deterministic worst-case strategy and we can consider it as a function that maps each possible challenge $\lambda \in \mathbb{F}$ to a polynomial $\tilde{p}_\lambda \in \mathbb{F}^n[X]$ with $\deg(\tilde{p}_\lambda) \leq k - 1$. Analogously,

T* implements a deterministic function by assumption. Thus, for each combination of λ and x it is fixed whether R finally rejects or not. This defines a relation between challenges λ and oracle inputs x. It can be represented as a bipartite graph, where a left-hand vertex λ is adjacent to a right-hand vertex x if R does not reject the corresponding protocol run. Our proof now boils down to showing that there exists a subset of "bad" edges E' such that

1. uniformly random λ and x are adjacent via a "bad" edge only with negligible probability, namely $|E'|/|\mathbb{F}|^2 \leq |\mathbb{F}|^{-\Omega(1)}$, and
2. after removal of all "bad" edges from the graph, T* implements on each neighborhood of a possible challenge λ a polynomial function of degree at most $k - 1$.

For the existence proof of E' see the full version [22]. The key observations used there are

- that T* implements a polynomial function of low degree on the common neighborhood $\mathcal{N}(\lambda) \cap \mathcal{N}(\lambda')$ of any distinct challenges λ, λ' and
- that after removal of only a few edges, our graph decomposes into a disjoint collection of complete bipartite subgraphs.

Once E' is shown to exist, we finally need to argue that the following event has probability $|\mathbb{F}|^{-\Omega(1)}$:

- The receiver does not reject and
- one of the oracle inputs $x_1, \ldots, x_k$ sampled by Ext is adjacent via a "bad" edge to the challenge λ given by τ, or $x_i = x_j$ for some $i \neq j$.

This implies that $\rho' \leq |\mathbb{F}|^{-\Omega(1)}$. However, since $|E'|/|\mathbb{F}|^2 \leq |\mathbb{F}|^{-\Omega(1)}$, we already have with probability $1 - |\mathbb{F}|^{-\Omega(1)}$ (taken over the randomness of λ) that the given challenge λ is only adjacent to an $|\mathbb{F}|^{-\Omega(1)}$-fraction of all inputs $x \in \mathbb{F}$ or λ is adjacent to $|\mathbb{F}|^{\Omega(1)}$ edges of which only an $|\mathbb{F}|^{-\Omega(1)}$-fraction is "bad". It follows that the event above has the claimed negligible probability.   □

4   Bounded-Resettable Tamper-Proof Hardware

In this section we define and discuss the ideal functionality for bounded-resettable tamper-proof hardware (q.v. Figure 2). It is a slightly modified version of the $\mathcal{F}_{\text{wrap}}$-functionality introduced by [38]. The token sender provides a (w.l.o.g., deterministic) Turing machine and the receiver can then run it once on an input word of his choice, staying oblivious of any internal secrets. A malicious receiver can reset the token and query it repeatedly, until some bound q is reached and the functionality does not respond any more. The query bound q models an estimation for how often an adversary could reset a token that is meant to shut down for good after the first query. All our protocols rely on q being polynomially bounded in the security parameter and a smaller bound q implies better efficiency.

Functionality $\mathcal{F}_{\text{wrap}}^{\text{b-r}}$

Implicitly parametrized by a query bound q. The variable *resets_left* is initialized by *resets_left* $\leftarrow q - 1$.

Creation:

1. Await an input (create, $\mathcal{M}, b$) from the token issuer, where $\mathcal{M}$ is a deterministic Turing program and $b \in \mathbb{N}$. Then, store $(\mathcal{M}, b)$ and send (created) to the adversary.
2. Await a message (delivery) from the adversary. Then send (ready) to the token receiver.

Execution:

3. Await an input (run, w) from the receiver. Run $\mathcal{M}$ on input w. When $\mathcal{M}$ halts without generating output or b steps have passed, send a special symbol $\perp$ to the receiver; else send the output of $\mathcal{M}$.

Reset (adversarial receiver only):

4. Upon receiving a message (reset) from a corrupted token receiver, verify that *resets_left* > 0. If so, decrease *resets_left* by 1 and go back to Step 3; otherwise ignore that message.

Fig. 2. The wrapper functionality by which we model bounded-resettable tamper-proof hardware. The runtime bound b is merely needed to prevent malicious token senders from providing a perpetually running program code $\mathcal{M}$; it will be omitted throughout the rest of the paper.

We stress that tokens are not actually required to contain a state that counts the number of queries. Our definition of $\mathcal{F}_{\text{wrap}}^{\text{b-r}}$ is just the most general way to model *any* kind of token for which an upper bound of resets can be derived. E.g., it suffices that $(1 - \rho)^q$ is negligible, where ρ is an lower bound for the probability that the token successfully self-destructs after a query. As well, the token could try to delete its program $\mathcal{M}$ or make it inaccessible but an adversarial receiver could slow down that process or interrupt the deletion before it is complete, so that several queries are possible before $\mathcal{M}$ becomes finally out of reach for him. One can also imagine that security is only needed for some limited time (which is usually the case for the binding property of commitments) and hence it suffices to estimate the number of queries within this time. The latter seems particularly feasible, because it relies on the minimum possible response time of an honestly generated token.

Further note that our definition can be canonically extended to tokens that can be queried more than once also by honest users. However, our approach has the advantage to be trivially secure against tokens that maliciously change their functionality depending on the input history.

Our model is weaker than the stateful-token model in the sense that no previously known protocol with stateful tokens can tolerate even a single reset. They would be all completely broken. Therefore, none of the known positive results for stateful tokens does carry over to our model (unless $q = 1$). In turn, bounded-resettable tokens can be trivially implemented from unresettable stateful tokens. So, our results are strictly stronger than the corresponding results for stateful tokens. On the other hand, bounded-resettable tokens are strictly more powerful than arbitrarily resettable (i.e., standard stateless) tokens, since non-interactive commitments and statistically secure OT are possible with the former but impossible with the latter.

4.1 Commitments from the Token Sender to the Token Receiver

The basic idea how the token issuer can commit himself to some secret s is quite simple. He stores a random degree-q polynomial p on the token and sends the token together with $r := s + p(0)$ to the receiver. The token lets the receiver evaluate p on arbitrary challenges x, except for $x = 0$. To unveil s, the sender sends a description of p. The scheme is perfectly hiding, because even a corrupted receiver can query the token on at most q inputs, receiving only randomness that is statistically independent of $p(0)$. The scheme is statistically binding, because for any two distinct unveil messages p, p' and a uniformly random token input x it holds with overwhelming probability (namely at least $1 - \frac{q}{|\mathbb{F}|-1}$, where $\mathbb{F}$ is the finite field in which all computations take place) that $p(x) \neq p'(x)$ and thus at least one unveil message will be inconsistent with the receiver's view.

Unfortunately, the scheme as stated above is not UC-secure against a corrupted sender. The reason for this is that the sender simulator must be able to extract the secret s from the token program and the commit message r. If the token is issued honestly and thus implements a degree-q polynomial p, the simulator can evaluate the token code on $q + 1$ different inputs, then reconstruct p, and compute $s = r - p(0)$. However, a maliciously issued token can implement an arbitrarily complicated function, which behaves like a degree-q polynomial only on a vanishing but still non-negligible fraction of inputs. It is at the very least unclear if one can extract the correct polynomial from such a token efficiently. Therefore, we employ our oracle validation scheme from Section 3 to make the token extractable. See Figure 3 for the resulting commitment protocol, which even implements many commitments using only one token.

Lemma 2. *The protocol $\Pi_{\text{COM}}^{\text{s-o}}$ (q.v. Figure 3) UC-implements $\mathcal{F}_{\text{COM}}^{\text{s-o}}$ (q.v. Appendix A.2).*

Proof (Sketch). We start with the case of a corrupted receiver. The main issue in this case is that the simulator has to equivocate commitments in the unveil phase. He can do so by picking polynomials $\hat{p}, \hat{p}' \in \mathbb{F}_{2^\ell}^n[X]$ such that

- $(\hat{p}(x), \hat{p}'(x)) = (p(x), p'(x))$ for all inputs x on which the simulated token was queried so far,
- $\lambda \cdot \hat{p} + \hat{p}' = \tilde{p}$, $\deg(\hat{p}, \hat{p}') \leq q$, and
- $\hat{p}_I(0) = \hat{s}_I$, where $\hat{s}_I$ is the desired result of the equivocation,

and reprogramming the token such that on input x it now outputs $(\hat{p}(x), \hat{p}'(x))$. The unveil message for equivocating the commitment to $\hat{s}_I$ is just $(I, \hat{p}_I)$. Since the corrupted receiver can query the token at most q times, this is in his view perfectly indistinguishable from a proper commitment.

Now we show security against a corrupted sender. The simulator has to extract commitments in the unveil phase. He can do so by running the extractor Ext from Lemma 1 on the transcript of the setup phase and with rewindable access to the token code T^*. The extracted polynomial p allows the simulator to reconstruct the committed secret s from the corrupted sender's commit message r as $s = r - p(0)$. Note that Ext may have exponential runtime, but only needs to be run if by the end of the commit phase it is not already clear that the receiver will reject anyway. Therefore, the simulator must first check that $\lambda \cdot y + y' = \tilde{p}(x)$ and then run Ext only if the check is passed. Since Ext has complexity $(\ell \cdot |T^*|)^{O(1)} \cdot \rho^{-1}$, where ρ is just the probability that this check is passed, we end up with an expected simulation complexity of $(\ell \cdot |T^*|)^{O(1)}$. $\qquad\square$

Remark 1. The commitment scheme $\Pi_{COM}^{s\text{-}o}$ is statistically binding, even if λ is fixed and known to the sender. This yields a statistically secure non-interactive commitment scheme in the bounded-resettable hardware model, which was proven impossible in the stateless-token model [28].

4.2 Commitments from the Token Receiver to the Token Sender

For a commitment from the token receiver to the token sender we need a slightly more sophisticated approach. As in our previous commitment scheme, the token implements a random degree-q polynomial p. The token receiver can then commit to some secret s by inputting a random x into the token, thus learning $p(x)$, and announcing a commit message that consists of

- a fraction of bits of $p(x)$, say the first quarter of its bit-string representation,
- a 2-universal hash function h, and
- $m := s + h(x)$.

To unveil s, he just needs to announce the used token input x. We briefly sketch now why this scheme is hiding and binding. We start with the latter. Due to the query-bound q, the token acts just like a perfectly random function. Thus, a corrupted commitment sender may only with negligible probability find two distinct unveil messages x, x' such that $p(x)$ and $p(x')$ agree on the first quarter of their bit-string representation. This establishes the binding property. The token issuer, however, learns only several bits of information about x during the commit phase, so that from his view x has still linear entropy afterwards. Since

---

Protocol $\Pi_{\text{COM}}^{\text{s-o}}$

Implicitly parametrized by a token query bound q, a commitment number n, and a commitment length ℓ. The security parameter is ℓ. For any vector $v = (v_1, \ldots, v_n)$ and $I \subseteq \{1, \ldots, n\}$ let $v_I := (v_i)_{i \in I}$.

Setup phase:

1. Sender: Pick two uniformly random polynomials $p, p' \in \mathbb{F}_{2^\ell}^n[X]$ of degree at most q. Program a token T which on input $x \in \mathbb{F}_{2^\ell} \setminus \{0\}$ outputs $(p(x), p'(x))$ and ignores input $x = 0$. Send T to the receiver.
2. Receiver: Pick $\lambda \in \mathbb{F}_{2^\ell}$ uniformly at random and send it to the sender.
3. Sender: Compute $\tilde{p} := \lambda \cdot p + p'$ and send it to the receiver.

Commit phase:

4. Sender: Let $s := (s_1, \ldots, s_n) \in \mathbb{F}_{2^\ell}^n$ be the sender's input. Send $r := s + p(0)$ to the receiver.
5. Receiver: Input a uniformly random $x \in \mathbb{F}_{2^\ell} \setminus \{0\}$ into T; let (y, y') denote the response.

Unveil phase:

6. Sender: Let $I \subseteq \{1, \ldots, n\}$ indicate the commitments to be opened. Send (I, p_I) to the receiver.
7. Receiver: If $\deg(\tilde{p}) \leq q$ and $\lambda \cdot y + y' = \tilde{p}(x)$ and $p_I(x) = y_I$, output $\hat{s}_I := r_I - p_I(0)$; else reject.

Fig. 3. Statistically UC-secure commitments where the sender is the token issuer

h is a 2-universal hash function, this means that he cannot predict $h(x)$ and thus the commitment is hiding. Still, we need to employ our oracle validation scheme from Section 3 again to make the token extractable, as otherwise we have no UC-security against a corrupted commitment receiver. See Figure 4 for the resulting protocol.

Lemma 3. *The protocol $\Pi_{\text{COM}}^{\text{rev}}$ (q.v. Figure 4) implements $\mathcal{F}_{\text{COM}}$ (q.v. Appendix A.1) UC-secure against a corrupted commitment sender.*

Proof (Sketch). We just have to exploit that the simulator sees all token inputs. As the number of token queries by the commitment sender is upper bounded by q, the token acts from his views like a perfectly random function. Hence, with overwhelming probability his announcement of $\tilde{y}$ in the commit phase either corresponds to a unique input x already sent to the token or he is caught cheating in the unveil phase. In the former case, the simulator can find x just by scanning through the token's input history, compute the correct secret $s = m - \langle h \mid \sigma(x) \rangle$ and send it to the ideal commitment functionality $\mathcal{F}_{\text{COM}}$. In the other case, the simulator can just send anything to the ideal functionality, because only with negligible probability he might need to unveil it later. $\square$

Protocol $\Pi_{\text{COM}}^{\text{rev}}$

Implicitly parametrized by a token query bound q and a commitment length ℓ. The security parameter is ℓ. Let $\sigma : \mathbb{F}_{2^{4\ell}} \to \mathbb{F}_{2^\ell}^4$, $x \mapsto (\sigma_1(x), \ldots, \sigma_4(x))$ be the canonical $\mathbb{F}_{2^\ell}$-vector space isomorphism.

Setup phase:

1. Receiver: Pick two uniformly random polynomials $p, p' \in \mathbb{F}_{2^{4\ell}}[X]$ of degree at most q and program a token T which on input $x \in \mathbb{F}_{2^{4\ell}}$ outputs $(p(x), p'(x))$. Send T to the sender.
2. Sender: Pick $\lambda \in \mathbb{F}_{2^{4\ell}}$ uniformly at random and send it to the receiver.
3. Receiver: Compute $\tilde{p} := \lambda \cdot p + p'$ and send it to the sender.

Commit phase:

4. Sender: Let $s \in \mathbb{F}_{2^\ell}$ be the sender's input. Input a uniformly random $x \in \mathbb{F}_{2^{4\ell}}$ into the token T; let (y, y') denote the response. If $\lambda \cdot y + y' = \tilde{p}(x)$ and $\deg(\tilde{p}) \leq q$, pick a uniformly random $h \in \mathbb{F}_{2^\ell}^4$ and compute $m := s + \langle h \,|\, \sigma(x) \rangle$ and $\tilde{y} := \sigma_1(y)$ and send $(m, h, \tilde{y})$ to the receiver; otherwise abort.

Unveil phase:

5. Sender: Send x to the receiver.
6. Receiver: Verify that $\tilde{y} = \sigma_1(p(x))$. If so, output $\hat{s} := m - \langle h \,|\, \sigma(x) \rangle$; otherwise reject.

Fig. 4. Statistically UC-secure commitment where the receiver is the token issuer

Proving UC-security against a corrupted commitment receiver, i.e. providing a simulator that equivocates commitments, is more challenging. Note that even after extracting a polynomial p that approximates the token functionality, it is still nontrivial to find a token input $\hat{x}$ such that the first quarter of bits of $p(\hat{x})$ matches the given commit message $(m, h, \tilde{y})$ while $m - h(x) = \hat{s}$ for a new secret $\hat{s}$. This problem can be expressed as a polynomial equation system. Here the efficient algorithm of [9] for sampling random solutions comes into play. (See Appendix B for a brief explanation that all preconditions of [9, Theorem 1.1] are met.) In addition, the simulator has to make sure that the sampled solution $\hat{x}$ is actually possible in the real model: He has to (re)sample $\hat{x}$ until $p(\hat{x})$ agrees with the token functionality and the consistency check in Step 4 of the commit phase of $\Pi_{\text{COM}}^{\text{rev}}$ is passed. See Figure 5 for the detailed simulator description. The resampling of $\hat{x}$ imposes some extra difficulty for the runtime estimation, but we refer to the full version [22] for the technical calculation. Next, we show that our scheme is statistically hiding. This is needed for the UC proof and has further application later in our construction of resettable zero-knowledge.

Lemma 4. *The protocol $\Pi_{\text{COM}}^{\text{rev}}$ is statistically hiding, even if λ is fixed.*

Simulator for a corrupted token issuer that receives commitments

Setup phase: Simulated straightforwardly, using a simulated version of the complete real model where the simulated adversary is wired to the ideal model's environment in the canonical way. Store $(\lambda, \tilde{p})$ and the token program T^* sent by the corrupted commitment receiver to the simulated functionality $\mathcal{F}_{\mathrm{wrap}}^{\mathrm{b\text{-}r}}$.

Commit phase: Simulated straightforwardly, with random sender input s. Store $(m, h, \tilde{y})$.

Unveil phase: If the simulated commitment sender has already aborted, do nothing. Otherwise, upon receiving $(\mathbf{opened}, \hat{s})$ from $\mathcal{F}_{\mathrm{COM}}$ replace the stored unveil information x in the simulated sender's memory with $\hat{x}$, computed by the following equivocation program, and let him then proceed with the protocol.

1. Setup the extractor Ext from Lemma 1 with parameters $\mathbb{F} := \mathbb{F}_{2^{4\ell}}$, $n := 1$, and $k := q + 1$. Provide Ext with the transcript $\tau := (\lambda, \tilde{p})$ and rewindable access to the token code T^*.
2. Start Ext. If Ext queries T^* more than 2^ℓ times, give up; otherwise let p denote Ext's output.
3. Compute the unique polynomial $p_1 \in \mathbb{F}_{2^\ell}[X_1, \ldots, X_4]$ such that $\deg(p_1) \leq \deg(p)$ and $\sigma_1 \circ p = p_1 \circ \sigma$, where "$\circ$" denotes the function composition operator. Then pick a uniformly random solution $\hat{x} \in \mathbb{F}_{2^{4\ell}}$ of the following polynomial equation system, using the efficient algorithm of [9]:

$$p_1(\sigma(\hat{x})) = \tilde{y}$$
$$\langle h \mid \sigma(\hat{x}) \rangle = m - \hat{s}$$

   Resample $\hat{x}$ until $p(\hat{x}) = t(\hat{x})$ and $\tilde{p}(\hat{x}) = \lambda \cdot t(\hat{x}) + t'(\hat{x})$, where $t, t' : \mathbb{F}_{2^{4\ell}} \to \mathbb{F}_{2^{4\ell}}$ such that $\mathsf{T}^*(\hat{x}) = (t(\hat{x}), t'(\hat{x}))$. Give up, if more than $2^{\sqrt{\ell}}$ iterations are required.
4. Replace x in the simulated sender's memory by $\hat{x}$.

Fig. 5. Simulator for a corrupted token issuer in the protocol $\Pi_{\mathrm{COM}}^{\mathrm{rev}}$ (q.v. Figure 4)

Proof. Let λ and $\tilde{p}$ be arbitrary but fixed. Let $t, t' : \mathbb{F}_{2^{4\ell}} \to \mathbb{F}_{2^{4\ell}}$ represent the (possibly) corrupted token functionality in the sense that the token maps $x \mapsto (t(x), t'(x))$. Moreover, let $Z := \mathbb{F}_{2^\ell} \cup \{\bot\}$ and for each $z \in Z$ let M_z denote the set of all token inputs x that lead to a commit message $(m, h, \tilde{y})$ with $\tilde{y} = z$. I.e., $M_z = \{x \in \mathbb{F}_{2^{4\ell}} \mid \lambda \cdot t(x) + t'(x) = \tilde{p}(x) \wedge \sigma_1(t(x)) = z\}$ for $z \in \mathbb{F}_{2^\ell}$ and $M_\bot = \{x \in \mathbb{F}_{2^{4\ell}} \mid \lambda \cdot t(x) + t'(x) \neq \tilde{p}(x)\}$. For uniformly random $x \in \mathbb{F}_{2^{4\ell}}$ and the corresponding $\tilde{y}$ (meaning that $\tilde{y} = \sigma_1(t(x))$ if $\lambda \cdot t(x) + t'(x) = \tilde{p}(x)$ and else $\tilde{y} = \bot$) it holds:

$$\max_{e: Z \to \mathbb{F}_{2^{4\ell}}} \Pr[x = e(\tilde{y})] = \mathrm{E}(|M_{\tilde{y}}|^{-1}) = \sum_{z \in Z} \Pr[x \in M_z] \cdot |M_z|^{-1}$$

$$= \sum_{z \in Z} \frac{1}{|\mathbb{F}_{2^{4\ell}}|} = 2^{-3\ell} + 2^{-4\ell}$$

Hence, for uniformly random $u \in \mathbb{F}_{2^\ell}$ we can conclude by the Generalized left-over hash lemma [20, Lemma 2.4]:

$$\Delta\big(\big(\langle h \mid \sigma(x)\rangle, h, \tilde{y}\big), (u, h, \tilde{y})\big) \le \frac{1}{2}\sqrt{\max_{e: Z \to \mathbb{F}_{2^{4\ell}}} \Pr[x = e(\tilde{y})] \cdot |\mathbb{F}_{2^\ell}|}$$

$$= \frac{1}{2}\sqrt{2^{-2\ell} + 2^{-3\ell}} < 2^{-\ell}.$$

It directly follows now that the statistical distance between a commitment on any secret s and a commitment on uniform randomness is also upper bounded by $2^{-\ell}$. □

Corollary 1. *The protocol $\Pi_{\text{COM}}^{\text{rev}}$ implements $\mathcal{F}_{\text{COM}}$ (q.v. Appendix A.1) UC-secure against a corrupted receiver. The simulation depicted in Figure 5 is indistinguishable from the real model.*

Proof. Consider the following sequence of experiments.

Experiment 1: This is the real model.

Experiment 2: The same as Experiment 1, except that the commitment sender commits to pure randomness in the commit phase and runs in the unveil phase a complete search over all token inputs to equivocate the commitment to his real input (which requires to reset the token exponentially many times).

Experiment 3: The same as Experiment 2, except that the complete search in the equivocation step is only over token inputs x on which the token functionality $x \mapsto \big(t(x), t'(x)\big)$ coincides with the mapping $x \mapsto \big(p(x), t'(x)\big)$, where p denotes the polynomial computed by Ext from the token program and the transcript of the setup phase.

Experiment 4: The ideal model, conditioned on the event that the simulator does not give up.

Experiment 5: This is the ideal model.

Experiment 1 and Experiment 2 are indistinguishable, because the commitment is statistically hiding (Lemma 4). Indistinguishability between Experiment 2 and Experiment 3 follows from the negligibility of Ext's failure probability ρ' (Lemma 1). Experiment 3 and Experiment 4 are indistinguishable by construction of the simulator—here we need that by [9] one finds solutions for a polynomial equation system that are statistically close to random solutions (cf. Appendix B). Experiment 4 and Experiment 5 are indistinguishable, since the simulator has polynomial expected runtime complexity (see full version [22]) and thus gives up only with negligible probability. □

5   Multiple OT from $O(1)$ Tokens

5.1   Multiple OT with Combined Abort

We adapt and enhance a protocol idea by [28] for a single OT instance in the stateless-token model. It works as follows. The OT-receiver first commits to his

---

Protocol $\Pi_{\mathrm{MOT}}^{\mathrm{c\text{-}ab}}$

Implicitly parametrized by the number n of single OTs to be implemented. Based upon our commitment schemes $\Pi_{\mathrm{COM}}^{\mathrm{s\text{-}o}}$ and $\Pi_{\mathrm{COM}}^{\mathrm{rev}}$ and a statistically secure message authentication scheme MAC, e.g. from [44].

1. Sender: Let $(s_0^{(1)}, s_1^{(1)}), \ldots, (s_0^{(n)}, s_1^{(n)})$ be the sender's n OT-inputs. Sample a key k for the message authentication scheme MAC. Commit to the $2n$ values $s_0^{(i)}, s_1^{(i)}$ via $\Pi_{\mathrm{COM}}^{\mathrm{s\text{-}o}}$ and prepare a hardware token T_{OT} with the following functionality and send it to the receiver:
   - On input (c, w, τ, σ), verify that $c \in \{0,1\}^n$, $\sigma = \mathsf{MAC}_k(\tau)$, and w is a correct $\Pi_{\mathrm{COM}}^{\mathrm{rev}}$-unveil of c with commit phase transcript τ. If so, return the $\Pi_{\mathrm{COM}}^{\mathrm{s\text{-}o}}$-unveil messages for $s_{c_1}^{(1)}, \ldots, s_{c_n}^{(n)}$.
2. Receiver: Let $c = (c_1, \ldots, c_n)$ be the receiver's choice bits. Commit to c via $\Pi_{\mathrm{COM}}^{\mathrm{rev}}$.
3. Sender: Take the message transcript τ of Step 2, compute $\sigma = \mathsf{MAC}_k(\tau)$, and send σ to the receiver.
4. Receiver: Let w be the $\Pi_{\mathrm{COM}}^{\mathrm{rev}}$-unveil message for c. Input (c, w, τ, σ) into T_{OT}; let $(r_1, \ldots, r_n)$ denote the response. Verify that $r_1, \ldots, r_n$ are correct unveil messages for the corresponding $\Pi_{\mathrm{COM}}^{\mathrm{s\text{-}o}}$-commitments from Step 1 indexed by c. If so, output the unveiled values; otherwise abort.

Fig. 6. Reduction of multiple OT with combined abort to our commitment protocols

choice bit. The OT-sender then programs a token T_{OT} and provides it with all his random coins and the message transcript of the commitment protocol. The token implements the following functionality. Upon receiving an unveil message for a bit c, the token checks if the unveil is correct; if so, it will provide an OT output s_c. The token T_{OT} is transferred to the receiver, he unveils to it his choice bit and learns the corresponding OT output.

Since the commitments of [28] in the stateless-token model require the commitment receiver to access some token in the unveil phase, they need the OT-sender to encapsulate tokens into each other. We can circumvent this by our commitment scheme $\Pi_{\mathrm{COM}}^{\mathrm{rev}}$, where the commitment receiver does not access any tokens at all. So far, we can implement one OT instance with two tokens. Now, if we implement many OT instances in parallel the straightforward way, i.e. letting the receiver unveil all his choice bits to the token T_{OT}, we run into trouble: Each of the many OT outputs by T_{OT} can arbitrarily depend on all choice bits. Therefore, we let the sender first commit to the OT outputs via our construction $\Pi_{\mathrm{COM}}^{\mathrm{s\text{-}o}}$. The token T_{OT} then merely unveils the requested OT outputs. Still, T_{OT} can abort depending on all the choice bits, but we are fine with this for the moment and deal with it in the next section. Thus, our OT construction implements a flawed version of the ideal multiple-OT functionality, where a corrupted sender can additionally upload a predicate that decides whether the receiver's choice bits are accepted (cf. Appendix A.3). A similar level of security was achieved by [33] in the context of non-interactive secure computation.

There is one further refinement of the protocol, by which we achieve that all tokens can be issued independently of the parties' OT inputs. So far, the program code of T_{OT} depends on the message sent by the OT-receiver for the $\Pi_{\mathrm{COM}}^{\mathrm{rev}}$-commitment on his choice bits. Instead, the token sender can give the receiver an information-theoretic MAC for this message, the receiver can input it together with the unveil message into T_{OT}, and the code of T_{OT} thus needs to depend only on the MAC-key—note that by construction of $\Pi_{\mathrm{COM}}^{\mathrm{s-o}}$, the unveil messages that T_{OT} outputs are independent of the committed secrets. The complete OT protocol is given in Figure 6.

Lemma 5. *The protocol $\Pi_{\mathrm{MOT}}^{\mathrm{c-ab}}$ (q.v. Figure 6) UC-implements $\mathcal{F}_{\mathrm{MOT}}^{\mathrm{c-ab}}$ (q.v. Appendix A.3).*

Proof (Proof-sketch). We first show UC-security against a corrupted OT-receiver. In this case, the simulator can fake a real protocol run, exploiting extractability of $\Pi_{\mathrm{COM}}^{\mathrm{rev}}$-commitments and equivocality of $\Pi_{\mathrm{COM}}^{\mathrm{s-o}}$-commitments. The simulation basically works as follows. Step 1 of $\Pi_{\mathrm{MOT}}^{\mathrm{c-ab}}$ is simulated straightforwardly with random input for the simulated sender. In Step 2, the corrupted receiver's choice bits $(c_1, \ldots, c_n)$ can be extracted (using the sender simulator for $\Pi_{\mathrm{COM}}^{\mathrm{rev}}$) and sent to the ideal functionality $\mathcal{F}_{\mathrm{MOT}}^{\mathrm{c-ab}}$. Then, Step 3 again is simulated straightforwardly. Finally, in Step 4, the unveil messages output by the simulated token T_{OT} are replaced (using the receiver simulator for $\Pi_{\mathrm{COM}}^{\mathrm{s-o}}$) such that the commitments from Step 1 are equivocated to the OT-outputs $\hat{s}_{c_1}^{(1)}, \ldots, \hat{s}_{c_n}^{(n)}$ received from $\mathcal{F}_{\mathrm{MOT}}^{\mathrm{c-ab}}$. Indistinguishability of the simulation from the real model follows from the UC-security of $\Pi_{\mathrm{COM}}^{\mathrm{s-o}}$ and $\Pi_{\mathrm{COM}}^{\mathrm{rev}}$ and the unforgeability of the message authentication scheme MAC.

Next, we show UC-security against a corrupted OT-sender. The simulator works as follows. In Step 1 of $\Pi_{\mathrm{MOT}}^{\mathrm{c-ab}}$, the corrupted sender's OT inputs $(s_0^{(1)}, s_1^{(1)}), \ldots, (s_0^{(n)}, s_1^{(n)})$ can be extracted (using the sender simulator for $\Pi_{\mathrm{COM}}^{\mathrm{s-o}}$). Step 2 and Step 3 of $\Pi_{\mathrm{MOT}}^{\mathrm{c-ab}}$ are simulated straightforwardly with random input for the simulated receiver. Then the simulator has to send the extracted OT inputs $(s_0^{(1)}, s_1^{(1)}), \ldots, (s_0^{(n)}, s_1^{(n)})$ together with an abort predicate Q to the ideal functionality $\mathcal{F}_{\mathrm{MOT}}^{\mathrm{c-ab}}$. The predicate Q is defined by the following program, parametrized with the $\Pi_{\mathrm{COM}}^{\mathrm{s-o}}$-commitments and the token code $\mathsf{T}_{\mathrm{OT}}^*$ obtained in Step 1 of $\Pi_{\mathrm{MOT}}^{\mathrm{c-ab}}$, the transcript τ of Step 2, and σ from Step 3:

1. Upon input $c \in \{0,1\}^n$, use the receiver simulator for $\Pi_{\mathrm{COM}}^{\mathrm{rev}}$ to obtain an unveil message $\widehat{w}$ that equivocates τ to c.
2. Run $\mathsf{T}_{\mathrm{OT}}^*$ on input $(c, \widehat{w}, \tau, \sigma)$; let $(r_1, \ldots, r_n)$ denote the response.
3. Simulate the check in Step 4 of $\Pi_{\mathrm{MOT}}^{\mathrm{c-ab}}$, i.e., verify that $r_1, \ldots, r_n$ are correct unveil messages for the corresponding $\Pi_{\mathrm{COM}}^{\mathrm{s-o}}$-commitments indexed by c. If so, accept; otherwise reject.

Indistinguishability of the simulation from the real model just follows from the UC-security of $\Pi_{\mathrm{COM}}^{\mathrm{s-o}}$ and $\Pi_{\mathrm{COM}}^{\mathrm{rev}}$. $\qquad\square$

Remark 2. Though stated as a three-token construction, our protocol $\Pi_{\mathrm{MOT}}^{\mathrm{c-ab}}$ can as well be implemented with two tokens, if one allows a token to be queried twice. In particular, the token T_{OT} gets with w a complete transcript of the messages sent to the token used in the subprotocol Π_{COM}^{rev} anyway. Hence, even if maliciously issued tokens can keep a complex state, it does not compromise security if these two tokens are combined into one query-twice token.

5.2   How to Get Rid of the Combined-Abort Flaw

The question of how to implement ideal oblivious transfer from the flawed version $\mathcal{F}_{MOT}^{c\text{-}ab}$ is closely related to the research field of OT combiners. However, an OT combiner needs access to *independent* OT instances, some of which may be corrupted. In contrast, $\mathcal{F}_{MOT}^{c\text{-}ab}$ leaks a predicate over the receiver's *joint* inputs for the multiple OT instances. Therefore we need an OT extractor, as defined in [34], rather than an OT combiner. However, the scope of [34] is skew to ours. They consider semi-honest parties, which follow the protocol, and only the leakage function is chosen maliciously. In this setting they aim at a constant extraction rate. In contrast, we consider malicious parties that may try to cheat in the extraction protocol, but we do not care much about the rate. For our purpose it suffices to implement n ideal OT instances from $n^{O(1)}$ flawed instances.

Our solution follows the basic idea of [36] to take an *outer protocol* with many parties and emulate some of the parties by an *inner protocol*, such that the security features of both protocols complement each other. However, before we describe our solution, we briefly sketch why a more classic OT combiner based on 2-universal hashing would be insecure in our case. Such combiners are usually built such that the receiver's ideal-OT choice bits are basically 2-universal hash values of his flawed-OT inputs, which are uniformly random, and similarly for the outputs. In the $\mathcal{F}_{MOT}^{c\text{-}ab}$-hybrid model, such an approach is susceptible to the following generic attack. The sender just follows the protocol, except that he randomly chooses two of his $\mathcal{F}_{MOT}^{c\text{-}ab}$-input tuples, say $\left(\tilde{s}_0^{(i)}, \tilde{s}_1^{(i)}\right)$ and $\left(\tilde{s}_0^{(j)}, \tilde{s}_1^{(j)}\right)$, and flips the bits of $\tilde{s}_1^{(i)}$ and $\tilde{s}_1^{(j)}$. Furthermore, he defines the abort predicate Q such that it rejects the receiver's choice bits $\tilde{c} := (\tilde{c}_1, \tilde{c}_2, \ldots)$ if and only if $\tilde{c}_i = \tilde{c}_j = 1$. This attack has the following effect. With non-negligible probability, the 2-universal hash functions are chosen such that

- the receiver's i-th flawed-OT input-output tuple $(\tilde{c}_i, \tilde{r}_i)$ influences the calculation of an ideal-OT input-output tuple (c_k, r_k), but not (c_l, r_l), where l is an index such that
- the receiver's j-th flawed-OT input-output tuple $(\tilde{c}_j, \tilde{r}_j)$ influences (c_l, r_l), but not (c_k, r_k).

In such a case, it happens with probability $\frac{1}{2}$ that (c_k, r_k) is affected by the bit-flip of $\tilde{s}_1^{(i)}$, namely if $\tilde{c}_i = 1$. Likewise, (c_l, r_l) is affected by the bit-flip of $\tilde{s}_1^{(j)}$ if $\tilde{c}_j = 1$. Both events are statistically independent of each other, but by definition of the abort predicate Q it will never happen that the receiver produces regular output while (c_k, r_k) and (c_l, r_l) are both affected by the attack. This correlation between the joint distribution of the receiver's ideal-OT inputs and outputs and the event of an abort is not simulatable with an ideal OT.

Our construction is at an abstract level very similar to the OT combiner of [32]. We believe that the constructions of [32] can also be proven secure when based on $\mathcal{F}_{\mathrm{MOT}}^{\mathrm{c\text{-}ab}}$, but we prefer to present a simple combination of results from the literature as opposed to tampering with the proof details. Our final OT construction consists of three ingredients:

1. Our $\mathcal{F}_{\mathrm{MOT}}^{\mathrm{c\text{-}ab}}$ implementation,
2. a construction for general, statistically UC-secure two-party computation in the OT-hybrid model, e.g. from [36], and
3. a statistically UC-secure protocol for multiple OT based on a single untrusted stateful tamper-proof hardware token, which we take from [21].

We take the token functionality from [21] and implement it by secure two-party computation from [36], based on $\mathcal{F}_{\mathrm{MOT}}^{\mathrm{c\text{-}ab}}$ instead of ideal OT. Note that OT can be stored and reversed [4,45] and therefore it suffices to query $\mathcal{F}_{\mathrm{MOT}}^{\mathrm{c\text{-}ab}}$ just once in the beginning with the token receiver being also the OT-receiver. The "emulated token" then replaces all token queries in the otherwise unchanged protocol of [21]. Now, any specification of the abort predicate in $\mathcal{F}_{\mathrm{MOT}}^{\mathrm{c\text{-}ab}}$ directly corresponds to a maliciously programmed token that stops functioning depending on its inputs. Since the construction of [21] is UC-secure against any malicious token behavior, we finally obtain UC-secure OT.

Remark 3. Notice that this directly provides an impossibility result for commitments in the stateless-token model where the unveil phase consists only of a single message from the sender to the receiver and local computations (without accessing any tokens) by the receiver. Otherwise our OT construction could be implemented in the stateless-token model (without encapsulation), contradicting the impossibility result for OT given in [28].

6 Bounded-Resettable Zero-Knowledge Proofs of Knowledge

We modify the constant-round zero-knowledge protocol of [26] for 3-COLOR such that the prover becomes resettable and only two tokens have to be sent to the verifier. In the protocol of [26], the verifier first commits to his challenge (the edges determining the vertices that are to be revealed), then the prover commits to permutations of the colored vertices. The verifier then reveals the challenge and the prover opens the specified commitments. The main problem imposed by a *resettable* prover is that a malicious verifier could try to run the same protocol several times, each time with different challenges, and hence step by step learn the prover's witness. The standard technique to deal with this is to let the prover's color permutations depend on the verifier's commitment in a pseudorandom way. Since we aim for *statistical* zero-knowledge, we cannot use a pseudorandom function, but need to replace it by a random polynomial of sufficient degree.

For our construction, we replace the computational commitments in [26] with the statistical commitments presented in the previous sections. Though, our

---

Protocol $\Pi_{\text{SZK}}^{\text{b-r}}$

Implicitly parametrized by a simple 3-colorable graph $G = (V, E)$ and a query bound q in the sense that a malicious verifier can reset the prover at most $q - 1$ times. Let $n := |V|$, $t := n \cdot |E|$, and $V := \{1, \ldots, n\}$.

Auxiliary input for prover: A 3-coloring of G, denoted $\varphi : V \to \{1, 2, 3\}$.

Setup phase:

– Prover: Select a random degree-q polynomial $f \in \mathbb{F}_{2^l}[X]$, where l is the number of random bits needed for token generation in $\Pi_{\text{COM}}^{\text{s-o}}$ for $n \cdot t$ commitments. Further, select a random degree-q polynomial $g \in \mathbb{F}_{2^k}[X]$, where k is the number of random bits needed to generate t random permutations over $\{1, 2, 3\}$. W.l.o.g., l and k are larger than the commit message length in $\widetilde{\Pi}_{\text{COM}}^{\text{rev}}$. Create two tokens T^{rev} and $\mathsf{T}^{\text{s-o}}$ with the following functionalities and send them to the receiver:
  - T^{rev}: Just implement the token functionality of $\widetilde{\Pi}_{\text{COM}}^{\text{rev}}$.
  - $\mathsf{T}^{\text{s-o}}$: Upon input (x, c^{rev}), simulate the token generation procedure of $\Pi_{\text{COM}}^{\text{s-o}}$ with randomness $f(c^{\text{rev}} \| 0 \ldots 0)$, evaluate the generated token program on input x, and output the result.

Proof phase:

1. Verifier: Uniformly and independently select a random value $\lambda^{\text{s-o}}$ according to the setup phase of $\Pi_{\text{COM}}^{\text{s-o}}$ and a t-tuple of edges $\bar{E} = (\{u_1, v_1\}, \ldots, \{u_t, v_t\})$ as a challenge for the zero-knowledge proof. Use $\widetilde{\Pi}_{\text{COM}}^{\text{rev}}$ to commit to $(\bar{E}, \lambda^{\text{s-o}})$ and send the corresponding commit message c^{rev} to the prover.
2. Prover: Compute $r = f(c^{\text{rev}} \| 0 \ldots 0)$ and $r' = g(c^{\text{rev}} \| 0 \ldots 0)$. Use r' to select t random permutations $\pi_1, \ldots, \pi_t$ over $\{1, 2, 3\}$ and set $\phi_i(v) = \pi_i(\varphi(v))$ for each $v \in V$ and $i \in \{1, \ldots, t\}$. Use r to simulate the token generation of $\Pi_{\text{COM}}^{\text{s-o}}$ and compute the corresponding $\Pi_{\text{COM}}^{\text{s-o}}$-commit message $c^{\text{s-o}}$ to commit to $\phi_i(v)$ for all $v \in V$ and $i \in \{1, \ldots, t\}$. Send $c^{\text{s-o}}$ to the verifier.
3. Verifier: Send c^{rev} and the corresponding $\widetilde{\Pi}_{\text{COM}}^{\text{rev}}$-unveil message to the prover, thus unveiling $(\bar{E}, \lambda^{\text{s-o}})$.
4. Prover: If the unveil was not correct, abort. Else, compute $r = f(c^{\text{rev}} \| 0 \ldots 0)$ and simulate the token generation of $\Pi_{\text{COM}}^{\text{s-o}}$ as in Step 2. Compute the response $\tilde{p}^{\text{s-o}}$ for $\lambda^{\text{s-o}}$ according to the setup phase of $\Pi_{\text{COM}}^{\text{s-o}}$. Let $w^{\text{s-o}}$ be the $\Pi_{\text{COM}}^{\text{s-o}}$-unveil message for the commitments indexed by $\bar{E}$. Send $(\tilde{p}^{\text{s-o}}, w^{\text{s-o}})$ to the verifier.
5. Verifier: Check the unveiled commitments according to the unveil phase of $\Pi_{\text{COM}}^{\text{s-o}}$. Also verify for each edge $\{u_i, v_i\} \in \bar{E}$ that $\phi_i(u_i) \neq \phi_i(v_i)$. If all checks are passed, accept the proof; if not, reject.

Fig. 7. Construction of a bounded-resettable statistical zero-knowledge proof of knowledge

commitment schemes have an interactive setup phase and become insecure if the token issuer is resettable. However, by fixing λ in the setup phase of $\Pi_{\text{COM}}^{\text{rev}}$, the resulting commitment scheme $\widetilde{\Pi}_{\text{COM}}^{\text{rev}}$ becomes resettable and remains statistically hiding (cf. Lemma 4). Making all the prover's random choices dependent on the verifier's first $\widetilde{\Pi}_{\mathrm{COM}}^{\mathrm{rev}}$-commit message c^{rev} is the lever we use to obtain resettability. This particularly has to include the randomness used in $\Pi_{\mathrm{COM}}^{\mathrm{s\text{-}o}}$ for token generation. Therefore, we need to use a modified token in $\Pi_{\mathrm{COM}}^{\mathrm{s\text{-}o}}$ with input domain $X \times C$, where C is the set of all possible commit messages c^{rev} in $\widetilde{\Pi}_{\mathrm{COM}}^{\mathrm{rev}}$ and X is the input space for the original token program in $\Pi_{\mathrm{COM}}^{\mathrm{s\text{-}o}}$. On input (x, c^{rev}), the modified $\Pi_{\mathrm{COM}}^{\mathrm{s\text{-}o}}$-token first simulates the token generation of $\Pi_{\mathrm{COM}}^{\mathrm{s\text{-}o}}$ with randomness c^{rev} and then runs the generated token program on input x. See Figure 7 for all further details.

Lemma 6. *The protocol $\Pi_{\mathrm{SZK}}^{\mathrm{b\text{-}r}}$ UC-implements $\mathcal{F}_{\mathrm{ZK}}$ (q.v. Appendix A.4).*

Proof (Sketch). We first show UC-security against a corrupted verifier, i.e., the simulator must fake a protocol run without knowing a witness. In Step 1 of $\Pi_{\mathrm{SZK}}^{\mathrm{b\text{-}r}}$, we exploit that $\widetilde{\Pi}_{\mathrm{COM}}^{\mathrm{rev}}$ is still UC-secure against a corrupted commitment sender and thus the challenge $\bar{E}$ can be extracted. Then, in Step 2, the simulated prover can commit to different colorings for each challenged vertex pair $\{u_i, v_i\} \in \bar{E}$ and to arbitrary colorings otherwise. The remaining protocol is just simulated straightforwardly. Indistinguishability from a real protocol run follows, because $\Pi_{\mathrm{COM}}^{\mathrm{s\text{-}o}}$ is statistically hiding.

We move on to show UC-security against a corrupted prover, i.e., the simulator has to extract a witness. The complete simulation just follows the real protocol. If in the end the simulated verifier accepts, the sender simulator for $\Pi_{\mathrm{COM}}^{\mathrm{s\text{-}o}}$ (provided with the corresponding message transcript and the token code $\mathsf{T}^{\mathrm{s\text{-}o}}$) is used to extract the commitments from Step 2 of $\Pi_{\mathrm{SZK}}^{\mathrm{b\text{-}r}}$, which yields t colorings for the graph G. If none of them is a valid 3-coloring, the simulator gives up; otherwise he sends a valid one to the ideal functionality $\mathcal{F}_{\mathrm{ZK}}$. It remains to show that the simulator gives up only with negligible probability. However, if none of the committed colorings is a valid 3-coloring, then the proof is accepted by the simulated verifier at most with the following probability (abstracting from the negligible case that some commitment is successfully broken by the corrupted prover):

$$\left(1 - \tfrac{1}{|E|}\right)^t = \left(1 - \tfrac{1}{|E|}\right)^{n \cdot |E|} = \exp\!\left(n \cdot |E| \cdot \log\!\left(1 - \tfrac{1}{|E|}\right)\right)$$
$$\leq \exp\!\left(n \cdot |E| \cdot \left(-\tfrac{1}{|E|}\right)\right) = \exp(-n) \qquad \square$$

Remark 4. Furthermore, $\Pi_{\mathrm{SZK}}^{\mathrm{b\text{-}r}}$ is bounded-resettably zero-knowledge. The resettability of the prover follows from two facts. Firstly, the prover's randomness (r, r') depends deterministically but otherwise unpredictable by the verifier on his first message c^{rev}. Secondly, by the binding property of $\widetilde{\Pi}_{\mathrm{COM}}^{\mathrm{rev}}$, a corrupted verifier cannot cheat in Step 3 of $\Pi_{\mathrm{SZK}}^{\mathrm{b\text{-}r}}$ other than switch to another instance of the zero-knowledge protocol with unrelated prover randomness (r, r').

Remark 5. Our construction $\Pi_{\mathrm{SZK}}^{\mathrm{b\text{-}r}}$ can directly be used to obtain a *non-interactive* zero-knowledge proof of knowledge scheme in the bounded-resettable hardware model by storing the prover functionality in another token (or two other tokens, if each token should be queried only once).

References

1. Arora, S., Safra, S.: Probabilistic checking of proofs; A new characterization of NP. In: Foundations of Computer Science - Proceedings of FOCS 1992, pp. 2–13. IEEE Computer Society (1992)
2. Barak, B., Goldreich, O., Goldwasser, S., Lindell, Y.: Resettably-sound zero-knowledge and its applications. In: Foundations of Computer Science - Proceedings of FOCS 2001, pp. 116–125. IEEE Computer Society (2001)
3. Barak, B., Lindell, Y., Vadhan, S.P.: Lower bounds for non-black-box zero knowledge. J. Comput. Syst. Sci. 72(2), 321–391 (2006)
4. Beaver, D.: Precomputing oblivious transfer. In: Coppersmith, D. (ed.) Advances in Cryptology - CRYPT0 1995. LNCS, vol. 963, pp. 97–109. Springer, Heidelberg (1995)
5. Bitansky, N., Paneth, O.: On the impossibility of approximate obfuscation and applications to resettable cryptography. In: Boneh, D., Roughgarden, T., Feigenbaum, J. (eds.) Symposium on Theory of Computing - Proceedings of STOC 2013, pp. 241–250. ACM (2013)
6. Canetti, R.: Universally composable security: A new paradigm for cryptographic protocols. In: Foundations of Computer Science - Proceedings of FOCS 2001, pp. 136–145. IEEE Computer Society (2001), revised full version online available at http://eprint.iacr.org/2000/067
7. Canetti, R., Goldreich, O., Goldwasser, S., Micali, S.: Resettable zero-knowledge (extended abstract). In: Yao, F.F., Luks, E.M. (eds.) Symposium on Theory of Computing - Proceedings of STOC 2000, pp. 235–244. ACM (2000)
8. Chandran, N., Goyal, V., Sahai, A.: New constructions for UC secure computation using tamper-proof hardware. In: Smart, N.P. (ed.) EUROCRYPT 2008. LNCS, vol. 4965, pp. 545–562. Springer, Heidelberg (2008)
9. Cheraghchi, M., Shokrollahi, A.: Almost-uniform sampling of points on high-dimensional algebraic varieties. In: Albers, S., Marion, J.Y. (eds.) Symposium on Theoretical Aspects of Computer Science - Proceedings of STACS 2009. LIPIcs, vol. 3, pp. 277–288. Schloss Dagstuhl - Leibniz-Zentrum für Informatik, Germany (2009)
10. Cho, C., Ostrovsky, R., Scafuro, A., Visconti, I.: Simultaneously resettable arguments of knowledge. In: Cramer, R. (ed.) TCC 2012. LNCS, vol. 7194, pp. 530–547. Springer, Heidelberg (2012)
11. Choi, S.G., Katz, J., Schröder, D., Yerukhimovich, A., Zhou, H.-S. (Efficient) universally composable oblivious transfer using a minimal number of stateless tokens. In: Lindell, Y. (ed.) TCC 2014. LNCS, vol. 8349, pp. 638–662. Springer, Heidelberg (2014)
12. Chung, K.M., Ostrovsky, R., Pass, R., Visconti, I.: Simultaneous resettability from one-way functions. In: Foundations of Computer Science - Proceedings of FOCS 2013, pp. 60–69. IEEE Computer Society (2013)
13. Chung, K.M., Pass, R., Seth, K.: Non-black-box simulation from one-way functions and applications to resettable security. In: Boneh, D., Roughgarden, T., Feigenbaum, J. (eds.) Symposium on Theory of Computing - Proceedings of STOC 2013, pp. 231–240. ACM (2013)
14. Di Crescenzo, G., Persiano, G., Visconti, I.: Constant-round resettable zero knowledge with concurrent soundness in the bare public-key model. In: Franklin, M. (ed.) CRYPTO 2004. LNCS, vol. 3152, pp. 237–253. Springer, Heidelberg (2004)

15. Di Crescenzo, G., Persiano, G., Visconti, I.: Improved setup assumptions for 3-round resettable zero knowledge. In: Lee, P.J. (ed.) ASIACRYPT 2004. LNCS, vol. 3329, pp. 530–544. Springer, Heidelberg (2004)
16. Damgård, I., Scafuro, A.: Unconditionally secure and universally composable commitments from physical assumptions. In: Sako, K., Sarkar, P. (eds.) ASIACRYPT 2013, Part II. LNCS, vol. 8270, pp. 100–119. Springer, Heidelberg (2013)
17. Deng, Y., Feng, D., Goyal, V., Lin, D., Sahai, A., Yung, M.: Resettable cryptography in constant rounds - the case of zero knowledge. In: Lee, D.H., Wang, X. (eds.) ASIACRYPT 2011. LNCS, vol. 7073, pp. 390–406. Springer, Heidelberg (2011)
18. Deng, Y., Goyal, V., Sahai, A.: Resolving the simultaneous resettability conjecture and a new non-black-box simulation strategy. In: Foundations of Computer Science - Proceedings of FOCS 2009, pp. 251–260. IEEE Computer Society (2009)
19. Deng, Y., Lin, D.: Resettable zero knowledge with concurrent soundness in the bare public-key model under standard assumption. In: Pei, D., Yung, M., Lin, D., Wu, C. (eds.) Inscrypt 2007. LNCS, vol. 4990, pp. 123–137. Springer, Heidelberg (2008)
20. Dodis, Y., Ostrovsky, R., Reyzin, L., Smith, A.: Fuzzy extractors: How to generate strong keys from biometrics and other noisy data. SIAM J. Comput. 38(1), 97–139 (2008)
21. Döttling, N., Kraschewski, D., Müller-Quade, J.: Unconditional and composable security using a single stateful tamper-proof hardware token. In: Ishai, Y. (ed.) TCC 2011. LNCS, vol. 6597, pp. 164–181. Springer, Heidelberg (2011), extended full version available at http://eprint.iacr.org/2012/135
22. Döttling, N., Kraschewski, D., Müller-Quade, J., Nilges, T.: General statistically secure computation with bounded-resettable hardware tokens. IACR Cryptology ePrint Archive 2014, Report 555 (2014), http://eprint.iacr.org/2014/555
23. Döttling, N., Mie, T., Müller-Quade, J., Nilges, T.: Implementing resettable UC-functionalities with untrusted tamper-proof hardware-tokens. In: Sahai, A. (ed.) TCC 2013. LNCS, vol. 7785, pp. 642–661. Springer, Heidelberg (2013)
24. Dziembowski, S., Kazana, T., Wichs, D.: Key-evolution schemes resilient to space-bounded leakage. In: Rogaway, P. (ed.) CRYPTO 2011. LNCS, vol. 6841, pp. 335–353. Springer, Heidelberg (2011)
25. Feige, U., Goldwasser, S., Lovász, L., Safra, S., Szegedy, M.: Approximating clique is almost NP-complete (preliminary version). In: Foundations of Computer Science - Proceedings of FOCS 1991, pp. 2–12. IEEE Computer Society (1991)
26. Goldreich, O., Kahan, A.: How to construct constant-round zero-knowledge proof systems for NP. J. Cryptology 9(3), 167–190 (1996)
27. Goldwasser, S., Kalai, Y.T., Rothblum, G.N.: One-time programs. In: Wagner, D. (ed.) CRYPTO 2008. LNCS, vol. 5157, pp. 39–56. Springer, Heidelberg (2008)
28. Goyal, V., Ishai, Y., Mahmoody, M., Sahai, A.: Interactive locking, zero-knowledge PCPs, and unconditional cryptography. In: Rabin, T. (ed.) CRYPTO 2010. LNCS, vol. 6223, pp. 173–190. Springer, Heidelberg (2010)
29. Goyal, V., Ishai, Y., Sahai, A., Venkatesan, R., Wadia, A.: Founding cryptography on tamper-proof hardware tokens. In: Micciancio, D. (ed.) TCC 2010. LNCS, vol. 5978, pp. 308–326. Springer, Heidelberg (2010)
30. Goyal, V., Maji, H.K.: Stateless cryptographic protocols. In: Ostrovsky, R. (ed.) Foundations of Computer Science - Proceedings of FOCS 2011, pp. 678–687. IEEE (2011)
31. Goyal, V., Sahai, A.: Resettably secure computation. In: Joux, A. (ed.) EURO-CRYPT 2009. LNCS, vol. 5479, pp. 54–71. Springer, Heidelberg (2009)

32. Harnik, D., Ishai, Y., Kushilevitz, E., Nielsen, J.B.: OT-combiners via secure computation. In: Canetti, R. (ed.) TCC 2008. LNCS, vol. 4948, pp. 393–411. Springer, Heidelberg (2008)

33. Ishai, Y., Kushilevitz, E., Ostrovsky, R., Prabhakaran, M., Sahai, A.: Efficient non-interactive secure computation. In: Paterson, K.G. (ed.) EUROCRYPT 2011. LNCS, vol. 6632, pp. 406–425. Springer, Heidelberg (2011)

34. Ishai, Y., Kushilevitz, E., Ostrovsky, R., Sahai, A.: Extracting correlations. In: Foundations of Computer Science - Proceedings of FOCS 2009, pp. 261–270. IEEE Computer Society (2009)

35. Ishai, Y., Mahmoody, M., Sahai, A.: On efficient zero-knowledge PCPs. In: Cramer, R. (ed.) TCC 2012. LNCS, vol. 7194, pp. 151–168. Springer, Heidelberg (2012)

36. Ishai, Y., Prabhakaran, M., Sahai, A.: Founding cryptography on oblivious transfer - efficiently. In: Wagner, D. (ed.) CRYPTO 2008. LNCS, vol. 5157, pp. 572–591. Springer, Heidelberg (2008)

37. Kalai, Y.T., Raz, R.: Interactive PCP. In: Aceto, L., Damgård, I., Goldberg, L.A., Halldórsson, M.M., Ingólfsdóttir, A., Walukiewicz, I. (eds.) ICALP 2008, Part II. LNCS, vol. 5126, pp. 536–547. Springer, Heidelberg (2008)

38. Katz, J.: Universally composable multi-party computation using tamper-proof hardware. In: Naor, M. (ed.) EUROCRYPT 2007. LNCS, vol. 4515, pp. 115–128. Springer, Heidelberg (2007)

39. Katz, J., Vaikuntanathan, V.: Signature schemes with bounded leakage resilience. In: Matsui, M. (ed.) ASIACRYPT 2009. LNCS, vol. 5912, pp. 703–720. Springer, Heidelberg (2009)

40. Kilian, J., Petrank, E.: Concurrent and resettable zero-knowledge in poly-loalgorithm rounds. In: Vitter, J.S., Spirakis, P.G., Yannakakis, M. (eds.) Symposium on Theory of Computing - Proceedings of STOC 2001, pp. 560–569. ACM (2001)

41. Kilian, J., Petrank, E., Tardos, G.: Probabilistically checkable proofs with zero knowledge. In: Leighton, F.T., Shor, P.W. (eds.) Symposium on Theory of Computing - Proceedings of STOC 1997, pp. 496–505. ACM (1997)

42. Micali, S., Reyzin, L.: Min-round resettable zero-knowledge in the public-key model. In: Pfitzmann, B. (ed.) EUROCRYPT 2001. LNCS, vol. 2045, pp. 373–393. Springer, Heidelberg (2001)

43. Moran, T., Segev, G.: David and Goliath commitments: UC computation for asymmetric parties using tamper-proof hardware. In: Smart, N.P. (ed.) EUROCRYPT 2008. LNCS, vol. 4965, pp. 527–544. Springer, Heidelberg (2008)

44. Wegman, M.N., Carter, L.: New hash functions and their use in authentication and set equality. J. Comput. Syst. Sci. 22(3), 265–279 (1981)

45. Wolf, S., Wullschleger, J.: Oblivious transfer is symmetric. In: Vaudenay, S. (ed.) EUROCRYPT 2006. LNCS, vol. 4004, pp. 222–232. Springer, Heidelberg (2006)

46. Yung, M., Zhao, Y.: Generic and practical resettable zero-knowledge in the bare public-key model. In: Naor, M. (ed.) EUROCRYPT 2007. LNCS, vol. 4515, pp. 129–147. Springer, Heidelberg (2007)

47. Zhao, Y., Deng, X., Lee, C.H., Zhu, H.: Resettable zero-knowledge in the weak public-key model. In: Biham, E. (ed.) EUROCRYPT 2003. LNCS, vol. 2656, pp. 123–139. Springer, Heidelberg (2003)

A   Ideal Functionalities

In this section we provide the ideal functionalities for our security proofs in the UC-framework. For better readability, we omit session identifiers and cover only the two-party case for each protocol.

A.1   Ideal Functionality for a Single Commitment

Functionality $\mathcal{F}_{\mathrm{COM}}$

Implicitly parametrized by a domain of secrets S.

Commit phase:

1. Await an input (commit, s) with $s \in S$ from the sender. Then, store s and send (committed) to the adversary.
2. Await a message (notify) from the adversary. Then send (committed) to the receiver.

Unveil phase:

3. Await an input (unveil, $\hat{s}$) with $\hat{s} \in S$ from the sender. Then, store $\hat{s}$ and send (opened) to the adversary.
4. Await a message (output) from the adversary. Then, if $\hat{s} = s$, send $\hat{s}$ to the receiver; otherwise, send a special reject message $\perp$.

A.2   Ideal Functionality for Commitments with Selective Opening

---

Functionality $\mathcal{F}_{\mathrm{COM}}^{\mathrm{s\text{-}o}}$

Implicitly parametrized by a domain of secrets S and the number n of commitments to be implemented.

Commit phase:

1. Await an input (commit, s) with $s = (s_1, \ldots, s_n) \in S^n$ from the sender. Then, store s and send (committed) to the adversary.
2. Await a message (notify) from the adversary. Then send (committed) to the receiver.

Unveil phase:

3. Await an input (unveil, $I, \hat{s}$) with $I \subseteq \{1, \ldots, n\}$ and $\hat{s} = (\hat{s}_i)_{i \in I} \in S^{|I|}$ from the sender. Then, store $(I, \hat{s})$ and send (opened) to the adversary.
4. Await a message (output) from the adversary. Then, if $\hat{s} = (s_i)_{i \in I}$, send $(I, \hat{s})$ to the receiver; otherwise, send a special reject message $\perp$.

A.3   Ideal Functionality for Multiple Oblivious Transfer with Combined Abort

---

Functionality $\mathcal{F}_{\text{MOT}}^{\text{c-ab}}$

Implicitly parametrized by a sender input domain S and the number n of single OTs to be implemented.

- Upon input $\left(\texttt{create}, (s_0^{(1)}, s_1^{(1)}), \ldots, (s_0^{(n)}, s_1^{(n)})\right)$ with $(s_0^{(i)}, s_1^{(i)}) \in S \times S$ from the sender, verify that the sender is uncorrupted; otherwise ignore that input. Next, store $(s_0^{(0)}, s_1^{(0)}), \ldots, (s_0^{(n)}, s_1^{(n)})$, send (**sent**) to the adversary, and henceforth ignore any further input from the sender.
- Upon input $\left(\texttt{mal\_create}, (s_0^{(1)}, s_1^{(1)}), \ldots, (s_0^{(n)}, s_1^{(n)}), Q\right)$ with $(s_0^{(i)}, s_1^{(i)}) \in S \times S$ and a predicate $Q : \{0,1\}^n \to \{\texttt{accept}, \texttt{reject}\}$ from the sender, verify that the sender is corrupted; otherwise ignore that input. Next, store $(s_0^{(0)}, s_1^{(0)}), \ldots, (s_0^{(n)}, s_1^{(n)})$ and Q, send (**sent**) to the adversary, and henceforth ignore any further input from the sender.
- Upon input $(\texttt{choice}, c)$ with $c = (c_1, \ldots, c_n) \in \{0,1\}^n$ from the receiver, store c, send (**chosen**) to the adversary, and henceforth ignore any further input from the receiver.
- Upon receiving a message (**output**) from the adversary, check that there are stored inputs $(s_0^{(1)}, s_1^{(1)}), \ldots, (s_0^{(n)}, s_1^{(n)})$ from the sender and c from the receiver; else ignore this message. If the sender is corrupted, compute $Q(c)$ and abort if $Q(c) = \texttt{reject}$. Next, send $(s_{c_1}^{(1)}, \ldots, s_{c_n}^{(n)})$ to the receiver and ignore any further (**output**)-messages from the adversary.
- Upon receiving a message (**notify**) from the adversary, check that there are stored inputs $(s_0^{(1)}, s_1^{(1)}), \ldots, (s_0^{(n)}, s_1^{(n)})$ from the sender and c from the receiver; else ignore this message. Next, send an empty output to the sender and ignore any further (**notify**)-messages from the adversary.

A.4   Ideal Functionality for Zero-Knowledge

Functionality $\mathcal{F}_{\text{ZK}}$

Implicitly parametrized with an *NP*-language L and a corresponding *NP*-problem instance x.

1. Await an input w from the sender. Then, store w and send (**sent**) to the adversary.
2. Await a message (**verify**) from the adversary. Then, if w is a witness for $x \in L$, send (**accept**) to the verifier; else send (**reject**).

B   Sampling Uniformly from Varieties of Constant Codimension

We briefly state the main theorem of [9], which is used in the proof of Corollary 1.

Theorem 1 ([9, Theorem 1.1]). *Let $k > 0$ be a constant integer, $n > k$ and $d > 0$ be integers, let p^ℓ be a sufficiently large prime power and $\epsilon > 0$ be an arbitrarily small constant. Suppose that $f_1, \ldots, f_k \in \mathbb{F}_{p^\ell}[x_1, \ldots, x_n]$ are polynomials, each of total degree at most d, and let*

$$V = V(f_1, \ldots, f_k) = \{\xi \in \mathbb{F}_{p^\ell}^n \mid f_1(\xi) = \ldots = f_k(\xi) = 0\}$$

be the variety defined by $f_1, \ldots, f_k$. There exists a randomized algorithm that, given the description of $f_1, \ldots, f_k$ as a list of their nonzero monomials, outputs a random point $v \in \mathbb{F}_{p^\ell}^n$ such that the distribution of v is $\frac{6}{p^{\ell(1-\epsilon)}}$-close to the uniform distribution on V. The worst-case runtime complexity of this algorithm is polynomial in $n, d, \ell \log(p)$ and the description of $f_1, \ldots, f_k$.

Concretely, in Corollary 1 the field is $\mathbb{F}_{2^\ell}$ and $n = 4$, as elements of $\mathbb{F}_{2^{4\ell}}$ are interpreted as 4-dimensional vectors over $\mathbb{F}_{2^\ell}$. The variety V is given by the polynomials (in $x = (x_1, \ldots, x_4)$)

$$p_1(x) - \tilde{y} = 0$$
$$\langle h \mid x \rangle - m + \hat{s} = 0$$

where $p_1(x) - \tilde{y} \in \mathbb{F}_{2^\ell}[x_1, \ldots, x_4]$ is a polynomial of degree q and $\langle h \mid x \rangle - m + \hat{s} \in \mathbb{F}_{2^\ell}[x_1, \ldots, x_4]$ is trivially a polynomial of degree 1. Thus the parameters are $k = 2$, $n = 4$, $p = 2$, and $d = q$. We can set $\epsilon = \frac{1}{2}$ and Theorem 1 yields an efficient algorithm that samples $\frac{6}{2^{\ell/2}}$-close to uniform from V.

Resettably Sound Zero-Knowledge Arguments from OWFs - The (Semi) Black-Box Way

Rafail Ostrovsky[1,*], Alessandra Scafuro[2,**],
and Muthuramakrishnan Venkitasubramanian[3]

[1] UCLA, USA
[2] Boston University and Northeastern University, USA
[3] University of Rochester, USA

Abstract. We construct a constant round resettably-sound zero knowledge argument of knowledge based on black-box use of any one-way function. Resettable-soundness was introduced by Barak, Goldreich, Goldwasser and Lindell [FOCS 01] and is a strengthening of the soundness requirement in interactive proofs demanding that soundness should hold even if the malicious prover is allowed to "reset" and "restart" the verifier. In their work they show that resettably-sound ZK arguments require non-black-box simulation techniques, and also provide the first construction based on the breakthrough simulation technique of Barak [FOCS 01]. All known implementations of Barak's non-black-box technique required non-black-box use of a collision-resistance hash-function (CRHF).

Very recently, Goyal, Ostrovsky, Scafuro and Visconti [STOC 14] showed an implementation of Barak's technique that needs only black-box access to a collision-resistant hash-function while still having a non-black-box simulator. (Such a construction is referred to as *semi black-box*.) Plugging this implementation in the compiler due to Barak et al. yields the first resettably-sound ZK arguments based on black-box use of CRHFs.

However, from the work of Chung, Pass and Seth [STOC 13] and Bitansky and Paneth [STOC 13], we know that resettably-sound ZK arguments can be constructed from non-black-box use of any one-way function (OWF), which is the *minimal* assumption for ZK arguments.

Hence, a natural question is whether it is possible to construct resettably-sound zero-knowledge arguments from black-box use of any OWF only. In this work we provide a positive answer to this question thus closing the gap between black-box and non-black-box constructions for resettably-sound ZK arguments.

* Work supported in part by NSF grants 09165174, 1065276, 1118126 and 1136174, US-Israel BSF grant 2008411, OKAWA Foundation Research Award, IBM Faculty Research Award, Xerox Faculty Research Award, B. John Garrick Foundation Award, Teradata Research Award, and LockheedMartin Corporation Research Award; and DARPA under Contract N00014 -11 -1-0392. The views expressed are those of the author and do not reflect the official policy or position of the Department of Defense or the U.S. Government.
** Work done while working at UCLA.

Y. Dodis and J.B. Nielsen (Eds.): TCC 2015, Part I, LNCS 9014, pp. 345–374, 2015.

1   Introduction

Zero-knowledge (ZK) proofs [13] allow a prover to convince a verifier of the validity of a mathematical statement of the form "$x \in L$" without revealing any additional knowledge to the verifier besides the fact that the theorem is true. This requirement is formalized using a simulation paradigm: for every malicious verifier there exists a simulator that having a "special" access to the verifier (special in the sense that the access granted to the simulator is not granted to the prover) but no witness, is able to reproduce the view that the verifier would obtain interacting with an honest prover (who knows the witness). The simulator has two special accesses to the verifier: black-box access, it has the power of resetting the verifier during the simulation; non-black-box access, it obtains the actual code of the verifier. While providing no additional knowledge, the proof must also be sound, i.e. no malicious prover should be able to convince a verifier of a false statement.

In this work we consider a stronger soundness requirement where the prover should not be able to convince the verifier even when having the power of resetting the verifier's machine (namely, having black-box access to the verifier). This notion of soundness, referred to as *resettable soundness*, was first introduced by Barak, Goldwasser, Goldreich and Lindell (BGGL) in [3], and is particularly relevant for cryptographic protocols being executed on embedded devices such as smart cards. Barak et al. in [3] prove that, unless **NP** $\subseteq$ BPP, interactive proofs for **NP** cannot admit a black-box zero knowledge simulator and be resettably-sound at same time. (Indeed, a resetting prover has the same special access to the verifier as a black-box simulator and it can therefore break soundness just by running the simulator's strategy.) Then, they provide the first resettably-sound zero-knowledge arguments for NP based on the non-black-box zero-knowledge protocol of [1] and on the existence of collision-resistant hash-functions (CRHFs). Recently, Chung, Pass and Seth (CPS) [9] showed that the minimal assumption for non-black-box zero-knowledge is the existence of one-way functions (OWFs). In their work they provide a new way of implementing Barak's non-black-box simulation strategy which requires only OWFs. Independently, Bitansky and Paneth [5] also showed that OWFs are sufficient by using a completely new approach based on the impossibility of approximate obfuscation.

Common to all the above constructions [3,9,5], beside the need of non-black-box simulation, is the need of non-black-box use of the underlying cryptographic primitives. Before proceeding, let us explain the meaning of black-box versus non-black-box use of a cryptographic primitive. A protocol makes black-box use of a cryptographic primitive if it only needs to access the input/output interface of the primitive. On the other hand, a protocol that relies on the knowledge of the implementation (e.g., the circuit) of the primitive is said to rely on the underlying primitive in a non-black-box way. A long line of work [19,17,28,6,31,14,22,15] starting from the seminal work of Impagliazzo and Rudich [18] aimed to understand the power of the non-black-box access versus the black-box access to a cryptographic primitive. Besides strong theoretical motivation, a practical reason is related to efficiency. Typically, non-black-box constructions are inefficient and as such, non-black-box constructions are used merely to demonstrate "feasibility" results. A first step towards making these constructions efficient is to obtain a construction that makes only black-box use of the underlying primitives.

In the resettable setting non-black-box simulation is necessary. In this work we are interested in understanding if non-black-box use of the underlying primitive is necessary as well. Very recently, [16] constructed a public-coin ZK argument of knowledge based on CRHFs in a black-box manner. They provided a non black-box simulator but a black-box construction based on CRHFs. Such a reduction is referred to as a *semi black-box* construction (see [29] for more on different notions of reductions). By applying the [3] transformation, their protocol yields the first (semi) black-box construction of a resettably-sound ZK argument that relies on CRHFs. In this paper, we address the following open question:

Can we construct resettably-sound ZK arguments under the minimal assumption of the existence of a OWF where the OWF is used in a black-box way?

1.1 Our Results

We resolve this question positively. Formally, we prove the following theorem.

Theorem 1 (Informal). *There exists a (semi) black-box construction of an O(1)-round resettably-sound zero-knowledge argument of knowledge for every language in NP based on one-way functions.*

It might seem that achieving such result is a matter of combining techniques from [16], which provides a "black-box" implementation of Barak's non-black-box simulation and [9], which provides an implementation of Barak's technique based on OWFs. However, it turns out that the two works have conflicting demands on the use of the underlying primitive which make the two techniques "incompatible".

More specifically, the construction presented in [16] crucially relies on the fact that a collision-resistance hash-function is publicly available. Namely, in [16] the prover (and the simulator) should be able to evaluate the hash function *on its own* on any message of its choice at any point of the protocol execution. In contrast, the protocol proposed in [9] replaces the hash function with digital signatures (that can be constructed from one-way functions), and requires that the signature key is hidden from the prover: the only way the prover can obtain a signature is through a "signature slot". Consequently in [9], in contrast with [16], the prover cannot compute signatures on its own, cannot obtain signatures at any point of the protocol (but only in the signature slot), and cannot obtain an arbitrary number of signatures.

Next, we explain the prior works in detail.

1.2 Previous Techniques and Their Limitations

We briefly review the works [16] and [9] in order to explain why they have conflicting demands. As they are both based on Barak's non black-box simulation technique, we start by describing this technique.

Barak's Non-Black-Box Zero Knowledge [1]. Barak's ZK protocol for an NP language L is based on the following idea: a verifier is convinced if one of the two statements is true: (1) the prover knows the verifier's next-message-function, or (2) $x \in L$. By definition a non-black-box simulator knows the code of the next-message-function of the verifier V^* which is a witness for statement 1, while the honest prover has the witness for statement 2. Soundness follows from the fact that no adversarial prover can predict the next-message-function of the verifier. Zero-knowledge can be achieved by employing a witness-indistinguishable (WI) proof for the above statements. The main bottleneck in translating this beautiful idea into a concrete construction is the size of statement 1. Since zero-knowledge demands simulation of arbitrary malicious non-uniform PPT verifiers, there is no *a priori* bound on the size of the verifier's next-message circuit and hence no strict-polynomial bound on the size of the witness for statement 1. Barak and Goldreich [2] in their seminal work show how to construct a WI argument that can *hide* the size of the witness. More precisely, they rely on Universal Arguments (UARG) that can be constructed based on collision-resistant hash-functions (CRHFs) via Probabilistically Checkable Proofs (PCPs) and Merkle hash trees (based on [21]). PCPs allow rewriting of proofs of NP statements in such a way that the verifier needs to check only a few bits to be convinced of the validity of the statement. Merkle hash-trees [23], on the other hand, allow committing to strings of arbitrary length with the additional property that one can selectively open some bits of the string by revealing only a small *fixed* amount of decommitment information. More precisely, a Merkle hash-tree is constructed by arranging the bits of the string on the leaves of a binary tree, and setting each internal node as the hash of its two children: a Merkle tree is a *hash chain*. The commitment to the string corresponds to the commitment to the root of the tree. The decommitment information required to open a single bit of the string is the path from the corresponding leaf in the tree to the root along with their siblings. This path is called *authentication* path. To verify the decommitment, it is sufficient to perform a *consistency check* on the path with the root previously committed. Namely, for each node along the path check if the node corresponds to the hash of the children, till the last node that must correspond to the root. Merkle trees allow for committing strings of super-polynomial length with trees of poly-logarithmic depth. Thus, one can construct a universal argument by putting PCP and Merkle tree together as follows. First, the prover commits to the PCP-proof via a Merkle hash-tree. The verifier responds with a sequence of locations in the proof that it needs to check and the prover opens the bits in the respective locations along with their authentication paths. While this approach allows constructing arguments for statements of arbitrary polynomial size, it is not witness indistinguishable because the authentication paths reveal the size of the proof, and therefore the witness used (this is because the length of the path reveals the depth of the tree). To obtain witness indistinguishability Barak's construction prevents the prover from revealing the actual values on any path. Instead, the prover commits to the paths padded to a fixed length, and then proves that the opening of the commitments corresponds to *consistent* paths leading to *accepting* PCP answers. A standard ZK is sufficient for this purpose as the size of the statement is strictly polynomial (is fixed as the depth of the Merkle tree). Such ZK proofs, however, need the code of the CRHFs used to build the Merkle-hash tree.

Black-box Implementation of Barak's Non-black-box Simulation Strategy [16]. Recently, Goyal, Ostrovsky, Scafuro and Visconti in [16] showed how to implement Barak's simulation technique using the hash function in a black-box manner.

They observe that in order to use a hash function in a black-box manner, the prover cannot prove the consistency of the paths by giving a proof, but instead it should reveal the paths and let the verifier recompute the hash values on its own and verify the consistency with the root of the tree. The problem is that to pass the consistency check an honest prover can open paths that are at most as long as the real tree, therefore revealing its size. Instead we need to let the verifier check the path while still keeping the size of the real tree hidden.

To tackle this, they introduce the concept of an *extendable* Merkle tree, where the honest prover will be able to extend any path of the tree *on the fly*, if some conditions are met. Intuitively, this solves the problem of hiding the size of the tree — and therefore the size of the proof— because a prover can show a path for any possible proof length.

To implement this idea they construct the tree in a novel manner, as a sequence of *"LEGO"* nodes, namely nodes that can be connected to the tree on the fly. More concretely, observe that checking the consistency of a path amounts to checking that each node of the path corresponds to the hash of the children. This is a *local* check that involves only 3 nodes: the node that we want to check, say A, and its two children say leftA, rightA, and the check passes if A is the hash of leftA, rightA. The LEGO idea of [16] consists of giving the node the following structure: a node A now has two fields: A = [*label*, *encode*], where *label* is the hash of the children leftA, rightA, while *encode* is some suitable encoding of the *label* that allows for black-box proof. Furthermore, *label* is not the hash of the *label* part of leftA, rightA, but it is the hash of the *encode* part. Thus there is not direct connection between the hash values, and the hash chain is broken. The second ingredient to guarantee binding, is to append to each node a proof of the fact that *"encode* is an encoding of *label"*. Note that now the tree so constructed is not an hash chain anymore, the chain is broken at each step. However, it is still binding because there is a proof that connects the hash with its encoding. More importantly note that, if one can cheat in the proof appended to a node, then one can replace/add nodes. Thus, provided that we introduce a *trapdoor* for the honest prover (and the simulator) to cheat in this proof, the tree is *extendable*. Therefore, now we can let the verifier check the hash value by itself, and still be able to hide the depth of the committed tree.

For the honest prover, which *does not* actually compute the PCP proof, the trapdoor is the witness for the theorem "$x \in L$". For the simulator, which is honestly computing the PCP proof and hence the Merkle tree, the trapdoor is the depth of the real tree, i.e., the depth, say d^*, at which the leaves of the real

Merkle tree lies. To guarantee binding, the depth d^* is committed at the very beginning. The simulator is allowed to cheat for all PCP queries associated to trees that are not of the committed depth.

The final piece of their construction deals with committing the code of V^* which also can be of an arbitrary polynomial size. The problem is that to verify a PCP proof, the standard PCP verifier needs to read the entire statement (in this case the code of V^*) or at least the size of the statement to process the queries. But revealing the size will invalidate ZK again. [16] get around this problem by relying on Probabilistically-Checkable Proof of Proximity (PCPP) [4] instead of PCPs which allow the verifier to verify any proof and arbitrary statement by querying only a few bits of the statement as well as the proof. For convenience of the reader, we provide a very simplified version of [16]'s protocol in Fig. 1.

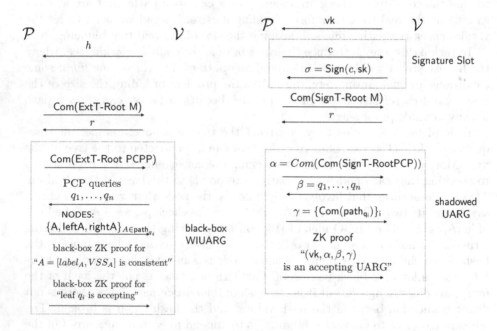

Fig. 1. Semi black-box resettably-sound ZK AoK from CRHF [16]

Fig. 2. Non-black-box resettably-sound ZK AoK from OWFs [9]

Summing up, some of the key ideas that allow [16] for a black-box use of the hash function are: (1) the prover unfolds the paths of the Merkle tree so that the verifier can directly check the hash consistency, (2) the prover/simulator can arbitrarily extend a path on the fly by computing fake LEGO nodes and cheat in the proof of consistency using their respective trapdoors. The work of [16] retains all the properties of Barak's ZK protocol, namely, is public-coin and constant-round (therefore resettable sound due to [3]), and relies on the underlying CRHF in a black-box manner.

Non black-box simulation using OWFs [9]. Chung, Pass and Seth [9] showed how to implement the non-black-box simulation technique of Barak using OWFs.

The main idea of CPS is to notice that digital signature schemes — which can be constructed from one-way functions — share many of the desirable properties of CRHFs, and to show how to appropriately instantiate (a variant of) Barak's protocol using signature schemes instead of using CRHFs. More precisely, CPS show that by relying on strong fixed-length signature schemes, one can construct a *signature tree* analogous to the Merkle hash-tree that allows compression of arbitrary length messages into fixed length commitments and additionally satisfies an analogue collision-resistance property. The soundness of such a construction will have to rely on the unforgeability (i.e. collision-resistance) of the underlying signature scheme. Hence it must be the case that is the verifier the one generating the secret key and the signatures for the prover. Towards this, CPS adds a signature slot at the beginning of the protocol. More precisely, first the verifier generates a signature key-pair vk, sk and sends only the verification key vk to the prover. Next, in a "signature slot", the prover sends a commitment c to the verifier, and the verifier returns a valid signature σ of c (using sk). The simulator constructs the signature tree by rewinding the (malicious) verifier, and then succeeds in the WIUARG proof as in Barak's protocol. While the simulator can use the signature slot to construct the signature tree, we cannot require the honest prover to construct any tree since it is not allowed to rewind the verifier. To address this, CPS uses a variant of Barak's protocol due to Pass and Rosen [26], which relies on a special-purpose WIUARG, in which the honest prover never needs to perform any hashing. The idea here is that since there exist public-coin UARGs, the prover can first engage in a shadowed UARG where the prover merely commits to its messages and then in a second phase proves using a witness-indistinguishable proof that either $x \in L$ or the messages it committed to constitute a valid UARG. This will allow the honest prover to "do nothing" in the shadowed UARG and use the witness corresponding to x in the second phase. The simulator instead is able to compute a valid signature-tree by rewinding the verifier, and it commits to valid messages in the shadowed UARG.

The resulting protocol is not public-coin, nevertheless [9] shows that it suffices to apply the PRF transformation of BGGL to obtain a protocol that is resettably-sound. We provide a very informal pictorial description of the CPS protocol in Fig. 2. It seems inherent that the CPS protocol needs a shadowed UARG, and hence proving anything regarding this shadowed argument needs to use the underlying OWF in a non-black-box manner.

Competing requirements of [16] and   [9]. In summary, in [16], in order to use the CRHF in a black-box manner, the prover is required to open the paths of the Merkle Tree corresponding to the PCPP queries and let the verifier check their consistency. To preserve size-hiding, the prover needs the ability to arbitrarily extend the paths of the tree by *privately* generating new nodes and this is possible because the prover can compute the hash function on its own. In contrast, in [9] the prover cannot compute nodes on its own, but it needs to get signatures from the verifier. Therefore the protocol in [9] is designed so that the prover never has to use the signatures.

In this work we show a technique for using the benefits of the signature while relying on the underlying OWF in a black-box manner and we explain this in the next section.

1.3 Our Techniques

Our goal is to implement the resettably-sound ZK protocol based on Barak's non-black-box simulation technique using OWFs in a black-box manner. We have illustrated the ideas of [16] to implement Barak's ZK protocol based on extendable Merkle hash-tree that uses a CRHF only as a black-box, and the ideas of [9] that show how to compute a Merkle signature-tree based on (non-black-box use of) OWFs.

The natural first step to reach our goal is then to take the protocol of [16] and implement the extendable Merkle tree with signatures instead of CRHF. As mentioned earlier, this replacement cannot work because the crucial property required to extend the tree is that the prover computes nodes on its own. If we replace CRHF with signatures, then the prover needs to ask signatures from the verifier for every node. This means that any path computed by the prover is already known by the verifier (even the concept of "opening a path" does not seem to make much sense here as the verifier needs to participate in the computation of the path). But instead we need the ability to commit to a tree and then extend it *without* the verifier knowing that we are creating new nodes.

We are able to move forward using the following facts underlying [16]'s protocol. First, although the prover is required to extend paths and prove consistency, it does so by cheating in every proof of consistency of the nodes. Indeed, recall that a node is a pair (*label, encode*), the consistency between *label* and *encode* is proved via ZK, and the prover cheats in this proof using the witness for "$x \in L$" as a trapdoor.

Under closer inspection we notice that the prover does not need to compute any tree; it just needs to compute the paths corresponding to the PCPP queries on-the-fly when it is asked for it. Indeed, an equivalent version of [16]'s protocol would be the following. The prover commits to the root by just committing to a random string.[1] Then, when it sees the PCPP queries $q_1, \ldots, q_n$ it computes *on-the-fly* the paths for leaves in the corresponding positions, and is able to prove consistency of the paths with the previously committed root by cheating in the proofs. Finally, we point out that the hash function is required only to compute the *label* part of the node, while the *encode* part can be computed by the prover on its own.

Armed with the above observations, we present our idea. As in [9] we place a signature slot at the very beginning of the protocol. This signature slot enables

---

[1] In [16] the prover actually commits to the hash of two random nodes. In our paper instead we use instance-dependent trapdoor commitments, computed on the instance x. The prover, that knows the witness w, can just commit to a random string and then equivocate later accordingly.

the simulator to get unbounded number of signatures by rewinding the verifier, and ultimately to construct the extendable Merkle trees. After the signature slot, the prover commits, using an instance-dependent trapdoor commitment scheme, to the roots of the extendable Merkle trees, one tree for the hidden statement of the PCPP (the code of the verifier) and one tree for the PCPP proof. Such trapdoor commitment allows the possessor of the witness to equivocate any commitment. Therefore, the roots committed by the prover – who knows the witness – are not binding. Next, when the prover receives the set of PCPP queries $q_1, \ldots, q_n$ (more specifically, one set of queries for each possible depth of the tree), it computes the paths on-the-fly with the help of the verifier. Namely, for each node along the paths for leaves $q_1, \ldots, q_n$, the prover computes the *encoding* part (which are equivocal commitments), and sends them to the verifier who computes the *label* part by "hashing" the encodings, namely, by computing their signature (in a proper order). Therefore, deviating from [9], we introduce a second signature slot where the prover gets the signatures required to construct the paths dictated by the PCPP queries. Once the labels/ signature for each node have been computed by the verifier, the paths are finally complete. Now the prover can proceed with proving that the paths just computed are *consistent* with the roots previously committed and lead to accepting PCPP answers (such black-box proof follows the ideas of [16]). Interestingly in this game the verifier knows that the prover is cheating and the paths cannot be possibly consistent, but is nevertheless convinced because the only way the prover can cheat in the consistency proof is by using the witness for $x \in L$, which guarantees the validity of x.

Remark 1. We remark that the prover does not compute any PCPP proof. In the entire protocol it just commits to random values and only in the last step it equivocates the commitments in such a manner that will convince the verifier in the proofs of consistency.

Now, let's look at the simulation strategy. The simulator honestly computes extendable Merkle signature-trees to commit to the machine V^* and to the PCPP, using the first signature slot. Then, when the verifier sends PCPP queries the simulator answers in the following way. For the queries that do not concern the real tree (recall that virtually there are polynomially many possible trees and the PCPP proof can lie in any of those, thus the verifier will provide queries $q_1, \ldots, q_n$ for each possible depth of the tree and expects correct answer only for the queries associated to the depth committed at the beginning) the simulator sends *encode* parts which are commitments of random strings. Later on it will cheat in the proof of consistency for such queries by using its *trapdoor* (as in [16] the trapdoor for the simulator is the size of the real tree). For the queries that hit the real tree, the simulator sends the same commitments that it sent in the first signature slot and that were used to compute the real tree. Indeed, for these nodes the simulator must prove that they are truly consistent, and it can do so by forcing in the view of the verifier the same paths that it already computed for the real tree.

Thus, for the simulation to go through it should be the case that the signatures that the simulator is collecting in the second signature slot, match the ones that it obtained in the first signature slot and that were used to compute the real tree.

Unfortunately, with the protocol outlined above we do not have such guarantee. This is due to two issues. First, the verifier can answer with some probability in the first slot and with another probability in the second slot, therefore skewing the distribution of the output of the simulator. Second, the verifier might not compute the signature deterministically: in this case the signatures obtained in the first slot and used to compute the real tree will not match the signatures obtained in the second slot, where the paths are "re-computed", and thus the simulator cannot prove that the nodes are consistent. We describe the two issue in details and we show how we change the construction to fix each issue.

Issue 1. V^* aborts in the two signature slots with different probabilities. We describe this issue with the following example. Consider a malicious verifier that aborts on all messages that start with bit 0 in the first signature slot and aborts on all sets of messages in the second slot if $3/4$ of them start with bit 1. The honest prover will succeed with probability close to a $1/2$ since the verifier will abort in the first message with probability $1/2$ and not abort with high probability in the second slot.[2] The simulator on the other hand can only obtain signatures of commitments that start with bit 1 in the first slot and has to use the same commitments in the second slot. This means that all the commitments sent by the simulator in the second slot will begin with the bit 1 and the verifier will always abort. Hence the simulator can never generate a view. The way out is to come up with a simulation strategy ensuring that the distribution fed in the second slot is indistinguishable to the honest prover's messages.

Fixing for issue 1. We first amplify the probability that the verifier gives a signature in the first signature slot by requesting the verifier to provide signatures for $T = O(n^{c+1})$ random commitments instead of just one. Using a Yao-type hardness-amplification, we can argue that if the verifier provides valid signatures with non-negligible probability then we can obtain signatures for at least $1 - \frac{1}{n^c}$ fraction of random tapes for the commitment. Lets call these random tapes good. Now if $k << n^c$ commitments are sent in the second slot, with probability at least $1 - \frac{k}{n^c}$ over random commitments made in the second slot, all of them will be good. This is already promising since the verifier at best will try to detect good messages in the second slot and with high probability all of them are good. However there is still a non-negligible probability (i.e. $\frac{k}{n^c}$) that the verifier could abort in this simulation strategy. To fix the next issue, we will have the verifier use several keys to sign and we will leverage that to handle this non-negligible fraction of bad messages.

Issue 2. V^* Does not Compute Signatures Deterministically. Our protocol requires the verifier to compute the signature deterministically, namely, the randomness used to compute the signatures must be derived from a PRG

---

[2] We assume here that random commitments are equally likely to have their first bits 0 or 1.

whose seed is sampled along with the parameters vk, sk and is part of the secret key. However, in the construction that we outlined before we cannot enforce a malicious verifier from signing deterministically. As mentioned before, if the verifier gives different (correct) signatures for the same node in the first and second slot, the simulator cannot proceed with the proof of consistency. The only way to catch the verifier is to demand from the verifier a proof that the signatures are computed deterministically. Because we need to use the cryptographic primitives in a black-box manner, we cannot solve the problem by attaching a standard proof of consistency.

Fixing for issue 2. We force the verifier to be honest using a cut-and-choose mechanism. At high level, we require the verifier to provide signatures for n different keys instead of just one in the signature slots. Together with sending the verification key vk_i, the verifier will also append the commitment to the randomness used in the key generation algorithm (the randomness determines the PRG seed that is used to deterministically sign the messages).

After the first signature slot, the prover asks the verifier to reveal $n/2$ keys (by requiring the verifier to decommit to the randomness used to generate the keys) and checks if, for the revealed keys, the signatures obtained so far were computed honestly. This verification requires the use of OWF only in a black-box manner. The prover proceeds with the protocol using the remaining half of the keys, namely by committing to $n/2$ roots, and obtaining $n/2$ sets of PCPP queries from the verifier. Later, after the second signature slot is completed, the prover will ask to open half of the remaining keys, namely $n/4$ keys, and checks again the consistency of the signatures obtained so far. If all checks pass, the prover is left with paths for $n/4$ trees/PCPP queries for which he has to prove consistency. Due to the cut-and-choose, we know that most of the remaining signatures were honestly generated, but not all of them. Therefore, at this point we will not ask the prover to prove consistency of all remaining $n/4$ executions. Instead, we allow the prover to choose one coordinate among the $n/4$ left, and to prove that the tree in this coordinate is consistent with the paths.

Allowing the prover to choose the tree for which to prove consistency, does not give any advantage compared to the original solution where there was only one tree. On the other hand, having a choice in the coordinate allows the simulator to choose the tree for which it received only consistent signatures and for which it will be able to provide a consistency proof. One can see this coordinate as another *trapdoor*. Namely, in the previous construction, the simulator commits to the depth of the real tree, so that in the consistency proof it can cheat in answering all queries that do not hit the committed value. We follow exactly the same concept by adding the commitment of the coordinate. The simulator commits to the coordinate for which it wants to provide the proof, and it is allowed to cheat in all the remaining proofs. Finally, because we are in the resettable setting, we require the prover to commit in advance to the coordinates that he wants the verifier to open. If this was not the case then the prover can rewind the verifier and get all the secret keys.

It only remains to argue that there is a strategy so that the simulator can always find one good key. Recall that the simulator rewinds the verifier several times in the first signature slot to obtain the signature on any message. The adversary has to sign deterministically in most coordinates because of the cut-and-choose mechanism. However, it can cheat in a few of them and in different rewindings it can choose to cheat in different keys. To ensure that the signatures obtained are the ones that are deterministically signed, we will make the simulator to obtain n signatures on a commitment and take that signature that occurs more than half of the times. However, there could be messages for which there will be no majority among the n signatures or worse the wrong signature in majority. This issue is quite subtle and combining the Yao-amplification and the cut-and-choose mechanism we argue that for all but small fraction of the messages, the simulator will obtain the deterministically-signed signature for most of the unopened keys. As with the first issue, we are still left with a small fraction of messages for which the simulator could receive a bad signature.

Handling Bad Messages and Bad Signatures. The fix for Issue 1 results in the simulator using a small fraction ($\approx \frac{1}{poly(n)}$) of commitments in the actual Merkle Tree whose signatures are not good messages w.r.t the first Signature Slot. If such a message is fed in the second Signature Slot, the Verifier can detect it. The fix for Issue 2 results in the simulator using a small fraction of commitment with bad signatures (i.e., not deterministically signed). If a message for which a bad signature was obtained is fed in the second Signature Slot then the signature obtained in the second slot will be different from the bad signature that the simulator obtained. However, we need to guarantee that the simulator will be successful in placing the commitments used in first signature slot to compute the Merkle tree, into the second signature slot, for at least one index among the $n/4$ trees remaining from the cut-and-choose. It will "test the waters" first: More precisely, before feeding the actual commitments in the second Signature Slot, for every unopened key, it will first generate random commitments to be sent in the second Signature Slot for that key and check if the random commitments are good messages w.r.t the first signature slot. More precisely, it will check if the first slot yields a signature for these commitments. A key is considered good if all these random commitments turn out to be good. For good keys, the simulator will swap the commitments with the actual commitments used to generate the Merkle Trees. It can be shown that the distribution induced by this swap is not skewed because a set of random good messages are swapped with the real commitments which are also random good commitments, since they yielded signatures in the first Signature Slot. This will help us handle bad messages as long as we can show there will be good keys. Arguing this turns out to be subtle and our proof will show that there will be only few bad keys. An analogous argument can be made to show there will be only few bad keys for which there is some commitment among the ones to be sent in the second Signature Slot by the simulator with a bad signature. If we start with sufficiently many ($O(n)$) keys then we will be able to show that there will be at least one good key that survives from bad messages and bad signatures.

2   Definitions

In this section we provide the definitions of some of the tools that we use in our construction. We refer the reader to the full version [25] for more details and for the definition of more standard tools, like instance-dependent equivocal commitment, that we omit in this section. We assume familiarity with interactive arguments and argument of knowledge.

Definition 1 (Zero-knowledge [13]). *An interactive protocol* (P, V) *for a language* L *is* zero-knowledge *if for every PPT adversarial verifier* V^* *and auxiliary input* $z \in \{0, 1\}^*$, *there exists a PPT simulator* S *such that the following ensembles are computationally indistinguishable over* $x \in L$:

$$\{\mathsf{View}_{V^*}\langle P, V^*(z)\rangle(x)\}_{x \in L, z \in \{0,1\}^*} \approx \{S(x, z)\}_{x \in L, z \in \{0,1\}^*}$$

Definition 2 (Resettably-sound Arguments [3]). *A* resetting attack *of a cheating prover* P^* *on a resettable verifier* V *is defined by the following two-step random process, indexed by a security parameter* n.

1. *Uniformly select and fix* $t = poly(n)$ *random-tapes, denoted* $r_1, \ldots, r_t$, *for* V, *resulting in deterministic strategies* $V^{(j)}(x) = V_{x, r_j}$ *defined by* $V_{x, r_j}(\alpha) = V(x, r_j, \alpha),$[3] *where* $x \in \{0, 1\}^n$ *and* $j \in [t]$. *Each* $V^{(j)}(x)$ *is called an incarnation of* V.
2. *On input* 1^n, *machine* P^* *is allowed to initiate* $poly(n)$-*many interactions with the* $V^{(j)}(x)$'s. *The activity of* P^* *proceeds in rounds. In each round* P^* *chooses* $x \in \{0, 1\}^n$ *and* $j \in [t]$, *thus defining* $V^{(j)}(x)$, *and conducts a complete session with it.*

Let (P, V) *be an interactive argument for a language* L. *We say that* (P, V) *is a* resettably-sound argument *for* L *if the following condition holds:*

- Resettable-soundness*: For every polynomial-size resetting attack, the probability that in some session the corresponding* $V^{(j)}(x)$ *has accepted and* $x \notin L$ *is negligible.*

Similarly to [9,8] we consider the following weaker notion of resettable soundness, where the statement to be proven is fixed, and the verifier uses a single random tape (that is, the prover cannot start many independent instances of the verifier).

Definition 3 (Fixed-input Resettably-sound Arguments [27]). *An interactive argument* (P, V) *for a* **NP** *language* L *with witness relation* $\mathcal{R}_L$ *is* fixed-input resettably-sound *if it satisfies the following property: For all non-uniform polynomial-time adversarial prover* P^*, *there exists a negligible function* $\mu(\cdot)$ *such that for every all* $x \notin L$,

$$\Pr[\mathsf{ran} \leftarrow \{0, 1\}^\infty; (P^{*V_{\mathsf{ran}}(x, \mathsf{pp})}, V_{\mathsf{ran}})(x) = 1] \leq \mu(|x|)$$

---

[3] Here, $V(x, r, \alpha)$ denotes the message sent by the strategy V on common input x, random-tape r, after seeing the message-sequence α.

This is sufficient because it was shown in [9] that any zero-knowledge *argument of knowledge* satisfying the weaker notion can be transformed into one that satisfies the stronger one, while preserving zero-knowledge (or any other secrecy property against malicious verifiers.

Claim. Let (P, V) be a fixed-input resettably sound zero-knowledge (resp. witness indistinguishable) argument of knowledge for a language $L \in \mathbf{NP}$. Then there exists a protocol (P', V') that is a (full-fledged) resettably-sound zero-knowledge (resp. witness indistinguishable) argument of knowledge for L.

Strong Deterministic Signature. In this section we define strong, fixed-length, deterministic secure signature schemes that we rely on in our construction. Recall that in a strong signature scheme, no polynomial-time attacker having oracle access to a signing oracle can produce a valid message-signature pair, unless it has received this pair from the signing oracle. The signature scheme being fixed-length means that signatures of arbitrary (polynomial-length) messages are of some fixed polynomial length. Deterministic signatures do not use fresh randomness in the signing process once the signing key has been chosen. In particular, once a signing key has been chosen, a message m will always be signed the same way.

Definition 4 (Strong Signatures). *A strong, length-ℓ, signature scheme* SIG *is a triple* (Gen, Sign, Ver) *of* PPT *algorithms, such that*

1. for all $n \in \mathcal{N}, m \in \{0,1\}^*$,

$$\Pr[(\mathsf{sk}, \mathsf{vk}) \leftarrow \mathsf{Gen}(1^n), \sigma \leftarrow \mathsf{Sign}_{\mathsf{sk}}](m); \mathsf{Ver}_{\mathsf{vk}}(m, \sigma) = 1 \wedge |\sigma| = \ell(n)] = 1$$

2. for every non-uniform PPT *adversary A, there exists a negligible function $\mu(\cdot)$ such that for all $(\mathsf{sk}, \mathsf{vk}) \leftarrow \mathsf{Gen}(1^n)$ it holds:*

$$\Pr[(m, \sigma) \leftarrow A^{\mathsf{Sign}_{\mathsf{sk}}(\cdot)}(1^n); \mathsf{Ver}_{\mathsf{vk}}(m, \sigma) = 1 \wedge (m, \sigma) \notin L] \le \mu(n),$$

where L denotes the list of query-answer pairs of A's queries to its oracle.

Strong, length-ℓ, **deterministic** signature schemes with $\ell(n) = n$ are known based on the existence of OWFs; see [24,30,12] for further details. In the rest of this paper, whenever we refer to signature schemes, we always means strong, length-n signature schemes.

Let us first note that signatures satisfy a "collision-resistance" property.

Claim. Let SIG = (Gen, Sign, Ver) be a strong (length-n) signature scheme. Then, for all non-uniform PPT adversaries A, there exists a negligible function $\mu(\cdot)$ such that for every $n \in \mathcal{N}$, for all $(\mathsf{sk}, \mathsf{vk}) \leftarrow \mathsf{Gen}(1^n)$ it holds:

$$Pr[(m_1, m_2, \sigma) \leftarrow A^{\mathsf{Sign}_{\mathsf{sk}}(\cdot)}(1^n, \mathsf{vk}); \mathsf{Ver}_{\mathsf{vk}}(m_1, \sigma) = \mathsf{Ver}_{\mathsf{vk}}(m_2, \sigma) = 1] \le \mu(n)$$

Verifiable Secret Sharing (VSS). A verifiable secret sharing scheme (VSS for short) [7] is a two-stage protocol run among $n + 1$ players. In the first stage, called $\mathsf{Share}(s)$, a special player, referred to as dealer, distributes a string s among the n players so that any t players (where $t = n/c$ for some constant $c > 3$) colluding cannot reconstruct the secret. The output of the Share phase is a set of VSS views $\mathsf{S}_1, \ldots, \mathsf{S}_n$ that we call VSS shares. In the second stage, called $\mathsf{Recon}(\mathsf{S}_1, \ldots, \mathsf{S}_n)$, any $(n - t)$ players can reconstruct the secret s by exchanging their VSS shares. The scheme guarantees that if at most t players are corrupted the Share stage is hiding, moreover a dishonest dealer is caught at the end of the Share phase through an accusation mechanism that disqualifies the dealer (this property is called t-privacy). A VSS scheme can tolerate errors on malicious dealer and players distributing inconsistent or incorrect shares, indeed the critical property is that even in case the dealer is dishonest but has not been disqualified, still the second stage always reconstructs the same string among the honest players.

MPC-in-the-head. MPC-in-the-head is a breakthrough technique introduced by Ishai at al. in [20] to construct a black-box zero-knowledge protocol. Let $\mathcal{F}_{ZK}$ be the zero-knowledge functionality for an NP language L, that takes as public input x and one share from each player P_i, and outputs 1 iff the secret reconstructed from the shares is a valid witness. Let MPCZK be a perfect (t, n)-secure MPC protocol implementing $\mathcal{F}_{ZK}$.

Very roughly, the "MPC-in-the-head" idea is the following. The prover runs *in his head* an execution of a (t, n)-secure MPCZK protocol among n imaginary players, each one participating in the protocol with a share of the witness. Then it commits to the view of each player separately. The verifier obtains t randomly chosen views, and checks that such views are consistent with an honest execution of the protocol and accepts if the output of every player is 1. Clearly P^* decides the randomness and the input of each player so it can cheat at any point and make players output 1. However, the crucial observation is that in order to do so, it must be the case that a constant fraction of the views committed are not consistent (this property is called t-robustness). Thus by selecting the t views at random, V will catch inconsistent views whp.

One can extend this technique further (as in [15]), to prove a general predicate ϕ about arbitrary values. Namely, one can consider the functionality $\mathcal{F}_\phi$ in which every player i participates with an input that is a view of a VSS player S_i. $\mathcal{F}_\phi$ collects all such views, and outputs 1 if and only if $\phi(\mathsf{Recon}(\mathsf{S}_1, \ldots \mathsf{S}_n)) = 1$. This idea is crucially used in [16].

Probabilistically Checkable Proofs. Informally, a PCP [2] system for a language L consists of a proof π written in a redundant form for a statement "$x \in L$", and a PPT verifier, which is able to decide the truthfulness of the statement by reading only few bits of the proof.

A PCP verifier V can be decomposed into a pair of algorithms: the query algorithm $\mathsf{Q}_{\mathsf{pcp}}$ and the decision algorithm $\mathsf{D}_{\mathsf{pcp}}$. $\mathsf{Q}_{\mathsf{pcp}}$ on input x and random tape r, outputs positions $q_1 = \mathsf{Q}_{\mathsf{pcp}}(x, r, 1), q_2 = \mathsf{Q}_{\mathsf{pcp}}(x, r, 2), \ldots, q_n = \mathsf{Q}_{\mathsf{pcp}}(x, r, p(|x|))$, for some polynomial p, and the prover answers with $b_i = \pi[q_i]$. V accepts if

$D_{pcp}(x, r, b_1, \ldots, b_{p(|x|)})$ outputs 1. For later, it is useful to see algorithm D_{pcp} as a predicate defined over a string π which is tested on few positions.

Probabilistically Checkable Proofs of Proximity. The standard PCP verifier decides whether to accept the statement $x \in L$ by probing few bits of the proof π and reading the entire statement x. A "PCP of proximity" (PCPP) [4] is a relaxation of PCP in which the verifier is able to make a decision without even reading the entire statement, but only few bits of it. More specifically, in a PCPP the theorem is divided in two parts (a, y). A public string a, which is read entirely by the verifier, a private string y, for which the verifier has only *oracle* access. Consequently, PCPP is defined for pair languages $L \subset \{0,1\}^* \times \{0,1\}^*$. For every $a \in \{0,1\}^*$, we denote $L_a = \{y \in \{0,1\}^* : (a, y) \in L\}$. The PCP Verifier can be seen as a pair of algorithms (Q_{pcpx}, D_{pcpx}), where $Q_{pcpx}(a, r, i)$ outputs a pair of positions (q_i, p_i): q_i denotes a position in the theorem y, p_i denotes a position in the proof π. D_{pcpx} decides whether to accept (a, y) by looking at the public theorem a, and at positions $y[q_i]$, $\pi[p_i]$. For later, it is useful to see algorithm D_{pcpx} as a predicate defined over two strings y, π, testing few positions of each string.

Definition 1 (PCPP verifier for a pair language). *For functions s, δ : $\mathcal{N} \to [0,1]$, a verifier V is a probabilistically checkable proof of proximity (PCPP) system for a pair language L with proximity parameter δ and soundness error s, if the following two conditions hold for every pair of strings (a, y):*

- *Completeness: If $(a, z) \in L$ then there exists π such that $V(a)$ accepts oracle $y \circ \pi$ with probability 1. Formally:*

$$\exists \pi, \Pr_{(Q,D) \leftarrow V(a)}[D((y \circ \pi)|_Q) = 1] = 1.$$

- *Soundness: If y is $\delta(|a|)$-far from $L(a)$, then for every π, the verifier $V(a)$ accepts oracle $y \circ \pi$ with probability strictly less than $s(|a|)$. Formally:*

$$\forall \pi, \Pr_{(Q,D) \leftarrow V(a)}[D((y \circ \pi)|_Q) = 1] < s(|a|).$$

Note that the query complexity of the verifier depends only on the public input a [10].

3 Protocol

Overview. The protocol starts with a signature slot, where the prover sends T commitments, and the verifier signs all of them. Then, as per Barak's construction, the prover sends a message z, which is the commitment to a machine M, the verifier sends a random string r, and finally the prover sends a commitment to a PCP of Proximity proof for the theorem: "the machine M committed in z is such that $M(z) = r$ in less then $n^{\log n}$ steps". M is the hidden theorem for the PCP of Proximity and the verifier has only oracle access to it. The above commitments are commitment to the roots of (extendable) Merkle signature-trees (that we will describe in details later). Next, the verifier sends the PCPP queries. As the verifier does not know the length of the PCPP proof and thus the depth of the Merkle tree, it will send a set of PCPP queries for each possible depth. Each PCPP query is a pair of indices, one index for the hidden theorem M and one for the PCPP proof. The verifier expects to see a path for each index.

At this point the prover needs to compute such paths. As we mentioned in the introduction, in a signature tree a path cannot be computed by the prover only: each nodes consists of two parts, the signature of the children, called *label*, and the encoding of the label, that we called *encode*. Thus, the prover continues as follows. For each path that must be provided for a query, he computes on-the-fly the *encode* parts of the nodes belonging to such path. It then sends all these "half" nodes to the verifier. The verifier computes the *label* parts for each node by signing the *encode* part and send them back to the prover. This is the second signature slot. Once all paths are completed, the prover starts the proof stage. Using a ZK protocol he proves that: (1) the paths are consistent with the root committed before, (2) the leaves of the paths open to accepting PCPP answers, in a black-box manner. How does the prover pass the proof stage? The prover computes all commitments using instance-dependent equivocal commitments and later cheats in the opening. How does the simulator pass the proof stage? The simulator computes consistent trees for the machine V^* and the PCPP proof by rewinding the verifier in the first signature slot, and committing to their depth at the beginning. On top of this outlined construction we use cut-and-choose to force the verifier to compute the signatures deterministically, so that the simulator can use the tree computed in the first signature slot. This concludes the high-level description of the protocol.

Now we need to show concretely how to compute the proofs of consistency using the signature and the commitment scheme only as *black-box*. This requires to go into the details of the (extendable) Merkle signature-tree and the mechanism for size hiding introduced in [16]. We provide such details in the next section.

3.1 Ingredients of the Construction

We present the ingredients of our construction in this section. Some of the ideas are adapted from [16].

String Representation. To allow black-box proofs for a committed string, the first ingredient is to represent the string with a convenient encoding that enables to give black-box proofs on top. For this purpose, following [16,15] we use a (t, n)-secure VSS scheme, defined in Sec. 2. To commit to any string s, the prover first runs, in his head, an execution of a *perfectly* (t, n)-secure VSS among $n + 1$ imaginary players where the dealer is sharing s, obtaining n views: $S[1], \ldots, S[n]$. Then it commits to each share separately using a statistically binding commitment.

Black-box Proof of a Predicate. With the VSS representation of strings, now the prover can commit to the string and prove any predicate about the committed string, using MPC-in-the-head as follow. Let $[S[1], \ldots, S[n]]$ be the VSS shares that reconstruct to a string s and let ϕ be a predicate. The prover wants to prove that $\phi(s)$ is true without revealing s. Define $\mathcal{F}_\phi$ as the n-party functionality that takes in input one VSS share $S[p]$ from each player p, and outputs $\phi(\text{Recon}(S[1],\ldots,S[n]))$. To prove $\phi(s)$, the prover runs a (t,n)-perfectly secure MPC-in-the-head among n players for the functionality $\mathcal{F}_\phi$. Each player participates to the protocol with input a VSS share of the string s. Then the prover commits to each view of the MPC-in-the-head so computed. The verifier checks the proof by observing t randomly chosen views of *both* the VSS and the MPC protocol, and checking that such views are consistent with an honest execution of the VSS and MPC protocols. Zero-knowledge follows from the t-privacy and soundness follows from the t-robustness of the MPC/VSS protocols, where t-robustness roughly means that, provided that the predicate to be proved is false and that the prover does not know in advance which views will be opened, corrupting only t players is not sufficient to convince the verifier with consistent views. On the other hand, by corrupting more than t players, the prover is caught whp.

(Extendable) Merkle Signature-Tree. As we discussed in the introduction, a node in an extendable Merkle tree is a pair $[label, encode]$. In our signature Merkle tree, the field *label* is a vector of signatures (computed by the verifier), and the field *encode* is a vector of commitments of VSS shares of *label*. Specifically, let γ be any node, let $\gamma 0 = [label^{\gamma 0}, \{\text{Com}(S^{\gamma 0}[i])\}_{i \in n}]$ be its left child, and $\gamma 1 = [label^{\gamma 1}, \{\text{Com}(S^{\gamma 1}[i])\}_{i \in n}]$ be its right child. Node γ is computed as follows. The label part is $label^\gamma = \{\text{Sign}_{\text{sk}}(\text{Com}(S^{\gamma b}[i]))\}_{b \in \{0,1\}, i \in n}$. The *encode* part is computed in two steps: First, compute shares $S_1^\gamma, \ldots, S_n^\gamma \leftarrow \text{Share}(label)$; next commit to each share separately. At **leaf level**, the $label^\gamma = s[\gamma]$, namely the γ-th bit that we want to commit.

Hiding the Size of the Tree. The size of the string committed, and hence the depth of the corresponding tree, is not known to the verifier and it must remain hidden. Specifically, the verifier should not know the size of the machine M and of the PCPP proof. Hence, the verifier will send a set of PCPP queries for each possible depth of the tree for the PCPP. Namely, for each possible depth $j \in [\log^2 d]$, V sends $\{q_{i,j}, p_{i,j}\}_{i \in k}{}^4$, where k is the soundness parameter.

Note that the prover (actually, the simulator) commits to one tree and is therefore able to correctly answer only the queries lying on the depth of the committed tree, and we want this to be transparent to the verifier. This is done by adding the commitment to the depth of the real tree at the beginning, and then proving for each query that either the query is correctly answered or the query refers to a depth that is different from the one committed. The commitment of the depth needs to be in the same format of the *encode* part of the nodes (i.e., it will be a commitment of VSS shares) because it will be used in the black-box proofs of consistency.

---

[4] We assume that the length PCPP proof is a power of 2. Also, for the sake of simplifying the notation we use the same index j for the queries to the machine $q_{i,j}$ and the proof $p_{i,j}$, even though the machine M and the corresponding PCPP proof π might lie on different depths.

Black-box Proofs about the Leaves of the Tree. As it should be clear by now, the proof consists in a sequence of paths, one for each PCPP query, and a sequence of proofs (one for every node and one for every possible depth) claiming that (1) the paths are consistent with the one committed root; (2) the paths open to accepting PCPP answers. This is done as follows. Attached to each path, there is a proof of consistency. This proof serves to convince the verifier that the path is consistent with the previously committed root. Attached to each set of paths (there is a set of paths for each depth $j = 1, \dots, \log^2 n$), there is a proof of acceptance. This proof serves to convince the verifier that those paths open to bits of PCPP proof/hidden theorem M that are accepting.

Both the prover and the simulator will cheat in this proof, but in different ways. The prover cheats in all proofs by equivocating the commitments (using the witness as trapdoor). The simulator cheats in all proofs concerning paths that do not match the depth of the real tree, by using the commitments of the depth as a trapdoor. Namely, the simulator will prove that either the path is consistent/accepting, or the depth of such path does not match the depth committed (note that there will exist one path that will match the committed depth). For the paths of the real tree, the simulator will honestly compute the proof. We now describe each step of the proof in more details.

Proof that a path is consistent. Let $p_{i,j}$ be a PCPP query for a tree of depth j. Associated to this query there is a path. Proving consistency of a path for $p_{i,j}$ amounts to prove consistency of each node along the path. For each node γ along the path for $p_{i,j}$, there the γ is $[label^\gamma, \{\mathsf{Com}(\mathsf{S}^\gamma[p])\}_{p \in n}]$, the prover proves that $\mathsf{Recon}(\mathsf{S}^\gamma[1], \dots, \mathsf{S}^\gamma[n]) = label^\gamma$. This is done via an MPC-in-the-head protocol, for a functionality $\mathcal{F}_{\mathsf{innode}}$ that takes in input the share $\mathsf{S}^\gamma[p]$ of the label, the share $\mathsf{S}_{\mathsf{depth}}[p]$ of the committed depth, and the string $label^\gamma$ and outputs 1 to all players if either $\mathsf{Recon}(\mathsf{S}^\gamma[1], \dots, \mathsf{S}^\gamma[n]) = label^\gamma$ or $\mathsf{Recon}(\mathsf{S}_{\mathsf{depth}}[1], \dots, \mathsf{S}_{\mathsf{depth}}[n]) \neq j$. Where $\{\mathsf{S}_{\mathsf{depth}}[1], \dots, \mathsf{S}_{\mathsf{depth}}[n]\} \leftarrow \mathsf{Share}(depth)$ are the share of the depth of the real tree and were committed at the beginning. The actual proof consists in the commitment of the views of the MPC players. The verifier verifies the proof by opening t views checking their consistency.

Proof that a set of paths is accepting. For each level j, the verifier sends queries $\{p_{i,j}, q_{i,j}\}_{i \in [k]}$ and the prover opens a path for each query. To prove that these queries are accepting, the prover computes a proof that involves the leaves of the paths of depth j. The prover runs an MPC-in-the-head for a functionality, that we call $\mathcal{F}_{\mathsf{VerPCPP}}$, that will check that the values in those positions will be accepted by a PCPP verifier. $\mathcal{F}_{\mathsf{VerPCPP}}$ takes in input shares: $\mathsf{S}^{p_{i,j}}[p]$, $\mathsf{S}^{q_{i,j}}[p]$, $\mathsf{S}_{\mathsf{depth}}[p]$ (with $p = 1, \dots, n$) and the public theorem; it then reconstructs the bits of the PCPP proof $\pi_{i,j} = \mathsf{Recon}(\mathsf{S}^{p_{i,j}}[1], \dots, \mathsf{S}^{p_{i,j}}[n])$ and of the hidden theorem $m_{i,j} = \mathsf{Recon}(\mathsf{S}^{q_{i,j}}[1], \dots, \mathsf{S}^{p_{i,j}}[n])$. It finally outputs 1 to all players iff: either the PCPP verifier accepts the reconstructed bits, i.e., $\mathsf{D}_{\mathsf{pcpx}}(m_{i,j}, \pi_{i,j}, q_{i,j}, p_{j,i}) = 1$, or if $\mathsf{Recon}(\mathsf{S}_{\mathsf{depth}}[1], \dots, \mathsf{S}_{\mathsf{depth}}[n]) \neq j$. The actual proof consists of the commitment of the views of the MPC players for $\mathcal{F}_{\mathsf{VerPCPP}}$. The verifier verifies the proof by checking the consistency of t views.

Verification of the proof. The verifier receives the commitments of all such views and ask the prover to open t of them. The prover will then decommits t views for *all* MPC/VSS protocol committed before as follow: first, it computes accepting MPC views by running the simulator granted by the t-security of the MPC-in-the-head protocol, then it open to such views by equivocating the corresponding commitments.

The Cut-and-choose. The mechanism described above is repeated n times: the verifier provides n signature keys, and the prover will compute n trees. During the protocol $3/4n$ of the secret keys will be revealed, so for those indexes the prover will not proceed to the proof phase. In fact, the prover will prove consistency of only one tree among the $1/4n$ trees left (but of course, we want the verifier to be oblivious about the tree that the prover is using). Thus, the last ingredient of our construction is to ask the prover to commit to an index J among the remaining $1/4n$ indexes. This commitment is again done via VSS of J and then committing to the view[5]. As expected, such VSS will be used in the computation of $\mathcal{F}_{innode}$ and $\mathcal{F}_{VerPCPP}$. The functionalities now will first check if the nodes are part of the J-th tree. If not, it means that the tree is not the one that must be checked, in such a case the functionality outputs 1 to all players regardless of whether any condition is satisfied.

3.2 The Construction

We now put everything together and provide the description of the final protocol. Some details are omitted for simplicity, a full specification of our protocol can be found in the full version [25]. We remark that in any step of the protocol the randomness used by the verifier to compute its messages is derived by the output of a PRF computed on the entire transcript computed so far. *Common Input:* An instance x of a language $L \in \mathbf{NP}$ with witness relation $\mathbf{R}_L$. *Auxiliary input to P:* A witness w such that $(x, w) \in \mathbf{R}_L$.

Cut-and-choose 1

- P_0: Randomly pick two disjoint subsets of $\{1, \ldots, n\}$ that we denote by J_1, J_2, with $|J_1| = n/2$ and $|J_2| = n/4$. Commit to J_1, J_2 using the equivocal commitment scheme.

- V_0: Run $(\mathsf{sk}_\kappa, \mathsf{vk}_\kappa) \leftarrow \mathsf{Gen}(1^n, r_\kappa)$ for $\kappa = 1, \ldots, n$. Send $\mathsf{vk}_\kappa, \mathsf{Com}(r_\kappa)$ for $\kappa = 1, \ldots, n$ to P.

- **Signature Slot 1.** P_1: Send $\mathsf{T} = O(n^c)$ commitments using the equivocal commitment scheme to 0^v (for some constant c and for v being as the size of the *encode* part). V_1: Signs each commitment.
  Check Signature Slot 1. P opens set J_1. V send sk_κ and decommitment to r_κ, for $\kappa \in J_1$. P checks that all signatures verified under key vk_κ are consistent with $\mathsf{sk}_\kappa, r_\kappa$. If not, abort.

---

[5] Attached to the VSS there will be also a proof that proves that $J \in \{1, \ldots, n\}/\{J_1 \cup J_2\}$.

Commitment to the Machine

- P_2: Send equivocal commitment to the *encode* part of the root of the (extendable) Merkle signature-tree for M, and equivocal commitment to the depth of the tree, this is done for each $\kappa \in \{1, \ldots, n\}/J_1$.

- V_2: Send a random string $r \in \{0,1\}^n$. Let (r, t) be the public theorem for the PCPP for the language: $\mathcal{L}_\mathcal{P} = \{(a = (r, \mathsf{t}), (Y)), \exists M \in \{0,1\}^*$ s.t. $Y \leftarrow \mathsf{ECC}(M)$, $M(z) \to r$ within t steps$\}$ (where $\mathsf{ECC}(\cdot)$ is a binary error correcting code tolerating a constant fraction $\delta > 0$ of errors, and δ is the proximity factor of PCPP).

Commitment to the PCPP proof

- P_3: Send equivocal commitment to the *encode* part of the root of the (extendable) Merkle signature-tree for the PCPP proof and the commitment to the depth of such tree. This is done for each $\kappa \in \{1, \ldots, n\}/J_1$.

- V_3: Send the random tapes for the PCPP queries. V and P obtain queries $(q_{i,j}, p_{i,j})$ for $i \in [k]$ and with $j = 1, \ldots, \log^2 n$.

- P_4: Send paths for $p_{i,j}, q_{i,j}$. Namely, send the *encode* part for each node along the paths $p_{i,j}, q_{i,j}$. This is done for each tree $\kappa \in \{1, \ldots, n\}/J_1$, previously committed.

Signature Slot 2 V_4: Sign the *encode* parts received from P.

Cut-and-choose 2. P opens the set J_2. V sends sk_κ and decommitment to r_κ, for $\kappa \in J_2$. P checks that all signatures verifier under key vk_κ are consistent with $\mathsf{sk}_\kappa, r_\kappa$. If not, abort.

Proof

- P_5: Commit to a random index $J \in \{1, \ldots, n\}/\{J_1 \cup J_2\}$.

  Then compute the proof of consistency of each path, and the proof of acceptance for each set of queries (as explained earlier), for each of the remaining trees. The proofs (i.e., the views of the MPC-in-the-head) are committed using an *extractable* equivocal commitment scheme.

- V_5: Select t players to check: Send indexes $p_1, \ldots, p_t$.

- P_6: Compute the t VSS shares and the t views of the MPC protocols that will make the verifier accept the proof. This is done by running the simulator guaranteed by the perfect t-security of the MPC protocols for the proofs. P then equivocates the previously committed views so that they open to this freshly computed views.

- V_6: Accept if all the opened views are consistent and output 1.

4   Security Proof

In this section we sketch the proof of the following theorem (for the formal proof the reader is referred to the full version of this work [25]).

Theorem 2. *There exists a (semi) black-box construction of a resettably-sound zero-knowledge argument of knowledge based on one-way functions.*

Resettable Soundness. We prove *fixed-input* resettable-soundness of the protocol without loss of generality. Assume for contradiction, there exists a PPT adversary P^*, sequences $\{x_n\}_{n \in \mathcal{N}} \subseteq \{0,1\}^*/L$, $\{z_n\}_{n \in \mathcal{N}} \subseteq \{0,1\}^*$ and polynomial $p(\cdot)$ such that for infinitely many n, it holds that P^* convinces V on common input $(1^n, x_n)$ and private input z_n with probability at least $\frac{1}{p(n)}$. Fix an n, for which this happens.

First, we consider a hybrid experiment HYB, where we run the adversary P^* on input (x_n, z_n) by supplying the messages of the honest verifier with a small modification. For all the randomness used by the verifier in the protocol via a PRF applied on the transcript, we instead supply truly random strings. In particular, the signature keys generated in the first message, the challenge string in message V_2, the random strings in V_3 and random t indices in V_5 are sampled truly randomly. By the pseudo-randomness of the PRF, we can conclude that P^* convinces the emulated verifier in this hybrid with probability at least $\frac{1}{p(n)} - \nu(n) > \frac{1}{2p(n)}$ for some negligible function $\nu(\cdot)$.

The high-level idea is that using P^*, we construct an oracle adversary A that violates the collision-resistance property of the underlying signature scheme. In more details, A is an oracle-aided PPT machine that on input vk, n and oracle access to a signing oracle $\mathsf{SIG}_{\mathsf{sk}}(\cdot)$ proceeds as follows: It internally incorporates the code of P^* and begins emulating the hybrid experiment HYB by providing the verifier messages. Recall that P^* is a resetting prover and can open arbitrary number of sessions by rewinding the verifier. A selects a random session i.

- For all the unselected sessions, A simply emulates the honest verifier.
- For the selected session, A proceeds as follows.
  (1) In the first message of the protocol, $A^{\mathsf{SIG}_{\mathsf{sk}}(\cdot)}$ chooses a random index $f \in [n]$ and places the vk in that coordinate. More precisely, it sends $(\mathsf{vk}_f, = \mathsf{vk}, c)$ where c is a commitment to the 0 string using Com. Note that since A does not have the secret key or the randomness used to generate $(\mathsf{sk}_f, \mathsf{vk}_f)$, it commits to the 0 string as randomness. It then emulates the protocol honestly. In either of the signature slots, whenever A needs to provide a signature under sk_f for any message m, A queries its oracle on m. For all the other keys, it possesses the signing key and can generate signatures on its own. If the sets J_1 or J_2 revealed in $P_{1.2}$ and $P_{4.2}$ contains f, then A simply halts.
  (2) If P^* fails to convince the verifier in the selected session, A halts. If it succeeds, then A stalls the emulation. Let C_0 contain the messages exchanged in session i. A then rewinds the prover to the message V_3. Let τ be the partial transcript of the messages exchanged until message V_3 in session i occurs.
  (3) Next, A uses fresh randomness to generate V_3, namely, the PCPP queries [6]. It then continues the execution from τ until P^* either convinces the verifier

---

[6] Here we use the reverse sampling property of the underlying PCPP of proximity.

in that session or aborts. If it aborts then A halts. Let C_1 be the transcript obtained from the second continuation of τ.

(4) Using the values revealed by P^* in the two continuations from the point τ, A will try to extract a collision if one exists and halts otherwise.

We describe below how the adversary A obtains a collision from two convincing transcripts starting from τ and argue that it can do so with non-negligible probability. Thus, we arrive at a contradiction to collision-resistance property of the signature scheme and will conclude the proof of resettable-soundness.

First, we consider a hybrid experiment HYB$'$, where a hybrid adversary A is provided with the actual commitment c to the randomness used to generate the signing key whose signing oracle it has access to. Besides that A' proceeds identically to A. By construction the internal emulation by A in HYB$'$ is identical to that of HYB. Below we analyze A's success in hybrid experiment HYB$'$. Finally, we claim that A will succeed with probability close to hybrid experiment HYB$'$ because the commitment scheme is hiding.

In a convincing session, for every PCPP query and every unopened signature key f, there is an associated set of paths that the prover reveals. More precisely, for every node γ in the paths, the prover provides $encode^\gamma$ which is the vector of commitments: $\{\mathsf{Com}(\mathsf{S}^\gamma[i])\}_{i \in [n]}$ of shares, which are supposed to reconstruct to valid signatures. If a node γ is supposed to be the child of the node γ', then it must be the case that $label^{\gamma'}$ contains the valid signatures of $\{\mathsf{Com}(\mathsf{S}^\gamma[i])\}_{i \in [n]}$.

Suppose that, in two convincing continuations from τ, for some pair of parent and child node γ' and γ, $label^{\gamma'}$ associated with γ' is the same in both the continuations but the commitments in the $encode$ part of γ are different. This means that there is at least one signature in the $label^{\gamma'}$ that verified correctly on two different values of $encode^\gamma$. Therefore, if A finds one such pair of nodes for the key sk_f, then it obtains a collision. We show below that with non-negligible probability, A will obtain a collision in this manner.

Next, we analyze the set of bad events when A receives two convincing continuations from τ. These bad events prevent A from finding a collision. So we bound the probabilities of these events happening. Fix a τ for which a random continuation yields a convincing session i with non-negligible probability.

B_1: P^* **equivocates the commitments.** If P^* equivocates the commitments then it can always compute t accepting views on the fly (as the honest prover does), and A's strategy fails. However, given that $x_n \notin L$, the commitment scheme used in the construction is statistically binding. Thus this event can happen only with negligible probability.

B_2: P^* **commits to a machine M that predicts V's next message.** Here P^* computes consistent trees and convinces the verifier using the same algorithm of the simulator. However, because the string r is chosen at random, the probability that this case happens is close to 2^{-n}, therefore is negligible.

B_3: P^* **cheats in the proofs of consistency.** Given that $x_n \notin L$, in step P_5 the prover convinces the verifier by proving that there exists an index J in which it constructed a consistent tree. P convinces the verifier by running MPC-in-the-head that proves the consistency of nodes and that the leaves are accepted by a PCPP verifier. Since B_1, B_2 happen with negligible probability, P^* cannot convince the verifier by neither equivocating the commitment nor by using a legitimate witness. Thus, it must be the case that P^* convinces the verifier computing an accepting MPC-in-the-head for a false predicate (i.e., the nodes are not consistent/not accepting). However, due to the t-robustness of the MPC protocols, this event happens with negligible probability.

Let J, J' be the coordinates that are chosen by P^* in the first and in the second continuation from τ generated by A. (Recall that J is the coordinate for which P^* will be required to provide an accepting proof. For any other coordinate P^* is not required to construct an accepting tree or provide any proof.) We now estimate the probability that $J = J' = f$, where f is the coordinate chosen by A. Let p be the probability that for a random continuation from τ, B_1, B_2 and B_3 do not occur. Since C_0 and C_1 are random continuations, it holds with probability at least p^2 that $J = J'$. The index f is chosen uniformly at random by A and is completely hidden. Hence with probability at least $\frac{p^2}{n}$, $J = J' = f$. Whenever this happens, we claim that A can find two nodes γ' and γ that will yield two strings with the same signature under key sk_f. This is because from B_2 we know that P^* cannot compute an accepting PCPP (as the simulator) and from B_3 we know that it cannot cheat in the proof. Hence, following similar arguments as in [2,16], we have that a prover P^* that convinces V must be able to open a random leaf of the tree as the value 0 and 1 with non-negligible probability. Given that the root and depth of the PCPP are fixed in τ, we have that in two random continuations C_0, C_1 there is a non negligible probability that P^* opens the same leaf as two different values. If this event happens that it must be the case that P^* has found a collision (conditioned on B_1, B_2, B_3 not occurring). Thus, A will find a collision with a polynomially related probability therefore contradicting the collision-resistance property of the underlying signature scheme.

Proof Sketch of Argument of Knowledge: Proving argument of knowledge will essentially follow from the same approach as the proof of soundness. Assume P^* convinces a verifier on a statement x. In the soundness proof, we crucially relied on the fact that the prover cannot equivocate any of the commitments and they were statistically binding. While proving argument of knowledge, this does not hold as x is in L. Instead, we observe that if the commitments were binding then P^* can find collisions to the signature scheme and that will happen only with negligible probability. Hence, it must be the case that P^* is equivocating the commitments. Since the commitments are instance based-commitments, if a commitment is opened to two different values, a witness can be extracted. The only witness for all the instance-based commitments used in the protocol is the witness to the input statement x. Since P^* equivocates with non-negligible probability, we can extract a witness with non-negligible probability. The extractor essentially runs the same algorithm as A does from the proof of soundness with the exception that it tries to find equivocated commitments as opposed to collision in the signatures.

Zero-Knowledge. The simulator Sim internally emulates the code of V^* and proceeds as follows: Sim proceeds as the honest prover through the first signature slot until V^* reveals the keys and randomness in J_1. If the set of signatures are not valid or consistent with what is revealed then the simulation halts immediately and outputs the transcript up to that point. Otherwise, Sim stalls the current execution and repetitively queries V^* at the Signature Slot 1 with fresh commitments to 0^v (where v is the size of a VSS share), until it obtains $2n$ sets of valid signatures (i.e. $2n$ rewindings where the verifier did not abort in the signing slot or return any invalid signature). Let θ be the number of queries Sim makes. If $\theta \geq 2^{n/2}$, then Sim aborts outputting fail_1. Next, it sets up a sub-procedure GetSign which is described below: Let c be a commitment to a message $m \in \{0,1\}^v$ and $\kappa \in [n]$.

Procedure GetSign(c, m, κ): Initialize $SIGS = \phi$ to be an empty list. For $i = 1$ to $2n^2\theta T^2$

> Choose i uniformly at random from $[T]$. Select $T - 1$ commitments to m, $c_1, \ldots, c_{i-1}, c_{i+1}, \ldots c_T$ where each of them are obtained by running Com^x on message m. Set $c_i = c$. Rewind V^* to message P_1 and feed $(c_1, \ldots, c_T)$ (the first Signature Slot). If V^* replies a valid set of signatures $(\sigma'_1, \ldots, \sigma'_T)$ under verification key vk_κ, then σ'_i is a valid signature of $c_i = c$. Add σ'_i to the list $SIGS$.

If fewer than n signatures are in the list $SIGS$, then return $\perp$. Let σ^* be that signature that occurs more than half the time in $SIGS$. Return σ^*. If no such signature exists return $\perp$.

Next, the simulator proceeds from where it stalled. It creates an (extendable) Merkle signature-tree for $\mathsf{ECC}(V^*)$ for every unopened signing key $\kappa \in [n]/J_1$. To obtain signature on a message m (to be used in the tree construction), it calls GetSign(c, m, κ) procedure T times, each time with a fresh commitment c to m. If any of the T trials succeed then it uses that commitment and signature for m. For every key $\kappa \in [n]/J_1$ it feeds the values stored in the encode part of the root to V^*. If for any message m and key $\kappa \in [n]/J_1$, GetSign fails to return a signature in T tries, then the simulator adds κ to the list BadKeys and just commits to dummy values for the roots of those trees by generating VSS shares of the all 0-string. It also commits to the depth of the tree, i.e. $\log |V^*|$ by generating its VSS and committing to each share. Sim receives the challenge r from V^* and proceeds similarly to generate and commit to the root and depth of the (extendable) Merkle-signature tree corresponding to the PCP of Proximity proof and as before, it appends BadKeys with keys for which the simualator fails to obtain the required signatures. After receiving the randomness required to generate the PCPP queries from V^*, Sim prepares the responses to get signatures in the second slot. For the keys $\kappa \in$ BadKeys, the simulator simply presents commitments to 0^v. For every $\kappa \in [n]/(J_1 \cup J_2 \cup \mathsf{BadKeys})$, the simulator proceeds as follows. Let N be the total number of commitments that it has to send for every key. Denote the commitments by $c_1, \ldots, c_N$. Define $\mathsf{VS} \subset [N]$ to be the subset of indexes that contain the commitments that the verifier expects to see for q_{i^*,j^*} where i^*, j^* are the actual depths of the two trees. The commitments for a key κ are generated as follows:

GENERATE(κ): Sim first generates $\tilde{c}_1, \ldots, \tilde{c}_N$ where each $\tilde{c}_i$ is a commitment to 0^v using $\mathsf{Com}^{\mathsf{x}}$. For every $\beta \in \mathsf{VS}$, run $\mathsf{GetSign}(\tilde{c}_\beta, 0^v, \kappa)$ and see if it returns a valid signature. If it receives a signature for all commitments, then replace all commitments $\tilde{c}_\beta$ by c'_β for all $\beta \in \mathsf{VS}$. Return $\tilde{c}_1, \ldots, \tilde{c}_N$ to be used for the key κ in the second slot. If for some commitment a signature was not obtained we say that GENERATE failed for κ and add κ to BadKeys.

After receiving the signatures, if for some key the signatures obtained for the commitments in VS were different in the first and second slot, the key is added to BadKeys. If BadKeys contains all keys not in $J_1 \cup J_2$, i.e. $[n] = \mathsf{BadKeys} \cup J_1 \cup J_2$, the simulator halts outputting fail_2. Otherwise, Sim proceeds to complete the execution by using some key $\kappa^* \in [n]/(J_1 \cup J_2 \cup \mathsf{BadKeys})$ to complete the simulation.

Running Time of the Simulator. First we analyze the running time of Sim, Let $p(m)$ be the probability that $\tilde{V}^*$ on query a random commitment $c = \mathsf{Com}(m, \tau)$ of $m \in \{0,1\}^l$ at the Signature Slot 1, returns a valid signature of c. Let $p = p(0^l)$.

We first argue that the simulator runs in expected polynomial time. To start, note that Sim aborts at the end of the Signature Slot 1 with probability $1 - p$, and in this case, Sim runs in polynomial time. With probability p, Sim emulates V^* only a strictly polynomial number of times and size of V^* is bounded by $T_{\tilde{V}^*}$. Thus, Sim runs in some $T' = poly(T_{\tilde{V}^*})$ time and makes at most T queries to its GetSign procedure, which in turn runs in time $\theta \cdot poly(n)$ to answer each query. Also note that Sim runs in time at most 2^n, since Sim aborts when $\theta \geq 2^{n/2}$. Now, we claim that $\theta \leq 10n/p$ with probability at least $1 - 2^{-n}$, and thus the expected running time of Sim is at most

$$(1 - p) \cdot poly(n) + p \cdot T' \cdot (10n/p) \cdot poly(n) + 2^{-n} \cdot 2^n \leq poly(T_{\tilde{V}^*}, n).$$

To see that $\theta \leq 10n/p$ with overwhelming probability, let $X_1, \ldots, X_{10n/p}$ be i.i.d. indicator variables on the event that V^* returns valid signatures for a random commitments to 0^s. If $\theta \leq 10n/p$ then via a standard Chernoff bound, we can conclude that $\sum_i X_i \leq 2n$ happens with probability at most 2^{-n}. Using a Markov argument, this also proves that the probability of fail_1 occurring is negligible.

Indistinguishability of Simulation. To prove indistinguishability, we analyze a hybrid simulator that has the witness and proceeds exactly as Sim with the exception that for every commitment it uses the equivocal commitment $\mathsf{EQCom}^{\mathsf{x}}$ scheme instead of $\mathsf{Com}^{\mathsf{x}}$. Indistinguishability of the simulation will follow using a standard hybrid argument and the indistinsguishability of the commitment scheme (analogous to [11]). Conditioned on the hybrid simulator not ending in one of the fail events, we can argue that the output of the hybrid simulator is identical to the real view. Notice that all messages until the second slot will be prepared by the simulator identical to the real simulator. Recall that the messages in the second signature slot are replaced with good commitments according to the GENERATE procedure. However, since we are replacing one random good commitment with another, the distribution of the simulator will be identical to the real provers message. Finally the rest of the messages only reveal t-views of all the MPC protocols and by the perfect t-privacy of the MPC protocols these messages will also be identically distributed. Therefore to prove correctness, it suffices to argue that all the fail events occur with negligible probability.

Claim. Except with negligible probability, there are at least $n/2 - n/10$ keys in $[n]/J_1$ for which the hybrid simulator obtains the deterministically signed signatures with probability at least $1 - \log^2 /n$.

Proof. Let $s = \frac{\log^2 n}{2}$. For every key, we define a good set of random tapes G_n. Now, we say that a key is good if GetSign fails to return a signature for at most $2s/n$ fraction of the good tapes. First, we show that on a good tape the GetSign procedure obtains n signatures with high-probability. Next, we show that the probability of a random tape being good is at least $1 - n/T$. Recall that it returns a signature only if it obtains n signatures and there exists a majority. We show that there exists at least $n/2 - n/10$ keys that are good. Suppose these claims were true, then it holds that there are at least $n/2 - n/10$ keys for which the probability that a random tape is good with probability least $1 - n/T - s/n = 1 - 2s/n$. Since the simulator calls GetSign T times for any message, it will yield a signature for every message in the good $n/2 - n/10$ keys with high-probability.

Defining G_n. Recall that, in the run-time analysis we showed that $n/p \leq \theta \leq 2^{n/2}$ with probability at least $1 - 2^{-\Omega(n)}$. Now, for every $m \in \{0,1\}^s$, $p(m) \geq p - \nu \geq p/2$ implies that $\theta \geq n/2p(m)$. Fix a message m and a key. Define G_n to be the set of random tapes τ such that the probability that V^* returns a signature on (c, c_{-i}) where $i \leftarrow [n]$, $c = \mathsf{Com}^x(m; \tau)$ and c_{-i} are random $T - 1$ commitments for m is at least $\frac{1}{2\theta T^2}$. This means that, for any $\tau \in G_n$, in $2n\theta T^2$ tries, the probability that GetSign fails to return a single signature is at most e^{-n}. Since GetSign makes $n(2n\theta T^2)$ attempts, it obtains n signatures except with negligible probability.

Probability of a Good Tape. We argue that a random τ is in G_n with probability at least $1 - \frac{n}{T}$. Assume for contradiction the fraction of tapes in G_n was smaller than $1 - \frac{n}{T}$. We now estimate the probability that V^* returns a signature on random commitments. There are two cases: (1) At least one commitment among the T commitments is not in G_n. Conditioned on this event, the probability that V^* honestly provides signatures is at most $\frac{T^2}{2\theta T^2}$. (2) All commitments are in G_n. The probability this occurs is at most $(1 - \frac{n}{T})^T \leq e^{-n}$. Overall the probability that V^* answers is at most $\frac{1}{2t} + e^{-n} < \frac{1}{\theta} < p(m)$ which is a contradiction. Therefore G_n must contain at least $1 - \frac{n}{T}$ fraction of the tapes.

Number of Good Keys. We need to argue that for most messages there will be a majority when n signatures are obtained. We will show that for most messages the n signatures have a majority and the popular signature will be the one that is computed deterministically. At the end of the first signature slot, the verifier opens the randomness and signing key used in the half the coordinates, i.e. those in J_1. Since J_1 is committed using an equivocal commitment in this hybrid, it is statistically hiding. So, given the commitments sent by the prover, J_1 is completely hidden in Hybrid H_2. Therefore, we can conclude that the probability that the verifier gives signatures that were not deterministically signed by more than s keys is at most $2^{-O(s)}$. Using an averaging argument, it holds that the probability that there exists more than $n/10$ keys such that the probability that the verifier gives "incorrect" signatures in those coordinates with probability bigger than $\frac{s}{10n}$ over messages sent in P_1 is negligible. This means that, for the remaining $n/2 - n/10$ keys, at most $s/10n$ fraction of possible messages (in the first slot) yield incorrect signatures. We argue next that GetSign gives the wrong signature for a random commitment to a message m in any of these $n/2 - n/10$ keys with probability at most s/n. Assuming this holds, it holds that for any message m, the probability that GetSign returns a deterministically signed signature for any of the $n/2 - n/10$ keys is at least $1 - \frac{s}{n}$. We now proceed to prove this claim.

The intuition is that if fewer than $1 - s/n$ fraction of the commitments yielded the correct signature with majority, then there will only be a small fraction of T-tuple of messages containing such commitments. Recall that, at least $1 - s/10n$ fraction of all T-tuple of messages are signed deterministically. Hence with probability at least $1 - s/n$ more than half of the T commitments must be deterministically signed and these commitments yield correct signatures with majority. Now suppose that fewer than $1 - s/n$ fraction of commitments yielded the deterministically generated signatures in majority. Then it must hold that $(1 - s/n)^{T/2} \geq (1 - s/n)p(m)$. Since $T = O(n^c)$, we can set c sufficiently large ($c = 5$ will suffice) to arrive at a contradiction since $p(m) > \frac{n}{2^{n/2}}$. This concludes proof of the claim.

Claim. The probability that the simulator outputs fail$_3$ is negligible.

Proof. We need to show that after the second signature slot BadKeys does not contain all the keys in $[n]/J_1 \cup J_2$. First we show that, GENERATE swaps the commitments for at least $n/20$ keys in $[n]/J_1 \cup J_2$. For the particular depths i^*, j^*, at most $\log^e(n)$ ($= |VS|$) commitments are sent (for some constant e). From the previous claim, we know that for at least $n/2 - n/10$ keys in $[n]/J_1$, (and therefore $n/4 - n/10$ keys in $[n]/J_1 \cup J_2$) , GetSign returns a signature on a commitment with probability at least $1 - \log^2 n/n$. A key fails in GENERATE if for some commitment among the commitments with index in VS does not yield a signature through GetSign. This happens with probability at most $\frac{\log^{e+2}(n)}{n}$. Since there are $n/4 - n/10$ keys in $[n]/J_1 \cup J_2$, the probability that more than $n/10$ keys fail in GENERATE is $(\frac{\log^{e+2}(n)}{n})^{O(n)}$, i.e. negligible. This means that for at least $n/20$ keys GENERATE successfully swaps. Next, we need to show that there exists at least one key for which the signature obtained by the simulator in the first and second are the same. Recall that GetSign in these keys returns the deterministic signatures on $1 - \log^2 n/n$ fraction of the commitments. Since we are concerned only about $\log^e(n)$ commitments, using the same argument, we can conclude that, the probability that there are more than $n/40$ keys for which the signature of some commitment among the $\log^e(n)$ commitments obtained in the first slot is not the one deterministically signed is negligible. In other words, there must be at least $n/20 - n/40 = n/40$ keys in $[n]/J_1 \cup J_2 \cup \mathsf{BadKeys}$ for which the simulator obtained deterministic signatures for the $\log^e(n)$ commitments in the first slot and GENERATE inserted those commitments in the second slot. Finally, from the second cut-and-choose, it will follow that there is at least one key among those for which it receives a deterministically signed signature for all the $\log^e(n)$ commitments in the second slot, and hence the same signature obtained from the first slot.

Acknowledgments. We thank the anonymous FOCS's reviewers for pointing out an issue with using digital signatures based on one-way functions in a previous version of our work. We thank Kai-Min Chung, Vipul Goyal, Huijia (Rachel) Lin, Rafael Pass and Ivan Visconti for valuable discussions.

References

1. Barak, B.: How to go beyond the black-box simulation barrier. In: FOCS, pp. 106–115. IEEE Computer Society (2001)
2. Barak, B., Goldreich, O.: Universal arguments and their applications. In: Computational Complexity, pp. 162–171 (2002)
3. Barak, B., Goldreich, O., Goldwasser, S., Lindell, Y.: Resettably-sound zero-knowledge and its applications. In: FOCS 2001, pp. 116–125 (2001)
4. Ben-Sasson, E., Goldreich, O., Harsha, P., Sudan, M., Vadhan, S.P.: Robust pcps of proximity, shorter pcps, and applications to coding. SIAM J. Comput. 36(4), 889–974 (2006)
5. Bitansky, N., Paneth, O.: On the impossibility of approximate obfuscation and applications to resettable cryptography. In: STOC, pp. 241–250 (2013)
6. Choi, S.G., Dachman-Soled, D., Malkin, T., Wee, H.: Simple, black-box constructions of adaptively secure protocols. In: Reingold, O. (ed.) TCC 2009. LNCS, vol. 5444, pp. 387–402. Springer, Heidelberg (2009)
7. Chor, B., Goldwasser, S., Micali, S., Awerbuch, B.: Verifiable Secret Sharing and Achieving Simultaneity in the Presence of Faults (Extended Abstract). In: Proceedings of the 26th Annual IEEE Symposium on Foundations of Computer Science, FOCS 1985, pp. 383–395 (1985)
8. Chung, K.M., Ostrovsky, R., Pass, R., Venkitasubramaniam, M., Visconti, I.: 4-round resettably-sound zero knowledge. In: TCC. pp. 192–216 (2014)
9. Chung, K.M., Pass, R., Seth, K.: Non-black-box simulation from one-way functions and applications to resettable security. In: STOC (2013)
10. Dachman-Soled, D., Kalai, Y.T.: Securing circuits against constant-rate tampering. In: Safavi-Naini, R., Canetti, R. (eds.) CRYPTO 2012. LNCS, vol. 7417, pp. 533–551. Springer, Heidelberg (2012)
11. Dachman-Soled, D., Malkin, T., Raykova, M., Venkitasubramaniam, M.: Adaptive and concurrent secure computation from new adaptive, non-malleable commitments. In: Sako, K., Sarkar, P. (eds.) ASIACRYPT 2013, Part I. LNCS, vol. 8269, pp. 316–336. Springer, Heidelberg (2013)
12. Goldreich, O.: Foundations of Cryptography — Basic Tools. Cambridge University Press (2001)

13. Goldwasser, S., Micali, S., Rackoff, C.: The knowledge complexity of interactive proof-systems (extended abstract). In: STOC, pp. 291–304 (1985)
14. Goyal, V.: Constant round non-malleable protocols using one way functions. In: Fortnow, L., Vadhan, S.P. (eds.) STOC, pp. 695–704. ACM (2011)
15. Goyal, V., Lee, C.K., Ostrovsky, R., Visconti, I.: Constructing non-malleable commitments: A black-box approach. In: FOCS, pp. 51–60. IEEE Computer Society (2012)
16. Goyal, V., Ostrovsky, R., Scafuro, A., Visconti, I.: Black-box non-black-box zero knowledge. In: STOC (2014)
17. Haitner, I.: Semi-honest to malicious oblivious transfer—the black-box way. In: Canetti, R. (ed.) TCC 2008. LNCS, vol. 4948, pp. 412–426. Springer, Heidelberg (2008)
18. Impagliazzo, R., Rudich, S.: Limits on the provable consequences of one-way permutations. In: Goldwasser, S. (ed.) Advances in Cryptology - CRYPT0 1988. LNCS, vol. 403, pp. 8–26. Springer, Heidelberg (1990)
19. Ishai, Y., Kushilevitz, E., Lindell, Y., Petrank, E.: Black-box constructions for secure computation. In: Proceedings of the 38th Annual ACM Symposium on Theory of Computing, Seattle, WA, USA, May 21-23, pp. 99–108. ACM (2006)
20. Ishai, Y., Kushilevitz, E., Ostrovsky, R., Sahai, A.: Zero-knowledge from secure multiparty computation. In: Johnson, D.S., Feige, U. (eds.) STOC, pp. 21–30. ACM (2007)
21. Kilian, J.: A note on efficient zero-knowledge proofs and arguments (extended abstract). In: Kosaraju, S.R., Fellows, M., Wigderson, A., Ellis, J.A. (eds.) STOC, pp. 723–732. ACM (1992)
22. Lin, H., Pass, R.: Black-box constructions of composable protocols without setup. In: Safavi-Naini, R., Canetti, R. (eds.) CRYPTO 2012. LNCS, vol. 7417, pp. 461–478. Springer, Heidelberg (2012)
23. Merkle, R.C.: A certified digital signature. In: Brassard, G. (ed.) CRYPTO 1989. LNCS, vol. 435, pp. 218–238. Springer, Heidelberg (1990)
24. Naor, M., Yung, M.: Universal one-way hash functions and their cryptographic applications. In: STOC 1989, pp. 33–43 (1989)
25. Ostrovsky, R., Scafuro, A., Venkitasubramaniam, M.: Resettably sound zero-knoweldge arguments from owfs - the (semi) black-box way. Cryptology ePrint Archive, Report 2014/284 (2014), http://eprint.iacr.org/
26. Pass, R., Rosen, A.: New and improved constructions of non-malleable cryptographic protocols. In: STOC 2005, pp. 533–542 (2005)
27. Pass, R., Tseng, W.-L.D., Wikström, D.: On the composition of public-coin zero-knowledge protocols. In: Halevi, S. (ed.) CRYPTO 2009. LNCS, vol. 5677, pp. 160–176. Springer, Heidelberg (2009)
28. Pass, R., Wee, H.: Black-box constructions of two-party protocols from one-way functions. In: Reingold, O. (ed.) TCC 2009. LNCS, vol. 5444, pp. 403–418. Springer, Heidelberg (2009)
29. Reingold, O., Trevisan, L., Vadhan, S.P.: Notions of reducibility between cryptographic primitives. In: Naor, M. (ed.) TCC 2004. LNCS, vol. 2951, pp. 1–20. Springer, Heidelberg (2004)
30. Rompel, J.: One-way functions are necessary and sufficient for secure signatures. In: Proceedings of the 22nd Annual ACM Symposium on Theory of Computing, Baltimore, Maryland, USA, May 13-17, pp. 387–394. ACM (1990)
31. Wee, H.: Black-box, round-efficient secure computation via non-malleability amplification. In: FOCS, pp. 531–540. IEEE Computer Society (2010)

A Rate-Optimizing Compiler for Non-malleable Codes Against Bit-Wise Tampering and Permutations

Shashank Agrawal[1,*], Divya Gupta[2,3,*,**], Hemanta K. Maji[2,3,**], Omkant Pandey[1,3,*,***], and Manoj Prabhakaran[1,*]

[1] University of Illinois Urbana-Champaign, USA
{sagrawl2,mmp}@illinois.edu
[2] University of California Los Angeles, USA
divyag@cs.ucla.edu,hemanta.maji@gmail.com,omkant@gmail.com
[3] Center for Encrypted Functionalities, USA

Abstract. A non-malleable code protects messages against a class of tampering functions. Informally, a code is non-malleable if the effect of applying any tampering function on an encoded message is to either retain the message or to replace it with an unrelated message. Two main challenges in this area – apart from establishing the feasibility against different families of tampering – are to obtain *explicit constructions* and to obtain *high-rates* for such constructions.

In this work, we present a compiler to transform low-rate (in fact, zero rate) non-malleable codes against certain class of tampering into an optimal-rate – i.e., rate 1 – non-malleable codes against the same class. If the original code is explicit, so is the new one.

When applied to the family of bit-wise tampering functions, this subsumes (and greatly simplifies) a recent result of Cheraghchi and Guruswami (TCC 2014). Further, our compiler can be applied to non-malleable codes against the class of bit-wise tampering and bit-level permutations. Combined with the rate-0 construction in a companion work,

---

* Research supported in part by NSF grant 1228856.
** Research supported in part from a DARPA/ONR PROCEED award, NSF Frontier Award 1413955, NSF grants 1228984, 1136174, 1118096, and 1065276, a Xerox Faculty Research Award, a Google Faculty Research Award, an equipment grant from Intel, and an Okawa Foundation Research Grant. This material is based upon work supported by the Defense Advanced Research Projects Agency through the U.S. Office of Naval Research under Contract N00014-11- 1-0389. The views expressed are those of the author and do not reflect the official policy or position of the Department of Defense, the National Science Foundation, or the U.S. Government.
*** Research supported in part from a DARPA/ONR PROCEED award, NSF Frontier Award 1413955, NSF grants 1228984, 1136174, 1118096. This material is based upon work supported by the Defense Advanced Research Projects Agency through the U.S. Office of Naval Research under Contract N00014-11- 1-0389. The views expressed are those of the author and do not reflect the official policy or position of the Department of Defense, the National Science Foundation, or the U.S. Government.

Y. Dodis and J.B. Nielsen (Eds.): TCC 2015, Part I, LNCS 9014, pp. 375–397, 2015.

this yields the first explicit rate-1 non-malleable code for this family of tampering functions.

Our compiler uses a new technique for boot-strapping non-malleability by introducing errors, that may be of independent interest.

Keywords: Non-malleable Codes, Explicit Construction, Information Theoretic, Rate-Optimizing Compiler, Rate 1.

1   Introduction

Non-Malleable Codes have emerged as an object of fundamental interest, at the intersection of coding theory and cryptography. Informally, a code is non-malleable if the message contained in a codeword that has been tampered with is either the original message, or a completely unrelated value. As a relatively new problem, several basic questions are still open. In particular, two main challenges in this area – apart from establishing the feasibility against different families of tampering – are to obtain *explicit constructions* and to obtain *high-rates*[1] for such constructions.

While existential results have been obtained for rate-1 non-malleable codes for very broad classes of tampering functions [8,17], explicit constructions have turned out to be much harder, in such generality. For the relatively simple class of bit-wise tampering functions introduced in [16], it was only recently that an explicit rate-1 construction was obtained [9]. For the more general class of "split-state" tampering functions, the first construction in [15] encoded only a single bit; in a break-through result, an explicit scheme (of rate 0) was proposed for arbitrary length messages by [1], and more recently, a constant rate construction (for 10 states) was provided in [6].

All the above explicit results relied on the tampering functions being "compartmentalized" — i.e., the codeword is partitioned *a priori* into separate blocks and each block is tampered independently of the others — In a companion paper [3], we presented the first instance of an explicit non-malleable code against a class of non-compartmentalized tampering functions. This class consists of functions which can *permute* the bits of a codeword, as well as tamper each bit independently. Apart from being of interest as a natural non-compartmentalized class, non-malleable codes against this class have direct cryptographic applications: in [3] it is used to obtain non-malleable string-commitments from non-malleable bit-commitments in a hardware token-based model, with information-theoretic security (or in the standard model under computational assumptions). This application also highlighted the need for explicit constructions, even if randomized constructions are efficient, since the latter calls for a trusted party to carry out the randomized construction.

---

[1] Rate refers to the asymptotic ratio of the length of a message to the length of its encoding (in bits), as the message length increases to infinity. The best rate possible is 1; if the length of the encoding is super-linear in the length of the message, the rate is 0.

The construction in [3] has 0 rate. In this paper, we present a simple but powerful compiler to transform such a non-malleable code into a rate-1 non-malleable code against the same family. In fact, our compiler is general enough that it can be applied to non-malleable codes against bit-wise tampering too, to improve their rate from 0 to 1. This subsumes (and greatly simplifies) a result of [9].

1.1   Our Contribution

Let $\mathcal{F}^*$ be the class of tampering functions $f : \{0,1\}^N \to \{0,1\}^N$ of the form $f(x) = f_\pi(f_1(x_1), \cdots, f_N(x_N))$, where f_π permutes the indices of its input according to a permutation $\pi : [N] \hookrightarrow [N]$, and each $f_i : \{0,1\} \to \{0,1\}$ is one of the four possible binary functions over $\{0,1\}$.

Our main technical result is the following.

Informal Theorem 1. *There exists a black-box compiler that takes a non-malleable code NMC_0 secure against $\mathcal{F}^*$, which may have a polynomial blowup in size during encoding (and hence rate 0), and defines a rate-1 non-malleable code NMC_1 secure against $\mathcal{F}^*$. The encoding and decoding algorithms of NMC_1 make only black-box calls to the respective functions of NMC_0 (on much smaller inputs).*

In fact, we present our compiler as consisting of two black-box components: NMC_0 and a rate-1 binary error-correcting code. (We require the error-correcting code to also have an easy to satisfy privacy requirement.) The encoding and decoding algorithms of NMC_1 make only black-box calls to the encoding, decoding and error-correcting functions of this error-correcting code. An error-correcting code with the requisite properties is easily instantiated using (low-distance) Reed-Solomon codes over a field of characteristic 2.

Instantiating NMC_0 with the non-malleable code of [3] (which has rate 0, as the codewords are super-linear in the length of the messages), we get our main result.

Corollary 1. *There exists an explicit and efficient rate-1 non-malleable code against $\mathcal{F}^*$.*

We point out that the above result has immediate implications for the class of all bit-wise tampering functions $\mathcal{F}_{\mathsf{BIT}}$ [16]. Non-malleable codes for this family has been studied by [16,9]. We note that $\mathcal{F}_{\mathsf{BIT}}$ is a subset of $\mathcal{F}^*$ in which π is restricted to be the identity permutation. As a consequence, we reproduce a result of [9] as a simple corollary to Corollary 1:

Corollary 2. *There exists an explicit and efficient rate-1 non-malleable code against $\mathcal{F}_{\mathsf{BIT}}$.*

In fact, Theorem 1 continues to hold true, without altering the compiler, if $\mathcal{F}^*$ is replaced by $\mathcal{F}_{\mathsf{BIT}}$ (so that NMC_0 and NMC_1 are both secure only against $\mathcal{F}_{\mathsf{BIT}}$). This provides a much simpler alternative to a compiler in [9], and proves Corollary 2 without relying on the recent construction from [3].

1.2 Prior Work

Cramer et al. [12] introduced the notion of arithmetic manipulation detection (AMD) codes, which is a special case of non-malleable codes against tampering functions with a simple algebraic structure; explicit AMD codes with optimal (second order) parameters have been recently provided by [13]. Dziembowski et al. motivated and formalized the more general notion of non-malleable codes in [16]. They showed existence of a constant rate non-malleable code against the class of all bit-wise independent tampering functions.

The existence of rate-1 non-malleable codes against various classes of tampering functions is known. For example, existence of such codes with rate $(1-\alpha)$ was shown against any tampering function family of size $2^{2^{\alpha n}}$; but this scheme has inefficient encoding and decoding [8]. For tampering functions of size $2^{\text{poly}(n)}$, rate-1 codes (with efficient encoding and decoding) exist, and can be obtained efficiently with overwhelming probability [18].

However, explicit constructions of non-malleable codes have remained elusive, except for some well structured tampering function classes. For the setting where the codeword is partitioned into separate blocks and each block can be tampered arbitrarily but independently, an encoding scheme was proposed in [10]. In the most general such compartmentalized model of tampering, where there are only two compartments (known as the split-state model), an explicit encoding scheme for bits was proposed by [15]. Recently, in a break-through result, an explicit scheme (of rate 0) was proposed for arbitrary length messages by [1]. A constant rate construction for 10 states was provided in [6].

In the computational setting, there has been a sequence of works on improving the rate of error-correcting codes [26,28,29,23,22,5] and constructing non-malleable codes and its variants [27,17].

An explicit rate 1 code for the class of bit-wise independent tampering function was proposed by [9]. Note that a tampering function in this class tampers each bit independently and is subsumed by our work and a companion paper [3]. In the construction of [9], they exhaustively search for an encoding scheme (which is guaranteed by [18]) for messages with logarithmic length. This is a complex procedure (and intuitively obscure) and the compiler which extends the non-malleability to long messages is also complicated. We, on the other hand, begin with a rate 0 code of [3] against a more general class of (non-compartmentalized) tampering functions and apply our compiler to obtain rate 1 non-malleable code against the more general class itself.

We remark that preliminary results leading to this work appear in [2]. The results of this paper and [3] together subsume and significantly extend the results in [2].[2]

---

[2] In [2] a weaker class of tampering functions was considered, which did not contain all bit-wise tampering functions. While a rate-amplification approach for this class was presented there, it was more complicated, and relied on the specifics of the rate-0 construction there. The approach for rate-amplification there breaks down when all bit-wise functions are allowed.

1.3   Technical Overview

Improved Efficiency via Hybrid Encoding. A recurring theme in cryptographic constructions for improving efficiency (in our setting, efficiency refers to the rate of the code) is a "boot-strapping" or "hybrid" approach. It takes a scheme with strong security (but low efficiency), and combines it with an efficient scheme (with a weak form of security) to obtain an efficient scheme with strong security. Perhaps the most well-known example of this approach in cryptography is that of "hybrid encryption," which improves the efficiency of a (non-malleable) public-key encryption scheme by using it to encrypt a short key for a symmetric-key encryption scheme, and then using the latter to encrypt the actual message (e.g., see [14,25]).

The high-level approach in this work, as well as in many works on improving the rate of error-correcting codes and non-malleable codes [22,9], fits this template. A basic idea for non-malleable codes in these works involves encoding the message using a high-rate (randomized) code and appending to it a tag that is encoded using an inefficient non-malleable encoding $\mathsf{NMC_0}$. That is, the final codeword has the form $(c, \mathsf{NMC_0}(\tau))$, where c is a (malleable) encoding of the message, and τ consists of some information about c that "binds" c to τ.

Intuitively, the short tag τ should encode some information about the much longer c in a way that makes it hard to change c without changing τ as well, and since the latter is encoded using a non-malleable code, one could hope that the over all code is non-malleable. One such choice of the tag, used in a preliminary version of this result [2], is $\tau = (h, h(c))$, where h is a randomly chosen hash function with a short description from a (statistical) collision-resistant hash function family. As shown in [2], this suffices for the class of attacks involving permutations (but not allowing the adversary to set/reset the bits). However, when the class of attacks allowed for the adversary includes the possibility of the adversary creating an entirely new tag τ^* (and a new purported codeword $\widetilde{c}$ obtained by mauling the original codeword c), this approach fails. This is because, it raises the possibility that the adversary can pick h such that $h(\widetilde{c})$ can be predicted with non-negligible probability, even though the adversary may have some uncertainty about $\widetilde{c}$. While it is plausible that most of the functions h in a function family would lead to unpredictable values of $h(\widetilde{c})$, it appears difficult to rule out there being no such h, or to provide an efficient algorithm for detecting them.

Our Approach: Adding Errors for Non-Malleability. We introduce a novel approach to boot-strapping non-malleability. We first motivate our approach using a loose analogy. Consider a student plagiarising a homework solution, by copying it from an original source, and blindly making a few alterations (without actually comprehending the original solution). The student would try to remove various pieces of identifying information (e.g., change variable names, reorder sentences etc.) and even introduce minor typographical errors, while hoping to make it look approximately correct to the grader. If there is not much variability in correct solutions, then, even if confronted with the original solution, the student will have plausible deniability that she came up with the solution on her own. However, if the original source happened to contain several minor random errors itself (which a grader would have recognized as minor, and ignored), then the chances are that many of them would make their way into the plagiarized solution as well. In this case, it will be unlikely that the student could have introduced those errors on her own, and this will be a strong indication of plagiarization.

While our problem of non-malleability is different, our solution follows the above intuition. Our encoding has the form $\mathsf{Enc}(s) = (\mathsf{ECSS}(s) \oplus e_R, \mathsf{NMC}_0(\tau))$, where now ECSS is a light-weight (rate-1) encoding of s, R is an appropriately sized random subset of indices (say, $|R| = n^\delta$ bits, where $|\mathsf{ECSS}(s)| = n$), e_R is a sparse n-bit vector, with zeros outside of R and uniformly randomly chosen bits in R. The tag τ is a succinct representation of the bits of $\mathsf{ECSS}(s) \oplus e_R$ at the positions in R. Note that $|\tau|$ is much shorter than n (e.g., $O(n^\delta \log n)$), so that (for an appropriate choice of δ), $\mathsf{NMC}_0(\tau)$ will be $o(n)$-bits long. The property we will need from ECSS is that it is an "error-correcting secret-sharing scheme" which is an error-correcting encoding that also behaves as a (ramp) secret-sharing scheme, so that any small subset of the bits in an encoding has values independent of the message it encodes. (Such a code can be readily instantiated using, for example, any linear error-correcting code of appropriate (sub-linear) distance and dual distance.)

To decode (c, σ), the following consistency check is carried out: apply error-correction to obtain a codeword $\hat{c}$ from c; also decode σ using the decoding of NMC_0 to obtain τ; then ensure that at the locations recorded in τ, c matches the recorded bits, and everywhere else c matches $\hat{c}$.

In other words, our encoding amounts to adding random errors to the efficiently encoded messages (while allowing error-correction); further, we require this to be accompanied by an "errata" (encoded using a non-malleable code) which lists *all the errors* in the first part. Now, intuitively, if the adversary chooses to create an errata on its own, but creates $\hat{c}$ by tampering c (i.e., using significantly many bits from c), then it is unlikely that the new errata matches $\hat{c}$. On the other hand, if the adversary retains the errata from a given codeword, then any significant tampering on c will result in a mismatch; instead, if $\hat{c}$ is obtained by only lightly tampering c, then, due to the distance of the code the only possibility to obtain a valid encoding is to have $\hat{c} = c$, and by a simple privacy requirement on the code, the probability of this happening when only a small number of bits in c are involved, is independent of the message.

Formal Analysis. We present a modular proof that the above construction is indeed a non-malleable code against $\mathcal{F}^*$, the family of permutations and bit-wise tampering attacks, by relying on the security of NMC_0 as well as the error-correction and privacy properties of ECSS in a black-box manner. The proof is somewhat simpler, if instead of considering $\mathcal{F}^*$, we considered only $\mathcal{F}_{\mathsf{BIT}}$, the family of bit-wise tampering functions, as in [9] (in which case, NMC_0 need be secure only against this class of tampering functions). Below we sketch this simpler proof, and indicate in footnotes the main points of departure for the full proof.

Formally, we need to argue that for any message s and any admissible attack f, for a randomly constructed codeword $\mathsf{Enc}(s)$, the outcome of $\mathsf{Dec}(f(\mathsf{Enc}(s)))$ is almost identically distributed as a simulated outcome which probabilistically maps f to the original message s, a fixed message distribution M_f, or $\perp$ (with M_f and the probabilities depending only on f). The simulated outcome is defined as follows:

Simulating $\mathsf{Dec}(f(\mathsf{Enc}(s)))$ (Given only f).

Let $\mathcal{L}$ and $\mathcal{R}$ denote the set of indices in our code corresponding to c and $\mathsf{NMC}_0(\tau)$, respectively. Given f, we proceed as follows.

- Define attacks $f^{(1)}$ and $f^{(2)}$ obtained by restricting respectively to $\mathcal{L}$ and $\mathcal{R}$.[3]
- Then, $f^{(2)}$ is simply a bit-wise tampering attack on $\mathsf{NMC}_0(\tau)$. By the security guarantee of NMC_0, we can sample, based only on $f^{(2)}$ (independent of τ), the outcome of $\mathsf{Dec}(f^{(2)}(\mathsf{NMC}_0(\tau)))$ as $\perp$, some string τ^*, or same* (i.e., τ itself).
- Case simulated outcome of $\mathsf{Dec}(f^{(2)}(\mathsf{NMC}_0(\tau)))$ is $\perp$: Set the simulated outcome of decoding $\mathsf{Dec}_1(f(\mathsf{ECSS}(s), \mathsf{NMC}_0(\tau)))$ to be $\perp$.
- Case simulated outcome of $\mathsf{Dec}(f^{(2)}(\mathsf{NMC}_0(\tau)))$ is τ^*: Let $c' = f^{(1)}(c)$. We consider two sub-cases, depending on the number of bits in c' that are not fixed by the attack $f^{(1)}$ (i.e., the number of bits that depend on the original bit at that position).
  - If the number of bits of c that influence c' is "small," then they can be sampled independent of the message s, by relying on the fact that ECSS is a (ramp) secret-sharing scheme. Then the simulated outcome is obtained by error-correcting c', decoding it and checking for consistency with τ^*.
  - If the number of bits of c' that depend on c is not small, then (following the analogy of plagiarism from above), there is an overwhelming probability that this set of bits contain several random bits and the probability that τ^* correctly records them is negligible. In this case, the simulated outcome is $\perp$.
- Case simulated outcome of $\mathsf{Dec}(f^{(2)}(\mathsf{NMC}_0(\tau)))$ is same*: In this case, τ remains unchanged. Again we consider two sub-cases, this time depending on the number of untampered bits in the attack $f^{(1)}$.
  - If there are only a small number of tampered bits in $f^{(1)}$, then $f^{(1)}$ ($\mathsf{ECSS}(s)$) results in a valid codeword iff $f^{(1)}$ has the effect of not altering $\mathsf{ECSS}(s)$. This depends only on the values of the bits of $\mathsf{ECSS}(s)$ which are tampered by $f^{(1)}$ (and τ), which in turn is independent of the message s, due to ECSS being a secret-sharing scheme. Hence we can sample τ

---

[3] When permutations are allowed, this is no more possible. Instead, we follow a more elaborate argument in which $f^{(2)}$ is defined after sampling the value of the bits from $\mathcal{L}$ that are moved to $\mathcal{R}$ by the permutation attack. To be able to do this independent of the encoded message, we rely on the error-correcting scheme ECSS being a ramp secret-sharing scheme.

and these bits, independent of s. This is used to simulate the outcome being same* or $\perp$.

- On the other hand, if there are several tampered bits, then we simulate the outcome to be $\perp$.

To argue that the last step results in only a negligible statistical error, we follow an argument similar to the plagiarism argument, but this time relying on the fact that τ is retained as it is, and will have a record of all the actual errors. Consider the set of bits that were tampered by $f^{(1)}$. Except with negligible probability, a significant number of bits in this set would have been recorded in τ. For each such bit, the probability that the tampered bit does not match the bit recorded in τ is at least $\frac{1}{2}$ (it is 1 if the tampering function is a bit-flip, and $\frac{1}{2}$ if it is a set/reset), independent of the other bits. Hence the probability that τ matches all of those bits is negligible.[4] Thus indeed, the outcome of the actual decoding would be $\perp$, except with negligible probability.

2   Preliminaries

We denote the set $\{1, \ldots, n\}$ by $[n]$. Probability distributions are represented by capital letters. The distribution U_S represents a uniform distribution over the set S. Given a distribution X, $x \sim X$ represents that x is sampled according to the distribution X. We shall often use the convention that a realization of a random variable denoted as X will be represented by the variable x.

For a joint variable $X = (X_1, \ldots, X_n)$ and $S = \{i_1, \ldots, i_{|S|}\} \subseteq [n]$, we define the random variable $X_S = (X_{i_1}, \ldots, X_{i_{|S|}})$, where $|S|$ denotes the size of the set S. We use a similar notation for vectors as well, for example x_S represents the vector restricted to indices in the set S. For a function $f(\cdot)$, the random variable $Y = f(X)$ represents the following distribution: sample $x \sim X$ and output $f(x)$. For a randomized algorithm A, we write $A(z)$ to denote the distribution of the output of A on an input z.

The statistical distance between two distributions S and T over a finite sample space I is defined as:

$$\mathrm{SD}\,(S, T) := \frac{1}{2} \sum_{i \in I} |\Pr_{x \sim S}[x = i] - \Pr_{x \sim T}[x = i]|.$$

The hamming distance between two vectors $c, c' \in \{0, 1\}^m$ is given by

$$\mathrm{HD}(c, c') := |\{i \in [m] | c_i \neq c'_i\}|.$$

---

[4] When we allow permutations, some amount of correlation can exist between the different tampered bits. For example, if two bits that are recorded got swapped with each other, the probability of not having an error is $\frac{1}{2}$ and not $\frac{1}{4}$. But this is the most extreme example: if k bits recorded in τ have been tampered with, we show that the error probability is at least $1 - (\frac{1}{2})^{k/2}$.

A function $f : \mathbb{N} \to \mathbb{R}^+$ is negligible if for every positive polynomial $\text{poly}(\cdot)$ and all sufficiently large n, $f(n) \leqslant 1/\text{poly}(n)$. We use $\text{negl}(M)$ to denote an (unspecified) negligible function in M.

Lastly, all logarithms in this paper would be to the base 2.

2.1 Classes of Tampering Functions

We shall consider the following basic tampering function classes.

1. Family of Permutations. Let $\mathcal{S}_N$ denote the set of all permutations $\pi : [N] \to [N]$. Given an input codeword $x_{[N]} \in \{0,1\}^N$, tampering with function $\pi \in \mathcal{S}_N$ yields the codeword: $x_{\pi^{-1}(1)} \cdots x_{\pi^{-1}(N)}$.
2. Family of Bit-Wise Tampering Functions. This class, represented by $\mathcal{F}_{\mathsf{BIT}}$, contains the following four functions, for a single bit input: a) $f(x) \mapsto x$, b) $f(x) \mapsto 1 \oplus x$, c) $f(x) \mapsto 0$, and d) $f(x) \mapsto 1$. These functions are, respectively, called *forward*, *toggle*, *reset* and *set* functions.

We define a more complex tampering function class $\mathcal{F}_{\mathsf{BIT}} \circ \mathcal{S}_N$ to consist of tampering functions of the form $f = (f_1, \ldots, f_N, \pi)$, where $\pi \in \mathcal{S}_N$ and $f_i \in \mathcal{F}_{\mathsf{BIT}}$, and

$$f(x_{[N]}) = f_{\pi^{-1}(1)}(x_{\pi^{-1}(1)}) \cdots f_{\pi^{-1}(N)}(x_{\pi^{-1}(N)}).$$

That is, to apply f to x, first we apply f_i to each position x_i, and apply the permutation π to the resulting string. Our main result provides an efficient rate-1 non-malleable code against this class.

2.2 Error-Correcting Secret-Sharing Scheme

In this section, we define error-correcting secret-sharing schemes that will be used in our construction.

Definition 1 (Error-Correcting Secret-Sharing Scheme (ECSS)). *Let $S = (X_0, X_1, \ldots, X_M)$ be a joint distribution over $\Lambda \times \{0,1\}^M$, such that the support of X_0 is all of Λ. (The random variable X_0 represents the secret being shared and X_i for $i \in [M]$ represents the i-th share.)*

We say that S is an $[M, L, T, D]$-error-correcting secret-sharing scheme if $\log|\Lambda| = L$, and the following conditions hold:

1. *T-privacy: $\forall\, s, s' \in \Lambda$, $\forall\, J \subseteq [M]$ such that $|J| \leqslant T$, we have*

$$\mathsf{SD}\left((X_J|X_0 = s), (X_J|X_0 = s')\right) = 0.$$

2. *D-error-correction: For any two distinct $c, c' \in \mathsf{Supp}(X_{[M]})$, the hamming distance between them $\mathsf{HD}(c, c') > 2d$, where $\mathsf{Supp}(X_{[M]})$ denotes the support of distribution $X_{[M]}$.*

3. Reconstruction: For any $s, s' \in \Lambda$ such that $s \neq s'$, we have

$$\mathrm{SD}\left((X_{[M]}|X_0 = s), (X_{[M]}|X_0 = s')\right) = 1.$$

In the remainder of the paper, by an ECSS scheme, we shall implicitly refer to a family of ECSS schemes indexed by M, i.e., $[M, L(M), T(M), D(M)]$-ECSS schemes for each positive integer M. We define the *rate* of such a scheme to be $\lim_{M \to \infty} \frac{L(M)}{M}$. We will be interested in *efficient* ECSS schemes. For this, we define three algorithms associated with such a scheme.

- $\mathsf{Enc}_{\mathsf{ECSS}}(s)$: This is a randomized algorithm that takes $s \in \Lambda$ as input and outputs a sample from the distribution $(X_{[M]}|X_0 = s)$.
- $\mathsf{ECorr}_{\mathsf{ECSS}}(\tilde{c})$: This algorithm takes a $\tilde{c} \in \{0,1\}^M$ as input, and outputs a $c \in \mathsf{Supp}(X_{[M]})$ such that $\mathrm{HD}(c, \tilde{c}) \leqslant D$. If such a c does not exist, it outputs $\perp$.
- $\mathsf{Rec}_{\mathsf{ECSS}}(c)$: This algorithm takes a $c \in \{0,1\}^M$ as input, and outputs a secret $s \in \Lambda$ such that $c \in \mathsf{Supp}(X_{[M]}|X_0 = s)$. If such a secret does not exist, it outputs $\perp$.

Note that the uniqueness of the output of algorithms $\mathsf{ECorr}_{\mathsf{ECSS}}$ and $\mathsf{Rec}_{\mathsf{ECSS}}$ is guaranteed by the D-error-correction and reconstruction properties respectively. An ECSS scheme is said to be *efficient* if the three algorithms defined above run in time bounded by a polynomial in M.

2.3   Non-malleable Codes

In Fig. 1 we present the definition of an $[N, L, \nu]$-non-malleable code against a family of tampering functions $\mathcal{F}$.

2.4   Concentration Bound

The following concentration bound will be useful in our proof. Below, we write $a \neq b \pm \varepsilon$ to mean $a \notin [b - \varepsilon, b + \varepsilon]$.

Lemma 1 (Tail Inequality for Hypergeometric Distribution [24,11]).
Let $c \in (0, 1/2)$ be a constant, $m, n \in \mathbb{N}$ and $m \in [cn, (1-c)n]$. Let X be a random variable distributed uniformly over all n-bit strings with exactly m 1s. For every $t \in \mathbb{N}$, we have

$$\Pr_{x \sim X}\left[\sum_{i \in [t]} x_i \neq t\left(\frac{m}{n} \pm \varepsilon\right)\right] \leqslant 2\exp\left(-\mathrm{D_{KL}}\left(\frac{m}{n} + \varepsilon, \frac{m}{n}\right) \cdot t\right) \leqslant 2\exp(-\varepsilon^2 t/3),$$

where $\mathrm{D_{KL}}(\alpha, \beta) := \alpha \ln \frac{\alpha}{\beta} + (1 - \alpha) \ln \frac{1-\alpha}{1-\beta}$.

Let $\mathcal{F}$ be a set of functions of the form $f : \{0,1\}^N \to \{0,1\}^N$. Consider two mappings $\mathsf{Enc} : \{0,1\}^L \to \{0,1\}^N$ (possibly randomized) and $\mathsf{Dec} : \{0,1\}^N \to \{0,1\}^L \cup \{\bot\}$.

For $f \in \mathcal{F}$ and $s \in \{0,1\}^L$, define a random variable $\mathsf{Tamper}_f^{(s)}$ over $\{0,1\}^L \cup \{\bot\}$ as follows:
$$\mathsf{Tamper}_f^{(s)} = \mathsf{Dec}(f(\mathsf{Enc}(s))).$$

Let Sim be a map from $\mathcal{F}$ to distributions over the sample space $\{0,1\}^L \cup \{\mathsf{same}^*, \bot\}$. For $f \in \mathcal{F}$ and $s \in \{0,1\}^L$, define the random variable $\mathsf{Copy}_{\mathsf{Sim}(f)}^{(s)}$ as follows.
$$\mathsf{Copy}_{\mathsf{Sim}(f)}^{(s)} = \begin{cases} s & \text{if } \mathsf{Sim}(f) = \mathsf{same}^* \\ \mathsf{Sim}(f) & \text{otherwise.} \end{cases}$$

The simulation error (or, advantage) is defined to be:
$$\mathsf{adv}_{\mathsf{Enc},\mathsf{Dec},\mathcal{F}} := \inf_{\mathsf{Sim}} \max_{\substack{s \in \{0,1\}^L \\ f \in \mathcal{F}}} \mathrm{SD}\left(\mathsf{Tamper}_f^{(s)}, \mathsf{Copy}_{\mathsf{Sim}(f)}^{(s)}\right)$$

$(\mathsf{Enc}, \mathsf{Dec})$ is called an $[N, L, \nu]$-*non-malleable code against* $\mathcal{F}$ if the following conditions hold:

- Correctness: $\forall s \in \{0,1\}^L$, $\Pr[\mathsf{Dec}(\mathsf{Enc}(s)) = s] = 1$.
- Non-Malleability: $\mathsf{adv}_{\mathsf{Enc},\mathsf{Dec},\mathcal{F}} \leqslant \nu(N)$.

We say that the coding scheme is *efficient* if Enc and Dec run in time bounded by a polynomial in N.

Fig. 1. Definition of Non-Malleable Codes

3 Construction and Proof

In this section, we shall prove our main theorem:

Theorem 1 (Compiler). *Suppose there exists a* $[t', t, \nu_0]$ *non-malleable code* NMC_0 *against the tampering class* $\mathcal{F}_{\mathsf{BIT}} \circ \mathcal{S}_{t'}$, *with* $t' \leqslant t^d$ *for some constant* $d \geqslant 1$, *and an* $[M, L, T, D]$ *binary error-correcting secret-sharing scheme* ECSS. *Then there exists an* $[N, L, \nu_1]$ *non-malleable code* NMC_1 *against the tampering class* $\mathcal{F}_{\mathsf{BIT}} \circ \mathcal{S}_N$ *with* $N \leqslant M + M^{d/(d+1)} \log^{2d} M$ *and* $\nu_1(N) = \mathsf{negl}(M) + \nu_0(N - M)$, *if* $T, D \geqslant 2M^{d/(d+1)} \log^{2d} M$. *Further, if* NMC_0 *and* ECSS *are efficient, then* NMC_1 *is also efficient.*

Note that above $N = M(1 + o(1))$. If $M = L(1 + o(1))$, then the code is a rate-1 code. Also, $\mathsf{poly}(N) < N - M < N$, and hence if ν_0 is a negligible function, so is ν_1.

Our compiler is described in Fig. 2. When properly instantiated (see Section 3.2) we can obtain our main results Corollary 1 and Corollary 2.

Ingredients:

1. An $[M, L, T, D]$ binary error correcting secret sharing scheme ECSS with encoding, error-correcting and reconstruction algorithms $\mathsf{Enc}_{\mathsf{ECSS}}, \mathsf{ECorr}_{\mathsf{ECSS}}$ and $\mathsf{Rec}_{\mathsf{ECSS}}$ respectively.
2. A $[t', t, \nu_0]$ non-malleable coding scheme NMC_0 against $\mathcal{F}_{\mathsf{BIT}} \circ \mathcal{S}_{t'}$ with encoding and decoding algorithms $\mathsf{Enc}_{\mathsf{NMC}_0}$ and $\mathsf{Dec}_{\mathsf{NMC}_0}$ respectively, with $t' \leqslant t^d$ for some constant $d \geqslant 1$. We require that $T, D \geqslant 2M^{d/(d+1)} \log^{2d} M$.

$\mathsf{Enc}(s \in \{0,1\}^L)$:

1. Let $N^{(1)} = M$, $p_e = (N^{(1)})^{-d/(d+1)}$ and $B = 2p_e N^{(1)}$. Sample a random subset $E = \{i_1, \ldots, i_B\}$ of $[N^{(1)}]$ of size B.
2. Draw $a^{(1)} \sim \mathsf{Enc}_{\mathsf{ECSS}}(s)$. For $i \in [N^{(1)}]$, define

$$c_i^{(1)} = \begin{cases} a_i^{(1)} & \text{if } i \notin E \\ a_i^{(1)} + e_i & \text{otherwise,} \end{cases}$$

where $e_i \sim U_{\{0,1\}}$.
3. Let $\tau = (E, c_E^{(1)})$, where $c_E^{(1)} = (c_{i_1}^{(1)}, \cdots, c_{i_B}^{(1)})$.
4. Draw $c^{(2)} \sim \mathsf{Enc}_{\mathsf{NMC}_0}(\tau)$. Note that $N^{(2)} := |c^{(2)}| \leqslant |\tau|^d$.
5. Output $(c^{(1)}, c^{(2)})$ as the codeword.

$\mathsf{Dec}(\tilde{c} \in \{0,1\}^N)$:

1. Interpret $\tilde{c}$ as $(\tilde{c}^{(1)}, \tilde{c}^{(2)})$ of length $N^{(1)}$ and $N^{(2)}$, respectively.
2. Let $\tau^* = \mathsf{Dec}_{\mathsf{NMC}_0}(\tilde{c}^{(2)})$.
3. $\mathsf{Dec}^*(\tilde{c}^{(1)}, \tau^*)$:
   (a) If $\tau^* = \bot$, output $\bot$ and halt.
   (b) Let $a^* = \mathsf{ECorr}_{\mathsf{ECSS}}(\tilde{c}^{(1)})$. If $a^* = \bot$, output $\bot$ and halt.
   (c) Let $\tilde{s} = \mathsf{Rec}_{\mathsf{ECSS}}(a^*)$. If $\tilde{s} = \bot$, output $\bot$ and halt.
   (d) Interpret τ^* as (E^*, r^*), where positions in r^* are indexed by elements in E^*. More formally, if $E^* = \{i_1, \ldots, i_B\}$, then $r^* = (r_{i_1}^*, \ldots, r_{i_B}^*)$. Let c^* be defined as follows: for $i \in [N^{(1)}]$, $c_i^* = a_i^*$ if $i \notin E^*$, and otherwise $c_i^* = r_i^*$. If $c^* \neq \tilde{c}^{(1)}$, output $\bot$; else, output $\tilde{s}$.

Fig. 2. Compiler for Rate-1 Non-Malleable Code

Discussion. Note that our compiler is a fully black-box compiler, in that both the components are used in fully black-box manner and the security of our compiler directly reduces to the security of its components in a black-box manner. Further, the output code is explicit if both ECSS and NMC_0 are explicit; and encoding and decoding of the new code is efficient if both its components are efficient.

3.1   Proof of Main Theorem

To achieve the parameters stated in the proof, we shall use the following parameters in our construction in Fig. 2: $T, D \geqslant 2M^{d/(d+1)} \log^{2d} M$, $N^{(1)} = M$, and $p_e =$

$M^{-d/(d+1)}$. From the construction, we get $|\tau| \leqslant B(\log M + 1) = 2p_e M \log 2M$. Using this we get, $N^{(2)} \leqslant |\tau|^d \leqslant 2^d M^{d/(d+1)} \log^d 2M < M^{d/(d+1)} \log^{2d} M$.

We shall interpret the codeword produced by Fig. 2 as a two-part codeword. The *left-part* is a share-packing based on the ECSS and the *right-part* has the non-malleable encoding of τ using $\mathsf{NMC_0}$. For ease of notation, let $\mathcal{L} = [N^{(1)}]$ and $\mathcal{R} = [N] \setminus \mathcal{L}$.

Consider the $[t', t, \nu_0]$ non-malleable coding scheme $\mathsf{NMC_0}$ against $\mathcal{F}' = \mathcal{F}_{\mathsf{BIT}} \circ \mathcal{S}_{t'}$ with encoding and decoding algorithms $\mathsf{Enc_{NMC_0}}$ and $\mathsf{Dec_{NMC_0}}$ respectively. For $\tau \in \{0,1\}^t$ and $f' \in \mathcal{F}'$, the random variable $\mathsf{Tamper}_{f',\mathsf{NMC_0}}^{(\tau)}$ over $\{0,1\}^t \cup \{\bot\}$ with respect $\mathsf{NMC_0}$ is given by

$$\mathsf{Tamper}_{f',\mathsf{NMC_0}}^{(\tau)} = \mathsf{Dec_{NMC_0}}(f'(\mathsf{Enc_{NMC_0}}(\tau))).$$

The non-malleability of $\mathsf{NMC_0}$ guarantees that there exists a map $\mathsf{Sim_0}$ from $\mathcal{F}'$ to distributions over the sample space $\{0,1\}^t \cup \{\mathsf{same}^*, \bot\}$ such that

$$\mathsf{SD}\left(\mathsf{Tamper}_{f',\mathsf{NMC_0}}^{(\tau)}, \mathsf{Copy}_{\mathsf{Sim_0}(f')}^{(\tau)}\right) \leqslant \nu_0(t').$$

for all $\tau \in \{0,1\}^t$ and $f' \in \mathcal{F}'$. See Fig. 1 for the definition of non-malleable codes. We use an additional subscript in the notation of Tamper function to distinguish it from the one we define below for $\mathsf{NMC_1}$.

Fix a tampering function $f = (f_1, \ldots, f_N, \pi) \in \mathcal{F}^* = \mathcal{F}_{\mathsf{BIT}} \circ \mathcal{S}_N$ and a message s. The random variable $\mathsf{Tamper}_{f,\mathsf{NMC_1}}^{(s)}$ over $\{0,1\}^L \cup \{\bot\}$ for the non-malleable coding scheme $\mathsf{NMC_1}$ described in Fig. 2 (with encoding and decoding functions Enc and Dec respectively) is given by

$$\mathsf{Tamper}_{f,\mathsf{NMC_1}}^{(s)} = \mathsf{Dec}(f(\mathsf{Enc}(s))).$$

Our goal is to show that there exists a map $\mathsf{Sim_1}$ from $\mathcal{F}^*$ to distributions over the sample space $\{0,1\}^L \cup \{\mathsf{same}^*, \bot\}$ such that

$$\mathsf{SD}\left(\mathsf{Tamper}_{f,\mathsf{NMC_1}}^{(s)}, \mathsf{Copy}_{\mathsf{Sim_1}(f)}^{(s)}\right) \leqslant \nu_1(N).$$

(It is easy to see that $\mathsf{NMC_1}$ is correct.)

We provide a description of $\mathsf{Sim_1}$ in the next section and show that $\mathsf{Copy}_{\mathsf{Sim_1}(f)}^{(s)}$ is statistically close to $\mathsf{Tamper}_{f,\mathsf{NMC_1}}^{(s)}$ after that.

Description of Simulator. Given $f = (f_1, \ldots, f_N, \pi) \in \mathcal{F}^*$, we define the following set of indices in $[N]$ which will be used in the description of the simulator (and later in the proof):

- Let $X \subseteq \mathcal{L}$ be the set of all indices which move from left to right as a result of applying the permutation π, i.e., $X = \{i \in \mathcal{L} \mid \pi(i) \in \mathcal{R}\}$.
- Similarly, let $Y = \{i \in \mathcal{R} \mid \pi(i) \in \mathcal{L}\}$ be the indices which move from right to left. Note that $|X| = |Y|$.

- Let $\overline{X} = \mathcal{L} \setminus X$.
- Let $J \subseteq \overline{X}$ such that for all $i \in J$, $f_i \in \{f_{\text{forward}}, f_{\text{toggle}}\}$. In other words, J is the set of all indices on the left which are mapped into the left codeword using f_{forward} or f_{toggle}.
- Let $V \subseteq \overline{X}$ such that for all $i \in V$, either $\pi(i) \neq i$ or $f_i \neq f_{\text{forward}}$.
- Let $\overline{V} = \overline{X} \setminus V$. In other words, $\overline{V}$ denotes the set of indices on the left which are not tampered.

Observe that only the bits at indices X in $c^{(1)}$ affect the tampered right codeword $\tilde{c}^{(2)}$. Hence, given $c_X^{(1)}$, we can construct a tampering function $f^{(2)} \in \mathcal{F}_{\text{BIT}} \circ \mathcal{S}_{t'}$ which acts on the right codeword. Let $\rho : Y \to X$ be an arbitrary bijection from Y to X. The function $f^{(2)} = (f_1^{(2)}, \ldots, f_{N^{(2)}}^{(2)}, \pi^{(2)})$ is given by

$$
f_{i-N^{(1)}}^{(2)} = \begin{cases} f_i & \text{if } i \in \mathcal{R} \setminus Y \\ f_{\text{reset}} & \text{if } i \in Y \text{ and } f_{\rho(i)}(c_{\rho(i)}^{(1)}) = 0 \\ f_{\text{set}} & \text{if } i \in Y \text{ and } f_{\rho(i)}(c_{\rho(i)}^{(1)}) = 1 \end{cases}
$$

$$
\pi^{(2)}(i - N^{(1)}) = \begin{cases} \pi(i) - N^{(1)} & \text{if } i \in \mathcal{R} \setminus Y \\ \pi(\rho(i)) - N^{(1)} & \text{otherwise} \end{cases}
$$

(1)

for $i \in \mathcal{R}$.

Let $\mathcal{D}_{\text{ECSS}}$ be the distribution of $\text{Enc}_{\text{ECSS}}(s')$ for a random $s' \in \{0,1\}^L$. We are now ready to describe how the distribution $\text{Sim}_1(f)$ is generated:

(1) Set τ, $c_X^{(1)}$ and $f^{(2)}$ as follows:
    First sample E as described in Step 1 of Fig. 2. Let $\tau = (E, r)$ where $r \sim U_{\{0,1\}^B}$ (i.e., r is a random bit-string of length B). The bits in r are indexed by the indices in E. Also, sample $a^{(1)} \sim \mathcal{D}_{\text{ECSS}}$. Together τ and $a^{(1)}$ is used to determine $c_X^{(1)}$: for $i \in X$, let $c_i^{(1)} = a_i^{(1)}$ if $i \notin E$; otherwise, $c_i^{(1)} = r_i$. Finally, define $f^{(2)}$ using $c_X^{(1)}$, as described in Equation 1. (If the number of error indices in X, i.e. $|X \cap E|$, is larger than $O(\log^2 M)$, output $\perp$ and stop. Using Lemma 1, we can show that this happens with only $\text{negl}(M)$ probability.)
(2) Draw $\theta \sim \text{Sim}_0(f^{(2)})$. Let $\tau^* = \text{Copy}_\theta^{(\tau)}$.
(3) Set $c^{(2)}$ as follows:
    $c^{(2)}$ is drawn from the output distribution of $\text{Enc}_{\text{NMC}_0}(\tau)$ conditioned on $\text{Dec}_{\text{NMC}_0}(f^{(2)}(c^{(2)})) = \tau^*$ [5]. If no such codeword exists, then the simulator fails.
(4) θ obtained in Step 2 could be an element in $\{0,1\}^t$, same* or $\perp$. Let $\text{Event}_{\text{fix}}$, $\text{Event}_{\text{same}^*}$ and $\text{Event}_\perp$ denote the corresponding events. Do the following based on which event takes place:
    (a) Case $\text{Event}_\perp$: Output $\perp$ and stop.

---

[5] It may not be possible to do this reverse sampling step efficiently. However, an efficient simulation is implied by the existence of an inefficient simulation; see Remark 1 in [9].

(b) Case $\mathsf{Event_{fix}}$: We have the following two sub cases, based on the size of the set J. Let $\alpha = M^{d/(d+1)} \log^2 M$.

    i. Case $|J| > \alpha$: Output $\perp$ and stop.

    ii. Case $|J| \leqslant \alpha$: Extend the definition of $c_X^{(1)}$ using $a^{(1)}, r$ from Step 1 to $c_J^{(1)}$ as well: i.e., for $i \in J$, define $c_i^{(1)} = a_i^{(1)}$ if $i \notin E$; otherwise, $c_i^{(1)} = r_i$. Combined with $c^{(2)}$ defined in the previous step, this defines a unique $\tilde{c}^{(1)}$ because for any $i \in \mathcal{L}$, $\tilde{c}_i^{(1)}$ is either 0 or 1 or has $\pi^{-1}(i) \in J \cup Y$. Output $\mathsf{Dec}^*(\tilde{c}^{(1)}, \tau^*)$ and stop (where Dec^* is as defined in Step 3 of the description of Dec, see Fig. 2).

(c) Case $\mathsf{Event_{same^*}}$: Define $n_{\text{non-id}} := |V|$. We have the following two cases based on $n_{\text{non-id}}$. Let $\beta = N^{(2)} + M^{d/(d+1)} \log^2 M$.

    i. Case $n_{\text{non-id}} > \beta$: Output $\perp$ and stop.

    ii. Case $n_{\text{non-id}} \leqslant \beta$: Extend the definition of $c_X^{(1)}$ using $a^{(1)}, r$ from Step 1 to $c_V^{(1)}$ as well: i.e., for $i \in V$, define $c_i^{(1)} = a_i^{(1)}$ if $i \notin E$; otherwise, $c_i^{(1)} = r_i$. Combined with $c^{(2)}$ defined in the previous step, this defines a unique $\tilde{c}_{V \cup X}^{(1)}$ because for any $i \in V \cup X$, $\pi^{-1}(i) \in V \cup Y$. Output same^* if $\tilde{c}_{V \cup X}^{(1)} = c_{V \cup X}^{(1)}$, else output $\perp$.

Hybrids and Their Indistinguishability. Recall that we want to show that the distributions $\mathsf{Tamper}_{f,\mathsf{NMC_1}}^{(s)}$ and $\mathsf{Copy}_{\mathsf{Sim_1}(f)}^{(s)}$ are statistically close to each other. Towards this, we first define four intermediate hybrids, (described below, and summarized, for quick reference, in Fig. 3), and show that for every $f \in \mathcal{F}^*$ and $s \in \{0,1\}^L$,

$$\mathsf{Tamper}_{f,\mathsf{NMC_1}}^{(s)} \equiv \mathsf{Hyb0}_f^{(s)} \equiv \mathsf{Hyb1}_f^{(s)} \equiv \mathsf{Hyb2}_f^{(s)} \approx \mathsf{Hyb3}_f^{(s)} \approx \mathsf{Copy}_{\mathsf{Sim_1}(f)}^{(s)}.$$

Of these, the first equality is because the experiment defining $\mathsf{Hyb0}_f^{(s)}$ is identical to that defining $\mathsf{Tamper}_{f,\mathsf{NMC_1}}^{(s)}$, but restated in a convenient form for comparison with the following hybrids. The second equality relies on the privacy of the ECSS code. The third equality follows since in defining $\mathsf{Hyb2}_f^{(s)}$ we merely change the order in which two random variables are sampled in $\mathsf{Hyb1}_f^{(s)}$, taking care to not change their distributions. The statistical difference between $\mathsf{Hyb2}_f^{(s)}$ and $\mathsf{Hyb3}_f^{(s)}$ will be bounded by $\nu_0(N - M)$ using the non-malleability of $\mathsf{NMC_0}$. Finally, we upper bound the statistical difference between $\mathsf{Hyb3}_f^{(s)}$ and $\mathsf{Copy}_{\mathsf{Sim_1}(f)}^{(s)}$ by $\mathsf{negl}(M)$, relying on the privacy and distance of ECSS and the "errata" technique. This will show that the statistical distance between $\mathsf{Tamper}_{f,\mathsf{NMC_1}}^{(s)}$ and $\mathsf{Copy}_{\mathsf{Sim_1}(f)}^{(s)}$ is at most $\mathsf{negl}(M) + \nu_0(N - M) = \nu_1(N)$, thus proving Theorem 1. We now discuss the above steps in detail.

Tamper vs. Hybrid 0: Our first claim, which is easy to verify, is that $\mathsf{Tamper}_f^{(s)}$ is identically distributed to $\mathsf{Hyb0}_f^{(s)}$ defined below.

Hybrid 0:

$$\tau$$

(1) $c^{(1)} \mid \tau, s$

   $f^{(2)} \mid c^{(1)}, f$

(2) $c^{(2)} \mid \tau$

(3) $\tau^* = \mathsf{Dec}_{\mathsf{NMC}_0}(f^{(2)}(c^{(2)}))$

(4) $\tilde{c}^{(1)} \mid c^{(1)}, c^{(2)}, f$

   $\mathsf{Hyb0}_f^{(s)} = \mathsf{Dec}^*(\tilde{c}^{(1)}, \tau^*)$

Hybrid 1:

$$\tau$$

(1) $(\star)$ $c_X^{(1)} \mid \tau$

   $(\star)$ $f^{(2)} \mid c_X^{(1)}, f$

(2) $c^{(2)} \mid \tau$

(3) $\tau^* = \mathsf{Dec}_{\mathsf{NMC}_0}(f^{(2)}(c^{(2)}))$

   $(\star)$ $c^{(1)} \mid c_X^{(1)}, \tau, s$

(4) $\tilde{c}^{(1)} \mid c^{(1)}, c^{(2)}, f$

   $\mathsf{Hyb1}_f^{(s)} = \mathsf{Dec}^*(\tilde{c}^{(1)}, \tau^*)$

Hybrid 2:

$$\tau$$

(1) $c_X^{(1)} \mid \tau$

   $f^{(2)} \mid c_X^{(1)}, f$

(2)

   $(\star)$ $\tau^* = \mathsf{Dec}_{\mathsf{NMC}_0}(f^{(2)}(\mathsf{Enc}_{\mathsf{NMC}_0}(\tau)))$

(3)

   $(\star)$ $c^{(2)} \mid \tau, \mathsf{Dec}_{\mathsf{NMC}_0}(f^{(2)}(c^{(2)})) = \tau^*$

   $c^{(1)} \mid c_X^{(1)}, \tau, s$

(4) $\tilde{c}^{(1)} \mid c^{(1)}, c^{(2)}, f$

   $\mathsf{Hyb2}_f^{(s)} = \mathsf{Dec}^*(\tilde{c}^{(1)}, \tau^*)$

Hybrid 3:

$$\tau$$

(1) $c_X^{(1)} \mid \tau$

   $f^{(2)} \mid c_X^{(1)}, f$

   $(\star)$ $\theta \sim \mathsf{Sim}_0(f^{(2)})$

(2) $(\star)$ $\tau^* = \mathsf{Copy}_\theta^{(\tau)}$

(3)

   $c^{(2)} \mid \tau, \mathsf{Dec}_{\mathsf{NMC}_0}(f^{(2)}(c^{(2)})) = \tau^*$

   $c^{(1)} \mid c_X^{(1)}, \tau, s$

(4) $\tilde{c}^{(1)} \mid c^{(1)}, c^{(2)}, f$

   $\mathsf{Hyb3}_f^{(s)} = \mathsf{Dec}^*(\tilde{c}^{(1)}, \tau^*)$

Fig. 3. For each hybrid experiment, the order in which the random variables are sampled is defined below. Notation $a \mid b$ denotes that a is sampled conditioned on b. The definition of the random variables is as in the description of the simulator. All the hybrids are parametrized by a function $f \in \mathcal{F}^*$ and a message $s \in \{0,1\}^L$. Items with a $(\star)$ before them indicate differences from the previous hybrid.

1. Sample E as described in Step 1 of Fig. 2. Let $\tau = (E, r)$ where $r \sim U_{\{0,1\}^B}$. (The bits in r are indexed by the indices in E.) Sample $a^{(1)} \sim \mathsf{Enc}_{\mathsf{ECSS}}(s)$. For $i \in \mathcal{L}$, let $c_i^{(1)} = a_i^{(1)}$ if $i \notin E$; otherwise, $c_i^{(1)} = r_i$. With $c^{(1)}$ defined, construct a tampering function $f^{(2)}$ as described before.
2. Sample $c^{(2)} \sim \mathsf{Enc}_{\mathsf{NMC}_0}(\tau)$. Let $\tilde{c}^{(2)} = f^{(2)}(c^{(2)})$.
3. Let $\tau^* = \mathsf{Dec}_{\mathsf{NMC}_0}(\tilde{c}^{(2)})$.
4. For $i \in \overline{X} \cup Y$, let $\tilde{c}_{\pi(i)}^{(1)} = f_i(c_i)$, where $\overline{X} = \mathcal{L} \setminus X$. Note that this completely defines $\tilde{c}^{(1)}$ because $\pi(\overline{X} \cup Y) = \mathcal{L}^6$. Output $\mathsf{Dec}^*(\tilde{c}^{(1)}, \tau^*)$.

Note that in sampling $\mathsf{Tamper}_{f, \mathsf{NMC}_1}^{(s)} = \mathsf{Dec}(f(\mathsf{Enc}(s)))$, the same steps as above are carried out, but in a different order: first $(c^{(1)}, c^{(2)})$ are sampled, then $(\tilde{c}^{(1)}, \tilde{c}^{(2)})$ are obtained, then τ^* is generated, and finally $\mathsf{Dec}^*(\tilde{c}^{(1)}, \tau^*)$ is output.

Hybrid 0 vs. Hybrid 1: Recall that $\mathcal{D}_{\mathsf{ECSS}}$ is the distribution of $\mathsf{Enc}_{\mathsf{ECSS}}(s')$ for a random $s' \in \{0,1\}^L$. In $\mathsf{Hyb1}_f^{(s)}$, Steps 1 and 4 change as described below (and the other two steps stay the same):

1. Sample $\tau = (E, r)$ in the same way as before. Sample $a^{(1)} \sim \mathcal{D}_{\mathsf{ECSS}}$. For $i \in X$, define $c_i^{(1)} = a_i^{(1)}$ if $i \notin E$; otherwise, $c_i^{(1)} = r_i$. Define $f^{(2)}$ using $c_X^{(1)}$.
4. Compute rest of $c^{(1)}$ consistent with $c_X^{(1)}$, τ and s. More formally, sample $b^{(1)}$ from the distribution $\mathsf{Enc}_{\mathsf{ECSS}}(s)$ conditioned on bits at indices X being $a_X^{(1)}$. For $i \in \overline{X}$, define $c_i^{(1)} = b_i^{(1)}$ if $i \notin E$; otherwise, $c_i^{(1)} = r_i$. Compute $\tilde{c}^{(1)}$ as described before and output $\mathsf{Dec}^*(\tilde{c}^{(1)}, \tau^*)$.

Effectively, $\mathsf{Hyb1}_f^{(s)}$ splits the sampling of $c^{(1)}$ into two parts. Only bits at indices in X are computed in Step 1, which is enough to define $f^{(2)}$. In order to show that $\mathsf{Tamper}_f^{(s)}$ and $\mathsf{Hyb1}_f^{(s)}$ are identically distributed, we must prove that $c_X^{(1)}$ can be sampled without knowledge of s as described in modified Step 1. This is indeed the case, since $|X| \leqslant N^{(2)} \leqslant T$, the privacy parameter of ECSS.

Hybrid 1 vs. Hybrid 2: We now define $\mathsf{Hyb2}_f^{(s)}$ which is a slightly different way of interpreting $\mathsf{Hyb1}_f^{(s)}$. Observe that the Steps 2 and 3 in $\mathsf{Hyb1}_f^{(s)}$ generate a τ^* from the distribution $\mathsf{Tamper}_{f^{(2)}, \mathsf{NMC}_0}^{(\tau)}$. In $\mathsf{Hyb2}_f^{(s)}$, we use this distribution to modify the following steps (rest of the steps remain unchanged):

2. Let $\tau^* \sim \mathsf{Tamper}_{f^{(2)}, \mathsf{NMC}_0}^{(\tau)}$. Recall that $\mathsf{Tamper}_{f^{(2)}, \mathsf{NMC}_0}^{(\tau)} = \mathsf{Dec}_{\mathsf{NMC}_0}(f^{(2)}$ $(\mathsf{Enc}_{\mathsf{NMC}_0}(\tau)))$.
3. Sample a random codeword $c^{(2)} \sim \mathsf{Enc}_{\mathsf{NMC}_0}(\tau)$ such that $\mathsf{Dec}_{\mathsf{NMC}_0}(f^{(2)}(c^{(2)}))$ $= \tau^*$.

It is clear that $\mathsf{Hyb1}_f^{(s)}$ is identically distributed to $\mathsf{Hyb2}_f^{(s)}$.

---

[6] Here there is a slight abuse of notation: $\pi(S)$ for any $S \subseteq [N]$ should be interpreted as the set $\{\pi(i) \mid i \in S\}$.

Hybrid 2 vs. Hybrid 3: The next hybrid – $\mathsf{Hyb3}_f^{(s)}$ – is same as $\mathsf{Hyb2}_f^{(s)}$ but with one difference:

– In Step 2, draw $\theta \sim \mathsf{Sim}_0(f^{(2)})$. Let $\tau^* = \mathsf{Copy}_\theta^{(\tau)}$.

In order to show that $\mathsf{Hyb2}_f^{(s)}$ and $\mathsf{Hyb3}_f^{(s)}$ are statistically close, we use the non-malleability of NMC_0. Consider an adversary $\mathcal{A}$ who, when given f and s as inputs, runs Steps 1 of $\mathsf{Hyb2}_f^{(s)}$ (or $\mathsf{Hyb3}_f^{(s)}$) to obtain τ and $f^{(2)}$, and sends them to a challenger. The challenger replies back with either $\mathsf{Tamper}_{f^{(2)},\mathsf{NMC}_0}^{(\tau)}$ or $\mathsf{Copy}_{\mathsf{Sim}_0(f^{(2)})}^{(\tau)}$; let τ^* denote this response. When adv receives τ^*, it runs Steps 3 and 4 of $\mathsf{Hyb2}_f^{(s)}$ (or $\mathsf{Hyb3}_f^{(s)}$), and outputs whatever the output of Step 4 is. It is easy to see that if challenger responds with $\mathsf{Tamper}_{f^{(2)},\mathsf{NMC}_0}^{(\tau)}$, then output of $\mathcal{A}$ is identically distributed to $\mathsf{Hyb2}_f^{(s)}$, and otherwise it is identically distributed to $\mathsf{Hyb3}_f^{(s)}$. We know that for all $f^{(2)} \in \mathcal{F}'$ and $\tau \in \{0,1\}^t$,
$$\mathsf{SD}\left(\mathsf{Tamper}_{f^{(2)},\mathsf{NMC}_0}^{(\tau)}, \mathsf{Copy}_{\mathsf{Sim}_0(f^{(2)})}^{(\tau)}\right) \leqslant \nu_0(t').$$
Hence, $\mathsf{SD}\left(\mathsf{Hyb2}_f^{(s)}, \mathsf{Hyb3}_f^{(s)}\right) \leqslant \nu_0(t') = \nu_0(N - M)$.

Hybrid 3 vs. Simulator: The final step in the proof is to show that the distributions $\mathsf{Hyb3}_f^{(s)}$ and $\mathsf{Copy}_{\mathsf{Sim}_1(f)}^{(s)}$ are statistically close. $\mathsf{Copy}_{\mathsf{Sim}_1(f)}^{(s)}$ is generated by the simulator as defined in Section 3.1, except that in Step 4-(c)-ii, "output same*" is replaced with "output s." Then we note that the first three steps in the experiment defining $\mathsf{Hyb3}_f^{(s)}$ and that defining $\mathsf{Copy}_{\mathsf{Sim}_1(f)}^{(s)}$ (or $\mathsf{Sim}_1(f)$) are identical[7].

Before proceeding further, we restate Step 4 of $\mathsf{Hyb3}_f^{(s)}$ here for the convenience of the reader:

– Sample $c_{\overline{X}}^{(1)}$ consistent with $c_X^{(1)}$, τ and s. Compute $\tilde{c}^{(1)}$ as follows: for $i \in \overline{X} \cup Y$, let $\tilde{c}_{\pi(i)}^{(1)} = f_i(c_i)$. Combined with $c^{(2)}$ from the previous step, this completely defines $\tilde{c}^{(1)}$ because $\pi(\overline{X} \cup Y) = \mathcal{L}$. Output $\mathsf{Dec}^*(\tilde{c}^{(1)}, \tau^*)$.

Consider the following case analysis based on the events defined by the value of θ. For each case, we show that the output of $\mathsf{Copy}_{\mathsf{Sim}_1(f)}^{(s)}$ is same as that of $\mathsf{Hyb3}_f^{(s)}$, except with probability $\mathsf{negl}(M)$.

1. *Case* $\mathsf{Event}_\perp$: In this case $\tau^* = \perp$, so $\mathsf{Hyb3}_f^{(s)}$ outputs $\perp$, just like $\mathsf{Copy}_{\mathsf{Sim}_1(f)}^{(s)}$ does.

2. *Case* $\mathsf{Event}_{\mathsf{fix}}$: We have the following two cases based on $|J|$. Recall that $J = \{i \in \mathcal{L} \mid \pi(i) \in \mathcal{L} \text{ and } f_i \in \{f_{\mathsf{forward}}, f_{\mathsf{toggle}}\}\}$.

---

[7] There is one minor difference in Step 1 though: $\mathsf{Sim}_1(f)$ outputs $\perp$ if the number of error indices in X is large, whereas $\mathsf{Hyb3}_f^{(s)}$ doesn't. However, this only adds a negligible amount of error. (See description of $\mathsf{Sim}_1(f)$ for more details.)

- Case $|J| > \alpha$: In this case, we show that $\Pr[\text{Hyb3}_f^{(s)} \neq \perp] \leqslant \text{negl}(M)$ over randomly chosen $\tau = (E, r)$ (conditioned on $c_X^{(1)}$). First, we define some notation. Let $Z \subseteq \mathcal{L}$ be set of indices $i \in \mathcal{L}$ such that $\pi^{-1}(i) \in \mathcal{R}$, i.e., Z is the set of indices on the left which come from the right. Let $\tau^* = (E^*, r^*)$, where the bits in r^* are indexed by the indices in E^*. (We know that $\tau^* \in \{0,1\}^t$ since $\theta \in \{0,1\}^t$). Let F_τ be the indices $i \in \mathcal{L}$ such that $\pi(j) = i$ and $j \in E$, i.e., j is an erroneous index according to original tag τ.

Fix any ECSS codeword $a^{(1)}$ for the left consistent with $c_X^{(1)}$. We first show that, irrespective of the value of τ, there exists at most one ECSS codeword a^* such that the tampered left codeword $\tilde{c}^{(1)}$ is consistent with a^* and τ^*. More precisely, for any two possible values of τ — τ_1 and τ_2 — let $\hat{c}^{\tau_1}$ and $\hat{c}^{\tau_2}$ be the corresponding values of $\tilde{c}^{(1)}$. Suppose $\text{Dec}^*(\hat{c}^{\tau_1}, \tau^*) \neq \perp$ and $\text{Dec}^*(\hat{c}^{\tau_2}, \tau^*) \neq \perp$. Let $\text{ECorr}_{\text{ECSS}}(\hat{c}^{\tau_1}) = a_1^*$ and $\text{ECorr}_{\text{ECSS}}(\hat{c}^{\tau_2}) = a_2^*$. Then, we argue that there is an element $c \in \{0,1\}^M$ such that $\text{HD}(a_1^*, c), \text{HD}(a_2^*, c) \leqslant D$. Hence, by the correctness of $\text{ECorr}_{\text{ECSS}}$, we have $a_1^* = a_2^*$.

To prove this, we shall let $c = \hat{c}^{\tau_1}$. Then $\text{HD}(a_1^*, c) \leqslant B < D$, since $\text{Dec}^*(c, \tau^*) \neq \perp$ (note that in the last step of Dec^* this is ensured). Similarly, $\text{HD}(a_2^*, \hat{c}^{\tau_2}) \leqslant B$. Then,

$$\text{HD}(a_2^*, c) \leqslant \text{HD}(a_2^*, \hat{c}^{\tau_2}) + \text{HD}(c, \hat{c}^{\tau_2}) \leqslant B + 2B + N^{(2)},$$

where we used the fact that, since $\hat{c}_{\mathcal{L}\backslash Z}^{\tau_1}$ and $\hat{c}_{\mathcal{L}\backslash Z}^{\tau_2}$ are derived from the same value of $a^{(1)}$ by adding at most B errors each, followed by applying a tampering function that cannot increase the hamming distance, their hamming distance is at most $2B$, and $|Z| \leqslant N^{(2)}$. By the choice of our parameters, $3B + N^{(2)} \leqslant D$, as required.

Thus, $\text{Hyb3}_f^{(s)} \neq \perp$ only if $\text{ECorr}_{\text{ECSS}}(\tilde{c}^{(1)}) = a^*$, for a value a^* which is fixed independent of τ. Further, since $\tau^* = (E^*, r^*)$ is fixed, there is a unique c^* as in the description of Dec^* in Fig. 2, and it must be the case that $\tilde{c}^{(1)} = c^*$. Next, we show that over randomness of τ (conditioned on $a^{(1)}, c_X^{(1)}$), the probability that tampered codeword $\tilde{c}^{(1)} = c^*$ is negligible in M.

Over randomness of τ, we know that number of error indices in J w.r.t. τ are $\Omega(\log^2 M)$ with $1 - \text{negl}(M)$ probability (see Lemma 1 and recall that the number of error indices in X is only $O(\log^2 M)$). Let these error indices be G.

Note that for a random τ, the value of erroneous indices outside X is uniform. This implies that each bit in $c_G^{(1)}$ is independent uniform bit (even after fixing $a^{(1)}$, $c_X^{(1)}$ and all other bits of $c_G^{(1)}$). Moreover, by definition of set J, for any $i \in G$, we have $\pi(i) \in \mathcal{L}$ (since $J \cap X = \emptyset$) and the bit $\tilde{c}_{\pi(i)}^{(1)} = f_i(c_i^{(1)})$ is a uniform random bit (since $f_i \in \{f_{\text{forward}}, f_{\text{toggle}}\}$). Hence $\Pr[\tilde{c}_{\pi(i)}^{(1)} = c_{\pi(i)}^*] = 1/2$, even conditioned on $\tilde{c}_{\pi(j)}^{(1)} = c_{\pi(j)}^*$ for $j \in G \setminus \{i\}$. Therefore, probability that $\tilde{c}_{\pi(G)}^{(1)} = c_{\pi(G)}^*$ is at most $2^{-\Omega(\log^2 M)} = \text{negl}(M)$.

- Case $|J| \leqslant \alpha$: In this case, $\text{Copy}_{\text{Sim}_1(f)}^{(s)}$ and $\text{Hyb3}_f^{(s)}$ behave in the same way except the manner in which they generate $c_J^{(1)}$ (rest of $c^{(1)}$ is not important).

In the former case, $c_J^{(1)}$ is sampled from $\mathcal{D}_{\mathsf{ECSS}}$ (conditioned on $c_X^{(1)}$), while in the latter case it is sampled from $\mathsf{Enc}_{\mathsf{ECSS}}(s)$ (again conditioned on $c_X^{(1)}$). This, however, makes no difference because T, the privacy parameter of ECSS, is at least $N^{(2)} + \alpha \geqslant |X| + |J|$.

3. *Case* $\mathsf{Event}_{\mathsf{same}^*}$: Since $\theta = \mathsf{same}^*$, we know that $\tau^* = \tau$. We have the following two cases based on $n_{\mathsf{non\text{-}id}} = |V|$. Recall that $V = \{i \in \mathcal{L} \mid \pi(i) \in \mathcal{L} \text{ and } (\pi(i) \neq i \text{ or } f_i \neq f_{\mathsf{forward}})\}$.

- Case $n_{\mathsf{non\text{-}id}} > \beta$: In this case, we show that $\Pr[\mathsf{Hyb3}_f^{(s)} \neq \bot] \leqslant \mathsf{negl}(M)$ over a random choice of $\tau = (E, r)$ (conditioned on $c_X^{(1)}$). In fact, we will show that for any ECSS codeword $a^{(1)}$ for the left (consistent with $c_X^{(1)}$), $\mathsf{Hyb3}_f^{(s)} = \bot$ with $1 - \mathsf{negl}(M)$ probability. Fix an ECSS codeword $a^{(1)}$ for the left consistent with $c_X^{(1)}$. Rest of the analysis will be over the randomness of τ, conditioned on $c_X^{(1)}$ and $a^{(1)}$.

  Recall that while sampling τ in Step 1, we begin by sampling a random set of error indices E. We consider the size of the set $V \cap E \setminus Z$, where $Z = \{i \in \mathcal{L} \mid \pi^{-1}(i) \in \mathcal{R}\}$. Over the random choice of E, we have $|V \cap E \setminus Z| = \Omega(\log^2 M)$ with $1 - \mathsf{negl}(M)$ probability. This follows from Lemma 1, because $|V \setminus Z| \geqslant \beta - N^{(2)} = M^{d/(d+1)} \log^2 M$ and the number of error indices in X is only $O(\log^2 M)$. We identify a set $U \subseteq V \cap E \setminus Z$, such that $|U| \geqslant |V \cap E \setminus Z|/2$ and for each $i \in U$, $\Pr[\tilde{c}_i^{(1)} = c_i^{(1)}] \leqslant 1/2$, even conditioned on $\tilde{c}_{i'}^{(1)} = c_{i'}^{(1)}$, for any set of $i' \in U \setminus \{i\}$. We build U iteratively: initialize $U = \emptyset$ and $W = V \cap E \setminus Z$. Pick any index $i \in W$, and let $j = \pi^{-1}(i)$. Update U to $U \cup \{i\}$ and W to $W \setminus \{i, j\}$, and repeat this until W is empty. Clearly, $|U| \geqslant |V \cap E \setminus Z|/2$. To verify the other property of U, note that at any step, when we pick $i \in W$, and let $j = \pi^{-1}(i)$, we have either 1) $j = i$ and $f_i \in \{f_{\mathsf{toggle}}, f_{\mathsf{reset}}, f_{\mathsf{set}}\}$ or 2) $j \neq i$ (since $i \in V$) but $j \in \mathcal{L}$ (since $i \notin Z$). Thus $\tilde{c}_i^{(1)} = f_i(c_i^{(1)})$ for $f_i \in \{f_{\mathsf{toggle}}, f_{\mathsf{reset}}, f_{\mathsf{set}}\}$, or $\tilde{c}_i^{(1)} = f_j(c_j^{(1)})$ for $j = \pi^{-1}(i) \notin U$. Since $i \in E$, each bit $c_i^{(1)}$ is uniformly random and hence $\Pr[\tilde{c}_i^{(1)} = c_i^{(1)}] \leqslant 1/2$. (This probability is 0 if $\pi(i) = i$ and $f_i = f_{\mathsf{toggle}}$; otherwise it is exactly $1/2$.) Further, the conditions $\tilde{c}_i^{(1)} = c_i^{(1)}$ for $i \in U$ are independent of each other, since each such condition involves either $c_i^{(1)}$ alone or a pair $(c_i^{(1)}, c_j^{(1)})$ such that $c_j^{(1)}$ never occurs in any other condition. Thus, $\Pr[\tilde{c}_E^{(1)} = c_E^{(1)}] \leqslant \Pr[\tilde{c}_U^{(1)} = c_U^{(1)}] \leqslant 1/2^{|V \cap E \setminus Z|/2} \leqslant \mathsf{negl}(M)$. But to be a valid left codeword, it has to be the case that $\tilde{c}_E^{(1)} = c_E^{(1)}$, since $\tau^* = \tau$, which records $c_E^{(1)}$. Hence, $\Pr[\mathsf{Hyb3}_f^{(s)} \neq \bot] \leqslant \mathsf{negl}(M)$.

- Case $n_{\mathsf{non\text{-}id}} \leqslant \beta$: We know that most of the indices on the left have been copied identically to the tampered codeword. We use this to argue that if the tampered codeword $\tilde{c}^{(1)}$ is valid w.r.t. τ then $\tilde{c}^{(1)} = c^{(1)}$, and that the probability of this happening depends only on a small number of bits in $c^{(1)}$ and hence independent of the message s. More formally, we have the following.

Let $a^{(1)}$ be an ECSS encoding of s consistent with $c_X^{(1)}$. The maximum number of errors in the tampered codeword $\tilde{c}^{(1)}$ could be $|\tilde{V}| + |X| + |E| \leqslant \beta + N^{(2)} + 2p_e N^{(1)} \leqslant D$, where D is the error-correction radius of the ECSS code. Hence, $a^* = \mathsf{ECorr_{ECSS}}(\tilde{c}^{(1)})$, computed in the algorithm Dec^* by $\mathsf{Hyb3}_f^{(s)}$, must be equal to $a^{(1)}$. Further, since $\tau^* = \tau$, $c^* = c^{(1)}$. Therefore, unless $c^{(1)} = \tilde{c}^{(1)}$, the output of $\mathsf{Hyb3}_f^{(s)}$ would be $\bot$.

Since indices in $\overline{V}$ have been copied identically, $c^{(1)} = \tilde{c}^{(1)}$ iff $c_{\overline{V} \cup X}^{(1)} = \tilde{c}_{\overline{V} \cup X}^{(1)}$. This is the exact check which $\mathsf{Sim_1}$ performs, by sampling $c_{\overline{V} \cup X}^{(1)}$ itself. Note that $\mathsf{Sim_1}$ generates $c_{\overline{V} \cup X}^{(1)}$ from the same distribution as in the experiment for $\mathsf{Hyb3}_f^{(s)}$, since the privacy parameter T of ECSS is at least $N^{(2)} + \beta \geqslant |X| + |V|$. Hence, in this case, $\mathsf{Sim_1}$ outputs same^* with the same probability as $\mathsf{Hyb3}_f^{(s)} = s$; otherwise, both the random variables are $\bot$.

This completes the proof.

3.2   Instantiations

We shall use the following results:

1. There exists a $[t', t, \nu_0]$ non-malleable code $\mathsf{NMC_0}$ against $\mathcal{F}_{\mathsf{BIT}} \circ \mathcal{S}_{t'}$, where $t' \leqslant t \cdot \mathrm{polylog}\, t$ and $\nu_0(\cdot)$ is a suitable negligible function. This follows from the work of [3]. For simplicity, we set $d = 2$ so that we have $t' < t^d$.
2. Using Shamir's secret-sharing scheme [30], with a standard share-packing technique [4,19], we can obtain an efficient $[M, L, T, D]$ ECSS such that: $M = n\varphi$, $L = \ell\varphi$ and $T = D = (n - \ell)/2$, where $\varphi = 2\log n$. We can choose $n - \ell = n^{3/4}$ so that $T, D = \widetilde{\Theta}(M^{3/4}) \geqslant 2M^{d/(d+1)} \log^{2d} M$, for $d = 2$.

   The secret sharing scheme is formally described as follows. We choose a field $\mathbb{F}$ of characteristic 2 such that $\varphi = |\mathbb{F}| \geqslant n^2$. Pick arbitrary elements in the field $\mathbb{F}$ and name them $S = \{-\ell, \ldots, -1, 1, \ldots, n\}$. The secret sharing scheme is constructed by picking a random $\ell + (n - \ell)/2$ degree polynomial $p(\cdot)$ and evaluating it at elements in S. The share X_0 is the concatenation of the evaluations of $p(\cdot)$ at $\{-\ell, \ldots, -1\}$ and interpreting it as a binary string. The share X_i, for $i \in [M]$, is the ith bit in the concatenation of evaluations of $p(\cdot)$ at $\{1, \ldots, n\}$.

   Note that we can also use Algebraic Geometric code [21,20] based share-packing techniques [7] over constant size field with characteristic 2. But we forgo this optimization for ease of presentation and simplicity of the resulting code.

Using the above choices for $\mathsf{NMC_0}$ and ECSS, we get the following rate of $\mathsf{NMC_1}$ in Theorem 1.

$$\frac{L}{N} \geqslant 1 - N^{-1/4} \mathrm{polylog}\, N.$$

This directly yields Corollary 1 and Corollary 2.

References

1. Aggarwal, D., Dodis, Y., Lovett, S.: Non-malleable codes from additive combinatorics. In: STOC, pp. 774–783 (2014)
2. Agrawal, S., Gupta, D., Maji, H.K., Pandey, O., Prabhakaran, M.: Explicit non-malleable codes resistant to permutations. Cryptology ePrint Archive, Report 2014/316 (2014), http://eprint.iacr.org/
3. Agrawal, S., Gupta, D., Maji, H.K., Pandey, O., Prabhakaran, M.: Explicit non-malleable codes resistant to permutations and perturbations. Cryptology ePrint Archive, Report 2014/841 (2014), http://eprint.iacr.org/
4. Blakley, G., Meadows, C.: Security of ramp schemes. In: Blakely, G.R., Chaum, D. (eds.) CRYPTO 1984. LNCS, vol. 196, pp. 242–268. Springer, Heidelberg (1985), http://dx.doi.org/10.1007/3-540-39568-7_20
5. Chandran, N., Kanukurthi, B., Ostrovsky, R.: Locally updatable and locally decodable codes. In: Lindell, Y. (ed.) TCC 2014. LNCS, vol. 8349, pp. 489–514. Springer, Heidelberg (2014)
6. Chattopadhyay, E., Zuckerman, D.: Non-malleable codes against constant split-state tampering. Electronic Colloquium on Computational Complexity, Report 2014/102 (2014), http://eccc.hpi-web.de/
7. Chen, H., Cramer, R.: Algebraic geometric secret sharing schemes and secure multi-party computations over small fields. In: Dwork, C. (ed.) CRYPTO 2006. LNCS, vol. 4117, pp. 521–536. Springer, Heidelberg (2006)
8. Cheraghchi, M., Guruswami, V.: Capacity of non-malleable codes. In: Naor, M. (ed.) ITCS, pp. 155–168. ACM (2014)
9. Cheraghchi, M., Guruswami, V.: Non-malleable coding against bit-wise and split-state tampering. In: Lindell, Y. (ed.) TCC 2014. LNCS, vol. 8349, pp. 440–464. Springer, Heidelberg (2014)
10. Choi, S.G., Kiayias, A., Malkin, T.: BiTR: Built-in tamper resilience. In: Lee, D.H., Wang, X. (eds.) ASIACRYPT 2011. LNCS, vol. 7073, pp. 740–758. Springer, Heidelberg (2011)
11. Chvátal, V.: The tail of the hypergeometric distribution. Discrete Mathematics 25(3), 285–287 (1979), http://www.sciencedirect.com/science/article/pii/0012365X79900840
12. Cramer, R., Dodis, Y., Fehr, S., Padró, C., Wichs, D.: Detection of algebraic manipulation with applications to robust secret sharing and fuzzy extractors. In: Smart, N.P. (ed.) EUROCRYPT 2008. LNCS, vol. 4965, pp. 471–488. Springer, Heidelberg (2008)
13. Cramer, R., Padró, C., Xing, C.: Optimal algebraic manipulation detection codes (2014), http://eprint.iacr.org/2014/116
14. Cramer, R., Shoup, V.: Design and analysis of practical public-key encryption schemes secure against adaptive chosen ciphertext attack. SIAM J. Comput. 33(1), 167–226 (2003), http://dx.doi.org/10.1137/S0097539702403773
15. Dziembowski, S., Kazana, T., Obremski, M.: Non-malleable codes from two-source extractors. In: Canetti, R., Garay, J.A. (eds.) CRYPTO 2013, Part II. LNCS, vol. 8043, pp. 239–257. Springer, Heidelberg (2013)
16. Dziembowski, S., Pietrzak, K., Wichs, D.: Non-malleable codes. In: Yao, A.C.C. (ed.) ICS, pp. 434–452. Tsinghua University Press (2010)
17. Faust, S., Mukherjee, P., Nielsen, J.B., Venturi, D.: Continuous non-malleable codes. In: TCC. pp. 465–488 (2014)
18. Faust, S., Mukherjee, P., Venturi, D., Wichs, D.: Efficient non-malleable codes and key-derivation for poly-size tampering circuits. In: Nguyen, P.Q., Oswald, E. (eds.) EUROCRYPT 2014. LNCS, vol. 8441, pp. 111–128. Springer, Heidelberg (2014)

19. Franklin, M.K., Yung, M.: Communication complexity of secure computation (extended abstract). In: Kosaraju, S.R., Fellows, M., Wigderson, A., Ellis, J.A. (eds.) Proceedings of the 24th Annual ACM Symposium on Theory of Computing, Victoria, British Columbia, Canada, May 4-6, pp. 699–710. ACM (1992), http://doi.acm.org/10.1145/129712.129780
20. Garcia, A., Stichtenoth, H.: On the asymptotic behaviour of some towers of function fields over finite fields. Journal of Number Theory 61(2), 248–273 (1996)
21. Goppa, V.D.: Codes on algebraic curves. In: Soviet Math. Dokl. pp. 170–172 (1981)
22. Guruswami, V., Smith, A.: Codes for computationally simple channels: Explicit constructions with optimal rate. In: FOCS, pp. 723–732. IEEE Computer Society (2010)
23. Hemenway, B., Ostrovsky, R.: Public-key locally-decodable codes. In: Wagner, D. (ed.) CRYPTO 2008. LNCS, vol. 5157, pp. 126–143. Springer, Heidelberg (2008)
24. Hoeffding, W.: Probability inequalities for sums of bounded random variables. Journal of the American Statistical Association 58(301), 13–30 (1963), http://www.jstor.org/stable/2282952
25. Kurosawa, K.: Hybrid encryption. In: Encyclopedia of Cryptography and Security, 2nd edn., pp. 570–572 (2011), http://dx.doi.org/10.1007/978-1-4419-5906-5_321
26. Lipton, R.J.: A new approach to information theory. In: Enjalbert, P., Mayr, E.W., Wagner, K.W. (eds.) STACS 1994. LNCS, vol. 775, pp. 699–708. Springer, Heidelberg (1994)
27. Liu, F.H., Lysyanskaya, A.: Tamper and leakage resilience in the split-state model. In: Safavi-Naini, R., Canetti, R. (eds.) CRYPTO 2012. LNCS, vol. 7417, pp. 517–532. Springer, Heidelberg (2012)
28. Micali, S., Peikert, C., Sudan, M., Wilson, D.A.: Optimal error correction against computationally bounded noise. In: Kilian, J. (ed.) TCC 2005. LNCS, vol. 3378, pp. 1–16. Springer, Heidelberg (2005)
29. Ostrovsky, R., Pandey, O., Sahai, A.: Private locally decodable codes. In: Arge, L., Cachin, C., Jurdziński, T., Tarlecki, A. (eds.) ICALP 2007. LNCS, vol. 4596, pp. 387–398. Springer, Heidelberg (2007)
30. Shamir, A.: How to share a secret. Communications of the ACM 22(11) (November 1979)

Leakage-Resilient Non-malleable Codes*

Divesh Aggarwal[1], Stefan Dziembowski[2], Tomasz Kazana[2],
and Maciej Obremski[2]

[1] Department of Computer Science, EPFL, Switzerland
[2] Institute of Informatics, University of Warsaw, Poland

Abstract. A recent trend in cryptography is to construct cryptosystems
that are secure against physical attacks. Such attacks are usually divided
into two classes: the *leakage* attacks in which the adversary obtains some
information about the internal state of the machine, and the *tampering*
attacks where the adversary can modify this state. One of the popular
tools used to provide tamper-resistance are the *non-malleable codes* introduced by Dziembowski, Pietrzak and Wichs (ICS 2010). These codes
can be defined in several variants, but arguably the most natural of them
are the information-theoretically secure codes in the *k-split-state model*
(the most desired case being $k = 2$).

Such codes were constucted recently by Aggarwal et al. (STOC 2014).
Unfortunately, unlike the earlier, computationally-secure constructions
(Liu and Lysanskaya, CRYPTO 2012) these codes are not known to be
resilient to leakage. This is unsatisfactory, since in practice one always
aims at providing resilience against *both* leakage and tampering (especially considering tampering without leakage is problematic, since the
leakage attacks are usually much easier to perform than the tampering
attacks).

In this paper we close this gap by showing a non-malleable code in
the 2-split state model that is secure against leaking almost a 1/12-th
fraction of the bits from the codeword (in the bounded-leakage model).
This is achieved via a generic transformation that takes as input any non-
malleable code (Enc, Dec) in the 2-split state model, and constructs out
of it another non-malleable code (Enc', Dec') in the 2-split state model
that is additionally leakage-resilient. The rate of (Enc', Dec') is linear in
the rate of (Enc, Dec). Our construction requires that Dec is *symmetric*,
i.e., for all x, y, it is the case that $\mathsf{Dec}(x, y) = \mathsf{Dec}(y, x)$, but this property holds for all currently known information-theoretically secure codes
in the 2-split state model. In particular, we can apply our transformation
to the code of Aggarwal et al., obtaining the first leakage-resilient code
secure in the split-state model. Our transformation can be applied to
other codes (in particular it can also be applied to a recent code of Aggarwal, Dodis, Kazana and Obremski constructed in the work subsequent
to this one).

---

* This work was supported by the WELCOME/2010-4/2 grant founded within the
framework of the EU Innovative Economy (National Cohesion Strategy) Operational
Programme.

Y. Dodis and J.B. Nielsen (Eds.): TCC 2015, Part I, LNCS 9014, pp. 398–426, 2015.

1 Introduction

Several attacks on cryptographic devices are based on exploiting physical weaknesses in their implementations. Such "physical attacks" are usually based on the *side-channel information* about the internals of the cryptographic device that the adversary can obtain by measuring its running-time, electromagnetic radiation, power consumption (see e.g. [24]), or on active tampering (see e.g. [4,24]). A recent trend in theoretical cryptography, initiated by [37,33,32], is to design schemes that are provably-secure even if their implementations can be attacked.

One of the tools used in this area are the so-called *non-malleable codes* introduced by Dziembowski, Pietrzak and Wichs in 2010 [23]. Informally, a code (Enc : $\mathcal{M} \to \mathcal{C}$, Dec : $\mathcal{C} \to \mathcal{M}$), where Enc is a randomized encoding function, and Dec is a partial decoding function, is *non-malleable* if an adversary that learns $C = \mathsf{Enc}(M)$ is not able to produce $C' = h(C)$ such that $\mathsf{Dec}(C')$ is not equal to M, but is "related" to it. The precise meaning of "not being related" is a little tricky to define but intuitively, what we require is that C' does not depend on C in any non-trivial way. For example: C' equal to C with the first bit set to zero, or C' equal to C with every bit negated, are obviously "related" to C, but a uniformly random C', or a constant C' are unrelated to C. It is easy to see that in order to construct such codes, one needs to restrict in some way the set of possible "manipulation functions" h that the adversary can use in order to compute C' from C. This is because otherwise the adversary could simply let h compute M from C (using the decoding function Dec), compute M' that is "related to M" (by say, negating all the bits of M), and then output $C' = \mathsf{Enc}(M')$. Therefore the non-malleable codes are always defined with respect to a family $\mathcal{H}$ of manipulation functions h that the adversary is allowed to use to compute C' from C.

The main application for this notion is the protection against tampering attacks. Imagine $C = \mathsf{Enc}(M)$ is stored on some device that the adversary can tamper with, and hence he can substitute C with some $C' \neq C$. Suppose (Enc, Dec) is non-malleable with respect to the set of manipulation functions that the adversary is able to induce by tampering. Then the only thing that the adversary can achieve is that C' will either decode to the same M, or to some M' that is totally unrelated to M. This is useful, since many practical attacks on cryptographic schemes are based on the so-called "related key attacks" [5], where the adversary is able to break a scheme $\mathsf{S}(K)$ (where K is the secret key) by having access to a device $\mathsf{S}(K')$ for some K' that is related to K. Clearly, storing K in an encoded form $\mathsf{Enc}(K)$ provides protection against such attacks. In [23] the authors describe also other applications of non-malleable codes. In particular they show how to use them in combination with the *algorithmic tamper proof security* framework of Gennaro et al. [30]. Recently, Faust et al. used the non-malleable codes to construct Random Access Machines secure against tampering and leakage attacks [27], and Coretti et al. [13] have shown how to use the non-malleable codes to construct public-key encryption schemes.

Since the invention of the non-malleable codes, there has been a significant effort to construct codes that would be secure against interesting classes of families. In [23] the authors show a construction of efficient codes secure against bitwise tapering, i.e. when every bit of the codeword is manipulated independently (this is achieved using the algebraic manipulation detection codes of Cramer et al. [14]). They also provide an existential result that for every sufficiently small family $\mathcal{H}$ of manipulation functions there exists a (not necessarily efficient) non-malleable code secure against it. This immediately gives a construction of non-malleable codes secure in the random oracle model [6].

Previous constructions of non-malleable codes in the split-state model. A very attractive and natural family of manipulation functions can be defined using the so-called *split-state model*. Assume that C is represented as a sequence of blocks $C = (C_1, \cdots, C_k)$. Then $\mathcal{H}$ is a family of k-*split state manipulation functions* if every $h \in \mathcal{H}$ manipulates each element $C_1, \ldots, C_k$ independently, i.e., for every h there exist functions $\{h_i\}_{i=1}^{k}$ such that $h(C_1, \ldots, C_k) = (h_1(C_1), \ldots, h_k(C_k))$. A practical justification for such a model comes from an observation that it may be easy to achieve in real life, by simply placing every C_i on a separate chip. Of course, the fewer parts are needed, the stronger the model is, and in particular the most desirable case is $k = 2$. In the sequel, we will sometimes refer to the 2-split state model simply as the "split model".

The aforementioned existential result of [23] implies that there exist non-malleable codes in the 2-split state model. The problem of showing an efficient construction of such codes was left open in [23]. The first step towards solving it was made by Liu and Lysyanskaya [36], who showed a construction of non-malleable codes computationally secure in the 2-split state model. Dziembowski, Kazana and Obremski [20] provided an efficient construction of information-theoretically non-malleable codes that works only for messages of length 1. The problem of constructing information-theoretically secure codes for messages of arbitrary length was finally solved by Aggarwal, Dodis and Lovett [3]. Their construction is based on the methods from additive combinatorics, including the so-called *Quasi-polynomial Freiman-Ruzsa Theorem*, and involves a substantial blow-up in the size of the codeword ($|C| = \tilde{O}((|M| + \kappa)^7)$, where κ is the security parameter). Very recently Chattopadhyay and Zuckerman [9] have shown a construction of non-malleable codes in 10-split state model that achieves linear blow-up.

A subsequent construction of non-malleable codes in the split-state model. In a paper subsequent to this one, Aggarwal et al. [2] show a general transformation of any k-split state model secure non-malleable code into one secure in the 2-split state model, that involves a linear blow-up in the codeword size. This, together with the result of [9] gives a construction of a non-malleable code in the 2-split state model with codeword of length linear in $|M|$. Their construction uses some of the techniques developed in this work. In particular, one of the steps of their construction which can be seen as a generalization of our reduction is a reduction from the 2-split-state tampering family to the so called 2-part t-lookahead tampering family.

However, the result of [2] does not consider leakage-resilience, and considering the number of levels of encoding required by their result, it is unlikely that their construction is resilient to any significant leakage.

Leakage-resilience of non-malleable codes. The ultimate goal of the "physically secure cryptography" is to provide both tampering- and leakage-resilience. The basic definition of the non-malleable codes does not consider any type of leakage information that the adversary can obtain about the codeword through the side channels. This may be considered unrealistic, as in practice it may be often relatively easy for the adversary to obtain such information (probably easier than to perform the tampering attacks). Therefore, it would be desirable to include also such attacks in this definition. The first paper that considered *leakage-resilient* non-malleable codes was [36] (in the computational settings). Our definition essentially follows their ideas, except that we consider the information-theoretic settings.

Let us now explain informally the concept of leakage-resilient non-malleable codes in the 2-split state model (the formal definition appears in Section 3). First consider the question what would be the most natural definition of leakage and tampering attacks. To be as general as possible we should give to the adversary right to simultaneously tamper the codeword $C = (L, R)$ and leak information from it. The leakage will be modeled by allowing the adversary to choose functions Leak_i^L and Leak_i^R and learn $\mathsf{Leak}_i^L(L)$ and $\mathsf{Leak}_i^R(R)$ (respectively). The entire process should happen in several rounds, and the adversary should be adaptive (i.e. his behavior in round i should depend on what he learned in the previous rounds). The functions will be arbitrary, except that we will have a bound on the total number of bits leaked from L and R, where the "number of leaked bits" is measured in terms of the total out size of the $\mathsf{Leak}_i^L(L)$ and $\mathsf{Leak}_i^L(R)$ functions. This is essentially the *independent leakage model* first considered in [22] (inspired by the "only computation leaks" paradigm of Micali and Reyzin [37]) and then in a sequence of papers (see, e.g.: [16,25,31,19,7]). It makes particular sense to use it in our context, as it is also motivated by the assumption that L and R are stored on two separate memory parts. Observe also that if we allowed joint leakage from L and R then we would need to have some additional restrictions on the leakage functions, as otherwise the adversary could choose a leakage function that first computes $M = \mathsf{Dec}(L, R)$ and then outputs the first bit $M[1]$ of M. This, in turn, would allow him to choose tampering functions that simply overwrite the original encoding with (L', R') such that $\mathsf{Dec}(L', R')$ is equal to $(M[1], 0, \ldots, 0)$ (such M' is obviously "related" to M and with overwhelming probability it is not equal to M). For similar reasons it is obvious that we always need some sort of restriction on the leakage functions Leak_i^L and Leak_i^R, since if the adversary learns the entire L (say) then he can then easily choose a leakage function Leak_i^R that first computes $\mathsf{Dec}(L, R)$ and then outputs $M[1]$.

It is also easy to see that without loss of generality we can restrict the adversary to choose deterministic leakage functions (since we can always convert a random function to a deterministic one by fixing its random input). Another natural observation is that it is enough to consider the case when all the leakage happens before the tampering functions are chosen. This is because the result of any leakage Leak_i from a tampered codeword $f(C)$ can be computed by a function $\mathsf{Leak}'_i = \mathsf{Leak}_i \circ f$ that is applied directly to C. The formal definition of our model appears in Section 3.

1.1   Our Contribution

As argued above, leakage-resilence is an important property for may applications of the non-malleable codes. Unfortunately, this aspect of these codes has been ignored in many recent papers on this topic. In particular, the authors of [3,9] do not consider leakage at all. Leakage-resilience was considered in [20], but their construction works only for messages of length 1. To summarize: until now, no construction of leakage-resilient 2-split state non-malleable codes for messages longer than 1 was known. Providing such a construction is the main contribution of this work.

In fact, our contribution is much more general. We show a generic transformation that takes as input any non-malleable code $(\mathsf{Enc}, \mathsf{Dec})$ in the 2-split state model, and constructs out of it another non-malleable code $(\mathsf{Enc}', \mathsf{Dec}')$ in the 2-split state model that additionally is secure against leaking a constant fraction of the bits from the codeword. The rate of $(\mathsf{Enc}', \mathsf{Dec}')$ is linear in the rate of $(\mathsf{Enc}, \mathsf{Dec})$. The only thing that we require is that $(\mathsf{Enc}, \mathsf{Dec})$ is *symmetric*, i.e., $\mathsf{Dec}(L, R) = \mathsf{Dec}(R, L)$. Since the code of [3] has this property, thus, combining this result with ours, we obtain a leakage-resilient 2-split state non-malleable code.

Let us also note that the code from the subsequent work of [2] (built by applying a reduction from 10-split to 2-split state to the code of [9]) is also symmetric, and therefore we can instantiate $(\mathsf{Enc}, \mathsf{Dec})$ with this construction, obtaining that $(\mathsf{Enc}', \mathsf{Dec}')$ is a constant rate and can tolerate leakage of a linear size.

Our key technical argument is contained in Theorem 2 that can be of independent interest. Informally, in this theorem we consider a "parallel composition of the inner product encodings", i.e., we consider encoding of a pair of messages $x_1, x_2 \in \mathbb{F}$ as random elements $L_1, R_1, L_2, R_2 \in \mathbb{F}^n$ such that $\langle L_1, R_1 \rangle = x_1$, and $\langle L_2, R_2 \rangle = x_2$. We show that it is partly resilient to tampering in the following sense. If (L_1, L_2) and (R_1, R_2) are independently tampered to obtain (L'_1, L'_2) and (R'_1, R'_2), and then we decode to get $x'_1 = \langle L'_1, R'_1 \rangle$, and $x'_2 = \langle L'_2, R'_2 \rangle$, then x'_1, x'_2 can only have a limited dependence on x_1, x_2. The proof of this result is done using a careful combinatorial argument that among other techniques, makes extensive use of the two-source extractor property of the inner-product, and Vazirani's XOR Lemma.

1.2   Comparison with Prior Work

Techniques. The proof technique of our main technical result (Theorem 2) has some similarity with the results of prior work [20,3,9]. We analyze the joint distribution

$$\phi_{f,g}(L,R) := \langle L_1, R_1 \rangle, \langle L_2, R_2 \rangle, \langle f_1(L_1, L_2), g_1(R_1, R_2) \rangle, \{f_2(L_1, L_2), g_2(R_1, R_2)\}$$

where (L_1, L_2, R_1, R_2) is uniform over a set $\mathcal{L} \times \mathcal{R} \subset \mathbb{F}^{4n}$, for some finite field $\mathbb{F}$, and $f_1, f_2 : \mathcal{L} \mapsto \mathbb{F}^n$, $g_1, g_2 : \mathcal{R} \mapsto \mathbb{F}^n$ are arbitrary functions.

The analysis of this result, and the main technical result of [20,3,9] proceeds by partitioning the ambient space (which is $\mathcal{L} \times \mathcal{R}$ in our result) into appropriate subsets, and analyzing the required distribution over these subsets and then combining using (an equivalent of) Lemma 7. However, the partitioning procedure for our work, in our opinion, is relatively more delicate which helps us obtain the result with good parameters without relying on advanced techniques from additive combinatorics as was the case in [3,9].

Leakage-resilience. The main reason we are able to make our codes resilient to leakage is because, in Theorem 2, we analyze the distribution of $\phi_{f,g}(L,R)$, conditioned on L, R being uniform in a *subset* of $\mathbb{F}^{2n}$, and not the entire $\mathbb{F}^{2n}$. This immediately leads to the question whether such an analysis over a subset of the ambient space would make the schemes of [3,9] also leakage-resilient. However, the proof of [3], in particular Lemma 8, which considers the case when f is far from linear and g is far from constant, will go through only if (in their notation) the subset has size bigger than p^{2n-t}, where $t = \Theta(n^{1/6}/\log p)$. This will lead to the code being resilient to only $\Theta(t/n) = \Theta(n^{-5/6}/\log p)$ fraction of leakage, as compared to a constant fraction in our result. On the contrary, the proof of [9] will work even if we restrict to a subset of the ambient space, but this will likely lead to a very small leakage rate. Also, their code is non-malleable in the 10-split-state model, and hence is less practical than the 2-split-state model that we consider, and for this reason we did not pursue this approach.

1.3   Other Related Work

The notion of non-malleability was introduced in cryptography by Dwork et al. [18]. Formal treatment of the tampering attacks was initiated in [32,30]. The non-malleable codes were also studied by Cheraghchi and Guruswami, who in [11] show improved constructions of the non-malleable codes secure against bit-wise tampering and show a connection between the non-malleable codes and the seed-less non-malleable extractors (which is a new notion that they introduce). In [10] the same authors study the problem of the capacity of non-malleable codes secure against different (non-split-state) families. Extensions of non-malleable codes to the case of *continuous* tampering were studied in [26]. Non malleable codes secure against tampering functions coming from restricted complexity classes were studied in [28], and secure against the linear tampering functions were considered in [8].

Simultaneous leakage and tampering attacks were also considered in [35] (who consider a more restricted type of leakage, called the "probing attacks"), in [34] who construct tamper- and leakage-resilient encryption and signature schemes, and in [15] who show a general way to transform any cryptographic functionality into one that is secure against tampering with individual bits, and leaking a logarithmic amount of information. Probabilistic tampering attacks on boolean circuits (where the adversary can tamper each wire with a certain probability) were also considered in [29].

2   Preliminaries

For a set T, let U_T denote a uniform distribution over T, and, for an integer ℓ, let U_ℓ denote uniform distribution over ℓ bit strings. The *statistical distance* between two random variables A, B is defined by $\Delta(A, B) = \frac{1}{2} \sum_v |\Pr[A = v] - \Pr[B = v]|$. We use $A \approx_\varepsilon B$ as shorthand for $\Delta(A, B) \le \varepsilon$.

Lemma 1. *For any (randomized) function* α, *if* $\Delta(A, B) \le \varepsilon$, *then*

$$\Delta(\alpha(A), \alpha(B)) \le \varepsilon .$$

The *min-entropy* of a random variable W is $\mathbf{H}_\infty(W) \overset{\text{def}}{=} -\log(\max_w \Pr[W = w])$, and the *conditional* min-entropy of W given Z is: $\mathbf{H}_\infty(W|Z) \overset{\text{def}}{=} -\log\left(\mathbb{E}_{z \leftarrow Z} \max_w \Pr[W = w | Z = z]\right).$[1]

Definition 1. *We say that an efficient function* $\mathsf{Ext} : \{0,1\}^n \times \{0,1\}^n \to \{0,1\}^m$ *is an* (n, k, m, ε)-*two-source extractor [12] if for all independent sources* $X, Y \in \{0,1\}^n$ *such that min-entropy* $\mathbf{H}_\infty(X) + \mathbf{H}_\infty(Y) \ge k$, *we have*

$$(Y, \mathsf{Ext}(X, Y)) \approx_\varepsilon (Y, U_m), \quad and \quad (X, \mathsf{Ext}(X, Y)) \approx_\varepsilon (X, U_m) .$$

For n being an integer multiple of m, and interpreting elements of $\{0,1\}^m$ as elements from $\mathbb{F}_{2^m}$ and those in $\{0,1\}^n$ to be from $(\mathbb{F}_{2^m})^{n/m}$, we have that the inner product function defined as $\langle (a_1, \ldots, a_{n/m}), (b_1, \ldots, b_{n/m}) \rangle := a_1 b_1 + \cdots + a_{n/m} b_{n/m}$ is a good 2-source extractor (cf. eg. [12,38]).

Lemma 2. *For all positive integers* m, n *such that* n *is a multiple of* m, *and for all* $\varepsilon > 0$, *there exists an efficient* $\left(n, n + m + 2\log\left(\frac{1}{\varepsilon}\right), m, \varepsilon\right)$ 2-*source extractor.*

We will need the following results. The proofs of these results can be found in Appendix A for completeness. The following is a simple result from [3].

Lemma 3. *Let* $X_1, Y_1 \in \mathcal{A}_1$, *and* $X_2, Y_2 \in \mathcal{A}_2$ *be random variables such that* $\Delta((X_1, X_2) ; (Y_1, Y_2)) \le \varepsilon$. *Then, for any non-empty set* $\mathcal{A}' \subseteq \mathcal{A}_1$, *we have*

$$\Delta(X_2 \mid X_1 \in \mathcal{A}' ; Y_2 \mid Y_1 \in \mathcal{A}') \le \frac{2\varepsilon}{\Pr(X_1 \in \mathcal{A}')} .$$

---

[1] Note that we use the variant of conditional entropy where the logarithm is taken *after* $\max_w$ is determined. This definition was introduced in [17] (it was called an *average min-entropy* there).

A *leakage oracle* is a machine Ω that takes as input $(L, R) \in \{0,1\}^n \times \{0,1\}^n$ and then answers the *leakage queries* of a type (L, f) and (R, g), where $f, g :$ $\{0,1\}^n \to \{0,1\}^*$. Each query (L, f_i) (resp.: (R, g_i)) is answered with $f_i(L)$ (resp.: $g_i(R)$) or $\perp$. An interactive machine $\mathcal{A}$ that issues the leakage queries is called a *leakage adversary*. Let $\mathsf{Leak}_\mathsf{L}^\mathcal{A}(L)$ (resp.: $\mathsf{Leak}_\mathsf{R}^\mathcal{A}(R)$) denote the concatenation of all the non-$\perp$ answers to the (L, f_i) (resp.: (R, g_i)) queries of $\mathcal{A}$. Moreover, let $\mathsf{Leak}^\mathcal{A}(L, R) := (\mathsf{Leak}_\mathsf{L}^\mathcal{A}(L), \mathsf{Leak}_\mathsf{R}^\mathcal{A}(R))$. The oracle Ω is m-bounded if it gives the non-$\perp$ answers to the (L, f_i) queries as long as $|\mathsf{Leak}_\mathsf{L}^\mathcal{A}(L)| \leq m$ and the non-$\perp$ answers to (R, f_i) queries as long as $|\mathsf{Leak}_\mathsf{R}^\mathcal{A}(R)| \leq m$. The following result follows easily Lemma 4 of [21] and Lemma 2.2 of [17].

Lemma 4. *Let $\tilde{L}$ and $\tilde{R}$ be two independent random variables uniformly distributed respectively on sets $\mathcal{L} \subseteq \{0,1\}^n$ and $\mathcal{R} \subseteq \{0,1\}^n$ and let $\mathcal{A}$ be an arbitrary leakage adversary interacting with an m-bounded oracle $\Omega(\tilde{L}, \tilde{R})$. Then $\tilde{L}$ and $\tilde{R}$ are independent given $\mathsf{Leak}^\mathcal{A}(\tilde{L}, \tilde{R})$. Moreover, for every $\delta > 0$ we have*

$$\Pr\left(\mathbf{H}_\infty(\tilde{L} \mid \mathsf{Leak}^\mathcal{A}(\tilde{L}, \tilde{R}) = a) \leq \log|\mathcal{L}| - m - \log(1/\delta)\right) \leq \delta$$

(where $a := \mathsf{Leak}^\mathcal{A}(\tilde{L}, \tilde{R})$), and

$$\Pr\left(\mathbf{H}_\infty(\tilde{R} \mid \mathsf{Leak}^\mathcal{A}(\tilde{L}, \tilde{R}) = b) \leq \log|\mathcal{L}| - m - \log(1/\delta)\right) \leq \delta,$$

(where $b := \mathsf{Leak}^\mathcal{A}(\tilde{L}, \tilde{R})$).

We say that $(\mathsf{Enc}_\mathsf{LR} : \mathcal{M} \to \mathcal{L} \times \mathcal{R}, \mathsf{Dec}_\mathsf{LR} : \mathcal{L} \times \mathcal{R} \to \mathcal{M})$ is an ε-*leakage-resilient encoding in the split-state model* [16] if for every $M_0, M_1 \in \mathcal{M}$ we have that

$$\Delta(\mathsf{Leak}(\mathsf{Enc}_\mathsf{LR}(M_0)) \, ; \, \mathsf{Leak}(\mathsf{Enc}_\mathsf{LR}(M_1)) \leq \varepsilon.$$

Such encodings can be easily constructed from the 2-source extractors [16]. The following is a generalization of the Vazirani's XOR Lemma (it is proven in Appendix A).

Lemma 5. *Let $X = (X_1, \ldots, X_t) \in \mathbb{F}^t$ be a random variable, where $\mathbb{F}$ is a finite field of order q. Assume that for all $a_1, \ldots, a_t \in \mathbb{F}^t$ not all zero, $\Delta(\sum_{i=1}^t a_i X_i \, ; \, U) \leq \varepsilon$, where U is uniform in $\mathbb{F}$. Then $\Delta(X_1, \ldots, X_t \, ; \, U_1, \ldots, U_t) \leq \varepsilon q^t$, where $U_1, \ldots, U_t$ are independent and uniform in $\mathbb{F}^t$.*

3   The Defintion of the Leakage Resilient Non-malleable Codes

In this section we present the defintion of the leakage resilient non-malleable codes in the split-state model. We first recall the definition of non-malleable codes in the split-state model from [23,3]. As discussed already informally the introduction in this model we assume that the codeword is split into two parts which are tampered independently.

Definition 2. *A* coding scheme in the split-state model *consists of two functions: a randomized encoding function* $\mathsf{Enc} : \mathcal{M} \mapsto \mathcal{L} \times \mathcal{R}$, *and a deterministic decoding function* $\mathsf{Dec} : \mathcal{L} \times \mathcal{R} \mapsto \mathcal{M} \cup \{\bot\}$ *such that, for each* $M \in \mathcal{M}$, $\Pr(\mathsf{Dec}(\mathsf{Enc}(M)) = M) = 1$ *(over the randomness of the encoding algorithm). Suppose* $\mathcal{L} = \mathcal{R}$ *and denote* $\mathcal{C} := \mathcal{L}(= \mathcal{R})$. *A coding scheme* $(\mathsf{Enc}, \mathsf{Dec})$ *is symmetric if for every* $(L, R) \in \mathcal{C}$ *we have that* $\mathsf{Dec}(L, R) = \mathsf{Dec}(R, L)$.

Definition 3. *Let* $(\mathsf{Enc} : \mathcal{M} \to \mathcal{C}^2, \mathsf{Dec} : \mathcal{C}^2 \to \mathcal{M})$ *be a coding scheme in a split state model. For tampering functions* $f, g : \mathcal{C} \to \mathcal{C}$, *and* $m \in \mathcal{M}$, *define the tampering-experiment*

$$\mathsf{Tamper}_m := \left\{ \begin{array}{c} (L, R) \leftarrow \mathsf{Enc}(m), \\ (\tilde{L}, \tilde{R}) := (f(L), g(R)) \\ \tilde{m} = \mathsf{Dec}(\tilde{L}, \tilde{R}) \\ \textit{Output: } \tilde{m}, \end{array} \right\}$$

which is a random variable over the randomness of the encoding function Enc. *We say that a coding scheme* $(\mathsf{Enc}, \mathsf{Dec})$ *is* ε-*non-malleable w.r.t.* $\mathcal{F}$ *if for each* $f \in \mathcal{F}$, *there exists a distribution (corresponding to the simulator)* D *over* $\mathcal{M} \cup \{\bot, \mathsf{same}\}$, *such that, for all* $m \in \mathcal{M}$, *we have that the statistical distance between* Tamper_m *and*

$$\mathsf{Sim}_m := \left\{ \begin{array}{c} \tilde{m} \leftarrow D \\ \textit{Output: } m \textit{ if } \tilde{m} = \mathsf{same}, \textit{ and } \tilde{m}, \textit{ otherwise} \end{array} \right\}$$

is at most ε.

We now define the notion of non-malleability against the leakage adversaries (which was first formulated by [36]). As explained in the introduction it is enough to consider the scenario where the adversary first learns some bounded information about the codeword (via the leakage oracle), and then chooses the tampering functions. Formally we have the following.

Definition 4. *Let* $\mathsf{Enc}, \mathsf{Dec}$ *be a coding scheme from* $\{0, 1\}^k$ *to* $\{0, 1\}^n \times \{0, 1\}^n$, *and let* $\gamma \in [0, 1)$ *be a parameter. Let* $\mathcal{A}$ *be any adversary that has oracle access to a* γn-*bounded leakage oracle* $\Omega(L, R)$ *(cf. 2), where* $L, R \in \{0, 1\}^n$, *and outputs functions* (f, g) *such that* $f, g : \{0, 1\}^n \to \{0, 1\}^n$. *Let* $m \in \{0, 1\}^k$ *be a message. Consider the following tampering experiment.*

$$\mathsf{Tamper}_m^\gamma := \left\{ \begin{array}{c} (L, R) \leftarrow \mathsf{Enc}(m), \\ (f, g) = \mathcal{A}(\Omega(L, R)), \\ (\tilde{L}, \tilde{R}) := (f(L), g(R)), \\ \tilde{m} = \mathsf{Dec}(\tilde{L}, \tilde{R}) \\ \textit{Output: } \tilde{m}, \end{array} \right\}$$

which is a random variable over the randomness of the encoding function Enc. *We say that a coding scheme* $(\mathsf{Enc}, \mathsf{Dec})$ *is* γ-*leakage resilient* ε-*non-malleable code if for each* $\mathcal{A}$, *there exists a distribution* D *over* $\{0, 1\}^k \cup \{\bot, \mathsf{same}\}$, *such that, for all* $m \in \{0,1\}^k$, *we have that the statistical distance between* Tamper_m^γ *and*

$$Sim_m := \left\{ \begin{array}{c} \tilde{m} \leftarrow D \\ \textit{Output: } m \textit{ if } \tilde{m} = \textit{same, and } \tilde{m}, \textit{ otherwise.} \end{array} \right\}$$

is at most ε.

Of course, every ε-non-malleable code is also a 0-leakage resilient ε-non-malleable code. On the other hand, it is easy to find codes that are ε-non-malleable but are not ξ-leakage resilient ε-non-malleable for an arbitrarily small ξ. For example, consider an ε-non-malleable code $(\mathsf{Enc}, \mathsf{Dec})$, and construct another code $(\mathsf{Enc}', \mathsf{Dec}')$ as follows. Let

$$\mathsf{Enc}'(M) = (\underbrace{(L, \cdots, L)}_{\lceil 1/\xi \rceil \text{ times}}, \underbrace{(R, \cdots, R)}_{\lceil 1/\xi \rceil \text{ times}}),$$

where $(L, R) \leftarrow \mathsf{Enc}(M)$. The decoding function for

$$(L', R') = ((L_1, \ldots, L_{\lceil 1/\xi \rceil}), ((R_1, \ldots, R_{\lceil 1/\xi \rceil}))$$

is defined as: $\mathsf{Dec}'(L', R') = \bot$ if for some i, j we have $L_i \neq L_j$ or $R_i \neq R_j$, and $\mathsf{Dec}'(L', R') = \mathsf{Dec}(L_1, R_1)$ otherwise. It is easy to show that $(\mathsf{Enc}', \mathsf{Dec}')$ is also ε-non-malleable. On the other hand, clearly, it is *not* ξ-leakage resilient ε-non-malleable, since the adversary can simply leak L_i and R_i (since $|L_i|/|L'| = |R_i|/|R'| = 1/(\lceil 1/\xi \rceil) \leq \xi$) and hence he can compute M before he chooses the tampering functions. Actually, it would even be enough for the adversary to leak L_i from one part (L, say), since in this case he could make the tampering function g fully dependent on M (since $M = \mathsf{Dec}(L_1, R_1)$), and, e.g., tamper with R only if $M = 0$ (which obviously means that the code is malleable).

4   Our Construction

This section contains the main construction of our paper. Let $(\mathsf{Enc} : \mathcal{M} \to \mathcal{X} \times \mathcal{X}, \mathsf{Dec} : \mathcal{X} \times \mathcal{X} \to \mathcal{M})$ be a (not leakage resilient) symmetric ε-non-malleable code in the split state model. Let $\gamma \in [0, 1/12)$ be a parameter. We are going to construct a γ-leakage resilient 3ε-non-malleable code $(\mathsf{Enc}', \mathsf{Dec}')$ in the split state mode (the code $(\mathsf{Enc}', \mathsf{Dec}')$ will also be symmetric).

The first obvious idea for constructing such a code could be to define $\mathsf{Enc}'(m)$ as follows: first compute $(x_1, x_2) = \mathsf{Enc}(m)$, and then "encode" both x_1 and x_2 using the leakage-resilient encoding (cf. Section 2). To be more concrete choose an encoding of [16] that is based on the inner product function[2]. Let (ℓ_1, r_1) and (ℓ_2, r_2) be such leakage-resilient encodings of x_1 and x_2, respectively. It is clear that given $(\ell_1, r_1, \ell_2, r_2)$ one can easily compute m as $\mathsf{Dec}(\langle \ell_1, r_1 \rangle, \langle \ell_2, r_2 \rangle)$.

---

[2] In this encoding in order to encode a message x one chooses random vectors ℓ and r (in some $\mathbb{F}^n$) such that $\langle \ell, r \rangle = x$, and to decode the message one simply computes $\langle \ell, r \rangle$.

Of course, what remains to be defined is how the variables ℓ_1, r_1, ℓ_2, and r_2 are represented in the final encoding, or, in other words: on which memory part one would store each of these variables. One option, of course, would be to move to the 4-split state model and say that the result of the encoding is $(\ell_1, r_1, \ell_2, r_2)$ (i.e. each of these variables can leak and be tampered independently). This approach can be proven secure, but it is clearly suboptimal since it increases to 4 the number of memory parts needed to implement the scheme.[3]

If we restrict ourselves to the 2-split state model then we could simply think of putting some of the ℓ_1, ℓ_2, r_1, r_2 variables on one part of the encoding and the remaining ones on the other part. It is easy to see that ℓ_1 and r_1 cannot be put together on one memory part (a symmetric argument works for ℓ_2 and r_2). This is because if the leakage function can be applied directly to ℓ_1 and r_1 then the adversary choose a function that it simply internally decodes x_1 from (ℓ_1, r_1) and leaks directly from x_1. Hence, the code would need to be non-malleable even if the adversary can choose the tampering function g (that is applied to x_2) after learning x_1, which is impossible (cf. the discussion at the end of Section 3).

Hence, the only option that has chances to work is to define the encoding of m to be equal to $((\ell_1, \ell_2), (r_1, r_2))$. Observe that the adversary can now obviously "swap" x_1 and x_2 by changing the encoding to $((\ell_2, \ell_1), (r_2, r_1))$. This is ok for us, since we assumed that our encoding is symmetric. He can also "copy" elements, and produce an encoding $((\ell_1, \ell_1), (r_1, r_1))$ (or $((\ell_2, \ell_2), (r_2, r_2))$). In this case the decoded value (x_1', x_2') will be equal to (x_1, x_1) (resp.: (x_2, x_2)). It turns out that this is also ok, since every non-malleable code in the 2-split state is essentially also a 2-out-of-2 secret sharing function (we prove this fact in Lemma 6), and thus x_1 and x_2 individually do not provide any significant information about the encoded message m (which, of course, implies that $\mathsf{Dec_{NM}}(x_1, x_1)$ is unrelated to m). Of course these attacks can be combined with tampering attacks applied to each of ℓ_1, ℓ_2, r_1 and r_2 individually. For example the adversary can transform the encoding to $((\ell_1, c \cdot \ell_1), (r_1, r_1))$, which (from the linearity of the inner product) would decode to $(x_1, c \cdot x_1)$. This, however, is not a problem, since such individual tampering is obviously tolerated by every non-malleable code.

Unfortunately, it turns out the the adversary can launch some more sophisticated attacks. Observe that in our encoding the values (ℓ_1, ℓ_2) and (r_1, r_2) can be treated as vectors of length $2n$. Suppose the adversary permutes both of them with the same random permutation σ, and let $(\ell_1', \ell_2') = \sigma(\ell_1, \ell_2)$ and $(r_1', r_2') = \sigma(r_1, r_2)$. Then the inner product of the $2n$-long vectors remains unchanged (i.e. $\langle (\ell_1', \ell_2'), (r_1', r_2') \rangle = \langle (\ell_1, \ell_2), (r_1, r_2) \rangle$), and therefore $\langle \ell_1, r_1 \rangle + \langle \ell_2, r_2 \rangle = \langle \ell_1', r_1' \rangle + \langle \ell_2', r_2' \rangle$, which means that $x_1' + x_2' = x_1 + x_2$ (where x_1' and x_2' are the results of decoding the manipulated encodings). Since $\langle \ell_1', r_1' \rangle$ is uniformly random, thus one can think of this attack as $(x_1', x_2') := (x_1 + Z, x_2 - Z)$ for some random Z. Fortunately, the non-malleable codes are obliviously secure against

---

[3] One can be tempted to say that in this case we can apply the transformation from the subsequent paper [2] to reduce the number of parts from 4 to 2, but for this approach to work one would need to show that the construction of [2] preserves the leakage-resilience which seems highly non-trivial.

attacks that add and subtract constants to the different parts of the encoding. Nevertheless, this example indicates that analyzing all possible strategies of the adversary may be non-trivial.

In Section 5 we characterize all such strategies by dividing them into classes. Very roughly speaking, it turns out that the attacks described above are examples of attacks from each of these classes. Namely: the adversary can either make x_1' depend only on x_1 and x_2' on x_2 (this case is denoted $\mathcal{D}_{id}$), or x_1' depend on x_2 and x_2' on x_1 (case $\mathcal{D}_{swap}$), or make both x_1' and x_2' depend *only* on x_1 (case $\mathcal{D}_{forget,2}$) or only on x_2 (case $\mathcal{D}_{forget,1}$). He can also make x_1' depend in an arbitrary way on x_1, x_2 and x_2' (case $\mathcal{D}_{unif,2}$). This comes at a cost of making x_1, x_2 and x_2' uniform and independent (note that the "$(x_1', x_2') := (x_1 + Z, x_2 - Z)$" attack falls into this category). Symmetrically, he can make x_2' depend in an arbitrary way on x_1, x_2 and x_1' (case $\mathcal{D}_{unif,1}$).

What remains is to prove security of the non-malleable codes when the adversary can perform the attacks from classes $\mathcal{D}_{id}, \mathcal{D}_{swap}, \mathcal{D}_{forget,1}, \mathcal{D}_{forget,2}, \mathcal{D}_{unif,1}$, and $\mathcal{D}_{unif,2}$. This is done in Section 6. In order to handle the cases $\mathcal{D}_{unif,1}$, and $\mathcal{D}_{unif,2}$ we need to modify our construction slightly. Namely, we make the leakage-resilient encoding "sparse" in the sense that a decoding of random codeword with overwhelming probability yields $\bot$ (this is slightly reminiscent of the construction of [20] where a similar technique was used to construct the non-malleable codes for 1-bit messages). This is achieved by requiring that the decoded value has to be in some sparse subset of $\mathbb{F}$ of size $q' \ll |\mathbb{F}|$. It will be convenient to define this set as the set of all x's such that $\psi(x) \le q'$, where $\psi : \mathbb{F} \to [q]$ is an arbitrary bijection. The technical details follow.

Our construction. Let n be an integer, and let $\mathbb{F} = \mathbb{F}_q$ be a finite field, and let $q' < q$ be an integer. Let $\prec$ be a total order on $\mathbb{F}$, and let $\psi : \mathbb{F} \to [q]$ be a bijection such that $a \prec b$ if and only $\psi(a) < \psi(b)$, for all $a, b \in \mathbb{F}$. Define the "leakage-resilient" decoding function $\mathsf{Dec_{LR}} : \mathbb{F}^{2n} \times \mathbb{F}^{2n} \to [q'] \times [q'] \cup \{\bot\}$ as follows:

$$\mathsf{Dec_{LR}}\left((\ell_1, \ell_2),\ (r_1, r_2)\right) = \begin{cases} \bot & \text{if } \psi(\langle \ell_1, r_1 \rangle) > q', \\ & \text{or } \psi(\langle \ell_2, r_2 \rangle) > q' \\ (\psi(\langle \ell_1, r_1 \rangle),\ \psi(\langle \ell_2, r_2 \rangle)) & \text{otherwise.} \end{cases}$$

We then define $\mathsf{Enc_{LR}} : [q'] \times [q'] \to \mathbb{F}^{2n} \times \mathbb{F}^{2n}$ as follows: for any $x \in [q'] \times [q']$, $\mathsf{Enc_{LR}}(x)$ is a random element y in $\mathbb{F}^{2n} \times \mathbb{F}^{2n}$, such that $\mathsf{Dec_{LR}}(y) = x$. The following is our main result. The proof is given in Section 6.

Theorem 1. *Let $(\mathsf{Enc_{NM}}, \mathsf{Dec_{NM}})$ be an ε-non-malleable code from $\{0,1\}^k$ to $[q'] \times [q']$ in the split-state model (for any $\varepsilon \in (0, 1/10)$) where $\mathsf{Dec_{NM}}$ is a symmetric function. Let $(\mathsf{Enc_{LR}}, \mathsf{Dec_{LR}})$ be as above, with $q \ge \frac{q'}{\varepsilon}$. Then for any $\gamma < \frac{1}{12}$ the encoding scheme $(\mathsf{Enc_{LR}} \circ \mathsf{Enc_{NM}}, \mathsf{Dec_{NM}} \circ \mathsf{Dec_{LR}})$ is an efficient 3ε-non-malleable γ-leakage resilient code in the split-state model from $\{0,1\}^k$ to $\mathbb{F}_q^n \times \mathbb{F}_q^n$, where $n = O\left(\frac{1}{1/12 - \gamma}\right)$.*

Thus, using the result of [3,1], we get the following result.

Corollary 1. *For any $\gamma < \frac{1}{12}$ and any $\varepsilon \in (0, 3/10)$, there exists an efficient γ-leakage-resilient ε-non-malleable code in the split-state model from k-bit messages to $\Theta\left(\frac{(k+\log(1/\varepsilon))^7}{1/12-\gamma}\right)$-bit codewords.*

Also, we get the following stronger result by combining Theorem 1 with an upcoming result [2], which gives a constant-rate non-malleable code in the split-state model.

Corollary 2. *For any $\gamma < \frac{1}{12}$ and any $\varepsilon \in (0, 3/10)$, there exists an efficient γ-leakage-resilient ε-non-malleable code in the split-state model from k-bit messages to $\Theta\left(\frac{k+\log(1/\varepsilon)}{1/12-\gamma}\right)$-bit codewords.*

5   The Joint Distribution of $\phi_{f,g}(L, R)$

Let $\mathbb{F} = \mathbb{F}_q$ be a finite field. Let $(L_1, L_2), (R_1, R_2)$ be independent and distributed uniformly over $\mathcal{L}, \mathcal{R} \subseteq \mathbb{F}^{2n}$, respectively. Let $f_1, g_1, f_2, g_2 : \mathbb{F}^n \times \mathbb{F}^n \to \mathbb{F}^n$ be a pair of functions. We consider the following family of distributions in $\mathbb{F}^4$,

$$\phi_{f,g}(L, R) := \langle L_1, R_1 \rangle, \langle L_2, R_2 \rangle, \langle f_1(L_1, L_2), g_1(R_1, R_2) \rangle, \langle f_2(L_1, L_2), g_2(R_1, R_2) \rangle ,$$

In this section, we analyze the possible joint distribution of $\phi_{f,g}(L, R)$ over $\mathbb{F}^4$ for arbitrary functions f_1, g_1, f_2, g_2. First, define the following set of distributions (which were already informally discussed in Section 4).

- $\mathcal{D}_{\mathsf{id}} := \{(U_1, U_2, h_1(U_1, Z), h_2(U_2, Z))\}$, where h_1, h_2 are functions from $\mathbb{F} \times \mathcal{Z}$ to $\mathbb{F}$, U_1, U_2 are independent and uniform in $\mathbb{F}$, and $Z \in \mathcal{Z}$ is some random variable independent of U_1, U_2.
- $\mathcal{D}_{\mathsf{swap}} := \{(U_1, U_2, h_1(U_2, Z), h_2(U_1, Z))\}$, where h_1, h_2 are functions from $\mathbb{F} \times \mathcal{Z}$ to $\mathbb{F}$, U_1, U_2 are independent and uniform in $\mathbb{F}$, and $Z \in \mathcal{Z}$ is some random variable independent of U_1, U_2.
- $\mathcal{D}_{\mathsf{unif},1} := \{(U_1, U_2, U_3, W)\}$, where U_1, U_2, U_3 are independent and uniform in $\mathbb{F}$, and W is a random variable over $\mathbb{F}$ arbitrarily correlated to U_1, U_2, U_3.
- $\mathcal{D}_{\mathsf{unif},2} := \{(U_1, U_2, W, U_3)\}$, where U_1, U_2, U_3 are independent and uniform in $\mathbb{F}$, and W is a random variable over $\mathbb{F}$ arbitrarily correlated to U_1, U_2, U_3.
- $\mathcal{D}_{\mathsf{forget},1} := \{(U_1, U_2, h_1(U_2, Z), h_2(U_2, Z))\}$, where h_1, h_2 are functions from $\mathbb{F} \times \mathcal{Z}$ to $\mathbb{F}$, U_1, U_2 are independent and uniform in $\mathbb{F}$, and $Z \in \mathcal{Z}$ is some random variable independent of U_1, U_2.
- $\mathcal{D}_{\mathsf{forget},2} := \{(U_1, U_2, h_1(U_1, Z), h_2(U_1, Z))\}$, where h_1, h_2 are functions from $\mathbb{F} \times \mathcal{Z}$ to $\mathbb{F}$, U_1, U_2 are independent and uniform in $\mathbb{F}$, and $Z \in \mathcal{Z}$ is some random variable independent of U_1, U_2.

Define $\mathcal{D}$ to be the family of convex combinations of

$$\mathcal{D}_{\mathsf{id}} \cup \mathcal{D}_{\mathsf{swap}} \cup \mathcal{D}_{\mathsf{unif},1} \cup \mathcal{D}_{\mathsf{unif},2} \cup \mathcal{D}_{\mathsf{forget},1} \cup \mathcal{D}_{\mathsf{forget},2} .$$

We show that for any f, g, the value of $\phi_{f,g}(L, R)$ is statistically close to some distribution in $\mathcal{D}$ if $\mathcal{L}, \mathcal{R}$ have size at least $q^{2n(1-\gamma)}$.

Theorem 2. *Let* $\mathbb{F} = \mathbb{F}_q$ *be a finite field with* $q \geq 4$, $n \geq 48$ *be an integer, and* $\gamma \in [0, 1/12)$. *Let* $L = (L_1, L_2)$, *and* $R = (R_1, R_2)$ *be distributed uniformly at random in sets* $\mathcal{L}, \mathcal{R}$ *of size at least* $q^{2n(1-\gamma)}$. *For any* $f_1, f_2, g_1, g_2 : \mathbb{F}^{2n} \to \mathbb{F}^n$, *there exists a distribution* $D \in \mathcal{D}$ *such that*

$$\Delta(\phi_{f,g}(L, R) \; ; \; D) \leq 7^2 \cdot 2^{-s}.$$

for any $s \leq ((\frac{1}{12} - \gamma)n - \frac{5}{4}) \log q - \frac{5}{6}$.

We give a proof of this theorem in Section 8.

6 Concluding Theorem 1 from Theorem 2

Before we present our proof, we state a result showing that non-malleable codes in 2−split state model also have the secret sharing property, i.e., that given any one part of the encoding of m, it is impossible to guess the message m. The proof of this can be found in Appendix B.

Lemma 6. *Let* Dec : $\mathcal{X} \times \mathcal{X} \to \mathcal{M}$, *and* Enc : $\mathcal{M} \to \mathcal{X} \times \mathcal{X}$ *be* ε−*non-malleable code in* 2−*split state model for some* $\varepsilon < \frac{1}{2}$. *For any pair of messages* $m_0, m_1 \in \mathcal{M}$, *let* $(X_1^0, X_2^0) \leftarrow$ Enc(m_0), *and let* $(X_1^1, X_2^1) \leftarrow$ Enc(m_1). *Then* $\Delta(X_0 \; ; \; X_1) \leq 2\varepsilon$.

Proof (Proof of Theorem 1). Let $n = \left\lceil \frac{6}{1/12-\gamma} \right\rceil$. Fix the message $m \in \mathcal{M}$, and let Enc$_{\mathsf{NM}}(m) = (X_1, X_2)$, and Enc$_{\mathsf{LR}}(X_1, X_2) = (L, R)$. Furthermore, fix manipulation functions $f, g : \mathbb{F}^n \times \mathbb{F}^n \to \mathbb{F}^n \times \mathbb{F}^n$ and the parameter $\gamma \in [0, \frac{1}{12})$. We need to analyze the distribution Tamper$_m^\gamma$ as in Definition 4. Before this, consider the following. Choose $\gamma' = \frac{1/12+\gamma}{2}$, so that $(1/12 - \gamma')n = (\gamma' - \gamma)n \geq 3$. Let $\tilde{L}, \tilde{R}$ be uniform in $\mathbb{F}^n$. From Lemma 4 we know that the min-entropies of $\tilde{L}$ and $\tilde{R}$ conditioned on the knowledge of Leak$^{\mathcal{A}}(\tilde{L}, \tilde{R})$ are at least $2n \log q(1 - \gamma')$ with probability at least $1 - 2 \cdot q^{2n(\gamma-\gamma')}$. So, at the cost of at most $2 \cdot q^{2n(\gamma-\gamma')}$ in the adversary's success probability, we can restrict ourselves to the case where $\tilde{L}$ and $\tilde{R}$ are distributed over uniformly over a set of size at least $q^{2n(1-\gamma')}$. Consider the joint distribution

$$\langle \tilde{L}_1, \tilde{R}_1 \rangle, \langle \tilde{L}_2, \tilde{R}_2 \rangle, \langle f_1(\tilde{L}_1, \tilde{L}_2), g_1(\tilde{R}_1, \tilde{R}_2) \rangle, \langle f_2(\tilde{L}_1, \tilde{L}_2), g_2(\tilde{R}_1, \tilde{R}_2) \rangle \;,$$

conditioned on the knowledge of Leak$^{\mathcal{A}}(\tilde{L}, \tilde{R})$. By Theorem 2, this has statistical distance at most

$$49 \cdot 2^{5/6} \cdot q^{5/4 - (1/12 - \gamma')n} + 2 \cdot q^{2n(\gamma-\gamma')} \leq 100 q^{-7/4} \leq 100 \left(\frac{\varepsilon^2}{2^k}\right)^{7/4} \leq \frac{\varepsilon}{2^{k+1}}$$

from some distribution D in $\mathcal{D}$. Here we used that $\varepsilon < 1/10$, and $q' \geq \frac{2^k}{\varepsilon}$. This holds for any non-malleable code in the 2-split-state model. Note that for any known non-malleable code in this model (in particular, those used in Corollary 1, and Corollary 2), we have that $q' \gg \frac{2^k}{\varepsilon}$. Since D is a convex combination (depending on f, g, and $\mathcal{A}$) of distributions in $\mathcal{D}_{\mathsf{forget},1}$, $\mathcal{D}_{\mathsf{forget},2}$, $\mathcal{D}_{\mathsf{unif},1}$, $\mathcal{D}_{\mathsf{unif},2}$, $\mathcal{D}_{\mathsf{id}}$, and $\mathcal{D}_{\mathsf{swap}}$, without loss of generality we will analyze the distribution Tamper_m^γ under the assumption that D belongs to one of these sets. Hence, we consider the following cases.

$D \in \mathcal{D}_{\mathsf{forget},1}$: In this case, D is of the form $U_1, U_2, h_1(U_2, Z), h_2(U_2, Z)$, where U_1, U_2 are independent and uniform in $\mathbb{F}$, and Z is independent of U_1, U_2. Of course, in our case L_1, R_1, L_2 and R_2 are not entirely uniform and independent, as they are a random encoding of a fixed message m. To take it into account we use Lemma 3, that states that in this case the statistical distance gets mutiplied by 2 divided by the probability that a random message M is equal to m (which is equal to 2^{-k}). Hence, we get that Tamper_m^γ has statistical distance at most

$$2^{k+1} \cdot \frac{\varepsilon}{2^{k+1}} = \varepsilon$$

from

$$V_1 = \mathsf{Dec}_{\mathsf{NM}}\left(\psi\left(h_1\left(\psi^{-1}(X_2), Z\right)\right), \psi\left(h_2\left(\psi^{-1}(X_2), Z\right)\right)\right),$$

where $\mathsf{Enc}_{\mathsf{NM}}(m) = (X_1, X_2)$, and Z is independent of X_1, X_2. Since V_1 is independent of X_1, using Lemma 6 we get the desired result (as X_2 cannot carry enough information to make V_1 dependent on m.

$D \in \mathcal{D}_{\mathsf{forget},2}$: This case is similar to the previous one.

$D \in \mathcal{D}_{\mathsf{unif},1}$: In this case, D is of the form U_1, U_2, U_3, W, where U_1, U_2, U_3 are independent and uniform in $\mathbb{F}$, and $W \in \mathbb{F}$ is arbitrarily correlated to U_1, U_2, U_3. Again, by Lemma 3, this implies that Tamper_m^γ has statistical distance at most

$$2^{k+1} \cdot \frac{\varepsilon}{2^{k+1}} = \varepsilon$$

from

$$V_2 = \mathsf{Dec}_{\mathsf{NM}}\left(\psi(U_3), W'\right),$$

where $\mathsf{Enc}_{\mathsf{NM}}(m) = (X_1, X_2)$, U_3 is uniform and independent in $\mathbb{F}$, W' is arbitrarily correlated to X_1, X_2, U_3. Note that $\psi(U_3) = \bot$ with probability $1 - q'/q$. Thus, Tamper_m^γ has statistical distance at most $\varepsilon + q'/q \le 2\varepsilon$ from $\bot$.

$D \in \mathcal{D}_{\mathsf{unif},2}$: This case is similar to the previous one.

$D \in \mathcal{D}_{\mathsf{id}}$: In this case, D is of the form $U_1, U_2, h_1(U_1, Z), h_2(U_2, Z)$, where U_1, U_2 are independent and uniform in $\mathbb{F}$, and Z is independent of U_1, U_2. By Lemma 3, this implies that Tamper_m^γ has statistical distance at most

$$2^{k+1} \cdot \frac{\varepsilon}{2^{k+1}} = \varepsilon$$

from

$$V_3 = \mathsf{Dec}_{\mathsf{NM}}\left(\psi\left(h_1\left(\psi^{-1}(X_1), Z\right)\right), \psi\left(h_2\left(\psi^{-1}(X_2), Z\right)\right)\right),$$

where $\mathsf{Enc}_{\mathsf{NM}}(m) = (X_1, X_2)$, and Z is independent of X_1, X_2. It is easy to see that V_3 is ε-close to Sim_m by the ε-non-malleability of $(\mathsf{Enc}_{\mathsf{NM}}, \mathsf{Dec}_{\mathsf{NM}})$.

$D \in \mathcal{D}_{\mathsf{swap}}$: Using the fact that the decoding function is symmetric, this case is similar to the previous one.

$\square$

7   Existential Result Using [11]

Cheraghchi and Guruswami [11] introduced the notion of seedless non-malleable extractors as a step towards constructing non-malleable codes defined as follows.

Definition 5. *A function* NMExt : $\{0,1\}^n \times \{0,1\}^n \to \{0,1\}^k$ *is a two-source non-malleable* (m, ε)*-extractor if, for every pair of independent random variables* X, Y *over* $\{0,1\}^n$ *such that* $\mathbf{H}_\infty(X) \geq m$, *and for any functions* $f, g : \{0,1\}^n \to \{0,1\}^n$, *there exists a distribution* D *over* $\{0,1\}^k \cup \{\mathsf{same}\}$, *such that*

$$\Delta(\mathsf{NMExt}(X, Y), \mathsf{NMExt}(f(X), g(Y)) ; U_k, \mathsf{copy}(D, U_k)) \leq \varepsilon ,$$

where U_k *is uniformly random in* $\{0,1\}^k$, *and* $\mathsf{copy}(D, U_k) = U_k$ *if* $D = \mathsf{same}$, *and* D, *otherwise.*

It was shown in [11] that assuming the existence of non-malleable extractors with $m = n$ immediately gives non-malleable codes with good rate. We observe that their proof easily extends to show that non-malleable extractors with small m implies non-malleable codes with good rate that also tolerate large amount of leakage.

Theorem 3. *Let* NMExt : $\{0,1\}^n \times \{0,1\}^n \to \{0,1\}^k$ *be a two-source non-malleable* (m, ε)*-extractor. Define a coding scheme* (Enc, Dec) *with message length* k *and block length* $2n$ *as follows. The decoder* Dec *is defined as* $\mathsf{Dec}(x, y) := \mathsf{NMExt}(x, y)$.

The encoder, given a message s, *outputs a uniformly random element* (X, Y) *in* $\{0,1\}^n \times \{0,1\}^n$ *such that* $\mathsf{Dec}(X, Y) = s$. *Then the pair* (Enc, Dec) *is* $(\varepsilon \cdot (2^k + 1)$*-non-malleable,* $1 - \frac{m}{n})$*-leakage-resilient code against split-state tampering.*

We now mention the result from [11] showing the existence of non-malleable codes.

Theorem 4. *Let* NMExt : $\{0,1\}^n \times \{0,1\}^n \to \{0,1\}^k$ *be a random function. For any* $\varepsilon, \delta > 0$, *and* $m \leq n$, *with probability at least* $1 - \delta$, *the function* NMExt *is a two-source non-malleable extractor provided that*

$$m \geq \max(k + \frac{3}{2} \cdot \log 1/\varepsilon + \frac{1}{2} \log \log(1/\delta) , \ \log n + \log \log(1/\delta) + O(1)) .$$

Combining Theorem 3 and Theorem 4 gives us the following corollary.

Corollary 3. *Let* $n = \mathsf{poly}(k)$, *and let* NMExt : $\{0,1\}^n \times \{0,1\}^n \to \{0,1\}^k$ *be a random function. With probability at least* $1 - \frac{1}{2^{2^k}}$, *the scheme* (NMExt^{-1}, NMExt) *is* 2^{-k}*-non-malleable,* $1 - \frac{5k}{n}$*-leakage-resilient.*

This implies that with probability very close to 1, a random function is an excellent leakage-resilient non-malleable code, and we can arbitrarily increase the amount of leakage (upto the total length of each part) at the cost of increasing the length of the codeword.

8   Proof of Theorem 2

8.1   The General Strategy

Our strategy is to divide $\mathcal{L} \times \mathcal{R}$ into several disjoint parts, and prove that $\phi_{f,g}(L, R)$ is close to some distribution in $\mathcal{D}$ for each of these parts separately. Let $L = (L_1, L_2)$, and $R = (R_1, R_2)$. Also, let $|\mathbb{F}| = q$, and $\tau = 1 - \gamma$. The following simple lemma shows that it suffices to bound the statistical distance between $\phi_{f,g}(L, R)$ and some distribution in $\mathcal{D}$ for (L, R) restricted to partitions of $\mathcal{L} \times \mathcal{R}$. This was shown in [3].

Lemma 7. *Let $\mathcal{S} \subseteq \mathcal{L} \times \mathcal{R}$. Let $\mathcal{S}_1, \dots, \mathcal{S}_k$ be a partition of $\mathcal{S}$. Also, let $D_1, \dots, D_k$ be some distribution in $\mathcal{D}$. Assume that for all $1 \leq i \leq k$,*

$$\Delta\left(\phi_{f,g}(L, R)|_{(L,R)\in\mathcal{S}_i} \ ; \ D_i\right) \leq \varepsilon_i \ .$$

Then there exists a distribution $D \in \mathcal{D}$ such that

$$\Delta\left(\phi_{f,g}(L, R)|_{(L,R)\in\mathcal{S}} \ ; \ D\right) \leq \sum \varepsilon_i \frac{|\mathcal{S}_i|}{|\mathcal{S}|} \ .$$

We construct a partition of $\mathcal{L} \times \mathcal{R}$ in Section 8.2. Then in Sections 8.3—8.3 we analyze the behavior of $\phi_{f,g}(L, R)$ on the constructed parts. Finally, in Section 8.4 we show how to combine the facts proven in previous sections in order to obtain the statement of the theorem.

8.2   Partitioning the Set $\mathcal{L} \times \mathcal{R}$.

We next define a partitioning of $\mathcal{L} \times \mathcal{R}$ based on f and g to which we will apply Lemma 7. This is done independently for $\mathcal{L}$ and $\mathcal{R}$ and hence we will focus only on $\mathcal{L}$ (the partitioning of $\mathcal{R}$ is done analogously). On a high level, our partitioning is constructed as follows: first we partition $\mathcal{L}$ into sets $\mathcal{L}_{\mathsf{ffb},1}, \mathcal{L}_{\mathsf{ffb},2}$, and $\mathcal{L}_1$. Then we partition $\mathcal{L}_1$ into $\mathcal{L}_{\mathsf{mix},1}, \mathcal{L}_{\mathsf{mix},2}$, and $\mathcal{L}_2$. Finally, we partition $\mathcal{L}_2$ into $\mathcal{L}_{\mathsf{id}}, \mathcal{L}_{\mathsf{swap}}$, and $\mathcal{L}_{\mathsf{rem}}$ (the meaning of the acronyms in the subscripts should become clear when the sets are defined). Altogether, we partition $\mathcal{L}$ into 7 sets $\mathcal{L}_{\mathsf{ffb},1}, \mathcal{L}_{\mathsf{ffb},2}, \mathcal{L}_{\mathsf{mix},1}, \mathcal{L}_{\mathsf{mix},2}, \mathcal{L}_{\mathsf{id}}, \mathcal{L}_{\mathsf{swap}}$, and $\mathcal{L}_{\mathsf{rem}}$.

Let $\beta_1 = \frac{1}{3}n \log q - 4(\tau - \frac{11}{12})n \log q + 4 \log q + 4s + 4$, and let $\beta_2 = \frac{1}{3}n \log q - 4(\tau - \frac{11}{12})n \log q + 6 \log q + 3s + 2$. We first partition $\mathcal{L}$ into $\mathcal{L}_{\mathsf{ffb},1}, \mathcal{L}_{\mathsf{ffb},2}$, and $\mathcal{L}_1$. Recall that the elements of $\mathcal{L}$ are pairs $(\ell_1, \ell_2) \in \mathbb{F}^n \times \mathbb{F}^n$. Intutively $\mathcal{L}_{\mathsf{ffb},i}$ (for $i \in \{1, 2\}$) will consist of the elements of $\mathcal{L}$ on which the function f is "far from a bijection", by which we mean that it "glues" at least $2^{\beta_1/2}$ elements on the ith component (cf. Steps 2 and 3 below). The set $\mathcal{L}_1$ will consists of the remaining elements (i.e. those that are "close to the bijection"). Since we want $\mathcal{L}_{\mathsf{ffb},1}, \mathcal{L}_{\mathsf{ffb},2}$, and $\mathcal{L}_1$ to be a partition, thus $\mathcal{L}_{\mathsf{ffb},1}$ and $\mathcal{L}_{\mathsf{ffb},2}$ have to be disjoint. Hence we first construct $\mathcal{L}_{\mathsf{ffb},1}$, and then, using the same method we construct $\mathcal{L}_{\mathsf{ffb},2}$, but in this construction we consider only ℓ's that belong to $\mathcal{L}^* := \mathcal{L} \setminus \mathcal{L}_{\mathsf{ffb},1}$. The set $\mathcal{L}_1$ consits of the elements of $\mathcal{L}$ that were not included in $\mathcal{L}_{\mathsf{ffb},1}$ or $\mathcal{L}_{\mathsf{ffb},2}$. To avoid repetition, we present the procedures for constructing $\mathcal{L}_{\mathsf{ffb},1}$ and $\mathcal{L}_{\mathsf{ffb},2}$ as one algorithm, whose behaviour depends on i. This algorithm, presented below, is executed first for $i = 1$ and then for $i = 2$.

1. Initialize $\mathcal{L}_{\text{ffb},i}$ to be empty, and let $\mathcal{L}^* := \mathcal{L}$ if $i = 1$, and $\mathcal{L}^* := \mathcal{L} \setminus \mathcal{L}_{\text{ffb},1}$, otherwise.
2. Let $\mathcal{W}$ be a largest subset of $\mathcal{L}^*$ such that for any two $\ell, \ell' \in \mathcal{W}$ it holds that $\ell_i \neq \ell'_i$, and $f(\ell) = f(\ell')$.
3. If $|\mathcal{W}| \geq 2^{\beta_1/2}$, then set $\mathcal{L}^* = \mathcal{L}^* \setminus \mathcal{W}$, set $\mathcal{L}_{\text{ffb},i} = \mathcal{L}_{\text{ffb},i} \cup \mathcal{W}$, and go to Step 2.
4. Return $\mathcal{L}_{\text{ffb},i}$.

The set $\mathcal{L}_1$ is defined to be

$$\mathcal{L}_1 = \mathcal{L} \setminus (\mathcal{L}_{\text{ffb},1} \cup \mathcal{L}_{\text{ffb},2}) .$$

The justification for this choice is that for $\tilde{L}$ chosen uniformly at random from $\mathcal{L}_{\text{ffb},i}$, we have

$$\mathbf{H}_\infty(\tilde{L}_i | f(\tilde{L}) = y) \geq \frac{\beta_1}{2} ,$$

where $y := f(\tilde{L})$ (this is because the number of pre-images of y under f, projected on the ith component, is at least $2^{\beta_1/2}$). Also, we have that for any $y \in \mathbb{F}^n$, the total number of elements $\ell \in \mathcal{L}_1$ such that $f(\ell) = y$ is at most $\left(2^{\frac{\beta_1}{2}}\right)^2 = 2^{\beta_1}$.

Recall that every $f : \mathbb{F}^n \times \mathbb{F}^n \to \mathbb{F}^n \times \mathbb{F}^n$ can be represened as a pair of functions $f_1, f_2 : \mathbb{F}^n \times \mathbb{F}^n \to \mathbb{F}^n$ defined as $(f_1(\ell), f_2(\ell)) := f(\ell)$. We further partition the set $\mathcal{L}_1$ depending on how $f_1(\ell), f_2(\ell)$ depend on ℓ_1, ℓ_2 for $\ell \in \mathcal{L}_1$. We will now define partitioning of $\mathcal{L}_1$. Before we do it, let us state the following auxiliary definition (note that it is defined for $\mathcal{L}$, not for $\mathcal{L}_1$).

Definition 6. *Define* $T^{i \to j} \subset \mathcal{L}$ *for* $i, j \in \{1, 2\}$, *as the set of all elements* $\ell \in \mathcal{L}$ *such that*

$$\left|\left\{\ell^* \in \mathcal{L} \;\middle|\; \ell_i = \ell_i^* \text{ and } f_j(\ell) = f_j(\ell^*)\right\}\right| \geq \frac{q^n}{2^{\beta_2}} .$$

Let us now prove the following simple result justifying the definition of $T^{i \to j}$. Intuitively, this result shows that for every $\ell \in T^{i \to j}$, the value of $f_j(\ell)$ can be computed given ℓ_i and a little more information.

Lemma 8. *Let* $\ell \in T^{i \to j}$ *for some* $i, j \in [t]$. *Then there exists some functions* $a_{i,j} : T^{i \to j} \mapsto \{0, 1\}^{\beta_2}$ *and* $\psi_{i,j} : \mathbb{F}^n \times \{0, 1\}^{\beta_2} \mapsto \mathbb{F}^n$ *such that for all* $\ell \in T^{i \to j}$,

$$f_j(\ell) = \psi_{i,j}(\ell_i, a_{i,j}(\ell)) .$$

Proof. Given $\ell \in T^{i \to j}$, let $T' = \{\ell^* \in T^{i \to j} \mid \ell_i^* = \ell_i\}$. Then, clearly $|T'| \leq |\mathbb{F}^n| = q^n$. Consider a partition of T' into sets $T'_1, \ldots, T'_m$ according to the value of function f_j. More formally, for any $u, v \in [m]$, and any $\ell' \in T'_u$, $\ell'' \in T'_v$, we have that $f_j(\ell') = f_j(\ell'')$ if and only if $u = v$. By definition of $T^{i \to j}$, we have that $|T'_u| \geq \frac{q^n}{2^{\beta_2}}$ for all $u \in [m]$. Thus

$$m \leq \frac{|T'| \cdot 2^{\beta_2}}{q^n} \leq 2^{\beta_2} .$$

We define $a_{i,j}(\ell)$ as the binary representation of k such that $\ell \in T'_k$. Now, it is easy to see that we can determine $f_j(\ell)$ given ℓ_i and $a_{i,j}(\ell)$. $\square$

We now define disjoint subsets $\mathcal{L}_{\text{mix},1}, \mathcal{L}_{\text{mix},2} \subseteq \mathcal{L}_1$ as follows.

$$\mathcal{L}_{\text{mix},1} := \left\{ \ell \in \mathcal{L}_1 \mid \ell \notin T^{1 \to 1} \cup T^{2 \to 1} \right\} ,$$

$$\mathcal{L}_{\text{mix},2} := \left\{ \ell \in \mathcal{L}_1 \setminus \mathcal{L}_{\text{mix},1} \mid \ell \notin T^{1 \to 2} \cup T^{2 \to 2} \right\} .$$

Informally speaking, $\ell \in \mathcal{L}_{\text{mix},j}$ implies that $f_j(\ell)$ depends on both ℓ_1 and ℓ_2. Now, let

$$\mathcal{L}_2 := \mathcal{L}_1 \setminus (\mathcal{L}_{\text{mix},1} \cup \mathcal{L}_{\text{mix},2}) .$$

We denote $\mathcal{T}(\ell, i)$ to be the set of $j \in \{1, 2\}$ such that $\ell \in T^{i \to j}$. Note that by the definition of $\mathcal{L}_{\text{mix},j}$, for any $\ell \in \mathcal{L}_2$ we have that $\mathcal{T}(\ell, 1) \cup \mathcal{T}(\ell, 2) = \{1, 2\}$. We further partition $\mathcal{L}_2$ into $\mathcal{L}_{\text{id}}$, $\mathcal{L}_{\text{swap}}$, and $\mathcal{L}_{\text{rem}}$ as follows.

$$\mathcal{L}_{\text{id}} := \left\{ \ell \in \mathcal{L}_2 \mid \mathcal{T}(\ell, 1) = \{1\}, \ \mathcal{T}(\ell, 2) = \{2\} \right\} ,$$

$$\mathcal{L}_{\text{swap}} := \left\{ \ell \in \mathcal{L}_2 \setminus \mathcal{L}_{\text{id}} \mid \mathcal{T}(\ell, 1) = \{2\}, \ \mathcal{T}(\ell, 2) = \{1\} \right\} ,$$

and

$$\mathcal{L}_{\text{rem}} := \mathcal{L}_2 \setminus (\mathcal{L}_{\text{id}} \cup \mathcal{L}_{\text{swap}}) .$$

The partitioning of $\mathcal{R}$ is defined similarly. We will later consider the partitions of $\mathcal{L} \times \mathcal{R}$ to be the product of individual partitions of $\mathcal{L}$ and $\mathcal{R}$ (hence, at the end $\mathcal{L} \times \mathcal{R}$ is partitioned into 7^2 parts).

8.3   Analyzing the Parts

We will argue that for any part, either its probability is small, or $\phi_{f,g}(L, R)$ conditioned on (L, R) belonging to it, is close to some distribution in $\mathcal{D}$. We then apply Lemma 7 to obtain a proof of Theorem 2.

Case: "f or g is far from bijection"

Lemma 9. *For $i = 1, 2$, and $\mathcal{R}^* \subset \mathcal{R}$, if $|\mathcal{L}_{\text{ffb},i} \times \mathcal{R}^*| \geq q^{4\tau n} \cdot 2^{-s}$ then there exists a distribution D that is a convex combination of distributions in $\mathcal{D}_{\text{forget},i}$ such that*

$$\Delta(\phi_{f,g}(L, R)|_{(L,R) \in \mathcal{L}_{\text{ffb},i} \times \mathcal{R}^*} \ ; D) \leq 2^{-s} .$$

Proof. Without loss of generality, let $i = 1$, and let $|\mathcal{L}_{\text{ffb},1} \times \mathcal{R}| \geq q^{4\tau n} \cdot 2^{-s}$. Let $\tilde{L}, \tilde{R}$ be distributed uniformly over $\mathcal{L}_{\text{ffb},1}$ and $\mathcal{R}^*$ respectively. Note that by the assumption we have that $|\mathcal{L}_{\text{ffb},1}| \geq q^{2\tau n} \cdot 2^{-s}$ and $|\mathcal{R}^*| \geq q^{2\tau n} \cdot 2^{-s}$. Thus,

$$\mathbf{H}_\infty(\tilde{L}_1) \geq (2\tau - 1)n \log q - s \quad \text{and} \quad \mathbf{H}_\infty(\tilde{L}_2|\tilde{L}_1) \geq (2\tau - 1)n \log q - s , \quad (8.1)$$

and

$$\mathbf{H}_\infty(\tilde{R}_1) \geq (2\tau - 1)n \log q - s \quad \text{and} \quad \mathbf{H}_\infty(\tilde{R}_2|\tilde{R}_1) \geq (2\tau - 1)n \log q - s . \quad (8.2)$$

Denote $\langle \tilde{L}_k, \tilde{R}_k \rangle$ by X_k, and $\langle f_k(\tilde{L}), g_k(\tilde{R}) \rangle$ by X'_k for $k = 1, 2$. We have that

$$\mathbf{H}_\infty(\tilde{L}_1 | X_2, f(\tilde{L})) \geq \frac{\beta_1}{2} - \log q \ .$$

Also, using Lemma 4, we get that $\tilde{L}$ and $\tilde{R}$ (and hence $\tilde{L}_1$ and $\tilde{R}_1$) are independent given $f(\tilde{L})$, $\tilde{R}_2$, and X_2. Thus, using the fact that Ext is a strong two-source extractor, we have that

$$X_1, \tilde{R}_1, X_2, f(\tilde{L}), \tilde{R}_2 \approx_{2^{-(s+1)}} U_1, \tilde{R}_1, X_2, f(\tilde{L}), \tilde{R}_2 \ ,$$

where U_1 is uniformly random in $\mathbb{F}$. This implies that

$$X_1, X_2, X'_1, X'_2 \approx_{2^{-(s+1)}} U_1, X_2, X'_1, X'_2 \ . \tag{8.3}$$

Now, X'_1, X'_2 are independent of U_1, and can be seen as a randomized function of X_2. Let

$$U_1, X_2, X'_1, X'_2 \equiv U_1, X_2, h_1(X_2, Z), h_2(X_2, Z) \ ,$$

for Z independent of U_1, X_2. Using Equation 8.1 and 8.2, and that Ext is a two-source randomness extractor, we have that X_2 is $2^{-(s+1)}$-close to uniform. Thus, using triangle inequality, we get that

$$X_1, X_2, X'_1, X'_2 \approx_{2^{-s}} U_1, U_2, h_1(U_2, Z), h_2(U_2, Z) \ ,$$

which implies the result. □

In a symetric way we can prove the following.

Lemma 10. *For $i = 1, 2$, and $\mathcal{L}^* \subset \mathcal{L}$, if $|\mathcal{L}^* \times \mathcal{R}_{\mathsf{ffb},i}| \geq q^{4\tau n} \cdot 2^{-s}$ then there exists a distribution D that is a convex combination of distributions in $\mathcal{D}_{\mathsf{forget},i}$ such that*

$$\Delta(\phi_{f,g}(L, R)|_{(L,R) \in \mathcal{L}^* \times \mathcal{R}_{\mathsf{ffb},i}} \ ; D) \leq 2^{-s} \ .$$

Case: "Output of f or g is Mixed"

Lemma 11. *For $j = 1, 2$, and $\mathcal{R}^* \subset \mathcal{R}$, if $|\mathcal{L}_{\mathsf{mix},j} \times \mathcal{R}^*| \geq q^{4\tau n} \cdot 2^{-s}$ then there exists a distribution D that is a convex combination of distributions in $\mathcal{D}_{\mathsf{unif},j}$, such that*

$$\Delta(\phi_{f,g}(L, R)|_{(L,R) \in \mathcal{L}_{\mathsf{mix},j} \times \mathcal{R}^*} \ ; D) \leq 2^{-s} \ .$$

Proof. Let us assume that $|\mathcal{L}_{\mathsf{mix},1} \times \mathcal{R}^*| \geq q^{4\tau n} \cdot 2^{-s}$. (Case $j = 2$ is analogous.) From the assumptions we have that $|\mathcal{L}_{\mathsf{mix},1}| \geq q^{2\tau n} \cdot 2^{-s}$ and $|\mathcal{R}^*| \geq q^{2\tau n} \cdot 2^{-s}$. Let $\tilde{L}, \tilde{R}$ be distributed uniformly over $\mathcal{L}_{\mathsf{mix},1}$ and $\mathcal{R}^*$ respectively. Denote $\mathsf{Ext}(\tilde{L}_k, \tilde{R}_k)$ by X_k, and $\langle f_k(\tilde{L}), g_k(\tilde{R}) \rangle$ by X'_k for $k = 1, 2$. Reasoning similarly as in Lemma 9, we have that X_1 is $q^{-3} \cdot 2^{-(s+1)}$-close to uniform, and also X_2 is $q^{-3} \cdot 2^{-(s+1)}$-close to uniform given $\tilde{L}_1, \tilde{R}_1$, and hence using the hybrid argument, we have that

$$X_1, X_2 \approx_{2^{-s} q^{-3}} U_1, U_2 \ , \tag{8.4}$$

where U_1, U_2, are independent and uniformly distributed in $\mathbb{F}$. We now give a lower bound for $\mathbf{H}_\infty(\tilde{L}_i | f_1(\tilde{L}))$ for $i = 1, 2$ using the definition of $\mathcal{L}_{\mathsf{mix},1}$.

$$
\begin{aligned}
\mathbf{H}_\infty(\tilde{L}_i | f_j(\tilde{L})) &= -\log\left(\sum_{y \in \mathbb{F}^n} \max_{\ell_i \in \mathbb{F}^n} \Pr(\tilde{L}_i = \ell_i \wedge f_j(\tilde{L}) = y) \right) \\
&\geq -\log\left(\sum_{y \in \mathbb{F}^n} \frac{q^n 2^{-\beta_2}}{|\mathcal{L}_{\mathsf{mix},1}|} \right) \\
&\geq -\log\left(\frac{q^{2n} 2^{-\beta_2}}{q^{2\tau n} 2^{-s}} \right) \geq \beta_2 - 2(1 - \tau) n \log q - s .
\end{aligned}
$$

Thus, we have that for $i = 1, 2$, X_i is $2^{-s-1} q^{-3}$-close to uniform given $f_1(\tilde{L}), \tilde{R}$, and hence,

$$
\Delta\left(X_i, \mathsf{Ext}(f_1(\tilde{L}), g_1(\tilde{R})) \; ; \; U_i, \mathsf{Ext}(f_1(\tilde{L}), g_1(\tilde{R})) \right) \leq q^{-3} 2^{-s-1} .
$$

Also, since $\mathcal{L}_{\mathsf{mix},1}$ and $\mathcal{R}^*$ are in the complement of $\mathcal{L}_{\mathsf{ffb}}$, and $\mathcal{R}_{\mathsf{ffb}}$, respectively, we have that

$$
\mathbf{H}_\infty(f_j(\tilde{L})) \geq \mathbf{H}_\infty(f(\tilde{L})) - n \log q \geq (2\tau - 1) n \log q - \beta_1 - s ,
$$

and

$$
\mathbf{H}_\infty(g_j(\tilde{R})) \geq (2\tau - 1) n \log q - \beta_1 - s .
$$

This implies that

$$
\Delta(X_i, X_1' \; ; \; U_i, U_1') \leq q^{-3} \cdot 2^{-s} , \tag{8.5}
$$

where U_1' is uniform in $\mathbb{F}$.

Now, we claim that

$$
\Delta(X_1, X_2, X_1' \; ; \; U_1, U_2, U_1') \leq 2^{-s} . \tag{8.6}
$$

If not, then by the XOR Lemma, there exist a_1, a_2, a_3, not all zero such that $a_1 X_1 + a_2 X_2 + a_3 X_1'$ is *not* $2^{-s} \cdot q^{-3}$ close to uniform. By Equation 8.4, we have that $a_3 \neq 0$, and by equation 8.5, we have that $a_1, a_2 \neq 0$. Consider two sources in $\mathbb{F}^3$ as $(a_1 \tilde{L}_1, a_2 \tilde{L}_2, a_3 f_1(\tilde{L}))$ and $(\tilde{R}_1, \tilde{R}_2, g_1(\tilde{R}))$. Applying Ext to these two sources gives $a_1 X_1 + a_2 X_2 + a_3 X_1'$. The two sources have min-entropy at least $2\tau n \log q - s$, and hence $\sum_{i=1}^t a_i X_i + a_{t+1} \mathsf{Ext}(f_j(\tilde{L}), g_j(\tilde{R}))$ is $2^{-s} q^{-3}$-close to uniform, which is a contradiction. $\qquad\square$

Symmetrically, we get that

Lemma 12. *For $j = 1, 2$, and $\mathcal{L}^* \subset \mathcal{L}$, if $|\mathcal{L}^* \times \mathcal{R}_{\mathsf{mix},j}| \geq q^{4\tau n} \cdot 2^{-s}$ then*

$$
\Delta(\phi_{f,g}(L, R)|_{(L,R) \in \mathcal{L}^* \times \mathcal{R}_{\mathsf{mix},j}} \; ; D) \leq 2^{-s} ,
$$

for some D that is a convex combination of distributions in $\mathcal{D}_{\mathsf{unif},j}$.

Case "$\tilde{L} \in \mathcal{L}_{\mathsf{id}} \cup \mathcal{L}_{\mathsf{swap}}$ and $\tilde{R} \in \mathcal{R}_{\mathsf{id}} \cup \mathcal{R}_{\mathsf{swap}}$"

Lemma 13. *If $|\mathcal{L}_{\mathsf{id}} \times \mathcal{R}_{\mathsf{swap}}| \geq q^{4\tau n} \cdot 2^{-s}$ then there exists a distribution D that is a convex combination of distributions in $\mathcal{D}_{\mathsf{forget},1}$ such that*

$$\Delta(\phi_{f,g}(L,R)|_{(L,R)\in\mathcal{L}_{\mathsf{id}}\times\mathcal{R}_{\mathsf{swap}}} ; D) \leq 2^{-s} .$$

Proof. This proof is almost identical to that of Lemma 9, except that it makes crucial use of Lemma 8. Let $|\mathcal{L}_{\mathsf{id}} \times \mathcal{R}_{\mathsf{swap}}| \geq q^{4\tau n} \cdot 2^{-s}$. Let $\tilde{L}, \tilde{R}$ be distributed uniformly over $\mathcal{L}_{\mathsf{id}}$ and $\mathcal{R}_{\mathsf{swap}}$ respectively. Note that by the assumption we have that $|\mathcal{L}_{\mathsf{id}}| \geq q^{2\tau n} \cdot 2^{-s}$ and $|\mathcal{R}_{\mathsf{swap}}| \geq q^{2\tau n} \cdot 2^{-s}$. Thus, for $k = 1, 2$,

$$\mathbf{H}_{\infty}(\tilde{L}_k|\tilde{L}_{k-1}) \geq (2\tau - 1)n \log q - s , \qquad (8.7)$$

and

$$\mathbf{H}_{\infty}(\tilde{R}_k|\tilde{R}_{k-1}) \geq (2\tau - 1)n \log q - s , \qquad (8.8)$$

where $\tilde{L}_0 = \tilde{R}_0 \equiv 0$. Denote $\langle \tilde{L}_k, \tilde{R}_k \rangle$ by X_k, and $\langle f_k(\tilde{L}), g_k(\tilde{R}) \rangle$ by X_k' for $k = 1, 2$. By Lemma 8, there exists maps $a_{1,1}, a_{2,2}$ from $\mathcal{L}_{\mathsf{id}}$ to $\{0,1\}^{\beta_2}$, and $b_{1,2}, b_{2,1}$ from $\mathcal{R}_{\mathsf{swap}}$ to $\{0,1\}^{\beta_2}$, such that $f_1(\tilde{L}), f_2(\tilde{L}), g_1(\tilde{R}), g_2(\tilde{R})$ are determined uniquely given $(\tilde{L}_1, a_{1,1}(\tilde{L})), (\tilde{L}_2, a_{2,2}(\tilde{L})), (\tilde{R}_2, b_{2,1}(\tilde{R})), (\tilde{R}_1, b_{1,2}(\tilde{R}))$, respectively.

Also, using Lemma 4, we get that $\tilde{L}$ and $\tilde{R}$ (and hence $\tilde{L}_1$ and $\tilde{R}_1$) are independent given $X_2 = \langle \tilde{L}_2, \tilde{R}_2 \rangle$, X_1', and X_2'. Thus, using the fact that Ext is a strong two-source extractor, we have that

$$X_1, \tilde{R}_1, X_2, f(\tilde{L}), \tilde{R}_2 \approx_{2^{-(s+1)}} U_1, \tilde{R}_1, X_2, f(\tilde{L}), \tilde{R}_2 ,$$

where U_1 is uniformly random in $\mathbb{F}$. This implies that

$$X_1, X_2, X_1', X_2' \approx_{2^{-(s+1)}} U_1, X_2, X_1', X_2' .$$

Now, X_1', X_2' are independent of U_1, and can be seen as a randomized function of X_2. Let

$$U_1, X_2, X_1', X_2' \equiv U_1, X_2, h_1(X_2, Z), h_2(X_2, Z) ,$$

for Z independent of U_1, X_2. Using Equation 8.7 and 8.8, and that Ext is a two-source randomness extractor, we have that X_2 is $2^{-(s+1)}$-close to uniform. Thus, using equation triangle inequality, we get that

$$X_1, X_2, X_1', X_2' \approx_{2^{-s}} U_1, U_2, h_1(U_2, Z), h_2(U_2, Z) ,$$

which implies the result. □

Symmetrically, we get the following:

Lemma 14. *If $|\mathcal{L}_{\mathsf{swap}} \times \mathcal{R}_{\mathsf{id}}| \geq q^{4\tau n} \cdot 2^{-s}$ then there exists a distribution D that is a convex combination of distributions in $\mathcal{D}_{\mathsf{forget},1}$ such that*

$$\Delta(\phi_{f,g}(L,R)|_{(L,R)\in\mathcal{L}_{\mathsf{swap}}\times\mathcal{R}_{\mathsf{id}}} ; D) \leq 2^{-s} .$$

We now look at the case when L, R are restricted to $\mathcal{L}_{\mathsf{id}}$, and $\mathcal{R}_{\mathsf{id}}$, respectively.

Lemma 15. *If* $|\mathcal{L}_{\mathsf{id}} \times \mathcal{R}_{\mathsf{id}}| \geq q^{4\tau n} \cdot 2^{-s}$ *then there exists a distribution D that is a convex combination of distributions in $\mathcal{D}_{\mathsf{id}}$ such that*

$$\Delta(\phi_{f,g}(L, R)|_{(L,R) \in \mathcal{L}_{\mathsf{id}} \times \mathcal{R}_{\mathsf{id}}} \,; D) \leq 2^{-s} .$$

Proof. Let $|\mathcal{L}_{\mathsf{id}} \times \mathcal{R}_{\mathsf{id}}| \geq q^{4\tau n} \cdot 2^{-s}$. Let $\tilde{L}, \tilde{R}$ be distributed uniformly over $\mathcal{L}_{\mathsf{id}}$ and $\mathcal{R}_{\mathsf{id}}$ respectively. Note that by the assumption we have that $|\mathcal{L}_{\mathsf{id}}| \geq q^{2\tau n} \cdot 2^{-s}$ and $|\mathcal{R}_{\mathsf{swap}}| \geq q^{2\tau n} \cdot 2^{-s}$. Denote $\langle \tilde{L}_k, \tilde{R}_k \rangle$ by X_k, and $\langle f_k(\tilde{L}), g_k(\tilde{R}) \rangle$ by X'_k for $k = 1, 2$.

By Lemma 8, there exists maps $a_{1,1}, a_{2,2}$ from $\mathcal{L}_{\mathsf{id}}$ to $\{0,1\}^{\beta_2}$, and $b_{1,1}$, $b_{2,2}$ from $\mathcal{R}_{\mathsf{id}}$ to $\{0,1\}^{\beta_2}$, such that $f_1(\tilde{L}), f_2(\tilde{L}), g_1(\tilde{R}), g_2(\tilde{R})$ are determined uniquely given $(\tilde{L}_1, a_{1,1}(\tilde{L})), (\tilde{L}_2, a_{2,2}(\tilde{L})), (\tilde{R}_1, b_{1,1}(\tilde{R})), (\tilde{R}_2, b_{2,2}(\tilde{R}))$, respectively. We define the random variable Y as

$$Y := \tilde{L}_1, \ \tilde{R}_2, \ a_{1,1}(\tilde{L}), \ a_{2,2}(\tilde{L}), \ b_{1,1}(\tilde{R}), \ b_{2,2}(\tilde{R}) .$$

Note that X'_1 is a deterministic function of Y and $\tilde{R}_1$. Similarly, X'_2 is a deterministic function of Y and $\tilde{L}_2$. Let W_1 be independent randomness used to sample $\tilde{R}_1$ given Y and X_1, and let W_2 be independent randomness used to sample $\tilde{L}_2$ given Y and X_2. Note that W_1, W_2 are independent from each other and from X_1, X_2, Y. Therefore, we have that

$$X_1, X_2, X'_1, X'_2 \equiv X_1, X_2, h_1(X_1, Y, W_1), h_2(X_2, Y, W_2) , \tag{8.9}$$

for some functions h_1, h_2. Also, using Lemma 2, we have that

$$X_1, X_2, Y, W_1, W_2 \approx_{2^{-(s+1)}} U_1, X_2, Y, W_1, W_2$$
$$\approx_{2^{-(s+1)}} U_1, U_2, Y, W_1, W_2 .$$

This implies the desired result using equation 8.9, and Lemma 1. $\qquad \square$

Symmetrically, we get the following:

Lemma 16. *If* $|\mathcal{L}_{\mathsf{swap}} \times \mathcal{R}_{\mathsf{swap}}| \geq q^{4\tau n} \cdot 2^{-s}$ *then there exists a distribution D that is a convex combination of distributions in $\mathcal{D}_{\mathsf{swap}}$ such that*

$$\Delta(\phi_{f,g}(L, R)|_{(L,R) \in \mathcal{L}_{\mathsf{swap}} \times \mathcal{R}_{\mathsf{swap}}} \,; D) \leq 2^{-s} .$$

Remaining Cases

Lemma 17. $|\mathcal{L}_{\mathsf{rem}}| \leq q^{2\tau n} 2^{-s}$, *and* $|\mathcal{R}_{\mathsf{rem}}| \leq q^{2\tau n} 2^{-s}$

Proof. Consider any $\ell \in \mathcal{L}_{\mathsf{rem}}$. Since $\ell \notin \mathcal{L}_{\mathsf{mix},j}$, we have that $\mathcal{T}(\ell, 1) \cup \mathcal{T}(\ell, 2) = \{1, 2\}$. Also, since $\ell \notin \mathcal{L}_{\mathsf{id}} \cup \mathcal{L}_{\mathsf{swap}}$, there exists some $k \in \{1, 2\}$, such that $\mathcal{T}(\ell, k) = \{1, 2\}$. Thus, using Lemma 8, we have that $f(\ell)$ can be determined given ℓ_k, $a_{k,1}, a_{k,2}$. This implies that $f(\ell)$ can be determined given at most $\tau n \log q + 2\beta_2$ bits. Therefore

$$|\mathcal{L}_{\mathsf{rem}}| \leq q^{\tau n} 2^{2\beta_2} \leq q^{2\tau n} 2^{-s} .$$

Symmetrically, $|\mathcal{R}_{\mathsf{rem}}| \leq q^{2\tau n} 2^{-s}$.

$\qquad \square$

8.4 Finishing the Proof

We partitioned $\mathcal{L} \times \mathcal{R}$ into following cases.

- $\mathcal{L}_{\mathsf{ffb},i} \times \mathcal{R}^*$ (see Lemma 9)
- $\mathcal{L}^* \times \mathcal{R}_{\mathsf{ffb},i}$ (see Lemma 10)
- $\mathcal{L}_{\mathsf{mix},j} \times \mathcal{R}^*$ (see Lemma 11)
- $\mathcal{L}^* \times \mathcal{R}_{\mathsf{mix},j}$ (see Lemma 12)
- $\mathcal{L}_{\mathsf{id}} \times \mathcal{R}_{\mathsf{swap}}$ (see Lemma 13)

- $\mathcal{L}_{\mathsf{swap}} \times \mathcal{R}_{\mathsf{id}}$ (see Lemma 14)
- $\mathcal{L}_{\mathsf{id}} \times \mathcal{R}_{\mathsf{id}}$ (see Lemma 15)
- $\mathcal{L}_{\mathsf{swap}} \times \mathcal{R}_{\mathsf{swap}}$ (see Lemma 16)
- $\mathcal{L}_{\mathsf{rem}} \times \mathcal{R}^*$ and $\mathcal{L}^* \times \mathcal{R}_{\mathsf{rem}}$ (see Lemma 17)

We showed that in every case for partition $\mathcal{L}^* \times \mathcal{R}^*$ we get either $\frac{|\mathcal{L}^* \times \mathcal{R}^*|}{|\mathcal{L} \times \mathcal{R}|} \leq 2^{-s}$ or there exists D' from $\mathcal{D}$ such that

$$\Delta(\phi_{f,g}(L,R)|_{L,R \in \mathcal{L}^* \times \mathcal{R}^*} \; ; \; D') \leq 2^{-s},$$

where $\mathcal{D}$ is a convex combination of distributions $\mathcal{D}_{\mathsf{id}} \cup \mathcal{D}_{\mathsf{swap}} \cup \mathcal{D}_{\mathsf{unif},1} \cup \mathcal{D}_{\mathsf{unif},2} \cup \mathcal{D}_{\mathsf{forget},1} \cup \mathcal{D}_{\mathsf{forget},2}$. We partitioned both $\mathcal{L}$ and $\mathcal{R}$ each into 7 subsets, thus by Lemma 7 we obtain that there exists a distribution D in $\mathcal{D}$ such that

$$\Delta(\phi_{f,g}(L,R) \; ; \; D) \leq 7^2 \cdot 2^{-s}.$$

This finishes the proof.   □

9 Conclusions and Open Problems

Our main result is a generic transformation from non-malleable codes in the 2-split-state model to non-malleable codes in the 2-split-state model that is resilient to leakage of length upto $1/12$-th of the length of the codeword. Combining with the best known non-malleable codes in the 2-split state model achieved by a subsequent work [2], we get constant-rate $1/12$-leakage resilient non-malleable codes.

We also observe in Section 7 that the result of [11] implies that we can achieve non-malleable codes resilient to a fraction of leakage arbitrarily close to 1.

Thus, our work can be viewed as initiating the study and achieving a constant factor leakage resilience for information-theoretically secure non-malleable codes. In view of the existential result, the main open question is whether we can give an efficient construction of non-malleable codes that can achieve leakage-resilience larger than a $1/12$-th fraction, and thereby make non-malleable codes practically more useful.

References

1. Aggarwal, D.: Affine-evasive sets modulo a prime. Cryptology ePrint Archive, Report 2014/328 (2014), http://eprint.iacr.org/
2. Aggarwal, D., Dodis, Y., Kazana, T., Obremski, M.: Non-malleable reductions and applications (2014) (unpublished manuscript)

3. Aggarwal, D., Dodis, Y., Lovett, S.: Non-malleable codes from additive combinatorics. In: Shmoys, D.B. (ed.) 46th ACM STOC, New York, NY, USA, May 31-June 3, pp. 774–783. ACM Press (2014)
4. Anderson, R., Kuhn, M.: Tamper resistance — a cautionary note. In: The Second USENIX Workshop on Electronic Commerce, pp. 1–11 (November 1996)
5. Bellare, M., Kohno, T.: A theoretical treatment of related-key attacks: RKA-PRPs, RKA-PRFs, and applications. In: Biham, E. (ed.) EUROCRYPT 2003. LNCS, vol. 2656, pp. 491–506. Springer, Heidelberg (2003), Full version available at http://www-cse.ucsd.edu/users/tkohno/papers/RKA/
6. Bellare, M., Rogaway, P.: Random oracles are practical: A paradigm for designing efficient protocols. In: Ashby, V. (ed.) ACM CCS 1993, Fairfax, Virginia, USA, November 3-5, pp. 62–73. ACM Press (1993)
7. Bitansky, N., Dachman-Soled, D., Lin, H.: Leakage-tolerant computation with input-independent preprocessing. In: Garay, J.A., Gennaro, R. (eds.) CRYPTO 2014, Part II. LNCS, vol. 8617, pp. 146–163. Springer, Heidelberg (2014)
8. Chabanne, H., Cohen, G., Patey, A.: Secure network coding and non-malleable codes: Protection against linear tampering. In: 2012 IEEE International Symposium on Information Theory Proceedings (ISIT), pp. 2546–2550 (July 2012)
9. Chattopadhyay, E., Zuckerman, D.: Non-malleable codes against constant split-state tampering. In: FOCS (2014)
10. Cheraghchi, M., Guruswami, V.: Capacity of non-malleable codes. In: Naor, M. (ed.) ITCS 2014, Princeton, NJ, USA, January 12-14, pp. 155–168. ACM (2014)
11. Cheraghchi, M., Guruswami, V.: Non-malleable coding against bit-wise and split-state tampering. In: Lindell, Y. (ed.) TCC 2014. LNCS, vol. 8349, pp. 440–464. Springer, Heidelberg (2014)
12. Chor, B., Goldreich, O.: Unbiased bits from sources of weak randomness and probabilistic communication complexity. SIAM Journal on Computing 17(2), 230–261 (1988)
13. Coretti, S., Maurer, U., Tackmann, B., Venturi, D.: From single-bit to multi-bit public-key encryption via non-malleable codes. Cryptology ePrint Archive, Report 2014/324 (2014), http://eprint.iacr.org/
14. Cramer, R., Dodis, Y., Fehr, S., Padró, C., Wichs, D.: Detection of algebraic manipulation with applications to robust secret sharing and fuzzy extractors. In: Smart, N.P. (ed.) EUROCRYPT 2008. LNCS, vol. 4965, pp. 471–488. Springer, Heidelberg (2008)
15. Dachman-Soled, D., Kalai, Y.T.: Securing circuits against constant-rate tampering. In: Safavi-Naini, R., Canetti, R. (eds.) CRYPTO 2012. LNCS, vol. 7417, pp. 533–551. Springer, Heidelberg (2012)
16. Davì, F., Dziembowski, S., Venturi, D.: Leakage-resilient storage. In: Garay, J.A., De Prisco, R. (eds.) SCN 2010. LNCS, vol. 6280, pp. 121–137. Springer, Heidelberg (2010)
17. Dodis, Y., Ostrovsky, R., Reyzin, L., Smith, A.: Fuzzy extractors: How to generate strong keys from biometrics and other noisy data. SIAM J. Comput. 38(1), 97–139 (2008)
18. Dolev, D., Dwork, C., Naor, M.: Non-malleable cryptography (1998) (manuscript)
19. Dziembowski, S., Faust, S.: Leakage-resilient circuits without computational assumptions. In: Cramer, R. (ed.) TCC 2012. LNCS, vol. 7194, pp. 230–247. Springer, Heidelberg (2012)
20. Dziembowski, S., Kazana, T., Obremski, M.: Non-malleable codes from two-source extractors. In: Canetti, R., Garay, J.A. (eds.) CRYPTO 2013, Part II. LNCS, vol. 8043, pp. 239–257. Springer, Heidelberg (2013)

21. Dziembowski, S., Pietrzak, K.: Intrusion-resilient secret sharing. In: FOCS, pp. 227–237. IEEE Computer Society (2007)
22. Dziembowski, S., Pietrzak, K.: Leakage-resilient cryptography. In: 49th Symposium on Foundations of Computer Science, Philadelphia, PA, USA, October 25-28, pp. 293–302. IEEE Computer Society (2008)
23. Dziembowski, S., Pietrzak, K., Wichs, D.: Non-malleable codes. In: Yao, A.C.-C. (ed.) ICS 2010, Beijing, China, January 5-7, pp. 434–452. Tsinghua University Press (2010)
24. ECRYPT. Side channel cryptanalysis lounge, http://www.crypto.ruhr-uni-bochum.de/en_sclounge.html (last accessed: August 26, 2009)
25. Faust, S., Kiltz, E., Pietrzak, K., Rothblum, G.N.: Leakage-resilient signatures. In: Micciancio, D. (ed.) TCC 2010. LNCS, vol. 5978, pp. 343–360. Springer, Heidelberg (2010)
26. Faust, S., Mukherjee, P., Nielsen, J., Venturi, D.: Continuous non-malleable codes. In: TCC 2014. Springer, Heidelberg (2014)
27. Faust, S., Mukherjee, P., Nielsen, J.B., Venturi, D.: A tamper and leakage resilient random access machine. Cryptology ePrint Archive, Report 2014/338 (2014), http://eprint.iacr.org/2014/338
28. Faust, S., Mukherjee, P., Venturi, D., Wichs, D.: Efficient non-malleable codes and key-derivation for poly-size tampering circuits. In: Nguyen, P.Q., Oswald, E. (eds.) EUROCRYPT 2014. LNCS, vol. 8441, pp. 111–128. Springer, Heidelberg (2014)
29. Faust, S., Pietrzak, K., Venturi, D.: Tamper-proof circuits: How to trade leakage for tamper-resilience. In: Aceto, L., Henzinger, M., Sgall, J. (eds.) ICALP 2011, Part I. LNCS, vol. 6755, pp. 391–402. Springer, Heidelberg (2011)
30. Gennaro, R., Lysyanskaya, A., Malkin, T., Micali, S., Rabin, T.: Algorithmic tamper-proof (ATP) security: Theoretical foundations for security against hardware tampering. In: Naor, M. (ed.) TCC 2004. LNCS, vol. 2951, pp. 258–277. Springer, Heidelberg (2004)
31. Goldwasser, S., Rothblum, G.N.: How to compute in the presence of leakage. In: 53rd FOCS, New Brunswick, NJ, USA, October 20-23, pp. 31–40. IEEE Computer Society Press (2012)
32. Ishai, Y., Prabhakaran, M., Sahai, A., Wagner, D.: Private circuits II: Keeping secrets in tamperable circuits. In: Vaudenay, S. (ed.) EUROCRYPT 2006. LNCS, vol. 4004, pp. 308–327. Springer, Heidelberg (2006)
33. Ishai, Y., Sahai, A., Wagner, D.: Private circuits: Securing hardware against probing attacks. In: Boneh, D. (ed.) CRYPTO 2003. LNCS, vol. 2729, pp. 463–481. Springer, Heidelberg (2003)
34. Kalai, Y.T., Kanukurthi, B., Sahai, A.: Cryptography with tamperable and leaky memory. In: Rogaway, P. (ed.) CRYPTO 2011. LNCS, vol. 6841, pp. 373–390. Springer, Heidelberg (2011)
35. Liu, F.-H., Lysyanskaya, A.: Algorithmic tamper-proof security under probing attacks. In: Garay, J.A., De Prisco, R. (eds.) SCN 2010. LNCS, vol. 6280, pp. 106–120. Springer, Heidelberg (2010)
36. Liu, F.-H., Lysyanskaya, A.: Tamper and leakage resilience in the split-state model. In: Safavi-Naini, R., Canetti, R. (eds.) CRYPTO 2012. LNCS, vol. 7417, pp. 517–532. Springer, Heidelberg (2012)
37. Micali, S., Reyzin, L.: Physically observable cryptography (extended abstract). In: Naor, M. (ed.) TCC 2004. LNCS, vol. 2951, pp. 278–296. Springer, Heidelberg (2004)

424   D. Aggarwal et al.

38. Rao, A.: An exposition of bourgain 2-source extractor. In: Electronic Colloquium on Computational Complexity (ECCC), vol. 14, p. 034 (2007)

A   Proofs of Lemmata from Section 2

Lemma 3 *Let $X_1, Y_1 \in \mathcal{A}_1$, and $X_2, Y_2 \in \mathcal{A}_2$ be random variables such that $\Delta((X_1, X_2) ; (Y_1, Y_2)) \leq \varepsilon$. Then, for any non-empty set $\mathcal{A}' \subseteq \mathcal{A}_1$, we have*

$$\Delta(X_2 \mid X_1 \in \mathcal{A}' ; Y_2 \mid Y_1 \in \mathcal{A}') \leq \frac{2\varepsilon}{\Pr(X_1 \in \mathcal{A}')} .$$

Proof. Let δ denote $\Delta(X_2 \mid X_1 \in \mathcal{A}' ; Y_2 \mid Y_1 \in \mathcal{A}')$. Then

$$\delta = \frac{1}{2} \sum_{x \in \mathcal{A}_2} \left| \Pr(X_2 = x \mid X_1 \in \mathcal{A}') - \Pr(Y_2 = x \mid Y_1 \in \mathcal{A}') \right|$$

$$\leq \frac{1}{2} \sum_{x \in \mathcal{A}_2} \left(\left| \frac{\Pr(X_2 = x \wedge X_1 \in \mathcal{A}')}{\Pr(X_1 \in \mathcal{A}')} - \frac{\Pr(Y_2 = x \wedge Y_1 \in \mathcal{A}')}{\Pr(X_1 \in \mathcal{A}')} \right| \right.$$

$$\left. + \Pr(Y_2 = x \wedge Y_1 \in \mathcal{A}') \left| \frac{1}{\Pr(Y_1 \in \mathcal{A}')} - \frac{1}{\Pr(X_1 \in \mathcal{A}')} \right| \right)$$

$$\leq \frac{\varepsilon}{\Pr(X_1 \in \mathcal{A}')} + \frac{\varepsilon \cdot \sum_{x \in \mathcal{A}_2} \Pr(Y_1 \in \mathcal{A}' \wedge Y_2 = x)}{\Pr(Y_1 \in \mathcal{A}') \cdot \Pr(X_1 \in \mathcal{A}')}$$

$$= \frac{2\varepsilon}{\Pr(X_1 \in \mathcal{A}')} .$$

$\square$

Lemma 5 *Let $X = (X_1, \ldots, X_t) \in \mathbb{F}^t$ be a random variable, where $\mathbb{F}$ is a finite field of order q. Assume that for all $a_1, \ldots, a_t \in \mathbb{F}^t$ not all zero, $\Delta(\sum_{i=1}^{t} a_i X_i ; U) \leq \varepsilon$, where U is uniform in $\mathbb{F}$. Then $\Delta(X_1, \ldots, X_t ; U_1, \ldots, U_t) \leq \varepsilon q^{(t+2)/2}$, where $U_1, \ldots, U_t$ are independent and uniform in $\mathbb{F}^t$.*

Proof. The proof uses basic Fourier analysis. Assume $\mathbb{F}$ has characteristic p. Let $\omega = e^{2\pi i/p}$ be a primitive p-th root of unity. Let $\mathrm{Tr} : \mathbb{F} \to \mathbb{F}_p$ denote the trace operator from $\mathbb{F}$ to $\mathbb{F}_p$. The additive characters of $\mathbb{F}$ are given by $\{\chi_a(x) : \mathbb{F} \to \mathbb{C} : a \in \mathbb{F}\}$ defined as

$$\chi_a(x) = \omega^{\mathrm{Tr}(ax)}.$$

The additive characters of $\mathbb{F}^t$ are given by $\chi_{a_1,\ldots,a_t}(x_1, \ldots, x_t) = \Pi_{i=1}^{t} \chi_{a_i}(x_i)$ for $a_1, \ldots, a_t \in \mathbb{F}$. First, we bound the Fourier coefficients of the distribution of $X = (X_1, \ldots, X_t)$. The $(a_1, \ldots, a_t)$ Fourier coefficient, for all non-zero $(a_1, \ldots, a_t)$, is given by

$$\mathbb{E}[\chi_{a_1,\ldots,a_t}(X_1,\ldots,X_t)] = \mathbb{E}[\omega^{\mathrm{Tr}(\sum_{i=1}^{t} a_i X_i)}] = \sum_{b \in \mathbb{F}} \omega^{\mathrm{Tr}(b)} \Pr[\sum_{i=1}^{t} a_i X_i = b]$$

$$= \sum_{b \in \mathbb{F}} \omega^{\mathrm{Tr}(b)} \left(\Pr_X[\sum_{i=1}^{t} a_i X_i = b] - \frac{1}{|\mathbb{F}|} \right),$$

where we used the fact that $\sum_{b \in \mathbb{F}} \omega^{\mathrm{Tr}(b)} = 0$. Hence for all non-zero $(a_1, \ldots, a_t)$,

$$\left| \mathbb{E}[\chi_{a_1,\ldots,a_t}(X_1,\ldots,X_t)] \right| \le \sum_{b \in \mathbb{F}} \left| \Pr[\sum_{i=1}^{t} a_i X_i = b] - \frac{1}{|\mathbb{F}|} \right| \le 2\varepsilon \cdot |\mathbb{F}| .$$

Let $p_{a_1,\ldots,a_t} = \Pr[(X_1,\ldots,X_t) = (a_1,\ldots,a_t)]$. By Parseval's identity,

$$\sum_{a_1,\ldots,a_t \in \mathbb{F}} \left(p_{a_1,\ldots,a_t} - \frac{1}{|\mathbb{F}|} \right)^2 = \sum_{(a_1,\ldots,a_t) \ne 0} \mathbb{E}[\chi_{a_1,\ldots,a_t}(X_1,\ldots,X_t)]^2 \le 4\varepsilon^2 |\mathbb{F}|^{t+2} ,$$

$\square$

B   Proof of Lemma 6

We will need the following fact.

Fact B.1 *Let* $\mathsf{Dec} : \mathcal{X} \times \mathcal{X} \to \mathcal{M} \cup \{\bot\}$, *and* $\mathsf{Enc} : \mathcal{M} \to \mathcal{X} \times \mathcal{X}$ *be* ε-*non-malleable scheme in* $2-$*split state model for some* $\varepsilon < \frac{1}{2}$. *For any two messages* $m_0, m_1 \in \mathcal{M}$, *there exist* $x_1^0, x_1^1, x_2 \in \mathcal{X}$ *such that*

- $\mathsf{Dec}(x_1^0, x_2) = m_0$ - $\mathsf{Dec}(x_1^1, x_2) = m_1$

Proof. By contradiction, let us assume there exists $m_0, m_1 \in \mathcal{M}$ such that $\forall_{x_2 \in \mathcal{X}} |\mathsf{Dec}(\mathcal{X}, x_2) \cap \{m_0, m_1\}| = 1$. Let us define sets $\mathcal{X}_2^0, \mathcal{X}_2^1$ as follows

$$\mathcal{X}_2^0 = \{x_2 \in \mathcal{X} : \mathsf{Dec}(\mathcal{X}, x_2) \cap \{m_0, m_1\} = \{m_0\}\}$$
$$\mathcal{X}_2^1 = \{x_2 \in \mathcal{X} : \mathsf{Dec}(\mathcal{X}, x_2) \cap \{m_0, m_1\} = \{m_1\}\}$$

Fix arbitrary $x_2^0 \in \mathcal{X}_2^0$, $x_2^1 \in \mathcal{X}_2^1$, and let

$$\mathcal{X}_1^0 = \{x \in \mathcal{X} : \mathsf{Dec}(x, x_2^0) = m_0\}$$
$$\mathcal{X}_1^1 = \{x \in \mathcal{X} : \mathsf{Dec}(x, x_2^1) = m_1\}$$

Consider tampering functions $h_1 : \mathcal{X} \to \mathcal{X}$, and $h_2 : \mathcal{X} \to \mathcal{X}$ as follows. Let $h_2(\mathcal{X}_2^0) := x_2^1$ and $h_2(\mathcal{X}_2^1) := x_2^0$, and h_2 is defined arbitrarily in $\mathcal{X} \setminus (\mathcal{X}_2^0 \cup \mathcal{X}_2^1)$. Also, if $\mathcal{X}_1^0 \cap \mathcal{X}_1^1$ is non-empty, then fix some $x \in \mathcal{X}_1^0 \cap \mathcal{X}_1^1$, and let $h_1(c) = x$ for all $c \in \mathcal{X}$. Otherwise choose arbitrary $x_1^0 \in \mathcal{X}_1^0$, $x_1^1 \in \mathcal{X}_1^1$, and let $h_1(\mathcal{X}_1^0) := x_1^1$ and $h_1(\mathcal{X}_1^1) := x_1^0$, and h_1 is defined arbitrarily in $\mathcal{X} \setminus (\mathcal{X}_1^0 \cup \mathcal{X}_1^1)$.

Then, we have that $\mathsf{Tamper}_{m_0} = m_1$, and $\mathsf{Tamper}_{m_1} = m_0$ with probability 1, where Tamper_{m_0} and Tamper_{m_1} are as in Definition 3. This implies that there exists a distribution D over $\mathcal{M} \cup \{\bot\}$ such that $\Pr(D = m_0) \geq 1 - \varepsilon$, and $\Pr(D = m_1) \geq 1 - \varepsilon$, which is a contradiction. $\qquad\square$

Lemma 6. *Let* $\mathsf{Dec} : \mathcal{X} \times \mathcal{X} \to \mathcal{M}$, *and* $\mathsf{Enc} : \mathcal{M} \to \mathcal{X} \times \mathcal{X}$ *be* $\varepsilon-non$-*malleable scheme in* $2-split$ *state model for some* $\varepsilon < \frac{1}{2}$. *For any pair of messages* $m_0, m_1 \in \mathcal{M}$, *let* $(X_1^0, X_2^0) \leftarrow \mathsf{Enc}(m_0)$, *and let* $(X_1^1, X_2^1) \leftarrow \mathsf{Enc}(m_1)$. *Then* $\Delta(X_1^0 ; X_1^1) \leq 2\varepsilon$.

Proof. By contradiction assume that $\Delta(X_1^0 ; X_1^1) > 2\varepsilon$. Then there exists distinguisher $\mathcal{A} : \mathcal{X} \to \{0, 1\}$ such that

$$\Pr(\mathcal{A}(X_1^0) = 1) - \Pr(\mathcal{A}(X_1^1) = 1) > 2\varepsilon . \tag{B.2}$$

By Fact B.1 we have $x_1^0, x_1^1, x_2 \in \mathcal{X}$ such that $\mathsf{Dec}(x_1^0, x_2) = m_0$ and $\mathsf{Dec}(x_1^1, x_2) = m_1$. Now let us choose following tampering functions:

$$h_2(r) = x_2, \quad \text{and} \quad h_1(\ell) = x_{\mathcal{A}(\ell)} .$$

Consider Tamper_{m_0} and Tamper_{m_1} as in Definiton 3. By Equation B.2, we have that

$$\Pr[\mathsf{Tamper}_{m_0} = m_1] - \Pr[\mathsf{Tamper}_{m_1} = m_1] > 2\varepsilon . \tag{B.3}$$

From Definition 3, we have that there exists a distribution D such that

$$|\Pr[\mathsf{Tamper}_{m_0} = m_1] - \Pr(D = m_1)| \leq \varepsilon, \quad \text{and}$$

$$|\Pr[\mathsf{Tamper}_{m_1} = m_1] - \Pr(D = m_1) - \Pr(D = \mathsf{same})| \leq \varepsilon .$$

By triangle inequality, this implies,

$$\Pr[\mathsf{Tamper}_{m_0} = m_1] - \Pr[\mathsf{Tamper}_{m_1} = m_1] + \Pr(D = \mathsf{same}) \leq 2\varepsilon ,$$

which contradicts Equation B.3. $\qquad\square$

Locally Decodable and Updatable Non-malleable Codes and Their Applications

Dana Dachman-Soled[1], Feng-Hao Liu[2], Elaine Shi[2], and Hong-Sheng Zhou[3]

[1] University of Maryland, USA
danadach@ece.umd.edu
[2] University of Maryland, USA
{fenghao,elaine}@cs.umd.edu
[3] Virginia Commonwealth University, USA
hszhou@vcu.edu

Abstract. Non-malleable codes, introduced as a relaxation of error-correcting codes by Dziembowski, Pietrzak and Wichs (ICS '10), provide the security guarantee that the message contained in a tampered codeword is either the same as the original message or is set to an unrelated value. Various applications of non-malleable codes have been discovered, and one of the most significant applications among these is the connection with tamper-resilient cryptography. There is a large body of work considering security against various classes of tampering functions, as well as non-malleable codes with enhanced features such as *leakage resilience*.

In this work, we propose combining the concepts of *non-malleability*, *leakage resilience*, and *locality* in a coding scheme. The contribution of this work is three-fold:

1. As a conceptual contribution, we define a new notion of *locally decodable and updatable non-malleable code* that combines the above properties.
2. We present two simple and efficient constructions achieving our new notion with different levels of security.
3. We present an important application of our new tool – securing RAM computation against memory tampering and leakage attacks. This is analogous to the usage of traditional non-malleable codes to secure implementations in the *circuit* model against memory tampering and leakage attacks.

1 Introduction

The notion of non-malleable codes was defined by Dziembowski, Pietrzak and Wichs [22] as a relaxation of error-correcting codes. Informally, a coding scheme is **non-malleable** against a tampering function if by tampering with the codeword, the function can either keep the underlying message unchanged or change it to an unrelated message. Designing non-malleable codes is not only an interesting mathematical task, but also has important implications in cryptography; for example, Coretti et al. [11] showed an efficient construction of a mulit-bit CCA

Y. Dodis and J.B. Nielsen (Eds.): TCC 2015, Part I, LNCS 9014, pp. 427–450, 2015.

secure encryption scheme from a single-bit one via non-malleable codes. Agrawal et al. [3] showed how to use non-malleable codes to build non-malleable commitments. Most notably, the notion has a deep connection with security against so-called *physical attacks*; indeed, using non-malleable codes to achieve security against physical attacks was the original motivation of the work [22]. Due to this important application, research on non-malleable codes has become an important agenda, and drawn much attention in both coding theory and cryptography.

Briefly speaking, physical attacks target implementations of cryptographic algorithms beyond their input/output behavior. For example, researchers have demonstrated that leaking/tampering with sensitive secrets such as cryptographic keys, through timing channel, differential power analysis, and various other attacks, can be devastating [2, 4, 5, 32, 39, 40, 45], and therefore the community has focused on developing new mechanisms to defend against such strong adversaries [12, 13, 15–19, 21, 26–28, 30, 31, 34–36, 38, 43, 44, 46]. Dziembowski, Pietrzak and Wichs [22] showed a simple and elegant mechanism to secure implementations against memory tampering attacks by using non-malleable codes – instead of storing the secret (in the clear) on a device, one instead stores an encoding of the secret. The security of the non-malleable code guarantees that the adversary cannot learn more than what can be learnt via black box access to the device, even though the adversary may tamper with memory.

In a subsequent work, Liu and Lysyanskaya [42] extended the notion to capture **leakage resilience** as well – in addition to non-malleability, the adversary cannot learn anything about the underlying message even while obtaining partial leakage of the codeword. By using the approach outlined above, one can achieve security guarantees against both tampering and leakage attacks. In recent years, researchers have been studying various flavors of non-malleable codes; for example some work has focused on constructions against different classes of tampering functions, some has focused on different additional features, (e.g. continual attacks, rates of the scheme, etc), and some focused on other applications [1, 3, 7–9, 20, 23, 25].

In this paper, we focus on another important feature inspired from the field of coding theory – **locality**. More concretely, we consider a coding scheme that is locally decodable and updatable. As introduced by Katz and Trevisan [37], local decodability means that in order to retrieve a portion of the underlying message, one does not need to read through the whole codeword. Instead, one can just read a few locations at the codeword. Similarly, local updatability means that in order to update some part of the underlying messages, one only needs to update some parts of the codeword. Locally decodable codes have many important applications in private information retrieval [10] and secure multi-party computation [33], and have deep connections with complexity theory; see [49]. Achieving local decodability and updatability simultaneously makes the task more challenging. Recently, Chandran et al. [6] constructed a locally decodable and updatable code in the setting of *error-correcting* codes. They also show an application to dynamic proofs of retrievability. Motivated by the above results, we further ask the following intriguing question:

Can we build a coding scheme enjoying all three properties, i.e., non-malleability, leakage resilience, and locality? If so, what are its implications in cryptography?

Our Results. In light of the above questions, our contribution is three-fold:

- **(Notions).** We propose new notions that combine the concepts of non-malleability, leakage resilience, and locality in codes. First, we formalize a new notion of *locally decodable and updatable non-malleable codes* (against one-time attacks). Then, we extend this new notion to capture leakage resilience under continual attacks.
- **(Constructions).** We present two simple constructions achieving our new notions. The first construction is highly efficient—in order to decode (update) one block of the encoded messages, only *two* blocks of the codeword must be read (written)—but is only secure against *one-time* attacks. The second construction achieves security against *continual* attacks, while requiring $\log(n)$ number of reads (writes) to perform one decode (update) operation, where n is the number of blocks of the underlying message.
- **(Application).** We present an important application of our new notion – achieving tamper and leakage resilience in the random access machine (RAM) model. We first define a new model that captures tampering and leakage attacks in the RAM model, and then give a generic compiler that uses our new notion as a main ingredient. The compiled machine will be resilient to leakage and tampering on the random access memory. This is analogous to the usage of traditional non-malleable codes to secure implementations in the *circuit* model.

1.1 Techniques

In this section, we present a technical overview of our results.

Locally Decodable Non-malleable Codes. Our first goal is to consider a combination of concepts of non-malleability and local decodability. Recall that a coding scheme is non-malleable with respect to a tampering function f if the decoding of the tampered codeword remains the same or becomes some unrelated message. To capture this idea, the definition in the work [22] requires that there exists a simulator (with respect to such f) who outputs same* if the decoding of the tampered codeword remains the same as the original one, or he outputs a decoded message, which is unrelated to the original one. In the setting of local decodability, we consider encodings of blocks of messages $M = (m_1, m_2, \ldots, m_n)$, and we are able to retrieve m_i by running $\text{DEC}^{\text{ENC}(M)}(i)$, where the decoding algorithm gets oracle access to the codeword.

The combination faces a subtlety that we cannot directly use the previous definition: suppose a tampering function f only modifies one block of the codeword, then it is likely that DEC remains unchanged for most places. (Recall a DEC will only read a few blocks of the codeword, so it may not detect the modification.)

In this case, the (overall) decoding of $f(C)$ (i.e. $(\text{DEC}^{f(C)}(1), \ldots, \text{DEC}^{f(C)}(n))$) can be highly related to the original message, which intuitively means it is highly malleable.

To handle this issue, we consider a more fine-grained experiment. Informally, we require that for any tampering function f (within some class), there exists a simulator that computes a vector of decoded messages $\boldsymbol{m}^*$, a set of indices $\mathcal{I} \subseteq [n]$. Here $\mathcal{I}$ denotes the coordinates of the underlying messages that have been tampered with. If $\mathcal{I} = [n]$, then the simulator thinks that the decoded messages are $\boldsymbol{m}^*$, which should be unrelated to the original messages. On the other hand, if $\mathcal{I} \subsetneq [n]$, the simulator thinks that all the messages not in $\mathcal{I}$ remain unchanged, while those in $\mathcal{I}$ become $\perp$. This intuitively means the tampering function can do only one of the following cases:

1. It destroys a block (or blocks) of the underlying messages while keeping the other blocks unchanged, or
2. If it modifies a block of the underlying messages to some unrelated string, then it must have modified all blocks of the underlying messages to encodings of unrelated messages.

Our construction of locally decodable non-malleable code is simple – we use the idea similar to the *key encapsulation mechanism/data encapsulation mechanism (KEM/DEM) framework*. Let NMC be a regular non-malleable code, and $\mathcal{E}$ be a secure (symmetric-key) authenticated encryption. Then to encode blocks of messages $M = (m_1, \ldots, m_n)$, we first sample a secret key sk of $\mathcal{E}$, and then output $(\text{NMC.ENC(sk)}, \mathcal{E}.\text{Encrypt}_{\text{sk}}(m_1, 1), \ldots, \mathcal{E}.\text{Encrypt}_{\text{sk}}(m_n, n))$. The intuition is clear: if the tampering function does not change the first block, then by security of the authenticated encryption, any modification of the rest will become $\perp$. (Note that here we include a tag of positions to prevent permutation attacks). On the other hand, if the tampering function modified the first block, it must be decoded to an unrelated secret key sk'. Then by semantic security of the encryption scheme, the decoded values of the rest must be unrelated. The code can be updated locally: in order to update m_i to some m_i', one just need to retrieve the 1st and $(i+1)^{\text{st}}$ blocks. Then he just computes a *fresh* encoding of NMC.ENC(sk) and the ciphertext $\mathcal{E}.\text{Encrypt}_{\text{sk}}(m_i')$, and writes back to the same positions.

Extensions to Leakage Resilience against Continual Attacks. We further consider a notion that captures leakage attacks in the continual model. First we observe that suppose the underlying non-malleable code is also leakage resilient [42], the above construction also achieves one-time leakage resilience. Using the same argument of Liu and Lysyanskaya [42], if we can refresh the whole encoding, we can show that the construction is secure against continual attacks. However, in our setting, refreshing the whole codeword is not counted as a solution since this is in the opposite of the spirit of our main theme – *locality*. The main challenge is how to refresh (update) the codeword locally while maintaining tamper and leakage resilience.

To capture the local refreshing and continual attacks, we consider a new model where there is an updater $\mathcal{U}$ who reads the whole underlying messages and decides how to update the codeword (using the local update algorithm). The updater is going to interact with the codeword in a continual manner, while the adversary can launch tampering and leakage attacks between two updates. To define security we require that the adversary cannot learn anything of the underlying messages via tampering and leakage attacks from the interaction.

We note that if there is no update procedure at all, then no coding scheme can be secure against continual leakage attacks if the adversary can learn the whole codeword bit-by-bit. In our model, the updater and the adversary take turns interacting with the codeword – the adversary tampers with and/or gets leakage of the codeword, and then the updater *locally* updates the codeword, and the process repeats. See Section 2 for the formal model.

Then we consider how to achieve this notion. First we observe that the construction above is not secure under continual attacks: suppose by leakage the adversary can get a full ciphertext $\mathcal{E}.\mathsf{Encrypt}_{\mathsf{sk}}(m_i, i)$ at some point, and then the updater updates the underlying message to m_i'. In the next round, the adversary can apply a *rewind attack* that modifies the codeword back with the old ciphertext. Under such attack, the underlying messages have been modified to some related messages. Thus the construction is not secure.

One way to handle this type of rewind attacks is to tie all the blocks of ciphertexts together with a "time stamp" that prevents the adversary from replacing the codeword with old ciphertexts obtained from leakage. A straightforward way is to hash all the blocks of encryptions using a collision resistant hash function and also encode this value into the non-malleable code, i.e., $C = (\mathsf{NMC}.\mathrm{ENC}(\mathsf{sk}, v), \mathcal{E}.\mathsf{Encrypt}(1, m_1), \ldots, \mathcal{E}.\mathsf{Encrypt}(n, m_n))$, where $v = h(\mathcal{E}.\mathsf{Encrypt}(1, m_1), \ldots, \mathcal{E}.\mathsf{Encrypt}(n, m_n))$. Intuitively, suppose the adversary replaces a block $\mathcal{E}.\mathsf{Encrypt}(i, m_i)$ by some old ciphertexts, then it would be caught by the hash value v unless he tampered with the non-malleable code as well. But if he tampers with the non-malleable code, the decoding will be unrelated to sk, and thus the rest of ciphertexts become "un-decryptable". This approach prevents the rewind attacks, yet it does not preserve the local properties, i.e. to decode a block, one needs to check the consistency of the hash value v, which needs to read all the blocks of encryptions. To prevent the rewind attacks while maintaining local decodability/updatability, we use the Merkle tree technique, which allows local checks of consistency. The final encoding outputs the following:

$$(\mathsf{NMC}.\mathrm{ENC}(\mathsf{sk}, v), \mathcal{E}.\mathsf{Encrypt}(1, m_1), \ldots, \mathcal{E}.\mathsf{Encrypt}(n, m_n), T)$$

where T is the Merkle tree of $(\mathcal{E}.\mathsf{Encrypt}(1, m_1), \ldots, \mathcal{E}.\mathsf{Encrypt}(n, m_n))$, and v is its root (it can also be viewed as a hash value). To decode a position i, the algorithm reads the 1st, and the $(i+1)^{\mathrm{st}}$ blocks together with a path in the tree. If the path is inconsistent with the root, then output $\bot$. To update, one only needs to re-encode the first block with a new root, and update the $(i+1)^{\mathrm{st}}$ block and the tree. We note that Merkle tree allows local updates: if there is only one single change at a leaf, then one can compute the new root given only a path passing through the leaf and the root. So the update of the codeword can be done locally by reading the 1st, the $(i+1)^{st}$ blocks and the path. We provide a detailed description and analysis in Section 3.2.

Concrete Instantiations. In our construction above, we rely on an underlying non-malleable code NMC against some class of tampering functions $\mathcal{F}$ and leakage resilient against some class of leakage functions $\mathcal{G}$. The resulting encoding scheme is a locally decodable and updatable coding scheme which is continual non-malleable against some class $\overline{\mathcal{F}}$ of tampering functions and leakage resilient against some class $\overline{\mathcal{G}}$ of leakage functions, where the class $\overline{\mathcal{F}}$ is determined by $\mathcal{F}$ and the class $\overline{\mathcal{G}}$ is determined by $\mathcal{G}$. In order to understand the relationship between these classes, it is helpful to recall the structure of the output of the final encoding scheme. The final encoding scheme will output $2n + 1$ blocks $x_1, \ldots, x_{2n+1}$ such that the first block x_1 is encoded using the underlying non-malleable code NMC. As a first attempt, we can define $\overline{\mathcal{F}}$ to consist of tampering functions $f(x_1, \ldots, x_{2n+1}) = (f_1(x_1), f_2(x_2, \ldots, x_{2n+1}))$, where $f_1 \in \mathcal{F}$ and f_2 is any polynomial-sized circuit. However, it turns out that we are resilient against an even larger class of tampering functions! This is because the tampering function f_1 can actually depend on all the values $x_2, \ldots, x_{2n+1}$ of blocks $2, \ldots, (2n+1)$. Similarly, for the class of leakage functions, as a first attempt, we can define $\overline{\mathcal{G}}$ to consist of leakage functions $g(x_1, \ldots, x_{2n+1}) = (g_1(x_1), g_2(x_2, \ldots, x_{2n+1}))$, where $g_1 \in \mathcal{G}$ and g_2 is any polynomial-sized circuit. However, again we can achieve even more because the tampering function g_1 can actually depend on all the values $x_2, \ldots, x_{2n+1}$. For a formal definition of the classes of tampering and leakage functions that we handle, see Theorem 3.

Finally, we give a concrete example of what the resulting classes look like using the NMC construction of Liu and Lysyanskaya [42] as the building block. Recall that their construction achieves both tamper and leakage resilience for split-state functions. Thus, the overall tampering function f restricted in the first block (i.e. f_1) can be any (poly-sized) split-state function. On the other hand f restricted in the rest (i.e. f_2) can be any poly-sized function. The overall leakage function g restricted in the first block (i.e. g_1) can be a (poly-sized) length-bounded split-state function; g on the other hand, can leak all the other parts. See Section 3.3 for more details.

Application to Tamper and Leakage Resilient RAM Model. Whereas regular non-malleable codes yield secure implementations against memory tampering in the circuit model, our new tool yields secure implementations against memory tampering (and leakage) in the RAM model.

In our RAM model, the data and program to be executed are stored in the random access memory. Through a CPU with a small number of (non-persistent) registers[1], execution proceeds in clock cycles: In each clock-cycle memory addresses

---

[1] These non-persistent registers are viewed as part of the circuitry that stores some transient states while the CPU is computing at each cycle. The number of these registers is small, and the CPU needs to erase the data in order to reuse them, so they cannot be used to store a secret key that is needed for a long term of computation.

are read and stored in registers, a computation is performed, and the contents of the registers are written back to memory. In our attack model, we assume that the CPU circuitry (including the non-persistent registers) is secure – the computation itself is not subject to physical attacks. On the other hand, the random access memory, and the memory addresses are prone to leakage and tampering attacks. We remark that if the CPU has *secure* persistent registers that store a secret key, then the problem becomes straightforward: Security can be achieved using encryption and authentication together with oblivious RAM [29]. We emphasize that in our model, persistent states of the CPU are stored in the memory, which are prone to leakage and tampering attacks. As our model allows the adversary to learn the access patterns the CPU made to the memory, together with the leakage and tampering power on the memory, the adversary can somewhat learn the messages transmitted over the *bus* or tamper with them (depending on the attack classes allowed on the memory). For simplicity of presentation, we do not define attacks on the bus, but just remark that these attacks can be implicitly captured by learning the access patterns and attacking the memory[2].

In our formal modeling, we consider a next instruction function Π, a database D (stored in the random access memory) and an internal state (using the non-persistent registers). The CPU will interact (i.e., read/write) with the memory based on Π, while the adversary can launch tamper and leakage attacks during the interaction.

Our compiler is very simple, given the ORAM technique and our new codes as building blocks. Informally speaking, given any next instruction function Π and database D, we first use ORAM technique to transform them into a next instruction function $\widetilde{\Pi}$ and a database $\widetilde{D}$. Next, we use our local non-malleable code (ENC, DEC, UPDATE) to encode $\widetilde{D}$ into $\widehat{D}$; the compiled next instruction function $\widehat{\Pi}$ does the following: run $\widetilde{\Pi}$ to compute the next "virtual" read/write instruction, and then run the local decoding or update algorithms to perform the physical memory access.

Intuitively, the inner ORAM protects leakage of the address patterns, and the outer local non-malleable codes prevent an attacker from modifying the contents of memory to some *different* but *related* value. Since at each cycle the CPU can only read and write at a small number of locations of the memory, using regular non-malleable codes does not work. Our new notion of locally decodable and updatable non-malleable codes exactly solves these issues!

1.2 Related Work

Different flavors of non-malleable codes were studied [1, 3, 7–9, 20, 22, 23, 25, 42]. We can use these constructions to secure implementations against memory attacks in the circuit model, and also as our building block for the locally decodable/updatable non-malleable codes. See also Section 3.3 for further exposition.

---

[2] There are some technical subtleties to simulate all leakage/tampering attacks on the values passing the bus using memory attacks (and addresses). We defer the rigorous treatment to future work.

Securing circuits or CPUs against physical attacks is an important task, but out of the scope of this paper. Some partial results can be found in previous work [12, 13, 16–19, 21, 26, 27, 30, 31, 34–36, 38, 41, 43, 44, 46–48].

In an independent and concurrent work, Faust et al. [24] also considered securing RAM computation against tampering and leakage attacks. We note that both their model and techniques differ considerably from ours. In the following, we highlight some of these differences. The main focus of [24] is constructing RAM compilers for keyed functions, denoted $\mathcal{G}_K$, to allow secure RAM emulation of these functions in the presence of leakage and tampering. In contrast, our work focuses on construcing compilers that transform any dynamic RAM machine into a RAM machine secure against leakage and tampering. Due to this different perspective, our compiler explicitly utilizes an underlying ORAM compiler, while they assume that the memory access pattern of input function $\mathcal{G}$ is independent of the secret state K (e.g., think of $\mathcal{G}$ as the circuit representation of the function). In addition to the split-state tampering and leakage attacks considered by both papers, [24] do not assume that memory can be overwritten or erased, but require the storage of a tamper-proof program counter. With regard to techniques, they use a stronger version of non-malleable codes in the split-state setting (called continual non-malleable codes [23]) for their construction. Finally, in their construction, each memory location is encoded using an expensive non-malleable encoding scheme, while in our construction, non-malleable codes are used only for a small portion of the memory, while highly efficient symmetric key authenticated encryption is used for the remainder.

2   Locally Decodable and Updatable Non-malleable Codes

In this section, we first review the concepts of non-malleable (leakage resilient) codes. Then we present our new notion that combines non-malleability, leakage resilience, and locality.

2.1   Preliminary

Definition 1 (Coding Scheme). *Let* $\Sigma, \hat{\Sigma}$ *be sets of strings, and* $\kappa, \hat{\kappa} \in \mathbb{N}$ *be some parameters. A coding scheme consists of two algorithms* (ENC, DEC) *with the following syntax:*

- *The encoding algorithm (*perhaps randomized*) takes input a block of message in* Σ *and outputs a codeword in* $\hat{\Sigma}$.
- *The decoding algorithm takes input a codeword in* $\hat{\Sigma}$ *and outputs a block of message in* Σ.

We require that for any message $m \in \Sigma$, $\Pr[\text{DEC}(\text{ENC}(m)) = m] = 1$, *where the probability is taken over the choice of the encoding algorithm. In binary settings, we often set* $\Sigma = \{0,1\}^{\kappa}$ *and* $\hat{\Sigma} = \{0,1\}^{\hat{\kappa}}$.

Definition 2 (Non-malleability [22]). *Let* k *be the security parameter,* $\mathcal{F}$ *be some family of functions. For each function* $f \in \mathcal{F}$, *and* $m \in \Sigma$, *define the tampering experiment:*

$$\textbf{Tamper}_m^f \overset{\text{def}}{=} \left\{ \begin{array}{c} c \leftarrow \text{ENC}(m), \tilde{c} := f(c), \tilde{m} := \text{DEC}(\tilde{c}). \\ Output : \tilde{m}. \end{array} \right\},$$

where the randomness of the experiment comes from the encoding algorithm. We say a coding scheme (ENC, DEC) *is non-malleable with respect to* $\mathcal{F}$ *if for each* $f \in \mathcal{F}$, *there exists a* PPT *simulator* $\mathcal{S}$ *such that for any message* $m \in \Sigma$, *we have*

$$\textbf{Tamper}_m^f \approx \textbf{Ideal}_{\mathcal{S},m} \overset{\text{def}}{=} \left\{ \begin{array}{c} \tilde{m} \cup \{\textsf{same}^*\} \leftarrow \mathcal{S}^{f(\cdot)}. \\ Output : m \; if \; \textsf{same}^*; \; otherwise \; \tilde{m}. \end{array} \right\}$$

Here the indistinguishability can be either computational or statistical.

We can extend the notion of non-malleability to leakage resilience (simultaneously) as the work of Liu and Lysyanskaya [42].

Definition 3 (Non-malleability and Leakage Resilience [42]). *Let* k *be the security parameter,* $\mathcal{F}, \mathcal{G}$ *be some families of functions. For each function* $f \in \mathcal{F}$, $g \in \mathcal{G}$, *and* $m \in \Sigma$, *define the tamper-leak experiment:*

$$\textbf{TamperLeak}_m^{f,g} \overset{\text{def}}{=} \left\{ \begin{array}{c} c \leftarrow \text{ENC}(m), \tilde{c} := f(c), \tilde{m} := \text{DEC}(\tilde{c}). \\ Output : (\tilde{m}, g(c)). \end{array} \right\},$$

where the randomness of the experiment comes from the encoding algorithm. We say a coding scheme (ENC, DEC) *is non-malleable and leakage resilience with respect to* $\mathcal{F}$ *and* $\mathcal{G}$ *if for any* $f \in \mathcal{F}$, $g \in \mathcal{G}$, *there exists a* PPT *simulator* $\mathcal{S}$ *such that for any message* $m \in \Sigma$, *we have*

$$\textbf{TamperLeak}_m^{f,g} \approx \textbf{Ideal}_{\mathcal{S},m} \overset{\text{def}}{=} \left\{ \begin{array}{c} (\tilde{m} \cup \{\textsf{same}^*\}, \ell) \leftarrow \mathcal{S}^{f(\cdot),g(\cdot)}. \\ Output : (m, \ell) \; if \; \textsf{same}^*; \; otherwise \; (\tilde{m}, \ell). \end{array} \right\}$$

Here the indistinguishability can be either computational or statistical.

2.2 New Definitions – Codes with Local Properties

In this section, we consider coding schemes with extra *local* properties – decodability and updatability. Intuitively, this gives a way to encode blocks of messages, such that in order to decode (retrieve) a single block of the messages, one only needs to read a small number of blocks of the codeword; similarly, in order to update a single block of the messages, one only needs to update a few blocks of the codeword.

Definition 4 (Locally Decodable and Updatable Code). *Let* $\Sigma, \hat{\Sigma}$ *be sets of strings, and* $n, \hat{n}, p, q$ *be some parameters. An* $(n, \hat{n}, p, q)$ *locally decodable and updatable coding scheme consists of three algorithms* (ENC, DEC, UPDATE) *with the following syntax:*

- *The encoding algorithm* ENC *(perhaps randomized) takes input an n-block (in Σ) message and outputs an $\hat{n}$-block (in $\hat{\Sigma}$) codeword.*
- *The (local) decoding algorithm* DEC *takes input an index in $[n]$, reads at most p blocks of the codeword, and outputs a block of message in Σ. The overall decoding algorithm simply outputs $(\mathrm{DEC}(1), \mathrm{DEC}(2), \ldots, \mathrm{DEC}(n))$.*
- *The (local) updating algorithm* UPDATE *(perhaps randomized) takes inputs an index in $[n]$ and a string in $\Sigma \cup \{\epsilon\}$, and reads/writes at most q blocks of the codeword. Here the string ϵ denotes the procedure of refreshing without changing anything.*

Let $C \in \hat{\Sigma}^{\hat{n}}$ be a codeword. For convenience, we denote DEC^C, UPDATE^C as the processes of reading/writing individual block of the codeword, i.e. the codeword oracle returns or modifies individual block upon a query. Here we view C as a random access memory where the algorithms can read/write to the memory C at individual different locations.

Remark 1. Throughout this paper, we only consider non-adaptive decoding and updating, which means the algorithms DEC and UPDATE compute all their queries at the same time before seeing the answers, and the computation only depends on the input i (the location). In contrast, an adaptive algorithm can compute a query based on the answer from previous queries. After learning the answer to such query, then it can make another query. We leave it as an interesting open question to construct more efficient schemes using adaptive queries.

Then we define the requirements of the coding scheme.

Definition 5 (Correctness). *An $(n, \hat{n}, p, q)$ locally decodable and updatable coding scheme (with respect to $\Sigma, \hat{\Sigma}$) satisfies the following properties. For any message $M = (m_1, m_2, \ldots, m_n) \in \Sigma^n$, let $C = (c_1, c_2, \ldots, c_{\hat{n}}) \leftarrow \mathrm{ENC}(M)$ be a codeword output by the encoding algorithm. Then we have:*

- *for any index $i \in [n]$, $\Pr[\mathrm{DEC}^C(i) = m_i] = 1$, where the probability is over the randomness of the encoding algorithm.*
- *for any update procedure with input $(j, m') \in [n] \times \Sigma \cup \{\epsilon\}$, let C' be the resulting codeword by running $\mathrm{UPDATE}^C(j, m')$. Then we have $\Pr[\mathrm{DEC}^{C'}(j) = m'] = 1$, where the probability is over the encoding and update procedures. Moreover, the decodings of the other positions remain unchanged.*

Remark 2. The correctness definition can be directly extended to handle any sequence of updates.

Next, we define several flavors of security about non-malleability and leakage resilience.

One-time Non-malleability. First we consider one-time non-malleability of locally decodable codes, i.e., the adversary only tampers with the codeword once. This extends the idea of the non-malleable codes (as in Definition 2). As discussed in the introduction, we present the following definition to capture the idea that the tampering function can only do either of the following cases:

- It destroys a block (or blocks) of the underlying messages while keeping the other blocks unchanged, or
- If it modifies a block of the underlying messages to some unrelated string, then it must have modified all blocks of the underlying messages to encodings of unrelated messages.

Definition 6 (Non-malleability of Locally Decodable Codes). *An $(n, \hat{n}, p, q)$-locally decodable coding scheme with respect to $\Sigma, \hat{\Sigma}$ is non-malleable against the tampering function class $\mathcal{F}$ if for all $f \in \mathcal{F}$, there exists some simulator S such that for any $M = (m_1, \ldots, m_n) \in \Sigma^n$, the experiment $\mathbf{Tamper}_M^f$ is (computationally) indistinguishable to the following ideal experiment $\mathbf{Ideal}_{S,M}$:*

- $(\mathcal{I}, \boldsymbol{m}^*) \leftarrow S(1^k)$, *where $\mathcal{I} \subseteq [n]$, $\boldsymbol{m}' \in \Sigma^n$. (Intuitively $\mathcal{I}$ means the coordinates of the underlying message that have been tampered with).*
- *If $\mathcal{I} = [n]$, define $\boldsymbol{m} = \boldsymbol{m}^*$; otherwise set $\boldsymbol{m}|_{\mathcal{I}} = \bot, \boldsymbol{m}|_{\bar{\mathcal{I}}} = M|_{\bar{\mathcal{I}}}$, where $\boldsymbol{x}|_{\mathcal{I}}$ denotes the coordinates $\boldsymbol{x}[v]$ where $v \in \mathcal{I}$, and the bar denotes the complement of a set.*
- *The experiment outputs $\boldsymbol{m}$.*

Remark 3. Here we make two remarks about the definition:

1. In the one-time security definition, we do not consider the update procedure. In the next paragraph when we define continual attacks, we will handle the update procedure explicitly.
2. One-time leakage resilience of locally decodable codes can be defined in the same way as Definition 3.

Security against Continual Attacks. In the following, we extend the security to handle continual attacks. Here we consider a third party called *updater*, who can read the underlying messages and decide how to update the codeword. Our model allows the adversary to learn the location that the updater updated the messages, so we also allow the simulator to learn this information. This is without loss of generality if the leakage class $\mathcal{G}$ allows it, i.e. the adversary can query some $g \in \mathcal{G}$ to figure out what location was modified. On the other hand, the updater does not tell the adversary what content was encoded of the updated messages, so the simulator needs to simulate the view without such information. We can think of the updater as an honest user interacting with the codeword (read/write). The security intuitively means that even if the adversary can launch tampering and leakage attacks when the updater is interacting with the codeword, the adversary cannot learn anything about the underlying encoded messages (or the updated messages during the interaction).

Our continual experiment consists of rounds: in each round the adversary can tamper with the codeword and get partial information. At the end of each round, the updater will run UPDATE, and the codeword will be somewhat updated and refreshed. We note that if there is no refreshing procedure, then no coding scheme can be secure against continual leakage attack even for one-bit leakage at a time[3],

---

[3] If there is no refreshing procedure, then the adversary can eventually learn the whole codeword bit-by-bit by leakage. Thus he can learn the underlying message.

so this property is necessary. Our concept of "continuity" is different from that of Faust et al. [23], who considered continual attacks on the same original codeword (the tampering functions can be chosen adaptively). Our model does not allow this type of "resetting attacks." Once a codeword has been modified to $f(C)$, the next tampering function will be applied on $f(C)$.

We remark that the one-time security can be easily extended to the continual case (using a standard hybrid argument) if the update procedure re-encodes the *whole* underlying messages (c.f. see the results in the work [42]). However, in the setting above, we emphasize on the *local property*, so this approach does not work. How to do a local update while maintaining tamper and leakage resilience makes the continual case challenging!

Definition 7 (Continual Tampering and Leakage Experiment). *Let k be the security parameter, $\mathcal{F}, \mathcal{G}$ be some families of functions. Let $(\text{ENC}, \text{DEC}, \text{UPDATE})$ be an $(n, \hat{n}, p, q)$-locally decodable and updatable coding scheme with respect to $\Sigma, \hat{\Sigma}$. Let $\mathcal{U}$ be an updater that takes input a message $M \in \Sigma^n$ and outputs an index $i \in [n]$ and $m \in \Sigma$. Then for any blocks of messages $M = (m_1, m_2, \ldots, m_n) \in \Sigma^n$, and any (non-uniform) adversary $\mathcal{A}$, any updater $\mathcal{U}$, define the following continual experiment $\mathbf{CTamperLeak}_{\mathcal{A},\mathcal{U},M}$:*

- *The challenger first computes an initial encoding $C^{(1)} \leftarrow \text{ENC}(M)$.*
- *Then the following procedure repeats, at each round j, let $C^{(j)}$ be the current codeword and $M^{(j)}$ be the underlying message:*
  - *$\mathcal{A}$ sends either a tampering function $f \in \mathcal{F}$ and/or a leakage function $g \in \mathcal{G}$ to the challenger.*
  - *The challenger replaces the codeword with $f(C^{(j)})$, or sends back a leakage $\ell^{(j)} = g(C^{(j)})$.*
  - *We define $\boldsymbol{m}^{(j)} \stackrel{\text{def}}{=} \left(\text{DEC}^{f(C^{(j)})}(1), \ldots, \text{DEC}^{f(C^{(j)})}(n) \right)$.*
  - *Then the updater computes $(i^{(j)}, m) \leftarrow \mathcal{U}(\boldsymbol{m}^{(j)})$ for the challenger.*
  - *Then the challenger runs $\text{UPDATE}^{f(C^{(j)})}(i^{(j)}, m)$ and sends the index $i^{(j)}$ to $\mathcal{A}$.*
  - *$\mathcal{A}$ may terminate the procedure at any point.*
- *Let t be the total number of rounds above. At the end, the experiment outputs*

$$\left(\ell^{(1)}, \ell^{(2)}, \ldots, \ell^{(t)}, \boldsymbol{m}^{(1)}, \ldots, \boldsymbol{m}^{(t)}, i^{(1)}, \ldots, i^{(t)} \right).$$

Definition 8 (Non-malleability and Leakage Resilience against Continual Attacks). *An $(n, \hat{n}, p, q)$-locally decodable and updatable coding scheme with respect to $\Sigma, \hat{\Sigma}$ is continual non-malleable against $\mathcal{F}$ and leakage resilient against $\mathcal{G}$ if for all PPT (non-uniform) adversaries $\mathcal{A}$, and PPT updaters $\mathcal{U}$, there exists some PPT (non-uniform) simulator $\mathcal{S}$ such that for any $M = (m_1, \ldots, m_n) \in \Sigma^n$, $\mathbf{CTamperLeak}_{\mathcal{A},\mathcal{U},M}$ is (computationally) indistinguishable to the following ideal experiment $\mathbf{Ideal}_{\mathcal{S},\mathcal{U},M}$:*

- *The experiment proceeds in rounds. Let $M^{(1)} = M$ be the initial message.*
- *At each round j, the experiment runs the following procedure:*

- *At the beginning of each round, $\mathcal{S}$ outputs $(\ell^{(j)}, \mathcal{I}^{(j)}, \boldsymbol{w}^{(j)})$, where $\mathcal{I}^{(j)} \subseteq [n]$.*
- *Define*

$$\boldsymbol{m}^{(j)} = \begin{cases} \boldsymbol{w}^{(j)} & \text{if } \mathcal{I}^{(j)} = [n] \\ \boldsymbol{m}^{(j)}|_{\mathcal{I}^{(j)}} := \perp, \boldsymbol{m}^{(j)}|_{\bar{\mathcal{I}}^{(j)}} := M^{(j)}|_{\bar{\mathcal{I}}^{(j)}} & \text{otherwise,} \end{cases}$$

where $\boldsymbol{x}|_{\mathcal{I}}$ denotes the coordinates $\boldsymbol{x}[v]$ where $v \in \mathcal{I}$, and the bar denotes the complement of a set.
- *The updater runs $(i^{(j)}, m) \leftarrow \mathcal{U}(\boldsymbol{m}^{(j)})$ and sends the index $i^{(j)}$ to the simulator. Then the experiment updates $M^{(j+1)}$ as follows: set $M^{(j+1)} := M^{(j)}$ for all coordinates except $i^{(j)}$, and set $M^{(j+1)}[i^{(j)}] := m$.*
- Let t be the total number of rounds above. At the end, the experiment outputs

$$\left(\ell^{(1)}, \ell^{(2)}, \ldots, \ell^{(t)}, \boldsymbol{m}^{(1)}, \ldots, \boldsymbol{m}^{(t)}, i^{(1)}, \ldots, i^{(t)} \right).$$

3 Our Constructions

In this section, we present two constructions. As a warm-up, we first present a construction that is one-time secure to demonstrate the idea of achieving non-malleability, local decodability and updatability simultaneously. Then in the next section, we show how to make the construction secure against continual attacks. Due to space limit, please find in the full version of the paper [14] the security proofs.

3.1 A First Attempt – One-time Security

Construction. Let $\mathcal{E} = (\text{Gen}, \text{Encrypt}, \text{Decrypt})$ be a symmetric encryption scheme, NMC = (ENC, DEC) be a coding scheme. Then we consider the following coding scheme:

- ENC(M): on input $M = (m_1, m_2, \ldots, m_n)$, the algorithm first generates the encryption key $\text{sk} \leftarrow \mathcal{E}.\text{Gen}(1^k)$. Then it computes $c \leftarrow \text{NMC.ENC}(\text{sk})$, $e_i \leftarrow \mathcal{E}.\text{Encrypt}_{\text{sk}}(m_i, i)$ for $i \in [n]$. The algorithm finally outputs a codeword $C = (c, e_1, e_2, \ldots, e_n)$.
- DEC^C(i): on input $i \in [n]$, the algorithm reads the first block and the $(i+1)$-st block of the codeword to retrieve (c, e_i). Then it runs $\text{sk} := \text{NMC.DEC}(c)$. If the decoding algorithm outputs $\perp$, then it outputs $\perp$ and terminates. Else, it computes $(m_i, i^*) = \mathcal{E}.\text{Decrypt}_{\text{sk}}(e_i)$. If $i^* \neq i$, or the decryption fails, the algorithm outputs $\perp$. If all the above verifications pass, the algorithm outputs m_i.
- UPDATE(i, m'): on inputs an index $i \in [n]$, a block of message $m' \in \Sigma$, the algorithm runs DEC^C(i) to retrieve (c, e_i) and (sk, m_i, i). If the decoding algorithm returns $\perp$, the algorithm writes $\perp$ to the first block and the $(i+1)$-st block. Otherwise, it computes a fresh encoding $c' \leftarrow \text{NMC.ENC}(\text{sk})$, and a fresh ciphertext $e'_i \leftarrow \mathcal{E}.\text{Encrypt}_{\text{sk}}(m', i)$. Then it writes back the first block and the $(i + 1)$-st block with (c', e'_i).

To analyze the coding scheme, we make the following assumptions of the parameters in the underlying scheme for convenience:

1.   The size of the encryption key is k (security parameter), i.e. $|\mathsf{sk}| = k$.

2.   Let Σ be a set, and the encryption scheme supports messages of length $|\Sigma| + \log n$. The ciphertexts are in the space $\hat{\Sigma}$.

3.   The length of $|\mathsf{NMC.ENC}(\mathsf{sk})|$ is less than $|\hat{\Sigma}|$.

Then clearly, the above coding scheme is an $(n, n + 1, 2, 2)$-locally updatable and decodable code with respect to $\Sigma, \hat{\Sigma}$. The correctness of the scheme is obvious by inspection. The rate (ratio of the length of messages to that of codewords) of the coding scheme is $1 - o(1)$.

Theorem 1. *Assume $\mathcal{E}$ is a symmetric authenticated encryption scheme, and NMC is a non-malleable code against the tampering function class $\mathcal{F}$. Then the coding scheme presented above is one-time non-malleable against the tampering class*

$$\bar{\mathcal{F}} \overset{\text{def}}{=} \left\{ \begin{array}{l} f : \hat{\Sigma}^{n+1} \to \hat{\Sigma}^{n+1} \text{ and } |f| \leq \text{poly}(k), \text{ such that :} \\ f = (f_1, f_2), \ f_1 : \hat{\Sigma}^{n+1} \to \hat{\Sigma}, \ f_2 : \hat{\Sigma}^n \to \hat{\Sigma}^n, \\ \forall (x_2, \ldots, x_{n+1}) \in \hat{\Sigma}^n, f_1(\cdot, x_2, \ldots, x_{n+1}) \in \mathcal{F} \\ f(x_1, x_2, \ldots, x_{n+1}) = (f_1(x_1, x_2, \ldots, x_{n+1}), f_2(x_2, \ldots, x_{n+1})) . \end{array} \right\} .$$

We have presented the intuition in the introduction. Before giving the detailed proof, we make the following remark.

Remark 4. The function class $\bar{\mathcal{F}}$ may look complex, yet the intuition is simple. The tampering function restricted in the first block (the underlying non-malleable code) falls into the class $\mathcal{F}$ – this is captured by $f_1 \in \mathcal{F}$; on the other hand, we just require the function restricted in the rest of the blocks to be polynomial-sized – this is captured by $|f_2| \leq |f| \leq \text{poly}(k)$.

For our construction, it is inherent that the function f_2 cannot depend on x_1 arbitrarily. Suppose this is not the case, then f_2 can first decode the non-malleable code, encrypt the decoded value and write the ciphertext into x_2, which breaks non-malleability. However, if the underlying coding scheme is non-malleable and also leakage resilient to $\mathcal{G}$, then we can allow f_2 to get additional information $g(x_1)$ for any $g \in \mathcal{G}$. Moreover, the above construction is one-time leakage resilient.

We present the above simpler version for clarity of exposition, and give this remark that our construction actually achieves security against a broader class of tampering attacks.

3.2   Achieving Security against Continual Attacks

As discussed in the introduction, the above construction is not secure if continual tampering and leakage is allowed – the adversary can use a rewind attack to modify the underlying message to some old/related messages. We handle this challenge using a technique of Merkle tree, which preserves local properties of the above scheme. We present the construction in the following:

Definition 9 (Merkle Tree). *Let* $h : \mathcal{X} \times \mathcal{X} \to \mathcal{X}$ *be a hash function that maps two blocks of messages to one.*[4] *A Merkle Tree* $\mathsf{Tree}_h(M)$ *takes input a message* $M = (m_1, m_2, \ldots, m_n) \in \mathcal{X}^n$. *Then it applies the hash on each pair* (m_{2i-1}, m_{2i}), *and resulting in* $n/2$ *blocks. Then again, it partitions the blocks into pairs and applies the hash on the pairs, which results in* $n/4$ *blocks. This is repeated* $\log n$ *times, resulting a binary tree with hash values, until one block remains. We call this value the root of Merkle Tree denoted* $\mathsf{Root}_h(M)$, *and the internal nodes (including the root) as* $\mathsf{Tree}_h(M)$. *Here* M *can be viewed as leaves.*

Theorem 2. *Assuming* h *is a collision resistant hash function. Then for any message* $M = (m_1, m_2, \ldots, m_n) \in \mathcal{X}^n$ *and any polynomial time adversary* $\mathcal{A}$,
$$\Pr\left[(m_i', p_i) \leftarrow \mathcal{A}(M, h) : m_i' \neq m_i, p_i \text{ is a consistent path with } \mathsf{Root}_h(M)\right] \leq \mathsf{negl}(k).$$

Moreover, given a path p_i *passing the leaf* m_i, *and a new value* m_i', *there is an algorithm that computes* $\mathsf{Root}_h(M')$ *in time* $\mathsf{poly}(\log n, k)$, *where* $M' = (m_1, \ldots, m_{i-1}, m_i', m_{i+1}, \ldots, m_n)$.

Construction. Let $\mathcal{E} = (\mathsf{Gen}, \mathsf{Encrypt}, \mathsf{Decrypt})$ be a symmetric encryption scheme, $\mathrm{NMC} = (\mathrm{ENC}, \mathrm{DEC})$ be a non-malleable code, H is a family of collision resistance hash functions. Then we consider the following coding scheme:

- $\mathrm{ENC}(M)$: on input $M = (m_1, m_2, \ldots, m_n)$, the algorithm first generates encryption key $\mathsf{sk} \leftarrow \mathcal{E}.\mathsf{Gen}(1^k)$ and $h \leftarrow H$. Then it computes $e_i \leftarrow \mathcal{E}.\mathsf{Encrypt}_{\mathsf{sk}}(m_i)$ for $i \in [n]$, and $T = \mathsf{Tree}_h(e_1, \ldots, e_n)$, $R = \mathsf{Root}_h(e_1, \ldots, e_n)$. Then it computes $c \leftarrow \mathrm{NMC}.\mathrm{ENC}(\mathsf{sk}, R, h)$, The algorithm finally outputs a codeword $C = (c, e_1, e_2, \ldots, e_n, T)$.

- $\mathrm{DEC}^C(i)$: on input $i \in [n]$, the algorithm reads the first block, the $(i + 1)$-st block, and a path p in the tree (from the root to the leaf i), and it retrieve (c, e_i, p). Then it runs $(\mathsf{sk}, R, h) = \mathrm{NMC}.\mathrm{DEC}(c)$. If the decoding algorithm outputs $\perp$, or the path is not consistent with the root R, then it outputs $\perp$ and terminates. Else, it computes $m_i = \mathcal{E}.\mathsf{Decrypt}_{\mathsf{sk}}(e_i)$. If the decryption fails, output $\perp$. If all the above verifications pass, the algorithm outputs m_i.

- $\mathrm{UPDATE}(i, m')$: on inputs an index $i \in [n]$, a block of message $m' \in \Sigma$, the algorithm runs $\mathrm{DEC}^C(i)$ to retrieve (c, e_i, p). Then the algorithm can derive $(\mathsf{sk}, R, h) = \mathrm{NMC}.\mathrm{DEC}(c)$. If the decoding algorithm returns $\perp$, the update writes $\perp$ to the first block, which denotes failure. Otherwise, it computes a fresh ciphertext $e_i' \leftarrow \mathcal{E}.\mathsf{Encrypt}_{\mathsf{sk}}(m')$, a new path p' (that replaces e_i by e_i') and a new root R', which is consistent with the new leaf value e_i'. (Note that this can be done given only the old path p as Theorem 2.) Finally, it computes a fresh encoding $c' \leftarrow \mathrm{NMC}.\mathrm{ENC}(\mathsf{sk}, R', h)$. Then it writes back the first block, the $(i + 1)$-st block, and the new path blocks with (c', e_i', p').

To analyze the coding scheme, we make the following assumptions of the parameters in the underlying scheme for convenience:

1. The size of the encryption key is k (security parameter), i.e. $|\mathsf{sk}| = k$ and the length of the output of the hash function is k.

---

[4] Here we assume $|\mathcal{X}|$ is greater than the security parameter.

2.   Let Σ be a set, and the encryption scheme supports messages of length $|\Sigma|$. The ciphertexts are in the space $\hat{\Sigma}$.

3.   The length of $|\mathsf{NMC.ENC}(\mathsf{sk}, v)|$ is less than $|\hat{\Sigma}|$, where $|v| = k$.

Clearly, the above coding scheme is an $(n, 2n + 1, O(\log n), O(\log n))$-locally updatable and decodable code with respect to $\Sigma, \hat{\Sigma}$. The correctness of the scheme is obvious by inspection. The rate (ratio of the length of messages to that of codewords) of the coding scheme is $1/2 - o(1)$.

Theorem 3. *Assume $\mathcal{E}$ is a semantically secure symmetric encryption scheme, and NMC is a non-malleable code against the tampering function class $\mathcal{F}$, and leakage resilient against the function class $\mathcal{G}$. Then the coding scheme presented above is non-malleable against continual attacks of the tampering class*

$$\bar{\mathcal{F}} \overset{\text{def}}{=} \left\{ \begin{array}{l} f : \hat{\Sigma}^{2n+1} \to \hat{\Sigma}^{2n+1} \text{ and } |f| \leq \mathrm{poly}(k), \text{ such that :} \\ f = (f_1, f_2), \ f_1 : \hat{\Sigma}^{2n+1} \to \hat{\Sigma}, \ f_2 : \hat{\Sigma}^{2n} \to \hat{\Sigma}^{2n}, \\ \forall (x_2, \ldots, x_{2n+1}) \in \hat{\Sigma}^n, f_1(\ \cdot\ , x_2, \ldots, x_{2n+1}) \in \mathcal{F}, \\ f(x_1, x_2, \ldots, x_{2n+1}) = (f_1(x_1, x_2, \ldots, x_{2n+1}), f_2(x_2, \ldots, x_{2n+1})). \end{array} \right\},$$

and is leakage resilient against the class

$$\bar{\mathcal{G}} \overset{\text{def}}{=} \left\{ \begin{array}{l} g : \hat{\Sigma}^{2n+1} \to \mathcal{Y} \text{ and } |g| \leq \mathrm{poly}(k), \text{ such that :} \\ g = (g_1, g_2), \ g_1 : \hat{\Sigma}^{2n+1} \to \mathcal{Y}', \ g_2 : \hat{\Sigma}^{2n} \to \hat{\Sigma}^{2n}, \\ \forall \ (x_2, \ldots, x_{2n+1}) \in \hat{\Sigma}^n, g_1(\ \cdot\ , x_2, \ldots, x_{2n+1}) \in \mathcal{G}. \end{array} \right\}.$$

The intuition of this construction can be found in the introduction. Before giving the detailed proof, we make a remark.

Remark 5. Actually our construction is secure against a broader class of tampering functions. The f_2 part can depend on $g'(x_1)$ as long as the function $g'(\cdot)$ together with the leakage function $g_1(\cdot, x_2, \ldots, x_{2n+1})$ belong to $\mathcal{G}$. That is, the tampering function $f = (f_1, f_2, g')$ and the leakage function $g = (g_1, g_2)$ satisfy the constraint $g'(\cdot) \circ g_1(\cdot, x_2, \ldots, x_{2n+1}) \in \mathcal{G}$ (Here we use $\circ$ to denote concatenation). For presentation clarity, we choose to describe the simpler but slightly smaller class of functions.

3.3 Instantiations

In this section, we describe several constructions of non-malleable codes against different classes of tampering/leakage functions. To our knowledge, we can use the explicit constructions (of the non-malleable codes) in the work [1, 3, 9, 22, 23, 25, 42].

First we overview different classes of tampering/leakage function allowed for these results: the constructions of [22] work for bit-wise tampering functions, and split-state functions in the random oracle model. The construction of Choi et al. [9] works for small block tampering functions. The construction of Liu and Lysyanskaya [42] achieves both tamper and leakage resilience against split-state functions in the common reference string (CRS) model. The construction of

Dziembowski et al. [20] achieves information theoretic security against split-state tampering functions, but their scheme can only support encoding for bits, so it cannot be used in our construction. The subsequent construction by Aggarwal et al. [1] achieves information theoretic security against split-state tampering without CRS. We believe that their construction also achieves leakage resilience against some length bounded split-state leakage yet their paper did not claim it. The construction by Faust et al. [25] is non-malleable against small-sized tampering functions. Another construction by Faust et al. [23] achieves both tamper and leakage resilience in the split-state model with CRS. The construction of Aggarwal et al. [3] is non-malleable against permutation functions.

Then we remark that actually there are other non-explicit constructions: Cheraghchi and Guruswami [8] showed the relation non-malleable codes and non-malleable two source extractors (but constructing a non-malleable two-source extractor is still open), and in another work Cheraghchi and Guruswami [7] showed the existence of high rate non-malleable codes in the split-state model but did not give an explicit (efficient) construction.

Finally, we give a concrete example of what the resulting class looks like using the construction of Liu and Lysyanskaya [42] as building block. Recall that their construction achieves both tamper and leakage resilience for split-state functions. Our construction has the form:

$$(\mathsf{NMC.ENC}(\mathsf{sk}, h, T), \mathsf{Encrypt}(m_1), \ldots, \mathsf{Encrypt}(m_n), T)$$

So the overall leakage function g restricted in the first block (i.e. g_1) can be a (poly-sized) length-bounded split-state function; g on the other hand, can leak all the other parts. For the tampering, the overall tampering function f restricted in the first block (i.e. f_1) can be any (poly-sized) split-state function. On the other hand f restricted in the rest (i.e. f_2) can be just any poly-sized function. We also remark that f_2 can depend on a split-state leakage on the first part, say g_1, as we discussed in the previous remark above.

4   Tamper and Leakage Resilient RAM

In this section, we first introduce the notations of the Random Access Machine (RAM) model of computation in presence of tampering and leakage attacks in Section 4.1. Then we define the security of tamper and leakage resilient RAM model of computation in Section 4.2, and then give a construction in Section 4.3. Due to space limit, please find in the full version of the paper [14], the building block Oblivious RAM (ORAM), more detailed construction, and the security analysis.

4.1   Random Access Machines

We consider RAM programs to be interactive stateful systems $\langle \Pi, \mathsf{state}, D \rangle$, where Π denotes a next instruction function, state the current state stored in registers, and D the content of memory. Upon state and an input value d, the next instruction function outputs the next instruction I and an updated state state'.

The initial state of the RAM machine, state, is set to (start, *). For simplicity we often denote RAM program as $\langle \Pi, D \rangle$. We consider four ways of interacting with the system:

- Execute(x): A user can provide the system with Execute(x) queries, for $x \in \{0, 1\}^u$, where u is the input length. Upon receiving such query, the system computes $(y, t, D') \leftarrow \langle \Pi, D \rangle(x)$, updates the state of the system to $D := D'$ and outputs (y, t), where y denotes the output of the computation and t denotes the time (or number of executed instructions). By Execute$_1(x)$ we denote the first coordinate of the output of Execute(x).

- doNext(x): A user can provide the system with doNext(x) queries, for $x \in \{0, 1\}^u$. Upon receiving such query, if state = (start, *), set state := (start, x), and $d := 0^r$; Here $\rho = |\text{state}|$ and $r = |d|$. The system does the following until termination:

  1. Compute $(I, \text{state}') = \Pi(\text{state}, d)$. Set state := state$'$.
  2. If $I = (\text{wait})$ then set state := 0^ρ, $d := 0^r$ and terminate.
  3. If $I = (\text{stop}, z)$ then set state := (start, *), $d := 0^r$ and terminate with output z.
  4. If $I = (\text{write}, v, d')$ then set $D[v] := d'$.
  5. If $I = (\text{read}, v, \perp)$ then set $d := D[v]$.

  Let $I_1, \ldots, I_\ell$ be the instructions executed by doNext(x). All memory addresses of executed instructions are returned to the user. Specifically, for instructions I_j of the form $(\text{read}, v, \perp)$ or (write, v, d'), v is returned.

- Tamper(f): We also consider tampering attacks against the system, modeled by Tamper(f) commands, for functions f. Upon receiving such command, the system sets $D := f(D)$.

- Leak(g): We finally consider leakage attacks against the system, modeled by Leak(g) commands, for functions g. Upon receiving such command, the value of $g(D)$ is returned to the user.

Remark 6. A doNext(x) instruction groups together instructions performed by the CPU in a single clock cycle. Intuitively, a (wait) instruction indicates that a clock cycle has ended and the CPU waits for the adversary to increment the clock. In contrast, a (stop, z) instruction indicates that the entire execution has concluded with output z. In this case, the internal state is set back to the start state.

We require that each doNext(x) operation performs exactly $\ell = \ell(k) = \text{poly}(k)$ instructions $I_1, \ldots, I_\ell$ where: The final instruction is of the form $I_\ell = (\text{stop}, \cdot)$ or $I_\ell = (\text{wait})$. For fixed $\ell_1 = \ell_1(k), \ell_2 = \ell_2(k)$ such that $\ell_1 + \ell_2 = \ell - 1$, we have that the first ℓ_1 instructions are of the form $I_\ell = (\text{read}, \cdot, \perp)$ and the next ℓ_2 instructions are of the form $I_\ell = (\text{write}, v, d')$. We assume that ℓ, ℓ_1, ℓ_2 are implementation-specific and public. The limitations on space are meant to model the fact that the CPU has a limited number of registers and that no persistent state is kept by the CPU between clock cycles.

Remark 7. We note that Execute(x) instructions are used by the ideal world adversary—who learns only the input-output behavior of the RAM machine and the run time—as well as by the real world adversary. The real world adversary may also use the more fine-grained $\mathsf{doNext}(x)$ instruction. We note that given access to the $\mathsf{doNext}(x)$ instruction, the behavior of the $\mathsf{Execute}(x)$ instruction may be simulated.

Remark 8. We note that our model does not explicitly allow for leakage and tampering on instructions I. E.g. when an instruction $I = (\mathsf{write}, v, d')$ is executed, we do not directly allow tampering with the values v, d' or leakage on d' (note that v is entirely leaked to the adversary). Nevertheless, as discussed in the introduction, since we allow full leakage on the addresses, the adversary can use the tampering and leakage attacks on the memory to capture the attacks on the instructions. We defer a rigorous treatment and analysis of such attacks to future work. In this work, for simplicity of presentation we assume these instructions are not subject to direct attacks.

4.2   Tamper and Leakage-Resilient (TLR) RAM

A tamper and leakage resilient (TLR) RAM compiler consists of two algorithms (CompMem, CompNext), which transform a RAM program $\langle \Pi, D \rangle$ into another program $\langle \widehat{\Pi}, \widehat{D} \rangle$ as follows: On input database D, CompMem initializes the memory and internal state of the compiled machine, and generates the transformed database $\widehat{D}$; On input next instruction function Π, CompNext generates the next instruction function of the compiled machine.

Definition 10. *A TLR compiler* (CompMem, CompNext) *is tamper and leakage simulatable w.r.t. function families* $\mathcal{F}, \mathcal{G}$, *if for every RAM next instruction function* Π, *and for any* PPT *(non-uniform) adversary* $\mathcal{A}$ *there exists a* PPT *(non-uniform) simulator* $\mathcal{S}$ *such that for any initial database* $D \in \{0,1\}^{\mathrm{poly}(k)}$ *we have*

$$\mathbf{TamperExec}(\mathcal{A}, \mathcal{F}, \mathcal{G}, \langle \mathsf{CompNext}(\Pi), \mathsf{CompMem}(D) \rangle) \approx \mathbf{IdealExec}(\mathcal{S}, \langle \Pi, D \rangle)$$

where **TamperExec** *and* **IdealExec** *are defined as follows:*

- **TamperExec**$(\mathcal{A}, \mathcal{F}, \mathcal{G}, \langle \mathsf{CompNext}(\Pi), \mathsf{CompMem}(D) \rangle)$: *The adversary* $\mathcal{A}$ *interacts with the system* $\langle \mathsf{CompNext}(\Pi), \mathsf{CompMem}(D) \rangle$ *for arbitrarily many rounds of interactions where, in each round:*

  1. *The adversary can "tamper" by executing a* $\mathsf{Tamper}(f)$ *command against the system, for some* $f \in \mathcal{F}$.
  2. *The adversary can "leak" by executing a* $\mathsf{Leak}(g)$ *command against the system, and receiving* $g(D)$ *in return.*
  3. *The adversary requests a* $\mathsf{doNext}(x)$ *command to be executed by the system. Let* $I_1, \ldots, I_\ell$ *be the instructions executed by* $\mathsf{doNext}(x)$. *If* I_ℓ *is of the form* (stop, z) *then output* z *is returned to the adversary. Moreover, all memory addresses corresponding to instructions* $I_1, \ldots, I_{\ell-1}$ *are returned to the adversary.*

The output of the game consists of the output of the adversary $\mathcal{A}$ at the end of the interaction, along with (1) all input-output pairs $(x_1, y_1), (x_2, y_2), \ldots,$ (2) all responses to leakage queries $\ell_1, \ell_2, \ldots$ (3) all outputs of $\mathsf{doNext}(x_1),$ $\mathsf{doNext}(x_2), \ldots.$

- **IdealExec$(\mathcal{S}, \langle \Pi, D \rangle)$:** *The simulator interacts with the system $\langle \Pi, D \rangle$ for arbitrarily many rounds of interaction where, in each round, it runs an $\mathsf{Execute}(x)$ query for some $x \in \{0, 1\}^u$ and receives output (y, t). The output of the game consists of the output of the simulator $\mathcal{S}$ at the end of the interaction, along with all of the execute-query inputs and outputs.*

For simplicity of exposition, we assume henceforth that the next instruction function Π to be compiled is the universal RAM next instruction function. In other words, we assume that the program to be executed is stored in the initial database D.

4.3   TLR-RAM Construction

Here we first give a high-level description of our construction and then state our theorem. More detailed construction and the security proof will be given in the full version of the paper [14].

High-level Description of Construction Let D be the initial database and let $\mathsf{ORAM} = (\mathsf{oCompMem}, \mathsf{oCompNext})$ be an ORAM compiler. Let $\mathsf{NMCode} = (\mathrm{ENC}, \mathrm{DEC}, \mathrm{UPDATE})$ be a locally decodable and updatable code. We present the following construction $\mathsf{TLR\text{-}RAM} = (\mathsf{CompMem}, \mathsf{CompNext})$ of a tamper and leakage resilient RAM compiler. In order to make our presentation more intuitive, instead of specifying the next message function $\mathsf{CompNext}(\Pi)$, we specify the pseudocode for the $\mathsf{doNext}(x)$ instruction of the compiled machine. We note that $\mathsf{CompNext}(\Pi)$ is implicitly defined by this description.

TLR-RAM takes as input an initial database D and a next instruction function Π and does the following:

- **CompMem:** On input security parameter k and initial database D, CompMem does:
    - Compute $\widetilde{D} \leftarrow \mathsf{oCompMem}(D)$, and output $\widehat{D} \leftarrow \mathrm{ENC}(\widetilde{D})$.
    - Initialize the ORAM state $\mathsf{state}_{\mathsf{ORAM}} := (\mathsf{start}, *)$ and $d_{\mathsf{ORAM}} := 0^r$, where $r = |d_{\mathsf{ORAM}}|$.
- **$\mathsf{doNext}(x)$:** On input x, do the following until termination:
    1. If $d_{\mathsf{ORAM}} = \bot$ then abort.
    2. Compute $(I, \mathsf{state}'_{\mathsf{ORAM}}) \leftarrow \mathsf{oCompNext}(\Pi)(\mathsf{state}_{\mathsf{ORAM}}, d_{\mathsf{ORAM}})$ and set $\mathsf{state}_{\mathsf{ORAM}} := \mathsf{state}'_{\mathsf{ORAM}}.$
    3. If $I = (\mathsf{wait})$ then set $\mathsf{state}_{\mathsf{ORAM}} := 0^\rho$ and $d_{\mathsf{ORAM}} := 0^r$ and terminate. Here $\rho = |\mathsf{state}_{\mathsf{ORAM}}|$ and $r = |d_{\mathsf{ORAM}}|$.
    4. If $I = (\mathsf{stop}, z)$ then set $\mathsf{state}_{\mathsf{ORAM}} := (\mathsf{start}, *)$, $d := 0^r$ and terminate with output z.
    5. If $I = (\mathsf{write}, v, d')$ then run $\mathrm{UPDATE}^{\widehat{D}}(v, d')$.

6. If $I = (\text{read}, v, \perp)$ then set $d_{\text{ORAM}} := \text{DEC}^{\widehat{D}}(v)$.

We are now ready to present the main theorem of this section:

Theorem 4. *Assume* ORAM = (oCompMem, oCompNext) *is an ORAM compiler which is access-pattern hiding and assume* NMCode = (ENC, DEC, UPDATE) *is a locally decodable and updatable code which is continual non-malleable against* $\mathcal{F}$ *and leakage resilient against* $\mathcal{G}$. *Then* TLR-RAM = (CompMem, CompNext) *presented above is tamper and leakage simulatable w.r.t. function families* $\mathcal{F}, \mathcal{G}$.

Acknowledgement. We thank Yevgeniy Dodis for helpful discussions. This research was funded in part by an NSF grant CNS-1314857, a subcontract from the DARPA PROCEED program, a Sloan Research Fellowship, and Google Faculty Research Awards. The views and conclusions contained herein are those of the authors and should not be interpreted as representing funding agencies.

References

1. Aggarwal, D., Dodis, Y., Lovett, S.: Non-malleable codes from additive combinatorics. In: STOC (2014), http://eprint.iacr.org/2013/201

2. Agrawal, D., Archambeault, B., Rao, J.R., Rohatgi, P.: The EM side-channel(s). In: Kaliski Jr., B.S., Koç, Ç.K., Paar, C. (eds.) CHES 2002. LNCS, vol. 2523, pp. 29–45. Springer, Heidelberg (2003)

3. Agrawal, S., Gupta, D., Maji, H.K., Pandey, O., Prabhakaran, M.: Explicit non-malleable codes resistant to permutations. In: Cryptology ePrint Archive, Report 2014/316 (2014)

4. Biham, E., Shamir, A.: Differential fault analysis of secret key cryptosystems. In: Kaliski Jr., B.S. (ed.) CRYPTO 1997. LNCS, vol. 1294, pp. 513–525. Springer, Heidelberg (1997)

5. Boneh, D., DeMillo, R.A., Lipton, R.J.: On the importance of eliminating errors in cryptographic computations. Journal of Cryptology 14(2), 101–119 (2001)

6. Chandran, N., Kanukurthi, B., Ostrovsky, R.: Locally updatable and locally decodable codes. In: Lindell, Y. (ed.) TCC 2014. LNCS, vol. 8349, pp. 489–514. Springer, Heidelberg (2014)

7. Cheraghchi, M., Guruswami, V.: Capacity of non-malleable codes. In: Naor, M. (ed.) ITCS 2014, pp. 155–168. ACM (January 2014)

8. Cheraghchi, M., Guruswami, V.: Non-malleable coding against bit-wise and split-state tampering. In: Lindell, Y. (ed.) TCC 2014. LNCS, vol. 8349, pp. 440–464. Springer, Heidelberg (2014)

9. Choi, S.G., Kiayias, A., Malkin, T.: BiTR: Built-in tamper resilience. In: Lee, D.H., Wang, X. (eds.) ASIACRYPT 2011. LNCS, vol. 7073, pp. 740–758. Springer, Heidelberg (2011)

10. Chor, B., Kushilevitz, E., Goldreich, O., Sudan, M.: Private information retrieval. J. ACM 45(6), 965–981 (1998)

11. Coretti, S., Maurer, U., Tackmann, B., Venturi, D.: From single-bit to multi-bit public-key encryption via non-malleable codes. In: Cryptology ePrint Archive, Report 2014/324 (2014)

12. Dachman-Soled, D., Kalai, Y.T.: Securing circuits against constant-rate tampering. In: Safavi-Naini, R., Canetti, R. (eds.) CRYPTO 2012. LNCS, vol. 7417, pp. 533–551. Springer, Heidelberg (2012)
13. Dachman-Soled, D., Kalai, Y.T.: Securing circuits and protocols against $1/\mathrm{poly}(k)$ tampering rate. In: Lindell, Y. (ed.) TCC 2014. LNCS, vol. 8349, pp. 540–565. Springer, Heidelberg (2014)
14. Dachman-Soled, D., Liu, F.-H., Shi, E., Zhou, H.-S.: Locally decodable and updatable non-malleable codes and their applications. Cryptology ePrint Archive, Report 2014/663 (2014), http://eprint.iacr.org/2014/663
15. Damgård, I., Faust, S., Mukherjee, P., Venturi, D.: Bounded tamper resilience: How to go beyond the algebraic barrier. In: Sako, K., Sarkar, P. (eds.) ASIACRYPT 2013, Part II. LNCS, vol. 8270, pp. 140–160. Springer, Heidelberg (2013)
16. Dodis, Y., Pietrzak, K.: Leakage-resilient pseudorandom functions and side-channel attacks on Feistel networks. In: Rabin, T. (ed.) CRYPTO 2010. LNCS, vol. 6223, pp. 21–40. Springer, Heidelberg (2010)
17. Duc, A., Dziembowski, S., Faust, S.: Unifying leakage models: From probing attacks to noisy leakage. In: Nguyen, P.Q., Oswald, E. (eds.) EUROCRYPT 2014. LNCS, vol. 8441, pp. 423–440. Springer, Heidelberg (2014)
18. Dziembowski, S., Faust, S.: Leakage-resilient cryptography from the inner-product extractor. In: Lee, D.H., Wang, X. (eds.) ASIACRYPT 2011. LNCS, vol. 7073, pp. 702–721. Springer, Heidelberg (2011)
19. Dziembowski, S., Faust, S.: Leakage-resilient circuits without computational assumptions. In: Cramer, R. (ed.) TCC 2012. LNCS, vol. 7194, pp. 230–247. Springer, Heidelberg (2012)
20. Dziembowski, S., Kazana, T., Obremski, M.: Non-malleable codes from two-source extractors. In: Canetti, R., Garay, J.A. (eds.) CRYPTO 2013, Part II. LNCS, vol. 8043, pp. 239–257. Springer, Heidelberg (2013)
21. Dziembowski, S., Pietrzak, K.: Leakage-resilient cryptography. In: 49th FOCS, pp. 293–302. IEEE Computer Society Press (October 2008)
22. Dziembowski, S., Pietrzak, K., Wichs, D.: Non-malleable codes. In: Yao, A.C.-C. (ed.) ICS 2010, pp. 434–452. Tsinghua University Press (January 2010)
23. Faust, S., Mukherjee, P., Nielsen, J.B., Venturi, D.: Continuous non-malleable codes. In: Lindell, Y. (ed.) TCC 2014. LNCS, vol. 8349, pp. 465–488. Springer, Heidelberg (2014)
24. Faust, S., Mukherjee, P., Nielsen, J.B., Venturi, D.: A tamper and leakage resilient random access machine. In: Cryptology ePrint Archive, Report 2014/338 (2014)
25. Faust, S., Mukherjee, P., Venturi, D., Wichs, D.: Efficient non-malleable codes and key-derivation for poly-size tampering circuits. In: Nguyen, P.Q., Oswald, E. (eds.) EUROCRYPT 2014. LNCS, vol. 8441, pp. 111–128. Springer, Heidelberg (2014)
26. Faust, S., Pietrzak, K., Venturi, D.: Tamper-proof circuits: How to trade leakage for tamper-resilience. In: Aceto, L., Henzinger, M., Sgall, J. (eds.) ICALP 2011, Part I. LNCS, vol. 6755, pp. 391–402. Springer, Heidelberg (2011)
27. Faust, S., Rabin, T., Reyzin, L., Tromer, E., Vaikuntanathan, V.: Protecting circuits from leakage: the computationally-bounded and noisy cases. In: Gilbert, H. (ed.) EUROCRYPT 2010. LNCS, vol. 6110, pp. 135–156. Springer, Heidelberg (2010)
28. Gennaro, R., Lysyanskaya, A., Malkin, T., Micali, S., Rabin, T.: Algorithmic tamper-proof (ATP) security: Theoretical foundations for security against hardware tampering. In: Naor, M. (ed.) TCC 2004. LNCS, vol. 2951, pp. 258–277. Springer, Heidelberg (2004)

29. Goldreich, O., Ostrovsky, R.: Software protection and simulation on oblivious rams. Journal of the ACM 43(3), 431–473 (1996)
30. Goldwasser, S., Rothblum, G.N.: Securing computation against continuous leakage. In: Rabin, T. (ed.) CRYPTO 2010. LNCS, vol. 6223, pp. 59–79. Springer, Heidelberg (2010)
31. Goldwasser, S., Rothblum, G.N.: How to compute in the presence of leakage. In: 53rd FOCS, pp. 31–40. IEEE Computer Society Press (October 2012)
32. Halderman, J.A., Schoen, S.D., Heninger, N., Clarkson, W., Paul, W., Calandrino, J.A., Feldman, A.J., Appelbaum, J., Felten, E.W.: Lest we remember: Cold boot attacks on encryption keys. In: USENIX Security Symposium, pp. 45–60 (2008)
33. Ishai, Y., Kushilevitz, E.: On the hardness of information-theoretic multiparty computation. In: Cachin, C., Camenisch, J.L. (eds.) EUROCRYPT 2004. LNCS, vol. 3027, pp. 439–455. Springer, Heidelberg (2004)
34. Ishai, Y., Prabhakaran, M., Sahai, A., Wagner, D.: Private circuits II: Keeping secrets in tamperable circuits. In: Vaudenay, S. (ed.) EUROCRYPT 2006. LNCS, vol. 4004, pp. 308–327. Springer, Heidelberg (2006)
35. Ishai, Y., Sahai, A., Wagner, D.: Private circuits: Securing hardware against probing attacks. In: Boneh, D. (ed.) CRYPTO 2003. LNCS, vol. 2729, pp. 463–481. Springer, Heidelberg (2003)
36. Juma, A., Vahlis, Y.: Protecting cryptographic keys against continual leakage. In: Rabin, T. (ed.) CRYPTO 2010. LNCS, vol. 6223, pp. 41–58. Springer, Heidelberg (2010)
37. Katz, J., Trevisan, L.: On the efficiency of local decoding procedures for error-correcting codes. In: 32nd ACM STOC, pp. 80–86. ACM Press (May 2000)
38. Kiayias, A., Tselekounis, Y.: Tamper resilient circuits: The adversary at the gates. In: Sako, K., Sarkar, P. (eds.) ASIACRYPT 2013, Part II. LNCS, vol. 8270, pp. 161–180. Springer, Heidelberg (2013)
39. Paul, C.: Timing attacks on implementations of Diffie-Hellman, RSA, DSS, and other systems. In: Koblitz, N. (ed.) CRYPTO 1996. LNCS, vol. 1109, pp. 104–113. Springer, Heidelberg (1996)
40. Kocher, P., Jaffe, J., Jun, B.: Differential power analysis. In: Wiener, M. (ed.) CRYPTO 1999. LNCS, vol. 1666, pp. 388–397. Springer, Heidelberg (1999)
41. Lie, D., Thekkath, C.A., Mitchell, M., Lincoln, P., Boneh, D., Mitchell, J.C., Horowitz, M.: Architectural support for copy and tamper resistant software. In: ASPLOS, pp. 168–177 (2000)
42. Liu, F.-H., Lysyanskaya, A.: Tamper and leakage resilience in the split-state model. In: Safavi-Naini, R., Canetti, R. (eds.) CRYPTO 2012. LNCS, vol. 7417, pp. 517–532. Springer, Heidelberg (2012)
43. Micali, S., Reyzin, L.: Physically observable cryptography (extended abstract). In: Naor, M. (ed.) TCC 2004. LNCS, vol. 2951, pp. 278–296. Springer, Heidelberg (2004)
44. Pietrzak, K.: A leakage-resilient mode of operation. In: Joux, A. (ed.) EUROCRYPT 2009. LNCS, vol. 5479, pp. 462–482. Springer, Heidelberg (2009)
45. Ristenpart, T., Tromer, E., Shacham, H., Savage, S.: Hey, you, get off of my cloud: Exploring information leakage in third-party compute clouds. In: Al-Shaer, E., Jha, S., Keromytis, A.D. (eds.) ACM CCS 2009, pp. 199–212. ACM Press (November 2009)
46. Rothblum, G.N.: How to compute under $\mathcal{AC}^0$ leakage without secure hardware. In: Safavi-Naini, R., Canetti, R. (eds.) CRYPTO 2012. LNCS, vol. 7417, pp. 552–569. Springer, Heidelberg (2012)

47. Suh, G.E., Clarke, D.E., Gassend, B., van Dijk, M., Devadas, S.: AEGIS: architecture for tamper-evident and tamper-resistant processing. In: Proceedings of the 17th Annual International Conference on Supercomputing, ICS 2003, pp. 160–171 (2003)
48. Vasudevan, A., McCune, J.M., Newsome, J., Perrig, A., van Doorn, L.: CARMA: A hardware tamper-resistant isolated execution environment on commodity x86 platforms. In: Youl Youm, H., Won, Y. (eds.) ASIACCS 2012, pp. 48–49. ACM Press (May 2012)
49. Yekhanin, S.: Locally decodable codes. Foundations and Trends in Theoretical Computer Science 6(3), 139–255 (2012)

Tamper Detection and Continuous Non-malleable Codes

Zahra Jafargholi and Daniel Wichs*

Northeastern University, USA

Abstract. WeN consider a public and keyless code (Enc, Dec) which is used to encode a message m and derive a codeword $c = \mathsf{Enc}(m)$. The codeword can be adversarially tampered via a function $f \in \mathcal{F}$ from some "tampering function family" $\mathcal{F}$, resulting in a tampered value $c' = f(c)$. We study the different types of security guarantees that can be achieved in this scenario for different families $\mathcal{F}$ of tampering attacks.

Firstly, we initiate the general study of *tamper-detection codes*, which must detect that tampering occurred and output $\mathsf{Dec}(c') = \bot$. We show that such codes exist for any family of functions $\mathcal{F}$ over n bit codewords, as long as $|\mathcal{F}| < 2^{2^n}$ is sufficiently smaller than the set of all possible functions, and the functions $f \in \mathcal{F}$ are further *restricted* in two ways: (1) they can only have a *few fixed points* x such that $f(x) = x$, (2) they must have *high entropy* of $f(x)$ over a random x. Such codes can also be made efficient when $|\mathcal{F}| = 2^{\mathsf{poly}(n)}$.

Next, we revisit *non-malleable codes*, which were introduced by Dziembowski, Pietrzak and Wichs (ICS '10) and require that $\mathsf{Dec}(c')$ either decodes to the original message m, or to some unrelated value (possibly $\bot$) that doesn't provide any information about m. We give a modular construction of non-malleable codes by combining tamper-detection codes and leakage-resilient codes. The resulting construction matches that of Faust et al. (EUROCRYPT '14) but has a more modular proof and improved parameters.

Finally, we initiate the general study of *continuous non-malleable codes*, which provide a non-malleability guarantee against an attacker that can tamper a codeword multiple times. We define several variants of the problem depending on: (I) whether tampering is *persistent* and each successive attack modifies the codeword that has been modified by previous attacks, or whether tampering is non-persistent and is always applied to the original codeword, (II) whether we can "*self-destruct*" and stop the experiment if a tampered codeword is ever detected to be invalid or whether the attacker can always tamper more. In the case of persistent tampering and self-destruct (weakest case), we get a broad existence results, essentially matching what's known for standard non-malleable codes. In the case of non-persistent tampering and no self-destruct (strongest case), we must further restrict the tampering functions to have few fixed points and high entropy. The two intermediate cases correspond to requiring only one of the above two restrictions.

---

* Supported by NSF grants 1347350, 1314722, 1413964.

Y. Dodis and J.B. Nielsen (Eds.): TCC 2015, Part I, LNCS 9014, pp. 451–480, 2015.

1   Introduction

Motivating Example. Consider a security-sensitive device such as a smart-card implementing a digital signature scheme. The user gives it messages as inputs and receives signatures as outputs. The computation relies on a secret signing key stored on the card. Moreover, the user's name, say "Eve", is stored on the card and the card only signs message that begin with the name "Eve". The security of signature schemes ensures that Eve cannot sign a message with any other name if she is given this card and uses it as a black-box. However, Boneh, DeMillo and Lipton [BDL01] show a surprising result: if the above is implemented using RSA signatures with Chinese remaindering, and Eve is able to simply flip a single bit of the signing key on the smart card and observe the resulting incorrectly generated signature, then she can factor the RSA modulus and completely recover the signing key. Alternatively, no matter which signature scheme is used, Eve may be able to flip a few bits of the name stored on the card (e.g., change the value from "Eve" to "Eva") without changing the signing key and then use the card to sign messages under a different name.

The above are examples of *tampering attacks*. By tampering with the internal state of a device (without necessarily knowing what it is) and then observing the outputs of the tampered device, an attacker may be able to learn additional sensitive information which would not be available otherwise. A natural approach to protecting against such attacks is to *encode* the data on the device in some way. For example, [BDL01] suggest using error-detection codes to thwart an attack that flips a small number of bits. This raises the question of what kind of codes are needed to achieve protection and what classes of tampering attacks can they protect against?

A Coding Problem. We can translate the above scenario into a coding problem. We would like to design a code (Enc, Dec) consisting of a possibly *randomized encoding* function and a *decoding* function with the correctness guarantee that $\mathsf{Dec}(\mathsf{Enc}(m)) = m$. There are no secret keys and anybody can encode and decode. We model tampering as a family of functions $\mathcal{F}$ that an attacker can apply to modify codewords.[1] We consider a "tampering experiment" with some message m and function $f \in \mathcal{F}$. The experiment begins by probabilistically encoding $c \leftarrow \mathsf{Enc}(m)$, then tampers $c' = f(c)$ and finally outputs the decoded value $m' = \mathsf{Dec}(c')$. We consider different types of security guarantees on the outcome m' of the experiment.

1.1   Tamper Detection Codes

Perhaps the simplest property that we could ask for is that tampering can always be detected with overwhelming probability, meaning that the decoded value is

---

[1] This is a departure from standard coding theory problems by focusing on the process (family of functions) $\mathcal{F}$ that modifies a codeword rather than on some notion of distance between the original and modified codeword.

some special symbol $m' = \bot$ indicating an error. In other words, a tamper-detection code for a family $\mathcal{F}$ ensures that for any message m and any tampering function $f \in \mathcal{F}$ we have $\Pr[\mathsf{Dec}(f(c)) \neq \bot \, : \, c \leftarrow \mathsf{Enc}(m)]$ is negligible. We ask for which function families $\mathcal{F}$ do such codes exist.

Surprisingly, this natural problem has not been studied at this level of generality and relatively little is known beyond a few specific function families. Standard error-detection codes provide this guarantee for the family of all functions f such that the hamming distance between c and $f(c)$ is always small, but non-zero. The *algebraic manipulation detection* (AMD) codes of Cramer et al. [CDF+08], consider this type of guarantee for functions f that can flip an arbitrary number of bits of c, but the error pattern is independent of c. In other words, AMD codes consider the family $\mathcal{F}_{AMD} = \{f_\Delta(c) := c \oplus \Delta \mid \Delta \neq 0^n\}$.

In this work, we show that tamper-detection codes exist for any function family $\mathcal{F}$ over n-bit codewords as long as the size of the family is bounded $|\mathcal{F}| < 2^{2^{\alpha n}}$ for some constant $\alpha < 1$ and the functions $f \in \mathcal{F}$ satisfy two additional restrictions:

- *High Entropy:* For each $f \in \mathcal{F}$, we require that $f(c)$ has sufficiently high min-entropy when $c \sim \{0,1\}^n$ is uniformly random.
- *Few Fixed Points:* For each $f \in \mathcal{F}$, we require that there aren't too many fixed points c s.t. $f(c) = c$.

Moreover, we show that such codes can achieve a rate (defined as the ratio of message size to codeword size) of $(1 - \alpha)$. We also show that the restrictions on the tampering functions (high entropy, few fixed points) in the above result cannot be removed.

This existence result relies on a probabilistic method argument and, in general, such codes may not be efficient. However, when the size of the function family is $|\mathcal{F}| \leq 2^{s(n)}$ for some polynomial s, then we can use a probabilistic method argument with limited independence to get efficient codes. More precisely, this yields a family of efficient codes $(\mathsf{Enc}_h, \mathsf{Dec}_h)$ indexed by some hash function h from a t-wise independent function family (for a polynomial t depending on s), such that a random member of the family is a secure tamper-detection code for $\mathcal{F}$ with overwhelming probability. We can also think of this as a construction of efficient tamper-detection codes in the *common random string* (CRS) model, where the random choice of h is specified in the CRS and known to everyone. The construction of this code family is extremely simple and it matches a construction proposed by Faust et al. [FMVW14] in the context of non-malleable codes. However, our analysis is fairly delicate and departs significantly from that of Faust et al.

This result generalizes AMD codes which which provide tamper-detection for the family $\mathcal{F}_{AMD}$ of size $|\mathcal{F}_{AMD}| = 2^n - 1$ that has full entropy n, and no fixed points. For example, using this result, we can get efficient tamper-detection codes (in the CRS model) for the class $\mathcal{F}_{\mathsf{poly},d}$ of polynomials f over $\mathbb{F}_{2^n}$ of some bounded degree $d = \mathsf{poly}(n)$, as long as we exclude the identity polynomial $f(x) = x$ and the constant (degree 0) polynomials. Alternatively, we get also get such tamper-detection codes for the class $\mathcal{F}_{\mathsf{affine},r}^-$ of all affine functions when we interpret $x \in \{0,1\}^n$ as an (n/r)-dimensional vector over the field $\mathbb{F}_{2^r}$ with a sufficiently large r, and we exclude the identity and the constant functions.

1.2   Non-malleable Codes

The work of Dziembowski, Pietrzak and Wichs [DPW10] introduced the notion of *non-malleable codes*, which asks for a weaker guarantee on the outcome of the tampering experiment. Instead of insisting that the decoded value after tampering is always $m' = \perp$, non-malleable security requires that either $m' = m$ is equal to the original message, or m' is a completely unrelated value (possibly $\perp$) that contains no information about the original message m. Moreover, the choice of which of these two options occurs is also unrelated to m. For example, non-malleability ensures that it should not be possible to tamper c to $c' = f(c)$ in such a way as to just flip a bit of the encoded message. Alternatively, it shouldn't be possible to tamper the codeword in such a way as to get either $m' = m$ or $m' = \perp$ depending on (say) the first bit of m.

Non-malleable codes offer strong protection against tampering. By encoding the data on an device with a non-malleable code for $\mathcal{F}$, we ensure that an attacker cannot learn anything more by tampering with the data on the device via functions $f \in \mathcal{F}$ and interacting with the tampered device, beyond what could be learned given black-box access to the device without the ability to tamper. Moreover, such codes can also protect against *continuous* tampering attacks, where the attacker repeatedly tampers with the data on the device and observes its outputs. On each invocation, the device must first decode the secret data, run the underlying functionality with the decoded value to produce some output, and then freshly *re-encode* the decoded value. This last step is important for security and ensures that each tampering attack acts on a fresh codeword. On the downside, this step usually requires fresh randomness and also requires that device is now *stateful* even if the underlying functionality is stateless.

Prior Work on Non-Malleable Codes. The work of [DPW10], gives a broad *existence* result showing that for any family $\mathcal{F}$ of tampering functions over n-bit codewords having size $|\mathcal{F}| < 2^{2^{\alpha n}}$ for $\alpha < 1$ there exists a (possibly inefficient) non-malleable code for $\mathcal{F}$. This covers various complex types of tampering attacks. The work of Cheraghchi and Guruswami [CG13a] further show that such codes can achieve a rate of $1 - \alpha$, which is optimal. These results rely on a probabilistic method argument, where the code is defined via a completely random function.

The works of [CG13a, FMVW14] also give *efficient* scaled-down versions of the above existence results for function families of singly-exponential size $|\mathcal{F}| \leq 2^{s(n)}$ for some polynomial s. For example, $\mathcal{F}$ could be the class of all circuits of size at most $s(n)$. These results are derived using a probabilistic method argument with limited independence, where the code $(\mathsf{Enc}_h, \mathsf{Dec}_h)$ is parameterized by some efficient hash function h chosen from a t-wise independent family for some polynomial t depending on s. The code is secure with overwhelming probability over the choice of h, which we can think of as a common random string (CRS).

Several other works [DPW10, CKM11, LL12, DKO13, ADL13, CG13b, FMNV14] and [CZ14] construct explicit non-malleable codes for interesting families that are restricted through their *granularity*. In particular, these works envision that the codeword consists of several components, each of which can be tampered arbitrarily but independently of the other components. The strongest variant of this is the *split-state model*, where the codeword consists of just two components that are tampered independently of each other. Such codes were recently constructed in the information theoretic setting in the works of [DKO13, ADL13]. Another recent result [AGM+14] shows how to construct non-malleable codes against functions that can permute (and perturbe) the bits of the codeword.

Our Results. We show a general way of obtaining non-malleable codes for general function families by combining tamper-detection codes for restricted function families (with few fixed points and high entropy) and a certain type of *leakage-resilient codes*, defined by Davì, Dziembowski and Venturi [DDV10]. The resulting non-malleable code construction matches that of Faust et al. [FMVW14] but with an optimized modular analysis. We show that this construction can simultaneously achieve optimal rate $(1 - \alpha)$ for function families of size $|\mathcal{F}| = 2^{2^{\alpha n}}$ and also efficient encoding/decoding that scales polynomially in the security parameter and n when $|\mathcal{F}| = 2^{\text{poly}(n)}$. Previously, each of these properties was known to be achievable individually but by different constrictions shown in [CG13a] and [FMVW14] respectively.

1.3 Continuous Non-malleable Codes

As mentioned, standard non-malleable codes already provide protection against *continuous* tampering attacks on a device, if the device freshly re-encodes its state after each invocation to ensure that tampering is applied to a fresh codeword each time. However, this is undesirable since it requires that: (1) the device has access to fresh randomness on each invocation, and (2) the device is stateful and updates its state on each invocation. This is the case even if the underlying functionality that the device implements (e.g., a signature scheme) is deterministic and stateless. This brings up the natural question (posed as an open problem by [DPW10]) whether we can achieve security against continuous tampering of a single codeword *without* re-encoding. We consider four variants of such *continuous non-malleable codes* depending on:

- Whether tampering is *persistent*, meaning that each tampering attack is applied to the current version of the codeword that has been modified by previous attacks, and the original codeword is otherwise lost once tampered. Alternatively, we can consider non-persistent tampering, where tampering is always applied to the original codeword.
- Whether tampering to an invalid codeword (one that decodes to $\bot$) causes a "*self-destruct*" meaning that the experiment stops and the attacker cannot

gain any additional information, or whether the attacker can always continue tampering more. We can think of this as corresponding to a physical device "self-destructing" (e.g., by erasing all internal data) if it detects that it has been tampered. This in turn corresponds to a very limited form of state, where the data on the device can only be erased but not updated otherwise.

Note that persistent tampering and self-destruct is the weakest variant, non-persistent tampering and no self destruct is the strongest variant, and the remaining two variants lie in between and are incomparable to each other. All of these variants are already stronger than standard non-malleable codes.

The notion of *continuous non-malleable codes* was first introduced in the work of Faust et al. [FMNV14], which considered the case of non-persistent tampering and self-destruct. It focused on tampering in the split-state model, where two halves of a codeword are tampered independently of each other, and showed that although such codes cannot exist information theoretically, they can be constructed under computational assumptions. The reason for focusing on non-persistent tampering is that this models attacks that have access to some "auxiliary memory" on the device beyond the n bits of "active memory" used to store the codeword. The initial tampering attack can make a copy the original codeword onto this auxiliary memory. Each subsequent attack can then tamper the original codeword from the auxiliary memory and place the tampered codeword into the active memory. In the case of persistent tampering, we implicitly assume that there is no such auxiliary memory on the device.

In this work, we give a general definition of continuous non-malleable codes that captures all of the above variants. We initiate the comprehensive study of what type of tampering attacks we can protect against under each variant in the information theoretic setting.

Our Results. We use the same template that we employed for constructing standard non-malleable codes to also construct continuous non-malleable codes. Depending on which of the four variants of continuous non-malleable codes we consider, we show that our construction achieves security for different classes of tampering attacks.

In the setting of *persistent tampering and self-destruct* (weakest), we show broad existence results which essentially match the existence results of [DPW10] and [CG13a] for standard non-malleable codes. In particular, we show that such codes *exist* (inefficiently) for any family $\mathcal{F}$ of tampering functions over n-bits, whose size is $|\mathcal{F}| = 2^{2^{\alpha n}}$ for $\alpha < 1$. Moreover, such codes achieve a rate of $1 - \alpha$. For example, this result shows existence of such codes (albeit inefficient) in the split-state model with rate $1/2$. Furthermore, we give an *efficient* scaled-down versions of the above existence results for function families of singly-exponential size $|\mathcal{F}| = 2^{\mathsf{poly}(n)}$. Unfortunately, in this case the efficiency of the code also depends polynomially on the number of tampering attacks we want to protect against. We conjecture that this dependence can be removed, and leave this fascinating problem for future work.

In the setting of *non-persistent tampering and no self destruct* (strongest), we must place additional restrictions on the function family $\mathcal{F}$ to ensure that the functions have *high entropy* and *few fixed points*, as in tamper-detection codes. In fact, a tamper-detection code *is* continuous non-malleable in this setting since each tampering attack simply leaves the codeword invalid (decodes to $\perp$) and so the attacker does not learn anything. However, in contrast to tamper-detection codes, continuous non-malleable codes in this setting can also trivially tolerate the "identity function" $f_{\mathsf{id}}(x) = x$ and the "always constant" functions $f_c(x) = c$. Therefore, we (e.g.,) get continuous non-malleable codes in this setting for the class $\mathcal{F}_{\mathsf{poly},d}$ of *all* low-degree polynomials or the class $\mathcal{F}_{\mathsf{affine},r}$ of *all* affine functions over a sufficiently large field.

In the two intermediate settings, we only need to impose one of the two restrictions on high entropy and few fixed points. For the case of *persistent tampering and no self destruct* we only require that the functions have *few fixed points* but can have arbitrary entropy. For the case of *non-persistent tampering and self destruct* we only require that the functions have *high entropy* but can have many fixed points.

1.4   Other Related Work and RKA Security

There is a vast body of literature that considers tampering attacks using other approaches besides (non-malleable) codes.See, e.g., [BK03, GLM+04, IPSW06] [BC10, BCM11, FPV11] [KKS11, AHI11, GOR11, Pie12, Wee12, BPT12, DFMV13] [ABPP14, GIP+14] etc.

One highly relevant line of work called *related key attacks* (RKA) security [BK03, BC10, BCM11, Wee12, BPT12, ABPP14] considers tampering with the secret key of a cryptographic primitive such as a pseudorandom function or a signature scheme. The definitions in those works usually require the schemes to be stateless, consider non-persistent tampering (tampering is always applied to the original key), and do not allow for self-destruct. These works focus on giving clever constructions of specific primitives that satisfy RKA security based on specific computational assumptions. We note that our results on continuous non-malleable codes in the setting of non-persistant tampering and no self-destruct provide a generic way of achieving qualitatively similar results to these works for any cryptographic scheme. In particular, by encoding the key of the scheme with such codes, we can achieve protection against the types of attacks that were considered in the RKA literature (e.g., additive tampering, low-degree polynomials) as well as many other interesting families.

Nevertheless, we note that the RKA constructions from the literature maintain some advantages, such as having uniformly random secret keys, whereas we would require the secret key to be a structured codeword. Also, the exact definitions of RKA security in the literature vary from one primitive to another and sometimes impose additional properties that our constructions would not satisfy. For example, definitions of RKA security for PRFs usually require that outputs under both the original and the tampered keys remain pseudorandom, whereas our solution would only guarantee that outputs under the original key remain pseudorandom even given outputs under a tampered key, but the latter may not be pseudorandom (they may just be $\perp$). It is not clear whether these differences are important in the envisioned applications, and therefore we view our results as providing qualitatively similar (but not equivalent) security guarantees to those studied in the context of RKA security.

2   Preliminaries

Notation. For a positive integer n, we define the set $[n] := \{1, \ldots, n\}$. Let X, Y be random variables with supports $S(X), S(Y)$, respectively. We define their *statistical distance* by $\mathbf{SD}(X, Y) = \frac{1}{2} \sum_{s \in S(X) \cup S(Y)} |\Pr[X = s] - \Pr[Y = s]|$. We write $X \approx_\varepsilon Y$ and say that X and Y are ε-statistically close to denote that $\mathbf{SD}(X, Y) \leq \varepsilon$. We let U_n denote the uniform distribution over $\{0, 1\}^n$. We use the notation $x \leftarrow X$ to denote the process of sampling a value x according to the distribution X. For a set S, we write $s \leftarrow S$ to denote the process of sampling s uniformly at random from S.

Tail Bound. We recall the following lemma from [BR94] which gives us a Chernoff-type tail bound for limited independence.

Lemma 1 (Lemma 2.3 of [BR94]). *Let $t \geq 4$ be an even integer. Suppose $X_1, \ldots, X_n$ are t-wise independent random variables over $\{0, 1\}$. Let $X := \sum_{i=1}^n X_i$ and define $\mu := \mathbf{E}[X]$ be the expectation of the sum. Then, for any $A > 0$, $\Pr[|X - \mu| \geq A] \leq 8 \left(\frac{t\mu + t^2}{A^2}\right)^{t/2}$. In particular, if $A \geq \mu$ then $\Pr[|X - \mu| \geq A] \leq 8 \left(\frac{2t}{A}\right)^{t/2}$.*

It is easy to check by observing the proof of the lemma that it also holds even when $X_1, \ldots, X_n$ are not truly t-wise independent but for any $S \subseteq [n]$ of size $|S| \leq t$ they satisfy $\Pr[\bigwedge_{i \in S} \{X_i = 1\}] \leq \prod_{i \in S} \Pr[X_i = 1]$. This is because the only use of independence in the proof is to bound $\mathbb{E}[\prod_{i \in S} X_i] \leq \prod_{i \in S} \mathbb{E}[X_i]$ which holds under the above condition.

Coding Schemes. It will be useful to define the following general notion of a coding scheme.

Definition 1. *A (k, n)-coding scheme consists of two functions: a randomized encoding function $\mathsf{Enc} : \{0, 1\}^k \to \{0, 1\}^n$, and a deterministic decoding function $\mathsf{Dec} : \{0, 1\}^n \to \{0, 1\}^k \cup \{\perp\}$ such that, for each $m \in \{0, 1\}^k$, $\Pr[\mathsf{Dec}(\mathsf{Enc}(m)) = m] = 1$. For convenience, we also define $\mathsf{Dec}(\perp) = \perp$.*

2.1   Leakage-Resilient Codes

The following definition of leakage-resilient codes is due to [DDV10]. Intuitively, it allows us to encode a message m into a codeword c in such a way that learning $f(c)$ from some class of *leakage functions* $f \in \mathcal{F}$ will not reveal anything about the message m. For convenience, we actually insist on a stronger guarantee that $f(c)$ is indistinguishable from f applied to the uniform distribution over n bit strings.

Definition 2. *Let* $(\mathsf{lrEnc}, \mathsf{lrDec})$ *be a* (k, n)-*coding scheme. For a function family* $\mathcal{F}$, *we say that* $(\mathsf{lrEnc}, \mathsf{lrDec})$ *is* $(\mathcal{F}, \varepsilon)$-*leakage-resilient, if for any* $f \in \mathcal{F}$ *and any* $m \in \{0, 1\}^k$ *we have* $f(\mathsf{lrEnc}(m)) \approx_\varepsilon f(U_n)$ *where* U_n *is the uniform distribution over* n *bit strings.*

Construction. We recall the following construction from [DDV10, FMVW14]. Let $\mathcal{H}$ be a t-wise independent hash function family consisting of functions $h : \{0, 1\}^v \to \{0, 1\}^k$. For any $h \in \mathcal{H}$ we define the $(k, n = k + v)$-coding scheme $(\mathsf{lrEnc}_h, \mathsf{lrDec}_h)$ where: (1) $\mathsf{lrEnc}_h(m) := (r, h(r) \oplus m)$ for $r \leftarrow \{0, 1\}^v$; (2) $\mathsf{lrDec}_h((r, z)) := z \oplus h(r)$.

We give an improved analysis of this construction: to handle a leakage family $\mathcal{F}$ with ℓ-bits of leakage (i.e., the output size of $f \in \mathcal{F}$ is ℓ-bits), the best prior analysis in [FMVW14] required overhead (roughly) $v = \log \log |\mathcal{F}| + \ell$, whereas our improved analysis only requires overhead (roughly) $v = \max\{\log \log |\mathcal{F}|, \ell\}$. This can yield up to a factor of 2 improvement and will be crucial in getting optimal rate for our non-malleable and continuous non-malleable codes.

Theorem 1. *Fix any function family* $\mathcal{F}$ *consisting of functions* $f : \{0, 1\}^n \to \{0, 1\}^\ell$. *With probability* $1 - \rho$ *over the choice of a random* $h \leftarrow \mathcal{H}$ *from a* t-*wise independent family* $\mathcal{H}$, *the coding scheme* $(\mathsf{lrEnc}_h, \mathsf{lrDec}_h)$ *above is* $(\mathcal{F}, \varepsilon)$-*leakage-resilient as long as* $v \geq v_{\min}$ *and* $t \geq t_{\min}$, *when* $B := \log |\mathcal{F}| + k + \log (1/\rho)$ *and*

either $v_{\min} = \log(B + 2^\ell) + \log \ell + 2 \log (1/\varepsilon) + O(1)$ *and* $t_{\min} = O(B)$,

or $v_{\min} = \log(B + 2^\ell) + 2 \log (1/\varepsilon) + O(1)$ *and* $t_{\min} = O(B + 2^\ell)$.

In particular, if $\rho = \varepsilon = 2^{-\lambda}$ *for security parameter* λ *and,* $|\mathcal{F}| \geq \max\{2^k, 2^\lambda\}$, $\ell \leq 2^\lambda$ *we get:* $v_{\min} = \max\{\log \log |\mathcal{F}|, \ell\} + O(\lambda)$ *and* $t_{\min} = O(\log |\mathcal{F}|)$.

See the full version [JW15] for the proof.

3   Tamper Detection Codes

We begin by defining the notion of a *tamper-detection code*, which ensures that if a codeword is tampered via some function $f \in \mathcal{F}$ then this is detected and the modified codeword decodes to $\perp$ with overwhelming probability. We give two flavors of this definition: a default ("standard") version which guarantees security for a worst-case message m and a weak version which only guarantees security for a random message m.

Definition 3. *Let* $(\mathsf{Enc}, \mathsf{Dec})$ *be a* (k, n)-*coding scheme and let* $\mathcal{F}$ *be a family of functions of the form* $f : \{0, 1\}^n \to \{0, 1\}^n \cup \perp$. *We say that* $(\mathsf{Enc}, \mathsf{Dec})$ *is a:*

– $(\mathcal{F}, \varepsilon)$-*secure tamper detection code* *if for any function* $f \in \mathcal{F}$ *and* *any message* $m \in \{0,1\}^k$, *we have* $\Pr\left[\mathsf{Dec}(f(\mathsf{Enc}(m))) \neq \perp\right] \leq \varepsilon$, *where the probability is over the randomness of the encoding procedure.*
– $(\mathcal{F}, \varepsilon)$-*weak tamper detection code* *if for any function* $f \in \mathcal{F}$ *we have* $\Pr_{m \leftarrow U_k}\left[\mathsf{Dec}(f(\mathsf{Enc}(m))) \neq \perp\right] \leq \varepsilon$, *where the probability is over a random message* m *and the randomness of the encoding procedure.*

It is easy to see that there are some small function families for which tamper detection (or even weak tamper detection) is impossible to achieve. One example is the family consisting of a single identity function $f_{\mathsf{id}}(x) = x$. Another example is the family consisting of all constant function $\mathcal{F}_{const} = \{f_c(x) = c \; : \; c \in \{0,1\}^n\}$. This family is of size only $|\mathcal{F}_{const}| = 2^n$ but no matter what code is used there is some function $f_c \in \mathcal{F}_{const}$ corresponding to a valid codeword c which breaks the tamper detection guarantee. Of course, there are other "bad" functions such as ones which are close to identity or close to some constant function. But we will show that these are the *only* bad cases. We begin by defining two restrictions on functions which ensure that they are far from the above bad cases.

Definition 4. *A function* $f : \{0,1\}^n \to \{0,1\}^n \cup \perp$ *is a* φ-*few fix points,* μ-*entropy function if*

– $\Pr_{x \leftarrow U_n}\left[f(x) = x\right] \leq \varphi$.   *($\varphi$-few fixed points)*
– $\forall y \in \{0,1\}^n, \; \Pr_{x \leftarrow U_n}\left[f(x) = y\right] \leq 2^{-\mu}$.   *($\mu$-entropy)*

The first property restricts the number of fixed points $x \; : \; f(x) = x$ to be less than $\varphi \cdot 2^n$. This ensures that the function is sufficiently far from being an identity function. The second property is equivalent to saying that the min-entropy $\mathbf{H}_\infty(f(U_n)) \geq \mu$. This ensures that the function is far from being a constant function.

3.1 Weak Tamper-Detection Codes

We begin by constructing weak tamper-detection codes. We will then show how to use weak tamper-detection codes to also construct standard tamper-detection codes.

Construction. Our construction of weak tamper detection codes will have a deterministic encoding. Let $\{0,1\}^k$ be the message space and $\{0,1\}^n$ to be the codeword space where $n = k + w$ for some $w > 0$. For a function $h : \{0,1\}^k \to \{0,1\}^w$ define $(\mathsf{Enc}_h, \mathsf{Dec}_h)$ as follow, $\mathsf{Enc}_h(m) := (m, h(m))$ and $\mathsf{Dec}_h(c)$ checks whether $c = (m, z)$ where $z = h(m)$: if so, it outputs m and otherwise it outputs $\perp$.

Theorem 2. *Let* $\mathcal{F}$ *be any finite family of* φ-*few fix points,* μ-*entropy functions and* $\mathcal{H} = \{h \mid h : \{0,1\}^k \to \{0,1\}^w\}$ *be a family of* $2t$-*wise independent hash functions, then*

$$\Pr_{h \leftarrow \mathcal{H}}\left[(\mathsf{Dec}_h, \mathsf{Enc}_h) \text{ is a } (\mathcal{F}, \varepsilon)\text{-weak tamper detection code}\right] > 1 - \rho$$

when the parameters satisfy

$$t \geq t_{\min} \quad where \quad t_{\min} = \log|\mathcal{F}| + k + \log(1/\rho) + 5$$
$$\mu \geq \log(t_{\min}) + w + \log(1/\varepsilon) + 6$$
$$w \geq \log(1/\varepsilon) + 3$$
$$\varphi \leq \varepsilon/4$$

For example, if $\rho = \varepsilon = 2^{-\lambda}$ for security parameter λ and, $|\mathcal{F}| \geq \max\{2^k, 2^\lambda\}$ we get:

$$t = O(\log|\mathcal{F}|), \quad \mu = \log\log|\mathcal{F}| + 2\lambda + O(1), \quad w = \lambda + O(1), \quad \varphi = 2^{-(\lambda + O(1))}$$

See the full version [JW15] for the proof.

Proof. Define the event BAD to occur if $(\mathsf{Dec}_h, \mathsf{Enc}_h)$ is not a $(\mathcal{F}, \varepsilon)$-weak tamper detection code. Then, by the union bound:

$$\Pr_{h \leftarrow \mathcal{H}}[\mathsf{BAD}] = \Pr_{h \leftarrow \mathcal{H}}\left[\exists f \in \mathcal{F}, \Pr_{m \leftarrow U_k}[\mathsf{Dec}_h(f(\mathsf{Enc}_h(m))) \neq \bot] > \varepsilon\right]$$
$$\leq \sum_{f \in \mathcal{F}} \underbrace{\Pr_{h \leftarrow \mathcal{H}}\left[\Pr_{m \leftarrow U_k}[\mathsf{Dec}_h(f(\mathsf{Enc}_h(m))) \neq \bot] > \varepsilon\right]}_{P_f} \qquad (3.1)$$

Let's fix some particular function $f \in \mathcal{F}$ and find a bound on the value of P_f. Define the indicator random variables $\{X_m\}_{m \in \{0,1\}^k}$ as $X_m = 1$ if $\mathsf{Dec}_h(f(\mathsf{Enc}_h(m))) \neq \bot$, where the randomness is only over the choice of h. Let's also define the variables $\{Y_{m,z}\}_{(m,z) \in \{0,1\}^n}$ as $Y_{m,z} = 1$ if $h(m) = z$ and $h(m') = z'$ where $(m', z') = f(m, z)$. Then $X_m = \sum_z Y_{m,z}$ and therefore:

$$P_f = \Pr\left[\sum_{m \in \{0,1\}^k} X_m > 2^k \varepsilon\right] = \Pr\left[\sum_{(m,z) \in \{0,1\}^n} Y_{m,z} \geq 2^k \varepsilon\right]. \qquad (3.2)$$

We can ignore variables $Y_{m,z}$ for values (m, z) such that $f(m, z) = \bot$ since in that case $Y_{m,z} = 0$ always. Otherwise, we have $\Pr[Y_{m,z} = 1] = \Pr[h(m) = z \wedge h(m') = z'] \leq 2^{-2w}$ if $f(m, z) = (m', z') \neq (m, z)$ is not a fixed point and $\Pr[Y_{m,z} = 1] = \Pr[h(m) = z] = 2^{-w}$ if $f(m, z) = (m, z)$ is a fixed point.

We might hope that the random variables $Y_{m,z}$ are t-wise independent but they are not. For example, say $c = (m, z)$ and $c' = (m', z')$ are two values such that $f(c) = c'$ and $f(c') = c$ then $Y_{m,z} = 1 \Leftrightarrow Y_{m',z'} = 1$. In order to analyze the dependence between these random variables we represent the tampering function f as a directed graph. We define a graph G with vertices $V = \{0,1\}^n$, representing the codewords, and edges $E = \{(c, c') \mid c' = f(c), c = (m, z), m \in \{0,1\}^k, z \in \{0,1\}^w\}$, representing the tampering function. Note that every vertex has at most one outgoing edge, so we can label each edge $e_{m,z}$ with the value (m, z) of its unique origin vertex. Using this representation we can associate each $Y_{m,z}$ to the unique edge $e_{m,z}$.

Now consider any subset $E' \subseteq E$ of edges such that the origin $c = (m, z)$ and destination $c' = (m', z') = f(c)$ of each edge (c, c') in E' is disjoint from the origin or destination of all other edges in E'. We call such sets E' *vertex-disjoint*. Then for any $S \subseteq E'$ of size $|S| = t$ we have:

$$\Pr\left[\bigwedge_{(m,z)\in S} Y_{m,z} = 1\right] \leq \prod_{(m,z)\in S} \Pr[Y_{m,z} = 1] \tag{3.3}$$

This is because $Y_{m,z} = 1$ iff $h(m) = z \wedge h(m') = z'$ where $(m', z') = f(m, z)$. Let $\tilde{S} \subseteq V$ be all the vertices contained in either the origin or destination of edges in E (so $|\tilde{S}| = 2S$ since $S \subseteq E'$ is vertex-disjoint). If two vertices (m, z) and (m, z') in $\tilde{S}$ share the same m, then it cannot be the case that $h(m) = z$ and $h(m) = z'$ so $\Pr[\bigwedge_{(m,z)\in S} Y_{m,z} = 1] = 0$. Otherwise, if the values m contained in $\tilde{S}$ are all disjoint, then variables $Y_{m,z}$ are independent since h is $2t$-wise independent.

Now we will partition the set E of edges into smaller subsets each of which is vertex-disjoint. Firstly, let E_{sl} be the set of all self-loops (fixed point) edges $e_{(m,z)}$ such that $f(m, z) = (m, z)$. It is clear that E_{sl} is vertex-disjoint. Since f is a φ-few fix points, μ-entropy function, we know that $|E_{\mathsf{sl}}| \leq \varphi 2^n$. Furthermore, $\mathbf{E}[\sum_{e_{(m,z)}\in E_{\mathsf{sl}}} \sum Y_{m,z}] \leq \varphi 2^{n-w} \leq \varphi 2^k$. By applying Lemma 1, with some $t_{\min} \leq t$ we get:

$$\Pr\left[\sum_{e_{(m,z)}\in E_{\mathsf{sl}}} Y_{m,z} \geq \frac{\varepsilon}{2} 2^k\right] \leq \Pr\left[\sum_{e_{(m,z)}\in E_{\mathsf{sl}}} Y_{m,z} \geq \varphi 2^k + (\frac{\varepsilon}{2} - \varphi) 2^k\right]$$

$$\leq \Pr\left[\sum_{e_{(m,z)}\in E_{\mathsf{sl}}} Y_{m,z} \geq \varphi 2^k + (\frac{\varepsilon}{4}) 2^k\right] \tag{3.4}$$

$$\leq 8\left(\frac{2t_{\min}}{(\frac{\varepsilon}{4}) 2^k}\right)^{t_{\min}/2} \leq 8\left(\frac{1}{2}\right)^{t_{\min}} \tag{3.5}$$

where inequality 3.4 follow by requiring that $\varphi \leq \varepsilon/4$ and the right hand inequality of 3.5 follows by requiring that $(\varepsilon/4)2^k \geq 8t_{\min}$. This in turn follows by observing that $k \geq \mu \geq \log(t_{\min}) + w + \log(1/\varepsilon) + 5$.

Next we define $E_{\mathsf{nsl}} = E \setminus E_{\mathsf{sl}}$ to be the set of edges that are not self loops. Let G_{nsl} be the corresponding graph consisting of G with all self-loops removed. Since f is a φ-few fix points, μ-entropy function, we know that each vertex in graph G has in-degree of at most $2^{n-\mu}$ and out-degree 1, and therefore total degree at most $2^{n-\mu} + 1$. By a theorem of Shannon [Sha49] on edge-colorings, we can color the edges of such a graph using at most $q \leq (3/2)2^{n-\mu} + 1 \leq 2^{n-\mu+1}$ colors so that no two neighboring edges (ignoring direction) are colored with the same color. In other words, we can partition E_{nsl} into exactly $q := 2^{n-\mu+1}$ subsets $E_1, \ldots, E_q$ such that each E_i is vertex-disjoint (we can always add dummy colors to make this exact). Further, recall that $\mathbf{E}[Y_{m,z}] \leq 2^{-2w}$ for edges $e_{(m,z)} \in E_{\mathsf{nsl}}$ that are not self loops. This implies that:

$$\Pr\left[\sum_{e_{m,z} \in E_{\mathsf{nsl}}} Y_{m,z} \geq \left(\frac{\varepsilon}{2}\right) 2^k\right]$$

$$\leq \Pr\left[\exists i \in [q] : \sum_{e_{m,z} \in E_i} Y_{m,z} \geq \left(\frac{\varepsilon}{2}\right) 2^k \left(\frac{1}{2}\right) \left(\frac{|E_i|}{2^n} + \frac{1}{q}\right)\right] \quad (3.6)$$

$$\leq \sum_{i \in [q]} \Pr\left[\sum_{e_{m,z} \in E_i} Y_{m,z} \geq \left(\frac{\varepsilon}{4}\right) 2^{-w} \left(|E_i| + 2^{\mu-1}\right)\right] \quad (3.7)$$

$$\leq \sum_{i \in [q]} \Pr\left[\sum_{e_{m,z} \in E_i} Y_{m,z} \geq |E_i| 2^{-2w} + A\right] \quad (3.8)$$

where $A = \left(\frac{\varepsilon}{4}\right) 2^{k-n} \left(|E_i| + 2^{\mu-1}\right) - |E_i| 2^{-2w}$.

Inequality 3.6 follows by observing that $\sum_{i \in [q]} \left(\frac{|E_i|}{2^n} + \frac{1}{q}\right) \leq 2$, inequality 3.7 follows by substituting $q = 2^{n-\mu+1}$ and $w = n - k$, and inequality 3.8 follows by the union bound. We can also bound:

$$A = \left(\frac{\varepsilon}{4}\right) 2^{-w} \left(|E_i| + 2^{\mu-1}\right) - |E_i| 2^{-2w}$$

$$\geq |E_i| 2^{-w} \left(\frac{\varepsilon}{4} - 2^{-w}\right) + \frac{\varepsilon}{4} 2^{-w} 2^{\mu-1}$$

$$\geq |E_i| 2^{-2w} + \frac{\varepsilon}{4} 2^{-w} 2^{\mu-1} \quad (3.9)$$

where equation 3.9 follows by requiring that $2^{-w} \leq \frac{\varepsilon}{8} \Leftrightarrow w \geq \log(1/\varepsilon) + 3$ and therefore $\left(\frac{\varepsilon}{4} - 2^{-w}\right) \geq \frac{\varepsilon}{8} \geq 2^{-w}$. This shows that $A \geq |E_i| 2^{-2w}$. Continuing from equation 3.8 and applying Lemma 1 with $t_{\min} \leq t$ we get:

$$\Pr\left[\sum_{e_{m,z} \in E_{\mathsf{nsl}}} Y_{m,z} \geq \left(\frac{\varepsilon}{2}\right) 2^k\right] \leq q 8 \left(\frac{2 t_{\min}}{\frac{\varepsilon}{4} 2^{-w} 2^{\mu-1}}\right)^{t_{\min}/2}$$

$$\leq 2^{n-\mu+1} 8 \left(\frac{1}{2}\right)^{t_{\min}} \quad (3.10)$$

where 3.10 follows by requiring $\frac{\varepsilon}{4} 2^{-w} 2^{\mu-1} \geq 8 t_{\min} \Leftrightarrow \mu \geq \log(t_{\min}) + w + \log(1/\varepsilon) + 6$.

Finally, combining 3.1, 3.2, 3.5 and 3.10 we have:

$$\Pr[\text{BAD}] \leq \sum_{f \in \mathcal{F}} P_f = \sum_{f \in \mathcal{F}} \Pr\left[\sum_{(m,z) \in \{0,1\}^n} Y_{m,z} \geq 2^k \varepsilon \right]$$

$$\leq \sum_{f \in \mathcal{F}} \left(\Pr\left[\sum_{e_{m,z} \in E_{\text{sl}}} Y_{m,z} \geq \left(\frac{\varepsilon}{2}\right) 2^k \right] + \Pr\left[\sum_{e_{m,z} \in E_{\text{nsl}}} Y_{m,z} \geq \left(\frac{\varepsilon}{2}\right) 2^k \right] \right)$$

$$\leq |\mathcal{F}| \left(8 \left(\frac{1}{2}\right)^{t_{\min}} + 2^{n-\mu+1} 8 \left(\frac{1}{2}\right)^{t_{\min}} \right)$$

$$\leq 16 |\mathcal{F}| 2^{n-\mu+1} \left(\frac{1}{2}\right)^{t_{\min}} \leq \rho$$

where the last line follows by requiring that $t_{\min} \geq \log|\mathcal{F}| + n - \mu + \log(1/\rho) + 5$ where $n = k + w$. This proves the theorem.

3.2   Upgrading Weak Tamper-Detection Codes

We now show how to convert weak tamper-detection codes to (standard) tamper-detection codes with security for a worst-case message. We do so via a "composed code" construction following Figure 1, which will be useful throughout the paper. The idea of such composed constructions, is to choose the inner code according to the weaknesses of the outer code; in other words the inner code complements the outer code to achieve a stronger notion of security which is expected from the composed code. In our case, the composed code is obtained by composing an inner "leakage-resilient (LR) code" and an outer "weak tamper-detection (WTD) code". The weakness of the outer WTD code comes from the fact that it only guarantees security for a uniformly random message. On the other hand, the inner LR code ensures that one cannot tell between a LR encoding of some worst-case message and a uniformly random inner codeword in the context of the tampering experiment.

Definition 5. *For a function f and a coding scheme* (E, D) *define the binary leakage function of f on* (E, D), *denoted by* $\mathsf{bL}_f[(\mathsf{E}, \mathsf{D})]$ *as follows,*

$$\mathsf{bL}_f[(\mathsf{E}, \mathsf{D})](x) = \begin{cases} 1 & \text{if } \mathsf{D}(f(\mathsf{E}(x))) = \bot \\ 0 & \text{otherwise.} \end{cases}$$

For a family of functions $\mathcal{F}$, define binary leakage function family as $\mathcal{BL}_{\mathcal{F}}[(\mathsf{E}, \mathsf{D})]$ $= \{\mathsf{bL}_f[(\mathsf{E}, \mathsf{D})] \mid \forall f \in \mathcal{F}\}$. When the coding scheme is implicit we omit the index (E, D) and write $\mathsf{bL}_f, \mathcal{BL}_{\mathcal{F}}$.

Theorem 3. *Let $\mathcal{F}$ be a family of functions, let* $(\mathtt{wtdEnc}, \mathtt{wtdDec})$ *be a* (k', n)-*coding scheme which is* $(\mathcal{F}, \varepsilon)$-*weak tamper detection code and let* $(\mathtt{lrEnc}, \mathtt{lrDec})$ *be a* (k, k')-*coding scheme which is* $(\mathcal{BL}_{\mathcal{F}}[(\mathtt{wtdEnc}, \mathtt{wtdDec})], \gamma)$-*leakage resilient code. Then the composed code* $(\mathsf{Enc}, \mathsf{Dec})$ *of Figure 1 is a* (k, n) *coding scheme which is* $(\mathcal{F}, \gamma + \varepsilon)$-*secure tamper detection code.*

Let $(\texttt{lrEnc}, \texttt{lrDec})$ be a (k, k')-code such that $\texttt{lrDec}$ never outputs $\perp$.
Let $(\texttt{wtdEnc}, \texttt{wtdDec})$ be a deterministic (k', n)-code. We define the composed (k, n)-code $(\texttt{Enc}, \texttt{Dec})$ via:

- $\texttt{Enc}(m)$: Let $c_{in} \leftarrow \texttt{lrEnc}(m)$ and output $c = \texttt{wtdEnc}(c_{in})$.
- $\texttt{Dec}(c)$: Let $c_{in} = \texttt{wtdDec}(c)$. If $c_{in} = \perp$, output $\perp$ else output $m = \texttt{lrDec}(c_{in})$.

In particular, let $(\texttt{lrEnc}_{h_1}, \texttt{lrDec}_{h_1})$ and $(\texttt{wtdEnc}_{h_2}, \texttt{wtdDec}_{h_2})$ be given by:

$$\texttt{lrEnc}_{h_1}(m) = (r, h_1(r) \oplus m) : r \leftarrow U_{v_1} , \quad \texttt{lrDec}_{h_1}(r, x) = (h_1(r) \oplus x)$$
$$\texttt{wtdEnc}_{h_2}(c_{in}) = (c_{in}, h_2(c_{in})) , \quad \texttt{wtdDec}_{h_2}(c_{in}, z) = c_{in} \text{ if } z = h_2(c_{in}) \text{ and } \perp \text{ if not.}$$

where $h_1 : \{0,1\}^{v_1} \rightarrow \{0,1\}^k$, $h_2 : \{0,1\}^{k'} \rightarrow \{0,1\}^{v_2}$, $k' := k + v_1, n := k' + v_2$.
Then the composed code $(\texttt{Enc}_{h_1, h_2}, \texttt{Dec}_{h_1, h_2})$ is defined as:

$$\texttt{Enc}_{h_1, h_2}(m) = \left\{ \begin{array}{c} r \leftarrow U_{v_1}, \ x := m \oplus h_1(r), \\ z := h_2(r, x), \text{ output } (r, x, z) \end{array} \right\}$$

$$\texttt{Dec}_{h_1, h_2}(r, x, z) = \left\{ \begin{array}{c} \text{If } z \neq h_2(r, x), \text{ output } \perp \\ \text{otherwise output } x \oplus h_1(r). \end{array} \right\}$$

Fig. 1. Composed Code Construction

See the full version [JW15] for the proof. Combining the results of theorems 1, 2 and 3 gives us the following corollary.

Corollary 1. *Let* $(\texttt{Enc}_{h_1, h_2}, \texttt{Dec}_{h_1, h_2})$ *be the construction in Figure 1 where* h_1 *is chosen from a t-wise independent hash family* $\mathcal{H}_1$, *and* h_2 *is chosen from a t-wise independent hash family* $\mathcal{H}_2$. *Then, for any family of* φ-*few fix points,* μ-*entropy functions* $\mathcal{F}$, *the code* $(\texttt{Enc}_{h_1, h_2}, \texttt{Dec}_{h_1, h_2})$ *is an* $(\mathcal{F}, 2\varepsilon)$-*tamper detection code with probability* $1 - \rho$ *over the choice of* h_1 *and* h_2, *as long as:*

$$t \geq t_{\min}, \qquad\qquad t_{\min} := O(\log |\mathcal{F}| + k + v_1 + \log(1/\rho))$$
$$v_1 \geq \log(t_{\min}) + 2\log(1/\varepsilon) + O(1), \qquad v_2 \geq \log(1/\varepsilon) + O(1).$$
$$\mu \geq \log(t_{\min}) + v_2 + \log(1/\varepsilon) + O(1), \qquad \varphi \leq \varepsilon/4$$

For example, if $\rho = \varepsilon = 2^{-\lambda}$ *for security parameter* λ *and,* $|\mathcal{F}| \geq \max\{2^k, 2^\lambda\}$ *we get:*

$$t = O(\log |\mathcal{F}|) , \quad \mu = \log\log|\mathcal{F}| + 2\lambda + O(1) , \qquad \varphi = 2^{-(\lambda + O(1))}$$
$$v_1 = \log\log|\mathcal{F}| + 2\lambda + O(1) , \qquad\qquad v_2 = \lambda + O(1).$$

See the full version [JW15] for the proof. When $|\mathcal{F}| \leq 2^{2^{\alpha n}}$ for some constant $\alpha < 1$, then the overhead of the code is $n - k = v_1 + v_2 = \alpha n + O(\lambda)$ and therefore the rate of the code approaches $k/n \approx (1 - \alpha)$. The above codes can be made efficient when $|\mathcal{F}| \leq 2^{s(n)}$ for some polynomial s, where the efficiency of the code depends on s (and in this case, the rate approaches 1). In particular, we get an efficient family of codes indexed by hash functions h_1, h_2 such that a random member of the family is a tamper-detection code for $\mathcal{F}$ with overwhelming probability. We can also think of this as an efficient construction in the common random string (CRS) model, where h_1, h_2 are given in the CRS.

Example: Tampering via Polynomials. Let $\mathcal{F}^-_{poly,d}$ be the set of all polynomials $p(x)$ over the field $\mathbb{F}_{2^n}$ of degree d, excluding the identity polynomial $p(x) = x$ and the degree 0 polynomials $\{p(x) = c \; : \; c \in \mathbb{F}_{2^n}\}$. Then $\mathcal{F}^-_{poly,d}$ is an φ-few fix points, μ-entropy function where $\varphi = d/2^n$ and $\mu = n - \log d$. Furthermore $|\mathcal{F}^-_{poly,d}| = 2^{n(d+1)}$.

Using corollary 1, we see that there exist $(\mathcal{F}^-_{poly,d}, \varepsilon)$-tamper-detection codes for degrees up to $d = 2^{\alpha n}$, for constant $\alpha < \frac{1}{2}$, with security ε negligible in n. The rate of such codes approaches $(1 - \alpha)$. Furthermore, when the degree $d = d(n)$ is polynomial in n, then we get an efficient construction in the CRS model with a rate that approaches 1.

Example: Tampering via Affine Functions. Let $\mathcal{F}^-_{affine,r}$ be the set of all affine tampering functions when we identify $\{0,1\}^n$ as the vector space $\mathbb{F}^m_{2^r}$ with $m = n/r$. We exclude the identity and constant functions. In particular, $\mathcal{F}^-_{affine,r}$ consists of all functions $f_{A,b}(x) = Ax + b$ where $A \in \mathbb{F}^{m \times m}_{2^r}, b \in \mathbb{F}^m_{2^r}$, and (1) A is not the all 0 matrix, (2) if A is the identity matrix then $b \neq 0$ is a non-zero vector.

In particular the family $\mathcal{F}^-_{affine,r}$ is a φ-few fix points, μ-entropy function where $\varphi = 2^{-r}$ and $\mu = r$. Furthermore the size of the family is $|\mathcal{F}^-_{affine,r}| \leq 2^{n^2+n}$. The high-entropy requirement is guaranteed by (1) and the few-fixed points requirement is guaranteed by (2) as follows. If x is a fixed point then $Ax + b = x$ means that $(A - I_n)x = b$; if A is identity then this cannot happen since $b \neq 0$ and if A is not identity then this happens with probability at most 2^{-r} over a random x.

Using corollary 1, we get $(\mathcal{F}^-_{affine,r}, \varepsilon)$-tamper-detection codes where ε is negligible in r and the rate of the code approaches 1. Furthermore, we get efficient constructions of such codes in the CRS model.

A similar result would also hold if we considered all affine functions over the vector space $\mathbb{F}^n_2$, given by $f_{A,b}(x) = Ax + b$ where $A \in \mathbb{F}^{n \times n}_2, b \in \mathbb{F}^n_2$, but in this case we would need to add the additional requirement that rank of A is at least r (to ensure high entropy), and either b is not in the column-span of $(A - I_n)$ or the rank of $(A - I_n)$ is at least r (to ensure few fixed points).

3.3   Predictable Codes

So far, we saw that tamper-detection codes are (only) achievable for "restricted" function families with few fixed points and high entropy. However, we now observe that such tamper-detection codes for restricted families can also provide a meaningful security guarantee for function families that don't have the above restrictions. The idea is to consider how tamper-detection can "fail". Firstly, it is possible that the tampering function gets a codeword which is a fixed-point c such that $f(c) = c$. In that case, tampering does not change the codeword and therefore will not get detected. In some sense this failure is not too bad since the codeword did not change. Secondly, it is possible that $f(c) = c^*$ where c^* is some "heavy" value that has many pre-images (responsible for low-entropy of f) and is a valid codeword. Fortunately, there cannot be too many such heavy values c^*. In other words, we can essentially predict what will happen as a result of tampering: either the codeword will not change at all, or it will be tampered to an invalid value that decodes to $\perp$, or it will be tampered to one of a few "heavy" values c^*. We capture this via the following definition of a "predictable code" (a similar notion called "bounded malleable codes" was defined in [FMVW14]) which says that the outcome of tampering a codeword via some function f lies in some "small" set $\mathcal{P}(f)$ which only depends on f and not on the message that was encoded.

Definition 6 (Predictable Codes). *Let* (Enc, Dec) *be a* (k, n)-*coding scheme. For a tampering function* $f : \{0,1\}^n \to \{0,1\}^n$ *and message* $m \in \{0,1\}^k$ *consider a distribution* $\mathsf{tamper}_{f,m}$ *that chooses* $c \leftarrow \mathsf{Enc}(m)$, *sets* $c' := f(c)$, *and if* $c' = c$ *it outputs a special symbol* same, *if* $\mathsf{Dec}(c') = \perp$ *it outputs* $\perp$, *and otherwise it outputs* c'.

For a family of tampering functions $\mathcal{F}$ *and a predictor* $\mathcal{P} : \mathcal{F} \to$ *powerset*$(\{0,1\}^n \cup$ same $\cup \perp)$ *we say that the code is* $(\mathcal{F}, \mathcal{P}, \varepsilon)$-*predictable if for all* $f \in \mathcal{F}, m \in \{0,1\}^k$: $\Pr[\mathsf{tamper}_{f,m} \notin \mathcal{P}(f)] \leq \varepsilon$. *We say that the code is* $(\mathcal{F}, \ell, \varepsilon)$-*predictable if it is* $(\mathcal{F}, \mathcal{P}, \varepsilon)$-*predictable for some* $\mathcal{P}$ *such that for all* $f \in \mathcal{F}, |\mathcal{P}(f)| \leq 2^\ell$.

Let $\mathcal{F}$ be a function family consisting of functions $f : \{0,1\}^n \to \{0,1\}^n$ and $\mu \in [n], \varphi > 0$ be two parameters. We say that $c' \in \{0,1\}^n$ is μ-heavy if $\Pr[f(c) = c' : c \leftarrow \{0,1\}^n] > 1/2^\mu$. Define

$$H_f(\mu) := \{c : c \in \{0,1\}^n \text{ is } \mu\text{-heavy}\}.$$

Note that $|H_f(\mu)| \leq 2^\mu$. For any function $f \in \mathcal{F}$ define the *restricted function* f' by setting $f'(c) := f(c)$ *unless*

(I) if $f(c) \in H_f(\mu)$ then $f'(c) := \perp$.
(II) if $\Pr_{x \in \{0,1\}^n}[f(x) = x] > \varphi$ and $f(c) = c$ then we set $f'(c) := \perp$.

It is clear that f' is a φ-few fix points, μ-entropy function. Define the family $\mathcal{F}[restrict(\mu, \varphi)] = \{f' : f \in \mathcal{F}\}$.

Theorem 4. *For any function family* $\mathcal{F}$, *if* (Enc, Dec) *is an* $(\mathcal{F}[restrict(\mu, \varphi)], \varepsilon)$-*TDC then it is also an* $(\mathcal{F}, \mathcal{P}, \varepsilon)$-*predictable code where* $\mathcal{P}(f) = \{\perp, \mathsf{same}\} \cup H_f(\mu)$. *In particular, it is* $(\mathcal{F}, \mu + 1, \varepsilon)$-*predictable.*

Furthermore, if $\mathcal{F}$ *has* μ-*high entropy, then* $\mathcal{P}(f) = \{\perp, \mathsf{same}\}$. *If* $\mathcal{F}$ *has* φ-*few fixed points then* $\mathcal{P}(f) = \{\perp\} \cup H_f(\mu)$. *If* $\mathcal{F}$ *has* μ-*high entropy and* φ-*few fixed points then* $\mathcal{P}(f) = \{\perp\}$.

See the full version [JW15] for the proof. Combining the results of Theorem 4 and Corollary 1 gives us the following corollary.

Corollary 2. *Let* $(\mathsf{Enc}_{h_1,h_2}, \mathsf{Dec}_{h_1,h_2})$ *be the construction in Figure 1 where* h_1 *is chosen from a t-wise independent hash family* $\mathcal{H}_1$, *and* h_2 *is chosen from a t-wise independent hash family* $\mathcal{H}_2$. *For any family of functions,* $\mathcal{F}$, *the code* $(\mathsf{Enc}_{h_1,h_2}, \mathsf{Dec}_{h_1,h_2})$ *is an* $(\mathcal{F}, \ell, \varepsilon)$-*predictable code with probability* $1 - \rho$ *over the choice of* h_1 *and* h_2, *as long as:*

$$t > t_{\min} \quad \text{where} \quad t_{\min} = O(\log |\mathcal{F}| + k + v_1 + \log (1/\rho))$$
$$v_1 \geq \log(t_{\min}) + 2\log (1/\varepsilon) + O(1) \quad , \quad v_2 \geq \log (1/\varepsilon) + O(1).$$
$$\ell \geq \log(t_{\min}) + v_2 + \log (1/\varepsilon) + O(1)$$

For example, if $\rho = \varepsilon = 2^{-\lambda}$ *for security parameter* λ *and,* $|\mathcal{F}| \geq \max\left\{2^k, 2^\lambda\right\}$ *we get:*

$$t = O(\log |\mathcal{F}|), \qquad\qquad \ell = \log \log |\mathcal{F}| + 2\lambda + O(1),$$
$$v_1 = \log \log |\mathcal{F}| + 2\lambda + O(1), \qquad v_2 = \lambda + O(1).$$

Furthermore, if $\mathcal{F}$ *has* μ-*high entropy for some* $\mu \geq \log \log |\mathcal{F}| + 2\lambda + O(1)$, *then the code is* $(\mathcal{F}, \mathcal{P}, \varepsilon)$-*predictable with* $\mathcal{P}(f) = \{\bot, \mathsf{same}\}$ *and if* $\mathcal{F}$ *has* φ-*few fixed points for some* $\varphi \leq 2^{-(\lambda+O(1))}$ *then* $\mathsf{same} \notin \mathcal{P}(f)$. *If* $\mathcal{F}$ *has* μ-*high entropy and* φ-*few fixed points then* $\mathcal{P}(f) = \{\bot\}$.

A similar result to the above corollary was also shown in [FMVW14] for "bounded malleable codes" via a direct proof that did not go through tamper-detection codes. Here, we achieve some important improvements in parameters. Most importantly, our overhead is $v_1 + v_2 = \log \log |\mathcal{F}| + O(\lambda)$ whereas previously it was at least $3 \log \log |\mathcal{F}| + O(\lambda)$. In other words, when the function family is of size $|\mathcal{F}| = 2^{2^{\alpha n}}$ then we get a rate $\approx (1 - \alpha)$ whereas the result of [FMVW14] would get a rate of $(1 - 3\alpha)$. This improvement in parameters will also translate to our constructions of non-malleable and continuous non-malleable codes.

4   Basic Non-malleable Codes

We now review the definition of non-malleable codes. Several different definitions (standard, strong, super) were proposed in [DPW10, FMVW14] and here we will by default use the strongest of these which was called "super" non-malleable codes in [FMVW14]. This notion considers the tampering experiment $\mathsf{tamper}_{f,m}$ that we previously used to define predictable codes: it chooses $c \leftarrow \mathsf{Enc}(m)$, sets $c' := f(c)$, and if $c' = c$ it outputs a special symbol same, if $\mathsf{Dec}(c') = \bot$ it outputs $\bot$, and otherwise it outputs c'. Non-malleable codes ensure that the output of the experiment is independent of the message m: for any two messages m_0, m_1 the distributions tamper_{f,m_0} and tamper_{f,m_1} should be statistically close.[2]

---

[2] For a weaker definition in [DPW10], the experiment $\mathsf{tamper}_{f,m}$ either outputs same if $c' = f(c)$ or $\mathsf{Dec}(c')$. In contrast, in our definition it outputs c' in full when $\mathsf{Dec}(c') \neq \bot$ which provides more information and makes the definition stronger.

Definition 7 (Non-malleable Code). *Let* $(\mathsf{Enc}, \mathsf{Dec})$ *be a* (k, n)-*coding scheme and* $\mathcal{F}$ *be a family of functions* $f : \{0,1\}^n \to \{0,1\}^n$. *We say that the scheme is* $(\mathcal{F}, \varepsilon)$-*non-malleable if for any* $m_0, m_1 \in \{0,1\}^k$ *and any* $f \in \mathcal{F}$, *we have* $\mathsf{tamper}_{f,m_0} \approx_\varepsilon \mathsf{tamper}_{f,m_1}$ *where*

$$\mathsf{tamper}_{f,m} := \left\{ \begin{array}{c} c \leftarrow \mathsf{Enc}(m), \ c' := f(c) \\ output \ \mathsf{same} \ if \ c' = c, \ output \perp if \ \mathsf{Dec}(c') = \perp, else \ output \ c'. \end{array} \right\}$$

We will argue that the composed code construction in Figure 1 already achieves non-malleability. Intuitively, we will rely on two facts: (1) we already showed that the composed code is predictable meaning that the outcome of the tampering experiment can be thought of as providing at most small amount of leakage, (2) the inner code is leakage-resilient so the small amount of leakage cannot help distinguish between two messages m and m'.

To make the above intuition formal, we need to translate a class of tampering functions into a related class of leakage functions for which we need the inner code to be leakage resilient. This translation is given in the following definition.

Definition 8. *Let* $(\mathsf{wtdEnc}, \mathsf{wtdDec})$ *be a deterministic coding scheme, let* $\mathcal{F} = \{f : \{0,1\}^n \to \{0,1\}^n\}$ *be a family of functions, and let* $\mathcal{P}(f) : \mathcal{F} \to powerset$ $\{\{0,1\}^n \cup \mathsf{same} \cup \perp\}$ *be a predictor. For a tampering function* $f \in \mathcal{F}$, *define the the corresponding leakage function* $\mathsf{L}_f(x)$ *as:*

$$\mathsf{L}_f(x) := \left\{ \begin{array}{ll} \mathsf{R}_f(x) & if \ \mathsf{R}_f(x) \in \mathcal{P}(f) \\ \min \mathcal{P}(f) & otherwise. \end{array} \right.$$

where

$$\mathsf{R}_f(x) := \left\{ \begin{array}{ll} \perp & if \ \mathsf{wtdDec}(f(\mathsf{wtdEnc}(x))) = \perp \\ \mathsf{same} & else \ if \ f(\mathsf{wtdEnc}(x)) = \mathsf{wtdEnc}(x), \\ f(\mathsf{wtdEnc}(x)) & otherwise. \end{array} \right.$$

where $\min \mathcal{P}(f)$ *denotes the minimal value in* $\mathcal{P}(f)$ *according to some ordering. Define the leakage family* $\mathcal{L}_f [(\mathsf{wtdEnc}, \mathsf{wtdDec}), \mathcal{P}] := \{\mathsf{L}_f \mid \forall f \in \mathcal{F}\}$. *When the coding scheme and the predictor function are implicit we simply write* $\mathcal{L}_\mathcal{F}$.

Next we argue that the composed code construction in Figure 1 is non-malleable for $\mathcal{F}$ as long as the composed code is predictable for $\mathcal{F}$ and the inner code is leakage resilient for $\mathcal{L}_\mathcal{F}$. Intuitively, the outer predictable code ensures that tampering can only result in a few possible values, while the inner leakage-resilient code ensures that the choice of which of these few values is the actual outcome of tampering does not reveal anything about the underlying message.

Theorem 5. *Let* $\mathcal{F}$ *be a finite family of functions and let* $(\mathsf{Enc}, \mathsf{Dec})$ *be the composed coding scheme of Figure 1 constructed using an inner code* $(\mathsf{lrEnc}, \mathsf{lrDec})$ *and an outer code* $(\mathsf{wtdEnc}, \mathsf{wtdDec})$. *If* $(\mathsf{Enc}, \mathsf{Dec})$ *is* $(\mathcal{F}, \mathcal{P}, \varepsilon)$-*predictable and* $(\mathsf{lrEnc}, \mathsf{lrDec})$ *is a* $(\mathcal{L}_f [(\mathsf{wtdEnc}, \mathsf{wtdDec}), \mathcal{P}], \gamma)$-*leakage resilient for some predictor* $\mathcal{P}$, *then* $(\mathsf{Enc}, \mathsf{Dec})$ *is a* $(\mathcal{F}, 2(\gamma + \varepsilon))$-*non-malleable code.*

Proof. For all $m_0, m_1 \in \{0,1\}^k$ and for all $f \in \mathcal{F}$, we have:

$$\mathsf{tamper}_{m_0,f} \equiv \mathsf{R}_f(\mathsf{lrEnc}(m_0))$$

$$\approx_\varepsilon \mathsf{L}_f(\mathsf{lrEnc}(m_0)) \tag{4.1}$$

$$\approx_\gamma \mathsf{L}_f(U_{k'}) \tag{4.2}$$

$$\approx_\gamma \mathsf{L}_f(\mathsf{lrEnc}(m_1)) \tag{4.3}$$

$$\approx_\varepsilon \mathsf{R}_f(\mathsf{lrEnc}(m_1)) \tag{4.4}$$

$$\equiv \mathsf{tamper}_{m_1,f}$$

where $\equiv$ denotes distributional equivalence. Lines 4.1 and 4.4 follow from the fact that $\Pr[\mathsf{R}_f(\mathsf{lrEnc}(m)) \neq \mathsf{L}_f(\mathsf{lrEnc}(m))] \leq \Pr[\mathsf{tamper}_{f,m} \notin \mathcal{P}(f)] \leq \varepsilon$ since the combined code is $(\mathcal{F}, \mathcal{P}, \varepsilon)$-predictable. Lines 4.2 and 4.3 follow from the fact that $(\mathsf{lrEnc}, \mathsf{lrDec})$ is a $(\mathcal{L}_\mathcal{F}, \gamma)$-leakage resilient code.

Corollary 3. *Let* $(\mathsf{Enc}_{h_1,h_2}, \mathsf{Dec}_{h_1,h_2})$ *be the construction in Figure 1 where* h_1 *is chosen from a hash family* $\mathcal{H}_1$ *which is* t-*wise independent, and* h_2 *is chosen from a hash family* $\mathcal{H}_2$ *which is* t-*wise independent. For any family of functions* $\mathcal{F}$, *the composed code* $(\mathsf{Enc}_{h_1,h_2}, \mathsf{Dec}_{h_1,h_2})$ *is a* $(\mathcal{F}, \varepsilon)$-*non-malleable code with probability* $1 - \rho$ *over the choice of* h_1 *and* h_2, *as long as the following holds:* $v_2 = \log(1/\varepsilon) + O(1)$, $v_1 \geq v_{\min}$ *and* $t \geq t_{\min}$ *for* $B := \log|\mathcal{F}| + k + v_1 + \log(1/\rho) + O(1)$ *and*

either $t_{\min} = O(B)$ *and* $v_{\min} = \log(B) + \log\log(B/\varepsilon^2) + 4\log(1/\varepsilon) + O(1)$

or $t_{\min} = O(B/\varepsilon^2)$ *and* $v_{\min} = \log(B) + 4\log(1/\varepsilon) + O(1)$.

In particular, if $\rho = \varepsilon = 2^{-\lambda}$ *and* $|\mathcal{F}| \geq \{2^k, 2^\lambda\}$, $\lambda \geq \log\log\log|\mathcal{F}|$ *then we get:*

$$t_{\min} = O(\log|\mathcal{F}|) \qquad and \qquad v_{\min} = \log\log|\mathcal{F}| + O(\lambda)$$

See the full version [JW15] for the proof.

The above corollary tells us that, for any tampering function family $\mathcal{F}$ of size up to $|\mathcal{F}| = 2^{2^{\alpha n}}$ for $\alpha < 1$ there exist (inefficient) non-malleable codes for $\mathcal{F}$ with additive overhead $v_1 + v_2 = \alpha n + O(\lambda)$ and therefore rate approaching $(1 - \alpha)$. This matches the positive results on the rate of non-malleable codes of [CG13a] and is known to be optimal. Furthermore, if $|\mathcal{F}| = 2^{s(n)}$ for some polynomial $s(n)$, then we can get an efficient family of codes such that a random member of the family is an $(\mathcal{F}, 2^{-\lambda})$ non-malleable code for $\mathcal{F}$ with overwhelming probability, where the efficiency of the code is $\mathrm{poly}(s(n), n, \lambda)$ and the additive overhead of the code is $v_1 + v_2 = O(\log n + \lambda)$, and therefore the rate approaches 1. For example, $\mathcal{F}$ could be the family of all circuits of size at most $s(n)$ and we can view our construction as giving an efficient non-malleable code for this family in the CRS model, where the random choice of h_1, h_2 is specified by the CRS. This matches the positive results of [FMVW14]. Therefore, we get a single construction and analysis which achieves the "best of both worlds" depending on the setting of parameters.

5 Continuous Non-malleable Codes

Intuitively, a continuous non-malleable code guarantees security against an attacker that can continuously tamper with a codeword without the codeword being refreshed. We give a unified definition of four variants of such codes depending on two binary variables: (I) a "self-destruct" flag sd which indicates whether the game stops if a codeword is ever detected to be invalid (corresponding to a device self-destructing or erasing its internals) or whether the attacker can continue tampering no matter what, (II) a "persistent" flag prs indicating whether each successive attack modifies the codeword that has been modified by previous attacks, or whether tampering can also always be applied to the original codeword (corresponding to the case where a copy of the original codeword may remain in some auxiliary memory on the device).

In more detail, the definition of continuous non-malleable codes considers a tampering experiment where a codeword $c \leftarrow \mathsf{Enc}(m)$ is chosen in the beginning. The attacker repeatedly chooses tampering functions f_i and we set the i'th tampered codeword to either be $c_i = f_i(c)$ if the "persistent" flag prs is off or $c_i = f_i \circ f_{i-1} \circ \cdots \circ f_1(c)$ if prs is on. Just like the non-continuous definition, we give the attacker the special symbol same if the codeword remains unchanged $c_i = c$, we give the attacker $\perp$ if $\mathsf{Dec}(c_i) = \perp$ and we give the attacker the entire codeword c_i otherwise. In the case where the "self-destruct" flag sd is on, the experiment immediately stops if we give the attacker $\perp$ to capture that the device self-destructs and the attacker has no more opportunity to tamper. In the case where the "persistent" flag prs is on, we also stop the experiment if the attacker ever gets the entire tampered codeword c_i (and not same or $\perp$) in some period i. This is without loss of generality since any future tampering attempts can be answered using c_i alone and hence there is no point in continuing the experiment.

Definition 9. *Let* (Enc, Dec) *be a* (k, n)-*coding scheme and let* $\mathcal{F}$ *be a family of tampering functions* $f : \{0,1\}^n \rightarrow \{0,1\}^n$. *We define four types of continuous non-malleable codes (CNMC) parameterized by the flags* sd $\in \{0, 1\}$ *(self destruct) and* prs $\in \{0, 1\}$ *(persistent tampering). We say that the scheme is a* $(\mathcal{F}, T, \varepsilon)$-*CNMC[prs, sd] if for any two messages* $m_0, m_1 \in \{0, 1\}^k$, *for any* $\mathcal{F}$-*legal adversary* $\mathcal{A}$ *we have* $\mathsf{ConTamper}_{\mathcal{A}, T, m_0} \approx_\varepsilon \mathsf{ConTamper}_{\mathcal{A}, T, m_1}$ *where the experiment* $\mathsf{ConTamper}_{\mathcal{A}, T, m}$ *is defined in figure 2. For* prs $= 0$, *an adversary* $\mathcal{A}$ *is* $\mathcal{F}$-*legal, if the tampering functions chosen in each round* $i \in [T]$ *satisfy* $f_i \in \mathcal{F}$. *For* prs $= 1$, *an adversary* $\mathcal{A}$ *is* $\mathcal{F}$-*legal if the tampering functions* f_i *chosen in each round* i *satisfy* $(f_i \circ \ldots \circ f_1) \in \mathcal{F}$.

Tamper-Resilience via Continuous Non-malleable Codes. We note that the above definition of continuous non-malleable codes directly guarantees strong protection against continuous tampering attacks on a device implementing an arbitrary cryptographic scheme. We imagine that the secret key sk of the cryptographic scheme is encoded using such a code and the codeword is stored on the device; on each invocation, the device first decodes to recover the key and

$\mathsf{ConTamper}_{\mathcal{A},T,m}[\mathsf{prs},\mathsf{sd}]$

 $c \leftarrow \mathsf{Enc}(m)$

 $f_0 := identity$

 Repeat $i = 1, \ldots, T$

  $\mathcal{A}$ chooses a function f_i'

  if $\mathsf{prs} \overset{?}{=} 1 : f_i := f_i' \circ f_{i-1}$ **else** $f_i := f_i'$

  $c' := f_i(c)$

  if $c' \overset{?}{=} c$: $\mathcal{A}$ receives same

  else if $\mathsf{Dec}(c') \overset{?}{=} \bot$: $\{\ \mathcal{A}$ receives $\bot$,   **if** $\mathsf{sd} \overset{?}{=} 1$ experiment stops$\}$

  else : $\{\ \mathcal{A}$ receives c',   **if** $\mathsf{prs} \overset{?}{=} 1$ experiment stops$\}$

The output of the experiment is the view of $\mathcal{A}$.

Fig. 2. Continuous Non-Malleability Experiment

then executes the original scheme. The device never needs to update the codeword. In the setting of "self-destruct" we need the device to erase the stored codeword or simply stop functioning if on any invocation it detects that the codeword decodes to $\bot$. In the setting of non-persistent tampering, we assume the tampering attacks have access to the original codeword rather than just the current value on the device (e.g., because a copy of the original codeword may remain somewhere on the device in some auxiliary memory).

We guarantee that whatever an attacker can learn by continuously tampering with the codeword stored on the device and interacting with the tampered device, could be simulated given only black-box access to the original cryptographic scheme with the original secret key sk and without the ability to tamper. The main idea is that the information that the attacker can learn by interacting with a tampered device is completely subsumed by the information provided by the experiment $\mathsf{ConTamper}_{\mathcal{A},T,sk}$ and interaction with the original untampered device: if the tampering results in $\bot$ then the attacker learns nothing from the tampered device beyond that this happened, if it results in same the attacker gets the outputs of the original device, and if it results in a new valid codeword c' then the attacker at most learns c'. On the other hand, $\mathsf{ConTamper}_{\mathcal{A},T,sk}$ does not provide any additional information about sk by the security of the continuous non-malleable codes.

A similar connection between standard non-malleable codes and tamper-resilience was formalized by [DPW10] in the case where the device updates the stored codeword after each invocation and the above claim simply says that the connection extends to the setting where the device does not update the codeword after each invocation and we use a continuous non-malleable code. The formalization of the above claim and formal proof would be essentially the same as in [DPW10].

5.1   Tool: Repeated Leakage Resilience

Towards the goal of constructing continuous non-malleable codes, we will rely on a new notion of leakage-resilient codes which we call "repeated leakage resilience".

Consider a family $\mathcal{F}$ consisting of leakage functions $f : \{0,1\}^n \to \{0,1\}^\ell$ where each function $f \in \mathcal{F}$ has an associated unique "repeat" value $\mathsf{repeat}_f \in \{0,1\}^\ell$. An attacker can specify a vector of functions $(f_1, \ldots, f_T)$ with $f_i \in \mathcal{F}$. We apply the functions to the codeword c one-by-one: if the output if $f_i(c) = \mathsf{repeat}_{f_i}$ we continue to the next function until we get the first i for which $f_i(c) \neq \mathsf{repeat}_{f_i}$ in which case we output $f_i(c)$. As in the case of standard leakage-resilient codes, we don't want the attacker to be able to distinguish between an encoding of a particular message m versus a random codeword.

Definition 10. *Let* $(\mathsf{lrEnc}, \mathsf{lrDec})$ *be a* (k, n)-*coding scheme. Let* $\mathcal{F}$ *be a family of functions* $f : \{0,1\}^n \to \{0,1\}^\ell$ *where each function* $f \in \mathcal{F}$ *has an associated "repeat value"* $\mathsf{repeat}_f \in \{0,1\}^\ell$. *For an integer* $T \geq 1$, *we say that* $(\mathsf{lrEnc}, \mathsf{lrDec})$ *is* $(\mathcal{F}, T, \varepsilon)$-*repeated-leakage-resilient, if for any* $\bar{f} = (f_1, \ldots, f_T) \in \mathcal{F}^T$ *and any* $m \in \{0,1\}^k$ *we have* $\mathsf{RLeakage}(\bar{f}, \mathsf{lrEnc}(m)) \approx_\varepsilon \mathsf{RLeakage}(\bar{f}, U_n)$, *where we define* $\mathsf{RLeakage}(\bar{f}, c)$ *to output the value* $(i, f_i(c))$ *for the smallest* i *such that* $f_i(c) \neq \mathsf{repeat}_{f_i}$, *or* $\perp$ *if no such* i *exists.*

Notice that if we allow T to be as large as $T = 2^n$ then there is a simple family $\mathcal{F}$ of only 2^n leakage functions with just 1-bit output such $\mathsf{RLeakage}(\bar{f}, \mathsf{lrEnc}(m))$ completely reveals m. In particular consider the family $\mathcal{F}$ of functions $f_{c'} : \{0,1\}^n \to \{\mathsf{stop}, \mathsf{repeat}\}$ that have a hard-coded value $c' \in \{0,1\}^n$ and we define $f_{c'}(c) = \mathsf{stop}$ if $c' = c$ and $f_{c'}(c) = \mathsf{repeat}$ otherwise (1 bit output). Let $\bar{f}$ be a vector of all functions in $\mathcal{F}$. Then $\mathsf{RLeakage}(\bar{f}, c)$ completely reveals c.

In general, we can always view repeated-leakage resilience as a special case of standard leakage resilience. In particular, for any family of leakage functions $\mathcal{F}$ with associated repeat values repeat_f we can define the family $\mathcal{F}_{\mathsf{repeat},T} := \{\mathsf{RLeakage}(\bar{f}, \cdot) : \bar{f} \in \mathcal{F}^T\}$. It is clear that an $(\mathcal{F}_{\mathsf{repeat},T}, \varepsilon)$-leakage-resilient code is also $(\mathcal{F}, T, \varepsilon)$-repeated-leakage-resilient. If the function-family $\mathcal{F}$ consists of functions with ℓ-bits of leakage then the functions in $\mathcal{F}_{\mathsf{repeat},T}$ have $\leq \ell + \log T + 1$ bits of leakage (e.g., the range of $f \in \mathcal{F}_{\mathsf{repeat},T}$ is $[T] \times \{0,1\}^\ell \cup \{\perp\}$). In other words, the amount of leakage only increases by $\log T + 1$ which is not very large. Unfortunately, even though the *amount* of leakage is not much larger, the *quality* of the leakage measured as the size of the leakage family is substantially larger: $|\mathcal{F}_{\mathsf{repeat},T}| = |\mathcal{F}|^T$. This hurts our parameters and requires the efficiency of our codes to depend on and exceed T. Summarizing the above discussion with the parameters of standard leakage-resilient codes from Section 2.1: Theorems 1 we get the following.

Theorem 6. *Let* $\mathcal{F}$ *be a family of functions of the form* $f : \{0,1\}^n \to \{0,1\}^\ell$ *where each function* f *has an associated "repeat" value* $\mathsf{repeat}_f \in \{0,1\}^\ell$. *If* $(\mathsf{lrEnc}, \mathsf{lrDec})$ *is a* (k, n)-*coding scheme which is* $(\mathcal{F}_{\mathsf{repeat},T}, \varepsilon)$-*leakage-resilient then it is also*

$(\mathcal{F}, T, \varepsilon)$-*repeated-leakage-resilient. In particular let* $(\mathtt{lrEnc}_h, \mathtt{lrDec}_h)$ *be the family of* $(k, n = k + v)$-*codes defined in Section 2.1. Then with probability* $1 - \rho$ *over the choice of* $h \leftarrow \mathcal{H}$ *chosen from a* t-*wise independent hash function family, the code* $(\mathtt{lrEnc}_h, \mathtt{lrDec}_h)$ *is* $(\mathcal{F}, T, \varepsilon)$-*repeated-leakage-resilient as long as:*

$$t \geq t_{\min} := O(T \cdot \log |\mathcal{F}| + k + \log(1/\rho))$$

$$v \geq \log\left(t_{\min} + (2^{\ell+1} \cdot T)\right) + \log(\ell + \log T + 1) + 2\log(1/\varepsilon) + O(1)$$

In the natural setting where $\varepsilon = \rho = 2^{-\lambda}$ *and* $T, \ell \leq 2^{\lambda}$, $|\mathcal{F}| \geq \max\{2^{\lambda}, 2^{k}\}$, *for security parameter* λ, *we have*

$$t = O(T \cdot \log |\mathcal{F}|) \qquad\qquad v = \max\{\ell, \log\log|\mathcal{F}|\} + O(\lambda)$$

Recall that in standard leakage-resilient codes, if we have $|\mathcal{F}| = 2^{s(n)}$ for some polynomial $s(n)$, then our construction was efficient and we required polynomial independence $t = O(s(n))$. Unfortunately, for repeated leakage resilience, the above theorem requires us to now set $t = O(Ts(n))$ depending on T. In general, we would like to have a fixed efficient construction which is secure for any polynomial number of repeated leakage queries T, rather than the reverse quantifiers where for any polynomial T we have an efficient construction which depends on T. We believe that our analysis of repeated-leakage resilience is sub-optimal and that the above goal is possible. Intuitively, we believe that the sub-optimality of the analysis comes from the fact that it completely ignores the structure of the function family $\mathcal{F}_{\mathsf{repeat},T}$ and only counts the number of functions. However, the functions $\mathsf{RLeakage}(\bar{f}, \cdot)$ in the family have significantly restricted structure and are not much more complex than the functions $f \in \mathcal{F}$. Indeed, we can think of $\mathsf{RLeakage}(\bar{f}, \cdot)$ as first computing $f_1(c), \ldots, f_T(c)$ and then computing some shallow circuit on top of the T outputs. We put forward the following conjecture, and leave it as an interesting open problem to either prove or refute it.

Conjecture 1. Theorem 6 holds in the setting $\varepsilon = \rho = 2^{-\lambda}$ and $T, \ell \leq 2^{\lambda}$, $|\mathcal{F}| \geq \max\{2^{\lambda}, 2^{k}\}$, with $t = O(\lambda \log |\mathcal{F}|)$ instead of $t = O(T \log |\mathcal{F}|)$.

5.2 Abstract Construction of Continuous Non-malleable Codes

We now show how to construct continuous non-malleable codes for all four cases depending on self-destruct (yes/no) and persistent tampering (yes/no). Our constructions follow the same template as the construction of standard non-malleable codes. In particular, to get continuous non-malleability for some function family $\mathcal{F}$ we will use the composed code construction from Figure 1 and rely on the fact that the combined code is "predictable" for $\mathcal{F}$ with some predictor $\mathcal{P}$ satisfying certain properties and that the inner code is repeated leakage resilient for a corresponding leakage family. Intuitively, we want to ensure that if the predictor $\mathcal{P}(f)$ outputs a large set of possible values that might be the outcomes of any single tampering attempt via a function f, there is at most one value $\mathsf{repeat}_f \in \mathcal{P}(f)$ which causes the continuous tamper experiment in figure 2 to continue and not stop. Therefore, the adversary does no learn much information during the tampering experiment.

Theorem 7. *Let $\mathcal{F}$ be a family of tampering functions of the form $f : \{0,1\}^n \to \{0,1\}^n$. Let $(\mathsf{Enc}, \mathsf{Dec})$ be the composed code from Figure 1, constructed using an inner code $(\mathsf{lrEnc}, \mathsf{lrDec})$ and an outer code $(\mathsf{wtdEnc}, \mathsf{wtdDec})$. Assume $(\mathsf{Enc}, \mathsf{Dec})$ is $(\mathcal{F}, \mathcal{P}, \varepsilon_1)$-predictable code with some predictor $\mathcal{P}$ satisfying the conditions below, and $(\mathsf{lrEnc}, \mathsf{lrDec})$ a $(\mathcal{L}_\mathcal{F}, T, \varepsilon_2)$-repeated leakage resilient code for the leakage family $\mathcal{L}_\mathcal{F} = \mathcal{L}_\mathcal{F}[(\mathsf{wtdEnc}, \mathsf{wtdDec}), \mathcal{P}]$ (see Definition 8) and where each $\mathsf{L}_f \in \mathcal{L}_\mathcal{F}$ has an associated repeat value repeat_f as defined below. Then the composed code $(\mathsf{Enc}, \mathsf{Dec})$ is $(\mathcal{F}, T, \varepsilon)$-CNMC[sd, prs] with $\varepsilon = 2T \cdot \varepsilon_1 + 2\varepsilon_2$.*

- For case $\mathsf{sd} = 0, \mathsf{prs} = 0$ we require that for all $f \in \mathcal{F}$, $|\mathcal{P}(f)| = 1$. In this case, the leakage-resilience requirement is vacuous since the leakage family $\mathcal{L}_\mathcal{F}$ has 0-bit output.
- For case $\mathsf{sd} = 1, \mathsf{prs} = 0$ we require that for all $f \in \mathcal{F}$, either $|\mathcal{P}(f)| = 1$ or $\mathcal{P}(f) = \{\perp, \mathsf{same}\}$. For each $\mathsf{L}_f \in \mathcal{L}_\mathcal{F}$, we define $\mathsf{repeat}_f = \mathcal{P}(f)$ if $|\mathcal{P}(f)| = 1$ or $\mathsf{repeat}_f = \mathsf{same}$ otherwise.
- For case $\mathsf{sd} = 0, \mathsf{prs} = 1$ we require that for all $f \in \mathcal{F}$, either $|\mathcal{P}(f)| = 1$ or $\mathsf{same} \notin \mathcal{P}(f)$. For each $\mathsf{L}_f \in \mathcal{L}_\mathcal{F}$, we define $\mathsf{repeat}_f = \mathcal{P}(f)$ if $|\mathcal{P}(f)| = 1$ or $\mathsf{repeat}_f = \perp$ otherwise.
- For case $\mathsf{sd} = 1, \mathsf{prs} = 1$, we have no requirements on $\mathcal{P}(f)$. For each $\mathsf{L}_f \in \mathcal{L}_\mathcal{F}$, we define $\mathsf{repeat}_f = \mathsf{same}$.

See the full version [JW15] for the proof.

5.3 Construction Results and Parameters

We now summarize what we get when we instantiate the above construction with the code $(\mathsf{Enc}_{h_1,h_2}, \mathsf{Dec}_{h_1,h_2})$ from Figure 1. We characterize the type of function families for which we achieve the various cases of continuous non-malleable security in terms of whether the family has "few fixed points" and "high entropy".

Corollary 4. *Let $(\mathsf{Enc}_{h_1,h_2}, \mathsf{Dec}_{h_1,h_2})$ be the (k, n)-coding scheme construction in Figure 1 where h_1 is chosen from a t_1-wise independent hash family $\mathcal{H}_1$, and h_2 is chosen from a t_2-wise independent hash family $\mathcal{H}_2$.*

Let $\mathcal{F}$ be a family of φ-few fix points, μ-entropy functions and assume that $|\mathcal{F}| \geq \max\{2^k, 2^\lambda\}$. Then for any $T \leq 2^\lambda$ the code $(\mathsf{Enc}_{h_1,h_2}, \mathsf{Dec}_{h_1,h_2})$ is an $(\mathcal{F}, T, T \cdot 2^{-\lambda})$-CNMC[sd, prs] with probability $1 - 2^{-\lambda}$ over the choice of h_1 and h_2 with the following parameters for each of the four options, where $\mathsf{sd} = 1$ indicates self-destruct and $\mathsf{prs} = 1$ indicates persistent tampering:

[sd, prs]	$\varphi \leq$	$\mu \geq$	$t_1 =$	$t_2 =$
[0 , 0]	$2^{-(\lambda+O(1))}$	$\log\log\|\mathcal{F}\| + 2\lambda + O(1)$	$O(\log\|F\|)$	$O(\log\|F\|)$
[0 , 1]	$2^{-(\lambda+O(1))}$	No restriction	$O(T \cdot \log\|\mathcal{F}\|)$	$O(\log\|F\|)$
[1 , 0]	No restriction	$\log\log\|\mathcal{F}\| + 2\lambda + O(1)$	$O(T \cdot \log\|\mathcal{F}\|)$	$O(\log\|F\|)$
[1 , 1]	No restriction	No restriction	$O(T \cdot \log\|\mathcal{F}\|)$	$O(\log\|F\|)$

In all four cases, parameter $t_2 = O(\log\|\mathcal{F}\|)$ and parameter $v_1 + v_2 = \log\log\|\mathcal{F}\| + O(\lambda)$. All results in the table also hold if $\mathcal{F}$ includes the always identity function $f_\mathsf{id}(x) = x$ or the always constant functions $\{f_c(x) = c : c \in \{0,1\}^n\}$.

If Conjecture 1 holds then we can set $t_1 = O(\lambda \cdot \log |\mathcal{F}|)$ for the last three rows of the table.

Proof. Theorem 7 tells us how to build continuous non-malleable codes using predictable codes and repeated-leakage resilient codes. Corollary 2 gives us the parameters for predictable codes and the type of predictability they achieve. Furthermore, we can always add the identity and constant functions f to any family and maintain predictability with $|\mathcal{P}(f)| = 1$ for these functions. Theorem 6 gives us the parameters for repeated-leakage-resilient codes. By plugging in the parameters, the corollary follows.

We explore the consequences of the above corollary in each of the four settings.

No Self-Destruct, Non-Persistent Tampering (Strongest). We begin with the strongest setting, where we assume no self-destruct and non-persistent tampering. In this case, an $(\mathcal{F}, \varepsilon)$-tamper-detection code is also an $(\mathcal{F}, \mathcal{P}, \varepsilon)$-predictable code with $\mathcal{P}(f) = \perp$ for each $f \in \mathcal{F}$, and hence it is also an $(\mathcal{F}, T, 2T \cdot \varepsilon)$-continuous non-malleable code in this setting. In particular, we get such codes for families of functions with high entropy and few fixed points. However, we can also add the "always identity" function $f_{\mathsf{id}}(x) = x$ or the "always constant" functions $\{f_c(x) = c : c \in \{0,1\}^n\}$ to $\mathcal{F}$ and maintain "predictability" and therefore also continuous non-malleable security in this setting.

As an example, we can achieve continuous non-malleable codes in this setting for the function families discusses in Section 3 in the context of tamper-detection. By adding in the "always identity" and the "always constant" functions to these families we get such continuous non-malleable codes for:

- The family $\mathcal{F}_{\mathsf{poly},d}$ consisting of *all* polynomials $p(x)$ of degree d over the field $\mathbb{F}_{2^n}$. For inefficient codes, we can set the degree as high as $d = 2^{\alpha n}$, for constant $\alpha < \frac{1}{2}$ and get rate $(1 - \alpha)$. For efficient codes (in the CRS model), the degree must be set to some polynomial $d = d(n)$ and the rate approaches 1.
- The family $\mathcal{F}_{\mathsf{affine},r}$ consisting of *all* affine tampering functions when we identify $\{0,1\}^n$ as the vector space $\mathbb{F}_{2^r}^m$ with $m = n/r$: i.e., all functions $f_{A,b}(x) = Ax + b$ where $A \in \mathbb{F}_{2^r}^{m \times m}, b \in \mathbb{F}_{2^r}^m$. Such codes are efficient (in the CRS model) and their security is $2^{-r+O(1)}$.

The number of tampering attacks T that such codes protect against can even be set to exponential in the security parameter $T = 2^\lambda$ without hurting efficiency.

Self Destruct, Persistent Tampering (Weakest). In the weakest setting, where we assume self-destruct and persistent tampering, we show the existence of (inefficient) continuous non-malleable codes which essentially matches the known results for standard non-malleable codes. In particular, for a function family $\mathcal{F}$ of size $|\mathcal{F}| = 2^{2^{\alpha n}}$ we can protect against even an exponential number $T = 2^\lambda$ tampering attempts with security $\varepsilon = 2^{-\lambda}$ and get a rate which approaches $(1 - \alpha)$. For example, we show the existence of such (inefficient) codes in the *split-state model* where the n-bit codeword is divided into two halves and the attacker can tamper each half of the codeword arbitrarily but independently of the other half.

Furthermore, if we take a function family of size $|\mathcal{F}| = 2^s$ for some polynomial s and want to protect against some polynomial T number of tampering attempts, then we get efficient constructions (in the CRS model) where the efficiency is $\mathsf{poly}(s, T, \lambda)$. For example, the family $\mathcal{F}$ could be all circuits of size s. The rate of such codes approaches 1. We know that the dependence between the efficiency of the code and s is necessary - for example, if the tampering functions are all circuits of size s then we know that the circuit-size of the code must exceed $O(s)$ as otherwise the tampering function could simply decode, flip the last bit, and re-encode. However, we do not know if the dependence between the efficiency of the code and T is necessary. This dependence comes from the parameters of repeated leakage-resilient codes and, if we could improve the parameters there as conjectured (see conjecture 1) we would get security for up to $T = 2^\lambda$ tampering attempts with a code having efficiency $\mathsf{poly}(s, \lambda)$.

Intermediate Cases. The two intermediate cases lie between the strongest and the weakest setting and correspond to either (I) having persistent tampering but no self-destruct or (II) having self-destruct but non-persistent tampering. These cases are incomparable to each other.

In case (I) we can allow function families with arbitrarily low-entropy but we need to require that the function only have few fixed points. The few-fixed points requirement is necessary in this setting as highlighted by the following attack. Consider the family $\mathcal{F}_{\mathsf{nextbit}}$ consisting of n functions $\{f_i(x) \ : \ i \in [n]\}$ where $f_i(x)$ looks at the i'th bit of x and if it a 0 then keeps x as is and otherwise it flips the $(i+1)$'st bit (if $i = n$ then the first bit) of x. We also add the identity function to the family. This family has full entropy $\mu = n$ but the functions (even the non-identity ones) have many fixed points $\varphi = 1/2$. The attacker does the following for $i = 1, \ldots, n$: he applies the function f_i and sees whether he gets back same. If so, he learns the i'th bit of the original codeword is a 0 and otherwise he learns that it was a 1. Either way the attacker applies f_i again to set the codeword back to the original. This is a legal attacker since if we compose the functions chosen by the attacker we get $f_i \circ \cdots \circ f_1 \circ f_1$ which is either just f_i or identity and therefore in the family $\mathcal{F}_{\mathsf{nextbit}}$. After n iterations the attacker completely learns the codeword and therefore also the encoded message. Notice that this attack crucially relies on the fact that there is no self-destruct, since each tampering attempt could result in $\perp$.

In case (II) we can allow function families with many fixed points but require entropy. Intuitively, the high entropy requirement is necessary in this setting to prevent the following attack. Consider the family $\mathcal{F}_{2const}$ consisting of $n2^{2n}$ functions $\{f_{i,c_0,c_1}(x) \ : \ c_0, c_1 \in \{0, 1\}^n, i \in [n]\}$ which output c_0 if the ith bit of x is a 0 and c_1 if the ith bit of x is a 1. This family has $\varphi = 2^{-n+1}$ few fixed points but has low entropy $\mu = 1$. Given any code (Enc, Dec) the attacker simply chooses any two distinct valid codewords $c_0 \neq c_1$, $\mathsf{Dec}(c_0), \mathsf{Dec}(c_1) \neq \perp$. For $i = 1, \ldots, n$: he applies the tampering function f_{i,c_0,c_1} and depending on whether he gets back c_0 or c_1, he learns the i'th bit of the original codeword c. After n iterations the attacker completely learns the codeword. Notice that this attack crucially relies on the fact that tampering is non-persistent, since each tampering attempt completely overwrites the codeword but the next tampering attempt still tampers the original codeword.

The parameters of our continuous non-malleable codes in these settings matches that of the weakest case. In particular, for a function family $\mathcal{F}$ of size $|\mathcal{F}| = 2^{2^{\alpha n}}$ satisfying the appropriate entropy and fixed-point requirements, we can inefficiently protect against even an exponential number $T = 2^\lambda$ tampering attempts with security $\varepsilon = 2^{-\lambda}$ and get a rate which approaches $(1 - \alpha)$. Furthermore, if we take a function family of size $|\mathcal{F}| = 2^s$ for some polynomial s and want to protect against some polynomial T number of tampering attempts, then we get efficient constructions (in the CRS model) where the efficiency is $\mathsf{poly}(s, T, \lambda)$. Under conjecture 1, we would get security for up to $T = 2^\lambda$ tampering attempts with a code having efficiency $\mathsf{poly}(s, \lambda)$.

6   Conclusion

In this paper, we introduced several new notions of codes that offer protection against tampering attacks. Most importantly, we defined a general notion of tamper-detection codes and various flavors of continuous non-malleable codes and explored the question of what families of functions admit such codes. Although some of our constructions can be made efficient, all of the constructions are Monte-Carlo constructions, which can also be interpreted as constructions in the CRS model. It remains an open problem to construct explicit and efficient codes (without a CRS) for interesting families, such as low-degree polynomials or affine functions (etc.).

References

[ABPP14] Abdalla, M., Benhamouda, F., Passelègue, A., Paterson, K.G.: Related-key security for pseudorandom functions beyond the linear barrier. In: Garay, J.A., Gennaro, R. (eds.) CRYPTO 2014, Part I. LNCS, vol. 8616, pp. 77–94. Springer, Heidelberg (2014)

[ADL13] Aggarwal, D., Dodis, Y., Lovett, S.: Non-malleable codes from additive combinatorics. Electronic Colloquium on Computational Complexity (ECCC) 20, 81 (2013)

[AGM+14] Agrawal, S., Gupta, D., Maji, H.K., Pandey, O., Prabhakaran, M.: Explicit non-malleable codes resistant to permutations and perturbations. Cryptology ePrint Archive, Report 2014/841 (2014), http://eprint.iacr.org/

[AHI11] Applebaum, B., Harnik, D., Ishai, Y.: Semantic security under related-key attacks and applications. In: ICS, pp. 45–60 (2011)

[BC10] Bellare, M., Cash, D.: Pseudorandom functions and permutations provably secure against related-key attacks. In: Rabin, T. (ed.) CRYPTO 2010. LNCS, vol. 6223, pp. 666–684. Springer, Heidelberg (2010)

[BCM11]   Bellare, M., Cash, D., Miller, R.: Cryptography secure against related-key attacks and tampering. In: Lee, D.H., Wang, X. (eds.) ASIACRYPT 2011. LNCS, vol. 7073, pp. 486–503. Springer, Heidelberg (2011)

[BDL01]   Boneh, D., DeMillo, R.A., Lipton, R.J.: On the importance of eliminating errors in cryptographic computations. J. Cryptology 14(2), 101–119 (2001)

[BK03]   Bellare, M., Kohno, T.: A theoretical treatment of related-key attacks: RKA-PRPs, RKA-PRFs. In: Biham, E. (ed.) EUROCRYPT 2003. LNCS, vol. 2656, pp. 491–506. Springer, Heidelberg (2003)

[BPT12]   Bellare, M., Paterson, K.G., Thomson, S.: RKA security beyond the linear barrier: IBE, encryption and signatures. In: Wang, X., Sako, K. (eds.) ASIACRYPT 2012. LNCS, vol. 7658, pp. 331–348. Springer, Heidelberg (2012)

[BR94]   Bellare, M., Rompel, J.: Randomness-efficient oblivious sampling. In: FOCS, pp. 276–287. IEEE Computer Society (1994)

[CDF+08] Cramer, R., Dodis, Y., Fehr, S., Padró, C., Wichs, D.: Detection of algebraic manipulation with applications to robust secret sharing and fuzzy extractors. In: Smart, N.P. (ed.) EUROCRYPT 2008. LNCS, vol. 4965, pp. 471–488. Springer, Heidelberg (2008)

[CG13a]   Cheraghchi, M., Guruswami, V.: Capacity of non-malleable codes. Electronic Colloquium on Computational Complexity (ECCC) 20, 118 (2013)

[CG13b]   Cheraghchi, M., Guruswami, V.: Non-malleable coding against bit-wise and split-state tampering. IACR Cryptology ePrint Archive, 2013:565 (2013)

[CKM11]   Choi, S.G., Kiayias, A., Malkin, T.: BiTR: Built-in tamper resilience. In: Lee, D.H., Wang, X. (eds.) ASIACRYPT 2011. LNCS, vol. 7073, pp. 740–758. Springer, Heidelberg (2011)

[CZ14]   Chattopadhyay, E., Zuckerman, D.: Non-malleable codes against constant split-state tampering. Electronic Colloquium on Computational Complexity (ECCC) 21, 102 (2014)

[DDV10]   Davì, F., Dziembowski, S., Venturi, D.: Leakage-resilient storage. In: Garay, J.A., De Prisco, R. (eds.) SCN 2010. LNCS, vol. 6280, pp. 121–137. Springer, Heidelberg (2010)

[DFMV13] Damgård, I., Faust, S., Mukherjee, P., Venturi, D.: Bounded tamper resilience: How to go beyond the algebraic barrier. In: Sako, K., Sarkar, P. (eds.) ASIACRYPT 2013, Part II. LNCS, vol. 8270, pp. 140–160. Springer, Heidelberg (2013)

[DKO13]   Dziembowski, S., Kazana, T., Obremski, M.: Non-malleable codes from two-source extractors. In: Canetti, R., Garay, J.A. (eds.) CRYPTO 2013, Part II. LNCS, vol. 8043, pp. 239–257. Springer, Heidelberg (2013)

[DPW10]   Dziembowski, S., Pietrzak, K., Wichs, D.: Non-malleable codes. In: ICS, pp. 434–452 (2010)

[FMNV14] Faust, S., Mukherjee, P., Nielsen, J.B., Venturi, D.: Continuous non-malleable codes. In: TCC 2014. Springer, Heidelberg (2014)

[FMVW14] Faust, S., Mukherjee, P., Venturi, D., Wichs, D.: Efficient non-malleable codes and key-derivation for poly-size tampering circuits. In: Nguyen, P.Q., Oswald, E. (eds.) EUROCRYPT 2014. LNCS, vol. 8441, pp. 111–128. Springer, Heidelberg (2014), http://eprint.iacr.org/2013/702

[FPV11]   Faust, S., Pietrzak, K., Venturi, D.: Tamper-proof circuits: How to trade leakage for tamper-resilience. In: Aceto, L., Henzinger, M., Sgall, J. (eds.) ICALP 2011, Part I. LNCS, vol. 6755, pp. 391–402. Springer, Heidelberg (2011)

[GIP+14]   Genkin, D., Ishai, Y., Prabhakaran, M., Sahai, A., Tromer, E.: Circuits resilient to additive attacks with applications to secure computation. In: Shmoys, D.B. (ed.) Symposium on Theory of Computing, STOC 2014, New York, NY, USA, May 31-June 03, pp. 495–504. ACM Press (2014)

[GLM+04]   Gennaro, R., Lysyanskaya, A., Malkin, T., Micali, S., Rabin, T.: Algorithmic tamper-proof (ATP) security: Theoretical foundations for security against hardware tampering. In: Naor, M. (ed.) TCC 2004. LNCS, vol. 2951, pp. 258–277. Springer, Heidelberg (2004)

[GOR11]   Goyal, V., O'Neill, A., Rao, V.: Correlated-input secure hash functions. In: Ishai, Y. (ed.) TCC 2011. LNCS, vol. 6597, pp. 182–200. Springer, Heidelberg (2011)

[IPSW06]   Ishai, Y., Prabhakaran, M., Sahai, A., Wagner, D.: Private circuits II: Keeping secrets in tamperable circuits. In: Vaudenay, S. (ed.) EUROCRYPT 2006. LNCS, vol. 4004, pp. 308–327. Springer, Heidelberg (2006)

[JW15]   Jafargholi, Z., Wichs, D.: Tamper detection and continuous non-malleable codes [full version] (2015), http://eprint.iacr.org/2014/956

[KKS11]   Kalai, Y.T., Kanukurthi, B., Sahai, A.: Cryptography with tamperable and leaky memory. In: Rogaway, P. (ed.) CRYPTO 2011. LNCS, vol. 6841, pp. 373–390. Springer, Heidelberg (2011)

[LL12]   Liu, F.-H., Lysyanskaya, A.: Tamper and leakage resilience in the split-state model. In: Safavi-Naini, R., Canetti, R. (eds.) CRYPTO 2012. LNCS, vol. 7417, pp. 517–532. Springer, Heidelberg (2012)

[Pie12]   Pietrzak, K.: Subspace LWE. In: Cramer, R. (ed.) TCC 2012. LNCS, vol. 7194, pp. 548–563. Springer, Heidelberg (2012)

[Sha49]   Claude, E.: Shannon. A theorem on coloring the lines of a network. Journal of Mathematics and Physics / Massachusetts Institute of Technology 28, 148–151 (1949)

[Wee12]   Wee, H.: Public key encryption against related key attacks. In: Fischlin, M., Buchmann, J., Manulis, M. (eds.) PKC 2012. LNCS, vol. 7293, pp. 262–279. Springer, Heidelberg (2012)

Optimal Algebraic Manipulation Detection Codes in the Constant-Error Model

Ronald Cramer[1], Carles Padró[2], and Chaoping Xing[3]

[1] CWI, Amsterdam and Mathematical Institute, Leiden University, The Netherlands
[2] Universitat Politècnica de Catalunya, Barcelona, Spain
[3] School of Physical and Mathematical Sciences, Nanyang Technological University, Singapore

Abstract. Algebraic manipulation detection (AMD) codes, introduced at EUROCRYPT 2008, may, in some sense, be viewed as *keyless* combinatorial authentication codes that provide security in the presence of an *oblivious, algebraic* attacker. Its original applications included robust fuzzy extractors, secure message transmission and robust secret sharing. In recent years, however, a rather diverse array of additional applications in cryptography has emerged. In this paper we consider, for the first time, the regime of arbitrary positive constant error probability ϵ in combination with unbounded cardinality M of the message space. There are several applications where this model makes sense. Adapting a known bound to this regime, it follows that the binary length ρ of the tag satisfies $\rho \geq \log \log M + \Omega_\epsilon(1)$. In this paper, we shall call AMD codes meeting this lower bound *optimal*. Known constructions, notably a construction based on dedicated polynomial evaluation codes, are a multiplicative factor 2 *off* from being optimal. By a generic enhancement using error-correcting codes, these parameters can be further improved but remain suboptimal. Reaching optimality efficiently turns out to be surprisingly nontrivial. We propose a novel constructive method based on symmetries of codes. This leads to an explicit construction based on certain BCH codes that improves the parameters of the polynomial construction and to an efficient randomized construction of optimal AMD codes based on certain quasi-cyclic codes. In all our results, the error probability ϵ can be chosen as an arbitrarily small positive real number.

1 Introduction

Algebraic manipulation detection (AMD) codes, introduced at EUROCRYPT 2008 [5], may, in some sense, be viewed as *keyless* combinatorial authentication codes that provide security in the presence of an *oblivious, algebraic* attacker. Briefly, a systematic AMD encoding is a pair consisting of a message m and a tag τ. Given the message, the tag is sampled probabilistically from some given finite abelian group, according to a distribution depending on the details of the scheme. The attack model considers an adversary which substitutes an intercepted pair (m, τ) by a pair $(\widetilde{m}, \widetilde{\tau})$ with $\widetilde{m} \neq m$ such that it knows $\Delta := \widetilde{\tau} - \tau$ and such that Δ is independently distributed from τ. It may, however, depend on m. The

Y. Dodis and J.B. Nielsen (Eds.): TCC 2015, Part I, LNCS 9014, pp. 481–501, 2015.
© International Association for Cryptologic Research 2015

error probability ϵ of an AMD code upper bounds the success probability of the best strategy to have a substitution accepted as a valid encoding.[1]

The original applications [5] of AMD codes included robust fuzzy extractors, secure message transmission, and robust secret sharing. During the last few years, however, several interesting new applications have emerged. Namely, AMD codes play a role in topics such as construction of non-malleable codes [7], codes for computationally bounded channels [10], unconditionally secure multiparty computation with dishonest majority [3], complete primitives for fairness [9], and public key encryption resilient against related key attacks [13].

In this paper we consider, for the first time, the regime of arbitrarily small positive constant error probability ϵ in combination with unbounded cardinality M of the message space. This model makes sense for most of the known information-theoretic applications of AMD codes. This is the case for secure message transmission, robust secret sharing and robust fuzzy extractors [5], and also for non-malleable codes [7, Theorem 4.1], unconditionally secure multiparty computation with dishonest majority [3, Theorem 8.3], and codes for computationally simple channels [10].[2]

Adapting a known bound to the constant-error model, it follows that the binary length ρ of the tag τ satisfies

$$\rho \geq \log \log M + \Omega_\epsilon(1),$$

where the hidden constant is about $-2 \log \epsilon$. In this work, *optimal* AMD codes are those meeting this lower bound, i.e., their tag-length is $\log \log M + O_\epsilon(1)$. Known constructions, notably a construction based on dedicated polynomial evaluation codes [5], are a multiplicative factor 2 *off* from being optimal (Proposition 3). By a generic combination of these polynomial AMD codes with asymptotically good error-correcting codes, AMD codes with tag-length

$$\rho = \log \log M + \log \log \log M + O_\epsilon(1)$$

are obtained (Proposition 4), which is still suboptimal. Bridging the gap to optimality efficiently turns out to be surprisingly nontrivial.

Owing to our refinement of the mathematical perspective on AMD codes, which focuses on symmetries of codes, we propose novel constructive principles. As we show, this leads to the following results.

1. There is a straightforward Gilbert-Varshamov type *nonconstructive* proof of the existence of *optimal* AMD codes (Theorem 1).
2. There is an explicit construction of AMD codes based on cyclic codes (Theorem 2). A construction with equivalent parameters to the polynomial construction from [5] is retrieved immediately by instantiating the latter with Reed-Solomon codes. Instantiating it with narrow-sense primitive BCH codes, AMD codes with improved parameters are obtained (Theorem 4).

---

[1] The adversary is even allowed to dictate the original message m that occurs in the intercepted encoding.

[2] Nevertheless, other applications require negligible error probability.

3. There is an efficient randomized construction of *optimal* AMD codes, based on twists of asymptotically good quasi-cyclic codes of finite index (Theorem 3). As an aside, the hidden constant in this construction is actually quite small, namely about $-6 \log \epsilon$, which is roughly 3 times the hidden constant in the lower bound (Remark 3). Nevertheless, the dependence on the error probability ϵ is worse than in the polynomial construction in [5], for which the tag-length is roughly $2 \log \log M - 2 \log \epsilon$.

Note that in all our results, the error probability ϵ can be chosen as an arbitrarily small positive real number.

Related Work The reader is referred to the survey [6] for more information about known results, techniques and applications of AMD codes. A class of AMD codes with a stronger security requirement was recently introduced in [11,12]. Namely, *all* algebraic manipulations, even those that do not change the message m but only the tag τ, should be detected with high probability. This additional requirement is not needed in most of the applications of AMD codes. [3] Our novel constructions of AMD codes in this paper satisfy that stronger security requirement.[4] A variant of AMD codes achieving leakage resilience has been presented [1].

2   Best Previous Constructions

The following definition of systematic AMD code was introduced in [5,6]. A new, equivalent definition, which fits our refinement of the mathematical perspective on AMD codes, is given in Section 3.

Definition 1. *Let ϵ be a real number with $0 \leq \epsilon \leq 1$ and let M, n be integers with $M, n \geq 1$. A systematic (M, n, ϵ)-AMD code consists of a map $f : \mathcal{M} \times G \rightarrow V$, where $\mathcal{M}$ is a set and G, V are finite abelian groups such that $M = |\mathcal{M}|$ and $n = |G| \cdot |V|$, and*

$$|\{g \in G \ : \ f(m, g) + c = f(m', ge)\}| \leq \epsilon \cdot |G|.$$

for all $m, m' \in \mathcal{M}$ with $m \neq m'$ and for all $(e, c) \in G \times V$. The tag-length *of an (M, n, ϵ)-AMD code is the quantity $\rho = \log_2 n$.*

As discussed after Definition 3, a message $m \in \mathcal{M}$ is encoded by choosing $g \in G$ uniformly at random and adding the *tag* $\tau = (g, f(m, g)) \in G \times V$ to the message m.

A simple example of a systematic AMD code, the so-called *multiplication* AMD code, is given in Proposition 1. It is extracted from the robust secret sharing construction in [4]. The proof of this result is straightforward.

---

[3] It is nevertheless essential for the non-malleable secret sharing schemes introduced in [9].

[4] The only exceptions appearing in this paper are the *non-constructive* family in Corollary 3 and the (known) multiplication AMD code in Proposition 1. The AMD code from [5] also satisfies the stronger security requirement.

Proposition 1. *Let q be a positive prime power and k, ℓ positive integers with $k \geq \ell$, and take an embedding of $\mathbb{F}_q^\ell$ into $\mathbb{F}_{q^k}$. Then the map $f : \mathbb{F}_q^\ell \times \mathbb{F}_{q^k} \to \mathbb{F}_{q^k}$ given by $f(m, g) = mg$ (here the embedding of $\mathbb{F}_q^\ell$ into $\mathbb{F}_{q^k}$ is used to compute the product mg) defines a systematic $(q^\ell, q^{2k}, 1/q^k)$-AMD code.*

We present next the family of efficient AMD codes, with rather good parameters, that was introduced in [5]. The reader is referred to [5,6] for more details about this construction.

Proposition 2. *Let $\mathbb{F}_q$ be a finite field of characteristic p. Let $d > 0$ be an integer such that $d + 1 < q$ and p is not a divisor of $d + 2$. Then the function $f : \mathbb{F}_q^d \times \mathbb{F}_q \to \mathbb{F}_q$ defined by*

$$f((m_1, \ldots, m_d), g) = g^{d+2} + \sum_{i=1}^{d} m_i g^i$$

determines a systematic $(q^d, q^2, (d+1)/q)$-AMD code.

The following discussion, which is adapted from [6, Section 6], demonstrates the flexibility in the values of the parameters of this family of AMD codes. In addition, it proves Proposition 3.

Consider a prime p, a real number ϵ_0 with $0 < \epsilon_0 < 1$, and an integer $M_0 \geq 1/\epsilon_0$. Take the smallest integer d such that $d + 2$ is not divisible by p and $\log M_0 \leq d(\log(d+1) - \log \epsilon_0)$,

$$k = \left\lceil \frac{\log(d+1) - \log \epsilon_0}{\log p} \right\rceil,$$

and $q = p^k$. Then $M = q^d \geq M_0$ and $\epsilon = (d+1)/q \leq \epsilon_0$. Therefore, there exists in the family introduced in Proposition 2 an (M, p^{2k}, ϵ_0)-AMD code, which can be trivially transformed into an $(M_0, p^{2k}, \epsilon_0)$-AMD code, with tag-length

$$\rho = 2k \log p \leq -2 \log \epsilon_0 + 2 \log(d+1) + 2 \log p$$

$$\leq -2 \log \epsilon_0 + 2 \log \left(-\frac{\log M_0}{\log \epsilon_0} + 3 \right) + 2 \log p.$$

We have used here that $(k-1) \log p \leq \log(d+1) - \log \epsilon_0$ and $(d-2)(\log(d-1) - \log \epsilon_0) \leq \log M_0$. The following two propositions are direct consequences of this discussion.

Proposition 3. *For every fixed value of ϵ with $0 < \epsilon < 1$ and arbitrarily large values of M, there exist systematic (M, n, ϵ)-AMD codes in the family introduced in Proposition 2 such that the asymptotic behavior of the tag-length is $\rho = 2 \log \log M + O(1)$.*

When comparing the result in Proposition 3 with the asymptotic lower bound in Corollary 1, we observe that the construction of AMD codes in [5] is a multiplicative factor 2 off from being optimal.

Finally, we observe here that it is possible to obtain an almost optimal construction by combining the AMD codes above with an asymptotically good family of $\mathbb{F}_q$-linear error-correcting codes. The idea is to encode the message $x \in \mathcal{M}$ with an error-correcting code C of length s in the family, take the tag $(g, C_g(x))$, where g is chosen uniformly at random from G_s, the cyclic group of order s, and $C_g(x) \in \mathbb{F}_q$ is the g-th component of the codeword $C(x)$, and then encode the tag $(g, C_g(x))$ with a suitable AMD code.

Proposition 4. *For every fixed value of ϵ with $0 < \epsilon < 1$ and arbitrarily large values of M, there exist systematic (M, n, ϵ)-AMD codes such that the asymptotic behavior of the tag-length is $\rho = \log \log M + \log \log \log M + O(1)$.*

Proof. Consider a family of $\mathbb{F}_q$-linear codes with constant rate $R > 0$ and constant relative minimum distance $\delta \geq 1 - \epsilon$. That is, for arbitrarily large values of s there is in the family a code $C : \mathbb{F}_q^k \to \mathbb{F}_q^s$ with length s, dimension $k \geq Rs$ and minimum distance at least δs. For every h in G_s and $x \in \mathcal{M} = \mathbb{F}_q^k$, let $C_g(x) \in \mathbb{F}_q$ be the g-th component of the codeword $C(x)$. We have seen before that one can find for these values of s AMD codes $f' : \mathcal{M}' \times G' \to V'$ with message space $\mathcal{M}' = G_s \times \mathbb{F}_q$, error probability ϵ, and tag-length $\log \log sq + O(1) = \log \log s + O(1)$. The proof is concluded by considering the AMD code

$$f : \mathbb{F}_q^k \times (G_s \times G') \to \mathbb{F}_q \times V'$$

defined by $f(x, (g, g')) = (C_g(x), f'((g, C_g(x)), g'))$.

3   Overview of Our Results

To enable a bird's eye view on our main results, we first briefly sketch our refinement of the mathematical perspective on AMD codes. Let V and G be finite abelian groups. Define the finite abelian group

$$V[G] = \bigoplus_{g \in G} V,$$

together with the group action denoted by "·" that turns $V[G]$ into a so-called G-module by having G act on the coordinates. More precisely, if $x \in V[G]$ with "coordinates" $x(g) \in V$ $(g \in G)$, then

$$h \cdot x \in V[G]$$

is defined such that

$$(h \cdot x)(g) := x(-h + g),$$

for all $g \in G$.[5] In particular, the G-action permutes coordinates. A *G-submodule* C' is a subgroup of $V[G]$ that is invariant under the G-action, i.e.,

$$G \cdot C' = C',$$

---

[5] Note that, $(h' \cdot (h \cdot x))(g) = x(-h - h' + g) = (h' + h) \cdot x$, for all $h, h', g \in G$ and $x \in V$.

or equivalently, $h \cdot x \in C'$ for all $h \in G$, $x \in C'$. Let $\Gamma \subset V[G]$ denote the G-submodule of *constants*, i.e., it consists of the elements $x \in V[G]$ such that $x(g) = x(g')$ for all $g, g' \in G$. If $x \in V[G]$, then $G \cdot x$ is the G-*orbit* of x, i.e., it is the set of elements $\{h \cdot x \mid h \in G\}$. Note that, if $x \neq 0$, then this is *not* a G-submodule. Recall that, if A, B are subsets of an additive group, then $A + B$ is defined as $\{a + b : a \in A, b \in B\}$.

Definition 2. *For $x, y \in V[G]$, the* AMD-equivalence relation *in $V[G]$ is defined by*

$$x \sim y \text{ if and only if } x \in (G \cdot y + \Gamma).$$

For $x \in V[G]$, the equivalence class of x under the AMD-equivalence relation is denoted by $\mathrm{cl}(x)$.

Consider the set $V[G]/\sim$, i.e., $V[G]$ taken modulo this equivalence relation. Also consider the *induced* Hamming-distance $\bar{d}_H$, which defines the distance between classes $\mathrm{cl}(x), \mathrm{cl}(x') \in V[G]/\sim$ as the minimum of the (regular) Hamming-distance $d_H(y, y')$ taken over all $y \in \mathrm{cl}(x)$ and $y' \in \mathrm{cl}(x')$.[6] Observe that $\bar{d}_H(\mathrm{cl}(x), \mathrm{cl}(x')) = d_H(\{x\}, \mathrm{cl}(x'))$. For a subset $C \subset V[G]$, the image of C under reduction by the equivalence relation is denoted by $\overline{C} \subset V[G]/\sim$.

Our new perspective concerns the observation that "good" AMD codes correspond to codes $C \subset V[G]$ such that $|C| = |\overline{C}|$, the cardinality $|\overline{C}|$ is "large" and the minimum distance $\bar{d}_{\min}(\overline{C})$ of $\overline{C} \subset V[G]/\sim$ (i.e, in terms of the induced Hamming-distance) is "large" as well. Only systematic algebraic manipulation detection codes are considered in this paper. The reader is referred to [6] for additional definitions and results about this and other classes of algebraic manipulation detection codes. For completeness, we present in Appendix 2 the equivalent definition of asymptotic AMD code from [5].

Definition 3 (AMD Codes). *Let ϵ be a real number with $0 \leq \epsilon \leq 1$ and let M, n be integers with $M, n \geq 1$. A systematic (M, n, ϵ)-algebraic manipulation detection (AMD) code consists of finite abelian groups G, V and a subset $C \subset V[G]$ such that $|C| = |\overline{C}| = M$ and $|G| \cdot |V| = n$, and $\bar{d}_{\min}(\overline{C}) \geq (1 - \epsilon) \cdot |G|$. The* tag-length *of an (M, n, ϵ)-AMD code is the quantity $\rho = \log_2 n$.*

We prove in the following that Definitions 3 and 1 are equivalent. First assume that $C \subset V[G]$ is a systematic AMD code given in Definition 3. Take $\mathcal{M} = C$ and consider the map $f : C \times G \to V$ defined by $(x, g) \mapsto x(g)$. Then it is easy to verify that this coincides with Definition 1. On the other hand, assume that we have a systematic AMD code given in Definition 1. Consider the set $C := \{\sum_{g \in G} f(m, g)g : m \in \mathcal{M}\}$. Then it is straightforward to verify that C is a systematic AMD code given in Definition 3.

In applications, a bijection $\phi : \mathcal{M} \to C$ between the message space $\mathcal{M}$ and the code C is fixed. To encode a message $m \in \mathcal{M}$, take $x = \phi(m)$, select $h \in G$ uniformly at random and set

---

[6] The (regular) Hamming-distance between two elements of $V[G]$ is, of course, the number of non-zero coordinates in their difference.

$$\tau := (h, x(h)) \in G \times V$$

as the tag.

AMD codes are a relaxation of combinatorial authentication codes. Their purpose is similar, namely ensuring *message integrity*. However, AMD codes are keyless and security is only guaranteed against a *non-adaptive, algebraic* adversary that has a priori knowledge of m and effectively replaces (m, τ) by $(m', \tau') \in \mathcal{M} \times (G \times V)$, under the following restrictions:

- $m' \neq m$.
- Effectively selects an *offset* $(e, c) \in G \times V$ and sets $\tau' = (h + e, x(h) + c) \in G \times V$.
- The selection of (m', e, c) may only depend on the message m and independent randomness chosen by the adversary. In particular, this selection does not depend on h.

Then the adversary is successful if and only if $x'(h + e) = x(h) + c$, where $x' = \phi(m') \in C$. It follows that success is equivalent to $((-e) \cdot x')(h) - c = x(h)$. Since x and x' are in distinct equivalence classes and since $h \in G$ is uniformly random and independent of x, x', e, c, the success probability of the adversary is at most ϵ because

$$1 - \frac{\overline{d}_H(\mathrm{cl}(x), \mathrm{cl}(x'))}{|G|} \leq 1 - \frac{\overline{d}_{\min}(\overline{C})}{|G|} \leq \epsilon.$$

In several specialized situations the adversary is effectively reduced to non-adaptive, algebraic attack. Moreover, authentication codes are typically not an option there: the secret key is susceptible to the same attack. Interestingly, the choice of the groups V, G is typically immaterial in applications.[7]

These observations motivate the following novel approaches to show existence of good AMD codes. Suppose, for now, that $C' \subset V[G]$ is such that

1. C' is a G-submodule.
2. $\Gamma \subset C'$.

Suppose that $|C'|$ is "large" and that the (regular) minimum distance $d_{\min}(C')$ is "large." In order to get a good AMD code out of this, it now suffices to develop an (efficient) method to select a subset $C \subset C'$ such that for each distinct $x, x' \in C$, the intersection between the orbits $G \cdot x$ and $G \cdot x'$ is empty (*orbit avoidance*). This way, one potentially achieves an AMD code C such that

$$|C| \geq \frac{|C'|}{|V| \cdot |G|},$$

where the denominator upper bounds the cardinality of a class, and such that the error probability ϵ satisfies

---

[7] Except perhaps that it is sometimes convenient if neither $|V|$ nor $|G|$ has a small prime divisor.

$$\epsilon = 1 - \frac{d_{\min}(C')}{|G|}.$$

This discussion is summarized in the following result.

Lemma 1. *Suppose* $C' \subset V[G]$ *is a* G-submodule, $\Gamma \subset C'$, *and* $d_{\min}(C') \geq (1-\epsilon) \cdot$ $|G|$ *for some* ϵ *with* $0 < \epsilon < 1$. *Then there exists a systematic* $\left(\dfrac{|C'|}{|V| \cdot |G|}, |V| \cdot |G|, \epsilon \right)$- *AMD code* $C \subset C'$.

As we show, this approach immediately leads to a greedy, *non-constructive* proof of the existence of optimal AMD codes.[8]

Theorem 1. *For every real number* ϵ *with* $0 < \epsilon < 1$, *there exist AMD codes with unbounded message space cardinality* M *and error probability at most* ϵ *whose tag-length* ρ *satisfies*

$$\rho = \log \log M + O_\epsilon(1),$$

which is optimal.

Remark 1 (Locking trick). Suppose the condition that $\Gamma \subset C'$. This complicates the situation as the (regular) relative minimum distance of the code C' no longer gives a non-trivial upper bound on the error probability of the AMD code C, i.e., the code obtained after application of orbit avoidance. But if $|V|$ is *constant*, the situation can be reduced to the previous situation by means of our *locking trick*, without harming *asymptotic performance*: simply augment an AMD encoding with a standard AMD encoding (with appropriate error probability) of the value $x(h)$ in the tag $\tau = (x, x(h))$. This way, at the cost of an additive constant increase in tag-length, we may as well assume that the adversary does not change the V-component of the tag. This obviates the need for considerations involving the constants Γ. As a consequence, the relative minimum distance of C' once again governs the error probability ϵ of the AMD code C.[9]

Any AMD code with the suitable parameters can be used in the locking trick as, for example, the simple multiplication AMD code in Proposition 1.[10].

Hence, the remaining question is about effective construction. We apply the idea above to *cyclic* $\mathbb{F}_q$-*linear codes* and show an efficiently enforceable algebraic conditions on the generator polynomial to ensure orbit avoidance. If $V = K$ is a (finite) field, then $K[G]$ is a ring, where multiplication is defined from the G-action by convolution, and hence $K[G]$ is a K-algebra (since K is contained in a natural way). A cyclic $\mathbb{F}_q$-linear code is a G-submodule of $\mathbb{F}_q[G]$, where G

---

[8] In fact, a Gilbert-Varshamov style argument.

[9] Note that locking only makes sense if $|V|$ is very small compared to $|G|$; otherwise this is too costly!

[10] If an AMD code under the stronger security requirements from [11] is needed, then one should select for the lock an AMD code that also satisfies those requirements as, for instance, the polynomial AMD code from [5].

is a finite cyclic group. It is convenient, though, to work with the following more common, equivalent definition.

Let q be a positive prime power and let $\mathbb{F}_q$ be a finite field with q elements. Let s be a positive integer that is not a multiple of the characteristic p of $\mathbb{F}_q$. Set $V = \mathbb{F}_q$ and $G = G_s$, the cyclic group of order s. Let $\pi_q(s) = \operatorname{ord}_s^*(q)$ be the multiplicative order of $q \bmod s$. An $\mathbb{F}_q$-*linear cyclic code* C' of length s is an ideal of $\mathbb{F}_q[G_s] \simeq \mathbb{F}_q[X]/(X^s - 1)$, and hence it is generated by the class $\overline{a(X)} \in \mathbb{F}_q[X]/(X^s - 1)$ of some polynomial $a(X) \in \mathbb{F}_q[X]$ that divides $X^s - 1$. This polynomial is called the *generator* of the cyclic code C'.

Theorem 2. *Let $\mathbb{F}_q$ be a finite field and $s > 1$ an integer that is not a multiple of the characteristic p of $\mathbb{F}_q$. Let $C' \subset \mathbb{F}_q[X]/(X^s - 1)$ be an $\mathbb{F}_q$-linear cyclic code of length s with generator $a(X) \in \mathbb{F}_q[X]$. Let d be the minimum distance of C'. Suppose that the following conditions are satisfied.*

1. *The all-one vector is in C' or, equivalently, $(X - 1)$ does not divide $a(X)$.*
2. *There is a primitive s-th root of unity $\omega \subset \overline{\mathbb{F}}_q$ with $a(\omega) \neq 0$.*
3. *$\pi_q(s) < s - \deg a - 1$.*

Then there exists an explicit *construction of a $(q^{s - \deg a - \pi_q(s) - 1}, sq, (s - d)/s)$-AMD code $C \subset C'$.*

Notice that the conditions imply $\Gamma \subset C'$, so the locking trick from Remark 1 is not necessary. For every real number ϵ with $0 < \epsilon < 1$, instantiation with Reed-Solomon codes defined over a large enough finite field leads to an explicit construction of AMD codes with arbitrarily large message space cardinality M and tag-length $2 \log \log M + O_\epsilon(1)$, which is the same as in the explicit construction from [5] (see Appendix 2). Instantiation with narrow-sense primitive BCH codes defined over a large enough finite field leads to an explicit construction of *almost*-optimal AMD codes, i.e., achieving tag-length $(1 + \delta) \log \log M + O_\epsilon(1)$ where δ is an arbitrary real constant with $0 < \delta < 1$.

One quickly sees that achieving optimality along the lines of twists on cyclic codes as discussed above would require the existence of asymptotically good cyclic $\mathbb{F}_q$-linear codes, which is one of the central open problems in the theory of error correcting codes.[11] The remainder of our results is concerned with bypassing this major open problem.

Our final result is a randomized construction of optimal AMD codes.

---

[11] It would not even be enough if asymptotically good cyclic $\mathbb{F}_q$-linear codes exist for *some* value of q. Namely, to suit our purposes, such codes should exist for infinitely many values of q and, when sending q to infinity, the relative minimum distance achieved should tend to 1. Finally, a certain condition on the lengths of the codes should hold. Specifically, given such a value of q, the lengths $\ell(C')$ occurring must satisfy $\pi_q(\ell(C')) \leq \gamma \ell(C')$ for some absolute real constant $\gamma > 0$. Otherwise the orbit-avoidance eats away too many codewords, causing the rate to drop to 0. If the codes do not contain the all-one vector, the locking trick alluded below could be applied.

Theorem 3. *For every real number ϵ with $0 < \epsilon < 1$, there exist an* efficient, *randomized construction of explicit AMD codes with arbitrarily large message space cardinality M and tag-length $\rho = \log\log M + O_\epsilon(1)$, which is* optimal.

Relying on our AMD perspective as outlined above, it is achieved by a series of twists on a result in a beautiful paper by Bazzi and Mitter [2] on *asymptotically good quasi-cyclic codes of constant index* ℓ. Let $\ell \geq 2$ and define $V = \mathbb{F}_2^\ell$. One of their results (stated in our terminology here) is that there exists a randomized construction of G_s-submodules $C' \subset \mathbb{F}_2^\ell[G_s]$ of rate $\log |C'|/(\ell s) = 1/\ell$ achieving the Gilbert-Varshamov bound when s tends to infinity.[12] The error probability of this randomized construction is exponentially small in s if the lengths s are carefully selected.

We use four twists on their result to show our claim. First, we generalize it to work over *all* finite fields $\mathbb{F}_q$, with $\ell \geq 2$ an arbitrary integer constant. This ensures that relative minimum distance arbitrarily close to 1 can be achieved, and hence ϵ can be selected arbitrarily close to 0. This generalization is straightforward, using some results from [8]. Second, this time we need to resort to the locking trick from Remark 1. Third, we need an adaptation of the efficient orbit avoidance method alluded to above. This adaptation is necessary not only because of the shift from cyclic codes to quasi-cyclic ones, but also because of the probabilistic nature of the construction. Fourth, we need to craft the lengths s with additional care to ensure that the rate of the code drops by at most a multiplicative positive constant factor after application of orbit avoidance.

4 Nonconstructive Optimal AMD Codes

In this section, we present the proof of Theorem 1. We begin by presenting the asymptotic lower bound in Corollary 1, which is a consequence of the bound in Proposition 6. This bound is a refinement of similar bounds presented in [6,11]. We are going to use the following trivial result

Proposition 5. *Let ϵ be a real number with $0 < \epsilon < 1$ and let $C \subset V[G]$ be an (M, n, ϵ)-AMD code. Then $|G| \geq 1/\epsilon$ and $|V| \geq 1/\epsilon$. As a consequence, the tag-length ρ satisfies $\rho \geq -2\log \epsilon$.*

Proof. Consider $x, x' \in C$ with $x \neq x'$. Then there exists $c \in V$ such that the set $\{g \in G : x(g) - x'(g) = c\}$ has cardinality at least $|G|/|V|$.

$$|\{g \in G : x(g) - x'(g) = c\}| \geq \left\lceil \frac{|G|}{|V|} \right\rceil.$$

Therefore,

$$(1 - \epsilon) \cdot |G| \leq \overline{d}_H(\mathrm{cl}(x), \mathrm{cl}(x')) \leq |G| - \left\lceil \frac{|G|}{|V|} \right\rceil \leq \min\left\{|G| - 1, |G| - \frac{|G|}{|V|}\right\}$$

and the proof is concluded.

---

[12] i.e., the relative minimum distance $\delta > 0$ of these codes is such that $H_2(\delta) = 1/\ell$, where $H_2(\cdot)$ is the binary Shannon-entropy function.

Other lower bounds on the tag-length are obtained by applying some known classical bounds from coding theory, as in Proposition 6. As a corollary, we obtain a lower bound on the asymptotic behavior of the tag-length.[13]

Proposition 6. *Let ϵ be a real number with $0 < \epsilon < 1$. Suppose that $M \geq 1/\epsilon$. Then the tag-length ρ of a systematic (M, n, ϵ)-AMD code satisfies*

$$\rho \geq \log \log M - 2 \log \epsilon - \max\{0, \log(-\log \epsilon)\}.$$

Proof. Let $C \subset V[G]$ be an (M, n, ϵ)-AMD code. From the definition of AMD code, $C + \Gamma$ is a code of size $M \cdot |V|$, length $|G|$ and minimum distance at least $(1 - \epsilon)|G|$ over the alphabet V. Therefore, by the Singleton bound, $M \leq |V|^{\epsilon \cdot |G|}$, and hence

$$\log |G| \geq \log \log M - \log \epsilon - \log \log |V|.$$

By Proposition 5, $\log |G| \geq -\log \epsilon$ and $\log |V| \geq -\log \epsilon$. Take $x = \log |V|$, $y = \log |G|$, $A = \max\{1, -\log \epsilon\}$, and $B = \log \log M$. Since $B - \log A \geq 0$, the minimum value of $x + y$ under the constraints $x, y \geq A$ and $y \geq A + B - \log x$ is attained when $x = A$ and $y = A + B - \log A$. $\qquad\blacksquare$

Corollary 1. *For every real number ϵ with If $0 < \epsilon < 1$, the tag-length ρ of the AMD codes with arbitrarily large message space cardinality M and error probability at most ϵ satisfies*

$$\rho \geq \log \log M + \Omega_\epsilon(1).$$

Application of the Hamming and Plotkin bounds instead of the Singleton bound gives better results, but the asymptotic results are not improved. Next, we prove Theorem 1 by using a variation on the Gilbert-Varshamov bound.

Definition 4. *Consider a finite field $\mathbb{F}_q$, a real number ϵ with $1/q < \epsilon < 1$, and a positive integer s. Then the quantity $A'_q(s, \epsilon)$ is defined as the maximum cardinality M of an (M, n, ϵ)-AMD code $C \subset \mathbb{F}_q[G_s]$.*

Proposition 7. *With conditions as above,*

$$A'_q(s, \epsilon) \geq \left\lfloor \frac{q^s}{qs \cdot V_q(s, 1 - \epsilon)} \right\rfloor,$$

where $V_q(s, 1 - \epsilon)$ is the volume of a sphere in $\mathbb{F}_q[G_s]$ with radius $(1 - \epsilon)s$.

Proof. Suppose that the result is false and take an (M, n, ϵ)-AMD code $C \subset \mathbb{F}_q[G_s]$ with $M = A'_q(s, \epsilon)$. Observe that $|\operatorname{cl}(x)| \leq qs$ for every $x \in \mathbb{F}_q[G_s]$. Therefore, the number of elements $y \in \mathbb{F}_q[G_s]$ such that $d_H(\{y\}, \operatorname{cl}(x)) < (1 - \epsilon)s$

---

[13] The looser, but easier to prove, lower bound $\rho \geq \log \log M - \log \epsilon$ is enough to obtain the asymptotic lower bound in Corollary 1. Nevertheless, the bound in Proposition 6 provides a better description of the behavior of the tag-length ρ in relation to the error probability ϵ.

is at most $qs \cdot V_q(s, 1 - \epsilon)$. Since $|C| \cdot qs \cdot V_q(s, 1 - \epsilon) < q^s$ there exist a vector $y \in \mathbb{F}_q[G_s] \smallsetminus C$ such that

$$\overline{d}_H(\mathrm{cl}(y), \mathrm{cl}(x)) = d_H(\{y\}, \mathrm{cl}(x)) \geq (1 - \epsilon)s$$

for all $x \in C$. Therefore, C has not maximum cardinality among all codes with the required property, a contradiction.

Corollary 2. *Let q be a positive prime power, let ϵ be a real number with $1/q < \epsilon < 1$, and let s be a positive integer. Then there exists a systematic*

$$\left(\left\lfloor \frac{q^s}{qs \cdot V_q(s, 1 - \epsilon)} \right\rfloor, qs, \epsilon \right) \text{-}AMD \ code$$

with $V = \mathbb{F}_q$ and $G = G_s$, the cyclic group of order s.

Lemma 2. *With conditions as above,*

$$\lim_{s \to \infty} \frac{\log A'_q(s, \epsilon)}{s} \geq (1 - H_q(1 - \epsilon)) \log q,$$

where H_q is the q-ary entropy function.

Proof. The result follows immediately from Corollary 2 by taking limits, taking into account that, by coding theory, $\lim\limits_{s \to \infty} \dfrac{\log_q V_q(s, 1 - \epsilon)}{s} = H_q(1 - \epsilon)$.

Finally, Theorem 1 is an immediate consequence of the following result.

Corollary 3 (Non-constructive optimality). *For any real constant $c > 0$, fix a positive prime power q and a real number ϵ with $\epsilon = 1/q + 1/q^{1+c}$. Then there exist AMD codes with arbitrarily large message space cardinality M, error probability at most ϵ and tag-length*

$$\rho = \log \log |M| - (2 + c) \log \epsilon + O(1).$$

Note that the tag-length is minimal up to an additive constant.

5   An Explicit Construction from Cyclic Codes

This section is devoted to prove Theorem 2. We present here an effective method to select, from any given cyclic code, a number of codewords in different AMD-equivalence classes. By Lemma 1, this provides an effective construction of systematic AMD codes.

5.1   General Construction

Let $\mathbb{F}_q$ be a finite field. Let $s > 1$ be an integer that is not a multiple of the characteristic p of $\mathbb{F}_q$. Then $\mathbb{F}_q[G_s]$ is a ring, where the product is defined from the G_s-action by convolution. So, $\mathbb{F}_q[G_s]$ is an $\mathbb{F}_q$-algebra. Since $X^s - 1$ is separable, it follows by the Chinese Remainder Theorem that $\mathbb{F}_q[G_s] \simeq \mathbb{F}_q[X]/(X^s - 1)$ is a product of finite extension fields of $\mathbb{F}_q$. Let $\omega \in \overline{\mathbb{F}}_q$ be a primitive s-th root of unity. Then the degree of $\mathbb{F}_q(\omega)$ over $\mathbb{F}_q$ equals

$$\pi_q(s) = \mathrm{ord}_s^*(q),$$

the multiplicative order of $q \bmod s$. Equivalently, it equals the degree of the minimal polynomial of ω over $\mathbb{F}_q$. It is also the *largest degree* occurring among the irreducible factors in the factorization of $X^s - 1$ over $\mathbb{F}_q[X]$ as each of the s roots sits in some intermediate extension of $\mathbb{F}_q(\omega) \supset \mathbb{F}_q$.

Definition 5. *Let $a(X) \in \mathbb{F}_q[X]$ be a polynomial such that $a(X)$ divides $X^s - 1$ and let $\omega \in \overline{\mathbb{F}}_q$ be a primitive s-th root of unity with $a(\omega) \neq 0$. We define $D(a(X), \omega) \subset \mathbb{F}_q[X]$ as the set of all polynomials $f(X) \in \mathbb{F}_q[X]$ such that*

1. *$f(\omega) = 1$, and*
2. *$\deg f < s - \deg a - \delta$, where $\delta = 0$ if $(X - 1)$ divides $a(X)$ and $\delta = 1$ otherwise.*

An $\mathbb{F}_q$-*linear cyclic code* C' of length s is a G_s-submodule of $\mathbb{F}_q[G_s]$. Equivalently, it is an ideal $C' \subset \mathbb{F}_q[X]/(X^s - 1)$ generated by the class of some polynomial $a(X) \in \mathbb{F}_q[X]$ that divides $X^s - 1$. This polynomial is called the *generator* of the cyclic code C'. Then

$$C' = \left\{ \overline{a(X)f(X)} : f(X) \in \mathbb{F}_q[X] \text{ and } \deg f < s - \deg a \right\} \subset \mathbb{F}_q[X]/(X^s - 1).$$

Let C' be an $\mathbb{F}_q$-linear cyclic code with generator $a(X)$ and suppose that $a(\omega) \neq 0$ for some primitive s-th root of unity $\omega \in \overline{\mathbb{F}}_q$. Let $C \subset C'$ be the set of all codewords $\overline{a(X)f(X)} \in \mathbb{F}_q[X]/(X^s - 1)$ with $f \in D(a(X), \omega)$.

Lemma 3. *No two distinct elements in C are in the same AMD-equivalence class.*

Proof. Suppose that there exist two different polynomials $f, g \in D$ such that the corresponding codewords in C are in the same AMD-equivalence class. Then

$$X^\ell a(X)f(X) + \lambda(X^{s-1} + \cdots + X + 1) \equiv a(X)g(X) \pmod{X^s - 1}$$

for some $\lambda \in \mathbb{F}_q$ and ℓ with $0 \leq \ell < s$. Therefore, $\omega^\ell a(\omega)f(\omega) = a(\omega)g(\omega)$, and hence $\omega^\ell = 1$ by the definition of C. Since ω is a primitive root, $\ell = 0$. Consequently,

$$a(X)(f(X) - g(X)) \equiv -\lambda(X^{s-1} + \cdots + X + 1) \pmod{X^s - 1}. \tag{1}$$

Suppose that $(X - 1)$ divides $a(X)$. Then $\lambda \neq 0$ because $\deg(f - g) < s - \deg a$, but this implies that $a(X)$ divides $X^{s-1} + \cdots + X + 1$, a contradiction. Suppose now that $(X - 1)$ does not divide $a(X)$, and hence $\delta = 1$ and $\deg(f - g) < s - \deg a - 1$. But then (1) is impossible because $\deg(a(f - g))$ is too small, a contradiction again.

Lemma 4. *Suppose that $\pi_q(s) < s - \deg a - \delta$. Then $|C| = q^{s - \deg a - \pi_q(s) - \delta}$.*

Proof. Take $h = s - \deg a - \delta$ and let $\mathbb{F}_q[X]_{<h}$ be the $\mathbb{F}_q$-vector space of the polynomials in $\mathbb{F}_q[X]$ with degree at most $h - 1$. Since $\pi_q(s) < h$, application of Lemma 7 implies that the kernel of the $\mathbb{F}_q$-linear map $\mathbb{F}_q[X]_{<h} \to \mathbb{F}_q[\omega]$, $f \mapsto f(\omega)$ has dimension $h - \pi_q(s)$.

The proof of Theorem 2 is now straightforward from Lemmas 1, 3 and 4.

5.2   Instantiations

Applying Theorem 2 to Reed-Solomon codes provides, for every real number ϵ with $0 < \epsilon < 1$, an effective construction of (M, n, ϵ)-AMD codes with unbounded message space cardinality M and tag-length $2 \log \log M + O_\epsilon(1)$, which is the same as in the polynomial construction from [5] (see Section 2). Indeed, consider a prime power q, a primitive element α of $\mathbb{F}_q^*$, an integer k with $1 \leq k \leq q - 1$, and the polynomial $a(X) = (X - \alpha)(X - \alpha^2) \cdots (X - \alpha^{q-k-1})$. By applying Theorem 2 to the $\mathbb{F}_q$-linear cyclic code with length $q - 1$ generated by $a(X)$, which is a Reed-Solomon code with minimum distance $d = q - k$, one obtains an effective AMD code with parameters $(q^{k-2}, q(q - 1), (k - 1)/(q - 1))$. The proof of our claim is concluded by using a similar argument as for the polynomial construction from [5] (see Section 2).

The instantiation to narrow-sense BCH codes is not so immediate. We refer to Appendix B for the background on BCH codes.

Let $e \geq 1$ be an integer. Let $s = q^e - 1$. Choose an element α of $\mathbb{F}_{q^e}$ of order $s = q^e - 1$. Let $m^{(i)}(x) \in \mathbb{F}_q[x]$ denote the minimal polynomial of α^i with respect to $\mathbb{F}_q$. For $0 < \epsilon < 1$, put $d = (1 - \epsilon)s$ and consider the BCH code B of length $s = q^e - 1$ with the generator polynomial $a(X) := \mathrm{lcm}\{m^{(1)}(X), m^{(2)}(X), \ldots, m^{(d-1)}(X)\}$. Then the minimum distance of B is at least d. Let $f_{a_i,j}(X)$ be the polynomials defined in (2) of Appendix B.

By Lemma 11, the dimension $s - \deg a$ of B is equal to the dimension of the $\mathbb{F}_q$-span $V_{s-d} = V_{\epsilon s}$ of $\{f_{a_i,j}(X) : 1 \leq i \leq t, 1 \leq j \leq s_{a_i}, \deg f_{a_i,j} \leq \epsilon n\}$. Hence, $\dim(B) = s - \deg a \geq (\epsilon(q - 1) + 1)^e \approx e + 1 + (\epsilon q)^e$. Note that in this case $\pi_q(s) = e$. Applying Theorem 2, we obtain the following AMD codes.

Theorem 4. *For any $\epsilon \in (0, 1)$, any integer $e \geq 1$ and prime power q, there exists an effective $(q^{(\epsilon q)^e}, (q^e - 1)q, \epsilon)$-AMD code. Thus, the tag-length equals to*

$$\frac{e + 1}{e} \log \log M - (e + 1) \log \epsilon + O(1).$$

Proof. Note that the message size $M = q^{(\epsilon q)^e}$ satisfies $\log \log M \approx e \cdot \log \epsilon + e \cdot \log q$. The tag-length satisfies $\log q + \log(q^e - 1) \leq (e+1) \log q \leq \dfrac{e+1}{e} \log \log M - (e+1) \log \epsilon$. This completes the proof.

Remark 2. When $e = 1$ in Theorem 4, we get almost the same result as in the one in [5]. If we choose $e = (\log \log M)^{0.5}$ in Theorem 4, then the tag-length is $\log \log M + O((\log \log M)^{0.5})$.

6   Monte-Carlo Construction of Optimal AMD Codes

In this section we prove Theorem 3. Namely, we present an efficient randomized construction of explicit optimal AMD codes. We proceed as follows. We begin by presenting in Theorem 5 a randomized construction of G_s-submodules $C' \subset \mathbb{F}_q^{\ell}[G_s]$. By considering the codes C' over the alphabet $\mathbb{F}_q$, they have rate $\log_q |C'|/(\ell s) = 1/\ell$ and minimum relative distance δ arbitrarily close to 1 achieving the Gilbert-Varshamov bound when s tends to infinity. This is an extension of the corresponding result by Bazzi and Mitter [2] for the case $q = 2$. This extension is based on some results from [8]. The error probability of this randomized construction is exponentially small in s if the lengths s are carefully selected. Then we apply the general method derived from Lemma 1 to those G_s-submodules $C' \subset \mathbb{F}_q^{\ell}[G_s]$. Since $\Gamma \not\subset C'$, we have to use the locking trick in Remark 1. Furthermore, we have to adapt orbit avoidance to this probabilistic scenario involving quasi-cyclic codes. In addition, we need to craft the lengths s with additional care to ensure that the rate of the AMD code remains positive after application of orbit avoidance. Finally, in Remark 3, we use a simple modification to reduce the size of the hidden constant in the tag-length.

Let $\mathbb{F}_q$ be a finite field and $s > 1$ an integer such that the characteristic p of $\mathbb{F}_q$ does not divide s. As before, G_s denotes the cyclic group of order s. Recall that, if $\omega \in \overline{\mathbb{F}}_q$ is a primitive s-th root of unity, then the degree of $\mathbb{F}_q(\omega)$ over $\mathbb{F}_q$ equals $\pi_q(s) = \mathrm{ord}_s^*(q)$, the multiplicative order of q mod s. The smallest degree of an extension of $\mathbb{F}_q$ containing some (not necessarily primitive) s-th root of unity different from 1 equals

$$\alpha_q(s) = \min_{p'|s} \mathrm{ord}_{p'}^*(q) = \min_{p'|s} \pi_q(p'),$$

where the minimum ranges over all prime divisors p' of s. Equivalently, this equals the *smallest degree* occurring among the irreducible factors in the factorization of $X^{s-1} + \cdots + X + 1$ over $\mathbb{F}_q[X]$.

Let $\mathbb{F}_q$ be a finite field and let s, ℓ be positive integers such that s is coprime with q. An $\mathbb{F}_q$-linear (s, ℓ)-*quasi-cyclic code* C' is of the form $C' = \{(fa_1, \ldots, fa_\ell) : f \in \mathbb{F}_q[G_s]\} \subset (\mathbb{F}_q[G_s])^{\ell}$, for some fixed $a_1, \ldots, a_\ell \in \mathbb{F}_q[G_s]$. In particular, C' is an $\mathbb{F}_q$-linear code of length $s\ell$.

Let $R \subset \mathbb{F}_q[G_s]$ be the set formed by all $a \in \mathbb{F}_q[G_s]$ with $\sum_{g \in G_s} a(g) = 0$. Equivalently, R is the set of all $\overline{a(X)} \in \mathbb{F}_q[X]/(X^s - 1)$ with $a(1) = 0$. Recall that H_q denotes the q-ary entropy function. The following theorem is a consequence of the results in [2,8].

Theorem 5. *For a finite field* $\mathbb{F}_q$, *an integer* $\ell > 1$, *and an integer* s *that is not a multiple of the characteristic* p *of* $\mathbb{F}_q$, *consider the randomized construction of quasi-cyclic codes*

$$C' = \{(fa_1, \ldots, fa_\ell) \,:\, f \in \mathbb{F}_q[G_s]\} \subset (\mathbb{F}_q[G_s])^\ell,$$

where $a_1, \ldots, a_\ell$ *are selected uniformly at random from* R. *Now consider* C' *as an* $\mathbb{F}_q$-*linear code of length* $s\ell$. *If* δ *is a real number with* $0 < \delta < 1 - 1/q$ *and*

$$H_q(\delta) \leq 1 - \frac{1}{\ell} - \frac{\log_q s}{\ell \alpha_q(s)},$$

Then the probability that the relative minimum distance of the code C' *is below* δ *or the rate of* C' *is below* $\dfrac{1}{\ell} - \dfrac{1}{\ell s}$ *is at most* $q^{-\beta}$, *where*

$$\beta = \ell \alpha_q(s) \left(1 - \frac{1}{\ell} - H_q(\delta) \right) - (\ell + 2) \log_q s - \ell(1 + \log_q \ell)$$

As a consequence, for fixed values of δ, q *and* ℓ, *if* $\alpha_q(s)$ *grows asymptotically faster than* $\log s$, *this code achieves the Gilbert-Varshamov (GV) bound for rate* $1/\ell$ *with high probability.*

There is a natural identification between $\mathbb{F}_q^\ell[G_s]$ and $(\mathbb{F}_q[G_s])^\ell$. Indeed, every element $\mathbf{x} \in \mathbb{F}_q^\ell[G_s]$ is of the form $(\mathbf{x}(g))_{g \in G_s}$, where $\mathbf{x}(g) = (x_1(g), \ldots, x_\ell(g)) \in \mathbb{F}_q^\ell$ for every $g \in G_s$. Then $\mathbf{x} \in \mathbb{F}_q^\ell[G_s]$ can be identified with $(x_1, \ldots, x_\ell) \in (\mathbb{F}_q[G_s])^\ell$. By this identification, every $\mathbb{F}_q$-linear (s, ℓ)-quasi-cyclic code C' is a G_s-submodule of $\mathbb{F}_q^\ell[G_s]$.

We proceed next with the detailed description of our efficient randomized construction of explicit optimal AMD codes. Given a real number ϵ with $0 < \epsilon < 1$, take a large enough prime power q such that $1/q < \epsilon$ and a large enough integer ℓ such that $1/\ell < 1 - H_q(1 - \epsilon)$. Note that this means that if an $\mathbb{F}_q$-linear code is on the GV-bound and it has rate $1/\ell$, then its relative minimum distance is at least $1 - \epsilon$.

Next, we select arbitrarily large values of s such that the following conditions are satisfied.

1. The characteristic p of $\mathbb{F}_q$ does not divide s.
2. The value $\alpha_q(s)$ grows asymptotically faster than $\log s$. By Theorem 5, this ensures that the relative minimum distance or the code C' is at least $1 - \epsilon$, except with exponentially small (in s) probability.
3. Finally, $\pi_q(s) \leq s/(\ell + 1)$. This condition is needed to ensure that the rate of the code drops by at most a multiplicative positive constant factor after application of orbit avoidance.

We describe next how to efficiently select arbitrarily large values of s satisfying those conditions. Take s a product of 2 distinct odd primes, In addition, we require that these primes are different from the characteristic p of $\mathbb{F}_q$, they have roughly the same size, and they satisfy $\pi_q(p') > \log^2 p'$. Then $\alpha_q(s)$ grows asymptotically faster than $\log s$. Indeed, since the primes p' are of similar size, $\alpha_q(s) = \Omega(\log^2 s)$. By the Prime Number Theorem, a random prime satisfies $\pi_q(p') > \log^2 p'$ with quite high probability. We can efficiently check that the condition is satisfied by simply factoring $p' - 1$ over a factor basis consisting of the primes up to $\log^2 p'$ (brute-force suffices as the factor basis is so small). Moreover, by the Chinese Remainder Theorem, it is straightforward to verify that the exponent of the group $(\mathbb{Z}/s\mathbb{Z})^*$ is at most $s/2$ if s is the product of 2 distinct odd primes. Therefore, $\pi_q(s) \leq s/2$.

Given a large enough integer s sampled as above, take a primitive s-th root of unity ω and a code $C' = \{(fa_1, \ldots, fa_\ell) : f \in \mathbb{F}_q[G_s]\} \subset \mathbb{F}_q^\ell[G_s]$ such that $a_1, \ldots, a_\ell \in R$ are selected independently and uniformly at random. By Theorem 5, the relative minimum distance of C' is at least $1 - \epsilon$ except with probability exponentially small in s. In addition, we require that $a_i(\omega) \neq 0$ for every $i = 1, \ldots, \ell$ and that there is no s-th root of unity $\eta \neq 1$ with $a_i(\eta) = 0$ for every $i = 1, \ldots, \ell$. The first property is used in Lemma 5 and the second property is used in Lemma 6. By using a similar argument as in the proof of Lemma 4, the probability that $a_i(\omega) = 0$ for some $i = 1, \ldots, \ell$ is at most $\ell q^{-\pi_q(s)}$. The probability that there is some s-th root of unity different from 1 that is a root of each $a_i(X)$ is at most $(s - 1)q^{-\ell\pi_q(s)}$. Therefore, these two additional requirements do not substantially decrease the success probability (use union bound) if $\alpha_q(s)$ is much larger than $\log s$.

Let $D \subset \mathbb{F}_q[X]$ be the subset of polynomials $f(X) \in \mathbb{F}_q[X]$ such that $\deg f < s - 1$ and $f(\omega) = 1$. The code $C \subset C'$ is now formed by the codewords $\overline{(f(X)a_1(X), \ldots, f(X)a_\ell(X))} \in C'$ such that $f(X) \in D$

The following two lemmas are conditioned on the "bad events" described above not happening.

Lemma 5. $G_s \cdot \mathbf{x}$ and $G_s \cdot \mathbf{x}'$ have empty intersection for every $\mathbf{x}, \mathbf{x}' \in C$ with $\mathbf{x} \neq \mathbf{x}'$.

Proof. Assume that the result is false. Then there exist polynomials $f(X), f'(X) \in D$ such that $X^i \cdot f(X) \cdot a_j(X) \equiv f'(X) \cdot a_j(X) \mod (X^s - 1)$ for some integers i, j with $1 \leq i < s$ and $1 \leq j \leq \ell$. Then this implies the identity $\omega^i = 1$, which is nonsense since ω is a primitive s-th root of unity.

Lemma 6. $|C| \geq q^{s-1-\ell\pi_q(s)} = q^{\Omega(s)}$.

Proof. Consider the map $\phi : \mathbb{F}_q[X]_{<s} \to (\mathbb{F}_q[X]/(X^s - 1))^\ell$ defined by $\phi(f) = (fa_1, \ldots, fa_\ell)$. Then the kernel of this map is spanned by the polynomial $X^{s-1} + \cdots + X + 1$. Since the degrees of the polynomials in D are smaller than $s - 1$, it follows that $|D| = |\phi(D)| = |C|$. It now suffices to lower bound $|D|$. By Lemma 7, the kernel of the map $\psi : \mathbb{F}_q[X]_{<s} \to \mathbb{F}_q$, $f \mapsto f(\omega)$ has dimension $s - \pi_q(s)$. Hence, $|D| \geq q^{s-1-\pi_q(s)}$. The claim follows since $\pi_q(s) \leq s/2$ by hypothesis.

The final ingredient in our construction is the locking trick in Remark 1, that is, we use the multiplication AMD code described in Proposition 1 to encode

$\mathbf{x}(g) \in \mathbb{F}_q^\ell$. Since $\epsilon > 1/q$, we can take $k = \ell$, and hence we add to the tag two elements from $\mathbb{F}_{q^\ell}$.[14] This increases the tag-length by an additive constant.

This concludes the proof of Theorem 3.

Remark 3 (Achieving a smaller hidden constant). Even though this randomized construction of AMD codes is optimal, the hidden constant is very large because so is the value of ℓ. This drawback can be avoided with a simple modification to our construction. Namely, instead of the tag $(g, \mathbf{x}(g)) \in G_s \times \mathbb{F}_q^\ell$ with a lock for $\mathbf{x}(g)$, use the tag $(g, h, x_h(g)) \in G_s \times G_\ell \times \mathbb{F}_q$ with locks for h and $x_h(g)$. In this way, the tag-length is reduced from $\log s + 3\ell \log q$ to $\log s + 3 \log \ell + 3 \log q$, which is around $\log s - 6 \log \epsilon$.

References

1. Ahmadi, H., Safavi-Naini, R.: Detection of Algebraic Manipulation in the Presence of Leakage. In: Padró, C. (ed.) ICITS 2013. LNCS, vol. 8317, pp. 238–258. Springer, Heidelberg (2014)
2. Bazzi, L.M.J., Mitter, S.K.: Some Randomized Code Constructions from Group Actions. IEEE Trans. Inf. Theory 52, 3210–3219 (2006)
3. Broadbent, A., Tapp, A.: Information-theoretic security without an honest majority. In: Kurosawa, K. (ed.) ASIACRYPT 2007. LNCS, vol. 4833, pp. 410–426. Springer, Heidelberg (2007), http://arxiv.org/abs/0706.2010
4. Cabello, S., Padró, C., Sáez, G.: Secret sharing schemes with detection of cheaters for a general access structure. Des. Codes Cryptogr. 25, 175–188 (2002)
5. Cramer, R., Dodis, Y., Fehr, S., Padró, C., Wichs, D.: Information-theoretic security without an honest majority. In: Smart, N.P. (ed.) EUROCRYPT 2008. LNCS, vol. 4965, pp. 471–488. Springer, Heidelberg (2008)
6. Cramer, R., Fehr, S., Padró, C.: Algebraic manipulation detection codes. Sci. China Math. 56, 1349–1358 (2013)
7. Dziembowski, S., Pietrzak, K., Wichs, D.: Non-Malleable Codes. In: Innovations in Computer Science, ICS 2010, pp. 434–452 (2010)
8. Fan, Y., Lin, L.: Thresholds of random quasi-abelian codes (2013), http://arxiv.org/pdf/1306.5377.pdf
9. Gordon, D., Ishai, Y., Moran, T., Ostrovsky, R., Sahai, A.: On Complete Primitives for Fairness. In: Micciancio, D. (ed.) TCC 2010. LNCS, vol. 5978, pp. 91–108. Springer, Heidelberg (2010)
10. Guruswami, V., Smith, A.: Codes for Computationally Simple Channels: Explicit Constructions with Optimal Rate. In: FOCS 2010, pp. 723–732 (2010), Full version available at arXiv.org, arXiv:1004.4017 [cs.IT]
11. Karpovski, M., Wang, Z.: Algebraic Manipulation Detection Codes and Their Applications for Design of Secure Communication or Computation Channels (2011) (manuscript), http://mark.bu.edu/papers/226.pdf
12. Wang, Z., Karpovsky, M.: Algebraic manipulation detection codes and their applications for design of secure cryptographic devices. In: IEEE 17th International On-Line Testing Symposium, IOLTS 2011, pp. 234–239 (2011)
13. Wee, H.: Public Key Encryption against Related Key Attacks. In: Fischlin, M., Buchmann, J., Manulis, M. (eds.) PKC 2012. LNCS, vol. 7293, pp. 262–279. Springer, Heidelberg (2012)

---

[14] If the additional security requirement introduced in [11] is required, a polynomial AMD code [5] with suitable parameters can be used instead.

A   A Generalization of Lagrange's Interpolation Theorem

It is convenient to recall a simple extension of the usual version of Lagrange Interpolation.

Lemma 7. *Let K be a field. Fix an algebraic closure $\overline{K}$ of K. Suppose $\alpha_1, \ldots, \alpha_m \in \overline{K}$ satisfy the property that if $m > 1$ then their respective minimal polynomials $h_i(X) \in K[X]$ are pair-wise distinct. Equivalently, α_i, α_j are not Galois-conjugate over K if $i \neq j$. For $i = 1, \ldots, m$, define*

$$\delta_i = \deg h_i \ (= \dim_K K(\alpha_i)).$$

Moreover, define

$$M = \sum_{i=1}^{m} \delta_i.$$

Let $K[X]_{\leq M-1}$ denote the K-vector space of polynomials $f(X) \in K[X]$ such that $\deg f \leq M - 1$.

Then the evaluation map

$$\mathcal{E} : K[X]_{\leq M-1} \longrightarrow \bigoplus_{i=1}^{m} K(\alpha_i)$$

$$f(X) \mapsto (f(\alpha_i))_{i=1}^{m}$$

is an isomorphism of K-vector spaces.

B   On BCH Codes

Let q be a prime power and let $e \geq 1$ be a positive integer. Put $s = q^e - 1$.

For any $a \in \mathbb{Z}_s$, we define a q-cylotomic coset modulo s

$$S_a := \{a \cdot q^i \bmod s : i = 0, 1, 2, \ldots\}.$$

It is a well-know fact that all q-cyclotomic cosets partition the set $\mathbb{Z}_s$. Let $S_{a_1}, S_{a_2}, \ldots, S_{a_t}$ stand for all distinct q-cyclotomic cosets modulo s. Then, we have that $\mathbb{Z}_s = \cup_{i=1}^{t} S_{a_i}$ and $s = \sum_{i=1}^{t} |S_{a_i}|$. We denote by s_a the size of the q-cyclotomic coset S_a. The following fact can be easily derived.

Lemma 8. *For every $a \in \mathbb{Z}_s$, the size s_a of S_a divides e which is the order of q modulo s.*

Proof. It is clear that s_a is the smallest positive integer such that $a \equiv aq^{s_a} \bmod s$, i.e, s_a is the smallest positive integer such that $s/\gcd(s,a)$ divides $q^{s_a} - 1$. Since $s/\gcd(s,a)$ also divides $q^e - 1$, we have $e \equiv 0 \bmod s_a$ by applying the long division.

Now for each S_a, we form s_a polynomials in the following way. Let $\alpha_1, \ldots, \alpha_{s_a}$ be an $\mathbb{F}_q$-basis of $\mathbb{F}_{q^{s_a}}$ (note that $\mathbb{F}_{q^{s_a}}$ is a subfield of $\mathbb{F}_{q^e}$). Define the polynomials

$$f_{a,j}(X) := \sum_{i=0}^{s_a-1} (\alpha_j X^a)^{q^i} \tag{2}$$

for $j = 1, 2, \ldots, s_a$.

Lemma 9. *For every $a \in \mathbb{Z}_s$, we have the following facts.*

(i) *The polynomials $f_{a,j}(X)$ for $j = 1, 2, \ldots, s_a$ are linearly independent over $\mathbb{F}_q$.*

(ii) *$f_{a,j}(\beta)$ belongs to $\mathbb{F}_q$ for all $\beta \in \mathbb{F}_{q^e}$.*

Proof. The first statement is clear since the coefficients of X^a in $f_{a,j}(X)$ are α_j and $\alpha_1, \alpha_2, \ldots, \alpha_{s_a}$ form an $\mathbb{F}_q$-basis of $\mathbb{F}_{q^{s_a}}$. To prove (ii), it is sufficient to prove that $(f_{a,j}(\beta))^q = f_{a,j}(\beta)$ for every $\beta \in \mathbb{F}_{q^e}$. Consider

$$(f_{a,j}(\beta))^q = \left(\sum_{i=0}^{s_a-1} (\alpha_j \beta^a)^{q^i} \right)^q = \sum_{i=0}^{s_a-1} (\alpha_j \beta^a)^{q^{i+1}}$$

$$= \sum_{i=1}^{s_a-1} (\alpha_j \beta^a)^{q^i} + \alpha_j^{q^{s_a}} \beta^{aq^{s_a}} = \sum_{i=1}^{s_a-1} (\alpha_j \beta^a)^{q^i} + \alpha_j \beta^a = f_{a,j}(\beta).$$

This completes the proof.

Lemma 10. *The following properties hold.*

(i) *The set $\{f_{a_i,j}(X) : j = 1, 2, \ldots, s_{a_i}, i = 1, 2, \ldots, t\}$ is linearly independent over $\mathbb{F}_q$.*

(ii) *Let V be the $\mathbb{F}_q$-span of the set $\{f_{a_i,j}(X) : j = 1, 2, \ldots, s_{a_i}, i = 1, 2, \ldots, t\}$. Then the map*

$$\pi : V \to \mathbb{F}_q^s; \quad f(X) \mapsto ((f(\alpha))_{\alpha \in \mathbb{F}_{q^e}^*} \tag{3}$$

is an $\mathbb{F}_q$-isomorphism.

Proof. (i) The degrees of $f_{a_{i_1},j_1}(X)$ and $f_{a_{i_2},j_2}(X)$ are distinct for any $i_1 \neq i_2$. Thus, the desired result follows from Lemma 9(ii).

Since both V and $\mathbb{F}_q^s$ have the same dimension, it is sufficient to prove that π is injective. This is clear since all polynomials in V has degree at most $s - 1$.

Choose an element α of $\mathbb{F}_{q^e}$ of order $s = q^e - 1$. Let $m^{(i)}(X) \in \mathbb{F}_q[X]$ denote the minimal polynomial of α^i with respect to $\mathbb{F}_q$. For $1 \leq d \leq s$, consider the BCH code B of length $s = q^e - 1$ with the generator polynomial l.c.m$\{m^{(1)}(X), m^{(2)}(X), \ldots, m^{(d-1)}(X)\}$. Then the minimum distance of B is at least d.

Lemma 11. *With notations defined above, we have $B = \pi(V_{n-d})$, where V_{n-d} is the $\mathbb{F}_q$-span of the set $\{f_{a_i,j}(X) : \deg(f_{a_i,j}(X)) \leq n - d\}$.*

Proof. It is clear that $f(X) = \sum_{i=0}^{s-1} f_i X^i \in \mathbb{F}_q[X]/(X^s - 1)$ belongs to B if and only if $f(\alpha^i) = 0$ for $i = 1, 2, \ldots, d - 2$. This means that $(f_0, f_1, \ldots, f_{s-1})$ belongs to the dual code of the following code

$$\{(a(1), a(\alpha), \ldots, a(\alpha^{s-1})) : a(X) \in \mathbb{F}_q[X]; \ 1 \le \deg(a(X)) \le d - 1\}.$$

On the other hand, the dual of the above code is in fact the generalized Reed-Solomon code

$$GRS(s - d) := \{(a(1), a(\alpha), \ldots, a(\alpha^{s-1})) : a(X) \in \mathbb{F}_q[X]; \ \deg(a(X)) \le s - d\}.$$

This means that $B = \mathbb{F}_q^s \cap GRS(s - d)$. The desired result follows from Lemma 10(ii).

Non-malleable Condensers for Arbitrary Min-entropy, and Almost Optimal Protocols for Privacy Amplification

Xin Li*

Department of Computer Science
Johns Hopkins University
Baltimore, MD 21218, USA
lixints@cs.jhu.edu

Abstract. Recently, the problem of privacy amplification with an active adversary has received a lot of attention. Given a shared n-bit weak random source X with min-entropy k and a security parameter s, the main goal is to construct an explicit 2-round privacy amplification protocol that achieves entropy loss $O(s)$. Dodis and Wichs [1] showed that optimal protocols can be achieved by constructing explicit *non-malleable extractors*. However, the best known explicit non-malleable extractor only achieves $k = 0.49n$ [2] and evidence in [2] suggests that constructing explicit non-malleable extractors for smaller min-entropy may be hard. In an alternative approach, Li [3] introduced the notion of a non-malleable condenser and showed that explicit non-malleable condensers also give optimal privacy amplification protocols.

In this paper, we give the first construction of non-malleable condensers for arbitrary min-entropy. Using our construction, we obtain a 2-round privacy amplification protocol with optimal entropy loss for security parameter up to $s = \Omega(\sqrt{k})$. This is the first protocol that simultaneously achieves optimal round complexity and optimal entropy loss for arbitrary min-entropy k. We also generalize this result to obtain a protocol that runs in $O(s/\sqrt{k})$ rounds with optimal entropy loss, for security parameter up to $s = \Omega(k)$. This significantly improves the protocol in [4]. Finally, we give a better non-malleable condenser for linear min-entropy, and in this case obtain a 2-round protocol with optimal entropy loss for security parameter up to $s = \Omega(k)$, which improves the entropy loss and communication complexity of the protocol in [2].

Keywords: privacy amplification, non-malleable, extractor, condenser.

1   Introduction

Modern cryptographic applications rely heavily on the use of randomness. Indeed, true randomness are provably necessary and key ingredients in even basic

---

* Most work was done while the author was a Simons postdoctoral fellow at University of Washington.

Y. Dodis and J.B. Nielsen (Eds.): TCC 2015, Part I, LNCS 9014, pp. 502–531, 2015.

tasks such as bit commitment and encryption. However, most of these applications require uniform random bits, yet real world random sources are rarely uniformly distributed. In addition, even initially uniform secret keys could be damaged by side channel attacks of an adversary. Naturally, the random sources we can use become imperfect, and it is therefore important to study how to run cryptographic applications using imperfect randomness. In [5], Dodis et. al showed that even slightly imperfect random sources cannot be used directly in many important cryptographic applications, thus we have to find a way to convert the imperfect random sources into nearly uniform random bits first.

In this general context, Bennett, Brassard, and Robert [6] introduced the basic cryptographic question of *privacy amplification*. The setting is as follows. Consider the simple model where two parties (Alice and Bob) share an n-bit secret key X, which is weakly random. They also have access to local (non-shared) uniform private random bits and share a public channel which is monitored by an adversary Eve. The goal now is for Alice and Bob to communicate over the channel to transform X into a nearly uniform secret key, so that Eve has negligible information about it. To measure the randomness in X, we use the standard min-entropy.

Definition 1. *The* min-entropy *of a random variable X is*

$$H_\infty(X) = \min_{x \in \text{supp}(X)} \log_2(1/\Pr[X = x]).$$

For $X \in \{0,1\}^n$, we call X an $(n, H_\infty(X))$-source, and we say X has entropy *rate $H_\infty(X)/n$.*

This problem arises naturally in several situations when two parties want to communicate with each other secretly (e.g., one-time pad). We note that shared randomness is an important resource and is often harder to obtain than local randomness. More importantly the quality of shared randomness generally may be much weaker than local randomness, thus it makes sense in the privacy amplification problem to assume that the parties have local uniform random bits and try to boost the quality of the shared weak random source.

Following [6], we assume the adversary Eve has unlimited computational power. If Eve is passive (i.e., can only see the messages but cannot change them), then this problem can be solved by using a well-studied combinatorial object called "strong extractor".

Notation. We let $[s]$ denote the set $\{1, 2, \ldots, s\}$. For ℓ a positive integer, U_ℓ denotes the uniform distribution on $\{0,1\}^\ell$, and for S a set, U_S denotes the uniform distribution on S. When used as a component in a vector, each U_ℓ or U_S is assumed independent of the other components.

Definition 2 (statistical distance). *Let W and Z be two distributions on a set S. Their* statistical distance *(variation distance) is*

$$\Delta(W, Z) =: \max_{T \subseteq S}(|W(T) - Z(T)|) = \frac{1}{2}\sum_{s \in S}|W(s) - Z(s)|.$$

We say W is ε-close to Z, denoted $W \approx_\varepsilon Z$, if $\Delta(W, Z) \leq \varepsilon$. For a distribution D on a set S and a function $h : S \to T$, let $h(D)$ denote the distribution on T induced by choosing x according to D and outputting $h(x)$.

Definition 3. *A function* $\mathsf{Ext} : \{0,1\}^n \times \{0,1\}^d \to \{0,1\}^m$ *is a strong* (k, ε)- *extractor if for every source* X *with min-entropy* k *and independent* Y *which is uniform on* $\{0,1\}^d$,

$$(\mathsf{Ext}(X, Y), Y) \approx_\varepsilon (U_m, Y).$$

Suppose we have a strong extractor Ext, we can then have Alice sample a fresh random string Y from her local random bits and send it to Bob. They then both compute $R = \mathsf{Ext}(X, Y)$. Since Eve only sees Y, the property of the strong extractor guarantees that the output is close to uniform even given this information.

However, if Eve is active (i.e., can arbitrarily change, delete and reorder messages), then the problem becomes much harder and the above simple solution fails. In this case, while one can show the task is still possible, the main goal is to try to use as few rounds as possible, and achieve a secret nearly uniform random string R that has length as close to $H_\infty(X)$ as possible. There has been a lot of effort in trying to achieve optimal parameters [7,8,1,9,10,4,11,12,3,2]. More specifically, [7] gave the first non-trivial protocol which takes one-round and works when the entropy rate of X is bigger than $2/3$. [8] later improved this to work for entropy rate bigger than $1/2$, yet both these results suffer from the drawback that the final secret key R is significantly shorter than the min-entropy of X. [1] showed that it is impossible to construct one-round protocol for if the entropy rate of X is less than $1/2$. Moreover, one can show that the final output R has to be at least $O(s)$ shorter than $H_\infty(X)$, where s is the security parameter of the protocol (A protocol has security parameter s if Eve cannot predict with advantage more than 2^{-s} over random. When Eve is active, we also require that Eve cannot make Alice and Bob output different secrets and not abort with probability more than 2^{-s}.). This difference is call the *entropy loss* of the protocol. Thus in general the optimal protocol should take 2 rounds and have entropy loss $O(s)$.

The first protocol which works for entropy rate below $1/2$ appeared in [9], which was simplified by [10] and shown to run in $O(s)$ rounds and achieve entropy loss $O(s^2)$. [1] improved the number of rounds to 2 but the entropy loss remains $O(s^2)$. [4] improved the entropy loss to $O(s)$ but the number of rounds increases to $O(s)$. The natural open question is therefore whether there is an explicit 2-round protocol with entropy loss $O(s)$. In the special case where X has entropy rate bigger than $1/2$, [11,12,3] gave 2-round protocols with entropy loss $O(s)$. For any constant $0 < \delta < 1$, [11] also gave a protocol for the case where X has entropy rate δ, which runs in $\mathsf{poly}(1/\delta)$ rounds with entropy loss $\mathsf{poly}(1/\delta)s = O(s)$. Recently, [2] gave an improved protocol for the case of entropy rate δ, which runs in 2 rounds and achieves optimal entropy loss $2^{\mathsf{poly}(1/\delta)}s = O(s)$, although the hidden constant can be quite large.

In [1], Dodis and Wichs introduced the notion of a "non-malleable extractor" and showed that such an object can be used to construct 2-round privacy amplification protocols with optimal entropy loss.

Definition 4. [1] *A function* nmExt : $\{0,1\}^n \times \{0,1\}^d \to \{0,1\}^m$ *is a* (k,ε)-*non-malleable extractor if, for any source X with $H_\infty(X) \geq k$ and any function $\mathcal{A} : \{0,1\}^d \to \{0,1\}^d$ such that $\mathcal{A}(y) \neq y$ for all y, the following holds. When Y is chosen uniformly from $\{0,1\}^d$ and independent of X,*

$$(\mathsf{nmExt}(X,Y), \mathsf{nmExt}(X,\mathcal{A}(Y)), Y) \approx_\varepsilon (U_m, \mathsf{nmExt}(X,\mathcal{A}(Y)), Y).$$

Dodis and Wichs showed that non-malleable extractors exist when $k > 2m + 3\log(1/\varepsilon) + \log d + 9$ and $d > \log(n - k + 1) + 2\log(1/\varepsilon) + 7$. However, they only constructed weaker forms of non-malleable extractors. The first explicit construction of non-malleable extractors appeared in [11], which works for entropy $k > n/2$. Later, various improvements appeared in [12,3,13]. However, the entropy requirement remains $k > n/2$. Recently, Li [2] gave the first explicit non-malleable extractor that breaks this barrier, which works for $k = (1/2 - \delta)n$ for some constant $\delta > 0$. [2] also showed a connection between non-malleable extractors and two-source extractors, which suggests that constructing explicit non-malleable extractors for smaller entropy may be hard.

Given the above background, an alternative approach seems promising. This is the notion of a non-malleable condenser introduced in [3]. While a non-malleable extractor requires the output to be close to uniform, a non-malleable condenser only requires the output to have enough min-entropy.

Definition 5. [2] *A* (k, k', ϵ) *non-malleable condenser is a function* nmCond : $\{0,1\}^n \times \{0,1\}^d \to \{0,1\}^m$ *such that given any (n, k)-source X, an independent uniform seed $Y \in \{0,1\}^d$, and any (deterministic) function $\mathcal{A} : \{0,1\}^d \to \{0,1\}^d$ such that $\forall y, \mathcal{A}(y) \neq y$, we have that with probability $1 - \epsilon$ over the fixing of $Y = y$,*

$$\Pr_{z' \leftarrow \mathsf{nmCond}(X,\mathcal{A}(y))}[\mathsf{nmCond}(X,y)|_{\mathsf{nmCond}(X,\mathcal{A}(y))=z'} \text{ is } \epsilon\text{-close to an } (m, k') \text{ source}] \geq 1 - \epsilon.$$

As can be seen from the definition, a non-malleable condenser is a strict relaxation of a non-malleable extractor and thus it may be easier to construct. In [3], Li showed that non-malleable condensers can also be used to construct 2-round privacy amplification protocols with optimal entropy loss. Thus to give optimal privacy amplification protocols for smaller min-entropy, one can hope to first construct explicit non-malleable condensers for smaller min-entropy.

1.1 Our Results

In this paper, we indeed succeed in the above approach. We construct explicit non-malleable condensers for essentially any min-entropy. Our first theorem is as follows.

---

[1] Following [11], we define worst case non-malleable extractors, which is slightly different from the original definition of average case non-malleable extractors in [1]. However, the two definitions are essentially equivalent up to a small change of parameters.

Theorem 1. *There exists a constant $C > 0$ such that for any $n, k \in \mathbb{N}$ and $s > 0$ with $k \geq C(\log n + s)^2$, there is an explicit $(k, s, 2^{-s})$-non-malleable condenser with seed length*
$d = O(\log n + s)^2$ *and output length $m = O(\log n + s)^2$.*

Combining this theorem with the protocol in [3], we immediately obtain a 2-round privacy amplification protocol with optimal entropy loss for any security parameter up to $\Omega(\sqrt{k})$. This is the first explicit protocol that simultaneously achieves optimal parameters in both round complexity and entropy loss, for arbitrary min-entropy.

Theorem 2. *There exists a constant C such that for any $\epsilon > 0$ with $k \geq C(\log n + \log(1/\epsilon))^2$, there exists an explicit 2-round privacy amplification protocol for (n, k) sources with security parameter $\log(1/\epsilon)$, entropy loss $O(\log n + \log(1/\epsilon))$ and communication complexity $O(\log n + \log(1/\epsilon))^2$.*

We note that except the protocol in [4], all previous results that work for arbitrary min-entropy k only achieve security parameter up to $s = \Omega(\sqrt{k})$ like our protocol and all of them have entropy loss $\Omega(s^2)$. In this paper, we finally manage to reduce the entropy loss to $O(s)$. Thus, for this range of security parameter, ignoring the communication complexity, we essentially obtain optimal privacy amplification protocols.

For the special case where $k = \delta n$ for some constant $0 < \delta < 1$, we can do better. Here we have the following theorem.

Theorem 3. *For any constant $0 < \delta < 1$ and $k = \delta n$ there exists a constant $C = 2^{\mathsf{poly}(1/\delta)}$ such that given any $0 < s \leq k/C$, there is an explicit $(k, s, 2^{-s})$-non-malleable condenser with seed length $d = \mathsf{poly}(1/\delta)(\log n + s)$ and output length $m = 2^{\mathsf{poly}(1/\delta)}(\log n + s)$.*

Combined with the protocol in [3], this theorem yields:

Theorem 4. *There exists an absolute constant $C_0 > 1$ such that for any constant $0 < \delta < 1$ and $k = \delta n$ there exists a constant $C_1 = 2^{\mathsf{poly}(1/\delta)}$ such that given any $\epsilon > 0$ with $C_1 \log(1/\epsilon) \leq k$, there exists an explicit 2-round privacy amplification protocol for (n, k) sources with security parameter $\log(1/\epsilon)$, entropy loss $C_0(\log n + \log(1/\epsilon))$ and communication complexity $\mathsf{poly}(1/\delta)(\log n + \log(1/\epsilon))$.*

Note that for security parameter s, the 2-round protocol for $k = \delta n$ in [2] has entropy loss $2^{\mathsf{poly}(1/\delta)}s$ and communication complexity $2^{\mathsf{poly}(1/\delta)}s$. Here, we improve the entropy loss to $C_0 s$ for an absolute constant $C_0 > 1$ and the communication complexity to $\mathsf{poly}(1/\delta)s$.

Finally, one can ask what if for arbitrary min-entropy k, we want to achieve security parameter bigger than $\sqrt{k}$, as in [4]. Using our techniques combined with some techniques from [4], we obtain the following theorem.

Theorem 5. *There exists a constant $C > 1$ such that for any $n, k \in \mathbb{N}$ with $k \geq \log^4 n$ and any $\epsilon > 0$ with $k \geq C(\log(1/\epsilon))$ there exists an explicit $O((\log n + \log(1/\epsilon))/\sqrt{k})$ round privacy amplification protocol for (n, k) sources with security parameter $\log(1/\epsilon)$, entropy loss $O(\log n + \log(1/\epsilon))$ and communication complexity $O((\log n + \log(1/\epsilon))\sqrt{k})$.*

Thus, we can essentially achieve security parameter up to $s = \Omega(k)$ with optimal entropy loss, at the price of increasing the number of rounds to $O(s/\sqrt{k})$. Note that the protocol in [4], though also achieving optimal entropy loss, runs in $\Omega(s)$ rounds. Thus our protocol improves their round complexity by a $\sqrt{k}$ factor. For large k this is a huge improvement, especially in practice.

Table 1 summarizes our results compared to some previous results, assuming the security parameter is s.

Table 1. Summary of Results on Privacy Amplification with an Active Adversary

Construction	Entropy of X	Security parameter	Rounds	Entropy loss
Optimal non-explicit	$k > \log n$	$s \leq \Omega(k)$	2	$\Theta(s + \log n)$
[7]	$k > 2n/3$	$s = \Theta(k)$	1	$(n - k)$
[8]	$k > n/2$	$s = \Theta(k)$	1	$(n - k)$
[9,10]	$k \geq \text{polylog}(n)$	$s \leq \Omega(\sqrt{k})$	$\Theta(s + \log n)$	$\Theta((s + \log n)^2)$
[1]	$k \geq \text{polylog}(n)$	$s \leq \Omega(\sqrt{k})$	2	$\Theta((s + \log n)^2)$
[4]	$k \geq \text{polylog}(n)$	$s \leq \Omega(k)$	$\Theta(s + \log n)$	$\Theta(s + \log n)$
[11]	$k \geq \delta n$	$s \leq k/\text{poly}(1/\delta)$	$\text{poly}(1/\delta)$	$\text{poly}(1/\delta)(s + \log n)$
[2]	$k \geq \delta n$	$s \leq k/2^{\text{poly}(1/\delta)}$	2	$2^{\text{poly}(1/\delta)}(s + \log n)$
This work	$k \geq \text{polylog}(n)$	$s \leq \Omega(\sqrt{k})$	2	$\Theta(s + \log n)$
This work	$k \geq \text{polylog}(n)$	$s \leq \Omega(k)$	$\Theta((s + \log n)/\sqrt{k})$	$\Theta(s + \log n)$
This work	$k \geq \delta n$	$s \leq k/2^{\text{poly}(1/\delta)}$	2	$\Theta(s + \log n)$

Subsequent Work. After the first version of this paper appeared online, Aggarwal et. al [14] made several improvements to our protocols to make them satisfy further security properties, such as *post-application robustness* and *source privacy*, at the cost of one or two extra rounds. In addition, they also applied techniques in our paper to the case of local computability and Bounded Retrieval Model [15,16].

2   Overview of the Constructions and Techniques

Here we give an informal overview of our constructions and the technique used. To give a clear description, we shall be imprecise sometimes.

2.1   Non-malleable Condenser for Arbitrary Min-entropy

For an (n, k) source X, our non-malleable condenser uses a uniform seed $Y = (Y_1, Y_2)$, where Y_2 has a bigger size than Y_1, say $|Y_1| = d$ and $|Y_2| = 10d$.

Consider now any function $\mathcal{A}(Y) = Y' = (Y_1', Y_2')$. In the following we will use letters with prime to denote variables produced with Y'. Since $Y' \neq Y$, we have two cases: $Y_1 = Y_1'$ or $Y_1 \neq Y_1'$. The output of our non-malleable condenser will be $Z = \mathsf{nmCond}(X, Y) = (V_1, V_2)$. Intuitively, V_1 handles the case where $Y_1 = Y_1'$ and V_2 handles the case where $Y_1 \neq Y_1'$. We now describe the two cases separately.

If $Y_1 = Y_1'$, then we take a strong extractor Ext and compute $W = \mathsf{Ext}(X, Y_1)$. Note that $W' = \mathsf{Ext}(X, Y_1') = W$ since $Y_1 = Y_1'$. Note that $Y' \neq Y$, thus we must have $Y_2' \neq Y_2$. We now fix Y_1 (and Y_1'). Note that conditioned on this fixing, $W = W'$ is still (close to) uniform since Ext is a strong extractor, and now Y_2' is a deterministic function of Y_2. At this point, we can take any non-malleable extractor nmExt from [11,12,3] and compute $V_1 = \mathsf{nmExt}(W, Y_2)$. Since W is uniform, by the property of the non-malleable extractor we have that V_1 is (close to) uniform even conditioned on the fixing of V_1' and (Y_2, Y_2'). Now let the size of V_1 be bigger than the size of V_2, say $|V_1| \geq |V_2| + s$. Thus the further conditioning on the fixing of V_2' will still leave V_1 with entropy roughly s. This takes care of our first case.

If $Y_1 \neq Y_1'$, then we first fix (Y_1, Y_1'). Note that fixing Y_1' may cause Y_2 to lose entropy. However, since $|Y_2| = 10|Y_1|$, conditioned on this fixing Y_2 still has entropy rate roughly $9/10$, and now Y_2' is a deterministic function of Y_2. We further fix $W' = \mathsf{Ext}(X, Y_1')$, which is now a deterministic function of X. As long as the entropy of X is larger than the size of W, conditioned on this fixing X still has a lot of entropy. Note that after these fixings X and Y_2 are still independent. Now, we use X and Y_2 to perform an alternating extraction protocol. Specifically, take the first $3d$ bits of Y_2 to be S_0, we compute the following random variables: $R_0 = \mathsf{Raz}(S_0, X), S_1 = \mathsf{Ext}(Y_2, R_0), R_1 = \mathsf{Ext}(X, S_1), S_2 = \mathsf{Ext}(Y_2, R_1), R_2 = \mathsf{Ext}(X, S_2), \cdots, S_t = \mathsf{Ext}(Y_2, R_{t-1}), R_t = \mathsf{Ext}(X, S_t)$. Here Raz is the strong two source extractor in [17], which works as long as the first source has entropy rate $> 1/2$, and Ext is a strong extractor. We take $t = 4d$ and let each R_i output s bits. Note that in the first step S_0 roughly has entropy rate $2/3$, thus we need to use the two-source extractor Raz. In all subsequent steps S_i, R_i are (close to) uniform, thus it suffices to use a strong extractor. The alternating extraction protocol is shown in Figure 1.

In the above alternating extraction protocol, as long as the size of each (S_i, R_i) is relatively small, one can show that for any j, R_j is (close to) uniform conditioned on $\{R_i, R_i', i < j\}$ and (Y_2, Y_2') (recall $\{R_j'\}$ are the random variables produced by Y_2' instead of Y_2). The intuitive reason is that in each step X still has enough entropy conditioned on all previous random variables produced, and we use a strong extractor which guarantees that the output is uniform even conditioned on the seed. Next, we borrow some ideas from [1]. Specifically, there they showed an efficient map f from a string with d bits to a subset of $[4d]$, such that for any $\mu \in \{0,1\}^d$, $f(\mu)$ has $2d$ elements. Moreover, for any $\mu \neq \mu'$, there exists a $j \in [4d]$ such that $|f(\mu)^{\geq j}| > |f(\mu')^{\geq j}|$, where $f(\mu)^{\geq j}$ denotes the subset of $f(\mu)$ which contains all the elements $\geq j$. Now, let $R = (R_1, \cdots, R_t)$ be the t random variables R_i produced in the above alternating extraction protocol. As

Y_2, S_0		X
S_0	$\xrightarrow{\quad S_0 \quad}$	
	$\xleftarrow{\quad R_0 \quad}$	$R_0 = \mathsf{Raz}(S_0, X)$
$S_1 = \mathsf{Ext}(Y_2, R_0)$	$\xrightarrow{\quad S_1 \quad}$	
	$\xleftarrow{\quad R_1 \quad}$	$R_1 = \mathsf{Ext}(X, S_1)$
$S_2 = \mathsf{Ext}(Y_2, R_1)$	$\xrightarrow{\quad S_2 \quad}$	
	$\xleftarrow{\quad R_2 \quad}$	$R_2 = \mathsf{Ext}(X, S_2)$
	$\cdots$	
$S_t = \mathsf{Ext}(Y_2, R_{t-1})$	$\xrightarrow{\quad S_t \quad}$	
		$R_t = \mathsf{Ext}(X, S_t)$

Fig. 1. Alternating Extraction

in [1], we define a "look-ahead" MAC (message authentication code) laMAC that uses R as the key. For any $\mu \in \{0,1\}^d$, we define $\mathsf{laMAC}_R(\mu) = \{R_i\}_{i \in f(\mu)}$. Now our V_2 is computed as $V_2 = \mathsf{laMAC}_R(Y_1)$.

Note that since we have fixed (Y_1, Y_1'), we can now view them as two different strings in $\{0,1\}^d$. Thus, there exists a $j \in [4d]$ such that $|f(Y_1)^{\geq j}| > |f(Y_1')^{\geq j}|$. We will now show that V_2 has entropy at least s conditioned on V_2'. To show this, let $\bar{R}$ be the concatenation of those R_is in $f(Y_1)^{\geq j}$ and $\bar{R}'$ be the concatenation of those R_i's in $f(Y_1')^{\geq j}$, then the size of $\bar{R}$ is bigger than the size of $\bar{R}'$ by at least s. Moreover, $\bar{R}$ is (close to) uniform conditioned on the fixing of $\{R_i', i < j\}$ and (Y_2, Y_2'). Thus $\bar{R}$ roughly has entropy s even conditioned on the fixing of $(\bar{R}', \{R_i', i < j\})$ and (Y_2, Y_2'), which also determines V_2'. Since $\bar{R}$ is part of V_2, we have that V_2 has entropy at least s conditioned on V_2'. Since we have fixed W' before, $V_1' = \mathsf{nmExt}(W', Y_2')$ is also fixed. Thus we have that $Z = (V_1, V_2)$ has entropy roughly s even conditioned on the fixing of $Z' = (V_1', V_2')$ and (Y_2, Y_2'). This takes care of our second case.

Thus, we obtain a non-malleable condenser for any min-entropy. However, since in the alternating extraction protocol each R_i outputs s bits, and we need $d = \Omega(s)$ to achieve error 2^{-s}, the entropy of X has to be larger than $4ds = \Omega(s^2)$. Thus we can only achieve s up to $\Omega(\sqrt{k})$.

2.2 Privacy Amplification Protocol

Combined with the techniques in [2], our non-malleable condenser immediately gives a 2-round privacy amplification protocol with optimal entropy loss for any min-entropy, with security parameter s up to $\Omega(\sqrt{k})$. To better illustrate the key idea, we also give a slightly simpler 2-round protocol with optimal entropy loss, without using the non-malleable condenser. Assuming the security parameter we want to achieve is s, we now describe the protocol below.

In the first round, Alice samples 3 random strings (Y_1, Y_2, Y_3) from her local random bits and sends them to Bob, where Bob receives (Y_1', Y_2', Y_3'). We let $|Y_1| = d, |Y_2| = 10d, |Y_3| = 50d$. Take a strong extractor Ext, now Alice and Bob each computes $R_1 = \mathsf{Ext}(X, Y_1)$ and $R_1' = \mathsf{Ext}(X, Y_1')$ respectively. Let R_1, R_1' each output $4s$ bits. Next, Alice and Bob each uses (X, Y_2) and (X, Y_2') to perform the alternating extraction protocol we described above, where they compute $R_2 = (R_{21}, \cdots, R_{2t})$ and $R_2' = (R_{21}', \cdots, R_{2t}')$ respectively, with $t = 4d$. Finally, using R_2 and R_2' as the key, they compute $Z = \mathsf{laMAC}_{R_2}(Y_1)$ and $Z' = \mathsf{laMAC}_{R_2'}(Y_1')$ respectively as described before.

In the second round, Bob samples a random string W' from his local random bits and sends it to Alice, where Alice receives W. Together with W', Bob also sends two tags (T_1', T_2'), where Alice receives (T_1, T_2). For T_1', Bob takes the two-source extractor Raz and computes $T_1' = \mathsf{Raz}(Y_3', Z')$. Let T_1' output s bits. For T_2', Bob takes a standard message authentication code (MAC) and computes $T_2' = \mathsf{MAC}_{R_1'}(W')$, where R_1' is used as the key to authenticate the message W'. Bob then computes $R_B = \mathsf{Ext}(X, W')$ as the final output. When receiving (W, T_1, T_2), Alice will check whether $T_1 = \mathsf{Raz}(Y_3, Z)$ and $T_2 = \mathsf{MAC}_{R_1}(W)$. If either test fails, Alice rejects and aborts. Otherwise Alice computes $R_A = \mathsf{Ext}(X, W)$ as the final output. The protocol is shown in Figure 2.

As before, the analysis can be divided into two cases: $Y_1 = Y_1'$ and $Y_1 \neq Y_1'$. In the first case, we have $R_1 = R_1'$ and is (close to) uniform and private. Thus R_1 can be used in the MAC to authenticate W' to Alice. The MAC works by the property that if Eve changes W' to a different W, then with high probability Even cannot produce the correct tag $T_2 = \mathsf{MAC}_{R_1}(W)$ even given T_2'. This works except that here Eve also has additional information from T_1'. However, although T_1' may give some information about the MAC key R_1, note that R_1 has size $4s$ and T_1' has size s. Thus even conditioned on T_1', R_1 has entropy roughly $3s$. We note that the MAC works as long as the entropy rate of R_1 is bigger than $1/2$. Thus in this case Bob can indeed authenticate W' to Alice and they will agree on a uniform and private final output.

In the second case, again we can first fix (Y_1, Y_1') and R_1'. As before we have that after this fixing, Y_2 still has entropy rate roughly $9/10$, X still has a lot of entropy, and X is independent of (Y_2, Y_3). Now we can view (Y_1, Y_1') as two different strings and by the same analysis before, Z roughly has entropy s conditioned on the fixing of Z' and (Y_2, Y_2'). Note that after this fixing Y_3 still has entropy rate $> 1/2$, and Y_3' is a deterministic function of Y_3. Since Raz is a strong two-source extractor, we have that $\mathsf{Raz}(Y_3, Z)$ is (close to) uniform even given (Y_3', Z', R_1', W'), which also determines (T_1', T_2'). Thus, in this case Alice will reject with probability $1 - 2^{-s}$, since the probability that Eve guesses $\mathsf{Raz}(Y_3, Z)$ correctly is at most 2^{-s}.

We note that our protocol shares some similarities with the 2-round protocol in [1], as they both use the alternating extraction protocol and the "look-ahead" MAC. However, there is one important difference. The protocol in [1] uses the

Alice: X	Eve: E	Bob: X

Sample random $Y = (Y_1, Y_2, Y_3)$.
Compute $R_2 = (R_{21}, \cdots, R_{2t})$
by alternating extraction of (X, Y_2).
$Z = \mathsf{laMAC}_{R_2}(Y_1)$.
$R_1 = \mathsf{Ext}(X, Y_1)$ and output $4s$ bits.

$$(Y_1, Y_2, Y_3)$$
$$\longrightarrow$$

$$(Y_1', Y_2', Y_3')$$

Sample random W' with d bits.
Compute $R_2' = (R_{21}', \cdots, R_{2t}')$
by alternating extraction of (X, Y_2').
$Z' = \mathsf{laMAC}_{R_2'}(Y_1')$.
$R_1' = \mathsf{Ext}(X, Y_1')$ and output $4s$ bits.
$T_1' = \mathsf{Raz}(Y_3', Z')$ with s bits,
$T_2' = \mathsf{MAC}_{R_1'}(W')$.
Set final $R_B = \mathsf{Ext}(X, W')$.

$$(W', T_1', T_2')$$
$$\longleftarrow$$

$$(W, T_1, T_2)$$

If $T_1 \neq \mathsf{Raz}(Y_3, Z)$ or
$T_2 \neq \mathsf{MAC}_{R_1}(W)$ *reject.*
Set final $R_A = \mathsf{Ext}(X, W)$.

Fig. 2. 2-round Privacy Amplification Protocol

look-ahead MAC to authenticate the string W' that Bob sends to Alice in the second round. The look-ahead MAC has size $\Omega(s^2)$ and is revealed in the second round, which causes an entropy loss of $\Omega(s^2)$. Our protocol, on the other hand, uses the look-ahead MAC to authenticate the string Y_1 that Alice sends to Bob in the first round. Although in the protocol we do compute some variables that have size $\Omega(s^2)$ (namely (Z, Z')), they are computed locally by Alice and Bob, and are *never* revealed in the protocol to Eve. Instead, what is revealed to Eve is $T_1' = \mathsf{Raz}(Y_3', Z')$, which only has size $O(s)$. In other words, in the case where $Y_1 \neq Y_1'$, since we know that Z has entropy s conditioned on Z', we can apply another extractor Raz to Z and Z' respectively, such that the resulting variable T_1' only has size $O(s)$ and $\mathsf{Raz}(Y_3, Z)$ is (close to) uniform conditioned on T_1'. This is enough for the purpose of authentication, while bringing the entropy loss down to $O(s)$.

One might think that the same trick can also be applied to the protocol in [1] directly. However, this is not the case. The reason is that conditioned on (Y, Y'), all the random variables in our protocol that are used to authenticate W' are (R_1, T_1, R_1', T_1'), which are deterministic functions of X and have size $O(s)$. Thus in the case where Bob successfully authenticates W' to Alice, we can fix them and conditioned on the fixing, X and W are still independent so we can apply a strong extractor to obtain the final output $\mathsf{Ext}(X, W)$. This results in a protocol with optimal entropy loss. In the protocol in [1], conditioned on (Y, Y'), the random variables that are used to authenticate W' include the output of the look-ahead extractor, which has size $\Omega(s^2)$. Thus conditioning on this random variable will cause X to lose entropy $\Omega(s^2)$. On the other hand, we cannot simply apply another extractor to this MAC to reduce the output size; since then the output will be a function of W and X, and thus conditioned on the fixing of it, W and X will no longer be independent.

We now describe our protocol for security parameter $s > \sqrt{k}$. The very high level strategy is as follows. At the beginning of the protocol, Alice samples a random string Y from her local random bits with $d_1 = O(s)$ bits and sends it to Bob, where Bob receives Y'. They each compute $R = \mathsf{Ext}(X, Y)$ and $R' = \mathsf{Ext}(X, Y')$ respectively, by using a strong extractor Ext. At the end of the protocol, Bob samples a random string W' from his local random bits with d_1 bits and sends it to Alice, together with a tag $T = \mathsf{MAC}_{R'}(W')$. Alice receives (W, T). Bob will compute $R_B = \mathsf{Ext}(X, W')$ as his final output and Alice will check if $T = \mathsf{MAC}_R(W)$. If the test fails then Alice rejects. Otherwise she will compute $R_A = \mathsf{Ext}(X, W)$ as her final output. In the case where $Y = Y'$, again we will have that $R = R'$ and is uniform and private. Thus in this case Bob can authenticate W' to Alice by using a MAC and R' as the key. We will now modify the protocol to ensure that if $Y \neq Y'$, then with probability $1 - 2^{-s}$ either Alice or Bob will reject.

If $s < \sqrt{k}$ then we can use our 2-round protocol described above. However, we want to achieve $s > \sqrt{k}$ and X does not have enough entropy for the 2-round protocol. On the other hand, we note that we can still use the 2-round protocol to authenticate a substring of Y with $s' = \Theta(\sqrt{k})$ bits to Bob, such that if Eve changes this string, then with probability $1 - 2^{-s'}$ Alice will reject. The key observation now is that after running this 2-round protocol, conditioned on the transcript revealed to Eve, X only loses $O(s')$ entropy. Thus X still has entropy $k - O(\sqrt{k})$ in Eve's view. Therefore, we can run the 2-round protocol again, using fresh random strings sampled from Alice and Bob's local random bits. This will authenticate another substring of Y with $s' = \Theta(\sqrt{k})$ bits to Bob. As long as X has enough entropy, we can keep doing this and it will take us $O(s/\sqrt{k})$ rounds to authenticate the entire Y to Bob, while the entropy loss is $O(s')O(s/\sqrt{k}) = O(s)$. Thus as long as $k \geq Cs$ for a sufficiently large constant C, the above approach will work.

However, the simple idea described above is not enough. The reason is that to change Y, Eve only needs to change one substring, and she can succeed with probability $2^{-s'} \gg 2^{-s}$. To fix this, we modify the protocol to ensure that, if Eve changes Y to $Y' \neq Y$, then she has to change $\Omega(s/\sqrt{k})$ substrings, i.e., a constant fraction of the substrings. This is where we borrow some ideas from [4]. Specifically, instead of having Alice just authenticate substrings of Y to Bob, we will use an asymptotically good code for edit errors and have Alice authenticate substrings of the encoding of Y to Bob. More specifically, let $M = \mathsf{Edit}(Y)$ be the encoding of Y, which has size $O(d_1)$. At the beginning of the protocol, Alice will send Y to Bob, where Bob receives Y'. Next, our protocol will run in $L = O(s/\sqrt{k})$ phases, with each phase consisting of two rounds. In phase i, Alice will send the i'th substring M_i of M to Bob, where M_i has $d_2 = \Theta(\sqrt{k})$ bits. In the first round of phase i, Alice samples two random strings (Y_{i2}, Y_{i3}) from her local random bits and sends them to Bob, together with M_i. Bob receives (M_i', Y_{i2}', Y_{i3}'). We will let $|Y_{i3}| \geq 10|Y_{i2}|$. As in the previous 2-round protocol, Alice will use X and Y_{i2} to perform an alternating extraction protocol, where she computes $R_i = (R_{i1}, \cdots, R_{it})$ with $t = 4d_2$ and $Z_i = \mathsf{laMAC}_{R_i}(M_i)$, where laMAC is the look-ahead MAC described before. Correspondingly, Bob will compute R_i' and $Z_i' = \mathsf{laMAC}_{R_i'}(M_i')$, using X and Y_{i2}'. In the second round, Bob will send $T_i' = \mathsf{Raz}(Y_{i3}', Z_i')$ to Alice, where Alice receives T_i. Alice will now check if $T_i = \mathsf{Raz}(Y_{i3}, Z_i)$ and she rejects if the test fails. By the same analysis of the 2-round protocol, if Eve changes the substring M_i to $M_i' \neq M_i$, then with probability $1 - 2^{-\Omega(\sqrt{k})}$ Alice will reject.

One problem of the above approach is that Eve can first delay messages from Alice, send fake messages to Bob to get responses that contain additional information, and then resume execution with Alice. To avoid this problem, we need to synchronize between Alice and Bob. To achieve this, in the second round of phase i, we will also have Bob sample a fresh random string W_i' from his local random bits and send it as a challenge to Alice, together with T_i'. Alice will receive (W_i, T_i). Now if Alice does not reject, then she will also compute a response $V_i = \mathsf{Ext}(X, W_i)$ and send it back to Bob in the first round of phase $i + 1$. Bob will receive V_i' and then check if $V_i' = \mathsf{Ext}(X, W_i')$. If the test fails then he rejects. Otherwise he proceeds as before. At the end of the protocol, Bob will first check if the received codeword $M' = M_1' \circ \cdots \circ M_L'$ is indeed equal to $\mathsf{Edit}(Y')$. If the test fails he rejects. Otherwise he proceeds as before. This gives our whole protocol. The formal protocol appears in Section 6, Figure 4.

For the analysis, by the property of the code, if Eve wants to change $M = \mathsf{Edit}(Y)$ to $M' = \mathsf{Edit}(Y')$ with $Y' \neq Y$, then she has to make $\Omega(d_1)$ edit operations (insertion, deletion or altering). Since changing one substring costs at most $\sqrt{k}$ edit operations, Eve has to change at least $\Omega(s/\sqrt{k})$ substrings. As in [4], we then show that as long as X has an extra entropy of $O(s)$, for a constant fraction of these changes, conditioned on the event that Eve has successfully made all previous changes, the probability that Eve can make this change successfully is at most $2^{-\Omega(\sqrt{k})}$. Thus the overall probability that Eve can change M to M' without causing either Alice or Bob to reject is at most $(2^{-\Omega(\sqrt{k})})^{\Omega(s/\sqrt{k})} = 2^{-\Omega(s)}$. The round complexity is $O(s/\sqrt{k})$ and the communication complexity is $O(s\sqrt{k})$ since in each phase, the communication complexity is $O(k)$.

2.3 Non-malleable Condenser for Linear Min-entropy

Our non-malleable condenser for linear min-entropy is similar to the construction for arbitrary min-entropy, except we use a different alternating extraction protocol, namely that in [2]. Specifically, we will again use a seed $Y = (Y_1, Y_2)$, where $|Y_1| = d$ and $|Y_2| \geq 10d$. The output will also be $Z = (V_1, V_2)$. For any function $\mathcal{A}(Y) = Y' = (Y_1', Y_2')$, we use V_1 to take care of the case where $Y_1 = Y_1'$ and use V_2 to take care of the case where $Y_1 \neq Y_1'$.

If $Y_1 = Y_1'$, then again we take a strong extractor Ext and compute $W = \text{Ext}(X, Y_1)$ and $V_1 = \text{nmExt}(W, Y_2)$. By the same argument before, as long as $|V_1| \geq |V_2| + s$, we have that V_1 roughly has min-entropy s conditioned on (V_1', V_2'). This takes care of our first case.

If $Y_1 \neq Y_1'$, then again we first fix (Y_1, Y_1') and W'. Conditioned on this fixing Y_2 still has entropy rate roughly 9/10, and now Y_2' is a deterministic function of Y_2. Moreover X still has a lot of entropy (say δn for some constant $\delta > 0$) and is independent of Y_2. Now we use the alternating extraction protocol in [2]. More specifically, since X has min-entropy $k = \delta n$ we can apply a somewhere condenser in [18], [17], [19] to X and obtain $\bar{X} = (X_1, \cdots, X_C)$ with $C = \text{poly}(1/\delta)$ such that at least one X_i has entropy rate 0.9. In [2], Li showed that as long as $k \geq 2^{\text{poly}(1/\delta)} s$, one can use $X, \bar{X}, Y_2$ to perform an alternating extraction protocol and then use the output and Y_1 to obtain V_2 with size $2^{\text{poly}(1/\delta)} s$, such that whenever $Y_1 \neq Y_1'$, V_2 roughly has entropy s conditioned on the fixing of V_2' and (Y_2, Y_2'). Since we have fixed (Y_1, Y_1') and W' before, this means that Z roughly has entropy s conditioned on the fixing of Z' and (Y, Y').

Combined with the protocol in [3], we thus reduce the entropy loss of the protocol in [2] to $O(s)$ for an absolute constant $O(\cdot)$ and the communication complexity to $\text{poly}(1/\delta)s$.

Organization. in Section 3 we give the formal definition of the privacy amplification problem. We then give some preliminaries in Section 4 and define alternating extraction in Section 5. We give our non-malleable condenser for arbitrary min-entropy and the 2-round protocol in Section 6. The general multi-round protocol and non-malleable condenser for linear min-entropy are deferred to the full version. We conclude with some open problems in Section 7.

3   Privacy Amplification with an Active Adversary

In this section we formally define the privacy amplification problem. First we define average conditional min-entropy.

Definition 6. *The* average conditional min-entropy *is defined as*

$$\widetilde{H}_\infty(X|W) = -\log\left(\mathbb{E}_{w \leftarrow W}\left[\max_x \Pr[X = x|W = w]\right]\right)$$
$$= -\log\left(\mathbb{E}_{w \leftarrow W}\left[2^{-H_\infty(X|W=w)}\right]\right).$$

We will follow [11] and define a privacy amplification protocol (P_A, P_B). The protocol is executed by two parties Alice and Bob, who share a secret $X \in \{0, 1\}^n$. An active, computationally unbounded adversary Eve might have some partial information E about X satisfying $\widetilde{H}_\infty(X|E) \geqslant k$. Since Eve is unbounded, we can assume without loss of generality that she is deterministic.

We assume that Eve has full control of the communication channel between the two parties. This means that Eve can arbitrarily insert, delete, reorder or modify messages sent by Alice and Bob to each other. In particular, Eve's strategy P_E defines two correlated executions (P_A, P_E) and (P_E, P_B) between Alice and Eve, and Eve and Bob, called "left execution" and "right execution", respectively. Alice and Bob are assumed to have fresh, private and independent random bits Y and W, respectively. Y and W are not known to Eve. In the protocol we use $\perp$ as a special symbol to indicate rejection. At the end of the left execution $(P_A(X, Y), P_E(E))$, Alice outputs a key $R_A \in \{0, 1\}^m \cup \{\perp\}$. Similarly, Bob outputs a key $R_B \in \{0, 1\}^m \cup \{\perp\}$ at the end of the right execution $(P_E(E), P_B(X, W))$. We let E' denote the final view of Eve, which includes E and the communication transcripts of both executions $(P_A(X, Y), P_E(E))$ and $(P_E(E), P_B(X, W))$. We can now define the security of (P_A, P_B).

Definition 7. *An interactive protocol (P_A, P_B), executed by Alice and Bob on a communication channel fully controlled by an active adversary Eve, is a (k, m, ϵ)-privacy amplification protocol if it satisfies the following properties whenever $\widetilde{H}_\infty(X|E) \geq k$:*

1. *Correctness. If Eve is passive, then $\Pr[R_A = R_B \wedge R_A \neq \perp \wedge R_B \neq \perp] = 1$.*
2. *Robustness. We start by defining the notion of pre-application robustness, which states that even if Eve is active, $\Pr[R_A \neq R_B \wedge R_A \neq \perp \wedge R_B \neq \perp] \leqslant \epsilon$. The stronger notion of post-application robustness is defined similarly, except Eve is additionally given the key R_A the moment she completed the left execution (P_A, P_E), and the key R_B the moment she completed the right execution (P_E, P_B). For example, if Eve completed the left execution before the right execution, she may try to use R_A to force Bob to output a different key $R_B \notin \{R_A, \perp\}$, and vice versa.*
3. *Extraction. Given a string $r \in \{0, 1\}^m \cup \{\perp\}$, let $\mathsf{purify}(r)$ be $\perp$ if $r = \perp$, and otherwise replace $r \neq \perp$ by a fresh m-bit random string U_m: $\mathsf{purify}(r) \leftarrow U_m$. Letting E' denote Eve's view of the protocol, we require that*

$$\Delta((R_A, E'), (\mathsf{purify}(R_A), E')) \leq \epsilon \quad and \quad \Delta((R_B, E'), (\mathsf{purify}(R_B), E')) \leq \epsilon$$

Namely, whenever a party does not reject, its key looks like a fresh random string to Eve.

The quantity $k - m$ is called the entropy loss *and the quantity $\log(1/\epsilon)$ is called the* security parameter *of the protocol.*

Remark 1. Our protocol, as well as many others in [1], [9], [10], [4], [11], [12], [3], [2] only achieve *pre-application* robustness. Recently, Aggarwal et. al [14] gave a general transformation that can convert any privacy amplification protocol with pre-application robustness into another privacy amplification protocol with *post-application* robustness at the cost of one extra round. Thus, using their transformation, our protocol can be turned into a 3-round post-application robust privacy amplification protocol with optimal entropy loss, for security parameter up to $s = \Omega(\sqrt{k})$ (as Aggarwal et. al did in [14]); or a $O(s/\sqrt{k})$ round post-application robust privacy amplification protocol with optimal entropy loss, for security parameter up to $s = \Omega(k)$.

4   Preliminaries

We often use capital letters for random variables and corresponding small letters for their instantiations. Let $|S|$ denote the cardinality of the set S. All logarithms are to the base 2.

4.1   Somewhere Random Sources, Extractors and Condensers

Definition 8 (Somewhere Random sources). *A source* $X = (X_1, \cdots, X_t)$ *is* $(t \times r)$ *somewhere-random (SR-source for short) if each* X_i *takes values in* $\{0,1\}^r$ *and there is an* i *such that* X_i *is uniformly distributed.*

Definition 9. *An elementary somewhere-k-source is a vector of sources* $(X_1, \cdots, X_t)$, *such that some* X_i *is a k-source. A somewhere k-source is a convex combination of elementary somewhere-k-sources.*

Definition 10. *A function* $C : \{0,1\}^n \times \{0,1\}^d \to \{0,1\}^m$ *is a* $(k \to l, \epsilon)$-*condenser if for every k-source* X, $C(X, U_d)$ *is* ϵ-*close to some l-source. When convenient, we call* C *a rate-*$(k/n \to l/m, \epsilon)$-*condenser.*

Definition 11. *A function* $C : \{0,1\}^n \times \{0,1\}^d \to \{0,1\}^m$ *is a* $(k \to l, \epsilon)$-*somewhere-condenser if for every k-source* X, *the vector* $(C(X, y)_{y \in \{0,1\}^d})$ *is* ϵ-*close to a somewhere-l-source. When convenient, we call* C *a rate-*$(k/n \to l/m, \epsilon)$-*somewhere-condenser.*

Definition 12. *A function* TExt $: \{0,1\}^{n_1} \times \{0,1\}^{n_2} \to \{0,1\}^m$ *is a strong two source extractor for min-entropy* k_1, k_2 *and error* ϵ *if for every independent* (n_1, k_1) *source* X *and* (n_2, k_2) *source* Y,

$$|(\mathsf{TExt}(X,Y), X) - (U_m, X)| < \epsilon$$

and

$$|(\mathsf{TExt}(X,Y), Y) - (U_m, Y)| < \epsilon,$$

where U_m *is the uniform distribution on* m *bits independent of* (X, Y).

4.2   Average Conditional Min-entropy

Dodis and Wichs originally defined non-malleable extractors with respect to average conditional min-entropy. However, this notion is essentially equivalent to the standard (worst-case) min-entropy, up to a small loss in parameters.

Lemma 1 ([20]). *For any* $s > 0$, $\Pr_{w \leftarrow W}[H_\infty(X|W = w) \geq \tilde{H}_\infty(X|W) - s] \geq 1 - 2^{-s}$.

Lemma 2 ([20]). *If a random variable B has at most 2^ℓ possible values, then $\tilde{H}_\infty(A|B) \geq H_\infty(A) - \ell$.*

To clarify which notion of min-entropy and non-malleable extractor we mean, we use the term *worst-case non-malleable extractor* when we refer to our Definition 4, which is with respect to traditional (worst-case) min-entropy, and *average-case non-malleable extractor* to refer to he original definition of Dodis and Wichs, which is with respect to average conditional min-entropy.

Corollary 1. *A (k, ε)-average-case non-malleable extractor is a (k, ε)-worst-case non-malleable extractor. For any $s > 0$, a (k, ε)-worst-case non-malleable extractor is a $(k + s, \varepsilon + 2^{-s})$-average-case non-malleable extractor.*

Throughout the rest of our paper, when we say non-malleable extractor, we refer to the worst-case non-malleable extractor of Definition 4.

4.3   Prerequisites from Previous Work

One-time message authentication codes (MACs) use a shared random key to authenticate a message in the information-theoretic setting.

Definition 13. *A function family $\{\mathsf{MAC}_R : \{0,1\}^d \to \{0,1\}^v\}$ is a ϵ-secure one-time MAC for messages of length d with tags of length v if for any $w \in \{0,1\}^d$ and any function (adversary) $A : \{0,1\}^v \to \{0,1\}^d \times \{0,1\}^v$,*

$$\Pr_R[\mathsf{MAC}_R(W') = T' \wedge W' \neq w \mid (W', T') = A(\mathsf{MAC}_R(w))] \leq \epsilon,$$

where R is the uniform distribution over the key space $\{0,1\}^\ell$.

Theorem 6 ([10]). *For any message length d and tag length v, there exists an efficient family of $(\lceil \frac{d}{v} \rceil 2^{-v})$-secure MACs with key length $\ell = 2v$. In particular, this MAC is ε-secure when $v = \log d + \log(1/\epsilon)$.*
More generally, this MAC also enjoys the following security guarantee, even if Eve has partial information E about its key R. Let (R, E) be any joint distribution. Then, for all attackers A_1 and A_2,

$$\Pr_{(R,E)}[\mathsf{MAC}_R(W') = T' \wedge W' \neq W \mid W = A_1(E),$$

$$(W', T') = A_2(\mathsf{MAC}_R(W), E)] \leq \left\lceil \frac{d}{v} \right\rceil 2^{v - \tilde{H}_\infty(R|E)}.$$

(In the special case when $R \equiv U_{2v}$ and independent of E, we get the original bound.)

Remark 2. Note that the above theorem indicates that the MAC works even if the key R has average conditional min-entropy rate $> 1/2$.

Sometimes it is convenient to talk about average case seeded extractors, where the source X has average conditional min-entropy $\widetilde{H}_\infty(X|Z) \geq k$ and the output of the extractor should be uniform given Z as well. The following lemma is proved in [20].

Lemma 3 ([20]). *For any $\delta > 0$, if* Ext *is a (k, ϵ) extractor then it is also a $(k + \log(1/\delta), \epsilon + \delta)$ average case extractor.*

Theorem 7 ([18,17,19]). *For any constant $\beta, \delta > 0$, there is an efficient family of rate-$(\delta \to 1 - \beta, \epsilon = 2^{-\Omega(n)})$-somewhere condensers* Cond $: \{0,1\}^n \to (\{0,1\}^m)^D$ *where $D = O(1)$ and $m = \Omega(n)$.*

For a strong seeded extractor with optimal parameters, we use the following extractor constructed in [21].

Theorem 8 ([21]). *For every constant $\alpha > 0$, and all positive integers n, k and any $\epsilon > 0$, there is an explicit construction of a strong (k, ϵ)-extractor* Ext $: \{0,1\}^n \times \{0,1\}^d \to \{0,1\}^m$ *with $d = O(\log n + \log(1/\epsilon))$ and $m \geq (1-\alpha)k$. It is also a strong (k, ϵ) average case extractor with $m \geq (1-\alpha)k - O(\log n + \log(1/\epsilon))$.*

We need the following construction of strong two-source extractors in [17].

Theorem 9 ([17]). *For any n_1, n_2, k_1, k_2, m and any $0 < \delta < 1/2$ with*

$- n_1 \geq 6 \log n_1 + 2 \log n_2$
$- k_1 \geq (0.5 + \delta)n_1 + 3 \log n_1 + \log n_2$
$- k_2 \geq 5 \log(n_1 - k_1)$
$- m \leq \delta \min[n_1/8, k_2/40] - 1$

There is a polynomial time computable strong 2-source extractor Raz $: \{0,1\}^{n_1} \times \{0,1\}^{n_2} \to \{0,1\}^m$ *for min-entropy k_1, k_2 with error $2^{-1.5m}$.*

Theorem 10 ([11,12,3]). *For every constant $\delta > 0$, there exists a constant $\beta > 0$ such that for every $n, k \in \mathbb{N}$ with $k \geq (1/2 + \delta)n$ and $\epsilon > 2^{-\beta n}$ there exists an explicit (k, ϵ) non-malleable extractor with seed length $d = O(\log n + \log \epsilon^{-1})$ and output length $m = \Omega(n)$.*

The following theorem is proved in [3].

Theorem 11 ([3]). *There exists a constant $C > 1$ such that the following holds. For any $n, k \in \mathbb{N}$ and $\epsilon > 0$, assume that there is an explicit (k, k', ϵ)-non-malleable condenser with seed length d such that $k' \geq C(\log n + \log(1/\epsilon))$. Then there exists an explicit 2-round privacy amplification protocol for (n, k) sources with entropy loss $O(\log n + \log(1/\epsilon))$ and communication complexity $O(d + \log n + \log(1/\epsilon))$.*

The following standard lemma about conditional min-entropy is implicit in [22] and explicit in [7].

Lemma 4 ([7]). *Let X and Y be random variables and let $\mathcal{Y}$ denote the range of Y. Then for all $\epsilon > 0$, one has*

$$\Pr_Y \left[H_\infty(X|Y = y) \geq H_\infty(X) - \log |\mathcal{Y}| - \log \left(\frac{1}{\epsilon} \right) \right] \geq 1 - \epsilon.$$

We also need the following lemma.

Lemma 5. *Let (X, Y) be a joint distribution such that X has range $\mathcal{X}$ and Y has range $\mathcal{Y}$. Assume that there is another random variable X' with the same range as X such that $|X - X'| = \epsilon$. Then there exists a joint distribution (X', Y) such that $|(X, Y) - (X', Y)| = \epsilon$.*

Proof. First let (X'', Y) be the same probability distribution as (X, Y). For any $x \in \mathcal{X}$, let $p''_x = \Pr[X'' = x]$ and $p'_x = \Pr[X' = x]$. For any $y \in \mathcal{Y}$, let $p_y = \Pr[Y = y]$. Let $p''_{xy} = \Pr[X'' = x, Y = y]$. Let $W = \{x \in \mathcal{X} : p''_x > p'_x\}$ and $V = \{x \in \mathcal{X} : p''_x < p'_x\}$. Thus we have that $\sum_{x \in W} |p''_x - p'_x| = \sum_{x \in V} |p''_x - p'_x| = \epsilon$.

We now gradually change the probability distribution X'' into X', while keeping the distribution Y the same, as follows. While W is not empty or V is not empty, do the following.

1. Pick $x \in W \cup V$ such that $|p''_x - p'_x| = min\{|p''_x - p'_x|, x \in W \cup V\}$.
2. If $x \in W$, we decrease $\Pr[X'' = x]$ to p'_x. Let $\tau = p''_x - p'_x$. To ensure this is still a probability distribution, we also pick any $\bar{x} \in V$ and increase $\Pr[X'' = \bar{x}]$ to $\Pr[X'' = \bar{x}] + \tau$. To do this, we pick the elements $y \in \mathcal{Y}$ one by one in an arbitrary order and while $\tau > 0$, do the following. Let $\tau' = min(p''_{xy}, \tau)$, $\Pr[X'' = x, Y = y] = \Pr[X'' = x, Y = y] - \tau'$, $\Pr[X'' = \bar{x}, Y = y] = \Pr[X'' = \bar{x}, Y = y] + \tau'$ and $\tau = \tau - \tau'$. We then update the sets $\{p''_x\}$ and $\{p''_{xy}\}$ accordingly. Note that since $p''_x = \tau + p'_x \geq \tau$, this process will indeed end when $\tau = 0$ and now $\Pr[X'' = x] = p'_x$. Note that after this change we still have that $p''_{\bar{x}} \leq p'_{\bar{x}}$. Also, for any $y \in \mathcal{Y}$ the probability $\Pr[Y = y]$ remains unchanged. Finally, remove x from W and if $p''_{\bar{x}} = p'_{\bar{x}}$, remove $\bar{x}$ from V.
3. If $x \in V$, we increase $\Pr[X'' = x]$ to p'_x. Let $\tau = p'_x - p''_x$. To ensure that X'' is still a probability distribution, we also pick any $\bar{x} \in W$ and decrease $\Pr[X'' = \bar{x}]$ to $\Pr[X'' = \bar{x}] - \tau$. To do this, we pick the elements $y \in \mathcal{Y}$ one by one in an arbitrary order and while $\tau > 0$, do the following. Let $\tau' = min(p''_{\bar{x}y}, \tau)$, $\Pr[X'' = x, Y = y] = \Pr[X'' = x, Y = y] + \tau'$, $\Pr[X'' = \bar{x}, Y = y] = \Pr[X'' = \bar{x}, Y = y] - \tau'$ and $\tau = \tau - \tau'$. We then update the sets $\{p''_x\}$ and $\{p''_{xy}\}$ accordingly. Note that since $p''_{\bar{x}} \geq \tau + p'_{\bar{x}}$, this process will indeed end when $\tau = 0$ and we still have $p''_{\bar{x}} \geq p_{\bar{x}}$. Also, for any $y \in \mathcal{Y}$ the probability $\Pr[Y = y]$ remains unchanged. Finally, remove x from V and if $p''_{\bar{x}} = p_{\bar{x}}$, remove $\bar{x}$ from W.

Note that in each iteration, at least one element will be removed from $W \cup V$. Thus the iteration will end after finite steps. When it ends, we have that $\forall x, \Pr[x'' = x] = p'_x$, thus $X'' = X'$. Since in each step the probability $\Pr[Y = y]$ remains unchanged, the distribution Y remains the same. Finally, it is clear from the algorithm that $|(X'', Y) - (X, Y)| = \epsilon$.

Next we have the following lemma.

Lemma 6. *Let X and Y be random variables and let $\mathcal{Y}$ denote the range of Y. Assume that X is ϵ-close to having min-entropy k. Then for any $\epsilon' > 0$*

$$\Pr_{Y}\left[(X|Y = y) \text{ is } \epsilon'\text{-close to a source with min-entropy } k - \log|\mathcal{Y}| - \log\left(\frac{1}{\epsilon'}\right)\right]$$
$$\geq 1 - \epsilon' - \frac{\epsilon}{\epsilon'}.$$

Proof. Let $\mathcal{X}$ denote the range of X. Assume that X' is a distribution on $\mathcal{X}$ with min-entropy k such that $|X - X'| \leq \epsilon$. Then by lemma 5, there exists a joint distribution (X', Y) such that

$$|(X, Y) - (X', Y)| \leq \epsilon.$$

Now for any $y \in \mathcal{Y}$, let $\Delta_y = \sum_{x \in \mathcal{X}} |\Pr[X = x, Y = y] - \Pr[X' = x, Y = y]|$. Then we have

$$\sum_{y \in \mathcal{Y}} \Delta_y \leq \epsilon.$$

For any $y \in \mathcal{Y}$, the statistical distance between $X|Y = y$ and $X'|Y = y$ is

$$\delta_y = \sum_{x \in \mathcal{X}} |\Pr[X = x|Y = y] - \Pr[X' = x|Y = y]|$$
$$= \left(\sum_{x \in \mathcal{X}} |\Pr[X = x, Y = y] - \Pr[X' = x, Y = y]|\right)/(\Pr[Y = y]) = \Delta_y / \Pr[Y = y].$$

Thus if $\delta_y \geq \epsilon'$ then $\Delta_y \geq \epsilon' \Pr[Y = y]$. Let $B_Y = \{y : \delta_y \geq \epsilon'\}$ then we have

$$\epsilon' \Pr[y \in B_Y] = \sum_{y \in B_Y} \epsilon' \Pr[Y = y] \leq \sum_{y \in B_Y} \Delta_y \leq \sum_{y \in \mathcal{Y}} \Delta_y \leq \epsilon.$$

Thus $\Pr[y \in B_Y] \leq \frac{\epsilon}{\epsilon'}$. Note that when $y \notin B_y$ we have $|X|Y = y - X'|Y = y| < \epsilon'$. Thus by Lemma 4 we have the statement of the lemma.

Quentin: Q, S_0 Wendy: X

$$
\begin{array}{lcl}
S_0 & \xrightarrow{\quad S_0 \quad} & \\
 & \xleftarrow{\quad R_0 \quad} & R_0 = \mathsf{Raz}(S_0, X) \\
S_1 = \mathsf{Ext}_q(Q, R_0) & \xrightarrow{\quad S_1 \quad} & \\
 & \xleftarrow{\quad R_1 \quad} & R_1 = \mathsf{Ext}_w(X, S_1) \\
S_2 = \mathsf{Ext}_q(Q, R_1) & \xrightarrow{\quad S_2 \quad} & \\
 & \xleftarrow{\quad R_2 \quad} & R_2 = \mathsf{Ext}_w(X, S_2) \\
 & \cdots & \\
S_t = \mathsf{Ext}_q(Q, R_{t-1}) & \xrightarrow{\quad S_t \quad} & \\
 & & R_t = \mathsf{Ext}_w(X, S_t)
\end{array}
$$

Fig. 3. Alternating Extraction

5 Alternating Extraction Protocol and Look Ahead Extractor

Recall that, an important ingredient in our construction is the following alternating extraction protocol modified from that in [1].

Alternating Extraction. Assume that we have two parties, Quentin and Wendy. Quentin has a source Q, Wendy has a source X. Also assume that Quentin has a weak source S_0 with entropy rate $> 1/2$ (which may be correlated with Q). Suppose that (Q, S_0) is kept secret from Wendy and X is kept secret from Quentin. Let Ext_q, Ext_w be strong seeded extractors with optimal parameters, such as that in Theorem 8. Let Raz be the strong two-source extractor in Theorem 9. Let d be an integer parameter for the protocol. For some integer parameter $t > 0$, the *alternating extraction protocol* is an interactive process between Quentin and Wendy that runs in $t + 1$ steps.

In the 0'th step, Quentin sends S_0 to Wendy, Wendy computes $R_0 = \mathsf{Raz}(S_0, X)$ and replies R_0 to Quentin, Quentin then computes $S_1 = \mathsf{Ext}_q(Q, R_0)$. In this step R_0, S_1 each outputs d bits. In the first step, Quentin sends S_1 to Wendy, Wendy computes $R_1 = \mathsf{Ext}_w(X, S_1)$. She sends R_1 to Quentin and Quentin computes $S_2 = \mathsf{Ext}_q(Q, R_1)$. In this step R_1, S_2 each outputs d bits. In each subsequent step i, Quentin sends S_i to Wendy, Wendy computes $R_i = \mathsf{Ext}_w(X, S_i)$. She replies R_i to Quentin and Quentin computes $S_{i+1} = \mathsf{Ext}_q(Q, R_i)$. In step i, R_i, S_{i+1} each outputs d bits. Therefore, this process produces the following sequence:

$$
S_0, R_0 = \mathsf{Raz}(S_0, X), S_1 = \mathsf{Ext}_q(Q, R_0), R_1 = \mathsf{Ext}_w(X, S_1), \cdots,
$$
$$
S_t = \mathsf{Ext}_q(Q, R_{t-1}), R_t = \mathsf{Ext}_w(X, S_t).
$$

Look-Ahead Extractor. Now we can define our look-ahead extractor. Let $Y = (Q, S_0)$ be a seed, the look-ahead extractor is defined as

$$\mathsf{laExt}(X, Y) = \mathsf{laExt}(X, (Q, S_0)) =: R_1, \cdots, R_t.$$

Note that the look-ahead extractor can be computed by each party (Alice or Bob) alone in our final protocol. We now have the following lemma.

Lemma 7. *In the alternating extraction protocol, assume that X has n bits and Q has at most n bits. Let $\epsilon > 0$ be a parameter and $d = O(\log n + \log(1/\epsilon)) > \log(1/\epsilon)$ be the number of random bits needed in Theorem 8 to achieve error ϵ. Assume that X has min-entropy at least $12d^2$, Q has min-entropy at least $11d^2$ and S_0 is a $(40d, 38d)$ source. Let Ext_w and Ext_q be strong extractors in Theorem 8 that use d bits to extract d bits. Let $t = 4d$.*

Let (Q', S_0') be another distribution on the same support of (Q, S_0) such that (Q, S_0, Q', S_0') is independent of X. Now run the alternating extraction protocol with X and (Q', S_0') where in each step we obtain S_i', R_i'. For any $i, 0 \leq i \leq t - 1$, let $\overline{S_i} = (S_0, \cdots, S_i)$, $\overline{S_i'} = (S_0', \cdots, S_i')$, $\overline{R_i} = (R_0, \cdots, R_i)$ and $\overline{R_i'} = (R_0', \cdots, R_i')$. Then for any $i, 0 \leq i \leq t - 1$, we have

$$(R_i, \overline{S_{i-1}}, \overline{S_{i-1}'}, \overline{R_{i-1}}, \overline{R_{i-1}'}, S_i, S_i', Q, Q')$$
$$\approx_{(2i+2)\epsilon} (U_d, \overline{S_{i-1}}, \overline{S_{i-1}'}, \overline{R_{i-1}}, \overline{R_{i-1}'}, S_i, S_i', Q, Q').$$

Proof. We first prove the following claim.

Claim. In step 0, we have

$$(R_0, S_0, S_0', Q, Q') \approx_\epsilon (U_d, S_0, S_0', Q, Q')$$

and

$$(S_1, R_0, S_0, R_0', S_0') \approx_{3\epsilon} (U_d, R_0, S_0, R_0', S_0').$$

Moreover, conditioned on (S_0, S_0'), (R_0, R_0') are both deterministic functions of X; conditioned on (R_0, S_0, R_0', S_0'), (S_1, S_1') are deterministic functions of (Q, Q').

Proof (Proof of the claim.). Note that S_0 is a $(40d, 38d)$ source. Thus by Theorem 9 we have that

$$(R_0, S_0) \approx_\epsilon (U_d, S_0).$$

Since conditioned on S_0, R_0 is a deterministic function of X, which is independent of (Q, Q'), we also have that

$$(R_0, S_0, S_0', Q, Q') \approx_\epsilon (U_d, S_0, S_0', Q, Q').$$

Now we fix (S_0, S_0') and (R_0, R_0') are both deterministic functions of X. Since the size of (S_0, S_0') is at most $80d$, by Lemma 4 we have that with probability $1 - \epsilon$ over these fixings, Q is a source with entropy $10d^2$. Since R_0, R_0' are both deterministic functions of X, they are independent of Q. Therefore by Theorem 8 we have

$$(S_1, R_0, R_0') \approx_\epsilon (U_d, R_0, R_0').$$

Thus altogether we have that

$$(S_1, R_0, S_0, R_0', S_0') \approx_{3\epsilon} (U_d, R_0, S_0, R_0', S_0')$$

Moreover, conditioned on (R_0, S_0, R_0', S_0'), (S_1, S_1') are deterministic functions of (Q, Q').

Now we fix (R_0, S_0, R_0', S_0'). Note that after this fixing, S_1, S_1' are are deterministic functions of (Q, Q'). Note that with probability $1 - \epsilon$ over this fixing, Q has min-entropy at least $10d^2$.

We now prove the lemma. In fact, we prove the following stronger claim.

Claim. For any i, we have that

$$(R_i, \overline{S_{i-1}}, \overline{S_{i-1}'}, \overline{R_{i-1}}, \overline{R_{i-1}'}, S_i, S_i', Q, Q')$$
$$\approx_{(2i+2)\epsilon}(U_d, \overline{S_{i-1}}, \overline{S_{i-1}'}, \overline{R_{i-1}}, \overline{R_{i-1}'}, S_i, S_i', Q, Q')$$

and

$$(S_{i+1}, \overline{S_i}, \overline{S_i'}, \overline{R_i}, \overline{R_i'}) \approx_{(2i+3)\epsilon} (U_d, \overline{S_i}, \overline{S_i'}, \overline{R_i}, \overline{R_i'}).$$

Moreover, conditioned on $(\overline{S_{i-1}}, \overline{S_{i-1}'}, \overline{R_{i-1}}, \overline{R_{i-1}'}, S_i, S_i')$, (R_i, R_i') are both deterministic functions of X; conditioned on $(\overline{S_i}, \overline{S_i'}, \overline{R_i}, \overline{R_i'})$, (S_{i+1}, S_{i+1}') are deterministic functions of (Q, Q').

We prove the claim by induction on i. When $i = 0$, the statements are already proved in Claim 5. Now we assume that the statements hold for $i = j$ and we prove them for $i = j + 1$.

We first fix $(\overline{S_j}, \overline{S_j'}, \overline{R_j}, \overline{R_j'})$. Since now (S_{j+1}, S_{j+1}') are deterministic functions of (Q, Q'), they are independent of X. Moreover S_{j+1} is $(2j+3)\epsilon$-close to uniform. Note that the average conditional min-entropy of X is at least $12d^2 - 2d \cdot 4d \geq 4d^2$. Therefore by Theorem 8 we have that

$$(R_{j+1}, \overline{S_j}, \overline{S_j'}, \overline{R_j}, \overline{R_j'}, S_{j+1}, S_{j+1}') \approx_{(2j+4)\epsilon} (U_d, \overline{S_j}, \overline{S_j'}, \overline{R_j}, \overline{R_j'}, S_{j+1}, S_{j+1}').$$

Since (S_{j+1}, S_{j+1}') are deterministic functions of (Q, Q'), we also have

$$(R_{j+1}, \overline{S_j}, \overline{S_j'}, \overline{R_j}, \overline{R_j'}, S_{j+1}, S_{j+1}', Q, Q') \approx_{(2j+4)\epsilon} (U_d, \overline{S_j}, \overline{S_j'}, \overline{R_j}, \overline{R_j'}, S_{j+1}, S_{j+1}', Q, Q').$$

Moreover, conditioned on $(\overline{S_j}, \overline{S_j'}, \overline{R_j}, \overline{R_j'}, S_{j+1}, S_{j+1}')$, (R_{j+1}, R_{j+1}') are both deterministic functions of X.

Next, since conditioned on $(\overline{S_j}, \overline{S_j'}, \overline{R_j}, \overline{R_j'}, S_{j+1}, S_{j+1}')$, (R_{j+1}, R_{j+1}') are both deterministic functions of X, they are independent of (Q, Q'). Moreover R_{j+1} is

$(2j + 4)\epsilon$-close to uniform. Note that the average conditional min-entropy of Q is at least $10d^2 - 8d^2 = 2d^2$. Therefore by Theorem 8 we have that

$$(S_{j+2}, \overline{S_j}, \overline{S'_j}, \overline{R_j}, \overline{R'_j}, S_{j+1}, S'_{j+1}, R_{j+1}, R'_{j+1})$$
$$\approx_{(2j+5)\epsilon} (U_d, \overline{S_j}, \overline{S'_j}, \overline{R_j}, \overline{R'_j}, S_{j+1}, S'_{j+1}, R_{j+1}, R'_{j+1}).$$

Namely,

$$(S_{j+2}, \overline{S_{j+1}}, \overline{S'_{j+1}}, \overline{R_{j+1}}, \overline{R'_{j+1}}) \approx_{(2(j+1)+3)\epsilon} (U_d, \overline{S_{j+1}}, \overline{S'_{j+1}}, \overline{R_{j+1}}, \overline{R'_{j+1}}).$$

Moreover, conditioned on $(\overline{S_{j+1}}, \overline{S'_{j+1}}, \overline{R_{j+1}}, \overline{R'_{j+1}})$, (S_{j+2}, S'_{j+2}) are deterministic functions of (Q, Q').

6   Non-malleable Condensers for Arbitrary Min-entropy

In this section we give our construction of non-malleable condensers for arbitrary min-entropy.

First, we need the following definitions and constructions from [1].

Definition 14. *[1] Given $S_1, S_2 \subseteq \{1, \cdots, t\}$, we say that the ordered pair (S_1, S_2) is top-heavy if there is some integer j such that $|S_1^{\geq j}| > |S_2^{\geq j}|$, where $S^{\geq j} =: \{s \in S | s \geq j\}$. Note that it is possible that (S_1, S_2) and (S_2, S_1) are both top-heavy. For a collection Ψ of sets $S_i \subseteq \{1, \cdots, t\}$, we say that Ψ is pairwise top-heavy if every ordered pair (S_i, S_j) of sets $S_i, S_j \in \Psi$ with $i \neq j$, is top-heavy.*

Now, for any m-bit message $\mu = (b_1, \cdots, b_m)$, consider the following mapping of μ to a subset $S \subseteq \{1, \cdots, 4m\}$:

$$f(\mu) = f(b_1, \cdots, b_m) = \{4i - 3 + b_i, 4i - b_i | i = 1, \cdots, m\}$$

i.e., each bit b_i decides if to include $\{4i - 3, 4i\}$ (if $b_i = 0$) or $\{4i - 2, 4i - 1\}$ (if $b_i = 1$) in S.

We now have the following lemma.

Lemma 8. *[1] The above construction gives a pairwise top-heavy collection Ψ of 2^m sets $S \subseteq \{1, \cdots, t\}$ where $t = 4m$. Furthermore, the function f is an efficient mapping of $\mu \in \{0, 1\}^m$ to S_μ.*

Now we have the following construction.

Let $r \in (\{0, 1\}^d)^t$ be the output of the look-ahead extractor defined above, i.e., $r = (r_1, \cdots, r_t) = \mathsf{laExt}(X, (Q, S_0))$. Let $\Psi = \{S_1, \cdots, S_{2^m}\}$ be the pairwise top-heavy collection of sets constructed above. For any message $\mu \in \{0, 1\}^m$, define the function $\mathsf{laMAC}_r(\mu) =: [r_i | i \in S_\mu]$, indexed by r.

Now we can describe our construction of the non-malleable condenser.

Algorithm 12 (nmCond(x, y))

Input: ℓ–an integer parameter. x — a sample from an (n, k)-source with $k \geq 60d^2$. y–an independent random seed with $y = (y_1, y_2)$ such that y_1 has size $d = O(\log n + \ell) > 5\ell$ and y_2 has size $12d^2$.
Output: z — an m bit string.

Sub-Routines and Parameters:
Let nmExt be the non-malleable extractor from Theorem 10, with error $2^{-4\ell}$.
Let Ext be the strong extractor with optimal parameters from Theorem 8, with error $2^{-5\ell}$.
Let laExt be the look-ahead extractor defined above, using Ext as Ext_q and Ext_s. laExt is set up to extract from x using seed (q, s_0) such that $q = y_2$ and s_0 is the string that contains the first $40d$ bits of y_2, and output a string $r \in (\{0, 1\}^d)^t$ with $t = 4d$.
Let $\text{laMAC}_r(\mu)$ be the function defined above.

---

1. Compute $w = \text{Ext}(x, y_1)$ with output size $20d^2$ and $r = \text{laExt}(x, (q, s_0))$.
2. Output $z = (\text{nmExt}(w, y_2), \text{laMAC}_r(y_1))$ such that $\text{nmExt}(w, y_2)$ has size $8d^2$.

We can now prove the following theorem.

Theorem 13. *There exists a constant $C > 0$ such that given any $s > 0$, as long as $k \geq C(\log n + s)^2$, the above construction is a $(k, s, 2^{-s})$-non-malleable condenser with seed length $O(\log n + s)^2$ and output length $O(\log n + s)^2$.*

Proof. Let $\mathcal{A}$ be any (deterministic) function such that $\forall y \in \text{Supp}(Y), \mathcal{A}(y) \neq y$. We will show that for most y, with high probability over the fixing of $\text{nmCond}(X, \mathcal{A}(y))$, $\text{nmCond}(X, y)$ is still close to having min-entropy at least ℓ. Let $Y' = \mathcal{A}(Y)$. Thus $Y' \neq Y$. In the following analysis we will use letters with prime to denote the corresponding random variables produced with Y' instead of Y. Let $V_1 = \text{nmExt}(W, Y_2)$ and $V_2 = \text{laMAC}_R(Y_1)$. Thus $Z = (V_1, V_2)$. We have the following two cases.

Case 1: $Y_1 = Y_1'$. In this case, since $Y' \neq Y$, we must have that $Y_2 \neq Y_2'$. Now by Theorem 8 we have that

$$(W, Y_1) \approx_{2^{-5\ell}} (U, Y_1).$$

Therefore, we can now fix Y_1 (and thus Y_1'), and with probability $1 - 2^{-\ell}$ over this fixing, W is $2^{-4\ell}$-close to uniform. Moreover, after this fixing W is a deterministic function of X and thus is independent of Y_2. Note also that after this fixing, Y_2' is a deterministic function of Y_2. Thus by Theorem 10 we have that

$$(V_1, V_1', Y_2, Y_2') \approx_{O(2^{-4\ell})} (U_{8d^2}, V_1', Y_2, Y_2').$$

Therefore, we can now further fix Y_2 (and thus Y_2') and with probability at least $1 - O(2^{-\ell})$ over this fixing, (V_1, V_1') is $2^{-3\ell}$-close to (U_{8d^2}, V_1'). Thus we can further fix V_1', and with probability at least $1 - 2^{-\ell}$ over this fixing, V_1 is $2^{-2\ell}$-close to uniform. Now note that V_1 has size $8d^2$ and V_2' has size $2d^2$. Thus by Lemma 6, we can further fix V_2', and with probability at least $1 - 2 \cdot 2^{-\ell}$ over this fixing, V_1 is 2^ℓ-close to having min-entropy at least $8d^2 - 2d^2 - \ell \geq 5d^2$.

Thus in this case we have shown that, with probability $1 - O(2^{-\ell})$ over the fixing of Y, with probability $1 - O(2^{-\ell})$ over the fixing of Z', Z is $2^{-\ell}$-close to having min-entropy at least $5d^2 > 5\ell^2$.

Case 2: $Y_1 \neq Y_1'$. In this case, we first fix Y_1 and Y_1'. Note that after this fixing, W and W' are now deterministic functions of X. We now further fix W and W' and after this fixing, X and Y_2 are still independent. Since the total size of (W, W') is $40d^2$, by Lemma 4 we have that with probability $1 - 2^{-2\ell}$ over this fixing, X still has min-entropy at least $60d^2 - 40d^2 - 2\ell > 12d^2$. Note also that after this fixing, Y_2' is a deterministic function of Y_2. However, since Y_1' may be a function of Y_2, fixing Y_1' may cause Y_2 to lose entropy. Note that Y_1' only has size d, thus by Lemma 4, with probability $1 - 2 \cdot 2^{-2\ell}$ over the fixing of (Y_1, Y_1'), we have that Y_2 has min-entropy at least $12d^2 - d - 2\ell > 11d^2$ and S_0 has min-entropy at least $40d - d - 2\ell > 38d$.

Now assume that X has min-entropy at least $12d^2$, Y_2 has min-entropy at least $11d^2$ and S_0 has min-entropy at least $38d$. This happens with probability at least $1 - O(2^{-\ell})$. For any $i, 0 \leq i \leq t - 1$, let $\overline{S_i} = (S_0, \cdots, S_i)$, $\overline{S_i'} = (S_0', \cdots, S_i')$, $\overline{R_i} = (R_0, \cdots, R_i)$ and $\overline{R_i'} = (R_0', \cdots, R_i')$. Now by Lemma 7 (note that $Y_2 = (Q, S_0)$) we have that for any $i, 0 \leq i \leq t - 1$,

$$(R_i, \overline{S_{i-1}}, \overline{S_{i-1}'}, \overline{R_{i-1}}, \overline{R_{i-1}'}, S_i, S_i', Y_2)$$
$$\approx_{(2i+2)2^{-5\ell}} (U_d, \overline{S_{i-1}}, \overline{S_{i-1}'}, \overline{R_{i-1}}, \overline{R_{i-1}'}, S_i, S_i', Y_2).$$

Therefore, we have that for any i,

$$(R_i, \overline{R_{i-1}}, \overline{R_{i-1}'}, Y_2) \approx_{(2i+2)2^{-5\ell}} (U_d, \overline{R_{i-1}}, \overline{R_{i-1}'}, Y_2).$$

Thus, for any i, with probability $1 - 2^{-1.25\ell}$ over the fixing of Y_2, we have

$$(R_i, \overline{R_{i-1}}, \overline{R_{i-1}'}) \approx_{(2i+2)2^{-3.75\ell}} (U_d, \overline{R_{i-1}}, \overline{R_{i-1}'}).$$

By the union bound, we have that with probability $1 - t2^{-1.25\ell}$ over the fixing of Y_2, for any i,

$$(R_i, \overline{R_{i-1}}, \overline{R_{i-1}'}) \approx_{(2i+2)2^{-3.75\ell}} (U_d, \overline{R_{i-1}}, \overline{R_{i-1}'}).$$

Consider a typical fixing of Y_2. Now note that $V_2 = \mathsf{laMAC}_R(Y_1)$ and $V_2' = \mathsf{laMAC}_{R'}(Y_1')$. Let the two sets in Lemma 8 that correspond to Y_1 and Y_1' be H and H'. Since $Y_1 \neq Y_1'$, by definition there exists $j \in [4d]$ such that $|H^{\geq j}| > |H'^{\geq j}|$. Let $l = |H^{\geq j}|$. Thus $l \leq t$ and $|H'^{\geq j}| \leq l - 1$. Let R_H be the concatenation of $\{R_i, i \in H^{\geq j}\}$ and $R_{H'}'$ be the concatenation of $\{R_i', i \in H'^{\geq j}\}$.

By the above equation and the hybrid argument we have that

$$(R_H, \overline{R_{j-1}}, \overline{R'_{j-1}}) \approx_{3t^2 \cdot 2^{-3.75\ell}} (U_{ld}, \overline{R_{j-1}}, \overline{R'_{j-1}}).$$

Thus now we can first fix $\overline{R'_{j-1}}$, and with probability $1 - 2^{-1.25\ell}$ over this fixing, we have

$$R_H \approx_{3t^2 \cdot 2^{-2.5\ell}} U_{ld}.$$

We now fix $R'_{H'}$. Since $|H'^{\geq j}| \leq l-1$, the size of $R'_{H'}$ is at most $(l-1)d$. Thus by Lemma 6 we have that with probability at least $1 - (3t^2 + 1) \cdot 2^{-1.25\ell}$ over this fixing, R_H is $2^{-1.25\ell}$-close to having min-entropy $d - 1.25\ell > \ell$. Note that after we fix $\overline{R'_{j-1}}$ and $R'_{H'}$, we have also fixed V'_2. Since W' and Y'_2 are already fixed, V'_1 is also fixed. Thus Z' is fixed. Therefore altogether we have that with probability $1 - 2 \cdot 2^{-2\ell} - t2^{-1.25\ell} = 1 - O(2^{-\ell})$ over the fixings of Y, with probability $1 - 2^{-1.25\ell} - (3t^2 + 1) \cdot 2^{-1.25\ell} = 1 - O(2^{-\ell})$ over the fixings of Z', Z is $2^{-1.25\ell}$-close to having min-entropy ℓ.

Combining **Case 1** and **Case 2**, and notice that the fraction of "bad seeds" that an adversary can achieve is at most the sum of the fraction of bad seeds in both cases. Thus by choosing an appropriate $\ell = O(s)$ we have that the construction is a $(k, s, 2^{-s})$-non-malleable condenser with seed length $O(\log n + s)^2$.

Combining Theorem 11 and Theorem 13, we immediately get a 2-round privacy amplification protocol with optimal entropy loss for any (n, k) source.

Theorem 14. *There exists a constant C such that for any $\epsilon > 0$ with $k \geq C(\log n + \log(1/\epsilon))^2$, there exists an explicit 2-round privacy amplification protocol for (n, k) sources with security parameter $\log(1/\epsilon)$, entropy loss $O(\log n + \log(1/\epsilon))$ and communication complexity $O(\log n + \log(1/\epsilon))^2$.*

In fact, we have a slightly simpler protocol that uses the look-ahead extractor and MAC somewhat more directly, while achieving the same performance.

We assume that the shared weak random source has min-entropy k, and the error ϵ we seek satisfies $\epsilon < 1/n$ and $k > C(\log n + \log(1/\epsilon))^2$ for some constant $C > 1$. For convenience, in the description below we introduce an "auxiliary" security parameter s. Eventually, we will set $s = \log(C'/\epsilon) + O(1) = \log(1/\epsilon) + O(1)$, so that $C'/2^s < \epsilon$, for a sufficiently large constant C' related to the number of "bad" events we need to account for. We need the following building blocks:

- Let Ext be a $(k, 2^{-5s})$-extractor with optimal entropy loss and seed length $d = O(\log n + s) > 202s$, from Theorem 8. Assume that $k \geq 15d^2$.
- Let Raz be the two source extractor from Theorem 9.
- Let MAC be the ("leakage-resilient") MAC, as in Theorem 6, with tag length $v = 2s$ and key length $\ell = 2v = 4s$.
- Let laExt be the look-ahead extractor defined above, using Ext as Ext_q and Ext_s. laExt is set up to extract from x using seed (q, s_0) such that $q = y_2$ and s_0 is the string that contains the first $40d$ bits of y_2, and output a string $r \in (\{0,1\}^d)^t$ with $t = 4d$.

- Let $\mathsf{laMAC}_r(\mu)$ be the function defined above.
- In the protocol Alice will sample three random strings Y_1, Y_2, Y_3, with size d, $12d^2$ and $50d^2$ respectively.

Using the above building blocks, the protocol is given in Figure 4. To emphasize the adversary Eve, we use letters with 'prime' to denote all the variables seen or generated by Bob; e.g., Bob picks W', but Alice may see a different W.

Alice: X Eve: E Bob: X

Sample random $Y = (Y_1, Y_2, Y_3)$.
Compute $R_2 = \mathsf{laExt}(X, Y_2)$.
$Z = \mathsf{laMAC}_{R_2}(Y_1)$.
$R_1 = \mathsf{Ext}(X, Y_1)$ and output $4s$ bits.

$$(Y_1, Y_2, Y_3)$$
$$\longrightarrow$$
$$(Y_1', Y_2', Y_3')$$

Sample random W' with d bits.
Compute $R_2' = \mathsf{laExt}(X, Y_2')$.
$Z' = \mathsf{laMAC}_{R_2'}(Y_1')$.
$R_1' = \mathsf{Ext}(X, Y_1')$ and output $4s$ bits.
$T_1' = \mathsf{Raz}(Y_3', Z')$ with s bits,
$T_2' = \mathsf{MAC}_{R_1'}(W')$.
Set final $R_B = \mathsf{Ext}(X, W')$.

$$(W', T_1', T_2')$$
$$\longleftarrow$$
$$(W, T_1, T_2)$$

If $T_1 \neq \mathsf{Raz}(Y_3, Z)$ or
$T_2 \neq \mathsf{MAC}_{R_1}(W)$ *reject.*
Set final $R_A = \mathsf{Ext}(X, W)$.

Fig. 4. 2-round Privacy Amplification Protocol

Theorem 15. *Assume that $k > C(\log n + \log(1/\epsilon))^2$ for some constant $C > 1$. The above protocol is a privacy amplification protocol with security parameter $\log(1/\epsilon)$, entropy loss $O(\log(1/\epsilon))$ and communication complexity $O(\log(1/\epsilon)^2)$.*

Proof. The proof can be divided into two cases: whether the adversary changes Y_1 or not.

Case 1: The adversary does not change Y_1. In this case, note that $R_1 = R_1'$ and is 2^{-5s}-close to uniform in Eve's view (even conditioned on Y_1, Y_2, Y_3). Thus the property of the MAC guarantees that Bob can authenticate W' to Alice. However, one thing to note here is that Eve has some additional information, namely T_1' which can leak information about the MAC key. On the other hand, the size of T_1' is s, thus by Lemma 2 the average conditional min-entropy $H_\infty(R_1|T_1')$ is at least $3s$. Therefore by Theorem 6 the probability that Eve can change W' to a different W without causing Alice to reject is at most

$$\left\lceil \frac{d_1}{2s} \right\rceil 2^{2s - \tilde{H}_\infty(R_1 | T_1')} + 2^{-5s} \le O(2^{2s - 3s}) + 2^{-5s} \le O(2^{-s}).$$

When $W = W'$, by Theorem 8 $R_A = R_B$ and is 2^{-5s}-close to uniform in Eve's view.

Case 2: The adversary does change Y_1. Thus we have $Y_1 \ne Y_1'$. Here the proof is similar to the proof of the non-malleable condenser. We first fix Y_1 and Y_1'. Note that after this fixing, R_1 and R_1' are now deterministic functions of X. We now further fix R_1 and R_1' and after this fixing, X and (Y_2, Y_3) are still independent. Since the total size of (R_1, R_1') is $8s$, by Lemma 4 we have that with probability $1 - 2^{-2s}$ over this fixing, X still has min-entropy at least $15d^2 - 8s - 2s > 12d^2$. Note also that after this fixing, Y_2' is a deterministic function of (Y_2, Y_3). However, since Y_1' may be a function of Y_2, fixing Y_1' may cause Y_2 to lose entropy. Note that Y_1' only has size d, thus by Lemma 4, with probability $1 - 2 \cdot 2^{-2s}$ over the fixing of (Y_1, Y_1'), we have that Y_2 has min-entropy at least $12d^2 - d - 2s > 11d^2$ and S_0 has min-entropy at least $40d - d - 2s > 38d$.

Now assume that X has min-entropy at least $12d^2$, Y_2 has min-entropy at least $11d^2$ and S_0 has min-entropy at least $38d$. This happens with probability at least $1 - O(2^{-s})$. For any $i, 0 \le i \le t - 1$, let $\overline{S_i} = (S_0, \cdots, S_i)$, $\overline{S_i'} = (S_0', \cdots, S_i')$, $\overline{R_i} = (R_0, \cdots, R_i)$ and $\overline{R_i'} = (R_0', \cdots, R_i')$. Again by Lemma 7 we have that for any i,

$$(R_i, \overline{S_{i-1}}, \overline{S_{i-1}'}, \overline{R_{i-1}}, \overline{R_{i-1}'}, S_i, S_i', Y_2, Y_2')$$
$$\approx_{(2i+2)2^{-5s}} (U_d, \overline{S_{i-1}}, \overline{S_{i-1}'}, \overline{R_{i-1}}, \overline{R_{i-1}'}, S_i, S_i', Y_2, Y_2').$$

Thus for any i, we have

$$(R_i, \overline{R_{i-1}}, \overline{R_{i-1}'}, Y_2, Y_2') \approx_{(2i+2)2^{-5s}} (U_d, \overline{R_{i-1}}, \overline{R_{i-1}'}, Y_2, Y_2').$$

Now by the same analysis as in the proof of the non-malleable condenser (and recall that $Y_1 \ne Y_1'$), we have that with probability $1 - t2^{-1.25\ell}$ over the fixing of (Y_2, Y_2'), with probability at least $1 - (3t^2 + 1) \cdot 2^{-1.25s}$ over the fixing of Z', Z is $2^{-1.25s}$-close to having min-entropy $d - 1.25s > 200s$.

Note that we have now fixed (Y_1, Y_1', Y_2, Y_2') and (R_1, R_1', Z'). After all these fixings, Z is a deterministic function of X and is $2^{-1.25s}$-close to having min-entropy $200s$. Thus Z is independent of Y_3 (note that Z' is also a deterministic function of X, thus fixing Z' does not influence the independence of Z and Y_3). Note that after these fixings, Y_3' is a deterministic function of Y_3, and since the size of (Y_1', Y_2') is $d + 12d^2 < 13d^2$, by Lemma 4 Y_3 is 2^{-s}-close to having min-entropy $50d^2 - 13d^2 - s > 36d^2$. Thus by Theorem 9 we have

$$(\mathsf{Raz}(Y_3, Z), Y_3, Y_3') \approx_{O(2^{-s})} (U_s, Y_3, Y_3').$$

Since we already fixed (Y_1, Y_1', Y_2, Y_2') and (R_1, R_1', Z'), and W' is independent of all random variables above, this also implies that

$$(\mathsf{Raz}(Y_3, Z), R_1', Z', Y, Y', W') \approx_{O(2^{-s})} (U_s, R_1', Z', Y, Y', W').$$

Note that $T_1' = \mathsf{Raz}(Y_3', Z')$ and $T_2' = \mathsf{MAC}_{R_1'}(W')$. Thus we have

$$(\mathsf{Raz}(Y_3, Z), T_1', T_2', Y, Y', W') \approx_{O(2^{-s})} (U_s, T_1', T_2', Y, Y', W').$$

Therefore, the probability that the adversary can guess the correct T_1 is at most $2^{-s} + O(2^{-s}) = O(2^{-s})$. For an appropriately chosen $s = \log(1/\epsilon) + O(1)$ this is at most ϵ. Note that conditioned on the fixing of Y, the random variables that are used to authenticate W' are (R_1, T_1), which are deterministic functions of X and have size $O(s)$, thus the entropy loss of the protocol is $O(\log(1/\epsilon))$. The communication complexity can be easily verified to be $O(\log(1/\epsilon)^2)$.

7   Conclusions and Open Problems

In this paper we construct explicit non-malleable condensers for arbitrary min-entropy, and use them to give an explicit 2-round privacy amplification protocol with optimal entropy loss for arbitrary min-entropy k, with security parameter up to $s = \Omega(\sqrt{k})$. This is the first explicit protocol that simultaneously achieves optimal parameters in both round complexity and entropy loss, for arbitrary min-entropy.

We then generalize this result to give a privacy amplification protocol that runs in $O(s/\sqrt{k})$ rounds and achieves optimal entropy loss for arbitrary min-entropy k, with security parameter up to $s = \Omega(k)$. This significantly improves the protocol in [4]. In the special case where $k = \delta n$ for some constant $\delta > 0$, we give better non-malleable condensers and a 2-round privacy amplification protocol with optimal entropy loss for security parameter up to $s = \Omega(k)$, which improves the entropy loss and communication complexity of the 2-round protocol in [2].

Some open problems include constructing better non-malleable extractors or non-malleable condensers, and to construct optimal privacy amplification protocols for security parameter bigger than $\sqrt{k}$. Another interesting problem is to find other applications of non-malleable extractors or non-malleable condensers.

References

1. Dodis, Y., Wichs, D.: Non-malleable extractors and symmetric key cryptography from weak secrets. In: Proceedings of the 41st Annual ACM Symposium on Theory of Computing, pp. 601–610 (2009)
2. Li, X.: Non-malleable extractors, two-source extractors and privacy amplification. In: Proceedings of the 53nd Annual IEEE Symposium on Foundations of Computer Science (2012)
3. Li, X.: Design extractors, non-malleable condensers and privacy amplification. In: Proceedings of the 44th Annual ACM Symposium on Theory of Computing (2012)
4. Chandran, N., Kanukurthi, B., Ostrovsky, R., Reyzin, L.: Privacy amplification with asymptotically optimal entropy loss. In: Proceedings of the 42nd Annual ACM Symposium on Theory of Computing, pp. 785–794 (2010)

5. Dodis, Y., Ong, S.J., Prabhakaran, M., Sahai, A.: On the (im)possibility of cryptography with imperfect randomness. In: FOCS 2004, pp. 196–205 (2004)
6. Bennett, C., Brassard, G., Robert, J.M.: Privacy amplification by public discussion. SIAM Journal on Computing 17, 210–229 (1988)
7. Maurer, U.M., Wolf, S.: Privacy amplification secure against active adversaries. In: Kaliski Jr., B.S. (ed.) CRYPTO 1997. LNCS, vol. 1294, pp. 307–321. Springer, Heidelberg (1997)
8. Dodis, Y., Katz, J., Reyzin, L., Smith, A.: Robust fuzzy extractors and authenticated key agreement from close secrets. In: Dwork, C. (ed.) CRYPTO 2006. LNCS, vol. 4117, pp. 232–250. Springer, Heidelberg (2006)
9. Renner, R.S., Wolf, S.: Unconditional authenticity and privacy from an arbitrarily weak secret. In: Boneh, D. (ed.) CRYPTO 2003. LNCS, vol. 2729, pp. 78–95. Springer, Heidelberg (2003)
10. Kanukurthi, B., Reyzin, L.: Key agreement from close secrets over unsecured channels. In: Joux, A. (ed.) EUROCRYPT 2009. LNCS, vol. 5479, pp. 206–223. Springer, Heidelberg (2009)
11. Dodis, Y., Li, X., Wooley, T.D., Zuckerman, D.: Privacy amplification and non-malleable extractors via character sums. In: Proceedings of the 52nd Annual IEEE Symposium on Foundations of Computer Science (2011)
12. Cohen, G., Raz, R., Segev, G.: Non-malleable extractors with short seeds and applications to privacy amplification. In: Proceedings of the 27th Annual IEEE Conference on Computational Complexity (2012)
13. Dodis, Y., Yu, Y.: Overcoming weak expectations. Manuscript (September 2012)
14. Aggarwal, D., Dodis, Y., Jafargholi, Z., Miles, E., Reyzin, L.: Amplifying privacy in privacy amplification. In: Garay, J.A., Gennaro, R. (eds.) CRYPTO 2014, Part II. LNCS, vol. 8617, pp. 183–198. Springer, Heidelberg (2014)
15. Dziembowski, S.: Intrusion-resilience via the bounded-storage model. In: Proceedings of the 3rd Theory of Cryptography Conference (2006)
16. Crescenzo, G.D., Lipton, R.J., Walfish, S.: Perfectly secure pass- word protocols in the bounded retrieval model. In: Proceedings of the 3rd Theory of Cryptography Conference (2006)
17. Raz, R.: Extractors with weak random seeds. In: Proceedings of the 37th Annual ACM Symposium on Theory of Computing, pp. 11–20 (2005)
18. Barak, B., Kindler, G., Shaltiel, R., Sudakov, B., Wigderson, A.: Simulating independence: New constructions of condensers, Ramsey graphs, dispersers, and extractors. In: Proceedings of the 37th Annual ACM Symposium on Theory of Computing, pp. 1–10 (2005)
19. Zuckerman, D.: Linear degree extractors and the inapproximability of max clique and chromatic number. In: Theory of Computing, pp. 103–128 (2007)
20. Dodis, Y., Ostrovsky, R., Reyzin, L., Smith, A.: Fuzzy extractors: How to generate strong keys from biometrics and other noisy data. SIAM Journal on Computing 38, 97–139 (2008)
21. Guruswami, V., Umans, C., Vadhan, S.: Unbalanced expanders and randomness extractors from Parvaresh-Vardy codes. Journal of the ACM 56(4) (2009)
22. Nisan, N., Zuckerman, D.: Randomness is linear in space. Journal of Computer and System Sciences 52(1), 43–52 (1996)
23. Schulman, L.J., Zuckerman, D.: Asymptotically good codes correcting insertions, deletions, and transpositions. IEEE Transactions on Information Theory 45(7), 2552–2557 (1999)

From Single-Bit to Multi-bit Public-Key Encryption via Non-malleable Codes

Sandro Coretti[1], Ueli Maurer[1], Björn Tackmann[2,*], and Daniele Venturi[3]

[1] ETH Zurich, Switzerland
{corettis,maurer}@inf.ethz.ch
[2] UC San Diego, USA
btackmann@eng.ucsd.edu
[3] Sapienza University of Rome, Italy
venturi@di.uniroma1.it

Abstract. One approach towards basing public-key encryption (PKE) schemes on weak and credible assumptions is to build "stronger" or more general schemes generically from "weaker" or more restricted ones. One particular line of work in this context was initiated by Myers and shelat (FOCS '09) and continued by Hohenberger, Lewko, and Waters (Eurocrypt '12), who provide constructions of multi-bit CCA-secure PKE from single-bit CCA-secure PKE.

It is well-known that encrypting each bit of a plaintext string independently is not chosen-ciphertext secure—the resulting scheme is *malleable*. This paper analyzes the conceptually simple approach of applying a suitable non-malleable code (Dziembowski *et al.*, ICS '10) to the plaintext and subsequently encrypting the resulting codeword bit-by-bit. An attacker's ability to make multiple decryption queries requires that the underlying code be *continuously* non-malleable (Faust *et al.*, TCC '14). This flavor of non-malleable codes can only be achieved if the decoder is allowed to "self-destruct" when it processes an invalid encoding. The resulting PKE scheme inherits this property and therefore only achieves a weaker variant of chosen-ciphertext security, where the decryption becomes dysfunctional once the attacker submits an invalid ciphertext.

We first show that the above approach based on non-malleable codes indeed yields a solution to the problem of domain extension for public-key encryption where the decryption may self-destruct, provided that the underlying code is continuously non-malleable against a *reduced* form of bit-wise tampering. This statement is shown by combining a simple information-theoretic argument with the constructive cryptography perspective on PKE (Coretti *et al.*, Asiacrypt '13). Then, we prove that the code of Dziembowski *et al.* is actually already continuously non-malleable against (full) bit-wise tampering; this constitutes the first *information-theoretically* secure continuously non-malleable code, a technical contribution that we believe is of independent interest. Compared to the previous approaches to PKE domain extension, our scheme is more efficient and intuitive, at the cost of not achieving full CCA security. Our result is also one of the first applications of non-malleable codes in a context other than memory tampering.

---

* Work done while author was at ETH Zurich.

Y. Dodis and J.B. Nielsen (Eds.): TCC 2015, Part I, LNCS 9014, pp. 532–560, 2015.

1 Introduction

1.1 Overview

A public-key encryption (PKE) scheme enables a sender A to send messages to a receiver B confidentially if B can send a single message, the public key, to A authentically. A encrypts a message with the public key and sends the ciphertext to B via a channel that could be authenticated or insecure, and B decrypts the received ciphertext using the private key. Following the seminal work of Diffie and Hellman [22], the first formal definition of public-key encryption has been provided by Goldwasser and Micali [32], and to date numerous instantiations of this concept have been proposed, e.g., [50,25,17,29,33,36,51,49], for different security properties and based on various different computational assumptions.

One natural approach towards developing public-key encryption schemes based on weak and credible assumptions is to build "stronger" or more general schemes generically from "weaker" or less general ones. While the "holy grail"—generically building a chosen-ciphertext secure scheme based on any chosen-plaintext secure one—has so far remained out of reach, and despite negative results [31], various interesting positive results have been shown. For instance, Cramer *et al.* [16] build *bounded-query* chosen-ciphertext secure schemes from chosen-plaintext secure ones, Choi *et al.* [10] *non-malleable* schemes from chosen-plaintext secure ones, and Lin and Tessaro [38] show how the security of weakly chosen-ciphertext secure schemes can be amplified. A line of work started by Myers, Sergi, and shelat [47] and continued by Dachman-Soled [18] shows how to obtain chosen-ciphertext secure schemes from plaintext-aware ones. Most relevant for our work, however, are the results of Myers and shelat [48] and Hohenberger, Lewko, and Waters [34], who generically build a multi-bit chosen-ciphertext secure scheme from a single-bit chosen-ciphertext secure one.

A naïve attempt at solving this problem would be to encrypt each bit m_i of a plaintext $m = m_1 \cdots m_k$ under an independent public key pk_i of the single-bit scheme. Unfortunately, this simple approach does not yield chosen-ciphertext security. The reason is that the above scheme is *malleable*: given a ciphertext $e = (e_1, \ldots, e_k)$, where e_i is an encryption of m_i, an attacker can generate a new ciphertext $e' \neq e$ that decrypts to a related message, for instance by copying the first ciphertext component e_1 and replacing the other components by fresh encryptions of, say, 0.

The above malleability issue suggests the following natural "encode-then-encrypt-bit-by-bit" approach: first encode the message using a non-malleable code[1] (a concept introduced by Dziembowski *et al.* [24]) to protect its integrity, obtaining an n-bit codeword $c = c_1 \cdots c_n$; then encrypt each bit c_i of the codeword using public key pk_i as in the naïve protocol from above.

It turns out that non-malleable codes as introduced by [24] are not sufficient: Since they are only secure against a single tampering, the security of the

---

[1] Roughly, a code is non-malleable w.r.t. a function class $\mathcal{F}$ if the message obtained by decoding a codeword modified via a function in $\mathcal{F}$ is either the original message or a completely unrelated value.

resulting scheme would only hold with respect to a single decryption. *Continuously* non-malleable codes (Faust *et al.* [26]) allow us to extend this guarantee to multiple decryptions. However, such codes "self-destruct" once an attack has been detected, and, therefore, so must any PKE scheme built on top of them. This is a restriction that we prove to be unavoidable for this approach based on non-malleable codes.

The resulting scheme achieves a notion weaker than full CCA, which we term *self-destruct chosen-ciphertext security (SD-CCA)*. Roughly, SD-CCA security is CCA security with the twist that the decryption oracle stops working once the adversary submits an invalid ciphertext.

Our paper consists of two main parts: First, we prove that the above approach allows to build multi-bit SD-CCA-secure PKE from single-bit SD-CCA-secure PKE, provided that the underlying code is continuously non-malleable against a *reduced* form of bit-wise tampering. This proof is greatly facilitated by rephrasing the problem using the paradigm of constructive cryptography [40], since it follows almost immediately from the composition theorem. For comparison, the full version of this paper [14] also contains a purely game-based proof. Second, we show that a simplified variant of the code by Dziembowski *et al.* [24] is already continuously non-malleable against the aforementioned reduced bit-wise tampering and that the full variant of said code achieves continuous non-malleability against *full* bit-wise tampering. This constitutes the first *information-theoretically* secure continuously non-malleable code, a contribution that we believe is of independent interest, and forms the technical core of this paper.

1.2   Techniques and Contributions

Constructive cryptography [40].   Statements about the security of cryptographic schemes can be stated as *constructions* of a "stronger" or more useful desired resource from a "weaker" or more restricted assumed one. Two such construction steps can be composed, i.e., if a protocol π constructs a resource S from an assumed resource R, denoted by $R \overset{\pi}{\Longmapsto} S$, and, additionally, a protocol ψ assumes resource S and constructs a resource T, then the composition theorem of constructive cryptography (see Section 2.4) states that the composed protocol, denoted $\psi \circ \pi$, constructs resource T from R. The resources considered in this work are different types of communication channels between two parties A and B; a channel is a resource that involves three entities: the sender, the receiver, and a (potential) attacker E.

We use and extend the notation by Maurer and Schmid [44], denoting different types of channels by different arrow symbols. A *confidential* channel (later denoted $-\diamond\!\!\rightarrow\!\bullet$) hides the messages sent by A from the attacker E but potentially allows her to inject *independent* messages; an *authenticated* channel (later denoted $\bullet\!-\diamond\!\!\rightarrow$) is dual to the confidential channel in that it potentially leaks the message to the attacker but prevents modifications and injections; an insecure channel (later denoted $-\rightarrow\!\!\rightarrow$) protects neither the confidentiality nor the authenticity. In all cases, the double arrow head indicates that the channel can be used to transmit multiple messages. A single arrow head, instead, means that channels are single-use. All channels used within this work are described formally in Section 2.5.

Warm-up: Dealing with the Malleability of the One-time pad.   To illustrate the intuition behind our approach, consider the following simple example: The one-time pad allows to encrypt an n-bit message m using an n-bit shared key κ by computing the ciphertext $e = m \oplus \kappa$. If e is sent via an insecure channel, an attacker can replace it by a different ciphertext e', in which case the receiver will compute $m' = e' \oplus \kappa = m \oplus (e \oplus e')$. This can be seen, as described in previous work by Maurer *et al.* [43], as constructing from an insecure channel and a shared secret n-bit key an "XOR-malleable" channel, denoted $\longrightarrow\!\!\oplus\!\!\rightarrow\!\bullet$, which is confidential but allows the attacker to specify a mask $\delta \in \{0,1\}^n$ ($= e \oplus e'$) to be XORed to the transmitted message.

Non-malleable codes can be used to deal with the XOR-malleability. To transmit a k-bit message m, we encode m with a (k,n)-bit non-malleable code, obtaining an n-bit codeword c, which we transmit via the XOR-malleable channel $\longrightarrow\!\!\oplus\!\!\rightarrow\!\bullet$. Since by XORing a mask δ to a codeword transmitted via $\longrightarrow\!\!\oplus\!\!\rightarrow\!\bullet$ the attacker can influence the value of each bit of the codeword only independently, a code that is non-malleable w.r.t. the function class $\mathcal{F}_{\mathsf{bit}}$, which (in particular) allows to either "keep" or "flip" each bit of a codeword only individually, is sufficient. Indeed, the non-malleability of the code implies that the decoded message will be either the original message or a completely unrelated value, which is the same guarantee as formulated by the single-message confidential channel (denoted $\longrightarrow\!\bullet$), and hence using the code, one achieves the construction

$$\longrightarrow\!\!\oplus\!\!\rightarrow\!\bullet \quad \Longmapsto \quad \longrightarrow\!\bullet.$$

A more detailed treatment and a formalization of this example appears in the full version of this paper [14]; suitable non-malleable codes are described in [15,24,9].

Dealing with the Malleability of Multiple Single-bit Encryptions.   Intuitively, CCA encryption guarantees that an attacker, by modifying a particular ciphertext, can either leave the message contained therein intact or replace it by an independently created one. This intuition is formally captured by the confidential channel $\longrightarrow\!\!\diamond\!\!\rightarrow\!\!\rightarrow\!\bullet$: at the attacker interface E, it allows to either forward messages sent by A or to inject *independent* messages. In [13], it is shown how CCA-secure encryption can be used to construct a confidential channel $\longrightarrow\!\!\diamond\!\!\rightarrow\!\!\rightarrow\!\bullet$ from A to B from an authenticated channel $\longleftarrow\!\bullet$ from B to A and an insecure channel $-\rightarrow$ from A to B. As shown in Section 3, this and the composition theorem imply that using n independent single-bit PKE schemes, one can construct n (independent) instances of the single-bit confidential channel $\overset{\text{1-bit}}{\longrightarrow\!\!\diamond\!\!\rightarrow\!\!\rightarrow\!\bullet}$, written $[\overset{\text{1-bit}}{\longrightarrow\!\!\diamond\!\!\rightarrow\!\!\rightarrow\!\bullet}]^n$.

The remaining step is showing how to achieve the construction

$$[\overset{\text{1-bit}}{\longrightarrow\!\!\diamond\!\!\rightarrow\!\!\rightarrow\!\bullet}]^n \quad \Longmapsto \quad \overset{k\text{-bit}}{\longrightarrow\!\!\diamond\!\!\rightarrow\!\!\rightarrow\!\bullet} \tag{1}$$

for some $k > 1$. Then, by the composition theorem, plugging these two steps together yields a protocol m-pke that constructs a k-bit confidential channel from an authenticated channel and an insecure channel.

To achieve construction (1), we use non-malleable codes. The fact that the channels are multiple-use leads to two important differences to the one-time-pad example above: First, the attacker can fabricate *multiple* codewords, which are then decoded. Second, each bit of such a codeword can be created by combining *any* of the bits sent by A over the corresponding channel. These capabilities can be formally captured by a particular class $\mathcal{F}_{copy}$ of tampering functions. We prove in Section 3 that any code that is *continuously non-malleable* w.r.t. $\mathcal{F}_{copy}$ can be used to achieve (1).

Unfortunately, we show in Section 5 that any code, in order to satisfy the above type of non-malleability, has to "self-destruct" in the event of a decoding error. For the application in the setting of public-key encryption, this means that the decryption algorithm of the receiver B also has to deny processing any further ciphertext once the code self-destructs.

Self-destruct CCA Security. In Section 3 we show how the protocol m-pke can be seen as a PKE scheme Π that achieves *self-destruct CCA security (SD-CCA)* and show that the single-bit confidential channel $\xrightarrow{\text{1-bit}}$ can also be constructed using a single-bit SD-CCA scheme (instead of a CCA-secure one). Thus, overall we obtain a way to transform 1-bit SD-CCA-secure PKE into multi-bit SD-CCA-secure PKE. For comparison, the full version of this paper [14] also contains a direct, entirely game-based proof that combining a single-bit SD-CCA PKE scheme with a non-malleable code as above yields a multi-bit SD-CCA scheme.[2]

SD-CCA is a (weaker) CCA variant that allows the scheme to self-destruct in case it detects an invalid ciphertext. The standard CCA game can easily be extended to include the self-destruct mode of the decryption: the decryption oracle keeps answering decryption queries as long as no invalid ciphertext (i.e., a ciphertext upon which the decryption algorithm outputs an error symbol) is received; after such an event occurs, no further decryption query is answered.

The guarantees of SD-CCA are perhaps best understood if compared to the q-bounded CCA notion by [10]. While q-CCA allows an *a priori determined* number q of decryption queries, SD-CCA allows an *arbitrary* number of *valid* decryption queries and one invalid query. From a practical viewpoint, an attacker can efficiently violate the availability with a scheme of either notion. However, as long as no invalid ciphertexts are received, an SD-CCA scheme can run indefinitely, whereas a q-CCA scheme has to necessarily stop after q decryptions.

Subsequent work [12] shows that SD-CCA security can in fact be achieved from CPA security only, by generalizing a technique by Choi *et al.* [10]. The resulting scheme, however, is considerably less efficient than the one we provide in this paper (under reasonable assumptions about the plaintext size). In [12],

---

[2] In fact, PKE scheme Π is only *replayable* SD-CCA secure; cf. Section 3.4.

the authors also study the relation between SD-CCA and other standard security notions and discuss possible applications.

Continuous Non-malleability w.r.t. $\mathcal{F}_{\text{copy}}$. The class $\mathcal{F}_{\text{copy}}$ can be seen as a multi-encoding version of the function class $\mathcal{F}_{\text{set}}$, which consists of functions that tamper with every bit of an encoding individually and may either leave it unchanged or replace it by a fixed value. In Section 4 we build a continuously non-malleable code w.r.t. $\mathcal{F}_{\text{copy}}$; the code consists of a linear error-correcting secret sharing (LECSS) scheme and can be seen as a simplified version of the code in [24]. The security proof of the code proceeds in two steps: First, we prove that it is continuously non-malleable w.r.t. $\mathcal{F}_{\text{set}}$ against tampering with a single encoding; the main challenge in this proof is showing that by repeatedly tampering with an encoding, an attacker cannot infer (too much) useful information about it. Then, we show that if a code is continuously non-malleable w.r.t. $\mathcal{F}_{\text{set}}$ against tampering with a single encoding, then it is also *adaptively* continuously non-malleable w.r.t. $\mathcal{F}_{\text{copy}}$, i.e., against tampering with many encodings simultaneously. In addition, in the full version of this paper [14], we also show that the full version of the code by [24] is non-malleable against *full* bit-wise tampering (i.e., when additionally the tamper function is allowed to flip bits of an encoding). These are the main technical contributions of this work.

1.3 More Details on Related Work

The work of Hohenberger *et al.* [34]—building on the work of Myers and shelat [48]—describes a multi-bit CCA-secure encrytion scheme from a single-bit CCA-secure one, a CPA-secure one, and a 1-query-bounded CCA-secure one. Their scheme is rather sophisticated and has a somewhat circular structure, requiring a complex security proof. The public key is of the form $\mathsf{pk} = (\mathsf{pk}_{in}, \mathsf{pk}_A, \mathsf{pk}_B)$, where the "inner" public key pk_{in} is the public key of a so-called *detectable CCA (DCCA)* PKE scheme, which is built from the single-bit CCA-secure scheme, and the "outer" public keys pk_A and pk_B are, respectively, the public key of a 1-bounded CCA and a CPA secure PKE scheme. To encrypt a k-bit message m one first encrypts a tuple (r_A, r_B, m), using the "inner" public key pk_{in}, obtaining a ciphertext e_{in}, where r_A and r_B are thought as being the randomness for the "outer" encryption scheme. Next, one has to encrypt the inner ciphertext e_{in} under the "outer" public key pk_A (resp. pk_B) using randomness r_A (resp. r_B) and thus obtaining a ciphertext e_A (resp. e_B). The output ciphertext is $e = (e_A, e_B)$.

To use the above scheme, we have to instantiate the DCCA, 1-bounded CCA and CPA components. As argued in [34], all schemes can be instantiated using a single-bit CCA-secure PKE scheme yielding a fully black-box construction of a multi-bit CCA-secure PKE from a single-bit CCA-secure PKE. Let us denote with l_p (resp., l_e) the bit-length of the public key (resp., the ciphertext) for the single-bit CCA-secure PKE scheme. When we refer to the construction of [16] for the 1-bounded CCA component, we get a public key of size roughly $(3 + 16s) \cdot l_p$ for the public key and $(k + 2s) \cdot 4s \cdot l_e^2$ for the ciphertext, for security parameter s.[3]

In contrast, our scheme instantiated with the information-theoretic LECSS scheme of [24] has a ciphertext of length $\approx 5k \cdot l_e$ and a public key of length $k \cdot l_p$. Note that the length of the public key depends on the length of the message, as we need independent public keys for each encrypted bit (whereas the DCCA scheme can use always the same public key). However, we observe that when k is not too large, e.g. in case the PKE scheme is used as a key encapsulation mechanism, we would have $k \approx s$ yielding public keys of comparable size. On the negative side, recall that our construction needs to self-destruct in case an invalid ciphertext is processed, which is not required in [34], and thus our construction only achieves SD-CCA security and not full-blown CCA security.

Non-malleable Codes.   Beyond the constructions of [24,9,26], non-malleable codes exists against block-wise tampering [11], against bit-wise tampering and permutations [5,4], against split-state tampering—both information-theoretic [23,2,7,3,1] and computational [39,19]—and in a setting where the computational complexity of the tampering functions is limited [8,28,35]. We stress that the typical application of non-malleable codes is to protect cryptographic schemes against memory tampering (see, e.g., [30,24,20,21]). A further application of non-malleable codes has been shown by Agrawal *et al.* [4] (in concurrent and independent work). They show that one can obtain a non-malleable multi-bit commitment scheme from a non-malleable single-bit commitment scheme by encoding the value with a (specific) non-malleable code and then committing to the codeword bits. Despite the similarity of the approaches, the techniques applied in their paper differ heavily from ours. The class of tampering functions the code has to protect against is different, and we additionally need continuous non-malleability to handle multiple decryption queries (this is not required for the commitment case).

2   Preliminaries

2.1   Systems: Resources, Converters, Distinguishers, and Reductions

Resources and Converters.   We use the concepts and terminology of abstract [42] and constructive cryptography [40]. The *resources* we consider are different types of communication channels, which are systems with three interfaces labeled by A, B, and E. A *converter* is a two-interface system which is directed in that it has an *inside* and an *outside* interface. Converters model protocol engines that are used by the parties, and using a protocol is modeled by connecting the party's interface of the resource to the inside interface of the converter (which hides those two interfaces) and using the outside interface of the converter instead. We generally use upper-case, bold-face letters (e.g., $\mathbf{R}$, $\mathbf{S}$) or channel symbols (e.g., $\bullet\!\!-\!\!\diamond\!\!-\!\!\twoheadrightarrow$) to denote resources or single-interface systems

---

[3] For simplicity, we assumed that the random strings r_A, r_B are computed by stretching the seed (of length s) of a pseudo-random generator.

and lower-case Greek letters (e.g., α, β) or sans-serif fonts (e.g., enc, dec) for converters. We denote by Φ the set of all resources and by Σ the set of all converters.

For $I \in \{A, B, E\}$, a resource $\mathbf{R} \in \Phi$, and a converter $\alpha \in \Sigma$, the expression $\alpha^I \mathbf{R}$ denotes the composite system obtained by connecting the inside interface of α to interface I of $\mathbf{R}$; the outside interface of α becomes the I-interface of the composite system. The system $\alpha^I \mathbf{R}$ is again a resource (cf. Figure 5 on page 546). For two resources $\mathbf{R}$ and $\mathbf{S}$, $[\mathbf{R}, \mathbf{S}]$ denotes the parallel composition of $\mathbf{R}$ and $\mathbf{S}$. For each $I \in \{A, B, E\}$, the I-interfaces of $\mathbf{R}$ and $\mathbf{S}$ are merged and become the *sub-interfaces* of the I-interface of $[\mathbf{R}, \mathbf{S}]$.

Two converters α and β can be composed serially by connecting the inside interface of β to the outside interface of α, written $\beta \circ \alpha$, with the effect that $(\beta \circ \alpha)^I \mathbf{R} = \beta^I \alpha^I \mathbf{R}$. Moreover, converters can also be taken in parallel, denoted by $[\alpha, \beta]$, with the effect that $[\alpha, \beta]^I [\mathbf{R}, \mathbf{S}] = [\alpha^I \mathbf{R}, \beta^I \mathbf{S}]$. We assume the existence of an identity converter $\mathrm{id} \in \Sigma$ with $\mathrm{id}^I \mathbf{R} = \mathbf{R}$ for all resources $\mathbf{R} \in \Phi$ and interfaces $I \in \{A, B, E\}$ and of a special converter $\perp \in \Sigma$ with an inactive outside interface.

Distinguishers. A *distinguisher* $\mathbf{D}$ connects to all interfaces of a resource $\mathbf{U}$ and outputs a single bit at the end of its interaction with $\mathbf{U}$. The expression $\mathbf{DU}$ defines a binary random variable corresponding to the output of $\mathbf{D}$ when interacting with $\mathbf{U}$, and the *distinguishing advantage of a distinguisher* $\mathbf{D}$ *on two systems* $\mathbf{U}$ *and* $\mathbf{V}$ is defined as

$$\Delta^{\mathbf{D}}(\mathbf{U}, \mathbf{V}) := |P[\mathbf{DU} = 1] - P[\mathbf{DV} = 1]|.$$

The distinguishing advantage measures how much the output distribution of $\mathbf{D}$ differs when it is connected to either $\mathbf{U}$ or $\mathbf{V}$. Note that the distinguishing advantage is a pseudo-metric.[4]

Reductions. When relating two distinguishing problems, it is convenient to use a special type of system $\mathbf{C}$ that translates one setting into the other. Formally, $\mathbf{C}$ is a converter that has an *inside* and an *outside* interface. When it is connected to a system $\mathbf{S}$, which is denoted by $\mathbf{CS}$, the inside interface of $\mathbf{C}$ connects to the (merged) interface(s) of $\mathbf{S}$ and the outside interface of $\mathbf{C}$ is the interface of the composed system. $\mathbf{C}$ is called a *reduction system* (or simply *reduction*).

To reduce distinguishing two systems $\mathbf{S}, \mathbf{T}$ to distinguishing two systems $\mathbf{U}, \mathbf{V}$, one exhibits a reduction $\mathbf{C}$ such that $\mathbf{CS} \equiv \mathbf{U}$ and $\mathbf{CT} \equiv \mathbf{V}$. Then, for all distinguishers $\mathbf{D}$, we have $\Delta^{\mathbf{D}}(\mathbf{U}, \mathbf{V}) = \Delta^{\mathbf{D}}(\mathbf{CS}, \mathbf{CT}) = \Delta^{\mathbf{DC}}(\mathbf{S}, \mathbf{T})$. The last equality follows from the fact that $\mathbf{C}$ can also be thought of as being part of the distinguisher (which follows from the *composition-order independence* [42]).

---

[4] That is, for any $\mathbf{D}$, it is symmetric, satisfies the triangle inequality, and $\Delta^{\mathbf{D}}(\mathbf{R}, \mathbf{R}) = 0$ for all $\mathbf{R}$.

2.2 Discrete Systems

The behavior of systems can be formalized by random systems as in [46,41]: A random system $\mathbf{S}$ is a sequence $(\mathsf{p}^{\mathbf{S}}_{Y^i|X^i})_{i\geq 1}$ of conditional probability distributions, where $\mathsf{p}^{\mathbf{S}}_{Y^i|X^i}(y^i, x^i)$ is the probability of observing the outputs $y^i = (y_1, \ldots, y_i)$ given the inputs $x^i = (x_1, \ldots, x_i)$. If for two systems $\mathbf{R}$ and $\mathbf{S}$,

$$\mathsf{p}^{\mathbf{R}}_{Y^i|X^i} = \mathsf{p}^{\mathbf{S}}_{Y^i|X^i}$$

for all i and for all parameters where both are defined, they are called *equivalent*, denoted by $\mathbf{R} \equiv \mathbf{S}$. In that case, $\Delta^{\mathbf{D}}(\mathbf{R}, \mathbf{S}) = 0$ for all distinguishers $\mathbf{D}$.

A system $\mathbf{S}$ can be extended by a so-called *monotone binary output* (or *MBO*) $\mathcal{B}$, which is an additional one-bit output $B_1, B_2, \ldots$ with the property that $B_i = 1$ implies $B_{i+1} = 1$ for all i.[5] The enhanced system is denoted by $\hat{\mathbf{S}}$, and its behavior is described by the sequence $(\mathsf{p}^{\hat{\mathbf{S}}}_{Y^i, B_i|X^i})_{i\geq 1}$. If for two systems $\hat{\mathbf{R}}$ and $\hat{\mathbf{S}}$ with MBOs,

$$\mathsf{p}^{\hat{\mathbf{R}}}_{Y^i, B_i=0|X^i} = \mathsf{p}^{\hat{\mathbf{S}}}_{Y^i, B_i=0|X^i}$$

for all i, they are called *game equivalent*, which is denoted by $\hat{\mathbf{R}} \stackrel{g}{\equiv} \hat{\mathbf{S}}$. In such a case, $\Delta^{\mathbf{D}}(\mathbf{R}, \mathbf{S}) \leq \Gamma^{\mathbf{D}}(\hat{\mathbf{R}}) = \Gamma^{\mathbf{D}}(\hat{\mathbf{S}})$, where $\Gamma^{\mathbf{D}}(\hat{\mathbf{R}})$ denotes the probability that $\mathbf{D}$ provokes the MBO. For more details and a proof of this fact, consult [41].[6]

2.3 The Notion of Construction

We formalize the security of protocols via the notion of *construction*, introduced in [42]:

Definition 1. *Let Φ and Σ be as above, and let ε_1 and ε_2 be two functions mapping each distinguisher $\mathbf{D}$ to a real number in $[0, 1]$. A protocol $\pi = (\pi_1, \pi_2) \in \Sigma^2$ constructs resource $\mathbf{S} \in \Phi$ from resource $\mathbf{R} \in \Phi$ with distance $(\varepsilon_1, \varepsilon_2)$ and with respect the simulator $\sigma \in \Sigma$, denoted[7]*

$$\mathbf{R} \stackrel{\pi,\sigma,(\varepsilon_1,\varepsilon_2)}{\Longmapsto} \mathbf{S},$$

if for all distinguishers $\mathbf{D}$,

$$\begin{cases} \Delta^{\mathbf{D}}(\pi_1{}^A \pi_2{}^B \perp^E \mathbf{R}, \perp^E \mathbf{S}) \leq \varepsilon_1(\mathbf{D}) & (availability) \\ \Delta^{\mathbf{D}}(\pi_1{}^A \pi_2{}^B \mathbf{R}, \sigma^E \mathbf{S}) \leq \varepsilon_2(\mathbf{D}) & (security). \end{cases}$$

The availability condition captures that a protocol must correctly implement the functionality of the constructed resource in the absence of the attacker. The security condition models the requirement that everything the attacker can achieve in the setting with the assumed resource and the protocol, she can also accomplish in the setting with the constructed resource (using the simulator to translate the behavior).

---

[5] In other words, once the MBO is 1, it cannot return to 0.

[6] Intuitively, this means that in order to distinguish the two systems, $\mathbf{D}$ has to provoke the MBO.

[7] In less formal contexts, we sometimes drop the superscripts on $\Longmapsto$.

2.4 The Composition Theorem

The above construction notion composes in the following two ways:[8] First, if one (lower-level) protocol constructs the resource that is assumed by the other (higher-level) protocol, then the composition of those two protocols constructs the same resource as the higher-level protocol, but from the resources assumed by the lower-level protocol, under the assumptions that occur in (at least) one of the individual security statements. Second, the security of constructions is maintained in the presence of arbitrary resources taken in parallel.

To state the theorem, we make use of the special converter id (cf. Section 2.1). Furthermore, we assume the operation $[\cdot, \ldots, \cdot]$ to be left-associative; in this way we can simply express multiple resources using the single variable $\mathbf{U}$.

Theorem 1. *Let* $\mathbf{R}, \mathbf{S}, \mathbf{T}, \mathbf{U} \in \Phi$ *be resources. Let* $\pi = (\pi_1, \pi_2)$ *and* $\psi = (\psi_1, \psi_2)$ *be protocols,* σ_π *and* σ_ψ *be simulators, and* $(\varepsilon_\pi^1, \varepsilon_\pi^2)$, $(\varepsilon_\psi^1, \varepsilon_\psi^2)$ *such that*

$$\mathbf{R} \xrightarrow{(\pi, \sigma_\pi, (\varepsilon_\pi^1, \varepsilon_\pi^2))} \mathbf{S} \quad and \quad \mathbf{S} \xrightarrow{(\psi, \sigma_\psi, (\varepsilon_\psi^1, \varepsilon_\psi^2))} \mathbf{T}.$$

Then

$$\mathbf{R} \xrightarrow{(\alpha, \sigma_\alpha, (\varepsilon_\alpha^1, \varepsilon_\alpha^2))} \mathbf{T}$$

with $\alpha = (\psi_1 \circ \pi_1, \psi_2 \circ \pi_2)$, $\sigma_\alpha = \sigma_\pi \circ \sigma_\psi$, *and* $\varepsilon_\alpha^i(\mathbf{D}) = \varepsilon_\pi^i(\mathbf{D}\sigma_\psi^E) + \varepsilon_\psi^i(\mathbf{D}\pi_1^A\pi_2^B)$, *where* $\mathbf{D}\sigma_\psi^E$ *and* $\mathbf{D}\pi_1^A\pi_2^B$ *mean that* $\mathbf{D}$ *applies the converters at the respective interfaces. Moreover*

$$[\mathbf{R}, \mathbf{U}] \xrightarrow{([\pi, (\mathrm{id}, \mathrm{id})], [\sigma_\pi, \mathrm{id}], (\bar{\varepsilon}_\pi^1, \bar{\varepsilon}_\pi^2))} [\mathbf{S}, \mathbf{U}],$$

with $\bar{\varepsilon}_\pi^i(\mathbf{D}) = \varepsilon_\pi^i(\mathbf{D}[\cdot, \mathbf{U}])$, *where* $\mathbf{D}[\cdot, \mathbf{U}]$ *means that the distinguisher emulates* $\mathbf{U}$ *in parallel. (The analogous statement holds with respect to* $[\mathbf{U}, \mathbf{R}]$ *and* $[\mathbf{U}, \mathbf{S}]$.)

2.5 Channel Resources

From the perspective of constructive cryptography, the purpose of a public-key encryption scheme is to construct a confidential channel from non-confidential channels. A channel is a resource that involves a sender A, a receiver B, and—to model channels with different levels of security—an attacker E. The main types of channels relevant to this work are defined below with respect to interface set $\{A, B, E\}$. All channels are parametrized by a message space $\mathcal{M} \subseteq \{0, 1\}^*$, which is only made explicit in the confidential channel (see below), however.

---

[8] The composition theorem was first explicitly stated in [45], but the statement there was restricted to asymptotic settings. Later, in [37], the theorem was stated in a way that also allows to capture concrete security statements. The proof, however, still follows the same steps as the one in [45].

Insecure Multiple-use Channel. The insecure channel $-\rightarrow\!\!\!\twoheadrightarrow$ transmits multiple messages $m \in \mathcal{M}$ and corresponds to, for instance, communication via the Internet. If no attacker is present (i.e., in case $\perp^E - \rightarrow\!\!\!\twoheadrightarrow$), then all messages are transmitted from A to B faithfully. Otherwise (for $- \rightarrow\!\!\!\twoheadrightarrow$), the communication can be controlled via the E-interface, i.e., the attacker learns all messages input at the A-interface and chooses the messages to be output at the B-interface. The channel is described in more detail in Figure 1.

Channel $\perp^E - \rightarrow\!\!\!\twoheadrightarrow$	Channel $- \rightarrow\!\!\!\twoheadrightarrow$	

on m *at* A
  | output m *at* B
	on m *at* A
	  | output (msg, m) *at* E
		on (inj, m) *at* E
		  | output m *at* B

Fig. 1. Insecure, multiple-use communication channel from A to B

Authenticated (Unreliable) Single-use Channel. The (single-use) authenticated channel $\bullet\!\!\longrightarrow$, described in Figure 2, allows the sender A to transmit a single message to the receiver B authentically. That means, while the attacker (at the E-interface) can still read the transmitted message, the only influence allowed is delaying the message (arbitrarily, i.e., there is no guarantee that the message will ever be delivered). The channel guarantees that *if* a message is delivered to B, *then* this message was input by A before. There are different constructions that result in the channel $\bullet\!\!\longrightarrow$, based on, for instance, MACs or signature schemes.

Channel $\perp^E \bullet\!\!\longrightarrow$	Channel $\bullet\!\!\longrightarrow$	

on first m *at* A
  | output m *at* B
	on first m *at* A
	  | output (msg, m) *at* E
		on first dlv *at* E
		  | output m *at* B (if defined)

Fig. 2. Authenticated, single-use communication channel from A to B

Confidential Multiple-use Channel. The k-bit confidential channel $\xrightarrow{k\text{-bit}}\!\!\!\diamond\!\!\twoheadrightarrow\!\!\bullet$ allows to transmit multiple messages $m \in \{0,1\}^k$. If no attacker is present (i.e., in case $\perp^E \xrightarrow{k\text{-bit}}\!\!\!\diamond\!\!\twoheadrightarrow\!\!\bullet$), then all messages are transmitted from A to B faithfully. Otherwise (for $\xrightarrow{k\text{-bit}}\!\!\!\diamond\!\!\twoheadrightarrow\!\!\bullet$), all messages $m \in \{0,1\}^k$ input at the A-interface are stored in a buffer $\mathcal{B}$.[9] The attacker can then choose messages from the buffer $\mathcal{B}$ (by using an index) to be delivered at the B-interface, or inject messages from $\{0,1\}^k$ which are then also output at the B-interface. Note that E cannot inject messages that depend on those in $\mathcal{B}$, i.e., the confidential channel is non-malleable. It is described in more detail in Figure 3.

---

[9] The $\diamond$ in the symbol $\xrightarrow{k\text{-bit}}\!\!\!\diamond\!\!\twoheadrightarrow\!\!\bullet$ is to suggest the presence of $\mathcal{B}$.

Channel $\perp^E$ $\xrightarrow{k\text{-bit}}\diamond\!\!\!\rightarrow\!\!\bullet$	Channel $\xrightarrow{k\text{-bit}}\diamond\!\!\!\rightarrow\!\!\bullet$	
	init	
	$\quad i \leftarrow 0$	**on** (dlv, i') *at* A
	$\quad \mathcal{B} \leftarrow \emptyset$	$\quad$ **if** $\exists m : (i', m) \in \mathcal{B}$
on $m \in \{0,1\}^k$ *at* A		$\quad\quad$ **output** m *at* B
$\quad$ **output** m *at* B	**on** $m \in \{0,1\}^k$ *at* A	
	$\quad i \leftarrow i + 1$	**on** (inj, m') *at* A
	$\quad \mathcal{B} \leftarrow \mathcal{B} \cup \{(i, m)\}$	$\quad$ **output** m' *at* E
	$\quad$ **output** (msg, i) *at* E	

Fig. 3. Confidential, multiple-use k-bit channel from A to B

2.6 Public-Key Encryption Schemes

A public-key encryption (PKE) scheme with message space $\mathcal{M} \subseteq \{0,1\}^*$ and ciphertext space $\mathcal{E}$ is defined as three algorithms $\Pi = (K, E, D)$, where the key-generation algorithm K outputs a key pair $(\mathsf{pk}, \mathsf{sk})$, the (probabilistic) encryption algorithm E takes a message $m \in \mathcal{M}$ and a public key pk and outputs a ciphertext $e \leftarrow E_{\mathsf{pk}}(m)$, and the decryption algorithm takes a ciphertext $e \in \mathcal{E}$ and a secret key sk and outputs a plaintext $m \leftarrow D_{\mathsf{sk}}(e)$. The output of the decryption algorithm can be the special symbol $\diamond$, indicating an invalid ciphertext. A PKE scheme is correct if $m = D_{\mathsf{sk}}(E_{\mathsf{pk}}(m))$ (with probability 1 over the randomness in the encryption algorithm) for all messages m and all key pairs $(\mathsf{pk}, \mathsf{sk})$ generated by K.

We introduce security notions for PKE schemes as we need them.

2.7 Continuously Non-malleable Codes

Non-malleable codes, introduced in [24], are coding schemes that protect the encoded messages against certain classes of adversarially chosen modifications, in the sense that the decoding will result either in the original message or in an unrelated value.

Definition 2 (Coding scheme). *A (k, n)-coding scheme (Enc, Dec) consists of a randomized encoding function* Enc $: \{0,1\}^k \to \{0,1\}^n$ *and a deterministic decoding function* Dec $: \{0,1\}^n \to \{0,1\}^k \cup \{\diamond\}$ *such that* Dec(Enc(x)) = x *(with probability 1 over the randomness of the encoding function) for each $x \in \{0,1\}^k$. The special symbol $\diamond$ indicates an invalid codeword.*

In the original definition [24], the adversary is allowed to modify the codeword via a function of a specified class $\mathcal{F}$ only once. Continuous non-malleability, introduced in [26], extends this guarantee to the case where the adversary is allowed to perform multiple such modifications for a fixed target codeword. The notion of *adaptive* continuous non-malleability considered here is an extension of the one in [26] in that the adversary is allowed to adaptively specify messages and the functions may depend on multiple codewords. That is, the class $\mathcal{F}$ is actually a sequence $(\mathcal{F}^{(i)})_{i \geq 1}$ of function families with $\mathcal{F}^{(i)} \subseteq \{f \mid f : (\{0,1\}^n)^i \to \{0,1\}^n\}$, and after encoding i messages, the adversary chooses functions from $\mathcal{F}^{(i)}$. A similar adaptive notion has been already considered for continuous strong non-malleability in the split-state model [27].

System $\mathbf{S}_{\mathcal{F}}^{\mathsf{real}}$		System $\mathbf{S}_{\mathcal{F},\tau}^{\mathsf{simu}}$	
init $\quad$ $i \leftarrow 0$ **on** (encode, x) $\quad$ $i \leftarrow i + 1$ $\quad$ $c^{(i)} \leftarrow_\$ \mathrm{Enc}(x)$	**on** (tamper, f) *with* $f \in \mathcal{F}^{(i)}$ $\quad$ $c' \leftarrow f(c^{(1)}, \dots, c^{(i)})$ $\quad$ $x' \leftarrow \mathrm{Dec}(c')$ $\quad$ **if** $x' = \diamond$ $\quad\quad$ **self-destruct** $\quad$ **out** x'	**init** $\quad$ $i \leftarrow 0$ **on** (encode, x) $\quad$ $i \leftarrow i + 1$ $\quad$ $x^{(i)} \leftarrow_\$ x$	**on** (tamper, f) *with* $f \in \mathcal{F}^{(i)}$ $\quad$ $x' \leftarrow_\$ \tau(i, f)$ $\quad$ **if** $x' = \diamond$ $\quad\quad$ **self-destruct** $\quad$ **if** $x' = (\mathsf{same}, j)$ $\quad\quad$ $x' \leftarrow x^{(j)}$ $\quad$ **out** x'

Fig. 4. Systems $\mathbf{S}_{\mathcal{F}}^{\mathsf{real}}$ and $\mathbf{S}_{\mathcal{F},\tau}^{\mathsf{simu}}$ defining adaptive continuous non-malleability of (Enc, Dec). The command **self-destruct** has the effect that $\diamond$ is output and all future queries are answered by $\diamond$.

Formally, adaptive continuous non-malleability w.r.t. $\mathcal{F}$ is defined by comparing the two random systems $\mathbf{S}_{\mathcal{F}}^{\mathsf{real}}$ and $\mathbf{S}_{\mathcal{F},\tau}^{\mathsf{simu}}$ defined in Figure 4. Both systems process encode and tamper queries from a distinguisher $\mathbf{D}$, whose objective is to tell the two systems apart.

System $\mathbf{S}_{\mathcal{F}}^{\mathsf{real}}$ produces a random encoding $c^{(i)}$ of each message $x^{(i)}$ specified by $\mathbf{D}$ and allows $\mathbf{D}$ to repeatedly issue tampering functions $f \in \mathcal{F}^{(i)}$. For each such query, $\mathbf{S}_{\mathcal{F}}^{\mathsf{real}}$ computes the modified codeword $c' = f(c^{(1)}, \dots, c^{(i)})$ and outputs $\mathrm{Dec}(c')$. Whenever $\mathrm{Dec}(c') = \diamond$, the system enters a self-destruct mode, in which all further queries are replied to by $\diamond$.

The second random system, $\mathbf{S}_{\mathcal{F},\tau}^{\mathsf{simu}}$, features a simulator τ, which is allowed to keep state. The simulator repeatedly takes a tampering function and outputs either a message x', (same, v) for $v \in \{1, \dots, i\}$, or $\diamond$, where (same, v) is used by τ to indicate that (it believes that) the tampering results in an n-bit string that decodes to the vth message encoded. System $\mathbf{S}_{\mathcal{F},\tau}^{\mathsf{simu}}$ outputs whatever τ outputs, except that (same, v) is replaced by the vth message $x^{(v)}$ specified by $\mathbf{D}$. Moreover, in case of $\diamond$, $\mathbf{S}_{\mathcal{F},\tau}^{\mathsf{simu}}$ self-destructs.

For $\ell, q \in \mathbb{N}$, $\mathbf{S}_{\mathcal{F},\ell,q}^{\mathsf{real}}$ is the system that behaves as $\mathbf{S}_{\mathcal{F}}^{\mathsf{real}}$ except that only the first ℓ encode-queries and the first q tamper-queries are handled (and similarly for $\mathbf{S}_{\mathcal{F},\tau,\ell,q}^{\mathsf{simu}}$ and $\mathbf{S}_{\mathcal{F},\tau}^{\mathsf{simu}}$). Note that by setting $\ell = 1$, one recovers continuous non-malleability as defined in [26],[10] and by additionally setting $q = 1$ the original definition of non-malleability.

Definition 3 (Continuous non-malleability). *Let* $\mathcal{F} = (\mathcal{F}^{(i)})_{i \geq 1}$ *be a sequence of function families* $\mathcal{F}^{(i)} \subseteq \{f \mid f : (\{0,1\}^n)^i \to \{0,1\}^n\}$ *and let* $\ell, q \in \mathbb{N}$. *A coding scheme* (Enc, Dec) *is adaptively continuously* $(\mathcal{F}, \varepsilon, \ell, q)$-*non-malleable (or simply* $(\mathcal{F}, \varepsilon, \ell, q)$-*non-malleable) if there exists a simulator* τ *such that* $\Delta^{\mathbf{D}}(\mathbf{S}_{\mathcal{F},\ell,q}^{\mathsf{real}}, \mathbf{S}_{\mathcal{F},\tau,\ell,q}^{\mathsf{simu}}) \leq \varepsilon$ *for all distinguishers* $\mathbf{D}$.

3 From Single-Bit to Multi-bit Channels

In this section we examine the issue of domain extension for chosen-ciphertext-secure (CCA) public-key encryption (PKE). To that end, we employ the constructive cryptography paradigm and proceed in two main steps: First, we reuse

---

[10] Being based on *strong* non-malleability [24], the notion of [26] is actually stronger than ours.

a result by [13] saying that CCA-secure PKE can be used to construct a confidential channel from an authenticated and an insecure channel. This implies that using n independent copies of a single-bit CCA-secure PKE scheme, one obtains n parallel instances of a single-bit confidential channel. Second, using continuously non-malleable codes, we tie the n single-bit channels together and obtain a k-bit confidential channel. Combining these two steps using the composition theorem results in a protocol that constructs a k-bit confidential channel from an authenticated and an insecure channel.

The combined protocol can be seen as the following simple PKE scheme: first encode a k-bit message using a continuously non-malleable (k, n)-code to protect its integrity, obtaining an n-bit codeword c; then encrypt c bit-wise using n independent public keys for a single-bit CCA-secure PKE. As we show, continuously non-malleable codes only exist if the decoder is allowed to "self-destruct" once it processes an invalid codeword. This property translates to the resulting PKE scheme, which therefore only achieves a weaker form of CCA security, called *self-destruct CCA security (SD-CCA)*, where the decryption oracle stops working after the attacker submits an invalid ciphertext. Noting that SD-CCA security suffices to construct the single-bit confidential channels yields a domain-extension technique for SD-CCA-secure PKE schemes.

We stress that the need for self-destruct is not a limitation of the security proof of our code (cf. Section 4), as continuous non-malleability for the class of tampering functions required for the above transformation to work is impossible without the self-destruct property (cf. Section 5 for details).

3.1   Single-Bit PKE Viewed Constructively

Following the proof of [13, Theorem 2], one can show that a 1-bit SD-CCA-secure PKE scheme can be used to design a protocol that achieves the construction

$$[\longleftarrow\!\bullet, -\!-\!\twoheadrightarrow] \quad \overset{\text{1-bit}}{\Longmapsto} \quad -\!\diamond\!\!-\!\!\twoheadrightarrow\!\bullet, \tag{2}$$

where, in a nutshell, the receiver's protocol converter is responsible for key generation, decryption, as well as self-destructing, the sender's protocol converter for encryption, and where the authenticated channel $\longleftarrow\!\bullet$ is used for the transmission of the public key and the insecure channel $-\!-\!\twoheadrightarrow$ for sending ciphertexts. The constructed single-bit confidential channel $\overset{\text{1-bit}}{-\!\diamond\!\!-\!\twoheadrightarrow\!\bullet}$ hides all messages sent by the sender from the attacker and allows the attacker to either deliver already sent messages or to inject *independent* messages. This captures the intuitive (SD-)CCA guarantee that an attacker, by modifying a particular ciphertext, can either leave the message contained therein intact or replace it by an independently created one.

Using n independent copies of the single-bit scheme in parallel yields a protocol 1-pke that achieves:

$$[\longleftarrow\!\bullet, -\!-\!\twoheadrightarrow] \quad \overset{\text{1-pke}}{\Longmapsto} \quad [\overset{\text{1-bit}}{-\!\diamond\!\!-\!\twoheadrightarrow\!\bullet}]^n, \tag{3}$$

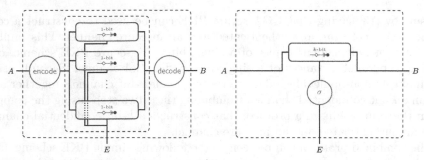

Fig. 5. Left: The assumed resource $[\overset{\text{1-bit}}{\multimap\!\!\rightarrow\!\!\bullet}]^n$ with protocol converters encode and decode attached to interfaces A and B, denoted $\text{encode}^A\text{decode}^B[\overset{\text{1-bit}}{\multimap\!\!\rightarrow\!\!\bullet}]^n$. Right: The constructed resource $\overset{k\text{-bit}}{\multimap\!\!\rightarrow\!\!\bullet}$ with simulator σ attached to the E-interface, denoted $\sigma^E \overset{k\text{-bit}}{\multimap\!\!\rightarrow\!\!\bullet}$. In particular, σ must simulate the E-interfaces of $[\overset{\text{1-bit}}{\multimap\!\!\rightarrow\!\!\bullet}]^n$. The protocol is secure if the two systems are indistinguishable.

which follows almost directly from the composition theorem. More details can be found in the full version of this paper [14].

3.2   Tying the Channels Together

We now show how to construct, using an adaptive continuously non-malleable (k,n)-code (cf. Section 2.7), a (single) k-bit confidential channel from the n independent single-bit confidential channels constructed in the previous section. This is achieved by having the sender encode the message with the non-malleable code and sending the resulting codeword over the 1-bit channels (bit-by-bit), while the receiver decodes all n-bit strings received via these channels. Additionally, due to the self-destruct property of continuously non-malleable codes, the receiver must stop decoding once an invalid codeword has been received.

More precisely, let (Enc, Dec) be a (k,n)-coding scheme and consider the following protocol $\text{nmc} = (\text{encode}, \text{decode})$: Converter encode encodes every message $m \in \{0,1\}^k$ input at its outside interface with fresh randomness, resulting in an n-bit encoding $c = c_1 \cdots c_n \leftarrow \text{Enc}(m)$. Then, for $i = 1, \ldots, n$, it outputs bit c_i to the ith channel at the inside interface. Converter decode, whenever it receives an n-bit string $c' = c'_1 \cdots c'_n$ (where the ith bit c'_i was received on the ith channel), it computes $m' \leftarrow \text{Dec}(c')$ and outputs m' at the outside interface. If $m' = \diamond$, it implements the self-destruct mode, i.e., it answers all future encodings received at the inside interface by outputting $\diamond$ at the outside interface.

The goal is now to show that protocol nmc achieves the construction

$$[\overset{\text{1-bit}}{\multimap\!\!\rightarrow\!\!\bullet}]^n \quad \overset{\text{nmc}}{\Longmapsto} \quad \overset{k\text{-bit}}{\multimap\!\!\rightarrow\!\!\bullet} . \tag{4}$$

The Required Non-malleability. By inspecting both sides of Figure 5, it becomes immediately apparent why adaptive continuously non-malleable codes are the proper choice to achieve construction (4): On the left-hand side, the distinguisher can repeatedly input messages $m^{(i)}$ at interface A, which results in encodings $c^{(i)}$ being input (bit-by-bit) into the single-bit channels. Using the E-interfaces of these channels, the distinguisher can repeatedly see the decoding of an n-bit string $c' = c'_1 \cdots c'_n$ at interface B, where each bit c'_j results from either forwarding one of the bits already in the jth channel or from injecting a fresh bit that is either 0 or 1.

Put differently, the distinguisher can effectively launch tampering attacks via functions from $\mathcal{F}_{\mathsf{copy}} := (\mathcal{F}_{\mathsf{copy}}^{(i)})_{i \geq 1}$, where $\mathcal{F}_{\mathsf{copy}}^{(i)} \subseteq \{f \mid f : (\{0,1\}^n)^i \to \{0,1\}^n\}$ and each function $f \in \mathcal{F}_{\mathsf{copy}}^{(i)}$ is characterized by a vector $\chi(f) = (f_1, \ldots, f_n)$ where $f_j \in \{\mathsf{zero}, \mathsf{one}, \mathsf{copy}_1, \ldots, \mathsf{copy}_i\}$, with the meaning that f takes as input i codewords $(c^{(1)}, \ldots, c^{(i)})$ and outputs an n-bit string $c' = c'_1 \cdots c'_n$ in which each bit c'_j is either set to 0 (zero), set to 1 (one), or copied from the jth bit in a codeword $c^{(v)}$ (copy_v) for $v \in \{1, \ldots, i\}$.

On the right-hand side, the distinguisher may again input messages $m^{(i)}$ at interface A, to the k-bit confidential channel. At interface E, this channel only allows to either deliver entire k-bit messages already sent by A or to inject independent messages. The simulator σ required to prove (4) needs to simulate the E-interfaces of the single-bit confidential channels at its outside interface and, based solely on what is input at these interfaces, decide whether to forward or inject a message, which corresponds exactly to the task of the simulator τ in the non-malleability experiment (cf. Section 2.7).

Theorem 2 below formalizes this correspondence; its proof is essentially a technicality: one merely needs to "translate" between the channel settings and the non-malleability experiment.

Theorem 2. *For any $\ell, q \in \mathbb{N}$, if $(\mathsf{Enc}, \mathsf{Dec})$ is $(\mathcal{F}_{\mathsf{copy}}, \varepsilon, \ell, q)$-continuously non-malleable, there exists a simulator σ such that*

$$\left[\overset{\mathsf{1\text{-}bit}, \ell, q}{\multimap\!\!\twoheadrightarrow\!\bullet} \right]^n \overset{(\mathsf{nmc}, \sigma, (0, \varepsilon))}{\Longrightarrow} \overset{k\text{-}\mathsf{bit}, \ell, q}{\multimap\!\!\twoheadrightarrow\!\bullet},$$

where the additional superscripts ℓ, q on a channel mean that it only processes the first ℓ queries at the A-interface and only the first q queries at the E-interface.

Proof. The availability condition holds by the correctness of the code.

Let $\mathcal{F} := \mathcal{F}_{\mathsf{copy}}$, $\mathbf{S}_{\mathcal{F}}^{\mathsf{real}} := \mathbf{S}_{\mathcal{F}, \ell, q}^{\mathsf{real}}$, and $\mathbf{S}_{\mathcal{F}, \tau}^{\mathsf{simu}} := \mathbf{S}_{\mathcal{F}, \tau, \ell, q}^{\mathsf{simu}}$ where τ is the simulator guaranteed to exist by Definition 3.

Consider the following simulator σ (based on τ), which simulates the E-sub-interfaces of the 1-bit confidential channels at its outside interface: When (msg, i) is received at the inside interface, it outputs (msg, i) at each outside sub-interface corresponding to a 1-bit confidential channel. Whenever σ receives one instruction to either deliver[11] $((\mathsf{dlv}, i')$ for $i' \in \mathbb{N})$ or inject $((\mathsf{inj}, m')$ for $m' \in \{0,1\})$ a

---

[11] For simplicity, assume that no deliver instruction (dlv, i') for some i' greater than the largest number i received via (msg, i) at the inside interface so far is input.

bit at each outside sub-interface corresponding to one of the confidential channels, it assembles these to a function f with $\chi(f) = (f_1, \ldots, f_n)$ as follows: For all $j = 1, \ldots, n$,

$$f_j := \begin{cases} \text{zero} & \text{if the instruction on the } j^{\text{th}} \text{ sub-interface is } (\text{inj}, 0), \\ \text{one} & \text{if the instruction on the } j^{\text{th}} \text{ sub-interface is } (\text{inj}, 1), \\ \text{copy}_{i'} & \text{if the instruction on the } j^{\text{th}} \text{ sub-interface is } (\text{dlv}, i'). \end{cases}$$

Then, σ invokes τ to obtain $x' \leftarrow_\$ \tau(i, f)$, where i is the number of instructions (msg, i) received at the inside interface so far. If $x' = (\text{same}, j)$, σ outputs (dlv, j) at the inside interface. Otherwise, it outputs (inj, x'). If $x' = \diamond$, σ outputs $(\text{inj}, \diamond)$ at the inside interface and implements the self-destruct mode, i.e., outputs $(\text{inj}, \diamond)$ at the inside interface for all future inputs to the simulated interfaces of the single-bit channels.

Consider the following reduction $\mathbf{C}$, which provides interfaces A, B, and E on the outside and expects to connect to either $\mathbf{S}_{\mathcal{F}}^{\text{real}}$ or $\mathbf{S}_{\mathcal{F}, \tau}^{\text{simu}}$ on the inside. When a message m is input at the A-interface, $\mathbf{C}$ outputs (encode, m) on the inside. Similarly to σ, it repeatedly collects instructions input at the E-sub-interfaces and uses them to form a tamper function f, which it outputs on the inside as (tamper, f). Then, it outputs the answer x' received on the inside at the B-interface. Additionally, if $x' = \diamond$, $\mathbf{C}$ implements the self-destruct mode, i.e., subsequently only outputs $\diamond$ at interface B.

One observes that

$$\mathbf{CS}_{\mathcal{F}}^{\text{real}} \equiv \text{encode}^A \text{decode}^B [\overset{\text{1-bit}}{-\diamond\!\!\!\twoheadrightarrow\!\bullet}]^n \quad \text{and} \quad \mathbf{CS}_{\mathcal{F}, \tau}^{\text{simu}} \equiv \sigma^E \overset{k\text{-bit}, \ell, q}{-\diamond\!\!\!\twoheadrightarrow\!\bullet} .$$

Thus, for all distinguishers $\mathbf{D}$,

$$\begin{aligned} \Delta^{\mathbf{D}} (\text{encode}^A \text{decode}^B [\overset{\text{1-bit}}{-\diamond\!\!\!\twoheadrightarrow\!\bullet}]^n, \sigma^E \overset{k\text{-bit}, \ell, q}{-\diamond\!\!\!\twoheadrightarrow\!\bullet}) &= \Delta^{\mathbf{D}} (\mathbf{CS}_{\mathcal{F}}^{\text{real}}, \mathbf{CS}_{\mathcal{F}, \tau}^{\text{simu}}) \\ &= \Delta^{\mathbf{DC}} (\mathbf{S}_{\mathcal{F}}^{\text{real}}, \mathbf{S}_{\mathcal{F}, \tau}^{\text{simu}}) \le \varepsilon. \end{aligned}$$

$\square$

3.3 Plugging It Together

The composition theorem of constructive cryptography (cf. Section 2.4) implies that the protocol m-pke = nmc $\circ$ 1-pke resulting from composing the protocols 1-pke and nmc for transformations (3) and (4), respectively, achieves

$$[\overset{}{\longleftarrow\!\bullet}, -\twoheadrightarrow] \overset{\text{m-pke}}{\Longrightarrow} \overset{k\text{-bit}}{-\diamond\!\!\!\twoheadrightarrow\!\bullet} . \tag{5}$$

3.4 SD-CCA Security

In this section we formally define the notion of SD-CCA security and show how protocol m-pke can be seen as a PKE scheme Π that achieves SD-CCA security;[12] a proof can be found in the full version of this paper [14]. There

---

[12] Actually, Π is only replayable SD-CCA secure; see below for details.

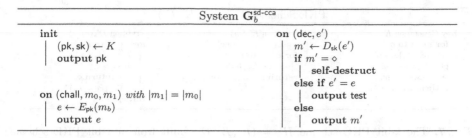

Fig. 6. System $\mathbf{G}_b^{\mathrm{sd\text{-}cca}}$, where $b \in \{0,1\}$, defining SD-CCA security of a PKE scheme $\Pi = (K, E, D)$. The command **self-destruct** causes the system to output $\diamond$ and to answer all future decryption queries by $\diamond$.

we also provide a direct game-based proof of the fact that combining single-bit SD-CCA-secure PKE with a non-malleable code as shown above yields a multi-bit SD-CCA-secure PKE scheme. That proof is a hybrid argument and is obtained by "unwrapping" the concatenation of the statements in this section. The modular nature and the intuitive simplicity of the proofs are lost, however.

Definition of SD-CCA. The only difference between the SD-CCA game and the standard game used to define CCA is that the decryption oracle self-destructs, i.e., it stops processing further queries once an invalid ciphertext is queried. Note that the self-destruct feature only affects the decryption oracle; the adversary is still allowed to get the challenge ciphertext after provoking a self-destruct. The game is phrased as a distinguishing problem between the two systems $\mathbf{G}_0^{\mathrm{sd\text{-}cca}}$ and $\mathbf{G}_1^{\mathrm{sd\text{-}cca}}$ described in Figure 6.

Definition 4. *A PKE scheme $\Pi = (K, E, D)$ is (t, q, ε)-SD-CCA secure if*

$$\Delta^{\mathbf{D}}(\mathbf{G}_0^{\mathrm{sd\text{-}cca}}, \mathbf{G}_1^{\mathrm{sd\text{-}cca}}) \leq \varepsilon$$

for all distinguishers $\mathbf{D}$ with running time at most t and making at most q decryption queries.

The PKE Scheme. The PKE scheme $\Pi = (K, E, D)$ corresponding to our protocol m-pke can be obtained as follows. The key generation algorithm K generates n independent key pairs of the 1-bit scheme. The encryption algorithm E first encodes a message using a non-malleable code and then encrypts each bit of the resulting encoding independently and outputs the n resulting ciphertexts. The decryption algorithm D first decrypts the n ciphertexts, decodes the resulting bitstring, and outputs the decoded message or the symbol $\diamond$, indicating an invalid ciphertext, if any of these steps fails. The scheme is described in more detail in Figure 7.

Security of Π. In the full version of this paper [14], we show that Π is *replayable* SD-CCA secure. The notion of *replayable* CCA security (RCCA) in

PKE Scheme $\Pi' = (K', E', D')$		
Key Generation K'	Encryption $E'_{pk}(m)$	Decryption $D'_{sk}(e)$
for $i \leftarrow 1$ **to** n	$c = c_1 \cdots c_n \leftarrow \text{Enc}(m)$	**for** $i \leftarrow 1$ **to** n
	$(pk_i, sk_i) \leftarrow_\$ K$	**for** $i \leftarrow 1$ **to** n		$c_i \leftarrow_\$ D_{sk_i}(e_i)$
$pk \leftarrow (pk_1, \ldots, pk_n)$		$e_i \leftarrow_\$ E_{pk_i}(c_i)$	**if** $c_i = \diamond$
$sk \leftarrow (sk_1, \ldots, sk_n)$	**return** $e = (e_1, \ldots, e_n)$		**return** $\diamond$
return (pk, sk)		$m \leftarrow \text{Dec}(c_1 \cdots c_n)$
		return m

Fig. 7. The k-bit PKE scheme $\Pi' = (K', E', D')$ built from a 1-bit PKE scheme $\Pi = (K, E, D)$ and a (k, n)-coding scheme (Enc, Dec)

general was introduced by Canetti *et al.* [6] to deal with the artificial strictness of full CCA security. Roughly, RCCA security weakens full CCA security by potentially allowing an attacker to maul a ciphertext into one that decrypts to the identical message. The SD-CCA game $\mathbf{G}_b^{\text{sd-cca}}$ can be easily modified to a new game $\mathbf{G}_b^{\text{sd-rcca}}$, which behaves as $\mathbf{G}_b^{\text{sd-cca}}$, except that it outputs test whenever $D_{sk}(e') \in \{m_0, m_1\}$ for a decryption query e'.

The reason that Π achieves only replayable SD-CCA security is that given any ciphertext e, an attacker can replace the first component of e by a fresh encryption of a randomly chosen bit and thereby obtain, with probability $1/2$, a ciphertext $e' \neq e$ that decrypts to the same message as e. In [6], the authors provide generic ways to achieve full CCA security from replayable CCA security. As shown in subsequent work [12] to this paper, these techniques can also be applied in the context of SD-CCA security.[13]

4 Continuous Non-malleability against $\mathcal{F}_{\text{copy}}$

In this section, we describe a code that is adaptively continuously non-malleable w.r.t. $\mathcal{F}_{\text{copy}}$. In the full version of this paper [14], we also provide a code secure w.r.t. to an extension $\mathcal{F}'_{\text{copy}}$ of $\mathcal{F}_{\text{copy}}$ that allows bit-flips as well.

The transition from continuous to *adaptive* continuous non-malleability w.r.t. $\mathcal{F}_{\text{copy}}$ is achieved generically:

Theorem 3. *If a coding scheme* (Enc, Dec) *is continuously* $(\mathcal{F}_{\text{copy}}, \varepsilon, 1, q)$-*non-malleable, it is also continuously* $(\mathcal{F}_{\text{copy}}, 2\ell\varepsilon + \frac{q\ell}{2^k}, \ell, q)$-*non-malleable, for all* $\ell, q \in \mathbb{N}$.

The proof of Theorem 3 also appears in the full version of this paper. It remains to construct a continuously non-malleable code that is secure against tampering with a single encoding, which we do below.

Continuous Non-malleability for Single Encoding. The code is based on a linear error-correcting secret-sharing (LECSS). The use of a LECSS is inspired by the work of [24], who proposed a (non-continuous) non-malleable

---

[13] SD-CCA is called IND-SDA security in [12].

code against bit-wise tampering based on a LECSS and, additionally, a so-called AMD-code, which essentially handles bit-flips. As we do not need to provide non-malleability against bit-flips, using only the LECSS is sufficient for our purposes. The following definition is taken from [24]:[14]

Definition 5 (LECSS code). *A* (k,n)-*coding scheme* (Enc, Dec) *is a* (d,t)-*linear error-correcting secret-sharing (LECSS) code if the following properties hold:*

- LINEARITY: *For all* $c \in \{0,1\}^n$ *such that* $\text{Dec}(c) \neq \bot$, *all* $\delta \in \{0,1\}^n$, *we have*

$$\text{Dec}(c \oplus \delta) = \begin{cases} \bot & \text{if } \text{Dec}(\delta) = \bot \\ \text{Dec}(c) \oplus \text{Dec}(\delta) & \text{otherwise.} \end{cases}$$

- DISTANCE d: *For all* $c' \in \{0,1\}^n$ *with Hamming weight* $0 < w_H(c') < d$, *we have* $\text{Dec}(c') = \bot$.
- SECRECY t: *For any fixed* $x \in \{0,1\}^k$, *the bits of* $\text{Enc}(x)$ *are individually uniform and* t-*wise independent (over the randomness in the encoding).*

It turns out that a LECSS code is already continuously non-malleable with respect to $\mathcal{F}_{\text{copy}}$:

Theorem 4. *Assume that* (Enc, Dec) *is a* (t,d)-*LECSS* (k,n)-*code for* $d > n/4$ *and* $d > t$. *Then* (Enc, Dec) *is* $(\mathcal{F}_{\text{copy}}, \varepsilon, 1, q)$-*continuously non-malleable for all* $q \in \mathbb{N}$ *and*

$$\varepsilon = 2^{-(t-1)} + \left(\frac{t}{n(d/n - 1/4)^2}\right)^{t/2}.$$

For brevity, we write $\mathcal{F}_{\text{set}}$ for $\mathcal{F}_{\text{copy}}^{(1)}$ below, with the idea that the tampering functions in $\mathcal{F}_{\text{copy}}^{(1)}$ only allow to keep a bit or to set it to 0 or to 1. More formally, a function $f \in \mathcal{F}_{\text{set}}$ can be characterized by a vector $\chi(f) = (f_1, \ldots, f_n)$ where $f_i \in \{\text{zero}, \text{one}, \text{keep}\}$, with the meaning that f takes as input a codeword c and outputs a codeword $c' = c'_1 \cdots c'_n$ in which each bit is either set to 0 (zero), set to 1 (one), or left unchanged (keep).

For the proof of Theorem 4, fix $q \in \mathbb{N}$ and some distinguisher **D**. For the remainder of this section, let $\mathcal{F} := \mathcal{F}_{\text{set}}$, $\mathbf{S}_{\mathcal{F}}^{\text{real}} := \mathbf{S}_{\mathcal{F},1,q}^{\text{real}}$ and $\mathbf{S}_{\mathcal{F},\tau}^{\text{simu}} := \mathbf{S}_{\mathcal{F},\tau,1,q}^{\text{simu}}$ (for a simulator τ to be determined). For a tamper query $f \in \mathcal{F}$ with $\chi(f) = (f_1, \ldots, f_n)$ issued by **D**, let $A(f) := \{i \mid f_i \in \{\text{zero}, \text{one}\}\}$, $B(f) := \{i \mid f_i \in \{\text{keep}\}\}$, and $a(f) := |A(f)|$. Moreover, let $\text{val}(\text{zero}) := 0$ and $\text{val}(\text{one}) := 1$. Queries f with $0 \leq a(f) \leq t$, $t < a(f) < n - t$, and $n - t \leq a(f) \leq n$ are called *low queries*, *middle queries*, and *high queries*, respectively.

Handling Middle Queries. Consider the hybrid system **H** that proceeds as $\mathbf{S}_{\mathcal{F}}^{\text{real}}$, except that as soon as **D** specifies a middle query f, **H** self-destructs, i.e., answers f and all subsequent queries by $\diamond$.

---

[14] The operator $\oplus$ denotes the bit-wise XOR.

Lemma 1. $\Delta^{\mathbf{D}}(\mathbf{S}_{\mathcal{F}}^{\text{real}}, \mathbf{H}) \leq \frac{1}{2^t} + \left(\frac{t}{n(d/n - 1/4)^2}\right)^{t/2}$.

Proof. Define a *successful* middle query to be a middle query that does not decode to $\diamond$. On both systems $\mathbf{S}_{\mathcal{F}}^{\text{real}}$ and $\mathbf{H}$, one can define an MBO $\mathcal{B}$ (cf. Section 2.2) that is provoked if and only if the *first* middle query is successful and the self-destruct has not been provoked up to that point.

Clearly, $\mathbf{S}_{\mathcal{F}}^{\text{real}}$ and $\mathbf{H}$ behave identically until MBO $\mathcal{B}$ is provoked, thus $\hat{\mathbf{S}}_{\mathcal{F}}^{\text{real}} \overset{g}{\equiv} \hat{\mathbf{H}}$, and

$$\Delta^{\mathbf{D}}(\mathbf{S}_{\mathcal{F}}^{\text{real}}, \mathbf{H}) \leq \Gamma^{\mathbf{D}}(\hat{\mathbf{S}}_{\mathcal{F}}^{\text{real}}).$$

Towards bounding $\Gamma^{\mathbf{D}}(\hat{\mathbf{S}}_{\mathcal{F}}^{\text{real}})$, note first that adaptivity does not help in provoking $\mathcal{B}$: For any distinguisher $\mathbf{D}$, there exists a *non-adaptive* distinguisher $\mathbf{D}'$ with

$$\Gamma^{\mathbf{D}}(\hat{\mathbf{S}}_{\mathcal{F}}^{\text{real}}) \leq \Gamma^{\mathbf{D}'}(\hat{\mathbf{S}}_{\mathcal{F}}^{\text{real}}). \tag{6}$$

$\mathbf{D}'$ proceeds as follows: First, it (internally) interacts with $\mathbf{D}$ only. Initially, it stores the message x output by $\mathbf{D}$ internally. Whenever $\mathbf{D}$ outputs a low query, $\mathbf{D}'$ answers with x. Whenever $\mathbf{D}$ outputs a high query $f = (f_1, \ldots, f_n)$, $\mathbf{D}'$ checks whether there exists a codeword c^* that agrees with f in positions i where $f_i \in \{\text{zero}, \text{one}\}$. If it exists, it answers with $\text{Dec}(c^*)$, otherwise with $\diamond$. As soon as $\mathbf{D}$ specifies a middle query, $\mathbf{D}'$ stops its interaction with $\mathbf{D}$ and sends x and all the queries to $\hat{\mathbf{S}}_{\mathcal{F}}^{\text{real}}$.

To prove (6), fix all randomness in experiment $\mathbf{D}'\mathbf{S}_{\mathcal{F}}^{\text{real}}$, i.e., the coins of $\mathbf{D}$ (inside $\mathbf{D}'$) and the randomness of the encoding (inside $\mathbf{S}_{\mathcal{F}}^{\text{real}}$). Suppose $\mathbf{D}$ would provoke $\mathcal{B}$ in the direct interaction with $\mathbf{S}_{\mathcal{F}}^{\text{real}}$. In that case all the answers by $\mathbf{D}'$ are equal to the answers by $\mathbf{S}_{\mathcal{F}}^{\text{real}}$. This is due to the fact that the distance of the LECSS is $d > t$; a successful low query must therefore result in the original message x and a successful high query in $\text{Dec}(c^*)$. Thus, whenever $\mathbf{D}$ provokes $\mathcal{B}$, $\mathbf{D}'$ provokes it as well.

It remains to analyze the success probability of non-adaptive distinguishers $\mathbf{D}'$. Fix the coins of $\mathbf{D}'$; this determines the tamper queries. Suppose there is at least one middle case, as otherwise $\mathcal{B}$ is trivially not provoked. The middle case's success probability can be analyzed as in [24, Theorem 4.1], which leads to

$$\Gamma^{\mathbf{D}'}(\hat{\mathbf{S}}_{\mathcal{F}}^{\text{real}}) \leq \frac{1}{2^t} + \left(\frac{t}{n(d/n - 1/4)^2}\right)^{t/2}$$

(recall that the MBO cannot be provoked after an unsuccessful first middle query). $\qquad\square$

Simulator. The final step of the proof consists of exhibiting a simulator τ such that $\Delta^{\mathbf{D}}(\mathbf{H}, \mathbf{S}_{\mathcal{F},\tau}^{\text{simu}})$ is small. The indistinguishability proof is facilitated by defining two hardly distinguishable systems $\mathbf{B}$ and $\mathbf{B}'$ and a wrapper system $\mathbf{W}$ such that $\mathbf{WB} \equiv \mathbf{H}$ and $\mathbf{WB}' \equiv \mathbf{S}_{\mathcal{F},\tau}^{\text{simu}}$.

System $\mathbf{B}$ works as follows: Initially, it takes a value $x \in \{0, 1\}^k$, computes an encoding $c_1 \cdots c_n \leftarrow_{\$} \text{Enc}(x)$ of it, and outputs λ (where the symbol λ indicates

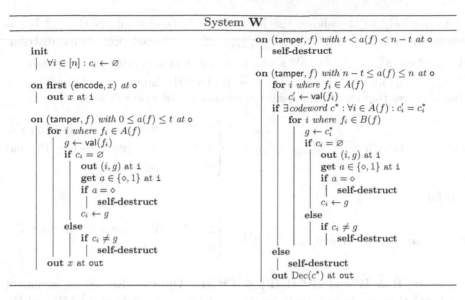

Fig. 8. The wrapper system **W**. The command **self-destruct** causes **W** to output $\diamond$ at o and to answer all future queries by $\diamond$. The symbol $\varnothing$ stands for "undefined."

an empty output). Then, it repeatedly accepts guesses $g_i = (j, b)$, where (j, b) is a guess b for c_j. If a guess g_i is correct, **B** returns $a_i = 1$. Otherwise, it outputs $a_i = \diamond$ and self-destructs (i.e., all future answers are $\diamond$). The system **B'** behaves as **B** except that the initial input x is ignored and the $c_1, \ldots, c_n$ are chosen uniformly at random and independently.

The behavior of **B** (and similarly the behavior of **B'**) is described by a sequence $(p^{\mathbf{B}}_{A^i|G^i})_{i \geq 0}$ of conditional probability distributions (cf. Section 2.2), where $p^{\mathbf{B}}_{A^i|G^i}(a^i, g^i)$ is the probability of observing the outputs $a^i = (\lambda, a_1, \ldots, a_i)$ given the inputs $g^i = (x, g_1, \ldots, g_i)$. For simplicity, assume below that g^i is such that no position is guessed twice (a generalization is straight-forward) and that a^i is of the form $\{\lambda\}\{1\}^*\{\diamond\}^*$ (as otherwise it has probability 0 anyway).

For system **B**, all i, and any g^i, $p^{\mathbf{B}}_{A^i|G^i}(a^i, g^i) = 2^{-(s+1)}$ if a^i has $s < \min(i, t)$ leading 1's; this follows from the t-wise independence of the bits of $\mathsf{Enc}(x)$. All remaining output vectors a^i, i.e., those with at least $\min(i, t)$ preceding 1's, share a probability mass of $2^{-\min(i,t)}$, in a way that depends on the code in use and on x. (It is easily verified that this yields a valid probability distribution.) The behavior of **B'** is obvious given the above (simply replace "t" by "n" in the above description).

Lemma 2. $\Delta^{\mathrm{D}}(\mathbf{B}, \mathbf{B}') \leq 2^{-t}$.

Proof. On both systems **B** and **B'**, one can define an MBO $\mathcal{B}$ that is zero as long as *less* than t positions have been guessed correctly. In the following, $\hat{\mathbf{B}}$ and $\hat{\mathbf{B}}'$ denote **B** and **B'** with the MBO, respectively.

Analogously to the above, the behavior of $\hat{\mathbf{B}}$ (and similarly that of $\hat{\mathbf{B}}'$) is described by a sequence $(\mathsf{p}^{\hat{\mathbf{B}}}_{A^i,B_i=0|G^i})_{i\geq 0}$ of conditional probability distributions, where $\mathsf{p}^{\hat{\mathbf{B}}}_{A^i,B_i=0|G^i}(a^i,g^i)$ is the probability of observing the outputs $a^i = (\lambda, a_1, \ldots, a_i)$ and $b_0 = b_1 = \ldots = b_i = 0$ given the inputs $g^i = (x, g_1, \ldots, g_i)$. One observes that due to the t-wise independence of $\mathrm{Enc}(x)$'s bits, for $i < t$,

$$\mathsf{p}^{\hat{\mathbf{B}}}_{A^i,B_i=0|G^i}(a^i,g^i) = \mathsf{p}^{\hat{\mathbf{B}}'}_{A^i,B_i=0|G^i}(a^i,g^i) = \begin{cases} 2^{-(s+1)} & \text{if } a^i \text{ has } s < i \text{ leading 1's,} \\ 2^{-i} & \text{if } a^i \text{ has } i \text{ leading 1's, and} \\ 0 & \text{otherwise,} \end{cases}$$

and for $i \geq t$,

$$\mathsf{p}^{\hat{\mathbf{B}}}_{A^i,B_i=0|G^i}(a^i,g^i) = \mathsf{p}^{\hat{\mathbf{B}}'}_{A^i,B_i=0|G^i}(a^i,g^i) = \begin{cases} 2^{-(s+1)} & \text{if } a^i \text{ has } s < t \text{ leading 1's,} \\ 0 & \text{otherwise.} \end{cases}$$

Therefore, $\hat{\mathbf{B}} \overset{g}{\equiv} \hat{\mathbf{B}}'$ and $\Delta^{\mathbf{D}}(\mathbf{B}, \mathbf{B}') \leq \Gamma^{\mathbf{D}}(\hat{\mathbf{B}}')$. Observe that by an argument similar to the one above, adaptivity does not help in provoking the MBO of $\hat{\mathbf{B}}'$. Thus, $\Gamma^{\mathbf{D}}(\hat{\mathbf{B}}') \leq 2^{-t}$, since an optimal non-adaptive strategy simply tries to guess distinct positions. □

Recall that the purpose of the wrapper system $\mathbf{W}$ is to emulate $\mathbf{H}$ and $\mathbf{S}^{\mathsf{simu}}_{\mathcal{F},\tau}$ using $\mathbf{B}$ and $\mathbf{B}'$, respectively. The key point is to note that low queries f can be answered knowing only the positions $A(f)$ of $\mathrm{Enc}(x)$, high queries knowing only the positions in $B(f)$, and middle queries can always be rejected. A full description of $\mathbf{W}$ can be found in Figure 8. It has an outside interface o and an inside interface i; at the latter interface, $\mathbf{W}$ expects to be connected to either $\mathbf{B}$ or $\mathbf{B}'$.

Lemma 3. $\mathbf{WB} \equiv \mathbf{H}$.

Proof. Since the distance of the LECSS is $d > t$, the following holds: A low query results in same if all injected positions match the corresponding bits of the encoding, and in ◇ otherwise. Similarly, for a high query, there can be at most one codeword that matches the injected positions. If such a codeword c^* exists, the outcome is $\mathrm{Dec}(c^*)$ if the bits in the keep-positions match c^*, and otherwise ◇. By inspection, it can be seen that $\mathbf{W}$ acts accordingly. □

Consider now the system $\mathbf{WB}'$. Due to the nature of $\mathbf{B}'$, the behavior of $\mathbf{WB}'$ is independent of the value x that is initially encoded. This allows to easily design a simulator τ as required by Definition 3. A full description of τ can be found in Figure 9.

Lemma 4. *The simulator τ of Figure 9 satisfies $\mathbf{WB}' \equiv \mathbf{S}^{\mathsf{simu}}_{\mathcal{F},\tau}$.*

Proof. Consider the systems $\mathbf{WB}'$ and $\mathbf{S}^{\mathsf{simu}}_{\mathcal{F},\tau}$. Both internally choose uniform and independent bits $c_1, \ldots, c_n$. System $\mathbf{WB}'$ answers low queries with the value x

Simulator τ	
init $\quad \mid \forall i \in [n] : c_i \leftarrow_\$ \{0,1\}$ **on** $(1,f)$ *with* $0 \leq a(f) \leq t$ $\quad$ **if** $\forall i \in A(f) : \mathsf{val}(f_i) = c_i$ $\quad\quad \mid$ **return** same $\quad$ **else** $\quad\quad \mid$ **return** $\diamond$	**on** $(1,f)$ *with* $t < a(f) < n-t$ $\quad \mid$ **return** $\diamond$ **on** $(1,f)$ *with* $n-t \leq a(f) \leq n$ $\quad$ **for** i where $f_i \in A(f)$ $\quad\quad \mid c_i' \leftarrow \mathsf{val}(f_i)$ $\quad$ **for** i where $f_i \in B(f)$ $\quad\quad \mid c_i' \leftarrow c_i$ $\quad c' \leftarrow c_1' \cdots c_n'$ $\quad$ **return** $\mathsf{Dec}(c')$

Fig. 9. The simulator τ

initially encoded if all injected positions match the corresponding random bits and with $\diamond$ otherwise. Simulator τ returns same in the former case, which $\mathbf{S}_{\mathcal{F},\tau}^{\mathsf{simu}}$ replaces by x, and $\diamond$ in the latter case.

Note that the answer by $\mathbf{WB}'$ to a high query f always matches $\mathsf{Dec}(c_1' \cdots c_n')$, where for $i \in A(f)$, $c_i' = \mathsf{val}(f_i)$, and for $i \in B(f)$, $c_i' = c_i$: If no codeword c^* matching the injected positions exists, then $\mathsf{Dec}(c_1' \cdots c_n') = \diamond$, which is also what $\mathbf{WB}'$ outputs. If such c^* exists and $c_i^* = c_i$ for all $i \in B(f)$, the output of $\mathbf{WB}'$ is $\mathsf{Dec}(c_1' \cdots c_n')$. If there exists an $i \in B(f)$ with $c_i^* \neq c_i$, $\mathbf{WB}'$ outputs $\diamond$, and in this case $\mathsf{Dec}(c_1' \cdots c_n') = \diamond$ since the distance of the LECSS is $d > t$. $\quad\square$

The proof of Theorem 4 now follows from a simple triangle inequality.

Proof (of Theorem 4). From Lemmas 1, 2, 3, and 4, one obtains that for all distinguishers $\mathbf{D}$,

$$\Delta^{\mathbf{D}}(\mathbf{S}_{\mathcal{F}}^{\mathsf{real}}, \mathbf{S}_{\mathcal{F},\tau}^{\mathsf{simu}}) \leq \Delta^{\mathbf{D}}(\mathbf{S}_{\mathcal{F}}^{\mathsf{real}}, \mathbf{H}) + \underbrace{\Delta^{\mathbf{D}}(\mathbf{H}, \mathbf{WB})}_{=0}$$

$$+ \underbrace{\Delta^{\mathbf{D}}(\mathbf{WB}, \mathbf{WB}')}_{=\Delta^{\mathbf{DW}}(\mathbf{B},\mathbf{B}')} + \underbrace{\Delta^{\mathbf{D}}(\mathbf{WB}', \mathbf{S}_{\mathcal{F},\tau}^{\mathsf{simu}})}_{=0}$$

$$\leq 2^{-t} + \left(\frac{t}{n(d/n - 1/4)^2} \right)^{t/2} + 2^{-t}$$

$$\leq 2^{-(t-1)} + \left(\frac{t}{n(d/n - 1/4)^2} \right)^{t/2} .$$

$\square$

5   On the Necessity of Self-destruct

In this section we show that no (k,n)-coding scheme (Enc, Dec) can achieve (even non-adaptive, i.e. for $\ell = 1$) continuous non-malleability against $\mathcal{F}_{\mathsf{copy}}$ without self-destruct. This fact is reminiscent of the negative result by Gennaro et al. [30], and was already observed by Faust et al. [26] (without a proof) for the easier case of *strong* continuous non-malleability. The impossibility proof in this section assumes that Dec is deterministic and that $\text{Dec}(\text{Enc}(x)) = x$ with probability 1 for all $x \in \{0,1\}^k$ (cf. Definition 2). The distinguisher $\mathbf{D}$ provided by Theorem 5 is universal, i.e., it breaks any coding scheme (if given oracle access to its decoding algorithm).

For the remainder of this section, let $\mathcal{F} := \mathcal{F}_{\text{set}}$ (as defined in Section 4), $\mathbf{S}_{\mathcal{F}}^{\text{real}} := \mathbf{S}_{\mathcal{F},1,n}^{\text{real}}$, and $\mathbf{S}_{\mathcal{F},\tau}^{\text{simu}} := \mathbf{S}_{\mathcal{F},\tau,1,n}^{\text{simu}}$ (with some simulator τ). Moreover, both $\mathbf{S}_{\mathcal{F}}^{\text{real}}$ and $\mathbf{S}_{\mathcal{F},\tau}^{\text{simu}}$ are stripped of the self-destruct mode.

Theorem 5. *There exists a distinguisher $\mathbf{D}$ such that for all coding schemes* (Enc, Dec) *and all simulators τ,*

$$\Delta^{\mathbf{D}}(\mathbf{S}_{\mathcal{F}}^{\text{real}}, \mathbf{S}_{\mathcal{F},\tau}^{\text{simu}}) \geq 1 - \frac{n+1}{2^k}.$$

The corollary below states no pair of converters $(\text{encode}, \text{decode})$ can achieve the constructive statement corresponding to Theorem 2 without relying on the self-destruct feature.

Corollary 1. *For any protocol* $\text{nmc} := (\text{encode}, \text{decode})$ *and all simulators σ, if both converters are stateless and*

$$[\overset{1\text{-bit}}{-\diamond\!\!\!\twoheadrightarrow\!\!\bullet}]^n \overset{((\text{encode},\text{decode}),\sigma,(0,\varepsilon))}{\Longrightarrow} \overset{k\text{-bit}}{-\diamond\!\!\!\twoheadrightarrow\!\!\bullet},$$

then,

$$\varepsilon \geq 1 - \frac{n+1}{2^k}.$$

Proof. Note that the protocol achieves perfect availability and thus constitutes a perfectly correct (k,n)-coding scheme (since the converters are stateless and with perfect correctness, decode can w.l.o.g. be assumed to be deterministic). Consider an arbitrary simulator σ. It can be converted into a simulator τ as required by Definition 3 in a straight-forward manner. Similarly, there exists a straight-forward reduction $\mathbf{C}$ such that

$$\mathbf{C}(\text{encode}^A \text{decode}^B [\overset{1\text{-bit},1,n}{-\diamond\!\!\!\twoheadrightarrow\!\!\bullet}]^n) \equiv \mathbf{S}_{\mathcal{F}}^{\text{real}} \quad \text{and} \quad \mathbf{C}(\sigma^E \overset{k\text{-bit},1,n}{-\diamond\!\!\!\twoheadrightarrow\!\!\bullet}) \equiv \mathbf{S}_{\mathcal{F},\tau}^{\text{simu}}.$$

Thus, $\mathbf{DC}$ achieves advantage $1 - \frac{n+1}{2^k}$. $\qquad\square$

5.1 Proof of Theorem 5

Distinguisher $\mathbf{D} := \mathbf{D}_{\text{Ext}}$ uses an algorithm Ext that always extracts the encoded message when interacting with system $\mathbf{S}_{\mathcal{F}}^{\text{real}}$ and does so with small probability only when interacting with system $\mathbf{S}_{\mathcal{F},\tau}^{\text{simu}}$ (for any simulator).

The Extraction Algorithm. Consider the following algorithm Ext, which repeatedly issues tamper queries (tamper, f) with $f \in \mathcal{F}_{\text{set}}$, expects an answer in $\{0,1\}^k \cup \{\diamond, \text{same}\}$, and eventually outputs a value $x' \in \{0,1\}^k$: Initially, it initializes variables $f_1, \ldots, f_n \leftarrow \varnothing$ (where the value $\varnothing$ stands for "undefined"). Then, for $i = 1, \ldots, n$ it proceeds as follows: It queries (tamper, f) with $\chi(f) = (f_1, \ldots, f_{i-1}, \text{zero}, \text{keep}, \ldots, \text{keep})$. If the answer is same, it sets $f_i \leftarrow \text{zero}$ and otherwise $f_i \leftarrow \text{one}$. In the end Ext outputs $x' \leftarrow \text{Dec}(\text{val}(f_1) \cdots \text{val}(f_n))$.

The Distinguisher. Consider the following distinguisher $\mathbf{D}_{\mathsf{Ext}}$: Initially, it chooses $x \leftarrow \{0,1\}^k$ and outputs (encode, x) to the system it is connected to. Then, it lets Ext interact with that system, replacing an answer by same whenever it is x. When Ext terminates and outputs a value x', $\mathbf{D}_{\mathsf{Ext}}$ outputs 1 if $x' = x$ and 0 otherwise.

Lemma 5. $\mathsf{P}[\mathbf{D}_{\mathsf{Ext}}\mathbf{S}_{\mathcal{F}}^{\mathsf{real}} = 1] = 1$.

Proof. Assume that before the ith iteration of Ext, asking the query (tamper, f) with $\chi(f) = (f_1, \ldots, f_{i-1}, \mathsf{keep}, \mathsf{keep}, \ldots, \mathsf{keep})$ to $\mathbf{S}_{\mathcal{F}}^{\mathsf{real}}$ yields the answer x. From this it follows that either $(f_1, \ldots, f_{i-1}, \mathsf{zero}, \mathsf{keep}, \ldots, \mathsf{keep})$ or $(f_1, \ldots, f_{i-1}, \mathsf{one}, \mathsf{keep}, \ldots, \mathsf{keep})$ leads to the answer x; Ext sets f_i appropriately (the fact that the answer x is replaced by same plays no role here). Thus, in the end, computing $\mathsf{Dec}(\mathsf{val}(f_1) \cdots \mathsf{val}(f_n))$ yields x. □

In other words, Lemma 5 means that Ext always succeeds at recovering the value x chosen by $\mathbf{D}$. Showing that this happens only with small probability when $\mathbf{D}_{\mathsf{Ext}}$ interacts with $\mathbf{S}_{\mathcal{F},\tau}^{\mathsf{simu}}$ completes the proof.

Lemma 6. $\mathsf{P}[\mathbf{D}_{\mathsf{Ext}}\mathbf{S}_{\mathcal{F},\tau}^{\mathsf{simu}} = 1] \leq \frac{n+1}{2^k}$.

Proof. Consider the following modified distinguisher $\hat{\mathbf{D}}_{\mathsf{Ext}}$ that works as $\mathbf{D}_{\mathsf{Ext}}$ except that it does *not* modify the answers received by the system it is connected to. Moreover, let $\hat{\mathbf{S}}_{\mathcal{F},\tau}^{\mathsf{simu}}$ be the the system that ignores all encode-queries and handles queries (tamper, f) by invoking $\tau(1, f)$ and outputting τ's answer.

Note that in both experiments, Ext's view is identical unless it causes τ to output x (the value encoded by $\mathbf{D}$), which happens with probability at most $\frac{n}{2^k}$. Thus,

$$|\mathsf{P}^{\mathbf{D}_{\mathsf{Ext}}\mathbf{S}_{\mathcal{F},\tau}^{\mathsf{simu}}}[\mathsf{Ext} \text{ outputs } x] - \mathsf{P}^{\hat{\mathbf{D}}_{\mathsf{Ext}}\hat{\mathbf{S}}_{\mathcal{F},\tau}^{\mathsf{simu}}}[\mathsf{Ext} \text{ outputs } x]| \leq \frac{n}{2^k}.$$

Furthermore, in experiment $\hat{\mathbf{D}}_{\mathsf{Ext}}\hat{\mathbf{S}}_{\mathcal{F},\tau}^{\mathsf{simu}}$, Ext's view is independent of x, and therefore, x is output by Ext with probability $\frac{1}{2^k}$. The claim follows. □

Acknowledgments. We thank Yevgeniy Dodis for insightful discussions and for suggesting many improvements to the paper. We thank Joël Alwen and Daniel Tschudi for helpful discussions, in particular on the impossibility proof in Section 5. The work was supported by the Swiss National Science Foundation (SNF), project no. 200020-132794. Björn Tackmann is supported by the Swiss National Science Foundation (SNF).

References

1. Aggarwal, D., Dodis, Y., Kazana, T., Obremski, M.: Non-malleable reductions and applications. Cryptology ePrint Archive, Report 2014/821 (2014), http://eprint.iacr.org/
2. Aggarwal, D., Dodis, Y., Lovett, S.: Non-malleable codes from additive combinatorics. In: STOC, pp. 774–783 (2014)

3. Aggarwal, D., Dziembowski, S., Kazana, T., Obremski, M.: Leakage-resilient non-malleable codes. Cryptology ePrint Archive, Report 2014/807 (2014). To appear in TCC (2015), http://eprint.iacr.org/

4. Agrawal, S., Gupta, D., Maji, H.K., Pandey, O., Prabhakaran, M.: Explicit non-malleable codes resistant to permutations and perturbations. Cryptology ePrint Archive, Report 2014/841 (2014), http://eprint.iacr.org/

5. Agrawal, S., Gupta, D., Maji, H.K., Pandey, O., Prabhakaran, M.: Explicit optimal-rate non-malleable codes against bit-wise tampering and permutations. Cryptology ePrint Archive, Report 2014/842, 2014. To appear in TCC (2015), http://eprint.iacr.org/

6. Canetti, R., Krawczyk, H., Nielsen, J.B.: Relaxing chosen-ciphertext security. In: Boneh, D. (ed.) CRYPTO 2003. LNCS, vol. 2729, pp. 565–582. Springer, Heidelberg (2003)

7. Chattopadhyay, E., Zuckerman, D.: Non-malleable codes against constant split-state tampering. In: FOCS, pp. 306–315 (2014)

8. Cheraghchi, M., Guruswami, V.: Capacity of non-malleable codes. In: ITCS, pp. 155–168 (2014)

9. Cheraghchi, M., Guruswami, V.: Non-malleable coding against bit-wise and split-state tampering. In: Lindell, Y. (ed.) TCC 2014. LNCS, vol. 8349, pp. 440–464. Springer, Heidelberg (2014)

10. Choi, S.G., Dachman-Soled, D., Malkin, T., Wee, H.M.: Black-box construction of a non-malleable encryption scheme from any semantically secure one. In: Canetti, R. (ed.) TCC 2008. LNCS, vol. 4948, pp. 427–444. Springer, Heidelberg (2008)

11. Choi, S.G., Kiayias, A., Malkin, T.: BiTR: Built-in tamper resilience. In: Lee, D.H., Wang, X. (eds.) ASIACRYPT 2011. LNCS, vol. 7073, pp. 740–758. Springer, Heidelberg (2011)

12. Coretti, S., Dodis, Y., Tackmann, B., Venturi, D.: Self-destruct non-malleability. IACR Cryptology ePrint Archive 2014, 866 (2014)

13. Coretti, S., Maurer, U., Tackmann, B.: Constructing confidential channels from authenticated channels - public-key encryption revisited. In: Sako, K., Sarkar, P. (eds.) ASIACRYPT 2013, Part I. LNCS, vol. 8269, pp. 134–153. Springer, Heidelberg (2013)

14. Coretti, S., Maurer, U., Tackmann, B., Venturi, D.: From single-bit to multi-bit public-key encryption via non-malleable codes. IACR Cryptology ePrint Archive 2014, 324 (2014)

15. Cramer, R., Dodis, Y., Fehr, S., Padró, C., Wichs, D.: Detection of algebraic manipulation with applications to robust secret sharing and fuzzy extractors. In: Smart, N.P. (ed.) EUROCRYPT 2008. LNCS, vol. 4965, pp. 471–488. Springer, Heidelberg (2008)

16. Cramer, R., Hanaoka, G., Hofheinz, D., Imai, H., Kiltz, E., Pass, R., Shelat, A., Vaikuntanathan, V.: Bounded CCA2-secure encryption. In: Kurosawa, K. (ed.) ASIACRYPT 2007. LNCS, vol. 4833, pp. 502–518. Springer, Heidelberg (2007)

17. Cramer, R., Shoup, V.: A practical public key cryptosystem provably secure against adaptive chosen ciphertext attack. In: Krawczyk, H. (ed.) CRYPTO 1998. LNCS, vol. 1462, pp. 13–25. Springer, Heidelberg (1998)

18. Dachman-Soled, D.: A black-box construction of a CCA2 encryption scheme from a plaintext aware encryption scheme. In: Krawczyk, H. (ed.) PKC 2014. LNCS, vol. 8383, pp. 37–55. Springer, Heidelberg (2014)

19. Dachman-Soled, D., Liu, F.-H., Shi, E., Zhou, H.-S.: Locally decodable and up-datable non-malleable codes and their applications. Cryptology ePrint Archive, Report 2014/663, 2014. To appear in TCC (2015), http://eprint.iacr.org/

20. Damgård, I., Faust, S., Mukherjee, P., Venturi, D.: Bounded tamper resilience: How to go beyond the algebraic barrier. In: Sako, K., Sarkar, P. (eds.) ASIACRYPT 2013, Part II. LNCS, vol. 8270, pp. 140–160. Springer, Heidelberg (2013)
21. Damgård, I., Faust, S., Mukherjee, P., Venturi, D.: The chaining lemma and its application. Cryptology ePrint Archive, Report 2014/979 (2014), http://eprint.iacr.org/
22. Diffie, W., Hellman, M.E.: New directions in cryptography. IEEE Transactions on Information Theory 22(6), 644–654 (1976)
23. Dziembowski, S., Kazana, T., Obremski, M.: Non-malleable codes from two-source extractors. In: Canetti, R., Garay, J.A. (eds.) CRYPTO 2013, Part II. LNCS, vol. 8043, pp. 239–257. Springer, Heidelberg (2013)
24. Dziembowski, S., Pietrzak, K., Wichs, D.: Non-malleable codes. In: ICS, pp. 434–452 (2010)
25. El Gamal, T.: A public key cryptosystem and a signature scheme based on discrete logarithms. In: Blakely, G.R., Chaum, D. (eds.) CRYPTO 1984. LNCS, vol. 196, pp. 10–18. Springer, Heidelberg (1985)
26. Faust, S., Mukherjee, P., Nielsen, J.B., Venturi, D.: Continuous non-malleable codes. In: Lindell, Y. (ed.) TCC 2014. LNCS, vol. 8349, pp. 465–488. Springer, Heidelberg (2014)
27. Faust, S., Mukherjee, P., Nielsen, J.B., Venturi, D.: A tamper and leakage resilient von Neumann architecture. IACR Cryptology ePrint Archive, 2014:338, 2014. To appear in PKC (2015)
28. Faust, S., Mukherjee, P., Venturi, D., Wichs, D.: Efficient non-malleable codes and key-derivation for poly-size tampering circuits. In: Nguyen, P.Q., Oswald, E. (eds.) EUROCRYPT 2014. LNCS, vol. 8441, pp. 111–128. Springer, Heidelberg (2014)
29. Fujisaki, E., Okamoto, T., Pointcheval, D., Stern, J.: RSA-OAEP is secure under the RSA assumption. In: Kilian, J. (ed.) CRYPTO 2001. LNCS, vol. 2139, pp. 260–274. Springer, Heidelberg (2001)
30. Gennaro, R., Lysyanskaya, A., Malkin, T., Micali, S., Rabin, T.: Algorithmic tamper-proof (ATP) security: Theoretical foundations for security against hardware tampering. In: Naor, M. (ed.) TCC 2004. LNCS, vol. 2951, pp. 258–277. Springer, Heidelberg (2004)
31. Gertner, Y., Malkin, T., Myers, S.: Towards a separation of semantic and CCA security for public key encryption. In: Vadhan, S.P. (ed.) TCC 2007. LNCS, vol. 4392, pp. 434–455. Springer, Heidelberg (2007)
32. Goldwasser, S., Micali, S.: Probabilistic encryption. J. Comput. Syst. Sci. 28(2), 270–299 (1984)
33. Hofheinz, D., Kiltz, E.: Practical chosen ciphertext secure encryption from factoring. In: Joux, A. (ed.) EUROCRYPT 2009. LNCS, vol. 5479, pp. 313–332. Springer, Heidelberg (2009)
34. Hohenberger, S., Lewko, A., Waters, B.: Detecting dangerous queries: A new approach for chosen ciphertext security. In: Pointcheval, D., Johansson, T. (eds.) EUROCRYPT 2012. LNCS, vol. 7237, pp. 663–681. Springer, Heidelberg (2012)
35. Jafargholi, Z., Wichs, D.: Tamper detection and continuous non-malleable codes. Cryptology ePrint Archive, Report 2014/956, 2014. To appear in TCC (2015), http://eprint.iacr.org/
36. Kiltz, E., Pietrzak, K., Stam, M., Yung, M.: A new randomness extraction paradigm for hybrid encryption. In: Joux, A. (ed.) EUROCRYPT 2009. LNCS, vol. 5479, pp. 590–609. Springer, Heidelberg (2009)

37. Kohlweiss, M., Maurer, U., Onete, C., Tackmann, B., Venturi, D.: Anonymity-preserving public-key encryption: A constructive approach. In: De Cristofaro, E., Wright, M. (eds.) PETS 2013. LNCS, vol. 7981, pp. 19–39. Springer, Heidelberg (2013)

38. Lin, H., Tessaro, S.: Amplification of chosen-ciphertext security. In: Johansson, T., Nguyen, P.Q. (eds.) EUROCRYPT 2013. LNCS, vol. 7881, pp. 503–519. Springer, Heidelberg (2013)

39. Liu, F.-H., Lysyanskaya, A.: Tamper and leakage resilience in the split-state model. In: Safavi-Naini, R., Canetti, R. (eds.) CRYPTO 2012. LNCS, vol. 7417, pp. 517–532. Springer, Heidelberg (2012)

40. Maurer, U.: Constructive cryptography - a new paradigm for security definitions and proofs. In: Mödersheim, S., Palamidessi, C. (eds.) TOSCA 2011. LNCS, vol. 6993, pp. 33–56. Springer, Heidelberg (2012)

41. Maurer, U.: Conditional equivalence of random systems and indistinguishability proofs. In: 2013 IEEE International Symposium on Information Theory Proceedings (ISIT), pp. 3150–3154 (2013)

42. Maurer, U., Renner, R.: Abstract cryptography. In: ICS, pp. 1–21 (2011)

43. Maurer, U., Rüedlinger, A., Tackmann, B.: Confidentiality and integrity: A constructive perspective. In: Cramer, R. (ed.) TCC 2012. LNCS, vol. 7194, pp. 209–229. Springer, Heidelberg (2012)

44. Maurer, U., Schmid, P.: A calculus for security bootstrapping in distributed systems. Journal of Computer Security 4(1), 55–80 (1996)

45. Maurer, U., Tackmann, B.: On the soundness of authenticate-then-encrypt: formalizing the malleability of symmetric encryption. In: ACM Conference on Computer and Communications Security, pp. 505–515 (2010)

46. Maurer, U.M.: Indistinguishability of random systems. In: Knudsen, L.R. (ed.) EUROCRYPT 2002. LNCS, vol. 2332, pp. 110–132. Springer, Heidelberg (2002)

47. Myers, S., Sergi, M., Shelat, A.: Blackbox construction of a more than non-malleable CCA1 encryption scheme from plaintext awareness. In: Visconti, I., De Prisco, R. (eds.) SCN 2012. LNCS, vol. 7485, pp. 149–165. Springer, Heidelberg (2012)

48. Myers, S., Shelat, A.: Bit encryption is complete. In: FOCS, pp. 607–616 (2009)

49. Peikert, C., Waters, B.: Lossy trapdoor functions and their applications. SIAM J. Comput. 40(6), 1803–1844 (2011)

50. Rivest, R.L., Shamir, A., Adleman, L.M.: A method for obtaining digital signatures and public-key cryptosystems (reprint). Commun. ACM 26(1), 96–99 (1983)

51. Rosen, A., Segev, G.: Chosen-ciphertext security via correlated products. SIAM J. Comput. 39(7), 3058–3088 (2010)

Constructing and Understanding Chosen Ciphertext Security via Puncturable Key Encapsulation Mechanisms

Takahiro Matsuda and Goichiro Hanaoka

Research Institute for Secure Systems (RISEC),
National Institute of Advanced Industrial Science and Technology (AIST), Japan
{t-matsuda,hanaoka-goichiro}@aist.go.jp

Abstract. In this paper, we introduce and study a new cryptographic primitive that we call *puncturable key encapsulation mechanism* (PKEM), which is a special class of KEMs that satisfy some functional and security requirements that, combined together, imply chosen ciphertext security (CCA security). The purpose of introducing this primitive is to capture certain common patterns in the security proofs of the several existing CCA secure public key encryption (PKE) schemes and KEMs based on general cryptographic primitives which (explicitly or implicitly) use the ideas and techniques of the Dolev-Dwork-Naor (DDN) construction (STOC'91), and "break down" the proofs into smaller steps, so that each small step is easier to work with/verify/understand than directly tackling CCA security.

To see the usefulness of PKEM, we show (1) how several existing constructions of CCA secure PKE/KEM constructed based on general cryptographic primitives can be captured as a PKEM, which enables us to understand these constructions via a unified framework, (2) their connection to detectable CCA security (Hohenberger et al. EUROCRYPT'12), and (3) a new security proof for a KEM-analogue of the DDN construction from a set of assumptions: *sender non-committing encryption* (SNCE) and non-interactive witness indistinguishable proofs.

Then, as our main technical result, we show how to construct a PKEM satisfying our requirements (and thus a CCA secure KEM) from a new set of general cryptographic primitives: *SNCE* and *symmetric key encryption secure for key-dependent messages* (KDM secure SKE). Our construction realizes the "decrypt-then-re-encrypt"-style validity check of a ciphertext which is powerful but in general has a problem of the circularity between a plaintext and a randomness. We show how SNCE and KDM secure SKE can be used together to overcome the circularity. We believe that the connection among three seemingly unrelated notions of encryption primitives, i.e. CCA security, the sender non-committing property, and KDM security, to be of theoretical interest.

Keywords: public key encryption, puncturable key encapsulation mechanism, chosen ciphertext security, sender non-committing encryption, key-dependent message secure symmetric-key encryption.

1   Introduction

In this paper, we continue a long line of work studying the constructions of public key encryption (PKE) schemes and its closely related primitive called *key encapsulation*

Y. Dodis and J.B. Nielsen (Eds.): TCC 2015, Part I, LNCS 9014, pp. 561–590, 2015.

mechanism (KEM) that are secure against chosen ciphertext attacks (CCA) [53,57,24] from general cryptographic primitives. CCA secure PKE/KEM is one of the most important cryptographic primitives that has been intensively studied in the literature, due to not only its implication to strong and useful security notions such as non-malleability [24] and universal composability [16], but also its resilience and robustness against practical attacks such as Bleichenbacher's attack [12].

There have been a number of works that show CCA secure PKE/KEMs from general cryptographic primitives: These include trapdoor permutations [24,30,31] (with some enhanced property [32]), identity-based encryption [19] and a weaker primitive called tag-based encryption [43,40], lossy trapdoor function [56] and trapdoor functions with weaker functionality/security properties [59,49,41,61], PKE with weaker than but close to CCA security [38,42,21], a combination of chosen plaintext secure (CPA secure) PKE and a hash function with some strong security [48], and techniques from program obfuscation [60,47].

One of the ultimate goals of this line of researches is to clarify whether one can construct CCA secure PKE only from CPA secure one (and in fact, a partial negative result is known [29]). This problem is important from both theoretical and practical points of view. To obtain insights into this problem, clarifying new classes of primitives that serve as building blocks is considered to be important, because those new class of primitives can be a new target that we can try constructing from CPA secure PKE schemes (or other standard primitives such as one-way injective trapdoor functions and permutations).

Our Motivation. Although differing in details, the existing constructions of CCA secure PKE schemes and KEMs from general cryptographic primitives [24,56,59,61,47,48,21] often employ the ideas and techniques of the Dolev-Dwork-Naor (DDN) construction [24], which is the first construction of CCA secure PKE from general primitives. The security proofs of these constructions are thus similar in a large sense, and it is highly likely that not a few future attempts to constructing CCA secure PKE/KEMs from general cryptographic primitives will also follow the DDN-style construction and security proof. Therefore, it will be useful and helpful for future research and also for understanding the existing works of this research direction if we can extract and abstract the common ideas and techniques behind the security proofs of the original DDN and the existing DDN-like constructions, and formalize them as a cryptographic primitive with a few formal functionality and security requirements (rather than heuristic ideas and techniques), so that most of the existing DDN-style constructions as well as potential future constructions are captured/explained/understood in a unified way, and in particular these are more accessible and easier-to-understand.

Our Contributions. Based on the motivation mentioned above, in this paper, we introduce and study a new cryptographic primitive that we call *puncturable key encapsulation mechanism* (PKEM). This is a class of KEMs that has two kinds of decryption procedures, and it is required to satisfy three simple security requirements, *decapsulation soundness*, *punctured decapsulation soundness*, and *extended CPA security* which we show in Section 3.3 that, combined together, implies CCA security. The intuition of these security notions as well as their formal definitions are explained in Section 3.2. The purpose of introducing this primitive is to capture certain common patterns in the security proofs of the several existing CCA secure PKE schemes and KEMs based on general cryptographic primitives which (explicitly or implicitly) use the ideas and techniques of the DDN construction [24], and "break down" the proofs into smaller steps, so that each small step is easier to work with/verify/understand than directly tackling CCA security. Our formalization of PKEM is inspired (and in some sense can be seen an extension of) the notion of *puncturable tag-based encryption* [48] (which is in turn inspired by the notion of *puncturable pseudorandom function* [60]), and we explain the difference from [48] in the paragraph *"Related Work"* below.

To see the usefulness of our framework of PKEM, we show (1) how the KEM-analogue of the original DDN [24] and several existing DDN-like constructions (e.g. [56,59,61,47,48]) can be understood as a PKEM in Section 3.4, (2) its connection to detectable CCA security which is a weaker security notion than CCA security introduced by Hohenberger et al. [38] in Section 3.5, and (3) a new security proof for a KEM-analogue of the DDN construction from a set of assumptions that are different from the one used in its known security proof: *sender non-committing encryption* (SNCE, see below) and non-interactive witness indistinguishable proofs. (For the purpose of exposition, this last result is shown in Section 5.)

Then, as our main technical result, in Section 4 we show how to construct a PKEM satisfying our requirements (and thus a CCA secure KEM) from a new set of general cryptographic primitives: *SNCE* and *symmetric key encryption secure for key-dependent messages* (KDM secure SKE) [11]. Roughly speaking, a SNCE scheme is a special case of non-committing encryption [18] and is a PKE scheme which is secure even if the sender's randomness used to generate the challenge ciphertext is corrupted by an adversary. See Section 2.1 where we define SNCE formally, explain the difference among related primitives, and how it can be realized from the standard cryptographic assumptions such as the decisional Diffie-Hellman (DDH), quadratic residuosity (QR), and decisional composite residuosity (DCR). The function class with respect to which we require the building block SKE scheme to be KDM secure, is a class of efficiently computable functions whose running time is a-priori fixed. Due to Applebaum's result [1,3] (and its efficient variant [6, §7.2]) we can realize a KDM secure SKE scheme satisfying our requirement from standard assumptions such as DDH, QR, DCR. For more details on KDM secure SKE, see Section 2.2.

Our proposed PKEM has a similalriy with the "double-layered" construction of Myers and Shelat [51] and its variants [38,45,21], in which a plaintext is encrypted twice: firstly by the "inner" scheme, and secondly by "outer" scheme. Strictly speaking, however, our construction is not purely double-layered, but in some sense is closer to "hybrid encryption" of a PKE (seen as a KEM) and a SKE schemes, much similarly to the recent constructions by Matsuda and Hanaoka [47,48]. Furthermore, our construction realizes the "decrypt-then-re-encrypt"-style validity check of a ciphertext, which is a powerful approach that has been adopted in several existing constructions that construct CCA secure PKE/KEM from general cryptographic primitives [27,56,59,51,41,38,47,48,21]. In general, however, this approach has a problem of the circularity between a plaintext and a randomness, and previous works avoid such a circularity using a random oracle [27], a trapdoor function [56,59,41], a PKE scheme which achieves some security which is (weaker than but) close to CCA security [51,38,21], or a power of additional building blocks with (seemingly very strong) security properties [47,48]. We show how SNCE and KDM secure SKE can be used together to overcome the circularity. Compared with the structurally similar constructions [38,47,48,21], the assumptions on which our construction is based could be seen weak, in the sense that the building blocks are known to be realizable from fairly standard computational assumptions such as the DDH, QR, and DCR assumptions. We believe that the connection among three seemingly unrelated notions of encryption primitives, i.e. CCA security, the sender non-committing property, and KDM security, to be of theoretical interest.

Open Problems. We believe that our framework of PKEM is useful for constructing and understanding the current and the potential future constructions of CCA secure PKE/KEMs based on the DDN-like approach, and motivates further studies on it. Our work leaves several open problems. Firstly, our framework of PKEM actually does not capture the recent construction by Dachman-Soled [21] who constructs a CCA secure PKE scheme from a PKE scheme that satisfies (standard model) plaintext awareness and some simulatability property. The construction in [21] is similar to our proposed (P)KEM in Section 4 and the recent similar constructions [47,48]. (Technically, to capture it in the language of PKEM, slight relaxations of some of the security requirements will be necessary, due to its double-layered use of PKE schemes similarly to [51].)

Secondly and perhaps more importantly, it will be worth clarifying whether it is possible to construct a PKEM satisfying our requirements only from CPA secure PKE or (an enhanced variant of) trapdoor permutations in a black-box manner. Note that a negative answer to this question will also give us interesting insights, as it shows that to construct a CCA secure PKE/KEM from these standard primitives, we have to essentially avoid the DDN-like construction.

Finally, it would also be interesting to find applications of a PKEM other than CCA secure PKE/KEMs.

Related Work. The notion of CCA security for PKE was formalized by Naor and Yung [53] and Rackoff and Simon [57]. We have already listed several existing constructions of CCA secure PKE/KEMs from general primitives in the second paragraph of Introduction. In our understanding, the works [24,56,59,61,47,48,21] are based on the ideas and techniques from the DDN construction [24].

As mentioned above, our notion of PKEM is inspired by the notion of *puncturable tag-based encryption* (PTBE) that was recently introduced by Matsuda and Hanaoka [48]. Similarly to PKEM, PTBE is a special kind of tag-based encryption [43,40] with two modes of decryption. (Roughly, in PKEM, a secret key can be punctured by a ciphertext, but in PTBE, a secret key is punctured by a tag.) Matsuda and Hanaoka [48] introduced PTBE as an abstraction of the "core" structure that appears in the original DDN construction (informally, it is the original DDN construction without a one-time signature scheme and a non-interactive zero-knowledge proof), and they use it to mainly reduce the "description complexity" of their proposed construction [48] and make it easier to understand the construction. However, they did not study it as a framework for capturing and understanding the existing DDN-style constructions (as well as potential future constructions) in a unified manner as we do in this paper. We note that Matsuda and Hanaoka [48] also formalized the security requirement called eCPA *security* whose formalization is a

PTBE-analogue of eCPA security for a PKEM (and thus we borrow the name). However, they did not formalize the security notions for PTBE that correspond to *decapsulation soundness* and *punctured decapsulation soundness* for a PKEM.

Paper Organization. The rest of the paper is organized as follows: In Section 2 and (in Appendix A), we review the notation and definitions of cryptographic primitives. In Section 3, we introduce and study PKEM, where in particular we show its implication to CCA security and how some of the existing constructions of KEMs can be interpreted and explained as a PKEM. In Section 4, we show our main technical result: a PKEM from SNCE and KDM secure SKE, which by the result in Section 3 yields a new CCA secure KEM from general assumptions. In Section 5, we show the CCA security of the DDN-KEM based on SNCE and non-interactive witness indistinguishable arguments.

2 Preliminaries

In this section, we give the definitions for sender non-committing encryption (SNCE) and symmetric key encryption (SKE) and its key-dependent message (KDM) security that are used in our main result in Section 4. The definitions for standard cryptographic primitives are given in Appendix A, which include PKE, KEMs, signature schemes, non-interactive argument systems, and universal one-way hash functions (UOWHFs). (The reader familiar with them need not check Appendix A at the first read, and can do so when he/she wants to check the details of the definitions.)

Basic Notation. $\mathbb{N}$ denotes the set of all natural numbers, and for $n \in \mathbb{N}$, we define $[n] := \{1, \ldots, n\}$. "$x \leftarrow y$" denotes that x is chosen uniformly at random from y if y is a finite set, x is output from y if y is a function or an algorithm, or y is assigned to x otherwise. If x and y are strings, then "$|x|$" denotes the bit-length of x, "$x\|y$" denotes the concatenation x and y, and "$(x \stackrel{?}{=} y)$" is the operation which returns 1 if $x = y$ and 0 otherwise. "PPTA" stands for a *probabilistic polynomial time algorithm*. For a finite set S, "$|S|$" denotes its size. If $\mathcal{A}$ is a probabilistic algorithm then "$y \leftarrow \mathcal{A}(x; r)$" denotes that $\mathcal{A}$ computes y as output by taking x as input and using r as randomness. $\mathcal{A}^{\mathcal{O}}$ denotes an algorithm $\mathcal{A}$ with oracle access to $\mathcal{O}$. A function $\epsilon(\cdot) : \mathbb{N} \to [0, 1]$ is said to be *negligible* if for all positive polynomials $p(\cdot)$ and all sufficiently large $k \in \mathbb{N}$, we have $\epsilon(k) < 1/p(k)$. Throughout this paper, we use the character "k" to denote a security parameter.

2.1 Sender Non-committing Public Key Encryption

Roughly, a SNCE scheme is a PKE scheme that remains secure even against an adversary who may obtain sender's randomness used to generate the challenge ciphertext. This security is ensured by requiring that there be an algorithm that generates a "fake transcript" pk and c that denote a public key and a ciphertext, respectively, so that the pair (pk, c) can be later explained as a transcript of an arbitrary message m. Our syntax of SNCE loosely follows that of sender-equivocable encryption [26,39], but departs from it because we need perfect correctness (or at least almost-all-keys-perfect correctness [25]) so that error-less decryption is guaranteed, which cannot be achieved

$$
\begin{array}{l|l|l}
\underline{\text{Expt}^{\text{SNC-Real}}_{\Pi,\mathcal{A}}(k):} & \underline{\text{Expt}^{\text{SNC-Sim}}_{\Pi,\mathcal{A}}(k):} & \underline{\text{Expt}^{\text{OTKDM}}_{E,\mathcal{F},\mathcal{A}}(k):} \\
(m,\mathsf{st}) \leftarrow \mathcal{A}_1(1^k) & (m,\mathsf{st}) \leftarrow \mathcal{A}_1(1^k) & (f,\mathsf{st}) \leftarrow \mathcal{A}_1(1^k) \\
(pk,sk) \leftarrow \mathsf{PKG}(1^k) & (pk,c,\omega) \leftarrow \mathsf{Fake}(1^k) & K \leftarrow \mathcal{K}_k \\
r \leftarrow \mathcal{R}_k & r \leftarrow \mathsf{Explain}(\omega,m) & m_1 \leftarrow f(K); \ m_0 \leftarrow \mathcal{M}_k \\
c \leftarrow \mathsf{Enc}(pk,m;r) & b' \leftarrow \mathcal{A}_2(\mathsf{st},pk,c,r) & b \leftarrow \{0,1\} \\
b' \leftarrow \mathcal{A}_2(\mathsf{st},pk,c,r) & \text{Return } b'. & c^* \leftarrow \mathsf{SEnc}(K,m_b) \\
\text{Return } b'. & & b' \leftarrow \mathcal{A}_2(\mathsf{st},c^*) \\
& & \text{Return } (b' \stackrel{?}{=} b).
\end{array}
$$

Fig. 1. Security experiments for defining the SNC security of a SNCE scheme (left and center) and that for the $\mathcal{F}$-OTKDM security of a SKE scheme (right)

by sender-equivocable encryption. We also note that recently, Hazay and Patra [35] introduced (among other notions) the notion that they call *NCE for the Sender* (NCES), which is a notion very close to SNCE we consider here. We will discuss the correctness and the difference between our definition and that of [35] later in this subsection.

Formally, a sender non-committing (public key) encryption (SNCE) scheme Π consists of the five PPTAs (PKG, Enc, Dec, Fake, Explain) where (PKG, Enc, Dec) constitutes a PKE scheme (where definitions for ordinary PKE can be found in Appendix A), and Fake and Explain are the simulation algorithms with the following syntax:

Fake: This is the "fake transcript" generation algorithm that takes 1^k as input, and outputs a "fake" public key/ciphertext pair (pk,c) and a corresponding state information ω (that will be used in the next algorithm).

Explain: This is the (deterministic) "explanation" algorithm that takes a state information ω (where ω is computed by $(pk,c,\omega) \leftarrow \mathsf{Fake}(1^k)$) and a plaintext m as input, and outputs a randomness r that "explains" the transcript (pk,c) corresponding to ω. Namely, it is required that $\mathsf{Enc}(pk,m;r) = c$ hold.

SNC *Security.* For a SNCE scheme $\Pi = (\mathsf{PKG},\mathsf{Enc},\mathsf{Dec},\mathsf{Fake},\mathsf{Explain})$ (where the randomness space of Enc is $\mathcal{R} = (\mathcal{R}_k)_{k\in\mathbb{N}}$) and an adversary $\mathcal{A} = (\mathcal{A}_1,\mathcal{A}_2)$, we define the SNC-Real experiment $\text{Expt}^{\text{SNC-Real}}_{\Pi,\mathcal{A}}(k)$ and the SNC-Sim experiment $\text{Expt}^{\text{SNC-Sim}}_{\Pi,\mathcal{A}}(k)$ as in Fig. 1 (left and center, respectively).

Definition 1. *We say that a SNCE scheme Π is* SNC *secure if for all PPTAs $\mathcal{A}$,* $\text{Adv}^{\text{SNC}}_{\Pi,\mathcal{A}}(k) := |\Pr[\text{Expt}^{\text{SNC-Real}}_{\Pi,\mathcal{A}}(k) = 1] - \Pr[\text{Expt}^{\text{SNC-Sim}}_{\Pi,\mathcal{A}}(k) = 1]|$ *is negligible.*

The Difference among Non-committing Encryption and Related Primitives. The original definition of non-committing encryption by Canetti et al. [18] ensures security under both the sender and receiver's corruption. This is ensured by requiring that the "explaining" algorithm output not only the sender's randomness but also receiver's (i.e. randomness used to generate public/secret keys). The original definition in [18] (and several works [23,28]) allows multi-round interaction between a sender and a receiver (and even the multi-party case), but in this paper we only consider the public-key case (equivalently, the one-round two-party protocol case). A SNCE scheme is a non-committing encryption scheme that only takes care of the sender's side corruption.

Sender-equivocable encryption [26,39] is a special case of a SNCE scheme in which a sender can, under an honestly generated public key, generate a fake ciphertext that can be later explained as an encryption of an arbitrary message (while a SNCE scheme allows that even a public key is a fake one).

Deniable encryption [17,54,10,60] has an even stronger property in which an honestly generated ciphertext under an honestly generated public key can be later explained as an encryption of an arbitrary message. For details on deniable encryption, we refer the reader to the papers [54,10].

The difference among these primitives is very important in our paper, as we explain below.

On Correctness of SNCE Schemes. In this paper, unlike most of the papers that treat (sender) non-committing encryption schemes and related primitives such as sender-equivocable encryption and deniable encryption, we require a SNCE scheme satisfy perfect correctness or at least almost-all-keys perfect correctness [25]. This is because our proposed constructions follow the Dolev-Dwork-Naor-style construction [24] which requires error-less decryption (under all but negligible fraction of key pairs) for a building block PKE scheme. Here, the non-committing property and (perfect or almost-all-keys perfect) correctness might sound contradicting. This is indeed the case for ordinary (i.e. bi-) and "receiver" non-committing encryption, sender-equivocable encryption, and deniable encryption, and thus we cannot use these primitives in our proposed constructions. However, "sender" non-committing encryption can avoid such an incompatibility, because the fake transcript generation algorithm Fake can generate (pk, c) such that pk is *not* in the range of the normal key generation algorithm PKG. Moreover, as we will see below, SNC secure SNCE schemes with perfect correctness (and even practical efficiency) can be realized from standard assumptions.

Concrete Instantiations of SNCE Schemes. Bellare et al. [8] formalized the notion of *lossy encryption* [8], which is a PKE scheme that has the "lossy key generation" algorithm. It outputs a "lossy public key" which is indistinguishable from a public key generated by the ordinary key generation algorithm, and an encryption under a lossy public key statistically hides the information of a plaintext. Bellare et al. [8] also introduced an additional property for lossy encryption called *efficient openability*, in which the lossy key generation algorithm outputs a trapdoor in addition to a lossy public key, and by using the trapdoor, an encryption under the lossy public key can be efficiently "explained" as a ciphertext of any plaintext.

We note that any lossy encryption with efficient openability yields a SNC secure SNCE scheme: the algorithm Fake generates a lossy public key pk as well as an encryption c of some plaintext, and keeps the trapdoor corresponding to pk as ω.; the algorithm Explain on input ω and a plaintext m outputs a randomness r that explains that $c = \mathsf{Enc}(pk, m; r)$ holds. Hence, we can use the existing lossy encryption schemes with efficient openability that are based on standard assumptions. These include the scheme based on the quadratic residuosity (QR) assumption [8, § 4.4] (which is essentially the multi-bit version of the Goldwasser-Micali scheme [33]), the scheme based on the decisional Diffie-Hellman (DDH) assumption [9, § 5.4] (which is the "bit-wise" encryption version of the DDH-based lossy encryption scheme [8, § 4.1]), and the scheme based on the decisional composite residuosity (DCR) assumption [36] (which shows that the original Paillier scheme [55] and the Damgård-Jurik scheme [22] can be extended to lossy encryption with efficient openability). In particular, the DCR-based schemes [55,22,36] have a compact ciphertext whose size does not grow linearly in the length of plaintexts.

On the Difference from the Formalization of "NCE for the Sender" in [35]. The definition of NCE for the Sender in [35] explicitly requires that the scheme have the "fake" key generation algorithm that outputs a "fake" public key together with a trapdoor, with which one can "equivocate" (or in our terminology, "explain") any ciphertext as an encryption of arbitrary plaintext m. Therefore, it seems to us that their formalization is close to lossy encryption with efficient openability [8]. On the other hand, our formalization requires that only a pair (pk, c) of public key and a ciphertext (or a "transcript" in a one-round message transmission protocol between two parties) be explained. We can construct a SNCE scheme in our formalization from NCE for the Sender of [35] (in essentially the same manner as we do so from lossy encryption with efficient openability), while we currently do not know if the converse implication can be established. Therefore, in the sense that currently an implication of only one direction is known, our formalization is weaker.

Some Useful Facts. For our result in Section 4, it is convenient to consider the so-called "repetition construction," in which a plaintext is encrypted multiple times by independently generated public keys.

More specifically, given a SNCE scheme $\Pi = (\mathsf{PKG}, \mathsf{Enc}, \mathsf{Dec}, \mathsf{Fake}, \mathsf{Explain})$, the n-wise repetition construction $\Pi^n = (\mathsf{PKG}^n, \mathsf{Enc}^n, \mathsf{Dec}^n, \mathsf{Fake}^n, \mathsf{Explain}^n)$ is defined as follows: The key generation algorithm PKG^n runs $(pk_i, sk_i) \leftarrow \mathsf{PKG}$ for $i \in [n]$ and returns public key $PK = (pk_i)_{i \in [n]}$ and secret key $SK = (sk_i)_{i \in [n]}$.; The encryption algorithm Enc^n, on input PK and a plaintext m, runs $c_i \leftarrow \mathsf{Enc}(pk_i, m; r_i)$ for $i \in [n]$ (where each r_i is an independently chosen randomness), and outputs a ciphertext $C = (c_i)_{i \in [n]}$.; The decryption algorithm Dec^n, on input SK and C, runs $m_i \leftarrow \mathsf{Dec}(sk_i, c_i)$ for $i \in [n]$, and returns m_1 if every m_i is equal or $\perp$ otherwise.; The fake transcript generation algorithm Fake^n runs $(pk_i, c_i, \omega_i) \leftarrow \mathsf{Fake}(1^k)$ for $n \in [n]$, and returns $PK = (pk_i)_{i \in [n]}$, $C = (c_i)_{i \in [n]}$, and a state information $W = (\omega_i)_{i \in [n]}$.; The explanation algorithm $\mathsf{Explain}^n$, on input W and m, runs $r_i \leftarrow \mathsf{Explain}(\omega_i, m)$ for $i \in [n]$, and returns the randomness $R = (r_i)_{i \in [n]}$ that explains that $C = \mathsf{Enc}^n(PK, m; R)$.

By a straightforward hybrid argument, we can show that for any polynomial $n = n(k) > 0$, if the underlying scheme Π is SNC secure, then so is the n-wise repetition construction Π^n. (It is also a well-known fact that if Π is CPA secure, then so is Π^n.)

We also note that the plaintext space of an SNCE scheme can be easily extended by considering the straightforward "concatenation construction," in which plaintext $m = (m_1, \ldots, m_n)$ is encrypted block-wise by independently generated public keys.

More formal statements regarding the repetition construction and the concatenation constructions are given in the full version.

2.2 Symmetric Key Encryption

A symmetric key encryption (SKE) scheme E with key space $\mathcal{K} = \{\mathcal{K}_k\}_{k \in \mathbb{N}}$ and plaintext space $\mathcal{M} = \{\mathcal{M}_k\}_{k \in \mathbb{N}}$[1] consists of the following two PPTAs ($\mathsf{SEnc}, \mathsf{SDec}$):

---

[1] In this paper, for simplicity, we assume that the key space $\mathcal{K}$ and plaintext space $\mathcal{M}$ of a SKE scheme satisfy the following conditions: For each $k \in \mathbb{N}$, (1) every element in $\mathcal{K}_k$ has

SEnc: The encryption algorithm that takes a key $K \in \mathcal{K}_k$ and a plaintext $m \in \mathcal{M}_k$ as input, and outputs a ciphertext c.

SDec: The (deterministic) decryption algorithm that takes $K \in \mathcal{K}_k$ and c as input, and outputs a plaintext m which could be the special symbol $\perp$ (which indicates that c is an invalid ciphertext under K).

Correctness. We require for all $k \in \mathbb{N}$, all keys $K \in \mathcal{K}_k$, and all plaintexts $m \in \mathcal{M}_k$, it holds that $\mathsf{SDec}(K, \mathsf{SEnc}(K, m)) = m$.

One-Time Key-Dependent Message Security. Let $E = (\mathsf{SEnc}, \mathsf{SDec})$ be a SKE scheme with key space $\mathcal{K} = \{\mathcal{K}_k\}_{k \in \mathbb{N}}$ and plaintext space $\mathcal{M} = \{\mathcal{M}_k\}_{k \in \mathbb{N}}$. Let $\mathcal{F} = \{\mathcal{F}_k\}_{k \in \mathbb{N}}$ be an ensemble (which we call *function ensemble*) where for each k, $\mathcal{F}_k$ is a set of efficiently computable functions with their domain $\mathcal{K}_k$ and range $\mathcal{M}_k$.

For the SKE scheme E, the function ensemble $\mathcal{F}$, and an adversary $\mathcal{A} = (\mathcal{A}_1, \mathcal{A}_2)$, we define the $\mathcal{F}$-OTKDM experiment $\mathsf{Expt}^{\mathsf{OTKDM}}_{E, \mathcal{F}, \mathcal{A}}(k)$ as in Fig. 1 (right). In the experiment, it is required that $f \in \mathcal{F}_k$.

Definition 2. *We say that a SKE scheme E is* OTKDM *secure with respect to $\mathcal{F}$ ($\mathcal{F}$-OTKDM secure, for short) if for all PPTAs $\mathcal{A}$,* $\mathsf{Adv}^{\mathsf{OTKDM}}_{E, \mathcal{F}, \mathcal{A}}(k) := 2 \cdot | \Pr[\mathsf{Expt}^{\mathsf{OTKDM}}_{E, \mathcal{F}, \mathcal{A}}(k) = 1] - 1/2|$ *is negligible.*

We would like to remark that our definition of OTKDM security is considerably weak: it is a single instance definition that need not take into account the existence of other keys, and an adversary is allowed to make a KDM encryption query (which is captured by f) only once.

Concrete Instantiations of OTKDM *Secure SKE Schemes.* In our proposed construction in Section 4, the class of functions with respect to which a SKE scheme is OTKDM secure needs to be rich enough to be able to compute the algorithm Explain in a SNCE scheme multiple (an a-priori bounded number of) times. Fortunately, Applebaum [1] showed how to generically convert any SKE scheme which is many-time KDM secure (i.e. secure for many KDM encryption queries) with respect to "projections" (i.e. functions each of whose output bit depends on at most one bit of inputs) into a SKE scheme which is many-time KDM secure (and thus OTKDM secure), with respect to a family of functions computable in a-priori fixed polynomial time. (We can also use a more efficient construction shown by Bellare et al. [6, §7.2].) This notion is sufficient for our proposed construction. Since most SKE and PKE schemes KDM secure with respect to the class of affine functions can be interpreted as (or easily converted to) "projection"-KDM secure SKE schemes [3, §A], we can use the existing (many-time) "affine"-KDM secure SKE schemes as a building block, and apply Applebaum's conversion (or that of [6, §7.2]). Therefore, for example, one can realize a OTKDM secure SKE scheme with respect to fixed poly-time computable functions, based on the DDH assumption [13], the QR assumption [15], the DCR assumption [15,44], the learning with errors (LWE) assumption [4], and the learning parity with noise (LPN) assumption [4,2]. Very recently,

---

the same length, (2) every element in $\mathcal{M}_k$ has the same length, (3) both $\mathcal{K}_k$ and $\mathcal{M}_k$ are efficiently recognizable, and (4) we can efficiently sample a uniformly random element from both $\mathcal{K}_k$ and $\mathcal{M}_k$.

Bellare et al. [5] introduced a notion of a family of hash function called *universal computational extractor* (UCE) which is seemingly quite strong (almost random oracle-like) but a standard model assumption, and then they showed (among many other things) how to construct a SKE scheme which is non-adaptively KDM secure (in which encryption queries have to be made in parallel) with respect to any efficiently computable functions. OTKDM security is the special case of non-adaptive KDM security, and hence we can also use the result of [5] in our proposed construction.

3   Chosen Ciphertext Security from Puncturable KEMs

In this section, we introduce the notion of a *puncturable KEM* (PKEM) and show several results on it.

This section is organized as follows: In Sections 3.1 and 3.2, we define the syntax and the security requirements of a PKEM, respectively. Then in Sections 3.3 and 3.5, we show the implication of a PKEM to a CCA secure KEM and a DCCA secure detectable KEM, respectively. We also explain how a wide class of the existing constructions of CCA secure KEMs can be understood via a PKEM in Section 3.4.

3.1   Syntax

Informally, a PKEM is a KEM that has additional procedures for "puncturing secret keys according to a ciphertext" and "punctured decapsulation." In a PKEM, one can generate a "punctured" secret key $\widehat{sk}_{c^*}$ from an ordinary sk and a ciphertext c^* via the "puncturing" algorithm Punc. Intuitively, although an ordinary secret key sk defines a map (via Decap) whose domain is the whole of the ciphertext space, $\widehat{sk}_{c^*}$ only defines a map whose domain is the ciphertext space that has a "hole" produced by the puncture of the ciphertext c^*. This "punctured" secret key $\widehat{sk}_{c^*}$ can be used in the "punctured" decapsulation algorithm PDecap to decapsulate all ciphertexts that are "far" from c^* (or, those that are not in the "hole" produced by c^*), while $\widehat{sk}_{c^*}$ is useless for decapsulating ciphertexts that are "close" to c^* (or, those that are in the "hole" including c^* itself), where what it means for a ciphertext to be close to/far from c^* is decided according to a publicly computable predicate F, which is also a part of a PKEM.

Formally, a puncturable KEM consists of the six PPTAs (KKG, Encap, Decap, F, Punc, PDecap), where (KKG, Encap, Decap) constitute a KEM, and the latter three algorithms are deterministic algorithms with the following interface:

F: The predicate that takes a public key pk (output by $KKG(1^k)$) and two ciphertexts c and c' as input, where c has to be in the range of $Encap(pk)$ (but c' need not), and outputs 0 or 1.

Punc: The "puncturing" algorithm that takes a secret key sk (output by $KKG(1^k)$) and a ciphertext c^* (output by $Encap(pk)$) as input, and outputs a punctured secret key $\widehat{sk}_{c^*}$.

PDecap: The "punctured" decapsulation algorithm that takes a punctured secret key $\widehat{sk}_{c^*}$ (output by $Punc(sk, c^*)$) and a ciphertext c as input, and outputs a session-key K which could be the special symbol $\bot$ (meaning that "c cannot be decapsulated by $\widehat{sk}_{c^*}$").

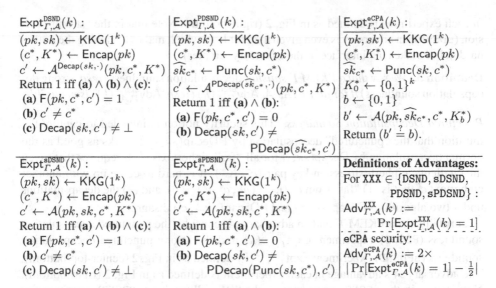

Fig. 2. Security experiments for a PKEM and the definition of an adversary's advantage in each experiment

The predicate F is used to define *decapsulation soundness* and *punctured decapsulation soundness*, which we explain in the next subsection. Its role is very similar to the predicate used to define DCCA security and unpredictability of detectable PKE in [38]. As mentioned above, intuitively, the predicate $F(pk, c^*, \cdot)$ divides the ciphertext space into two classes: ciphertexts that are "close" to c^* and those that are "far" from c^*, and for each of the classes, we expect the decapsulation algorithms Decap and PDecap to work "appropriately," as we will see below.

3.2 Security Requirements

For a PKEM, we consider the three kinds of security notions: *decapsulation soundness*, *punctured decapsulation soundness*, and *extended CPA security*. The intuition for each of the security notions as well as formal definitions are explained below. Furthermore, for the first two notions, we consider two flavors: the ordinary version and the strong version (where the latter formally implies the former). We only need the ordinary notions for showing the CCA security of a PKEM, while the strong notions are usually easier to test/prove.

Decapsulation Soundness. This security notion is intended to capture the intuition that the only valid ciphertext which is "close" to c^* is c^* itself: It requires that given the challenge ciphertext/session-key pair (c^*, K^*), it is hard to come up with another ciphertext $c' \neq c^*$ that is (1) "close" to c^* (i.e. $F(pk, c^*, c') = 1$), and (2) valid (i.e. $\mathsf{Decap}(sk, c') \neq \perp$).

Formally, for a PKEM Γ and an adversary $\mathcal{A}$, consider the decapsulation soundness (DSND) experiment $\mathsf{Expt}^{\mathsf{DSND}}_{\Gamma,\mathcal{A}}(k)$ and the strong decapsulation soundness (sDSND) experiment $\mathsf{Expt}^{\mathsf{sDSND}}_{\Gamma,\mathcal{A}}(k)$ defined as in Fig. 2 (left-top/bottom). The adversary $\mathcal{A}$'s advantage in each experiment is defined as in Fig. 2 (right-bottom). Note that in the "strong" version (sDSND), an adversary is even given a secret key (which makes achieving the notion harder, but makes the interface of the adversary simpler).

Definition 3. *We say that a PKEM Γ satisfies* decapsulation soundness *(resp.* strong decapsulation soundness*) if for all PPTAs $\mathcal{A}$, $\mathsf{Adv}_{\Gamma,\mathcal{A}}^{\mathrm{DSND}}(k)$ (resp. $\mathsf{Adv}_{\Gamma,\mathcal{A}}^{\mathrm{sDSND}}(k)$) is negligible.*

Punctured Decapsulation Soundness. This security notion is intended to capture the intuition that the "punctured" decapsulation by $\mathsf{PDecap}(\widehat{sk}_{c^*}, \cdot)$ works as good as the normal decapsulation by $\mathsf{Decap}(sk, \cdot)$ for all "far" ciphertexts c': It requires that given the challenge ciphertext/session-key pair (c^*, K^*), it is hard to come up with another ciphertext c' that is (1) "far" from c^* (i.e. $\mathsf{F}(pk, c^*, c') = 0$), and (2) the decapsulations under two algorithms $\mathsf{Decap}(sk, c')$ and $\mathsf{PDecap}(\widehat{sk}_{c^*}, c')$ disagree.

Formally, for a PKEM Γ and an adversary $\mathcal{A}$, consider the punctured decapsulation soundness (PDSND) experiment $\mathsf{Expt}_{\Gamma,\mathcal{A}}^{\mathrm{PDSND}}(k)$ and the strong punctured decapsulation soundness (sPDSND) experiment $\mathsf{Expt}_{\Gamma,\mathcal{A}}^{\mathrm{sPDSND}}(k)$ defined as in Fig. 2 (center-top/bottom). The adversary $\mathcal{A}$'s advantage in each experiment is defined as in Fig. 2 (right-bottom). Note that as in the sDSND experiment, in the "strong" version (sPDSND), an adversary is even given a secret key (which makes achieving the notion harder, but makes the interface of the adversary simpler).

Definition 4. *We say that a PKEM Γ satisfies* punctured decapsulation soundness *(resp.* strong punctured decapsulation soundness*) if for all PPTAs $\mathcal{A}$, $\mathsf{Adv}_{\Gamma,\mathcal{A}}^{\mathrm{PDSND}}(k)$ (resp. $\mathsf{Adv}_{\Gamma,\mathcal{A}}^{\mathrm{sPDSND}}(k)$) is negligible.*

Extended CPA Security: CPA security in the presence of a punctured secret key. *Extended CPA security (eCPA security, for short) requires that the CPA security hold even in the presence of the punctured secret key $\widehat{sk}_{c^*}$ corresponding to the challenge ciphertext c^*.*

Formally, for a PKEM Γ and an adversary $\mathcal{A}$, consider the eCPA experiment $\mathsf{Expt}_{\Gamma,\mathcal{A}}^{\mathrm{eCPA}}(k)$ defined as in Fig. 2 (right-top). We define the advantage of an adversary as in Fig. 2 (right-bottom).

Definition 5. *We say that a PKEM Γ is* eCPA secure *if for all PPTAs $\mathcal{A}$, $\mathsf{Adv}_{\Gamma,\mathcal{A}}^{\mathrm{eCPA}}(k)$ is negligible.*

3.3 CCA Secure KEM from a Puncturable KEM

Here, we show that a PKEM satisfying all security notions introduced in Section 3.2 yields a CCA secure KEM. (The formal proof is given in the full version.)

Theorem 1. *Let $\Gamma = (\mathsf{KKG}, \mathsf{Encap}, \mathsf{Decap}, \mathsf{F}, \mathsf{Punc}, \mathsf{PDecap})$ be a PKEM satisfying decapsulation soundness, punctured decapsulation soundness, and eCPA security. Then, $\Gamma^* = (\mathsf{KKG}, \mathsf{Encap}, \mathsf{Decap})$ is a CCA secure KEM.*

Specifically, for any PPTA $\mathcal{A}$ that attacks the CCA security of Γ^* and makes in total $Q = Q(k) > 0$ decapsulation queries, there exist PPTAs $\mathcal{B}_{\mathsf{d}}$, $\mathcal{B}_{\mathsf{a}}$, and $\mathcal{B}_{\mathsf{e}}$ such that

$$\mathsf{Adv}_{\Gamma^*,\mathcal{A}}^{\mathrm{CCA}}(k) \leq 2 \cdot \mathsf{Adv}_{\Gamma,\mathcal{B}_{\mathsf{d}}}^{\mathrm{DSND}}(k) + 2Q \cdot \mathsf{Adv}_{\Gamma,\mathcal{B}_{\mathsf{a}}}^{\mathrm{PDSND}}(k) + \mathsf{Adv}_{\Gamma,\mathcal{B}_{\mathsf{e}}}^{\mathrm{eCPA}}(k). \tag{1}$$

Proof Sketch of Theorem 1. Let $\mathcal{A}$ be any PPTA adversary that attacks the KEM Γ^* in the sense of CCA security. Consider the following sequence of games:

Game 1: This is the CCA experiment $\mathsf{Expt}^{\mathsf{CCA}}_{\Gamma^*,\mathcal{A}}(k)$ itself.

Game 2: Same as Game 1, except that all decapsulation queries c satisfying $\mathsf{F}(pk, c^*, c)$ $= 1$ are answered with $\perp$.

Game 3: Same as Game 2, except that all decapsulation queries c satisfying $\mathsf{F}(pk, c^*, c)$ $= 0$ are answered with $\mathsf{PDecap}(\widehat{sk}_{c^*}, c)$, where $\widehat{sk}_{c^*} = \mathsf{Punc}(sk, c^*)$.

For $i \in [3]$, let Succ_i denote the event that in Game i, $\mathcal{A}$ succeeds in guessing the challenge bit (i.e. $b' = b$ occurs). We will show that $|\Pr[\mathsf{Succ}_i] - \Pr[\mathsf{Succ}_{i+1}]|$ is negligible for each $i \in [2]$ and that $|\Pr[\mathsf{Succ}_3] - 1/2|$ is negligible, which proves the theorem.

Firstly, note that Game 1 and Game 2 proceed identically unless $\mathcal{A}$ makes a decapsulation query c satisfying $\mathsf{F}(pk, c^*, c') = 1$ and $\mathsf{Decap}(sk, c) \neq \perp$, and hence $|\Pr[\mathsf{Succ}_1] - \Pr[\mathsf{Succ}_2]|$ is upperbounded by the probability of $\mathcal{A}$ making such a query in Game 1 or Game 2. Recall that by the rule of the CCA experiment, $\mathcal{A}$'s queries c must satisfy $c \neq c^*$. But $\mathsf{F}(pk, c^*, c') = 1$, $c \neq c^*$, and $\mathsf{Decap}(sk, c) \neq \perp$ are exactly the conditions of violating the decapsulation soundness, and the probability of $\mathcal{A}$ making a query satisfying these conditions is negligible.

Secondly, note that Game 2 and Game 3 proceed identically unless $\mathcal{A}$ makes a decapsulation query c satisfying $\mathsf{F}(pk, c^*, c) = 0$ and $\mathsf{Decap}(sk, c) \neq \mathsf{PDecap}(\widehat{sk}_{c^*}, c)$, where $\widehat{sk}_{c^*} = \mathsf{Punc}(sk, c^*)$. Hence $|\Pr[\mathsf{Succ}_2] - \Pr[\mathsf{Succ}_3]|$ is upperbounded by the probability of $\mathcal{A}$ making such a query in Game 2 or Game 3. However, since these conditions are exactly those of violating the punctured decapsulation soundness, the probability of $\mathcal{A}$ making a query satisfying the above conditions is negligible.

Finally, we can upperbound $|\Pr[\mathsf{Succ}_3] - 1/2|$ to be negligible directly by the eCPA security of the PKEM Γ. More specifically, any eCPA adversary $\mathcal{B}_{\mathrm{e}}$, which receives $(pk, \widehat{sk}_{c^*}, c^*, K_b^*)$ as input, can simulate Game 3 for $\mathcal{A}$, where $\mathcal{A}$'s decapsulation oracle in Game 3 is simulated perfectly by using $\widehat{sk}_{c^*}$, so that $\mathcal{B}_{\mathrm{e}}$'s eCPA advantage is exactly $2 \cdot |\Pr[\mathsf{Succ}_3] - 1/2|$. This shows that $|\Pr[\mathsf{Succ}_3] - 1/2|$ is negligible.   □

On the Tightness of the Reduction. In the equation (1) of the above proof, the reason why we have the factor Q (the number of a CCA adversary $\mathcal{A}$'s decapsulation queries) in front of the advantage $\mathsf{Adv}^{\mathsf{PDSND}}_{\Gamma,\mathcal{B}_{\mathrm{a}}}(k)$ of the reduction algorithm $\mathcal{B}_{\mathrm{a}}$ attacking punctured decapsulation soundness, is that the reduction algorithm $\mathcal{B}_{\mathrm{a}}$ cannot check whether a ciphertext c' satisfies the condition (b) of violating punctured decapsulation soundness, i.e. $\mathsf{Decap}(sk, c') \neq \mathsf{PDecap}(\widehat{sk}_{c^*}, c')$, and thus $\mathcal{B}_{\mathrm{a}}$ picks one of $\mathcal{A}$'s decapsulation queries randomly. However, if we instead use a PKEM with *strong* punctured decapsulation soundness, then, when proving security, a reduction algorithm attacking *strong* punctured decapsulation soundness is given the secret key sk as input, which enables it to check whether the condition $\mathsf{Decap}(sk, c') \neq \mathsf{PDecap}(\widehat{sk}_{c^*}, c')$ is satisfied. Therefore, the reduction algorithm need not pick one of the decapsulation queries randomly, but can find a ciphertext c' that violates the conditions of strong punctured decapsulation soundness whenever the adversary $\mathcal{A}$ asks such a ciphertext as a decapsulation query, which leads to a tight security reduction. We will explain this in more details in the full version.

3.4   Understanding the Existing Constructions of CCA Secure KEMs via Puncturable KEM

To see the usefulness of a PKEM and the result in Section 3.3, here we demonstrate how the existing constructions of CCA secure KEMs can be understood via a PKEM.

The Dolev-Dwork-Naor KEM. We first show how a security proof of the KEM version of the DDN construction [24], which we call the *DDN-KEM*, can be understood via a PKEM. This is the KEM obtained from the original DDN construction (which is a PKE scheme) in which we encrypt a random value and regard it as a session-key.

Let $\Pi = (\mathsf{PKG}, \mathsf{Enc}, \mathsf{Dec})$ be a PKE scheme whose plaintext space is $\{0,1\}^k$ and whose randomness space (for security parameter k) is $\mathcal{R}_k$. Consider the NP language $L = \{L_k\}_{k \in \mathbb{N}}$ where each L_k is defined as follows:

$$L_k := \left\{ \left. ((pk_i)_{i \in [k]}, (c_i)_{i \in [k]}) \;\right|\; \begin{array}{c} \exists ((r_i)_{i \in [k]}, K) \in (\mathcal{R}_k)^k \times \{0,1\}^k \text{ s.t.} \\ \forall i \in [k] : \mathsf{Enc}(pk_i, K; r_i) = c_i \end{array} \right\}.$$

Let $\mathcal{P} = (\mathsf{CRSG}, \mathsf{Prove}, \mathsf{PVer})$ be a non-interactive argument system for the language L. Moreover, let $\Sigma = (\mathsf{SKG}, \mathsf{Sign}, \mathsf{SVer})$ and $\mathcal{H} = (\mathsf{HKG}, \mathsf{H})$ be a signature scheme and a UOWHF, respectively. (The definitions of an ordinary PKE scheme, a signature scheme, a UOWHF, and a non-interactive argument system can be found in Appendix A.) Then we construct the PKEM $\Gamma_{\mathsf{DDN}} = (\mathsf{KKG}_{\mathsf{DDN}}, \mathsf{Encap}_{\mathsf{DDN}}, \mathsf{Decap}_{\mathsf{DDN}}, \mathsf{F}_{\mathsf{DDN}}, \mathsf{Punc}_{\mathsf{DDN}}, \mathsf{PDecap}_{\mathsf{DDN}})$, which is based on the DDN-KEM, as in Fig. 3. The original DDN-KEM Γ^*_{DDN} is $(\mathsf{KKG}_{\mathsf{DDN}}, \mathsf{Encap}_{\mathsf{DDN}}, \mathsf{Decap}_{\mathsf{DDN}})$.

For the PKEM Γ_{DDN}, the three security requirements are shown as follows:

Lemma 1. *If $\mathcal{H}$ is a UOWHF and Σ is a SOT secure signature scheme, then the PKEM Γ_{DDN} satisfies strong decapsulation soundness.*

Lemma 2. *If the non-interactive argument system $\mathcal{P}$ satisfies adaptive soundness, then the PKEM Γ_{DDN} satisfies strong punctured decapsulation soundness.*

Lemma 3. *If the PKE scheme Π is CPA secure and the non-interactive argument system $\mathcal{P}$ is ZK secure, then the PKEM Γ_{DDN} is eCPA secure.*

The formal proofs of these lemmas are given in the full version, and here we give some intuitions below.

The first two lemmas are almost trivial. Specifically, let $C^* = (vk^*, (c_i^*)_i, \pi^*, \sigma^*)$ be the challenge ciphertext, and let $C' = (vk', (c_i')_i, \pi', \sigma')$ be a ciphertext output by an adversary in the sDSND experiment or the sPDSND experiment (recall that the interface of an adversary in these experiments is the same). Then, a simple observation shows that if C' is a successful ciphertext that violates strong decapsulation soundness, then C' must satisfy one of the following two conditions: (1) $\mathsf{H}_\kappa(vk^*) = \mathsf{H}_\kappa(vk')$ and $vk^* \neq vk'$, or (2) $\mathsf{SVer}(vk', ((c_i')_i, \pi'), \sigma') = \top$, $((c_i^*)_i, \pi^*, \sigma^*) \neq ((c_i')_i, \pi', \sigma')$, and $vk^* = vk'$. However, a ciphertext with the first condition is hard to find due to the security of the UOWHF $\mathcal{H}$, and a ciphertext with the second condition is hard to find due to the SOT security of the signature scheme Σ. Similarly, again a simple observation shows that in order for C' to be a successful ciphertext that violates strong punctured

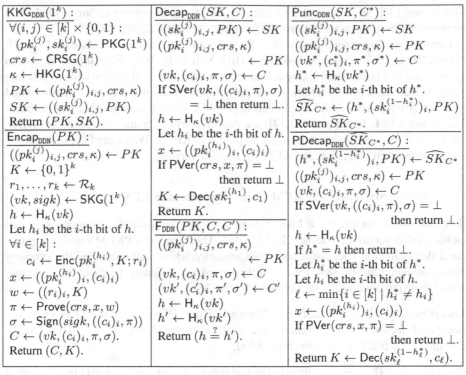

$\mathsf{KKG}_{\mathrm{DDN}}(1^k):$	$\mathsf{Decap}_{\mathrm{DDN}}(SK, C):$	$\mathsf{Punc}_{\mathrm{DDN}}(SK, C^*):$
$\forall (i,j) \in [k] \times \{0,1\}:$	$((sk_i^{(j)})_{i,j}, PK) \leftarrow SK$	$((sk_i^{(j)})_{i,j}, PK) \leftarrow SK$
$\quad (pk_i^{(j)}, sk_i^{(j)}) \leftarrow \mathsf{PKG}(1^k)$	$((pk_i^{(j)})_{i,j}, crs, \kappa)$	$((pk_i^{(j)})_{i,j}, crs, \kappa) \leftarrow PK$
$crs \leftarrow \mathsf{CRSG}(1^k)$	$\qquad\qquad\qquad\quad \leftarrow PK$	$(vk^*, (c_i^*)_i, \pi^*, \sigma^*) \leftarrow C$
$\kappa \leftarrow \mathsf{HKG}(1^k)$	$(vk, (c_i)_i, \pi, \sigma) \leftarrow C$	$h^* \leftarrow \mathsf{H}_\kappa(vk^*)$
$PK \leftarrow ((pk_i^{(j)})_{i,j}, crs, \kappa)$	If $\mathsf{SVer}(vk, ((c_i)_i, \pi), \sigma)$	Let h_i^* be the i-th bit of h^*.
$SK \leftarrow ((sk_i^{(j)})_{i,j}, PK)$	$\quad = \bot$ then return $\bot$.	$\widehat{SK}_{C^*} \leftarrow (h^*, (sk_i^{(1-h_i^*)})_i, PK)$
Return (PK, SK).	$h \leftarrow \mathsf{H}_\kappa(vk)$	Return $\widehat{SK}_{C^*}$.
$\underline{\mathsf{Encap}_{\mathrm{DDN}}(PK):}$	Let h_i be the i-th bit of h.	$\underline{\mathsf{PDecap}_{\mathrm{DDN}}(\widehat{SK}_{C^*}, C):}$
$((pk_i^{(j)})_{i,j}, crs, \kappa) \leftarrow PK$	$x \leftarrow ((pk_i^{(h_i)})_i, (c_i)_i)$	$(h^*, (sk_i^{(1-h_i^*)})_i, PK) \leftarrow \widehat{SK}_{C^*}$
$K \leftarrow \{0,1\}^k$	If $\mathsf{PVer}(crs, x, \pi) = \bot$	$((pk_i^{(j)})_{i,j}, crs, \kappa) \leftarrow PK$
$r_1, \ldots, r_k \leftarrow \mathcal{R}_k$	$\qquad$ then return $\bot$	$(vk, (c_i)_i, \pi, \sigma) \leftarrow C$
$(vk, sigk) \leftarrow \mathsf{SKG}(1^k)$	$K \leftarrow \mathsf{Dec}(sk_1^{(h_1)}, c_1)$	If $\mathsf{SVer}(vk, ((c_i)_i, \pi), \sigma) = \bot$
$h \leftarrow \mathsf{H}_\kappa(vk)$	Return K.	$\qquad\qquad\qquad$ then return $\bot$.
Let h_i be the i-th bit of h.	$\underline{\mathsf{F}_{\mathrm{DDN}}(PK, C, C'):}$	$h \leftarrow \mathsf{H}_\kappa(vk)$
$\forall i \in [k]:$	$((pk_i^{(j)})_{i,j}, crs, \kappa)$	If $h^* = h$ then return $\bot$.
$\quad c_i \leftarrow \mathsf{Enc}(pk_i^{(h_i)}, K; r_i)$	$\qquad\qquad\qquad\quad \leftarrow PK$	Let h_i^* be the i-th bit of h^*.
$x \leftarrow ((pk_i^{(h_i)})_i, (c_i)_i)$	$(vk, (c_i)_i, \pi, \sigma) \leftarrow C$	Let h_i be the i-th bit of h.
$w \leftarrow ((r_i)_i, K)$	$(vk', (c_i')_i, \pi', \sigma') \leftarrow C'$	$\ell \leftarrow \min\{i \in [k] \mid h_i^* \neq h_i\}$
$\pi \leftarrow \mathsf{Prove}(crs, x, w)$	$h \leftarrow \mathsf{H}_\kappa(vk)$	$x \leftarrow ((pk_i^{(h_i)})_i, (c_i)_i)$
$\sigma \leftarrow \mathsf{Sign}(sigk, ((c_i)_i, \pi))$	$h' \leftarrow \mathsf{H}_\kappa(vk')$	If $\mathsf{PVer}(crs, x, \pi) = \bot$
$C \leftarrow (vk, (c_i)_i, \pi, \sigma)$.	Return $(h \stackrel{?}{=} h')$.	$\qquad\qquad\qquad$ then return $\bot$.
Return (C, K).		Return $K \leftarrow \mathsf{Dec}(sk_\ell^{(1-h_\ell^*)}, c_\ell)$.

Fig. 3. The PKEM Γ_{DDN} based on a PKE scheme Π and a non-interactive argument system $\mathcal{P}$. In the figure, "$(r_i)_i$" and "$(pk_i^{(j)})_{i,j}$" are the abbreviations of "$(r_i)_{i \in [k]}$" and "$(pk_i^{(j)})_{i \in [k], j \in \{0,1\}}$", respectively, and we use a similar notation for other values.

decapsulation soundness, C' has to satisfy $\mathsf{PVer}(crs, x', \pi') = \top$ and $x' \notin L_k$ where $x' = ((pk_i^{(h_i')})_i, (c_i')_i)$, and hence the adaptive soundness of the non-interactive argument system $\mathcal{P}$ guarantees that the probability that an adversary coming up with such a ciphertext in the sPDSND experiment is negligible. The eCPA security is also easy to see. Specifically, we can first consider a modified experiment in which crs and π are respectively generated by using the simulation algorithms SimCRS and SimPrv which exist by the ZK security of $\mathcal{P}$. By the ZK security, an eCPA adversary cannot notice this change. Then, the CPA security of the underlying PKE scheme directly shows that the information of a session-key does not leak, leading to the eCPA security.

Capturing Other Existing Constructions. Our framework with a PKEM can explain other existing constructions that, explicitly or implicitly, follow a similar security proof to the DDN construction. For example, the Rosen-Segev construction based on an injective trapdoor function (TDF) secure under correlated inputs [59], the Peikert-Waters construction [56] based on a lossy TDF and an all-but-one lossy TDF (ABO-TDF) in which the ABO-TDF is instantiated from a lossy TDF (see this construction in [56, §2.3]). Moreover, the construction based on CPA secure PKE and an obfuscator for point functions (with multi-bit output) by Matsuda and Hanaoka [47] and one based on

CPA secure PKE and a hash function family satisfying the strong notion (called UCE security [5]) from the same authors [48] can also be captured as a PKEM.

Furthermore, our framework with a PKEM can also capture KEMs based on *all-but-one extractable hash proof systems* (ABO-XHPS) by Wee [61] (and its extension by Matsuda and Hanaoka [46]), by introducing some additional property for underlying ABO-XHPS. Although the additional property that we need is quite subtle, it is satisfied by most existing ABO-XHPS explained in [61,46]. Since a number of recent practical CCA secure KEMs (e.g. [14,20,34,37]) are captured by the framework of ABO-XHPS, our result is also useful for understanding practical KEMs. We expand the explanation for capturing ABO-XHPS-based KEMs in the full version.

3.5 DCCA Secure Detectable KEM from a Puncturable KEM

Here, we show that even if a PKEM does not have decapsulation soundness, it still yields a DCCA secure detectable KEM [38,45]. Therefore, if a PKEM satisfying punctured decapsulation soundness and eCPA security additionally satisfies the property called *unpredictability* [38,45] it can still be used as a building block in the constructions [38,45] to obtain fully CCA secure PKE/KEM.[2]

Theorem 2. *Let* Γ = (KKG, Encap, Decap, F, Punc, PDecap) *be a PKEM satisfying punctured decapsulation soundness and* eCPA *security. Then,* $\Gamma^\dagger$ = (KKG, Encap, Decap, F) *is a* DCCA *secure detectable KEM.*

Proof Sketch of Theorem 2. The proof of this theorem is straightforward given the proof of Theorem 1 (it is only simpler), and thus we omit a formal proof. The reason why we do not need decapsulation soundness is that an adversary in the DCCA experiment is not allowed to ask a decapsulation query c with $F(pk, c^*, c) = 1$, and we need not care the behavior of Decap for "close" ciphertexts. Thus, as in the proof of Theorem 1, the punctured decapsulation soundness guarantees that $\mathsf{PDecap}(\widehat{sk}_{c^*}, \cdot)$ works as good as $\mathsf{Decap}(sk, \cdot)$ for all "far" ciphertexts c with $F(pk, c^*, c) = 0$, and then the eCPA security guarantees the indistinguishability of a real session-key K_1^* and a random K_0^*.   □

4 Puncturable KEM from Sender Non-committing Encryption and KDM Secure SKE

In this section, we show our main technical result: a PKEM that uses a SNCE scheme and a OTKDM secure SKE scheme (with respect to efficiently computable functions). By Theorem 1, this yields a CCA secure KEM. Therefore, this result clarifies a new set of general cryptographic primitives that implies CCA secure PKE/KEM.

The construction of the proposed PKEM is as follows: Let Π = (PKG, Enc, Dec, Fake, Explain) be a SNCE scheme such that the plaintext space is $\{0, 1\}^n$ (for some polynomial $n = n(k) > 0$) and the randomness space of Enc is $\mathcal{R}_k$. Let E = (SEnc,

---

[2] We note that the DDN-KEM reviewed in Section 3.4 and our proposed KEM in Section 4 achieve strong unpredictability (based on the security of the building blocks), which we show in the full version.

SDec) be a SKE scheme whose key space and plaintext space (for security parameter k) are $\mathcal{K}_k$ and $\mathcal{M}_k$, respectively. We require $\mathcal{K}_k \subseteq \{0,1\}^n$ and $(\mathcal{R}_k)^{k+1} \times \{0,1\}^k \subseteq \mathcal{M}_k$. Furthermore, let $\mathcal{H} = (\mathsf{HKG}, \mathsf{H})$ be a hash function family (which is going to be assumed to be a UOWHF). Then we construct a PKEM $\widehat{\varGamma} = (\widehat{\mathsf{KKG}}, \widehat{\mathsf{Encap}}, \widehat{\mathsf{Decap}}, \widehat{\mathsf{F}}, \widehat{\mathsf{Punc}}, \widehat{\mathsf{PDecap}})$ as in Fig. 4.

Function Ensemble for OTKDM *Security.* For showing the eCPA security of $\widehat{\varGamma}$, we need to specify a function ensemble $\mathcal{F} = \{\mathcal{F}_k\}_{k \in \mathbb{N}}$ with respect to which E is OTKDM secure. For each $k \in \mathbb{N}$, define a set $\mathcal{F}_k$ of efficiently computable functions as follows:

$$
\mathcal{F}_k := \left\{ \begin{array}{l} f_z : \mathcal{K}_k \to \mathcal{M}_k \text{ given by} \\ f_z(\alpha) := ((\mathsf{Explain}(\omega_i, \alpha))_{i \in [k+1]}, K) \end{array} \middle| \begin{array}{l} z = ((\omega_i)_{i \in [k+1]}, K) \text{ where } K \in \{0,1\}^k \\ \text{and each } \omega_i \text{ is output from } \mathsf{Fake}(1^k) \end{array} \right\}
$$

Note that each function in $\mathcal{F}_k$ is parameterized by z, and is efficiently computable.

Security of $\widehat{\varGamma}$. The three security requirements of the PKEM $\widehat{\varGamma}$ can be shown as follows: (The formal proofs of Lemmas 4, 5, and 6 are given in Appendices B.1, B.2, and B.3, respectively.)

$\widehat{\mathsf{KKG}}(1^k)$:	$\widehat{\mathsf{Decap}}(SK, C)$:	$\widehat{\mathsf{Punc}}(SK, C^*)$:
$\forall (i,j) \in [k] \times \{0,1\}$:	$((sk_i^{(j)})_{i,j}, PK) \leftarrow SK$	$((sk_i^{(j)})_{i,j}, PK) \leftarrow SK$
$\quad (pk_i^{(j)}, sk_i^{(j)}) \leftarrow \mathsf{PKG}(1^k)$	$((pk_i^{(j)})_{i,j}, pk_{k+1}, \kappa) \leftarrow PK$	$(h^*, (c_i^*)_i, \widetilde{c}^*) \leftarrow C^*$
$(pk_{k+1}, sk_{k+1}) \leftarrow \mathsf{PKG}(1^k)$	$(h, (c_i)_i, \widetilde{c}) \leftarrow C$	Let h_i^* be the i-th bit of h^*.
$\kappa \leftarrow \mathsf{HKG}(1^k)$	Let h_i be the i-th bit of h.	$\widehat{SK}_{C^*} \leftarrow$
$PK \leftarrow ((pk_i^{(j)})_{i,j}, pk_{k+1}, \kappa)$	$\alpha \leftarrow \mathsf{Dec}(sk_1^{(h_1)}, c_1)$	$\quad\quad (h^*, (sk_i^{(1-h_i^*)})_i, PK)$
$SK \leftarrow ((sk_i^{(j)})_{i,j}, PK)$	If $\alpha = \bot$ then return $\bot$.	Return $\widehat{SK}_{C^*}$.
Return (PK, SK).	$\beta \leftarrow \mathsf{SDec}(\alpha, \widetilde{c})$	
	If $\beta = \bot$ then return $\bot$.	$\widehat{\mathsf{PDecap}}(\widehat{SK}_{C^*}, C)$:
$\widehat{\mathsf{Encap}}(PK)$:	$((r_i)_{i \in [k+1]}, K) \leftarrow \beta$	$(h^*, (sk_i^{(1-h_i^*)})_i, PK)$
$((pk_i^{(j)})_{i,j}, pk_{k+1}, \kappa) \leftarrow PK$	$c_{k+1} \leftarrow \mathsf{Enc}(pk_{k+1}, \alpha; r_{k+1})$	$\quad\quad\quad\quad \leftarrow \widehat{SK}_{C^*}$
$\alpha \leftarrow \mathcal{K}_k$; $K \leftarrow \{0,1\}^k$	If (a) $\wedge$ (b) then return K	$(h, (c_i)_i, \widetilde{c}) \leftarrow C$
$r_1, \ldots, r_{k+1} \leftarrow \mathcal{R}_k$	$\quad\quad\quad\quad\quad$ else return $\bot$:	If $h^* = h$ then return $\bot$.
$\beta \leftarrow ((r_i)_{i \in [k+1]}, K)$	$\quad$ (a) $\mathsf{H}_\kappa(c_{k+1} \| \widetilde{c}) = h$	Let h_i^* be the i-th bit of h^*.
$\widetilde{c} \leftarrow \mathsf{SEnc}(\alpha, \beta)$	$\quad$ (b) $\forall i \in [k]$:	Let h_i be the i-th bit of h.
$c_{k+1} \leftarrow \mathsf{Enc}(pk_{k+1}, \alpha; r_{k+1})$	$\quad\quad \mathsf{Enc}(pk_i^{(h_i)}, \alpha; r_i) = c_i$	$\ell \leftarrow \min\{i \in [k] \mid h_i^* \neq h_i\}$
$h \leftarrow \mathsf{H}_\kappa(c_{k+1} \| \widetilde{c})$		$\alpha \leftarrow \mathsf{Dec}(sk_\ell^{(1-h_\ell^*)}, c_\ell)$
Let h_i be the i-th bit of h.	$\widehat{\mathsf{F}}(PK, C, C')$:	Run exactly as $\widehat{\mathsf{Decap}}$ from
$\forall i \in [k]$:	$(h, (c_i)_i, \widetilde{c}) \leftarrow C$	$\quad$ the sixth step and return
$\quad c_i \leftarrow \mathsf{Enc}(pk_i^{(h_i)}, \alpha; r_i)$	$(h', (c_i')_i, \widetilde{c}') \leftarrow C'$	$\quad\quad\quad\quad\quad$ the result.
$C \leftarrow (h, (c_i)_i, \widetilde{c})$.	Return $(h \stackrel{?}{=} h')$.	
Return (C, K).		

Fig. 4. The PKEM $\widehat{\varGamma}$ based on a SNCE scheme Π and a SKE scheme E. In the figure, "$(r_i)_i$" and "$(pk_i^{(j)})_{i,j}$" are the abbreviations of "$(r_i)_{i \in [k]}$" and "$(pk_i^{(j)})_{i \in [k], j \in \{0,1\}}$", respectively, and we use similar notation for other values.

Lemma 4. *If $\mathcal{H}$ is a UOWHF, then the PKEM $\widehat{\Gamma}$ satisfies strong decapsulation soundness.*

Lemma 5. *The PKEM $\widehat{\Gamma}$ satisfies strong punctured decapsulation soundness (even against computationally unbounded adversaries) unconditionally.*

Lemma 6. *If the SNCE scheme Π is SNC secure and the SKE scheme E is $\mathcal{F}$-OTKDM secure, then the PKEM $\widehat{\Gamma}$ is eCPA secure.*

Here, we explain high-level proof sketches for each lemma. Regarding strong decapsulation soundness (Lemma 4), recall that in the sDSND experiment, in order for a ciphertext $C' = (h', (c'_i)_i, \widetilde{c}')$ to violate (strong) decapsulation soundness, it must satisfy $\widehat{\mathsf{F}}(PK, C^*, C') = 1$ (which implies $h^* = h'$), $C' \neq C^*$, and $\widehat{\mathsf{Decap}}(SK, C') \neq \perp$, which (among other conditions) implies $h^* = \mathsf{H}_\kappa(c^*_{k+1}\|\widetilde{c}^*) = \mathsf{H}_\kappa(c'_{k+1}\|\widetilde{c}') = h'$, where the values with asterisk are those related to the challenge ciphertext $C^* = (h^*, (c^*_i)_i, \widetilde{c}^*)$ and c'_{k+1} is the intermediate value calculated during the computation of $\widehat{\mathsf{Decap}}(SK, C')$. On the other hand, a simple observation shows that the above conditions also imply another condition $(c^*_{k+1}, \widetilde{c}^*) \neq (c'_{k+1}, \widetilde{c}')$. This means that a successful ciphertext that violates (strong) decapsulation soundness leads to a collision for the UOWHF $\mathcal{H}$, which is hard to find by the security of the UOWHF $\mathcal{H}$.

Regarding punctured decapsulation soundness (Lemma 5), we show that for any (possibly invalid) ciphertext $C' = (h', (c'_i)_i, \widetilde{c}')$, if $h' \neq h^*$, then it always holds that $\widehat{\mathsf{Decap}}(SK, C') = \widehat{\mathsf{PDecap}}(\widehat{SK}_{C^*}, C')$. This can be shown due to the correctness of the building block SNCE scheme Π and the validity check by re-encryption performed at the last step of $\widehat{\mathsf{Decap}}$ and $\widehat{\mathsf{PDecap}}$. In particular, the validity check by re-encryption works like a non-interactive proof with perfect soundness in the DDN construction, and hence for any adversary, its sPDSND advantage is zero.

Finally, we explain how the eCPA security (Lemma 6) is proved. Let $\mathcal{A}$ be any eCPA adversary. Consider the following sequence of games:

Game 1: This is the eCPA experiment itself. To make it easier to define the subsequent games, we change the ordering of the operations as follows (note that this does not change $\mathcal{A}$'s view):

$\alpha^* \leftarrow \mathcal{K}_k$;
For $i \in [k+1]$:
  $(pk'_i, sk'_i) \leftarrow \mathsf{PKG}(1^k)$;
  $r_i^* \leftarrow \mathcal{R}_k$;
  $c_i^* \leftarrow \mathsf{Enc}(pk'_i, \alpha^*; r_i^*)$;
End For
$K_1^* \leftarrow \{0,1\}^k$;
$\beta^* \leftarrow ((r_i^*)_{i \in [k+1]}, K_1^*)$;
$\widetilde{c}^* \leftarrow \mathsf{SEnc}(\alpha^*, \beta^*)$;
$\kappa \leftarrow \mathsf{HKG}(1^k)$;
$h^* = (h_1^*\|\ldots\|h_k^*) \leftarrow \mathsf{H}_\kappa(c^*_{k+1}\|\widetilde{c}^*)$;
(Continue to the right column $\nearrow$)

For $i \in [k]$:
  $pk_i^{(h_i^*)} \leftarrow pk'_i$;
  $(pk_i^{(1-h_i^*)}, sk_i^{(1-h_i^*)}) \leftarrow \mathsf{PKG}(1^k)$;
End For
$PK \leftarrow ((pk_i^{(j)})_{i,j}, pk'_{k+1}, \kappa)$;
$C^* \leftarrow (h^*, (c_i^*)_i, \widetilde{c}^*)$;
$\widehat{SK}_{C^*} \leftarrow (h^*, (sk_i^{(1-h_i^*)})_i, PK)$;
$K_0^* \leftarrow \{0,1\}^k$;
$b \leftarrow \{0,1\}$;
$b' \leftarrow \mathcal{A}(PK, \widehat{SK}_{C^*}, C^*, K_b^*)$

Game 2: Same as Game 1, except that we generate each tuple $(pk_i^{(h_i^*)}, c_i^*, r_i^*)$ and $(pk_{k+1}, c_{k+1}^*, r_{k+1}^*)$ by using the simulation algorithms Fake and Explain of the SNCE scheme Π. More precisely, in this game, the step with the underline in Game 1 is replaced with: "$(pk_i', c_i^*, \omega_i^*) \leftarrow \mathsf{Fake}(1^k); r_i^* \leftarrow \mathsf{Explain}(\omega_i^*, \alpha^*)$."

Game 3: Same as Game 2, except that the information of $\beta^* = ((r_i^*)_{i \in k+1}, K_1^*)$ is erased from $\widetilde{c}^*$. More precisely, in this game, the step "$\widetilde{c}^* \leftarrow \mathsf{SEnc}(\alpha^*, \beta^*)$" in Game 2 is replaced with the steps "$\beta' \leftarrow \mathcal{M}_k; \widetilde{c}^* \leftarrow \mathsf{SEnc}(\alpha^*, \beta')$."

For $i \in [3]$, let Succ_i be the event that $\mathcal{A}$ succeeds in guessing the challenge bit (i.e. $b' = b$ occurs). We will show that $|\Pr[\mathsf{Succ}_i] - \Pr[\mathsf{Succ}_{i+1}]|$ is negligible for each $i \in [2]$, and that $\Pr[\mathsf{Succ}_3] = 1/2$, which proves the eCPA security of the PKEM $\widehat{\Gamma}$.

Firstly, we can show that $|\Pr[\mathsf{Succ}_1] - \Pr[\mathsf{Succ}_2]|$ is negligible due to the SNC security of the $(k + 1)$-repetition construction Π^{k+1}, which in turn follows from the SNC security of the underlying SNCE scheme Π by a standard hybrid argument (see the explanation in the last paragraph of Section 2.1).

Secondly, we can show that $|\Pr[\mathsf{Succ}_2] - \Pr[\mathsf{Succ}_3]|$ is negligible due to the $\mathcal{F}$-OTKDM security of the SKE scheme E. Here, the key idea is that we view the plaintext $\beta^* = ((r_i^*)_{i \in [k+1]}, K_1^*) = ((\mathsf{Explain}(\omega_i^*, \alpha^*)_{i \in [k+1]}, K^*)$ which will be encrypted under the key α^* as a "key-dependent message" of the key α^*. More specifically, in the full proof we show how to construct a OTKDM adversary $\mathcal{B}_e$ that uses the KDM function $f \in \mathcal{F}_k$ defined by $f(\alpha^*) = ((\mathsf{Explain}(\omega_i^*, \alpha^*)_{i \in [k+1]}, K^*)$ (where $(\omega_i^*)_{i \in [k+1]}$ and K_1^* are viewed as fixed parameters hard-coded in f) for the challenge KDM query, and depending on $\mathcal{B}_e$'s challenge bit, $\mathcal{B}_e$ simulates Game 2 or Game 3 perfectly for $\mathcal{A}$ so that $\mathsf{Adv}_{E, \mathcal{F}, \mathcal{B}_e}^{\mathsf{OTKDM}}(k) = |\Pr[\mathsf{Succ}_2] - \Pr[\mathsf{Succ}_3]|$.

Finally, observe that in Game 3, the challenge ciphertext C^* is independent of K_1^*, and the input $(PK, \widehat{SK}_{C^*}, C^*, K_b^*)$ to $\mathcal{A}$ is distributed identically for both $b \in \{0, 1\}$. This implies $\Pr[\mathsf{Succ}_3] = 1/2$.

Our construction of the PKEM $\widehat{\Gamma}$, and the combination of Lemmas 4 to 6 and Theorem 1 lead to our main result in this paper:

Theorem 3. *If there exist a SNC secure SNCE scheme and a SKE scheme that is OTKDM secure with respect to efficiently computable functions, then there exist a CCA secure PKE scheme/KEM.*

Finally, it would be worth noting that our construction of a CCA secure PKE (via a PKEM) is black-box, in the sense that the construction uses the building blocks in a black-box manner, while our security reductions of the eCPA security is non-black-box, in the sense that our reduction algorithm needs to use the description of the Explain algorithm as a KDM encryption query. Such a situation was encountered in [50,21] where these constructions use the building block PKE scheme in a black-box manner, while the security proof (reduction) is non-black-box because they need to rely on plaintext awareness.

5   Dolev-Dwork-Naor KEM Revisited

In this section, we show that the eCPA security of the DDN-PKEM Γ_{DDN} (Fig. 5) that we reviewed in Section 3.4 can be shown from different assumptions on the PKE scheme

Π and the non-interactive argument system $\mathcal{P}$. More specifically, we show that if Π is a SNC secure SNCE scheme and $\mathcal{P}$ is WI secure, then we can still show that the PKEM Γ_{DDN} is eCPA secure. We emphasize that this change of assumptions does *not* affect the other assumptions used for decapsulation soundness and punctured decapsulation soundness, and thus we see that this result is a concrete evidence of the usefulness of "breaking down" the steps in a security proof into small separate steps. By Theorem 1, we obtain a new CCA security proof for the DDN-KEM based on a SNCE scheme and a non-interactive witness indistinguishable argument system (in the common reference string model).

We believe this new proof for the classical construction with different set of assumptions to be theoretically interesting, and another qualitative evidence of the usefulness of SNCE in the context of constructing CCA secure PKE/KEM. In particular, compared with the original DDN-KEM, our result here shows a trade-off among assumptions on building blocks: a stronger assumption on a PKE scheme and instead a weaker assumption on a non-interactive argument system. Our result shows that the difference between a CPA secure PKE scheme and a SNC secure SNCE scheme is as large/small as the difference between the ZK security and WI security of a non-interactive argument system.

Lemma 7. *If Π is a SNC secure SNCE scheme and the non-interactive argument system $\mathcal{P}$ is WI secure, then the PKEM Γ_{DDN} is eCPA secure.*

The formal proof is given in the full version, and here we give a proof sketch. Recall that in the proof of Lemma 3, we first use the ZK security of $\mathcal{P}$ to "cut" the relation between the components $(c_i^*)_i$ and the proof π^*, and then use the CPA security of the k-repetition construction Π^k (which in turn follows from the CPA security of Π) to "hide" the information of the challenge bit. The proof of Lemma 7 uses the properties of the building blocks in the reversed order.

Proof Sketch of Lemma 7. Let $\mathcal{A}$ be any PPTA adversary that attacks the eCPA security of Γ_{DDN}. Consider the following sequence of games:

Game 1: This is the eCPA experiment itself. To make it easier to define the subsequent games, we change the ordering of the operations as follows (note that this does not change $\mathcal{A}$'s view):

$K_1^* \leftarrow \{0,1\}^k$;
For $i \in [k]$:
$\quad (pk_i', sk_i') \leftarrow \mathsf{PKG}(1^k)$;
$\quad \underline{r_i^* \leftarrow \mathcal{R}_k}$;
$\quad \underline{c_i^* \leftarrow \mathsf{Enc}(pk_i', K_1^*; r_i^*)}$;
End For
$x^* \leftarrow ((pk_i')_i, (c_i^*)_i)$;
$w^* \leftarrow ((r_i^*)_i, K_1^*)$;
$crs \leftarrow \mathsf{CRSG}(1^k)$;
$\pi^* \leftarrow \mathsf{Prove}(crs, x^*, w^*)$;
$(vk^*, sigk^*) \leftarrow \mathsf{SKG}(1^k)$;
$\sigma^* \leftarrow \mathsf{Sign}(sigk^*, ((c_i^*)_i, \pi^*))$;
(Continue to the right column $\nearrow$)

$\kappa \leftarrow \mathsf{HKG}(1^k)$;
$h^* = (h_1^* \| \ldots \| h_k^*) \leftarrow \mathsf{H}_\kappa(vk^*)$;
For $i \in [k]$:
$\quad pk_i^{(h_i^*)} \leftarrow pk_i'$;
$\quad (pk_i^{(1-h_i^*)}, sk_i^{(1-h_i^*)}) \leftarrow \mathsf{PKG}(1^k)$;
End For
$PK \leftarrow ((pk_i^{(j)})_{i,j}, crs, \kappa)$;
$C^* \leftarrow (vk^*, (c_i^*)_i, \pi^*, \sigma^*)$;
$\widehat{SK}_{C^*} \leftarrow (h^*, (sk_i^{(1-h_i^*)})_i, PK)$;
$K_0^* \leftarrow \{0,1\}^k$;
$b \leftarrow \{0,1\}$;
$b' \leftarrow \mathcal{A}(PK, \widehat{SK}_{C^*}, C^*, K_b^*)$

Game 2: Same as Game 1, except that we generate each tuple $(pk_i^{(h_i^*)}, c_i^*, r_i^*)$ by using the simulation algorithms Fake and Explain of the SNCE scheme Π. More

precisely, in this game, the step with the <u>underline</u> in Game 1 is replaced with:
"$(pk'_i, c^*_i, \omega^*_i) \leftarrow \mathsf{Fake}(1^k)$ and $r^*_i \leftarrow \mathsf{Explain}(\omega^*_i, K^*_1)$."

Game 3: Same as Game 2, except that the information of K^*_1 is erased from the witness w^*. More precisely, in this game, the steps "$r^*_i \leftarrow \mathsf{Explain}(\omega^*_i, K^*_1)$" and "$w^* \leftarrow ((r^*_i)_i, K^*_1)$" in Game 2 are replaced with the steps "$r'_i \leftarrow \mathsf{Explain}(\omega^*_i, 0^k)$" and "$w' \leftarrow ((r'_i)_i, 0^k)$," respectively.

For $i \in [3]$, let Succ_i be the event that $\mathcal{A}$ succeeds in guessing the challenge bit (i.e. $b' = b$ occurs). We will show that $|\Pr[\mathsf{Succ}_i] - \Pr[\mathsf{Succ}_{i+1}]|$ is negligible for each $i \in [2]$ and that $\Pr[\mathsf{Succ}_3] = 1/2$, which proves the eCPA security of the PKEM Γ_{DDN}.

Firstly, we can show that $|\Pr[\mathsf{Succ}_1] - \Pr[\mathsf{Succ}_2]|$ is negligible due to the SNC security of the k-repetition construction Π^k, which in turn follows from the SNC security of the underlying SNCE scheme Π (see the explanation in the last paragraph of Section 2.1).

Secondly, we can show that $|\Pr[\mathsf{Succ}_2] - \Pr[\mathsf{Succ}_3]|$ is negligible due to the WI security of the non-interactive argument system $\mathcal{P}$. Note that in Game 2 (and Game 3), every pair $(pk_i^{(h^*_i)}, c^*_i)$ is generated by the simulation algorithm Fake, and hence can be explained as an encryption of an arbitrary plaintext (by using $\mathsf{Explain}$). This in particular means that there are many witnesses for the statement $x^* = ((pk_i^{(h^*_i)})_i, (c^*_i)_i) \in L_k$, and we exploit this fact. Specifically, for $i \in [n]$, let ω_i be the state information corresponding to $(pk_i^{(h^*_i)}, c^*_i)$, and let $w_1 = (K^*_1, (r^*_i)_i)$ (resp. $w_0 = (0^k, (r'_i)_i)$) be a witness for the fact that "each c^*_i encrypts K^*_1 (resp. 0^k)," where each r^*_i (resp. r'_i) is computed by $r^*_i = \mathsf{Explain}(\omega_i, K^*_1)$ (resp. $r'_i = \mathsf{Explain}(\omega_i, 0^k)$). We can construct a reduction algorithm that attacks the WI security of $\mathcal{P}$ so that it uses the above witnesses w_1 and w_0 as its challenge, simulates Game 2 and Game 3 for $\mathcal{A}$ depending on its challenge bit, and has advantage exactly $|\Pr[\mathsf{Succ}_2] - \Pr[\mathsf{Succ}_3]|$.

Finally, observe that in Game 3, the challenge ciphertext C^* is independent of K^*_1, and the input $(PK, \widehat{SK}_{C^*}, C^*, K^*_b)$ to $\mathcal{A}$ is distributed independently for both $b \in \{0, 1\}$. This implies $\Pr[\mathsf{Succ}_3] = 1/2$. $\square$

References

1. Applebaum, B.: Key-dependent message security: Generic amplification and completeness. In: Paterson, K.G. (ed.) EUROCRYPT 2011. LNCS, vol. 6632, pp. 527–546. Springer, Heidelberg (2011)
2. Applebaum, B.: Garbling XOR gates "for free" in the standard model. In: Sahai, A. (ed.) TCC 2013. LNCS, vol. 7785, pp. 162–181. Springer, Heidelberg (2013)
3. Applebaum, B.: Key-dependent message security: Generic amplification and completeness. J. of Cryptology 27(3), 429–451 (2014)
4. Applebaum, B., Cash, D., Peikert, C., Sahai, A.: Fast cryptographic primitives and circular-secure encryption based on hard learning problems. In: Halevi, S. (ed.) CRYPTO 2009. LNCS, vol. 5677, pp. 595–618. Springer, Heidelberg (2009)
5. Bellare, M., Hoang, V.T., Keelveedhi, S.: Instantiating random oracles via UCEs. In: Canetti, R., Garay, J.A. (eds.) CRYPTO 2013, Part II. LNCS, vol. 8043, pp. 398–415. Springer, Heidelberg (2013)
6. Bellare, M., Hoang, V.T., Rogaway, P.: Foundations of garbled circuits, Full version of [7] (2012), http://eprint.iacr.org/2012/265
7. Bellare, M., Hoang, V.T., Rogaway, P.: Foundations of garbled circuits. In: CCS 2012, pp. 784–796 (2012)

8. Bellare, M., Hofheinz, D., Yilek, S.: Possibility and impossibility results for encryption and commitment secure under selective opening. In: Joux, A. (ed.) EUROCRYPT 2009. LNCS, vol. 5479, pp. 1–35. Springer, Heidelberg (2009)

9. Bellare, M., Yilek, S.: Encryption schemes secure under selective opening attack. This is an updated full version of a preliminary version with Hofheinz [8]. Available at eprint.iacr.org/2009/101

10. Bendlin, R., Nielsen, J.B., Nordholt, P.S., Orlandi, C.: Lower and upper bounds for deniable public-key encryption. In: Lee, D.H., Wang, X. (eds.) ASIACRYPT 2011. LNCS, vol. 7073, pp. 125–142. Springer, Heidelberg (2011)

11. Black, J., Rogaway, P., Shrimpton, T.: Encryption-scheme security in the presence of key-dependent messages. In: Nyberg, K., Heys, H.M. (eds.) SAC 2002. LNCS, vol. 2595, pp. 62–75. Springer, Heidelberg (2003)

12. Bleichenbacher, D.: Chosen ciphertext attacks against protocols based on the RSA encryption standard PKCS #1. In: Krawczyk, H. (ed.) CRYPTO 1998. LNCS, vol. 1462, pp. 1–12. Springer, Heidelberg (1998)

13. Boneh, D., Halevi, S., Hamburg, M., Ostrovsky, R.: Circular-secure encryption from decision Diffie-Hellman. In: Wagner, D. (ed.) CRYPTO 2008. LNCS, vol. 5157, pp. 108–125. Springer, Heidelberg (2008)

14. Boyen, X., Mei, Q., Waters, B.: Direct chosen ciphertext security from identity-based techniques. In: CCS 2005, pp. 320–329 (2005)

15. Brakerski, Z., Goldwasser, S.: Circular and leakage resilient public-key encryption under subgroup indistinguishability - (or: Quadratic residuosity strikes back). In: Rabin, T. (ed.) CRYPTO 2010. LNCS, vol. 6223, pp. 1–20. Springer, Heidelberg (2010)

16. Canetti, R.: Universally composable security: A new paradigm for cryptographic protocols. In: FOCS 2001, pp. 136–145 (2001)

17. Canetti, R., Dwork, C., Naor, M., Ostrovsky, R.: Deniable encryption. In: Kaliski Jr., B.S. (ed.) CRYPTO 1997. LNCS, vol. 1294, pp. 90–104. Springer, Heidelberg (1997)

18. Canetti, R., Feige, U., Goldreich, O., Naor, M.: Adaptively secure multi-party computation. In: STOC 1996, pp. 639–648 (1996)

19. Canetti, R., Halevi, S., Katz, J.: Chosen-ciphertext security from identity-based encryption. In: Cachin, C., Camenisch, J.L. (eds.) EUROCRYPT 2004. LNCS, vol. 3027, pp. 207–222. Springer, Heidelberg (2004)

20. Cash, D.M., Kiltz, E., Shoup, V.: The twin Diffie-Hellman problem and applications. In: Smart, N.P. (ed.) EUROCRYPT 2008. LNCS, vol. 4965, pp. 127–145. Springer, Heidelberg (2008)

21. Dachman-Soled, D.: A black-box construction of a CCA2 encryption scheme from a plaintext aware (sPA1) encryption scheme. In: Krawczyk, H. (ed.) PKC 2014. LNCS, vol. 8383, pp. 37–55. Springer, Heidelberg (2014)

22. Damgård, I., Jurik, M.: A generalization, a simplification and some applications of Paillier's probabilistic public-key system. In: Kim, K.-c. (ed.) PKC 2001. LNCS, vol. 1992, pp. 119–136. Springer, Heidelberg (2001)

23. Damgård, I.B., Nielsen, J.B.: Improved non-committing encryption schemes based on a general complexity assumption. In: Bellare, M. (ed.) CRYPTO 2000. LNCS, vol. 1880, pp. 432–450. Springer, Heidelberg (2000)

24. Dolev, D., Dwork, C., Naor, M.: Non-malleable cryptography. In: STOC 1991, pp. 542–552 (1991)

25. Dwork, C., Naor, M., Reingold, O.: Immunizing encryption schemes from decryption errors. In: Cachin, C., Camenisch, J.L. (eds.) EUROCRYPT 2004. LNCS, vol. 3027, pp. 342–360. Springer, Heidelberg (2004)

26. Fehr, S., Hofheinz, D., Kiltz, E., Wee, H.: Encryption schemes secure against chosen-ciphertext selective opening attacks. In: Gilbert, H. (ed.) EUROCRYPT 2010. LNCS, vol. 6110, pp. 381–402. Springer, Heidelberg (2010)
27. Fujisaki, E., Okamoto, T.: Secure integration of asymmetric and symmetric encryption schemes. J. of Cryptology 26(1), 80–101 (2013)
28. Garay, J.A., Wichs, D., Zhou, H.-S.: Somewhat non-committing encryption and efficient adaptively secure oblivious transfer. In: Halevi, S. (ed.) CRYPTO 2009. LNCS, vol. 5677, pp. 505–523. Springer, Heidelberg (2009)
29. Gertner, Y., Malkin, T., Myers, S.: Towards a separation of semantic and CCA security for public key encryption. In: Vadhan, S.P. (ed.) TCC 2007. LNCS, vol. 4392, pp. 434–455. Springer, Heidelberg (2007)
30. Goldreich, O.: Foundations of Cryptography - Volume 1. Cambridge University Press (2001)
31. Goldreich, O.: Foundations of Cryptography - Volume 2. Cambridge University Press (2004)
32. Goldreich, O.: Basing non-interactive zero-knowledge on (enhanced) trapdoor permutations: The state of the art. In: Goldreich, O. (ed.) Studies in Complexity and Cryptography. LNCS, vol. 6650, pp. 406–421. Springer, Heidelberg (2011)
33. Goldwasser, S., Micali, S.: Probabilistic encryption. J. of Computer and System Sciences 28(2), 270–299 (1984)
34. Hanaoka, G., Kurosawa, K.: Efficient chosen ciphertext secure public key encryption under the computational Diffie-Hellman assumption. In: Pieprzyk, J. (ed.) ASIACRYPT 2008. LNCS, vol. 5350, pp. 308–325. Springer, Heidelberg (2008)
35. Hazay, C., Patra, A.: One-sided adaptively secure two-party computation. In: Lindell, Y. (ed.) TCC 2014. LNCS, vol. 8349, pp. 368–393. Springer, Heidelberg (2014)
36. Hemenway, B., Libert, B., Ostrovsky, R., Vergnaud, D.: Lossy encryption: Constructions from general assumptions and efficient selective opening chosen ciphertext security. In: Lee, D.H., Wang, X. (eds.) ASIACRYPT 2011. LNCS, vol. 7073, pp. 70–88. Springer, Heidelberg (2011)
37. Hofheinz, D., Kiltz, E.: Practical chosen ciphertext secure encryption from factoring. In: Joux, A. (ed.) EUROCRYPT 2009. LNCS, vol. 5479, pp. 313–332. Springer, Heidelberg (2009)
38. Hohenberger, S., Lewko, A., Waters, B.: Detecting dangerous queries: A new approach for chosen ciphertext security. In: Pointcheval, D., Johansson, T. (eds.) EUROCRYPT 2012. LNCS, vol. 7237, pp. 663–681. Springer, Heidelberg (2012)
39. Huang, Z., Liu, S., Qin, B.: Sender-equivocable encryption schemes secure against chosen-ciphertext attacks revisited. In: Kurosawa, K., Hanaoka, G. (eds.) PKC 2013. LNCS, vol. 7778, pp. 369–385. Springer, Heidelberg (2013)
40. Kiltz, E.: Chosen-ciphertext security from tag-based encryption. In: Halevi, S., Rabin, T. (eds.) TCC 2006. LNCS, vol. 3876, pp. 581–600. Springer, Heidelberg (2006)
41. Kiltz, E., Mohassel, P., O'Neill, A.: Adaptive trapdoor functions and chosen-ciphertext security. In: Gilbert, H. (ed.) EUROCRYPT 2010. LNCS, vol. 6110, pp. 673–692. Springer, Heidelberg (2010)
42. Lin, H., Tessaro, S.: Amplification of chosen-ciphertext security. In: Johansson, T., Nguyen, P.Q. (eds.) EUROCRYPT 2013. LNCS, vol. 7881, pp. 503–519. Springer, Heidelberg (2013)
43. MacKenzie, P.D., Reiter, M.K., Yang, K.: Alternatives to non-malleability: Definitions, constructions and applications. In: Naor, M. (ed.) TCC 2004. LNCS, vol. 2951, pp. 171–190. Springer, Heidelberg (2004)
44. Malkin, T., Teranishi, I., Yung, M.: Efficient circuit-size independent public key encryption with KDM security. In: Paterson, K.G. (ed.) EUROCRYPT 2011. LNCS, vol. 6632, pp. 507–526. Springer, Heidelberg (2011)

45. Matsuda, T., Hanaoka, G.: Achieving chosen ciphertext security from detectable public key encryption efficiently via hybrid encryption. In: Sakiyama, K., Terada, M. (eds.) IWSEC 2013. LNCS, vol. 8231, pp. 226–243. Springer, Heidelberg (2013)
46. Matsuda, T., Hanaoka, G.: Key encapsulation mechanisms from extractable hash proof systems, revisited. In: Kurosawa, K., Hanaoka, G. (eds.) PKC 2013. LNCS, vol. 7778, pp. 332–351. Springer, Heidelberg (2013)
47. Matsuda, T., Hanaoka, G.: Chosen ciphertext security via point obfuscation. In: Lindell, Y. (ed.) TCC 2014. LNCS, vol. 8349, pp. 95–120. Springer, Heidelberg (2014)
48. Matsuda, T., Hanaoka, G.: Chosen ciphertext security via UCE. In: Krawczyk, H. (ed.) PKC 2014. LNCS, vol. 8383, pp. 56–76. Springer, Heidelberg (2014)
49. Mol, P., Yilek, S.: Chosen-ciphertext security from slightly lossy trapdoor functions. In: Nguyen, P.Q., Pointcheval, D. (eds.) PKC 2010. LNCS, vol. 6056, pp. 296–311. Springer, Heidelberg (2010)
50. Myers, S., Sergi, M., shelat, a.: Blackbox construction of a more than non-malleable CCA1 encryption scheme from plaintext awareness. In: Visconti, I., De Prisco, R. (eds.) SCN 2012. LNCS, vol. 7485, pp. 149–165. Springer, Heidelberg (2012)
51. Myers, S., Shelat, A.: Bit encryption is complete. In: FOCS 2009, pp. 607–616 (2009)
52. Naor, M., Yung, M.: Universal one-way hash functions and their cryptographic applications. In: STOC 1989, pp. 33–43 (1989)
53. Naor, M., Yung, M.: Public-key cryptosystems provably secure against chosen ciphertext attacks. In: STOC 1990, pp. 427–437 (1990)
54. O'Neill, A., Peikert, C., Waters, B.: Bi-deniable public-key encryption. In: Rogaway, P. (ed.) CRYPTO 2011. LNCS, vol. 6841, pp. 525–542. Springer, Heidelberg (2011)
55. Paillier, P.: Public-key cryptosystems based on composite degree residuosity classes. In: Stern, J. (ed.) EUROCRYPT 1999. LNCS, vol. 1592, pp. 223–238. Springer, Heidelberg (1999)
56. Peikert, C., Waters, B.: Lossy trapdoor functions and their applications. In: STOC 2008, pp. 187–196 (2008)
57. Rackoff, C., Simon, D.R.: Non-interactive zero-knowledge proof of knowledge and chosen ciphertext attack. In: Feigenbaum, J. (ed.) CRYPTO 1991. LNCS, vol. 576, pp. 433–444. Springer, Heidelberg (1992)
58. Rompel, J.: One-way functions are necessary and sufficient for secure signatures. In: STOC 1990, pp. 387–394 (1990)
59. Rosen, A., Segev, G.: Chosen-ciphertext security via correlated products. In: Reingold, O. (ed.) TCC 2009. LNCS, vol. 5444, pp. 419–436. Springer, Heidelberg (2009)
60. Sahai, A., Waters, B.: How to use indistinguishability obfuscation: deniable encryption, and more. In: STOC 2014, pp. 475–484 (2014)
61. Wee, H.: Efficient chosen-ciphertext security via extractable hash proofs. In: Rabin, T. (ed.) CRYPTO 2010. LNCS, vol. 6223, pp. 314–332. Springer, Heidelberg (2010)

A   Basic Cryptographic Primitives

Public Key Encryption. A public key encryption (PKE) scheme Π consists of the three PPTAs $(\mathsf{PKG}, \mathsf{Enc}, \mathsf{Dec})$ with the following interface:

Key Generation:	**Encryption:**	**Decryption:**
$(pk, sk) \leftarrow \mathsf{PKG}(1^k)$	$c \leftarrow \mathsf{Enc}(pk, m)$	$m \text{ (or } \perp) \leftarrow \mathsf{Dec}(sk, c)$

where Dec is a deterministic algorithm, (pk, sk) is a public/secret key pair, and c is a ciphertext of a plaintext m under pk. We require for all $k \in \mathbb{N}$, all (pk, sk) output by $\mathsf{PKG}(1^k)$, and all m, it holds that $\mathsf{Dec}(sk, \mathsf{Enc}(pk, m)) = m$.

$\mathsf{Expt}^{\mathsf{CPA}}_{\Pi,\mathcal{A}}(k):$	$\mathsf{Expt}^{\mathsf{ATK}}_{\Gamma,\mathcal{A}}(k):$	$\mathsf{Expt}^{\mathsf{SOT}}_{\Sigma,\mathcal{A}}(k):$
$(pk, sk) \leftarrow \mathsf{PKG}(1^k)$	$(pk, sk) \leftarrow \mathsf{KKG}(1^k)$	$(vk, sigk) \leftarrow \mathsf{SKG}(1^k)$
$(m_0, m_1, \mathsf{st}) \leftarrow \mathcal{A}_1(pk)$	$(c^*, K_1^*) \leftarrow \mathsf{Encap}(pk)$	$(m, \mathsf{st}) \leftarrow \mathcal{A}_1(vk)$
$b \leftarrow \{0, 1\}$	$K_0^* \leftarrow \{0, 1\}^k$	$\sigma \leftarrow \mathsf{Sign}(sigk, m)$
$c^* \leftarrow \mathsf{Enc}(pk, m_b)$	$b \leftarrow \{0, 1\}$	$(m', \sigma') \leftarrow \mathcal{A}_2(\mathsf{st}, \sigma)$
$b' \leftarrow \mathcal{A}_2(\mathsf{st}, c^*)$	$b' \leftarrow \mathcal{A}^{\mathsf{Decap}(sk, \cdot)}(\mathsf{st}, c^*, K_b^*)$	Return 1 iff **(a)** $\wedge$ **(b)** :
Return $(b' \stackrel{?}{=} b)$.	Return $(b' \stackrel{?}{=} b)$.	**(a)** $\mathsf{SVer}(vk, m', \sigma') = \top$
		(b) $(m', \sigma') \neq (m, \sigma)$

Fig. 5. The CPA security experiment for a PKE scheme Π (left), the ATK security experiment (with $\mathsf{ATK} \in \{\mathsf{CCA}, \mathsf{DCCA}, \mathsf{CPA}\}$) for a (detectable) KEM Γ (center), and the SOT security experiment (right)

We say that a PKE scheme Π is CPA secure if for all PPTAs $\mathcal{A} = (\mathcal{A}_1, \mathcal{A}_2)$, $\mathsf{Adv}^{\mathsf{CPA}}_{\Pi,\mathcal{A}}(k) := 2 \cdot | \Pr[\mathsf{Expt}^{\mathsf{CPA}}_{\Pi,\mathcal{A}}(k) = 1] - 1/2|$ is negligible, where the experiment $\mathsf{Expt}^{\mathsf{CPA}}_{\Pi,\mathcal{A}}(k)$ is defined as in Fig. 5 (left). In the experiment, it is required that $|m_0| = |m_1|$.

(Detectable) Key Encapsulation Mechanism. A key encapsulation mechanism (KEM) Γ consists of the three PPTAs (KKG, Encap, Decap) with the following interface:

Key Generation:	**Encapsulation:**	**Decapsulation:**
$(pk, sk) \leftarrow \mathsf{KKG}(1^k)$	$(c, K) \leftarrow \mathsf{Encap}(pk)$	$K \text{ (or } \bot) \leftarrow \mathsf{Decap}(sk, c)$

where Decap is a deterministic algorithm, (pk, sk) is a public/secret key pair, and c is a ciphertext of a session-key $K \in \{0, 1\}^k$ under pk. We require for all $k \in \mathbb{N}$, all (pk, sk) output by $\mathsf{KKG}(1^k)$, and all $(c, K) \leftarrow \mathsf{Encap}(pk)$, it holds that $\mathsf{Decap}(sk, c) = K$.

A tuple of PPTAs $\Gamma = (\mathsf{KKG}, \mathsf{Encap}, \mathsf{Decap}, \mathsf{F})$ is said to be a *detectable* KEM if the tuple (KKG, Encap, Decap) constitutes a KEM, and F is a predicate that takes a public key pk and two ciphertexts c, c' as input and outputs either 0 or 1. (The interface is exactly the same as that of the predicate F of a PKEM introduced in Section 3.) The predicate F is used to define *detectable CCA (DCCA) security* (and another notion *unpredictability*) for a detectable KEM.[3]

For $\mathsf{ATK} \in \{\mathsf{CCA}, \mathsf{DCCA}, \mathsf{CPA}\}$, we say that a (detectable) KEM Γ is ATK secure if for all PPTAs $\mathcal{A}$, $\mathsf{Adv}^{\mathsf{ATK}}_{\Gamma,\mathcal{A}}(k) := 2 \cdot | \Pr[\mathsf{Expt}^{\mathsf{ATK}}_{\Gamma,\mathcal{A}}(k) = 1] - 1/2|$ is negligible, where the ATK experiment $\mathsf{Expt}^{\mathsf{ATK}}_{\Gamma,\mathcal{A}}(k)$ is defined as in Fig. 5 (center). In the experiment, $\mathcal{A}$ is not allowed to submit "prohibited" queries that are defined based on ATK: If $\mathsf{ATK} = \mathsf{CCA}$, then the prohibited query is c^*.; If $\mathsf{ATK} = \mathsf{DCCA}$, then the prohibited queries are c such that $\mathsf{F}(pk, c^*, c) = 1$.; If $\mathsf{ATK} = \mathsf{CPA}$, then $\mathcal{A}$ is not allowed to submit any query.

Signature. A signature scheme Σ consists of the three PPTAs (SKG, Sign, SVer) with the following interface:

Key Generation:	**Signing:**	**Verification:**
$(vk, sigk) \leftarrow \mathsf{SKG}(1^k)$	$\sigma \leftarrow \mathsf{Sign}(sigk, m)$	$\top \text{ or } \bot \leftarrow \mathsf{SVer}(vk, m, \sigma)$

---

[3] In this proceedings version we do not recall *unpredictability* of a detectable KEM. For its formal definition, see the full version (or the papers [38,45]).

$\mathsf{Expt}^{\mathsf{Sound}}_{\mathcal{P},\mathcal{A}}(k):$	$\mathsf{Expt}^{\mathsf{WI}}_{\mathcal{P},\mathcal{A}}(k):$	$\mathsf{Expt}^{\mathsf{ZK\text{-}Real}}_{\mathcal{P},\mathcal{A}}(k):$	$\mathsf{Expt}^{\mathsf{ZK\text{-}Sim}}_{\mathcal{P},\mathcal{S},\mathcal{A}}(k):$
$crs \leftarrow \mathsf{CRSG}(1^k)$	$(x, w_0, w_1, \mathsf{st})$	$(x, w, \mathsf{st}) \leftarrow \mathcal{A}_1(1^k)$	$(x, w, \mathsf{st}) \leftarrow \mathcal{A}_1(1^k)$
$(x, \pi) \leftarrow \mathcal{A}(crs)$	$\qquad\qquad \leftarrow \mathcal{A}_1(1^k)$	$crs \leftarrow \mathsf{CRSG}(1^k)$	(crs, td)
Return 1 iff (a) $\wedge$ (b):	$crs \leftarrow \mathsf{CRSG}(1^k)$	$\pi \leftarrow \mathsf{Prove}(crs, x, w)$	$\qquad \leftarrow \mathsf{SimCRS}(1^k)$
(a) $x \notin L_k$	$b \leftarrow \{0, 1\}$	$b' \leftarrow \mathcal{A}_2(\mathsf{st}, crs, \pi)$	$\pi \leftarrow \mathsf{SimPrv}(td, x)$
(b) $\mathsf{PVer}(crs, x, \pi)$	$\pi \leftarrow \mathsf{Prove}(crs, x, w_b)$	Return b'.	$b' \leftarrow \mathcal{A}_2(\mathsf{st}, crs, \pi)$
$\qquad = \top$	$b' \leftarrow \mathcal{A}_2(\mathsf{st}, crs, \pi)$		Return b'.
	Return $(b' \overset{?}{=} b)$.		

Fig. 6. Security experiments for a non-interactive argument system

where SVer is a deterministic algorithm, $(vk, sigk)$ is a verification/signing key pair, and σ is a signature on a message m under the key pair $(vk, sigk)$. The symbol $\top$ (resp. $\perp$) indicates "accept" (resp. "reject"). We require for all $k \in \mathbb{N}$, all $(vk, sigk)$ output by $\mathsf{SKG}(1^k)$, and all m, it holds that $\mathsf{SVer}(vk, m, \mathsf{Sign}(vk, m)) = \top$.

We say that a signature scheme Σ is strongly one-time secure (SOT secure, for short) if for all PPTAs $\mathcal{A} = (\mathcal{A}_1, \mathcal{A}_2)$, $\mathsf{Adv}^{\mathsf{SOT}}_{\Sigma, \mathcal{A}}(k) := \Pr[\mathsf{Expt}^{\mathsf{SOT}}_{\Sigma, \mathcal{A}}(k) = 1]$ is negligible, where the experiment $\mathsf{Expt}^{\mathsf{SOT}}_{\Sigma, \mathcal{A}}(k)$ is defined as in Fig. 5 (right).

A SOT secure signature scheme can be built from any one-way function [52,58].

Universal One-Way Hash Function. We say that a pair of PPTAs $\mathcal{H} = (\mathsf{HKG}, \mathsf{H})$ is a universal one-way hash function (UOWHF) if the following two properties are satisfied: (1) On input 1^k, HKG outputs a hash-key κ. For any hash-key κ output from $\mathsf{HKG}(1^k)$, H defines an (efficiently computable) function of the form $\mathsf{H}_\kappa : \{0, 1\}^* \rightarrow \{0, 1\}^k$. (2) For all PPTAs $\mathcal{A} = (\mathcal{A}_1, \mathcal{A}_2)$, $\mathsf{Adv}^{\mathsf{UOW}}_{\mathcal{H}, \mathcal{A}}(k) := \Pr[\mathsf{Expt}^{\mathsf{UOW}}_{\mathcal{H}, \mathcal{A}}(k) = 1]$ is negligible, where the experiment is defined as follows:

$$\mathsf{Expt}^{\mathsf{UOW}}_{\mathcal{H}, \mathcal{A}}(k) : [\, (m, \mathsf{st}) \leftarrow \mathcal{A}_1(1^k); \ \kappa \leftarrow \mathsf{HKG}(1^k); \ m' \leftarrow \mathcal{A}_2(\mathsf{st}, \kappa);$$
$$\text{Return 1 iff } \mathsf{H}_\kappa(m') = \mathsf{H}_\kappa(m) \wedge m' \neq m. \,].$$

A UOWHF can be built from any one-way function [52,58].

Non-interactive Argument Systems. Let $L = \{L_k\}_{k \in \mathbb{N}}$ be an NP language (for simplicity, we assume that L consists of sets L_k parameterized by the security parameter k). A non-interactive argument system $\mathcal{P}$ for L consists of the three algorithms (CRSG, Prove, PVer) with the following interface:

CRS Generation:	**Proving:**	**Verification:**
$crs \leftarrow \mathsf{CRSG}(1^k)$	$\pi \leftarrow \mathsf{Prove}(crs, x, w)$	$\top$ or $\perp \leftarrow \mathsf{PVer}(crs, x, \pi)$

where PVer is a deterministic algorithm, crs is a common reference string (CRS), x is a statement, w is a witness for the fact that $x \in L_k$, and π is a proof string (that is supposed to prove that $x \in L_k$). The symbol $\top$ (resp. $\perp$) indicates "accept" (resp. "reject"). We require for all $k \in \mathbb{N}$, all crs output by $\mathsf{CRSG}(1^k)$, and all statement/witness pairs $(x, w) \in L_k \times \{0, 1\}^*$ (where w is a witness for the fact that $x \in L_k$), it holds that $\mathsf{PVer}(crs, x, \mathsf{PVer}(crs, x, w)) = \top$.

We say that a non-interactive argument system $\mathcal{P}$ for a language L satisfies *adaptive soundness* if for all PPTAs $\mathcal{A}$, $\mathsf{Adv}^{\mathtt{Sound}}_{\mathcal{P},\mathcal{A}}(k) := \Pr[\mathsf{Expt}^{\mathtt{Sound}}_{\mathcal{P},\mathcal{A}}(k) = 1]$ is negligible, where the Sound experiment $\mathsf{Expt}^{\mathtt{Sound}}_{\mathcal{P},\mathcal{A}}(k)$ is defined as in Fig. 6 (leftmost).

We say that a non-interactive argument system $\mathcal{P}$ for an NP language L satisfies *witness indistinguishability* (WI security, for short) if for all PPTAs $\mathcal{A} = (\mathcal{A}_1, \mathcal{A}_2)$, $\mathsf{Adv}^{\mathtt{WI}}_{\mathcal{P},\mathcal{A}}(k) := 2 \cdot | \Pr[\mathsf{Expt}^{\mathtt{WI}}_{\mathcal{P},\mathcal{A}}(k) = 1] - 1/2|$ is negligible, where the WI experiment $\mathsf{Expt}^{\mathtt{WI}}_{\mathcal{P},\mathcal{A}}(k)$ is defined as in Fig. 6 (second-left), and it is required that $x \in L_k$, and both w_0 and w_1 are witnesses for the fact that $x \in L_k$ in the WI experiment. [4]

Finally, we recall the definition of the *zero-knowledge property* (ZK security, for short). We say that a non-interactive argument system $\mathcal{P}$ for an NP language L satisfies the *zero-knowledge* property (ZK secure, for short) if there exists a pair of PPTAs $\mathcal{S} = (\mathsf{SimCRS}, \mathsf{SimPrv})$ satisfying the following properties:

- **(Syntax:)** SimCRS is the "simulated common reference string" generation algorithm that takes 1^k as input, and outputs crs and a corresponding trapdoor td.; SimPrv is the "simulated proof" generation algorithm that takes td (output by SimCRS) and a statement $x \in \{0,1\}^*$ (which may not belong to L_k) as input, and outputs a "simulated proof" π.

- **(Zero-Knowledge:)** For all PPTAs $\mathcal{A} = (\mathcal{A}_1, \mathcal{A}_2)$, $\mathsf{Adv}^{\mathtt{ZK}}_{\mathcal{P},\mathcal{S},\mathcal{A}}(k) := | \Pr[\mathsf{Expt}^{\mathtt{ZK\text{-}Real}}_{\mathcal{P},\mathcal{A}}(k) = 1] - \Pr[\mathsf{Expt}^{\mathtt{ZK\text{-}Sim}}_{\mathcal{P},\mathcal{S},\mathcal{A}}(k) = 1]|$ is negligible, where the ZK-Real experiment $\mathsf{Expt}^{\mathtt{ZK\text{-}Real}}_{\mathcal{P},\mathcal{A}}(k)$ and the ZK-Sim experiment $\mathsf{Expt}^{\mathtt{ZK\text{-}Sim}}_{\mathcal{P},\mathcal{S},\mathcal{A}}(k)$ are defined as in Fig. 6 (second-right and rightmost, respectively), and furthermore it is required that $x \in L_k$ and w is a witness for the fact that $x \in L_k$ in both of the experiments.

B   Postponed Proofs

B.1   Proof of Lemma 4: Strong Decapsulation Soundness of $\widehat{\varGamma}$

Let $\mathcal{A}$ be a PPTA sDSND adversary. Let (PK, SK, C^*, K^*) be a tuple that is input to $\mathcal{A}$ in the sDSND experiment, where $PK = ((pk_i^{(j)})_{i,j}, pk_{k+1}, \kappa)$, $SK = ((sk_i^{(j)})_{i,j}, PK)$, and $C^* = (h^*, (c_i^*)_i, \widetilde{c}^*)$.

Let us call $\mathcal{A}$'s output $C' = (h', (c_i')_i, \widetilde{c}')$ in the sDSND experiment *successful* if C' satisfies the conditions that make the experiment output 1, i.e. $\widehat{\mathsf{F}}(PK, C^*, C') = 1$ (which is equivalent to $h' = h^*$), $C' \neq C^*$, and $\widehat{\mathsf{Decap}}(SK, C') \neq \bot$. Below, we use asterisk (*) to denote the values generated/chosen during the generation of C^*, and prime (′) to denote the values generated during the calculation of $\widehat{\mathsf{Decap}}(SK, C')$.

We first confirm that a successful ciphertext C' must additionally satisfy $(c_{k+1}', \widetilde{c}') \neq (c_{k+1}^*, \widetilde{c}^*)$. To see this, assume the opposite, i.e. $(c_{k+1}', \widetilde{c}') = (c_{k+1}^*, \widetilde{c}^*)$. Here, $c_{k+1}' = c_{k+1}^*$ implies $\alpha' = \alpha^*$ (due to the correctness of the SNCE scheme $\varPi$). This and $\widetilde{c}' = \widetilde{c}^*$ imply $(r_i')_{i \in [k+1]} = (r_i^*)_{i \in [k+1]}$ (due to the correctness of the SKE scheme E), which in turn implies $(c_i')_i = (c_i^*)_i$. Hence, it holds that $C' = (h', (c_i')_i, \widetilde{c}') = (h^*, (c_i^*)_i, \widetilde{c}^*) = C^*$, but this contradicts $C' \neq C^*$.

---

[4] We note that unlike soundness, we do *not* need a version of the WI security in which a statement (and witnesses) may depend on a common reference string.

So far, we have seen that a successful ciphertext C' must satisfy $\mathsf{H}_\kappa(c'_{k+1}\|\widetilde{c}') = h' = h^* = \mathsf{H}_\kappa(c^*_{k+1}\|\widetilde{c}^*)$ and $(c'_{k+1}, \widetilde{c}') \neq (c^*_{k+1}, \widetilde{c}^*)$, which means that $(c'_{k+1}\|\widetilde{c}')$ and $(c^*_{k+1}\|\widetilde{c}^*)$ constitute a collision pair for H_κ. Using this fact, we can construct a PPTA $\mathcal{B}_\mathsf{h}$ whose advantage in the UOW experiment regarding $\mathcal{H}$ is exactly the probability that $\mathcal{A}$ outputs a successful ciphertext in the sDSND experiment, which combined with the security of the UOWHF $\mathcal{H}$, proves the lemma. Since the reduction algorithm is straightforward given the explanation here, we omit its description. (In the full version, we provide the details of the reduction algorithm.) $\square$

B.2   Proof of Lemma 5: Strong Punctured Decapsulation Soundness of $\widehat{\Gamma}$

Let (PK, SK) be a key pair output by $\widehat{\mathsf{KKG}}(1^k)$, where $PK = ((pk_i^{(j)})_{i,j}, pk_{k+1}, \kappa)$ and $SK = ((sk_i^{(j)})_{i,j}, PK)$. Let $C^* = (h^*, (c_i^*)_i, \widetilde{c}^*)$ be any ciphertext output by $\widehat{\mathsf{Encap}}(PK)$, and let $\widehat{SK}_{C^*} = (h^*, (sk_i^{(1-h_i^*)})_i, PK)$ be the punctured secret key generated by $\widehat{\mathsf{Punc}}(SK, C^*)$. We show that for any ciphertext $C = (h, (c_i)_i, \widetilde{c})$ (which might be outside the range of $\widehat{\mathsf{Encap}}(PK)$) satisfying $\widehat{\mathsf{F}}(PK, C^*, C) = 0$ (i.e. $h \neq h^*$), it holds that $\widehat{\mathsf{Decap}}(SK, C) = \widehat{\mathsf{PDecap}}(\widehat{SK}_{C^*}, C)$. Note that this implies that there exists no ciphertext that violates (strong) punctured decapsulation soundness of the PKEM $\widehat{\Gamma}$, and thus for any (even computationally unbounded) sPDSND adversary $\mathcal{A}$, $\mathsf{Adv}_{\widehat{\Gamma}, \mathcal{A}}^{\mathsf{sPDSND}}(k) = 0$, which will prove the lemma.

To show the above, fix arbitrarily a ciphertext $C = (h, (c_i)_i, \widetilde{c})$ satisfying $\widehat{\mathsf{F}}(PK, C^*, C) = 0$ (and hence $h^* \neq h$) and let $\ell = \min\{i \in [k] \mid h_i^* \neq h_i\}$, where each of h_i and h_i^* are the i-th bit of h and h^*, respectively. For notational convenience, let $\alpha_1 = \mathsf{Dec}(sk_1^{(h_1)}, c_1)$ and $\alpha_\ell = \mathsf{Dec}(sk_\ell^{(1-h_\ell^*)}, c_\ell) = \mathsf{Dec}(sk_\ell^{(h_\ell)}, c_\ell)$, where the latter equality is because $h_\ell^* \neq h_\ell$ implies $1 - h_\ell^* = h_\ell$. We consider the following two cases, and show that the results from both of the algorithms $\widehat{\mathsf{Decap}}$ and $\widehat{\mathsf{PDecap}}$ always agree.

Case $\alpha_1 = \alpha_\ell$: Both $\widehat{\mathsf{Decap}}$ and $\widehat{\mathsf{PDecap}}$ proceed identically after they respectively compute α_1 and α_ℓ, and thus the outputs from these algorithms agree.

Case $\alpha_1 \neq \alpha_\ell$: In this case, both $\widehat{\mathsf{Decap}}$ and $\widehat{\mathsf{PDecap}}$ return $\bot$. Specifically, $\alpha_1 \neq \alpha_\ell$ and the correctness of the SNCE scheme Π imply that there does not exist r_ℓ such that $\mathsf{Enc}(pk_\ell^{(h_\ell)}, \alpha_1; r_\ell) = c_\ell$, and thus $\widehat{\mathsf{Decap}}$ returns $\bot$ in its last step at the latest (it may return $\bot$ earlier if $\alpha_1 = \bot$ or $\mathsf{SDec}(\alpha_1, \widetilde{c}) = \bot$). Symmetrically, there does not exist r_1 such that $\mathsf{Enc}(pk_1^{(h_1)}, \alpha_\ell; r_1) = c_1$, and thus $\widehat{\mathsf{PDecap}}$ returns $\bot$ in its last step at the latest (it may return $\bot$ earlier as above).

This completes the proof of Lemma 5. $\square$

B.3   Proof of Lemma 6: eCPA Security of $\widehat{\Gamma}$

Let $\mathcal{A}$ be any PPTA adversary that attacks the eCPA security of $\widehat{\Gamma}$. For this $\mathcal{A}$, we consider the sequence of games described in the explanation in Section 4. Here, we only show that $|\Pr[\mathsf{Succ}_1] - \Pr[\mathsf{Succ}_2]|$ and $|\Pr[\mathsf{Succ}_2] - \Pr[\mathsf{Succ}_3]|$ are negligible, which should be sufficient for the proof of Lemma 6, given the intuitive explanation in Section 4.

Claim 1. *There exists a PPTA $\mathcal{B}_p$ such that $\mathrm{Adv}_{\Pi^{k+1},\mathcal{B}_p}^{\mathrm{SNC}}(k) = |\Pr[\mathrm{Succ}_1] - \Pr[\mathrm{Succ}_2]|$.*

Proof of Claim 1. We show how to construct a PPTA adversary $\mathcal{B}_p$ that attacks the SNC security of the $(k+1)$-repetition construction Π^{k+1} of the SNCE scheme with the claimed advantage. The description of $\mathcal{B}_p = (\mathcal{B}_{p1}, \mathcal{B}_{p2})$ as follows:

$\mathcal{B}_{p1}(1^k)$: $\mathcal{B}_{p1}$ picks $\alpha^* \in \mathcal{K}_k$ uniformly at random, and sets $\mathrm{st}_\mathcal{B} \leftarrow (\mathcal{B}_{p1}$'s entire view). Then $\mathcal{B}_{p1}$ terminates with output $(\alpha^*, \mathrm{st}_\mathcal{B})$ (where α^* is regarded as $\mathcal{B}_p$'s challenge message).

$\mathcal{B}_{p2}(\mathrm{st}_\mathcal{B}, PK' = (pk_i')_{i \in [k+1]}, C'^* = (c_i^*)_{i \in [k+1]}, R'^* = (r_i^*)_{i \in [k+1]})$: $\mathcal{B}_{p2}$ picks $K_1^* \leftarrow \{0,1\}^k$ uniformly at random, sets $\beta^* \leftarrow ((r_i^*)_{i \in [k+1]}, K_1^*)$, and runs $\widetilde{c}^* \leftarrow \mathrm{SEnc}(\alpha^*, \beta^*)$, $\kappa \leftarrow \mathrm{HKG}(1^k)$, and $h^* = (h_1^* \| \dots \| h_k^*) \leftarrow \mathrm{H}_\kappa(c_{k+1}^* \| \widetilde{c}^*)$. For each $i \in [k]$, $\mathcal{B}_{p2}$ sets $pk_i^{(h_i^*)} \leftarrow pk_i'$ and runs $(pk_i^{(1-h_i^*)}, sk_i^{(1-h_i^*)}) \leftarrow \mathrm{PKG}(1^k)$. Next $\mathcal{B}_{p2}$ sets $PK \leftarrow ((pk_i^{(j)})_{i,j}, pk_{k+1}', \kappa)$, $C^* \leftarrow (h^*, (c_i^*)_i, \widetilde{c}^*)$, and $\widehat{SK}_{C^*} \leftarrow (h^*, (sk_i^{(1-h_i^*)})_i, PK)$. Then $\mathcal{B}_{p2}$ picks $K_0^* \in \{0,1\}^k$ and $b \in \{0,1\}$ uniformly at random, runs $b' \leftarrow \mathcal{A}(PK, \widehat{SK}_{C^*}, C^*, K_b^*)$, and terminates with output $(b' \overset{?}{=} b)$.

The above completes the description of $\mathcal{B}_p$. Note that $\mathcal{B}_{p2}$ outputs 1 only when $b' = b$ occurs. $\mathcal{B}_p$'s SNC advantage can be estimated as follows:

$$\mathrm{Adv}_{\Pi^{k+1},\mathcal{B}_p}^{\mathrm{SNC}}(k) = |\Pr[\mathrm{Expt}_{\Pi^{k+1},\mathcal{B}_p}^{\mathrm{SNC-Real}}(k) = 1] - \Pr[\mathrm{Expt}_{\Pi^{k+1},\mathcal{B}_p}^{\mathrm{SNC-Sim}}(k) = 1]|$$
$$= |\Pr[\mathrm{Expt}_{\Pi^{k+1},\mathcal{B}_p}^{\mathrm{SNC-Real}}(k) : b' = b] - \Pr[\mathrm{Expt}_{\Pi^{k+1},\mathcal{B}_p}^{\mathrm{SNC-Sim}}(k) : b' = b]|.$$

Consider the case when $\mathcal{B}_p$ runs in $\mathrm{Expt}_{\Pi^{k+1},\mathcal{B}_p}^{\mathrm{SNC-Real}}(k)$. It is easy to see that in this case, $\mathcal{B}_p$ perfectly simulates Game 1 for $\mathcal{A}$. In particular, every $pk_i^{(j)}$ and pk_{k+1} in PK are generated honestly by running $\mathrm{PKG}(1^k)$, and every c_i^* in C^* is generated as $c_i^* \leftarrow \mathrm{Enc}(pk_i^{(h_i^*)}, \alpha^*; r_i^*)$ where $\alpha^* \in \mathcal{K}_k$ and each of $r_i^* \in \mathcal{R}_k$ are chosen uniformly at random, as done in Game 1. Under this situation, the probability that $b' = b$ occurs is exactly the same as the probability that $\mathcal{A}$ succeeds in guessing its challenge bit in Game 1, i.e., $\Pr[\mathrm{Expt}_{\Pi^{k+1},\mathcal{B}_p}^{\mathrm{SNC-Real}}(k) : b' = b] = \Pr[\mathrm{Succ}_1]$.

When $\mathcal{B}_p$ runs in $\mathrm{Expt}_{\Pi^{k+1},\mathcal{B}_p}^{\mathrm{SNC-Sim}}(k)$, on the other hand, each of pairs $(pk_i^{(h_i^*)}, c_i^*)$ and each r_i^* are generated by using the simulation algorithms Fake and Explain of the underlying SNCE scheme Π, in such a way that the plaintext corresponding to c_i^* is "explained" as $\alpha^* \in \mathcal{K}_k$ that is chosen uniformly at random, as done in Game 2. The rest of the procedures remains unchanged from the above case. Therefore, the probability that $b' = b$ occurs is exactly the same as the probability that $\mathcal{A}$ succeeds in guessing its challenge bit in Game 2, i.e., $\Pr[\mathrm{Expt}_{\Pi^{k+1},\mathcal{B}_p}^{\mathrm{SNC-Sim}}(k) : b' = b] = \Pr[\mathrm{Succ}_2]$.

In summary, we have $\mathrm{Adv}_{\Pi^{k+1},\mathcal{B}_p}^{\mathrm{SNC}}(k) = |\Pr[\mathrm{Succ}_1] - \Pr[\mathrm{Succ}_2]|$. This completes the proof of Claim 1. $\qquad\square$

Claim 2. *There exists a PPTA $\mathcal{B}_e$ such that $\mathrm{Adv}_{E,\mathcal{F},\mathcal{B}_e}^{\mathrm{OTKDM}}(k) = |\Pr[\mathrm{Succ}_2] - \Pr[\mathrm{Succ}_3]|$.*

Proof of Claim 2. We show how to construct a PPTA adversary $\mathcal{B}_e$ that attacks the $\mathcal{F}$-OTKDM security of the underlying SKE scheme E with the claimed advantage. The description of $\mathcal{B}_e = (\mathcal{B}_{e1}, \mathcal{B}_{e2})$ is as follows:

$\mathcal{B}_{e1}(1^k)$: For every $i \in [k+1]$, $\mathcal{B}_e$ runs $(pk'_i, c^*_i, \omega^*_i) \leftarrow \mathsf{Fake}(1^k)$. Then, $\mathcal{B}_{e1}$ picks $K^*_1 \in \{0,1\}^k$ uniformly at random. Next, $\mathcal{B}_{e1}$ specifies the function $f : \mathcal{K}_k \to \mathcal{M}_k$ which is used as an encryption query in the OTKDM experiment, defined by: $\alpha \overset{f}{\mapsto} (\mathsf{Explain}(\omega^*_i, \alpha)_{i \in [k+1]}, K^*_1)$, where each ω^*_i and K^*_1 are treated as fixed parameters hard-coded in f. (Note that $f \in \mathcal{F}_k$.) Finally, $\mathcal{B}_{e1}$ sets $\mathsf{st}_\mathcal{B} \leftarrow (\mathcal{B}_{e1}\text{'s entire view})$, and terminates with output $(f, \mathsf{st}_\mathcal{B})$.

$\mathcal{B}_{e2}(\mathsf{st}_\mathcal{B}, \widetilde{c}^*)$: $\mathcal{B}_{e2}$ runs $\kappa \leftarrow \mathsf{HKG}(1^k)$ and $h^* = (h^*_1 \| \dots \| h^*_k) \leftarrow \mathsf{H}_\kappa(c^*_{k+1} \| \widetilde{c}^*)$. Next, for every $i \in [k]$, $\mathcal{B}_{e2}$ sets $pk_i^{(h^*_i)} \leftarrow pk'_i$ and runs $(pk_i^{(1-h^*_i)}, sk_i^{(1-h^*_i)}) \leftarrow \mathsf{PKG}(1^k)$. Then, $\mathcal{B}_{e2}$ sets $PK \leftarrow ((pk_i^{(j)})_{i,j}, pk'_{k+1}, \kappa)$, $C^* \leftarrow (h^*, (c^*_i)_i, \widetilde{c}^*)$, and $\widehat{SK}_{C^*} \leftarrow (h^*, (sk_i^{(1-h^*_i)})_i, PK)$. $\mathcal{B}_{e2}$ picks $K^*_0 \in \{0,1\}^k$ and $b \in \{0,1\}$ uniformly at random, runs $b' \leftarrow \mathcal{A}(PK, \widehat{SK}_{C^*}, C^*, K^*_b)$, and terminates with output $\gamma' \leftarrow (b' \overset{?}{=} b)$.

The above completes the description of $\mathcal{B}_e$. Let $\gamma \in \{0,1\}$ be $\mathcal{B}_e$'s challenge bit. $\mathcal{B}_e$'s $\mathcal{F}$-OTKDM advantage is estimate as follows:

$$\mathsf{Adv}^{\mathsf{OTKDM}}_{E,\mathcal{F},\mathcal{B}_e}(k) = 2 \cdot \left| \Pr[\gamma' = \gamma] - \frac{1}{2} \right| = |\Pr[\gamma' = 1 | \gamma = 1] - \Pr[\gamma' = 1 | \gamma = 0]|$$
$$= |\Pr[b' = b | \gamma = 1] - \Pr[b' = b | \gamma = 0]|.$$

Let $\alpha^* \in \mathcal{K}_k$ be the key, and $M_1 = f(\alpha^*)$ and $M_0 \in \mathcal{M}_k$ be the plaintexts calculated/chosen in $\mathcal{B}_e$'s OTKDM experiment. Consider the case when $\gamma = 1$, i.e. $\widetilde{c}^*$ is an encryption of $M_1 = f(\alpha^*) = ((r^*_i)_{i \in [k+1]}, K^*_1)$. Note that by the definition of the experiment $\mathsf{Expt}^{\mathsf{OTKDM}}_{E,\mathcal{F},\mathcal{B}_e}(k)$, if we regard the key $\alpha^* \in \mathcal{K}_k$ and $M^*_1 = f(\alpha^*)$ in $\mathsf{Expt}^{\mathsf{OTKDM}}_{E,\mathcal{F},\mathcal{B}_e}(k)$ as α^* and β^* in Game 2, then each r^*_i is generated by $r^*_i \leftarrow \mathsf{Explain}(\omega^*_i, \alpha^*)$, so that the plaintext corresponding to each c^*_i is α^*, which is how these values are generated in Game 2. Moreover, the public key PK, the values $(c^*_i)_{i \in [k+1]}$ used in the challenge ciphertext C^*, and the punctured secret key $\widehat{SK}_{C^*}$ are distributed identically to those in Game 2. Hence, $\mathcal{B}_e$ simulates Game 2 perfectly for $\mathcal{A}$. Under this situation, the probability that $b' = b$ occurs is exactly the same as the probability that $\mathcal{A}$ succeeds in guessing the challenge bit in Game 2, i.e. $\Pr[b' = b | \gamma = 1] = \Pr[\mathsf{Succ}_2]$.

Next, consider the case when $\gamma = 0$. In this case, $\widetilde{c}^*$ is an encryption of a random message $M_0 \in \mathcal{M}_k$ that is independent of any other values. Then, if we regard the key α^* and the random message M_0 in $\mathsf{Expt}^{\mathsf{OTKDM}}_{E,\mathcal{B}_e}(k)$ as α^* and β' in Game 3, respectively, then $\mathcal{A}$'s challenge ciphertext C^* is generated in such a way that they are distributed identically to those in Game 3, and thus $\mathcal{B}_e$ simulates Game 3 perfectly for $\mathcal{A}$. Therefore, with a similar argument to the above, we have $\Pr[b' = b | \gamma = 0] = \Pr[\mathsf{Succ}_3]$.

In summary, we have $\mathsf{Adv}^{\mathsf{OTKDM}}_{E,\mathcal{F},\mathcal{B}_e}(k) = |\Pr[\mathsf{Succ}_2] - \Pr[\mathsf{Succ}_3]|$. This completes the proof of Claim 2. $\square$

Due to our assumptions on the building blocks, and the SNC security of the $(k+1)$-repetition construction Π^{k+1} (see the explanation in Section 2.1), we can conclude that $|\Pr[\mathsf{Succ}_1] - \Pr[\mathsf{Succ}_2]|$ and $|\Pr[\mathsf{Succ}_2] - \Pr[\mathsf{Succ}_3]|$ are negligible. Combined with the intuitive explanations given in Section 4, this completes the proof of Lemma 6. $\square$

Non-committing Encryption from Φ-hiding

Brett Hemenway[1], Rafail Ostrovsky[2,*], and Alon Rosen[3,**]

[1] University of Pennsylvania, USA
fbrett@cis.upenn.edu
[2] UCLA, USA
rafail@cs.ucla.edu
[3] Herzliya Interdisciplinary Center, Israel
alon.rosen@idc.ac.il

Abstract. A multiparty computation protocol is said to be adaptively secure if it retains its security even in the presence of an adversary who can corrupt participants as the protocol proceeds. This is in contrast to the static corruption model where the adversary is forced to choose which participants to corrupt before the protocol begins.

A central tool for constructing adaptively secure protocols is non-committing encryption (Canetti, Feige, Goldreich and Naor, STOC '96). The original protocol of Canetti et al. had ciphertext expansion that was quadratic in the security parameter, and prior to this work, the best known constructions had ciphertext expansion that was linear in the security parameter.

In this work, we present the first non-committing encryption scheme that achieves ciphertext expansion that is logarithmic in the message length. Our construction has optimal round complexity (2-rounds), where (just as in all previous constructions) the first message consists of a public-key of size $\tilde{\mathcal{O}}(n\lambda)$ where n is the message length and λ is the security parameter. The second message consists of a ciphertext of size $\mathcal{O}(n\log n + \lambda)$. The security of our scheme is proved based on the Φ-hiding problem.

1 Introduction

Secure multiparty computation (MPC) allows a group of players to compute any joint function of their inputs while maintaining the privacy of each individual

* The work of R. Ostrovsky was supported in part by NSF grants CCF-0916574, IIS-1065276, CCF-1016540, CNS-1118126, CNS-1136174; US-Israel BSF grant 2008411; OKAWA Foundation Research Award; IBM Faculty Research Award; Xerox Faculty Research Award; B. John Garrick Foundation Award; Teradata Re- search Award; and Lockheed-Martin Corporation Research Award. This material is also based upon work supported by the Defense Advanced Research Projects Agency through the U.S. Office of Naval Research under Contract N00014-11-1-0392.

** Work supported by ISF grant no. 1255/12 and by the European Research Council under the European Union's Seventh Framework Programme (FP/2007-2013) / ERC Grant Agreement n. 307952. Part of this work done while visiting UCLA.

Y. Dodis and J.B. Nielsen (Eds.): TCC 2015, Part I, LNCS 9014, pp. 591–608, 2015.

player's input. When defining security of an MPC protocol, it is important to consider how and when players are corrupted. Traditionally security has been considered in two adversarial models, *static* and *adaptive*. In the static model, the corrupted players are fixed before the protocol begins, while in the adaptive model players may become corrupted at any point in the protocol. In particular, the adversary may choose to corrupt players adaptively (up to a certain threshold) based on messages sent in earlier rounds of the protocol. The adaptive security model, which more accurately reflects conditions in the real-world, is widely recognized to be "the right" model for security in the MPC setting. Unfortunately, analysis of protocols in the adaptive model can be difficult, and many protocols are only proven secure in the weaker static corruption model.

The Goldreich, Micali, Wigderson protocol [14] provides computational security in the static model, and it was observed in [11] that the original GMW proof of security does not hold in the adaptive corruption model. The works of Ben-Or, Goldwasser and Wigderson [3] and Chaum, Crépeau and Damgård [7] take an alternate approach to MPC and create protocols with information theoretic security. These protocols rely on a majority on private channels between the players, and require a majority of players to be honest. Like the GMW protocol, the BGW and CCD protocols were only proven secure in the static corruption model.

1.1   Non-committing Encryption

In [5], Canetti, Feige, Goldreich and Naor introduced the notion of *non-committing encryption*, and showed that if the private channels in the BGW and CCD protocols are replaced by non-committing encryption, then the MPC protocols can achieve adaptive security. Subsequent work has made similar use of non-committing encryption in order to construct cryptographic protocols achieving security against adaptive adversaries [1,6].

Despite its utility, previous constructions of non-committing encryption have been much less efficient than standard (IND-CPA secure) cryptosystems. The original construction of Canetti et. al. created non-committing encryption from *common-domain* trapdoor permutations, which in turn could be realized from either the CDH or RSA assumption. The resulting scheme requires $\mathcal{O}(\lambda)$ ciphertexts (each of bit length $\mathcal{O}(\lambda)$) to encrypt a single bit plaintext with security parameter λ.

Beaver [1] constructed a non-committing key-exchange protocol based on the DDH assumption, but his protocol requires 3 rounds. Beaver's construction provided an *interactive* encryption protocol with ciphertext expansion $\mathcal{O}(\lambda)$. Damgård and Nielsen [11] demonstrated a generalization of Beaver's scheme that could be implemented under general assumptions. The protocol of Damgård and Nielsen, like that of Beaver, requires three rounds and has ciphertext expansion $\mathcal{O}(\lambda)$. The construction of Damgård and Nielsen was later improved, reducing the round complexity to two rounds, by Choi, Dachman-Soled, Malkin and Wee [8]. The construction of [8] gives a two-round protocol with ciphertext expansion $\mathcal{O}(\lambda)$ and public keys of size $\mathcal{O}(n\lambda)$ for n-bit messages. Achieving non-interactive (2-round) non-committing encryption with optimal rate (i.e., $\mathcal{O}(1)$ ciphertext expansion) remained an open question.

1.2 Our Contributions

We construct the first non-committing encryption scheme with logarithmic ciphertext expansion. Our scheme is 2-round, which is optimal in terms of round complexity. The security of our construction rests on the Φ-hiding assumption [4]. Our construction improves on the ciphertext size of previous constructions, with only a modest cost in key size (our public key sizes are $\tilde{\mathcal{O}}(n\lambda)$, compared to $\mathcal{O}(n\lambda)$ in [8]).

All prior constructions of non-committing encryption are loosely based on the following paradigm. The receiver will generate a set of n public-keys for which he only knows the decryption keys corresponding to a small subset, I_R. The sender will first encode his message using an error-correcting code of block-length n, then the sender will choose a small subset of keys, I_S, and encrypt one symbol of the codeword under keys in I_S and sample ciphertexts obliviously for the remaining keys. The parameters then need to be tuned so that if the sender's and receiver's sets are chosen independently, the intersection $I_R \cap I_S$ will be large enough to recover the message with high probability. On the other hand, the simulator (who knows decryption keys for all public-keys, and encryption randomness for all ciphertexts) should be able to find a subset of secret keys, and a subset of encryption randomness, such that the ciphertexts in their intersection decode to any target message.

In these schemes, each of the underlying encryptions encrypts one symbol from an error-correcting encoding of the sender's message. Let Σ denote the alphabet of the error-correcting code. Then, even if the simulator knows decryptions for every one of the n ciphertexts, decryptions will only agree with a random vector in Σ^n in a $\frac{1}{|\Sigma|}$ fraction of coordinates. Thus if Σ is large, the simulator's job of finding an intersection $I_R \cap I_S$ that matches the target message will be impossible. On the other hand, if Σ is small then the scheme suffers from large ciphertext expansion because each codeword symbol is encrypted separately, and the semantic security of the underlying cryptosystem puts a lower bound on the size of each ciphertext. This has been a fundamental barrier to achieving non-committing encryption with good rate.

In order to overcome this barrier, we design a cryptosystem where randomness can be re-used across keys. Cryptosystems with this property are easy to construct, and even El-Gamal has this property. In fact, cryptosystems coming from smooth hash proof systems [10] have this property, as do cryptosystems based on the extended decisional diffie-hellman assumption [16]. This notion of randomness re-use across keys was a necessary ingredient in constructing lossy trapdoor functions [20] and multi-recipient cryptosystems [2].

Now, if we have n ciphertexts, but each uses the same randomness, the total length could be as small as $\mathcal{O}(n+\lambda)$ instead of $\mathcal{O}(n\lambda)$. Unfortunately, reusing randomness across ciphertexts makes the simulator's job much harder. Now, when the simulator generates encryption randomness, it must efficiently generate a *single* random element that is simultaneously valid randomness for all n ciphertexts. For a given cryptosystem, it can be difficult to determine whether such randomness exists, and more difficult still to devise an *efficient* scheme for finding it.

Although we can build randomness-reusable cryptosystems from almost all standard hardness assumptions, (e.g. DDH, DCR, QR) in order to achieve the efficient randomness sampling required by the simulator, we rely on the Φ-hiding assumption [4].

At a high level, our non-committing encryption behaves as follows. The receiver generates n public-keys for a randomness-reusable cryptosystem based on the Φ-hiding assumption, in such a way that he only knows the decryption keys for a small subset I_R. The sender encodes his message using a binary error-correcting code of block-length n, and chooses a subset of keys $I_S \subset [n]$. The sender generates a single piece of randomness and encodes the symbols of his codeword using this single random value for $i \in I_S$, and the sender samples ciphertexts obliviously otherwise.

Encryption under our scheme is conceptually similar to all previous constructions of non-committing encryption, but re-using randomness makes creating an efficient simulator significantly more difficult. On the positive side re-using randomness allows us to decrease the ciphertext size beyond the bounds achieved compared to all previously known constructions.

Reference	Rounds	Ciphertext Expansion	Assumption
[5]	2	$\mathcal{O}(\lambda^2)$	Common-Domain TDPs
[1]	3	$\mathcal{O}(\lambda)$	DDH
[11]	3	$\mathcal{O}(\lambda)$	Simulatable PKE
[8]	2	$\mathcal{O}(\lambda)$	Simulatable PKE
This work	2	$\mathcal{O}(\log |n|)$	Φ-hiding

Fig. 1. Comparison to prior work. The parameter λ denotes the security parameter, and $|n|$ denotes the message length. Notice that $\lambda \gg \log |n|$ since by assumption super-polynomial time calculations in λ are infeasible, while exponential-time calculation in $\log |n|$ are feasible.

2   Preliminaries

2.1   Notation

If A is a Probabilistic Polynomial Time (PPT) machine, then we use $a \xleftarrow{\$} A$ to denote running the machine A and obtaining an output, where a is distributed according to the internal randomness of A. If R is a set, we use $r \xleftarrow{\$} R$ to denote sampling uniformly from R.

We use the notation

$$\Pr[A(x, r) = c : r \xleftarrow{\$} R, x \xleftarrow{\$} X],$$

to denote the probability that A outputs c when x is sampled uniformly from X and r is sampled uniformly from R. We define the statistical distance between two distributions X, Y to be

$$\Delta(X, Y) = \frac{1}{2} \sum_x |\Pr[X = x] - \Pr[Y = x]|$$

If X and Y are families of distributions indexed by a security parameter λ, we use $X \approx_s Y$ to mean the distributions X and Y are statistically close, i.e., for all polynomials p and sufficiently large λ, we have $\Delta(X, Y) < \frac{1}{p(\lambda)}$. We use $X \approx_c Y$ to mean X and Y are computationally close, i.e., for all PPT adversaries A, for all polynomials p, then for all sufficiently large λ, we have $|\Pr[A^X = 1] - \Pr[A^Y = 1]| < 1/p(\lambda)$.

2.2   Non-committing Encryption

In this section, we review the notion of non-committing encryption [5]. Essentially, a non-committing encryption scheme is one where there is an efficient simulator, that can generate public keys and ciphertexts, such that, at a later point, a corresponding secret key and encryption randomness can be generated which "opens" the ciphertext to any given message. For security, we require that the distribution on simulated keys and ciphertexts is computationally indistinguishable from the distribution on keys and ciphertexts generated by the real encryption protocol. The formal definition is given in Definition 1.

Definition 1 (Non-Committing Encryption).
A cryptosystem $\mathcal{PKE} = (\mathsf{Gen}, \mathsf{Enc}, \mathsf{Dec})$ is called non-committing, if there exists a simulator Sim such that the following properties hold:

1. **Efficiency:** *The algorithms $\mathsf{Gen}, \mathsf{Enc}, \mathsf{Dec}$ and Sim are all probabilistic polynomial time.*
2. **Correctness:** *For any message $m \in \mathcal{M}$*

$$\Pr[\mathsf{Dec}(sk, c) = m : (pk, sk) \overset{\$}{\leftarrow} \mathsf{Gen}(1^\lambda), c \overset{\$}{\leftarrow} \mathsf{Enc}(pk, m)] = 1$$

3. **Simulatability:** *The simulator $\mathsf{Sim} = (\mathsf{Sim}_1, \mathsf{Sim}_2)$ For any message $m \in \mathcal{M}$, define the distributions:*

$$\Lambda_m^{\mathrm{Sim}} = \{(m, pk, sk, r_1, r_2) : (pk, c, t) \overset{\$}{\leftarrow} \mathsf{Sim}_1(1^\lambda), (sk, r_1, r_2) \overset{\$}{\leftarrow} \mathsf{Sim}_2(m, t)\}$$

and

$$\Lambda_m^{\mathrm{Real}} = \{(m, pk, sk, r_1, r_2) : (pk, sk) \overset{\$}{\leftarrow} \mathsf{Gen}(1^\lambda; r_1), c \overset{\$}{\leftarrow} \mathsf{Enc}(pk, m; r_2)\}$$

We require that $\Lambda_m^{\mathrm{Sim}} \approx_c \Lambda_m^{\mathrm{Real}}$.

We do not explicitly require that the system be semantically secure, this is implied by the simulatability of the system.

2.3   The Φ-hiding Assumption

The security of our construction rests on the Φ-hiding assumption [4]. Informally, the Φ-hiding assumption asserts that it is hard to determine whether a given integer, e, divides the size of the group $\mathbb{Z}_N^*$ when $N = pq$ is an RSA modulus.

Let $\mathsf{PRIMES}(\lambda)$ denote the set of primes of bit-length λ, and let $\mathsf{RSA}(\lambda)$ denote the set of RSA moduli of bit-length λ, i.e.,

$$\mathsf{RSA}(\lambda) = \{N : N = pq, p, q \in \mathsf{PRIMES}(\lambda/2), \gcd(p-1, q-1) = 2\}$$

Let $\mathsf{RSA}_e(\lambda)$ denote the set of RSA moduli of length λ that Φ-hide e.

$$\mathsf{RSA}_e(\lambda) = \{N \in \mathsf{RSA}(\lambda) : e \text{ divides } p-1\}$$

The Φ-hiding assumption states that a random sample from $\mathsf{RSA}_e(\lambda)$ is computationally indistinguishable from a random sample from $\mathsf{RSA}(\lambda)$.

Definition 2 (The Φ-hiding assumption).
For all $\epsilon > 0$, and for all $e \in \mathbb{Z}$ such that $3 < e < 2^{\lfloor \log(\lambda/4) \rfloor - \epsilon}$ and for all PPT distinguishers, A

$$\left| \Pr[A(N, e) : N \xleftarrow{\$} \mathsf{RSA}(\lambda)] - \Pr[A(N, e) : N \xleftarrow{\$} \mathsf{RSA}_e(\lambda)] \right| < \nu(\lambda)$$

For some negligble function $\nu(\cdot)$.

The original Φ-hiding assumption stated that $\mathsf{RSA}_e(\lambda)$ and $\mathsf{RSA}_{e'}(\lambda)$ are indistinguishable for all $3 < e \le e' < 2^{\lambda/5}$. For simplicity, we use a slightly modified assumption (i.e., that $\mathsf{RSA}_e(\lambda) \approx_c \mathsf{RSA}(\lambda)$). All of our constructions go through essentially unchanged under the original Φ-hiding assumption, except that the key generator would have to choose an e to sample from $\mathsf{RSA}_e(\lambda)$ and then ignore e throughout the protocol.

The bound $e < 2^{\lfloor \log(\lambda/4) \rfloor - \epsilon}$ is necessary to avoid Coppersmith's attack [9]. The Φ-hiding assumption was originally used to build efficient Private Information Retrieval (PIR) protocols [4,13]. In [17] it was observed that the Φ-hiding assumption turns the RSA map $x \mapsto x^e \bmod N$ into a lossy trapdoor function (LTF) [20]. Kiltz, O'Neill and Smith leveraged the lossiness to remove the random oracle from the proof of security of RSA-OAEP. More recently, the regularity of the lossy RSA map has been explored [18].

3   Smooth Hash Proofs from Φ-hiding

Our work builds on the notion of smooth hash proof systems as defined by Cramer and Shoup [10]. A hash proof system for a language, L, is a set of keyed hash functions H_k, along with "public" and "private" evaluation algorithms for H_k. A hash function H_k (for $k \in K$) takes $X \to \Pi$.

The public algorithm takes $x \in L$, along with a witness w, and the "projected key" $\alpha(k)$, and outputs $H_k(x)$. The private evaluation algorithm simply takes x, k and outputs $H_k(x)$. While the public and private evaluations must match on $x \in L$, if $x \notin L$, the smoothness property says that $H_k(x)$ should be almost uniformly distributed (even conditioned on $\alpha(k)$). Formally, we have

Definition 3 (Hash Proof Systems [10]). *The set* $(H, K, X, L, \Pi, S, \alpha)$ *is a hash proof system if, for all* $k \in K$, *the action of* H_k *on the subset* L *is completely determined by* $\alpha(k)$.

Thus for a hash proof system the projection key $\alpha(k)$ determines the action of H_k on L. We also require that given $x \in L$, and a witness w, along with the projection key $\alpha(k)$ then the hash value $H_k(x)$ can be *efficiently* computed. This is called the *public evaluation algorithm*.

A hash proof system is *smooth* if the projected key $\alpha(k)$ encodes almost no information about the action of $H_k(\cdot)$ outside L.

Definition 4 (Smooth Hash Proof Systems [10]). *Let* $(H, K, X, L, \Pi, S, \alpha)$ *be a hash proof system, and define two distributions* Z_1, Z_2 *taking values on the set* $X \setminus L \times S \times \Pi$. *For* Z_1, *we sample* $k \leftarrow K$, $x \leftarrow X \setminus L$, *and set* $s = \alpha(k)$, $\pi = H_k(x)$, *for* Z_2 *we sample* $k \leftarrow K$, $x \leftarrow X \setminus L$, *and* $\pi \leftarrow \Pi$, *and set* $s = \alpha(k)$. *The hash proof system is called* ν-*smooth if* $\Delta(Z_1, Z_2) < \nu$.

We will call an infinite family of hash proof systems smooth if the family is ν-smooth for some negligible function ν. For a language, L with no polynomial-time distinguisher, smooth hash proof systems immediately imply IND-CPA secure encryption by letting k be the secret key, $\alpha(k)$ be the public key, and to encrypt a message $m \in \Pi$ you sample $x \in L$ along with a witness and use the public evaluation algorithm to output $E(m) = (x, H_k(x) + m)$.

We begin by describing a simple smooth hash proof system from the Φ-hiding assumption. Building a hash proof system based on Φ-hiding is straightforward, because Φ-hiding essentially implies that the discrete-log problem is hard[1], and then we can used the framework for the original discrete-log based hash proof systems from [10].

- **Key Generation:** Choose a strong $(\tau, 2^{-\lambda})$-extractor[2] $\mathrm{Ext} : \{0,1\}^d \times \mathbb{Z}_N \rightarrow \{0,1\}$ Fix a prime p with $2^\tau < p < N^{\frac{1}{4}}$. Sample $N \xleftarrow{\$} \mathrm{RSA}_p(\lambda)$ Sample $r \xleftarrow{\$} \mathbb{Z}_N$. Set $g = r^p \in \mathbb{Z}_N$.
  Sample an extractor seed, $\mathfrak{s} \xleftarrow{\$} \{0,1\}^d$.
  The language will be the set of pth powers in $\mathbb{Z}_N$.[3]
  Sample $k \xleftarrow{\$} \mathbb{Z}_N$ Set the projection key, $\alpha(k) = g^k \mod N$.

---

[1] Suppose g is a generator for the largest cyclic component of $\mathbb{Z}_N^*$ and let $h = g^p$ mod N. If N Φ-hides p, then h most elements of $\mathbb{Z}_N^*$ will *not* be powers of h. On the other hand, if $\gcd(p, \varphi(N)) = 1$, then h will generate the largest cyclic component of $\mathbb{Z}_N^*$. If we had an oracle that could find the discrete-logarithm (with base h) in $\mathbb{Z}_N^*$, then we could distinguish these two cases. For the same reason, the Φ-hiding assumption also implies that the determining membership in the subgroup of pth powers is hard.

[2] In other words, Ext, if X is a random variable supported on $\mathbb{Z}_N$ with min-entropy at least τ, then if $\mathfrak{s} \xleftarrow{\$} \{0,1\}^d$ the statistical distance between $(\mathfrak{s}, \mathrm{Ext}_\mathfrak{s}(X))$ and $(\mathfrak{s}, U_{\{0,1\}})$ is less than $2^{-\lambda}$.

[3] Note that if N Φ-hides p then $\mathbb{Z}_N^*$ has a cyclic subgroup of order p, and so at most $\varphi(N)/p$ residues in $\mathbb{Z}_N^*$ that are pth powers, otherwise all elements in $\mathbb{Z}_N^*$ will be pth powers.

– **Public Evaluation:** Given $x = g^w \mod N$, and a witness, w,

$$H_k(x) = \text{Ext}_{\mathfrak{s}}\left((g^k)^w \mod N\right)$$

– **Private Evaluation:** Given the secret key, k, and a point x,

$$H_k(x) = \text{Ext}_{\mathfrak{s}}\left(x^k \mod N\right)$$

Correctness is easy to verify. To see smoothness, suppose N Φ-hides p. Then, $\mathbb{Z}_N^*$ has a subgroup of order p, but $g = r^p \mod N$ has no component in that subgroup. In particular, $h = g^k \mod N$, only depends on $k \mod \varphi(N)/p$.

Now, if x has full order in the largest cyclic subgroup[4] of $\mathbb{Z}_N^*$ then, assuming $\gcd(p, w) = 1$, the value $(x^w)^k \mod N$ depends on all of k, and in particular has $\log p$ bits of entropy even conditioned on $k \mod \varphi(N)/p$ (and hence the same holds conditioned on g^x). Thus the extractor produces an almost uniform value, even conditioned on the projected key $g^k \mod N$.

3.1   A Compact Hash Proof System from the Φ-hiding Assumption

In this section, we build a simple IND-CPA secure cryptosystem that will be the foundation of our full non-committing construction.

– **Key Generation:** Let ℓ be the (bit) length of messages that will be encrypted. Choose a strong $(\tau, 2^{-\lambda})$-extractor[5] $\text{Ext} : \{0,1\}^d \times \mathbb{Z}_N \to \{0,1\}$
Let $p_i = (i+2)$nd prime,

$$(p_1, p_2, p_3, \ldots) = (5, 7, 11, \ldots)$$

Let $e_i = \lceil \tau \log_{p_i} 2 \rceil$ for $i = 1, \ldots, \ell$. Let $q_i = p_i^{e_i}$, thus $q_i > 2^\tau$.
Sample $N \xleftarrow{\$} \text{RSA}(\lambda)$.
Sample an extractor seed, $\mathfrak{s} \xleftarrow{\$} \{0,1\}^d$. Sample $a_1, \ldots, a_\ell \xleftarrow{\$} \mathbb{Z}_N$. Set

$$h_i = \left(g^{\prod_{j \neq i} q_i}\right)^{a_i} \mod N$$

for $i = 1, \ldots, \ell$.
– **Encryption:** Given a message $m = (m_1, \ldots, m_\ell) \in \{0,1\}^\ell$

Pick $w \xleftarrow{\$} \mathbb{Z}_N$

$$c = (g^w \mod N, \mathfrak{s}, c_1, \ldots, c_\ell)$$

where

$$c_i = \text{Ext}_{\mathfrak{s}}\left((h_i)^w\right) \oplus m_i$$

---

[4] If we choose N to be the product of two primes, N_1, N_2 and $\gcd(N_1 - 1, N_2 - 1) = 2$, then we have that $\mathbb{Z}_N^*$ has a cyclic subgroup of order $\varphi(N)/2 = (N_1 - 1)(N_2 - 1)/2$, and we can work in this group.

[5] The explicit extractors of [15] give seed length $d = \mathcal{O}(\log \log N + \log 1/\epsilon)$ and can extract $\omega(k)$ bits where k is the min-entropy of the source. In our construction, we will only need to extract one bit.

– **Decryption:** Notice that

$$(h_i)^w = \left((g^w)^{\Pi_{j \neq i} q_j}\right)^{a_i}$$

Thus the receiver, who has a_i, can recover m_i by setting

$$m_i = \text{Ext}_s \left(\left((g^w)^{\Pi_{j \neq i} q_j}\right)^{a_i}\right) \oplus c_i$$

The IND-CPA security of this scheme follows immediately from the security of the underlying hash proof system.

Remark 1: Since messages must be polynomial in the security parameter, λ, we have that $\ell \log \ell < \sqrt[5]{N} \approx 2^{\lambda/5}$, so the conditions of the Φ-hiding assumption are satisfied.

Remark 2: Although the security follows from the Φ-hiding assumption, the honest receiver does not need to choose a modulus that Φ-hides anything.

Remark 3: Notice that the factorization of N is never used, and need not be known for decryption.

4 Non-committing Encryption from Φ-hiding

The high level idea is as follows: to encrypt a message m the sender encodes the message m using a standard binary error correcting code. The sender will then flip a small number of bits of the codeword. The sender then encrypts each bit of this corrupted codeword separately (using different public-keys, but the same encryption randomness).

The receiver will know only a subset of the total set of possible secret keys. The receiver will decode only the bits corresponding to the keys he knows. With high probability, the overlap between the receiver's decryption keys and the bits of the codeword flipped by the sender will be small enough to be handled by the error correcting code. The formal scheme is given in Figure 2.

4.1 Probability of Decryption Error

First, we examine the probability of decryption error. It is a simple calculation to verify that the bits in $I_R \cap I_S \subset [n]$ will be decrypted correctly. Since I_R is uniformly chosen, and the bits in $I_R \cap I_S$ are decrypted correctly while the others will be in error with probability $\frac{1}{2}$, we can bound the probability of decryption error.

Lemma 1. *The probability the receiver decrypts more than δn bits incorrectly is bounded by*

$$\Pr[\text{ more than } \delta n \text{ bits are decrypted incorrectly}] \leq e^{-\frac{1}{2}c_r^2 c_s n} + e^{-2\left(\frac{c_r c_s}{2} + \delta - \frac{1}{2}\right)^2 n / \left(1 - \frac{c_r c_s}{2}\right)}$$

Proof. First, we lower-bound $|I_R \cap I_S|$ (recall these bits are always decrypted correctly).

- **Key Generation:**
  Choose an error correcting code $\mathbf{ECC} : \{0,1\}^\ell \to \{0,1\}^n$ such that $n = \mathcal{O}(\ell)$ that is recoverable from a $\frac{1}{2} - \delta$ fraction of errors.[a] Sample $N \xleftarrow{\$}$ RSA$(\min(\lambda, \mathcal{O}(\ell \log \ell))$. Choose a strong $(\tau, 2^{-\lambda})$ extractor $\mathrm{Ext} : \{0,1\}^d \times \mathbb{Z}_N \to \{0,1\}$. Let $q_i = p_i^{e_i}$ where p_i is the $(i+2)nd$ prime and e_i is the smallest integer such that $q_i > 2^\tau$.
  Chooses a random subset $I_R \subset [n]$, with $|I_R| = c_r n$. Then

  - for $i \in I_R$, the receiver chooses $a_i \xleftarrow{\$} \mathbb{Z}_N$ and sets $h_i = \left(g^{\Pi_{j \neq i} q_i}\right)^{a_i}$ mod N.
  - For $i \in [n] \setminus I_R$, The receiver chooses $h_i \xleftarrow{\$} \mathbb{Z}_N$.

  The receiver will know the decryption key, k_i, for the h_i with $i \in I_R$. The receiver will not be able to decrypt messages sent under public-keys h_i for $I \in [n] \setminus I_R$. As in the basic scheme, the factorization of N is *not* part of the secret key, and will not be used for decryption.

  $$\text{public key: } \{h_i\}_{i=1}^n \qquad \text{secret key: } \{a_i\}_{i \in I_R}$$

- **Encryption:**
  Given a message $m = m_1, \ldots, m_\ell \in \{0,1\}^\ell$, the sender encodes m as $(y_1, \ldots, y_n) = \mathbf{ECC}(m)$, then chooses a random $w \xleftarrow{\$} \mathbb{Z}_N$, and sets $x = g^w$ mod N. The sender chooses a random extractor seed, $\mathfrak{s} \xleftarrow{\$} \{0,1\}^d$. Then the sender chooses a random subset $I_S \subset [n]$, with $|I_S| = c_s n$ and

  - for $i \in I_S$, the sender sets

    $$c_i = \mathrm{Ext}_{\mathfrak{s}}\left((h_i)^w \mod N\right) \oplus y_i$$

  - for $i \in [n] \setminus I_S$, the sender chooses b_i uniformly from $\{0,1\}$, and sets $c_i = b_i$.

  The encryption randomness is $\{I_S, w, \mathfrak{s}, \{b_i\}_{i \in [n] \setminus I_S}\}$. The sender sends the ciphertext

  $$C = (x, \mathfrak{s}, c_1, \ldots, c_n) \in \mathbb{Z}_N \times \{0,1\}^d \times \{0,1\}^n$$

- **Decryption:**
  Given a ciphertext $C = (x, \mathfrak{s}, c_1, \ldots, c_n)$ and a secret key $\{k_i\}_{i \in I_R}$ the receiver decrypts as follows:

  - if $i \in I_R$ the receiver calculates

    $$y_i' = \mathrm{Ext}_{\mathfrak{s}}\left(\left(x^{\Pi_{j \neq i} q_j}\right)^{a_i}\right) \oplus c_i = \mathrm{Ext}_{\mathfrak{s}}\left(\left((g^w)^{\Pi_{j \neq i} q_j}\right)^{a_i}\right) \oplus c_i = \mathrm{Ext}_{\mathfrak{s}}\left(\left((g^{a_i})^{\Pi_{j \neq i} q_j}\right)^w\right) \oplus c_i$$

  - if $I \in [n] \setminus I_R$ the receiver sets $y_i' = \bot$.

  Given $y_1', \ldots, y_n'$ the receiver decodes using the decoding algorithm for $\mathbf{ECC}$ to obtain a message $m_1', \ldots, m_\ell'$.

---

[a] Constant rate binary codes that can efficiently recover from a $\frac{1}{2} - \delta$ fraction of errors exist in the bounded channel model [19].

Fig. 2. Non-committing encryption from Φ-hiding

Since I_R and I_S are both chosen uniformly at random, we can imagine fixing I_R and choosing I_S at random. Now, let X_i be the indicator variable indicating whether the ith element of I_S is in I_R. Let $X = \sum_{i=1}^{|I_S|} X_i$. Thus $X = |I_R \cap I_S|$. Then $\Pr[X_i = 1] = c_r$, so $E(X) = c_r|I_S| = c_r c_s n$.

Since the X_i are negatively correlated we can apply a Chernoff bound ([12] Thms 1.1 and 4.3) to obtain

$$\Pr[X < E(X) - t] \le e^{-2t^2/|I_S|}$$

taking $t = \frac{E(X)}{2} = \frac{c_r c_s n}{2}$ we have

$$\Pr\left[X < \frac{c_r c_s n}{2}\right] < e^{-\frac{1}{2}c_r^2 c_s n}$$

Now, for the elements in $[n] \setminus I_R \cap I_S$, these have a half chance of decrypting correctly. Let Y_i be the indicator random variable that indicates whether the ith element of $[n] \setminus I_R \cap I_S$ decrypts correctly. Thus $E(Y_i) = \frac{1}{2}$. If we let $Y = \sum_{i=1}^{n-|I_R \cap I_S|} Y_i$, we have $E(Y) = \frac{n-X}{2} = (n - |I_R \cap I_S|)/2$.

Now X is the size of $|I_R \cap I_S|$ and Y is the number of bits outside of $I_R \cap I_S$ that are decrypted correctly. So the probability that more than δn of the bits are decrypted incorrectly is

$$
\begin{aligned}
&\Pr\left[X + Y < (1 - \delta)n\right] \\
&= \Pr\left[Y < (1 - \delta)n - X\right] \\
&= \Pr\left[Y < E(Y) - (E(Y) - (1 - \delta)n + X)\right] \\
&= \Pr\left[Y < E(Y) - \left(\frac{n - X}{2} - (1 - \delta)n + X\right)\right] \\
&= \Pr\left[Y < E(Y) - \left(\frac{X}{2} - \left(\frac{1}{2} - \delta\right)n\right)\right] \\
&= \Pr\left[Y < E(Y) - \left(\frac{X}{2} - \left(\frac{1}{2} - \delta\right)n\right) \,\middle|\, X < \frac{c_r c_s n}{2}\right] \Pr\left[X < \frac{c_r c_s n}{2}\right] \\
&\quad + \Pr\left[Y < E(Y) - \left(\frac{X}{2} - \left(\frac{1}{2} - \delta\right)n\right) \,\middle|\, X \ge \frac{c_r c_s n}{2}\right] \Pr\left[X \ge \frac{c_r c_s n}{2}\right] \\
&\le e^{-\frac{1}{2}c_r^2 c_s n} + \Pr\left[Y < E(Y) - \left(\frac{X}{2} - \left(\frac{1}{2} - \delta\right)n\right) \,\middle|\, X \ge \frac{c_r c_s n}{2}\right] \\
&\le e^{-\frac{1}{2}c_r^2 c_s n} + \Pr\left[Y < E(Y) - \left(\frac{c_r c_s}{2} + \delta - \frac{1}{2}\right)n \,\middle|\, X \ge \frac{c_r c_s n}{2}\right] \\
&\le e^{-\frac{1}{2}c_r^2 c_s n} + e^{-2\left(\left(\frac{c_r c_s}{2} + \delta - \frac{1}{2}\right)n\right)^2 / \left((1 - \frac{c_r c_s}{2})n\right)} \quad (X \ge \frac{c_r c_s n}{2} \text{ and Chernoff}) \\
&\le e^{-\frac{1}{2}c_r^2 c_s n} + e^{-2\left(\frac{c_r c_s}{2} + \delta - \frac{1}{2}\right)^2 n / (1 - \frac{c_r c_s}{2})}
\end{aligned}
$$

In the second to last line, recall that X is the number of summands in Y so the Chernoff bound for Y gets better as X increases.

Note that to apply Chernoff, we need $\frac{c_r c_s}{2} + \delta - \frac{1}{2} > 0$, in other words $\delta > \frac{1}{2} - \frac{c_r c_s}{2}$. Thus we will need a code **ECC** that can correct from an error-rate of $\frac{1}{2} - \frac{c_r c_s}{2}$. This can be achieved, and in fact constant-rate codes that can recover from a δ fraction of errors exist for any $\delta < \frac{1}{2}$ in the bounded channel model [19].

5   The Simulator

The simulation game in a non-committing encryption proceeds as follows: the simulator generates a public-key and a ciphertext. The adversary then chooses a target message, and the simulator must produce a secret key and encryption randomness such that the entire ensemble of public-key, secret key, ciphertext, encryption randomness and target message is indistinguishable from a valid run of the key-generation and encryption algorithms encrypting the target message.

- **Key Generation:** As in the real scheme the simulator makes use of an error correcting code **ECC** : $\{0,1\}^\ell \to \{0,1\}^n$, and an extractor Ext : $\{0,1\}^d \times \mathbb{Z}_N \to \{0,1\}$.

  Note that unlike the valid sender and receiver, the simulator *will* need the factorization of the RSA modulus N (details below).

  - The simulator chooses a random extractor seed $\mathfrak{s} \xleftarrow{\$} \{0,1\}^d$.
  - The simulator uniformly chooses a subset $I_{\text{good}} \subset [n]$ of size $c_g n$, and sets $I_{\text{bad}} = [n] \setminus I_{\text{good}}$. Let $c_b = \frac{|I_{\text{bad}}|}{n} = 1 - c_g$.
  - The simulator generates an Φ-hiding modulus, N, such that N Φ-hides q_j for $j \in I_{\text{bad}}$.[6]
  - The simulator chooses $r \xleftarrow{\$} \mathbb{Z}_N$ and sets $g = r^{\prod_{j \in I_{\text{bad}}} q_j} \mod N$.
  - The simulator then chooses keys $a_i \xleftarrow{\$} \mathbb{Z}_N$ for $i = 1, \ldots, n$.
  - The simulator then outputs:

  $$\text{public parameters: } N, g \qquad \text{Public key:} \{h_i\}_{i=1}^n$$

The public-key will be the collection $\{h_i\}_{i=1}^n$. A crucial point of our scheme is that the simulator knows the decryption keys for all h_i for all i, and will be able to decrypt for all $i \in I_{\text{good}}$, unlike in the real scheme, where the receiver only knows the keys for some smaller subset I_R.

- **Ciphertext Generation:**

  - The simulator generates $w \xleftarrow{\$} \mathbb{Z}_n$
  - The simulator partitions I_{good} into two random disjoint sets A_1 and A_0, with $|A_1| \approx |A_0| \approx c_g n/2$, and $A_1 \cup A_0 = I_{\text{good}}$.

---

[6] This means that $N \geq \left(\prod_{j \in I_{\text{bad}}} q_j\right)^5$ to avoid Coppersmith's attack. Since $|I_{\text{bad}}| = \mathcal{O}(n)$, and the $p_i \approx i \log i$ by the Prime Number Theorem, we have $N \approx (n \log n)^{\mathcal{O}(n)}$, which implies $\log N = \mathcal{O}(n \log n)$.

- For $i \in A_1$, the simulator sets $c_i = \text{Ext}_{\mathfrak{s}}(h_i^w \mod N) \oplus 1$ and for $i \in A_0$, the simulator sets $c_i = \text{Ext}_{\mathfrak{s}}(h_i^w \mod N)$.
- For $i \in I_{\text{bad}}$ the simulator chooses a random bit b_i and sets $c_i = b_i$.

The simulator outputs the ciphertext

$$C = (x, \mathfrak{s}, c_1, \ldots, c_n) \in \mathbb{Z}_N \times \{0,1\}^d \times \{0,1\}^n$$

- **Simulating Encryption Randomness:** After generating a public-key and a ciphertext, the simulator is given a target message m^*. The simulator calculates $y^* = \mathbf{ECC}(m^*)$.

  The receiver chooses the set I_S as follows: Let $M_1 = \{i : y_i^* = 1\}$, and $M_0 = \{i : y_i^* = 0\}$.

  If $|M_1 \cap A_1 \cup M_0 \cap A_0| < c_s n$ the simulator outputs $\bot$.

  We show in Section 5.1 that this happens with negligible probability.

  Otherwise, the simulator samples $c_s n$ elements uniformly from $M_1 \cap A_1 \cup M_0 \cap A_0$ and calls this set I_S.

- **Generating the Secret Key:** If $|M_1 \cap A_1 \cup M_0 \cap A_0| < c_r n$ the simulator outputs $\bot$.

  We show in Section 5.1 that this happens with negligible probability.

  Otherwise, the simulator samples $c_r n$ elements uniformly from $|M_1 \cap A_1 \cup M_0 \cap A_0|$ and calls this set I_R. Since $A_1 \cup A_0 = I_{\text{good}}$, and the receiver knows all the secret keys corresponding to the indices in I_{good} the simulator can output a_i for $i \in I_R$.

5.1   Good Randomness and Secret Keys Exists with Overwhelming Probability

The simulator outputs $\bot$ if $|M_1 \cap A_1 \cup M_0 \cap A_0| < \max(c_r, c_s)n$, thus we need to show that this happens with only negligible probability. Recall that M_i is the set of indices where the (encoded) target message has value i, whereas A_i is the set of indices where the ciphertext encrypts value i. The sets A_i were chosen uniformly at random, and the sets M_i were adversarially chosen (since they are determined by the adversary's choice of message).

We begin by showing that $|M_1 \cap A_1 \cup M_0 \cap A_0|$ will be large if M_0, M_1 were chosen independently of A_0, A_1. Then we will argue that the adversary cannot do much better than choosing M_0, M_1 independently.

If M_0 and M_1 were chosen independently of A_0, and A_1, then the probability that the simulator outputs $\bot$ can be bounded by Chernoff. In particular, we have $E(|M_i \cap A_i|) = \frac{|A_i||M_i|}{n} = \frac{c_g|M_i|}{2}$ (since $|A_i| = \frac{c_g}{2}n$).

$$\Pr\left[|M_i \cap A_i| < \frac{c_g|M_i|}{2} - \frac{c_g|M_i|}{4}\right] < e^{-2\frac{c_g|M_i|^2}{8n}}$$

Since $|M_0| + |M_1| = n$, at least one of the $|M_i|$ will have $|M_i| \geq \frac{n}{2}$, thus

$$\Pr\left[|M_1 \cap A_1 \cup M_0 \cap A_0| < \frac{c_g n}{8}\right] < e^{-2\frac{c_g n}{32}}$$

which is negligible in n. Thus we have the desired result whenever $\max(c_r, c_s) < \frac{c_g}{8}$.

Now, the sets M_0, M_1 are not independent of A_0, A_1 because they were generated adversarially after seeing the ciphertext C (which encodes A_0, A_1). Since the adversary is polynomially-bounded, and the sets A_0 and A_1 are hidden by the semantic security of the underlying Φ-hiding based encryption, a standard argument shows that if the adversary could cause $|M_1 \cap A_1 \cup M_0 \cap A_0| > \min(c_r, c_s)n$ with non-neglible probability, then this adversary could be used to break the Φ hiding assumption. (Note that the simulator does *not* need the factorization of N to determine the size $|M_1 \cap A_1 \cup M_0 \cap A_0|$.)

5.2   Simulating the Choice of w

Simulating w. Most of the technical difficulty in this construction lies in finding a suitable w^* for the simulator to output. Since $g^w \bmod N$ is part of the ciphertext, $g^{w^*} = g^w \bmod N$.

Since g is a power of q_j for all $j \in I_{\text{bad}}$ this means $w = w^* \bmod \varphi(N)/$ $(\prod_{j \in I_{\text{bad}}} q_j)$. If we simply output $w^* = w$, the adversary can distinguish the simulator's output from a true output.

In a valid encryption for $i \in [n] \setminus I_s$ about half of the c_i should equal $\text{Ext}_s(h_i^w \bmod N) + y_i$, but if the simulator simply outputs $w^* = w$, by using w^* to "decrypt" c_i for $i \in [n] \setminus I_S$ and the adversary will find that many fewer than half the locations $i \in [n] \setminus I_s$ match the encoded message. In the next section we will provide an algorithm for choosing w^*.

Finding a Target set T. The simulator will have revealed secret keys for all $i \in I_R \subset M_1 \cap A_1 \cup M_0 \cap A_0 \subset I_{\text{good}}$. The simulator cannot lie about values whose indices are in I_{good}.

From the adversary's view, for $i \in [n] \setminus I_S$ the encryptions c_i will be encryptions of y^*. In an honest execution of the encryption algorithm, for $i \notin [n] \setminus I_S$, the encryptions, c_i should be uniformly random (in particular, they should be independent of y^*).

Thus the simulator needs to choose values in I_{bad} so that the c_i for $i \in [n] \setminus I_S$ encrypt values that are uniformly distributed on $\{0, 1\}$. This is because the honest encrypter will choose c_i at random for $i \notin I_S$.

Let $\alpha_b = \{i \in I_{\text{good}} \setminus I_S : \text{Ext}_s(g^w \bmod N) \oplus b = c_i\}$. Thus α_b denotes the number of indices that are encryptions of b (that the simulator cannot change).

In an honest execution, the number of encryptions of 0 that an adversary sees in $[n] \setminus I_S$ is binomially distributed with parameters $n - |I_S| = (1 - c_s)n$ and $\frac{1}{2}$.

The simulator will sample a random variable, X, whose distribution is a binomial, shifted by α_1.

$$\Pr[X = j] = \binom{(1 - c_s)n}{j + \alpha_1} \left(\frac{1}{2}\right)^{(1-c_s)n} \quad \text{for } j = -\alpha_1, \ldots, (1 - c_s n) - \alpha_1$$

If $X > c_b n$ or $X < 0$ then the simulator will return $\perp$.[7]

Otherwise, the simulator will choose a random subset of I_{bad} of size X, and set those X elements 1, and the remaining $c_b - X$ elements to 0.

If we denote $I_{bad} = \{i_1, \ldots, i_{c_b n}\}$, then the simulator has chosen $\{t_{i_j}\}_{j=1}^{c_b n}$ so that exactly X of the t_{i_j} are 1, and $c_b n - X$ of the t_{i_j} are 0.

Now, consider the distribution of encryptions of 1 in the set $[n] \setminus I_S$. There are exactly $\alpha_1 + X$ ones in that set, and assuming the simulator does not output $\perp$, the probability of this occurring is exactly $\binom{(1-c_s)n}{\alpha_1 + X} \left(\frac{1}{2}\right)^{(1-c_s)n}$ which is exactly the binomial distribution.

Calculating w^*. Denote $I_{bad} = \{i_1, \ldots, i_{c_b n}\}$. For $i \in I_{bad}$ the simulator has set $h_i = \left(g^{\prod_{j \neq i} q_j}\right)^{a_i} = \left(\left(r^{\prod_{j \in I_{bad}} q_j}\right)^{\prod_{j \neq i} q_j}\right)^{a_i} \mod N$. For the bad set I_{bad} the ciphertexts are "lossy", so the smoothness of the hash proof system makes the ciphertexts non-binding. Suppose the simulator has chosen target values $\{t_i\}_{i \in c_b n} \subset \{0,1\}^{c_b n}$ (using the methods in the previous section). The simulator's goal is for c_{i_j} to be an encryption of t_j for $j = 1, \ldots, c_b n$. We will show how the simulator can choose a w^* so that c_{i_j} looks like an encryption of t_j for $j = 1, \ldots, c_b n$. In particular, this means

$$\text{Ext}_s \left(h_{i_j}^{w^*} \mod N\right) = c_{i_j} \oplus t_j \quad \text{for } j = 1, \ldots, c_b n$$

Now, $w^* = w \mod \varphi(N) / \left(\prod_{j \in I_{bad}} q_j\right)$. The simulator then runs the following iterative algorithm (Algorithm 1) to find w^*. (Recall: $I_{bad} = \{i_1, \ldots, i_{c_b n}\}$.)

Algorithm 1. Calculating w^*

$w^* = w$

for $j = 1, \ldots, c_b n$ **do**

$\quad$ **while** $\text{Ext}_s \left(g^{w^*} \mod N\right) \neq c_{i_j} \oplus t_j$ **do**

$\quad\quad w^* = w^* + \varphi(N)/q_{i_j} \mod \varphi(N)$

$\quad$ **end while**

end for

First, notice that at every step $w^* = w \mod \varphi(N) / \prod_{j \in I_{bad}} q_j$. Second, notice that when processing target index $i_j \in I_{bad}$, the value of w^* does change modulo

---

[7] Chernoff says that this happens with negligible probability. In particular, by Chernoff

$$\Pr[X < 0] < e^{-2 \frac{\left(\frac{(1-c_s)n}{2} - \alpha_1\right)^2}{(1-c_s)n}}$$

and

$$\Pr[X < c_b n] < e^{-2 \frac{\left(c_b - \frac{(1-c_s)n}{2}\right)^2}{(1-c_s)n}}$$

(see for example [5] Claim 4.6)

q_{i_j}, but not for any other primes $q_k \in I_{bad} \setminus \{i_j\}$. Since g is *not* a q_{i_j}th power, the value g^{w^*} changes as w^* changes modulo q_{i_j}. Thus the expected running time of the while loop is only 2.

6   Efficiency and Parameters

A codeword in our cryptosystem is given by

$$C = (x, \mathfrak{s}, c_1, \ldots, c_n) \in \mathbb{Z}_N \times \{0,1\}^d \times \{0,1\}^n$$

Since $x \in \mathbb{Z}_N$, $\mathfrak{s} \in \{0,1\}^d$, and $c_i \in \{0,1\}$, the length of the codeword is $\log N + d + n$. Using the explicit extractors of [15], we can set the seed length d to be $\mathcal{O}(\log \log N + \log 1/\epsilon)$, so setting $\epsilon = 2^{-\lambda}$, we can choose $d = \mathcal{O}(\lambda)$, and extract bits that are negligibly close to uniform.

Now, the element $x \in \mathbb{Z}_N$, so $|x| = \log N$. Since the modulus, N, generated by the simulator must be able to Φ-hide all q_j for $j \in I_{bad} \subset [n]$. Thus $\sqrt[5]{N} \geq \prod_{j \in I_{bad}} q_j$ to avoid Coppersmith's attack. Since $|I_{bad}| = \mathcal{O}(n)$, and the $p_i \approx i \log i$ by the Prime Number Theorem so $N \approx (n \log n)^{\mathcal{O}(n)}$, which means that $\log N = \mathcal{O}(n \log n)$. This means the codewords are of length $\mathcal{O}(n \log n + \lambda + n) = \mathcal{O}(n \log n)$. In our scheme, the plaintext messages are of length ℓ, and $n = |\mathbf{ECC}(m)| = \mathcal{O}(\ell)$, so $\mathcal{O}(n \log n) = \mathcal{O}(\ell \log \ell)$. Thus our scheme achieves a logarithmic ciphertext expansion. By contrast, the best previously known schemes have a ciphertext expansion of λ [8]. By assumption, super-polynomial-time calculations in λ are infeasible while polynomial-time calculations in ℓ (and hence exponential-time calculations in $\log(\ell)$) are efficient. Thus $\lambda \gg \log \ell$, so our scheme achieves an improved ciphertext expansion.

Table 1. Summary of the parameters in our scheme

λ	Security Parameter
ℓ	Message length
n	Codeword length ($n = |\mathbf{ECC}(m)| = \mathcal{O}(\ell)$)
N	Φ-hiding modulus ($\mathcal{O}(n \log n)$ bits)
δ	fraction of errors tolerated by $\mathbf{ECC}$ ($\delta > \frac{1}{2} - \frac{c_r c_s}{2}$, so $\delta = \mathcal{O}(1)$)
c_r	fraction of indices where receiver can decrypt ($c_r = \mathcal{O}(1)$)
c_s	fraction of indices correctly encoded by sender ($c_s = \mathcal{O}(1)$)
c_g	fraction of "good" indices known by the simulator ($\max(c_r, c_s) < \frac{c_g}{8}$)
d	Length of extractor seed ($d = \mathcal{O}(\lambda)$)
$\log N + d + n$	Codeword length
$n \log N$	Public-key length
$c_r n \log N$	Private-key length

7   Conclusion

In this work, we constructed the first non-committing encryption scheme with messages of length $|m|$ and ciphertexts of length $\mathcal{O}(|m| \log |m|)$. This improves on the best prior constructions which had ciphertexts of length $\mathcal{O}(|m|\lambda)$ [8]. Like most previous protocols, a public key in our system consists of n public-keys for which the receiver has decryption keys for only a small subset. The sender then encodes the message using an error-correcting code, and encrypts each bit of the encoded message using a different key. Since each bit of the encoded message is encrypted as a separate ciphertext, a blowup of $\mathcal{O}(\lambda)$ seems unavoidable. The main technical contribution of our work is to show that by using the Φ-hiding assumption, we can encode each bit of the encoded message using a different key, but the same randomness, thus eliminating the factor $\mathcal{O}(\lambda)$ ciphertext blowup. Many cryptosystems allow randomness re-use across different keys, (indeed El-Gamal has this property, and this is the basis of most Lossy-trapdoor function constructions [20,16]). The difficulty arises in generating a system where false randomness can be *efficiently* generated when the randomness is re-used across multiple keys. To achieve this, we turn to the Φ-hiding assumption [4].

It remains an interesting open question whether this protocol can be generalized, e.g. implemented with any smooth projective hash proof system, or whether the ciphertext blowup can be reduced to $\mathcal{O}(1)$. It is also open to construct a non-committing encryption scheme with n-bit messages and public-keys of size $\mathcal{O}(n + \lambda)$.

References

1. Beaver, D.: Plug and play encryption. In: Kaliski Jr., B.S. (ed.) CRYPTO 1997. LNCS, vol. 1294, pp. 75–89. Springer, Heidelberg (1997)
2. Bellare, M., Boldyreva, A., Kurosawa, K., Staddon, J.: Multirecipient Encryption Schemes: How to Save on Bandwidth and Computation Without Sacrificing Security. IEEE Transactions on Information Theory 53(11), 3927–3943 (2007)
3. Ben-Or, M., Goldwasser, S., Wigderson, A.: Completeness theorems for non-cryptographic fault-tolerant distributed computation. In: STOC 1988, pp. 1–10. ACM, New York (1988)
4. Cachin, C., Micali, S., Stadler, M.: Computationally Private Information Retrieval with Polylogarithmic Communication. In: Stern, J. (ed.) EUROCRYPT 1999. LNCS, vol. 1592, pp. 402–414. Springer, Heidelberg (1999)
5. Canetti, R., Feige, U., Goldreich, O., Naor, M.: Adaptively secure multi-party computation. In: STOC 1996: Proceedings of the Twenty-Eighth Annual ACM Symposium on Theory of Computing, pp. 639–648. ACM, New York (1996)
6. Canetti, R., Lindell, Y., Ostrovsky, R., Sahai, A.: Universally Composable Two-Party and Multi-Party Secure Computation. In: STOC 2002 (2002)
7. Chaum, D., Crépeau, C., Damgård, I.: Multiparty Unconditionally Secure Protocols. In: STOC, pp. 11–19 (1988)
8. Choi, S.G., Dachman-Soled, D., Malkin, T., Wee, H.: Improved Non-committing Encryption with Applications to Adaptively Secure Protocols. In: Matsui, M. (ed.) ASIACRYPT 2009. LNCS, vol. 5912, pp. 287–302. Springer, Heidelberg (2009)

9. Coppersmith, D.: Small Solutions to Polynomial Equations, and Low Exponent RSA Vulnerabilities. Journal of Cryptology 10(4), 233–260 (1997)
10. Cramer, R., Shoup, V.: Universal Hash Proofs and a Paradigm for Adaptive Chosen Ciphertext Secure Public-Key Encryption. In: Knudsen, L.R. (ed.) EUROCRYPT 2002. LNCS, vol. 2332, pp. 45–64. Springer, Heidelberg (2002)
11. Damgård, I., Nielsen, J.B.: Improved Non-committing Encryption Schemes Based on a General Complexity Assumption. In: Crypto 2000: Proceedings of the 20th Annual International Cryptology Conference on Advances in Cryptology, pp. 432–450. Springer, Heidelberg (2000)
12. Dubhashi, D.P., Panconesi, A.: Concentration of Measure for the Analysis of Randomized Algorithms. Cambridge University Press (March 2012)
13. Gentry, C., Ramzan, Z.: Single-Database Private Information Retrieval with Constant Communication Rate. In: Caires, L., Italiano, G.F., Monteiro, L., Palamidessi, C., Yung, M. (eds.) ICALP 2005. LNCS, vol. 3580, pp. 803–815. Springer, Heidelberg (2005)
14. Goldreich, O., Micali, S., Wigderson, A.: How to Play any Mental Game. In: STOC 1987, pp. 218–229 (1987)
15. Guruswami, V., Umans, C., Vadhan, S.: Unbalanced Expanders and Randomness Extractors from Parvaresh–Vardy Codes. J. ACM 56(4) (July 2009)
16. Hemenway, B., Ostrovsky, R.: Extended-DDH and Lossy Trapdoor Functions. In: Fischlin, M., Buchmann, J., Manulis, M. (eds.) PKC 2012. LNCS, vol. 7293, pp. 627–643. Springer, Heidelberg (2012),
http://eccc.hpi-web.de/report/2009/127/
17. Kiltz, E., O'Neill, A., Smith, A.: Instantiability of RSA-OAEP under chosen-plaintext attack. In: Proceedings of the 30th annual conference on Advances in cryptology, Santa Barbara, CA, USA. LNCS, pp. 295–313. Springer, Heidelberg (2010)
18. Lewko, M., O'Neill, A., Smith, A.: Regularity of Lossy RSA on Subdomains and Its Applications. In: Johansson, T., Nguyen, P. (eds.) Advances in Cryptology EUROCRYPT 2013. LNCS, vol. 7881, pp. 55–75. Springer, Heidelberg (2013)
19. Micali, S., Peikert, C., Sudan, M., Wilson, D.A.: Optimal Error Correction Against Computationally Bounded Noise. In: Kilian, J. (ed.) TCC 2005. LNCS, vol. 3378, pp. 1–16. Springer, Heidelberg (2005)
20. Peikert, C., Waters, B.: Lossy trapdoor functions and their applications. In: STOC 2008: Proceedings of the 40th Annual ACM Symposium on Theory of Computing, pp. 187–196. ACM, New York (2008)

On the Regularity of Lossy RSA
Improved Bounds and Applications
to Padding-Based Encryption

Adam Smith* and Ye Zhang

Computer Science and Engineering Department
Pennsylvania State University, University Park, PA 16802, USA
{asmith,yxz169}@cse.psu.edu

Abstract. We provide new bounds on how close to regular the map $x \mapsto x^e$ is on arithmetic progressions in $\mathbb{Z}_N$, assuming $e|\Phi(N)$ and N is composite. We use these bounds to analyze the security of natural cryptographic problems related to RSA, based on the well-studied Φ-Hiding assumption. For example, under this assumption, we show that RSA PKCS #1 v1.5 is secure against chosen-plaintext attacks for messages of length roughly $\frac{\log N}{4}$ bits, whereas the previous analysis, due to [19], applies only to messages of length less than $\frac{\log N}{32}$.

In addition to providing new bounds, we also show that a key lemma of [19] is incorrect. We prove a weaker version of the claim which is nonetheless sufficient for most, though not all, of their applications.

Our technical results can be viewed as showing that exponentiation in $\mathbb{Z}_N$ is a deterministic extractor for every source that is uniform on an arithmetic progression. Previous work showed this type of statement only on average over a large class of sources, or for much longer progressions (that is, sources with much more entropy).

1 Introduction

Cryptographic schemes based on the RSA trapdoor permutation [23] are ubiquitous in practice. Many of the schemes, are simple, natural and highly efficient. Unfortunately, their security is often understood only in the random oracle model [3], if at all.[1] When can the security of natural constructions be proven under well-defined and thoroughly studied assumptions? For example, consider the "simple embedding" RSA-based encryption scheme (of which RSA PKCS #1 v1.5, which is still in wide use, is a variant): given a plaintext x, encrypt it as $(x\|R)^e \bmod N$, where R is a random string of appropriate length and '$\|$' denotes string concatenation. Until recently [19], there was no proof of security for

---

* A.S. and Y.Z. were supported by National Science Foundation awards #0747294 and #0941553. as well as a Google research award. Part of this work was done while A.S. was on sabbatical at Boston University's Hariri Institute for Computing.
[1] There are many RSA-based constructions without random oracles, e.g., [5,14,15], but they are less efficient and not currently widely used.

Y. Dodis and J.B. Nielsen (Eds.): TCC 2015, Part I, LNCS 9014, pp. 609–628, 2015.

this scheme under a well-understood assumption. The security of this scheme under chosen plaintext attacks is closely related to another fundamental question, namely, whether many physical bits of RSA are simultaneously hardcore [2,1].

Indistinguishability of RSA on Arithmetic Progressions. Both of these questions are related to the hardness of a basic computational problem, which we dub *RSA-AP*. Consider a game in which a distinguisher is first given an RSA public key (N, e) and a number K. The distinguisher then selects the description of an arithmetic progression (abbreviated "AP") $P = \{\sigma i + \tau \mid i = 0, \ldots, K - 1\}$ of length K. Finally, the distinguisher gets a number $Y \in \mathbb{Z}_N$, and must guess whether Y was generated as $Y = X^e \bmod N$, where X is uniform in the AP P, or Y was drawn uniformly from $\mathbb{Z}_N$. We say RSA-AP is hard for length K (where K may depend on the security parameter) if no polynomial-time distinguisher can win this game with probability significantly better than it could by random guessing.

Hardness statements for the RSA-AP problem have important implications. For example, in the "simple embedding" scheme above, the input to the RSA permutation is $x\|R$, which is distributed uniformly over the AP $\{x2^\rho + i \mid i = 0 \ldots, 2^\rho - 1\}$ where ρ is the bit length of R. If RSA-AP is hard for length 2^ρ, then $(x\|R)^e \bmod N$ is indistinguishable from uniform for all messages x and so simple embedding is CPA secure.

In this paper, we show that RSA-AP is hard under well-studied assumptions, for much shorter lengths K than was previously known. From this, we draw conclusions about classic problems (the CPA security of PKCS #1 v1.5 and the simultaneous hardcoreness of many physical bits of RSA) that were either previously unknown, or for which previous proofs were incorrect.

Φ-Hiding, Lossiness and Regularity. The Φ-Hiding assumption, due to [7], states that it is computationally hard to to distinguish standard RSA keys— that is, pairs (N, e) for which $\gcd(e, \Phi(N)) = 1$—from *lossy* keys (N, e) for which $e \mid \Phi(N)$. Under a lossy key, the map $x \mapsto x^e$ is not a permutation: if $N = pq$ where p, q are prime, e divides $p - 1$ and $\gcd(e, q - 1) = 1$, then $x \mapsto x^e$ is e-to-1 on $\mathbb{Z}_N^*$. We consider two variants for the lossy mode: one where p and q are chosen to have the same bit length, and one where their bit lengths differ by a specified difference θ (see Section 2).

The Φ-Hiding assumption has proven useful since under it, statements about *computational* indistinguishability in the real world (with regular keys) may be proven by showing the *statistical* indistinguishability of the corresponding distributions in the "lossy world" (where $e \mid \Phi(N)$) [18,16,19].

Specifically, [19] showed that under Φ-Hiding, the hardness of RSA-AP for length K is implied by the approximate *regularity* of the map $x \mapsto x^e$ on arithmetic progressions when $e \mid \phi(N)$. Recall that a function is regular if it as the same number of preimages for each point in the image. For positive integers

e, N and K, let $Reg(N, e, K, \ell_1)$ denote the maximum, over arithmetic progressions P of length K, of the statistical difference between $X^e \bmod N$, where $X \leftarrow_\$ P$, and a uniform e-th residue in $\mathbb{Z}_N$. That is,

$$Reg(N, e, K, \ell_1) \stackrel{def}{=}$$
$$\max \left\{ SD(X^e \bmod N; U^e \bmod N) \;\middle|\; \begin{array}{l} \sigma \in \mathbb{Z}_N^*, \tau \in \mathbb{Z}_N, \\ X \leftarrow_\$ \{\sigma i + \tau \mid i = 0, \ldots, K-1\}, \\ U \leftarrow_\$ \mathbb{Z}_N \end{array} \right\}$$

Note that the maximum is taken over the choice of the AP parameters σ and τ. We can restrict our attention, w.l.o.g., to the case where $\sigma = 1$ (see Section 2); the maximum is thus really over the choice of τ.

Lewko et al. [19] observed that if $Reg(N, e, K, \ell_1)$ is negligible for the lossy keys (N, e), then Φ-Hiding implies that RSA-AP is hard for length K. Motivated by this, they studied the regularity of lossy exponentiation on arithmetic progressions. They claimed two types of bounds: average-case bounds, where the starting point τ of the AP is selected uniformly at random, and much weaker worst-case bounds, where τ is chosen adversarially based on the key (N, e).

1.1   Our Contributions

We provide new, worst-case bounds on the regularity of lossy exponentiation over $\mathbb{Z}_N$. These lead directly to new results on the hardness of RSA-AP, the CPA-security of simple padding-based encryption schemes, and the simultaneous hardcoreness of physical RSA bits. In addition, we provide a corrected version of the incorrect bound from [19] which allows us to recover some, though not all, of their claimed results.

Notice that in order to get any non-trivial regularity for exponentiation, we must have $K \geq N/e$, since there are at least N/e images. If the e-th powers of different elements were distributed uniformly and independently in $\mathbb{Z}_N$, then in fact we would expect statistical distance bounds of the form $\sqrt{\frac{N}{eK}}$. The e-th powers are of course not randomly scattered, yet we recover this type of distance bound under a few different conditions.

Our contributions can be broken into three categories:

Worst-case Bounds (Section 3). We provide a new worst-case bound on the regularity of exponentiation for integers with an unbalanced factorization, where $q > p$. We show that

$$Reg(N, e, K, \ell_1) = O\left(\frac{p}{q} + \sqrt{\frac{N}{eK}}\right). \tag{1}$$

When q is much larger than p, our bound scales as $\sqrt{\frac{N}{eK}}$. This bound is much stronger than the analogous worst-case bound from [19], which is

$\tilde{O}(\sqrt{\frac{N}{eK}} \cdot \sqrt{\frac{N}{K}} \cdot \sqrt[8]{pe})$ (where $\tilde{O}(\cdot)$ hides polylogarithmic factors in N).[2] In particular, we get much tighter bounds on the security of padding-based schemes than [19] (see "Applications", below).

Applying our new bounds requires one to assume a version of the Φ-Hiding assumption in which the "lossy" keys are generated in such a way that $q \gg p$ (roughly, $\log(q) \geq \log(p) + \lambda$ for security parameter λ). We dub this variant the *unbalanced* Φ-hiding assumption.

Perhaps surprisingly, the proof of our worst-case bounds in $\mathbb{Z}_N$ uses average-case bounds (for the smaller ring $\mathbb{Z}_p$), described next.

Average-case Bounds (Section 4). We can remove the assumption that lossy keys have different-length factors if we settle for an average-case bound, where the average is taken over random translations of an arithmetic progression of a given length. We show that if X is uniform over an AP of length K, then

$$\mathop{\mathbb{E}}_{c \leftarrow_\$ \mathbb{Z}_N} \left(SD\left((c + X)^e \bmod N \; ; \; U^e \bmod N \right) \right) = O\left(\sqrt{\frac{N}{eK}} + \frac{p+q}{N} \right),$$

where U is uniform in $\mathbb{Z}_N^*$. The expectation above can also be written as the distance between the pairs $(C, (C + X)^e \bmod N)$ and $(C, U^e \bmod N)$, where $C \leftarrow_\$ \mathbb{Z}_N$. This average-case bound is sufficient for our application to simultaneous hardcore bits.

This result was claimed in [19] for *arbitrary* random variables X that are uniform over a set of size K. The claim is false in general (to see why, notice that exponentiation by e does not lose any information when working modulo q, and so $X \bmod q$ needs to be close to uniform in $\mathbb{Z}_q$). However, the techniques from our worst-case result can be used to prove the lemma for arithmetic progressions (and, more generally, for distributions X which are high in min-entropy and are distributed uniformly modulo q).

Applications (Section 5). Our bounds imply that, under Φ-Hiding, the RSA-AP problem is hard roughly as long as $K > \frac{N}{e}$. This, in turn, leads to new results on the security of RSA-based cryptographic constructions.

1. Simple encryption schemes that pad the message with a random string before exponentiating (including PKCS #1 v1.5) are semantically secure under unbalanced Φ-hiding as long as the random string is more than $\log(N) - \log(e)$ bits long (and hence the message is roughly $\log(e)$ bits). In contrast, the results of [19] only apply when the message has length at most $\frac{\log(e)}{16}$ bits.[3]

---

[2] The bound of [19] relies on number-theoretic estimates of *Gauss sums*. Under the best known estimates [12], the bound has the form above. Even under the most optimistic number-theoretic conjecture on Gauss sums (the "MVW conjecture" [21]), the bounds of [19] have the form $\tilde{O}(\sqrt{\frac{N}{eK}} \cdot \sqrt{\frac{N}{K}})$ and are consequently quite weak in the typical setting where $K \ll N$.

[3] Even under the MVW conjecture (see footnote 2), one gets security for messages of at most $\frac{\log(e)}{8}$ bits.

Known attacks on Φ-Hiding fail as long as $e \ll \sqrt{p}$ (see "Related Work", below). Thus, we can get security for messages of length up to $\frac{\log(N)}{4}$, as opposed to $\frac{\log(N)}{16}$. For example, when N is 8192 bits long, our analysis supports messages of 1735 bits with 80-bit security, as opposed to 128 bits [19].

2. Under Φ-hiding, the $\log(e)$ most (or least) significant input bits of RSA are simultaneously hardcore. This result follows from both types of bounds we prove (average- and worst-case). If we assume only that RSA is hard to invert, then the best known reductions show security only for a number of bits proportional to the security parameter (e.g., [1]), which is at most $O(\sqrt[3]{\log N})$.

Lewko et al. [19] claimed a proof that *any* contiguous block of about $\log(e)$ physical bits of RSA is simultaneously hardcore. Our corrected version of their result applies on to the most or least significant bits, however. Proving security of other natural candidate hardcore functions remains an interesting open problem.

Techniques. The main idea behind our new worst-case bounds is to lift an average-case bound over the smaller ring $\mathbb{Z}_p$ to a worst-case bound on the larger ring $\mathbb{Z}_N$. First, note that we can exploit the product structure of $\mathbb{Z}_N \equiv \mathbb{Z}_q \times \mathbb{Z}_p$ to decompose the problem into mod p and mod q components. The "random translations" lemma of [19] *is* correct over $\mathbb{Z}_p$ (for p prime), even though it is false over $\mathbb{Z}_N$. The key observation is that, when the source X is drawn from a long arithmetic progression, the mod q component (which is close to uniform) acts as a random translation on the mod p component of X.

More specifically, let $V = [X \bmod q]$ denote the mod q component of X (drawn from an arithmetic progression of length much greater than q) and, for each value $v \in \mathbb{Z}_q$, let X_v denote the conditional distribution of X given $V = v$. Then

$$X_v \approx X_0 + v.$$

That is, X_v is statistically close to a translation of the shorter but sparser AP X_0 (namely, elements of the original AP which equal 0 modulo q). In the product ring $\mathbb{Z}_q \times \mathbb{Z}_p$, the random variable X is thus approximated by the pair

$$(\underbrace{V}_{\in \mathbb{Z}_q}, \underbrace{X_0 + V}_{\in \mathbb{Z}_p}).$$

Since V is essentially uniform in $\mathbb{Z}_q$, its reduction modulo p is also close to uniform in $\mathbb{Z}_p$ when $q \gg p$. This allows us to employ the random translations lemma in $\mathbb{Z}_p$ [19] to show that $X^e \bmod N$ is close to $U^e \bmod N$.

Discussion. Our worst-case bounds can be viewed as stating that multiplicative homomorphisms in $\mathbb{Z}_N$ (all of which correspond to exponentiation by a divisor of $\phi(N)$) are deterministic extractors for the class of sources that are uniform on arithmetic progressions of length roughly the number of images of the homomorphism. This is in line with the growing body of work in additive combinatorics that seeks to understand how additive and multiplicative structure interact. Interestingly, our proofs are closely tied to the product structure of $\mathbb{Z}_N$. The Gauss-sums-based results of [19] remain the best known for analogous questions in $\mathbb{Z}_p$ when p is prime.

1.2 Related Work

Lossy trapdoor functions, defined by [22], are one of many concepts that allow us to deploy information-theoretic tools to analyze computationally-secure protocols. Lossiness is mostly used in the literature as a tool for designing new cryptographic systems. However, as mentioned above, [18] and [19] showed that the concept also sheds light on existing constructions since, under the Φ-hiding assumption, the RSA permutation is lossy.

The Φ-hiding assumption predates those works considerably— introduced by [7], it is the basis for a number of efficient protocols [7,6,10,13]. Following [18], Kakvi and Kiltz [16] showed that lossiness of RSA under ΦA is also useful to understand security of a classical RSA-based *signatures*. The best known cryptanalytic attack on Φ-hiding uses Coppersmith's technique for solving low-degree polynomials [8,20] and applies when e is close to $p^{1/2}$ (the attack has a ratio of running time to success probability of at least $\sqrt{p}/e$, which implies that we should take $\log(e) \leq \frac{\log p}{2} - \lambda$ for security parameter λ). Other attacks [24] are for moduli of a special form that do not arise in the applications to RSA.

The security of the specific constructions we analyze has also been studied considerably. [4] gave *chosen-ciphertext* attacks on PKCS #1. [9] gave *chosen-plaintext* attacks for instantiations of PKCS #1 v1.5 encryption which pad the plaintext with a very short random string. In contrast, our security proofs require a large random string of at least $\frac{3}{4} \log N$ bits (though this is still shorter than the $\frac{7}{8} \log N$ random bits needed by the analysis of [19]). [17, p. 363] mention that PKCS #1 v1.5 is believed to be CPA-secure for appropriate parameters, but no proof is known.

The "large hardcore bit conjecture" for RSA and the security of the simple embedding scheme are mentioned as important open problems by [11]. Assuming that RSA is hard to invert implies only that λ bits are simultaneously hardcore, where 2^λ is the time needed to invert (see, e.g., [2,1]). Prior progress was made by [25], who showed that the $1/2 - 1/e - \epsilon - o(1)$ least significant bits of RSA are simultaneously hardcore under a computational problem related to the work of [8]. (This result does not apply directly to PKCS #1 v1.5 because the latter does not use the full RSA domain—some bits are fixed constants.)

2 Preliminaries

We denote by $SD(A; B)$ the statistical distance between the distributions of random variables A and B taking values in the same set. We write $A \approx_\epsilon B$ as shorthand for $SD(A; B) \leq \epsilon$. We consider adversaries that are restricted to probabilistic polynomial time (PPT), and let $negl(k)$ be a negligible function in

k, that is, one that decreases faster than the reciprocal of any polynomial. We write $A \leftarrow_\$ B$ to indicate that the random variable A is generated by running (randomized) algorithm B using fresh random coins, if B is an algorithm, or that A is distributed uniformly in B, if B is a finite set.

Given an integer $I \in \mathbb{Z}^+$, we write $[I]$ for the set $\{0, 1, 2, \ldots, I - 1\}$. Thus, an arithmetic progression ("AP") of length K can be written $P = \sigma[K] + \tau$ for some $\sigma, \tau \in \mathbb{Z}$.

Let Primes_t denote the uniform distribution of t-bit primes, and let $\mathsf{Primes}_t[\cdots]$ be shorthand the uniform distribution over t-bit primes that satisfy the condition in brackets. Let RSA_k denote the usual modulus generation algorithm for RSA which selects $p, q \leftarrow_\$ \mathsf{Primes}_{\frac{k}{2}}$ and outputs (N, p, q) where $N = pq$. Note that k is generally taken to be $\Omega(\lambda^3)$, where λ is the security parameter, so that known algorithms take 2^λ expected time to factor $N \leftarrow_\$ \mathsf{RSA}_k$.

The RSA-AP Problem. The RSA-AP problem asks an attacker to distinguish $X^e \bmod N$ from $U^e \bmod N$, where $X \leftarrow_\$ P$ is drawn from an arithmetic progression and $U \leftarrow \mathbb{Z}_N$. We allow the attacker to choose the arithmetic progression based on the public key; this is necessary for applications to CPA security. We define RSA-AP$(1^k, K)$ to be the assumption that the two following distributions are computationally indistinguishable, for any PPT attacker $\mathcal{A}$:

Experiment RSA-AP$(1^k, K)$:
$(N, p, q) \leftarrow_\$ \mathsf{RSA}_k$
$(\sigma, \tau) \leftarrow \mathcal{A}(N, e)$ where $\sigma \in \mathbb{Z}_N^*$ and $\tau \in \mathbb{Z}$
$X \leftarrow_\$ \{\sigma i + \tau : i = 0, \ldots K - 1\}$
Return (N, e, X)

Experiment RSA-Unif$(1^k, K)$:
$(N, p, q) \leftarrow_\$ \mathsf{RSA}_k$
$(\sigma, \tau) \leftarrow \mathcal{A}(N, e)$
$U \leftarrow_\$ \mathbb{Z}_N$
Return (N, e, U)

Note that without loss of generality, we may always take $\sigma = 1$ in the above experiments, since given the key (N, e) and the element $X^e \bmod N$ where X is uniform in $P = \{\sigma i + \tau : i = 0, \ldots K - 1\}$, one can compute $(\sigma^{-1} X)^e \bmod N$ where σ^{-1} is an inverse of σ modulo N. The element $\sigma^{-1} X$ is uniform in $P' = \{i + \sigma^{-1} \tau : i = 0, \ldots K - 1\}$, while the element $\sigma^{-1} U$ will still be uniform in $\mathbb{Z}_N$. Hence, a distinguisher for inputs drawn from P can be used to construct a distinguisher for elements drawn from P', and vice-versa.

Φ-Hiding Assumption. Let θ be an even integer and $c \in (0, 1)$ be a constant. We define two alternate parameter generation algorithms for RSA keys:

Algorithm $RSA_{c,\theta}^{\mathrm{inj}}(1^k)$:
$e \leftarrow_\$ \mathsf{Primes}_{ck}$
$(N, p, q) \leftarrow_\$ \mathsf{RSA}_k$
Return (N, e)

Algorithm $RSA_{c,\theta}^{\mathrm{loss}}(1^k)$
$e \leftarrow_\$ \mathsf{Primes}_{ck}$
$p \leftarrow_\$ \mathsf{Primes}_{\frac{k}{2} - \frac{\theta}{2}}[p = 1 \bmod e]$
$q \leftarrow_\$ \mathsf{Primes}_{\frac{k}{2} + \frac{\theta}{2}}$
Return (pq, e)

Definition 1 ((c, θ)-Φ-Hiding Assumption (ΦA)). *Let θ, c be parameters that are functions of the modulus length k, where $\theta \in \mathbb{Z}^+$ is even and $c \in (0, 1)$.*

For any probabilistic polynomial-time distinguisher $\mathcal{D}$,

$$Adv^{\Phi A}_{c,\theta,\mathcal{D}}(k) = \left| \Pr[\mathcal{D}(RSA^{inj}_{c,\theta}(1^k)) = 1] - \Pr[\mathcal{D}(RSA^{loss}_{c,\theta}(1^k)) = 1] \right| \le negl(k).$$

where $negl(k)$ is a negligible function in k.

As mentioned in the introduction, the regularity of lossy exponentiation on AP's of length K implies, under Φ-hiding, that RSA-AP is hard:

Observation 1. *Suppose that $Reg(N, e, K, \ell_1) \le \epsilon$ for a $1 - \delta$ fraction of outputs of $RSA^{loss}_{c,\theta}(1^k)$. Then the advantage of an attacker $\mathcal{D}$ at distinguishing* RSA-AP$(1^k, K)$ *from* RSA-Unif(1^k) *is at most $Adv^{\Phi A}_{c,\theta,\mathcal{D}}(k) + \epsilon + \delta$.*

Though the definitions above are stated in terms of asymptotic error, we state our main results directly in terms of a time-bounded distinguisher's advantage, to allow for a concrete security treatment.

3   Improved ℓ_1-Regularity Bounds for Arithmetic Progressions

Let $\mathcal{P} = \sigma[K] + \tau$ be an arithmetic progression where $K \in \mathbb{Z}^+$. In this section, we show that if X is uniformly distributed over an arithmetic progression, then $X^e \bmod N$ is statistically close to a uniformly random e-th residue in $\mathbb{Z}_N$.

Theorem 2. *Let $N = pq$ (p, q primes) and we assume $q > p$ and $\gcd(\sigma, N) = 1$. Let $\mathcal{P}$ be* AP *where $\mathcal{P} = \sigma[K] + \tau$ and assume that $K > q$. Let e be such that $e | p - 1$ and $\gcd(e, q - 1) = 1$. Then,*

$$SD(X^e \bmod N, U^e \bmod N) \le \frac{3q}{K} + \frac{2p}{q - 1} + \frac{2}{p - 1} + \sqrt{\frac{N}{eK}}$$

where $X \leftarrow_\$ \mathcal{P}$ and $U \leftarrow_\$ \mathbb{Z}_N^*$.

Recall, from Section 2, that it suffices to prove the Theorem for $\sigma = 1$. The main idea behind the proof is as follows: For any $v \in \mathbb{Z}_q$ and a set $\mathcal{P} \subset \mathbb{Z}_N$, we define $\mathcal{P}_v = \{x \in \mathcal{P} | x \bmod q = v\}$. First, we observe that $SD(X^e \bmod N, U^e \bmod N) \approx \mathbb{E}_{v \in \mathbb{Z}_q^*}(SD(X_v^e \bmod p, U_p^e \bmod p))$ (Lemma 1) where $U_p \leftarrow_\$ \mathbb{Z}_p^*$ and for any $v \in \mathbb{Z}_q^*$, $X_v \leftarrow_\$ \mathcal{P}_v$. Second, we show that $\mathcal{P}_v$ is almost identical to $\mathcal{P}_0 + v$ (that is, the set $\mathcal{P}_0$ shifted by $v \in \mathbb{Z}_q$) (Lemma 2). Therefore, we can replace $\mathbb{E}_{v \leftarrow_\$ \mathbb{Z}_q^*}(SD(X_v^e \bmod p, U_p^e \bmod p))$ with $\mathbb{E}_{v \leftarrow_\$ \mathbb{Z}_q^*}(SD((Y + v')^e \bmod p, U_p^e \bmod p))$ where $Y \leftarrow_\$ \overline{\mathcal{P}}$. The last term can be bounded via hybrid arguments and a similar technique to [19, Lemma 3] (our Lemma 4).

In order to prove this theorem, we need the following lemmas (whose proof will be given at the end of this section):

Lemma 1. *Let $N = pq$ (p, q primes). Let $\mathcal{P}$ be an* AP *where $\mathcal{P} = [K] + \tau$ and assume that $K > q$. Let e be such that $e | p - 1$ and $\gcd(e, q - 1) = 1$. Then,*

$$SD(X^e \bmod N, U^e \bmod N) \le \frac{q}{K} + \mathop{\mathbb{E}}_{v \leftarrow_\$ \mathbb{Z}_q^*}(SD(X_v^e \bmod p, U_p^e \bmod p))$$

where $X \leftarrow_\$ \mathcal{P}$, $U \leftarrow_\$ \mathbb{Z}_N^*$, $U_p \leftarrow_\$ \mathbb{Z}_p^*$ and for any $v \in \mathbb{Z}_q^*$, $X_v \leftarrow_\$ \mathcal{P}_v$.

Lemma 2. *Let $N = pq$ (p, q primes). Let $\mathcal{P}$ be an AP where $\mathcal{P} = [K] + \tau$. For any $v \in \mathbb{Z}_q^*$, $|\mathcal{P}_v \triangle (\mathcal{P}_0 + v)| \leq 2$ where $\triangle$ denotes symmetric difference.*

Lemma 3. *Let $N = pq$ (p, q primes) and assume $q > p$. Let e be such that $e | p - 1$ and $\gcd(e, q - 1) = 1$. Let $\overline{\mathcal{K}} \subset \mathbb{Z}_N$ be an arbitrary subset (not necessarily an AP):*

$$SD((C \bmod p, (C + R)^e \bmod p), (C \bmod p, U_p^e \bmod p))$$

$$\leq SD((V_p, (V_p + R)^e \bmod p), (V_p, U_p^e \bmod p)) + \frac{2p}{q - 1}.$$

where $C \leftarrow_{\$} \mathbb{Z}_q^*, V_p, U_p \leftarrow_{\$} \mathbb{Z}_p^*$ and $R \leftarrow_{\$} \overline{\mathcal{K}}$.

Notice that in this lemma, the random variable C is chosen from $\mathbb{Z}_q^*$ but always appears reduced modulo p. Roughly speaking, Lemma 3 says that if $[I]$ ($I \in \mathbb{Z}^+$; e.g., $I = q - 1$) is large enough ($I > p$), we can replace $Q \bmod p$ with V_p, where $Q \leftarrow_{\$} [I]$ and $V_p \leftarrow_{\$} \mathbb{Z}_p^*$. Then, we can apply the random translations lemma [19] over $\mathbb{Z}_p^*$ to show Lemma 4.

We should point out that the mistake in the proof of [19] does not apply to Lemma 4. Specifically, the mistake in [19] is due to the fact that $\omega - 1$ may not be invertible in $\mathbb{Z}_N$ where $N = pq$, $\omega^e = 1 \bmod N$ and $\omega \neq 1$ (refer Section 4 for more detailed explanation). However, $\omega - 1$ is invertible in $\mathbb{Z}_p$, (since p is prime) which is the ring used in Lemma 4. Specifically, we apply the following corrected version of [19, Lemma 3]:

Lemma 4 (Random Translations Lemma, adapted from [19]). *Let $N = pq$ (p, q primes). Let $V_p, U_p \leftarrow_{\$} \mathbb{Z}_p^*$. Let $R \leftarrow_{\$} \overline{\mathcal{K}}$ where $\overline{\mathcal{K}} \subset \mathbb{Z}_N$ and $|\overline{\mathcal{K}}| = \overline{K}$.*

$$SD((V_p, (V_p + R)^e \bmod p), (V_p, U_p^e \bmod p)) \leq \frac{2}{p - 1} + \sqrt{\frac{p - 1}{e\overline{K}}}.$$

The proof of Lemma 4 is given in Appendix A. We can now prove our main result, Theorem 2:

Proof (of Theorem 2). Let $X \leftarrow_{\$} \mathcal{P}$, $U \leftarrow_{\$} \mathbb{Z}_N^*$, $U_p \leftarrow_{\$} \mathbb{Z}_p^*$. For any $v \in \mathbb{Z}_q$, let $X_v \leftarrow_{\$} \mathcal{P}_v$ (recall $\mathcal{P}_v$ is a set $\{x \in \mathcal{P} | x \bmod q = v\}$). By Lemma 1, we have:

$$SD(X^e \bmod N, U^e \bmod N) \leq \frac{q}{K} + \mathop{\mathbb{E}}_{v \leftarrow_{\$} \mathbb{Z}_q^*} (SD(X_v^e \bmod p, U_p^e \bmod p)).$$

Let $Y \leftarrow_{\$} \mathcal{P}_0$. By the triangle inequality:

$$\mathop{\mathbb{E}}_{v \leftarrow_{\$} \mathbb{Z}_q^*} SD(X_v^e \bmod p, U_p^e \bmod p)$$

$$\leq \mathop{\mathbb{E}}_{v \leftarrow_{\$} \mathbb{Z}_q^*} (SD(X_v, Y + v) + SD((Y + v)^e \bmod p, U_p^e \bmod p)).$$

Note that $SD(A^e \bmod p, B^e \bmod p) \leq SD(A, B)$ for any A and B. By Lemma 2, for every $v \in \mathbb{Z}_q^*$, we have $|\mathcal{P}_v \triangle (\mathcal{P}_0 + v)| \leq 2$. Therefore, we have $SD(X_v, Y + v) = \frac{|\mathcal{P}_v \triangle (\mathcal{P}_0 + v)|}{|\mathcal{P}_0|} \leq \frac{2}{|\mathcal{P}_0|} \leq \frac{2q}{K}$. Then,

$$\mathop{\mathbb{E}}_{v \leftarrow \$ \, \mathbb{Z}_q^*} \left(SD(X_v, Y+v) + SD((Y+v)^e \bmod p, U_p^e \bmod p) \right)$$

$$\leq \mathop{\mathbb{E}}_{v \leftarrow \$ \, \mathbb{Z}_q^*} SD(X_v, Y+v) + \mathop{\mathbb{E}}_{v \leftarrow \$ \, \mathbb{Z}_q^*} SD \left((Y+v)^e \bmod p, U_p^e \bmod p \right)$$

$$\leq \frac{2q}{K} + \mathop{\mathbb{E}}_{v \leftarrow \$ \, \mathbb{Z}_q^*} SD \left((Y+v)^e \bmod p, U_p^e \bmod p \right).$$

First, note that only the reduced value of $v \bmod p$ affects the statistical distance $SD((Y+v)^e \bmod p, U_p^e \bmod p)$. so the expression above can be rewritten as:

$$\mathop{\mathbb{E}}_{v \leftarrow \$ \, \mathbb{Z}_q^*} SD(((Y+v)^e \bmod p, U_p^e \bmod p))$$

$$= \mathop{\mathbb{E}}_{v \leftarrow \$ \, \mathbb{Z}_q^*; w \leftarrow \$ \, v \bmod p} SD \left((Y+w)^e \bmod p, U_p^e \bmod p \right).$$

Let $U_q \leftarrow \$ \, \mathbb{Z}_q^*$. The expectation above can be written as the distance between two pairs:

$$\mathop{\mathbb{E}}_{v \leftarrow \$ \, \mathbb{Z}_q^*; w \leftarrow \$ \, v \bmod p} SD((Y+w)^e \bmod p, U_p^e \bmod p)$$

$$= SD \left(U_q \bmod p, (Y+U_q)^e \bmod p, (U_q \bmod p, U_p^e \bmod p) \right).$$

By Lemma 3 and 4, $SD \left(U_q \bmod p, (R+U_q)^e \bmod p, (U_q \bmod p, U_p^e \bmod p) \right)$ is less than $\frac{2p}{q-1} + \frac{2}{p-1} + \sqrt{\frac{p-1}{e|\overline{\mathcal{K}}|}}$ where $\overline{\mathcal{K}} \subset \mathbb{Z}_N$ and $R \leftarrow \$ \, \overline{\mathcal{K}}$. We apply the inequality with $\overline{\mathcal{K}} = \mathcal{P}_0$:

$$SD((U_q \bmod p, (Y+U_q)^e \bmod p), (U_q \bmod p, U_p^e \bmod p))$$

$$\leq \frac{2p}{q-1} + \frac{2}{p-1} + \sqrt{\frac{p-1}{e|\mathcal{P}_0|}} \leq \frac{2p}{q-1} + \frac{2}{p-1} + \sqrt{\frac{N}{eK}}.$$

since $|\mathcal{P}_0| = \{\lfloor \frac{K}{q} \rfloor, \lceil \frac{K}{q} \rceil\}$.

3.1   Proofs of Lemmas

We now prove the technical lemmas from previous section.

Proof (of Lemma 1). The proof is done via hybrid arguments. By the Chinese Remainder Theorem, the mapping $a \mapsto (a \bmod p, a \bmod q)$ is an isomorphism from $\mathbb{Z}_N \to \mathbb{Z}_p \times \mathbb{Z}_q$. Therefore, we can rewrite $SD(X^e \bmod N, U^e \bmod N)$ as $SD((X^e \bmod p, X^e \bmod q), (U_p^e \bmod p, U_q^e \bmod q))$ where $U \leftarrow \$ \, \mathbb{Z}_N^*, U_p \leftarrow \$ \, \mathbb{Z}_p^*$ and $U_q \leftarrow \$ \, \mathbb{Z}_q^*$. Furthermore, as $\gcd(e, q-1) = 1$, $a \to a^e \bmod q$ is a 1-to-1 mapping over $\mathbb{Z}_q^*$. Therefore,

$$SD((X^e \bmod p, X^e \bmod q), (U_p^e \bmod p, U_q^e \bmod q))$$
$$= SD((X^e \bmod p, X \bmod q), (U_p^e \bmod p, U_q \bmod q)).$$

Now, we define $T_0 = (X \bmod q, X^e \bmod p)$, $T_1 = (U_q, X^e_{U_q} \bmod p)$ and $T_2 = (U_q, U^e_p \bmod p)$ where X_{U_q} is the random variable that chooses $v \leftarrow_\$ \mathbb{Z}^*_q$ and then $X_{U_q} \leftarrow_\$ \mathcal{P}_v$. By the triangle inequality (hybrid arguments),

$$SD(T_0, T_2) \le SD(T_0, T_1) + SD(T_1, T_2)$$

where we have $SD(T_1, T_2) = \mathbb{E}_{v \in \mathbb{Z}^*_q} SD((X^e_v \bmod p, U^e_p \bmod p).$

Now, we bound $SD(T_0, T_1)$. Define $T'_0 = (W \bmod q, X^e_{W \bmod q} \bmod p)$ where $W \leftarrow_\$ [K]$ (recall that $|\mathcal{P}| = K$). We claim that $SD(T_0, T_1) = SD(T'_0, T_1)$. Specifically,

$$SD(T_0, T_1)$$

$$= \frac{1}{2} \sum_{a \in \mathbb{Z}^*_q} \left| \Pr_{(\ell + \tau) \leftarrow_\$ \mathcal{K}}[\ell + \tau = a \bmod q] - \Pr_{x \leftarrow_\$ \mathbb{Z}^*_q}[x = a \bmod q] \right|$$

$$= \frac{1}{2} \sum_{a \in \mathbb{Z}^*_q} \left| \Pr_{\ell \leftarrow_\$ [K]}[\ell = (a - \tau) \bmod q] - \Pr_{x \leftarrow_\$ \mathbb{Z}^*_q}[x = a \bmod q] \right|$$

$$= \frac{1}{2} \sum_{a \in \mathbb{Z}^*_q} \left| \Pr_{\ell \leftarrow_\$ [K]}[\ell = (a - \tau) \bmod q] - \Pr_{x \leftarrow_\$ \mathbb{Z}^*_q}[x = (a - \tau) \bmod q] \right|$$

$$= SD(T'_0, T_1).$$

Now, we bound $SD(T'_0, T_1)$:

$$SD(T'_0, T_1) = SD((W \bmod q, X^e_{W \bmod q} \bmod p), (U_q, X^e_{U_q} \bmod p))$$
$$\le SD(W \bmod q, U_q).$$

Let $r = K \bmod q$. Then,

$$SD(W \bmod q, U_q) = \frac{1}{2} \sum_{a \in \mathbb{Z}^*_q} \left| \Pr_{x \leftarrow_\$ [K]}[x \bmod q = a] - \Pr_{x \leftarrow_\$ \mathbb{Z}^*_q}[x = a] \right|$$
$$= r \left(\frac{(K - r)/q + 1}{K} - \frac{1}{q - 1} \right).$$

Note that $\frac{(K-r)/q+1}{K} \le (1 + \frac{q-r}{K}) \frac{1}{q-1}$ and we have:

$$r \left(\frac{(K-r)/q+1}{K} - \frac{1}{q-1} \right) \le \frac{r}{(q-1)} \frac{q-r}{K} \le \frac{q}{K}$$

as $0 \le r \le q - 1$. To conclude,

$$SD(T_0, T_2) \le SD(T'_0, T_1) + SD(T_1, T_2)$$
$$\le \frac{q}{K} + \mathbb{E}_{v \leftarrow_\$ \mathbb{Z}^*_q} SD((X^e_v \bmod p, U^e_p \bmod p)).$$

Proof (of Lemma 2). Let $u \in \mathbb{Z}_q$, we have

$$\begin{aligned}
\mathcal{P}_u &= \{x \in \mathcal{P} | x \bmod q = u\} \\
&= \{\ell + \tau | \ell \leq K \wedge \ell = u - \tau \bmod q\} \\
&= \{(u - \tau) \bmod q + qk + \tau | 0 \leq k \leq \frac{K - (u - \tau) \bmod q}{q}\}.
\end{aligned}$$

Specifically, we have $\mathcal{P}_0 = \{qk - \tau \bmod q + \tau | 0 \leq k \leq \frac{K + \tau \bmod q}{q}\}$. Recall that $v < q$ ($v \in \mathbb{Z}_q^*$), we have:

$$\mathcal{P}_v = \begin{cases} \{qk - \tau \bmod q + \tau + q + v | 0 \leq k \leq \frac{K - v + \tau \bmod q)}{q} - 1\} & v < \tau \bmod q; \\ \{qk - \tau \bmod q + \tau + v | 0 \leq k \leq \frac{K - v + \tau \bmod q)}{q}\} & \text{otherwise.} \end{cases}$$

Therefore, for any $v \in \mathbb{Z}_q^*$, $|\mathcal{P}_v \triangle (\mathcal{P}_0 + v)| \leq 2$ where $\triangle$ denotes symmetric difference.

Proof (of Lemma 3). The proof is done via hybrid arguments. Let $T_0 = (C \bmod p, (C \bmod p + R)^e \bmod p)$, $T_1 = (V_p, (V_p + R)^e \bmod p)$, $T_2 = (V_p, U_p^e \bmod p)$ and $T_3 = (C \bmod p, U_p^e \bmod p)$. Then,

$$SD(T_0, T_3) \leq SD(T_0, T_1) + SD(T_1, T_2) + SD(T_2, T_3).$$

Via the similar technique (to show $SD(W \bmod q, U_q)$) in Lemma 1, we have:

$$SD(T_0, T_1) = SD(T_2, T_3) = SD(C \bmod p, U_p)$$
$$\leq \frac{p}{|C|} = \frac{p}{q - 1}.$$

4   Average-Case Bounds over Random Translations

In this section, we point out a mistake in the proof of Lemma 4 from [19]. We give a counter example to the lemma, explain the error in the proof and prove a corrected version of the lemma which still implies the main conclusions from [19]. First, we restate their lemma:

Incorrect Claim 1 (Lemma 4 [19]). *Let $N = pq$ and e be such that $e | p - 1$ and $\gcd(e, q - 1) = 1$. Let $\mathcal{K} \subset \mathbb{Z}_N$ such that $|\mathcal{K}| \geq \frac{4N}{e\alpha^2}$ for some $\alpha \geq \frac{4(p + q - 1)}{N}$. Then,*

$$SD((C, (C + X)^e \bmod N), (C, U^e \bmod N) \leq \alpha$$

where $C, U \leftarrow_\$ \mathbb{Z}_N$ and $X \leftarrow_\$ \mathcal{K}$.

4.1   Counterexample to Lemma 4 in LOS

The problem with this lemma, as stated, is that raising numbers to the e-th power is a permutation in $\mathbb{Z}_q$, and so exponentiation does not erase any information (statistically) about the value of the input mod q. (It may be that information is lost computationally when p, q are secret, but the claim is about statistical distance.)

Adding a publicly available random offset does not help, since the composition of translation and exponentiation is still a permutation of $\mathbb{Z}_q$. Hence, if $X \bmod q$ is not close to uniform, then $(C, (C + X)^e \bmod q)$ is not close to uniform in $\mathbb{Z}_N \times \mathbb{Z}_q$, and so $(C, (C + X)^e \bmod N)$ is not close to uniform in $\mathbb{Z}_N^2$.

To get a counterexample to the claimed lemma, let $\mathcal{K} = \{x \in \mathbb{Z}_N : x \bmod q \in \{0, ..., \frac{q-1}{2}\}\}$ (the subset of $\mathbb{Z}_N$ with mod q component less than $q/2$). $\mathcal{K}$ is very large (size about $N/2$) but the pair $C, (X + C)^e \bmod q$ will never be close to uniform when $X \leftarrow_{\$} \mathcal{K}$.

The above attack was motivated by the discovery of a mistake in the proof of Lemma 4 from [19]. Specifically, the authors analyze the probability that $(C+X)^e = (C+Y)^e$ by decomposing the event into events of the form $(C+X) = \omega(C + Y)$ where ω is an e-th root of unity. The problem arises because

$$\Pr[(C + X) = \omega(C + Y)] \neq \Pr[C = (\omega - 1)^{-1}(\omega Y - X)]$$

since $\omega - 1$ is not invertible in $\mathbb{Z}_N^*$ (it is 0 mod q).

4.2   Corrected Translation Lemma

It turns out that distinguishability mod q is the only obstacle to the random translation lemma. We obtain the following corrected version:

Lemma 5. *Let* $N = pq$ *and* e *be such that* $e|p - 1$ *and* $\gcd(e, q - 1) = 1$. *Let* $\mathcal{K} \subset \mathbb{Z}_N$ *be an arithmetic progression. Specifically, let* $\mathcal{K} = \sigma[K] + \tau$ *with* $K > q$. *Then,*

$$SD((C, (C + X)^e \bmod N), (C, U^e \bmod N)) \leq \frac{3}{p-1} + \sqrt{\frac{N}{eK}} + \frac{q}{K}.$$

where $C, U \leftarrow_{\$} \mathbb{Z}_N$ *and* $X \leftarrow_{\$} \mathcal{K}$.

Proof. Applying the same idea in Lemma 1, let $U_p \leftarrow_{\$} \mathbb{Z}_p, U_q \leftarrow_{\$} \mathbb{Z}_q$, we have:

$$SD((C, (C + X)^e \bmod N), (C, U^e \bmod N))$$
$$= \mathop{\mathbb{E}}_{c \leftarrow_{\$} \mathbb{Z}_N} (SD((c + X)^e \bmod N, U^e \bmod N))$$
$$= \mathop{\mathbb{E}}_{c \leftarrow_{\$} \mathbb{Z}_N} (SD(((c + X)^e \bmod p, (c + X) \bmod q), (U_p^e \bmod p, U_q))).$$

Notice that the mod$q$ components are not raised to the e-th power. This is because exponentiation is a permutation of $\mathbb{Z}_q^*$ as $\gcd(e, q - 1) = 1$. For any $c \in \mathbb{Z}_N$, let $T_0(c) = ((c + X)^e \bmod p, (c + X) \bmod q), T_1(c) = ((c + X)_{U_q}^e \bmod p, U_q)$, $T_2 = (U_p^e \bmod p, U_q)$. We rewrite $\mathbb{E}_{c \leftarrow_{\$} \mathbb{Z}_N} (SD(((c + X)^e \bmod p, (c + X) \bmod q), (U_p^e \bmod p, U_q)))$ as $\mathbb{E}_{c \leftarrow_{\$} \mathbb{Z}_N} SD(T_0(c), T_2)$. By the triangle inequality, we have:

$$\mathop{\mathbb{E}}_{c \leftarrow_{\$} \mathbb{Z}_N} SD(T_0(c), T_2) \leq \mathop{\mathbb{E}}_{c \leftarrow_{\$} \mathbb{Z}_N} (SD(T_0(c), T_1(c)) + SD(T_1(c), T_2)).$$

For each $c \in \mathbb{Z}_N$:

$$SD(T_0(c), T_1) = SD\Big(((c+X)^e \bmod p, (c+X) \bmod q), ((c+X)^e_{U_q} \bmod p, U_q)\Big)$$
$$\leq SD((c+X) \bmod q, U_q) \leq SD(X \bmod q, U_q).$$

The last equality holds because translation by c is a permutation of $\mathbb{Z}_q$. We have:

$$SD(T_1(c), T_2) = SD\Big(((c+X)^e_{U_q} \bmod p, U_q), (U^e_p \bmod p, U_q)\Big)$$
$$= \mathop{\mathbb{E}}_{v \leftarrow^\$ \mathbb{Z}_q} SD((X+c)^e_v \bmod p, U^e_p \bmod p).$$

Recall that for any $v \in \mathbb{Z}_q$, $(c+X)_v$ denotes the random variable $c+X$ conditioned on the event that $c+X \bmod q = v$. To sum up,

$$\mathop{\mathbb{E}}_{c \leftarrow^\$ \mathbb{Z}_N} ((c+X)^e \bmod N, U^e \bmod N)$$
$$\leq SD(X \bmod q, U_q) + \mathop{\mathbb{E}}_{v \leftarrow^\$ \mathbb{Z}_q} \mathop{\mathbb{E}}_{c \leftarrow^\$ \mathbb{Z}_N} SD\left((X+c)^e_v \bmod p, U^e_p \bmod p\right).$$

Note that only the value of $c \bmod p$ affects $SD((X+c)^e_v \bmod p, U^e_p \bmod p)$. We can replace $c \leftarrow^\$ \mathbb{Z}_N$ with $V_p \leftarrow^\$ \mathbb{Z}^*_p$. Specifically, let **BAD** be the event that $\gcd(c, p) \neq 1$. As $c \leftarrow^\$ \mathbb{Z}_N$, we have $\Pr[\textbf{BAD}] = \Pr_{c \leftarrow^\$ \mathbb{Z}_N}[\gcd(c, p) \neq 1] = \frac{1}{p}$. Therefore, for any $v \in \mathbb{Z}_q$,

$$\mathop{\mathbb{E}}_{c \leftarrow^\$ \mathbb{Z}_N} SD((X+c)^e_v \bmod p, U^e_p \bmod p)$$
$$\leq \Pr[\textbf{BAD}] \cdot 1 + 1 \cdot \mathop{\mathbb{E}}_{c \leftarrow^\$ \mathbb{Z}^*_p} SD((X+c)^e_v \bmod p, U^e_p \bmod p)$$
$$\leq \frac{1}{p} + \mathop{\mathbb{E}}_{V_p \leftarrow^\$ \mathbb{Z}^*_p} SD((X+V_p)^e_v \bmod p, U^e_p \bmod p)$$

as $\Pr[\textbf{BAD}] < 1$ and statistical distance $SD(\cdot, \cdot) < 1$.

By Lemma 4, we have $\mathbb{E}_{V_p \leftarrow^\$ \mathbb{Z}^*_p} \left((X+V_p)^e_v \bmod p, U^e_p \bmod p\right) \leq \frac{2}{p-1} + \sqrt{\frac{N}{eK}}$. Thus,

$$SD_{C \leftarrow^\$ \mathbb{Z}_N}((C, (C+X)^e \bmod N), (C, U^e \bmod N))$$
$$\leq \frac{1}{p} + \frac{2}{p-1} + \sqrt{\frac{N}{eK}} + SD(X \bmod q, U_q) \leq \frac{1}{p} + \frac{2}{p-1} + \sqrt{\frac{N}{eK}} + \frac{q}{K}.$$

5   Applications

In this section, we apply the above results to understanding the IND-CPA security of PKCS #1 v1.5 and to showing that the most/least $\log e - 3 \log \frac{1}{\epsilon} + o(1)$ significant RSA bits are simultaneously hardcore. To illustrate our results, we show that our bounds imply improvements to the concrete security parameters from [19].

5.1   IND-CPA Security of PKCS #1 v1.5

Below, a_{16} denotes the 16-bit binary representation of a two-symbol hexadecimal number $a \in \{00, ..., FF\}$. Let $PKCS(x; r) = x||00_{16}||r$. The ciphertext for message x under PKCS #1 v1.5 [4] is then $(00_{16}||02_{16}||PKCS(x; r))^e \bmod N$, where r is chosen uniformly random from $\{0, 1\}^\rho$.

Theorem 3 (CPA security of PKCS #1 v1.5). *Let λ be the security parameter, $k = k(\lambda) \in \mathbb{Z}^+$ and $\epsilon(\lambda), c(\lambda) > 0$. Suppose ΦA holds for c and $\theta \geq 4 + \log \frac{1}{\epsilon}$. Let Π_{PKCS} be the PKCS #1 v1.5 encryption scheme. Assume that $\rho \geq \log N - \log e + 2 \log(1/\epsilon) + 4$ and $\theta \geq 4 + \log \frac{1}{\epsilon}$. Then for any IND-CPA adversary $\mathcal{A}$ against Π_{PKCS}, there exists a distinguisher $\mathcal{D}$ for Φ-Hiding with $time(\mathcal{D}) \leq time(\mathcal{A}) + O(k^3)$ such that for all $\lambda \in \mathbb{N}$:*

$$Adv^{ind-cpa}_{\Pi_{PKCS}, \mathcal{A}}(\lambda) \leq Adv^{\Phi A}_{c, \theta, \mathcal{D}}(\lambda) + \epsilon(\lambda).$$

Proof. Define **Game$_0$** be the original IND-CPA security game with the adversary $\mathcal{A}$. Let **Game$_1$** be identical to **Game$_0$** except that (N, e) is generated via lossy RSA key generation (Section 2, Φ-Hiding Assumption), such that $e|p - 1$ and $gcd(e, q - 1) = 1$. **Game$_2$** is identical to **Game$_1$** except that the challenge ciphertext $c^* = (00_{16}||02_{16}||PKCS(x^*, r^*))^e \bmod N$ is replaced with $U^e \bmod N$ where $U \leftarrow_\$ \mathbb{Z}_N^*$.

An adversary who performs differently in **Game$_0$** and **Game$_1$** can be used to attack the ΦA assumption; the increase in running time is the time it takes to generate a challenge ciphertext, which is at most $O(k^3)$. The difference between **Game$_1$** and **Game$_2$** is $SD((00_{16}||02_{16}||PKCS(x^*, r^*))^e \bmod N, (\mathbb{Z}_N^*)^e \bmod N)$ (information theoretically) where x^* is the challenge plaintext and r^* is the encryption randomness. Specifically, given the challenge plaintext x^* that may depend on $pk = (N, e)$, $00_{16}||02_{16}||PKCS(x^*, \cdot) = \{r + x^* 2^{\rho+8} + 2^{\rho+8+|x|} | r \in \{0, 1\}^\rho\}$ is an arithmetic progression with length 2^ρ. By Theorem 2,

$$SD(0002_{16}||PKCS(x^*, r^*)^e \bmod N, (\mathbb{Z}_N^*)^e \bmod N)$$

$$\leq \frac{1}{p-1} + \frac{2p}{q-1} + \frac{3q}{2^{\rho+1}} + \sqrt{\frac{N}{e2^\rho}} \leq 2 \left(\frac{2p}{q-1} + \sqrt{\frac{N}{e2^\rho}} \right) < \epsilon,$$

where we have $\frac{2p}{q-1} < \frac{\epsilon}{4}$ when $\theta \geq 4 + \log \frac{1}{\epsilon}$, and $\sqrt{\frac{N}{e2^\rho}} < \frac{\epsilon}{4}$ when $\rho \geq \log N - \log e - 2 \log \epsilon + 4$. Note that the advantage of $\mathcal{A}$ in **Game$_2$** is 0.

Achievable Parameters. To get a sense of the parameters for which our analysis applies, recall the best known attack on Φ-Hiding (using Coppersmith's algorithm) has a tradeoff of time to success probability of at least 2^λ when $p < q$ and $\log(e) = \frac{\log(p)}{2} - \lambda$. We therefore select this value of e (that is, $e = \sqrt{p}/2^\lambda$) for security parameter λ.

---

[4] RFC2313, http://tools.ietf.org/html/rfc2313

For a message of length m, PKCS #1 v1.5 uses a random string of length $\rho = \log N - m - 48$ (since six bytes of the padded string are fixed constants). To apply Theorem 3, we need two conditions. First, we need $\rho \geq \log N - \log e + 2 \log(1/\epsilon) + 4$; for this, it suffices have a message of length $m \leq \log(e) - 2 \log(1/\epsilon) - 52$. Second, we need $\theta = \log q - \log p \geq 4 + \log(1/\epsilon)$. Setting p to have length $\log(p) = \frac{\log(N)}{2} - \frac{\log(1/\epsilon) + 4}{2}$ satisfies this condition.

Using the value of e based on Coppersmith's attack, and setting $\epsilon = 2^{-\lambda}$ in Theorem 3, we get CPA security for messages of length up to

$$m = \tfrac{1}{4} \log N - \tfrac{13}{4} \lambda - 53. \tag{2}$$

with security parameter λ.

In contrast, the analysis of [19] proves security for messages of length only $m = \frac{\log N}{16} - \Theta(\lambda)$. Even under the most optimistic number-theoretic conjecture (the MVW conjecture on Gauss sums), their analysis applies to messages of length only $m = \frac{\log N}{8} - \Theta(\lambda)$. [5] Their proof methodology cannot go beyond that bound. Our results therefore present a significant improvement over the previous work.

Concrete Parameters: Take the modulus length $k = \log N = 8192$ as an example. We will aim for $\lambda = 80$-bit security. We get CPA security for messages of length up to

$$m = \frac{\log N}{4} - \frac{13}{4}\lambda - 53 = 1735 \text{ (bits)}.$$

This is improves over the 128 bit messages supported by the analysis of [19] by a factor of 13. (That said, we do not claim to offer guidance for setting parameters in practice, since our results require an exponent e much larger than the ones generally employed.)

5.2 (Most/Least Significant) Simultaneously Hardcore Bits for RSA

Let λ be the security parameter and let $k = \log N$ be the modulus length. For $1 \leq i < j \leq k$, we want to show that the two distributions $(N, e, x^e \bmod N, x[i, j])$ and $(N, e, x^e \bmod N, r)$ are computationally indistinguishable, where $x \leftarrow_\$ \mathbb{Z}_N^*$, $r \leftarrow_\$ \{0, 1\}^{j-i-1}$, and $x[i : j]$ denotes bits i through j of the binary representation of x.

In this section, we apply Theorem 2 to show the most and least $\log e - O(\log \frac{1}{\epsilon})$ significant bits of RSA functions are simultaneously hardcore (Theorem 4). We should note that we can apply the corrected random translations lemma (our Lemma 5) to this problem, which yields an essentially identical result. For brevity, we omit its proof.

Theorem 4. *Let λ be the security parameter, $k = k(\lambda) \in \mathbb{Z}^+$ and $\epsilon(\lambda), c(\lambda) > 0$. Suppose ΦA holds for c and $\theta > 4 + \log \frac{1}{\epsilon}$. Then, the most (or least) $\log e - 2 \log \frac{1}{\epsilon} - 2$ significant bits of RSA are simultaneously hardcore. Specifically, for any*

---

[5] Even under MVW, the result of [19] requires that $\rho \geq \log N - \frac{1}{2} \log(e) + \lambda + O(1)$. Combined with the requirement that $\log(e) \leq \frac{1}{2} \log(p) - \lambda$, we get a message of length $m = \log(N) - \rho - O(1) \leq \frac{1}{8} \log(N) - \frac{3}{2}\lambda - O(1)$.

distinguisher $\mathcal{D}$, there exists a distinguisher $\overline{\mathcal{D}}$ running in time $time(\mathcal{D}) + O(k^3)$ such that

$$\left| \Pr[\mathcal{D}(N, e, x^e \bmod N, x[i:j]) = 1] - \Pr[\mathcal{D}(N, e, x^e \bmod N, r[i:j]) = 1] \right|$$

$$\leq Adv_{c,\theta,\overline{\mathcal{D}}}^{\Phi A}(\lambda) + 2\epsilon.$$

where $r \leftarrow_\$ \mathbb{Z}_N$, $|j - i| \leq \log e - 2\log\frac{1}{\epsilon} - 2$ and either $i = 1$ or $j = k$. Furthermore, the distribution of $r[i; j]$ is 2^{k-j}-far from uniform on $\{0,1\}^{j-i+1}$.

It's important to note that the theorem is stated in terms of the distinguishability between bits i through j of the RSA input, and bits i through j of a random element r of $\mathbb{Z}_N$. The string $r[i:j]$ is not exactly uniform – indeed, when $j = k$, it is easily distinguishable from uniform unless N happens to be very close to a power of 2.

Depending on the application, it may be important to have $x[i:j]$ indistinguishable from a truly uniform string. In that case, one may either set $i = 1$ (use the least significant bits) or, in the case $j = k$, ignore the top $\log(1/\epsilon)$ bits of $r[i; k]$ (effectively reducing the number of hardcore bits to about $\log(e) - 3\log(1/\epsilon)$ bits).

Proof (of Theorem 4). We define two games. Let $U \leftarrow_\$ \mathbb{Z}_N^*$. **Game$_0$** is to distinguish $(N, e, x^e \bmod N, x[i,j])$ and $(N, e, U^e \bmod N, x[i,j])$; **Game$_1$** is to distinguish $(N, e, U^e \bmod N, x[i,j])$ and $(N, e, U^e \bmod N, r)$. Since x is chosen uniform randomly from $\mathbb{Z}_N^*$, the advantage in **Game$_1$** is at most 2^{j-k} (since k is the bit length). Let $\mathcal{D}$ be any distinguisher, and let $\overline{\mathcal{D}}$ be the distinguisher for the Φ-Hiding game that prepares inputs to $\mathcal{D}$ using a challenge public key and uses $\mathcal{D}$'s output as its own. We have

$$Adv_{\mathcal{D}}^{\mathbf{Game}_0}(1^\lambda) = \left| \Pr[\mathcal{D}(N, e, x^e \bmod N, x[i,j]) = 1] \right.$$
$$\left. - \Pr[\mathcal{D}(N, e, (\mathbb{Z}_N^*)^e \bmod N, x[i,j]) = 1] \right|$$
$$\leq Adv_{c,\theta,\overline{\mathcal{D}}}^{\Phi A}(\lambda) + SD(\mathcal{P}^e \bmod N, U^e \bmod N)$$

where $\mathcal{P}$ is the set of integers with bits i through j set to $x[i:j]$.

The structure of $\mathcal{P}$ depends on the integers i and j. In general, when $j < k$ and $i > 1$, $\mathcal{P}$ may not be well-approximated by an arithmetic progression. However, if $j = k$, then $\mathcal{P}$ is the arithmetic progression $\mathcal{P} = \{x[i,j] \cdot 2^{i-1} + a \mid a = 0, \ldots, 2^{i-1} - 1\}$. If $i = 1$, then the set $\mathcal{P}$ is more complicated, but it is closely approximated by an AP. Specifically, let $\mathcal{P}' = \{x[i,j] + b \cdot 2^j \mid b = 0, ..., N_j\}$, where $N_j \overset{def}{=} N$ div 2^j is the integer obtained by consider only bits $j+1$ through k of the binary representation of the modulus N. Then the uniform distribution on $\mathcal{P}$ is at most 2^{k-j}-far from the uniform distribution on $\mathcal{P}'$.

As Theorem 2 applies to arithmetic progressions, we can apply it in the cases $i = 1$ and $j = k$. By Theorem 2,

$$Adv_{\mathcal{D}}^{\mathbf{Game}_0}(1^\lambda) \leq 2\left(\frac{2p}{q-1} + \sqrt{\frac{N}{e2^{k-|j-i|}}}\right) < 2\epsilon.$$

The last inequality uses the hypotheses that $\theta = \log q - \log p \geq 4 + \log \frac{1}{\epsilon}$ and $|j - i| < \log e - 2 \log \frac{1}{\epsilon} - 2$.

Concrete Parameters: Let λ denote the security parameter. As in the calculations for PKCS in the previous section, we require $\log(e) \leq \frac{\log p}{2} - \lambda$ (for Coppersmith's attack to be ineffective) and $\epsilon = 2^{-\lambda}$. To apply Theorem 4, we require that $\theta \geq 4 + \log \frac{1}{\epsilon} = 4 + \lambda$, and therefore $\log e \leq \frac{k-\theta}{4} - \lambda \leq \frac{k-5\lambda}{4} - 1$. Theorem 4 then proves security for a run of bits with length $\log e - 2\lambda - 2 = \frac{1}{4}k - \frac{13}{4}\lambda - 3$. For example, for a modulus of length $k = 2048$ bits and security parameter $\lambda = 80$, we get that the 249 least significant bits are simultaneously hardcore. Alternatively, our analysis shows that the 169 bits in positions $k - 249$ through $k - 169$ are simultaneously hardcore (see the discussion immediately after Theorem 4).

References

1. Akavia, A., Goldwasser, S., Safra, S.: Proving hard-core predicates using list decoding. In: 44th Annual Symposium on Foundations of Computer Science, pp. 146 –159. IEEE Computer Society Press (October 2003)
2. Alexi, W., Chor, B., Goldreich, O., Schnorr, C.: RSA and Rabin functions: Certain parts are as hard as the whole. SIAM Journal on Computing 17(2), 194–209 (1988)
3. Bellare, M., Rogaway, P.: Random oracles are practical: A paradigm for designing efficient protocols. In: Ashby, V. (ed.) ACM CCS 1993: 1st Conference on Computer and Communications Security, pp. 62–73. ACM Press (November 1993)
4. Bleichenbacher, D.: Chosen ciphertext attacks against protocols based on the RSA encryption standard PKCS #1. In: Krawczyk, H. (ed.) CRYPTO 1998. LNCS, vol. 1462, pp. 1–12. Springer, Heidelberg (1998)
5. Blum, M., Goldwasser, S.: An efficient probabilistic public-key encryption scheme which hides all partial information. In: Blakely, G.R., Chaum, D. (eds.) CRYPTO 1984. LNCS, vol. 196, pp. 289–302. Springer, Heidelberg (1985)
6. Cachin, C.: Efficient private bidding and auctions with an oblivious third party. In: ACM CCS 1999: 6th Conference on Computer and Communications Security, pp. 120–127. ACM Press (November 1999)
7. Cachin, C., Micali, S., Stadler, M.A.: Computationally private information retrieval with polylogarithmic communication. In: Stern, J. (ed.) EUROCRYPT 1999. LNCS, vol. 1592, pp. 402–414. Springer, Heidelberg (1999)
8. Coppersmith, D.: Small solutions to polynomial equations, and low exponent RSA vulnerabilities. Journal of Cryptology 10(4), 233–260 (1997)
9. Coron, J.-S., Joye, M., Naccache, D., Paillier, P.: New attacks on PKCS #1 v1.5 encryption. In: Preneel, B. (ed.) EUROCRYPT 2000. LNCS, vol. 1807, pp. 369–381. Springer, Heidelberg (2000)
10. Gentry, C., Mackenzie, P.D., Ramzan, Z.: Password authenticated key exchange using hidden smooth subgroups. In: Atluri, V., Meadows, C., Juels, A. (eds.) ACM CCS 2005: 12th Conference on Computer and Communications Security, pp. 299–309. ACM Press (November 2005)
11. Goldreich, O.: Foundations of Cryptography: Basic Applications, vol. 2. Cambridge University Press, Cambridge (2004) (hardback) ISBN 0-521- 83084-2
12. Heath-Brown, D., Konyagin, S.: New bounds for gauss sums derived from kth powers, and for heilbronns exponential sum. The Quarterly Journal of Mathematics 51(2), 221–235 (2000)

13. Hemenway, B., Ostrovsky, R.: Public-key locally-decodable codes. In: Wagner, D. (ed.) CRYPTO 2008. LNCS, vol. 5157, pp. 126–143. Springer, Heidelberg (2008)
14. Hofheinz, D., Kiltz, E.: Practical chosen ciphertext secure encryption from factoring. In: Joux, A. (ed.) EUROCRYPT 2009. LNCS, vol. 5479, pp. 313–332. Springer, Heidelberg (2009)
15. Hohenberger, S., Waters, B.: Short and stateless signatures from the RSA assumption. In: Halevi, S. (ed.) CRYPTO 2009. LNCS, vol. 5677, pp. 654–670. Springer, Heidelberg (2009)
16. Kakvi, S.A., Kiltz, E.: Optimal security proofs for full domain hash, revisited. In: Pointcheval, D., Johansson, T. (eds.) EUROCRYPT 2012. LNCS, vol. 7237, pp. 537–553. Springer, Heidelberg (2012)
17. Katz, J., Lindell, Y.: Introduction to Modern Cryptography. Chapman and Hall/CRC Press (2007)
18. Kiltz, E., O'Neill, A., Smith, A.: Instantiability of RSA-OAEP under chosen-plaintext attack. In: Rabin, T. (ed.) CRYPTO 2010. LNCS, vol. 6223, pp. 295–313. Springer, Heidelberg (2010)
19. Lewko, M., O'Neill, A., Smith, A.: Regularity of lossy RSA on subdomains an its applications. In: Johansson, T., Nguyen, P.Q. (eds.) EUROCRYPT 2013. LNCS, vol. 7881, pp. 55–75. Springer, Heidelberg (2013)
20. May, A.: Using LLL-reduction for solving RSA and factorization problems. In: Phong, B.V., Nguyen, Q. (eds.) The LLL Algorithm: Survey and Applications, pp. 315–348. Springer (2010)
21. Montgomery, H., Vaughan, R., Wooley, T.: Some remarks on gauss sums associated with kth powers. In: Mathematical Proceedings of the Cambridge Philosophical Society, vol. 118, pp. 21–33. Cambridge Univ. Press (1995)
22. Peikert, C., Waters, B.: Lossy trapdoor functions and their applications. In: Dwork, C. (ed.) STOC, pp. 187–196. ACM (2008) ISBN 978-1-60558- 047-0
23. Rivest, R.L., Shamir, A., Adleman, L.M.: A method for obtaining digital signatures and public-key cryptosystems. Commun. ACM 21(2), 120–126 (1978)
24. Schridde, C., Freisleben, B.: On the validity of the phi-hiding assumption in cryptographic protocols. In: Pieprzyk, J. (ed.) ASIACRYPT 2008. LNCS, vol. 5350, pp. 344–354. Springer, Heidelberg (2008)
25. Steinfeld, R., Pieprzyk, J., Wang, H.: On the provable security of an efficient RSA-based pseudorandom generator. In: Lai, X., Chen, K. (eds.) ASIACRYPT 2006. LNCS, vol. 4284, pp. 194–209. Springer, Heidelberg (2006)

A   Lemma 4

Proof (of Lemma 4). This proof is observed by [19]. However, in [19], $\omega - 1$ may not be invertible in $\mathbb{Z}_N$ (recall that $\omega \in \{x | x^e \bmod N = 1\}$) but $\omega - 1$ is invertible in $\mathbb{Z}_p$ as $e | p - 1$.

Let $\mathcal{Q}$ be the distribution of $(V, (V + X)^e \bmod p)$ and $\mathcal{T}$ be the distribution of $(V, U^e \bmod p)$. $\mathcal{Q}_0$ is identical to $\mathcal{Q}$ except that the event $(V + X)^e \bmod p = 0$ occurs; $\mathcal{T}_0$ is identical to $\mathcal{T}$ except that the event $U^e \bmod p = 0$ occurs. Similarly, $\mathcal{Q}_1$ is defined to be identical to $\mathcal{Q}$ except that $(V + X)^e \bmod p \neq 0$; $\mathcal{T}_1$ is identical to $\mathcal{T}$ except that $U^e \bmod p \neq 0$. Then, we have:

$$SD(\mathcal{Q}, \mathcal{T}) = SD(\mathcal{Q}_0, \mathcal{T}_0) + SD(\mathcal{Q}_1, \mathcal{T}_1).$$

$$SD(\mathcal{Q}_0, \mathcal{T}_0) \leq\ <1, \mathcal{Q}_0> + <1, \mathcal{T}_0>$$

$$\leq \frac{1}{p-1} + \frac{1}{p-1} \leq \frac{2}{p-1}.$$

$$SD(\mathcal{Q}_1, \mathcal{T}_1) \leq \sqrt{supp(\mathcal{Q}_1 - \mathcal{T}_1)\|\mathcal{Q}_1\|_2^2 - 1}$$

$$\leq \sqrt{\frac{(p-1)^2}{e}\|\mathcal{Q}_1\|_2^2 - 1}.$$

Where,

$$\|\mathcal{Q}_1\|_2^2 = \Pr[(V, (V+X)^e \bmod p) = (V', (V'+Y) \bmod p)]$$

$$= \frac{1}{p-1}\Pr[(V+X)^e \bmod p = (V+Y)^e \bmod p]$$

$$= \frac{1}{p-1} \sum_{\omega \in \{x | x^e \bmod p = 1\}} \Pr[(V+X) = \omega(V+Y) \bmod p]$$

$$= \frac{1}{p-1}(\Pr[X = Y \bmod p] + \sum_{\omega \neq 1} \Pr[V = (\omega-1)^{-1}(X - \omega Y) \bmod p])$$

$$= \frac{1}{p-1}(\Pr[X = Y \bmod p] + \frac{e-1}{p-1})$$

$$\leq \frac{1}{p-1}(\frac{1}{p} + \frac{1}{\overline{K}} + \frac{e-1}{p-1})$$

$$\leq \frac{1}{p-1}(\frac{e}{p} + \frac{1}{\overline{K}}).$$

Therefore, we have:

$$SD(\mathcal{Q}, \mathcal{T})$$

$$\leq \frac{2}{p-1} + \sqrt{\frac{p-1}{e}(e/p) - 1 + \frac{p-1}{e\overline{K}}}$$

$$\leq \frac{2}{p-1} + \sqrt{\frac{p-1}{e\overline{K}}}.$$

Tightly-Secure Authenticated Key Exchange*

Christoph Bader[1], Dennis Hofheinz[2], Tibor Jager[1], Eike Kiltz[1], and Yong Li[1]

[1] Horst Görtz Institute for IT Security, Ruhr-University Bochum, Germany
{christoph.bader,tibor.jager,eike.kiltz,yong.li}@rub.de
[2] Karlsruhe Institute of Technology, Germany
dennis.hofheinz@kit.edu

Abstract. We construct the first Authenticated Key Exchange (AKE) protocol whose security does not degrade with an increasing number of users or sessions. We describe a three-message protocol and prove security in an enhanced version of the classical Bellare-Rogaway security model.

Our construction is modular, it can be instantiated efficiently from standard assumptions (such as the SXDH or DLIN assumptions in pairing-friendly groups). For instance, we provide an SXDH-based protocol with only 14 group elements and 4 exponents communication complexity (plus some bookkeeping information).

Along the way we develop new, stronger security definitions for digital signatures and key encapsulation mechanisms. For instance, we introduce a security model for digital signatures that provides existential unforgeability under chosen-message attacks in a *multi-user setting* with *adaptive corruptions of secret keys*. We show how to construct efficient schemes that satisfy the new definitions with *tight* security proofs under standard assumptions.

1 Introduction

Authenticated Key Exchange (AKE) protocols allow two parties to establish a cryptographic key over an insecure channel. Secure AKE protects against strong active attackers that may for instance read, alter, drop, replay, or inject messages, and adaptively corrupt parties to reveal their long-term or session keys. This makes such protocols much stronger (and thus harder to construct) than simpler passively secure key exchange protocols like e.g. [19].

Probably the most prominent example of an AKE protocol is the TLS Handshake [16,17,18], which is widely used for key establishment and authentication on the Internet. The widespread use of TLS makes AKE protocols one of the most widely-used cryptographic primitives. For example, the social network Facebook.com reports 802 million *daily* active users on average in September 2013. This makes more than 2^{29} executions of the TLS Handshake protocol *per day* only

---

* A public version of this paper has been posted to the ePrint Archive at http://eprint.iacr.org/2014/.

Y. Dodis and J.B. Nielsen (Eds.): TCC 2015, Part I, LNCS 9014, pp. 629–658, 2015.

on this single web site.[1] The wide application of AKE protocols makes it necessary and interesting to study their security in large-scale settings with many millions of users.

Provably-secure AKE and tight reductions. A reduction-based security proof describes an algorithm, the *reduction*, which turns an efficient attacker on the protocol into an efficient algorithm solving an assumed-to-be-hard computational problem. The quality of such a reduction can be measured by its efficiency: the running time and success probability of the reduction running the attacker as a subroutine, relative to the running time and success probability of the attacker alone. Ideally the reduction adds only a minor amount of computation and has about the same success probability as the attacker. In this case the reduction is said to be *tight*.

The existence of tight security proofs has been studied for many cryptographic primitives, like, e.g., digital signatures [8,34,28,2], public-key encryption [4,25,31], or identity-based encryption [15,11]. However, there is no example of an authenticated key exchange protocol that comes with tight security proof under a standard assumption, not even in the Random Oracle Model [6].

Known provably secure AKE protocols come with a reduction which loses a factor that depends on the number μ of users and the number ℓ of sessions per user. The loss of the reduction ranges typically between $1/(\mu \cdot \ell)$ (if the reduction has to guess only one party participating in a particular session) and $1/(\mu \cdot \ell)^2$ (if the reduction has to guess both parties participating in a particular session). This may become significant in large-scale applications. We also consider tight reductions as theoretically interesting in their own right, because it is challenging to develop new proof strategies that avoid guessing. We will elaborate on the difficulty of constructing tightly secure AKE ini the next paragraph.

The difficulty of Tightly-Secure AKE. There are two main difficulties with proving tight security of an AKE protocol, which we would like to explain with concrete examples.

To illustrate the first, let us think of an AKE protocol where the long-term key pair (pk_i, sk_i) is a key pair for a digital signature scheme. Clearly, at some point in the security proof the security of the signature scheme must be used as an argument for the security of the AKE protocol, by giving a reduction from forging a signature to breaking the AKE protocol. Note that the attacker may use the Corrupt-query to learn the long-term secret of *all* parties, except for communication partner P_j of the Test-oracle. The index j might be chosen at random by the attacker.

A standard approach in security proofs for AKE protocols is to let the reduction, which implements the challenger in order to take advantage of the attacker, *guess* the index j of party P_j. The reduction generates all key pairs (pk_i, sk_i) with $i \neq j$ on its own, and thus is able to answer Corrupt-queries to party P_i for

---

[1] Figure obtained from http://newsroom.fb.com/Key-Facts on May 26, 2014. We assume that each active user logs-in once per day.

all $i \neq j$. In order to use the security of the signature scheme as an argument, a challenge public-key pk^* from the security experiment of the signature scheme is embedded as $pk_j := pk^*$.

Note that this strategy works *only if* the reduction guesses the index $i \in [\ell]$ correctly, which leads to a loss factor of $1/\ell$ in the success probability of the reduction. It is not immediately clear how to avoid this guessing: a reduction that avoids it would be required to be able to reveal *all* long-term secret key at any time in the security experiment, while simultaneously it needs to use the security of the signature scheme as an argument for the security of the AKE protocol. It turns out that we can resolve this seeming paradox by combining two copies of a signature scheme with a non-interactive proof system in a way somewhat related to the Naor-Yung paradigm [33] for public-key encryption.

To explain the second main difficulty, let us consider signed-DH protocol as an example. Let us first sketch this protocol. We stress that we leave out many details for simplicity, to keep the discussion on an intuitive level. In the sequel let $\mathcal{G}$ be a cyclic group of order p with generator g. Two parties P_i, P_j exchange a key as follows.

1. $\mathcal{P}_i$ chooses $x \xleftarrow{\$} \mathbb{Z}_p$ at random. It computes g^x and a digital signature σ_i over g^x, and sends (g^x, σ_i) to P_j.
2. If $\mathcal{P}_j$ receives (g^x, σ_i). It verifies σ_i, chooses $y \xleftarrow{\$} \mathbb{Z}_p$ at random, computes g^y and a digital signature σ_j over g^y, and sends (g^y, σ_j) to P_i. Moreover, P_j computes the key as $K = (g^x)^y$.
3. If $\mathcal{P}_i$ receives (g^y, σ_j), and σ_j is a valid signature, then P_i computes the key as $K = (g^y)^x$.

The security of this protocol can be proved [13] based on the (assumed) security of the signature scheme and the hardness of the decisional Diffie-Hellman problem, which asks for a given a vector $(g, g^x, g^y, g^w) \in \mathcal{G}$ to determine whether $w = xy$ or w is random. However, even though the DDH problem is randomly self-reducible [4], it seems impossible to avoid guessing at least one oracle participating in the Test-session.

To explain this, consider an attacker in the AKE security model from Section 4.1. Assume that the attacker asks $\mathsf{Send}(i, s, (\top, j))$ to an (uncorrupted) oracle π_i^s. According to the protocol specification, the oracle has to respond with (g^x, σ_i). At some point in the security proof the security of the protocol is reduced to the hardness of the DDH problem, thus, the challenger of the AKE security experiment has to decide whether it embeds (a part of) the given DDH-instance in g^x. Essentially, there are two options:

– The challenger decides that it embeds (a part of) the given DDH-instance in g^x. In this case, there exists an attacker which makes the simulation fail (with probability 1) if oracle π_i^s does not participate in the Test-session. This attacker proceeds as follows.

    1. It corrupts some unrelated party P_j to learn sk_j.

    2. It computes g^y for $y \xleftarrow{\$} \mathbb{Z}_p$ along with a signature σ_j under sk_j, and asks $\mathsf{Send}(i, s, (g^y, \sigma_j))$ to send (g^y, σ_j) to π_i^s.

3. Finally it asks Reveal(i, s) to learn the session key k_i^s computed by π_i^s, and checks whether $k_i^s = (g^x)^y$.

A challenger interacting with this attacker faces the problem that it needs to be able to compute $k_i^s = (g^y)^x$, *knowing neither x or y*. Note that the challenger can not answer with an incorrect k_i^s, because the attacker knows y and thus is able check whether k_i^s is computed correctly.

– The challenger decides that it does not embed (a part of) the given DDH-instance in g^x. If now the attacker asks Test(i, s), then the challenger is not able to take advantage of the attacker, because the DDH-challenge is not embedded in the Test-session.

The only way we see to circumvent this technical issue is to let the challenger guess in advance (at least) one oracle that participates in the Test-session, which however leads to a loss of $1/(\mu\ell)$ in the reduction.

The challenge with describing a tightly-secure AKE protocol is therefore to come up with a proof strategy that that avoids guessing. This requires to apply a strategy where essentially an instance of a hard computational problem is embedded into *any* protocol session, while at the same time the AKE-challenger is always able to compute the same keys as the attacker.

Our contribution. We construct the first AKE protocols whose security does not degrade in the number of users and instances. Following [5] we consider a very strong security model, which allows adaptive corruptions of long-term secrets, adaptive reveals of session keys, and multiple adaptive Test queries.

Our model provides *perfect forward secrecy* [9,29]: the corruption of a long-term secret does not foil the security of previously established session keys. In addition to that, we prevent *key-compromise impersonation* attacks [10,23]: in our security model, an attacker may introduce maliciously-generated keys. On the other hand, we do not allow reveals of internal states or intermediate results of computations, as considered in the (extended) Canetti-Krawczyk model [13,30]. The existence of a tightly secure construction in such a model is an interesting open problem.

While our approach is generic and modular, we give efficient instantiations from standard assumptions (such as the SXDH or DLIN assumptions in pairing-friendly groups). Specifically, we propose an SXDH-based AKE protocol with a communication complexity of only 14 group elements and 4 exponents (plus some bookkeeping information). The security reduction to SXDH loses a factor of κ (the security parameter), but does not depend on the number of users or instances. (Using different building blocks, this reduction loss can even be made constant, however at a significant expense of communication complexity.)

Our approach. At a very high level, our AKE protocol follows a well-known paradigm: we use a public-key encryption scheme to transport shared keys, and a digital signature scheme to authenticate exchanged messages. Besides, we use one-time signature scheme to provide a session-specific authentication, and thus to guarantee a technical "matching conversations" property.[2] The combination of these building blocks in itself is fairly standard; the difficulty in our case is to construct suitable buildings blocks that are *tightly and adaptively* secure.

More specifically, we require, e.g., a signature scheme that is tightly secure in face of adaptive corruptions. Specifically, it should be hard for an adversary $\mathcal{A}$ to forge a new signature in the name of *any* so far uncorrupted party in the system, even though $\mathcal{A}$ may corrupt arbitrary other parties adaptively. While regular signature security implies adaptive security in this sense, this involves a (non-tight) guessing argument. In fact, currently, no adaptively tightly secure signature scheme is known: while, e.g., [25] describe a tightly secure signature scheme, their analysis does not consider adaptive corruptions, and in particular no release whatsoever of signing keys. (The situation is similar for the encryption scheme used for key transport.)

How we construct adaptively secure signatures. Hence, while we cannot directly use existing building blocks, we can use the (non-adaptively) tightly secure signature scheme of [25] as a basis to construct adaptively and tightly secure components. In a nutshell, our first (less efficient but easier-to-describe) scheme adapts the "double encryption" technique of Naor and Yung [33] to the signature setting. A little more concretely, our scheme uses two copies of an underlying signature scheme SIG (that has to be tightly secure, but not necessarily against adaptive corruptions). A public key in our scheme consists of two public keys pk_1, sk_2 of SIG; however, our secret key consists only of one (randomly chosen) secret key sk_b of SIG. Signatures are (non-interactive, witness-indistinguishable) proofs of knowledge of *one* signature σ_i under one sk_i.

During the security proof, the simulation will know one valid secret key sk_b for each scheme instance.[3] This allows to plausibly reveal secret keys upon corruptions. However, the witness-indistinguishability of the employed proof system will hide *which* of the two possible keys sk_i are known for each user until that user is corrupted. Hence, an adversary $\mathcal{A}$ who forges a signature for an uncorrupted user will (with probability about $1/2$) forge a signature under a secret key which is unknown to the simulation. Hence, the simulation will lose only about a factor of 2 relative to the success probability of $\mathcal{A}$.

Of course, this requires using a suitable underlying signature scheme and proof system. For instance, the tightly secure (without corruptions) signature scheme from [25,1] and the Groth-Sahai non-interactive proof system [24] will be suitable DLIN-based building blocks.

---

[2] Intuitively, the matching conversations property, introduced by Bellare and Rogaway [5] establishes the notion of a "session" between two communication partners (essentially as the transcript of exchanged messages itself). Such a notion is essential in a model without explicit session identifiers (such as the one of Canetti and Krawczyk [13,14]) that separate different protocol instances.

[3] This can be seen as a variation of the approach of "two-signatures" approach of [21]. Concretely, [21] construct a signature scheme in which the simulation – by cleverly programming a random oracle – knows one out of two possible signatures for each message.

Efficient adaptively secure signatures. The signature scheme arising from the generic approach above is not overly efficient. Hence, we also construct a very optimized scheme that is not as modularly structured as the scheme above, but has extremely compact ciphertexts (of only 3 group elements). In a nutshell, this compact scheme uses the signature scheme that arises out of the recent almost-tightly secure MAC of [11] as a basis. Instead of Groth-Sahai proofs, we use a more implicit consistency proof reminiscent of hash proof systems. Security can be based on a number of computational assumptions (including SXDH and DLIN), and the security reduction loses a factor of κ (the security parameter), independently of the number of users or generated signatures. We believe that this signature scheme can be of independent interest.

Adaptively secure PKE and AKE schemes. A similar (generic) proof strategy allows to construct adaptively (chosen-plaintext) secure public-key encryption schemes using a variation of the Naor-Yung double encryption strategy [33]. (In this case, the simulation will know one out of two possible decryption keys. Furthermore, because we only require chosen-plaintext security, no consistency proof will be necessary.) Combining these tightly and adaptively secure building blocks with the tightly secure one-time signature scheme from [25] finally enables the construction of a tightly secure AKE protocol. As already sketched, our signature scheme ensures authenticated channels, while our encryption scheme is used to exchange session keys. (However, to achieve perfect forward secrecy – i.e., the secrecy of finished sessions upon corruption –, we generate PKE instances freshly for each new session.)

Notation. The symbol $\emptyset$ denotes the empty set. Let $[n] := \{1, 2, \ldots, n\} \subset \mathbb{N}$ and let $[n]^0 := [n] \cup \{0\}$. If A is a set, then $a \xleftarrow{\$} A$ denotes the action of sampling a uniformly random element from A. If A is a probabilistic algorithm, then we denote by $a \xleftarrow{\$} A$ that a is output by A using fresh random coins. If an algorithm A has black-box access to an algorithm $\mathcal{O}$, we will write $A^{\mathcal{O}}$.

2   Digital Signatures in the Multi-user Setting with Corruptions

In this section we define digital signature schemes and their security in the multi-user setting. Our strongest definition will be *existential unforgeability under adaptive chosen-message attacks in the multi-user setting with adaptive corruptions*. We show how to construct a signature scheme with tight security proof, based on a combination of a non-interactive witness indistinguishable proof of knowledge with a signature scheme with weaker security properties.

2.1   Basic Definitions

Definition 1. *A (one-time) signature scheme* SIG *consists of four probabilistic algorithms:*

- $\Pi \xleftarrow{\$} \mathsf{SIG.Setup}(1^{\kappa})$: *The parameter generation algorithm on input a security parameter 1^{κ} returns public parameters Π, defining the message space $\mathcal{M}$, signature space $\mathcal{S}$, and key space $\mathcal{VK} \times \mathcal{SK}$.*
- $\mathsf{SIG.Gen}(\Pi)$: *On input Π the key generation algorithm ouputs a key pair $(vk, sk) \in \mathcal{VK} \times \mathcal{SK}$.*
- $\mathsf{SIG.Sign}(sk, m)$: *On input a private key sk and a message $m \in \mathcal{M}$, the signing algorithm outputs a signature σ.*
- $\mathsf{SIG.Vfy}(vk, m, \sigma)$: *On input a verification key vk, a message m, and a purported signature σ, the verification algorithm returns $b \in \{0, 1\}$.*

We note that our security definition below assumes a trusted setup of public parameters (using $\mathsf{SIG.Setup}$). Moreover, throughout the paper, we will assume signature schemes with message space $\{0, 1\}^{*}$ for simplicity. It is well-known that such a scheme can be constructed from a signature scheme with arbitrary message space $\mathcal{M}$ by applying a collision-resistant hash function $H : \{0, 1\}^{*} \rightarrow \mathcal{M}$ to the message before signing.

Security Definitions. The standard security notion for signature schemes in the single user setting is *existential unforgeability under chosen-message attacks*, as proposed by Goldwasser, Micali and Rivest [22]. We consider natural extensions of this notion to the multi-user setting with or without adaptive corruptions.

Consider the following game between a challenger $\mathcal{C}$ and an adversary $\mathcal{A}$, which is parametrized by the number of public keys μ.

1. For each $i \in [\mu]$, $\mathcal{C}$ runs $(vk^{(i)}, sk^{(i)}) \leftarrow \mathsf{SIG.Gen}(\Pi)$, where Π are public parameters. Furthermore, the challenger initializes a set $\mathcal{S}^{\mathsf{corr}}$ to keep track of corrupted keys, and μ sets $\mathcal{S}_1, \ldots, \mathcal{S}_{\mu}$, to keep track of chosen-message queries. All sets are initially empty. Then it outputs $(vk^{(1)}, \ldots, vk^{(\mu)})$ to $\mathcal{A}$.
2. $\mathcal{A}$ may now issue two different types of queries. When $\mathcal{A}$ outputs an index $i \in [\mu]$, then $\mathcal{C}$ updates $\mathcal{S}^{\mathsf{corr}} := \mathcal{S}^{\mathsf{corr}} \cup \{i\}$ and returns sk_i. When $\mathcal{A}$ outputs a tuple (m, i), then $\mathcal{C}$ computes $\sigma := \mathsf{SIG.Sign}(sk_i, m)$, adds (m, σ) to $\mathcal{S}_i$, and responds with σ.
3. Eventually $\mathcal{A}$ outputs a triple $(i^{*}, m^{*}, \sigma^{*})$.

Now we can derive various security definitions from this generic experiment. We start with existential unforgeability under chosen-message attacks in the multi-user setting with corruptions.

Definition 2. *Let $\mathcal{A}$ be an algorithm that runs in time t. We say that $\mathcal{A}$ (t, ϵ, μ)-breaks the $\mathrm{MU\text{-}EUF\text{-}CMA}^{\mathsf{Corr}}$-security of SIG, if in the above game it holds that*

$$\Pr\left[(m^{*}, i^{*}, \sigma^{*}) \leftarrow \mathcal{A}^{\mathcal{C}} : \begin{array}{l} i^{*} \notin \mathcal{S}^{\mathsf{corr}} \wedge (m^{*}, \cdot) \notin \mathcal{S}_{i^{*}} \\ \wedge \mathsf{SIG.Vfy}(vk^{(i^{*})}, m^{*}, \sigma^{*}) = 1 \end{array}\right] \geq \epsilon$$

In order to construct an $\mathrm{MU\text{-}EUF\text{-}CMA}^{\mathsf{Corr}}$-secure signature scheme, we will also need the following weaker definition of EUF-CMA security in the multi-user setting *without* corruptions. We note that this definition was also considered in [32].

Definition 3. *Let $\mathcal{A}$ be an algorithm that runs in time t. We say that $\mathcal{A}$ (t, ϵ, μ)-breaks the MU-EUF-CMA-security of SIG, if in the above game it holds that*

$$\Pr\left[(m^*, i^*, \sigma^*) \leftarrow \mathcal{A}^{\mathcal{C}} : \begin{array}{l} \mathcal{S}^{corr} = \emptyset \wedge (m^*, \cdot) \notin \mathcal{S}_{i^*} \\ \wedge \mathsf{SIG.Vfy}(vk^{(i^*)}, m^*, \sigma^*) = 1 \end{array}\right] \geq \epsilon$$

Note that both MU-EUF-CMA^{Corr} and MU-EUF-CMA security notions are polynomially equivalent to the standard (single user) EUF-CMA security notion for digital signatures. However, the reduction is not tight.

Finally, we need *strong* existential unforgeability in the multi-user setting without corruptions for *one-time signatures*.

Definition 4. *Let $\mathcal{A}$ be an algorithm that runs in time t. We say that $\mathcal{A}$ (t, ϵ, μ)-breaks the MU-sEUF-1-CMA-security of SIG, if in the above game it holds that*

$$\Pr\left[(m^*, i^*, \sigma^*) \leftarrow \mathcal{A}^{\mathcal{C}} : \begin{array}{l} \mathcal{S}^{corr} = \emptyset \wedge |\mathcal{S}_i| \leq 1, \forall i \wedge (m^*, \sigma^*) \notin \mathcal{S}_{i^*} \\ \wedge \mathsf{SIG.Vfy}(vk^{(i^*)}, m^*, \sigma^*) = 1 \end{array}\right] \geq \epsilon$$

2.2 MU-EUF-CMA^{Corr}-Secure Signatures from General Assumptions

In this section we give a generic construction of a MU-EUF-CMA^{Corr}-secure signature scheme, based on a MU-EUF-CMA-signature scheme and a non-interactive witness-indistinguishable proof of knowledge that allows a tight security proof. The main purpose of this construction is to resolve the "paradox" explained in the introduction.

NIWI Proofs of Knowledge. Let R be a binary relation. If $(x, w) \in R$, then we call x the *statement* and w the *witness*. R defines a language $\mathcal{L}_R := \{x : \exists w : (x, w) \in R\}$. A *non-interactive proof system* NIPS $=$ (NIPS.Gen, NIPS.Prove, NIPS.Vfy) for R consists of the following efficient algorithms.

- Algorithm NIPS.Gen takes as input the security parameter and ouputs a *common reference string* CRS $\xleftarrow{\$}$ NIPS.Gen(1^κ).
- Algorithm NIPS.Prove takes as input the CRS, a statement x and a witness w, and outputs a proof $\pi \xleftarrow{\$}$ NIPS.Prove(CRS, x, w).
- The verification algorithm NIPS.Vfy(CRS, x, π) $\in \{0, 1\}$ takes as input the CRS, a statement x, and a purported proof π. It outputs 1 if the proof is accepted, and 0 otherwise.

Definition 5. *We call NIPS a witness indistinguishable proof of knowledge (NIWI-PoK) for R, if the following conditions are satisfied:*
Perfect Completeness. *For all $(x, w) \in R$, $\kappa \in \mathbb{N}$, CRS $\xleftarrow{\$}$ NIPS.Gen(1^κ), and all proofs π computed as $\pi \xleftarrow{\$}$ NIPS.Prove(CRS, x, w) holds that*

$$\Pr\left[\mathsf{NIPS.Vfy}(\mathsf{CRS}, x, \pi) = 1\right] = 1$$

Perfect Witness Indistinguishability. *For all* CRS $\overset{\$}{\leftarrow}$ NIPS.Gen(1^κ), *for all* (x, w_0, w_1) *such that* $(x, w_0) \in R$ *and* $(x, w_1) \in R$, *and all algorithms* $\mathcal{A}$ *it holds that*

$$\Pr[\mathcal{A}(\pi_0) = 1] = \Pr[\mathcal{A}(\pi_1) = 1] \tag{1}$$

where $\pi_0 \overset{\$}{\leftarrow}$ NIPS.Prove(CRS, x, w_0) *and* $\pi_1 \overset{\$}{\leftarrow}$ NIPS.Prove(CRS, x, w_1).

Simulated CRS. *There exists an algorithm* $\mathcal{E}_0$, *which takes as input* κ *and outputs a simulated common reference string* CRS$_{sim}$ *and a trapdoor* τ.

Perfect Knowledge Extraction on Simulated CRS. *There exists an algorithms* $\mathcal{E}_1$ *such that for all* (CRS$_{sim}, \tau$) $\overset{\$}{\leftarrow} \mathcal{E}_0(1^\kappa)$ *and all* $(\pi, x) \leftarrow \mathcal{A}$ *such that* NIPS.Vfy(CRS$_{sim}, x, \pi$) = 1

$$\Pr\left[w \overset{\$}{\leftarrow} \mathcal{E}_1(\text{CRS}_{sim}, \pi, x, \tau) : (x, w) \in R\right] = 1$$

Security Definition for NIWI-PoK. *An algorithm* $(t, \epsilon_{\text{CRS}})$-*breaks the security of a NIWI-PoK if it runs in time* t *and for all* $\kappa \in \mathbb{N}$, CRS$_{real}$ $\overset{\$}{\leftarrow}$ NIPS.Gen(1^κ), *all* (CRS$_{sim}, \tau$) $\overset{\$}{\leftarrow} \mathcal{E}_0(1^\kappa)$, *it holds that*

$$\Pr\left[\mathcal{A}(\text{CRS}_{real}) = 1)\right] - \Pr\left[\mathcal{A}(\text{CRS}_{sim}) = 1\right] \geq \epsilon_{\text{CRS}}$$

We note that *perfect* witness indistinguishability is preserved if the algorithm $\mathcal{A}$ sees more than one proof. That is, let $\mathcal{O}_b^q(x, w_0, w_1)$ denote an oracle which takes as input (x, w_0, w_1) with $(x, w_0) \in R$ and $(x, w_1) \in R$, and outputs NIPS.Prove(CRS, x, w_b) for random $b \in \{0, 1\}$. Consider an algorithm $\mathcal{A}$ which asks $\mathcal{O}_b^q$ at most q times. The following is proven in the full version of our paper [3].

Lemma 1. *Equation 1 implies for all* $q \in \mathbb{N}$:

$$\Pr\left[\mathcal{A}^{\mathcal{O}_1^q} = 1 : \text{CRS} \overset{\$}{\leftarrow} \text{NIPS.Gen}(1^\kappa)\right] = \Pr\left[\mathcal{A}^{\mathcal{O}_0^q} = 1 : \text{CRS} \overset{\$}{\leftarrow} \text{NIPS.Gen}(1^\kappa)\right]$$

Generic Construction. In this section we show how to generically construct an MU-EUF-CMA^{Corr}-secure (Definition 2) signature scheme SIG$_{\text{MU}}$ from a signature scheme SIG that is MU-EUF-CMA-secure (Definition 3) and a NIWI-PoK.

In the sequel let NIPS = (NIPS.Gen, NIPS.Prove, NIPS.Vfy) denote a NIWI-PoK for relation

$$R := \left\{ ((\text{vk}_0, \text{vk}_1, m), (\sigma_0, \sigma_1)) : \begin{array}{l} \text{SIG.Vfy}(\text{vk}_0, m, \sigma_0) = 1 \\ \vee \text{SIG.Vfy}(\text{vk}_1, m, \sigma_1) = 1 \end{array} \right\}.$$

That is, R consists of statements of the form $(\text{vk}_0, \text{vk}_1, m)$, where $(\text{vk}_0, \text{vk}_1)$ are verification keys for signature scheme SIG, and m is a message. Witnesses are tuples (σ_0, σ_1) such that either σ_0 is a valid signature for m under vk_0, or σ_1 is a valid signature for m under vk_1, or both.

The new signature scheme SIG$_{\text{MU}}$ = (SIG.Setup$_{\text{MU}}$, SIG.Gen$_{\text{MU}}$, SIG.Sign$_{\text{MU}}$, SIG.Vfy$_{\text{MU}}$) works as follows:

- $\Pi_{\mathsf{SIG_{MU}}} \xleftarrow{\$} \mathsf{SIG.Setup_{MU}}(1^\kappa)$: The setup algorithm runs $\Pi_{\mathsf{SIG}} \xleftarrow{\$} \mathsf{SIG.Setup}(1^\kappa)$ and $\mathsf{CRS} \xleftarrow{\$} \mathsf{NIPS.Gen}(1^\kappa)$. It outputs $\Pi_{\mathsf{SIG_{MU}}} := (\Pi_{\mathsf{SIG}}, \mathsf{CRS})$.
- $\mathsf{SIG.Gen_{MU}}(\Pi_{\mathsf{SIG_{MU}}})$: The key generation algorithm generates two key pairs by running the key generation algorithm of SIG twice: For $i \in \{0,1\}$, it runs $(\mathsf{vk}_i, \mathsf{sk}_i) \xleftarrow{\$} \mathsf{SIG.Gen}(\Pi_{\mathsf{SIG}})$. Then it flips a random coin $\delta \xleftarrow{\$} \{0,1\}$ and returns $(vk, sk) = ((\mathsf{vk}_0, \mathsf{vk}_1), (\mathsf{sk}_\delta, \delta))$. Observe that $\mathsf{sk}_{1-\delta}$ is discarded.
- $\mathsf{SIG.Sign_{MU}}(sk, m)$: The signing algorithm generates a SIG-signature $\sigma_\delta \xleftarrow{\$} \mathsf{SIG.Sign}(\mathsf{sk}_\delta, m)$. Then it defines a witness w as

$$w := \begin{cases} (\sigma_\delta, \bot), & \text{if } \delta = 0, \\ (\bot, \sigma_\delta), & \text{if } \delta = 1, \end{cases}$$

  where $\bot$ is an arbitrary constant (e.g., a fixed element from the signature space). Note that $((\mathsf{vk}_0, \mathsf{vk}_1, m), w) \in R$. Finally it returns a signature as $\sigma = \pi \xleftarrow{\$} \mathsf{NIPS.Prove}(\mathsf{CRS}, (\mathsf{vk}_0, \mathsf{vk}_1, m), w)$.
- $\mathsf{SIG.Vfy_{MU}}(vk, m, \sigma)$: The verification algorithm parses vk as $(\mathsf{vk}_0, \mathsf{vk}_1)$ and returns whatever $\mathsf{NIPS.Vfy}(\mathsf{CRS}, (\mathsf{vk}_0, \mathsf{vk}_1, m), \sigma)$ returns.

Theorem 1. *Let $\mathsf{SIG_{MU}}$ be as described above. From any attacker $\mathcal{A}_{\mathsf{SIG_{MU}}}$ that (t, ϵ, μ)-breaks the MU-EUF-CMA^{Corr}-security (with corruptions) of $\mathsf{SIG_{MU}}$, we can construct algorithms $\mathcal{B}_{\mathsf{NIPS}}$ and $\mathcal{B}_{\mathsf{SIG}}$ such that either $\mathcal{B}_{\mathsf{NIPS}}$ $(t_{\mathsf{CRS}}, \epsilon_{\mathsf{CRS}})$-breaks the security of NIWI-PoK or $\mathcal{B}_{\mathsf{SIG}}$ $(t_{\mathsf{SIG}}, \epsilon_{\mathsf{SIG}}, \mu)$-breaks the MU-EUF-CMA-security (without corruptions) of SIG, where*

$$\epsilon < 2 \cdot \epsilon_{\mathsf{SIG}} + \epsilon_{\mathsf{CRS}}$$

We have $t_{\mathsf{CRS}} = t + t'_{\mathsf{CRS}}$ and $t_{\mathsf{SIG}} = t + t'_{\mathsf{SIG}}$, where t'_{CRS} and t'_{SIG} correspond to the respective runtimes required to provide $\mathcal{A}_{\mathsf{SIG_{MU}}}$ with the simulated experiment as described below.

Proof. We proceed in a sequence of games. The first game is the real game that is played between an attacker $\mathcal{A}$ and a challenger $\mathcal{C}$, as described in Section 2.1. We denote by χ_i the event that $\mathcal{A}_{\mathsf{SIG_{MU}}}$ outputs (m^*, i^*, σ^*) such that $\mathsf{SIG.Vfy}(vk^{(i^*)}, m^*, \sigma^*) \wedge i^* \notin \mathcal{S}^{\mathsf{corr}} \wedge (m^*, \cdot) \notin \mathcal{S}_{i^*}$ in Game i.

GAME 0. This is the real game that is played between $\mathcal{A}$ and $\mathcal{C}$. We set

$$\Pr[\chi_0] = \epsilon.$$

GAME 1. In this game we change the way keys are generated and chosen-message queries are answered by the challenger.

When generating a key pair by running $\mathsf{SIG.Gen_{MU}}$, the challenger does not discard $\mathsf{sk}_{1-\delta}$ but keeps it. However, corruption queries by the attacker are still answered by responding only with sk_δ. Therefore this change is completely oblivious to $\mathcal{A}$.

To explain the second change, recall that a $\mathsf{SIG_{MU}}$-signature in Game 0 consists of a proof $\pi \xleftarrow{\$} \mathsf{NIPS.Prove}(\mathsf{CRS}, (\mathsf{vk}_0, \mathsf{vk}_1, m), w)$, where either $w = (\sigma_\delta, \bot)$

or $w = (\bot, \sigma_\delta)$ for $\sigma_\delta \xleftarrow{\$} \mathsf{SIG.Sign}(\mathsf{sk}_\delta, m)$. In Game 1 the challenger now defines w as follows. It first computes two signatures $\sigma_0 \xleftarrow{\$} \mathsf{SIG.Sign}(\mathsf{sk}_0, m)$ and $\sigma_1 \xleftarrow{\$} \mathsf{SIG.Sign}(\mathsf{sk}_1, m)$, and sets $w := (\sigma_0, \sigma_1)$. Then it proceeds as before, by computing π as $\pi \xleftarrow{\$} \mathsf{NIPS.Prove}(\mathsf{CRS}, (\mathsf{vk}_0, \mathsf{vk}_1, m), w)$. Thus, in Game 1 *two* valid signatures are used as witnesses. Due to the *perfect* witness indistinguishability property of NIPS we have:

$$\Pr[\chi_0] = \Pr[\chi_1]$$

GAME 2. This game is very similar to the previous game, except that we change the way the CRS is generated. Now, we run $(\mathsf{CRS}_{\mathsf{sim}}, \tau) \xleftarrow{\$} \mathcal{E}_0$ and all proofs are generated with respect to $\mathsf{CRS}_{\mathsf{sim}}$. Since the contrary would allow $\mathcal{B}_{\mathsf{NIPS}}$ to break the $(t, \epsilon_{\mathsf{CRS}})$-security of NIPS we have

$$|\Pr[\chi_1] - \Pr[\chi_2]| < \epsilon_{\mathsf{CRS}}$$

GAME 3. This game is similar to Game 2 except for the following. We abort the game (and $\mathcal{A}$ loses) if the forgery (i^*, m^*, σ^*) returned by $\mathcal{A}$ is valid, i.e., satisfies $\mathsf{SIG.Vfy}_{\mathsf{MU}}\left(vk^{(i^*)}, m^*, \sigma^*\right) = 1$, but the extractor $\mathcal{E}_1$ is not able to extract a witness (s_0, s_1) from σ^*. Due to the *perfect* knowledge extraction property of NIPS on a simulated CRS we have:

$$\Pr[\chi_2] = \Pr[\chi_3]$$

GAME 4. In this game we raise event $\mathsf{abort}_{\delta^{(i^*)}}$ and abort (and $\mathcal{A}$ loses) if $\mathcal{A}$ outputs a forgery (i^*, m^*, σ^*) such that the following holds.

Given (i^*, m^*, σ^*), the challenger first runs the extractor $(s_0, s_1) \xleftarrow{\$} \mathcal{E}_1(\tau, \sigma^*)$. Then it checks whether

$$\mathsf{SIG.Vfy}\left(\mathsf{vk}^{(i^*)}_{1-\delta^{(i^*)}}, m^*, s_{1-\delta^{(i^*)}}\right) = 0.$$

Recall here that $\delta^{(i^*)}$ denotes the random bit chosen by the challenger for the generation of the long-term secret of user i^*. If this condition is satisfied, then the game is aborted. Putting it differently, the challenger aborts, if the witness $s_{1-\delta^{(i^*)}}$ is *not* a valid signature for m^* under $\mathsf{vk}^{(i^*)}_{1-\delta^{(i^*)}}$.

Since $\mathcal{A}$ is not allowed to corrupt the secret key of user i^*, and the adversary sees only proofs which use *two* valid signatures (s_0, s_1) as witnesses (cf. Game 1), the random bit $\delta^{(i^*)}$ is information-theoretically perfectly hidden from $\mathcal{A}$. Therefore, we have $\Pr[\mathsf{abort}_{\delta^{(i^*)}}] \leq 1/2$ and

$$\Pr[\chi_3] \leq 2 \cdot \Pr[\chi_4]$$

Claim. For any attacker $\mathcal{A}_{\mathsf{SIG}_{\mathsf{MU}}}$ that breaks the $(t, \Pr[\chi_4], \mu)$-MU-EUF-CMA^{Corr}-security of $\mathsf{SIG}_{\mathsf{MU}}$ in Game 4 there exists an attacker $\mathcal{B}_{\mathsf{SIG}}$ that breaks the $(t_{\mathsf{SIG}}, \epsilon_{\mathsf{SIG}}, \mu)$-MU-EUF-CMA-security of SIG with $t_{\mathsf{SIG}} \approx t$ and $\epsilon_{\mathsf{SIG}} \geq \Pr[\chi_4]$.

Given the above claim, we can conclude the proof of Theorem 1. In summary we have $\epsilon \leq \epsilon_{\mathsf{CRS}} + 2 \cdot \epsilon_{\mathsf{SIG}}$.

Proof of 4. Attacker $\mathcal{B}_{\mathsf{SIG}}$ simulates the challenger for an adversary $\mathcal{A}_{\mathsf{SIG}_{\mathsf{MU}}}$ in Game 4. We show that any successful forgery that is output by $\mathcal{A}_{\mathsf{SIG}_{\mathsf{MU}}}$ can be used by $\mathcal{B}_{\mathsf{SIG}}$ to win the SIG security game.

$\mathcal{B}_{\mathsf{SIG}}$ receives μ public verification keys $\mathsf{vk}^{(i)}, i \in [\mu]$, and public parameters Π_{SIG} from the SIG challenger. Next, it samples μ key pairs $(\mathsf{vk}^{(i)}, \mathsf{sk}^{(i)}) \xleftarrow{\$}$ $\mathsf{SIG.Gen}(\Pi_{\mathsf{SIG}}), i \in \{\mu+1, \ldots, 2\mu\}$. Moreover, it chooses a random vector $\delta = (\delta^{(1)}, \ldots, \delta^{(\mu)}) \in \{0,1\}^{\mu}$. It sets

$$(vk^{(i)}, sk^{(i)}) \leftarrow \left(\left(\mathsf{vk}^{(\delta^{(i)}\mu + i)}, \mathsf{vk}^{((1-\delta^{(i)})\mu + i)} \right), \left(\mathsf{sk}^{\mu+i}, 1 - \delta^{(i)} \right) \right).$$

Note that now each $\mathsf{SIG}_{\mathsf{MU}}$-verification key contains one SIG-verification key that $\mathcal{A}_{\mathsf{SIG}}$ has obtained from its challenger, and one that was generated by $\mathcal{B}_{\mathsf{SIG}}$. We note further that, given $vk^{(i)}$, $sk^{(i)}$ is distributed correctly and may be returned by $\mathcal{B}_{\mathsf{SIG}}$ when $\mathcal{A}_{\mathsf{SIG}_{\mathsf{MU}}}$ issues a corrupt query (since it is generated by $\mathcal{B}_{\mathsf{SIG}}$ itself).

Over that $\mathcal{B}_{\mathsf{SIG}}$ generates a "simulated" CRS for the NIWI-PoK along with a trapdoor by running $(\mathsf{CRS}_{\mathsf{sim}}, \tau) \xleftarrow{\$} \mathcal{E}_0$. $\mathcal{A}_{\mathsf{SIG}_{\mathsf{MU}}}$ receives as input $\{vk^{(i)} : i \in [\mu]\}$, Π_{SIG} and CRS.

Now, when asked to sign a message m under public key $vk^{(i)}$, $\mathcal{A}_{\mathsf{SIG}}$ proceeds as follows. Let $\delta^{(i)} = 0$ without loss of generality. Then it computes $\sigma_1 = \mathsf{SIG.Sign}(\mathsf{sk}^{(\mu+i)}, m)$. Moreover it requests a signature for public key $\mathsf{vk}^{(i)}$ and message m from its SIG-challenger. Let σ_0 be the response. $\mathcal{A}_{\mathsf{SIG}}$ computes the signature for m using both signatures $w = (\sigma_0, \sigma_1)$ as witnesses. Note that this is a perfect simulation of Game 4.

If Game 4 is not aborted, then any valid forgery of $\mathcal{A}_{\mathsf{SIG}_{\mathsf{MU}}}$ can be used by $\mathcal{B}_{\mathsf{SIG}}$ as a forgery in the SIG security game. The claim follows. $\square$

(Somewhat Inefficient) Instantiation From Existing Building Blocks.

The generic construction $\mathsf{SIG}_{\mathsf{MU}}$ above can be instantiated conveniently from existing building blocks:

- Suitable tightly secure MU-EUF-CMA-secure signatures can be found in [26,1] (based on the DLIN assumption in pairing-friendly groups).
- A suitable tightly MU-sEUF-1-CMA-secure one-time signature scheme is described in [26, Section 4.2]. Its security is based on the discrete logarithm assumption.
- Finally, a compatible NIWI-PoK is given by Groth-Sahai proofs [24]. (In a Groth-Sahai proof system, there exist "hiding" and "binding" CRSs. These correspond to our honestly generated, resp. simulated CRSs.) The security of Groth-Sahai proofs can be based on a number of assumptions, including the DLIN assumption in pairing-friendly groups.

When used in our generic construction, this yields a signature scheme whose MU-EUF-CMA^{Corr} security can be tightly (i.e., with a small constant loss) reduced to the DLIN assumption in pairing-friendly groups. However, we note that the resulting scheme is not overly efficient. In particular, the scheme suffers from public keys and signatures that contain a linear – in the security parameter – number of group elements.

Thus, in the next section, we offer an optimized, significantly more efficient MU-EUF-CMA^{Corr}-secure signature scheme.

2.3   Efficient and Almost Tightly MU-EUF-CMA^{Corr}-Secure Signatures

Here, we present a very efficient signature scheme whose MU-EUF-CMA^{Corr} security can be almost tightly (i.e., with a reduction loss that is linear in the security parameter) reduced to a number of standard assumptions in cyclic groups. In fact, we prove security under any *matrix assumption* [20], which encompasses, e.g., the SXDH, DLIN, and k-Linear assumptions. The following definitions are taken from [11].

Pairing Groups and Matrix Diffie-Hellman Assumption. Let GGen be a probabilistic polynomial time (PPT) algorithm that on input 1^κ returns a description $\mathcal{G} = (\mathbb{G}_1, \mathbb{G}_2, \mathbb{G}_T, q, g_1, g_2, e)$ of asymmetric pairing groups where $\mathbb{G}_1, \mathbb{G}_2, \mathbb{G}_T$ are cyclic groups of order q for a κ-bit prime q, g_1 and g_2 are generators of $\mathbb{G}_1$ and $\mathbb{G}_2$, respectively, and $e : \mathbb{G}_1 \times \mathbb{G}_2$ is an efficiently computable (non-degenerated) bilinear map. Define $g_T := e(g_1, g_2)$, which is a generator in $\mathbb{G}_T$.

We use implicit representation of exponents by group elements as introduced in [20]. For $s \in \{1, 2, T\}$ and $a \in \mathbb{Z}_q$ define $[a]_s = g_s^a \in \mathbb{G}_s$ as the *implicit representation* of a in $\mathbb{G}_s$. More generally, for a matrix $\mathbf{A} = (a_{ij}) \in \mathbb{Z}_q^{n \times m}$ we define $[\mathbf{A}]_s$ as the implicit representation of $\mathbf{A}$ in $\mathbb{G}_s$:

$$[\mathbf{A}]_s := \begin{pmatrix} g_s^{a_{11}} & \cdots & g_s^{a_{1m}} \\ & & \\ g_s^{a_{n1}} & \cdots & g_s^{a_{nm}} \end{pmatrix} \in \mathbb{G}_s^{n \times m}$$

We will always use this implicit notation of elements in $\mathbb{G}_s$, i.e., we let $[a]_s \in \mathbb{G}_s$ be an element in $\mathbb{G}_s$. Note that under the discrete logarithm assumption in $\mathbb{G}_s$ it is hard to compute a from $[a]_s \in \mathbb{G}_s$. Further, from $[b]_T \in \mathbb{G}_T$ it is hard to compute the value $[b]_1 \in \mathbb{G}_1$ and $[b]_2 \in \mathbb{G}_2$ (pairing inversion problem). Obviously, given $[a]_s \in \mathbb{G}_s$ and a scalar $x \in \mathbb{Z}_q$, one can efficiently compute $[ax]_s \in \mathbb{G}_s$. Further, given $[a]_1, [a]_2$ one can efficiently compute $[ab]_T$ using the pairing e. For $\mathbf{a}, \mathbf{b} \in \mathbb{Z}_q^k$ define $e([\mathbf{a}]_1, [\mathbf{b}]_2) := [\mathbf{a}^\top \mathbf{b}]_T \in \mathbb{G}_T$.

We recall the definition of the Matrix Diffie-Hellman (MDDH) assumption [20].

Definition 6 (Matrix Distribution). *Let* $k \in \mathbb{N}$. *We call* $\mathcal{D}_k$ *a matrix distribution if it outputs matrices in* $\mathbb{Z}_q^{(k+1) \times k}$ *of full rank* k *in polynomial time.*

For $\mathbf{B} \in \mathbb{Z}_q^{(k+1) \times n}$, we define $\overline{\mathbf{B}} \in \mathbb{Z}_q^{k \times n}$ as the first k rows of $\mathbf{B}$ and $\underline{\mathbf{B}} \in \mathbb{Z}_q^{1 \times n}$ as the last row vector of $\mathbf{B}$. Without loss of generality, we assume the first k rows $\overline{\mathbf{A}}$ of $\mathbf{A} \overset{\$}{\leftarrow} \mathcal{D}_k$ form an invertible matrix.

The $\mathcal{D}_k$-Matrix Diffie-Hellman problem is to distinguish the two distributions $([\mathbf{A}], [\mathbf{A}\mathbf{w}])$ and $([\mathbf{A}], [\mathbf{u}])$ where $\mathbf{A} \overset{\$}{\leftarrow} \mathcal{D}_k$, $\mathbf{w} \overset{\$}{\leftarrow} \mathbb{Z}_q^k$ and $\mathbf{u} \overset{\$}{\leftarrow} \mathbb{Z}_q^{k+1}$.

Definition 7 ($\mathcal{D}_k$-Matrix Diffie-Hellman Assumption $\mathcal{D}_k$-MDDH). *Let $\mathcal{D}_k$ be a matrix distribution and $s \in \{1, 2, T\}$. We say that $\mathcal{A}$ (ϵ, t)-breaks the $\mathcal{D}_k$-Matrix Diffie-Hellman ($\mathcal{D}_k$-MDDH) Assumption relative to GGen in group $\mathbb{G}_s$ if it runs in time at most t and*

$$\Pr[\mathcal{A}(\mathcal{G}, [\mathbf{A}]_s, [\mathbf{Aw}]_s) = 1] - \Pr[\mathcal{A}(\mathcal{G}, [\mathbf{A}]_s, [\mathbf{u}]_s) = 1]| \leq \epsilon,$$

where the probability is taken over $\mathcal{G} \leftarrow \mathsf{GGen}(1^\lambda)$, $\mathbf{A} \leftarrow \mathcal{D}_k, \mathbf{w} \overset{\$}{\leftarrow} \mathbb{Z}_q^k, \mathbf{u} \overset{\$}{\leftarrow} \mathbb{Z}_q^{k+1}$ and the random coins of $\mathcal{A}$.

The Construction and Its Security. Let GGen be a pairing group generator and let $\mathcal{D}_k$ be a matrix distribution. The new signature scheme $\mathsf{SIG}_\mathsf{C} = (\mathsf{SIG.Setup}_\mathsf{C}, \mathsf{SIG.Gen}_\mathsf{C}, \mathsf{SIG.Sign}_\mathsf{C}, \mathsf{SIG.Vfy}_\mathsf{C})$ for message $m \in \{0, 1\}^\ell$ is based on a tightly-secure signature scheme from [11]. Whereas [11] obtained their signature scheme from a tightly-secure single-user algebraic MAC, we implicitly construct a tightly-secure *multi-user* algebraic MAC. More precisely, the signatures consist of the algebraic MAC part (elements $[\mathbf{r}]_2, [u]_2$) plus a NIZK proof $[\mathbf{v}]_2$ showing that the MAC is correct with respect to the committed MAC secret key $[\mathbf{c}]_1$.

The scheme works as follows.

- $\Pi \overset{\$}{\leftarrow} \mathsf{SIG.Setup}_\mathsf{C}(1^\kappa)$: The parameter generation algorithm $\mathsf{SIG.Setup}_\mathsf{C}$ runs $\mathcal{G} \overset{\$}{\leftarrow} \mathsf{GGen}$, $\mathbf{A}, \mathbf{A}' \overset{\$}{\leftarrow} \mathcal{D}_k$ and defines $\mathbf{B} := \overline{\mathbf{A}'} \in \mathbb{Z}_q^{k \times k}$, the $k \times k$ matrix consisting of the k top rows of $\mathbf{A}'$. For $0 \leq i \leq \ell, 0 \leq b \leq 1$ it picks $\mathbf{x}_{i,b} \overset{\$}{\leftarrow} \mathbb{Z}_q^k$, $\mathbf{Y}_{i,b} \overset{\$}{\leftarrow} \mathbb{Z}_q^{k \times k}$, and defines $\mathbf{Z}_{i,b} = (\mathbf{Y}_{i,b}^\top \| \mathbf{x}_{i,b}) \cdot \mathbf{A} \in \mathbb{Z}_q^{k \times k}$. It outputs

$$\Pi := \left(\mathcal{G}, [\mathbf{A}]_1, [\mathbf{B}]_2, ([\mathbf{Z}_{i,b}]_1, [\mathbf{x}_{i,b}^\top \mathbf{B}]_2, [\mathbf{Y}_{i,b}\mathbf{B}]_2)_{1 \leq i \leq \ell, 0 \leq b \leq 1} \right).$$

For a message $m = (m_1, \ldots, m_\ell) \in \{0, 1\}^\ell$, define the following functions

$$\mathbf{x}(m) := \sum_{i=1}^{\ell} \mathbf{x}_{i,m_i}^\top \in \mathbb{Z}_q^{1 \times k}, \quad \mathbf{Y}(m) := \sum_{i=1}^{\ell} \mathbf{Y}_{i,m_i} \in \mathbb{Z}_q^{k \times k},$$

$$\mathbf{Z}(m) := \sum_{i=1}^{\ell} \mathbf{Z}_{i,m_i} = (\mathbf{Y}(m)^\top \| \mathbf{x}(m)^\top) \cdot \mathbf{A} \in \mathbb{Z}_q^{k \times k}. \tag{2}$$

- $\mathsf{SIG.Gen}_\mathsf{C}(\Pi)$: The key generation algorithm picks $a \overset{\$}{\leftarrow} \mathbb{Z}_q$, $\mathbf{b} \overset{\$}{\leftarrow} \mathbb{Z}_q^k$, and defines $\mathbf{c}^\top = (\mathbf{b}^\top \| a) \cdot \mathbf{A} \in \mathbb{Z}_q^{1 \times k}$. It returns $(\mathsf{vk}, \mathsf{sk}) = ([\mathbf{c}]_1, ([a]_2, [\mathbf{b}]_2)) \in \mathbb{G}_1^k \times \mathbb{G}_2^{k+1}$.

- $\mathsf{SIG.Sign}_\mathsf{C}(\Pi, \mathsf{sk}, m)$: The signing algorithm parses sk as $\mathsf{sk} = ([a]_2, [\mathbf{b}]_2)$. Next, it picks $\mathbf{r}' \overset{\$}{\leftarrow} \mathbb{Z}_q^k$ and defines

$$\mathbf{r} := \mathbf{B} \cdot \mathbf{r}' \in \mathbb{Z}_q^k, \quad u = a + \mathbf{x}(m) \cdot \mathbf{r} \in \mathbb{Z}_q, \quad \mathbf{v} = \mathbf{b} + \mathbf{Y}(m) \cdot \mathbf{r} \in \mathbb{Z}_q^k. \tag{3}$$

The signature for message m is $\sigma := ([\mathbf{r}]_2, [u]_2, [\mathbf{v}]_2) \in \mathbb{G}_2^{2k+1}$. Note that $[u]_2, [\mathbf{v}]_2$ can be computed from $\mathbf{r}'$ and Π.

– $\mathsf{SIG.Vfy}_\mathsf{C}(\Pi, \mathsf{vk} = [\mathbf{c}]_1, m, \sigma = ([\mathbf{r}]_2, [u]_2, [\mathbf{v}]_2))$: The verification algorithm picks $\mathbf{s} \xleftarrow{\$} \mathbb{Z}_q^k$ and returns 1 iff the equation

$$e([\mathbf{c}^\top \cdot \mathbf{s}]_1, [1]_2) = e([\mathbf{A} \cdot \mathbf{s}]_1, \begin{bmatrix} \mathbf{v} \\ u \end{bmatrix}_2) \cdot e([\mathbf{Z}(m) \cdot \mathbf{s}]_1, [\mathbf{r}]_2)^{-1} \tag{4}$$

holds, where $e([\mathbf{z}]_1, [\mathbf{z}']_2) := [\mathbf{z}^\top \cdot \mathbf{z}']_T$.

Instantiated under the SXDH assumption (i.e., $k = 1$ and DDH in $\mathbb{G}_1$ and $\mathbb{G}_2$) we obtain a signature scheme with $|\mathsf{vk}| = 1 \times \mathbb{G}_1$ and $|\sigma| = 3 \times \mathbb{G}_2$. Instantiated under the k-Lin assumption, we obtain a signature scheme with $\mathsf{vk}| = k \times \mathbb{G}_1$ and $|\sigma| = (2k + 1) \times \mathbb{G}_2$. In both cases the public parameters contain ℓk^2 group elements.

Theorem 2. *For any attacker $\mathcal{A}$ that (t, ϵ, μ)-breaks the $\mathsf{MU\text{-}EUF\text{-}CMA}^{\mathsf{Corr}}$-security of SIG_C, there exists an algorithm $\mathcal{B} = (\mathcal{B}_1, \mathcal{B}_2)$ such that $\mathcal{B}_1$ (t_1, ϵ_1)-breaks the $\mathcal{D}_k$-MDDH assumption in $\mathbb{G}_1$, and $\mathcal{B}_2$ (t_2, ϵ_2)-breaks the $\mathcal{D}_k$-MDDH assumption in $\mathbb{G}_2$ where $\epsilon < \epsilon_1 + 2\ell\epsilon_2 + 2/q$. We have $t_1 = t + t_1'$ and $t_2 = t + t_2$, where t_1' and t_2' correspond to the respective runtimes required to provide $\mathcal{A}_{\mathsf{SIG}_{\mathsf{MU}}}$ with the simulated experiment as described below.*

Proof. As before, we proceed in a sequence of games where the first game is the $\mathsf{MU\text{-}EUF\text{-}CMA}^{\mathsf{Corr}}$-security game that is played between an attacker $\mathcal{A}$ and a challenger $\mathcal{C}$, as described in Section 2.1. We denote by χ_i the event that $\mathcal{A}_{\mathsf{SIG}_{\mathsf{MU}}}$ outputs (m^*, i^*, σ^*) such that $\mathsf{SIG.Vfy}(vk^{(i^*)}, m^*, \sigma^*) \wedge i^* \notin \mathcal{S}^{\mathsf{corr}} \wedge (m^*, \cdot) \notin \mathcal{S}_{i^*}$ in Game i.

GAME 0. This is the real game that is played between $\mathcal{A}$ and $\mathcal{C}$. We use $(\mathsf{vk}_i, \mathsf{sk}_i) = ([\mathbf{c}_i]_1, ([a_i]_2, [\mathbf{b}_i]_2))$ to denote the verification/signing key of the i-th user. We have

$$\Pr[\chi_0] = \epsilon.$$

GAME 1. In this game we change the way the experiment treats the final forgery $\sigma^* = ([\mathbf{r}^*]_2, [u^*]_2, [\mathbf{v}^*]_2)$ for user i^* on message m^*. The experiment picks $\mathbf{s}^* \xleftarrow{\$} \mathbb{Z}_q^k$ and defines $\mathbf{t}^* = \mathbf{A} \cdot \mathbf{s}^*$. Next, it changes verification equation (4) and returns 1 iff equation

$$e([(\mathbf{b}_{i^*}^\top || a_{i^*}) \cdot \mathbf{t}^*]_1, [1]_2) = e([\mathbf{t}^*]_1, \begin{bmatrix} \mathbf{v}^* \\ u^* \end{bmatrix}_2) \cdot e([(\mathbf{Y}^\top(m^*) || \mathbf{x}(m^*)^\top) \cdot \mathbf{t}^*]_1, [\mathbf{r}^*]_2)^{-1} \tag{5}$$

holds. By equation (2) and by the definition of $\mathbf{c}_{i^*}^\top = (\mathbf{b}_{i^*}^\top || a_{i^*}) \cdot \mathbf{A}$, equations (4) and (5) are equivalent. Hence,

$$\Pr[\chi_1] = \Pr[\chi_0].$$

GAME 2. In this game, we again change the way the experiment treats the final forgery. Instead of defining $\mathbf{t}^* = \mathbf{A} \cdot \mathbf{s}^*$, we pick $\mathbf{t}^* \xleftarrow{\$} \mathbb{Z}_q^{k+1}$. Clearly, there exists an adversary $\mathcal{B}_1$ such that $\mathcal{B}_1$ (t_1, ϵ_1)-breaks the $\mathcal{D}_k$-MDDH assumption in $\mathbb{G}_1$ with $t \approx t_1$ and

$$\Pr[\chi_2] - \Pr[\chi_1] = \epsilon_1.$$

GAME 3. In this game, we make a change of variables by substituting all $\mathbf{Y}_{i,b}$ and $\mathbf{b}_i$ using the formulas

$$\mathbf{Y}_{i,b}^\top = (\mathbf{Z}_{i,b} - \mathbf{x}_{i,b} \cdot \underline{\mathbf{A}})\overline{\mathbf{A}}^{-1}, \quad \mathbf{b}_i^\top = (\mathbf{c}_i^\top - a_i \cdot \underline{\mathbf{A}})\overline{\mathbf{A}}^{-1}, \tag{6}$$

respectively. The concrete changes are as follows. First, the public parameters Π are computed by picking $\mathbf{Z}_{i,b}$ and $\mathbf{x}_{i,b}$ at random and then defining $\mathbf{Y}_{i,b}$ using (6). Second, the verification keys vk_i for user i are computed by picking $\mathbf{c}_i$ and a_i at random and then defining $\mathbf{b}_i$ using (6).

Third, on a signing query (m, i), the values $\mathbf{r}$ and u are computed as before, but the value $\mathbf{v}$ is computed as

$$\mathbf{v}^\top = (\mathbf{r}^\top \mathbf{Z}(m) + \mathbf{c}_i^\top - u \cdot \underline{\mathbf{A}}) \cdot \overline{\mathbf{A}}^{-1}. \tag{7}$$

Fourth, the verification query for message m^* and user i^* is answered by picking $h^* \xleftarrow{\$} \mathbb{Z}_q$ and $\overline{\mathbf{t}^*} \xleftarrow{\$} \mathbb{Z}_q^k$, defining $\underline{\mathbf{t}^*} = h^* + \underline{\mathbf{A}}\,\overline{\mathbf{A}}^{-1}\overline{\mathbf{t}^*}$ and changing equation (5) to

$$e([\mathbf{c}_{i^*}^\top \cdot \overline{\mathbf{A}}^{-1}\overline{\mathbf{t}^*} + a_{i^*} \cdot h^*]_1, [1]_2)$$
$$= e([\mathbf{t}^*]_1, \begin{bmatrix} \mathbf{v}^* \\ \mathbf{u}^* \end{bmatrix}_2) \cdot e([\mathbf{Z}(m^*)\overline{\mathbf{A}}^{-1}\overline{\mathbf{t}^*} + \mathbf{x}(m^*)h^*]_1, [\mathbf{r}^*]_2)^{-1} \tag{8}$$

By the substitution formulas for $\mathbf{Y}_{i,b}$ and $\mathbf{b}_i$ and be the definition of h and $\overline{\mathbf{t}^*}$, equations (3) and (7) and equations (5) and (8) are equivalent. Hence,

$$\Pr[\chi_3] = \Pr[\chi_2].$$

GAME 4. In this game, the answer $\sigma = ([\mathbf{r}]_2, [u]_2, [\mathbf{v}]_2)$ to a signing query (m, i) is computed differently. Concretely, the values $\mathbf{r}$ and $\mathbf{v}$ are computed as before, but the value u is chosen as $u \xleftarrow{\$} \mathbb{Z}_q$.

The remaining argument is purely information-theoretic. Note that in Game 4, the value a_{i^*} from sk_{i^*} only leaks through vk_{i^*} via $\mathbf{c}_{i^*}^\top = (\mathbf{b}_{i^*}^\top || a_{i^*}) \cdot \mathbf{A}$. As the uniform $\mathbf{t}^* \notin span(\mathbf{A})$ (except with probability $1/q$) the value $(\mathbf{b}_{i^*}^\top || a_{i^*}) \cdot \mathbf{t}^*$ from (5) (which is equivalent to (8)) is uniform and independent from $\mathcal{A}$'s view. Hence,

$$\Pr[\chi_4] = 2/q.$$

The following lemma which essentially proves that the underlying message authentication code is tightly secure in a multi-user setting with corruptions completes the proof of the Theorem 2. It follows [11,15].

Lemma 2. *There exists an adversary $\mathcal{B}_2$ such that $\mathcal{B}_2$ (t_2, ϵ_2)-breaks the $\mathcal{D}_k$-MDDH assumption in $\mathbb{G}_2$ with $t \approx t_1$ and*

$$\Pr[\chi_4] - \Pr[\chi_3] \leq 2\ell\epsilon_2.$$

To prove the lemma, we define the following hybrid games H_j, $0 \le j \le \ell$ that are played with an adversary $\mathcal{C}$. All variables are distributed as in Game 4. For $m \in \{0, 1\}^*$, define $m_{|j}$ as the j-th prefix of m. (By definition, $m_{|0}$ is the empty string ε.) Let $\mathsf{RF}_{i,j} : \{0, 1\}^j \to \mathbb{Z}_q$ be independent random functions. (For concreteness, one may think of $\mathsf{RF}_{i,0}(\varepsilon) := a_i$, the MAC secret key $\mathsf{sk}_{\mathsf{MAC}}$ of the i-th user. In each hybrid H_j, we will double the number of secret-keys used in answering the queries until each query uses an independent secret key.) In Hybrid H_j, adversary $\mathcal{C}$ first obtains the values $[\mathbf{B}]_2$ and $([\mathbf{x}_{i,b}^\top \mathbf{B}]_2)_{i,b}$, which can be seen as the public MAC parameters Π_{MAC}. Next, adversary $\mathcal{C}$ can make an arbitrary number of tagging and corruption queries, plus one forgery query. On a tagging query called with (m, i), hybrid H_j picks a random $\mathbf{r} \in \mathbb{Z}_q^k$, computes $u = \mathsf{RF}_{i,j}(m_{|j}) + \mathbf{x}(m) \cdot \mathbf{r}$ and returns $([\mathbf{r}]_2, [u]_2)$ (the MAC tag) to adversary $\mathcal{C}$. Note that the value $\mathbf{v}$ is not provided by the oracle. On a Corrupt query called with i, hybrid H_j returns $[a_i]_2 = [\mathsf{RF}_{i,j}(m(i)_{|j})]_2$ to $\mathcal{C}$, where $m(i)$ is the first message for which the tagging oracle was called for with respect to user i. (We make one dummy query if $m(i)$ is undefined.) Further, user i is added to the list of corrupted users. The adversary is also allowed to make one single forgery query (i^*, m^*) for an uncorrupted user i^* which is answered with $([h^*]_1, [h^* \cdot \mathsf{RF}_{i^*,j}(m_{|j}^*)]_1, [h^* \cdot \mathbf{x}(m^*)]_1)$, for $h^* \xleftarrow{\$} \mathbb{Z}_q$. Finally, hybrid H_j outputs whatever adversary $\mathcal{C}$ outputs.

Note that Game 3 can be perfectly simulated using the oracles provided by hybrid H_0. The reduction picks $\mathbf{A} \xleftarrow{\$} \mathcal{D}_k$, inputs $[\mathbf{B}]_2$ and $([\mathbf{x}_{i,b}\mathbf{B}]_2)_{i,b}$ from the hybrid game H_0, picks $\mathbf{Z}_{i,b}$ at random, and computes $[\mathbf{Y}_{i,b}\mathbf{B}]_2$ via (6). The public verification keys $\mathsf{vk}_i = [\mathbf{c}_i]_1$ are picked at random, without knowing $\mathsf{sk}_i = ([a_i]_2, [\mathbf{b}_i]_2)$. To simulate a signing query on (m, i), the reduction queries the tagging oracle to obtain $([\mathbf{r}]_2, [u]_2)$ and computes the value $[\mathbf{v}]_2$ as in Game 3 via (7). Forgery and Corrupt queries can be simulated the same way by defining $\mathsf{RF}_{i,0}(\varepsilon) =: a_i$. Hence $\Pr[\chi_3] = \Pr[H_0 = 1]$. Similarly, $\Pr[\chi_4] = \Pr[H_\ell = 1]$ as in hybrid H_ℓ are values $\mathbf{u} = \mathsf{RF}_{i,\ell}(m) + \mathbf{x}(m) \cdot \mathbf{r}$ are uniform.

We make the following claim:

Claim. $|\Pr[H_{j-1} = 1] - \Pr[H_j = 1]| \le 2\epsilon_2$, for a suitable adversary $\mathcal{B}_2$.

The proof of this claim essentially follows verbatim from Lemma B.3 of [11]. The reduction uses the fact that the $\mathcal{D}_k$-MDDH assumption is random self-reducible. There is a multiplicative loss of 2 since the reduction has to guess m_j^*, the j-th bit of the forgery m^*.

Fix $0 \le j \le \ell - 1$. Let Q be the maximal number of tagging queries. Adversary $\mathcal{B}_2$ inputs a Q-fold $\mathcal{D}_k$-MDDH challenge $([\mathbf{A}']_2, [\mathbf{H}]_2) \in \mathbb{G}_2^{(k+1) \times k} \times \mathbb{G}_2^{(k+1) \times Q}$ and has to distinguish $\mathbf{H} = \mathbf{A}'\mathbf{W}$ for $\mathbf{W} \in \mathbb{Z}_q^{k \times Q}$ from $\mathbf{H} \xleftarrow{\$} \mathbb{Z}_q^{(k+1) \times Q}$. The Q-fold $\mathcal{D}_k$-MDDH assumption has been proved tightly equivalent to the $\mathcal{D}_k$-MDDH assumption in [20].

Adversary $\mathcal{B}_2$ defines $\mathbf{B} := \overline{\mathbf{A}'}$ and picks a random bit α which is a guess for m_j^*, the j-th bit of m^*. We assume that this guess is correct, which happens with probability $1/2$. For each user i, define the random function $\mathsf{RF}_{i,j}(\cdot)$ via

$$\mathsf{RF}_{i,j}(m_{|j}) := \begin{cases} \mathsf{RF}_{i,j-1}(m_{|j-1}) & m_j = \alpha \\ \mathsf{RF}_{i,j-1}(m_{|j-1}) + R_{i,m_{|j}} & m_j = 1 - \alpha \end{cases}, \tag{9}$$

where $R_{i,m_{|j}} \xleftarrow{\$} \mathbb{Z}_q$. Let $\pi_{i,j} : \{0,1\}^j \to Q$ be arbitrary injective functions. Next, for $i = 1, \ldots, \ell$, $b = 0, 1$ with $(i, b) \neq (j, 1 - \alpha)$, $\mathcal{B}_2$ picks $\mathbf{x}_{i,b} \xleftarrow{\$} \mathbb{Z}_q^k$ and implicitly defines $\mathbf{x}_{j,1-\alpha}^\top \mathbf{B} := \mathbf{x}'^\top \mathbf{A}'$ for $\mathbf{x}' \xleftarrow{\$} \mathbb{Z}_q^{k+1}$. Note that $\mathbf{x}_{j,1-\alpha}$ is not known to $\mathcal{B}_2$, only $[\mathbf{x}_{j,1-\alpha}^\top \mathbf{B}]_2$. Adversary $\mathcal{B}_2$ returns the values $\Pi_{\mathsf{MAC}} = ([\mathbf{B}]_2, ([\mathbf{x}_{i,b}^\top \mathbf{B}]_2)_{i,b})$.

A signing query on (i, m) is simulated as follows. We distinguish two cases. Case 1, if $m_j = \alpha$, then pick random $\mathbf{r} \in \mathbb{Z}_q^k$ and define $u = \mathsf{RF}_{i,j-1}(m_{|j-1}) + \mathbf{x}(m) \cdot \mathbf{r}$. By (9), the value u has the same distribution in H_{j-1} and H_j. Case 2, if $m_j \neq \alpha$ (i.e., only $[\mathbf{x}_{j,m_j}^\top \mathbf{B}]_2$ is known, $\mathbf{x}_{j,m_j}$ not), then pick random $\mathbf{r}' \in \mathbb{Z}_q^k$, define $\mathbf{r} := \mathbf{B}\mathbf{r}' + \overline{\mathbf{H}}_\beta$ and $u := \mathsf{RF}_{i,j-1}(m_{|j-1}) + \sum_{l \neq j} \mathbf{x}_{l,m_l}^\top \cdot \mathbf{r} + \mathbf{x}'^\top (\mathbf{A}'\mathbf{r}' + \mathbf{H}_\beta)$. Here $\mathbf{H}_\beta$ is the β-th column of $\mathbf{H}$ and $\beta = \pi_{i,j}(m_{|j})$. Let $\mathbf{H}_\beta = \mathbf{A}'\mathbf{W}_\beta + \mathbf{R}_\beta$, where $\mathbf{R}_\beta = 0$ or $\mathbf{R}_\beta$ is uniform. Then $\mathbf{r} = \overline{\mathbf{A}'}(\mathbf{r}' + \mathbf{W}_\beta) + \mathbf{R}_\beta$ and

$$\begin{aligned} \mathbf{x}'^\top (\mathbf{A}'\mathbf{r}' + \mathbf{H}_\beta) &= \mathbf{x}'^\top \mathbf{A}'(\mathbf{r}' + \mathbf{W}_\beta) + \mathbf{x}'^\top \mathbf{R}_\beta \\ &= \mathbf{x}_{j,m_j}^\top \mathbf{B}(\mathbf{r}' + \mathbf{W}_\beta) + \mathbf{x}'^\top \mathbf{R}_\beta \\ &= \mathbf{x}_{j,m_j}^\top \mathbf{r} + \mathbf{x}'^\top \mathbf{R}_\beta \end{aligned}$$

such that $u = \mathsf{RF}_{i,j-1}(m_{|j-1}) + \sum_l \mathbf{x}_{l,m_l}^\top \cdot \mathbf{r} + \mathbf{x}'^\top \mathbf{R}_\beta$. Hence, if $\mathbf{H}$ comes from the Q-fold MDDH distribution, then $\mathbf{R}_\beta = 0$ and u is distributed as in H_{j-1}; if $\mathbf{H}$ comes from the uniform distribution, then u is distributed as in H_j with $R_{i,m_{|j}} := \mathbf{x}'^\top \mathbf{R}_\beta$.

A verification query on (i^*, m^*, σ^*) is answered with $([h^*]_1, [h^* \cdot \mathsf{RF}_{i^*,j}(m_{|j}^*)]_1, [h^* \cdot \mathbf{x}(m^*)]_1)$, for uniform h^*. Note that $\mathbf{x}(m^*)$ can be computed as all $\mathbf{x}_{l,m_l^*}$ are known to $\mathcal{B}_2$.

Finally, a Corrupt query for user i is answered with $[a_i]_2 = [\mathsf{RF}_{i,j}(m(i)_{|j})]_2$. Note that $[\mathsf{RF}_{i,j}(m_{|j})]_2$ can be computed for all m.

3   KEMs in the Multi-user Setting with Corruptions

In this section we will describe a generic construction of a key encapsulation mechanism (KEM) with tight MU-IND-CPA^{Corr}-security proof, based on any public-key encryption scheme with tight security proof in the multi-user setting *without* corruptions. Encryption schemes with the latter property were described in [4,25]. In particular, a tight security proof for the DLIN-based scheme from [12] is given in [25]. A similar scheme was generalized to hold under any MDDH-assumption [20].

Due to space limitations, we refer to the full version of our paper [3] for standard definitions of public key encryption (PKE) and KEMs.

Before we proceed let us first recall public key encryption and key encapsulation mechanisms.

3.1 Public-Key Encryption

A PKE scheme is a four-tuple of algorithms PKE = (PKE.Setup, PKE.KGen, PKE.Enc, PKE.Dec) with the following syntax:

- $\Pi \xleftarrow{\$} $ PKE.Setup(1^κ): The algorithm PKE.Setup, on input the security parameter 1^κ, outputs a set, Π, of system parameters. Π determines the message space $\mathcal{M}$, the ciphertext space $\mathcal{C}$, the randomness space $\mathcal{R}$, and the key space $\mathcal{PK} \times \mathcal{SK}$.
- $(sk, pk) \xleftarrow{\$} $ PKE.KGen(Π): This algorithm takes as input Π and outputs a key pair $(sk, pk) \in \mathcal{SK} \times \mathcal{PK}$.
- $c \xleftarrow{\$} $ PKE.Enc(pk, m): This probabilistic algorithm takes as input a public key and a message $m \in \mathcal{M}$, and outputs a ciphertext $c \in \mathcal{C}$.
- $m = $ PKE.Dec(sk, c): This deterministic algorithm takes as input a secret key sk and a ciphertext c, and outputs a plaintext $m \in \mathcal{M}$ or an error symbol, $\bot$.

Security. The standard security notions for public key encryption in the multi-user setting (without corruptions) go back to Bellare, Boldyreva and Micali [4]. Security is formalized by a game that is played between an attacker $\mathcal{A}$ and a challenger $\mathcal{C}$.

1. After running $\Pi \xleftarrow{\$} $ PKE.Setup(1^κ), $\mathcal{C}$ generates $\mu \cdot \ell$ key pairs $(sk_i^s, pk_i^s) \xleftarrow{\$} $ PKE.KGen(Π) for $(i, s) \in [\mu] \times [\ell]$, and chooses $b \xleftarrow{\$} \{0, 1\}$ uniformly at random.
2. $\mathcal{A}$ receives Π and $pk_1^1, \ldots, pk_\mu^\ell$, and may now adaptively query an oracle $\mathcal{O}_{\mathsf{Encrypt}}$, which takes as input (pk_i^s, m_0, m_1), computes $c \xleftarrow{\$} $ PKE.Enc(pk_i^s, m_b) and responds with c.
3. Eventually $\mathcal{A}$ ouputs a bit b'.

Definition 8. *We say that $\mathcal{A}$ (t, ϵ, μ, ℓ)-breaks the MU-IND-CPA security of PKE, if it runs in time t in the above security game and*

$$\Pr[b' = b] \geq 1/2 + \epsilon$$

3.2 Key Encapsulation Mechanisms

Definition 9. *A key encapsulation mechanism consists of four probabilistic algorithms:*

- $\Pi \xleftarrow{\$} $ KEM.Setup(1^κ): *The algorithm KEM.Setup, on input the security parameter 1^κ, outputs public parameters Π, which determine the session key space $\mathcal{K}$, the ciphertext space $\mathcal{C}$, the randomness space $\mathcal{R}$, and key space $\mathcal{SK} \times \mathcal{PK}$.*
- $(sk, pk) \xleftarrow{\$} $ KEM.Gen(Π): *This algorithm takes as input parameters Π and outputs a key pair $(sk, pk) \in \mathcal{SK} \times \mathcal{PK}$.*
- $(K, C) \xleftarrow{\$} $ KEM.Encap(pk) *takes as input a public key pk, and outputs a ciphertext $C \in \mathcal{C}$ along with a key $K \in \mathcal{K}$.*
- $K = $ KEM.Decap(sk, C) *takes as input a secret key sk and a ciphertext C, and outputs a key $K \in \mathcal{K}$ or an error symbol $\bot$.*

We require the usual correctness properties.

Multi User Security of KEMs. We extend the standard indistinguishability under chosen-plaintext attacks (IND-CPA) security for KEMs to a multi-user setting with $\mu \geq 1$ public keys and adaptive corruptions of secret keys. We will refer to this new notion as MU-IND-CPA^{Corr}-security.

Consider the following game played between a challenger $\mathcal{C}$ and an attacker $\mathcal{A}$.

1. At the beginning $\mathcal{C}$ generates parameters $\Pi \xleftarrow{\$} \mathsf{KEM.Setup}(1^\kappa)$. Then, for each $(i, s) \in [\mu] \times [\ell]$, it generates a key pair $(sk_i^s, pk_i^s) \xleftarrow{\$} \mathsf{KEM.Gen}(\Pi)$ and chooses an independently random bit $b_i^s \xleftarrow{\$} \{0, 1\}$. Finally, the challenger initializes a set $\mathcal{S}^{\mathsf{corr}} := \emptyset$ to keep track of corrupted keys. The attacker receives as input $(pk_1^1, \ldots, pk_\mu^\ell)$.
2. Now the attacker may adaptively query two oracles. $\mathcal{O}_{\mathsf{Corrupt}}$ takes as input a public key pk_i^s. It appends (i, s) to $\mathcal{S}^{\mathsf{corr}}$ and responds with sk_i^s. Oracle $\mathcal{O}_{\mathsf{Encap}}$ takes as input a public key pk_i^s. It generates a ciphertext-key-pair as $(C_i^s, K_{i,1}^s) \xleftarrow{\$} \mathsf{KEM.Encap}(pk_i^s)$ and chooses a random key $K_{i,0}^s$. It responds with $(C_i^s, K_{i,b_i^s}^s)$.
3. Finally, the attacker outputs a pair (i, s, b).

Definition 10 (MU-IND-CPA^{Corr}-security). *Algorithm $\mathcal{A}$ (t, ϵ, μ, ℓ)-breaks the MU-IND-CPA^{Corr}-security of the KEM, if it runs in time at most t and it holds that*

$$\Pr[b_i^s = b \wedge (i, s) \notin \mathcal{S}^{\mathsf{corr}}] \geq 1/2 + \epsilon$$

Remark 1. It is easy to see that security in the sense of Definition 10 can efficiently be reduced to standard IND-CPA security. However, the reduction incurs a loss of $1/(\mu \cdot \ell)$. We will describe a KEM with tight security proof.

3.3   Generic KEM Construction

Our KEM $\mathsf{KEM_{MU}}$ is based on a PKE-scheme $\mathsf{PKE} = (\mathsf{PKE.Setup}, \mathsf{PKE.KGen}, \mathsf{PKE.Enc}, \mathsf{PKE.Dec})$. It works as follows:

- $\Pi \xleftarrow{\$} \mathsf{KEM.Setup_{MU}}(1^\kappa)$: The parameter generation algorithm $\mathsf{KEM.Setup_{MU}}$ on input κ runs $\Pi_{\mathsf{PKE}} \xleftarrow{\$} \mathsf{PKE.Setup}(1^\kappa)$. The session key space $\mathcal{K}$ is set to $\mathcal{M}$, the message space of PKE that is determined by Π_{PKE}.
- $(sk, pk) \xleftarrow{\$} \mathsf{KEM.Setup_{MU}}(\Pi)$: The key generation algorithm generates two keys of the PKE scheme by running $(\mathsf{sk}_i, \mathsf{pk}_i) \xleftarrow{\$} \mathsf{PKE.KGen}(\Pi)$ for $i \in \{0, 1\}$. It furthermore flips a random coin $\delta \xleftarrow{\$} \{0, 1\}$ and returns $(sk, pk) = ((\mathsf{sk}_\delta, \delta), (\mathsf{pk}_0, \mathsf{pk}_1))$.
- $(K, C) \xleftarrow{\$} \mathsf{KEM.Encap_{MU}}(pk)$: On input $pk = (\mathsf{pk}_0, \mathsf{pk}_1)$ the encapsulation algorithm samples a random key $K \xleftarrow{\$} \mathcal{K}$, computes two ciphertexts (C_0, C_1) as $C_i \xleftarrow{\$} \mathsf{PKE.Enc}(\mathsf{pk}_i, K)$ for $i \in \{0, 1\}$, sets $C := (C_0, C_1)$, and outputs (K, C).
- $K \leftarrow \mathsf{KEM.Decap_{MU}}(sk, C)$: The decapsulation algorithm parses the secret key as $sk = (\mathsf{sk}_\delta, \delta)$ and $C = (C_0, C_1)$. It computes $K \leftarrow \mathsf{PKE.Dec}(\mathsf{sk}_\delta, C_\delta)$ and returns K.

Theorem 3. *Let* KEM$_{MU}$ *be as described above. For each attacker* $\mathcal{A}^{KEM}$ *that* $(\epsilon_{kem}, t_{kem}, \mu, \ell)$-*breaks the* MU-IND-CPA^{Corr}-*security of* KEM$_{MU}$ *there exists an attacker* $\mathcal{A}^{PKE}$ *that* $(\epsilon_{pke}, t_{pke}, \mu, \ell)$-*breaks the* MU-IND-CPA-*security of* PKE *with* $t_{kem} = t_{pke} + t'_{kem}$ *and* $\epsilon_{kem} \leq \epsilon_{pke}$. *Here* t'_{kem} *is the runtime required to provide* $\mathcal{A}^{KEM}$ *with the simulation described below.*

Due to space limitations we omit the proof of Theorem 3 here. It can be found in the full version of our paper [3]. □

4 A Tightly-Secure AKE Protocol

4.1 Secure Authenticated Key-Exchange

In this section we present a formal security model for authenticated key-exchange (AKE) protocols. We follow the approach of Bellare and Rogaway [5] and use oracles to model concurrent and multiple protocol executions within a party and the concept of *matching conversations* to define partnership between oracles.

Essentially our model is a strenghtened version of the AKE-security model of [27], which allows an additional RegisterCorrupt-query. Moreover, we let the adversary issue more than one Test-query, in order to achieve tightness also in this dimension.

Execution Environment. In our security model, we consider μ parties $P_1, \ldots, P_\mu$. In order to formalize several sequential and parallel executions of an AKE protocol, each party P_i is represented by a set of ℓ oracles, $\{\pi_i^1, \ldots, \pi_i^\ell\}$, where $\ell \in \mathbb{N}$ is the maximum number of protocol executions per party.

Each oracle π_i^s has access to the long-term key pair $(sk^{(i)}, pk^{(i)})$ of party P_i and to the public keys of all other parties. Let $\mathcal{K}$ be the session key space. Each oracle π_i^s maintains a list of internal state variables that are described in the following:

- Pid$_i^s$ stores the identity of the intended communication partner.
- $\Psi_i^s \in \{\texttt{accept}, \texttt{reject}\}$ is a boolean variable indicating wether oracle π_i^s succesfully completed the protocol execution.
- $k_i^s \in \mathcal{K}$ is used to store the session key that is computed by π_i^s.
- Γ_i^s is a variable that stores all messages sent and received by π_i^s in the order of appearance. We call Γ_i^s the transcript.

For each oracle π_i^s these variables are initialized as $(\text{Pid}_i^s, \Psi_i^s, k_i^s, \Gamma_i^s) = (\emptyset, \emptyset, \emptyset, \emptyset)$, where $\emptyset$ denotes the empty string. The computed session key is assigned to the variable k_i^s if and only if π_i^s reaches the accept state, i.e., if $\Psi_i^s = \texttt{accept}$.

Adversarial Model. The attacker $\mathcal{A}$ interacts with these oracles through oracle queries. We consider an active attacker that has full control over the communication network, i.e., $\mathcal{A}$ can schedule all sessions between the parties, delay, drop, change or replay messages at will and inject own generated messages of its choice. This is modeled by the Send-query defined below.

To model further real world capabilites of $\mathcal{A}$, such as break-ins, we provide further types of queries. The Corrupt-query allows the adversary to compromise the long-term key of a party. The Reveal-query may be used to obtain the session key that was computed in a previous protocol instance. The RegisterCorrupt enables the attacker to register maliciously-generated public keys. Note that we do not require the adversary to know the corresponding secret key. The Test-query does not correspond to a real world capability of $\mathcal{A}$, but it is used to evaluate the advantage of $\mathcal{A}$ in breaking the security of the key exchange protocol.

More formally, the attacker may ask the following queries:

- Send(i, s, m): $\mathcal{A}$ can use this query to send any message m of its choice to oracle π_i^s. The oracle will respond according to the protocol specification and depending on its internal state. If $m = (\top, j)$ is sent to π_i^s, then π_i^s will send the first protocol message to P_j.

  If Send(i, s, m) is the τ-th query asked by $\mathcal{A}$, and oracle π_i^s sets variable $\Psi_i^s = \mathrm{accept}$ after this query, then we say that π_i^s has τ-*accepted*.

- Corrupt(i): This query returns the long-term secret key sk_i of party P_i. If the τ-th query of $\mathcal{A}$ is Corrupt(P_i), then we call P_i τ-corrupted. If Corrupt(P_i) has never been issued by $\mathcal{A}$, then we say that party i is ∞-corrupted.

- RegisterCorrupt$(P_i, pk^{(i)})$: This query allows $\mathcal{A}$ to register a new party P_i, $i > \mu$, with public key $pk^{(i)}$. If the same party P_i is already registered (either via RegisterCorrupt-query or $i \in [\mu]$), a failure symbol $\perp$ is returned to $\mathcal{A}$. Otherwise, P_i is registered, the pair $(P_i, pk^{(i)})$ is distributed to all other parties, and the symbol $\top$ is returned.

  Parties registered by this query are called *adversarially-controlled*.

  All adversarially-controlled parties are defined to be 0-corrupted.

- Reveal(i, s): In response to this query π_i^s returns the contents of k_i^s. Recall that we have $k_i^s \neq \emptyset$ if and only if $\Psi_i^s = \mathrm{accept}$. If Reveal$(i, s)$ is the τ-th query issued by $\mathcal{A}$, we call π_i^s τ-revealed. If Reveal(i, s) has never been issued by $\mathcal{A}$, then we say that party i is ∞-revealed.

- Test(i, s): If $\Psi_i^s \neq \mathrm{accept}$, then a failure symbol $\perp$ is returned. Otherwise π_i^s flips a fair coin b_i^s, samples $k_0 \xleftarrow{\$} \mathcal{K}$ at random, sets $k_1 = k_i^s$, and returns $k_{b_i^s}$.

  If Test(i, s) is the τ-th query issued by $\mathcal{A}$, we call π_i^s τ-tested. If Test(i, s) has never been issued by $\mathcal{A}$, then we say that party i is ∞-tested.

  The attacker may ask many Test-queries to different oracles, but only once to each oracle.

Security Definitions. We recall the concept of *matching conversations* here that was first introduced by Bellare and Rogaway [5]. We adopt the refinement from [27].

Recall that Γ_i^s be the transcript of oracle π_i^s. By $|\Gamma_i^s|$ we denote the number of the messages in Γ_i^s. Assume that there are two transcripts, Γ_i^s and Γ_j^t, where $|\Gamma_i^s| = w$ and $|\Gamma_j^t| = n$. We say that Γ_i^s is a *prefix* of Γ_j^t if $0 < w \leq n$ and the first w messages in transcripts Γ_i^s and Γ_j^t are identical.

Definition 11 (Matching conversations). *We say that π_i^s has a matching conversation to oracle π_j^t, if*
- *π_j^t has sent all protocol messages and Γ_j^t is a prefix of Γ_i^s, or*
- *π_j^t has sent all protocol messages and $\Gamma_i^s = \Gamma_j^t$.*

We say that two oracles, π_i^s and π_j^t, have matching conversations if π_i^s has a matching conversation to process π_j^t and vice versa.

Definition 12 (Correctness). *We say that a two-party AKE protocol, Σ, is correct, if for any two oracles, π_i^s and π_j^t, that have matching conversations it holds that $\Psi_i^s = \Psi_j^t = \mathsf{accept}$, $\mathsf{Pid}_i^s = j$ and $\mathsf{Pid}_j^t = i$ and $k_i^s = k_j^t$.*

Security Game. Consider the following game that is played between an adversary, $\mathcal{A}$, and a challenger, $\mathcal{C}$, and that is parametrized by two numbers, μ (the number of honest identities) and ℓ (the maximum number of protocol executions per identity).

1. At the beginning of the game, $\mathcal{C}$ generates system parameters that are specified by the protocol and μ long-term key pairs $(sk^{(i)}, pk^{(i)}), i \in [\mu]$. Then $\mathcal{C}$ implements a collection of oracles $\{\pi_i^s : i \in [\mu], s \in [\ell]\}$. It passes to $\mathcal{A}$ all public keys, $pk^{(1)}, \ldots, pk^{(\mu)}$, and the public parameters.
2. Then $\mathcal{A}$ may adaptively issue Send, Corrupt, Reveal, RegisterCorrupt and Test queries to $\mathcal{C}$.
3. At the end of the game, $\mathcal{A}$ terminates with outputting a tuple (i, s, b') where π_i^s is an oracle and b' is its guess for b_i^s.

For a given protocol Σ by $\mathbb{G}_\Sigma(\mu, \ell)$ we denote the security game that is carried out with parameters μ, ℓ as described above and where the oracles implement protocol Σ.

Definition 13 (Freshness). *Oracle π_i^s is said to be τ-fresh if the following requirements satisfied:*
- *π_i^s has $\tilde{\tau}$-accepted, where $\tilde{\tau} \leq \tau$.*
- *π_i^s is $\hat{\tau}$-revealed, where $\hat{\tau} > \tau$.*
- *If there is an oracle, π_j^t, that has matching conversation to π_i^s, then π_j^t is ∞-revealed and ∞-tested.*
- *If $\mathsf{Pid}_i^s = j$ then P_j is $\tau^{(j)}$-corrupted with $\tau^{(j)} > \tau$ [4].*

Definition 14 (AKE Security). *We say that an attacker (t, μ, ℓ, ϵ)-breaks the security of a two-party AKE protocol, Σ, if it runs in time t in the above security game $\mathbb{G}_\Sigma(\mu, \ell)$ and it holds that:*

1. *Let $\mathcal{Q}$ denote the event that there exists a τ and a τ-fresh oracle π_i^s and there is no unique oracle π_j^t such that π_i^s and π_j^t have matching conversations. Then $Pr[\mathcal{Q}] \geq \epsilon$, or*
2. *When $\mathcal{A}$ returns (i, s, b') such that $\mathsf{Test}(\pi_i^s)$ was $\mathcal{A}s$ τ-th query and π_i^s is a τ-fresh oracle that is ∞-revealed throughout the security game then the probability that b' equals b_i^s is upper bounded by*

---

[4] We note that for any $P_i, i > \mu$, we have $\tau^{(i)} = 0$. Therefore for any $\tau \geq 1$, the intended partner of a τ- fresh oracle must not be adversarially controlled.

$$|\Pr[b_i^s = b'] - 1/2| \geq \epsilon.$$

We discuss and highlight properties of the model in the full version of our paper [3, Remark 2]

4.2   Our Tightly Secure AKE Protocol

Here, we construct an AKE-protocol AKE, which is based on three building blocks: a key encapsulation mechanism, a signature scheme, and a one-time signature scheme.

The protocol is a key transport protocol that needs three messages to authenticate both participants and to establish a shared session key between both parties. Informally, the key encapsulation mechanism guarantees that session keys are indistinguishable from random keys. The signature scheme is used to guarantee authentication: The long-term keys of all parties consist of verification keys of the signature scheme. Finally, the one-time signature scheme prevents oracles from accepting without having a (unique) partner oracle.

In the sequel let SIG and OTSIG be signature schemes and let KEM be a key-encapsulation mechanism. We will assume common parameters $\Pi_{\mathsf{SIG}} \xleftarrow{\$} \mathsf{SIG.Setup}(1^\kappa)$, $\Pi_{\mathsf{OTSIG}} \xleftarrow{\$} \mathsf{OTSIG.Setup}(1^\kappa)$, and $\Pi_{\mathsf{KEM}} \xleftarrow{\$} \mathsf{KEM.Setup}(1^\kappa)$.

Long-term secrets. Each party possesses a key pair $(vk, sk) \xleftarrow{\$} \mathsf{SIG.Gen}(\Pi_{\mathsf{SIG}})$ for signature scheme SIG. In the sequel let $(vk^{(i)}, sk^{(i)})$ and $(vk^{(j)}, sk^{(j)})$ denote the key pairs of parties P_i, P_j, respectively.

Protocol execution. In order to establish a key, parties P_i, P_j execute the following protocol.

1. First, P_i runs $(sk_{\mathsf{KEM}}^{(i)}, pk_{\mathsf{KEM}}^{(i)}) \xleftarrow{\$} \mathsf{KEM.Gen}(\Pi_{\mathsf{KEM}})$ and $(vk_{\mathsf{OTS}}^{(i)}, sk_{\mathsf{OTS}}^{(i)}) \xleftarrow{\$} \mathsf{OTSIG.Gen}(\Pi_{\mathsf{OTSIG}})$ and computes a signature $\sigma^{(i)} := \mathsf{SIG.Sign}(sk^{(i)}, vk_{\mathsf{OTS}}^{(i)})$. It defines $\mathsf{Pid} = j$ and $m_1 := (vk_{\mathsf{OTS}}^{(i)}, \sigma^{(i)}, pk_{\mathsf{KEM}}^{(i)}, \mathsf{Pid}, i)$ and transmits m_1 to P_j.

2. Upon receiving m_1, P_j parses m_1 as the tuple $(vk_{\mathsf{OTS}}^{(i)}, \sigma^{(i)}, pk_{\mathsf{KEM}}^{(i)}, \mathsf{Pid}, i)$. Then it checks whether $\mathsf{Pid} = j$ and $\mathsf{SIG.Vfy}(vk^{(i)}, vk_{\mathsf{OTS}}^{(i)}, \sigma^{(i)}) = 1$. If at least one of both check is not passed, then P_j outputs $\perp$ and rejects. Otherwise it runs $(vk_{\mathsf{OTS}}^{(j)}, sk_{\mathsf{OTS}}^{(j)}) \xleftarrow{\$} \mathsf{OTSIG.Gen}(\Pi_{\mathsf{OTSIG}})$, encapsulates a key as $(K, C) \xleftarrow{\$} \mathsf{KEM.Encap}(pk_{\mathsf{KEM}}^{(i)})$ and computes a signature $\sigma^{(j)} := \mathsf{SIG.Sign}(sk^{(j)}, vk_{\mathsf{OTS}}^{(j)})$. Then it sets $m_2 := (vk_{\mathsf{OTS}}^{(j)}, \sigma^{(j)}, C)$ and computes a one-time signature $\sigma_{\mathsf{OTS}}^{(j)} := \mathsf{OTSIG.Sign}(sk_{\mathsf{OTS}}^{(j)}, (m_1, m_2))$ and transmits the tuple $(m_2, \sigma_{\mathsf{OTS}}^{(j)})$ to P_i.

3. Upon receiving the message $(m_2, \sigma_{\mathsf{OTS}}^{(j)})$, P_i parses m_2 as $(vk_{\mathsf{OTS}}^{(j)}, \sigma^{(j)}, C)$ and checks whether $\mathsf{SIG.Vfy}(vk^{(j)}, vk_{\mathsf{OTS}}^{(j)}, \sigma^{(j)}) = 1$ and $\mathsf{OTSIG.Vfy}(vk_{\mathsf{OTS}}^{(j)}, (m_1, m_2), \sigma_{\mathsf{OTS}}^{(j)}) = 1$. If at least one of both check is not passed, then P_i outputs $\perp$ and rejects.

Otherwise it computes $\sigma_{\text{OTS}}^{(i)} := \text{OTSIG.Sign } (sk_{\text{OTS}}^{(i)}, (m_1, m_2))$ and sends $\sigma_{\text{OTS}}^{(i)}$ to P_j. P_i outputs the session key as $K_{i,j} := \text{KEM.Decap}(sk_{\text{KEM}}^{(i)}, C)$.

4. Upon receiving $\sigma_{\text{OTS}}^{(i)}$, P_j checks whether $\text{OTSIG.Vfy}(vk_{\text{OTS}}^{(i)}, (m_1, m_2), \sigma_{\text{OTS}}^{(i)})$ $= 1$. If this fails, then $\perp$ is returned. Otherwise P_j outputs the session key $K_{i,j} := K$.

In the full version of the paper [3], we elaborate on the efficiency of our protocol when it is instantiated with building blocks from the literatur.

4.3   Proof of Security

Theorem 4. *Let* AKE *be as described above. If there is an attacker* $\mathcal{A}_{\text{AKE}}$ *that* $(t, \mu, \ell, \epsilon_{\text{AKE}})$*-breaks the security of* AKE *in the sense of Definition 14 then there is an algorithm* $\mathcal{B} = (\mathcal{B}_{\text{KEM}}, \mathcal{B}_{\text{SIG}}, \mathcal{B}_{\text{OTSIG}})$ *such that either* $\mathcal{B}_{\text{KEM}}$ $(t', \mu \cdot \ell, \epsilon_{\text{KEM}})$*-breaks the* MU-IND-CPA^{Corr}*-security of* KEM *(Definition 10), or* $\mathcal{B}_{\text{SIG}}$ $(t', \epsilon_{\text{SIG}}, \mu)$*-breaks the* MU-EUF-CMA*-security of* SIG *(Definition 2), or* $\mathcal{B}_{\text{OTSIG}}$ $(t', \epsilon_{\text{OTSIG}}, \mu \cdot \ell)$*-breaks the* MU-sEUF-1-CMA*-security of* OTSIG *(Definition 4) where*

$$\epsilon_{\text{AKE}} \leq 4\epsilon_{\text{OTSIG}} + 2\epsilon_{\text{SIG}} + \epsilon_{\text{KEM}}.$$

Here, $t' = t + t''$ *where* t'' *corresponds to the runtime required to provide* $\mathcal{A}_{\text{AKE}}$ *with the simulated experiment as described below.*

Proof. We prove the security of the proposed protocol AKE using the sequence-of-games approach, following [35,7]. The first game is the original attack game that is played between a challenger and an attacker. We then describe a sequence of games where we modify the original game step by step. We show that the advantage of distinguishing between two successive games is negligible.

We prove Theorem 4 in two stages. First, we show that the AKE protocol is a secure authentication protocol except for probability ϵ_{Auth}. That is, the protocol fulfills security property 1.) of the AKE security definition Definition 14. Informally, the authentication property is guaranteed by the uniqueness of the transcript and the security of the MU-EUF-CMA secure signature scheme SIG and the security of the one-time signature scheme OTSIG. We show that for any τ and any τ-accepted oracle π_i^s with internal state $\Psi_i^s = \text{accept}$ and $\text{Pid}_i^s = j$ there exists an oracle, π_j^t, such that π_i^s and π_j^t have matching conversations. Otherwise the attacker $\mathcal{A}$ has forged a signature for either SIG or OTSIG.

In the next step, we show that the session key of the AKE protocol is secure except for probability ϵ_{Ind} in the sense of the Property 2.) of the AKE security Definition 14. The security of the authentication protocol guarantees that there can only be passive attackers on the test oracles, so that we can conclude the security for key indistinguishability from the security of the underlying KEM. We recall that μ denotes the number of honest identities and that ℓ denotes the maximum number of protocol executions per party. In the proof of Theorem 4, we consider the following two lemmas. Lemma 3 bounds the probability ϵ_{Auth} that an attacker breaks the authentication property of AKE and Lemma 4 bounds the probability ϵ_{Ind} that an attacker is able to distinguish real from random keys. It holds:

$$\epsilon_{\mathsf{AKE}} \leq \epsilon_{\mathsf{Auth}} + \epsilon_{\mathsf{Ind}}.$$

4.4   Authentication

Lemma 3. *For all attackers $\mathcal{A}$ that $(t, \mu, \ell, \epsilon_{\mathsf{Ind}})$-break the AKE protocol by breaking Property 1.) of Definition 14 there exists an algorithm $\mathcal{B} = (\mathcal{B}_{\mathsf{SIG}}, \mathcal{B}_{\mathsf{OTSIG}})$ such that either $\mathcal{B}_{\mathsf{SIG}}$ $(t', \mu, \epsilon_{\mathsf{SIG}})$-breaks the security of SIG or $\mathcal{B}_{\mathsf{OTSIG}}$ $(t', \epsilon_{\mathsf{OTSIG}}, \mu\ell)$-breaks the security of OTSIG where $t \approx t'$ and*

$$\epsilon_{\mathsf{Auth}} \leq \epsilon_{\mathsf{SIG}} + 2 \cdot \epsilon_{\mathsf{OTSIG}}.$$

Proof. Let $\mathsf{break}_\delta^{(\mathsf{Auth})}$ be the event that there exists a τ and a τ-fresh oracle π_i^s that has internal state $\Psi_i^s = \mathtt{accept}$ and $\mathsf{Pid}_i^s = j$, but there is no unique oracle π_j^t such that π_i^s and π_j^t have matching conversations, in Game δ. If $\mathsf{break}_\delta^{(\mathsf{Auth})}$ occurs, we say that $\mathcal{A}$ wins in Game δ.

GAME $\mathsf{G_0}$. This is the original game that is played between an attacker $\mathcal{A}$ and a challenger $\mathcal{C}$, as described in Section 4.1. Thus we have:

$$\Pr[\mathsf{break}_0^{(\mathsf{Auth})}] = \epsilon_{\mathsf{Auth}}$$

GAME $\mathsf{G_1}$. In this game, the challenger proceeds exactly like in the previous game, except that we add an abort rule. Let π_i^s be a τ-accepted oracle with internal state $\mathsf{Pid}_i^s = j$, where P_j is $\hat{\tau}$-corrupted with $\hat{\tau} > \tau$. We want to ensure that the OTSIG public key $vk_{\mathsf{OTSIG}}^{(j)}$ received by π_i^s was output by an oracle π_j^t (and not generated by the attacker).

Technically, we abort and raise the event $\mathsf{abort}_{\mathsf{SIG}}$, if the following condition holds:
- there exists a τ and a τ-fresh oracle π_i^s with internal state $\mathsf{Pid}_i^s = j$[5] and
- π_i^s received a signature $\sigma^{(j)}$ that satisfies $\mathsf{SIG.Vfy}(vk^{(j)}, vk_{\mathsf{OTS}}^{(j)}, \sigma^{(j)})$, but there exists no oracle π_j^t which has previously output a signature $\sigma^{(j)}$ over $vk_{\mathsf{OTS}}^{(j)}$.

Clearly we have

$$\left| \Pr[\mathsf{break}_0^{(\mathsf{Auth})}] - \Pr[\mathsf{break}_1^{(\mathsf{Auth})}] \right| \leq \Pr[\mathsf{abort}_{\mathsf{SIG}}].$$

Claim. $\Pr[\mathsf{abort}_{\mathsf{SIG}}] \leq \epsilon_{\mathsf{SIG}}$.

We refer to the full version of the paper [3] for a proof of the claim. $\square$

GAME $\mathsf{G_2}$. In this game, the challenger proceeds exactly like the challenger in Game 1, except that we add an abort rule. Let $\mathsf{abort}_{\mathsf{collision}}$ denote the event that

---

[5] Since π_i^s is τ-fresh it holds that P_j is $\hat{\tau}$-corrupted, where $\hat{\tau} > \tau$.

two oracles, π_i^s and π_j^t, sample the same verification key, vk_{OTS}, for the one-time signature scheme. More formally, let

$$\mathsf{abort}_{\mathsf{collision}} := \left\{ \exists (i,j) \in [\mu \cdot \ell]^2 : vk_{\mathsf{OTS}}^{(i)} = vk_{\mathsf{OTS}}^{(j)} \land i \neq j \right\}.$$

The simulator aborts if $\mathsf{abort}_{\mathsf{collision}}$ occurs and $\mathcal{A}$ loses the game. Clearly, we have

$$\left| \Pr[\mathsf{break}_1^{(\mathsf{Auth})}] - \Pr[\mathsf{break}_2^{(\mathsf{Auth})}] \right| \leq \Pr[\mathsf{abort}_{\mathsf{collision}}].$$

Claim. $\Pr[\mathsf{abort}_{\mathsf{collision}}] \leq \epsilon_{\mathsf{OTSIG}}$

We refer to the full version of the paper [3] for a proof of the claim.   □

GAME G_3. In this game, the challenger proceeds exactly like in the previous game, except that we add an abort rule. Let π_i^s be a τ-accepted oracle, for some τ, that received a one-time signature key, $vk_{\mathsf{OTS}}^{(j)}$, from an uncorrupted oracle, π_j^t. Informally, we want to make sure that if π_i^s accepts then π_j^t has previously output *the same* one-time signature $\sigma_{\mathsf{OTS}}^{(j)}$ over (m_1, m_2) that is valid under $vk_{\mathsf{OTS}}^{(j)}$. Note that in this case π_i^s confirms the "view on the transcript" of π_j^t.

Technically, we raise the event $\mathsf{abort}_{\mathsf{OTSIG}}$ and abort (and $\mathcal{A}$ loses), if the following condition holds:

- there exists a τ-fresh oracle π_i^s that has internal state $\mathsf{Pid}_i^s = j$ and
- π_i^s receives a valid one-time signature $\sigma_{\mathsf{OTS}}^{(j)}$ for (m_1, m_2) and accepts, but there is no unique oracle, π_j^t, which has previously output $\left((m_1, m_2), \sigma_{\mathsf{OTS}}^{(j)} \right)$.

Clearly we have

$$\left| \Pr[\mathsf{break}_2^{(\mathsf{Auth})}] - \Pr[\mathsf{break}_3^{(\mathsf{Auth})}] \right| \leq \Pr[\mathsf{abort}_{\mathsf{OTSIG}}].$$

Claim. $\Pr[\mathsf{abort}_{\mathsf{OTSIG}}] \leq \epsilon_{\mathsf{OTSIG}}$

We refer to the full version of the paper [3] for a proof of the claim.   □

Claim. $\Pr[\mathsf{break}_3^{(\mathsf{Auth})}] = 0$

Proof. Note that $\mathsf{break}_3^{(\mathsf{Auth})}$ occurs only if there exists a τ-fresh oracle π_i^s and there is no *unique* oracle π_j^t such that π_i^s and π_j^t have matching conversations.

Consider a τ-fresh oracle π_i^s. Due to Game 1 there exists (at least one) oracle π_j^t which has output the verification key $vk_{\mathsf{OTS}}^{(j)}$ received by π_i^s, along with a valid SIG-signature $\sigma^{(j)}$ over $vk_{\mathsf{OTS}}^{(j)}$, as otherwise the game is aborted. $vk_{\mathsf{OTS}}^{(j)}$ (and therefore also π_j^t) is unique due to Game 2.

π_i^s accepts only if it receives a valid one-time signature $\sigma_{\mathsf{OTS}}^{(j)}$ over the transcript (m_1, m_2) of messages. Due to Game 3 there must exist an oracle which has output this signature $\sigma_{\mathsf{OTS}}^{(j)}$. Since (m_1, m_2) contains $vk_{\mathsf{OTS}}^{(j)}$, this can only be π_j^t. Thus, if π_i^s accepts, then it must have a matching conversation to π_j^t.

Summing up we see that:

$$\epsilon_{\mathsf{Auth}} \leq \epsilon_{\mathsf{SIG}} + 2\epsilon_{\mathsf{OTSIG}}$$

4.5 Key Indistinguishability

Lemma 4. *For any attacker A that $(t, \mu, \ell, \epsilon_{Ind})$-break AKE by breaking Property 2.) of Definition 14 there exists an algorithm $B = (B_{KEM}, B_{SIG}, B_{OTSIG})$ such that either B_{KEM} $(t', \mu\ell, \epsilon_{KEM})$-breaks the security of KEM, or B_{SIG} $(t', \mu, \epsilon_{SIG})$- breaks the security of SIG or B_{OTSIG} $(t', \epsilon_{OTSIG}, \mu\ell)$-breaks the security of OTSIG where $t \approx t'$ and*

$$\epsilon_{Ind} \leq \epsilon_{SIG} + 2 \cdot \epsilon_{OTSIG} + \epsilon_{KEM}.$$

The proof of Lemma 4 can be found in the full version of the paper [3]. Summing up probabilities, we obtain that

$$\epsilon_{Ind} \leq \epsilon_{SIG} + 2 \cdot \epsilon_{OTSIG} + \epsilon_{KEM}$$

$\square$

References

1. Abe, M., David, B., Kohlweiss, M., Nishimaki, R., Ohkubo, M.: Tagged one-time signatures: Tight security and optimal tag size. In: Kurosawa, K., Hanaoka, G. (eds.) PKC 2013. LNCS, vol. 7778, pp. 312–331. Springer, Heidelberg (2013)
2. Bader, C.: Efficient signatures with tight real world security in the random oracle model. In: Gritzalis, D., Kiayias, A., Askoxylakis, I. (eds.) CANS 2014. LNCS, vol. 8813, pp. 370–383. Springer, Heidelberg (2014)
3. Bader, C., Hofheinz, D., Jager, T., Kiltz, E., Li, Y.: Tightly secure authenticated key exchange. Cryptology ePrint Archive, Report 2014/797 (2014) http://eprint.iacr.org/.
4. Bellare, M., Boldyreva, A., Micali, S.: Public-key encryption in a multi-user setting: Security proofs and improvements. In: Preneel, B. (ed.) EUROCRYPT 2000. LNCS, vol. 1807, pp. 259–274. Springer, Heidelberg (2000)
5. Bellare, M., Rogaway, P.: Entity authentication and key distribution. In: Stinson, D.R. (ed.) CRYPTO 1993. LNCS, vol. 773, pp. 232–249. Springer, Heidelberg (1994)
6. Bellare, M., Rogaway, P.: Random oracles are practical: A paradigm for designing efficient protocols. In: Ashby, V. (ed.) ACM CCS 1993 1st Conference on Computer and Communications Security, pp. 62–73. ACM Press (November 1993)
7. Bellare, M., Rogaway, P.: The security of triple encryption and a framework for code-based game-playing proofs, pp. 409–426 (2006)
8. Bernstein, D.J.: Proving tight security for Rabin-Williams signatures. In: Smart, N.P. (ed.) EUROCRYPT 2008. LNCS, vol. 4965, pp. 70–87. Springer, Heidelberg (2008)
9. Blake-Wilson, S., Johnson, D., Menezes, A.: Key agreement protocols and their security analysis. In: Darnell, M.J. (ed.) Cryptography and Coding 1997. LNCS, vol. 1355, pp. 30–45. Springer, Heidelberg (1997)
10. Blake-Wilson, S., Menezes, A.: Authenticated Diffie-Hellman key agreement protocols (invited talk). In: Tavares, S., Meijer, H. (eds.) SAC 1998. LNCS, vol. 1556, pp. 339–361. Springer, Heidelberg (1999)

11. Blazy, O., Kiltz, E., Pan, J. (Hierarchical) identity-based encryption from affine message authentication. In: Garay, J.A., Gennaro, R. (eds.) CRYPTO 2014, Part I. LNCS, vol. 8616, pp. 408–425. Springer, Heidelberg (2014)
12. Boneh, D., Boyen, X., Shacham, H.: Short group signatures. In: Franklin, M. (ed.) CRYPTO 2004. LNCS, vol. 3152, pp. 41–55. Springer, Heidelberg (2004)
13. Canetti, R., Krawczyk, H.: Analysis of key-exchange protocols and their use for building secure channels. In: Pfitzmann, B. (ed.) EUROCRYPT 2001. LNCS, vol. 2045, pp. 453–474. Springer, Heidelberg (2001)
14. Canetti, R., Krawczyk, H.: Universally composable notions of key exchange and secure channels. In: Knudsen, L.R. (ed.) EUROCRYPT 2002. LNCS, vol. 2332, pp. 337–351. Springer, Heidelberg (2002)
15. Chen, J., Wee, H.: Fully (almost) tightly secure IBE and dual system groups. In: Canetti, R., Garay, J.A. (eds.) CRYPTO 2013, Part II. LNCS, vol. 8043, pp. 435–460. Springer, Heidelberg (2013)
16. Dierks, T., Allen, C.: The TLS Protocol Version 1.0. RFC 2246 (Proposed Standard). Obsoleted by RFC 4346, updated by RFCs 3546, 5746 (January 1999)
17. Dierks, T., Rescorla, E.: The Transport Layer Security (TLS) Protocol Version 1.1. RFC 4346 (Proposed Standard). Obsoleted by RFC 5246, updated by RFCs 4366, 4680, 4681, 5746 (April 2006)
18. Dierks, T., Rescorla, E.: RFC 5246: The transport layer security (tls) protocol; version 1.2 (August 2008)
19. Diffie, W., Hellman, M.E.: New directions in cryptography. IEEE Transactions on Information Theory 22(6), 644–654 (1976)
20. Escala, A., Herold, G., Kiltz, E., Ràfols, C., Villar, J.: An algebraic framework for Diffie-Hellman assumptions. In: Canetti, R., Garay, J.A. (eds.) CRYPTO 2013, Part II. LNCS, vol. 8043, pp. 129–147. Springer, Heidelberg (2013)
21. Goh, E.-J., Jarecki, S., Katz, J., Wang, N.: Efficient signature schemes with tight reductions to the Diffie-Hellman problems. Journal of Cryptology 20(4), 493–514 (2007)
22. Goldwasser, S., Micali, S., Rivest, R.L.: A digital signature scheme secure against adaptive chosen-message attacks. SIAM Journal on Computing 17(2), 281–308 (1988)
23. Gorantla, M.C., Boyd, C., González Nieto, J.M.: Modeling key compromise impersonation attacks on group key exchange protocols. In: Jarecki, S., Tsudik, G. (eds.) PKC 2009. LNCS, vol. 5443, pp. 105–123. Springer, Heidelberg (2009)
24. Groth, J., Sahai, A.: Efficient non-interactive proof systems for bilinear groups. In: Smart, N.P. (ed.) EUROCRYPT 2008. LNCS, vol. 4965, pp. 415–432. Springer, Heidelberg (2008)
25. Hofheinz, D., Jager, T.: Tightly secure signatures and public-key encryption. In: Safavi-Naini, R., Canetti, R. (eds.) CRYPTO 2012. LNCS, vol. 7417, pp. 590–607. Springer, Heidelberg (2012)
26. Hofheinz, D., Jager, T.: Tightly secure signatures and public-key encryption. Cryptology ePrint Archive, Report 2012/311 (2012), http://eprint.iacr.org/
27. Jager, T., Kohlar, F., Schäge, S., Schwenk, J.: On the security of TLS-DHE in the standard model. In: Safavi-Naini, R., Canetti, R. (eds.) CRYPTO 2012. LNCS, vol. 7417, pp. 273–293. Springer, Heidelberg (2012)
28. Kakvi, S.A., Kiltz, E.: Optimal security proofs for full domain hash, revisited. In: Pointcheval, D., Johansson, T. (eds.) EUROCRYPT 2012. LNCS, vol. 7237, pp. 537–553. Springer, Heidelberg (2012)

29. Krawczyk, H.: HMQV: A high-performance secure Diffie-Hellman protocol. In: Shoup, V. (ed.) CRYPTO 2005. LNCS, vol. 3621, pp. 546–566. Springer, Heidelberg (2005)

30. LaMacchia, B.A., Lauter, K., Mityagin, A.: Stronger security of authenticated key exchange. In: Susilo, W., Liu, J.K., Mu, Y. (eds.) ProvSec 2007. LNCS, vol. 4784, pp. 1–16. Springer, Heidelberg (2007)

31. Libert, B., Joye, M., Yung, M., Peters, T.: Concise multi-challenge cca-secure encryption and signatures with almost tight security. In: Sarkar, P., Iwata, T. (eds.) ASIACRYPT 2014, Part II. LNCS, vol. 8874, pp. 1–21. Springer, Heidelberg (2014), https://eprint.iacr.org/2014/743.pdf

32. Menezes, A., Smart, N.P.: Security of signature schemes in a multi-user setting. Des. Codes Cryptography 33(3), 261–274 (2004)

33. Naor, M., Yung, M.: Public-key cryptosystems provably secure against chosen ciphertext attacks. In: 22nd Annual ACM Symposium on Theory of Computing, pp. 427–437. ACM Press (May 1990)

34. Schäge, S.: Tight proofs for signature schemes without random oracles. In: Paterson, K.G. (ed.) EUROCRYPT 2011. LNCS, vol. 6632, pp. 189–206. Springer, Heidelberg (2011)

35. Shoup, V.: Sequences of games: A tool for taming complexity in security proofs. Cryptology ePrint Archive, Report 2004/332 (2004), http://eprint.iacr.org/

Author Index